# Neurosciences - From Molecule to Behavior: A University Textbook

C. Giovanni Galizia • Pierre-Marie Lledo
Editors

# Neurosciences - From Molecule to Behavior: A University Textbook

Springer Spektrum

*Editors*
C. Giovanni Galizia
Department of Neurobiology
University of Konstanz
Konstanz
Germany

Pierre-Marie Lledo
Laboratoire Perception et Mémoire
Pasteur Institute and CNRS
Paris Cedex
France

Based on the 2nd edn. of the textbook "Neurowissenschaft" by Josef Dudel, Randolf Menzel and Robert F. Schmidt

Illustrations: Fotosatz-Service Köhler GmbH - Reinhold Schöberl, Würzburg

ISBN 978-3-662-51881-6 ISBN 978-3-642-10769-6 (eBook)
DOI 10.1007/978-3-642-10769-6
Springer Heidelberg New York Dordrecht London

Springer Spektrum

Softcover reprint of the hardcover 1st edition 2013


Printed on acid-free paper

Springer Spektrum is a brand of Springer DE. Springer DE is part of Springer Science+Business Media
www.springer-spektrum.de

# Preface

For several decades now, scientists have attempted to translate ancient beliefs on the nature of the human mind inspired by Greek philosophers, into a scientifically based understanding of the functions and processes occurring within the physical matter amassed in the 1.5 kg of tissue that make up the human brain. The goal of this modern science that we call *Neuroscience* is to understand how brains are able to retain for decades, and recreate at will, the color and taste of a ripe apple or the scent of a rose.

Taboos are vanishing now as scientists are collecting mounting data on the question of mind with the empirical language of biology, physiology, ecology, and ethology. Indeed, during the last years, *Neuroscience* has expanded into and converged with quite many domains – from anatomy to zoology, from physiology to computer science and many more. *Neuroscience* has received vast attention in the media and in public discourse, among others through our increasing knowledge of neurological diseases such as Alzheimer and Parkinson's, by addressing ethical questions such as brain death, or because of interdisciplinary convergences which in turn create new disciplines – even pedagogy, politics, and prosthetics now have their own specific "*neuro*" flavors.

This textbook covers *Neuroscience* from the cellular and molecular level, to behavior, and cognitive processing – thus giving every interested reader a solid basis of knowledge about this now well-established discipline. We also address evolution of the nervous system, computational neuroscience, neurophilosophy, and the history of *Neuroscience* as a discipline – to name but a few. The book provides the newest state-of-the-art knowledge about *Neuroscience* from across the animal kingdom, with particular emphasis on model species commonly used in neuroscience labs across the world: mouse, zebra fish, fruit fly, honeybee, and nematode worm. This choice has not only practical roots – it is based on the observation that all animals share molecular mechanisms and a fundamental design in their nervous systems. After all, the function of nervous systems in general is to perceive sensory information and trigger actions that benefit the organism.

Thus, comprehending the human brain, understanding where consciousness arises, finding treatments for neurological diseases, all is based on knowing how the nervous system works and how it evolved to become what it is today. Quite astounding but true: our brains bear many structural and functional similarities to those of our many animal relatives – and yet again, there are important differences! In the light of these interrelationships, studying animals can provide insights that may help us understand ourselves.

Considering all, it is no longer merely *Neuroscience*, but indeed has become the ***Neurosciences***! The approach of this book thus follows and expands on the tradition of its predecessor: the German textbook "Neurowissenschaft" edited by Dudel, Menzel, and Schmidt, published in two successful editions by Springer.

The editors and authors would like to thank the many people involved in such a major undertaking, most importantly in each of the authors' groups the many collaborators who have supported them by reading, correcting, adjusting, and fine-tuning the individual chapters. For graphical work, we would like to thank Michaela Baumann and Reinhold Schöberl from Fotosatz-Service Köhler GmbH in Würzburg, Germany, for their tremendous commitment. For attentive and devoted work on the project across all chapters we thank Mihaela Mihaylova and Brigitte Geiger in Konstanz. Important help and assistance was also granted by Christine Dittrich and many students of the University of Konstanz. Many thanks go to the Springer editorial team for their professional support, in particular to Stefanie Wolf as project coordinator and to Theodor C.H. Cole as copy editor.

Konstanz, Germany
Paris, France

C. Giovanni Galizia
Pierre-Marie Lledo

# Contents

# From Neurophysiology to Neuroscience: New Technologies and New Concepts in the Twentieth Century

**1**

François Clarac

Modern neurophysiology started with four major pioneers: C. Golgi (1843–1926) and S. Ramón y Cajal (1852–1934) in anatomy and histology, C. Sherrington (1857–1952) in physiology, and I. Pavlov (1849–1936) in behavior. The first roots of this discipline appeared in the nineteenth century and have been reviewed by an abundant literature on the history of neurology, neurophysiology, neuroanatomy; see [1–7].

At the onset of the twentieth century, three major contributions were presented on the central nervous system (CNS), the concept of the neuron, its implication in the regulation of the reflex, and the cartography of the brain. The neuronal histology was only accepted after great fighting between the **reticularists** and the **neuronists**. While Golgi was the first in 1873 to properly achieve the staining of nervous cells by impregnating fixed nervous tissue with potassium bichromate and silver nitrate, he thought the adjacent cells were in continuity with fused terminal connections making complex networks. It was Ramón y Cajal with his systematic histological analysis who demonstrated neuronal organization [8]. Cells are in contact but separated by a "synapse", the name given by Sherrington. However, the founder of the "*neuron theory*" in 1891 was the German anatomist and physiologist Wilhelm von Waldeyer (1836–1921) who coined a great number of biological terms such as chromosome and **neuron**. In fact, the definite proof was only given in 1954 by Palay and Palade [9] when the cleft of the synapse was made visible by electron microscopy. S.L. Palay (1918–2002), G.E. Palade (1912–2008), and mainly Eduardo de Robertis (1913–1988), in Argentina, blocked the effect of transmitters with antibodies against synaptosomes.

The reflex was the dominant concept with the very detailed work by Sherrington on the cat spinal cord [10]. He defined the proprioceptive regulations with the stretch of the muscle spindles and the neuronal organization of the myotactic reflex with two neurons, one sensory from the muscle itself and one motoneuron. With Liddell, in 1926, they studied an extensor reflex response demonstrating that it can be completely overridden by an inhibitory process due to a simultaneous action of an antagonist, a flexor. In 1943, Lloyd proved the monosynapticity of the reflex. It was Pavlov who defined in 1904 the conditioned reflex, demonstrating the possibility of learning and adaptation. For him the brain was composed of "*cellular analyzers*" able to detail the complexity of the external world, the brain being not independent from the external medium.

K. Brodmann (1868–1918) working in Berlin in the laboratory of Vogt started in 1901 a comparative cytoarchitectonics of the human brain. The main results were published between 1903 and 1908 in the *Journal für Psychologie und Neurologie*. He distinguished 47 areas depending upon their histological localizations, distinguishing sensory, motor, and associative areas.

As in any discipline in science, major breakthroughs in neuroscience have been driven by three distinct yet overlaping forces: technology, sociological, and political contexts. In this review we will focus on the first point and only mention some examples for the two others. The successive events of the century, the different wars, the political regime crises in various continents,

F. Clarac
CNRS, INT, Campus Timone-Santé,
27 Bd Jean Moulin, 13385 Marseille,
cedex 05, France
e-mail: francois.clarac@univ-amu.fr

C.G. Galizia, P.-M. Lledo (eds.), *Neurosciences - From Molecule to Behavior: A University Textbook*,
DOI 10.1007/978-3-642-10769-6_1, © Springer-Verlag Berlin Heidelberg 2013

the religious oppositions, and the ideological aggressions have several consequences on neuroscience. For example, the improvements brought on by electrophysiologists are linked to major achievements during World War II in the field of electronics.

This chapter cannot provide a full summary of the data obtained in the twentieth century – it can merely touch on some features that for the author seemed to have been crucial in the emergence of modern neuroscience. We will consider the data obtained during the first part of the twentieth century, the main changes resulting from this multidisciplinary concept, and the initiation of some new scientific fields of study.

## 1.1 Nervous System Explorations from 1900 to 1950

### 1.1.1 Electrophysiology and Nervous Conduction

After the first conclusions on animal electricity obtained in the nineteenth century from L. Galvani (1737–1798), C. Matteucci, (1811–1868), and E. Du Bois-Reymond, (1818–1896) the subject was further elaborated by J. Bernstein (1839–1917) who published in 1902 the first physical model of bio-electric events during nervous conduction; in 1868, he gave the first accurate description of the spatial distribution of that "*negative variation*". He described these action potentials using a sort of rheotome, a "*current slicer*" that can calculate precisely the conduction velocity. In 1902, he applied the Nernst equation to the cellular membrane. He limited his considerations to potassium ions and explained that nervous conduction was due to a changing permeability at the level of the membrane. These speculations were confirmed 50 years later by A.F. Huxley (1917–2012), A.L. Hodgkin (1914–1998), and B. Katz (1911–2003) in several papers published in 1952 (see Hodgkin et al. [45]). The experiments were performed on a giant axon from the mantle of the Atlantic squid (*Loligo pealei*), where it was possible to put two electrodes inside (see Chap. 7), one stimulating the fiber and the other clamping the current (voltage clamp). The internal solution could be changed giving the possibility to study separately the different ions. During the action potential, sodium penetrates through the membrane explaining the overshoot while potassium leaves the axon. The membrane potential depends upon the maintenance of ionic concentration gradients across it.

The great problem during that first period was the difficulty for recording the nervous electrical phenomena. G. Lippmann (1845–1921) invented the **capillary electrometer** that was routinely used despite the fact that it was often difficult to operate. When the biological electrical pulse arrives it changes the surface tension of the mercury and allows it to leap up a short distance in the capillary tube full of sulfuric acid. The other system was the **string galvanometer** promoted by W. Einthoven (1860–1927). This device used a very thin filament of conductive wire passing between very strong electromagnets. When a current passed through the filament, the electromagnetic field would cause the string to move. A light shining on the string would cast a shadow on a moving roll of photographic paper, thus forming a continuous curve showing the movement of the string. At that time, the electrophysiologists were also physicists and often built themselves their proper apparatus or modified them to improve their responses.

K. Lucas (1879–1916), with his pupil E. Adrian (1889–1977), established two fundamental laws on muscle contraction and on nervous conduction [11]. Using the *cutaneous dorsi* of the frog, they stimulated a small number of muscle fibers and demonstrated a discontinuous gradation in the contraction describing the excitatory "*all or none*" response. With a system of double shocks they showed that after a first firing, the neuron goes through an absolute refractory period, followed by a relative refractory period. During the 1920s, Adrian started new preparations where he isolated single sensory fibers and characterized their type of discharge. Action potentials have similar amplitudes and the intensity of a sensation is correlated with an increasing spike frequency.

BHC Matthews [12] in 1931 was able to demonstrate that the muscle spindle in the frog is stimulated by stretching the muscle and is released during contraction (Fig. 1.1). However, the best electrical recordings were obtained in 1921 by J. Erlanger (1874–1965) and H.S. Gasser (1888–1963). They used a cathode ray tube with a new type of amplifier. Their most significant contribution was that larger nerve fibers conducted electrical impulses faster than smaller ones, different nerve fibers being involved in different functions (Gasser and Erlanger 1927 [13]). H.K. Hartline (1903–1983) using minute electrodes

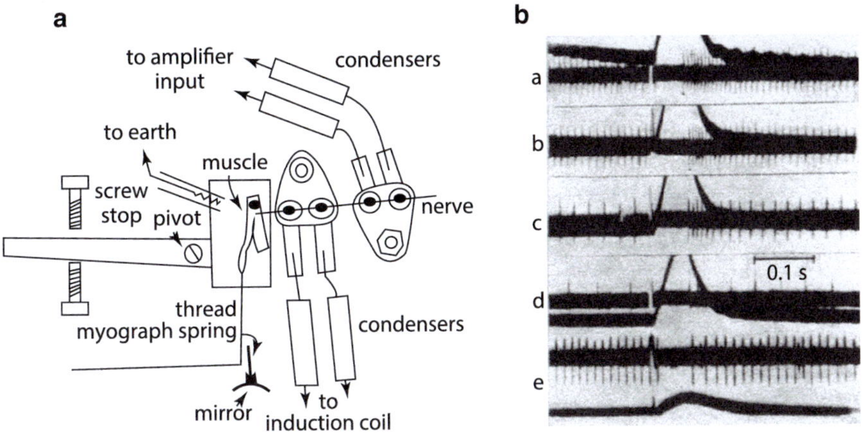


**Fig. 1.1** **The response of a muscle spindle during active contraction of a muscle.** (**a**) Presentation of the preparation with the muscle (an extensor of a frog's toe), the nerve that is recorded and the different dispositives. (**b**) Response in the nerve when the muscle twitches. *a–e*: decreasing contraction. There is silence during the twitch (From Matthews [12] with permission)

to study the eye of the horseshoe crab (*Limulus polyphemus*) obtained single electrical impulses from the optic nerve fiber, the receptors being stimulated by light (Hartline 1935 [14]). He demonstrated lateral inhibition between the receptors.

A new challenge was to understand the neuronal patterns of the different nervous centers. Thirty years later a more sophisticated coding was deciphered at the level of the cortex: V. Mountcastle studied the sensori-motor cortex of the cat and presented the columnar brain functional organization. When an electrode is inserted perpendicularly to the cortical surface, it recorded neurons that have nearly identical receptive fields (1978 [15]). There are 50–100 cortical minicolumns that form a hypercolumn, each comprising around 80 neurons. The columnar organization is functional by definition, and reflects the local connectivity of the cerebral cortex.

In parallel, Hubel and Wiesel in 1962 [16] described the visual cellular coding in anesthetized cat that had been studied previously by S.W. Kuffler (1913–1980) at the peripheral level (1953 [17]). Inserting a microelectrode into the primary visual cortex, they projected patterns of light and dark on a screen in front of the cat. They found neurons firing rapidly with presented lines at different angles; each cell had a maximum firing with a certain angle and was silent with an opposite direction. They called these neurons "*simple cells*". Other neurons, termed "*complex cells*", had identical responses to light and dark patterns. Later they described how signals from the eye are processed by the brain to generate main visual parameters, motion, color, or stereoscopic depth. By depriving kittens from using one eye, they demonstrated that cortical visual primary columns can receive inputs from the other eye taking over the areas that would normally receive input from the deprived eye. This demonstrated a "*critical period*" during development.

### 1.1.2 Acetylcholine (ACh) and Chemical Synapses

The neuronal theory was based on the notion of synapse that was more or less unknown at the onset of the twentieth century. At that particular level the major question was the nature of such conduction. Initially most of the scientists imagined only electrical connections: If the conduction was electrical in the axon, electrical mechanisms would exist at the synaptic level. The idea of different chemical substances was considered only for visceral regulation. J.N. Langley (1852–1925) analyzed the role of nicotine and of curare on the muscular contraction in chick. He established that both substances played opposite roles. Moreover using denervated frog gastrocnemius, he demonstrated that the substances act exactly at the nerve endings. In 1905, Thomas Elliott, student of Langley in Cambridge, claimed that the electrical stimulation of the cardiac sympathetic nerves increases its rhythm. This effect was mimicked by extracts of the adrenal glands that contained norepinephrin. In 1921, Otto Loewi (1873–1961) completed the pioneering observations of H.H. Dale (1875–1968). He stimulated the vagus nerve of an isolated heart diminishing its beating frequency. Taking a second heart perfused by the liquid of the first stimulated one, he observed the same slowing frequency. This was done without any electrical stimulation. He named this substance the **Vagusstoff** (vagal substance) and demonstrated that it was ACh.

During the 1930s, W. Feldberg (1900–1993) demonstrated the presence of cholinergic neurons in different regions of the nervous system and not only in visceral systems. D. Nachmanson (1899–1983) found that acetylcholine esterase was able to transform ACh in choline and in acetate. A. Fessard (1900–1982) a french neurophysiologist invited both at the marine station of Arcachon in 1939 where it was easy to collect a particular fish, *Torpedo marmorata* that possesses an electric organ that corresponds to a giant neuromuscular junction. Nachmanson found ACh enzyme to be highly concentrated. Eserin that inhibits the action of the enzyme rapidly induced fatigue and the electric organ stopped responding. Feldberg and Fessard in 1942 [18] made a detailed study of the electric organ demonstrating that there were 40–100 μg of ACh/g of fresh tissue. Stimulating the nerves produced ACh, but it was immediately destroyed by the enzyme. To find ACh, it was necessary to add some eserin. It was R. Couteaux (1909–1999) who, studying the presynaptic part of the synapse, demonstrated the presence of synaptic vesicles and pouches at the level of "*active zones*" of the frog neuromuscular junction (1970 [19]).

Later V.P. Whittaker and his group used fractionation of nerve tissue and obtained pure fractions of synaptic vesicles with ACh (synaptosomes: 140 nmol/g). The concept of receptors was introduced by R.P. Alquist (1948) who classified adrenoceptors into alpha and beta types based on differential sensitivity to epinephrine. Around 1950, only two substances seemed to be present at the synaptic level, ACh or Adrenaline, acting in opposition to regulate CNS mechanisms.

The debate to confirm GABA as an inhibitory neurotransmitter was quite long. Bazemore et al. [20] in 1956 demonstrated that GABA extracted from mammalian CNS blocked crustacean stretch receptor discharge. At the end of an international meeting in 1960, it was said that "*GABA entered the conference as a proud transmitter candidate and left it as a poor metabolite*!". Ernst Florey (1927–1997) played a great role in that discovery when he described in 1961 a substance I that was, in fact, GABA [21]. Kravitz et al. [22] in 1962 analyzed the peripheral nervous system of *Cancer borealis* and found different compounds like GABA, beta-alanine, taurine, and glutamate. Later they looked at the 60-μm-diameter fibers in the walking legs of the lobster, *Homarus americanus*. They were able to dissect out several meters of isolated axonal substance, which they separated into inhibitory and excitatory axons! They demonstrated that only the inhibitory axon contains an important quantity of GABA. Takeuchi and Takeuchi in 1965 using iontophoretic injections focused on the nerve-muscle junction in the crayfish leg. Using GABA, they mimicked inhibitory synaptic actions [23].

### 1.1.3 Behavioral Concepts

The reflex concept was also present at the integrated level when J.B. Watson (1878–1958) promoted "*behaviorism*" (1913 [24]). This theory proposed to consider the brain as a "*black box*" and put behavior at the center of psychology. The major idea was to abandon introspection and to ignore consciousness. This was in total opposition with the psychoanalytic approach by S. Freud (1856–1939). In behaviorism, the key element was to explain the different human reactions. It continued to be developed during the 1940s by B.F. Skinner (1904–1990) who promoted that **operant conditioning** operates on the environment and is maintained by its consequences which are either positive or negative, with reinforcement, punishment, and extinction (1938 [25]). Such vision was progressively criticized and several other concepts were presented. The notion of "**Gestalt**" was first introduced by C. von Ehrenfels (1859–1932) who explained that a given phenomenon must be considered in its entity. For example, conscious experience has to be considered in its globality and not by its different components. If during the first part of the century central psychological events were nearly abandoned, it can be said that the success of the "cognitive revolution" of the 1960s was a sort of revolt against behaviorism.

An important event occurred in 1924 but was not accepted immediately. The German neurologist H. Berger (1873–1941) made a systematic study of the electrical activity of human brain and developed the **electroencephalography** (**EEG**) (1929 [26]). Such activity was recorded in animals about 50 years before by the English Richard Caton (1842–1926) but had been curiously ignored. Using the EEG, Berger described the different waves or rhythms which were present in the normal and abnormal brain, such as the alpha wave rhythm (8–12 Hz). EEG became immediately useful in two domains, in sleep with the regulation of vigilance and in pathology with epilepsy.

During the 1940s it appeared that in the central axis of the CNS some complex structures control vigilance. Morruzzi and Magoun [27] in 1949 described that a high-frequency electrical stimulation of the reticular formation induces fast cortical rhythms corresponding

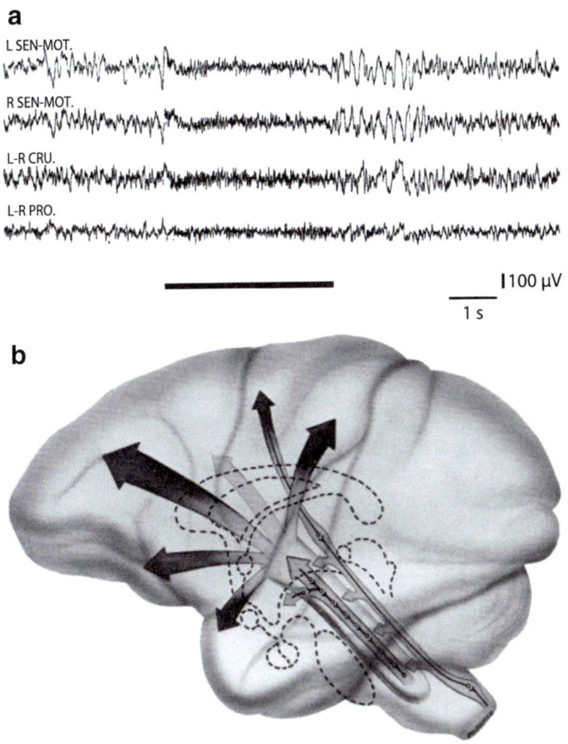


**Fig. 1.2** **The reticular formation and its control of the cortex.** The stimulation of the ascending reticular formation at high frequency induces an awareness and a low-voltage fast activity and a cessation of synchronized discharge in the cortex. (**a**) Four cortical recordings (left and right sensori-motor cortex, left to right cruciate gyrus, left to right gyrus proreus). *Black bar*: stimulation time. (**b**) Lateral view of the monkey's brain, with the ascending reticular activating system in the brain stem receiving different collaterals and projecting primarily to the associational areas of the hemisphere (From Moruzzi and Magoun [27] with permission)

to awareness (Fig. 1.2). For some decades this structure composed of numerous nuclei was extensively studied and considered as essential in the origin of the EEG activities. Penfield developed the idea of a "centrencephalon" where **consciousness** was due to a dialog between cortical areas, diencephalic and brain stem reticular formations (Delafresnay 1955 [28], see Fig. 1.2). The behaviorism was "over"!

In the rat brain some particular regions were found where the animal "wants to be electrically stimulated" (Olds and Milner 1954 [29]): Rats tested in Skinner boxes would stimulate themselves by pressing a lever. With electrodes permanently implanted and a 60-Hz stimulation, they received no other reward than the electrical stimulus during the experiments. The electrode was implanted in the lateral hypothalamus, where the medial forebrain bundle (MFB) interconnects the brainstem with most areas of the cerebrum. Such stimulation induces a dopamine production. The animal will press the lever and even forgets to eat.

Perhaps the most exciting experiment on rat learning was done by M.R. Rosenzweig (1922–2009) and his group in 1964 [30]. Rats that were raised as "pets" later solved tests better than rats raised in limited cages. Growing up in enriched environments affected not only learning behaviors but also cholinesterase activity in the cortex. Environmental enrichment increased cerebral cortex volume due to increased cerebral cortex thickness, greater synapses and glial cell numbers.

## 1.2 New Scientific Developments and the Foundation of Neuroscience

Neuroscience started around 1960. It was not a new discipline but a sort of another thinking manner, an ensemble of collaborations between disciplines and a new synthesis in these approaches. Its roots have to be searched in laboratories that studied the brain with multidisciplinary perspectives like the Institut für Hirnforschung (Institute for Brain Research) in Berlin created by O. Vogt (1870–1959) and C. Vogt-Meunier (1875–1862), the Montreal Neurological Institute, with W. Penfield (1891–1976), the Institute of Problems of Information Transmission in Moscow, with N. Bernstein (1896–1966) and after WWII, several laboratories in Cambrige (England), in France with A. Fessard and the Marey Institute, in Marseille with J. Paillard (1920–2006), the Institute of Neurophysiology and Psychophysiology in Pisa, Italy with G. Moruzzi (1910–1986), the Brain Research Institute in Zurich with K. Akert. In the U.S., several scientists proposed interdisciplinary projects like H.W. Magoun (1907–1991) in the Brain Institute at the University of California at Los Angeles. This led to the creation of IBRO (International Brain Research Organization). After several international EEG meetings, in particular after that in Moscow in 1958, a first society devoted to brain studies started in 1960, under the auspices of UNESCO. H.H. Jasper (1906–1999) was its first secretary and the first meeting was in Pisa in 1961. The main interest was to facilitate contacts between western and eastern countries that were politically separated by the iron curtain. This separation was very efficient. Some great Soviet scientists were ignored in the west and were recognized only later. Perhaps the best example is N. Bernstein (1896–1966), a great theoretician on motor control.

One of the main impulses to neuroscience was due to an MIT professor, Francis O. Schmitt (1903–1995), who organized meetings with scientists from different fields, from molecular biology to psychology, from mathematics to zoology. In 1961, he started a discussion group on «*The Men's Project*». In 1962, under his leadership a small group of committed scientists from diverse backgrounds but with a common interest in brain function began to meet regularly to share ideas about how the brain works. It was the beginning of the «*Neuroscience Research Program*» (NRP). The term "Neuroscience" became official. The consequences were three volumes published in 1967, 1970, and 1974 that proposed new trends in neuroscience (Worden et al. 1973 [31]). Schmitt proposed a *Neuroscience Research Foundation* to collect the necessary funds. The most characteristic laboratory of that new discipline in 1967 was Kuffler's at the Harvard Medical School. With D. Hubel, T. Wiesel, E. Furspan, and D. Potter, they were two "*brain boys*" and two "*membrane boys*"! The definitive consecration was the first meeting of the American Society of Neuroscience held with great success in 1971, October 27–30, in Washington DC. Soon after and progressively each country developed similar societies. A European Society was also established with a first meeting in Munich in 1972. The emergence of neuroscience concepts corresponded to new research types. We will consider two of them, the development of neurochemical processes and the use of simple preparations like those from invertebrates.

### 1.2.1 Neurotransmitters and Neurochemical Controls

Chemistry became integrated into the discipline very slowly. In 1946, Vittorio Esparmer (1909–1999) had demonstrated the presence of enteramine in intestinal chromaffine cells. In fact, he was investigating another neurotransmitter, 5-hydroxytryptamine (5-HT or **serotonin**). The characterization of the other amines (noradrenaline, dopamine) occurred in the 1950s. 3-Hydroxytyramine (**dopamine**) was identified as an intermediary in the synthesis of adrenaline from tyrosine. A. Carlson and his group found that the akinetic effects of reserpine treatment could be reversed by dopamine. It seemed that depletion of dopamine and not of noradrenaline or 5-HT was the cause of the akinetic state of reserpine. The same group reported that dopamine was mainly located in the striatum. This led to the hypothesis that dopamine is involved in Parkinson's disease. Cotzias published the first article on L-DOPA therapy in 1967 [32].

It was at that time possible to visualize a catecholamine neurotransmitter within nerve cells and to map anatomically the location of the various amines in the different brain nuclei with a new technique developed in 1962 by Falck and Hillarp (1916–1965). This microscopic method was based on the exposure of freeze-dried tissue to formaldehyde vapor allowing dopamine or noradrenaline to be converted to isoquinoline molecules that emit a yellow-green fluorescence. T. Hökfelt was from this group (Fig. 1.3). His first work was to identify cathecholamine by electron microscopy. Later, Hökfelt proved that the Dell principle elaborated in 1935 was wrong according to which

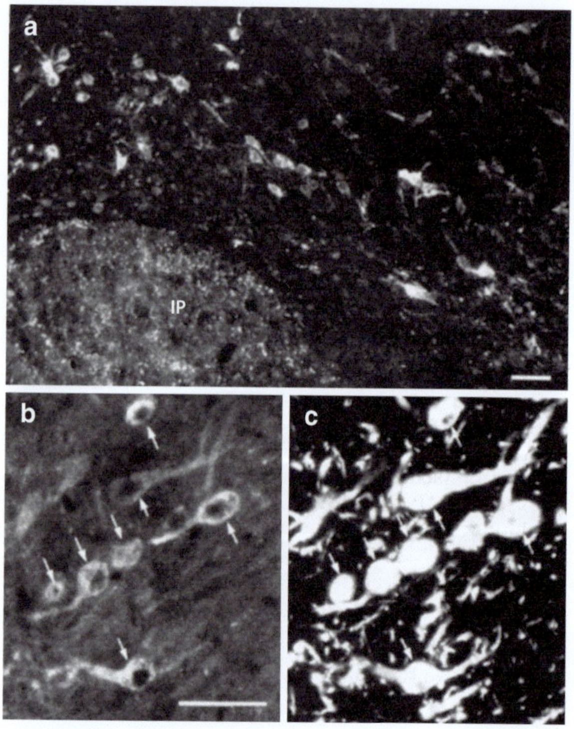


**Fig. 1.3 Neurochemical techniques: Presence at the same synapse of an amine and of a peptide**: Micrographs of the ventral tegmental area of colchicine-treated rat after incubation with antisera to gastrin/CCK (**a** and **b**) and tyrosine hydroxylase (**c**). (**b**) and (**c**) show the same section which after photography (**b**) and removal of the gastrin/cCCK antiserum has been reincubated with TH antiserum demonstrating the colocalization of dopamine and gastrin/CCK (*arrows*). *IP* interpeduncular nucleus (From Hökfelt et al. [56] with permission)

neurons only release a single transmitter. He demonstrated that at the same synapse a peptide and an amine can be colocalized.

Opioid neuropeptides were discovered simultaneously around 1975 by two groups: J. Hughes and H.W. Kosterlitz (1903–1996) isolated in the basal ganglia a substance of the brain of the pig that they called **enkephalin** (1975 [33]). R. Simantov and S.H. Snyder (1976) in the U.S. found a substance in the calf brain that they called **endorphin** for "*endogenous morphine*". Such studies were a direct consequence of the Vietnam War and the heroin used by the American soldiers. In an interview (2002) Snyder explained his motivation and how he found the opiate receptors.

Then, new maps describing the cortical geography of the different neurotranmitters, amines, ACh, peptides with their particular pathways and nuclei were elaborated. After the research of electrical pathways with electrophysiology the different circuits of the neurotransmitters were now established. A new technique with "*push-pull*" cannulas which both withdraw and inject fluid was developed to determine the effect of a certain chemical on specific groups of cells. Several other techniques like autoradiography or immunohistochemistry were developed to characterize the location of these substances in different types of tissues. The production of monoclonal antibodies around 1980 enabled researchers to examine very accurately protein expression within specific brain structures.

### 1.2.2 Invertebrate Preparations

In the 1950s, Eccles and his group recorded the first motoneuronal soma in the spinal cord of a cat, but most of the following intracellular results have been obtained in invertebrate preparations by A. Arvanitaki (1901–1983) and by L. Tauc (1926–1999) in different types of neurons of *Aplysia*. Synaptic mechanisms were studied with two electrodes inserted, one at the presynaptic and the other at the postsynpatic levels. In the squid giant synapse, Bullock and Hagiwara (1957) described a chemical synapse. In the crayfish giant motor synapse, Furshpan and Potter (1957) recorded the activity of an electrical synapse. In the former, the synaptic delay was of 1.2 ms; in the latter approx. 0.1 ms.

New staining techniques were discovered giving clear visualization of neuronal structures. Remler et al. (1968 [34]) characterized ipsilateral dendritic arborization from a contralateral soma by injection of the fluorescent Procion Yellow into the lateral giant fiber of crayfish. Pitman et al. (1972 [35]) injected cobalt chloride into an identified nerve cell body in an insect ganglion. This cell staining can be visualized both by light and electron microscopy. These techniques provided accurate locations for neurons in electrophysiological studies and enabled the construction of detailed maps of invertebrate ganglia mainly in the crayfish, *Aplysia*, or locust (Fig. 1.4). The concept of CPG (central pattern generator) emerged and corresponded to an ensemble of interneurons (or motoneurons) inducing a patterned activity regulated by modulary input and controlled by sensory afferents (Delcomyn 1980 [36]). This concept has been applied recently also to the cortex.

In parallel, W. McCulloch (1898–1969) and W. Pitts (1923–1969) considered the theoretical concept of a neuron and the idea of an artificial neuron was presented (1943 [37]). It became more elaborated with Rosenblatt (1928–1971) and his perceptron (1958). N. Wiener (1894–1964) a pioneer in the study of stochastic and noise processes wrote a book on cybernetics that described electronic engineering, electronic communication, and control systems (1948 [38]). Computational theories of mind promoted by artificial intelligence started at that time. This great intellectual activity in brain studies was made possible through the "Macy Conferences" which were a series of meetings of scholars from various fields held in New York by the initiative of Warren McCulloch and the Macy Foundation from 1946 to 1953. The principal purpose of these conferences was to set the foundations for a *general science of the workings of the human mind*.

## 1.3 Neuroscience in Progress: 1960–2000

When neuroscience emerged, the goal was to understand the CNS with an ensemble of classical disciplines like anatomy, physiology, behavior and new disciplines like molecular biology (Watson and Crick presented the hypothesis of the double helix of DNA in 1952), genetics (the first transgenic mouse was elaborated in 1980) and at the more elaborated level of complexity, cognition (the Royaumont meeting in 1975 illustrated by the discussion between Schomsky and Piaget (1896–1980) is considered as a start of that discipline).

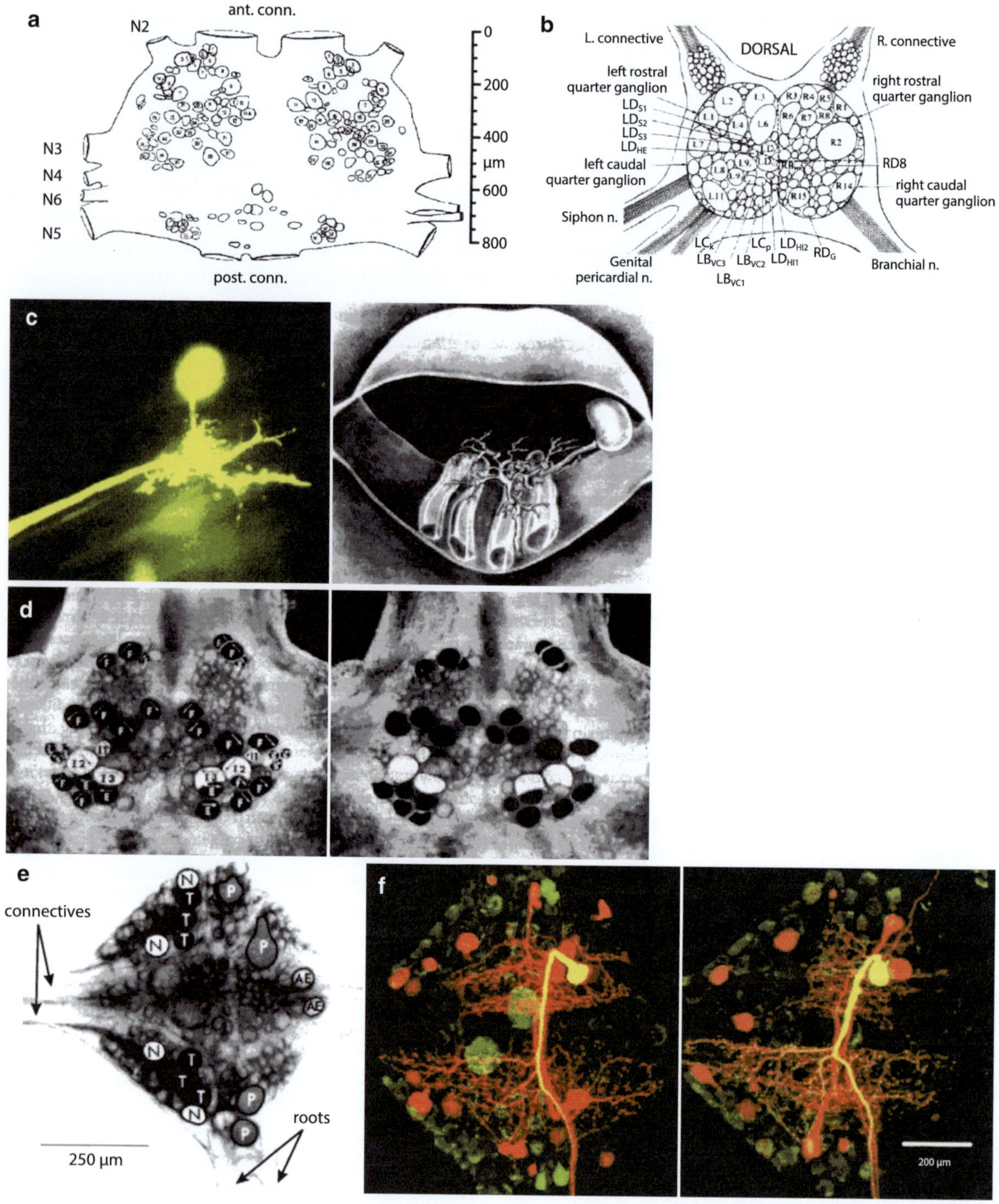


**Fig. 1.4 Mapping of cell localization in invertebrate ganglia**. (**a**) Dorsal view of the cockroach metathoracic ganglion (From Cohen and Jacklet [57]). (**b**) Dorsal view of the abdominal ganglion of *Aplysia* with right and left organization [58]. (**c**) *Left*, morphology of the fast flexor motoneuron in the third abdominal ganglion of the crayfish stained with lucifer yellow; *right*, drawing of the same neuron and its relation with the giant fibers [50]. (**d**) Chemical maps of lobster A3 ganglion; *left*, flexor and extensor motoneurons are represented; *right*, map of GABA containing cells [53]. (**e**) Ventral view of the leech segmental ganglion showing localization of sensory cells responding to touch (*T*), pressure (*P*), or noxious (*N*) stimulation of the skin [59]. (**f**) Confocal imaging of motoneuron-coupled cells in a leech ganglion. A single motoneuron is injected with two dyes but only one of these passes through gap junctions. This allows to identify all the coupled cells (in *red*) to the injected motoneuron (in *yellow*) [60] (This figure is derived from Fig. 4 in Clarac and Pearlstein [61]) (Figures from: (**a**) Cohen and Jacklet [57]; (**b**)Winlow and Kandel [58]; (**c**) Kennedy et al. [50]; (**d**) Otsuka et al. [53]; (**e**) Nicholls and Baylor [59]; (**f**) Fan et al., and Clarac and Pearlstein [60, 61] all with permission)

**Molecular biology** became central with specific techniques combined from genetics and biochemistry. This involved different studies of "mutants" (organisms which lack one or more functional components always compared with the "wild type" or normal phenotype). Transgenic mice, the worm *Caenorhabditis elegans* (a very simple animal with a fixed number of 302 neurons), the fruit fly *Drosophila melanogaster* as an insect, and the zebrafish *Danio rerio*, all became excellent models. They were used as transgenic animals to analyze molecular mechanisms and for studying human degenerative diseases.

At the integrated level, new technologies in cortical imagery gave the possibility to study the "*brain alive*": In the early 1980s, the development of radioligands allowed single photon emission computed tomography (SPECT) and positron emission tomography (PET) of the brain. During the 1980s, scientists demonstrated that the large blood flow changes measured by PET could also be made visible through magnetic resonance imaging (MRI). Functional magnetic resonance imaging (fMRI) was developed and is now the most widely used system in human brain mapping due to the absence of radiation exposure as in the PET method. fMRI allows images that reflect the particular structures in the brain that are activated during different tasks associated with different experimental paradigms.

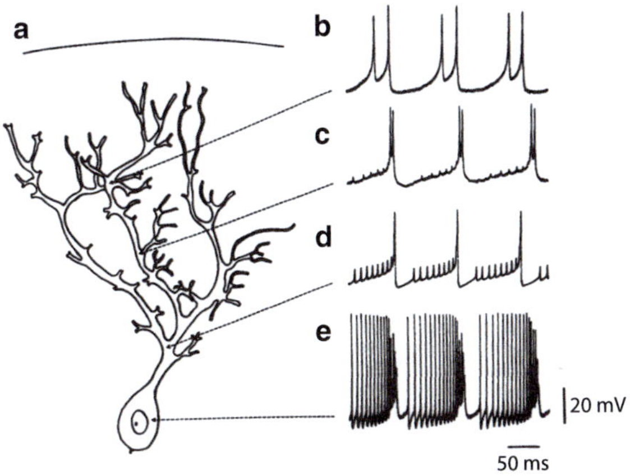


**Fig. 1.5** **Relationship between somatic and dendritic action potentials following DC depolarization through the recording electrode**. This composite picture shows four different recording sites: In (**a**), drawing of the neuron. In (**b–d**) are represented different locations of the different parts of the dendrites with their successive recordings. In (**e**) is the intracellular body recording. There is a shift in amplitude from the fast spike and the dendritic Ca-dependent potentials (From Llinas and Sugimori [40])

### 1.3.1 Molecular Mechanisms and "Simple" Preparations in Vertebrates

Acute experiments on cats seemed too complex and over the years raised substantial ethical concerns. The rat was considered as a nuisance, rising less ethical concerns and became an important model in electrophysiology. Simpler preparations were also developed, such as cutting brain slices. These new preparations were associated with new techniques such as **patch clamp** that allows to record the currents going through a single ion channel (see Chap. 7). To see the evolution of excitation, calcium imaging was developed.

Glutamate was discovered as the main neurotransmitter but due to its implication in the synthesis of proteins it took time to demonstrate its role. In 1961, Curtis and Watkins [39] showed that L-glutamate depolarized and excited central neurons, as expected for an excitatory transmitter. One year later, the synthesis of the first agonist (NMDA) was achieved. Several events during the 1980s confirmed the importance of glutamate: the role of magnesium during depolarization and the involvement of second messenger systems to stimulate the formation of inositol phosphates. This finding provided evidence for the existence of metabotropic glutamate receptors. In 1991, the first metabotropic glutamate receptor was cloned.

During that period, it was also demonstrated that spikes could be complex and could also be produced in dendrites [40] (Fig. 1.5). The conductances present in the different types of neurons were analyzed and it seemed that each cell had a diverse array of conductances giving them individual properties that can be modified or even completely changed in the presence of different neuromodulators. Recently, a new class of active substances has been added in neuronal metabolism, the so-called **transmitter transporters**, a class of proteins that span the cellular membrane of neurons. Their primary function is to carry neurotransmitters across membranes and to direct their further transport to specific intracellular locations. There are more than 20 types of neurotransmitter transporters for most of the classical neurotransmitters, except for ACh which is rapidly degraded. The CNS considered 50 years ago as a rigid system appeared progressively as a very *plastic system*.

### 1.3.2 Neural Networks – Substrate of Behaviors

The second part of the twentieth century was characterized by the development of new preparations trying to confirm the Hebb principle (see Chap. 26). In 1970, E. Kandel elaborated the gill-withdrawal reflex (GWR) paradigm from his work with *Aplysia*: a water current on the gills produced their retraction. He elaborated the circuit involved, from the tactile receptors in the skin to the monosynaptic and polysynaptic pathways that induced the contraction of the gills. Over a period of 30 years, Kandel and coworkers explained the cellular and molecular mechanisms of habituation and dishabituation. They demonstrated differential classical conditioning of the GWR of *Aplysia* and proposed activity-dependent presynaptic facilitation and presynaptic interaction between high intracellular $Ca^{2+}$ levels and 5-HT. Many reviews have been published on the preparation differentiating short- and long-term memory storage (see Chap. 26). Postsynaptic processes are also involved in memory mechanisms in invertebrates. Neurotransmitters, second messenger systems, protein kinases, ion channels, and transcription factors like CREB have been confirmed to function in both vertebrate and invertebrate learning and memory storage.

The second preparation was on the mammalian hippocampus, a very stereotyped structure composed of an S curve, the dentate gyrus, the subiculum, and the enthorinal cortex and the three parts, CA3, CA2, and CA1. This structure was known from EEG recordings for its characteristic 4–12 Hz oscillation, the theta rhythm associated with movement through space. The phenomenon of long-term potentiation (LTP) was initially observed in 1966 in the hippocampus of the rabbit by Terje Lomo, then working in the laboratory of Andersen. In 1973, he described the phenomenon in a seminal paper [41]. High frequency stimulation increased the excitatory response in the postsynaptic cell for 30 min to 10 h (Fig. 1.6). LTP was confirmed in hippocampus slices in 1975 by Schwartzkroin and Wester. At the beginning of the 1980s, the role of NMDA receptors and calcium influx at the postsynaptic cell became evident. Over the last years, evidence has accumulated to support the notion that dendritic spines are not passive structures but are involved in memory storage. Also, the hippocampus appeared to possess "*place cells*" that fire when the animal is in a specific place of a given environment. Following this discovery, O'Keefe and Nadel in 1978 [42] proposed that the primary function of the rat hippocampus is to form a "cognitive map" of the environment (see Chap. 28).

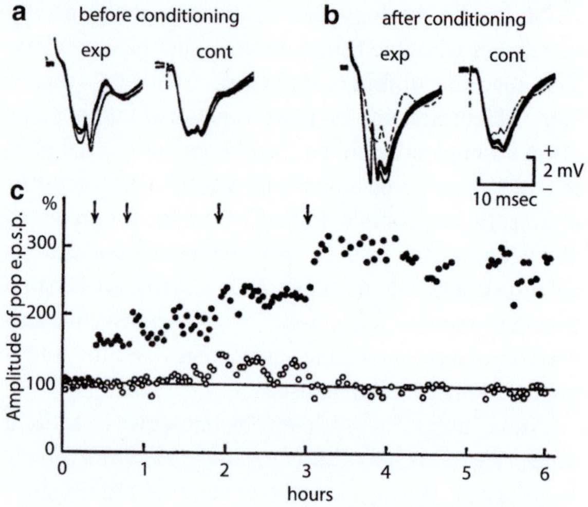


**Fig. 1.6 Synaptic plasticity in the hippocampus: long-term potentiation**. Repetitive stimulation (10–15 Hz for 10 s) of the perforant path fibers of the hippocampus of rabbits leads to LTP. The amplitude of the population EPSPs signaling the depolarization of the granule cells are increased for several hours. (**a**) Before stimulation. (**b**) 2.5 h after stimulations. The *dotted line* shows the average for comparison. (**c**) Graph showing the amplitude of the population EPSP for the experimental pathways (*black dots*) and the ipsilateral control pathway (*white dots*) that has not been stimulated (From Bliss and Lomo [41] with permission)

Invertebrates also perform complex neuronal activities. Learning in honeybees can be compared to learning in vertebrates in many respects, and the honeybee can be used as a model system to gain an understanding of the intermediate levels of complexity of cognitive functions and their neural substrates. The "*minibrain*" of the honeybee, with its 960,000 neurons, is now extensively explored; as Menzel said in a recent article "*Small brain, bright mind*!".

Analysis of the brain's cortical activities linked with a learning behavior was studied systematically in monkeys where different cortical units of different areas, sensory or motor, have been recorded in relation with a given behavior (see Sect. 1.5). In the late 1980s, it became possible to record several units simultaneously; with this one could analyze the temporal structure of activity patterns. It appeared that ongoing activity has predictive power and that self-generated fluctuations can sometimes synchronize in attention or during high-level activities [43]. The **mirror neurons** discovered by the group of Rizzolatti in 1996 have spectacular properties. These cells are both activated when the subject performs a motor action or when the subject observes someone in front performing the same action. Described first in the monkey, mirror neurons were confirmed in human. They have been considered as the substrate of several behaviors linked with action, intention, emotion, and empathy. Their dysfunction leads to impairment in social interactions as seen, for instance, in autism disorders; see [44].

### 1.3.3 Cognitive Neuroscience and Cortex Imaging

Two types of recordings were used: first the EEG and then cortical imaging. EEG was a very global activity; still several events seemed linked with cognition. In 1964, Walter et al. described the **contingent negative variation** (**CNV**) as a response linked to attention and vigilance. Sutton et al. in 1965 [45] described the "*P300* (*P3*)" wave, an event-related potential (ERP). Both were different, the CNV reflecting early arousal and attentional processes, P300 related to a decisional behavior involved in stimulus evaluation or categorization. Analyzed with other waves this showed that the brain produced some electrical signals linked with cognitive activities.

B. Libet (1916–2007) in the 1980s suggested that a readiness potential is formed in the cortex 500 ms before a voluntary movement [46]. Thus decisions made by a subject are first being made at a subconscious level and become conscious only afterwards. With new brain scanning technology, Soon et al. (2008 [47]) were even able to demonstrate that the delay could reach 10 s before a subject is conscious of his decision.

N. Geschwind (1826–1984) worked on aphasia and epilepsy, as well as dyslexias and on the neuroanatomy of cerebral lateral asymmetries. He coined the term **behavioral neurology** in the 1970s to describe the ensemble of elements composing higher cortical functions. Ungerleider and Mishkin (1982) proposed separate pathways for characterizing object shape and location in extrastriate visual cortex. Around the 1990s, brain imaging became extensively used to confirm the data previously collected in neuropsychology with clinical observations and post-mortem analysis (autopsy). The different primary areas, sensory and motor, were characterized much more precisely. When more complex tasks were analyzed, associative areas were added.

## 1.4 Computational Neuroscience

Computational neuroscience is the theoretical analysis and modeling of biological neural systems (see Chap. 30). Its goal is to propose models of biological neural networks in order to relate them to cognitive processes and behavior. They range from simple behavior with individual neurons, through models with learning and memory and short-term/long-term plasticity. Some of the models use architectures of Bayesian neural networks to include predictive capabilities and to use probabilities in order to characterize and process information.

## 1.5 History of "Animal Models"

Animal models had been employed from the early days of science. Perhaps the first was the frog by the Dutch biologist Jan Swammerdam (1637–1680) who analyzed the nerve-muscle preparation and demonstrated that the nervous conduction was not due to the "*animal's spirits*" as it was said by Galen (129–201 AD) and René Descartes (1596–1650), an old notion that considered the presence of a fluid necessary for muscle contractions. Swammerdam proved that when a muscle contracted, its volume did not change (see Chap. 22). Frogs were used routinely thereafter. Their great advantage was that the preparation stays alive for several hours at room temperature, as the animals are poikilotherms.

### 1.5.1 Invertebrates

In 1976, S.W. Kuffler and J. Nicholls published a book that was very popular for several generations of students: "*From Neuron to Brain*". To be clear and didactic, they used a great number of examples from invertebrates. They explained why they used them: "Fortunately, in the brains of all animals that have been studied there is apparent uniformity of principles for neurological signaling. We are convinced that behind each problem that appears extraordinary complex and insoluble there lies a simplifying principle that will lead to an unravelling of the events."

#### 1.5.1.1 First Models

During the nineteenth century, invertebrates were considered for some of their very particularly large neural structures, such as the giant fiber system of the earthworm, of the leech, or of the crab. The animal receiving most attention at the time was the crayfish that the British biologist T.H. Huxley (1825–1895) described in great detail in his book "The crayfish" (1879). In the preface he explained the goal of such models defending comparative anatomy with great emphasis: "I have desired, in fact, to show how the careful study of one of the commonest and most insignificant of animals, leads us, step by step, from every-day knowledge to the widest generalizations and the most difficult problems of zoology; and, indeed, of biological science in general."

During the twentieth century, invertebrates were used to find particular neuronal structures that could be analyzed easily in a particular animal. The best example is the famous squid axon that gave Hodgkin, Huxley, and Katz the possibility to understand nervous conduction (1952 [48]). The discovery of the giant axon has been described by J.S. Young. He had identified the long fiber at the Naples' marine research station and had initially thought that it was a blood vessel. In 1935, Young came to Woods Hole where squids were kept in very good shape in circular aquariums. He observed that an electrical stimulation of that giant fiber did not induce a muscle twitch. Young suggested that the best stimulus could be a chemical stimulus like $Na^+$ citrate. Then, he connected a fiber to an amplifier and a speaker and put some citrate on the end of it. "Out came – buzzzzzzzz – one of the best sounds I have ever heard" said Young several years later. It was the beginning of intense work on this nerve fiber on both sides of the Atlantic with Cole and Curtiss in Woods Hole, and the British in Plymouth and Cambridge.

#### 1.5.1.2 Arthropods

Crayfish and the larger crustaceans *Homarus* or *Palinurus* have been excellent models for characterizing neuromuscular innervation. The Dutch Cornelius Adrianus Gerrit Wiersma (1905–1979) at the University of California (UCLA) analyzed systematically the peripheral nerve-muscle system of crustaceans. Recording after stimulation of the different leg motor nerves, excitatory junction potentials with an Einthoven string galvanometer, he demonstrated the existence of separate excitatory and inhibitory motoneurons (MNs). The opener distal muscle was exceptional with only two motoneurons, one excitatory and one inhibitory. Insects, in particular crickets and cockroaches, presented the same advantages and were often used. In reflex activities it has been possible to characterize the proprioceptive control of each motoneuron of a given muscle (see Fig. 1.7).

#### 1.5.1.3 Mollusks

They were known to have large nervous cells. This was demonstrated by Angélique Arvanitaki (1901–1983) in Lyon, France, when working on the sea slug *Aplysia*. At the marine station of Tamaris (near Toulon) she discovered its very complex nervous system. She presented a general description of the different ganglia focusing on the visceral ganglion with the largest cells (up to 500 μm in diameter). Several electrical recordings were made. She observed spontaneous rhythmic activity in some cells (Arvanitaki and Cardot 1941 [49]). Her best intracellular recordings were carried out in 1955. L. Tauc at the Marey Institute in Paris analyzed the different intracellular potentials with two microelectrodes. Arvanitaki travelled to the U.S. in 1958 and reported there on her preparation in several leading laboratories. As a result, the procedure was adopted by the scientific community, in particular by Eric Kandel.

#### 1.5.1.4 Central Pattern Generators

*Aplysia* – as most invertebrate preparations – was studied because neural networks consist of only a small number of neurons. This has several advantages. It is possible to locate cell bodies inside ganglia and to produce an initial form of "neurogeography" ([50] see Fig. 1.4) showing a small number of interneurons able to control selected hierarchical behaviors. Cells were identified as contributing to restricted circuits supporting a stereotyped behavior through reflex modulation. It was "the golden age" for these preparations!

Miller and Selverston, studying the stomatogastric ganglion of Crustacea – perhaps the key preparation

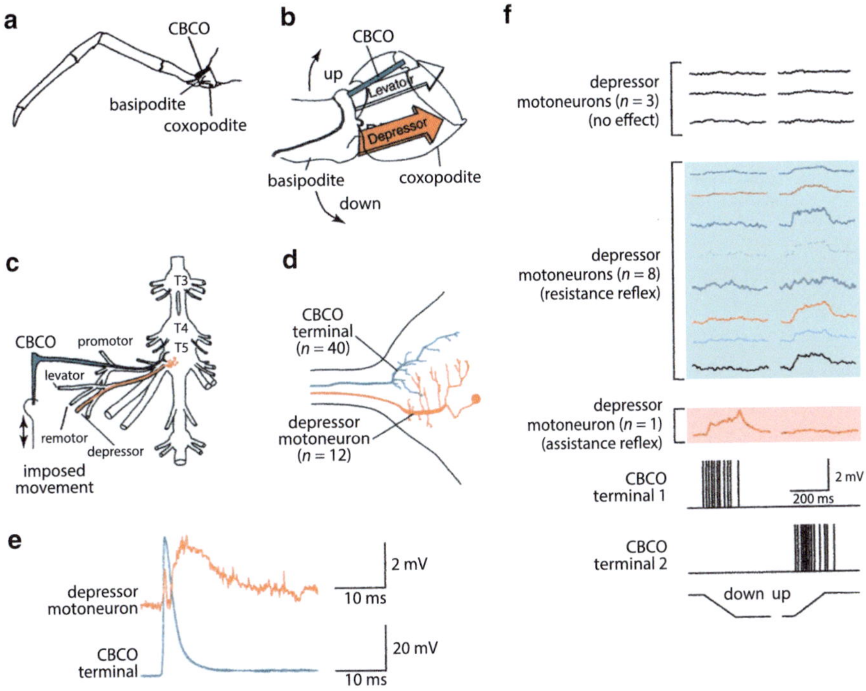


**Fig. 1.7 Different reflex controls on identified motoneurons in a depressor leg muscle by a mechanoreceptor, a chordotonal organ.** (**a**) and (**b**) Arrangement of the mechanoreceptor at the coxo-basipodite leg joint (CBCO) in the crayfish leg commanded by levator and depressor muscles. (**c**) *In vitro* preparation with the ventral nerve cord together with the motor and sensory nerves. The CBCO can be stimulated to mimic leg movements. (**d**) CBCO sensory terminal and a depressor motoneuron. (**e**) paired recordings from a CBCO terminal coding for upward leg movement with a PPSE from a depressor motoneuron. (**f**) Responses of proprioceptive neurons (two classes of CBCO terminals, 1 and 2) and of the 12 depressor motoneurons. 3 MNs do not respond, 8 are activated monosynaptically by upward movements and one by downward movements (From Clarac et al. [62] with permission)

for neural networks (see Chap. 23) – filled identified neurons with procyon yellow and killed them with blue light (1979 [51]). This allowed to study the network with and without a single neuron. It appeared around 1960, that a great number of isolated invertebrate ganglia presented a central rhythmical activity that was called a **central pattern generator** (**CPG**), a term first used in 1965 by Wilson and Wyman.

The pharmacological differences between invertebrates and vertebrates have also been used. Expression of the insect allatostatin receptor in a population of mouse neurons allows to activate them selectively by applying allatostatin, since there is no mammalian receptor that responds to this peptide (Fig. 1.8). The onset and reversal inactivation can be accomplished rapidly. For example, in the mouse spinal cord it was possible to characterize the role of V1 interneurons. The new field of "optogenetics" now develops several probes that when activated by light open channels or initiate second messenger cascades, increasing the toolbox for manipulative experiments.

## 1.5.2 Vertebrates and Mammals

### 1.5.2.1 Cat Models

In the eighteenth century, it was not possible to keep mammals alive for more than 10 min after surgical opening. Once they could be kept alive for several hours they were considered for neurobiological studies. Cats were employed by Sherrington for spinal cord experiments – his most famous experiment induced decerebrate rigidity after severance at the level of the brain stem: the cat was in complete extension due to the suppression of inhibition coming from the cortex. F. Bremer (1892–1982) proposed in 1935 the "isolated encephalon (*encéphale isolé*)" and the "isolated brain stem (*cerveau isolé*)" that were considered as key elements in the understanding of sleep. In the former

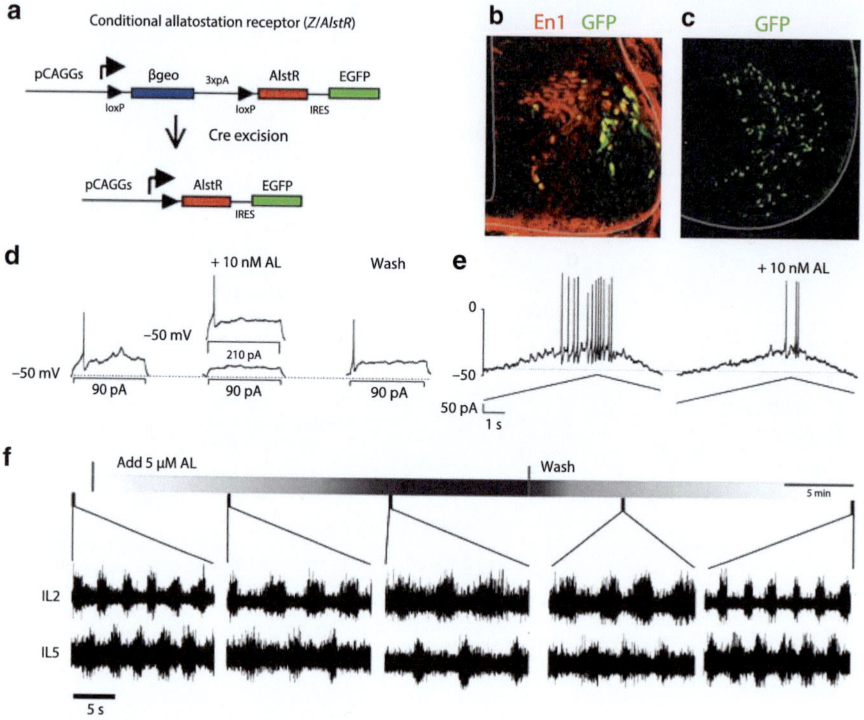


**Fig. 1.8** **Acute silencing of V1 neurons via allatostatin receptor causes a slowing of locomotor rhythmicity.** (**a**) Structural organization of the allatostatin receptor (Z/AlstR) transgene. Crossing mice that harbor this transgene with *En1*[cre] mouse excises the *loxP*-flanked *b-geo* sequences, leading to transcription of *AlstR* in En1-derived cells. (**b**) Coexpression of GFP (*green*) and En1 (*red*) in E11,5 *En1*[cre] (**c**) Spinal cord slice from a P0 *En1*[cre] (**d**) Recording from a V1 neuron expressing AlstR in a P1 *En1*[cre] (**e**) Recordings from an AlstR-expressing V1 neuron in response to a current ramp. Before allatostatin (AL) application, the cell begins to fire upon current injection of 105 pA. After allatostatin application (10 nM), the cell only fires when the injected current reaches 190 pA. (**f**) Electroneurogram recordings for two spinal roots (L2, *right* and *left*) just after birth (at P1). *AlstR192* mouse, showing the slowing of the rhythm due to the effect of allatostatin (5 μM) application and wash out on locomotor activity in the isolated spinal cord (From Gosgnach et al. [63] with permission)

case, the cat was awake and sleepy as normal, in the latter it slept all the time (Fig. 1.9).

The difficulty was to stimulate central nuclei or particular pathways. This was only possible with the apparatus proposed by Horsley-Clarke in the early twentieth century. Around 1890, the famous neurosurgeon Victor Horsley (1857–1916), who was professor in London at Brown Institute, met Robert-Henry Clarke (1853–1931) from Queens College of Cambridge and they became close friends. Around 1895, Clarke proposed to build an apparatus that fixed the head of the cat at the mouth, ears, and at the base of the orbits. The brain was mapped in a 3D system (three cartesian coordinates were taken, laterolateral ($x$), dorsoventral ($y$), and rostrocaudal ($z$)). In 1906, James Swift built the prototype in London. It was published in *Brain* in 1908 and it was quite useful for limited electrolytic lesions. The original apparatus is now in the Museum of the University College Hospital, London. But the two friends had very different characters, Clarke was a great sportsman, enjoying life but was very shy. Horsley was "very Victorian" and extrovert. Clarke did not accept the glory around his colleague and became paranoiac. The apparatus went to California and seemed abandoned. It was only used again 30 years later around 1940 with the electrical stimulations of the reticular formation. It was then used extensively in all physiological laboratories for more than 40 years.

#### 1.5.2.2 "Simple Models"

The "cat model"-period came to an end in the 1980s; cats were progressively abandoned for ethical reasons but also for their high cost and for the duration of the preparation. The rat appeared as a suitable alternative, it was much easier to explore, the anatomy and the pharmacology were well described. The success of the invertebrates had introduced some "simple

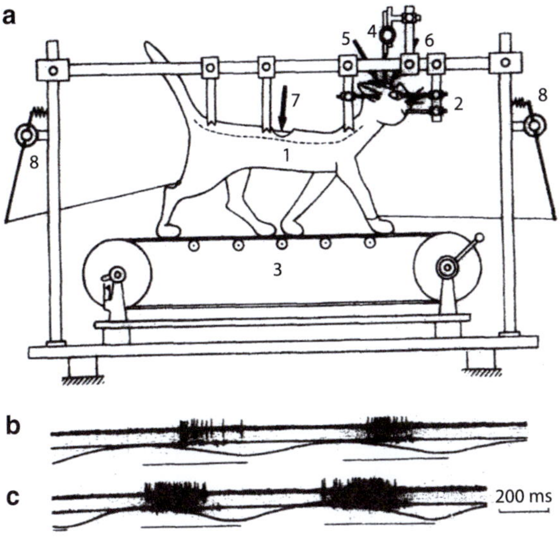


**Fig. 1.9 Chronic decerebrate walking cat above a treadmill.** Experimental arrangement for investigation of locomotion. (**a**) The cat has a severance between the inferior and the superior colliculus (mesencephalic cat). It is fixed in the Horsley-Clarke apparatus with electrodes for stimulating and recording. In (**b**, **c**) are represented the activity of two neurons in the dorsal spinal cerebellar tracts (DSCT) that responded during hindlimb locomotion. (**b**) Slow locomotion. (**c**) Fast locomotion with a characteristic increasing of the neuron discharge (From Arshavsky et al. [64] with permission)

vertebrates" like the lamprey, *Xenopus* eggs, or the tortoise. Lower vertebrate preparations were a good solution with even their simplicity, a similar anatomy and functional properties as they exist in all vertebrates. For example, S. Grillner analyzed in great detail the anguilliform mode of locomotion of the lamprey, and A. Roberts and K. Sillar the swimming of the developing *Xenopus* tadpole.

Two other very popular preparations that can be used today for *in vitro* or *in vivo* chronological recordings are the chick and the neonatal rat. The great interest is that both have an embryonic stage of 3 weeks and a postnatal life of similar duration, 3 weeks. Their development is quite different though: the precocial chick is able to walk at birth, while the altricial neonatal rat needs to remain in the nest, its development is gradual. A comparison of the two species is very useful for understanding maturation processses. Pioneering work on chicks was performed by Hamburger and Balaban (1963 [52]) and on rats by Otsuka and Konishi (1974 [53]). Recently, mice have become the most important laboratory animals because techniques have been developed to generate transgenic strains.

Two other methods have been used in neuroscience:

- Brain slices: This technique was developed in the late 1960s to study the hippocampus. Thick sections of living brain tissue are maintained *in vitro* in a solution of artificial cerebrospinal fluid that diffuses into all pieces of tissue. Used extensively and for almost all regions of mammalian central nervous system, slices preserve many circuits that can be finely analyzed.
- Cell cultures: Slices of CNS tissues that are prepared from young rodents are maintained in culture with a stable medium for many weeks. Nerve cells continue to differentiate and to develop a tissue organization that closely resembles that observed in situ. Long-term survival is achieved by culturing slices at the air/liquid interface, either by continuously rotating the preparation (roller-tube cultures, Gahwiler, 1981 [54]) or by culturing them on semiporous membranes (Stoppini et al. 1991 [55]).

#### 1.5.2.3 Behavioral Models

Several models are used depending on the behavioral sequences that can be recorded and the facility to obtain and reproduce them. Let us consider birds and their singing behaviors, for instance. They vary among species, and most singing takes place during the breeding season. These songs are learned and not inherited. The avian vocal organ is called the syrinx; it is a bony structure at the bottom of the trachea. The most common species studied are the zebra finch (*Taeniopygia guttata*), the chaffinch (*Fringilla coelebs*), the European starling, and the robin. Studies on singing nervous centers confirmed the complexity of its production and some correspondences with human speaking.

In mammals, most of the behavioral analysis is done on rats or mice. Monkeys are also used: the squirrel monkey (*Saïmiri sciureus*) is interesting for vision, vocalization, and olfactory communication, the baboon (genus *Papio*) for manual coordination. In *Papio papio* light-induced epilepsy was discovered by R. Naquet. With photonic stimulation at 20–25 Hz spontaneous bifrontal discharges (mainly spike and wave) occur. This epilepsy has been found also in humans.

The most useful monkeys for electrophysiology have been *Macacus* and *Rhesus mulatta* that survive well in captivity. In 1965, E.V. Evarts (1926–1985) was able to record single cortical neurons from the motor area of monkeys controlling the flexion or the extension of their wrist. He found that some units are activated about 100 ms before the movements. This was a start of several studies on these animals to analyze cortical organization.

#### 1.5.2.4 Animals for Genetic Experiments

With this new discipline, new models were created with two opposite considerations. One was to find the simplest system for particular questions, the second was to be near to humans to examine different pathologies.

#### 1.5.2.5 *Caenorhabditis elegans*

Sydney Brenner as a molecular biologist created the first computer matrix analysis of nucleic acids. He was then interested to find the most simple animal to analyze biological activities. He explained his goal in a letter to Max Perutz: (5 June 1963): "The experimental approach I would like to follow is to attempt to define the unitary steps of development using the techniques of genetic analysis… Our success with bacteria has suggested to me that we could use the same approach to study the specification and control of more complex processes in cells of higher organisms. As a first stage, I would like to initiate studies into the control of cell division in higher cells, in particular to try to find out what determines meiosis and mitosis. In this work there is a great need to "microbiologize" the material so that one can handle the cells as one handles bacteria and viruses. Hence, like in the case of replication and transcription, one wants a model system." He then decided to work on a small nematode worm, *Caenorhabditis briggsae* (in fact, it was *C. elegans* that was selected). Brenner has demonstrated very clearly that a scientific project is exclusively linked with the model used. At his Nobel lecture on December 2002, he proposed the following title to congratulate his modest nematode: "*Nature's Gift to Science*".

Investigations with *Drosophila melanogaster* started even earlier – Thomas Hunt Morgan (1866–1945) had employed it in his genetic studies. He was a highly competent zoologist who had first worked at Naples marine station in 1894. By 1904, when Morgan took a professorship in experimental zoology at Columbia University, he was becoming increasingly focused on the mechanisms of heredity and evolution. In 1900 Hugo De Vries (1848–1935), a Dutch botanist, proposed that new species are created by mutation. Following the American entomologist C.W. Woodworth (1865–1940) and the American geneticist W.E. Castle (1867–1962), Morgan in 1908 started working on the fruit fly *D. melanogaster*. In 1909, a series of heritable mutants appeared, some of which displaying Mendelian inheritance patterns. In 1910, Morgan found a white-eyed mutant male among the red-eyed wild types. In 1915, he published with his team the seminal book *The Mechanism of Mendelian Heredity* and he was later awarded the Nobel Prize (1933) for his genetic studies of this small fly.

The zebrafish *Danio rerio*, a tropical freshwater fish, is an ideal model organism for studies of vertebrate development and gene function for various reasons: the genetic code is fully sequenced, the fish is easily observable and testable for developmental behavior, the mutants are well-characterized, the embryonic development is rapid, it takes only 3 days from egg to larva. Embryos that develop outside the mother are large and transparent. Drugs may be administered directly to the tank.

Today, the most intensively used animals for genetic studies are mice. Work on this model is so dominant that we have sometimes the impression to only characterize the neuroscience of mice!

## References

1. Brazier MAB (1984) A history of neurophysiology in the 17th and 18th centuries: from concept to experiment. Raven, New York
2. Brazier MAB (1988) A history of neurophysiology in the 19th century. Raven, New York
3. Clarke E, Jacyna LS (1987) Nineteenth-century origins of neuroscientific concepts. University of California Press, Berkeley
4. Clarac F, Ternaux JP (2008) Encyclopédie Historique des Neurosciences. De Boeck Université, Publ, Bruxelles
5. Finger S (1994) Origins of neuroscience. Oxford University Press, Oxford/New York
6. Ochs S (2004) A history of nerve functions: from animal spirits to molecular mechanisms. Cambridge University Press, Cambridge/New York
7. Shepherd GM (1991) Foundations of the neuron doctrine. Oxford University Press, New York
8. Ramón y Cajal S (1911) Histologie du système nerveux de l'homme et des vertébrés (trans: Dr. Azoulay), 2 vols. Maloine, Paris

9. Palay SL, Palade GE (1954) Electron microscopic observations of interneuronal and neuromuscular synapses. Anat Rec 118:335–336
10. Sherrington CS (1906/1947) The integrative action of the nervous system, 5th edn. Cambridge University Press, Cambridge, p 433
11. Adrian ED (1914) The all-or-none principle in nerve. J Physiol 47:460–474
12. Matthews BHC (1931) The response of a muscle spindle during active contraction of a muscle. J Physiol 72: 153–174
13. Gasser HS, Erlanger J (1927) The role played by the sizes of the constituent fibers of a nerve trunk in determining the form of the action potential wave. Am J Physiol 80:522–547
14. Hartline H (1935) The response of single sense cells to light of different wave lengths. J Gen Physiol 18:917–931
15. Mountcastle V (1978) An organizing principle for cerebral function: the unit mödel and the distributed system. In: Edelman GM, Mountcastle VB (eds) The mindful brain. MIT Press, Cambridge, MA
16. Hubel DH, Wiesel TN (1962) Receptive fields, binocular interaction and functional architecture in the cat's visual cortex. J Physiol 160:106–154
17. Kuffler SW (1953) Discharge patterns and functional organisation of mammalian retina. J Neurophysiol 16:37–68
18. Feldberg W, Fessard A (1942) The cholinergic nature of the nerves to the electric organ of the Torpedo (*Torpedo marmorata*). J Physiol 101:200–216
19. Couteaux R, Pecot-Dechavassine M (1970) Vesicules synaptiques et poches au niveau des "zones actives" de la jonction neuromusculaire. CR Sci (Paris) 271:2346–2349
20. Bazemore A, Elliott KA et al. (1956) Factor I and gamma-aminobutyric acid. Nature 178:1052–1053
21. Florey E (1961) Comparative physiology – transmitter substances. Annu Rev Physiol 23:501
22. Kravitz EA, Potter DD et al. (1962) Gamma-aminobutyric acid and other blocking substances extracted from crab muscle. Nature 194:382
23. Takeuchi A, Takeuchi N (1965) Localized action of gamma-aminobutyric acid on crayfish muscle. J Physiol 177:225–238
24. Watson JB (1913) Psychology as the behaviorist views it. Psychol Rev 20:158–177
25. Skinner BF (1938) The behavior of organisms. Appleton, New York
26. Berger H (1929) Über das Elektroenzephalogramm des Menschen. Arch Psychiatr Nervenkr 87:527–570
27. Moruzzi G, Magoun HW (1949) Brain stem reticular formation and activation of the EEG. Electroencephalogr Clin Neuro 1:455–473
28. Delafresnay JD (1955) Brain mechanisms and consciousness. Blackwell, Oxford
29. Olds J, Milner PM (1954) Positive reinforcement produced by electrical stimulation of septal area and other regions of rat brain. J Comp Physiol Psychol 47:419–427
30. Diamond MC, Krech D, Rosenzweig MR (1964) The effects of an enriched environment on the histology of the rat cerebral cortex. J Comp Neurol 123:111–120
31. Worden FG, Swazey JP, Adelman G (1973/1992) The neurosciences: paths of discovery I, 2nd edn. Birkhäuser, Basel, Boston
32. Cotzias GC, van Woert MH, Schiffer LM (1967) Aromatic amino acids and modification of parkinsonism. N Engl J Med 276:374–379
33. Hughes J, Smith T, Kosterlitz H, Fothergill L, Morgan B, Morris H (1975) Identification of two related pentapeptides from the brain with potent opiate agonist activity. Nature 258:577–580
34. Remler MP, Selverston AI, Kennedy D (1968) Lateral giant fibers of crayfish: locations of somata by dye injection. Science 162:281–283
35. Pitman RM, Tweedle CD, Cohen MJ (1972) Branching of central neurons: intracellular cobalt injection for light and electron microscopy. Science 176:412–414
36. Delcomyn F (1980) Neural basis of rhythmic behavior in animals. Science 210:492–498
37. McCulloch WS, Pitts W (1943) A logical calculus of the ideas immanent in nervous activity. Bull Math Biophys 5:115–133
38. Wiener N (1948) Cybernetics, or control and communication in the animal and the machine. MIT Press/Wiley, Cambridge MA/New York
39. Curtis DR, Watkins JC (1961) Analogues of glutamic and gamma-amino-*n*-butyric acids having potent actions on mammalian neurones. Nature 191:1010–1011
40. Llinas RR, Sugimori M (1980) Electrophysiological properties of in vitro purkinje cell dendrites in mammalian cerebellar slices. J Physiol 305:197–213
41. Bliss TVP, Lomo T (1973) Long-lasting potentiation of synaptic transmission in the dentate area of the anesthetized rabbit following stimulation of the perforant path. J Physiol 232:331–356
42. O'Keefe J, Nadel L (1978) The hippocampus as a cognitive map. Clarendon, Oxford
43. Engel AK, Fries P, Singer W (2001) Dynamic predictions: oscillations and synchrony in top-down processing. Nat Rev Neurosci 2:704–716
44. Rizzolatti G, Fabbri-Destro M, Cattaneo L (2009) Mirror neurons and their clinical relevance. Nat Clin Pract Neurol 5:24–34
45. Sutton S, Braren M, Zubin J, John ER (1965) Evoked-potential correlates of stimulus uncertainty. Science 150:1187–1188
46. Libet B, Gleason CA, Wright EW, Pearl DK (1983) Time of conscious intention to act in relation of onset of cerebral activity (readiness-potential). The unconscious initiation of freely voluntary act. Brain 106:623–642
47. Soon C, Brass M, Heinze H, Haynes J (2008) Unconscious determinants of free decisions in the human brain. Nat Neurosci 11:543–545
48. Hodgkin AL, Huxley AF, Katz B (1952) Measurements of current voltage relations in the membrane of the giant axon of *Loligo*. J Physiol 116:424–448
49. Arvanitaki A, Cardot H (1941) Les caractéristiques de l'activité rythmique ganglionnaire "spontanée" chez l'Aplysie. CR Soc Biol 135:1207–1211
50. Kennedy D, Selverston AI, Remler MP (1969) Analysis of restricted neural networks. Science 164:1488–1496
51. Miller JP, Selverston AI (1979) Rapid killing of single neurons by irradiation of intrecellular injected dye. Science 206:702–704
52. Hamburger V, Balaban M (1963) Observations and experiments on spontaneous rhythmical behavior in the chick embryo. Dev Biol 7:533–545

53. Otsuka M, Kravitz EA, Potter DD (1967) Physiological and chemical architecture of a lobster ganglion with particular reference gamma-aminobutyrate and glutamate. J Neurophysiol 30:725–752
54. Gahwiler BH (1981) Organotypic monolayer cultures of nervous tissue. J Neurosci Methods 4:329–342
55. Stoppini L, Buchs PA, Muller D (1991) A simple method for organotypic cultures of nervous tissue. J Neurosci Methods 37:173–182
56. Hökfelt T, Rehfeld JF, Skirboll L, Ivemark B, Goldstein M, Markey K (1980) Evidence for coexistence of dopamine and CCK in mesolimbic neurons. Nature 285:476–478
57. Cohen MJ, Jacklet JWJ (1967) Functional organization of motor neurons in an insect ganglion. Philos Trans R Soc Lond B 252:561
58. Winlow W, Kandel ER (1976) The morphology of identified neurons in the abdominal ganglion of *Aplysia californica* Brain Res 112:221–249
59. Nicholls JG, Baylor DA (1968) Specific modalities and receptive fields of sensory neurons in CNS of leech. J Neurophysiol 31:740–756
60. Fan RJ, Marin-Burgin A, French KA, Friesen WO (2005) A dye mixture (Neurobiotin and Alexa 488) reveals extensive dye-coupling among neurons in leeches. J Comp Physiol A 191:1157–1171
61. Clarac F, Pearlstein E (2007) Invertebrate preparations and their contribution to neurobiology in the second half of the 20th century. Brain Res Rev 54:113–161
62. Clarac F, Cattaert D, Le Ray D (2000) Central control components of a 'simple' stretch reflex. Trends Neurosci 23: 199–208
63. Gosgnach S, Lanuza GM, Butt SJB, Saueressig H, Zhang Y, Velasquez T, Riethmacher D, Callaway EM, Kiehn O, Goulding M (2006) V1 spinal neurons regulate the speed of vertebrate locomotor outputs. Nature 440: 215–219
64. Arhavsky YI, Berkinblit MB, Fukson OI, Gelfand IM, Orlovsky GN (1972) Recording of neurones of the dorsal spinocerebellar tract during evoked locomotion. Brain Res 43:272–275

# Evolution of Nervous Systems and Brains

# 2

Gerhard Roth and Ursula Dicke

The modern theory of biological evolution, as established by Charles Darwin and Alfred Russel Wallace in the middle of the nineteenth century, is based on three interrelated facts: (i) **phylogeny** – the common history of organisms on earth stretching back over 3.5 billion years, (ii) **evolution** in a narrow sense – modifications of organisms during phylogeny and underlying mechanisms, and (iii) **speciation** – the process by which new species arise during phylogeny. Regarding the phylogeny, it is now commonly accepted that all organisms on Earth are derived from a common ancestor or an ancestral gene pool, while controversies have remained since the time of Darwin and Wallace about the major mechanisms underlying the observed modifications during phylogeny (cf. [1]).

The prevalent view of **neodarwinism** (or better "new" or "modern evolutionary synthesis") is characterized by the assumption that evolutionary changes are caused by a combination of two major processes, (i) heritable variation of individual genomes within a population by mutation and recombination, and (ii) natural selection, i.e., selective environmental or genomic forces leading to better adaptation of those bearing the mutation and as a consequence to their greater differential reproductive success. The **modern synthesis** holds that evolutionary changes are gradual in the sense that larger, macro-evolutionary, changes are the sum of smaller, micro-evolutionary, changes. Other experts argue that such a gradualistic view of evolution (including "sexual selection" and "genetic drift") is incomplete; they point to a number of other and perhaps equally important mechanisms such as (i) neutral gene evolution without natural selection, (ii) mass extinctions wiping out up to 90 % of existing species (such as the Cambrian, Devonian, Permian, and Cretaceous-Tertiary mass extinctions) and (iii) genetic and epigenetic-developmental ("**evo-devo**") self-canalization of evolutionary processes [2]. It remains uncertain as to which of these possible processes principally drive the evolution of nervous systems and brains.

G. Roth (✉) • U. Dicke
Brain Research Institute, University of Bremen,
28334 Bremen, Germany
e-mail: gerhard.roth@uni-bremen.de; dicke@uni-bremen.de

## 2.1 Reconstruction of the Evolution of Nervous Systems and Brains

In most cases, the reconstruction of the evolution of nervous systems and brains cannot be based on fossilized material, since their soft tissues decompose, but has to make use of the distribution of neural traits in extant species. This is usually done by means of the phylogenetic or cladistic method as originally developed by the German entomologist Willi Hennig [3]. This method is based on whether a given character (or trait) represents an ancestral, *plesiomorphic*, or a derived, *apomorphic* state. The result of such an analysis is a "tree" called *cladogram* reflecting shared derived characters called *synapomorphies* or *homologies*. Characters used for the construction of a *cladogram* can be of any nature, but mostly are anatomical or biochemical. The standard criterion between competing cladograms is the **principle of parsimony** stating that the most likely evolutionary hypothesis is that minimizing the number of independent (*convergent* or *homoplastic*) steps of modification of the character under consideration. If different species of animals

C.G. Galizia, P.-M. Lledo (eds.), *Neurosciences - From Molecule to Behavior: A University Textbook*,
DOI 10.1007/978-3-642-10769-6_2, © Springer-Verlag Berlin Heidelberg 2013

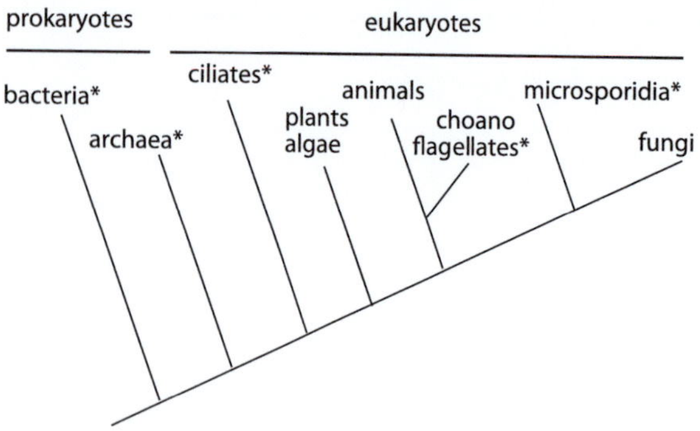


**Fig. 2.1** **"Tree of life"**. Lifeforms are classified into prokaryotes (unicellular, without a cell nucleus) and eukaryotes (cells with a nucleus), the latter being either unicellular (*marked by* *, only some examples shown) or multicellular (fungi, plants, and animals). The phylogeny of these groups remains an open field of research

coincide in the formation of a given character in detail (e.g., the structure of the inner ear or the genetic sequence of a membrane channel), then it appears more likely that these species are related and the coincidence of characters is due to common ancestry, i.e., that they are *homologous*. However, there is increasing evidence that convergent-homoplastic evolution is much more common than previously believed, which weakens the principle of parsimony.

In the reconstruction of the evolution of nervous systems and brains, however, the primary goal is not the construction of new cladograms, but an answer to the question whether within a group (taxon) of animals the presence or absence of a given character, e.g., the six-layered isocortex as found in all mammals, represents an ancestral (plesiomorphic) or derived (apomorphic) state. This requires the availability of well-established cladograms, which are based on non-neural characters in order to avoid circular conclusions. However, such well-established cladograms do not always exist, and therefore one often has to operate with competitive cladograms.

Figure 2.1 illustrates the present knowledge about the major groups of organisms – the "tree of life". According to most recent evidence, the earth was formed about 4.5 billion years ago (bya). First organisms appeared about 3.5 bya as prokaryotic bacteria and archaea (these are organisms without a cell nucleus). Unicellular eukaryotes (which bear a cell nucleus) originated 2.7–1.6 bya, simple multicellular animals (sponges) about 1 bya, and coelenterates about 700 million years ago (mya). First deuterostomes and first arthropods appeared 570 mya, cephalopods as well as first fishlike animals and proto-amphibians about 500 mya, insects about 400 mya, amphibians 360 mya, reptiles 300 mya, mammals 200 mya, and birds 150 mya. First human-like animals (australopithecines) appeared 4 mya, and modern humans (*Homo sapiens sapiens*) 200,000–150,000 years ago.

## 2.2 Organisms Without a Nervous System

Nervous systems and brains have a dual function, i.e., the maintenance of inner "vital" functions of the organism and the control of behavior of that organism within a given environment [4]. Unicellular organisms exert the same functions and exhibit remarkably complex behaviors, although they do not possess, by definition, a nervous system.

*Bacteria* sense nutritive substances (e.g., sugar) or toxins (e.g., heavy metal) in their environments through a large number of chemoreceptors as well as obstacles through mechanoreceptors [5]. This diverse information is integrated and, through a chain of complex chemical reactions, drives their flagella for movement. *Escherichia coli* has six flagella, each possessing a proton-driven motor, that are combined to one single superflagellum for forward propulsion ("run") as soon as receptors detect an increase in nutritive or a decrease in toxic substances. In the opposite case, the superflagellum disintegrates, and the single flagella move independently. As a consequence, *E. coli* starts "tumbling" and randomly changes its direction of movement until the receptors sense a new gradient, and a new "run" begins. The bacterium has a mini-memory, by which it can compare the incoming information with previous information, and this "knowledge" determines the behavior. Thus, even in these most primitive organisms we find the three basic components for adaptive control of behavior, i.e., a sensorium, a motorium and in between information storage and processing. Other bacteria or archaea like *Halobacterium* possess light-sensitive spots that make them swim toward sunlight.

Unicellular eukaryotes such as *Paramecium* or *Euglena*, despite their unicellular organization, exhibit a much more complex control of behavior than bacteria or archaea [6]. They possess either flagella composed of microtubules and attached to their front, which bend to perform a breakstroke, or cilia that cover the entire body and are able to exert coordinated movements, or pseudopodia. They gather information about their environment through voltage-gated, hyperpolarizing potassium and depolarizing calcium ion channels, while voltage-gated sodium channels are absent and found only from planarians, possibly cnidarians onwards. The calcium and potassium channels are used, among others, for the release of forward and backward movement. Many protozoans respond to chemical, tactile, temperature and visual stimuli. *Euglena* and *Paramecium* and other unicellular eukaryotes possess light-sensitive organelles for phototaxis. It is debated whether they already possess an intracellular system for central movement coordination.

Figure 2.2 illustrates the phylogeny of the Metazoa, i.e., multicellular organisms, which are divided into Nonbilateria and Bilateria. Porifera (sponges, 8,000 species, all aquatic) are the simplest metazoans. They possess "independent effectors" or *myocytes*, which have sensori-motor functions and directly respond to stimuli, but are not electrically excitable [7]. Sponges can regulate the water stream through openings (ostia) inside their body by modifying the diameters of these ostia. The inner surface of the ostia is covered with choanocytes carrying one flagellum each, and their movement can drive the water through the ostium in a coordinated fashion. The presence of true nerve cells is debated, because there is no convincing evidence for electrical signal conduction.

## 2.3 Nervous Systems in Eumetazoans

Eumetazoans comprise all metazoans except the Porifera. The evolutionary origin of the first nerve cells among eumetazoans is still a matter of debate (cf. [7, 8]). One assumption is that sensory and nerve cells originated from neuromuscular cells, while other authors postulate an independent origin of sensory, nerve, and muscle cells from epithelial cells. The "paraneuron" concept proposes the evolution of nerve cells from secretory cells. Even unicellular eukaryotes, plants, and non-neuronal cells display many features of nerve cells such as membrane potential, transmitters and other neuroactive substances, membrane receptors, ion channels, many chemical processes relevant for "neuronal information processing", and even action potentials – all of which being more than 1 billion years old and thus older than nerve cells and nervous systems. One remarkable exception is the voltage-gated sodium channel, which

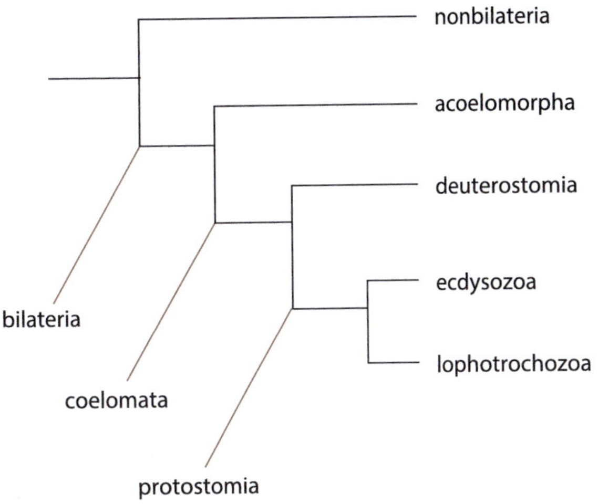


**Fig. 2.2** **Phylogeny of metazoans,** i.e., multicellular animals. Metazoans comprise non-bilaterally organized animals (such as sponges and coelenterates) and bilaterally organized animals, which include those *without* a secondary body cavity (or coelom) – the Acoelomorpha, and those *with* a coelom – the Coelomata; for further explanation see text

is found first in planarians, perhaps already in cnidarians.

### 2.3.1 Coelenterata (Cnidaria, Ctenophora)

Coelenterates are nonbilaterian eumetazoans. Today they are considered two independent phyla, Cnidaria and Ctenophora (together about 11,000 species) and exhibit both the simplest types of nervous system (nerve nets) as well as relatively complex forms, i.e., radially symmetric nervous systems (Fig. 2.3). A central nervous system is absent [7]. The phylum Cnidaria comprises the sessile Anthozoa (sea anemones, corals), Scyphozoa (jellyfish), Cubozoa (box jellies), and Hydrozoa (hydras). Epidermal nerve nets are found in sessile hydrozoans like the freshwater polyp *Hydra*. There is a concentration (nerve rings) around the mouth and the peduncle of this animal. Complex sense organs are absent, but *Hydra* responds to mechanical, chemical, visual, and temperature stimuli. The free-swimming medusa forms of scyphozoans, in contrast, possess complex circular nervous systems inside the rim of the umbrella – the inner *subumbrellar* nerve ring contains large bipolar "swim motor neurons" for synchronous umbrella contraction, and the outer *exumbrellar* nerve ring consists of small multipolar sensory cells which are in contact with light-sensitive cells in mouth and tentacles. Both nerve rings are

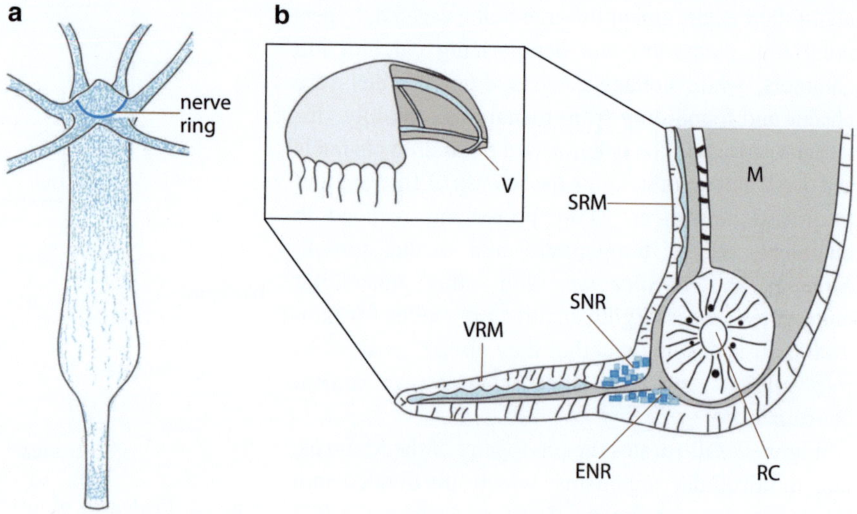


**Fig. 2.3 The nervous system of the polyp *Hydra* (a) and radial section through the umbrella of a hydromedusa (b).** *ENR* exumbrellar nerve ring, *M* mesogloea, *RC* ring canal, *SNR* subumbrellar nerve ring, *SRM* subumbrellar ring muscle, *V* velum, *VRM* velar ring muscle (After Satterlie and Spencer [9] with permission)

interconnected. Sensory organs are the ocelli (pigment spots, cup ocelli, or even "eyes" with biconvex lenses), statocystes (balance organs), and "rhopalia", i.e., complex clublike balance organs, often combined with photo- and chemoreceptors, which initiate the rhythmic contraction of the medusas. The nervous system of cnidarians is characterized by the dominance of electric synapses, although chemical synapses are likewise present. Chemical transmission is mostly exerted by a number of neuropeptides (e.g., FMRFamides and RFamides; cf. [10]), although there is evidence of cholingergic, serotonergic, dopaminergic, and glutamatergic transmissions in different cnidarians species.

## 2.3.2 Bilateria

Animals with bilateral symmetry ("Bilateria") comprise the three major groups of phyla, the Acoelomorpha, Protostomia, and Deuterostomia, the latter two together forming the "Coelomata", i.e., organisms with a secondary body cavity (see Fig. 2.2). However, the phylogeny of the Bilateria has not yet been firmly consolidated.

### 2.3.2.1 Acoelomorpha

Acoelomorpha include very small bilateral animals resembling flatworms and were previously assigned to the phylum Platyhelminthes (see below). They possess a diffuse subepidermal nerve net resembling that of *Hydra* and representing the simplest form of a bilateral nervous system. Since such diffuse subepidermal nerve nets are likewise found in other flatworm-like organisms, it is not certain whether this type has evolved independently, e.g., via secondary simplification of more complex types [11] or represents the ancestral form of all bilateral nervous systems [12].

### 2.3.2.2 Protostomia

According to molecular phylogeny (cf. Fig. 2.2), protostome phyla are grouped into the Lophotrochozoa – animals carrying a lophophor (a complex feeding organ) or possessing a trochophora larva, and the Ecdysozoa, i.e., with ecdysis (see below). Many phyla of the lophotrochozoans include small and often sessile organisms such as the Bryozoa (also called Ectoprocta), the Brachiopoda, Echiura, Entoprocta, or Nemertini – all with simple to very simple (possibly simplified) nervous systems consisting of two or more nerve cords extending from a supraesophageal ganglion through the elongated body, which are connected by a number of commissures. In the following, only the larger phyla are described.

#### Lophotrochozoa

Platyhelminthes (flatworms; 25,000–30,000 species), comprise a number of species previously called "Turbellaria" (whereas other "turbellarians" are now included in the Acoela, which are not considered to belong to the Platyhelminthes), and the endoparasitic tapeworm groups Cestoda (tapeworms; 3,500 species) and Trematoda (flukes; about 20,000 species). The phylogeny of platyhelminths remains unresolved, however. "Turbellarian" platyhelminths may possess very

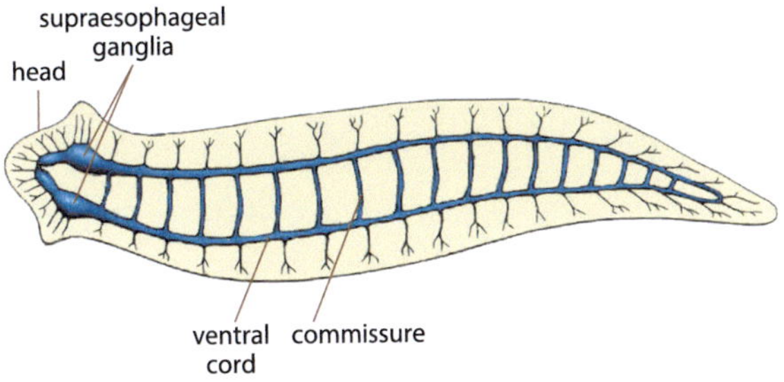


**Fig. 2.4** **Nervous system and brain of a flatworm.** For further information see text

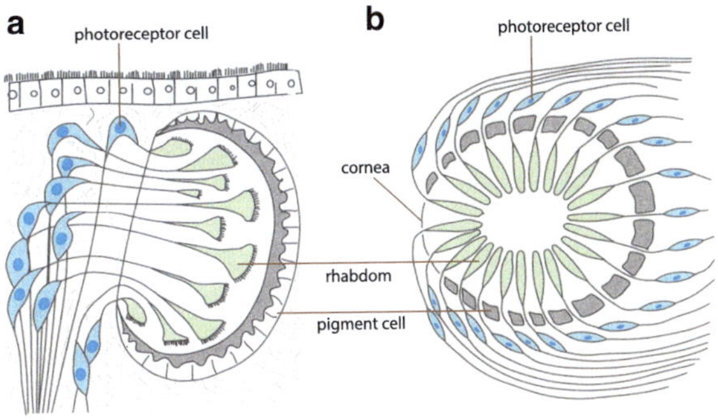


**Fig. 2.5** **Eyes of "turbellarian" flatworms.** (**a**) Inverse pigment cup eye of a freshwater planarian (After Bullock and Horridge [8] and Paulus [13]). (**b**) Everse eye of a land planarian (After Bullock and Horridge [8] with permission)

simple nervous systems resembling the subepidermal diffuse nerve net found in the Acoela. In other forms, there is a supraesophageal ganglion giving rise to dorsal and ventral longitudinal cords connected by commissural tracts. The longitudinal cords can either consist entirely of fibers or of fibers forming regularly arranged ganglia (Fig. 2.4). The most complex nervous systems are found in predatory flatworms (planarians) such as *Notoplana* and *Stylochoplana* with cerebral ganglia consisting of five different "brain masses". Flatworms possess a variety of sense organs such as tactile or chemoreceptors on the head and all over the body, statocysts and inverse or everse pigment pit eyes containing several hundred photoreceptors (Fig. 2.5). Some of them are found as a pair of ocelli on the head, other terrestrial flatworms have more than 1,000 ocelli. Due to their endoparasitic life style, trematodes and cestodes have a simplified nervous system consisting of a simple "brain" and a varying number of longitudinal fiber tracts.

Annelida (segmented worms; about 18,000 species) are another large lophotrochozoan group. They are divided into Polychaeta (with hairs, chetae, and leglike parapodia) and Clitellata, the latter comprising oligochaetes (earthworms) and hirudineans (leeches), both without hairs and parapodia. Annelids possess a paired "ladder-type" central nervous system (Fig. 2.6). In its simplest state this structure consists of a cerebral or supraesophageal nerve ring or ganglion giving rise to paired ventral cords with a pair of ganglia per body segment connected by transverse connectives (anastomoses). In annelids, the cerebral ganglion has variously undergone an increase in complexity; in predatory polychaetes it has developed into a three-partite brain resembling the proto-, deuto-, and tritocerebrum of the insect brain (see below). Here, but also in some oligochaetes, the first segments of the ventral nerve cord are often fused into a subesophageal ganglion. In the oligochaetes, we find a modest, and in hirudineans a massive simplification of this basic organization. Within the ventral nerve cord of most oligochaetes and some polychaetes there are giant fibers with very fast conduction velocity (three in oligochaetes) separated from the thinner fibers. Annelids possess a large variety of tactile and chemosensory organs, feelers or antennae, palps, and one ciliated "nuchal organ" possibly involved in light detection. Other light-sensitive organs range from very simple pigment spots and eye pits to compound eyes and lens eyes with accommodation mechanism in some predatory polychaetes, and have evolved independently of similar eye types in other animal groups (Fig. 2.7).

Molluska are the largest lophotrochozoan group (100,000 or more species). Their phylogenetic relationships are still unresolved. Besides several smaller groups, there are three large taxa, i.e., Gastropoda (snails and slugs, about 70,000 species), Bivalvia (clams, oysters, mussels, scallops; 10,000–20,000 species), and Cephalopoda (cuttlefish, squid, and octopods; about 800 species).

The molluskan nervous systems range from relatively simple (or simplified) forms resembling those found in acoelans to the most complex ones among invertebrates, in the cephalopods. The basic pattern is a *tetraneural* nervous system consisting of a cerebral ganglion, which gives rise to two dorsal pleurovisceral and two ventral pedal nerve cords. In the ancestral state, nerve cell bodies are not concentrated in ganglia, but are dispersed throughout the cords. The formation

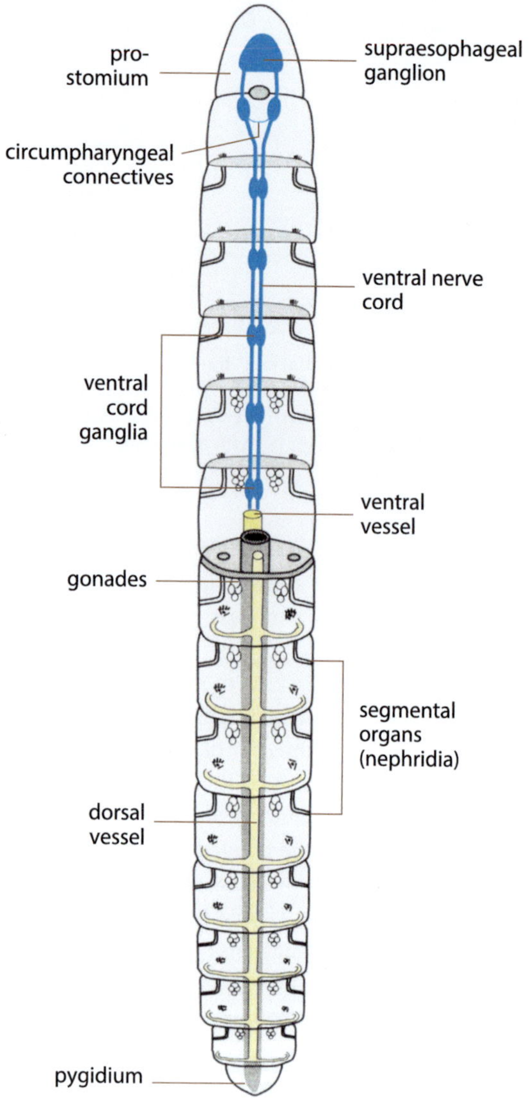


**Fig. 2.6** **Paired "ladder-type" central nervous system of annelids** (After Hennig [14] with permission)

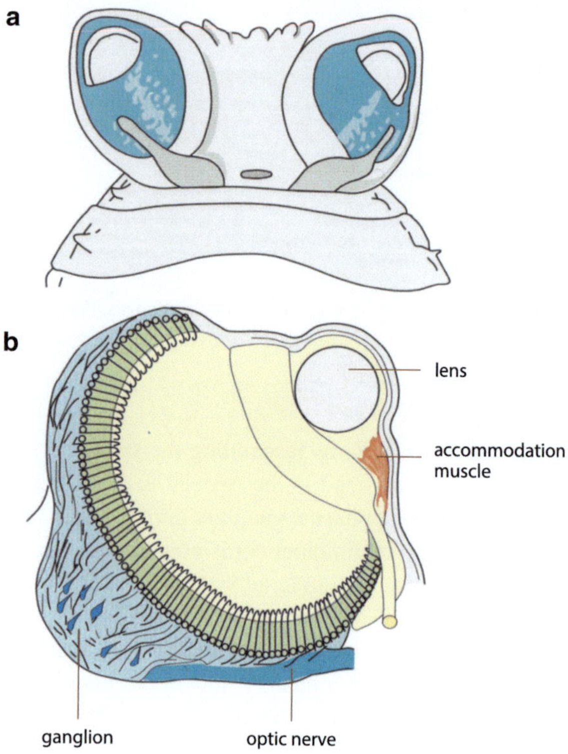


**Fig. 2.7** **Camera eye of the polychaete *Alciope*** with lens accommodation mechanism. (**a**) Ventral view. (**b**) Cross section through the optical axis (After Bullock and Horridge [8] with permission)

of ganglia in mollusks is a derived state that occurred independently of the formation of ganglia in other forms such as annelids and arthropods.

### Gastropoda

The nervous system of snails and slugs consists of four nerve cords – hence the term *tetraneural* nervous system – with maximally six pairs of ganglia and mostly one unpaired visceral ganglion (Fig. 2.8a). The nerve cords are mostly linked by commissures. The paired cerebral ganglia connected by a commissure are located around the esophagus and process information to and from the eyes, to statocysts, head tentacles, skin, and muscles of the lip, head, and sometimes penis region. One pair of buccal ganglia with a commissure is situated below the esophagus and innervates the pharynx, salivary glands, a nerve plexus of the esophagus and the stomach. One pair of pleural (i.e., lung membrane) ganglia without a commissure is connected by cords with the cerebral, buccal, and parietal-visceral ganglia. The pedal ganglia innervate feet muscles and skin. The cerebral, pleural, and pedal ganglia together form the "brain". The supra- and subintestinal ganglia innervate the gills, the "osphradium" (an olfactory organ) and parts of the mantle and skin; one pair of parietal ganglion (not present in all gastropods) innervates the lateral walls of the body. Finally, the unpaired visceral ganglion supplies the caudal region of the gut, anus, and neighboring regions of the skin and body wall, sexual organs, kidney, liver, and heart. It completes the "visceral loop", i.e., the chain of ganglia and cords from the pleural to the visceral ganglion.

A fusion of ganglia, mostly of the "visceral loop", is observed in many gastropods, e.g., in air-breathing landsnails. The most highly developed gastropod "brain" is found in *Helix pomatia* (the Roman or Burgundy snail). It consists of a protocerebrum with globuli (i.e., globe-like neuronal contact zones) and dense neuropils, a mesocerebrum, and a postcerebrum with pleural and pedal lobes. This organization is remarkably similar to that of other invertebrates with complex brains, but has presumably evolved independently.

Gastropods have chemoreceptive and mechanoreceptive sense organs distributed all over the body. Complex sense organs comprise statocysts, eyes ranging from widely open pit eyes (*Patella*), pinhole eyes (*Trochus*) to lense eyes (*Helix*), and chemosensitive osphradia in the mantle near the gills.

Some sea slugs have gained fame in modern neurobiology, e.g., the Californian "sea hare" *Aplysia californica*, which possesses some very large nerve cells that can be detected with the naked eye and are well-suited for studies of neuronal information processing and learning processes [15].

### Bivalvia

*Bivalves* have a secondarily simplified nervous system with only three pairs of ganglia with an emphasis on the visceral ganglion, which is often fused with the parietal ganglia. In most species, the rostralmost ganglion is a fused cerebral, pleural, and buccal ganglion. Some bivalves, e.g., the scallop *Pecten*, have eyes on the rim of the mantle, often with a complex anatomy (e.g., a distal and proximal retina).

### Cephalopoda

Cephalopods have highly developed nervous systems characterized by fusion of ganglia and subsequent development into lobes forming a complex brain around the esophagus. There are lobes that correspond to the cerebral, buccal, labial, pleural, and visceral ganglia of other mollusks, while innovative structures are the central optic-visual, olfactory, and peduncular ganglia as well as peripheral branchial and stellar ganglia (cf. [16, 17]).

The well-known *Nautilus* possesses a relatively simple brain without bulging supraesophageal lobes and with unfused subesophageal lobes, which probably represents the ancestral state of cephalopods. Members of the subclass Coleoidea comprising cuttlefish, squids, and octopods have much more complex brains. The most complex nervous system and brain of all invertebrates is that of the octopus (Fig. 2.8b, c). Its nervous system contains about 550 million neurons, 350 of which are located inside the eight arms, 120–180 million neurons in the giant optic lobes, and 42 million neurons in the brain. The latter encircles the esophagus and is composed of 38 lobes. The supraesophageal part is divided into 16 lobes and contains the mass of neurons. It has a ventral portion involved in the control of feeding and locomotion, and a dorsal portion exerting

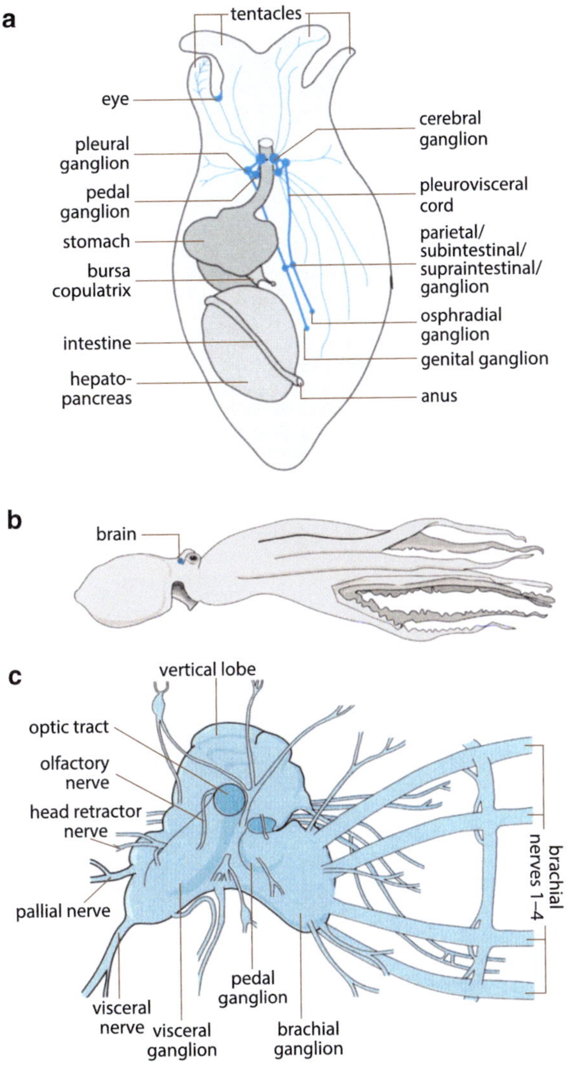


**Fig. 2.8** **Central nervous system of mollusks.** (**a**) Nervous system of the sea slug *Aplysia*. (**b**) Site of the brain of *Octopus*, (**c**) *Octopus* brain and nerves (After Bullock and Horridge [8] with permission)

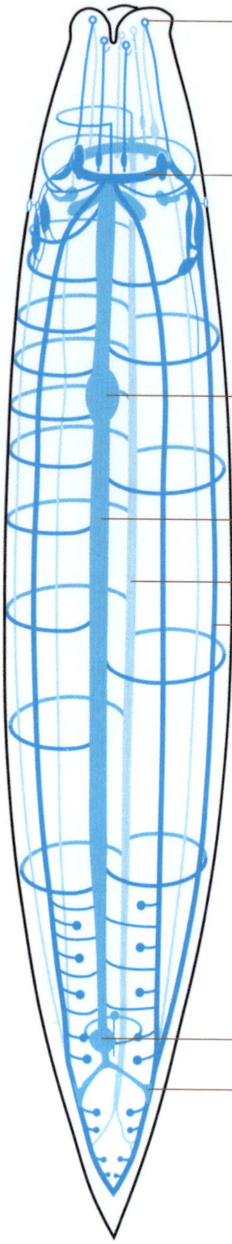


**Fig. 2.9** **Central nervous system of the nematode *Ascaris*,** ventral view (After Bullock and Horridge [8] with permission)

cognitive functions, mostly visual and tactile-chemosensory. The vertical lobe is considered the most complex part of the octopus brain. It is composed of five lobules and contains about 26 million neurons. The vertical lobe is closely connected to the subvertical lobe containing about 800,000 neurons, and the interaction of both lobes, processed by a spectacularly regular network of millions of fibers arranged in a rectangular fashion, is regarded as the neural basis of the astonishing learning and memory capacity of *Octopus* [18]. Likewise complex are the giant optic lobes exhibiting a laminar neuropil resembling the cortex of mammals. They process the visual information arriving from the large lens eyes. These eyes are capable of lens accommodation and pupil contraction achieved by muscles, and have a striking similarity to the vertebrate eye, although they are the product of convergent evolution.

## Ecdysozoa

The Ecdysozoa comprise all invertebrate animals that shed their exoskeleton – a process called "ecdysis". According to present taxonomy they include eight phyla, the largest of them being the Nematoda and the Arthropoda. The smaller groups mostly have relatively simple or simplified brains and simple sense organs.

### Nematoda

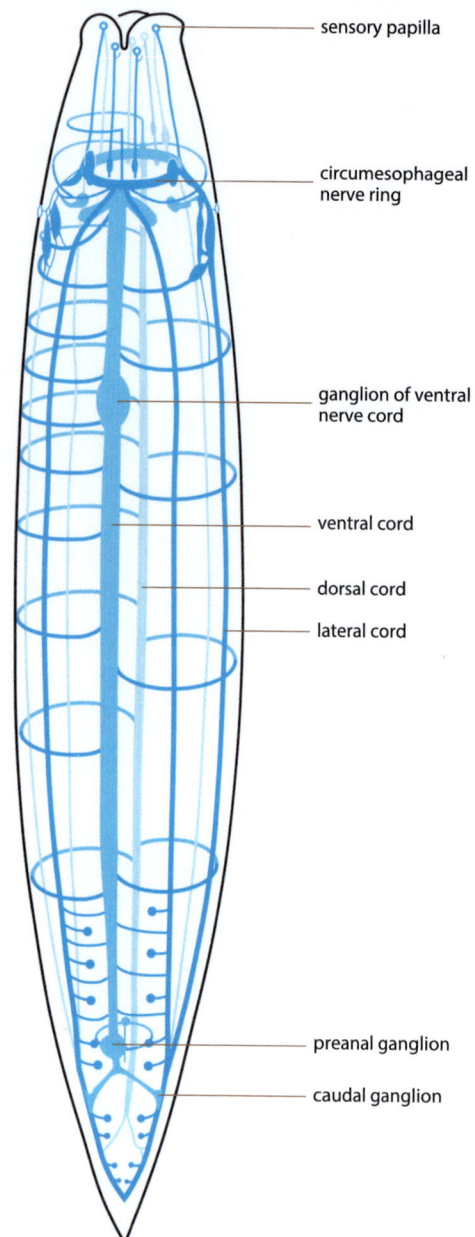

Nematodes (roundworms; about 28,000 species) are the most numerous multicellular animals on Earth. Probably due to their predominantly parasitic lifestyle, they have very simple nervous systems (Fig. 2.9) consisting of a nerve ring around the esophagus and a number of ganglia connected to this ring. Four to twelve ventral cords originate from the ring and are irregularly connected by half-sided commissures. Local ganglia and nerves are found in the caudal gut and anal region. Some nerves extend from the esophageal nerve ring to the sense organs in the "head" region such as sensory papillae and bristles. Other sense organs are chemoreceptive organs called "amphidia".

The tiny nematode *Caenorhabditis elegans* has become a model organism in molecular and developmental neurobiology by the work of the South African molecular neurobiologist Sydney Brenner and colleagues [19, 20], a reason being the fact that it has a very simple nervous system composed of exactly 302 neurons. The basic genetic features and the connectivity of this nervous system was completely mapped by those authors and subsequent studies explored the neural and molecular mechanisms responsible for a variety of behaviors shown by *C. elegans* [21].

### Arthropoda

Arthropods are by far the largest (about 1.2 million species described, more than ten million estimated) and most diverse group of animals. They are divided into protoarthropods (onychophorans, possibly tardigrades) and euarthropods (chelicerates, crustaceans, myriapods, and hexapods, the latter three taxa called "mandibulates"). Their taxonomy is not fully established.

Like annelids, arthropods have a ventral, regularly segmented, paired "ladder-type" nerve cord. Based on the new taxonomy of protostomes mentioned above, this organization either has evolved independently in the lophotrochozoans and ecdysozoans from an unsegmented "ur-bilaterian" nervous system or was ancestral and has been lost in many cases [11]. In all arthropods, the first ganglia have fused into a complex brain. In mandibulates there are three major brain divisions, i.e., a proto-, deuto-, and tritocerebrum. The protocerebrum is associated with the paired optic lobes, the deutocerebrum with the first and the tritocerebrum with the second pair of antennae. Mandibulates display a subesophageal ganglion having formed by fusion of the three first ventral ganglia. They supply the mouth region and mandibles, in crustaceans the first and second maxillae, in insects the maxillae, mandibles and labium. Caudal ganglia of the ventral cords exhibit a strong tendency to fuse and to form specialized abdominal structures [22].

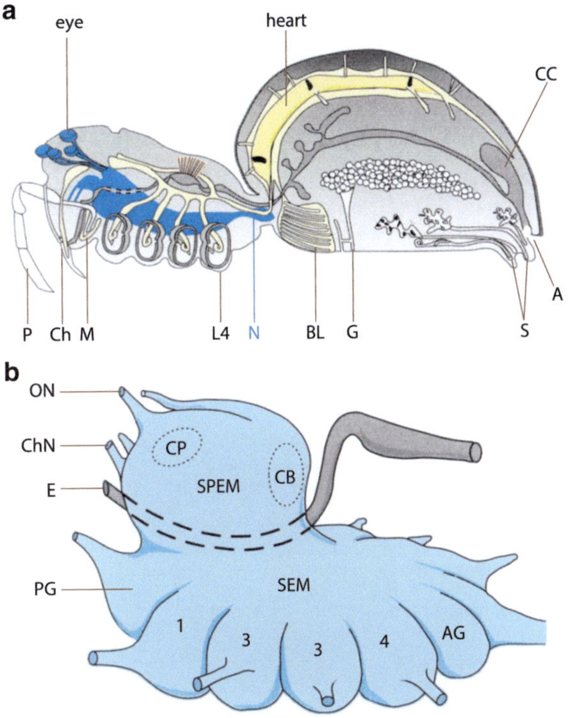


**Fig. 2.10** **Central nervous system of arachnids.** (**a**) Site of the CNS (*blue*) inside the body of the house spider *Tegenaria*, side view (After Kästner [23]). (**b**) Closer view of the CNS. Abbreviations: *1–4* leg ganglia, *A* anus, *AG* abdominal ganglia, *BL* book lung, *CB* central body, *Ch* chelicerae, *CP* corpora pedunculata, *CC* cloacal chamber, *ChN* cheliceral nerve, *E* esophagus, *G* aperture of gonads, *L4* insertion of leg 4, *M* mouth, *N* nerve to abdomen, *ON* optic nerve, *P* pedipalp, *PG* pedipalp ganglion, *S* spinneret, *SPEM* supraesophageal mass, *SEM* subesophageal mass (After Foelix [24] with permission)

### Chelicerata

Extant Chelicerata (about 100,000 species) comprise the Arachnida (spiders, scorpions, mites, and others) and Xiphosura (horseshoe crabs). They all possess specialized feeding appendages called chelicerae (claw horns), while lacking antennae. The CNS of the chelicerates is characterized by the absence of a deutocerebrum because of lack of antennae; the tritocerebrum supplies the chelicerae. In xiphosurans, scorpions, and araneans (spiders) there is an increasing tendency towards fusion of ganglia during ontogeny. In many species of these groups the entire chain of ventral ganglia forms a compact mass around the mouth, in the araneans below the brain.

The brain (supraesophageal ganglion) of arachnids consists of a protocerebrum and tritocerebrum. In the anterior median part of the protocerebrum, corpora pedunculata ("mushroom bodies") are found, which – in contrast to insects – are exclusively visual neuropils associated with the secondary eyes. A central body is found in the posterodorsal part and is probably an integrative center for visual information from the main eyes. The homology of both the corpora pedunculata and the central body of arachnids with those of insects remains uncertain [22]. The tritocerebrum is the ganglion linked with the chelicerae and is often fused with the subesophageal mass supplying the legs. This mass is found below the brain. It consists of a highly variable number of fused ventral ganglia (16 in araneans) (Fig. 2.10).

Arachnids have a large variety of sense organs. There are vibration-sensitive slit-like lyriform organs involved in the detection of vibration and in proprioception and hair sensilla called "trichobothria" on the legs and lateral and dorsal parts of the body, which are involved in the detection of airborne vibration and air currents. Species differ in number of main and secondary eyes. The main eyes are considered homologous to the ocelli, and the secondary eyes to the compound eyes of insects.

### Crustacea

Crustaceans (crabs, lobsters, crayfish, shrimp, krill, and barnacles; totaling >50,000 species) with the largest group Malacostraca (crabs, lobsters, crayfish) have

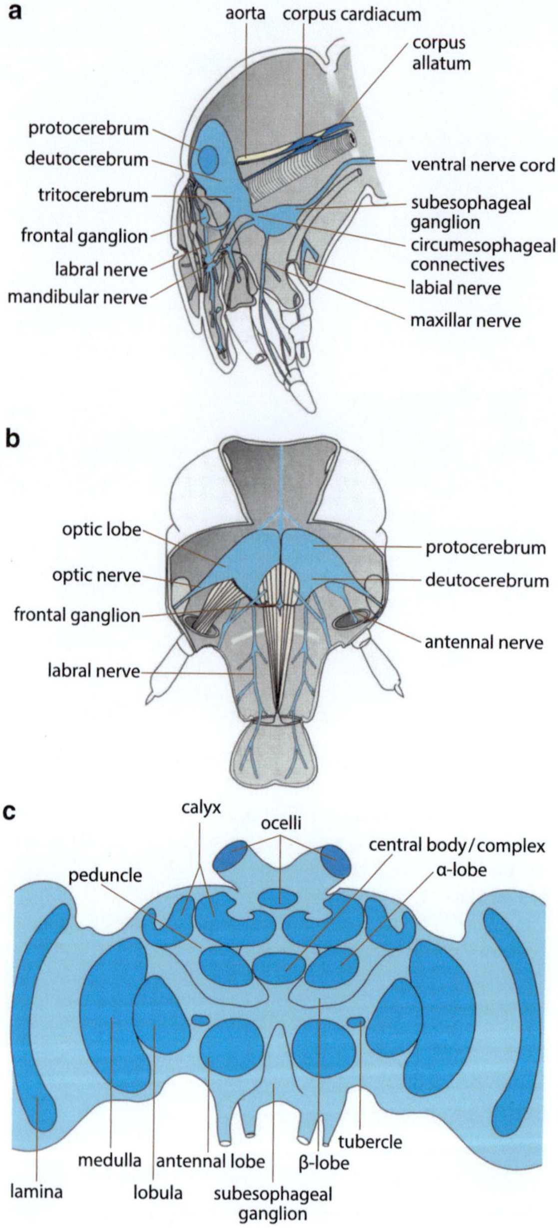


**Fig. 2.11** **Insect brain**. (**a**) Lateral view and (**b**) ventral view of the brain and nerves in the scorpionfly *Panorpa* (After Bullock and Horridge [8]). (**c**) Schematic of the brain of a honey bee (After Mobbs [25] with permission)

a typical paired "ladder-type" nerve cord as the ancestral form. The brain (supraesophageal ganglion) is linked via two connectives with the ventral nerve cords. The protocerebrum consists of two lateral optic lobes and the median protocerebrum containing the anterior and posterior optic neuropils, the protocerebral bridge, and the central body. Neuropils of the optic lobes are highly variable. In decapod crustaceans (e.g., crabs), there are additional visual neuropils within the optic lobes, i.e., a terminal medulla and the so-called hemiellipsoid bodies, which some authors conceive to be homologous to the insect mushroom bodies (see below). Both include a varying number of complex neuropils; most of them contain glomeruli. The hemiellipsoid bodies and some of the other neuropils have connections with the accessory and olfactory lobes of the deutocerebrum. The deutocerebrum contains the medial and lateral neuropils receiving vestibular and mechanosensory input from the first antennae, the olfactory and parolfactory lobes (the latter with unknown input), and the lateral glomeruli. The tritocerebrum receives information from the second antennae and sends motor nerves to them. There are strong differences in the degree of fusion of ventral cord ganglia. A subesophageal ganglion controlling mouth appendages is found in many malacostracans, and the fusion of ganglia is maximal in crabs.

Crustaceans have a large number of sense organs. These comprise proprioceptive mechanoreceptors of leg joints, the chordotonal organs. The surface including distal limbs and antennae is covered with mechano- and chemoreceptors possessing sensilla or setae. Only malacostracans have vestibular organs. An unpaired nauplius eye, frontal simple eyes and compound eyes are found, the latter are located either directly on the head or on eyes elevated on movable stalks. The compound eyes can consist of a few or several 1,000 ommatidia.

### Insecta (Hexapoda)

Insects are the largest group of arthropods, with an estimated 6–10 million species, most of which are terrestrial (as opposed to crustaceans). The nervous system of insects consists of a brain (supraesophageal ganglion) and ventral nerve cords (Fig. 2.11). The brain, formed by fusion of the first three ganglia, consists of a large protocerebrum, a smaller deutocerebrum, and a very small tritocerebrum. Fiber tracts connect the brain with the subesophageal ganglion, constituted by fusion of the first three ventral cord ganglia. The protocerebrum consists of two hemispheres, which are continuous with the lateral optic lobes receiving input from the compound eyes. Terminal fields of the nerves from the ocelli are found in the posterior median protocerebrum. The central complex and the *corpora pedunculata* or "mushroom bodies" (MB), are located in the median protocerebrum (see below). The MB receive olfactory input from the antennae via the antennal lobes situated in the deutocerebrum and the antennocerebral tract (ACT). In hymenopterans (bees, wasps, ants) the MB also receive visual projections from the optic lobes, which terminate in the calyces of the MB. The optic tubercle is also found in the median protocerebrum which receives visual input from the optic lobes. These structures are connected with the ventral cords via descending tracts. The smaller deutocerebrum is connected with the protocerebrum by a supraesophageal commissure. Mechanoreceptive fibers terminate in its dorsal lobe. Here, the antennal lobe is found as terminal field of olfactory afferents from the antennae. Projection neurons of the antennal lobe send axons to the MB and to the protocerebral lobe of the protocerebrum via ACT. The deutocerebrum gives rise to the sensory and motor antennal nerves. The small tritocerebrum is related to taste perception and origin of the frontal connectives and the labral nerves.

The chain of ventral cord ganglia consists of subesophageal, thoracic, and abdominal ganglia. The first innervates the mandibles, maxillae, and labium as well as the neck musculature. It is also involved in the innervations of the salivary glands, the corpora allata (endocrine glands producing the juvenile hormone), and the frontal ganglion and is considered a higher motor center for the initiation and control of behavior. Most insects have three thoracic ganglia: a pro-, meso-, and metathoracic ganglion supplying legs and wings, if present, with sensory and motor nerves. Abdominal ganglia (11 in the embryonic stage) are reduced and fused during development.

The visual system of insects comprises the retina of the compound eye and three optic neuropils, the lamina, medulla, and lobula complex, which in flies and butterflies is divided into a lobula and lobula plate. In addition to the compound eyes, insects have dorsal eyes, so-called ocelli, which are simple lens eyes and thought to exert steering functions during walking and flight.

Antennae bear mechanosensitive, olfactory, hygroreceptive, and temperature-sensitive receptors. The neuropil of the antennal lobe in the deutocerebrum contains a species-specific number of glomeruli, in which sensory afferents and interneurons make contacts. Macroglomeruli are found in some male insects related to sexual pheromone processing.

The MBs in hymenopterans are composed of one calyx or two calyces (a medial and a lateral one) and a peduncle consisting of two lobes, alpha and beta. The somata of neurons ("Kenyon cells", bees having around 300,000) together with their axons ("Kenyon fibers") form the peduncle; the neuronal somata are located in the outer rim of the calyx. The Kenyon fibers split up – one collateral enters the $\alpha$, and another the $\beta$ lobe. In the honeybee, the calyces exhibit three vertically arranged regions, the lip, collar, and basal ring region. Afferents from the antennal lobe terminate in the lip region, afferents from the medulla and lobula of the optic neuropils terminate in the collar region, and the basal ring region receives collaterals from both afferents as well as from the subesophageal ganglion. The $\alpha$ and $\beta$ lobes send fibers to the median protocerebrum between the two MBs, the protocerebral lobe lateral to the MB, the contralateral MB, the optic tubercle and back to their own calyces. In hymenopterans, the MB represents a highly complex multimodal center that forms the neural basis of processing and integrating olfactory/visual and mechanosensory information and enables learning (mostly olfactory and visual), complex cognitive functions, and complex behavior such as navigation [26]. Their output has sensory, movement-related, and sensorimotor functions. MBs differ substantially across species. For example, *Drosophila* only has a single cup, and spatially segregated $\alpha$, $\beta$, and $\lambda$ lobes.

The central complex of insects consists of four neuropils, i.e., the protocerebral bridge, an upper division (in *Drosophila* called fan-shaped body), a lower division (in *Drosophila* called ellipsoid body), and the paired nodules. It receives strong visual as well as mechanosensory input, but only weak input from the MB. Their precise function is still unclear, but the central complex has to do with premotor integration, orientation, and control of complex locomotion and path integration.

The homology of the MB and the central bodies (CB) or central complexes in arthropods is debated. The CB of insects and crustaceans are probably homologous, whereas homology with CB of chelicerates is controversial. The same holds for the MB in insects and crustaceans (here called hemiellipsoid body), on the one hand, and of chelicerates, on the other, partly because in the latter, the MB receive only visual input. Some authors place the MB of chelicerates closer to those of onychophorans [27]. Accordingly, the MB of

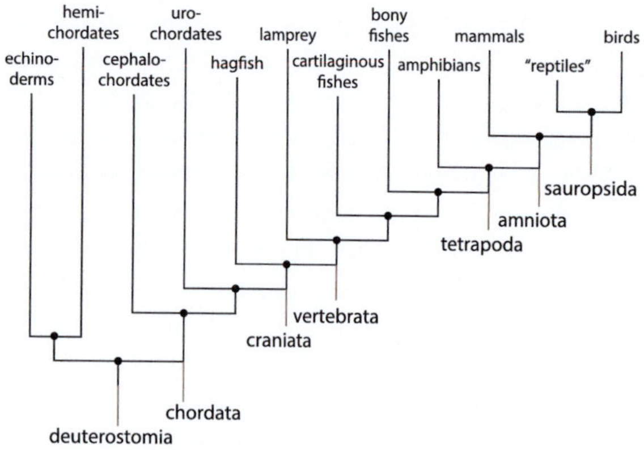


**Fig. 2.12** **Phylogeny of deuterostomes**

onychophorans and chelicerates would have evolved independently of those of the mandibulates or underwent substantial functional and structural changes in their ancestral organization.

### 2.3.2.3 Deuterostomia

As illustrated in Fig. 2.12, the superphylum Deuterostomia, i.e., animals with a "secondary mouth", comprise the phyla Echinodermata, Xenoturbellaria (not shown), Hemichordata, and Chordata. The split between protostomes and deuterostomes has happened in Precambrian time about 560 mya.

#### Echinodermata

The Echinodermata (starfish, sea urchins, sand dollars, brittle stars, sea cucumbers, and feather stars, together about 6,300 species) are radially symmetric (pentaradial) animals at the adult stage. The radial symmetry of body and nervous system appears to be secondary, since echinoderms most probably derive from bilaterally symmetric ancestors, as is reflected by their bilateral larva. The nervous system is composed of a sensory ectoneural nerve ring of ectodermal origin surrounding the mouth, and a motor hyponeural system of mesodermal origin, from which radial nerves enter the arms and the rest of the body coordinating the movement of the animal. The connection between the two systems is unclear.

#### Hemichordata

Hemichordates (with the classes Enteropneusta and Pterobranchia; 100 species) are wormlike or sessile marine animals with a primitive or secondarily simplified nervous system consisting of a dorsal and ventral nerve cord interpreted by some authors to be homologous to the spinal cord of chordates, while others consider it a result of independent evolution.

#### Chordata

The phylum Chordata consists of the Cephalochordata (lancelets), Urochordata (tunicates), and Craniata (cf. Fig. 2.12). They all possess, at least at some point of their life, a chorda dorsalis or notochord, i.e., a flexible cartilaginous rod, and a hollow dorsal nerve or "spinal" cord.

##### Cephalochordata and Urochordata

Cephalochordates (21 species, e.g., the lancelet *Branchiostoma*, previously called *Amphioxus*) possess a neural plate, but no neural crest or placodes and, as a consequence, no head. Recent studies based on neural gene expression patterns [28] reveal that the neural tube and its rostral "cerebral vesicle" are homologous with most parts of the vertebrate CNS, i.e., a spinal cord, a rhombencephalon, mesencephalon, and diencephalon and perhaps parts of a telencephalon, which is connected with the unpaired frontal eye. Urochordates (2,200 species) are sessile animals having a free-swimming larva and a very primitive nervous system, probably as a consequence of secondary simplification related to their sessile life style.

##### Craniata/Vertebrata

The group Craniata (i.e., animals with a skull) comprises the Myxinoidea and all members of the subphylum Vertebrata including the group Petromyzontida (lampreys) (cf. Fig. 2.12). Myxinoids and petromyzontids have no jaws and are, therefore, often called Agnatha, i.e., jawless fishes.

Myxinoids (hagfishes; about 60 species) are eel-like exoparasites with a well-developed olfactory and mechanosensory system – no lateral-line system as in all other aquatic vertebrates – and have degenerated eyes.

##### Vertebrata

As shown in Fig. 2.12, the subphylum Vertebrata comprises the classes Petromyzontida (about 50 species), Chondrichthyes (cartilaginous fishes, i.e., sharks, rays, skates and chimaeras, about 1,100 species), Osteichthyes (bony fishes, i.e., actinopterygian, brachiopterygian, and sarcopterygian fishes, the latter comprising lungfishes and crossopterygians; together more than 30,000 species), Amphibia (frogs, salamanders, caecilians, about 6,000 species), "Reptilia" (chelonians, i.e., turtles; rhynchocephalians, i.e., the tuatara; squamates, i.e., lizards and snakes; and crocodilians – together about 9,500 species), Aves (birds, about 10,000 species), and Mammalia (about 5,700 species). The former class "Reptilia" is now considered a paraphyletic taxon, i.e., one without a common ancestor, because crocodilians (and the extinct dinosaurs) are more closely related to birds than to other "reptiles". The representatives of the former class "Reptilia" and birds together form the superclass Sauropsida.

The class Mammalia comprises the three subclasses Prototheria (egg-laying mammals, with the single order Monotremata, 5 species), Metatheria or Marsupialia (i.e., "pouched" mammals, about 340 species), and Eutheria or Placentalia (i.e., mammals with a placenta, 5,300 species). According to recent taxonomy, the latter comprises the four supraorders Afrotheria (containing elephants, manatees, and species previously included in the now obsolete group "insectivores"), Xenarthra (anteaters, armadillos, sloths), Euarchontoglires (e.g., primates, rodents, hares), and Laurasiatheria (including bats, carnivores, ungulates, cetaceans, and certain species previously referred to "insectivores").

The CNS of craniates reveals a highly uniform organization [29, 30] (Figs. 2.13 and 2.14a–j). In its hypothetical ancestral form, it exhibits the "standard" organization into the three: prosencephalon, mesencephalon, and rhombencephalon. Whether such a tripartite organization, occurring in lophotrochozoans as well as ecdysozoans, is due to "deep homology" found in all bilaterians [11] or to convergent evolution, is debated. All extant craniates have brains consisting of a rhombencephalon composed of a myelencephalon or medulla oblongata, and a metencephalon including a cerebellum, a mesencephalon (midbrain) including an isthmic region, and a prosencephalon composed of a diencephalon (or "primary prosencephalon") and a telencephalon (or "secondary prosencephalon" – endbrain). Medulla oblongata and mesencephalon together form the "brainstem". It is now generally accepted that most parts of the brain like the spinal cord have a *segmental organization* (Fig. 2.15). The rhombencephalon consists of rhombomeres R1–7, the mesencephalon is composed of an isthmic neuromer and a mesencephalic neuromer proper, the diencephalon (as "primary prosencephalon") and at least ventral parts of the

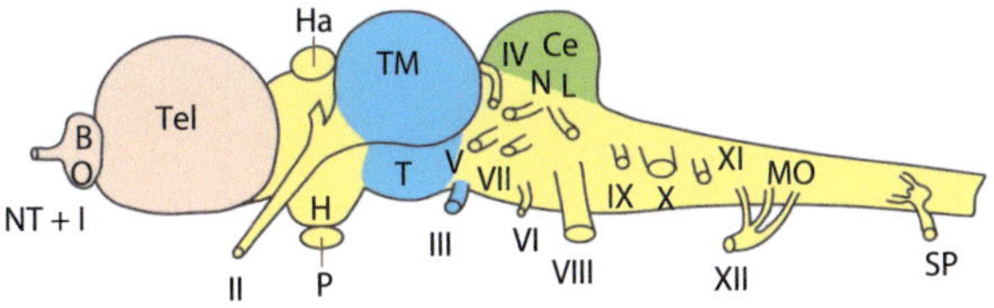


**Fig. 2.13 Basic organization of the vertebrate brain**. *BO* olfactory bulb, *Ce* cerebellum, *H* hypothalamus, *Ha* habenula, *MO* medulla oblongata, *NL* lateral nerves, *NT* terminal nerve, *P* hypophysis/pituitary, *SP* first spinal nerve, *T* tegmentum, *Tel* telencephalon, *TM* tectum mesencephali, *I–XII* cranial nerves

telencephalon (as "secondary prosencephalon") are segmented into six prosomeres P1–6. The exact segmentation of the telencephalon dorsal and rostral to P6 into pallial and subpallial regions has yet to be determined.

The *medulla spinalis* consists of an inner gray substance around the central canal consisting mostly of nerve cells covered by white substance containing dendrites and ascending and descending nerve fibers. The gray substance is divided into a dorsal somatosensory and viscerosensory region and a ventral visceromotor and somatomotor region. Nerve cells innervate the various parts of the body via spinal nerves in a segmental fashion.

The *medulla oblongata* reveals the same dorsoventral organization as the medulla spinalis and contains, in all vertebrates, from dorsal to ventral: somatosensory, viscerosensory, visceromotor, and sensorimotor areas and nuclei of the cranial nerves V to X. Tetrapod vertebrates (amphibians, "reptiles", birds, and mammals) also have the additional cranial nerves XI (*N. accessorius*) and XII (*N. hypoglossus*). The dorsal sensory roots of the cranial nerves include ganglia containing the somata of sensory neurons. The regions of sensory nuclei may undergo strong enlargement and complication, as, for instance, in goldfish with the gustatory vagal lobe and its highly evolved gustatory system. A mechano- and electroreceptive lateral-line system associated with cranial nerves is present in all vertebrates and was lost in some terrestrial amphibians and in all amniotes, i.e., in "reptiles", birds, and mammals.

The *reticular formation system* is found inside the medulla oblongata, the pons (only in mammals, though birds have evolved a somewhat similar structure independently), and in the tegmental midbrain. It is relatively uniform among vertebrates and contains

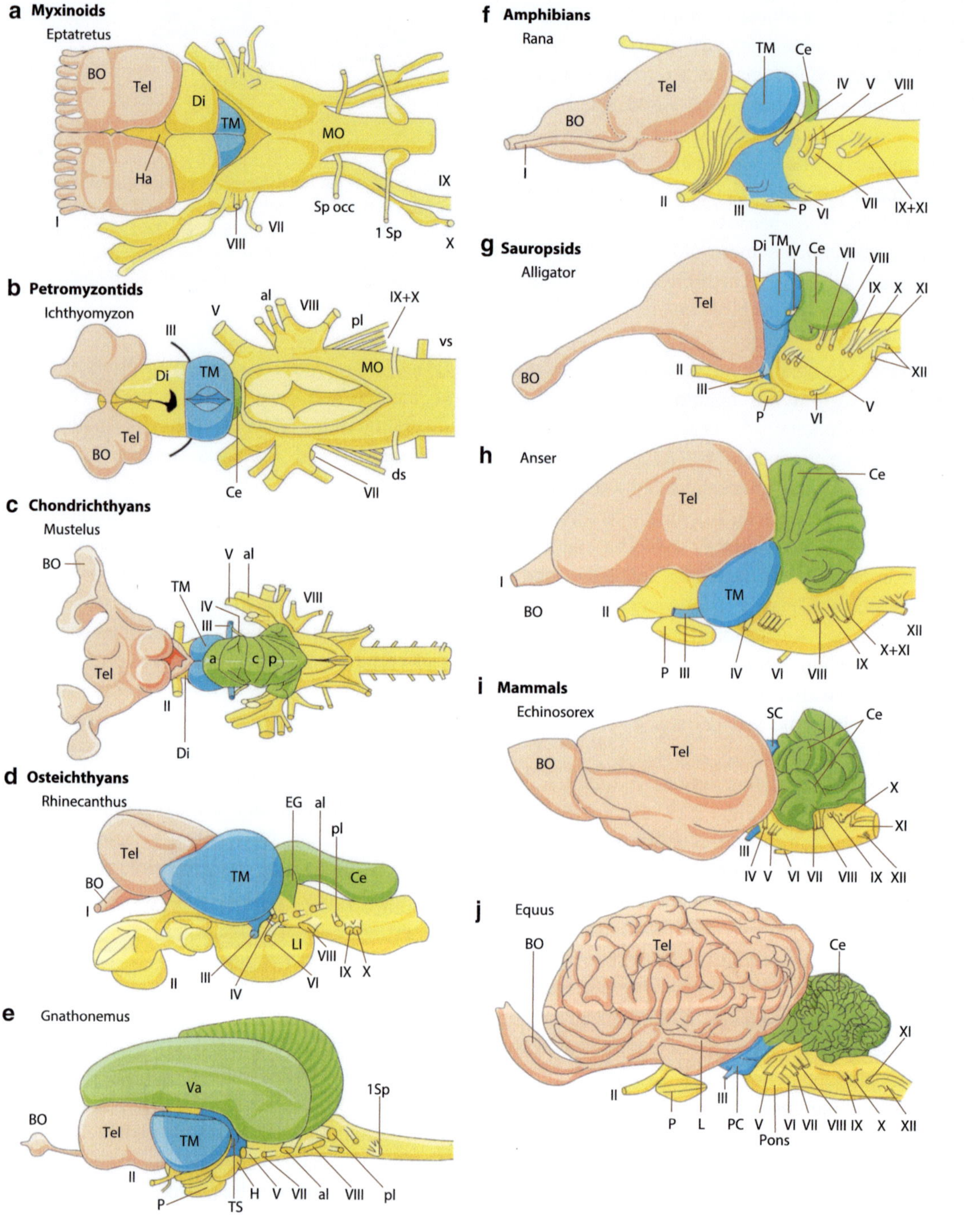


**Fig. 2.14** **Brains of representatives of major groups of craniates.** (**a**) Hagfish, dorsal view (After Northcutt [31]). (**b**) Lamprey, dorsal view (After Northcutt [31]). (**c**) Common smooth-hound, dorsal view (After Northcutt [32]). (**d**) Trigger fish, lateral view. (**e**) Elephantnose, lateral view. (**f**) Frog, lateral view. (**g**) Alligator, lateral view. (**h**) Goose, lateral view. (**i**) Moon rat. (**j**) Horse, (**d–j**) after Romer and Parson [33]). For abbreviations see Fig. 2.13; further abbreviations: *a* anterior cerebellar lobe, *al* anterior lateral nerve, *c* central cerebellar lobe, *Di* diencephalon, *ds* dorsal spinal nerve, *EG* eminentia granularis, *H* habenula, *P* pituitary, *LI* inferior lobe, *p* posterior cerebellar lobe, *pl* posterior lateral nerve, *SC* superior culliculus, *Sp occ* spino-occipital nerve, *1Sp* first spinal nerve, *TS* torus semicircircularis, *Va* valvula cerebella, *vs* ventral spinal nerve

important neuromodulator-producing centers such as the noradrenergic *locus coeruleus* and the serotonergic *raphe* nuclei. The reticular formation controls centers for breathing and cardiovascular activity and gives rise to an ascending activation system for vigilance, awareness, and consciousness. A *pons* ("bridge") is found only in mammals and is situated in the rostral medulla oblongata and caudal tegmentum. It contains relay nuclei of fiber bundles that connect the cerebral cortex and the cerebellum. However, similar pathways and nuclei are likewise found in birds, but most likely developed independently.

The *cerebellum* is a formation of the dorsal metencephalon and present in all vertebrates, but absent in myxinoids, perhaps due to secondary loss. In all vertebrates except petromyzontids, it exhibits a uniform three-layered organization, i.e., a deep small-celled granular layer, a large-celled layer of Purkinje cells, and a peripheral molecular layer. The vestibulolateral lobes of the cerebellum processing primary vestibular and – if present – mechano- and electroreceptive lateral-line information are present in all vertebrates. A *corpus cerebelli* is found in cartilaginous and bony fishes and terrestrial vertebrates and has undergone hypertrophy in some groups of cartilaginous fishes, and is strongly reduced in size in amphibians. In some actinopterygian fishes, predominantly in electric fish, there is a strong enlargement of parts of the cerebellum, the *valvula* [34]. Mammals have novel lateral cerebellar structures, the *cerebellar hemispheres*, receiving telencephalo-pontine input. Besides vestibular, somatosensory, and sensorimotor functions, the cerebellum of mammals, and perhaps of birds, is likewise involved in "higher" cognitive functions such as thinking and action planning as well as language in humans.

The *mesencephalon* consists, from dorsal to ventral, of the: *tectum* (in mammals called "colliculi superiores"), the *torus semicircularis* (in mammals called "colliculi inferiores"), and the *tegmentum*. In all craniates, except mammals, the tectum is the major visual processing and integration center. In its ancestral state, it exhibits a laminar organization consisting of alternating cellular and fiber layers. The absence of such lamination in some vertebrates such as salamanders, caecilians, and South American and African lungfishes is a consequence of secondary simplification [35]. Besides visual input, other sensory information such as auditory and – if present – mechano- and electrosensory information originating in the torus semicircularis terminate in deeper layers of the tectum and

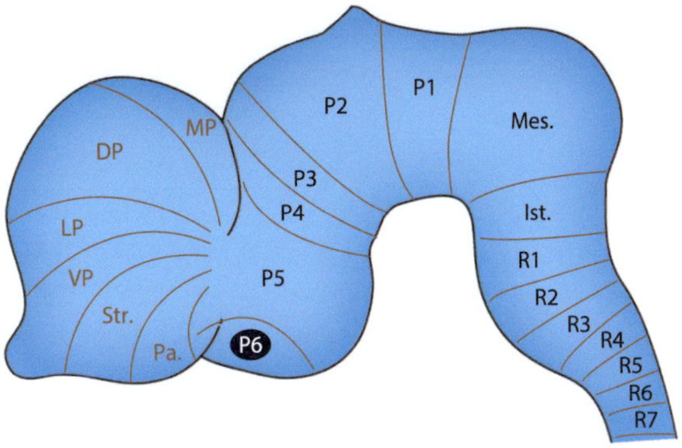


**Fig. 2.15** **Segmental organization of the craniate brain**. *DP* dorsal pallium, *Ist.* isthmic neuromer, *LP* lateral pallium, *Mes.* mesencephalic mesomer, *MP* medial pallium, *P1–6* prosomeres, *Pa.* pallidum, *R1–7* rhombomeres, *Str.* striatum, *VP* ventral pallium (After Striedter [30] with permission)

contribute to a multimodal representation for orientation in space. Likewise, telencephalic efferents terminate in the tectum of cartilaginous and bony fishes and of terrestrial vertebrates. Main tectal efferents descend to the medulla oblongata/pons and medulla spinalis (tectobulbar and tectospinal tracts, respectively).

The *torus semicircularis* is, in its plesiomorphic state, the midbrain relay station for auditory, mechano-, and electrosensitive projections ascending to the diencephalon and telencephalon (cf. Fig. 2.16d). Like the tectum, it is characterized by a laminar organization, which is most spectacular in electric fish [29] (cf. Fig. 2.17).

The *tegmentum* is involved in (pre)motor functions and exhibits a number of specialties. One ancient component is the nucleus of the oculomotor nerve (cranial nerve III), which is absent in myxinoids, but present in all vertebrates. In the tegmentum, massive fiber tracts descending from the cortex to the pons, medulla oblongata, and medulla spinalis are found (corticopontine tracts and pyramidal tract). The tegmentum contains the dopaminergic *substantia nigra* exhibiting reciprocal connections with the telencephalic striatum, and the premotor *nucleus ruber*. The latter receives crossed efferents from the cerebellum and gives rise to crossed descending motor tracts to the spinal cord (rubrospinal tracts).

The *diencephalon* is divided from dorsal to ventral into the epithalamus, thalamus, and hypothalamus (Fig. 2.16c). The *epithalamus* contains the *habenular nuclei*, which are important parts of the limbic system and present in all craniates. They project, via the

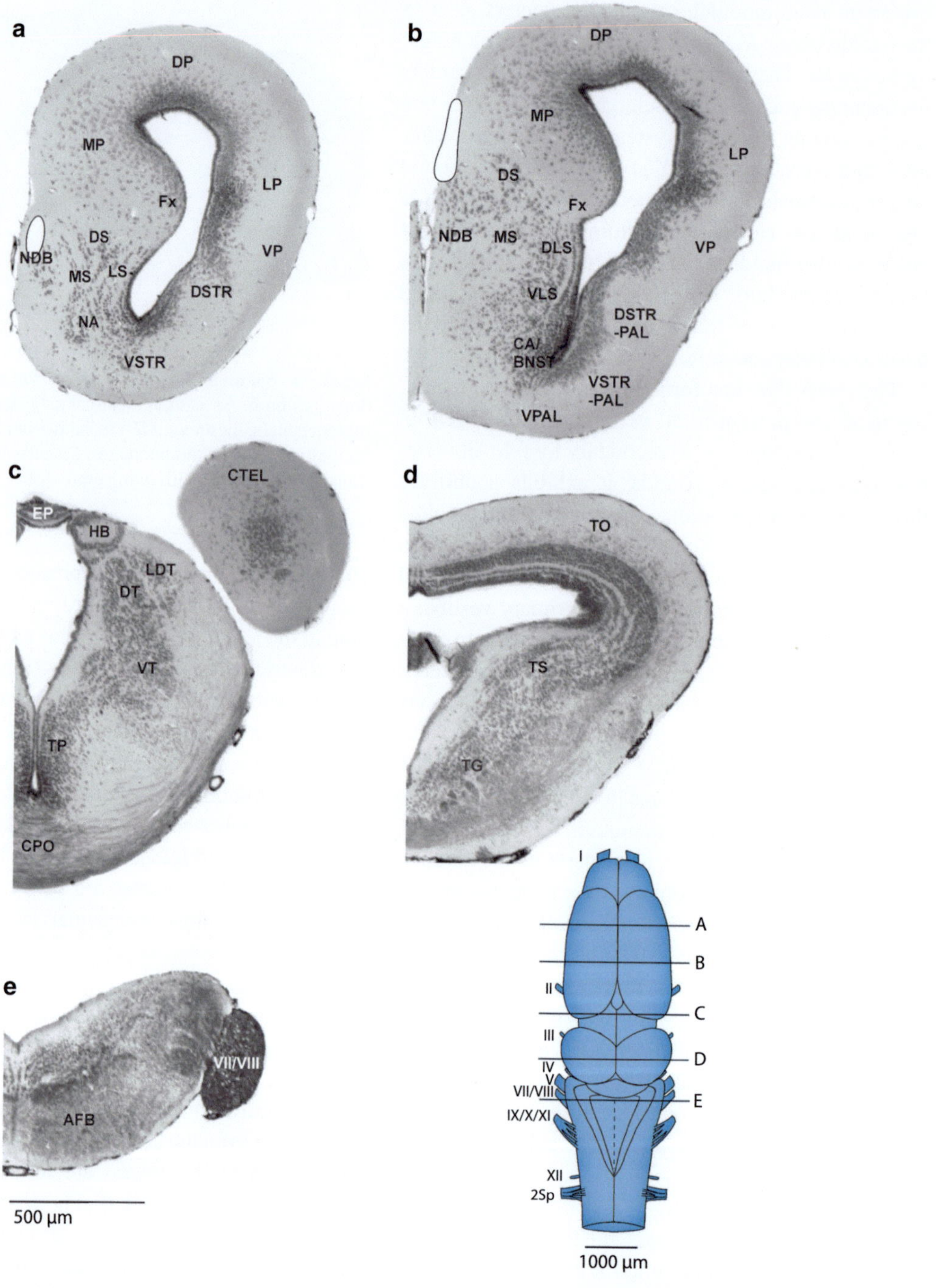


**Fig. 2.16** **Cross sections through the brain of the frog *Bombina orientalis*.** Levels of cross sections (**a**–**e**) are indicated in dorsal view of the brain at *lower right*. (**a**) Rostral telencephalon at the level of the nucleus accumbens. (**b**) Central telencephalon at the level of dorsal and ventral striatum. (**c**) Diencephalon at the level of the habenula and postoptic commissure. (**d**) Midbrain with optic tectum and torus semicircularis. (**e**) Rostral medulla oblongata at the level of entrance of cranial nerve VII. Abbreviations: *AFB* descending fiber bundles, *CA-BNST* central amygdala-nucleus interstitialis of the stria terminalis, *CPO* commissura postoptica, *CTEL* caudal telencephalon, *DLS* dorsal lateral septum, *DS* dorsal septum, *DP* dorsal pallium, *DSTR* dorsal striatum, *DSTR-PAL* dorsal striatopallidum, *DT* dorsal thalamus, *EP* epiphysis/pineal organ, *Fx* fornix, *HB* habenula, *LP* lateral pallium, *LS* lateral septum, *LDT* lateral dorsal thalamus, *MP* medial pallium, *MS* medial septum, *NA* nucleus accumbens, *NDB* nucleus of diagonal band of Broca, *TG* tegmentum, *TO* optic tectum, *TP* tuberculum posterius, *TS* torus semicircularis, *VLS* ventral lateral septum, *VSTR* ventral striatum, *VSTR-PAL* ventral striatopallidum, *VP* ventral pallium, *VT* ventral thalamus, *VII/VII* 7th/8th cranial nerve, *2SP* 2nd spinal nerve

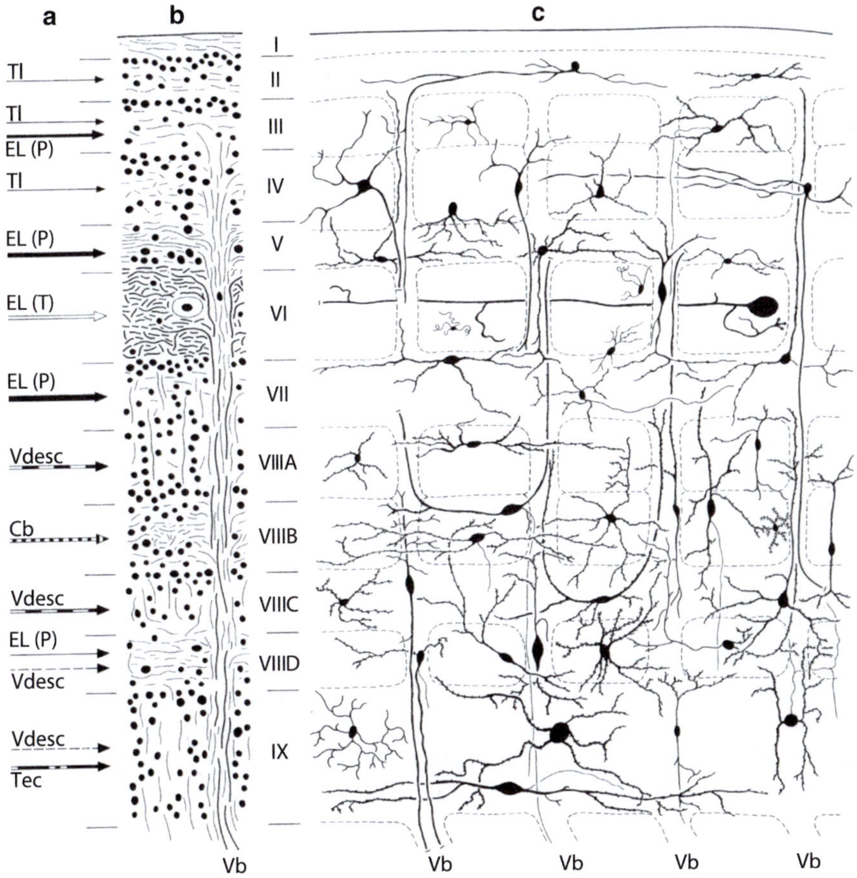


**Fig. 2.17 Anatomy of the torus semicircularis of the electric fish *Eigenmannia virescens*** exhibiting a spectacular laminar organization. (**a**) Afferents from different brain regions terminate in different layers of the torus. (**b**) Laminar organization of the torus in bodian staining. (**c**) Cytoarchitecture of the torus in golgi staining. Abbreviations: *Cb* cerebellum, *EL(P)/EL(T)* electrosensory P- and T-type afferents, *Vdesc* nucleus descendens of the trigeminal nerve, *Tec* tectum opticum, *Tl* torus longitudinalis, *Vb* vertical bundle (After Nieuwenhuys et al. [29], modified, with permission)

fasciculus retroflexus, to the midbrain tegmentum. In many craniates, the epithalamus carries the pineal organ or "epiphysis", a small endocrine gland releasing the hormone melatonin, which affects wake-sleep patterns and seasonal functions. The *thalamus* is composed of a dorsal and a ventral part and the posterior tuberculum. In all craniates, the *dorsal thalamus* receives visual, somatosensory, auditory, and gustatory information, either directly (e.g., from the eyes) or indirectly via brainstem relay centers, and sends them to telencephalic regions. Visual pathways terminate in the dorsal pallium of fishes, amphibians, "reptiles", and birds and in the occipital cortex of mammals. In "reptiles" and birds, auditory as well as visual and somatosensory pathways terminate in a special region of the lateral telencephalon called *dorsal ventricular ridge* (DVR) in "reptiles" and *mesonidopallium* in birds. The thalamus has undergone a strong enlargement in amniotes, particularly mammals and birds, and became parcellated into many nuclei related to sensory, cognitive, and motor as well as limbic functions. In mammals (Fig. 2.18), visual functions are relayed by the corpus geniculatum laterale, auditory functions by the corpus geniculatum mediale, both projecting to the primary visual and auditory cortex, respectively. Ray-finned fishes have independently developed a projection from the posterior tuberculum of the diencephalon to the pallium. The *ventral thalamus* and subthalamus (zona incerta) of mammals projects to telencephalic parts of the basal ganglia, i.e., corpus striatum and globus pallidus, and to the hippocampus. The *hypothalamus* and its appendage, the *pituitary* (*hypophysis*), are the main hormone-based control centers for basal homeostatic functions. Cartilaginous and bony fishes exhibit a hypertrophy of the lateral hypothalamus (*lobus inferior hypothalami*), with unknown functions.

The evolution of the *telencephalon* in craniates is not fully understood. In all craniates, it receives olfactory information from the olfactory bulb as the only direct sensory input. Comparative neuroanatomists, therefore, previously believed that in its ancestral state the telencephalon had to be considered the "olfactory brain". Later it was found that in all craniates the

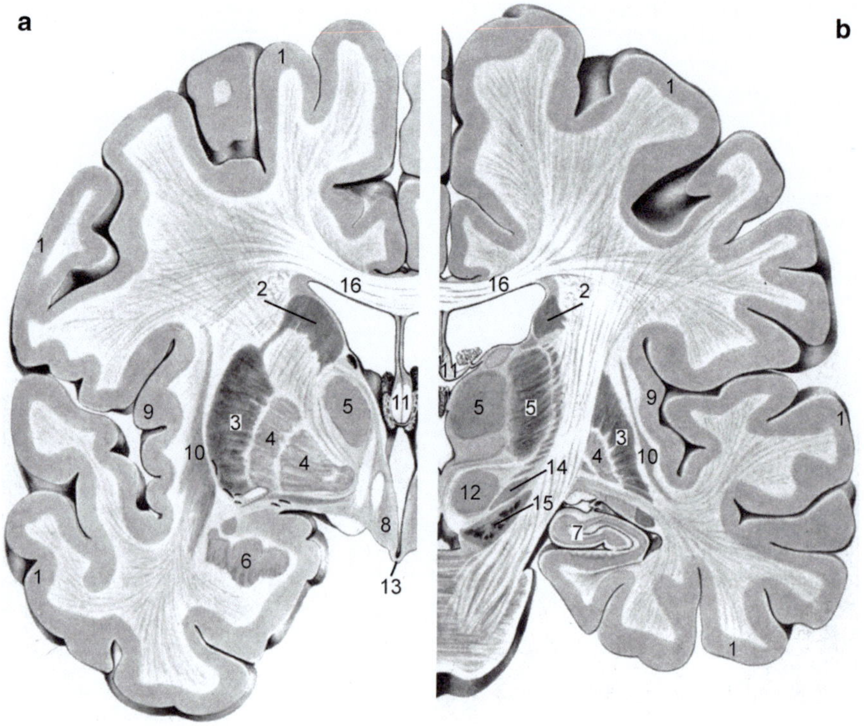


**Fig. 2.18** **Cross section through the human brain.** (**a**) At the level of hypothalamus, amygdala, and striatopallidum. (**b**) At the level of hippocampus and thalamus. Numbers: *1* cerebral cortex; *2* nucleus caudatus, *3* putamen, *4* globus pallidus, *5* thalamus, *6* amygdala, *7* hippocampus; *8* hypothalamus, *9* insular cortex, *10* claustrum, *11* fornix, *12* nucleus ruber, *13* infundibulum of pituitary, *14* nucleus subthalamicus, *15* substantia nigra, *16* corpus callosum (After Nieuwenhuys et al. [36] with permission)

telencephalon also receives information from other senses, e.g., visual, auditory, and mechanosensory via pathways ascending from the diencephalon (see above), and the telencephalon was regarded "multimodal" in its ancestral state. Recent studies, however, revealed that in all craniates except birds and mammals these nonolfactory sensory afferents to the telencephalon are either multimodal or do not form topographic representations. This would imply that the development of topographic representations of thalamic sensory afferents to pallial/cortical regions has happened independently in birds and mammals and would again strengthen the "olfactory brain" interpretation.

The telencephalon of all craniates is composed of a pallium and a ventral subpallium ("pallium" means "mantle") surrounding the paired telencephalic ventricles, which in its ancestral form of tetrapods is clearly visible in the amphibian brain ([35] cf. Fig. 2.19). With the exception of petromyzontids, the pallium is divided into a medial, dorsal, lateral, and ventral pallium, and the subpallium into a septal and a striato-pallidal region, the latter including parts of the amygdala involved in limbic-autonomic functions (in mammals the "central amygdala"). In "reptiles", pallial divisions are called medial, dorsal, and lateral cortex and dorsal ventricular ridge (see below). The medial pallium/cortex of amphibians and sauropsids corresponds to the hippocampal formation of mammals, the lateral pallium to the olfactory (in mammals "piriform") cortex, and the ventral pallium to the vomeronasal pallium.

The ontogeny of the telencephalon of actinopterygian (ray-finned) fishes deviates from that of other craniates [34]; Fig. 2.19. In the latter, the unpaired embryonic telencephalon extends laterally and forms two hemispheres through *evagination* resulting in an arrangement of a medial, dorsal, lateral, and ventral pallium, and a subpallium around the ventricles. In contrast, in actinopterygians two hemispheres are formed by *eversion* in such a way that parts of the brain which in the evagination type are found medially and dorsally, occupy a ventrolateral position. These differences in ontogeny contribute to the difficulty with homologizing parts of the telencephalon of actinopterygian fishes and other vertebrates.

In sauropsids, the lateral parts of the pallium have developed into a large structure, the *dorsal ventricular ridge*, DVR, which bulges in medial direction into the ventricles and for a long time was considered part of the corpus striatum as the major telencephalic

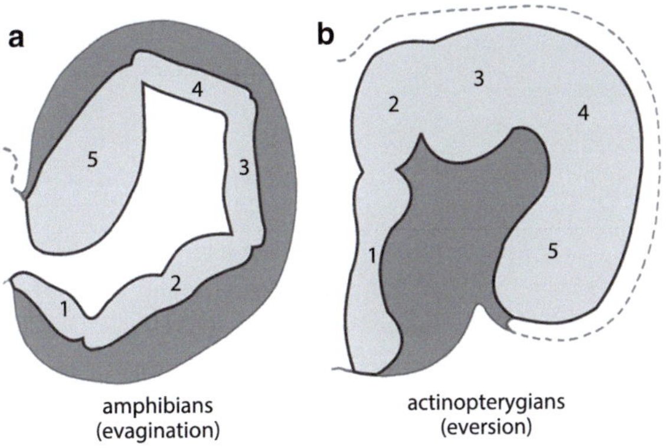


**Fig. 2.19 Differences in ontogenies of the telencephalon.** (**a**) Evaginated telencephalon as in most vertebrates (here amphibians). (**b**) Everted telencephalon as in actinopterygian bony fishes. *Numbers* indicate the major regions of the telencephalon: *1* ventromedial subpallium, *2* ventrolateral subpallium (striatopallidum), *3* lateral pallium, *4* dorsal pallium, *5* medial pallium. After Nieuwenhuys et al. [29], modified, with permission

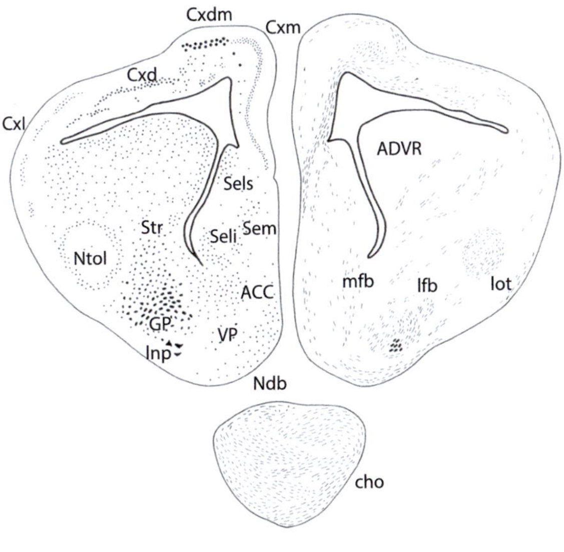


**Fig. 2.20 Cross section through the telencephalon of the lizard *Tupinambis teguixin*** at the level of the striatum. The anterior dorsal ventricular ridge (ADVR) bulges into the telencephalic ventricle. Abbreviations: *ACC* nucleus accumbens, *cho* chiasma opticum, *Cxd* cortex dorsalis, *Cxdm* cortex dorsomedialis, *Cxl* cortex lateralis, *Cxm* cortex medialis, *fx* fornix, *lfb* lateral forebrain bundle, *lot* lateral olfactory tract, *mfb* medial forebrain bundle, *Ndb* nucleus of the diagonal band of Broca, *Ntol* nucleus of the tuberculum olfactorium, *Seli* inferior lateral septum, *Sels* superior lateral septum, *Sem* medial septum, *Str* striatum, *VP* ventral pallidum (From Nieuwenhuys et al. [29], modified, with permission)

component of the basal ganglia (Fig. 2.20). Accordingly, parts of the DVR of birds were called *ectostriatum*, *neostriatum*, and *hyperstriatum*. Today it is generally accepted that the DVR is not homologous to the striatum, but is of pallial origin [37].

However, as illustrated in Fig. 2.21, there is a debate whether the DVR and the mesonidopallium of birds are homologous to the lateral cortex or the claustrum of mammals and ventral pallium of amphibians [38, 39]. In the latter case, different parts of the dorsal telencephalon would give rise to centers involved in intelligence and other mental abilities (see below).

The dorsal pallium of myxinoids exhibits a five-layered structure, which has developed independently of other lamination patterns found in craniates (see below). The pallium of cartilaginous and actinopterygian fishes is unlaminated. The medial and dorsal pallium of lungfishes (dipnoans) displays some lamination, while the pallium of amphibians is generally unlaminated despite extensive cell migration in medial and dorsal parts [29]; cf. Fig. 2.18a, b. In the medial, dorsomedial, and dorsal cortex of "reptiles" there is a three- to four-fold lamination, which however, is discontinuous and not comparable in its cytoarchitecture to the laminated cortex of mammals (cf. Fig. 2.20). In birds, the dorsally situated hyperpallium is considered homologous to the visual, auditory, and somatosensory cortex of mammals, while the meso-/nidopallium of birds is considered to be homologous either to the lateral (temporal) isocortex or a derivative of the ventral pallium of amphibian-reptilian ancestors. In birds, both the hyperpallium and meso-/nidopallium are unlaminated (Fig. 2.22).

Mammals in general possess a six-layered cortex called "isocortex" (Fig. 2.23) and a 3- to 5-layered "allocortex" or limbic cortex. The evolution of the mammalian cortex is unclear, but its laminar organization appears to have evolved independently of the lamination occurring in the pallia of other craniates and probably evolved from a three-layered olfactory cortex. A common organizational principle of the mammalian cortex is the parcellation into functionally different (sensory, motor, integrative) areas. In small mammalian brains, the number of such areas is low, having about 10 primary sensory and motor areas without signs of integrative-associative areas; cf. [41]. The number of cortical areas increases with cortex volume in most mammals and all primates. Concurrently,

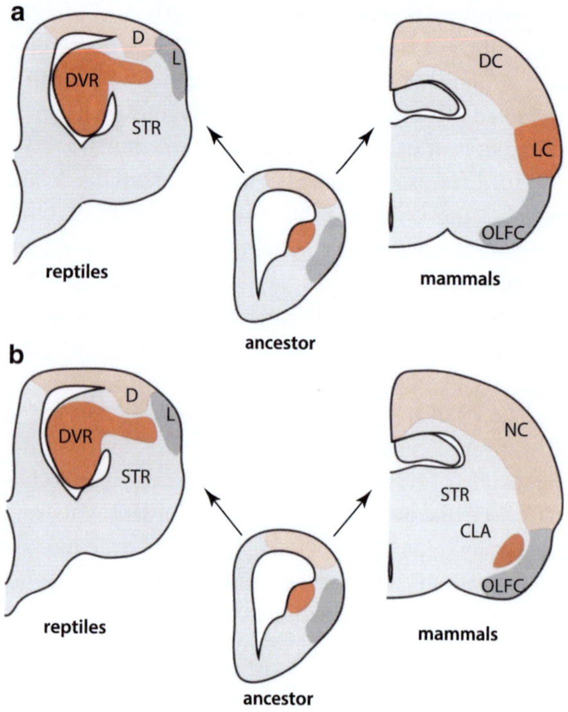


**Fig. 2.21** **Two hypotheses concerning the homology of the lateral mammalian cortex (LC) and the dorsal ventricular ridge (DVR) of "reptiles"**. (**a**) Hypothesis of "common origin" of the DVR and LC from the same embryonic material of the amniote ancestor. (**b**) Hypothesis of the de-novo formation of LC and DVR. For further information see text. Abbreviations: *CLA* claustrum, *D* dorsal cortex of reptiles, *L* lateral cortex of reptiles, *DC* dorsal cortex of mammals, *LC* lateral cortex of mammals, *NC* neocortex of mammals, *OLFC* olfactory cortex, *STR* striatum (From Striedter [30], modified, with permission)

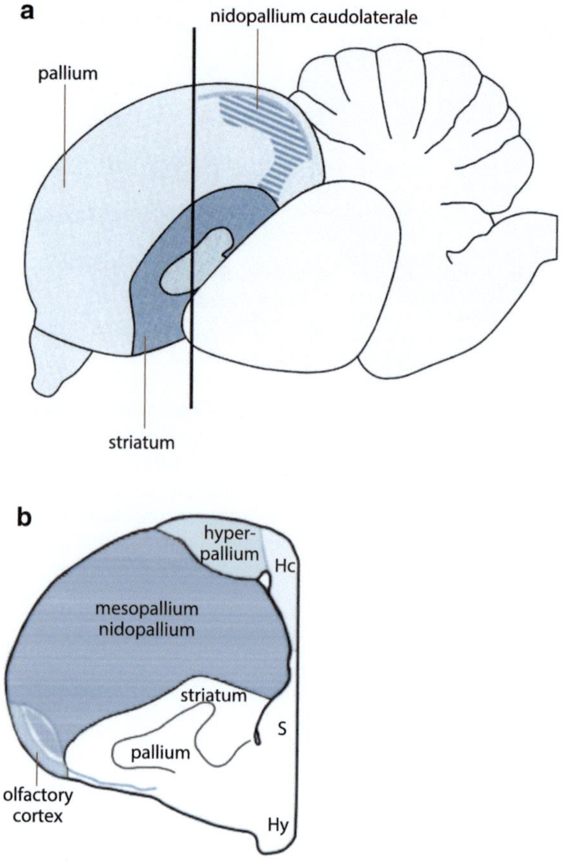


**Fig. 2.22** **Brain of a pigeon**. (**a**) Lateral view. The telencephalon is composed of a pallium and a striatopallidum. A special pallial region is the nidopallium caudolaterale; (**b**) Cross section through the telencephalon at the level indicated in (**a**). Most of the pallium consists of the mesopallium and nidopallium. The hyperpallium is situated mediodorsally. Striatum and pallidum are located below the meso-/nidopallium. Abbreviations: *Hc* hippocampus, *S* septum, *Hy* hypothalamus. For further information see text ((a) After Gunturkun [43], modified, with permission. (b) After Reiner et al. [38], modified, with permission)

the relative sizes of cortical areas are supposed to decrease. The human cortex is assumed to possess 150 areas and 60 connections per area resulting in 9,000 area-area connections [42].

The subpallium/subcortex of vertebrates consists of a large striatopallidum (*nucleus caudatus* and *putamen*, together forming the "corpus striatum", and *globus pallidus* in mammals; cf. Fig. 2.17) as the main component of the basal ganglia, a subcortical amygdalar complex, and a medially situated septal region. The amygdala of vertebrates consists of a portion exerting autonomic-limbic functions ("central amygdala") as well as olfactory and vomeronasal functions (cortical and medial amygdala of mammals, respectively). In addition, in mammals a basolateral amygdala has evolved, with strong reciprocal connections to the isocortex and limbic cortex as well as afferents from the sensory nuclei of the thalamus.

## 2.4 Major Evolutionary Changes of the Vertebrate Brain

The basic organization of the brain is surprisingly uniform across all vertebrate taxa, as described above. Besides the development of structures for the processing of specialized senses such as the vagal and facial

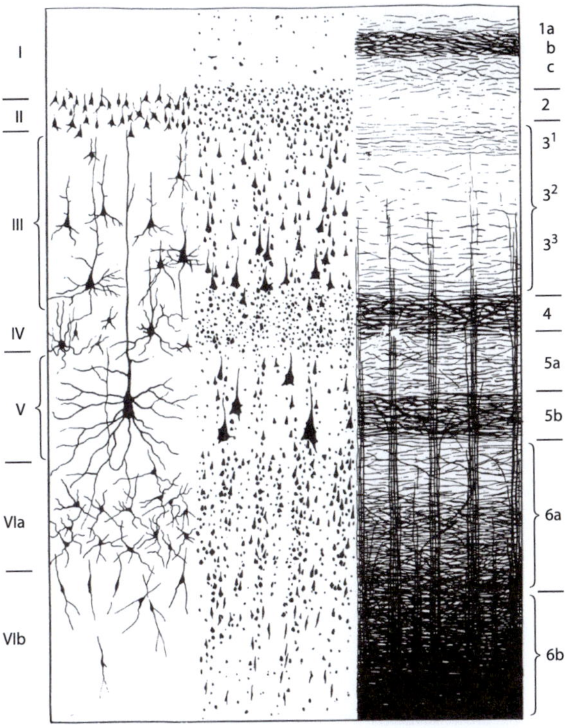


**Fig. 2.23 Cytoarchitecture of the six-layered mammalian isocortex.** The *left side* of the figure shows the distribution of nerve cells, predominantly pyramidal cells, in a Golgi staining. In the *middle*, the distribution of cell bodies is shown in a Nissl staining. The *right side* shows the distribution of myelinated fibers in a Weigert staining. *Roman numbers* to the *left* indicate the gross lamination of the cortex, *arabic numbers* to the *right* indicate the sublamination based on a Nissl staining (After Vogt and Brodmann from Creutzfeldt [40], modified, with permission)

lobe (taste) or the valvula cerebelli (electroreception) in teleost fishes, major changes are observed in the telencephalic roof, i.e., pallium or cortex. However, one of the most striking differences concerns absolute brain size (ABS), which in vertebrates varies from 1 mg (or $mm^3$) in miniaturized fishes and amphibians up to 8,000 g in the false killer whale, which is a range of almost seven orders of magnitude (Table 2.1).

The classes of craniates likewise differ markedly in brain size relative to body size (RBS) (Fig. 2.24). The smallest RBS is found in agnathans (myxinoids and petromyzontids), followed by bony fishes, amphibians, "reptiles", and cartilaginous fishes, the latter having surprisingly large brains. Mammals and birds, on average, have brains that are about ten times larger than those of the other groups of craniates of the same body size. In birds, parrots (Psittacidae) and corvids (Corvidae) have 6–10 times larger brains than other birds. Among mammals, primates (with the exception of prosimians) generally have larger brains than the other orders with the same body size. In primates, prosimians and tarsiers have relatively small brains with an average of 6.7 g, followed by New World monkeys with an average of 45 g and Old World monkeys at an average of 115 g, with the largest monkey brains found in baboons. Among apes, gibbons have brain sizes (88–105 g) which fall within the range of Old World monkeys, while the large apes (orangutans, gorillas, chimpanzees) have brain weights between 330 and 570 g (males) [47].

Brain size in vertebrates is mostly determined (more than 90 %) by body size [48]. However, brain size does not increase proportionally with body size, but "lags behind", i.e., with an exponent (or allometric coefficient) of 0.6–0.8, which is due to the fact that with an increase in body size brains become absolutely larger, but relatively smaller – this is called *negative brain allometry* [44]. As a consequence, in small mice or insectivores brain volume may constitute 10 % or more of body volume, while in the blue whale, the largest living animal, the brain makes up only 0.01 % or even less of body mass [49]; (Fig. 2.25). Primates, in general, have higher RBS than all other groups of mammals.

The human brain has a weight of 1,250–1,450 g on average and represents about 2 % of body mass. Although the human brain is neither exceptional in ABS or RBS, it is unusually large in terms of body size. This can be demonstrated using various statistical methods, e.g., the encephalization quotient EQ, which indicates the extent to which the brain size of a given species $E_a$ deviates from the expected brain size $E_e$;cf. [44]. Within primates, humans have the highest EQ of 7.4–7.8, meaning that the human brain is 7–8 times larger than that of an average mammal of the same body size (Table 2.1). They are followed by the monkeys *Cebus* and *Saimiri* with EQs of 4.8 and 2.8, respectively, while chimpanzees and orangutans have low (1.7 and 1.9, respectively) and gorillas very low EQs (1.5). Other mammals have EQs between 0.4 (rat) and 1.3 (elephant).

During mammalian brain evolution, all parts increased in size relative to the body, except the olfactory system. However, the telencephalon as well as the cerebellum underwent a faster growth than the other parts resulting in *positive allometry* (i.e., with an

**Table 2.1** Brain weight, encephalization quotient, and number of cortical neurons in selected mammals

| Animal taxa | Brain weight (in g)[a] | Encephalization quotient[b,c] | Number of cortical neurons (in millions)[d] |
|---|---|---|---|
| Whales | 2,600–9,000 | 1.8 | 10,500 |
| False killer whale | 7,650 | | |
| African elephant | 4,200 | 1.3 | 11,000 |
| *Homo sapiens* | 1,250–1,450[e] | 7.4–7.8 | 15,000 |
| Bottlenose dolphin | 1,350 | 5.3 | 5,800 |
| Walrus | 1,130 | 1.2 | |
| Camel | 762 | 1.2 | |
| Ox | 490 | 0.5 | |
| Horse | 510 | 0.9 | 1,200 |
| Gorilla | 430[e]–570 | 1.5–1.8 | 4,300 |
| Chimpanzee | 330–430[e] | 2.2–2.5 | 6,200 |
| Lion | 260 | 0.6 | |
| Sheep | 140 | 0.8 | |
| Old World monkeys | 41–122 | 1.7–2.7 | 840 |
| Rhesus monkey | 88 | 2.1 | |
| Gibbon | 88–105 | 1.9–2.7 | |
| Capuchin monkeys | 26–80 | 2.4–4.8 | 720 |
| White-fronted capuchin | 57 | 4.8 | |
| Dog | 64 | 1.2 | 160 |
| Fox | 53 | 1.6 | |
| Cat | 25 | 1.0 | 300 |
| Squirrel monkey | 23 | 2.3 | 450 |
| Rabbit | 11 | 0.4 | |
| Marmoset | 7 | 1.7 | |
| Opossum | 7.6 | 0.2 | 27 |
| Squirrel | 7 | 1.1 | |
| Hedgehog | 3.3 | 0.3 | 24 |
| Rat | 2 | 0.4 | 15 |
| Mouse | 0.3 | 0.5 | 4 |

[a]Data from Jerison, Haug, and Russell [44–46]

[b]Indicates the deviation of the brain size of a species from brain size expected on the basis of a "standard" species of the same taxon, in this case of the cat

[c]Data after Jerison and Russell [44, 46]

[d]Calculated using data from Haug [45]

[e]Basis for calculation of neuron number

exponent larger than 1; [50]). Inside the telencephalon, the cortex has grown faster both in surface and thickness. However, while the increase in surface was 10,000-fold (from 0.8 $cm^2$ in "insectivore"-like species to 7,400 $cm^2$ in whales), cortical thickness increased only slightly, from 0.5 mm in mice to 3 mm in *Homo sapiens*. Interestingly, mammals with much larger brains and consequently cortical surfaces such as elephants and cetaceans (whales and dolphins) have unusually thin cortices of 1.5 and 1 mm, respectively.

In all mammalian cortices, cortical cell density decreases with increasing cortex volume, but primates in general have higher densities (around 50,000 cells per $mm^3$) than expected compared to other mammals – elephants and cetaceans have the lowest cortical cell densities (6,000–7,000 cells per $mm^3$). Humans combine a relatively large brain, a very thick cortex with a relatively high cortical cellular density (about 30,000 cells per $mm^3$) which results in the fact that they have considerably more cortical neurons (about 15 billion) than elephants and cetaceans with their much bigger brains (11 and 10.5 billion, respectively), which is the highest number of cortical neurons found in animals [51, 52].

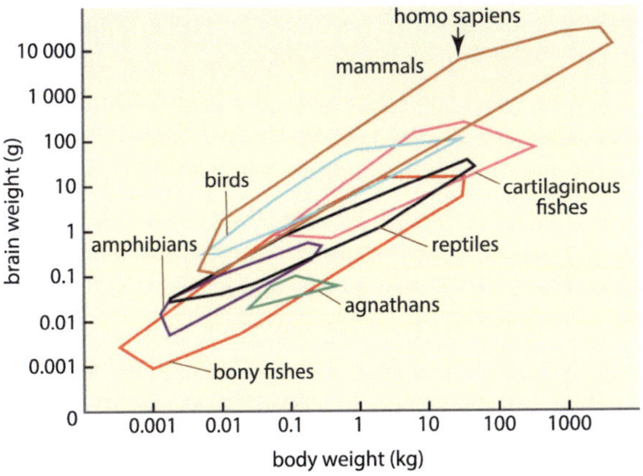


**Fig. 2.24** **The relationship between brain weight** (ordinate, gram) **and body weight** (abscissa, kilogram) in the vertebrate classes in a double-logarithmic presentation using the polygon method developed by Jerison. Mammals and birds generally have larger relative brain weights or volumes than agnathans, bony fishes, amphibians, and "reptiles". The brains of cartilaginous fishes are in between. The weight/volume of the human brain is on top of the distribution, when corrected for body size. Further information in text (After Jerison [44], modified, with permission)

### 2.4.1 Brain Evolution During Hominid Evolution

The earliest human-like primates, the australopithecines (such as "Lucy", *Australopithecus afarensis*) existed in East Africa at about 3–4 mya and had a brain volume of 400–450 cm$^3$, which is equal to or only slightly larger than that of present chimpanzees (see Fig. 2.26). A strong increase in brain volume occurred only with the appearance of *Homo habilis* about 2 mya having 700 cm$^3$. This means that brain size of our ancestors remained constant for 1.5 mya despite strong environmental changes. The next strong increase in brain size occurred 1.8 mya with the appearance of *Homo erectus*, who had a brain volume of 800–1,000 cm$^3$. The appearance of early forms of *Homo sapiens* about 400,000 years ago with brain volumes between 1,100 and 1,500 cm$^3$ represents the latest step of brain evolution in hominins. Importantly, not *Homo sapiens*, but *Homo neanderthalensis*, with 1,400–1,900 cm$^3$ had the largest brain of all hominins and primates. The reasons for this dramatic increase in brain size in a relatively short evolutionary time are unclear despite a large number of scenarios.

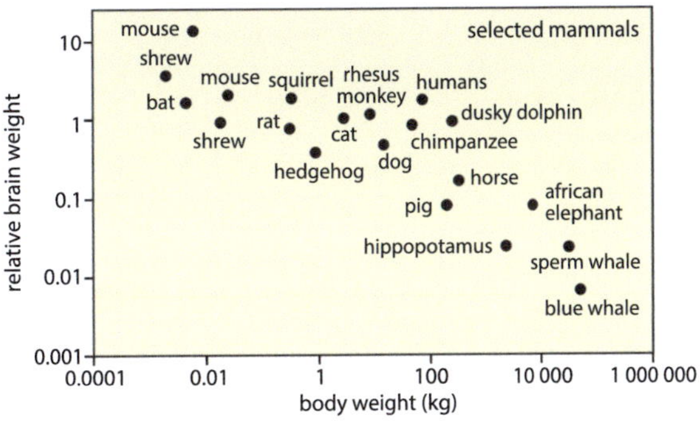


**Fig. 2.25** **Mammalian brain size in percent of body size.** The figure shows brain weight as a percentage of body weight for the same 20 mammalian species as in Fig. 2.2, again plotted in log-log coordinates. As can be seen, small mammals such as mice and shrews have much larger brains in relative terms (10 % or more of body weight) than cetaceans (less than 0.01 %). Humans, with a brain representing 2 % of body weight, have a much higher relative brain size than expected (i.e., around 0.3 %) (From Dongen [49], modified, with permission)

## 2.5 Brain and Intelligence

In humans, intelligence is commonly defined as mental capacities such as abstract thinking, understanding, communication, reasoning, learning, and memory formation, action planning, and problem solving [51]. Usually, human intelligence is measured by intelligence tests and expressed in intelligence quotient (IQ) values expressing different contents (e.g., visual-spatial, verbal, numerical). Evidently, such a definition and measurement of intelligence cannot be applied directly to nonhuman animals, because any test depending on verbalization is inapplicable. A number of comparative and evolutionary psychologists and cognitive ecologists converge on the view that mental or behavioral *flexibility* is a good measure of intelligence culminating in the appearance of novel solutions not part of the animal's normal repertoire [53, 54].

Intelligence defined in such a manner has developed several times independently during evolution, e.g., in cephalopods (e.g., *Octopus*), social insects (e.g., the honeybee), some teleost fishes (e.g., cichlids), some birds (corvids and parrots), and mammals. In all these cases, high intelligence is coupled with (*i*) larger to much larger brains as compared to less intelligent members of the respective taxon, (*ii*) specialized brain

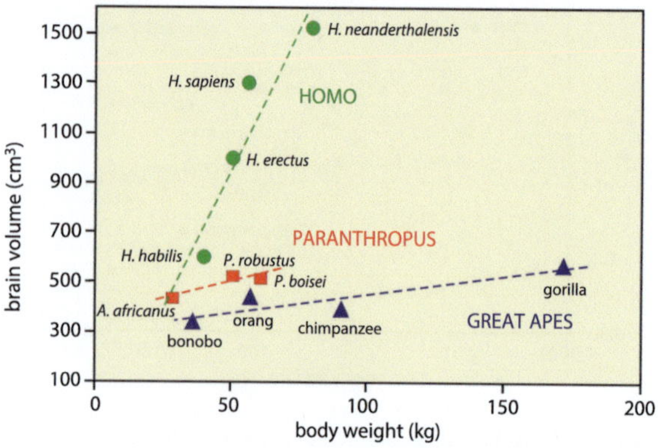


**Fig. 2.26 The relationship between body size and brain size or endocranial volume** (extinct species) in great apes (bonobo, chimpanzee, orangutan, gorilla), australopithecines (*Australopithecus africanus*, *Paranthropus robustus*, *P. boisei*), and the genus *Homo* (*Homo habilis*, *H. erectus*, *H. sapiens*, *H. neanderthalensis*) (Data from Jerison [44]). While in the great apes as well as in the extinct australopithecines brain/endocranial volume has increased only slightly with body size, in the genus *Homo* a steep increase in brain/endocranial volume has occurred during 2.5 mya culminating in the brain of the extinct *Homo neanderthalensis*, which with a volume of 1,200–1,750 ccm was considerably larger than that of modern *Homo sapiens*

centers with a high number of densely packed and interconnected neurons, and (*iii*) structures and mechanisms for fast information processing, e.g., the vertical and subvertical lobe of *Octopus*, the mushroom bodies in the honeybee, the mesonidopallium of corvids, and the isocortex of primates. Primates are, on average, more intelligent than other mammals and all other animals, with the great apes and finally humans at the far end. Because of higher relative cortex volume and higher neuron packing density (NPD), primates have considerably more cortical neurons than other mammals of the same brain size. Likewise, information processing capacity (IPC) is generally higher in primates due to short interneuronal distance and high axon conduction velocity. Finally, primate cortices exhibit extensive parcellation according to the principle of intense local and sparse global connectivity [52].

Across taxa, differences in intelligence correlates best with differences in number of pallial or cortical neurons and synapses plus processing speed. The human brain combines large cortical volume with reasonable NPD, high conduction velocity, and high cortical parcellation. Cetaceans and elephants have much larger brains than even humans, but less cortical neurons, because of much lower NPD. This could explain why cetaceans and elephants are not as intelligent as one would expect on the basis of brain size. The contrary is the case for corvid birds with very small brains, but high NPD and IPC, which could explain why these animals reveal an intelligence comparable to primates with much larger brains. All aspects of human intelligence are present at least in principle in nonhuman primates and in some other mammals and more distant vertebrates – except syntactical language. The latter can be regarded as a very potent "intelligence amplifier".

## 2.6 Convergence or "Deep Homologies"?

An unsolved question in the study of evolution, in general, and of nervous systems and brains, in particular, is the question whether the strikingly numerous cases of similar structures found in sense organs and brains are the result of true convergent evolution (*homoplasy*) or of the action of "deep" homologies (see below). Striking examples for convergent evolution in the traditional sense are the lens eye of *Octopus* and of vertebrates; the paired "ladder"-type ventral nerve cord of annelids and arthropods; and the tripartite brain in polychetes, arthropods, and (at least embryologically) craniates, among others.

The concept of "deep homologies" attempts to describe the role of very ancient genetic mechanisms governing growth and differentiation processes in metazoans [11]. This includes so-called *homeotic genes* that control the differentiation of organisms along their body axes, of sense organs, and nervous systems and brains. They are found to be the same in very distantly related animals such as the fruit fly *Drosophila* and the clawed toad *Xenopus*. In *Drosophila* (and all other insects studied) the formation of the ventral cords and the tritocerebrum, and in *Xenopus* (and all other chordates, including vertebrates) the development of the spinal cord and myelencephalon, is controlled by *Hox* genes. On the other hand, non*Hox* genes (*otd*/*Otx*) determine the formation of the proto- and deutocerebrum of insects and of the mes-, di-, and metencephalon in chordates. The zone in between these regions, the tritocerebrum in insects and the isthmic region in vertebrates, is controlled by so-called *Pax* genes in both groups [55]. These findings could now be extended to all bilaterians and even to phyla having lost their bilateral organization in the adult stage (echinoderms), which leads to the assumption that all bilaterally organized animals possess a common "bauplan" for a tripartite brain since about 600 mya [11]. Precursors of such genes have been found even in coelenterates and sponges.

Such a new concept radically simplifies many problems concerning the evolution of nervous systems and brains, e.g., explaining the large number of seemingly "convergent" or "homoplastic" steps in the evolution of neural and sensory structures, because they appear to be based on the same regulatory developmental genes – this, however, leads to new problems, because an uncountable number of cases of secondary simplification have to be assumed. Even if there is a "deep" homology of such genes, the parts of the nervous systems and brains controlled by them may have developed independently, and the same or very similar developmental genes may have undergone a change in function. Furthermore, it is still unclear how the same regulatory genes can lead to very different structures. Insect and vertebrate brains have hardly any resemblance, and the same is true for the insect compound eye and the vertebrate lens eye, despite similar regulatory genes. These may represent basic organizational commands such as "form a tripartite brain!" or "develop a light-sensitive organ!", and it is left up to epigenetic (i.e., gene-expression regulating) mechanisms to form a pigment spot, a compound eye, or a lens eye.

## 2.7 Summary – Major Trends in the Evolution of Nervous Systems and Brains

At the level of bacteria and unicellular eukaryotes we already find a fundamental division into sensorium and motorium, with some sort of information processing in between, which involves a short-term memory and represents the basic organization of "cognitive" functions. In the earliest eumetazoans diffuse "basi-epithelial" nerve nets exist similar to that found in *Hydra*. From there, two major evolutionary trends took their course. The first represents a sidetrack leading to the evolution of ring-shaped nerve systems in cnidarians and ctenophorans ("coelenterates"), the other is the main track leading to a bilaterally organized nervous system with a circumesophageal ganglion and ventral cords having originated already in Precambrian times 560 mya or earlier. Even planarians feature such a CNS, and at this level all extant neuronal mechanisms (including ion channels, synaptic mechanisms, and transmitters) were already present. Interestingly, both "coelenterates" and bilateral organisms possess very similar genes controlling the body plan – including the brain.

From the CNS of the most primitive bilaterally symmetric animals ("ur-bilateria"), again two major evolutionary trends originated – one in the protostomes, the other in the deuterostomes. The former split into two evolutionary lineages: the lophotrochozoan and the ecdysozoan schemes – in both lines we find the formation of complex sense organs and brains. Among lophotrochozoans, this is the case for predatory platyhelminths, polychaetes, and likewise predatory cephalopods. Here we find complex lens eyes and visual systems as well as a multilobed supraesophageal ganglia, and the brain of *Octopus* is regarded the most complex protostome brain. Among ecdysozoans, arthropods likewise exhibit highly complex sense organs and brains, and those of the flies and hymenopterans (wasps, bees, ants) are considered the most complex ones. The deuterostomes, with an unresolved phylogenetic origin, likewise exhibit two major developmental lineages, one leading to the Echinodermata with radially symmetric adult nervous systems superficially resembling those of the "coelenterates"; the other line leading via hemi-, uro-, and cephalochordates – all with very simple nervous systems – to the craniates and eventually to the vertebrates. The vertebrate brain evolved very early (about 500 mya) and since then remained relatively uniform in its basic organization. Major evolutionary changes concerned the specialization, complication, and enlargement (sometimes reduction) of the different brain regions and formation of novel sensory systems such as gustatory, electroreceptive, infrared, echolocation, and visual systems. The most dramatic changes occurred in the dorsal telencephalon (pallium, cortex) together with related changes in the dorsal thalamus projecting to this structure.

Birds and, independently, mammals evolved unimodal afferents to the pallium-cortex – in birds mostly to the dorsal ventricular ridge and in mammals to the isocortex forming topographically organized sensory and motor areas. In birds as well as in mammals there is a dramatic increase in brain size and in the size of the pallium and isocortex, in particular. Among birds the largest brains are found in psittacids and corvids and among mammals the largest brains occur in primates (in all cases corrected for body size) and this correlates roughly with the degree of cognitive abilities such as learning, problem solving, use and fabrication of tools, imitation, insight, thinking, action planning, and language. Humans have the largest brain of all animals (corrected for body size) and in addition they have the largest number of cortical neurons as well as the most efficient information-processing capacities among large-brained animals and so far appear to be the only lifeform with a grammatically and syntactically structured language.

## References


1. Futuyma DJ (2009) Evolution, 2nd edn. Sinauer, Sunderland
2. Raff R (1996) The shape of life. Genes, development, and the evolution of animal form. University of Chicago Press, Chicago
3. Hennig W (1966, 1979) Phylogenetic systematics. University of Illinois Press, Urbana
4. Ghysen A (2003) The origin and evolution of the nervous system. Int J Dev Biol 47:555–62
5. Berg HC (2000) Motile behavior of bacteria. Phys Today 53:24–9
6. Armus HL, Montgomery AR, Jellison JL (2006) Discrimination learning in paramecia (*Paramecium caudatum*). Psychol Rec 56:489–98
7. Lichtneckert R, Reichert H (2007) Origin and evolution of the first nervous systems. In: Kaas J, Bullock TH (eds) Evolution of nervous systems. A comprehensive review, vol 1, Theories, development, invertebrates. Academic (Elsevier), Amsterdam/Oxford, pp 289–315
8. Bullock TH, Horridge GA (1965) Structure and function in the nervous system of invertebrates. Freeman, San Francisco
9. Satterlie RA, Spencer AN (1983) Neuronal control of locomotion in hydrozoan medusae. J Comp Physiol 150: 195–206
10. Grimmelikhuijzen CJP, Carstensen K, Darmer D, McFarlane I, Moosler A, Nothacker HP, Reinscheid RK, Rinehart KL, Schmutzler C, Vollert H (1992) Coelenterate neuropeptides: structure, action and biosynthesis. Am Zool 32:1–12
11. Hirth F, Reichert H (2007) Basic nervous system types: one or many. In: Striedter GF, Rubenstein JL (eds) Evolution of nervous systems. Theories, development, invertebrates. Academic (Elsevier), Amsterdam/Oxford, pp 55–72
12. Moroz LL (2009) On the independent origins of complex brains and neurons. BBE 74:177–90
13. Paulus HF (1979) Eye structure and the monophyly of the arthropoda. In: Gupta AP (ed) Arthropod phylogeny. Van Nostrand Reinhold, New York
14. Hennig W (1972) Taschenbuch der speziellen Zoologie, Teil 2, Wirbellose II. H Deutsch, Frankfurt
15. Kandel ER (1976) Cellular basis of behavior – an introduction to behavioral neurobiology. WH Freeman, New York
16. Young JZ (1971) The anatomy of the nervous system of *Octopus vulgaris*. Clarendon, Oxford
17. Nixon M, Young JZ (2003) The brains and lives of cephalopods. Oxford Biology, Oxford
18. Shomrat T, Zarrella I, Fiorito G, Hochner B (2008) The octopus vertical lobe modulates short-term learning rate and uses LTP to acquire long-term memory. Curr Biol 18:337–42
19. Brenner S (1974) The genetics of *Caenorhabditis elegans*. Genetics 77:71–94
20. White JG, Southgate E, Thomson JN, Brenner S (1976) The structure of the ventral nerve cord of *Caenorhabditis elegans*. Phil T Roy Soc B 275:327–48
21. Schafer WR (2005) Deciphering the neural and molecular mechanisms of *C. elegans* behavior. Curr Biol 15:R723–9
22. Withington PM (2007) The evolution of arthropod nervous systems; insight from neural development in the Onychophora and Myriapoda. In: Kaas J, Bullock TH (eds) Evolution of nervous systems. A comprehensive review, vol 1, Theories, development, invertebrates. Academic (Elsevier), Amsterdam/Oxford, pp 317–36
23. Kästner A (1969) Lehrbuch der speziellen Zoologie, Band I, Wirbellose 1. Teil. G Fischer, Stuttgart
24. Foelix RF (2010) Biology of spiders, 3rd edn. Oxford University Press, Oxford/New York
25. Mobbs PG (1984) Neural networks in the mushroom bodies of the honeybee. J Insect Physiol 30:43–58
26. de Marco RJ, Menzel R (2008) Learning and memory in communication and navigation in insects. In: Byrne JH, Menzel R (eds) Learning theory and behavior, vol 4, Learning and memory: a comprehensive reference. Academic (Elsevier), Amsterdam/Oxford, pp 477–98
27. Strausfeld NJ, Strausfeld C, Loesel R, Rowell D, Stowe S (2006) Arthropod phylogeny: onychophoran brain organization suggests an archaic relationship with a chelicerate stem lineage. Proc Biol Sci 7:1857–66
28. Holland LZ, Short S (2008) Gene duplication, co-option and recruitment during the origin of the vertebrate brain from the invertebrate chordate brain. BBE 72:91–105
29. Nieuwenhuys R, ten Donkelaar HJ, Nicholson C (1998) The central nervous system of vertebrates, 3rd edn. Springer, Berlin/Heidelberg/New York
30. Striedter GF (2005) Brain evolution. Sinauer, Sunderland
31. Northcutt RG (1987) Brain and sense organs of the earliest vertebrates: reconstruction of a morphotype. In: Foreman RE, Gorbman A, Dodd JM, Olsson R (eds) Evolutionary biology of primitive fishes, vol 103, NATO ASI series, series A: life sciences. Plenum, New York
32. Northcutt RG (1989) Brain variation and phylogenetic trends in elasmobranch fishes. J Exp Zool Suppl 2:83–100
33. Romer AS, Parson TS (1986) The vertebrate body. Saunders College Publishing, Philadelphia
34. Wullimann MF, Vernier P (2007) Evolution of the nervous system in fishes. In: Kaas J, Bullock TH (eds) Evolution of nervous systems. A comprehensive review, vol 2, Non-mammalian vertebrates. Academic (Elsevier), Amsterdam/Oxford, pp 39–60
35. Dicke U, Roth G (2007) Evolution of the amphibian nervous system. In: Kaas JH, Bullock TH (eds) Evolution of nervous systems. A comprehensive review, vol 2, Non-mammalian vertebrates. Academic (Elsevier), Amsterdam/Oxford, pp 61–124

36. Nieuwenhuys R, Voogd J, van Huijzen C (1988) The human central nervous system. Springer, Berlin/Heidelberg/New York
37. Karten HJ (1991) Homology and evolutionary origins of the "neocortex". Brain Behav Evol 38:264–72
38. Reiner A, Yamamoto K, Karten HJ (2005) Organization and evolution of the avian forebrain. Anat Rec A 287A: 1080–102
39. Medina L (2007) Do birds and reptiles possess homologues of mammalian visual, somatosensory, and motor cortices. In: Kaas J, Bullock TH (eds) Evolution of nervous systems. A comprehensive review, vol 2, Non-mammalian vertebrates. Academic (Elsevier), Amsterdam/Oxford, pp 163–94
40. Creutzfeldt OD (1983) Cortex Cerebri. Leistung, strukturelle und funktionelle Organisation der Hirnrinde. Springer, Berlin/Heidelberg/New York
41. Kaas JH (2007) Reconstructing the organization of neocortex of the first mammals and subsequent modifications. In: Kaas JH, Krubitzer LA (eds) Evolution of nervous systems. A comprehensive review, vol 3, Mammals. Academic (Elsevier), Amsterdam/Oxford, pp 27–48
42. Changizi MA (2007) Scaling the brain and its connections. In: Kaas JH, Krubitzer LA (eds) Evolution of nervous systems. A comprehensive review, vol 3, Mammals. Academic (Elsevier), Amsterdam/Oxford, pp 167–80
43. Güntürkün O (2008) Wann ist ein Gehirn intelligent? Spektrum der Wissenschaft 11:124–132
44. Jerison HJ (1973) Evolution of the brain and intelligence. Academic, Amsterdam/Oxford
45. Haug H (1987) Brain sizes, surfaces, and neuronal sizes of the cortex cerebri: a stereological investigation of man and his variability and a comparison with some mammals (primates, whales, marsupials, insectivores, and one elephant). Am J Anat 180:126–42
46. Russell S (1979) Brain size and intelligence: a comparative perspective. In: Oakley DA, Plotkin HC (eds) Brain, behavior and evolution. Methuen, London, pp 126–53
47. Falk D (2007) Evolution of the primate brain. In: Henke W, Tattersall I (eds) Handbook of paleaanthropology. Primate evolution and human origins, vol 2. Springer, Berlin, pp 1133–62
48. Hofman MA (2000) Evolution and complexity of the human brain: some organizing principles. In: Roth G, Wullimann MF (eds) Brain, evolution and cognition. Spektrum Akademischer Verlag, Heidelberg, pp 501–21
49. van Dongen PAM (1998) Brain size in vertebrates. In: Niewenhuys R (ed) The central nervous system of vertebrates. Springer, Berlin/Heidelberg/New York, pp 2099–134
50. Pilbeam D, Gould SJ (1974) Size and scaling in human evolution. Science 186:892–901
51. Roth G, Dicke U (2005) Evolution of the brain and intelligence. Trends Cogn Sci 9:250–7
52. Roth G, Dicke U (2012) Evolution of the brain and intelligence in primates. Prog Brain Res 195:413–30
53. Byrne R (1995) The thinking ape. Evolutionary origins of intelligence. Oxford University Press, Oxford/New York
54. Gibson KR, Rumbaugh D, Beran M (2001) Bigger is better: primate brain size in relationship to cognition. In: Falk D, Gibson KR (eds) Evolutionary anatomy of the primate cerebral cortex. Cambridge University Press, Cambridge/New York, pp 79–97
55. Farris SM (2008) Evolutionary convergence of higher brain centers spanning the protostome-deuterostome boundary. Brain Behav Evol 72:106–22

# 3 Ontogeny of the Vertebrate Nervous System

Salvador Martínez, Eduardo Puelles, and Diego Echevarria

The vertebrate Central Nervous System (CNS) originates from the embryonic dorsal ectoderm. Differentiation of the neural plate epithelium from the ectoderm constitutes the first phase of complex processes called gastrulation and neurulation, which culminates in the formation of the neural tube. During these processes, molecular and cellular mechanisms are orchestrated to specify the body and CNS axes. This is a required step to generate topological landmarks, from were positional information will be distributed by morphogenetic signals controlling cell differentiation and organogenesis.

The common origin of the brain in the animal kingdom is an extensively studied question, the answer of which still remaining unclear to the present. Recent comparative studies on molecular regionalization of brain development in invertebrates (mainly in *Drosophila melanogaster*) and vertebrates are showing that homologous genes control similar biological processes, suggesting common ancestor molecules regulating developmental mechanisms, which have been conserved by evolution [1]. Divergent evolution is supported by the different localization of nerve cords in arthropods (ventral) and vertebrates (dorsal). A consensual explanation is that this difference is due to the inversion of the dorsoventral body axis in one of the two animal groups. Recent developmental genetic evidence supports the dorsoventral inversion theory at the level of molecular signals regulating patterning and subsequent molecular regionalization in presumptive neural tissues (see later).

At neural-plate and neural-tube stages local signaling centers in the neuroepithelium, known as organizers, refine the anteroposterior and dorsoventral specification of different neural territories by providing the nearby cells with the proper positional information. According to this positional information, a specific set of transcription factors will be activated to locally control histogenetic processes.

In this chapter we will describe the main steps of CNS development in birds and mammals, with some comparative references to amphibians and fish. We will start from early stages of embryogenesis (gastrulation and neurulation) leading up to the formation of a variety of different neural regions by the processes known as neuroepithelial regionalization and morphogenesis. These processes contribute to the structural and functional complexity of the brain. We will emphasize the cellular and molecular events and mechanisms involved in the regionalization of the neural tube and the key role played by localized signaling centers.

## 3.1 Induction and Regionalization of the Neural Plate: The Planar Map

A fundamental early step in neural development is the allocation of a group of ectodermal cells as precursors of the entire nervous system (Fig. 3.1a). This process involves an inductive interaction first demonstrated in amphibian embryos in the 1920s [2]. The according experiments, which involved the grafting of differently pigmented species of newt, established the concept of

S. Martínez (✉) • E. Puelles • D. Echevarria
Instituto de Neurociencias de Alicante,
Consejo Superior de Investigaciones Científicas–Universidad Miguel Hernandez (UMH-CSIC),
E-03550 San Juan de Alicante, Alicante, Spain
e-mail: smartinez@umh.es

C.G. Galizia, P.-M. Lledo (eds.), *Neurosciences - From Molecule to Behavior: A University Textbook*,
DOI 10.1007/978-3-642-10769-6_3, © Springer-Verlag Berlin Heidelberg 2013

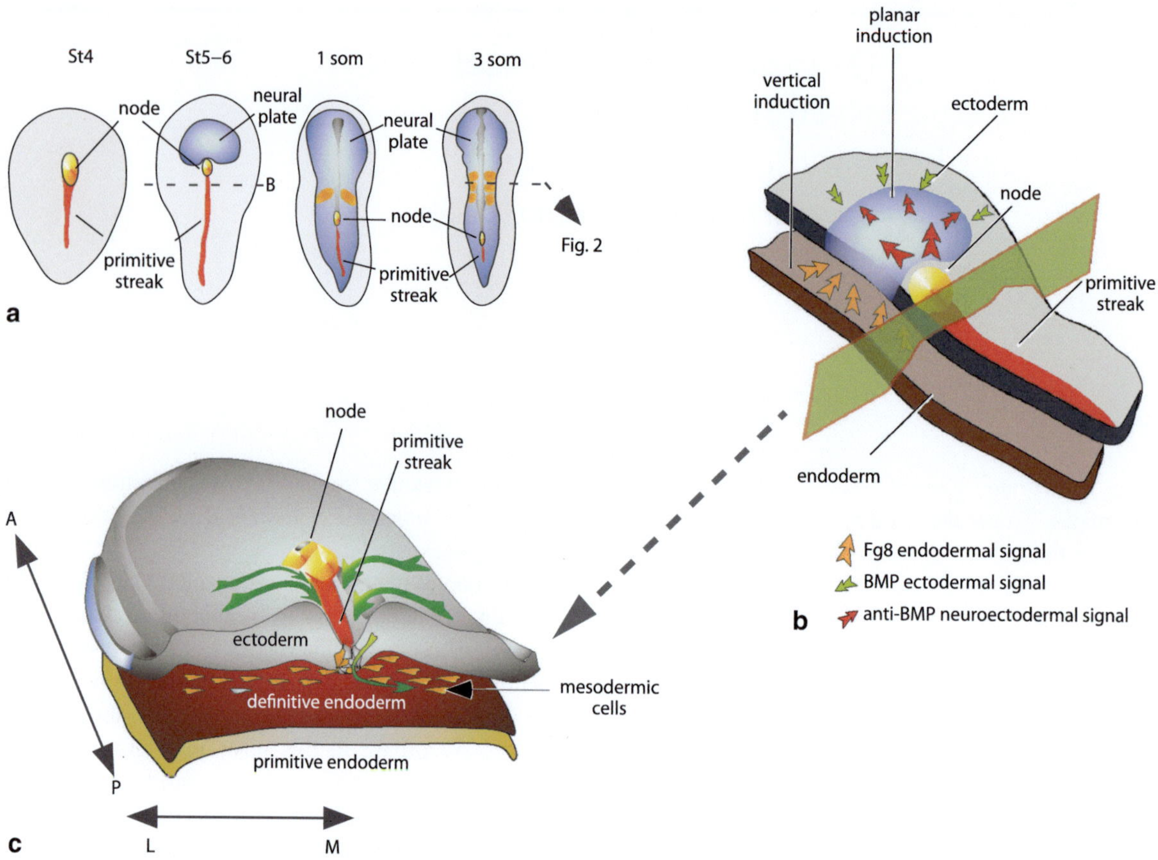


**Fig. 3.1 Neural induction and early gastrulation stage.** (**a**) Mammalian embryo drawings at different developmental stages in which the processes of neural induction and gastrulation occur. (**b**) 3D drawing of (**a**) at stages 4–6 in which vertical and planar induction of neural plates is made thus, vertical induction (*orange arrows*) from the underlying axial mesendoderm, together with planar induction from the Hensen's node (*red arrows*) and ectoderm (*green arrows*) regulate dorsoventral polarity and the initial steps of anteroposterior regionalization in the neuroepithelium. (**c**) During the process of gastrulation cells from the ectoderm detach from the ectodermal sheet at the primitive streak into the space between the ectoderm and the definitive endoderm forming the future axial and paraxial mesendodermal sheet

neural induction as an instructive interaction between the dorsal lip of the blastopore (the "**organizer**") and the neighboring ectoderm. The discovery of a neural organization center for the amphibian gastrula initiated a search for homologous structures in other vertebrates. Soon thereafter, the equivalent region was discovered in most vertebrate species, including the shield of teleost fishes. In birds and mammals, the region was named "**Hensen's node**" and "the node", respectively. When C.H. Waddington [3] transplanted the Hensen's node of a chick embryo, reproducing the initial experiment of Spemann and Mangold [2], he observed the induction of an ectopic neural plate or the formation of a partial new embryonic axis containing neural tube, notochord and somites. This demonstration provided the first evidence that in chick embryos, the nervous system is induced by signals from non-neural cells. Recent works demonstrated that the capacity of ectodermal cells to undergo neural differentiation represents their default state. In fact, neural differentiation must be suppressed in the lateral ectoderm by signals transmitted between neighboring cells, in order to develop as epidermis. These molecular signals are members of the bone morphogenetic protein (BMP) subclass of transforming growth factor-β (TGF-β)-related proteins (Fig. 3.1b).

Recent studies in chick have shown that neural induction really begins prior to the formation of the organizer region and thus must be initiated by signals derived from other cellular areas. Members of other families of signaling molecules, notably the fibroblast growth factors (FGFs), have now been proposed as early-acting factors, which initiate neural induction by a progressive sequence of molecular interactions. First, the presumptive neural plate area is established by Fgf8 activity coming from the primary endoderm (Figs. 3.1a, b). Subsequently, the suppression of BMP signaling maintains rather than initiates the process of neural differentiation [4].

These molecular interactions together with the participation of *Hox* genes regulate cellular inductive events during the process of gastrulation leading to the definition of the anteroposterior and dorsoventral axes of the embryo, as well as to the generation of the three blastodermal layers: ectoderm, mesoderm and endoderm (Fig. 3.1c). Thus, in the central area of the embryo (at its prospective dorsal region), ectodermal cells are induced to develop as neural plate cells as a result of these progressive cellular and molecular interactions, acting via planar and vertical inductions (Fig. 3.1b). Indeed, formation of the neural plate involves apicobasal thickening and pseudostratification of the ectoderm, resulting in the formation of a flat but thickened epithelial region which expresses a characteristic pattern of molecular markers.

The process of neurulation involves cell-shape changes and epithelial rearrangement generating the neural grove with the progressive apposition of its latent edges to form the neural tube (Fig. 3.2a), which represents the anlage of the CNS in sauropsids and mammals. In fishes neural-plate folding is somewhat less overt and the neural plate cells do not expose a space as they fold (neural keel and neural rod). They secondarily cavitate to form the central canal resulting in the neural tube.

During neurulation the neural plate lengthens along the anteroposterior axis and becomes narrower, so that subsequent bending will form a tube. Full anteroposterior formation and extension of the neural tube requires normal gastrulation movements and in particular, regression of the primitive streak (Figs. 3.1 and 3.2).

Experiments involving the labeling of individual cells or small groups of cells in order to analyze their fate during gastrulation and neurulation have been performed in different species. The resulting fate maps showed that the generation of the neural plate and tube involves similar morphogenetic programs in all vertebrates. During gastrulation, the bending of the neural plate, together with the intercalation of neuroectodermal precursors and regional differences in proliferation, transform the initial mediolateral arrangement of cells in the neural plate into the ventrodorsal organization of the neural tube (Fig. 3.2).

Gene expression patterns provide insights into the location, onset and developmental consequences of inductive processes, which generate regional specification within the developing brain. During the past two decades, researchers have identified many regulatory genes whose patterns of expression in the embryonic neural plate and tube have yielded important insights into brain regionalization. Interpretation of the expression patterns in terms of the topology of the neural plate axes provides a clearer picture of its molecular regionalization and also contributes to our understanding of how the expression of specific sets of genes in the neuroepithelium is related to brain histogenesis. Thus, longitudinal patterns reflect expressions that extend along parts of or the entire anteroposterior axis and may mark the longitudinal primordia of the floor, basal and alar plates (Fig. 3.2b).

## 3.2 Topologic and Topographic Patterning of the Early Brain: Dorsoventral (DV) Patterning and Anterior-Posterior (AP) Patterning

The specification of longitudinally and transversally aligned regions within the neuroepithelium involves patterning along the mediolateral (ML) and AP dimension of the neural plate.

The ML patterning in the neural plate is topologically equivalent to the DV patterning in the neural tube (Figs. 3.1 and 3.2). It has been well established that within the posterior neural plate, ML regional identities are specified in part by molecules produced by adjacent non-neural tissues. Both gain- and loss-of-function experiments have demonstrated that medial signaling is regulated by 'sonic hedgehog' (Shh) produced by the axial mesendoderm [5]. Shh is first produced by the notochord, and later its expression is induced in the overlying medial neural plate (Fig. 3.2a). Moreover, gain-of-function experiments and gene expression data support the idea that lateral signaling is coded by members of the TGF-β superfamily, such as BMP4 and BMP7 produced by non-neural ectoderm. A set of homologous genes are involved in the formation of DV regions of the developing CNS in insects and vertebrates. In *D. melanogaster* proneural clusters are arranged in three longitudinal columns on either side of the midline, similar to the three columns of neuroepithelial cell clusters that give rise to primary neurons at both sides of the floor plate in frog and zebrafish

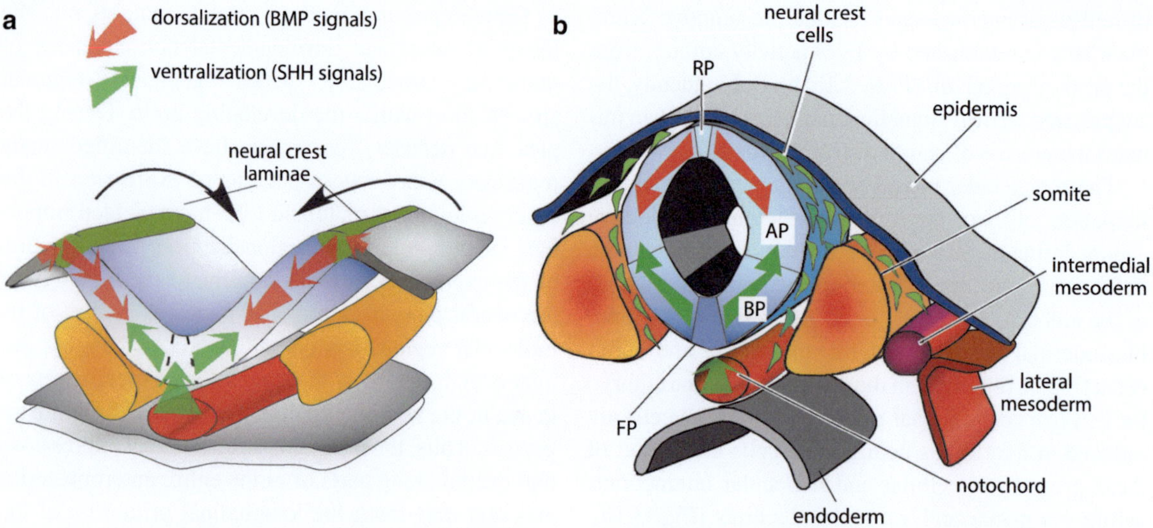


**Fig. 3.2** **Neurulation**. (**a**) During neurulation process, neural folds close at the dorsal midline forming the neural tube. Neural crest cells delaminate and migrate from the neural folds edges before closure and the neural groove become the lumen of the neural tube. Planar information from the ventral midline (*FP* floor plate, *green arrow*) and dorsal midline (*RP* roof plate, *red arrow*) plays a fundamental role in the establishment of definitive dorsoventral regionalization, using sonic hedgehog (Shh) and bone morphogenetic proteins (BMP) as signaling molecules. (**b**) As a consequence of these inductive events, the lateral wall of the neural tube is subdivided into two columnar domains: the basal plate (close to the floor plate) and the alar plate (close to the roof plate). *AP* alar plate, *BP* basal plate

embryos. Moreover, homologous genes of *Nkx2-vnd*, *Gsh-ind*, and *Msx-msh* families define and specify the medial, intermediate and lateral neurogenic columns in vertebrate and fly embryos, respectively.

Because the notochord does not underlie the anterior forebrain (the anterior end of the notochord ends at the level of the anterior diencephalon at the prethalamic basal plate), it is unclear whether patterning of the medial (ventral) forebrain is regulated by mechanisms distinct from those that occur in more posterior regions (see Figs. 3.3 and 3.4). Anterior to the notochord, an axial mesendodermal structure named the prechordal plate underlies the anterior part of the neural plate (Fig. 3.1c). Several lines of molecular and genetic evidence now suggest that ventral specification of the forebrain is regulated by the combinatory influences of notochord and prechordal plate, and involves molecular mechanisms similar to those in more posterior CNS regions. In fact, analysis of mice lacking a functional *Shh* gene has demonstrated that Shh is essential for medial patterning of the entire CNS [6]. The prechordal plate (and the dorsal foregut) induces medial (ventral) and represses lateral (dorsal) molecular properties in prosencephalic explants (Fig. 3.2b). The prechordal plate functions alone in the initial specification of the medial prosencephalon [7].

Thus, the prechordal plate and notochord share similar roles in medial neural plate specification and longitudinal neural patterning. However, they also exhibit distinct molecular properties, which may endow them with specific inductive abilities, which contribute to different properties in the overlying prechordal and epichordal neuroepithelium. In fact, epichordal diencephalic and prechordal hypothalamic regions have different inductive properties when ectopically grafted into the thalamus or telencephalon [8]. Due to this molecular and structural complexity the anterior pole of the basal plate in the hypothalamus has not yet been definitively established.

The AP patterning is the process that leads to the generation of distinct transverse domains at different axial positions. There is evidence that AP patterning begins during early gastrulation in the neural plate. Vertical signals from underlying tissues (mesoderm and endoderm) to the overlying dorsal ectoderm, and planar signals from the organizer (Hensen's node) acting through the plane of the ectodermal epithelium, contribute to the specification of AP regional differences (Fig. 3.1). The initial AP pattern is induced by the combined action of two signals produced by the dorsal (axial) mesoderm. The first signals that induce the neuroectoderm specify anterior neural fate (forebrain and midbrain). Candidate regional molecules to regulate this signal are *Lim1* and *Otx2*. Then, a graded second signal posteriorizes the neural plate, inducing hindbrain and spinal cord development. Candidate signals for the posteriorizing signal include retinoic acid, basic FGF and Wingless (WNT) signaling, regulating local expression of *Hox*-family genes. In mammals, this latter gene family comprises 39 closely related genes for homeodomain transcription factors, organized in 4 homologous clusters (A, B, C, and D) (Fig. 3.5a). The *Hox* genes have sharply defined anterior expression boundaries but their posterior boundaries are typically less clear and overlap with the expression of more posterior *Hox* genes (Fig. 3.5).

Furthermore, mesendodermal tissues underlying the anterior neural plate could also regulate regional patterns of genes, as *Orthodenticle homeobox 2* (*Otx2*) and *Engrailed* (*En1*), within the rostral brain (Fig. 3.3). Prechordal mesoderm is necessary to develop normal regionalization of the ventral prosencephalon in chick embryos. One protein that may regulate this process is named Cerberus. This secreted protein is expressed in a broad anterior domain flanking the expression of *Chordin* and *Lim1* in the prechordal plate. Thus, *Cerberus* may specify anterolateral structures, such as anterior axial mesendoderm (prechordal plate) and would then regulate medial specification within the prosencephalic neural plate. When *Cerberus* is ectopically expressed in *Xenopus* embryos, it induces nearly complete head structures [9]. Other genes such as *Noggin*, *Follistatin*, *Cripto*, and *Chordin* also induce anterior neural tissues, but these genes may not be essential for AP patterning.

We have already described that two homeodomain transcription factors, *Lim1* and *Otx2*, expressed in the tissues underlying the anterior neural plate, are essential for the development of anterior CNS structures. Loss-of-function mutants result in mouse embryos lacking forebrain and midbrain, suggesting that *Lim1* and *Otx2* have a role in early AP patterning. *Lim1* is expressed in the primitive streak and prechordal mesoderm. Because expression is not detected in the neural plate, the lack of forebrain and midbrain in *Lim1* mutants is evidence for an essential role of this mesoderm in anterior CNS development. Understanding the mechanisms underlying the *Otx2* phenotype is more

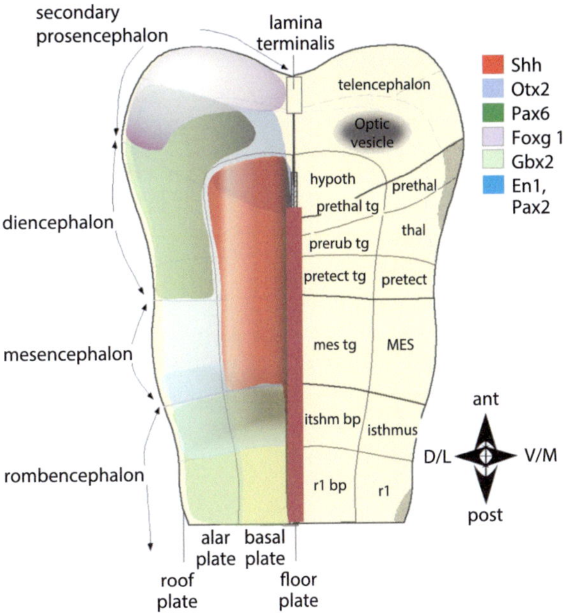


**Fig. 3.3 Gene expression pattern and fate map studies at neural plate stages**. Schematic representation of some gene expression domains at the neural plate stage and the fate map of these different brain regions. Different colors represent different genes (*gene symbol* and *color codes* are identified in the schematic diagram). Mediolateral (dorsoventral) and anteroposterior (rostrocaudal) regionalization is identified by the limits between expression domains. Longitudinal domains: *RP* roof plate, *AP* alar plate *BP* (*bp*) basal plate, *FP* floor plate. Abbreviations: *Hypoth* hypothalamus, *prethal* prethalamus tg:, *thal* thalamus, *prerub* prerubral, *pretect* pretectum, *MES* mesencephalon, *mes tg*, mesencephalic tegmentum, *r1* rhombomere 1

difficult due to the dynamics of its expression pattern and to the complexity of its molecular interactions.

A variety of evidences indicate that AP regionalization can generate transverse blocks of neuroepithelium that have distinct competence to respond to the same inductive signal. This phenomenon is clearly illustrated by inductive responses to Shh. This gene is expressed along the entire AP extent of the prechordal plate and the notochord (Fig. 3.3). Whereas Shh induces the expression of some genes (e.g., *Shh*, *HNF3β*, *Nkx2.2*) in all regions of the medial neural plate/ventral neural tube, other genes are induced within particular intervals along the AP axis. For instance, whereas *Nkx2.1* is expressed only in the prosencephalic neural plate, *Nkx6.1* is expressed in more posterior locations [10]. Thus, distinct gene expression programs at different AP positions could be due to intrinsic differences in competence of neural plate epithelium. *Fgf8* is another example of an inductive signal

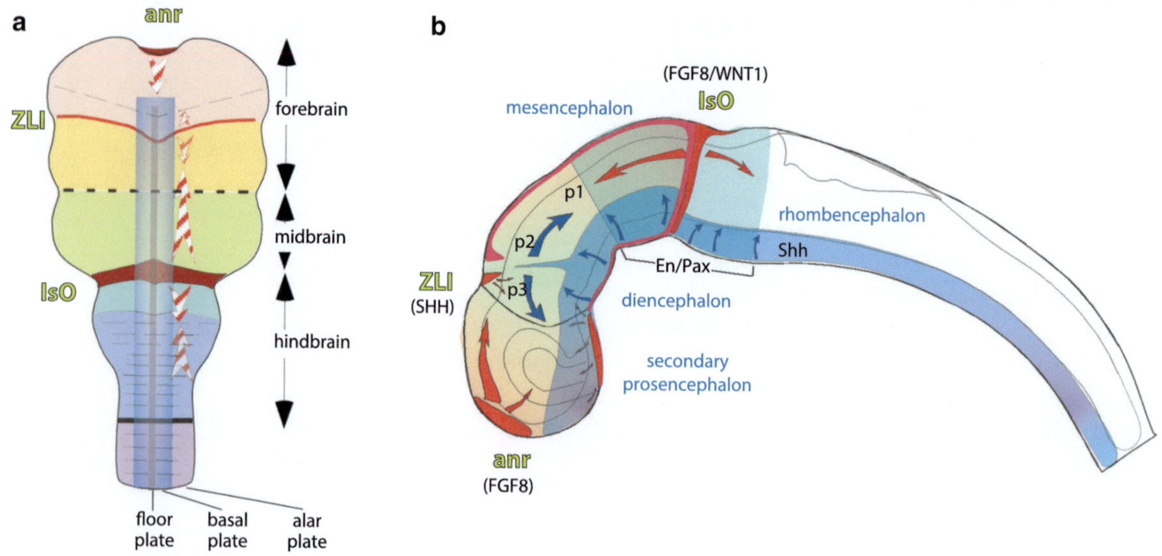


**Fig. 3.4 Secondary organizer specification and signals during neural tube regionalization.** (**a**) Schematic representation of secondary organizer specification in relation to brain domains (*color codes*) at neural plate stages. Many molecular interactions at the limits between these domains specify the development of morphogenetic organizers that generate a secondary wave of inductive signals (*dashed red arrows*) to regulate the correct patterning of structural properties in the surrounding neural tube regions. The presumptive epithelia of different brain regions have been identified, as well as the presumptive localization of the secondary organizers, in relation to precise boundaries between gene expression domains. Secondary organizers: *ANR* anterior neural ridge, *ZLI* zona limitans intrathalamica, *IsO* isthmic organizer. (**b**) Representation of a lateral view of an E9.5 mouse neural tube showing the main neuronal regions and the transverse segments of the neural tube in relation to the secondary organizers. Each secondary organizer (*written in green*) is represented and the inductive signal as *colored arrows*. In general the morphogenetic activity has a gradient effect from the source to the neighbor tissue. In the diencephalic alar plate the influence of Shh has a ventrodorsal and anteroposterior effect. In fact, Shh morphogenetic gradient activity from the basal plate and from the *ZLI* are represented by *graded blue colors*; both control the expression of diencephalic selector genes (see also Fig. 3.6c) and is fundamental for the correct specification of the diencephalic regions. *Arrows* indicate gradient morphogenetic activity by secreted molecule (signaling molecule)

that generates distinct molecular responses at different axial levels (Figs. 3.4a and 3.5). When Fgf8 is applied to prosencephalic and mesencephalic domains of neural plate explants, it induces distinct genes: in anterior epithelium it induces *FoxG1* (*Bf1*), whereas posteriorly, it induces *En2*.

## 3.3 Regionalization of the Neural Tube

Neurulation is a fundamental event of embryogenesis that culminates in the formation of the neural tube, which is the precursor of the brain and spinal cord. The bending of the neural plate involves the formation of hinge regions where the neural tube contacts surrounding tissues. Elevation of the neural folds establishes a trough-like space called the neural groove, which becomes the lumen of the primitive neural tube after closure. In addition, the neural folds will generate the specialized cells of the neural crest. The neural tube closes as the paired neural folds are brought together at the dorsal midline, establishing the roof plate of the neural tube (Fig. 3.2b).

The early neural tube is, in most vertebrates, a straight structure. However, even before the posterior portion of the tube has formed, the most anterior portion of the neural tube is undergoing drastic changes. In this region, the anterior tube balloons into three primary vesicles: the forebrain (prosencephalon), midbrain (mesencephalon) and hindbrain (rhombencephalon) (Fig. 3.4). By the time the posterior end of the neural tube closes, secondary bulges – the optic vesicles – have extended laterally from each side of the developing forebrain. At this early stage of development (three vesicle stage), the bending of the long axis, already observed at the late neural plate stages, increases considerably after neurulation, leading to the cephalic and cervical flexures of the neural tube

(Figs. 3.4b and 3.5b). Then, the prosencephalon becomes subdivided into the anterior secondary prosencephalon (telencephalon and hypothalamus) and the more caudal diencephalon.

The discovery that putative regulatory genes are expressed in regionally restricted patterns in the developing forebrain has provided new tools for defining histogenetic domains and their boundaries at higher resolution. Based on gene expression patterns as well as morphological information, two models have been used to interpret neural plate and tube regionalization: a regional-topographic model, largely aimed at saving the classic concept of sulcal division of the diencephalon into the 4 longitudinal columnar zones of Herrick and a segmental-topological model called the "prosomeric model" (Fig. 3.5b, [11]). The latter is more consistent with emergent morphological, molecular and experimental data, which cannot be satisfactorily rationalized in terms of the zones of Herrick.

The prosomeric model has been applied to explain the molecular regionalization of the neural tube in almost all vertebrate species. This topologic paradigm proposes that the embryonic forebrain is a neuromeric structure subdivided into a grid-like pattern of histogenetic longitudinal (columnar) and transverse (segmental) domains, as a result of the evolution of the Cartesian organization of the neural plate (Fig. 3.5b). The longitudinal boundaries segregate columns of cells with similar properties. These are specified by dorsoventral (DV) patterning mechanisms, which are equivalent to the lateromedial patterning mechanisms of the neural plate (Fig. 3.3). All these interactions, which occur during DV patterning, give rise to four longitudinal columnar territories, which from ventral to dorsal are the floor plate, basal plate, alar plate and roof plate (Figs. 3.2b and 3.4a).

Transverse boundaries subdivide the brain into segments (neuromeres). In the prosencephalon, these segments are called prosomeres (p1–p3 plus the secondary prosencephalon). In the rhombencephalon, the segments are termed rhombomeres (r1–r7) and pseudorhombomeres (r8–r11) (Figs. 3.4a and 3.5). The mesencephalon has not been subdivided into internal sub-segments so far, and consequently is considered as a single segmental unit. The prosomeric model has unveiled the morphological significance of numerous gene expression patterns in the forebrain, suggesting the existence of additional molecular subdivisions of the main AP and DV zones, representing histogenetically specified domains of neural precursors (Figs. 3.3 and 3.4a). Examination of this molecular-structural association has shown how the prosencephalic expression of particular genes is directly related to its specific morphogenetic and cytogenetic development. For instance, *Gbx2* expression is associated with the generation of thalamic neurons, which develop into the thalamocortical projection (Fig. 3.6, [12]), whereas *Nkx2.1* expression is associated with hypothalamic development (Fig. 3.7).

A detailed comparison of gene expression patterns and developmental neuroanatomy in vertebrates and urochordates (ascidians) has unveiled a common tripartite ground plan along the anteroposterior axis for the embryonic CNS. Anterior regions of brain primordium express *Otx* family genes, followed by the expression of *Pax* genes in midbrain and *Hox* gene family expression in the caudal neural tube (hindbrain and spinal cord anlage [13]). Recently, a similar tripartite of gene expression has been reported for arthropods and hemichordates, suggesting an evolutionary more ancient origin of this organization [1].

## 3.4 Genetic Regionalization According to Brain Subdivisions: The Importance of Secondary Organizers

Regionalization of the anterior neural plate appears to result from the superposition of multiple distinct patterning mechanisms. Initial events of AP patterning that we have described in Sect. 3.1 create transverse zones, each with a distinct histogenetic competence, while patterning along the ML axis generates longitudinally aligned domains. The combination of ML and AP patterning then generates a grid-like organization of distinct regional brain primordia (Fig. 3.5). Therefore, neural progenitors in the epithelium will establish their differentiation program under the control of positional information (as defined in a Cartesian map) transmitted by molecular signals (see also Fig. 3.4b).

Distinct neural and glial identities are acquired by neuroepithelial cells through progressive restriction of histogenetic potential under the influence of local environmental signals. Evidence for morphogenetic controlling processes at specific locations of the neural tube has led to the concept of secondary organizers, which regulate the identity and regional polarity of

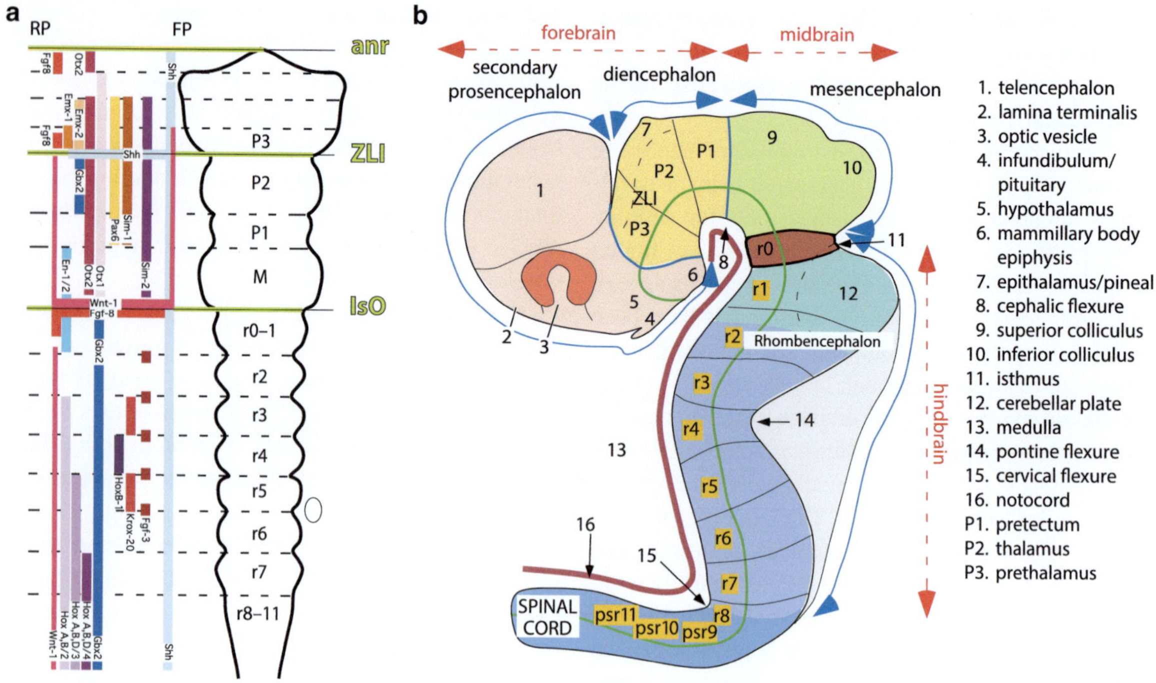


**Fig. 3.5** **Molecular pathways controlling brain regionalization.** (**a**) Planar reconstruction of gene pattern expression profile of several transcription factors and secreted molecules along the AP axis, showing the neural genetic territories influenced by the morphogenetic activity of secondary organizers (ANR *left, ZLI middle,* and IsO *right,* rostral is *up*). (**b**) Schematic representation of a midline sagittal section of an E10.5 mouse embryo showing the different transversal and longitudinal domains according to the postulated neuromeric model (the prosomeric model; Puelles and Rubenstein [11]) throughout the AP and DV axis. *Green lines* indicate the boundary between basal and alar plate form the most rostral parts of the anterior neural tube to the most caudal parts

neighboring neuroepithelial regions (Fig. 3.4). Thus, these organizers, secondary to those that operate throughout the embryo during gastrulation, usually develop within the previously broadly regionalized neuroectoderm at given genetic boundaries (frequently where cells expressing different transcription factors are juxtaposed) and their subsequent activity refines local neural identities along the AP or DV axes (Figs. 3.3, 3.4, and 3.5a).

Three regions in the neural plate and tube have been identified as putative secondary organizers (Fig. 3.4): the anterior neural ridge (ANR) at the anterior pole of the secondary prosencephalon, the zona limitans intrathalamica (ZLI) in the diencephalon, and the isthmic organizer (IsO) at the mid-hindbrain junction.

### 3.4.1 The Rhombencephalon: The Isthmic Organizer (IsO)

The rhombencephalon or hindbrain compromises the brain segment between the spinal cord and the mesencephalon. This region contains, from rostral to caudal: the isthmus, the pons (ventral), the cerebellum (dorsal), and the medulla oblongata. Homotopic and isochronic quail-chick grafting experiments performed in the late 1980s and early 1990s consistently showed that the caudal part of the early "midbrain vesicle" contains the rostral part of the prospective isthmocerebellum (Figs. 3.4 and 3.5). This caudal portion produced the isthmus, with the trochlear motor nucleus and other basal plate nuclei, as well as the velum medullare and median part of the presumptive cerebellum in its alar plate. Recent analysis of cell lineages at the mid-hindbrain limiting regions in the mouse disclosed that a clonal restriction boundary equivalent to that of interrhombomeric limits becomes established around E9.5–E10. This implies that the final fixed isthmic constriction, coincident with the fate-mapped isthmo–mesencephalic boundary, acquires intersegmental properties later than the more precocious interrhombomeric boundaries. The delay in fixing this boundary is related in part to the higher relative proliferative activity maintained in the caudal midbrain and isthmic areas.

The secondary organizer, the isthmic organizer (IsO), controls midbrain and anterior hindbrain regionalization is localized at the mid-hindbrain transition (the isthmus) (Fig. 3.4; [14]). Analysis of *Fgf8* expression in the mid-hindbrain region allows following the development of the IsO, which occurs at the expression limits of *Otx2* and *Gbx2* domains. At early neural stages, *Otx2* and *Gbx2* are, for a short period of time, co-expressed in a domain at their respective interfaces of expression, having an intracellular and/or intercellular molecular repressive interaction (Figs. 3.4b and 3.5a). *Otx2* is quickly down-regulated, being limited to the posterior part of the mesencephalon, whereas the expression of *Gbx2* is restricted to the isthmus and rhombencephalon. The isthmic organizer develops at the limit between *Otx2* and *Gbx2* expression domains (Fig. 3.5a). Loss of *Gbx2* function or reduction of *Otx2* in mice shows a re-patterning phenotype in the mid-hindbrain regions [15, 16].

An isthmic-like molecular region has been also identified in ascidian and *Drosophila* embryos, characterized by the expression of homologous genes: *Pax2/5/8* group, between *Otx-otd* genes expressed rostrally (in anterior neural anlages) and *Hox* genes expressed in the caudal regions [17].

Between the caudal and rostral limits of *Otx2* and *Gbx2*, respectively, *Fgf8* expression is induced (Fig. 3.5). This neuroepithelial region also dynamically expresses *Lmxb1*, which is required for *Fgf8* expression. Moreover, approximately at the end of gastrulation, *Pax2*, *Pax5*, *En1*, and *Wnt1* are all expressed in a transverse domain which coincides with the contact area between *Otx2* and *Gbx2*, and subsequently with the IsO, but in a domain which is broader than that of *Fgf8* (Fig. 3.5a).

The morphogenetic activity of the isthmic region was first suggested by loss-of-function experiments centered on the *Wnt1* gene. Expression of IsO signal *Wnt1* gets restricted to the caudal midbrain and *Fgf8* to the rostral R1 (Figs. 3.4b and 3.5b, [18, 19]). The early development of midbrain–hindbrain region requires both *Wnt1* and *Fgf8* signals. In *Wnt1* null mutant and *Fgf8* midbrain–R1 mutant embryos, cells in the midbrain and R1 die apoptotically around E8.5. Morphogenetic activity of the IsO was demonstrated by using quail-chick grafts in which the isthmic neuroepithelium in anterior mesencephalic and diencephalic regions induced caudal mesencephalic/rhombencephalic molecular expressions and subsequent histogenetic development [20]. FGF8 was reported to be the signal molecule associated with isthmic activity [21]. Ectopic FGF8 protein can induce IsO-characteristic and structural alterations in the diencephalic caudal prosomeres (p1–p2), the midbrain, and the hindbrain [14].

Experimental manipulations of these genes expressed in the IsO have demonstrated that both the presence of their protein products and their normal combined patterns of expression are required for the normal morphogenetic process. Mutant mice lacking *Wnt1*, *Pax2*, *En1*, *Gbx2*, or *Fgf8* do not develop isthmo-cerebellar structures [22]. In addition, experiments involving mutations of *En1/2*, *Pax2/5*, *Otx1/2* or a hypomorphic allele for the *Fgf8* gene, showed that the observed anatomical malformations in these models are due to the mismatch of the IsO [23].

Fgf8 is essential for cell survival, as evidenced by the finding that inactivation of *Fgf8* in the early neural plate causes extensive cell death throughout the mesencephalon and rostral hindbrain, resulting in full deletion of the midbrain and cerebellum. Interestingly, when *Fgf8* expression is modestly reduced, rather than eliminated, the rostral most portion of the midbrain is spared and appears normal, whereas the remaining dorsal midbrain, isthmus and cerebellum are absent. This suggests that there are regional differences in sensitivity to FGF signaling within the mesencephalon and/or rhombencephalon.

FGF signaling is mediated via receptor tyrosine kinases (RTKs). These transmembrane FGF receptors (FGFRs) activate signaling cascades including the phosphatidylinositol-3 kinase (PI3K) and Ras-ERK pathways (MAPK). The expression of molecules such as Mkp3/Sef and those belonging to the sprouty (*Spry*) family are induced by *Fgf8* expression in the organizers and may determine the spatial reduction of *Fgf8* activity in a gradient manner by interaction with the intracellular mechanism of the MAP kinase cascade. In particular, the negative feedback modulator of *Fgf8* signaling, *Mkp3*, selectively inactivates the ERK1/2 class of MAP kinases by diphosphorylation leading to catalytic inactivation. Thus *Mkp3* prevents translocation into the nucleus, resulting in inhibition of ERK1/2-dependent transcription.

Hence, FGF8 activate signal from the organizer regions and present a gradient-like distribution in the extracellular compartment. This gradient, acting through FGF receptors, activates intracellular transduction pathways which are required for cell-autonomous control of *Fgf8* expression and for the activation of expression of multiple genes necessary for a variety of developmental processes, including cell fate decisions, determination of axial polarity, and promotion of cell survival. The morphogenetic activity of the IsO is then a consequence

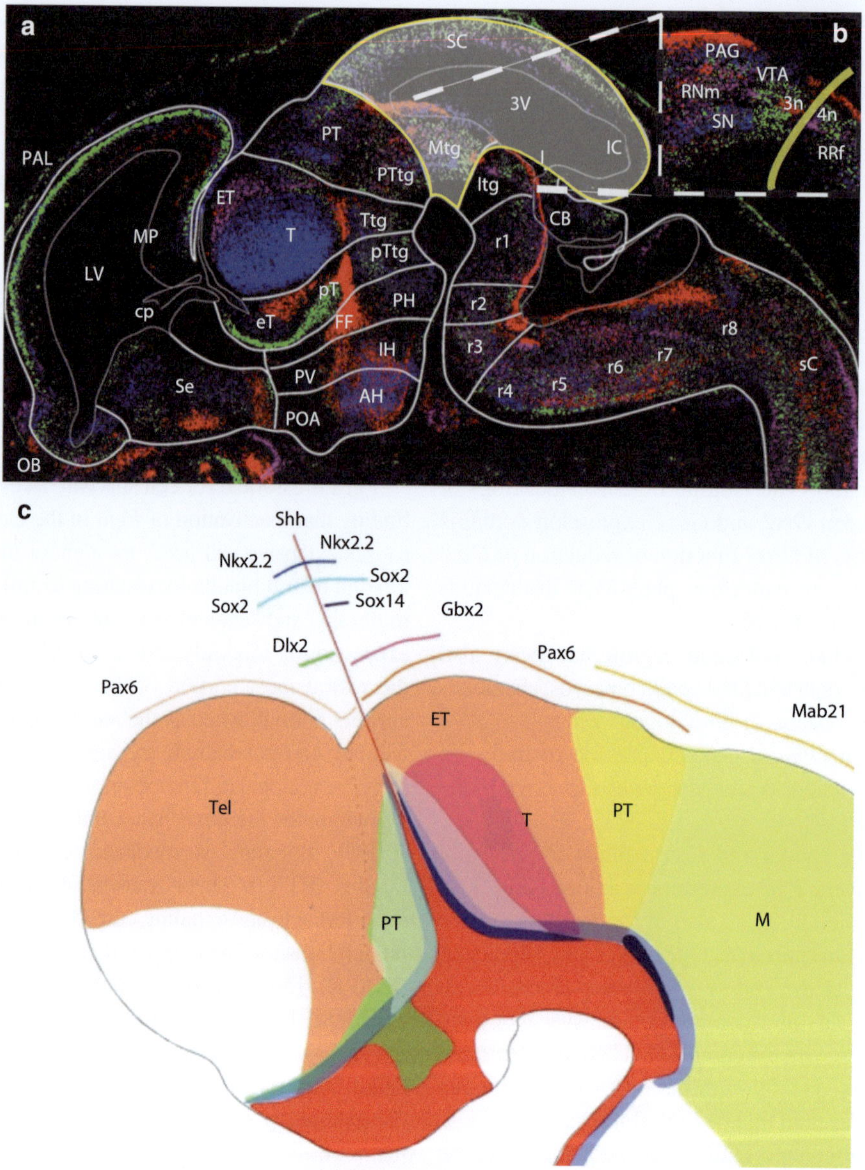


**Fig. 3.6 I-Molecular pathways controlling brain regionalization.** (**a**) Pseudocolor combined image showing the expressions of different genes (each color represent the expression of a gene) in a sagittal section of a E14.5 mouse brain. Front is to the *left* and dorsal is *up*. The predicted limits and neural regions by the prosemeric model (*white lines*) allow to easily interpret and localize the expression domains. The mesencephalon was identified between *yellow lines* and *shadowed* with a *white transparency*. (**b**) Mesencephalic tegmental (*basal*) region shown in high power to illustrate how specific genes are expressed in identifiable neuronal populations. (**c**) A representative scheme of diencephalic expressed genes: domains schematized by *colored areas* or *lines*, in relation to the *ZLI*, where *Shh* is expressed. The nested and complementary distribution of these domains at both sides of the *ZLI* suggests a mophogenetic role of this transversal band of epithelial cells. Abbreviations: *3n* third cranial nerve (external oculomotor nucleus); *3V* third ventricle, *4n* fourth cranial nerve (trochlear nerve nucleus), *AH* anterior hypothalamus, *CB* cerebellum, *Cp* caudal pallium, *eT* elimentia thalami, *ET* epithalamus, *FF* fore fields, *I* isthmus, *IC* inferior colliculus, *IH* intermediate hypothalamus, *Itg* isthmic tegmentum, *LV* lateral ventricle, *MP* medial pallium, *Mtg* mesencephalic tegmentum, *OB* olfactory bulb; *PAG* periaqueductal gray, *PAL* pallium, *PH* posterior hypothalamus, *POA* preoptic area, *PT* pretectum, *pT* prethalamus, *PTtg* prethalamic tegmentum, *PV* paraventricular hypothalamic region, *r1–r8* rhombomere 1–8, *RNm* red nucleus magnocelularis, *RRf* reticular rafe formation, *sC* spinal cord, *SC* superior colliculus, *Se* septum, *SN* substantia nigra, *T* thalamus, *Ttg* thalamic tegmentum, *VTA* ventral tegmental area

of specific molecular expression patterns, which regulate the differential specification of neuroepithelial territories. The decreasing *Fgf8* signal activity in the IsO is fundamental for cell survival and the differential development of cerebellar, isthmic, and mesencephalic structures. In the basal plate, FGF8 concentration gradients are the key players in cell survival and, together with Shh, regulate caudal serotoninergic and rostral dopaminergic fates in cellular progenitors (see below), as well as the localization and development of other basal derivatives, such as noradrenergic cells in the locus coeruleus (in the rhombencephalon) and the red nucleus (in the mesencephalic tegmentum).

### 3.4.2 The Mesencephalon

The mesencephalon or midbrain is a primary cerebral vesicle localized rostral to the isthmus with a clear dorsoventral organization (Figs. 3.4b, 3.5b, and 3.6a). The dorsal midbrain, also known generically as tectum, is divided into two main structures. The superior colliculus (SC) (rostrally) is a plurilaminated retinorecipient structure. The inferior colliculus (IC) (caudally) consists of a globular formation formed by a core and a shell, associated with the auditory pathway (Fig. 3.5b). Studies in chick embryos have identified two additional alar midbrain domains of smaller size: the tectal gray, rostral to the superior colliculus, and the pre-isthmus, caudal to the inferior colliculus (possible mammalian homologues are the misnamed posterior pretectal nucleus and the cuneiform area, respectively). This clear structural and functional division in the dorsal aspect of the midbrain cannot be traced ventrally into the basal plate (also called midbrain tegmentum), which has a rather homogenous organization along the anteroposterior length of the mesencephalon. This is the fundamental support for the idea that the midbrain consists of a single neural tube segment. The tegmentum contains some well-defined neuronal populations and other, more diffuse cellular areas (see below).

Inside the well-defined neuronal populations, the dopaminergic population is the most studied because of its relation with human Parkinson's disease. It has two main components: the ventral tegmental area (VTA) and the substantia nigra (SN) (Fig. 3.6b). The VTA is in the medial mantle zone, derived from the floor plate. Its projections innervate the nucleus accumbens and the limbic cerebral cortex, modulating cognitive processes [24]. The SN is positioned lateroventrally, mostly in the basal plate domain. It is divided into superficial (pars reticularis, or SNR) and deep (pars compacta, or SNC) portions. Neurodegeneration of these nuclei produces the motor syndrome Parkinson's disease in humans. The genetic cascade responsible for their differentiation program involves three groups of factors. One group is formed by extracellular signals inducing the dopaminergic phenotype, such as *Shh*, *Fgf8*, TGFs, and *Wnt1*. Another group, composed of *Nr4a2* (*Nurr1*), *Lmx1b*, *Pitx3*, *Otx1/2*, and *En1/2*, among others, is involved in the intrinsic cell differentiation program of this pre-specified cell type. Finally, there is a group of genes related to expression of all the enzymes, transporters, and receptors required for the proper synthesis, storage, release, and reuptake of dopamine by these neurons. Regarding signaling molecules, Shh is capable of inducing dopaminergic neurons by itself in the ventral midbrain territory, but to do so in other neural territories, it requires the presence of Fgf8. *Wnt1* plays a dual role in this mechanism, both in the establishment of the progenitor domain and in the terminal differentiation of these nuclei, clearly a direct result of the cephalic flexure of the neural tube axis; thus, all the congruently curved limits are longitudinal (Fig. 3.4). This combination of gene domains might be reflected in the dorsoventral organization of the periaqueductal gray (PAG) and reticular formation, conferring different functional roles to each cell subpopulation. Unfortunately, little is known about the genetic mechanisms involved in the development of these two populations. Thus *Shh* and *Fgf8* are the key signaling genes in the specification of all the ventral midbrain structures. Several of the transcription factors that confer the prospective mesencephalic territory with the ability and competence to receive, process, and interpret this signaling information have been identified. The choice of one differentiation program over others depends on the combination of those factors that coincide in space and time during early development of the midbrain.

### 3.4.3 The Diencephalon: The Zona Limitans Intrathalamica (ZLI)

The diencephalon is the central area of the developing brain that generates the thalamic nuclear complex and the pretectum in its alar plate, while its basal plate generates anterior tegmental structures. Morphological,

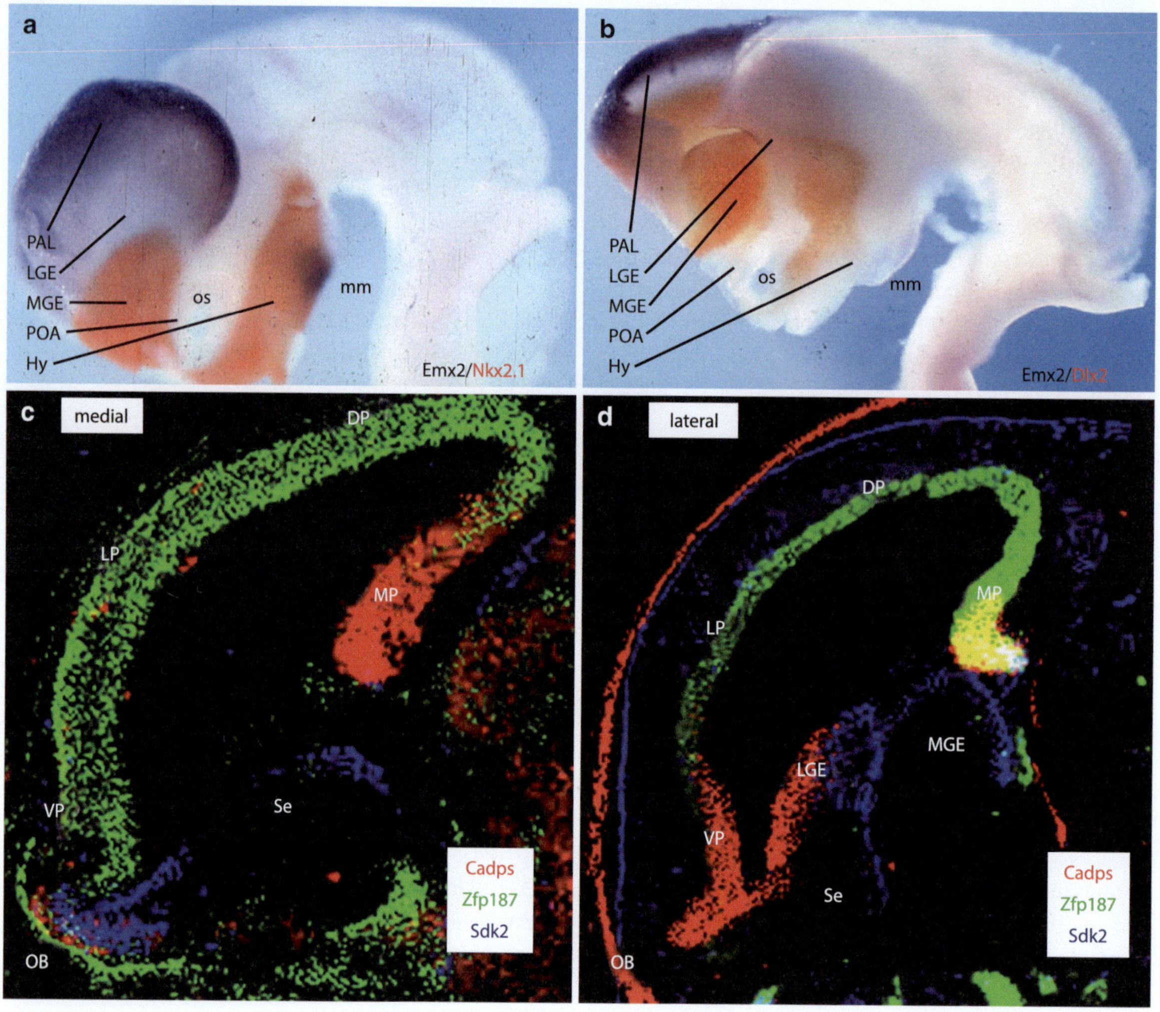


**Fig. 3.7 II-Molecular pathways controlling brain regionalization.** (**a**) Lateral view of an E10 mouse neural tube (anterior to the *left* and dorsal is *up*) processed to detect the expression of *Emx2* (*in blue*) and *Nkx2.1* (*in red*). (**b**) Lateral view of a E12.5 mouse neural tube processed to reveal *Emx1* (*blue color*) and *Dlx2* (*red color*). Note how a negative band of pallium is distributed between the subpallium and the pallium, which corresponds to the ventral pallium (*VP*). (**c**, **d**) Combination of gene expression patterns in the telencephalon, each gene expression domain colored by *Photoshop*, showing the presence of molecular regions where later in development will appear functional cortical areas. Abbreviations: *DP*: dorsal pallium, *Hy*: hypothalamus, *LGE*: lateral ganglionic eminence, *LP* lateral pallium, *MGE* medial ganglionic eminence, *mm* mammillar region *MP* medial pallium, *OB* olfactory bulb, *os* optic stalk, *POA* preoptic area, *Se* septum, *VP* ventral pallium

histogenetic, and molecular processes have showed that the diencephalon is subdivided into three transversal territories of a characteristic molecular identity, which can be described as prosomeres. In the mature brain, these prosomeres contain from caudal to rostral the following structures in their alar plates: the pretectum (in prosomere 1; p1), the dorsal thalamus and epithalamus (in prosomere 2; p2), the ventral thalamus and the eminentia thalami (in prosomere 3; p3). In their basal plates they contain structures that are classified as retromammillary, prerubral, and pretectal tegmentum, as shown by fate-mapping studies using experimental embryology or developmentally maintained gene expressions. The zona limitans intrathalamica (ZLI) is a boundary that appears as a transversal ventricular ridge at the neural tube stages, corresponding to a surface constriction, or furrow, between p2 and p3, separating the anterior (p3) from the posterior (p2 and p1) thalamic regions. This intrathalamic limit appears early at neural tube development and exhibits a unique molecular expression profile, which suggests an important role for this area as a secondary morphogenetic organizer in diencephalic histogenesis (Fig. 3.6c).

The cellular and molecular mechanisms which regulate positioning and specification of the ZLI are based on interactions between prechordal and epichordal neuroepithelia. This pre-epichordal planar interaction in the alar plate induces the expression of *Shh* in the ZLI and activates the morphogenetic properties of this organizer, specifying in turn the compartmentalization and cell fate of the different diencephalic prosomeres, through the control of specific gene expressions (Fig. 3.6c). The pattern of *Shh* expression in the ZLI is highly dynamic in both mouse and chick embryos. It starts by being limited to the diencephalic basal plate and then extends dorsally into the basal part of the presumptive ZLI epithelium, by a process of homogenetic induction. The importance of *Shh* expression in the ZLI is supported by the fact that mice with mutant *Shh* show defects in early development, with an important reduction in size of the diencephalic vesicle.

Cellular identity in the diencephalon may be under the control of genes whose expression is regulated by signaling cascades activated by ZLI-derived morphogenes (Fig. 3.6). Nested within the *Shh* expression domain, ZLI cells express the transcription factor *Sim1*. At both rostral and caudal sides of the ZLI, *Nkx2.2* and *Fgf15* are expressed. The *Gbx2* transcription factor is expressed caudal to the ZLI and serves as a marker for the thalamus. *Dlx2* and *Nkx2.1* are expressed in the alar and basal plate, respectively, just rostral to the ZLI (Figs. 3.7a, b). Also, the dorsal end of *Shh* expression in the ZLI is flanked by *Wnt1* caudally and *Fgf8* rostrally.

Shh signals from both the ZLI and the basal plate play an important role in the molecular patterning of the diencephalic alar plate (Figs. 3.4b and 3.6b; [8]). Basal plate signals could play an initial role in the specification of longitudinal territories in the thalamic area (ventrodorsal patterning), while posterior development of the ZLI represents an additional source of morphogenetic signals that superimpose anteroposterior information over thalamic epithelium (Fig. 3.6). The combinatory effects of these two types of information could contribute to the complexity of thalamic molecular regionalization and as a consequence, to its complex anatomy.

### 3.4.4 The Secondary Prosencephalon: Anterior Neural Ridge (ANR)

At early neural tube stages, the forebrain is formed simply by the primary prosencephalon. The prosencephalon subsequently develops into two AP parts: the secondary prosencephalon and the diencephalon proper (Figs. 3.4, 3.6, and 3.7). The main arguments for distinguishing these units are that the secondary prosencephalon is distinctly prechordal and participates in the formation of the eyes, the telencephalon and hypothalamus (entities which react together to developmental noxious stimuli to mainly generate holoprosencephaly), while the diencephalon is epichordal and lies caudal to the telencephalon (see below). Note that here the definition of the diencephalon differs notably from the usual textbook columnar formulation (and also from some previous neuromeric models), since the hypothalamus is excluded from the diencephalon and reinterpreted as a ventral part of the secondary prosencephalon.

The anterior secondary organizer, the ANR, was first described by Houart el al. [25] in zebrafish at the junction between the most rostral part of the neural plate, the anlage of the anterior commissure and non-neural ectoderm. Genes expressed in this region control others necessary for telencephalic regionalization. In particular, the *Fgf8* gene is expressed very early in ANR cells and has been shown to be crucial for the specification of the anterior areas of the forebrain and telencephalon. *Fgf8* expression in the ANR is necessary for the induction and/or maintenance of *FoxG1* (*Bf1*) expression, which in turn is essential for telencephalic precursor proliferation. In addition, implantation of FGF8 protein into the prospective area of the telencephalon in chick embryos generates changes in the patterns of gene expression and consequently a redistribution of telencephalic and optic derivatives. Ectopic expression of *Fgf8* in the caudal telencephalon of mouse embryos produces duplication of functional areas of the cortex. *Fgf8* regulates prosencephalic regionalization, at least in part, through inhibition of *Otx2* and *Emx2* expression and cooperation with *Bmp4*, *Wnt* and *Shh*. Other members of the FGF family, such as *Fgf15* and *Fgf17*, are also expressed in the ANR. Their expression domain are closely related to that of *Fgf8* and induced by FGF8.

Another signaling protein secreted near the ANR is Shh. Considerable evidence suggests that SHH is both necessary and sufficient for the specification of ventral structures throughout the nervous system, including the telencephalon. In fact, normal patterning in the telencephalon depends on the ventral repression of *Gli3* function by Shh and, conversely on the dorsal repression of Shh signaling by *Gli3*. Finally, the activity of transcription factor *Nkx2.1* (a homeodomain gene), required for the development of the hypothalamus and ventral forebrain, is also regulated by Shh (Fig. 3.7a, b).

#### 3.4.4.1 Telencephalon

The fundamental molecular and structural constituents of the telencephalon in mouse, chicken, frog, and fish embryos have been re-analyzed by means of abundant data on molecular regionalization. In concordance with classical descriptions the pallium and the subpallium are the most dorsal derivatives of the alar secondary prosencephalon. The pallio–subpallial boundary, defined molecularly by the interface between *Dlx* family genes expressed in the subpallium and *Tbr1* expressed in the pallium, runs all the way across the telencephalon, stretching from the amygdala, caudally, to the septum, rostromedially (Fig. 3.7b). This means that both amygdala and septum possess pallial and subpallial portions, comparable to central telencephalic parts in terms of primary genetic codes that define intratelencephalic domains: cortex, claustrum, and basal ganglia (from dorsal to ventral), independently of their secondarily specialized development (Fig. 3.7c, d). The subpallium can be divided into three parallel zones, identified as anterior entopeduncular/preoptic areas (AEP/POA), pallidum, and striatum, respectively (Fig. 3.7a, b). While all these areas express *Dlx* genes, only AEP/POA and pallidum express additionally *Nkx2.1*, and only AEP/POA expresses *Shh* in its ventricular zone (Fig. 3.7a). In the mammalian forebrain, the *Nkx2.1* gene is also expressed in the basal hypothalamus, which are severely malformed and reduced in size in *Nkx2.1*-knockout mice. As regards to the pallium, four parallel subdivisions can be postulated: the medial, dorsal, lateral, and ventral pallial portions (Figs. 3.7c, d). This implies one subdivision more than those in other pallial schemas, namely "ventral pallium", defined as an *Emx1/2* negative zone at the limit between the pallium and the subpalllium (defined by the expression of *Dlx2*; Fig. 3.7b). These results in mouse and chick embryos were reported also in the turtle and frog telencephalon, suggesting the maintenance of the four-part pallial model in tetrapods.

#### 3.4.4.2 Hypothalamus

The hypothalamus is defined classically as the ventral diencephalon, since the neural longitudinal axis was thought to follow the subthalamic sulcus into the telencephalic vesicles (Figs. 3.3 and 3.7). If we are aware of the topological organization of the neural tube, bent by cervical and cephalic flexures, and the distribution of intersegmental boundaries (Figs. 3.4 and 3.5), the hypothalamus is actually the ventral part of the secondary prosencephalon [11]. This region contains dorsal (alar) structures which form the telencephalic and optic peduncular areas, as well as midline derivatives (lamina terminalis and chiasmatic region) (Fig. 3.5a). In addition to the retinal fibers that enter and partially decussate to the contralateral side in the optic chiasm, other longitudinal tracts run through the ventral lining of the alar plate and across to the contralateral side, underneath the optic chiasm. The ventral region of the hypothalamus contains three regions: anterior, intermediate (tuberoinfundibular), and posterior (mammillary) hypothalamic area (Fig. 3.6a). The complexity of inductive interactions between prechordal and epichordal regions (discussed previously in Sect. 3.4.3), together with the induction of the tubero/pineal evagination by the endoderm of Radket's pouch makes specially complex the temporospatial patterns of gene expression and structural development.

## 3.5 Summary

The overall organization of the vertebrate CNS is largely due to the concerted action of morphogenetic signals acting during the early gastrula stage of embryonic development. Primary neural induction and fundamental anteroposterior or dorsoventral regionalization of the early neural tube is due to the activity of the "primary organizer". Slightly later in development, local signaling centers in the neuroepithelium, known as "secondary organizers", refine the anteroposterior specification of the three main domains in the brain primordium: forebrain, midbrain, and hindbrain. Additionally, the morphogenetic activity of these secondary organizers controls the polarity and the generation of neural subregions inside these main regions. Morphogenetic organizers develop in specific domains of the neuroepithelium as a consequence of interactions between two differently pre-specified zones. They confer positional identity by secreting a graded concentration of signal, which triggers concentration-specific genetic cascades, which are often conserved during the evolution in invertebrate and vertebrate brain embryogenesis. A complex combination of transcription factors refines the "inside" of the neural domains to form the adult brain as we know.

Due to space limitations, in this chapter we did not cover the development of invertebrate nervous systems. In addition to the cited references, good reviews and papers can be found for *Drosophila* in [26], for insects in general in [27], for mollusks in [28], and for nematodes in [29]. Very little is known from other animal phyla.

## References


1. Lichtneckert R, Reichert H (2005) Insights into urbilaterian brain: conserved genetic patterning mechanisms in insect and vertebrate brain development. Heredity 94:465–477
2. Spemann H, Mangold H (2001) Induction of embryonic primordia by implantation of organizers from a different species. 1923. Int J Dev Biol 45:13–38
3. Waddington CH (1936) Organizers in mammalian development. Nature 138:125
4. Stern CD (2005) Neural induction: old problem, new findings, yet more questions. Development 132:2007–2021
5. Tanabe Y, Jessell TM (1996) Diversity and pattern in the developing spinal cord. Science 274:1115–1123
6. Chiang C, Litingtung Y, Lee E, Young KE, Corden JL, Westphal H, Beachy PA (1996) Cyclopia and defective axial patterning in mice lacking sonic hedgehog gene function. Nature 383:407–413
7. Shimamura K, Rubenstein JL (1997) Inductive interactions direct early regionalization of the mouse forebrain. Development 124:2709–2718
8. Vieira C, Garcia-Lopez R, Martinez S (2006) Positional regulation of Pax2 expression pattern in mesencephalic and diencephalic alar plate. Neuroscience 137:7–11
9. Bouwmeester T, Kim S, Sasai Y, Lu B, De Robertis EM (1996) Cerberus is a head-inducing secreted factor expressed in the anterior endoderm of Spemann's organizer. Nature 382:595–601
10. Qiu M, Shimamura K, Sussel L, Chen S, Rubenstein JL (1998) Control of anteroposterior and dorsoventral domains of *Nkx-6.1* gene expression relative to other *Nkx* genes during vertebrate CNS development. Mech Dev 72:77–88
11. Puelles L, Rubenstein JL (2003) Forebrain gene expression domains and the evolving prosomeric model. Trends Neurosci 26:469–476
12. Miyashita-Lin EM, Hevner R, Wassarman KM, Martinez S, Rubenstein JL (1999) Early neocortical regionalization in the absence of thalamic innervation. Science 285:906–909
13. Holland LZ, Holland ND (1999) Chordate origins of the vertebrate central nervous system. Curr Opin Neurobiol 9:596–602
14. Martinez S, Crossley PH, Cobos I, Rubenstein JL, Martin GR (1999) FGF8 induces formation of an ectopic isthmic organizer and isthmocerebellar development via a repressive effect on *Otx2* expression. Development 126:1189–1200
15. Millet S, Campbell K, Epstein DJ, Losos K, Harris E, Joyner AL (1999) A role for *Gbx2* in repression of *Otx2* and positioning the mid/hindbrain organizer. Nature 401:161–164
16. Broccoli V, Boncinelli E, Wurst W (1999) The caudal limit of *Otx2* expression positions the isthmic organizer. Nature 401:164–168
17. Reichert H (2005) A tripartite organization of the urbilaterian brain: developmental genetic evidence form *Drosophila*. Brain Res Bull 66:491–494
18. MacMahon AP, Bradley A (1990) The *Wnt-1* (*int-1*) proto-oncogene is required for development of a large region of the mouse brain. Cell 62:1073–1085
19. Thomas KR, Capecchi MR (1990) Targeted disruption of the murine *int-1* proto-oncogene resulting in severe abnormalities in midbrain and cerebellar development. Nature 346:847–850
20. Martinez S, Wassef M, Alvarado-Mallart RM (1991) Induction of a mesencephalic phenotype in the 2-day-old chick prosencephalon is preceded by the early expression of the homeobox gene *en*. Neuron 6:971–981
21. Crossley PH, Martinez S, Martin GR (1996) Midbrain development induced by FGF8 in the chick embryo. Nature 380:66–68
22. Joyner AL (1996) *Engrailed*, *Wnt* and *Pax* genes regulate midbrain–hindbrain development. Trends Genet 12:15–20
23. Vieira C, Pombero A, Garcia-Lopez R, Gimeno L, Echevarria D, Martinez S (2010) Molecular mechanisms controlling brain development: an overview of neuroepithelial secondary organizers. Int J Dev Biol 54(1):7–20
24. Prakash N, Wurst W (2006) Genetic networks controlling the development of midbrain dopaminergic neurons. J Physiol 575:403–410
25. Houart C, Westerfield M, Wilson SW (1998) A small population of anterior cells patterns the forebrain during zebrafish gastrulation. Nature 391:788–792
26. Boyan GS, Reichert H (2011) Mechanisms for complexity in the brain: generating the insect central complex. TINS 34:247–250
27. Reichert H, Boyan G (1997) Building a brain: developmental insights in insects. Trends Neurosci 20:258–264
28. Lee PN, Callaerts P, de Couet HG, Martindale MK (2003) Cephalopod *Hox* code and the origin of morphological novelties. Nature 424:1061–1065
29. Hobert O (2010) Neurogenesis in the nematode *Caenorhabditis elegans*. WormBook 4:1–24

# 4 Diseases

Jean-Jacques Hauw, Marie-Anne Colle, and Danielle Seilhean

Diseases affecting the nervous system encompass a broad spectrum of disorders which induce, for example, abnormalities of motor, sensory, vision, taste, olfaction, sleep, memory, cognitive functions, or behavior. They may involve primarily the nervous system or may be part of a general disorder (infectious disease, cancer…). They can be of genetic, environmental, or mixed, often still partly obscure, mechanisms. Their diagnosis may be easily performed by the usual clinical examination (history taking and physical evaluation such as assessment of gait, muscle power testing, deep tendon reflexes evaluation…), but a number of modern investigations, such as neuroimaging, metabolic, or genetic studies are now widely used. These diseases benefit by the outburst of knowledge in the neurosciences. The heavy cost and high degree of disability that some of these conditions induce, along with their rising occurrence and associated increased life expectancy, in most countries induce a major public health concern. As an example, the prevalence (number of affected persons) of Alzheimer disease (see Sect. 4.2.10) is estimated to increase from 1.5 % in the age group 65–70 to 6 % in for people age 75–80, and over 25 % above 85 years of age.

J.-J. Hauw (✉)
Académie Nationale de Médecine, 16 rue Bonaparte, 75272 Paris, Cedex 06, France
e-mail: jjhauw@laposte.net

M.-A. Colle
Ecole Nationale Vétérinaire, Agroalimentaire et de l'Alimentation Nantes-Atlantique (Oniris), CS 40706, F-44307 Nantes Cedex 03, France
e-mail: marie-anne.colle@oniris-nantes.fr

D. Seilhean
Service de Neuropathologie, Groupe Hospitalier Pitié-Salpêtrière, AP-HP, CR-ICM U 975, UPMC-Sorbonne Universités, 75013 Paris, France
e-mail: danielle.seilhean@psl.aphp.fr

## 4.1 Main Mechanisms of Cell Pathology in the Nervous System

Injuries (inducing "lesions") of neurons generally effect changes in other cells of the central (CNS) or peripheral (PNS) nervous system in all animal species. In mammals these include glial (or neuroglial) cells, i.e., astrocytes and ependymal glial cells, oligodendrocytes, microglia, some stem cells (located especially in the paraventricular area, the hippocampus and the olfactory bulb) and the supporting structures (connective tissue including blood vessels, site of the blood-brain barrier, and meninges). In the PNS, neuronal processes are surrounded by Schwann cells (the oligodendrocyte counterpart), and by endoneural and perineural cells. Contacts with muscle cells are established through synaptic (neuromuscular) junctions.

In some pathologies, selected cells are preferentially disturbed: in degenerative disorders, neuronal degeneration and death are the main pathological processes; in multiple sclerosis and some peripheral neuropathies, oligodendrocytes or Schwann cells are more specifically affected [4].

### 4.1.1 Neuronal Degeneration and Death

Neuronal death may be acute (few minutes or hours) as in severe anoxia (lack of supply of oxygen to the brain), which leads to **necrosis** (Fig. 4.1a). In subacute or

C.G. Galizia, P.-M. Lledo (eds.), *Neurosciences - From Molecule to Behavior: A University Textbook*,
DOI 10.1007/978-3-642-10769-6_4, 

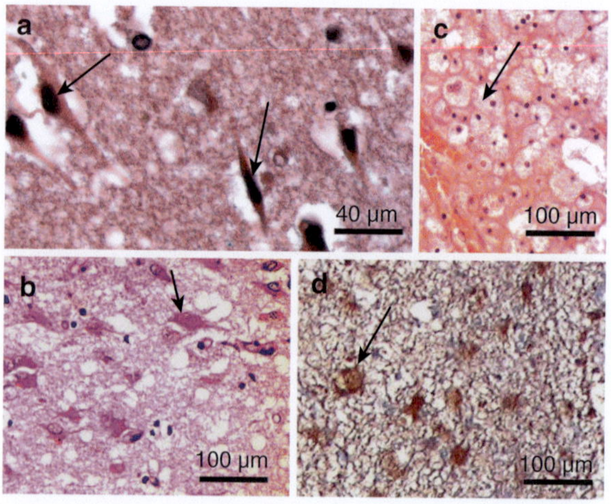


**Fig. 4.1** (**a**) Histological view of a human cerebral infarct with *neuronal necrosis* after ischemia, indicated by marked pink staining of the cytoplasm by eosin dye, shrinkage and excessive dark blue staining of the nucleus by hematein dye, with disappearance of the nucleolus (*arrow*). Necrotic neurons are surrounded by a clear area indicating *cytotoxic edema* of the adjoining astrocytes (see text). In (**b**), microglial cells have enlarged after phagocytosis of necrotic material (*arrow*). In (**c** and **d**), astrocytes proliferate and enlarge (*arrows*) ("*astrocytic gliosis*"). (**a–c**): hematein-eosin stain; (**d**): immunohistochemistry *for glial fibrillary acidic protein* (*GFAP*), the most widely used marker for astrocytes

chronic conditions, cell death often results from **apoptosis** or programmed cell death, an active, energy-consuming process. Various signals may trigger, inhibit, or reduce apoptosis which is controlled by very complex mechanisms. These include "extrinsic inducers" (toxins, hormones, growth factors, nitric oxide, or cytokines, mitochondrial induced apoptotic cascade, abortive induction of re-entry into the cell cycle of post-mitotic neurons) or "intrinsic inducers" of genetic or epigenetic origin. Some treatable mechanisms for chronic cell death are oxidative stress, NMDA glutamate receptor toxicity, chronic inflammatory processes. Neuronal death can lead to atrophy of the nervous tissue seen by neuroimaging techniques or post-mortem examination. Fragmentation of neuronal processes ("neurites") without nerve cell death can occur in acute conditions such as trauma. Before neuronal death, axonal degeneration and synaptic loss occurs in several chronic conditions. Some diseases affect specifically the synapses.

### 4.1.2 Glial Reactions

**Microglial cells**, the resident macrophages of the brain, scavenge the CNS for any pathological compound and when a target is found, they become "activated". They secrete cytotoxic and pro-inflammatory signaling molecules. They act as antigen-presenting cells activating the few T-cells which physiologically cross the blood-brain barrier; the blood lymphoid cells and monocyte/macrophages converge towards the inflammatory area, they proliferate and phagocyte foreign materials (Fig. 4.1c). In subacute inflammatory disorders, they are present around the vessels and through the brain tissue, together with lymphocytes of B and T lineages; in chronic inflammation, they are seen in a number of lesions such as plaques of Alzheimer's disease. In addition, in healing processes, microglia remove the damaged neurites and synapses from preserved neurons, promoting regeneration of injured neural circuits. **Astrocytes**, although more resistant than neurons to acute injuries, can also die under these conditions. In subacute or chronic, and even in some acute injuries of the central nervous system, they proliferate and enlarge ("astrocytic gliosis", Fig. 4.1b, d) in a few days. They are considered part of the reactive inflammatory process of the CNS, ultimately forming a glial scar which can inhibit axonal regeneration in pathological processes such as nervous system infarction or spinal cord trauma. **Oligodendrocytes** lose their myelin compacted process (myelin sheath) in axonal degeneration ("secondary demyelination"). They can be more selectively involved in some disorders such as leukodystrophies (delayed or defective myelination due to an abnormal metabolism) or multiple sclerosis (focal loss of apparently normal myelin by an immune reaction). This leads to so-called "primary demyelination", often called in simple terms "demyelination".

### 4.1.3 Cerebral Edema (Fig. 4.2)

In most acute and some subacute pathological processes, the CNS volume enlarges. Three mechanisms are responsible for this: (i) vasogenic edema, resulting from increased permeability of the blood-brain barrier and consequent raise of serum proteins in the extracellular space; (ii) cytotoxic edema, due to excessive amounts of water entering the intracellular compartments of the CNS, resulting from an impaired capacity of the cells to maintain ionic homeostasis by failure of ion pumps (e.g., ATP-dependent sodium pump); (iii) interstitial or hydrocephalic edema due to the obstruction of the pathways of cerebrospinal fluid produced by the choroid plexus (or, seldom, to an overproduction of the cerebrospinal fluid by a tumor); it accumulates in the extracellular spaces. Cerebral edema causes increased intracranial pressure which may induce herniation of the brain (shifting of some regions of the nervous tissue across some openings in the skull) and irreversible brain damage [4].

#### 4.1.3.1 Main Mechanisms of Tissue Pathology in the Brain

- Cerebral infarction (Fig. 4.2) is the consequence of ischemia (absence or drastic decrease in blood flow), usually due to occlusion (or marked narrowing) of one or several arteries. This leads to necrosis of the whole, or part of, the territory of the brain supplied by that artery.
- Infectious diseases are due to infection of more or less selected areas of the brain (and sometimes of meninges). The nervous system can be specifically (or mainly) involved (e.g., in the human prion disease called Creutzfeldt-Jakob disease). In other diseases (e.g., in AIDS), a number of other organs are also affected.
- Nutritional deficiencies such as vitamin B deficiencies can induce brain disorders.
- Immunopathologic disorders are involved in a number of diseases such as myasthenia gravis, narcolepsy or multiple sclerosis. The origin is often still poorly understood, but combination of genetic and remote infectious processes is often assumed.
- Tumors of the nervous system may be "primary" (issued from cells of the nervous system, the most frequent being of astrocytic, oligodendroglial, and meningeal lineages) or "secondary", i.e., metastases to the brain from primary tumors of lung or breast, for example.
- Degenerative diseases are due to progressive deterioration of neuronal systems due to various, often still vaguely understood, processes. Genetic and epigenetic mechanisms as well as environmental factors may interfere [4, 5].
- Other disorders, such as congenital malformations, epilepsy, trauma, and involvement of the brain in various conditions such as toxic or metabolic disorders will not be considered here.

Depending on the mechanism, the clinical onset of the disease is *acute* (large infarct involving areas of the nervous system which govern recognizable functions, some nutritional deficiencies), *subacute* (infectious diseases, tumors…), *remittent* (multiple sclerosis), or *chronic* (most nutritional deficiencies, degenerative diseases), the latter being particularly slow, with a preclinical stage lasting sometimes over 10 or 20 years.

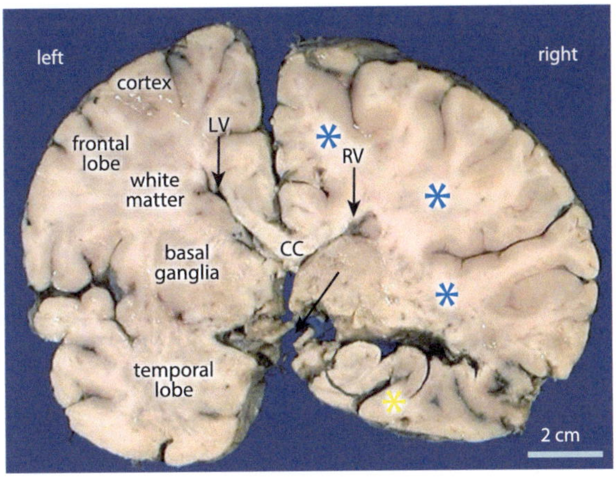


**Fig. 4.2** Macroscopic coronal brain section of a human brain with brain edema due to massive recent infarction of the right internal carotid artery territory (*blue stars*). Note the swelling of the brain hemisphere involving mostly the white matter, which spares the hemispheric territory of the right posterior artery territory supplied by the basilar artery (*yellow star*). The edema induces the collapsus of the ventricular cavities (including right and left lateral ventricles, *RV* and *LV*) and herniation of the brain (*long arrow*). Compare with Fig. 4.8

#### 4.1.3.2 Specificity of the Pathology of the Nervous System

The symptoms (abnormal feelings noticed by a patient) and signs (abnormal observations made by the examiner) of diseases of the nervous system are essentially linked to the affected areas or system(s) often governing primary functions (motor, sensory) or more complex tasks (language, problem solving, everyday jobs). Few pathologies involve a single function – an example being classic amyotrophic lateral sclerosis, a degenerative disease which affects only the motor system. Most disorders (infarction, tumor, infection) involve several systems, in an acute, subacute, or progressive way. The more we know, the more complex are the mechanisms that we learn to understand. Parkinson's disease, a degenerative disorder, was thought to be due solely to a single defect: loss of dopaminergic neurons of the substantia nigra, where abnormal inclusions, Lewy bodies, were noticed. The patients indeed improved when L-DOPA was administered, which is a dopamine precursor.

Today it is evident that a number of other systems are involved, and also that the main compound of Lewy bodies, α-synuclein, the hallmark of Parkinson's disease, is found in various regions, including the olfactory bulb and the enteric peripheral autonomic nervous system where it may be used for diagnostic purpose [7].

## 4.2 Representative Diseases

### 4.2.1 Human Myasthenia Gravis

Myasthenia gravis (MG) [14, 16, 19, 20] is a disorder of the neuromuscular junction of skeletal muscles. It is thought to be an autoimmune immunopathological polyclonal process implicating B and T cells. It involves the ionotropic acetylcholine receptor (AChR). Antibodies bind to different regions of the AChR, including the alpha subunit governing the opening of the gate. This impairs the neuromuscular transmission by various mechanisms: blockage of the binding of ACh to AChR, cross-linkage of adjacent AChR, increase of AChR internalization rate, triggering of complement-mediated destruction of the postsynaptic membrane, and therefore decrease in the number of AChRs with simultaneous reduction in new AChRs. The most consistent pathological changes in skeletal muscles are scattered degenerating fibers, lymphocytic infiltrates and reduced enzymatic activity for choline acetyltransferase and cholinesterase. Electron microscopy of end plate has shown the simplification of the postsynaptic regions and decreased areas of related nerve terminal, studies with radioactive α-bungarotoxin reduction of the number of junctional acetylcholine receptors. The nervous system *per se* is unaffected.

MG is a rare disease. The prevalence (number of cases in the population at a given time) is estimated to be 20 per 100,000. It is more frequent in women under 40, and people from 50 to 70, of either sex.

The main characteristic of MG is fatigability. Weakness increases progressively during activity and improves with rest. Extra-ocular muscles that control eye and eyelid movements, inducing ptosis (floppy eyelids), are usually the first to be affected, then muscles that influence facial expressions, talking, swallowing and breathing and, later, the muscles in charge of neck and limb movements. Often, the physical examination is within normal limits. The disease generally progresses over weeks to months, and affects hands and fingers ("generalized disease"). A sudden exacerbation ("myasthenic crisis") with paralysis of the respiratory muscles requiring assisted ventilation can occur at any time, often due to an intercurrent illness, stress or inappropriate medication.

MG is often associated with other autoimmune conditions (e.g., thyroid diseases, rheumatoid arthritis) and is found in rare families with these conditions. In 65 % of patients, it is linked to a hyperplasia (benign enlargement) of the thymus which plays an important role in the development of immunity related T cells, in 5 % to a thymic tumor (thymoma).

Laboratory studies currently performed are (*i*) *blood tests* for antibodies against the AChR, the MusK protein (a receptor tyrosine kinase needed for the formation of the neuromuscular junction), striated muscle, titin and ryanodine receptor (RyR) antibodies; (*ii*) *electrodiagnostic studies*, performed in a proximal muscle; single-fiber electromyography records abnormal variability of the interpotential interval between two or more single muscle fibers of the same motor unit; low-frequency repetitive nerve stimulation, by stimulating a nerve-muscle motor unit with short sequences of rapid, regular electrical impulses, before and after exercising the motor unit, measures the fatiguability of the muscle; (*iii*) other tests include *the "Edrophonium test*": intravenous administration of a short-acting AChE inhibitor temporarily reverts muscle weakness in patients with MG; (*iv*) *imaging*: a chest X-ray is frequently performed to eliminate alternative diagnoses (e.g., a Lambert-Eaton syndrome due to a lung tumor). Imaging studies are usually done to search for a thymoma; (*v*) a *muscle biopsy* is indicated only if the diagnosis is in doubt or for research purpose.

Medications consist of cholinesterase inhibitors and immunosuppressant drugs. For emergency treatment or in preparation for surgery, extracorporal treatment of plasma, plasma exchanges from donors or intravenous immunoglobulins can be used as a temporary measure to remove antibodies from the blood circulation. Thymectomy is an essential treatment option if a thymoma is present.

The overall life expectancy is normal or subnormal, with the exception of those patients with a malignant thymoma, whose prognosis is linked to that of the tumor. Quality of life depends on the severity of the disease. It usually reaches a steady state 3 years after the onset and may improve after this date.

> Myasthenia gravis is a rare human auto-immune disease affecting the cholinergic receptors of the neuromuscular junction, and occurring often in association with pathologies of the thymus which plays an important role in the development of immunity related T cells.

### 4.2.2 Vitamin $B_1$ (Thiamine) Deficiency in Cattle: Polioencephalomalacia

A progressive encephalopathy secondary to *thiamine deficiency* or a disturbance of thiamine metabolism is reported in a large spectrum of domestic animals, including ruminants and carnivores, and in humans (beriberi and Wernicke-Korsakoff encephalopathy). In cattle, sheep, and goats, "polioencephalopathy" (lesion of the gray matter) is a necrosis selectively affecting the pyramidal neurons of the cerebral cortex in association with a cytotoxic edema of neighboring astrocytes. It involves the parieto-occipital areas of the cerebrum (Fig. 4.3). In fact, similar lesions can result from other causes, such as excess dietary sulfur which inactivates thiamine, lead poisoning, or water deprivation. It classically occurs in young animals (peak incidence 9–12 months of age) well fed with high level grain diets for several weeks. Outbreaks can occur suddenly, affecting up to 25 % of groups of feeder cattle. Sporadic episodes affect older animal.

Clinical signs of the disorder are a sudden onset of blindness, muscle tremors, walking aimlessly, ataxia (lack of coordination of motor activity), and a progressive separation from the group. Thiamin and erythrocyte transketolase activity are difficult to interpret. A number of new techniques (capillary electrophoresis and in-capillary enzyme reaction methods) have emerged as potential alternative techniques for the determination of blood thiamin levels.

Acute polioencephalomalacia (PEM) is an emergency: when treatment by intravenous administration

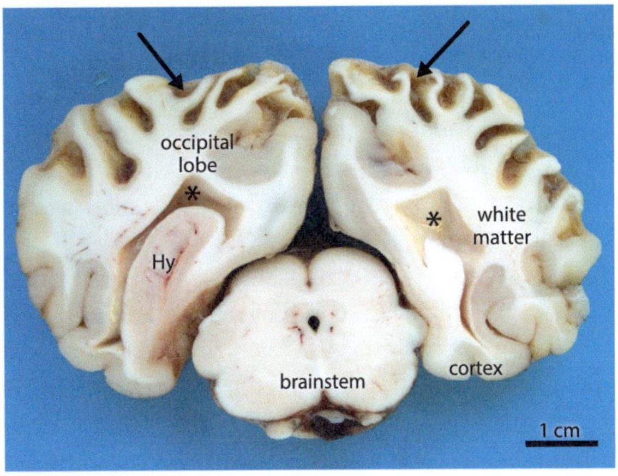


**Fig. 4.3** Chronic thiamine deficiency polioencephalopathy in a cow. Macroscopic coronal brain section reveals bilateral extensive loss of dorsal cerebral cortex in occipital areas (*arrows*), *Asteriscs* indicate the ventricles, *Hy* hippocampus

of thiamin hydrochloride is given within a few hours of the onset of signs, a beneficial response is common within 1–6 h and a complete clinical recovery can occur in 24 h.

> Polioencephalomalacia (PEM) in the cattle is a selective necrosis of the pyramidal neurons of the cerebral cortex due to deficiency or disturbance of the metabolism of vitamin $B_1$.

### 4.2.3 Human Narcolepsy

Narcolepsy is a chronic sleep disorder mainly due to a hormonal defect [14, 20]. It is classified into three subtypes:

*Narcoplexy with cataplexy*, defined by excessive daytime sleepiness occurring as sudden and uncontrollable episodes of sleep attacks, and whole body sudden loss of muscle tone (cataplexy). It could be associated with paralysis before falling asleep or upon awakening (sleep paralysis). Another symptom is hallucinations when falling asleep (hypnagogic hallucinations). Nearly all patients with idiopathic narcolepsy have a massive loss of orexin/hypocretin producing neurons and astrocytic gliosis of the hypothalamus, associated with very low levels of hypocretin 1 in the cerebrospinal fluid (CSF). These lesions have been attributed to a burnt out chronic inflammatory/autoimmune process, sometimes linked to infectious diseases or vaccinations. The gliosis extends to proven or supposed projections of the hypothalamic hypocretin system, most of them involved in sleep and especially arousal, while other areas involved in arousal, such as the thalamus, are not affected [6].

The diagnosis is confirmed by (*i*) nocturnal polysomnography (a comprehensive recording of the physiological changes occurring during sleep) showing the intrusion of rapid eye movement (REM) sleep during wakefulness, without the initial non-REM phases of sleep; (*ii*) Multiple sleep latency test, measuring the time to fall asleep in the day; (*iii*) Assessment of hypocretin level in the CSF. In more than 92 % of patients, an area of chromosome 6, the HLA complex, predisposes to narcolepsy. There are also associations with various regions of chromosomes 17, 4, and 21.

In *narcoplexy without cataplexy* only some patients will develop cataplexy. There are genuine narcolepsies and idiopathic hypersomnias to be distinguished from the more common consequence of sleep deprivation. Generally low levels of hypocretin 1 can be found, and in 10 % of patients levels are very low.

*Symptomatic narcolepsy* is often due to lesions located either in the hypothalamus or along the hypothalamic hypocretin projections. In some patients with Lewy body disease, a neurodegenerative disease related to Parkinson's disease, symptomatic narcolepsy is an early symptom.

There is today no specific treatment for narcolepsy. Stimulants, sodium oxybate, antidepressants, and other drugs that suppress REM sleep can be used but side effects may occur. Lifestyle changes are important, such as scheduling regular short naps (10–15 min) to help control excessive daytime sleepiness. Research studies include novel drugs and transplantation of hypocretin neurons.

> Human narcoplexy is a disorder of sleep organization manifesting by uncontrollable episodes of REM sleep (sleep attacks: cataplexy) due to loss of the hypothalamic neurons producing the hormone orexin/hypocretin.

### 4.2.4 Human Hereditary Cerebellar Ataxias

Ataxia [14, 20] is characterized by lack of coordination of motor activity inducing errors in rate, direction, force, timing. It may involve limbs, bulbar or eye muscles, affecting gait and hands, speech and eye movements. It can be due to three main mechanisms: dysfunction of cerebellum, sensory peripheral afferents or vestibular system. It is acute, when resulting, for instance, from a benign paroxysmal positional vertigo, subacute, for example in cerebellar tumors, has episodic course in multiple sclerosis, or has a chronic progressive progression in chronic alcoholism or in genetic spinocerebellar degenerations or hereditary ataxias [1].

The diagnosis requires: (*i*) *detecting the typical symptoms and signs*: poor coordination of movement and unsteady gait, often associated with dysarthria (motor speech disorder) and nystagmus (involuntary eye movement); (*ii*) *testing non-genetic causes*: alcoholism, vitamin deficiencies, multiple sclerosis, vascular disease, primary or metastatic tumors; (*iii*) *documenting the hereditary nature* by recognition of clinical associations of symptoms and signs characteristic of a genetic form, finding a family history of ataxia or identifying an ataxia-causing mutation. Inheritance can be autosomal dominant, autosomal recessive, X-linked, or mitochondrial.

With the exception of ataxia with vitamin E deficiency, there are no specific treatments for hereditary ataxias. Rehabilitation medicine, occupational and physical therapy are used. Genetic counseling for issues related to risk assessment and testing of at-risk relatives has to be performed in specialized centers.

> Hereditary cerebellar ataxias are a group of genetically induced diseases characterized by the lack of coordination of muscle movement by the cerebellar system, often associated with other disorders.

### 4.2.5 Human and Animal Rabies

Rabies [14, 16, 18, 20] is an endemic disease due to an enveloped RNA rhabdovirus. The disease is characterized by a lethal infection of PNS ("ganglionitis") and

CNS ("encephalomyelitis"). The virus is transmitted by bite inoculation. It can affect humans (55,000 people die of rabies every year), domestic and non domestic animals (e.g., dogs, foxes, bats). The incubation period before development of neurological signs usually is of 1–3 months but can be decreased by the anatomical location of the inoculation, viral strain or the quantity of inoculated virus. During the incubation period the virus can replicate locally in muscle cells or attach directly to nerve endings, then migrates along peripheral nerves towards the CNS by retrograde transport. From the brain and spinal cord, there is a third phase with centrifugal spread along peripheral nerves to many organs (including viscera, skin, salivary glands). Transmission occurs from neuron to neuron across synaptic junctions. When the virus reaches the CNS, there is a massive replication within neurons, many harboring in their cytoplasm characteristic ovoid "inclusion bodies", called Negri bodies, which primarily house the transcription and replication of the viral genome. They can be found anywhere in the CNS and in peripheral ganglia but, in dog, are more readily encountered in hippocampal neurons and in cerebellar Purkinje cells (Fig. 4.4). CNS lesions predominate in the gray matter ("polioencephalitis"). They consist in perivascular lymphocytes and microglial activation. Their severity varies largely between species and does not correlate with that of neurological signs.

The major clinical sign in rabid infection is an alteration of behavior which may be explained by the effects of the infection on the limbic system. For instance, unusual friendliness may be observed in wild animals such as foxes. On the contrary, other forms of rabies can be described as "furious" with episodes of aggressive and destructive behavior towards animals, humans, or objects. Various patterns of ataxia, muscle trembling and paralysis of jaw or tongue may also occur. Pharyngeal paralysis can lead to profuse salivation.

The diagnosis of rabies can be made by identification of antigen or viral RNA, by virus isolation, or post-mortem examination of the brain. The mortality is 100 % in rabid animals or nontreated humans once the neurological signs have appeared, although a few cases have survived after induced long-term coma. Prevention is thus paramount by elimination of infection in animal vectors and vaccination. Post-exposure prophylaxis (human rabbit immunoglobulins and vaccination) is an option as an emergency measure.

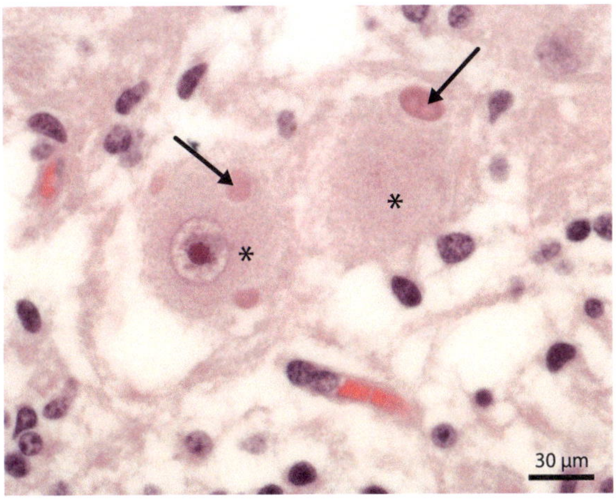


**Fig. 4.4** **Rabies**. Cytoplasmic eosinophilic (*red*) inclusions (Negri bodies indicated by *arrows*) in the Purkinje cells (*asterisks*) of the cerebellum; dog (stain: hematein-eosin)

> Rabies is an endemic viral disease transmitted by bite inoculation from infested animals. The virus migrates along peripheral nerves towards the central nervous system by retrograde transport, then, from the brain and spinal cord, spreads centrifugally.

### 4.2.6 Marek's Disease in Birds

Marek's disease consists of several distinct pathologic diseases due to infection by a distinct herpes virus (Marek Disease Virus, MDV) [11, 12, 15, 16]. T-cell lymphoid tumors are most frequently associated with Marek's disease, lymphoma (solid tumors) being the most common. Chicken are the most important natural host but other birds such as quail, turkeys, and pheasants are also susceptible to disease. Marek's disease is highly contagious and spreads by bird-to-bird contact. Its distribution is worldwide, affecting poultry-producing countries. Prior to the use of vaccines, losses were estimated to range from a few birds to 25 % in affected flocks. Chickens with Marek's disease lymphoma may exhibit signs related to peripheral nerve involvement: asymmetric progressive paresis (decrease in muscle strength), then paralysis of one leg. Bad coordination or stilted gait is usually the first observed sign. Pathologic changes consist mainly of nerve lesions and lymphomas in one or more organs

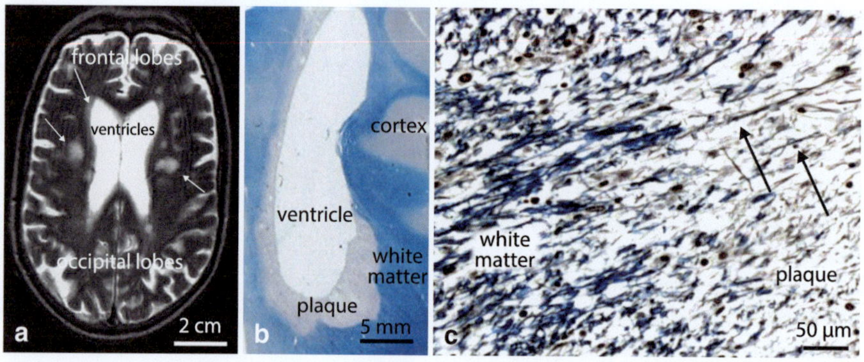


**Fig. 4.5** **Multiple sclerosis.** (**a**) Magnetic resonance imaging (MRI) T2 transverse section showing multiple plaques (hypodense areas) in the white matter, particularly in periventricular location (*arrows*). (**b**) Histological low magnification of one of these plaques. Note the *sharp border* of the demyelination area. Luxol fast blue stain for myelin. (**c**) Histological high magnification of the border of a plaque demonstrating the lack of myelin (stained in *blue* by the luxol stain) and the presence of axons (shown in *brown* by Bodian's silver stain, *arrow*)

such as ovaries, proventriculus, heart, and liver. Nerve lesions, often unilateral, are mainly characterized by enlarged peripheral nerves that show a gray or yellow discoloration and loss of striations. Microscopic examination shows two types of lesions in different nerves of the same bird or even in the different areas of the same nerve. Type A is characterized by marked cellular infiltration of proliferating lymphoblastic cells, whereas type B consists of edema, scattered infiltrating small and medium lymphocytes, and plasma cells.

The diagnosis of Marek's disease involves a combination of disease-specific criteria such as clinical, gross pathology, characterization of the proliferating cell population, epidemiology, and virological criteria. No curative treatment is available. Chicken can be vaccinated at the hatchery. While the vaccination prevents tumor formation, it does not prevent infection by the virus.

> Marek's disease in birds is a highly contagious disease due to a herpes virus which induces tumors of the lymphoid system affecting the peripheral nervous system and other organs.

### 4.2.7 Human Multiple Sclerosis

Multiple sclerosis (MS) [4, 14, 20] is predominantly due to "demyelination", an acquired loss of the myelin sheath made up of the compacted processes of oligodendrocytes, in brain and spinal cord white matter. This results in absence of conduction of the electric signal in the related axon. In MS, demyelination occurs in focal areas of the brain and spinal cord, called the "plaques" (Fig. 4.5a–c). In "active plaques", there is loss of oligodendrocytes, phagocytosis by microglial cells of myelin debris, astrocytic gliosis, and infiltration of T and B lymphocytes around the small vessels, characteristic of an immune pathology. The blood–brain barrier is impaired. In "chronic" plaques, there is only loss of oligodendrocytes and astrocytic gliosis, but once demyelination has occurred, a repair process, called "remyelination" can take place.

Peak age of onset is between 20 and 30. Women, especially when they lived in Northern countries before age 15, are more affected than men. Some genetic, such as *HLA–DRB1*, and environmental risk factors, such as Epstein-Barr virus infection and vitamin D deficiency, have been recognized.

Clinical presentation may be quite variable, depending on the plaque location. Limb weakness or bad coordination, spasticity, or sensory symptoms such as paresthesias (abnormal sensation such as tingling or numbness) may occur when the motor or the sensory systems, respectively, are primarily affected. Unilateral loss of vision by involvement of the optic nerve ("optic neuritis"), other visual difficulties due to lesions of the posterior optic pathways or some cranial nerves or areas of the brain stem which control eye motility are frequent. Other symptoms may be vertigo (involvement of brain stem areas which mismatch vestibular, visual and sensory inputs), bladder dysfunction, impotence (spinal cord lesions), or frequently, fatigue. The course of the disease is variable. The most frequent is the relapsing-remitting type, where acute periods are followed by remissions, then unpredictable worsening (called relapses) occur.

Acute relapses are treated by administration of high doses of intravenous corticosteroids (steroid hormones produced in the adrenal cortex regulating immune response). A number of disease-modifying agents including two interferons (proteins triggering the protective defenses of the immune system) are used in relapse-remitting courses. Other treatments have been tested, for example to reduce the migration of lymphocytes into the CNS but none is perfect and some may contribute to facilitate the development of viral infections.

> Multiple sclerosis is an autoimmune disease, mostly affecting myelinated fibers of the central nervous system, with both genetic and environmental origin.

### 4.2.8 "Brainworm": The Lancet Fluke

Some parasites can take over and control the behavior of their hosts, leading to positive returns for them. An example is given by the lancet fluke, *Dicrocoelium dendriticum*, a trematode worm [8, 9, 10]. This parasite is widespread in grazing ruminants (sheep and goats) in many countries of the world (Europe, North America, Asia), and occasionally can affect humans. Both ruminants and humans act as ultimate hosts. Adults of *D. dendriticum* live in the hepatic bile ducts and gall bladders of ruminants. Their embryonated eggs pass through the intestine and are excreted in the feces. Once eaten by snails, the first intermediate host, miracidia stages hatch from the eggs and penetrate the intestinal wall of the land mollusks. In the hepatopancreas, the miracidia become mother-sporocysts, producing large numbers of daughter-sporocysts with developing cercariae. The mature cercariae abandon the sporocyst and migrate to the respiratory chamber where they are released from the respiratory pore in small slime balls. Different species of ants (e.g., *Formica*) readily feed on the slime balls and become second intermediate host. The ingested cercariae penetrate the ant's crop, loose their tails, and distribute in the body cavity. Most of them encyst to metacercariae with a robust cyst wall, but one, or sometimes two migrate as "brainworms" to the subesophageal ganglion where they encyst with a soft cyst wall (Fig. 4.6 red arrow). The metacercaria of the brainworm manipulates the ant's behavior. In the evening, when temperatures drop below 15 °C, the parasitized animal does not return to the nest, but rather climbs up the leaves of grass and stays there for the rest of the night, attached with a tetanus of the mandibular muscles. As temperature rises in the morning, the ant returns to normal behavior and joins her sisters during the day. However, if the ant, together with the grass, is eaten by a ruminant, the parasite has reached its final host. "Plant-topping" of the infected ant was described as a "disorientation" mechanism. The mechanism of action of the brainworm in the suboesophageal ganglion remains mysterious. It has been suggested that an as yet unknown metabolic product would influence the nervous system of the ant, possibly acting via neuromodulators or neurotransmitters within the subesophageal ganglion. The pathway of pyrokinin/pheromone-biosynthesis-activating neuropeptide (PBAN) family of peptides (Fig. 4.6), which is required for diverse physiological functions, including muscle contraction and various behaviors, may be a promising research pathway. Ultimately, the ant's behavior comes to be the extended phenotype of the brainworm. Other parasites that affect the behavior of insects have also been described, including fungal infections that create "zombie ants". The fungus grows in the ant's brain, controls the ant's behavior, and kills the host once it has reached a location that is ideal for the fungus to develop, increasing the fungus' evolutionary fitness.

> Encephalic dicrocoeliosis in ants is a parasitic disease which induces an altered behavior of the infected insect called "brainworm tetania", blocking it on the tip of grass, thus promoting its ingestion by grazing ruminants. This allows the parasite to complete its life cycle.

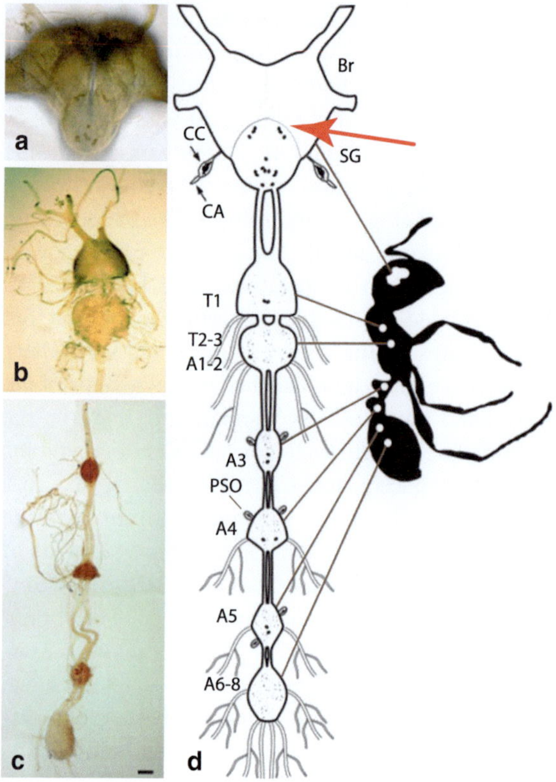


**Fig. 4.6** Ant nervous system. *Left*, photomicrographs of the brain and subesophageal ganglion (**a**), thoracic ganglia (**b**), and abdominal ganglia (**c**); (**d**) representation of CNS and cells that label for the peptide PBAN(PBAN-like immunoreactivity) in adult fire ant (*Br* brain, *SG* subesophageal ganglion, *CC* corpora cardiaca, *CA* corpora allata, *PSO* perisympathetic organ, *T1–T3* first to third thoracic ganglia, *A1–A8* first to eighth abdominal ganglia). Bar: 50 μm (From Choi et al. [2]). The *red arrow* points to the subesophageal ganglion where the brainworm migrates

### 4.2.9 Occlusion of the Posterior Cerebral Arteries

An infarction can be caused by the occlusion of the posterior cerebral arteries (PCA) [3] and induce cortical blindness, visual agnosia, and cortical visual impairment, among other symptoms. In *cortical blindness* [14], the loss of vision is caused by bilateral damage to the primary visual area (V1) in the calcarine cortex of the occipital lobe of the brain, whereas anterior visual pathways are normal. The patient does not see anything, although pupils normally dilate and contract with light exposure, since the pupillary-light reflex does not involve the posterior visual pathways. In *visual agnosia*, the brain is unable to make sense of, or make use of, some part of otherwise normal visual stimuli: for example, the person does not recognize familiar objects or faces. This is due to lesions of the visual association areas (V2–V5, which organize sensory information into a coherent representation). Still different is *cortical visual impairment* where the patient is not totally blind, but the field of view may be limited, imperfect visual processing being slow and requiring extreme effort.

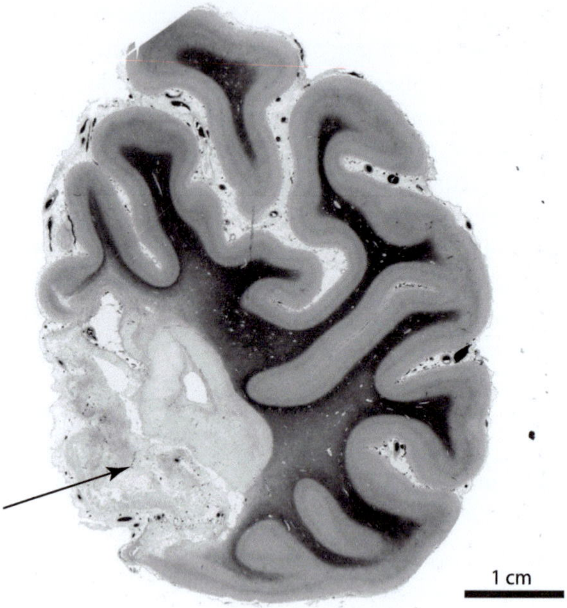


**Fig. 4.7** Infarct of the occipital territory of the posterior cerebral artery (*arrow*). Coronal section of the brain. Loyez stain for myelin

Two PCAs, issued from a single basilar artery supply each the occipital cortex (Fig. 4.7), the ventral part of the temporal cortex, and some basal ganglia of a cerebral hemisphere. This system can be occluded by a range of mechanisms, and the blood supply to the cortex reduced or abolished, leading to necrosis in the territory of one or both PCAs; the entire territory, or only parts thereof, can be damaged. The onset is usually acute, with sometimes a regressive course.

Bilateral occlusion of hemispheric territories induces cortical blindness, severe amnesia, and agitated delirium; in occlusion of the left PCA, there is visual agnosia, and also inability to name colors, right hemianopsia (blindness in half the right visual field of both eyes), and other deficits (on language and memory functions); in occlusion of the right PCA, left associated hemianopsia and/or visual neglect (deficit in attention to and awareness of one side of space), auditory and cognitive symptoms may be observed.

In the acute stroke, the patient must be seen in a stroke center within the first 3 h to set up the treatment which might include lysis of the thrombus, antiplatelet agents, or anticoagulant therapy to prevent a new occlusion.

Occlusion of the Posterior Cerebral Artery induces various neurological syndromes, often including vision and/or memory deficits, depending on the involved cerebral territory.

### 4.2.10 Human Alzheimer's Disease (AD)

The prevalence of this degenerative dementia is estimated to increase from 1.5 % at age 65–70 to 6 % at 75–70, and over 25 % above 85 years of age. AD [4, 14, 20] is due to a combination of risk factors, genetic (such as apolipoprotein E4, a lipid-carrying plasma protein encoded on chromosome 19) and environmental. There are rare autosomal dominant families linked to genes related to the metabolism of Aβ (see later), usually of early onset.

Two main brain lesions are seen (Fig. 4.8): (*i*) *amyloid plaques*, deposits of an extracellular neuronal peptide, called Aβ, the 1–42 amino acids form of which tends to aggregate into toxic β-pleated extracellular monomers; when mature "senile plaques" occur, the amyloid deposit of Aβ is surrounded by a crown of degenerating neurites filled with clusters of paired helical filaments composed of abnormally phosphorylated tau, a tubule-associated intraneuronal protein which governs axonal flow; (*ii*) *neurofibrillary tangles* are also clusters of paired helical filaments of abnormally phosphorilated and also β–pleated tau accumulated in the neuronal perikarya. They are preceded by intracellular tau storage called pre-tangle. These abnormal proteins are seen early in the brain, long before the disease is recognized (likely 10–20 years at least). The first to appear is the abnormally phosphorylated tau, which is found in the brain stem of children and young adults. Then, neurofibrillary tangles, tightly linked to the neuronal degeneration and synaptic loss, become visible in the allocortex, which includes the entorhinal olfactory cortex and the hippocampus. From there it develops into the olfactory bulb and the cerebral neocortex in demented patients. Some cholinergic nuclei of the basal forebrain projecting to the cortex (including Meynert basal nucleus) are involved very early

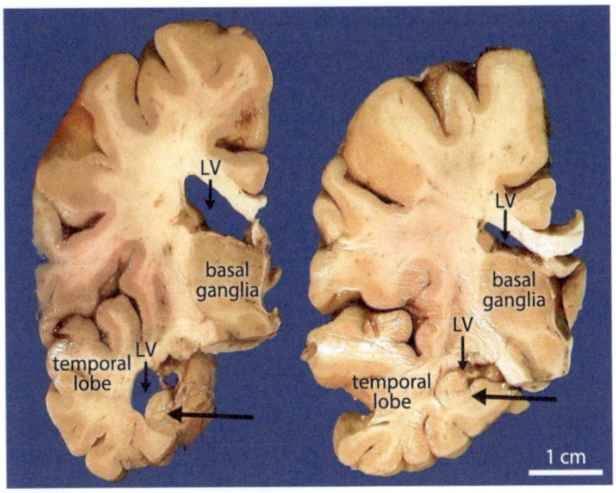


**Fig. 4.8** Macroscopic coronal brain sections of an Alzheimer's disease affected patient (*left*) and a control brain (*right*). Note the cerebral atrophy which affects principally the hippocampus (*arrow*) and the adjacent internal temporal regions. *LV* lateral ventricle

(before the hippocampus). The Aβ protein occurs later in the brain. It is first seen in the neocortex, then in the allocortex, the basal ganglia, basal forebrain and brain stem, and lastly in the cerebellum. The first senile plaques generally develop above the age of 40, culminating in the tenth decade (75 % of cases).

AD is a chronic (7–10 years) progressive disorder revealing between 40 and 95 but affecting mostly old people, and the most common cause of *dementia* (i.e., loss of cognitive abilities always including memory and at least one of the following: language, execution of purposeful movements, recognition of objects, persons, sounds, loss of planning and problem solving). Memory loss, which initially affects "episodic memory", is associated with disorientation to time and place, and can occur after other disabilities such as language or visuospatial deficits. There is progressive anosognosia (lack of awareness of cognitive deficit). In addition, psychotic symptoms and sleep disorders, mainly sleep-wake rhythm disturbances due to lack of clock genes expression are frequent. In contrast, physical examination is usually normal, although rigidity and low movements, and sometimes myoclonus (brief muscular contractions) may occur, especially in the late stages.

The definite diagnosis of AD, the most frequent degenerative dementia, is made by pathological brain examination. A highly likely diagnosis relies on clinical information documented by brief psychological tests confirmed by a formal neuropsychological testing and normal blood examination. Neuroimaging (brain scan and MRI) shows only cerebral atrophy (predominating on hippocampus), CSF examination decreased Aβ42 and increased tau protein. These are performed to corroborate dementia, and dismiss other causes (vascular dementia, Lewy body disease and other degenerative dementia, Creutzfeldt-Jakob disease). New diagnostic tests include blood tests (genomic and proteomic by mass spectroscopy) and, overall neuroimaging. Three techniques are increasingly used: PET (positron emission tomography) scan using compounds binding to Aβ, the first of which was PIB compound, can measure Aβ load; *tractography* (using special techniques of MRI), showing the defect of bundles of white matter, and *functional MRI*, which discloses signal modifications in the brain that are due to changing activity.

Pharmacologic approach of AD treatment includes cholinergic enhancement (cholinesterase inhibitors are the most widely used) and glutamate NMDA receptors antagonists, which have some therapeutic efficiency. Depending on the regulations and the practice of each country, other drugs (antioxidants, anti-inflammatory agents) may be used. Treatments in investigation include modulation of production, aggregation, clearance, or toxicity of Aβ and, more recently, of abnormally phosphorylated tau. Future research will also involve increase of synaptogenesis, nerve cell plasticity, and stem cells production and differentiation.

Alzheimer's disease is due to an abnormal metabolism of Aβ and tau proteins inducing degeneration and death of neurons in selective areas involved in memory and in higher cognitive functions. This causes, decades later, the most frequent degenerative dementia.

## References


1. Bird TD (1993) Hereditary ataxia overview. In: Pagon RA, Bird TD, Dolan CR, Stephens K, Adam MP (eds) GeneReviews™ [Internet]. University of Washington, Seattle [updated 2013 Jan 17]
2. Choi MY, Raina A, Vander Meer RK (2009) PBAN/pyrokinin peptides in the central nervous system of the fire ant, *Solencos invicta*. Cell Tissue Res 335:431–439
3. De Girolami U, Seilhean D, Hauw JJ (2009) Neuropathology of central nervous system arterial syndromes. Part I: the supratentorial circulation. J Neuropathol Exp Neurol 68: 113–124
4. Finsterer J, Papic L, Auer-Grumbach M (2011) Motor neuron, nerve, and neuromuscular junction disease. Curr Opin Neurol 24:469–474
5. Gray F, Duyckaerts C, De Girolami U (2013) Escourolle and Poirier's Manual of Basic Neuropathology. Oxford University Press, New York
6. Hauw JJ, Hausser-Hauw C, De Girolami U, Seilhean D (2011) Neuropathology of sleep disorders: a review. J Neuropathol Exp Neurol 70:243–252
7. Hawkes DP, Del Tredicy K, Braak H (2009) Parkinson's diseases: the dual hit theory revisited. Ann NY Acad Sci 1170:615–622
8. Hughes DP, Andersen SB, Hywel-Jones NL, Himaman W, Billen J, Boomsma J (2011) Behavioral mechanisms and morphological symptoms of zombie ants dying from fungal infection. Ecology 11:13–23
9. Manga-González MY, González-Lanza C (2005) Field and experimental studies on *Dicrocoelium dendriticum* and dicrocoeliasis in northern Spain. J Helminthol 79: 291–302
10. Martínez-Ibeas AM, Martínez-Valladares M, González-Lanza C, Miñambres B, Manga-González MY (2011) Detection of *Dicrocoelium dendriticum* larval stages in mollusc and ant intermediate hosts by PCR, using mitochondrial and ribosomal internal transcribed spacer (ITS-2) sequences. Parasitology 24:1–8
11. Mwangi WN, Smith LP, Baigent SJ, Beal RK, Nair V, Smith AL (2011) Clonal structure of rapid-onset MDV-driven CD4+ lymphomas and responding CD8+ T cells. PLoS Pathog 7(5):e1001337
12. Radostis OM, Gay C, Hinchcliff KW, Constable PD (eds) (2000) Veterinary medicine: a textbook of the diseases of cattle, horses, sheep, pigs and goats, 10th edn. Saunders, Philadelphia
13. Romi F (2011) Thymoma in myasthenia gravis. From diagnosis to treatment. Autoimmun Dis 2011:474512
14. Ropper AH, Samuels MA (2009) Adams and Victor principles of neurology, 8th edn. McGraw-Hill, New York
15. Saif YM, Fadly AM, Glisson JR, McDougald LR, Nolan LK, Swayne DL (2008) Diseases of poultry, 12th edn. Blackwell Publishing, London
16. Summers BA, Cummings J, de Lahunta A (1995) Veterinary neuropathology. Mosby, St. Louis
17. Thannickal TC, Nienhuis R, Siegek JM (2009) Localized loss of hypocretin (orexin) cells in narcolepsy without cataplexy. Sleep 32:993–998
18. Warrell MJ, Warrell DA (2004) Rabies and other lyssavirus diseases. Lancet 363:959–969
19. Yoshimura T, Motomura M, Tsujihata M (2011) Histochemical findings of and fine structural changes in motor endplates in diseases with neuromuscular transmission abnormalities. Brain Nerve 63:719–727 (article in Japanese)
20. Zaidat OO, Lerner A (2008) The little black book of neurology, Mobile medicine series. Saunders, Mosby/Elsevier, Philadelphia

# Neurophilosophy

# 5

Georg Northoff

## 5.1 Background: The History of Neurophilosophy

> Neurophilosophy stands for the investigation of philosophical questions in the context of a neuroscientific hypothesis.

Recent neuroscientific progress has led to the extension of neuroscience to apply and include also concepts like consciousness, free will, self, etc. that were originally discussed in philosophy. This has led to the recent emergence of a new field – neurophilosophy. The term "neurophilosophy" is often used either implicitly or explicitly for the characterization of an investigation of philosophical theories in relation to neuroscientific hypothesis. According to Breidbach [1], pp. 393–394, "neurophilosophy" had already been implicitly practiced at the turn of the last century by W. Wundt (1832–1920), for instance. Another neurophilosopher, though not named as such, was Schopenhauer who was probably the first philosopher to introduce the concept of the brain in the philosophical context. The French philosopher M. Merleau-Ponty (1908–1961) may also be considered a neurophilosopher since in his 'Phenomenology of perception' he explicitly introduces the brain and its neural organisation and links it to perception and other originally philosophical concepts.

Other important developments in this regard were put forward by the American philosopher W. von Orman Quine (1908–2000): He raised the question whether what we can know about ourselves and the world as usually dealt with in the philosophical discipline of epistemology can be traced back to nature itself and ultimately to evolution [2]. This was complemented by the collaboration between the philosopher K. Popper (1902–1994) and the neuroscientist J.C. Eccles (1903–1997) who discussed the relation between brain and mind from both perspectives, neuroscientifically and philosophically [3]. Finally, the term 'neurophilosophy' was explicitly coined by the American philosopher P. Churchland [4] in her book 'Neurophilosophy' where she discussed empirical results side by side with theoretical issues.

The current field of neurophilosophy covers mainly three different domains, 'Empirical Neurophilosophy', 'Practical Neurophilosophy', and 'Theoretical Neurophilosophy'. 'Empirical Neurophilosophy' describes the "application of neuroscientific concepts to traditional philosophical questions" [5], p. 1. Here concepts like consciousness, self, and free will (see below for details) that have traditionally been dealt with theoretically in philosophy are now investigated experimentally in neuroscience. Secondly, there is the field of 'Practical Neurophilosophy' that deals with ethical concepts like free will, moral judgment, and informed consent in the neural context of the brain. Thereby, as in empirical neurophilosophy, the philosophical-ethical concepts may also be extended from the originally purely human domain to animals, like

G. Northoff
Canada Research Chair for Mind, Brain Imaging, and Neuroethics, Michael Smith Chair for Neuroscience and Mental Health, University of Ottawa Institute of Mental Health Research,
Ottawa, Canada
e-mail: georg.northoff@theroyal.ca

C.G. Galizia, P.-M. Lledo (eds.), *Neurosciences - From Molecule to Behavior: A University Textbook*,
DOI 10.1007/978-3-642-10769-6_5, © Springer-Verlag Berlin Heidelberg 2013

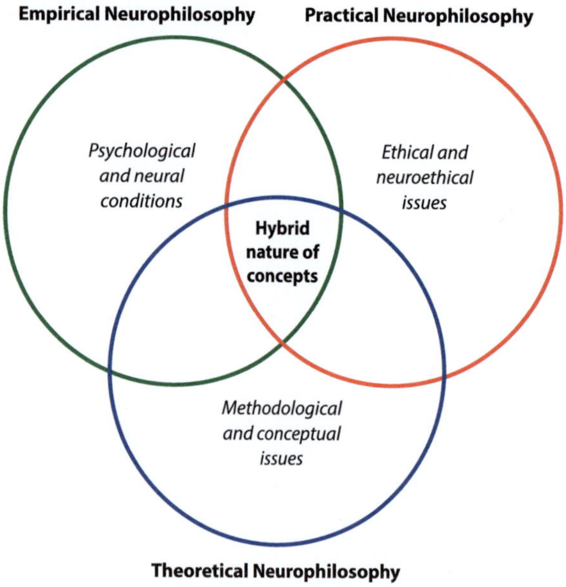


**Fig. 5.1** The figure illustrates the three main domains of neurophilosophy, empirical, theoretical, and practical. Empirical neurophilosophy is concerned with the search for the neural and psychological conditions of originally philosophical terms like self, consciousness, free will, etc. Theoretical neurophilosophy is about the methodological and conceptual issues when linking neuroscientific data/facts and philosophical concepts. Finally, practical neurophilosophy is about the linkage between neuroscience and ethics with ethical issues in neuroscience and neuroscientific mechanisms underlying ethical concepts

whether the latter have free will or not. Third, and finally, there is 'Theoretical Neurophilosophy' which focuses on methodological issues like how to link empirical data and theoretical concepts in neurophilosophical investigation (Fig. 5.1).

## 5.2 Empirical Neurophilosophy – Experimental Investigation of Philosophical Concepts

One of the main originally philosophical concepts investigated in neuroscience is consciousness. What is consciousness? **Consciousness** is often understood as the ability to detect, evaluate and report about the experience of a particular object or event in the environment or the own thoughts. Since detection, reporting, and evaluating requires access to the content in question, this form is often called 'access consciousness'.

'Access consciousness' must be distinguished from the experience itself which, following philosophers like Th. Nagel [6], can be characterized by a particular point of view, a stance in the world, from which we perceive and experience ourselves and others. That form of consciousness has been described as 'phenomenal consciousness'. The distinction between 'phenomenal and access consciousness' is considered by many a core distinction which has also aroused plenty of controvery. Some authors deny for instance that phenomenal consciousness can be distinguished from access consciousness. However, animals may have phenomenal consicousness while they may remain unable to report the contents of their consciousness thus lacking access consciousness.

Following Christoph Koch and Francis Crick [7] we need to identify what they call the 'neural correlates of consciousness' (NCC). The NCC describe the search for those minimally neuronal conditions that are jointly sufficient for any one specific conscious, i.e., phenomenal, percept that we can experience. Several neuronal mechanisms have been discussed as possible candidate mechanisms for the NCC. In the following I highlight some of the main and most popular suggestions.

G. Edelman [8] considers cyclic processing and thus circularity within the brain's neural organisation as central for constituting consciousness. Cyclic processing describes the re-entrance of neural activity in the same region after looping and circulating in so-called re-entrant (or feedback) circuits.

This is for instance the case in primary visual cortex (V1): The initial neural activity in V1 is transferred to higher visual regions such as the inferotemporal cortex (IT) in feedforward connections. From there it is conveyed to the thalamus which relays the information back to V1 and the other cortical regions implying thalamocortical re-entrant connections. Consciousness is assumed to be constituted on the basis of such feedback or re-entrant connections that allow for cyclic processing.

What is the exact mechanism of the feedback or re-entrant circuits? Re-entrant circuits integrate information. This leads Giulio Tononi to emphasize the integration of information as the central neuronal mechanism in yielding consciousness. He consecutively developed what he calls 'Integrated Information Theory' (IIT). We usually focus on the content that is selected to become conscious, i.e., 'what is perceived'. Instead, as the IIT claims, we may better search for the neuronal mechanisms that allow excluding content from becoming conscious, i.e., 'what is ruled out'. The information that is ruled out to become conscious may suffer from insufficient integration of information and remains therefore unconscious.

Tononi assumes the integration of information to be particularly related to the thalamo-cortical re-entrant connections: These re-entrant connections process all kinds of stimuli thus remaining unspecific with regard to the selected content. They make it possible to generate a particular point of view and an associated quality of experience (also called qualia) as hallmark feature of consciousness. Linkage of these qualia to the content processed via thalamo-cortical information integration may then allow these contents to become conscious. This distinguishes them from the unconscious contents that do not undergo such cyclic processing via the thalamus – and therefore the addition of the specific quality, the qualia, remains impossible.

Another suggestion for the neural correlate of consciousness comes from B. Baars [9, 10] and others like S. Dehaene. They assume global distribution of neural activity across many brain regions in a so-called global workspace to be central for yielding consciousness: The information and its contents processed in the brain must be globally distributed across the whole brain in order for them to become associated with consciousness.

When information is only processed locally within a particular region but not throughout the whole brain, it can not be associated with consciousness anymore. The main distinction between unconsciousness and consciousness is thus supposed to be manifest in the difference between local and global distribution of neural activity. Hence, the global distribution of neural activity is here considered a sufficient condition and thus neural correlate of consciousness.

Taken together, there are currently these neuroscientific suggestions for consciousness. Future research is needed though to further specify the neuronal mechanisms themselves and the features of consciousness itself. Consciousness may by itself not be as homogenous as it appears; instead, it may be characterized by different features as for instance a point of view (see above), a quality (see above), and a particular unity as unifying convergence point for different contents.

Another originally philosophical concept now hotly debated in neuroscience is the concept of the **self**. The question of the self has been one of the most salient problems throughout the history of philosophy and more recently also in psychology and neuroscience. For example, William James (1842–1910) distinguished between a physical self, a mental self, and a spiritual self. These distinctions seem to reappear in recent concepts of self as discussed in neuroscience. Damasio [11] and Panksepp [12] suggest a "proto-self" in the sensory and motor domains, respectively, which resembles James' description of the physical self. Similarly, what has been described as "minimal self" [13, 14] or "core or mental self" [11] might correspond more or less to James' concept of mental self. Finally, Damasio's "autobiographical self" and Gallagher's "narrative self" strongly rely on linking past, present, and future events with some resemblances to James' spiritual self.

These distinct selves are now related to distinct brain regions. For instance, the "proto-self" outlining one's body in strongly affective and sensory-motor terms is associated with subcortical regions like the periaqueductal gray, the colliculi, and the tectum. The "core or mental self" building upon the "proto-self" in mental terms is associated more with the thalamus and cortical regions like the ventromedial prefrontal cortex (see, for instance [11, 15]). Finally, the "autobiographical or extended self" that allows one to reflect upon one's "proto-self" and "core or mental self" is associated with cortical regions like the hippocampus and the cingulate cortex.

Humans show various cortical regions, predominantly the so-called cortical midline structures (CMS), to be involved in what is called self-related processing (SRP) that are integrated with subcortical processes to yield an integrated subcortical-cortical midline system (SCMS). The lowest regions of this distributed SCMS network include the periaqueductal gray, the superior colliculi, and the adjacent mesencephalic locomotor region as well as preoptic areas, the hypothalamus, and dorsomedial thalamus, while cortical regions include the ventro- and dorsomedial prefrontal cortex, the pre- and supragenual anterior cingulate cortex and the posterior cingulate cortex, and the medial parietal cortex. The association of the subcortical regions with a sense of self has led to the assumption that already animals may have a sense of self [16, 17] though most likely not as cognitively elaborated as the human self.

## 5.3 Theoretical Neurophilosophy – Methodology and Knowledge of the Linkage Between Brain Data and Philosophical Concepts

One of the main issues in neurophilosophy is the question for methodology. How can we link empirical data, so-called facts as obtained in neuroscience, to the concepts and their meaning as dealt with in philosophy?

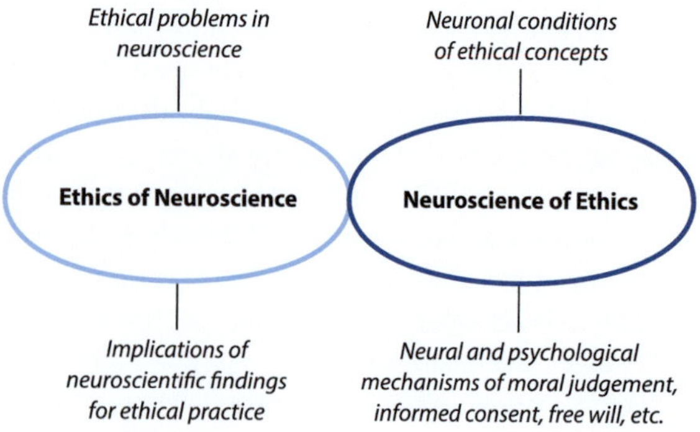


**Fig. 5.2** The figure illustrates the fields of practical neurophilosophy. Ethics of neuroscience concersn ethical problems in neuroscience like informed consent. While neuroscience of ethics refers to the neural and psychological conditions of ethical concepts

Data and facts do not require any definition and determination. They rely on observation and can in principle be obtained by anybody; they thus remain investigator-independent. This is different in the domain of concepts. Concepts carry a meaning, a semantic dimension, which may be closely related to the investigator and how he defines and uses the concept in question.

However, empirical-experimental investigation cannot do without concepts. For instance, in formulating the hypotheses of the experiment as starting point for developing an appropriate experimental design, concepts play a substantial role. And after obtaining the data they must be interpreted for which again concepts are necessary. This concerns only concepts within the natural world, the world we live in, and thus what philosophers call the 'natural conditions'. Such 'natural conditions' must be distinguished from 'logical conditions' that describe logically possible worlds which may or may not be realized within the context of our current natural world.

The neurophilosopher is thus confronted with the principal gap between data/facts and concepts in a twofold manner. First, there is the gap between data/facts and concepts within the domain of the natural world: How do certain data about, for instance, the reward system in animals stand to the concept of reward in general in both animals and humans? This is a gap the neuroscientists themselves already face which, due to the predominant experimental focus, is often neglected. Secondly, there is the gap between neuroscientific data/facts in the natural world and the concepts in the logical worlds of the philosophers: How can we infer from neuroscientific data about consciousness to the philosophical concept of consciousness and, vice versa, how can we translate the latter into experimental designs to test it empirically? This is a truly neurophilosophical gap which we need to bridge if neurophilosophy is to succeed in both methodology and knowledge.

## 5.4 Practical Neurophilosophy – Neuroethics and the Relevance of Ethical Concerns in Neuroscience

Practical neurophilosophy or neuroethics focuses, on the one hand, on the investigation of the psychological and neural conditions of ethical concepts like free will, decision making, moral judgment, and informed consent. This can be described as 'neuroscience of ethics'. At the same time, practical neurophilosophy also deals with ethical problems in neuroscience and thus with issues of validity of informed consent in psychiatric patients, enhancement of cognitive functions by neuroscientific interventions, coincidental findings in neuroimaging (Fig. 5.2). That amounts to an 'ethics of neuroscience' [18].

Do we have a free will or not? The **free will** is, for instance, manifest in our daily decisions if, for instance, we choose the red rather than the green apples in the supermarket. Recent neuroscience detected the neural mechanisms of decision making that seem to involve a number of different brain regions including those where reward is processed. The reward regions include the ventral tegmental area (VTA), the ventral striatum (VS), and the ventromedial prefrontal cortex (VMPFC). All these regions have their homologs in various animal species so that the same question for the free will may also be extended from humans to non-human animals.

The observation that the apparently free decision making is related to and, in fact, temporally preceded by neural activity specifically related to the decision in question has put the concept of free will in doubt. If the free will is pre-determined by the neural events in the brain, one can no longer speak of a free will. The free will is then no longer free but nothing but a mere illusion on our side with the brain determining our actions and decisions. Are we thus no longer free in our will? That obviously is an interpretation of the data and also depends on the definition of the concept of free will. If one, for instance, presupposes a narrow concept of free will that excludes any preceding changes, the present brain data may tell us that there is indeed no free will. Brain data and free will are then incompatible. Conversely, a wider concept of free will that does not exclude preceding neural activity changes may then be well compatible with the brain data.

The debate about free will pertains to a wider issue, the question of determinism versus indeterminism. Determinism assumes that all our decision and also what we call free will is determined completely and exclusively by the brain and its neural activity. Our person or our self, as presumably distinct from the brain, has then no say at all in our decision. Hence, it is then the brain rather than the self that makes the decision and has a 'neuronal will' rather than a 'free will'. That however is countered by indeterminism. Indeterminism argues that the brain itself and its neural activity changes does not determine completely and exclusively our decision making so that there are traces of free will left in our decisions. Who is right, determinism or indeterminism? As said above, it may strongly depend not only on the data but also on the conceptual definitions.

Besides such questions belonging to the 'neuroscience of ethics', the neuroscientific investigation of ethical concepts, there are also issues pertaining to ethical problems in neuroscience. One problem here is, for instance, the one of informed consent which subjects have to give when participating in experimental investigations. Being able to give informed consent may include a variety of different functions, cognitive, social, and affective, that are all ultimately brain-based. Does this mean that we have to exclude those subjects that suffer from impairments in these functions? Furthermore, recent research demonstrates that animals possess many of the cognitive and social functions originally attributed to humans only. Do we therefore need to develop more rules for animal participation in research by, for instance, considering that they can have consciousness, feel pain, and empathize with co-species?

## 5.5 Summary

Neurophilosophy is a young and novel field right at the intersection between neuroscience and philosophy. Unlike more established disciplines, it has not yet an established method that needs to be developed in the future as part of a 'theoretical neurophilosophy'. At the same time though neurophilosophy is a highly promising field which will be able to provide novel answers to questions discussed in philosophy for more than 3,000 years. This will not only enrich neuroscience and provide new ideas for experimental designs but will also change and reverberate in philosophy itself by allowing for a shift from the hitherto mind-based philosophy to a more brain-based neurophilosophy.

## References


1. Breidbach O (1997) Die Materialisierung des Ichs – Eine Geschichte der Hirnforschung im 19. und 20. Jahrhundert. Suhrkamp, Frankfurt aM
2. Quine WO (1969) Epistemology naturalized. In: Quine WO (ed) Ontological relativity and other essays. Columbia University Press, New York
3. Popper K, Eccles J (1989) Das Ich und sein Gehirn. Piper, München
4. Churchland P (1986) Neurophilosophy: toward a unified science of the mind-brain. MIT Press, Cambridge, MA
5. Bickle J, Mandik P, Landreth A (2006) The philosophy of neuroscience. Journal [serial on the Internet]. Available from: http://plato.stanford.edu/entries/neuroscience/
6. Nagel T (1979) What is it like to be a bat? Mortal questions. Cambridge University Press, New York, p 166
7. Koch C, Crick F (2001) The zombie within. Nature 411:893
8. Edelman GM (2003) Naturalizing consciousness: a theoretical framework. Proc Nat Acad Sci USA 100:5520–5524
9. Baars BJ (2005) Global workspace theory of consciousness: toward a cognitive neuroscience of human experience. Prog Brain Res 150:45–53
10. Baars BJ, Franklin S (2007) An architectural model of conscious and unconscious brain functions: Global Workspace Theory and IDA. Neural Netw 20:955–961
11. Damasio A (1999) How the brain creates the mind. Sci Am 281:112–117
12. Panksepp J (2003) At the interface of the affective, behavioral, and cognitive neurosciences: decoding the emotional feelings of the brain. Brain Cogn 52:4–14
13. Gallagher S (2000) Philosophical conception of the self: implications for cognitive science. Trends Cogn Sci 4:14–21
14. Gallagher HL, Frith CD (2003) Functional imaging of 'theory of mind'. Trends Cogn Sci 7:77–83
15. Damasio A (2003) Feelings of emotion and the self. Ann N Y Acad Sci 1001:253–261
16. Northoff G, Panksepp J (2008) The trans-species concept of self and the subcortical-cortical midline system. Trends Cogn Sci 12:259–264
17. Northoff G (2004) Philosophy of the brain. John Benjamins Publishing, Amsterdam
18. Roskies A (2002) Neuroethics for the new millenium. Neuron 35:21–23

# 6 Cellular and Molecular Basis of Neural Function

Herbert Zimmermann

Neurons and glia represent the fundamental building blocks orchestrating nervous systems. Their properties are determined by a highly complex cellular and molecular machinery that is subject to continued activity-dependent modification. Indeed, activity of the nervous system involves a constant reciprocal interplay between genes, the encoded proteins, and the resulting cellular activity patterns and neural circuitry that governs behavior. Individual cells are only grains of sand within the huge and highly structured nervous systems. This chapter briefly investigates principle cellular and molecular mechanisms underlying the cellular dynamics of the nervous system. It serves as a general introduction to the chapters following where individual mechanisms are elucidated in detail regarding specific physiological or pathophysiological processes.

## 6.1 Cell Membranes

Without a membrane capable of separating a reaction volume from the surrounding exterior no evolution of life would have occurred. Neural cells, like cells in other tissues, are surrounded by an approximately 5-nm-**thick lipid bilayer**, a barrier impermeable for most water-soluble molecules. This membrane separates the extracellular from the intracellular medium with all its metabolites and ions. Only nonpolar substances such as $O_2$, $CO_2$, and to a lesser extent small uncharged molecules such as urea and water, can diffuse through cell membranes. Similar membranes confine defined reaction volumes inside cell organelles. This allows a subcompartmentalization of metabolic activities within the cell. Such compartments include the nucleus, mitochondria, endoplasmic reticulum, Golgi apparatus, lysosomes, or secretory vesicles.

H. Zimmermann
Institute of Cell Biology and Neuroscience,
Goethe University, Max-von-Laue-Str. 13,
60438 Frankfurt am Main, Germany
e-mail: h.zimmermann@bio.uni-frankfurt.de

### 6.1.1 Amphipathic Lipid Molecules Build Up the Lipid Bilayer

The structure of the cell membrane is determined by the chemical nature of its individual lipid modules (Fig. 6.1). Phosphoglycerides, sphingolipids, and the sterol cholesterol are the building blocks of the lipid bilayer. These are amphipathic molecules consisting of a hydrophilic head group and a hydrophobic tail. Phospholipids are the most abundant membrane lipids. Phosphoglycerides carry the trivalent alcohol glycerol as a backbone. Two of its hydroxyl groups are linked through ester bonds with long-chain fatty acid molecules that form the hydrophobic tail. The remaining hydroxyl group connects via a phosphodiester bond to different types of charged head groups such as the amino alcohol choline (phosphatitylcholine) or an amino acid residue (for example, phosphatidylethanolamine or phosphatidylserine). In contrast, the phospholipid sphingomyelin is based on the long-chain unsaturated amino alcohol sphingosine with a fatty acid tail attached to its amino group (yielding ceramide) and a phosphodiester bond-linked choline group. These phospholipids provide the hydrophilic head groups that interact with the surrounding aqueous exterior. The equally ceramide-based and sugar

C.G. Galizia, P.-M. Lledo (eds.), *Neurosciences - From Molecule to Behavior: A University Textbook*,
DOI 10.1007/978-3-642-10769-6_6, © Springer-Verlag Berlin Heidelberg 2013

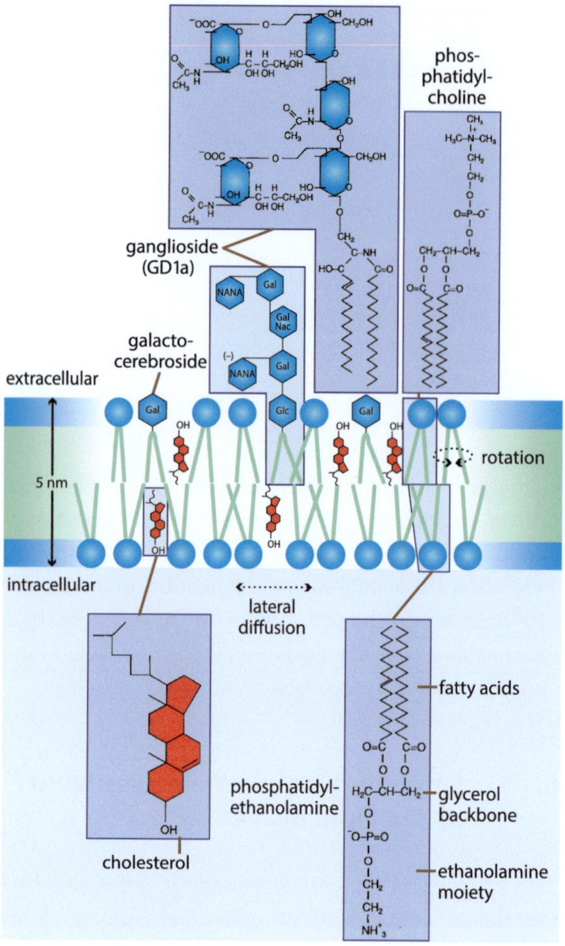


**Fig. 6.1** **Lipid structure of the plasma membrane.** Phospholipid molecules have a hydrophilic head (*blue*) and a lipophilic tail (*green*) and form a bilayer. Glycolipids are directed extracellularly. Only one (GD1a) of the many potential structures of gangliosides is depicted. Individual lipid components can rotate and diffuse laterally. *Gal* galactose, *GalNac* *N*-acetylgalactosamine, *Glc* glucose, *NANA* *N*-acetylneuraminic acid (sialic acid)

moieties-containing **glycolipids** reveal a comparable amphipathic structure. The sterol **cholesterol**, the final important lipid component, with its hydrophobic and rigid aromatic ring structure and single polar hydroxyl group, also displays a directed membrane insertion.

The spontaneous hydrophobic interactions between the tails of the lipid molecules result in the formation of two membrane leaflets, the **bilayer**. Individual lipid molecules are asymmetrically distributed between the inner and outer membrane leaflet of the lipid bilayer. The lipid bilayer behaves as a two-dimensional fluid allowing free diffusion of individual lipid molecules. The fluidity of the membrane depends on its lipid composition that varies between cells and even subdomains of the cell membrane. It decreases with the contribution of unsaturated fatty acids and increases with the contents in cholesterol molecules with their rigid plate-like steroid rings. Particularly high cholesterol contents have been found in the membrane stacks of myelin.

#### 6.1.1.1 Membrane Proteins: Morphological and Functional Diversity

The interaction and communication between the membrane-bound cellular compartments is mediated by membrane proteins. These allow the controlled transport of solutes such as ions, amino acids, sugars, or proteins. Membrane proteins are also of central importance for the transmission of signals into the cell interior and serve the interaction between membrane compartments inside the cell. The rapid formation and propagation of electrical signals at the plasma membrane represents one of the most prominent features of nerve cells and sensory cells (see Chap. 7). It results from the interplay between the charge-separating properties of the lipid bilayer and transmembrane proteins capable of inducing rapid and controlled changes in ion flux.

The lipid bilayer represents the backbone for the assembly of membrane proteins. Proteins take advantage of the physicochemical properties of the lipid bilayer for both hydrophobic and hydrophilic interactions (Fig. 6.2). The lipid bilayer in turn can affect the functional properties of proteins. Similar to lipids, the protein composition not only varies considerably between individual membrane compartments of the cell. Proteins are asymmetrically distributed over the cell membrane and many proteins can laterally diffuse within the membrane.

**Peripheral membrane proteins** bind via ionic or noncovalent interactions to proteins or lipids at the membrane surface. Interestingly, specific cellular signals can induce the transfer and specific binding of proteins to the inner leaflet of the plasma membrane. Both, lipid-binding motifs and specific motifs of membrane-associated proteins can be involved. For example, proteins with a PH (pleckstrin homology)-domain bind to phosphorylated inositol rings of phosphoinositides (phosphorylated derivatives of phosphatidylinositol) (compare Fig. 6.15). Proteins with an SH2 (Src-homology 2) domain bind to a distinct sequence of amino acids surrounding a phosphotyrosine residue as can be found in the cytosolic domain of phosphorylated receptor proteins (compare Fig. 6.14). As a result of these and other binding interactions, the inner surface of the plasma membrane represents a highly versatile metabolic subcompartment of the cell that plays a specific

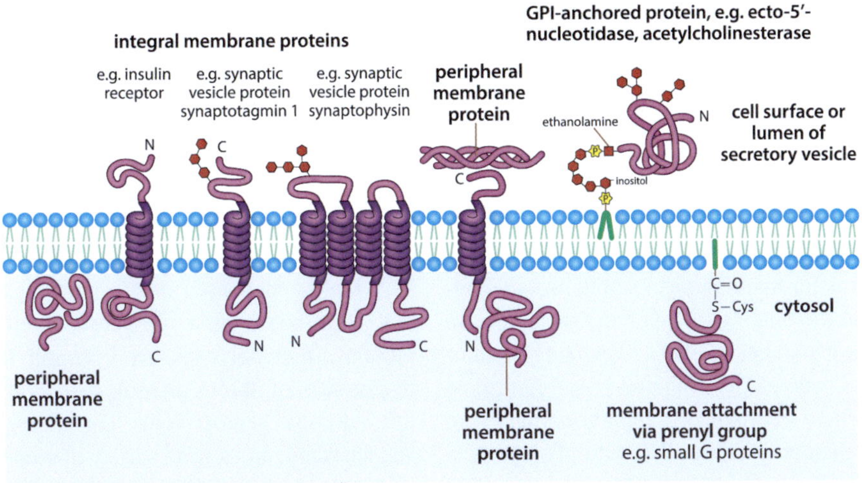


**Fig. 6.2 Examples of protein interactions with the lipid bilayer.** Glycoproteins are directed towards the lumen of intracellular organelles (ER, Golgi apparatus, secretory vesicles) or the extracellular medium (plasma membrane). Integral membrane proteins can permeate the lipid bilayer via one or multiple α-helices or form a barrel structure (mitochondrial porins, not shown). Other proteins are anchored to the membrane via a covalently attached lipid chain (fatty acid or prenyl group) or via a glycosylphosphatidyl inositol (GPI) anchor. Peripheral membrane proteins can directly bind to lipids (*left*) or interact with integral membrane proteins (*center*)

role in various cellular functions, in particular in receptor-mediated signal transfer. Similarly, extracellular proteins can bind to transmembrane proteins: for example, extracellular matrix proteins bind to integrins.

Typically **integral membrane proteins** contain extended α-helical domains that allow them to penetrate the lipid bilayer with direct hydrophobic interactions. A single polypeptide chain can traverse the lipid membrane several times forming a **multipass transmembrane protein**. Examples of proteins with multiple transmembrane helices include neurotransmitter receptors, ion channels, and neurotransmitter transporters. Membrane-spanning proteins are always polarized and their extracellular and intracellular domains may carry separate functional domains.

Other proteins can be anchored to the lipid bilayer via a covalently bound **lipid anchor** such as a fatty acyl group (e.g., myristate or palmitate) or a prenyl anchor. In some cases these anchors are dynamically attached or removed which allows signal-induced recruitment of proteins to the membrane. Among such lipid-anchored proteins are small GTPases of the Ras (from *ra*t *s*arcoma) family (compare Fig. 6.14) – including Rab (from *ras* related in *b*rain) – that are important for the targeting of synaptic vesicles to their specific fusion site [1] (compare Fig. 6.19). Glycosylphosphatidylinositol (GPI) anchors allow the attachment of proteins to the cell surface. In this case the protein is bound via a glycan to phosphatidylinositol which acts as a membrane anchor by insertion of its two fatty acyl groups into the outer membrane leaflet [2]. GPI-anchored proteins include some of the highly glycosylated and surface-located proteoglycans, cell adhesion molecules, or ectoenzymes that metabolize extracellular substrates such as subtypes of acetylcholinesterase or ecto-5′-nucleotidase, an AMP-hydrolyzing and extracellular adenosine-producing enzyme. The GPI anchor can be cleaved by phospholipase C releasing the protein into the extracellular space.

### 6.1.2 Glycolipids and Glycoproteins Are Functional Components of the Cell Surface

The exclusive surface orientation of glycosylated lipids and proteins implies that glycosylation is immediately linked to cell surface-specific functions, as for example, the binding of specific ligands or cell/cell interactions. Glycolipids are differentially distributed between cell types. For example, **galactocerebrosides** are particularly enriched in myelin. The negatively charged **gangliosides** carry varying numbers of sialic acid residues (*N*-acetylneuraminic acid) (Fig. 6.1). Varying the order and number of mainly sialic acid residues via specific glycosyltransferases creates manifold molecular structures reminiscent of the variability of protein structures resulting from differentially combined amino acid residues. Neuronal and particularly synaptic membranes are rich in complex gangliosides. The pattern of gangliosides varies during development of the nervous system. Gangliosides can interact with membrane proteins and thereby affect their function [3]. They also act as receptors for the cellular uptake of some serotypes of botulinum neurotoxin and of choleratoxin.

There is no unifying hypothesis for the nearly ubiquitous glycosylation of cell surface-oriented membrane proteins. **Protein glycosylation** appears to be equally relevant for the quality control or the intracellular sorting of proteins as for maintaining a stable protein conformation or for the binding of water and metal ions to membranes. In a number of well-investigated examples an involvement of glycosylated membrane proteins in cell/cell or cell/matrix interactions has been demonstrated. These include the sialoglycoprotein neural cell adhesion protein (N-CAM). The extent of sialylation controls the homophilic interaction between N-CAM molecules and thus the degree of cell adhesion. Polysialylation is observed particularly during embryonic development. The high density of negative charges and the resulting heavy hydration reduces the homophilic molecular interactions of N-CAMs and thus is thought to impair cell adhesion during the early stages of development.

In addition to membrane proteins, carbohydrates are also associated with glycoproteins or **proteoglycans** that have been secreted into the extracellular space. There they can become components of the highly glycosylated **extracellular matrix** that forms a boundary around cells to protect them against mechanical and chemical damage, amongst others. Cell bodies and proximal dendrites of some classes of neurons are additionally surrounded by a highly condensed matrix known as **perineuronal net**. These nets consist of glycosaminoglycans bound to proteins (proteoglycans) or unbound in the form of hyaluronan, of fibrous proteins (e.g., collagens and elastin), adhesive glycoproteins (e.g., fibronectin, laminin, and tenascin), and a wide variety of secreted growth factors and other molecules [4]. They are thought to be involved in various functions including synaptic stabilization and limitation of synaptic plasticity, the support of ion homeostasis around highly active types of neurons, and in neuroprotection. Proteoglycans can also be essential components of receptor activation, as in the case of the activation of the Wnt signaling pathway or the activation of the fibroblast growth factor (FGF) receptor by FGF complexed with heparin sulfate chains.

Technical Box 1

**The Movement of Integral Membrane Proteins Can Be Traced by Single-Molecule Tracking**

The diffusion properties of individual membrane proteins over time can be visualized by single-molecule tracking methods. Molecules may be traced after labeling with organic fluorophores or with fluorescent proteins. Recently, quantum dots have been successfully employed for the analysis of surface trafficking of neurotransmitter receptors and channels on viable cultured neurons and also for the analysis of endocytotic trafficking of cell surface receptors [5]. Quantum dots are small (1–10 nm) semiconductor nanocrystals. The surface of quantum dots is covered by a hydrophilic coating which enables conjugation of molecules of interest such as antibodies or small ligands (Fig. 6.3). This permits tagging of the quantum dot to the molecule of interest. The quantum dots can be traced by confocal laser scanning microscopy as single particles (bound to individual molecules) by the induced luminescent blinking and their trajectories can be analyzed. Quantum dots with different diameters can be designed that emit at different wavelengths after excitation at a single wavelength. This offers the opportunity of multichannel analysis.

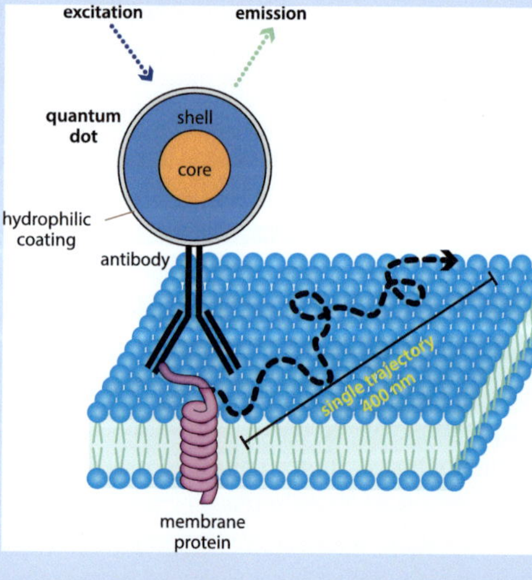


**Fig. 6.3 Quantum dot-tracking of single cell surface molecules.** The surface of the quantum dot is covered with a hydrophilic coating which enables conjugation to biological compounds such as antibodies. Life cells (e.g., cultured neurons) are first incubated with the labeled quantum dots to allow binding of the antibody to its surface-located antigen. The subsequent dot tracking by confocal laser scanning microscopy provides a 10 nm pointing accuracy which offers the possibility to construct a trajectory in submicrometer membrane domains (e.g., clusters of postsynaptic receptors). Excitation with laser light results in intermittent fluorescence emission (blinking) of the quantum dot

## 6.2 From Gene to Functional Protein

The pathways leading from DNA to functional protein and its final cellular destination are complex and regulated. They are highly relevant for understanding protein expression during development, the pattern of protein expression in individual cells or their progeny, alterations in protein expression following physiological or pathological stimuli, or also the cellular distribution of proteins.

### 6.2.1 Transcription and Translation Are Highly Regulated

The first step on the way to protein synthesis comprises the **transcription** of the DNA base sequence into that of **mRNA**. The transcription of the genetic information into mRNA is accomplished by RNA polymerase II. The enzyme attaches to the coding DNA strand and catalyzes the transcription of the DNA code into the complementary RNA code. Only one of the two DNA strands acts as a template. In the majority of cases transcription is regulated. The attachment of RNA polymerase to the DNA and the transcription of the DNA strand are strictly controlled and aided by **transcription factors**.

A **gene** consists of two basic functional units: a **coding region** that is transcribed into RNA and a **regulatory region** that controls transcription (Fig. 6.4). The regulatory regions comprise the **promoter** (the site of assembly of transcription factors and polymerase) and additional **regulatory sequences** (controlling the rate of transcription initiation, including the modification of chromatin structure) for binding of regulatory proteins. The regulatory sequences can be distant from the promoter region. The promoter region of many abundantly expressed proteins contains an AT-rich sequence, the **TATA box** that directs RNA polymerase II to transcription initiation. Specific proteins within the region of the TATA box and additional DNA segments regulate the exact positioning of the RNA polymerase. Depending on the regulatory proteins bound, transcription can be induced intermittently, permanently, or not at all. The termination of mRNA synthesis and the dissociation of the primary transcript mRNA is similarly complex and determined by specific nucleotide sequences. Transcription factors such as the homeodomain proteins or the basic helix-loop-helix proteins are highly relevant for the regulation of development, including neural development.

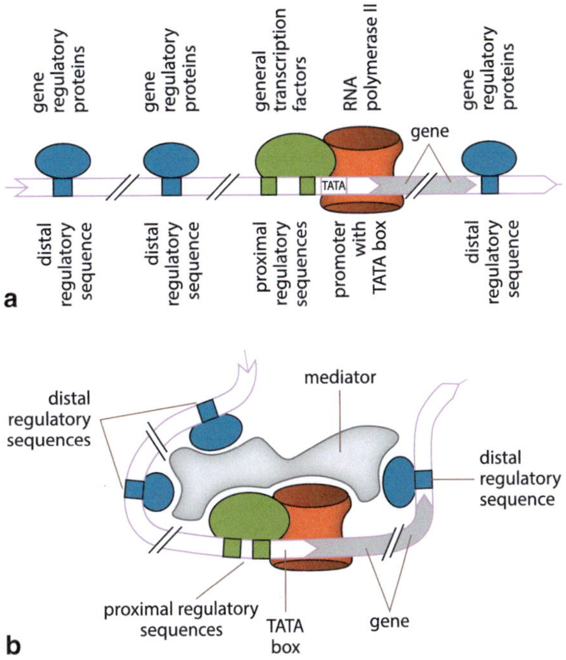


**Fig. 6.4 Simplified scheme of eukaryotic transcription control.** (**a**) Transcription initiation at the promoter requires the assembly of general transcription factors and RNA polymerase. Additional regulatory sequences can bind gene regulatory proteins depending on the state of cellular activity and control the protein assembly at the promoter. These can be situated in proximity to the promoter or far upstream or downstream and even within the protein encoding gene. (**b**) Interaction of the DNA-bound proteins requires DNA looping and can involve the mediator protein complex that facilitates the interaction of transcriptional activators with the RNA polymerase II transcription factor complex

The primary transcript (**pre-mRNA**) is modified co-transcriptionally in its nascent state (Fig. 6.5). A cap consisting of a modified guanine nucleotide is added to the 5′-end of the RNA molecule and the 3′-end becomes elongated by a poly-A tail, a sequence not encoded by the DNA. The cap and the poly-A tail are later important, amongst others, for the nuclear export of RNA and the stability of mRNA during translation. Before transcription has been terminated the silent **introns** are enzymatically removed and the coding **exons** spliced together to form the mature and translatable mRNA. **Alternative splicing** of pre-mRNAs containing several exons can lead to the synthesis of multiple proteins from one pre-mRNA or a single gene. For example, polymorphic forms of the enzyme acetylcholinesterase can be formed by alternative splicing and alternative promoter choices. The six different isoenzymes exist as soluble monomers or combine to form amphipathic dimers or tetramers that

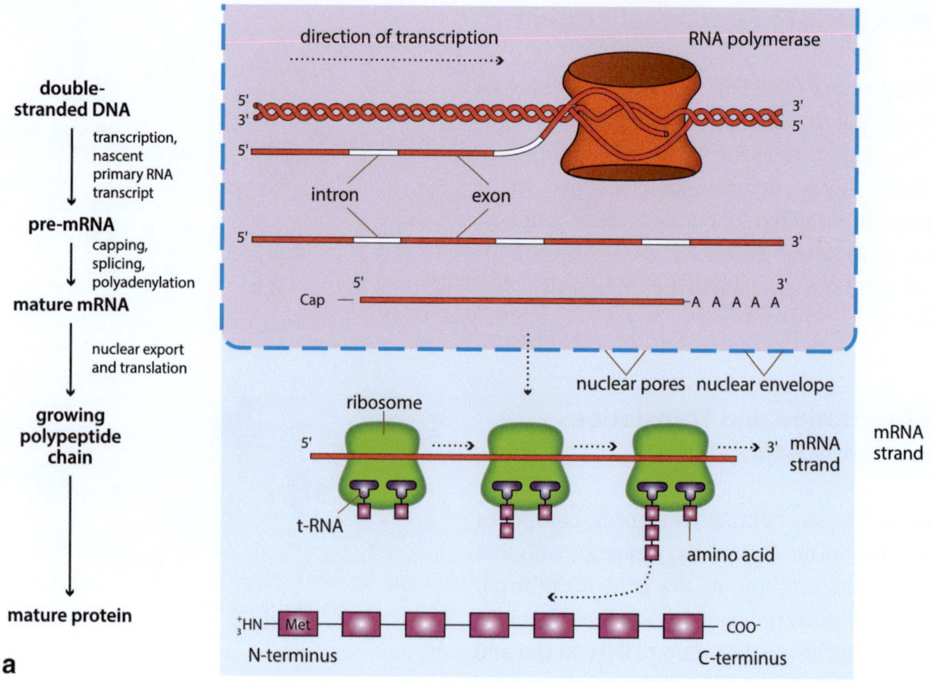


a

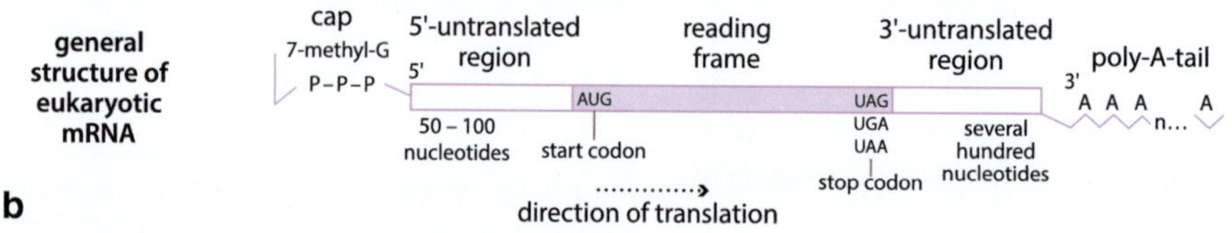


b

**Fig. 6.5 From gene to protein in eukaryotic cells, essential steps.** (**a**) Main nucleus- and cytosol-located processes. (**b**) Structural features of eukaryotic mRNA. The cap is enzymatically added to the nascent RNA transcript. Typically it consists of a 7-methylguanosine moiety that is connected via a triphosphate bridge to the first nucleotide at the 5′ end. After 3′ cleavage of the RNA the poly-A tail consisting of 100–200 adenine nucleotides is sequentially added by poly-A-polymerase. Following transcription and splicing the mRNA still contains nontranslated nucleotide sequences at its 5′ and 3′ end. The translation of the coding region is warranted by the start and stop condons

become associated with the plasma membrane or the extracellular matrix.

### 6.2.2 Epigenetic, Posttranscriptional, and Posttranslational Mechanisms

As revealed in recent years, the classical pathway from gene to protein can be modified by multiple epigenetic processes. These include DNA methylation, histone modifications, nucleosome repositioning, higher-order chromatin remodelling, noncoding RNAs, and RNA and DNA editing [6]. These mechanisms orchestrate a seemingly infinite variety of molecular and cellular processes essential for higher nervous system functioning. **RNA editing** is one way to modify the RNA message. This posttranslational mechanism equally occurs in *C. elegans* and *D. melanogaster* but its impact has revealed a dramatic increase during mammalian evolution with highest levels and most complex forms in hominids. In mammals, editing is accomplished by deamination of adenine to inosine and of cytosine to uracil. In mRNA, the alteration in codon sequence can result in alterations in pre-mRNA splicing or in the amino acid sequence of the resulting protein [7]. In *Drosophila* and mammals mRNA editing is particularly prominent in the nervous system. For example, three AMPA and two kainate receptors are subject to mRNA editing. This can result in the formation of receptors with different physiological properties or also in the modulation of receptor trafficking to the plasma membrane or in subunit assembly. Accordingly regulated mRNA editing can directly impact on multiple physiological processes including synaptic transmission.

Enzymatically controlled shortening of the poly-A tail typically controls the **stability and half-life of mRNA**s. Recently, another mechanism for posttranslational control has been uncovered whose physiological implications are just being deciphered. The DNA encoding the short non-protein-coding single-stranded **microRNA**s (miRNAs) is transcribed by either RNA polymerase II or RNA polymerase III into primary miRNA transcripts that can be further processed. They often are capped and polyadenlyated [8, 9]. Similar to mRNA, miRNA precursors can undergo RNA editing. In the brain these miRNAs are expressed at high levels. They target specific mRNAs and negatively regulate their stability and translation (Fig. 6.6a). In humans, several hundred miRNAs have been identified. A single miRNA can directly downregulate the production of hundreds of proteins [10]. Increasing evidence suggests that miRNAs have multiple and severe physiological and pathological implications for neural development, synaptic maturation and plasticity, learning behavior, and neurological disease. Mutation of a miRNA expressed in hair cells of the inner ear has been related to progressive hearing loss [11].

Epigenetic processes include DNA and histone modifications that can modulate gene expression patterns without modifying the actual nucleotide sequence and still yield profound effects on the cellular repertoire of expressed genes. **DNA methylation** affects protein binding to the DNA and is typically correlated with gene silencing. Similarly, the reversible **histone acetylation** and **methylation** can directly affect chromatin structure and the activity state of the DNA. Importantly, epigenetic information is often inherited within cell lineages and may last for multiple generations.

Technical Box 2

**Targeting mRNA for Inhibition of Protein Expression**

Novel techniques take advantage of endogenous cytoplasmic mechanisms that control the stability of mRNA or its translation into proteins. Suppressing the synthesis of a protein of interest at the mRNA level can provide important information regarding the functional role of this protein in a complex cellular context. In the traditional **antisense** approach, single-stranded DNA oligonucleotides complementary to a segment of the mRNA in question is introduced into the cell and used for hybridization, similarly to the in situ hybridization. As a result, the hybridized mRNA is degraded. The now widely used **RNA interference** (**RNAi**) takes advantage of endogenous cellular defense mechanisms against viruses and transposable elements. One way is to use small double-stranded RNA molecules 21–23 nucleotides in length (also referred to as small interfering RNA or silencing RNAs, **siRNAs**) that are introduced into the cell. The cellular mechanisms for processing microRNAs and of siRNAs reveal conspicuous parallels (Fig. 6.6b). The siRNA is processed in the cytosol and then hybridizes with the targeted mRNA. This directs degradation of the mRNA, a process also referred to as **siRNA knockdown**. Similarly, **small hairpin RNA** (**shRNA**, also short hairpin RNA) can be used to silence gene expression (Fig. 6.6b). shRNA denotes a sequence of RNA revealing a tight hairpin turn. The shRNA hairpin structure is cleaved into siRNA that in turn is processed and binds to its target mRNA. shRNA can be expressed from appropriate DNA vectors in cells of interest and passed on to daughter cells causing long-term silencing. Moreover, transgenic animals can be produced that express the RNAi under the control of a specific promoter. Depending on the type of promoter the RNA may be expressed in specific tissues or cell types or also during specific stages of development or stages of cell activation. This allows cell or **tissue-specific inactivation of gene function** and a detailed analysis of the function of the targeted protein in situ.

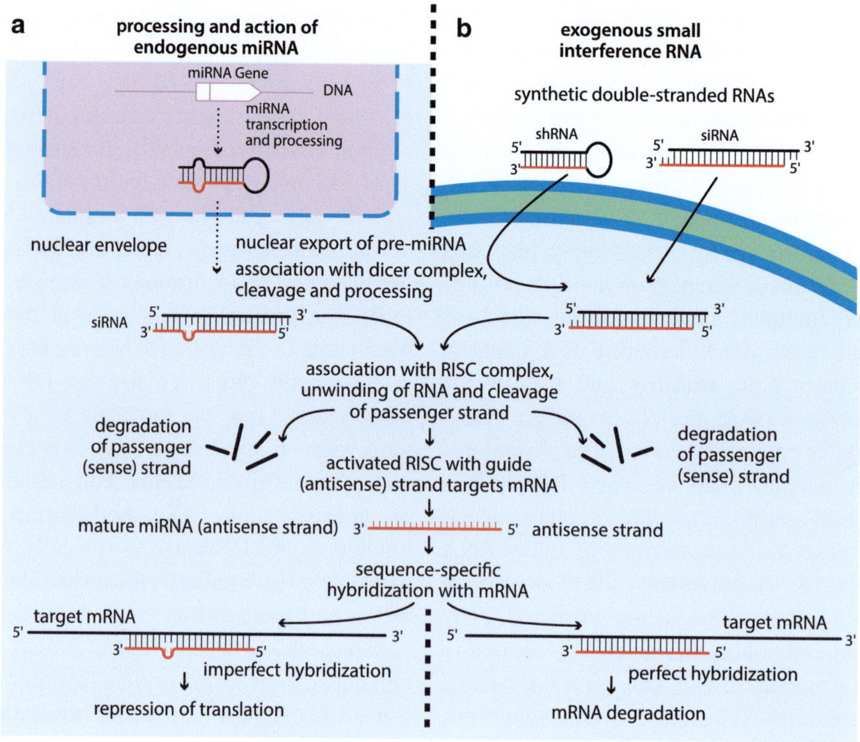


**Fig. 6.6** **Simplified scheme of gene inactivation by RNA interference in animals.** (**a**) Endogenous miRNA. The hairpin miRNA is encoded by DNA and generally transcribed by polymerase II. Processing involves cropping within the nucleus, export into the cytosol, and cleaving by the RNAse Dicer to produce double-stranded si (small interference) RNA, a 21–23 duplex. The siRNA then associates with the RISC (RNA-induced silencing) complex that unwinds the duplex via its helicase activity. Whereas the sense strand can be degraded, the antisense strand guides RISC to the target mRNA. miRNA hybridizes imperfectly with the mRNA and generally inhibits translation. (**b**) Exogenously provided synthetic siRNAs are converted into functional siRNAs and subsequently processed along the same cytoplasmic processing steps as miRNA. Note that sh (short hairpin) RNA and siRNA enter the pathway at different points. In contrast to shRNA, exogenous siRNA does not require Dicer processing and is directly incorporated into the RISC complex. Thus the pathways for exogenous synthetic double-stranded RNAs and endogenous miRNA converge at the Dicer and RISC complex, respectively, eventually resulting in gene silencing. siRNA with its perfect fit induces rapid mRNA degradation. shRNA can also be constructed in a plasmid backbone for cell transfection or packaged into a virus and transduced into target cells, resulting in the stable integration and expression of shRNA for long-term gene knockdown

### 6.2.3 The mRNA Carries Sequence Information for Subcellular Localization of the Protein

During **translation** the nucleotide sequence of mRNA is converted into the amino acid sequence of the protein. Of particular importance in the present context is the observation that proteins determined for the **cytosol** (for example, soluble enzymes or proteins of the cytoskeleton) are translated at free ribosomes. Following translation they are released from the ribosome and folded. In contrast, **membrane proteins** and **secretory proteins** are translated at ribosomes that attach to the rough endoplasmic reticulum (ER). Such proteins typically carry a hydrophobic N-terminal signal sequence that facilitates the attachment of the ribosome carrying the nascent polypeptide chain to receptors located at the ER. The hydrophobic N-terminus also acts as a start-transfer signal for the cotranslational transport of the polypeptide chain through the aqueous pore of a translocator into the ER lumen (Fig. 6.7). Following the insertion of the hydrophobic signal sequence into the lipid membrane, the polypeptide chain continues to grow into the ER lumen.

This process is interrupted when a stop codon of the mRNA has been reached and the polysome detaches. Specific protein sequences determine the final form of membrane incorporation of the protein into the lipid bilayer or its release into the ER lumen where it is properly folded.

### 6.2.4 Orientation, Folding, and Posttranslational Modification of Membrane Proteins Is Determined by the Primary Structure

**Secretory proteins** are generated in soluble form within the ER lumen. Following the synthesis of the **preproprotein**, the N-terminal hydrophobic signal sequence remains in the ER membrane. It is cleaved by a signal peptidase releasing the **proprotein** into the ER lumen (Fig. 6.7a). From there it can be transported via the Golgi apparatus to the lumen of secretory vesicles (compare Fig. 6.19). Within the Golgi apparatus or the secretory granules such proproteins can be proteolytically cleaved at specific sites into physiologically active secretory proteins or peptides. Many hormones are synthesized this way and one proprotein can give rise to several peptide hormones with differing physiological functions (see Chap. 11); in a similar way, proenzymes are cleaved to functional enzymes.

There are several ways the growing polypeptide chain of **membrane-spanning proteins** can be inserted into the ER membrane (Fig. 6.7b). One way to generate a single-pass transmembrane protein with a luminal N-terminus and a cytosolic C-terminus is by cleaving the signal sequence and interrupting further membrane translocation of the polypeptide chain by another hydrophobic sequence, a stop-transfer signal. In case of a luminal C-terminus and cytosolic N-terminus the polypeptide chain contains an internal signal sequence that initiates the translocation of the protein through the translocation apparatus. The signal sequence is not cleaved and anchors the protein to the membrane.

In many proteins that serve the transfer of ions or metabolites or the signal transfer into the cell, the polypeptide chain traverses the lipid membrane several times (multipass transmembrane protein). For example, the α subunit of the voltage-dependent $Na^+$ channel contains 24 membrane-spanning domains (see Chap. 7). The process of membrane insertion of such proteins is complex. It is assumed that the membrane orientation is accomplished by alternate start-transfer and stop-transfer sequences. Mostly, these membrane-spanning sequences reveal an **α-helical structure** resulting from a sequence of about 20 hydrophobic (uncharged) amino acid residues whose hydrophilic side chains are in the interior of the helix. When reaching the cell surface, the orientation of the ER proteins will be inverted as a result of transport vesicle exocytosis. Additional steps necessary for the production of functional proteins are located within the ER or at least are initiated there. These include protein folding and quality control, the formation of disulfide bonds, and protein glycosylation that is completed in the Golgi apparatus.

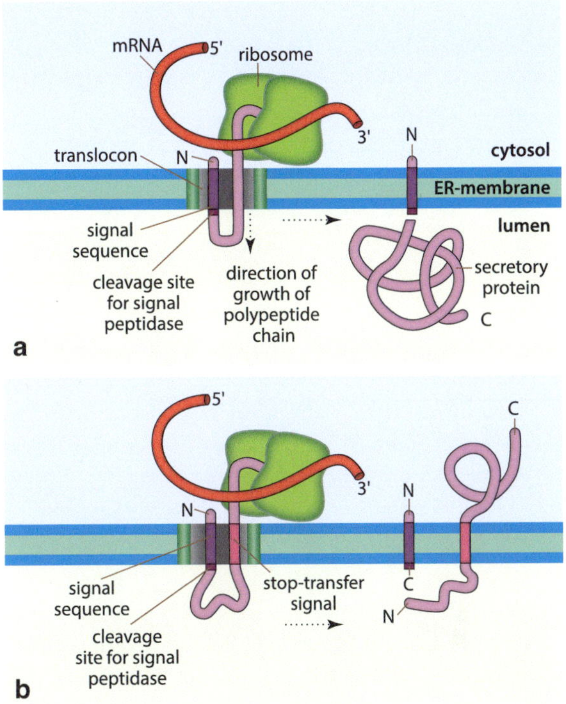


**Fig. 6.7 Import of the growing polypeptide chain into the lumen of the ER via a signal sequence.** (**a**) Binding of the hydrophobic N-terminal ER signal sequence (*magenta*) opens the translocon complex and initiates the transfer of the growing polypeptide chain through the aqueous channel across the lipid bilayer. Cleavage of the signal sequence by the signal peptidase creates a soluble protein in the lumen of the ER that is subsequently packaged into secretory vesicles via the Golgi apparatus. (**b**) Formation of a transmembrane protein involving a hydrophobic stop-transfer-signal (*light red*) and cleavage of the signal sequence (*magenta*). The remaining signal sequence is subsequently degraded

Technical Box 3

**Genetic Tools for Analyzing Protein Expression and Function**

Impressive methodological progress in molecular biology has created new tools for studying the structure, function, expression, and cellular distribution of proteins. These methods take advantage of the universality of the genetic code and the insight into the regulatory mechanisms governing transcription, translation, and recombination. For example, restriction endonucleases that are used by bacteria to destroy the foreign DNA of bacteriophages are employed for cutting DNA molecules at defined sites of their nucleotide sequence. Today several hundreds of such **restriction enzymes** are available. Similarly, with the help of **ligases** DNA fragments can be ligated to create new **recombinant DNA** molecules. **DNA polymerase I** can produce double-stranded DNA from single-stranded DNA also in the test tube, a process that normally takes place when DNA is replicated during the S phase of eukaryotic cell division. **Reverse transcriptase**, an RNA-dependent DNA polymerase of retroviruses permits the *in vitro* synthesis of DNA molecules complementary to isolated mRNA molecules. This complementary DNA (**cDNA**) is free of the noncoding intron sequences of the initial DNA.

Of particular experimental relevance is the possibility to insert genetic material into other cells. To achieve this, **vectors** are required as transport vehicles. The type of vector used depends on the experimental requirements. A vector not only contains the **passenger DNA** in question but also promoter sequences that assure that this DNA is amplified or that the corresponding protein is synthesized (Fig. 6.8). Using suitable expression vectors, proteins may be expressed in bacteria but also in yeast, insect, or mammalian cells. Heterologous expression can serve for analyzing, for example, the cell-specific localization or functional properties of proteins or the overproduction of a protein for subsequent high yield purification and structural analysis. Together this results in a broad spectrum of methodological approaches some of which are briefly characterized.

**DNA cloning**

DNA fragments can be amplified in a vector by a process referred to as DNA cloning. Several approaches may be taken. The fragment of a DNA of interest can be inserted into the DNA genome of a bacteriophage or a bacterial plasmid. The resulting **recombinant DNA** can be replicated in a bacterial host cell. The DNA of interest is often obtained from a collection of cloned DNA fragments (library). Such libraries may be produced from genomic DNA of a particular organism (**genomic library**) or from the total of mRNAs transcribed in a particular tissue or cell type (**cDNA library**).

Bacterial plasmids impose restrictions to the size of the DNA insert to be cloned. For cloning very large DNA molecules the linear **yeast artificial chromosomes** (**YACs**) are employed. More recently new plasmid vectors have become available that can be amplified in the more rapidly dividing bacteria. The **bacterial artificial chromosome** (**BAC**) accepts inserts of up to one million nucleotide pairs and thus the insertion of large segments of a genome. Also **human artificial chromosomes** (**HACs**) have been designed. They offer a promising system for delivery and expression of full-length human genes. Today BAC libraries are available that contain the entire genome of various organisms. BACs can carry both the coding region of a gene and its regulatory sequences that determine where and when the gene is expressed. BACs are also very useful for making transgenic animals with segments of DNAs. This allows the expression of large marker genes, recombination enzymes, toxins, tract-tracing proteins, or multiple combinations of these. For example, the pattern of the natural expression of a gene during brain development can be followed after replacing the coding region of the gene with a reporter sequence (for example, enhanced green fluorescent protein, EGFP, gene tagging, Fig. 6.8a) and subsequent tracing the pattern of EGFP expression in tissue sections.

**Site-directed mutagenesis for altering protein function**

If the primary structure of a protein is known, any amino acid residue can be exchanged via **site-directed mutagenesis**. This requires the synthesis of an

**Technical Box 3** (*continued*)

oligonucleotide complementary to the DNA sequence in question that contains the codon for an altered amino acid at the required position. The synthetically altered nucleotide can be hybridized to the original DNA forming a single mismatched nucleotide pair. Double strand completion is then accomplished by DNA polymerase and DNA ligase. Following replication and selection of the mutated strand the mutated DNA can be used to synthesize the mutated protein for functional analysis. Similarly, **protein chimeras** may be produced by replacing one segment of a protein by that of another protein or by fusing two proteins. This allows the design of proteins with arbitrarily altered primary structure. After expression in suitable cell systems the functional role of individual amino acids or of entire protein domains can be characterized. Or a short peptide sequence such as the myc-tag may be used for immunohistochemical detection of the expressed protein (epitope tagging, Fig. 6.8c).

**Transgenic animals for the study of cell-specific protein expression and function**

The methodological repertoire of molecular genetics permits the transfer of genes into individual cells (**cell transfection**) and the production of **transgenic animals** that possess external or modified genes including the **inactivation of genes**.

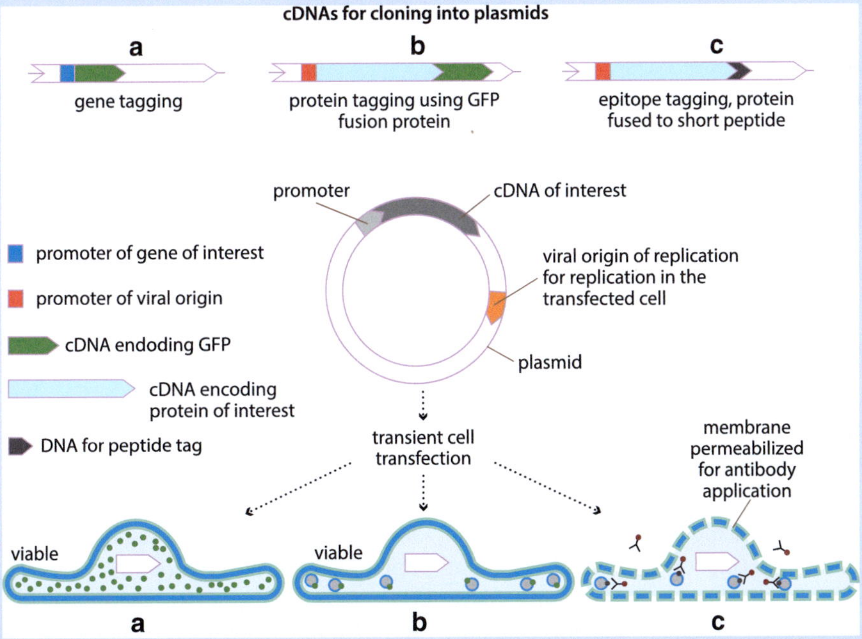


**Fig. 6.8** **Methodological approaches for gene tagging, protein tagging, and epitope tagging.** The scheme depicts examples with simple transient cell transfection using bacterial plasmids. (**a**) For gene tagging the cDNA encoding a reporter protein (for example, green fluorescent protein, GFP) is placed under the control of the endogenous promoter of interest. When the promoter is activated, the soluble GFP (*green dots*) distributes randomly in the cytosol of the viable cells and can be visualized by fluorescence microscopy. (**b**) In the case of protein tagging, the subcellular distribution of a protein can be visualized in viable cells by constructing a chimera of the endogenous protein and GFP. In this case and in (**c**) a strong promoter recognized by mammalian RNA polymerase is required to assure strong protein expression. In the example shown, the protein-GFP chimera (*green dots*) is targeted to the membrane of vesicles. (**c**) For epitope tagging, a short peptide sequence (such as the myc-tag consisting of 10 amino acids) for which commercially antibodies are available is fused to the protein of interest. In this case, the tagged protein can be visualized only by immunocytochemical methods requiring previous cell fixation and plasma membrane permeabilization. The epitope tag is incorporated into a vesicle protein (*black dot*) and detected with a fluorophore-conjugated (*red dot*) antibody. Corresponding techniques can be used for stable cell transfection or when creating transgenic animals

**Technical Box 3** (*continued*)

These technologies are extensively applied in all kinds of variants for the study of neural function.

By the use of specific targeting techniques genes can be altered or inactivated in mice. In **knockout mice** both copies of a gene are inactivated or deleted. Recently, also **knockout rats** have become available. These may become important new tools as rats more closely correspond with human physiology and disease than mice. A disadvantage of the knockout approach can be compensatory protein expression that becomes apparent during development of the knockout animal. In **conditional mutants** the deletion of the gene can be induced in a specific cell type or tissue at will, also in the adult stage. **Transgenic mouse lines** were originally generated by injection of linear DNA into the nucleus of an oocyte or transfected into embryonic stem cells. This approach results in random DNA integration into the genome and does not control copy number and may thus cause the inactivation of another gene at the site of integration. Methods have now become available for controlled and site-specific integration. This procedure achieves the integration of a single transgene by an induced and site-specific recombination process. The transgene constructs can utilize the endogenous promoter and gene sequences in BACs or YACs. These techniques also allow the expression of molecular **tags** such as the enhanced green fluorescent protein (**EGFP**) or other fluorescent proteins under a cell-specific promoter and the analysis of gene expression in specific cells. For example, neuron-specific expression is often induced via the Thy-1 promoter or expression in astrocytes via the promoter for glial fibrillary acidic protein (GFAP).

Similarly, **fate mapping** experiments are used to tag a progenitor cell with a genetic marker in order to later identify its descendants. In addition, fusion proteins consisting of the protein to be investigated and a fluorescent protein allows the mapping of the cellular expression and cellular distribution of individual proteins (protein tagging, Fig. 6.8b). Animals can be produced that simultaneously express proteins tagged with different fluorescent proteins (such as green, blue, yellow, or red fluorescent protein) in different cell types (multicolor transgenes). This allows the fluorescent imaging of the cellular dynamics of individual proteins equally well as of dynamic cell interactions.

## 6.3 Membrane Transport

Only very small and polar molecules can rapidly diffuse through membranes (Fig. 6.9a). Their distribution depends on the direction of their concentration gradient over the membrane. All other molecules require auxiliary mechanisms to effectively cross the membrane. Two principle mechanisms have evolved: direct membrane passage with the involvement of membrane proteins (transporters, ion channels) and endocytosis (compare Fig. 6.19). In the case of endocytosis, uptake of proteins and small molecular complexes is mediated via the budding of vesicles from the plasma membrane. All transporters and ion channels (Fig. 6.9) are integral membrane proteins with polypeptide chains passing the membrane multiple times. See Chap. 7 for the functional role of ion channels for the formation of membrane potentials and the propagation of action potentials.

### 6.3.1 Ion Channels

Ion channels serve the rapid membrane passage of ions. They form a hydrophilic aqueous pore structure (Fig. 6.9b) allowing the facilitated diffusion of ions through the membrane along their electrochemical gradient. With transporters they share their specificity for the solute to be transported. For example, some ion channels are permeable selectively for $Na^+$, $K^+$, $Ca^{2+}$, or $Cl^-$ whereas others are generally permeable for small metal cations. The ion specificity is determined by the diameter of the channel pore and its charge properties. All ion channels alternate between an open and closed state. Channels can be **gated** (opened) via a variety of mechanisms including changes in membrane potential, the binding of neurotransmitters or intracellular ligands including $Ca^{2+}$ ions, cyclic AMP (cAMP), or cyclic GMP (cGMP). Accordingly, they are referred to as voltage-gated, ligand-gated, nucleotide-gated etc.

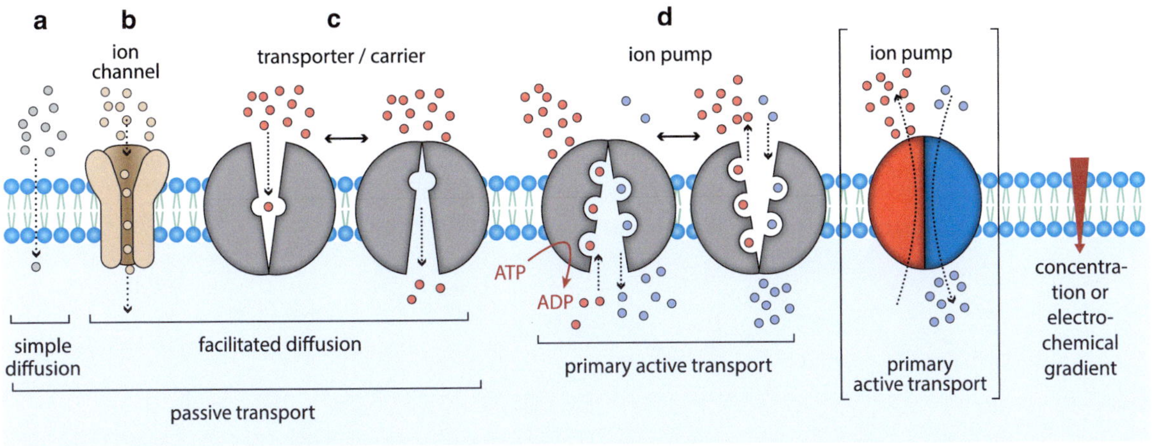


**Fig. 6.9** **Mechanisms of movement of solutes across membranes.** (**a**) Hydrophobic or very small molecules can diffuse directly through the lipid bilayer along their concentration gradient. (**b**) Ions can diffuse through the hydrophilic pore of an ion channel along their electrochemical gradient. (**c**) Transporters (carriers) facilitate the membrane passage of small lipid-insoluble solutes that otherwise would pass the lipid bilayer only very slowly. (**d**) Transport against a concentration or electrochemical gradient requires energy (active transport). In the case of primary active transport of ions (ion pump), the energy derived from the hydrolysis of ATP directly provides the driving force for the transport of ions against their electrochemical gradient. In the example depicted ($Na^+$-$K^+$ ATPase) $Na^+$ (*red*) and $K^+$ (*blue*) ions are pumped into opposite directions

ion channels. Voltage-gated ion channels are of central importance for neuronal excitability (see Chap. 7). Cyclic nucleotide-gated channels function in signal transduction in retinal photoreceptor cells (see Chap. 18) and in the sensory neurons of the olfactory epithelium (see Chap. 13). Other ion channels open when the membrane is mechanically distorted. They play an important role in signal transduction in the hair cells of the inner ear (see Chap. 17). Yet others, such as the resting $K^+$ channels (that are largely responsible for the generation of the resting potential) are nongated, and are not affected by specific stimuli.

### 6.3.2 Transporters

Differing functional principles apply to membrane proteins mediating the rapid transport through membranes. **Transporters** or **carriers** actively bind molecules and transport them through the membrane. The direction of transport is determined by the concentration gradient or (in the case of charged substances) the electrochemical gradient. This process is referred to as **passive transport** or facilitated diffusion (Fig. 6.9c). Energy is required for the transport of molecules against their concentration gradient (**active transport**) (Fig. 6.9d). The energy is derived either directly from the hydrolysis of ATP or indirectly from an electrochemical gradient generated by the hydrolysis of ATP. For example, the $Ca^{2+}$ stores of the endoplasmic reticulum are filled by an ATP-consuming $Ca^{2+}$ pump ($Ca^{2+}$ ATPase) for later release by a receptor-mediated process (compare Fig. 6.13). At the plasma membrane, $Na^+$ ions are transported to the cell exterior against their electrochemical gradient via the ATP-consuming $Na^+$ pump ($Na^+$-$K^+$ ATPase). This active transport is essential for maintaining the membrane potential. In addition, the electrochemical $Na^+$ gradient generated is utilized for the carrier-mediated cellular import of substances against their concentration gradient (Fig. 6.10a). Examples include the transporters mediating the reuptake of neurotransmitters such as glutamate, GABA, or noradrenaline or of the precursor substance choline. These substances are taken up against a considerable concentration gradient from the extracellular fluid together (**cotransport or symport**) with $Na^+$ ions that follow their electrochemical gradient [14]. Similarly protons are pumped into the lumen of synaptic vesicles by a transport ATPase (proton pump, $H^+$ ATPase). The proton gradient generated is then utilized for the vesicular uptake of neurotransmitters. In the depicted example (Fig. 6.10b) acetylcholine is taken up into synaptic vesicles in exchange (**antiport**) of the excess of protons in the vesicle lumen.

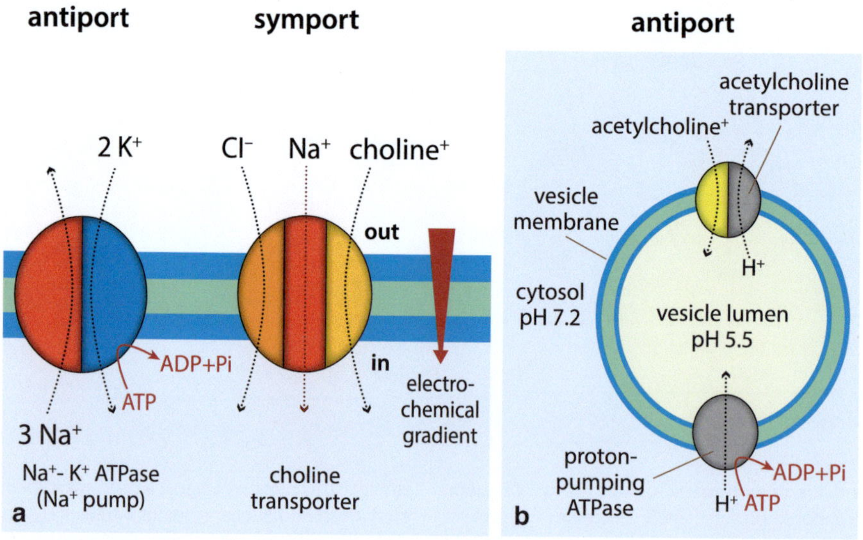


**Fig. 6.10** **Production of the driving force for the transport of molecules against their concentration gradient.** (**a**) The electrochemical gradient produced by the $Na^+$-$K^+$ ATPase is the driving force for the uptake of the neurotransmitter precursor choline via the high affinity choline transporter ($K_m$ approx. 1 μM) into cholinergic nerve terminals. The cotransport requires $Cl^-$ in addition to $Na^+$ ions [12]. (**b**) The proton-pumping ATPase (V-ATPase) creates an excess of protons in the synaptic vesicle lumen and acetylcholine is taken up in exchange of protons via an antiport (vesicular acetylcholine transporter) [13], leading to an approx. 100-fold concentration of luminal ACh as compared to the surrounding cytosol

## 6.4 Cell Communication

Multiple mechanisms of receptor-mediated cell communication have evolved. These include the type and cellular localization of receptors, the chemical nature and origin of the signaling substances, and the cellular mechanisms induced by receptor activation. Only cells in the immediate environment can communicate via **direct cell contact**. Communication with distant cells or between tissues requires secreted **extracellular signal molecules**. Of particular relevance in the nervous system are neurotransmitters, hormones, or growth factors. Signal molecules include low-molecular-weight substances such as acetylcholine, noradrenaline, or various amino acids but also peptides, lipids such as derivatives of

**Technical Box 4**

**Channelrhodopsins, Tools for Noninvasive Excitation**

Recently, **channelrhodopsins**, light-gated ion channels, have been introduced as powerful new tools allowing the temporally precise and noninvasive control of the activity of genetically defined subpopulations of neurons. The widely used channelrhodopsin-2 (ChR2) was isolated from the alga *Chlamydomonas reinhardtii*. Like rhodopsin (see Chap. 18) it is a seven-pass membrane protein and binds the chromophore all-*trans*-retinal. It reacts to illumination with blue light with the rapid opening of a channel pore permeable to $H^+$, $Na^+$, $K^+$, and $Ca^{2+}$ and can thus be employed for neural excitation. When fused to a fluorescent protein the ChR2-transfected cells can directly be visualized. This tool has now been applied to *C. elegans*, *Drosophila*, zebrafish, and mice. It can also be used in mammals for effective photostimulation using light targeted via fiberoptic-based systems to the ChR2-expressing neurons [15]. Other recently developed tools include halorhodopsin, a yellow light-activated chloride pump from Halobacteria that suppresses action potentials. When expressed together with ChR2 under the control of cell-specific promoters this allows wavelength-dependent multiple activation and silencing of neural activity with millisecond precision [16]. These novel tools permit the analysis of synaptic function equally well as cell type-specific neural circuit control and the analysis of the role of defined circuits in the control of behavior under the control of light. They open up the new field of **optogenetics** for the exploration by biomedical engineers.

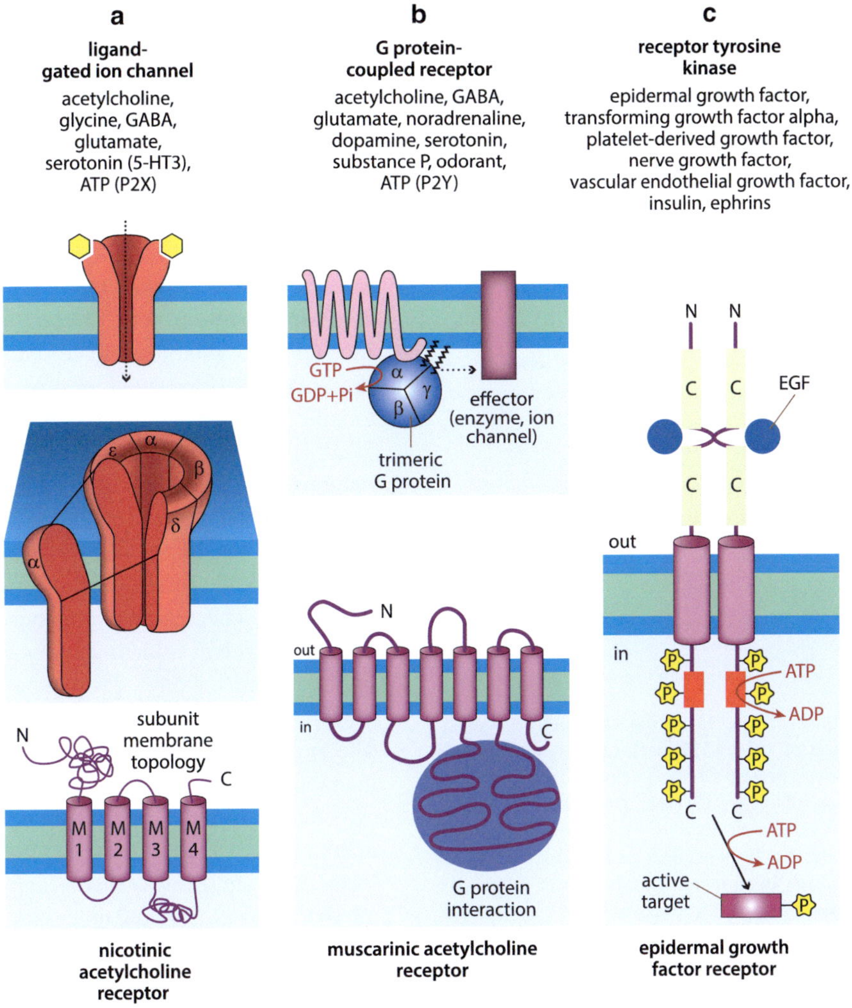


**Fig. 6.11 Three types of membrane-bound receptors, their membrane topology, and selection of ligands.** (**a**) Ligand-gated ion channels represent a group of receptors activated by multiple neurotransmitters. The nicotinic acetylcholine receptor is depicted as an example. In the adult mammal it consists of five subunits (2α, β, δ, ε) with identical membrane topology and four transmembrane α-helices (M1–M4) each. Whereas the receptors for glycine, GABA, and the serotonin (5-HT3) receptors belong to the same receptor superfamily, the ionotropic glutamate receptors and the P2X ATP receptors are members of different protein families with differing subunit composition and membrane topology. (**b**) G protein-coupled receptors represent a large protein family whose individual members are activated by multiple ligands including many neurotransmitters. They share the typical membrane topology with seven transmembrane α-helical regions and the cytosolic binding of a trimeric G protein. The muscarinic acetylcholine receptor is shown as an example. (**c**) Receptor tyrosine kinases are depicted as an example for catalytic receptors. They share an extracellular ligand-binding domain and an intracellular catalytic domain, in this case a tyrosine kinase domain. Binding of the ligand induces receptor dimerization and autophosphorylation at the intracellular C-terminal domain, followed by phosphorylation of target proteins. The ligand-activated and dimerized EGF receptor (ErbB1) is shown as an example. *C* cysteine-rich extracellular domains, *EGF* epidermal growth factor

arachidonic acid or phospholipids, or gaseous substances such as NO. The broad spectrum of chemical messengers faces a limited number of principal signaling mechanisms in the target cell. Water-soluble substances interact with receptors at the cell surface whereas lipophilic substances may pass the plasma membrane and bind to cytosolic receptors or receptors in the nucleus.

### 6.4.1 There Are Multiple Types of Receptors and Ligands

Multiple pathways have evolved for the transduction of extracellular to intracellular signals. Only some of the principal receptor types can be briefly introduced here (Figs. 6.11, 6.12, 6.13, 6.14, and 6.15). Individual cells may carry dozens of different receptor types.

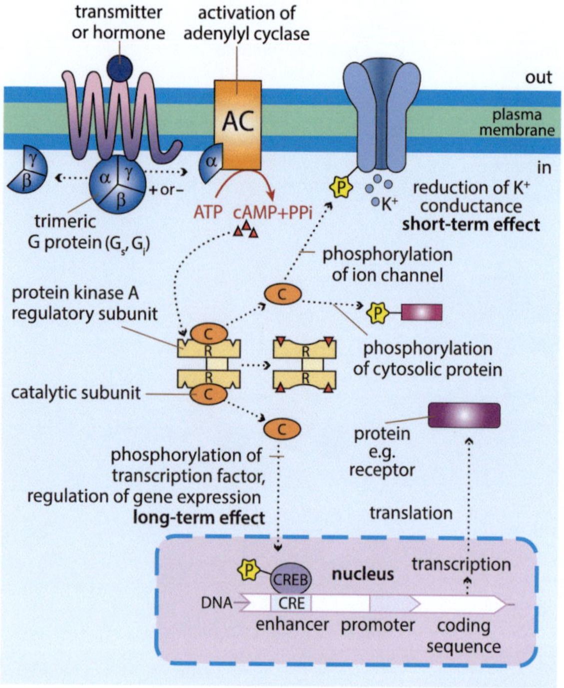


**Fig. 6.12** **Essentials of the G protein-coupled receptor-mediated activation of the cAMP pathway.** Ligand binding activates the trimeric G protein whose α subunit can either activate (+, $G_s$ type) or inhibit (–, $G_i$ type) the membrane-bound enzyme adenylyl cyclase. In case of activation, the cAMP formed stimulates the cAMP-dependent protein kinase (protein kinase A, PKA) which in turn induces short-term effects by phosphorylation of ion channels or cytosolic proteins or long-term effects by regulating transcription. Activated PKA enters the nucleus where it phosphorylates the transcription factor CREB (CRE-binding protein) that binds to the cAMP response element (CRE). Also the β,γ complex can directly activate membrane-bound effectors. cAMP can in addition directly gate ion channels in the plasma membrane, as, for example, in olfactory sensory cells (see Chap. 13)

General principles of receptor function include the following:

- The receptor is a ligand-gated ion channel. Ligand binding gates and thus opens the ion channel (**ionotropic receptor**). This is the typical pathway employed by fast neurotransmitters.
- The receptor contains an extracellular ligand-binding domain and an intracellular binding domain for a transducer protein (trimeric G protein). The ligand-induced binding of the trimeric G protein induces a series of intracellular reactions (**metabotropic receptor**). Many peptide hormones, but also other signal substances, utilize this pathway.
- The receptor is a ligand-activated enzyme (enzyme-coupled receptor, **catalytic receptor**). The ligand binds to the extracellular domain but the catalytic activity is produced by the cytosolic domain. The intrinsic catalytic function includes activity such as tyrosine protein kinase or serine/threonine protein kinase or also phosphatase activity. Typical ligands of receptor tyrosine kinases include growth factors but also the membrane-tethered protein boss, important for *Drosophila* ommatidium development, or the ephrins. Ligands of serine/threonine kinases include members of the transforming growth factor-β (TGFβ) superfamily (including bone morphogenetic protein) that are important morphogens during development.
- Binding of the ligand triggers the interaction of transmembrane proteins that in turn induces a cellular signal. This mechanism is typical for the Wnt and the hedgehog signaling pathways that are both important for many developmental processes, including brain development.
- The induction of a signaling event involves the interaction of transmembrane proteins of adjacent cells that in turn induces a signaling event in one or in both cells. An example of the former is the Notch/Delta signaling pathway that is highly relevant for a number of developmental processes. An example of the latter is the ephrin/Eph receptor signaling pathway. Ephrins are involved in various aspects of cell-cell communication during development, including axonal path finding and cell-cell interactions of the vascular endothelial cells.
- The receptor is an intracellular protein activated by lipophilic agonists such as gaseous substances or lipid hormones.

Different intracellular signaling pathways may overlap allowing for synergism or antagonism. In many cases these signaling pathways serve the amplification of the ligand-induced signal inside the cells and in addition the activation of transcription factors that regulate protein expression. Typical amplifiers include protein kinases. These may represent an intrinsic component of the receptor protein, be tightly bound to the receptor or also become activated via a cascade of protein interactions within the cytosol. In the following, four examples of principal receptor types are briefly discussed.

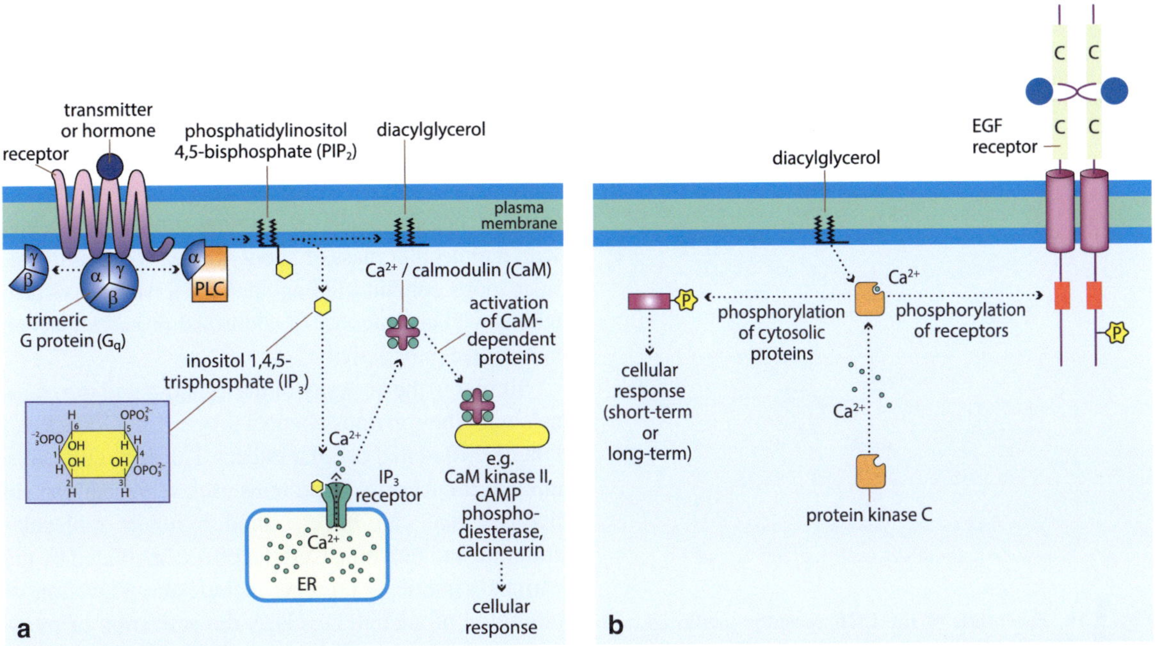


**Fig. 6.13** **Essentials of the G protein-coupled receptor-mediated phospholipase Cβ pathway.** Activation of phospholipase Cβ by the trimeric G protein (Gq type) leads to the production of two different second messengers from phosphatidylinositol 4,5-bisphosphate ($PIP_2$). (**a**) The water-soluble and diffusible inositol 1,4,5-trisphosphate ($IP_3$) elevates cytosolic $Ca^{2+}$ (*green dots*) via activation of the $IP_3$ receptor (an $IP_3$-gated $Ca^{2+}$ channel) in the membrane of the endoplasmic reticulum (ER). Binding of released $Ca^{2+}$ to calmodulin leads to activation of several proteins, including $Ca^{2+}$/calmodulin-dependent protein kinase II (CaM kinase II) that can phosphorylate multiple proteins, including neurotransmitter receptors. (**b**) The membrane-bound diacylglycerol activates protein kinase C (PKC) that is recruited to the plasma membrane by elevated $Ca^{2+}$. Activated PKC phosphorylates and regulates multiple cytosolic or membrane proteins including the EGF receptor

### 6.4.2 Ligand-Gated Ion Channels: A Class of Neurotransmitter Receptors

Ligand-gated ion channels combine the advantage of fast signal transduction (ion flux and change in membrane potential) with the activation via a select signal substance. They are generally activated by neurotransmitters and mainly located at synapses of the central and peripheral nervous system. Neurotransmitters taking advantage of this fast signaling pathway include acetylcholine, GABA, glycine, glutamate, serotonin, and ATP (Fig. 6.11a). All receptors are oligomers consisting of several subunits that assemble to form a functional ion channel. The type of subunits of a particular receptor can vary between tissues, e.g., between muscle, different brain regions, or during development.

The first membrane-bound receptor to be isolated and characterized in molecular terms was the **nicotinic acetylcholine receptor**. This receptor is a pentamer of four different subunits with molecular masses of 40–60 kDa encoded by four different genes [17]. Since the α subunit occurs twice, the receptor complex has the subunit stoichiometry $\alpha_2$, β, δ, ε (in the adult and $\alpha_2$, β, δ, γ in embryos) (Fig. 6.11a). Two acetylcholine molecules are necessary to gate the channel and the acetylcholine-binding sites are situated at subunit interfaces always involving an α subunit. Each subunit reveals four transmembrane helices and the ion channel of the muscle nicotinic receptor permits the passage of small metal cations, with decreasing permeability: $K^+ > Cs^+ > Na^+ > Li^+$. Under conditions of resting potential, channel opening results in a massive depolarizing influx of $Na^+$ ions and a small rectifying efflux of $K^+$ ions. The protein family includes a cation-permeable serotonin receptor and the $Cl^-$-permeable GABA and glycine receptors. The cation-permeable glutamate and ATP receptors form two additional protein families [18]. Like most other receptors, ligand-gated ion channels couple to scaffolding proteins at the inner membrane surface and connect to the cytoskeleton.

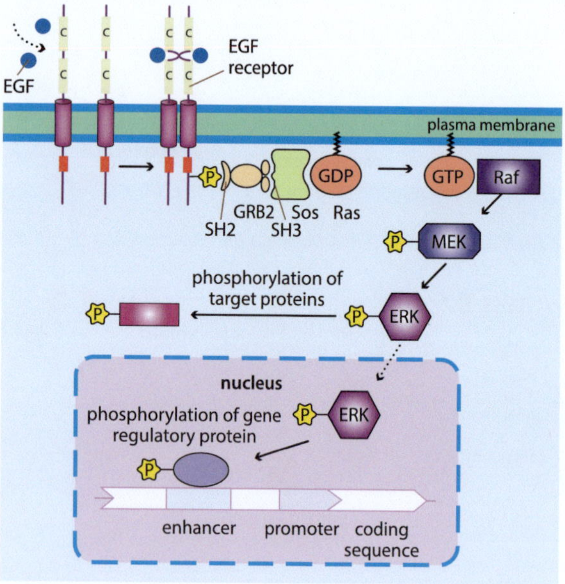


**Fig. 6.14** **Essentials of the EGF receptor-mediated activation of the MAK kinase pathway.** Ligand binding induces receptor dimerization via the interaction of two extracellular loop segments. This leads to activation of the intracellular tyrosine kinase domain and mutual phosphorylation of tyrosine residues. Via its phosphotyrosine-binding SH2 (src-homology 2) domain the adapter protein GRB2 binds to the C-terminal part of the receptor and recruits the nucleotide exchange factor Sos (son of sevenless) via its two SH3 domains (that bind to proline-rich sequences). Sos couples to the GDP-bound inactive Ras (linked to the membrane via a farnesyl anchor) and promotes the exchange of GDP for GTP. In mammalian cells activation of Ras leads to the subsequent phosphorylation in series of the three serine/threonine protein kinases Raf, MEK, and ERK (MAP kinase). MAP kinase can phosphorylate multiple cytosolic proteins and enter the nucleus to phosphorylate gene regulatory proteins. This leads to increased expression of multiple proteins (early-response genes such as c-fos)

### 6.4.3 G Protein-Coupled Receptors (GPCRs): A Family of Receptors Interacts with a Family of Trimeric G Proteins

G proteins bind and hydrolyze GTP. Trimeric G proteins are heteromers of the subunits α, β, and γ. G protein-coupled receptors belong to a superfamily of **seven-pass transmembrane proteins (7TM)** (Fig. 6.11b) derived from a common ancestor [19]. This group of receptors can be activated by a wide range of molecules, such as neurotransmitters, several nucleotides, peptide hormones, lipids such as sphingosine 1-phosphate or lysophosphatidic acid, a variety of chemokines, odor and taste molecules, or the protease thrombin. The protein family includes the retinal-binding protein opsin of the visual pigment rhodopsin. Contrary to the ligand-gated ion channels these receptors consist of a single polypeptide chain with a molecular mass of about 40 kDa. Their intracellular loops contain binding sites for a trimeric G protein as well as a number of additional proteins that can modify receptor activity.

Similar to the diversity of the ligand-gated ion channels with their various channel properties, GPCRs can induce differential cellular effects. The α, β, and γ subunits of the trimeric G proteins each represent protein families that can be classified by their molecular structure and function. Cellular effects mediated by the various trimeric G proteins include the activation or inhibition of adenylyl cyclase, the activation of phospholipase Cβ, the inhibition of cGMP-dependent phosphodiesterase in retinal photoreceptors, or the modulation of $K^+$ and $Ca^{2+}$ channels.

The trimeric G protein functions as a shuttle between membrane proteins. Both, the α and the γ subunits are fastened to the inner surface of the plasma membrane via a lipid anchor. The α subunit exerts the name-giving function. It carries the catalytic activity for the hydrolysis of GTP. When the activated seven-pass receptor binds and activates the G protein, the GDP bound to the α subunit is exchanged for GTP causing a conformational change of the protein that now adopts its active state. According to the original model, the α subunit dissociates from the trimeric complex and binds to and activates or inhibits an effector (a membrane-bound enzyme or an ion channel). Similarly the βγ complex can regulate the activity of target proteins. Recent evidence implies that – at least in some cases – the α subunit and the βγ subunits may interact with their targets without previous dissociation. The functional cycle becomes terminated by hydrolysis of GTP and dissociation of the G protein from its target protein.

Generally, the activated enzymes induce the formation of intracellular signal substances (**second messengers**) and thus an amplification of the signal initially induced by the ligand-binding to the receptor. The induced physiological function results from the properties of the targeted enzyme or ion channel and in addition from the cell context. Since many receptors employ the same G proteins, the resulting intracellular functions can be identical.

Interestingly, the neurotransmitters acetylcholine, GABA, glutamate, serotonin, and ATP – but not glycine – all make use of both ligand-gated ion channels and GPCRs (Figs. 6.11a, b), depending on the cellular and functional context necessary for the induction of fast or enduring signals, respectively.

### 6.4.4 Adenylyl Cyclase and Phospholipases C Generate Second Messengers

One of the functional consequences of GPCR activation is the activation or inhibition of adenylyl cyclase by trimeric G proteins of the $G_s$ and $G_i$ type, respectively (Fig. 6.12). The **cAMP** formed as a result of the stimulation of adenylyl cyclase and the subsequent activation of the cAMP-dependent protein kinase (protein kinase A, PKA) has multiple functional implications. Short-term effects include the phosphorylation of cytosolic proteins or ion channels. The latter induces alterations in channel properties such as the reduction of channel conductance. Via PKA-mediated phosphorylation of transcription factors (CREB) the ligand bound to the receptor directly impacts on the expression pattern of proteins in the target cell.

The consequences of the activation of phospholipase C-β (PLC-β) by the G protein $G_q$ are more complex. As a result of the hydrolysis of phosphatidylinositol 4,5-bisphosphate ($PIP_2$) two second messengers are formed, the hydrophilic and freely diffusible **inositol 1,4,5-trisphosphate ($IP_3$)** and the membrane-bound **1,2-diacylglycerol** (Fig. 6.13a). $IP_3$ activates a specific receptor in the membrane of the endoplasmic reticulum. This $IP_3$ receptor is a $Ca^{2+}$ channel allowing the release of $Ca^{2+}$ into the cytosol. Another type of $Ca^{2+}$ channel of the endoplasmic reticulum prominent in neurons is the ryanodine receptor (named after the plant alkaloid ryanodine) that can be activated by $Ca^{2+}$ ($Ca^{2+}$-activated $Ca^{2+}$ channel) and further amplifies the $Ca^{2+}$ signal. The $Ca^{2+}$ released from the lumen of the endoplasmic reticulum into the cytosol represents a third messenger. It binds to specific $Ca^{2+}$-binding proteins such as calmodulin that in turn exert diverse cellular functions. One possibility is the activation of $Ca^{2+}$/calmodulin-dependent protein kinase (CaM kinase). Within nerve endings important targets of this kinase include the synaptic vesicle-associated protein synapsin and cAMP phosphodiesterase that catalyzes the conversion of cAMP to AMP and thus terminates its function. CaM kinase is also a key enzyme modulating synaptic plasticity, learning, and memory and development [20].

Diacylglycerol in turn activates **protein kinase C** (PKC) (Fig. 6.13b). Activation of some of the PKC isoforms requires in addition $Ca^{2+}$. In analogy to the other protein kinases, PKC exerts multiple cellular effects. One prominent example is potentiating the exocytotic release of hormones or transmitters from glands and nerve cells, respectively.

### 6.4.5 Receptor Tyrosine Kinases Induce Multiple Intracellular Signaling Pathways

Receptor tyrosine kinases (**RTKs**) are single pass membrane proteins with an intracellular tyrosine kinase domain and an extracellular ligand-binding domain. Agonists include various growth factors, insulin, or ephrins (Fig. 6.11c). In many of the RTKs, ligand binding induces receptor dimerization. This in turn causes mutual cross-phosphorylation on multiple tyrosines of the intracellular segment of the polypeptide chain. The phosphorylated tyrosine residues serve as docking sites and scaffolding for intracellular signal-transduction proteins. These eventually amplify the ligand-induced signal and elicit the cellular reaction. Multiple signaling pathways have been uncovered that may mutually interact, only some of which are briefly discussed here.

Recruited proteins include docking proteins that lead to the activation of several intracellular targets. These include the **MAP kinase** (mitogen-activated protein kinase) signaling pathway, **phospholipase C-γ** that induces essentially the same signaling pathways as the trimeric G protein-activated phospholipase C-β (compare Fig. 6.13), the soluble and cytoplasmic tyrosine kinase Src (pronounced "sarc", from sarcoma), and **phosphoinositide 3-kinase (PI 3-kinase)**.

PI 3-kinase creates additional docking sites at the inner membrane surface by the phosphorylation of inositol phospholipids. Figures 6.14 and 6.15 depict three of the several signaling pathways elicited by the EGF receptor, the first receptor tyrosine kinase to be discovered. The EGF receptor (ErbB1) is activated by EGF and by a number of additional agonists including transforming growth factor alpha (TGFα), amphiregulin, heparin-binding EGF-like growth factor (HB-EGF), or betacellulin. By forming heterodimers it can interact with its three homologs ErbB2, ErbB3, and ErbB4, in a ligand-dependent fashion [21].

Essentially every RTK activates the **Ras/MAP kinase pathway**, one of the several MAP kinase cascades active in eukaryotic cells. The name-giving (but not unique) function is the stimulation of cell proliferation by RTK-activating growth factors (mitogens). In a first step this involves the activation of the small lipid-anchored G protein **Ras** by exchange of GDP for GTP, similar to the activation of the α subunit of trimeric G proteins. This is brought about by a phosphotyrosine-binding adaptor protein (**GRB2**) and a guanine nucleotide exchange factor (**Sos**, son of sevenless). Activated Ras induces a highly amplifying kinase cascade that begins with the recruitment of the serine/threonine protein kinase **Raf** to the inner surface of the plasma membrane. Activated Raf phosphorylates and activates the serine/threonine protein kinase **MEK** (MAP and ERK kinase) which in turn phosphorylates and activates **MAP kinase** (mitogen-activated protein kinase – also named **ERK**, extracellular signal-regulated kinase). MAP kinase finally phosphorylates and thus regulates the activity of proteins including multiple transcription factors. This leads to the altered transcription of genes that are important for activating the cell cycle.

PI 3-kinase can be recruited to the phosphotyrosines of RTKs via its SH2 domain (Fig. 6.15). The PI 3-kinase superfamily consists of several structurally related enzymes. Class-I PI 3-kinases can catalyze the formation of PI 3,4,5-trisphosphate ($PIP_3$) from PI 4,5-bisphosphate ($PIP_2$). The phosphorylated lipid then acts as a docking site and recruits proteins to the inner surface of the plasma membrane. The **phosphoinositide-dependent protein kinase-1** (**PDK1**) and the protein kinase **AKT** (also referred to as **protein kinase B**, **PKB**) can bind to these PI 3-phosphates via their **PH domain** (pleckstrin homology domain). The activated AKT dissociates from the plasma membrane and can phosphorylate and inactivate proteins involved in the synthesis of pro-apoptotic proteins. Via this mechanism AKT promotes cell survival. Amongst others, activation of PI 3-kinase has been implicated in the induction of long-term potentiation.

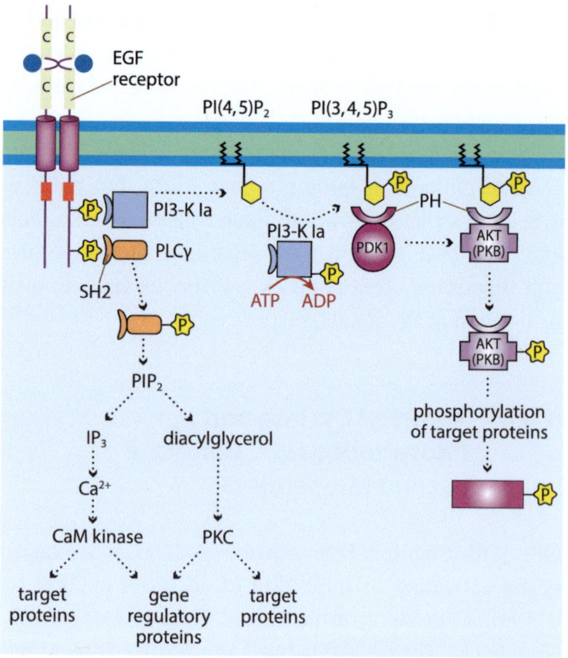


**Fig. 6.15 Essentials of EGF receptor-mediated signaling through phosphoinositide 3-kinase Ia and phospholipase Cγ.** Both enzymes are activated through docking to the cytosolic domain of the activated receptor via their SH2-domains. Of the three classes of typical PI 3-kinases, only class Ia is activated (phosphorylated) by tyrosine kinase receptors. The mode of coupling can vary between homo- and heterodimers of ErbB receptors. For example, direct activation of PI 3-kinase Ia (PI 3-K Ia) is typical for ErbB1/ErbB3 heterodimers whereby PI 3-K Ia binds to a phosphotyrosine residue in Erb3. Other modes of PI 3-K Ia activation via the EGF receptor are possible, including the activation via adaptor proteins. Activated PI 3-K Ia phosphorylates phosphatidylinositol 4,5-bisphosphate ($PI(4,5)P_2$, $PIP_2$) to phosphatidylinositol 3,4,5-trisphosphate ($PI(3,4,5)P_3$, $PIP_3$) that serves as a docking site for proteins with a pleckstrin homology (PH) domain. Phosphoinositide-dependent protein kinase I (PDK1) phosphorylates and activates the serine/threonine protein kinase AKT (named after a viral oncogene AKT8 and also referred to as protein kinase B, PKB). Activated AKT dissociates from the plasma membrane and phosphorylates proteins supporting cell survival and growth. Phospholipase Cγ is phosphorylated and activated by the EGF receptor and produces the two second messengers $IP_3$ and diacylglycerol as described for phospholipase Cβ (Fig. 6.13)

### 6.4.6 Steroid Hormones Are Regulators of Transcription

Steroid hormones comprise the sexual hormones, cortisol from the cortex of the adrenal glands, or the moulting hormone ecdysone of insects. These hormones can simply pass the plasma membrane. Their intracellular receptors represent a protein superfamily and function as ligand-regulated transcription factors. They show ligand specificity, reversible ligand binding and a very low binding constant (about $10^{-9}$ M). The cytosol-located steroid hormone receptors possess three typical functional domains: a hormone-binding domain, a DNA-binding domain, and an N-terminal domain involved in the regulation of transcription. The receptor hormone complex translocates to the nucleus and binds with high affinity to response elements. Binding of the hormone protein complex exerts enhancer or suppressor function.

### 6.4.7 Gaseous Transmitters Are Short-Ranged

Gaseous messenger substances such as NO or CO equally pass the plasma membrane. NO binds to the cytosolic protein guanylyl cyclase in adjacent cells. This enzyme contains the prosthetic group heme and thus (like hemoglobin) is capable of binding the gaseous molecule. This results in enzyme activation and the formation of cGMP from GTP as a second messenger. cGMP can modulate the permeability of ion channels or also activate cGMP-dependent protein kinases (protein kinase G).

## 6.5 Filamentous Cell Proteins Form the Cytoskeleton

Filamentous cell proteins include the **actin** filaments (microfilaments), **intermediate filaments**, and **microtubules** (Fig. 6.16). These filaments have differential but also interconnected cellular functions and together form the **cytoskeleton**. All three types of filaments are polymers of monomeric subunits. They interact with a vast spectrum of additional proteins and also with each other. This way a filamentous cytoplasmic network is formed that is anchored to the plasma membrane and to organelles. Generally, the individual filaments fulfill different requirements. A major function of the intermediate filaments is in the maintenance of cell structure and the cohesion of cellular assemblies via specific contact sites, the desmosomes. Microtubules and actin filaments participate in a large number of cellular processes. Actin filaments are important for locomotion. This includes muscle contraction and cell migration but also organelle transport. Microtubules have a central role in the positioning and movement of organelles and large molecular complexes. There exist a considerable number of tissue-specific but closely related actin and tubulin genes.

### 6.5.1 Actin Exists Largely in Polymerized Form

The 7–9-nm-thick actin filaments (F-actin, microfilaments) (Fig. 6.16a) are formed by polymerization from globular G-actin subunits (42 kDa, β and γ actin in non-muscle cells). Polymerization consumes ATP and the filaments formed are composed of two helically intertwined protein strands of less than 1 μm in length. The filament is polarized with preferential incorporation of subunits at the plus end. Actin filaments are highly dynamic structures. Polymerization and depolymerization of subunits as well as filament crosslinking, filament branching, or membrane attachment are targeted by a considerable number of signaling pathways and highly regulated by regulatory actin-binding proteins. Neuronal actin filaments are enriched at dendritic spines, the filopodia of growth cones, in stress fibers (contractile cellular filament bundles consisting of actin, crosslinking proteins, and myosin II motors) or at the plasma membrane. Thereby they participate in the formation of the cell cortex, a shell-like contractile meshwork beneath the plasma membrane for mechanical support and movement. Via the membrane-spanning integrins filamentous actin interacts with extracellular matrix proteins such as fibronectin or laminin. This way the cell skeleton contributes to cell migration and cell adhesion.

### 6.5.2 Microtubules Are Polarized

Microtubules are stiff tubes of 24 nm outer diameter and up to hundreds of micrometers in length

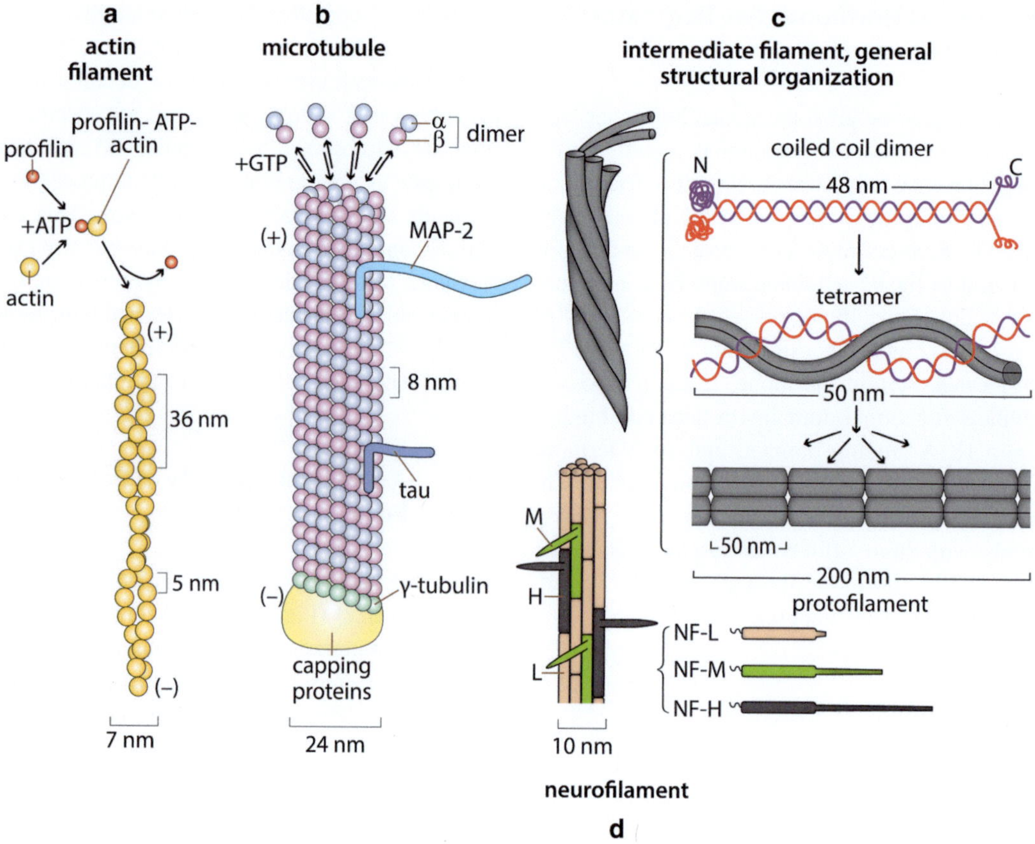


**Fig. 6.16 The three principal types of intermediate filaments.** (**a**) Actin filaments in nonmuscle cells are instable and filament stability is highly regulated by actin-binding proteins. For example, profilin catalyzes nucleotide exchange on actin monomers and promotes the assembly of actin monomers at the actin filaments. (**b**) In microtubules α, β-dimers assemble to form protofilaments, 13 of which arrange in parallel to form the tubule. Addition of dimers and disassembly take place mainly at the (+) end. The cartoon also exemplifies the dimensions of the microtubule-associated proteins MAP-2 and tau. (**c**) The stable intermediate filaments are generally formed by the parallel arrangement of helical monomers to coiled coil dimers and soluble antiparallel tetramers (one dimer is depicted in colors) that form protofilaments and in turn interact helically to form the filament. (**d**) Neurofilaments differ from other intermediate filaments by typically consisting of three assembly blocks of the neurofilament (NF) proteins NF-L (light), NF-M (medium) and NF-H (heavy), with NF-L subunit forming the filament core. Phosphorylated NF-H and NF-M protrude laterally from the filament backbone. The additional subunit α-internexin can associate with NF-L, NF-M, and NF-H (not shown)

(Fig. 6.16b). They are formed in the presence of GTP from tubulin dimers of a globular α and a globular β subunit that in turn assemble into 13 protofilaments arranged in parallel to form a hollow tube. A third tubulin isoform, γ-tubulin, functions as a template for the correct assembly of microtubules [22]. Microtubules are nucleated at a microtubule-organizing center and can subsequently be transported to their often far distant destiny. The two microtubule-composing tubulin monomers are encoded by closely related genes and exhibit both a molecular mass of 55 kDa. The structure of the microtubules is polar, whereby addition (elongation) and release (shortening) of tubulin dimers preferentially take place at the plus end. In general, microtubules are more stable structures than actin filaments. As for actin, microtubule-associated proteins (**MAPs**) are very highly relevant for the regulation of microtubule stability, microtubule spacing, bundling, and its interaction with other proteins. The high-molecular-weight MAPs form side arms protruding from the microtubule and control axon diameter. MAPs include the tau proteins that are important constituents of neurofibrillary tangles in the brain of Alzheimer patients.

### 6.5.3 Intermediate Filaments in Neurons and Glia and Neural Stem Cells

Intermediate filament proteins form a superfamily of five distinct classes and reveal a largely cell- and development-specific expression pattern. For this reason they are often used as cell type-specific markers in immunocytochemistry. All intermediate filament proteins share a common structural organization (Fig. 6.16c). They can self-assemble into flexible, nonpolar filaments (diameter approx. 10 nm). The molecular building blocks consist of dimers of two identical monomers with α-helical domains that form a coiled-coil structure. The dimers associate to form tetramers that assemble into long protofilaments that eventually combine in parallel to form a cylindrical filament. As compared to actin and myosin filaments intermediate filaments are much more stable and possess considerable mechanical strength.

Intermediate filaments of particular interest for the study of the nervous system include the intermediate filament of astrocytes consisting of glial fibrillary acidic protein (**GFAP**) and the **neurofilaments**. Neurofilaments (NF) are special as they are made up of several filament components (Fig. 6.16d). In CNS neurons, the primary protein constituents of neuronal intermediate filaments are the neurofilament triplet proteins and the more recently discovered α-internexin. The triplet proteins differ in molecular mass (light NF-L, medium NF-M, and heavy NF-H have molecular masses of 68, 160, and 205 kDa, respectively) and are encoded by three distinct genes. A fifth neuronal intermediate filament protein, peripherin, adds to the triplet proteins in the peripheral nervous system. Neurofilaments form parallel bundles. The NF-H and NF-M proteins possess lengthy C-terminal tail domains involved in the spacing between neighboring filaments. These side arms limit packing density and are important for determining axon caliber. Neurofilaments outnumber microtubules with which they form dynamic crosslinks. Abnormal accumulations of neurofilaments are a pathological hallmark of several human neurodegenerative diseases. These include Alzheimer's disease, Parkinson's disease, and amyotrophic lateral sclerosis [23]. Intermediate filaments composed of the protein **nestin** are typical for embryonic and adult neural stem cells and also vascular endothelial cells. Mature oligodendrocytes apparently lack intermediate filaments.

## 6.6 Molecular Motors and Axonal Transport

The targeted transport of organelles, secretory vesicles and their precursors, and of protein complexes is essential for the exchange of materials between different cytoplasmic compartments. This process requires microtubule-based motors: kinesins and cytoplasmic dyneins, and the actin-based myosin motors (Fig. 6.17). These motor proteins are excellent examples of "nanotechnology" in nature. They convert the chemical energy stored in ATP into mechanical work. They are highly relevant for every cell and in particular for the highly polarized neurons. Neurons require efficient mechanisms for the supply of dendrites and in particular of the extended axon and its arborizations with metabolites, macromolecules, and organelles. These include the enzymes for transmitter metabolism, synaptic vesicle precursors, granules, and mitochondria. Conversely, substances and organelles need to be transported back to the cell soma. Via endocytosis and retrograde transport of vesicular structures the neuron can further obtain information regarding the chemical status at the nerve ending. Also toxins and viruses use these pathways as free-riders.

### 6.6.1 Neuronal Polarity and Transport

The observation that axoplasmic constituents are transported from the cell body towards the nerve terminal was first made by ligating peripheral nerves. Today one differentiates between fast and slow axonal transport. **Fast axonal transport** serves the transport of membrane-bound organelles in an **anterograde** or **retrograde** manner (from the soma to the nerve terminal and vice versa). The components of the cytoskeleton and cytosolic proteins move towards the nerve terminal via **slow axonal transport**. The transport velocities as well as the biochemical

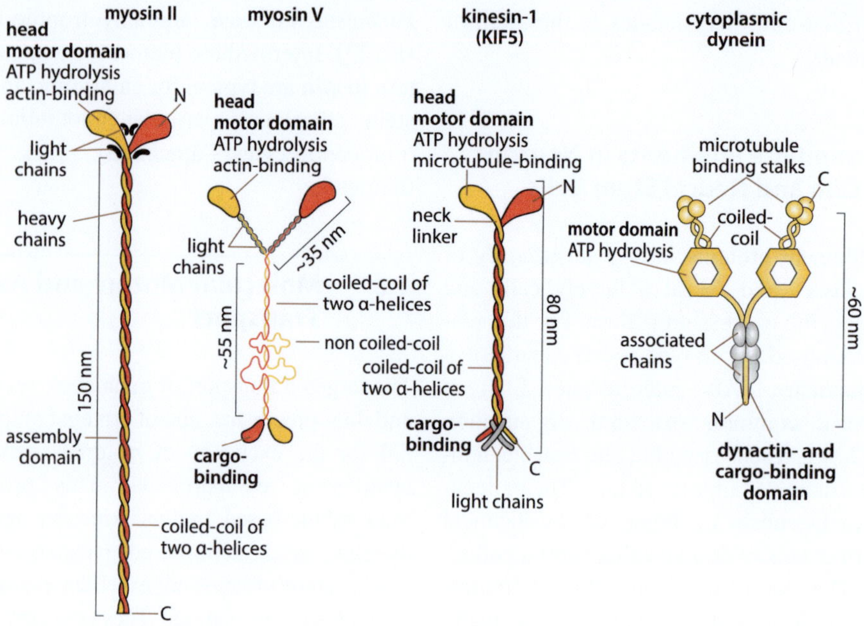


**Fig. 6.17 Structural models of myosin, kinesin, and cytoplasmic dynein.** The three cellular motors come in multiple forms. The four polar molecular complexes depicted share important structural and functional properties. They possess two motor domains that hydrolyze ATP and convert chemical energy into force and motion. They all are dimers with associated proteins. Myosin II in nonmuscle cells can produce contraction by sliding actin filaments relative to each other. Myosin V can transport organelles along actin filaments. Kinesin-1 (conventional kinesin, KIF5) and cytoplasmic dynein transport organelles along microtubules. The motor head formed by the dynein heavy chain has multiple ATP binding sites that are arranged within seven globular domains. Connection of dynein to the vesicular cargo requires the dynactin complex. The molecular complexes are depicted proportional to size

characteristics imply differential molecular mechanisms underlying the two transport mechanisms. Transport velocities range from 100 to 400 mm/day (~1–5 μm/s) for fast and ~0.2–6 mm/day for slow transport. Substances that interfere with microtubule structure selectively interrupt the fast transport. These include colchicin, the toxin of the meadow saffron (*Colchicum autumnale*) that attaches to the tubulin dimers and thus prevents polymerization.

The parallel arrangement and unidirectional orientation of the axonal microtubules is of central importance for axonal organelle transport (Fig. 6.18a). The minus end of axonal microtubules always points towards the cell soma and the growing plus end towards the nerve terminal. This allows the formation of two opposing and directional transport mechanisms. In contrast, microtubules in the cell soma and in dendrites are oriented in both directions. Also dendrites are supplied by motor-driven transport mechanisms. In the following, principle molecular mechanisms of cargo transport are briefly described.

### 6.6.2 Myosins, Kinesins, and Dyneins: Separate Classes of Motor Proteins

Using components of the cytoskeleton as tracks, myosins, kinesins, and dyneins (Fig. 6.17) are all involved in cargo transport throughout eukaryotic cells. They are members of larger protein families with specific structural and functional adaptions. They represent polar elongated molecular complexes composed of several subunits and use ATP as an energy source. The conformational changes they undergo during their catalytic cycle are coupled to net movement along fibers within cells. This general principle was first unravelled for muscle **myosin** (myosin II) and its interaction with filamentous actin. In addition to myosin II with its two globular heads, some 40 additional myosin genes have been identified in the human genome, consisting of 18 subfamilies. Besides the two-headed members of the muscle myosin II subfamily these include the two-headed myosins V and the one-headed myosins I. All types of myosins, except one, move towards the plus end of actin filaments. They are involved in cell migration and also in the movement of organelles. **Axonemal dyneins** are the motors required for the beating of cilia and flagella. Within cilia the dynein arms bridge between microtubules and initiate the bending of cilia. The ependymal cells of the vertebrate brain, for example, propagate the cerebrospinal fluid by way of their beating cilia. The related and ubiquitous **cytoplasmic dyneins** play an important role in the retrograde transport of cell organelles to the soma of neurons. The multiple **kinesins** comprise the third class of molecular motors. They play an important role in organelle movement from yeast onwards and are also involved in mitosis.

### 6.6.3 Multiplicity and Functional Diversity of Kinesin Motors

Kinesin superfamily proteins (kinesin family members, **KIFs**) come in multiple molecular forms and have been grouped into 14 families according to phylogenetic relationships. The families are numbered kinesin-1 to kinesin-14 and in mammals individual proteins are referred to as KIFs [24]. Individual KIFs vary regarding the number of heavy and associated chains, the number of head groups, the location of the motor domain within the molecule, and the direction of transport along microtubules. Some can form heteromers.

The well-investigated conventional kinesin and founding member of the superfamily (now kinesin-1, KIF5), moves towards the plus end of microtubules. It represents a dimer of two heavy chains with a long coiled-coil stalk to which a pair of globular head domains is connected via a flexible linker domain (Fig. 6.17). The light chain is attached to the tail of the molecule. While the two globular domains bind ATP and connect to microtubules, the linker domains provide motility and the tail domain interacts with the cargo (Fig. 6.18b, c). Individual kinesin 1 molecules were shown to walk down individual microtubules taking hundreds of steps. As for foot steps, one foot remains firmly attached to the rail as the motor molecule moves forward. Thereby the motor domains move foot over foot between β subunits of microtubule protofilaments involving a cycle of ATP hydrolysis (Fig. 6.18c). This was demonstrated by experiments measuring the steps of single fluorescence-labeled kinesin heads. The center-of-mass advance for one step is 8 nm whereby the rear head undergoes a displacement of 16 nm [25]. The dimensions should be kept in mind: The diameter of a synaptic vesicle is in the order of 50 nm. This is little more than half of the length of a kinesin molecule.

Interestingly, the principal transport mechanisms can be reconstituted *in vitro*. This requires the isolated molecular components and video-enhanced contrast microscopy. Isolated microtubules can be adsorbed to coverslips. Added latex beads are transported along microtubules in an energy-dependent manner when kinesin molecules are added. When the coverslips are coated with kinesin, added microtubules migrate along the glass surface when ATP is provided. Similarly, inhibition of kinesin synthesis by RNA interference blocks fast axonal transport.

Different classes of organelles are transported by different KIFs which can explain differences in axonal transport rates between organelles. KIFs interact with, e.g., mitochondria, vesicles at the Golgi apparatus, or diverse vesicular organelles transporting synaptic vesicle proteins but also with ribonucleoprotein particles containing mRNA. For example, the monomeric kinesin-3 family members KIF1A and KIF1Bβ act as motors for the transport of synaptic vesicle precursors containing synaptotagmin, synaptophysin, and synaptic vesicle protein 2 (SV2), whereas the monomeric KIF1Bα drives mitochondrial transport. The three variants of the conventional and dimeric kinesin-1 (KIF5A-C) play an essential role in axonal transport including SNARE protein-containing organelles (see Chap. 8) and also mitochondria whereas the kinesin-2 family member KIF17 drives the transport of NMDA-receptor-containing vesicles.

An intensively investigated issue concerns the mechanisms that ensure the cargo-specific association of individual kinesins. KIFs bind to cargoes through adaptors or scaffolding complexes including lipids that vary between individual motors (Fig. 6.18b) [26]. Cargo binding can be regulated by the phosphorylation and possibly other posttranslational modifications of motor complex and adaptor proteins. After release from their cargoes at the destination, motor proteins are supposed to be degraded. It is obvious that axonal transport defects will result in a variety of neurodegenerative diseases.

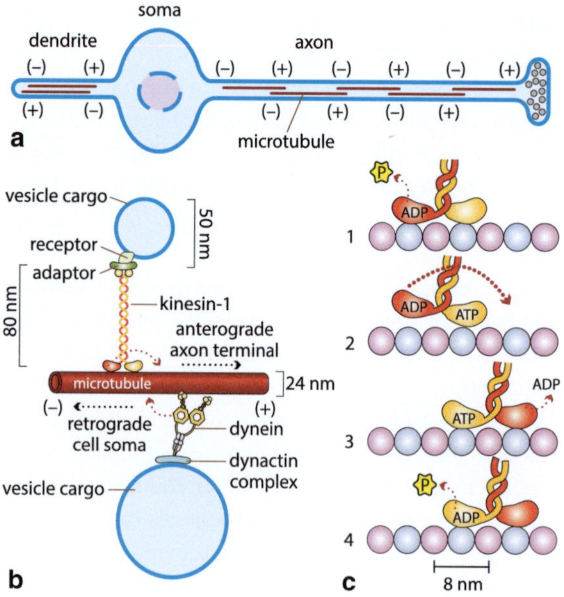


**Fig. 6.18 Model of the molecular mechanism of axonal transport with the motors kinesin-1 and dynein.** (**a**) Unlike in dendrites, in axons microtubules share the same polarity with the (+) ends pointing towards the axon terminal. (**b**) ATP-mediated conformational changes direct cargo transport of kinesin-1 to the (+) end and of dynein to the (–) end of microtubules. The scheme is roughly drawn to size. (**c**) Simplified motility cycle of the motor domain of kinesin-1 when walking along a microtubule protofilament. (*1*) Release of inorganic phosphate (P) loosens the binding of the lagging motor head (*red*) to the microtubule. (*2*) Binding of ATP to the leading motor head (*yellow*) induces a conformational change which throws the lagging head forward to the next tubulin binding site 16 nm away. (*3*) The new leading head docks to the microtubule and releases its ADP. (*4*) Hydrolysis of ATP by the now lagging head and release of phosphate initiates the next cycle. During one cycle the center of mass of the kinesin-1 molecule advances by 8 nm

### 6.6.4 Cytoplasmic Dynein: A Major Motor of Retrograde Transport

Cytoplasmic dyneins (as opposed to axonemal dyneins) represent the motors for fast retrograde transport. *In vitro* experiments revealed that the polarity of the dynein-mediated transport is directed towards the minus end of microtubules and thus opposite to that of most kinesins. Dyneins consist of a ~60-nm-long complex molecular structure with two heavy chains (Fig. 6.17) and a mass about ten times that of kinesins. Similar to KIF5 they possess two head domains containing the ATPase activity and a microtubule-binding domain at the end of a stalk region. Binding of cytosolic dyneins to their cargoes requires large complexes of microtubule-binding proteins, such as dynactin. The dynactin multiprotein complex contains a short actin-like filament composed of the actin-related protein 1 (Arp1) and is essential for linking cytoplasmic dynein to its cargo and for regulating its activity [27]. Whereas the motor domains of myosin and kinesin reveal structural similarity indicating a common evolutionary origin, the motor head formed by the dynein heavy chain has multiple ATP binding sites that are arranged within seven globular domains forming a ring-like structure around a central cavity. Dynein heavy chains belong to the AAA superfamily of mechanoenzymes (ATPases associated with diverse cellular activities). In common with the two-headed myosin and kinesin molecules, dynein motility requires two heads but the power stroke appears to differ [28]. Similar to kinesin-1 the center-of-mass advance is about 8 nm but variable advances have been reported (4–24 nm).

As viewed by time lapse microscopy, organelles transported in vital axons do not always simply move one way. They can be carried back and forth, implicating that they contain multiple binding sites for motor proteins and that there is a tug-of-war between anterograde and retrograde motors. *In vitro* experiments with extruded cytoplasm further demonstrate that organelles can be transported simultaneously in opposite directions and pass each other unhindered at an individual microtubule.

**Actin- and myosin-dependent transport.** Axoplasmic organelles obtained from the squid giant axon move on actin filaments at an average velocity of 1 μm/s, possibly involving myosin motors. However, at present the neural functions of myosins are not well defined. The dimeric myosins II and V can contribute to the delivery of secretory vesicles to the plasma membrane. Myosin V (Fig. 6.17) binds with its extended neck domain to specific receptors on the melanosomes in melanocytes of the skin and drives organelle movement on actin filaments. Similar to kinesins, myosin V moves hand over hand but takes considerably larger steps (36 nm). Myosin V is expressed in the brain and contributes to the transport of membrane compartments into dendritic spines, the docking of secretory granules and possibly also the short-distance transport along actin filaments within the cortical meshwork subjacent to the plasma membrane. It interacts with synaptic vesicles [29]. Mice lacking myosin V die due to neurological seizures.

### 6.6.5 Slow Axonal Transport

The velocity of slow axonal transport corresponds to that of a growing axon and is about 100 times slower than that of vesicular traffic [30]. Neurofilaments, tubulin, actin, and soluble cytosolic proteins are amongst the constituents of slow axonal transport. Experiments using GFP-tagged neurofilament proteins expressed in cultured neurons and real-time live cell microscopy revealed that the apparent slow transport is in fact the result of fast motions (up to 3 μm per second for neurofilaments) that are interspersed with long pauses. Neurofilament protein is transported primarily in a polymeric form, whereas tubulin and actin are transported as individual subunits or oligomers. As for fast axonal transport, microtubules and the kinesins provide the main transport substrate also for the components of the slow transport, whereby kinesins assemble with their cargo via scaffolding proteins. Transport velocity can further be affected by the phosphorylation state of transported proteins [31].

### 6.6.6 Transport of mRNA

The synthesis of soluble and membrane proteins within the cell body and their export into axons or dendrites has long been considered the single mechanism for protein supply to these distant cell compartments. There is now compelling evidence for novel mechanisms that ensure local targeting of mRNA into axons and in particular into dendrites, associated with the machinery for "on-site" translation and translational regulation. This is thought to be of profound functional importance not only during development for axon guidance, synaptogenesis, and dendritogenesis or in axonal regeneration but also for rapidly modulating synaptic plasticity and memory storage in the adult brain [32].

A subset of apparently dormant mRNA is transported into dendrites, complexed into large particles (mostly ribonucleoprotein particles, RNPs). Dendritic transport depends on specific targeting elements, usually located in the 3′ untranslated regions of the mRNA. RNPs have been estimated to have a diameter of 200 nm and to contain a large number of proteins including regulators of RNA transport and translation. In situ hybridization studies of hippocampal or cerebellar slices and biochemical analysis of dendrites from cultured neurons have identified mRNAs of synaptically relevant cytosolic, cytoskeletal, and integral membrane proteins including microtubule-associated protein 2 (MAP2), the α subunit of $Ca^{2+}$/calmodulin-dependent protein kinase II (CaMKIIα, that makes a substantial proportion of the postsynaptic density), brain-derived neurotrophic factor (BDNF), activity-regulated cytoskeleton-associated protein (Arc), tyrosine-related kinase B (TrkB) receptor, IP3 receptor, the AMPA receptor subunit GluR2, or the glycine receptor α subunit. In addition, mRNAs encoding proteins essential for translation have been identified suggesting that dendritic protein synthesis can be locally controlled. Acute extracellular signaling cues regulating translation of dendritic mRNAs include BDNF and glutamate. Of interest in the context of this chapter is the finding that RNPs containing mRNA for CaMKIIα, arc, or shank1 (a scaffolding protein of the postsynaptic density) were found to be driven by the KIF5 (kinesin-1) molecular motor [33, 34] at a velocity corresponding to that of fast axonal transport (up to 5 μm/s). The association of the molecular motor with RNA is assured by RNA-binding proteins and adaptor proteins.

Similar mechanisms are at work in the axon. Export of mRNAs and local protein synthesis are important for growth cone guidance and axonal regeneration. In immature axons tau and actin mRNAs are axonally targeted as RNPs involving molecular motors.

## 6.7 Membrane Trafficking, Exocytosis, and Endocytosis

### 6.7.1 The Secretory Pathway and Exocytosis

A major cellular function of neurons consists in the secretion of signal molecules (Fig. 6.19). Secretion occurs at structurally specialized domains of the plasma membrane via the fusion of storage organelles. Similar to other secretory cells, neurons use two principal pathways for exocytosis. **Constitutive exocytosis** serves the continuous release of secretory substances and the incorporation of membrane constituents into the plasma membrane. In contrast, **regulated exocytosis** is activated by specific cellular signals such as the influx of $Ca^{2+}$ ions into the nerve terminal. It is of note that fusion of secretory organelles with release of peptides and neurotransmitters

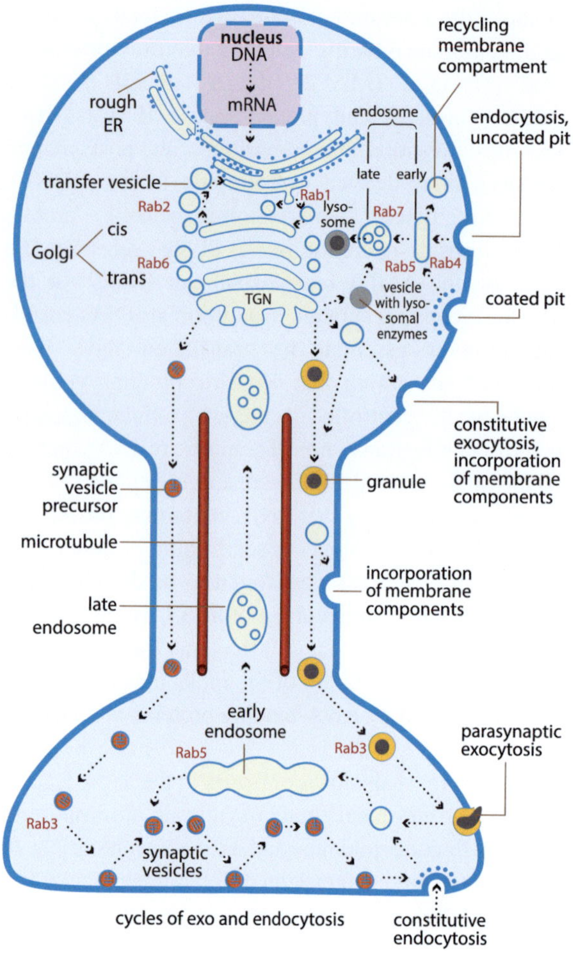


**Fig. 6.19** **Membrane traffic in the secretory nerve cell.** For simplicity the dendritic compartment and somatodendritic membrane targeting are not depicted. Neurons possess two types of vesicular organelles for regulated secretion of low-molecular messengers, the synaptic vesicles and the larger granules that in addition contain peptides and proteins. While the granules are fully generated at the somatic Golgi apparatus, synaptic vesicles are thought to largely derive from precursor vesicles that transport vesicular membrane constituents to the nerve terminal. Following cycles of exo- and endocytosis these become inserted into synaptic vesicles proper. Within the nerve terminal the synaptic vesicle membrane compartment undergoes multiple cycles of exo- and endocytosis and refilling with neurotransmitter that can involve a synaptic sorting endosome. Peptide granules typically fuse laterally to the active zone. They need to be replenished via the cell soma. The incorporation of lipids, receptors and other proteins into the plasma membrane of dendrites, soma and nerve terminal is equally important. Examples of the localization of members of the Rab family of small G proteins which are important for organelle targeting are depicted. *TGN* trans-Golgi network

can also occur from dendrites [35]. Exocytosis requires the fusion of the membrane of the secretory organelle and the plasma membrane. This process is protein-mediated. The release of gaseous substances such as NO is differently regulated – these substances are synthesized when required and leave the cell by diffusion without further control.

### 6.7.2 Endocytosis Serves in Membrane Recycling and Cellular Uptake

Endocytosis comprises the withdrawal of a plasma membrane patch into the cell interior. At the synapse it serves the recycling of membrane constituents or the cellular uptake of specific cargo (receptor-mediated endocytosis). Together with specific adaptor proteins **clathrin** molecules play an important role in the identification of the membrane segment and the formation of the endocytic vesicle. Vesicle fission involves the GTPase **dynamin** that (together with other proteins) polymerizes around the neck of the pinching-off vesicles and severs the membrane [36] (Fig. 6.20a). After endocytosis the clathrin coat rapidly dissociates from the vesicular surface. Similar to exocytosis, endocytosis can be constitutive or induced by a cellular signal. For example, binding of nerve growth factor to its receptor at the cell surface acts as a signal for the rapid endocytosis of the receptor-containing membrane segment. The endocytosed membrane compartment and its contents can follow different pathways [37]. In the simplest case they both undergo lysosomal degradation. Or they are transferred to a sorting compartment that ensures reuse of the membrane compartment while the transported cargo is released into the cell. For example, the cell surface-located receptors for lipid-transporting particles (low-density lipoprotein receptors) remain unmodified and are recycled for reuse to the plasma membrane. Recent evidence suggests that signaling from internalized receptors for neurotrophins can continue in endosomal compartments (signaling endosomes). When retrogradely transported they can convey survival signals from the nerve terminals to the cell body.

### 6.7.3 Rab Proteins and Vesicle Targeting

Membrane compartments destined for secretion, endocytosis, or also cellular degradation need to be transported to the right place and interact with the right target membrane. This not only requires the existence of directed transport mechanisms but also mechanisms for specific recognition. Membrane transport generally uses microtubule tracks. The Rab proteins, a subfamily of small G proteins (GTPases), play an important role in membrane targeting. In contrast to the trimeric G proteins associated with plasma membrane-located receptors, the small G proteins are monomers. They are grouped into several protein subfamilies of which the **Rab proteins** contribute more than 60 members. Rab proteins are involved in the regulation of membrane traffic from the ER via the Golgi apparatus to the plasma membrane or lysosomes and from endocytosis to the recycling of membrane vesicles or their lysosomal degradation [38] (Fig. 6.19). They bind to **Rab effector** complexes which support vesicle transport, membrane tethering, and membrane fusion [39]. Individual Rabs bind to specific organelles. For example, rab3A associates with synaptic vesicles and secretory granules, whereas Rab 5 assembles on early endosomal membranes (Fig. 6.19). As for other G proteins, functional activity requires the binding of GTP.

### 6.7.4 Regulated Exocytosis Occurs via Granules or Vesicles

Two principal types of secretory organelles, granules and synaptic vesicles, are involved in the secretion via regulated exocytosis from neurons. **Granules** store proteins, peptides, and in some cases also low-molecular-weight components. Proteins and peptides are packaged into secretory granules via the ER and Golgi apparatus (Fig. 6.19).

The smaller (50 nm) **synaptic vesicles** represent the second compartment for regulated exocytosis. This type of vesicles is found in neurons as well as in neuroendocrine cells and recent data suggest that similar vesicles also exist in astrocytes. In neurons, vesicles serve the fast release of low-molecular-weight neurotransmitters such as acetylcholine, $\gamma$-aminobutyric acid (GABA), glycine, glutamate, or ATP. In contrast to peptides, these substances are synthesized and taken up into synaptic vesicles via specific transporters within the nerve terminals (compare Fig. 6.10). Synaptic vesicles are densely packed with membrane proteins that are involved in many functions during the synaptic vesicles life cycle. Besides uptake of neurotransmitters, these include the controlled interaction with elements of the cytoskeleton during axonal transport (microtubules) or within the nerve terminal (actin filaments), or synaptic vesicle docking, fusion, and recycling.

Due to the extreme polarity of nerve cells, the secretory compartment is far distant from the cell soma with its capacity for synthesis and degradation. This requires rapid mechanisms not only of fast axonal transport of the synaptic vesicle membrane compartment. The nerve terminal possesses mechanisms for rapid exocytosis and subsequent local membrane recycling. This allows repeated cycles of reloading and reuse of the synaptic vesicle membrane compartment. This is best exemplified by *in vitro* experiments with **synaptosomes**, nerve terminals detached from their axons as a result of gentle brain homogenization. The plasma membrane of the synaptosomes reseals. As a consequence, synaptosomes can be stimulated to release neurotransmitter in the test tube for an extended period of time and the synaptic vesicle membrane compartment recycles without the aid of an adjacent axon. In contrast, granules cannot be reloaded with high-molecular-weight components within the nerve terminal. They need to be reformed via the Golgi apparatus.

### 6.7.5 The Molecular Control of Exocytosis Is Complex

Synaptic vesicles are clustered within the nerve terminal compartment. Resting vesicles are tied to the actin cytoskeleton via the surface-associated protein **synapsin**. An increase in intracellular $Ca^{2+}$ concentration induces the calmodulin-dependent phosphorylation of synapsin, alters its conformation, releases the vesicle from its anchorage at the cytoskeleton, and delivers vesicles to the active zone [40]. For exocytosis, synaptic vesicles first enter into a lose association with the presynaptic plasma membrane (Fig. 6.20a). The subsequent formation of the tight docking complex results from the interaction of three proteins, the vesicle-bound **synaptobrevin-2** and the two plasma membrane-located proteins **syntaxin** and **SNAP-25**. These three proteins are also referred to as **SNARE proteins** or SNAREs. They form the **SNARE complex** which consists of four intertwined $\alpha$-helical domains (Fig. 6.20b). Additional proteins associate with the SNARE complex and contribute to the regulation of

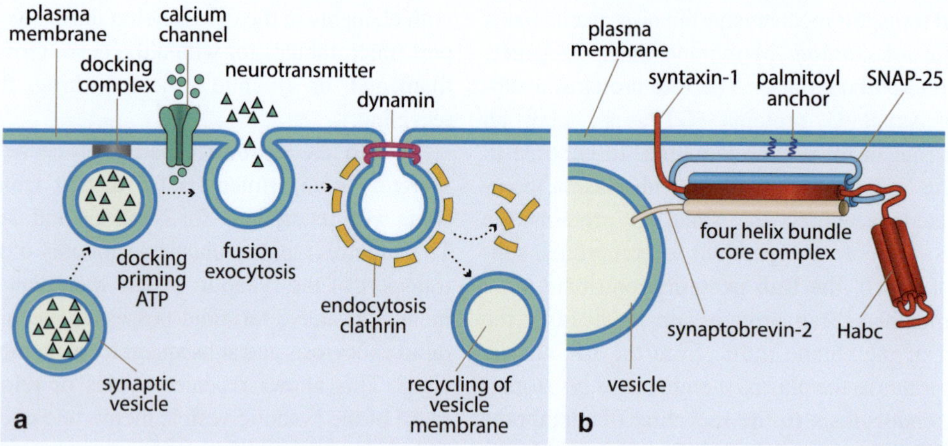


**Fig. 6.20 Principal steps involved in vesicular membrane fusion.** (**a**) Synaptic vesicles attach to specific binding sites at the plasma membrane. The four SNARE proteins synaptobrevin-2, SNAP-25, and syntaxin 1 form the core of the docking complex. ATP is required for priming the vesicles for fusion. Membrane fusion is elicited by influx of $Ca^{2+}$ ions (*blue*) through voltage-gated calcium channels positioned at the release sites. Recycling of the vesicular membrane compartment is thought to occur by clathrin-mediated endocytosis. (**b**) The SNARE complex plays a central role in vesicle docking and fusion. It represents a highly stable antiparallel bundle of four helices (the SNARE motifs, 60–70 amino acids in length). The vesicle protein synaptobrevin-2 and the plasma membrane integral protein syntaxin-1 contribute one helix each, the palmitoyl-anchored SNAP-25 contributes two helixes. Additional proteins (not shown) control the SNARE-mediated fusion process. The flexible N-terminal region of syntaxin 1 (Habc domain) folds into a three helix bundle and contributes to the regulation of SNARE complex assembly

membrane fusion [41]. Typically, the fusion of docked vesicles is initiated by depolarization of the presynaptic plasma membrane and the opening of voltage-gated $Ca^{2+}$ channels adjacent to the release sites. Thereby the synaptic vesicle protein **synaptotagmin** functions as the $Ca^{2+}$ sensor. One SNARE complex appears to be sufficient for membrane fusion [42]. Following fusion, the SNARE complex is dissolved involving ATP hydrolysis and the vesicular membrane compartment is recycled via clathrin-mediated endocytosis. There exists a multiplicity of related SNARE proteins and SNARE-mediated membrane fusion is not restricted to exocytosis. SNARE proteins are equally responsible for intracellular membrane fusion processes as, e.g., at the ER, Golgi apparatus, or at endosomes and lysosomes [43]. However, the control of membrane fusion via an increase in intracellular $Ca^{2+}$ is specific for regulated exocytosis.

The mechanism of fusion of the lipid bilayers and the structure, size, and stability of the fusion event are only partially characterized. Possibly a short opening of the fusion pore is sufficient for releasing synaptic vesicle contents (sometimes referred to as kiss-and-run mechanism). In this case, a complete collapse of the synaptic vesicle membrane into the plasma membrane would not be required [44].

The exocytosis of the larger granules is thought to be controlled by similar mechanisms. But synaptic location of the fusion site and $Ca^{2+}$ dependency of the fusion differ from those of synaptic vesicles. Granules fuse laterally to the active zone or even with the nonsynaptic plasma membrane (parasynaptic exocytosis) (Fig. 6.19). The release of the protein matrix and of peptides from granules is a slower process than the phasic release of fast neurotransmitters.

## 6.8 Summary

The chapter provides the reader with an overview and selective information regarding cellular and molecular mechanisms required for the understanding of more complex neural functions. Select reading will be required for a deeper understanding of molecular structures and mechanisms. Neurons and glia share with other cells their basic cell biological mechanisms including those of gene regulation. But there is also a multitude of genes specifically expressed in neural cells and neurons possess specific molecular machineries as a prerequisite for fast information processing, neural plasticity, and memory formation. While elucidating the cellular and molecular mechanisms of neural function remains a fascinating challenge, modern neurobiology equally integrates molecular biology, structural analysis, physiology, neural development, system function, and behavior to fully understand the working of nervous systems.

## References


1. ten Klooster JP, Hordijk PL (2007) Targeting and localized signalling by small GTPases. Biol Cell 99:1–12
2. Paulick MG, Bertozzi CR (2008) The glycosylphosphatidylinositol anchor: a complex membrane-anchoring structure for proteins. Biochemistry 47:6991–7000
3. Yu RK, Tsai YT, Ariga T, Yanagisawa M (2011) Structures, biosynthesis, and functions of gangliosides – an overview. J Oleo Sci 60:537–544
4. Kwok JC, Dick G, Wang D, Fawcett JW (2011) Extracellular matrix and perineuronal nets in CNS repair. Dev Neurobiol 71:1073–1089
5. Dahan M, Levi S, Luccardini C, Rostaing P, Riveau B, Triller A (2003) Diffusion dynamics of glycine receptors revealed by single-quantum dot tracking. Science 302:442–445
6. Mehler MF (2008) Epigenetics and the nervous system. Ann Neurol 64:602–617
7. Tang W, Fei Y, Page M (2012) Biological significance of RNA editing in cells. Mol Biotechnol 52:91–100
8. Winter J, Jung S, Keller S, Gregory RI, Diederichs S (2009) Many roads to maturity: microRNA biogenesis pathways and their regulation. Nat Cell Biol 11:228–234
9. Olde Loohuis NF, Kos A, Martens GJ, Van Bokhoven H, Nadif Kasri N, Aschrafi A (2012) MicroRNA networks direct neuronal development and plasticity. Cell Mol Life Sci 69:89–102
10. Selbach M, Schwanhäusser B, Thierfelder N, Fang Z, Khanin R, Rajewsky N (2008) Widespread changes in protein synthesis induced by microRNAs. Nature 455:58–63
11. Mencia A, Modamio-Hoybjor S, Redshaw N, Morin M, Mayo-Merino F, Olavarrieta L, Aguirre LA, del Castillo I, Steel KP, Dalmay T, Moreno F, Moreno-Pelayo MA (2009) Mutations in the seed region of human miR-96 are responsible for nonsyndromic progressive hearing loss. Nat Genet 41:609–613
12. Okuda T, Haga T, Kanai Y, Endou H, Ishihara T, Katsura I (2000) Identification and characterization of the high-affinity choline transporter. Nat Neurosci 3:120–125
13. Varoqui H, Erickson JD (1996) Active transport of acetylcholine by the human vesicular acetylcholine transporter. J Biol Chem 271:27229–27232
14. Gether U, Andersen PH, Larsson OM, Schousboe A (2006) Neurotransmitter transporters: molecular function of important drug targets. Trends Pharmacol Sci 27:375–383
15. Fenno L, Yizhar O, Deisseroth K (2011) The development and application of optogenetics. Annu Rev Neurosci 34:389–412
16. Liewald JF, Brauner M, Stephens GJ, Bouhours M, Schultheis C, Zhen M, Gottschalk A (2008) Optogenetic analysis of synaptic function. Nat Methods 5:895–902
17. Albuquerque EX, Pereira EF, Alkondon M, Rogers SW (2009) Mammalian nicotinic acetylcholine receptors: from structure to function. Physiol Rev 89:73–120
18. Collingridge GL, Olsen RW, Peters J, Spedding M (2009) A nomenclature for ligand-gated ion channels. Neuropharmacology 56:2–5
19. Katritch V, Cherezov V, Stevens RC (2012) Diversity and modularity of G protein-coupled receptor structures. Trends Pharmacol Sci 33:17–27
20. Wayman GA, Lee YS, Tokumitsu H, Silva AJ, Soderling TR (2008) Calmodulin-kinases: modulators of neuronal development and plasticity. Neuron 59:914–931
21. Yarden Y, Sliwkowski MX (2001) Untangling the ErbB signalling network. Nat Rev Mol Cell Biol 2:127–137
22. Conde C, Caceres A (2009) Microtubule assembly, organization and dynamics in axons and dendrites. Nat Rev Neurosci 10:319–332
23. Perrot R, Eyer J (2009) Neuronal intermediate filaments and neurodegenerative disorders. Brain Res Bull 80:282–295
24. Hirokawa N, Niwa S, Tanaka Y (2010) Molecular motors in neurons: transport mechanisms and roles in brain function, development, and disease. Neuron 68:610–638
25. Gennerich A, Vale RD (2009) Walking the walk: how kinesin and dynein coordinate their steps. Curr Opin Cell Biol 21:59–67
26. Verhey KJ, Kaul N, Soppina V (2011) Kinesin assembly and movement in cells. Annu Rev Biophys 40:267–288
27. Kardon JR, Vale RD (2009) Regulators of the cytoplasmic dynein motor. Nat Rev Mol Cell Biol 10:854–865
28. DeWitt MA, Chang AY, Combs PA, Yildiz A (2012) Cytoplasmic dynein moves through uncoordinated stepping of the AAA+ ring domains. Science 335:221–225
29. Hammer JA III, Sellers JR (2012) Walking to work: roles for class V myosins as cargo transporters. Nat Rev Mol Cell Biol 13:13–26
30. Miller KE, Heidemann SR (2008) What is slow axonal transport? Exp Cell Res 314:1981–1990
31. Shea TB, Chan WK (2008) Regulation of neurofilament dynamics by phosphorylation. Eur J Neurosci 27:1893–1901
32. Doyle M, Kiebler MA (2011) Mechanisms of dendritic mRNA transport and its role in synaptic tagging. EMBO J 30:3540–3552
33. Kanai Y, Dohmae N, Hirokawa N (2004) Kinesin transports RNA: isolation and characterization of an RNA-transporting granule. Neuron 43:513–525
34. Falley K, Schutt J, Iglauer P, Menke K, Maas C, Kneussel M, Kindler S, Wouters FS, Richter D, Kreienkamp HJ (2009) Shank1 MRNA: dendritic transport by kinesin and translational control by the 5′ untranslated region. Traffic 10:844–857
35. Ovsepian SV, Dolly JO (2011) Dendritic SNAREs add a new twist to the old neuron theory. Proc Natl Acad Sci USA 108:19113–19120
36. McMahon HT, Boucrot E (2011) Molecular mechanism and physiological functions of clathrin-mediated endocytosis. Nat Rev Mol Cell Biol 12:517–533
37. Schmidt MR, Haucke V (2007) Recycling endosomes in neuronal membrane traffic. Biol Cell 99:333–342
38. Stenmark H (2009) Rab GTPases as coordinators of vesicle traffic. Nat Rev Mol Cell Biol 10:513–525

39. Gandini MA, Felix R (2012) Functional interactions between voltage-gated $Ca^{2+}$ channels and Rab3-interacting molecules (RIMs): new insights into stimulus-secretion coupling. Biochim Biophys Acta 1818:551–558
40. Bykhovskaia M (2011) Synapsin regulation of vesicle organization and functional pools. Semin Cell Dev Biol 22:387–392
41. Rizo J, Rosenmund C (2008) Synaptic vesicle fusion. Nat Struct Mol Biol 15:665–674
42. van den Bogaart G, Holt MG, Bunt G, Riedel D, Wouters FS, Jahn R (2010) One SNARE complex is sufficient for membrane fusion. Nat Struct Mol Biol 17:358–364
43. Jahn R, Scheller RH (2006) SNAREs – engines for membrane fusion. Nat Rev Mol Cell Biol 7:631–643
44. Jackson MB, Chapman ER (2008) The fusion pores of $Ca^{2+}$-triggered exocytosis. Nat Struct Mol Biol 15:684–689

## Recommended Textbooks

Alberts B et al. (2008) Molecular biology of the cell, 5th edn. Garland Science, New York

Brady ST et al. (2012) Basic neurochemistry – molecular, cellular and medical aspects, 8th edn. Elsevier, Amsterdam

Lodish H et al. (2012) Molecular cell biology, 7th edn. W.H. Freemann, New York

# Electrical Activity in Neurons

# 7

Veronica Egger and Dirk Feldmeyer

To exchange and process information cells rely mostly on biochemical signaling pathways. Since these pathways are not well suited to rapidly transmit signals over larger distances, neurons use electrical activity in addition, exploiting the energy stored in the electrical gradients across cellular membranes. These gradients result from the uneven distribution of ions between the intra- and extracellular space caused by the ongoing activity of ion transporters and pumps and the fact that the membrane is semi-permeable for certain ions. The resulting membrane potential allows for fast flow of ions across cellular membranes via specialized classes of membrane-spanning proteins, namely ion channels. In neurons, some of these channels are voltage-gated, creating an electrical feedback loop that can generate stereotyped, regenerative depolarizing responses. Voltage-gated channels mediate the generation and conduction of action potentials and the release of transmitters at the synapse.

The first sections of this chapter explain the basic principles that govern ion channel structure and function. Special emphasis rests on the mechanisms of voltage-dependent gating because of its relevance for electrical signaling. Most voltage-gated channels are rather selective for a certain ion, thus there are voltage-gated potassium, sodium, calcium, and chloride ion channels. In many types of channels gating is controlled by specialized activating and inactivating processes. Recent findings allow to correlate these gating steps to certain movements in the channel structure, that is, conformational changes of the channel protein. We also discuss some evolutionary aspects of most of the presently known voltage-gated ion channels, including related, voltage-independent channels.

In the second half we focus on the generation and conduction of electrical signals within neurons. Two main forms of electrical signals serve either to receive information at the dendrites – the receptor potentials at sensors and synapses – or to transmit information to other neurons and effector organs via axons – the action potentials. In a nutshell, the classical action potential arises from the tight sequential activation of sodium and potassium voltage-gated ion channels, as first described by A.L. Hodgkin and A.F. Huxley: an inward current of sodium ions depolarizes the membrane, then quick repolarization is achieved by activation of an outward potassium current and inactivation of the sodium current. We also briefly cover regenerative calcium action potentials and the mechanisms underlying pacemaking activity. The conduction of electrical signals in dendrites and axons occurs passively and/or actively; in the latter case the signal is regenerated at the local membrane and there is no loss in amplitude. In vertebrates, a morphological specialization of the axon called myelination allows for fast conduction of action potentials across long distances in large bundles of axons – as required for large multicellular organisms. Action potentials may

V. Egger (✉)
Neurophysiology, Zoological Institute,
University of Regensburg, 93040 Regensburg, Germany
e-mail: veronica.egger@biologie.uni-regensburg.de

D. Feldmeyer
Function of Neuronal Microcircuits Group
Institute of Neuroscience and Medicine INM-2,
Leo-Brandt-Strasse, 52425 Juelich, Germany
e-mail: d.feldmeyer@fz-juelich.de

C.G. Galizia, P.-M. Lledo (eds.), *Neurosciences - From Molecule to Behavior: A University Textbook*,
DOI 10.1007/978-3-642-10769-6_7, © Springer-Verlag Berlin Heidelberg 2013

also propagate backwards into dendrites or even be generated there.

## 7.1 Ion Channel Function: General Principles

An ion channel is a loophole for ions across lipid membranes that are otherwise not permeable due to their strongly hydrophobic interior. Since ions carry charges and therewith electrical currents, ion channels provide the main pathway for electrical signaling in neurons.

Structurally, an ion channel is a large membrane-spanning protein with a pore in the center. The pore is formed by the transmembrane segments (TM) of the macromolecule, usually α-helices, in some cases β-sheets. Most ion channels are not constitutively open but are gated by external stimuli such as changes in the **membrane potential** $V_m$. The sensing and gating mechanisms involve conformational changes of the channel protein.

### 7.1.1 Gated Ion Channel Currents

The ease with which the current of ions flows across a channel is measured in terms of **conductance**, the inverse of resistance. Most ion channels, when open, have an approximately constant conductance that does not depend on $V_m$. Thus their current–voltage relationship follows Ohm's law: the single channel current rises linearly with voltage, its slope given by the single channel conductance $g_{SC}$, shown in Fig. 7.1a ($I_{SC} = g_{SC}\ \Delta V$). Characteristic conductance values of single channels are on the order of 2–100 pS (unit Siemens $S = A/V = 1/\Omega$). At the **reversal potential** $V_{rev}$ for an ion channel the net flow direction of ions flips and the single current amplitude is 0. The value of $V_{rev}$ depends on the individual conductances of the types of ions that the channel admits and their extracellular and intracellular concentration (see below Sects. 7.4.3 and 7.4.4; if the channel was perfectly selective for one type of ion that was equally distributed, $V_{rev}$ would equal 0 mV). The total amplitude of a single channel current is thus given by $g_{SC}$ multiplied by the difference between $V_{rev}$ and the actual membrane potential $V_m$ (Eq. 7.1, Fig. 7.1a). $I_{SC}$ is on the order of 1–30 pA, and the corresponding ion fluxes are ~$10^6$–$10^8$ ions per second and channel.

However, although at a certain $V_m$ the current amplitude has a fixed value $I_{SC}$, it does not follow automatically that a current $I_{SC}$ will indeed flow at this $V_m$ because the opening of a gated channel is a stochastic process. If an external gating stimulus is applied, the channel will not open instantly but only with a certain open probability $P_o$. This stochastic behavior can be described in terms of channel kinetics using rates or time constants for the transition between the open and the closed channel state. For example, if a channel has an exponential activation time constant of $\tau = 1$ ms for a certain stimulus, this channel will open with a probability of $(1 - 1/e) = 0.67$ within 1 ms following the onset of the stimulus. But we do not know when precisely the transition will happen. An equivalent approach is to consider a group of the same channels. If now the gating stimulus is applied, 67 % of the total number of channels will open within $\tau = 1$ ms – on average, since there is always a certain fluctuation. So repetitive measurements of the same process are required to reliably determine channel kinetics, whereas the conductance remains constant (see also Fig. 7.7a). Note that the stochastic nature of ion channel gating is nothing fancy but inherent to any molecular movements.

For voltage-dependent channels the ion channel open probability $P_o$ depends on the membrane potential $V_m$. In Fig. 7.1 the channel is activated by depolarizations and $P_o$ increases with rising $V_m$. If we now consider a neuron with many thousands of similar voltage-dependent channels, the macroscopic current $I$ is the product of the total number of ion channels $N$, their open probability $P_o(V_m)$ and the single channel current $I_{SC}$ (Eq. 7.2 in Fig. 7.1). Similarly, the macroscopic conductance $g$ is the product of $N$, the open probability $P_o(V_m)$, and the single channel conductance $g_{SC}$.

### 7.1.2 Methods to Investigate Ion Channel Structure and Function

Most of our knowledge on ion channels is obtained with the techniques shown in Fig. 7.2.

**a single channel current**

$$I_{SC} = g_{SC} \cdot (V_{rev} - V_m) \quad \text{eq. 7.1}$$

**b open probability $P_O$**

**c macroscopic current**

$$I = N \cdot P_O \cdot I_{SC} = N \cdot P_O \cdot g_{SC} \cdot (V_{rev} - V_m)$$
$$\text{eq. 7.2}$$

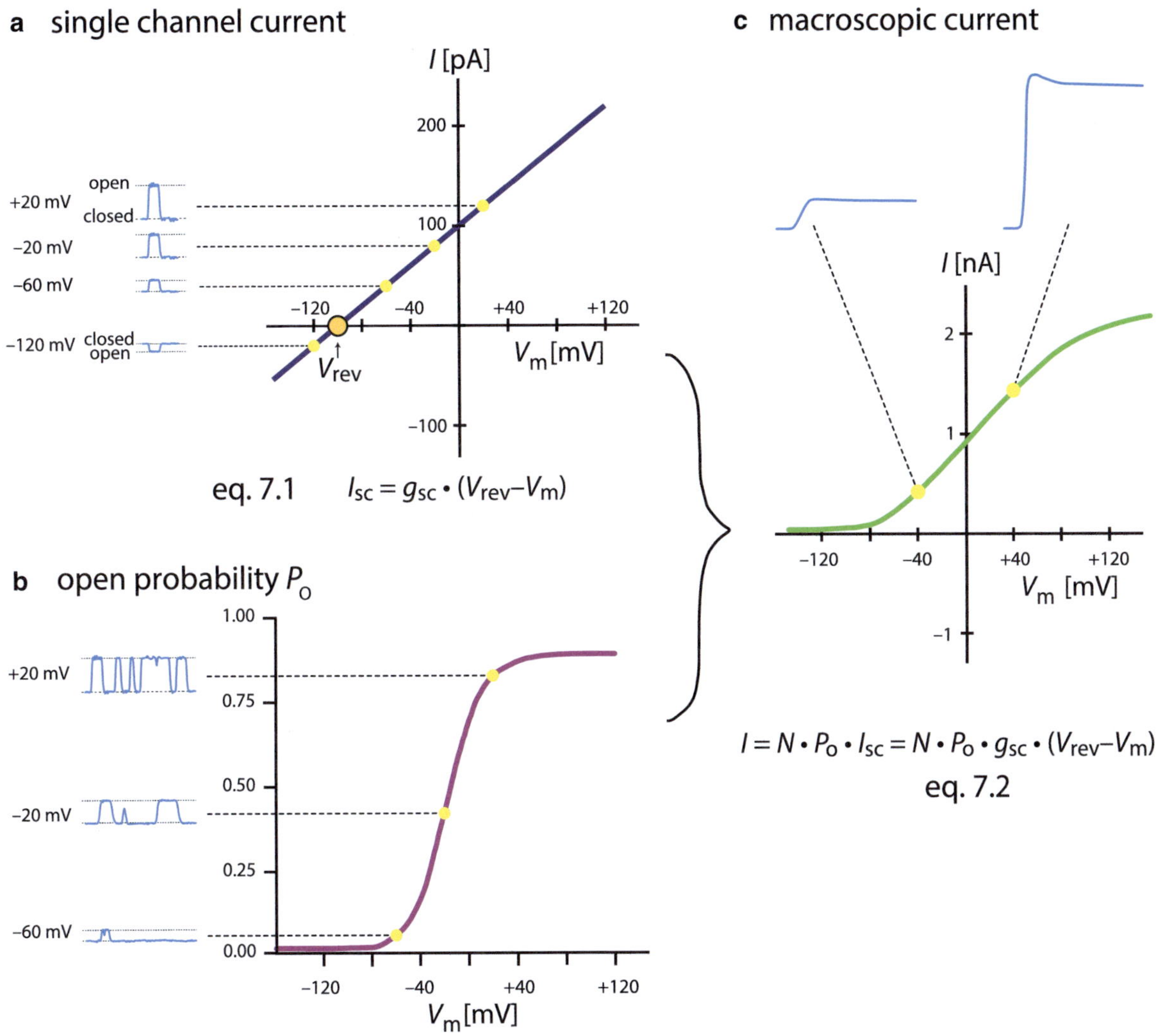

**Fig. 7.1 Microscopic and macroscopic ion channel currents** (see Sect. 7.1.1 for more details). (**a**) Exemplary current–voltage (*I–V*) relationship for an ohmic ion channel. The single channel conductance ($g_{SC}$) is 100 pS for this example. The reversal potential $V_{rev}$ for this ion channel is –100 mV. At more negative potentials, the single channel current $I_{SC}$ is negative, at more positive potentials it is positive, as seen in the single channel openings on the *left*. The total amplitude of $I_{SC}$ is given by Eq. 7.1. In electrophysiology, positive amplitudes of cation currents denote an outward movement of ions to the extracellular space. Thus if this channel was selective for K+ ions, the ions would flow out of the cell for all potentials more positive than $V_{rev}$. (**b**) Voltage-dependence of the ion channel open probability $P_O$. At membrane potentials

more negative than –80 mV, $P_O$ of this ion channel is almost 0. When $V_m$ becomes more depolarized, $P_O$ increases as can be seen on the *left*: while channel openings are rare at –60 mV they are rather frequent at +20 mV. Note that the maximal $P_O$ does not have to reach 1.0, as in this example. (**c**) Current–voltage relationship for a macroscopic current through 15,000 ion channels of the type shown in (a, b) (Eq. 7.2). This particular type of I–V curve is called a rectifier. Rectification means that the channel allows ion flow only in one direction. Again, if it were a K+ channel, it would allow only efflux of K+ ions. Such rectifying channels play an important role in repolarization after the action potential (AP) (see Sect. 7.5.1)

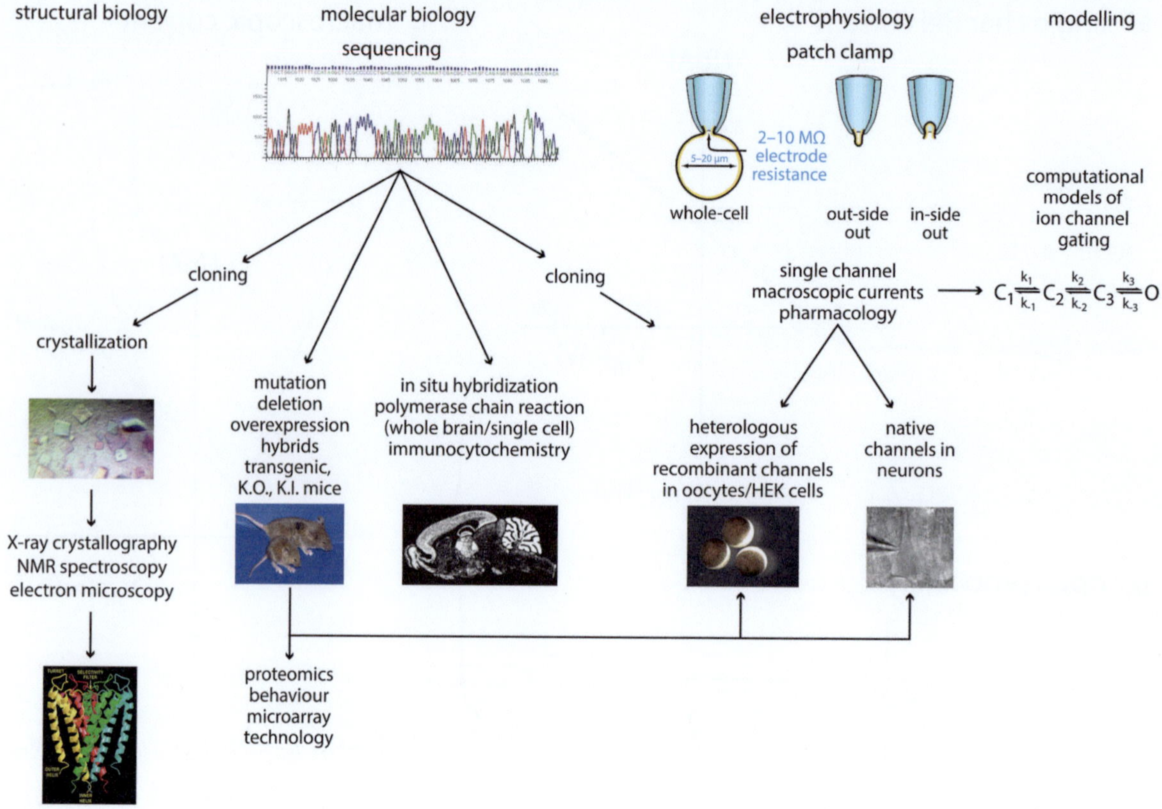


**Fig. 7.2 Technical approaches to study ion channels. Molecular biology** serves to sequence the DNA of an ion channel gene. Channel DNA or the transcribed mRNA is introduced into a host cell such as a *Xenopus laevis* (African clawed frog) oocyte via expression cloning, so that large amounts of a particular ion channel protein are expressed and its function can be investigated in isolation using the patch clamp technique (see below). Recombinant DNA technology (site-directed mutagenesis) allows precise point mutations to study functionally important sections of the ion channel gene product such as pore-lining domains or auxiliary subunits (Sect. 7.2.1). Genetic engineering techniques such as gene knockout (K.O., in which the ion channel gene or part of it is made inoperative), gene knock in (K.I., in which a cDNA sequence coding for, e.g., an additional ion channel protein is inserted in the animal's chromosome) or insertion of transgenes (i.e., genes that originated in different species) are used to test the function of ion channels in the whole organism. More refined versions of these techniques allow a so-called conditional knockout that is limited to certain brain areas and/or neuronal classes or start at certain time points well after birth. In situ hybridization is used to localize the DNA or RNA sequence of an ion channel in the tissue. Single cell polymerase chain reaction (PCR) allows to identify the specific ion channel RNA sequence in individual cell types; the RNA is harvested with a patch pipette from the intracellular solution of a single cell and subsequently amplified using PCR. Immunocytochemistry allows to detect ion channel proteins at the tissue, cellular and even subcellular level using specific, monoclonal antibodies. **Electrophysiological techniques** are used for functional studies of ion channels. Here the 'patch clamp' technique (Technical Box 1) allows to measure whole cell currents or single channel currents as exemplified in Fig. 7.1. Electrophysiological measurements can be performed using heterologous expression systems (see above) or native neurons in cell culture or from acute brain slices. Pharmacological agents help to isolate specific ion channels by blocking others. Ion channel characteristics have also been probed using ions of different size to gain information about the pore diameter. For **structural biology** see also Technical Box 2. If the amount of ion channel protein produced by expression cloning is sufficiently high, it may be feasible to grow a protein crystal. With X-ray crystallography it is then possible to localize nearly all of the atoms in the protein. Examples for structures determined by crystallography can be found in Figs. 7.5, 7.6, 7.7, and 7.8. Other techniques such as electron microscopy or nuclear magnetic resonance can complement these data. **Computational approaches** include modeling of ion channel gating to infer the different mechanisms that govern ion channel kinetics using rate models of the transitions between the channel states. Usually there are several activation steps necessary to reach the open state. Computer-based modeling is also used to predict the function and two- and three-dimensional structure of ion channel proteins from their amino acid sequence. Bioinformatics allows to construct evolutionary trees of protein families based on sequenced genomes (cf Fig. 7.4)

**Technical Box 1**

**Voltage Clamp and Patch Clamp**

The 'voltage clamp' is an electronic feed-back system to control the potential difference across the cellular membrane while simultaneously measuring the membrane current. This concept was invented by Cole and Marmount in the 1940s. The voltage clamp allows to manipulate $V_m$ independently of the ionic currents and thus to establish the current–voltage relationships of ion channels as Hodgkin and Huxley demonstrated in 1952 [15, 16]. In two-microelectrode voltage clamp one microelectrode measures $V_m$ while the other passes the current needed to keep $V_m$ at the desired value.

The patch clamp method was introduced to electrophysiology in the late 1970s by Neher and Sakmann [11]. It is a variant of the voltage clamp technique using a glass micropipette with a diameter of ~1 µm. The tip of the pipette is pressed against the cell membrane; subsequently suction is applied until the formation of a tight seal between patch pipette and membrane with an electrical resistance of 1–10 GΩ. A highly sensitive operational amplifier is used to control $V_m$ and measure current across the membrane patch. The high resistance allows the measurement even of single channel currents. Single channel measurements can be performed in the 'cell attached' recording mode, where the pipette is only pressed onto a small 'patch' (hence the name) of the cell membrane under investigation. Membrane patches can also be excised from cells, either in the 'outside-out' or 'inside-out' configuration, where the extra- or intracellular side of the membrane is facing the bathing solution, respectively. The 'whole cell' configuration allows to record from entire cells (see Fig. 7.2). Recently, these methods have been automated to some extent, allowing for high-throughput tests of the effects of new pharmacological substances on ion channels.

**Technical Box 2**

**X-Ray Crystallography**

In the past 15 years, X-ray crystallography has become a valuable tool for the structural analysis of ion channels because its resolution is high enough to visualize the channel structure. The method exploits the fact that X-rays are diffracted by crystals. Hence the technique requires the growth of ion channel protein crystals of a certain size (edge at least 0.1 mm) to achieve sufficient resolution. The fabrication of such crystals is the main obstacle for the structural analysis of membrane proteins since the proteins need to be separated from the lipids of their expression system, solubilized with detergents and crystallized without compromising their conformation.

X-rays are waves of electromagnetic radiation (wavelength $\lambda$ ~0.1 nm = $10^{-10}$ m) and crystals are regular arrays of atoms. When a beam of monochromatic X-rays penetrates a crystal, the electron cloud of the crystal's atoms will partially scatter the X-rays, producing a regular diffraction pattern that is related to the crystal structure through a mathematical operation called Fourier transform (please refer to physics textbooks for further information). The inverse Fourier transform of the diffraction pattern will yield the original function, which is the electron density of the crystal protein. Given the fact that protein crystals are somewhat irregular structures (e.g., in comparison to a salt crystal), the calculated models require reiterative refinements in order to arrive at the accurate molecular structure.

## 7.2 Voltage-Gated Cation Channels

This section explains general properties of voltage-gated cation channels whose function is at the core of electrical activity in neurons. We review the structural make-up and current kinetics of the most important types of cationic channels and describe the voltage-gated cation channel superfamily and its evolution. Then we illustrate the highly sophisticated selectivity mechanisms that make some ion channels permeable for exactly one type of ion with the famous example of the voltage-gated potassium channel that was explored mostly by R. MacKinnon and his group (the bacterial KscA channel; Nobel Prize in 2003). This selectivity filter of this ion channel mimicks the aqueous environment of $K^+$ ions in solution, but not the environment of other cations. Next we describe what is known about voltage sensitivity within the channel protein – the property that directly governs voltage-gating and therewith activation of the channels. Finally there is the process of inactivation that will close a channel even when the membrane depolarization remains high enough for activation, and thus limits the duration of currents via many types of voltage-gated channels.

### 7.2.1 Voltage-Gated Cation Channel Structure and Currents

The elementary building block of most voltage-gated cation channels is a sequence of 6 α-helical transmembrane segments (6TM). Different functions are attributed to the different segments (S): S1–S4 is often called the voltage-sensing domain with the main voltage-sensor located at S4, while the protein-aqueous pore interface is lined by the pore domains S5–S6 that govern the selectivity of the channel. Pore domains are also often associated with extracellular loops. Moreover, extracellular and intracellular loops and may provide binding sites for second messengers and the like, and phosphorylation and glycosylation sites that serve to modulate channel properties, e.g., conductance. Intracellular loops may be involved in special forms of channel gating such as the inactivation of $Na_v$ channels (see Sect. 7.2.5). As shown in Fig. 7.3, the 6TM building block is either assembled as a tetramer – the $K_v$ channel – or concatenated into a single protein with four similar 6TM domains – the $Na_v$ and $Ca_v$ channels. This pore-forming part of an ion channel is usually called its α-subunit.

Auxiliary subunits serve to modulate the trafficking of channels and their biophysical properties such as gating and activation kinetics. Some important representatives are shown in Fig. 7.3. Note that in the case of $Na_v$ channels the auxiliary β-subunits are transmembrane glycoproteins – unlike the intracellular β-subunits of $Ca_v$ and $K_v$ channels. $K_v$ channels have four β-subunits, one for each α-subunit.

Figure 7.3 also shows characteristic representatives of each type of ion channel current. In general, currents mediated by $Na_v$ channels follow a fast activation and inactivation time course compared to the $K^+$ and $Ca^{2+}$ currents. For $K^+$ channels the picture is more diverse (also reflected in the large number of family members, see Sect. 7.2.2). For example, the 'inwardly rectifying' $K_{ir}$ channels conduct $K^+$ ions at hyperpolarizing membrane potentials and are not voltage-gated. The classic $K_v$ channels are often of the delayed rectifier type (see Fig. 7.1). $Ca_v$ currents come in high-voltage activated ($Ca_v1$, $Ca_v2$) and low-voltage activated types ($Ca_v3$). Both activate quickly and display a relatively slow inactivation.

### 7.2.2 Molecular Relationships and Evolution of Ion Channels

The genetic sequence for voltage-dependent cation channels can be found in the genome of Eukarya, Bacteria, and Archaea as well as in that of some viruses. In multicellular organisms they are found in both excitable and non-excitable tissues. Most likely, all voltage-gated cation channel proteins and possibly all $K^+$ channels as well as several other channel types evolved from the same common ancestral gene, belonging to the transporter protein superfamily shown in Fig. 7.4. Most of the family members are 6TM channels, apart from the inward rectifier $K^+$ ($K_{ir}$) channels and the two-pore-domain background $K^+$ ($K_{2P}$) channels.

$K_{ir}$ channels fall into seven different subfamilies. Their molecular architecture is quite distinct since they lack the voltage-sensing domain S1–S4 and contain only two α-helical segments (2TM) and the P-loop, i.e., the pore-forming domain. Many bacteria have 2TM $K^+$ channels resembling $K_{ir}$ channels, only some have 6TM voltage-gated $K^+$ ($K_v$) channels. Thus, bacterial 2TM channels may represent phylogenetically early $K^+$ channels. The inward rectification is due to the voltage-dependent block of $K_{ir}$ channels by intracellular polyamines and $Mg^{2+}$. The activity of some $K_{ir}$ channel types is also regulated by the pH, G-proteins, and ATP.

The $K_{2P}$ channels are passive $K^+$-selective channels and therefore particularly relevant for the maintenance of the resting membrane potential (see Sect. 7.4.2). They are also involved in the respiratory chemoreception of the CNS. $K_{2P}$ channels are composed of two 2TM domains linked together (see Fig. 7.4) and channels are formed by the dimeric association of two of these. So far there are six known families of $K_{2P}$

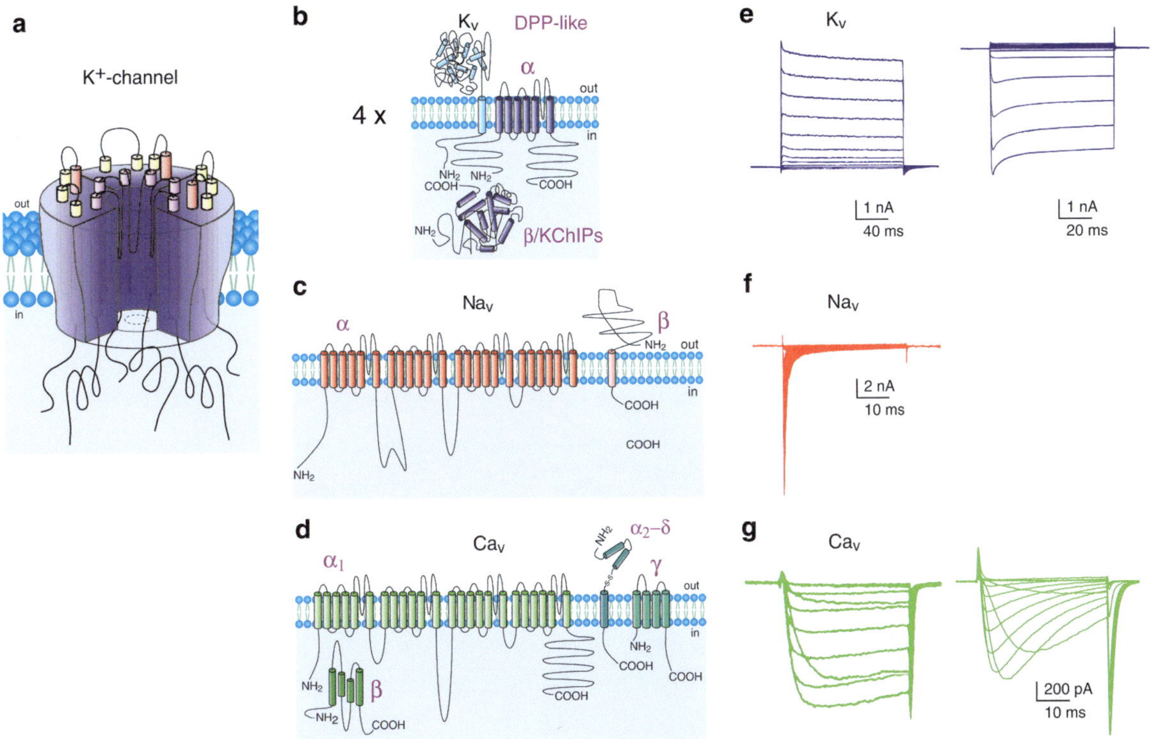


**Fig. 7.3 Voltage-gated cation channel structure and currents.** *Structure of a voltage-gated ion channel.* (**a**) Schematic image of the core tetrameric structure of a $K_v$ channel embedded in the cell membrane (modified from [10]). Each α-subunit has an amino- and a carboxyterminal end. One subunit is removed to expose the ion channel pore. For each subunit (in *blue*), the six transmembrane domains and the pore-loops are shown (*yellow*: S1–S3; *red*: voltage-sensor S4; *violet*: pore-lining S5 and S6). Recent data suggest that S4 is at the protein–lipid interface with most charged amino acids not exposed to the lipid. *Subunit compositions and two-dimensional membrane topologies of voltage-gated cation channels.* (**b**) The principal, pore-forming α-subunit and some auxiliary subunits of voltage-gated $K^+$ channels. Each $K_v$ channel is a homotetramer as shown in (a). Also shown is the auxiliary intracellular β-subunit, and the membrane bound dipeptidyl aminopeptidase-like (DPP-like) subunit. (**c**) The pore-forming α- and the β-auxiliary subunits of $Na_v$ channels. The α-subunits of $Na_v$ channels are heterotetrameric-unit structures with 24 transmembrane domains (4×6) and four pore-lining P-loops equivalent to the four α-subunits of a $K_v$ channel. (**d**) The principal $\alpha_1$-subunit and the auxiliary β-, γ-, $\alpha_2$-δ-subunits of $Ca_v$ channels. Like the α-subunits of $Na_v$ channels, the $\alpha_1$-subunits of $Ca_v$ channels are pore-forming heterotetrameric-unit structures. *Cation channel currents.* (**e**) *Left*, family of currents (i.e., responses to applied rectangular pulses of increasing membrane depolarization) through voltage-gated, 'delayed rectifying' $K^+$ channels [18] ($K_v$2.2 channels). Compare with Fig. 7.1: small depolarization leads to no current (many lines at the bottom), and with increasing depolarization the current increases (lines shift upwards). *Right*, family of currents through 'inwardly rectifying' $K^+$ channels [26] ($K_{ir}$2.1 channels). (**f**) Family of $Na^+$ currents ($Na_v$1.2 channels) [28]. Note the phasic response. (**g**) *Left*, family of neuronal L-type $Ca^{2+}$ currents (L for "long-lasting", also $Ca_v$1.2) [20]. *Right*, family of T-type $Ca^{2+}$ currents (T for "transient", also $Ca_v$3.1) [20]. Note the more complex temporal response structure (All with permission)

channels which are regulated by a variety of physicochemical factors such as extracellular pH, intracellular [$Ca^{2+}$], and lipid environment.

The 6TM $Ca^{2+}$-activated $K^+$ ($K_{Ca}$) channels are gated by voltage and the elevation of intracellular [$Ca^{2+}$]. In a subfamily of $K_{Ca}$ channels (the 'small conductance' or SK channels), $Ca^{2+}$ activates the channel by binding to intracellular calmodulin that is linked to the channel protein. A similar type of gating mechanism has also been reported for $Ca_v$1 channels.

The 6TM cyclic-nucleotide gated (CNG) and hyperpolarization-activated cyclic-nucleotide modulated (HCN) channel families are relatively nonselective cation channels that are regulated by cGMP and cAMP. These cyclic nucleotides activate the channels via binding sites at the C-terminal end. The well-known CNG channels in the retinal rod cells are opened by cGMP in the dark causing the so-called depolarizing 'dark current'; light reduces the cGMP concentration and thus leads to hyperpolarization (see Chap. 18).

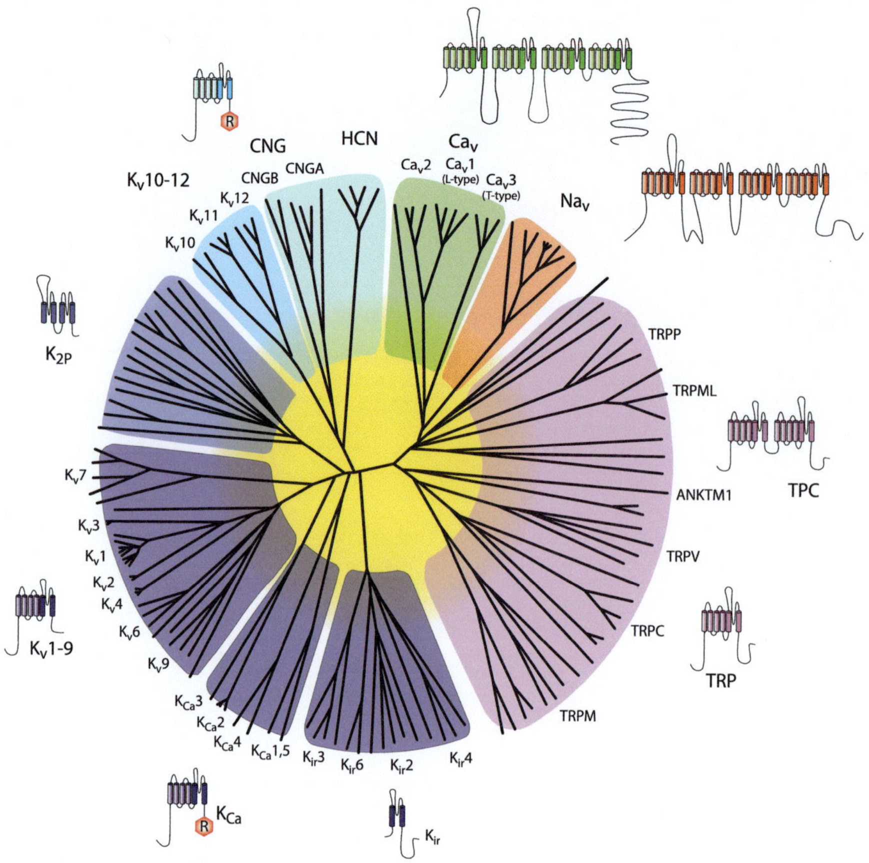


**Fig. 7.4 Evolutionary tree of voltage-gated and related cation channels.** Evolutionary relationship of the minimal pore regions of the voltage-gated ion channel superfamily according to their amino acid sequence, from the human genome. These structurally related ion channel genes form seven groups of ion channel families, shown with their respective membrane topologies. Background colors separate the ion channel proteins into related groups: *green*, $Ca_v$; *red*, $Na_v$; *violet*, TRP channels; *blue*, $K^+$ channels, except $K_v$10–12, which have a cyclic nucleotide binding domain; *cyan*, $K_v$10–12 channels and cyclic nucleotide-modulated CNG and HCN channels. Apparently all channels are variations on the same theme. However, for some channels subunits have formed oligomers of two or four (e.g., for $Ca_v$ and $Na_v$ channels and $K_{2P}$ and TPC channels, respectively) and the number of transmembrane domains can be as low as two per subunit (e.g., for $K_{ir}$ channels). Due to space limits not all individual channels are named (Modified after Yu et al. [30] with permission)

HCN channels are found in many neurons and may affect the time course of synaptic potentials. They also play important roles in pacemaking mechanisms of the heart and brain (see, e.g., Fig. 7.12).

Voltage-dependent $K^+$ channels ($K_v$) consist of tetramers of single 6TM domains (Figs. 7.3 and 7.4); their properties will be described in detail below. In contrast to $K_v$ channels, all voltage-dependent $Ca^{2+}$ and $Na^+$ ($Ca_v$ and $Na_v$) channel family members contain four linked homologous 6TM domains; therefore, the $Ca_v$ and $Na_v$ channel α-subunits are about four times as large as $K_v$ channel subunits. It is therefore highly probable that $Ca_v$ and $Na_v$ channels have evolved from $K_v$ channels.

In turn, the gene sequence and the functional properties of $Na_v$ channels are clearly less diverse than those of $Ca_v$ channels. Thus $Na_v$ channels may have evolved last, which is also supported by the fact that $Ca_v$ channels are found in single cell eukaryotes like *Paramecium* while $Na_v$ channels are not. Four domain $Na_v$ channels are first seen in early metazoans such as cnidarian jellyfish and appear to have evolved to allow for high frequency signaling and rapid signal propagation in larger organisms (for more details on $Na_v$ channels see also below).

Finally, the loosely related transient receptor potential (TRP) channel family members are not yet completely understood. They respond to temperature, touch, pain, osmolarity, pheromones, taste, and other stimuli. Although some of them have a preferential permeability for $Ca^{2+}$, TRP channels are generally non-selective cation channels with a weak voltage-dependence. Structurally, TRP channels resemble $K_v$, $Na_v$, and $Ca_v$ channels, but their amino acid sequence is less conserved. Two-pore channels (TPC) belong also to this group but have so far not been identified in neuronal cell types.

The anion or $Cl^-$ channels (ClC) described in more detail in Sect. 7.3 below belong to an entirely different protein family, the so-called 'Major Facilitator Superfamily'. This is a very old, large and diverse superfamily that comprises many transporter proteins such as the vesicular glutamate transporters but also some ion channels. ClC proteins are found in many Bacteria, Archaea, and a vast number of Eukarya. However, they are not present in several prokaryotes with small genomes.

### 7.2.3 The Ion Channel Pore and Ion Selectivity

The intra- and extracellular space contains a large variety of different cations and anions, among them $K^+$, $Na^+$, $Ca^{2+}$, $Cl^-$, and $HCO_3^-$ ions. Voltage-gated ion channels are, however, very selective for a particular ion species, in contrast to most ligand-gated channels. How is this remarkably high ion selectivity achieved?

In solution, ions are hydrated, i.e., they are surrounded by a hydration shell of six to eight water molecules which are electrically dipolar and arranged to shield the ion's charge. Thus, in the case of cations the electronegative oxygen atom of the water molecules is oriented towards the ion (e.g., $K^+$, see Fig. 7.5e), while the electropositive hydrogen atoms are turned towards anions.

When a hydrated $K^+$ ion moves through the $K_v$ channel pore from the cytoplasmic to the extracellular side it will enter the channel by the gate region on the intracellular side of the membrane. This region will alter its configuration from a 'closed' to an 'open' state during the activation of the channel (see Sect. 7.2.4), allowing hydrated ions to enter the channel cavity.

Having passed the gate region the $K^+$ ion will then enter the water-filled central cavity of the $K_v$ channel near the center of the pore. The gate region and the central cavity are lined with hydrophobic amino acid residues, so the hydration shell remains intact (Fig. 7.5b, c). The existence of this central cavity aids to reduce the energetic barrier for ion conduction through the pore.

To pass the narrowest region of the ion channel pore, the so-called selectivity filter, $K^+$ ions now have to shed their shell of water molecules because hydrated ions are too big to permeate the channel. In $K_v$ channels, this filter region is rather short and takes approximately one quarter of the ion conduction pathway within the transmembrane section (Fig. 7.5a, b). The selectivity filter is located on the extracellular side of the membrane and formed by the highly conserved amino acid residues threonine-valine-glycine-tyrosine-glycine (TVGYG, from prokaryotes to humans) in the P-loop from each subunit. In particular the two glycine residues are conserved that allow torsions of the backbone amino acids. The carbonyl oxygen atoms of the amino acids are oriented towards the filter pore (Fig. 7.5c) and thus form surrogate hydration shells for the $K^+$ ions in the pore: As in solution, each $K^+$ ion is surrounded by two groups of four oxygen atoms. The total energy required for shedding of the hydration shell at the entry of the filter and entering the surrogate shell within the filter is low, which allows to achieve a high flux of $10^7$–$10^8$ ions per second. The interaction of $K^+$ with the carbonyl oxygen helps also to overcome the repulsive electrostatic forces that act against the entry of further cations into the filter pore. Having entered the pore, $K^+$ ions will now permeate the filter in single file, separated from one another by single water molecules. When the ion leaves the filter the shielding hydration shell will be reformed by the water molecules in the extracellular solution.

Why is the selectivity filter highly selective for positive ions? The exclusion of anions is inherent in the make-up of the filter pore, because they are repelled by the negatively charged carbonyl oxygens lining the filter. Why are cations like $Na^+$ ions prevented from permeation? In this case the size of the ions matters: $Na^+$ ions are much smaller than $K^+$ ions. In water both ions are hydrated by a similar arrangement of water molecules, but in the $K_v$ channel filter the carbonyl oxygens are optimally positioned to form stable complexes only with the dehydrated $K^+$ ion. To form similarly stable complexes with the smaller dehydrated $Na^+$ ions, the filter pore diameter would have to shrink beyond what is structurally possible (Fig. 7.5d). The total energetic cost for the dehydration and the entry into the filter is therefore higher for $Na^+$ ions than for $K^+$ ions. Thus $Na^+$ ions are prevented from entering the

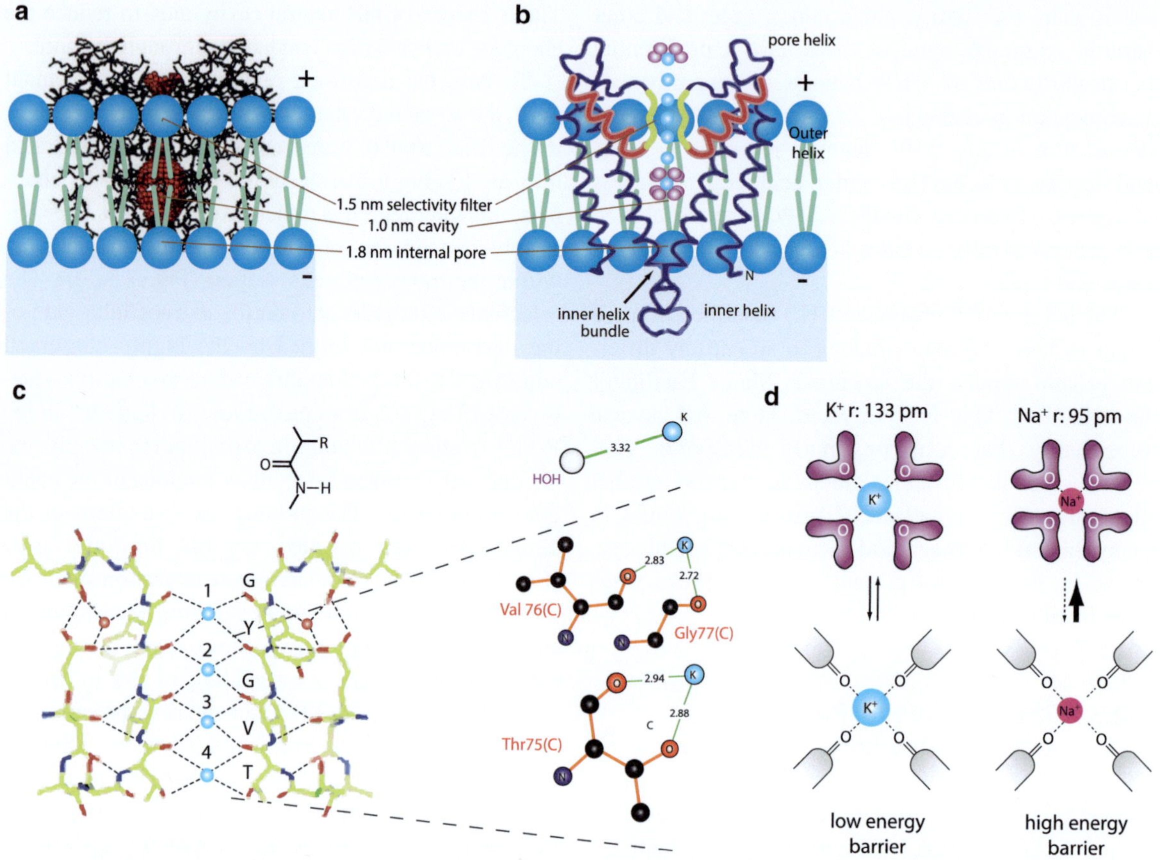


**Fig. 7.5** **Pore of a K⁺ channel.** (**a**) Three-dimensional stick model of a bacterial K⁺ channel, the KcsA channel, that is gated by pH. The channel pore is 4.5 nm long and about 1 nm wide. *Black*, amino acid backbone of the channel protein, *red*, actual size of the ion-conducting pore within the membrane. Each of the four subunits of the channel consists of two membrane-spanning domains and a pore loop that inserts into the membrane ([6] with permission). (**b**) The core of the bacterial K⁺ channel in cutaway showing, from bottom to top, the gate region (closed), a hydrated K⁺ ion (*magenta* with eight *violet* water molecules) in the inner cavity, five dehydrated K⁺ ions in the filter, and a partially hydrated K⁺ ion at the outer mouth of the filter. The length of the inner cavity is 1.0 nm, that of the selectivity filter 1.5 nm, and that of the hydrophilic internal pore 1.8 nm (Modified after MacKinnon [21] with permission). (**c**) Stick model of the selectivity filter. The backbone carbonyl oxygen atoms (*red*) represent four binding sites for K⁺ ions. These carbonyl oxygens replace the hydration shell of a K⁺ ion: the energy of binding the K⁺ ion in the selectivity filter is lower than that required for dehydration (Modified after MacKinnon [21] with permission). (**d**) *Top*, size of K⁺ and Na⁺ ions and the water molecules forming their hydration shell in solution. *Bottom*, the fixed structure of the K⁺ channel selectivity filter is fine-tuned to accommodate a K⁺ ion but it cannot shrink sufficiently to properly bind the smaller Na⁺ ions. The energetic cost for dehydration is therefore higher for Na⁺ ions than for K⁺ ions and thus selectivity is achieved (Modified after http://www.nobelprizes.org/nobel_prizes/chemistry/laureates/2003/popular.html with permission)

filter pore and the selectivity for a particular cation, K⁺, is achieved. For $Na_v$ and $Ca_v$ channels the mechanisms of selectivity are quite similar.

### 7.2.4 Activation and the Voltage Sensor of Ion Channels

The macroscopic conductance of voltage-gated ion channels like the $Na_v$, $Ca_v$, and $K_v$ channels discussed above is steeply dependent on the membrane potential. For $Na_v$ channels, conductance values can increase up to 150 fold for an increment of 10 mV in $V_m$. However, the conductance of the single open channel $g_{SC}$ is only weakly voltage-dependent; hence the steep voltage dependence of the ion channel current must result from the voltage dependence of the open probability $P_o$ (see Sect. 7.1.1). This high sensitivity to voltage is crucial because of the small size of neuronal voltage changes.

What is the mechanism that allows voltage-gated ion channels to detect minor changes in $V_m$ and open and close the channel pore in fractions of milliseconds? To sense changes in the electric field that result from changes in $V_m$, the transmembrane channel protein must contain mobile charged groups. Upon a change in $V_m$ such charged groups will rearrange and thereby trigger a change in the conformation of the entire channel protein, leading to an opening of the channel pore (depicted schematically in Fig. 7.6a for the extreme case of a reversal of $V_m$).

The relocation of the charged groups can be measured in form of a so-called 'gating current' or 'charge movement' at the onset and offset of a voltage pulse (Fig. 7.6b). These gating currents are a direct expression of the molecular rearrangements that occur during the operation (or gating) of the channel molecule. Because the charges are located at certain amino acids of the ion channel molecule, the charge moved in one direction during depolarization ($Q_{ON}$) is identical to that moved into the opposite direction upon repolarization to the resting membrane potential ($Q_{OFF}$) (Fig. 7.6b); this symmetry is in marked contrast to the current flowing through the channel pore which follows the electrochemical gradient (Sect. 7.4.3). In electrotechnical terms, gating currents are therefore capacitive currents while ionic currents through the channel pore are ohmic or resistive currents. The gating current always precedes the ionic current since it corresponds to the movement of charged residues of the channel protein that ultimately cause the opening of the channel pore (Fig. 7.6c). Electrophysiological measurements have shown that the gating current corresponds to the translocation of the equivalent of ~10 elementary charges per channel across the membrane, which accounts for the steep voltage dependence of voltage-gated channels.

When the subunits of voltage-gated cation channels such as $Na_v$, $Ca_v$, and $K_v$ channels were first cloned it was found that of their 6 TM segments the fourth segment (S4) stood out as a natural candidate for the **voltage sensor** that causes the ion channel to open and close. Depending on the type of voltage-gated channel, the S4 segment has between four and seven positive charges, mostly from arginine or lysine residues (Fig. 7.6d). Other techniques have provided evidence that S4 indeed plays an important role in voltage sensing, that it moves in response to $V_m$ changes and that some of its amino acids are alternately exposed on the internal or external face of the membrane, depending on $V_m$. Because the first three transmembrane segments (S1–S3) contain several negatively charged amino acids, they also belong to the voltage-sensing domain (S1–S4; Fig. 7.6d). The voltage sensor is coupled to the pore-forming unit via the so-called 'S4–S5 linker' sequence that mediates channel opening and closing upon changes in the membrane electric field.

Notably, the S4 segment is also the main voltage-sensing element in HCN channels, with similar positively charged residues and the same direction of movement in response to depolarization. The opposite effects of potential change ($K_v$ channels open with depolarization, HCN channels close) may be the result of a different attachment site of the linker between S4 and the activation gate.

What then are the structural rearrangements during voltage sensing that lead to the opening of the ion channel pore? This issue has not been fully resolved and data from different laboratories are controversial. It is generally accepted that an outward movement of the charges (i.e., towards the extracellular side of the membrane) occurs in all voltage-gated cation channels. However, the present models differ markedly in the location of the S4 segment and in the extent of transmembranal movement of the charged, voltage-sensing residues:

The 'conventional' model proposes a highly localized electric field (~1 nm), which is significantly shorter than the lipid bilayer (4–5 nm). In this model, the S4 segment responds to changes in $V_m$ with rather small movements while the S1–S3 segments remain more or less unperturbed. The S4 movements can involve one or more combinations of tilt, rotation, axial and lateral translation, and minor changes in secondary structure.

In contrast, X-ray cristallography of a $K_v$ channel of the archaebacterium *Aeropyrum pernix*, the $K_{vAP}$ channel, and the eukaryotic D-type $K_v1.2$ and the 'delayed rectifier' $K_v2.1$ channels suggests a different structure of the voltage sensor (Fig. 7.6e). According to this data, the S4 and the C-terminal half of the S3 segment are arranged like a 'paddle' close to the lipid interface. The voltage sensor operates according to a very straightforward mechanism: In response to changes in the electric field, the charges attached to the paddle will be transported from the intra- to the extracellular face of the membrane (i.e., over a large fraction of the transmembranal distance), thereby acting as levers that open and close the ion conduction pathway of the channel. Hence, changes in $V_m$ are sensed by the paddle and converted into mechanical work.

Very recently, two proteins containing voltage sensor domains have been identified that are not attached to an ion channel pore. These isolated S1–S4 segments showed unexpected biological functions. One was

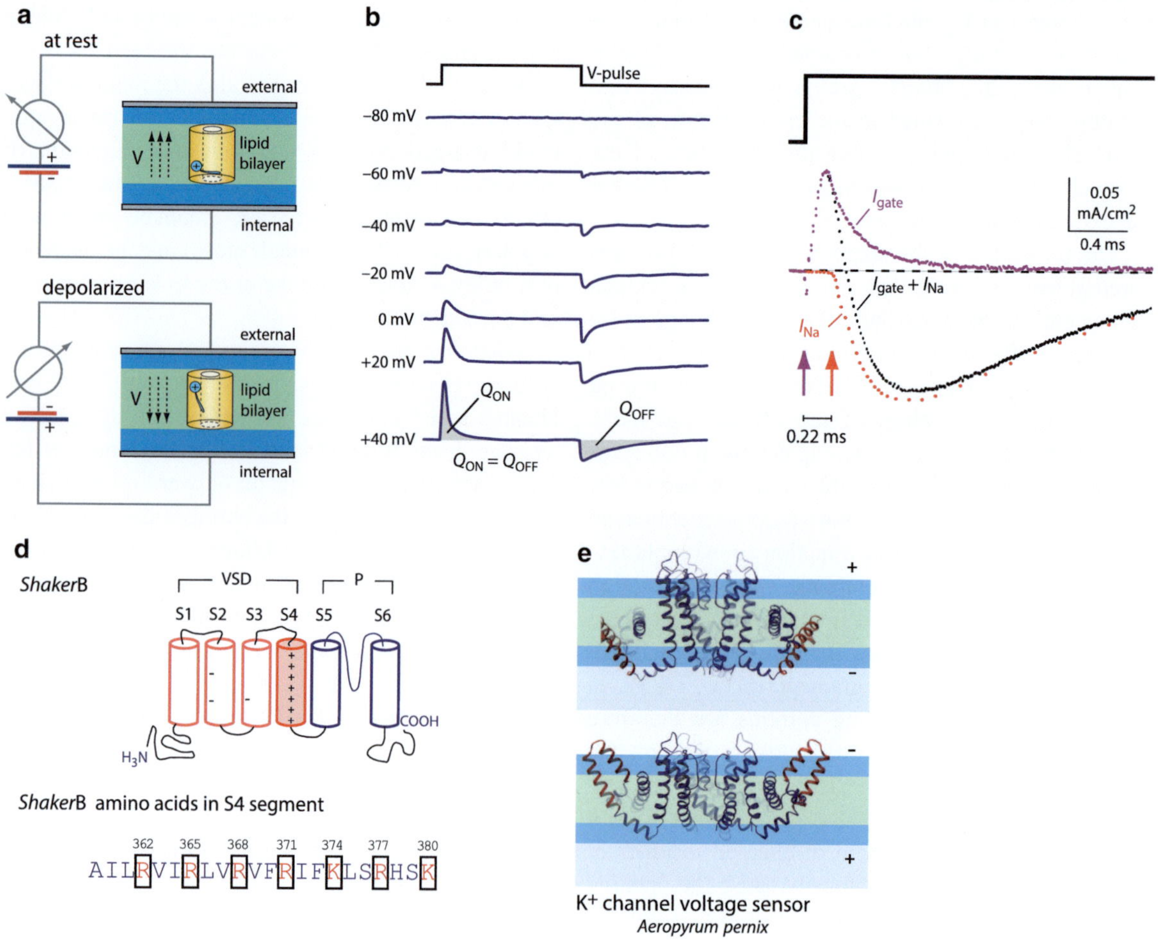


**Fig. 7.6 Voltage sensor of an ion channel.** (**a**) Schematic representation of a voltage-gated ion channel (*yellow*) in cell membrane (*gray*). A mobile, positively charged residue is linked to the 'gate' of the ion channel (*light blue*). When the membrane potential changes, the charged residue(s) in the channel molecule moves accordingly towards the negative side, thereby generating a capacitive current, i.e., a current in an electrical capacitance. This capacitance is shown in the equivalent circuit in *violet*. The movement of the electrical charge is linked to a conformational change in the channel molecule and results in pore opening. (**b**) Gating currents (charge movements) of $Na^+$ channels elicited by depolarizations from −80 mV to the membrane potentials given on the *left* of each trace. Note that the charge moved during the pulse onset ($Q_{ON}$) is similar to that moved at the pulse offset ($Q_{OFF}$), i.e., the charge is moved forth and back within the membrane. (**c**) Sequence of currents induced by a depolarization. The gating current (*violet*, $I_{gate}$) is generated by the voltage-driven movement of charged amino acid residues within the membrane that induces the change from the closed to the open conformation of the ion channel. It precedes the $Na^+$ current (*red*, $I_{Na}$) by about ~200 μs. The total current is shown in *black*. The gating current can be isolated by blocking the $Na^+$ channel pore with tetrodotoxin or by removing $Na^+$ from the external solution (Modified after Armstrong [2, 3] with permission). (**d**) *Top*, the architecture of a $K_v$ channel subunit, the ShakerB ($K_v$1.1). Cylinders represent membrane-spanning helical segments. Of the six transmembrane domains, the main voltage sensing domain is the S4 segment which contains seven positive amino acid residues. The S1 and S2 segments contain negatively charged residues. *Bottom*, five of the positive charged amino acids are arginine (R), two are lysine (K). Compare with Fig. 7.3a. (**e**) Three-dimensional 'paddle model' of two subunits of a $K_v$ channel from the bacterium *Aeropyrum pernix*. *Top*, in the resting state the S4 segment is near the intracellular membrane surface and the channel pore is closed; *red*, voltage-sensing S4 segment; *blue*, the other transmembrane domains. *Bottom*, depolarization causes a rotation of the voltage sensor towards the extracellular end of the membrane in a paddle-like movement. This results in a conformational change that opens the pore (Modified after Jiang et al. [17] with permission)

associated with a phosphatase that controls the hydrolysis of membrane phosphoinositides in a voltage-dependent manner. Another S1–S4 domain without an attached pore domain acted as a proton channel by itself – the proton permeation pathway was embedded in the S1–S4 domain. These findings indicate that the S1–S4 voltage sensor domain evolved separately. Association with pore-forming domains or enzymes enables it to exert a voltage-dependent control over ion channel gating or enzyme activity.

### 7.2.5 The Mechanism of Voltage-Dependent Ion Channel 'Inactivation'

When Hodgkin and Huxley first described the currents underlying the action potential in the squid giant axon (see Sect. 7.5.1) they noted that while the 'delayed rectifier' $K^+$ current remained largely constant during ongoing depolarization the $Na^+$ current underwent a rapid decrease. During sustained depolarization, single $Na_v$ channels displayed a few openings at the pulse onset but then remained closed despite the ongoing depolarization (Fig. 7.7a). This behavior which occurs within milliseconds or tens of milliseconds is called ion channel **inactivation**. Current inactivation has been observed for $Na_v$ and $Ca_v$ channels and also for some $K_v$ channels (the fast inactivating A-type or 'shaker' $K_v4$ and the more slowly inactivating D-type or $K_v1$ channels).

Electrophysiological studies in the early 1970s demonstrated that the structural basis for inactivation of the $Na_v$ channel is a cytoplasmic protein domain: When pronase (a mixture of proteolytic enzymes) was perfused in a squid giant axon, the inactivation of sodium channels was completely destroyed. Eventually, this leads to the proposal of the 'ball-and-chain' model of inactivation in which a positively charged inactivation particle (the ball) on a tether (the chain) can block the channel by binding to the channel pore. For $Na_v$ channels, the 'ball-and-chain' structure has been identified as the linker sequence between the third and fourth domain. Structural analysis predicts that this peptide sequence is an α-helical structure preceded by an isoleucine-phenylalanine-methionine (IFM) motif (Fig. 7.7b, c). The critical phenylalanine (F1489) blocks the conduction pathway of the channel by interacting with and binding to a site in the pore. To remove inactivation the membrane must be repolarized to the resting potential. The ensuing conformational change in the channel protein will slowly dissociate the inactivation sequence from its binding site, thereby restoring the activatable state of the channel.

Some $K_v$ channels show fast inactivation. The mechanism of inactivation in these $K_v$ channels is also of the 'ball-and-chain' type but structurally different from that found in $Na_v$ channels: Here, the inactivation peptide is located at the N-terminal end of each subunit. Ablation of the N-terminal peptide results in a loss of inactivation. In $K_v$ channels the 'ball-and-chain' inactivation mechanism is therefore also called 'N-type inactivation'. The N-terminal inactivation peptide of one of the four channel subunits enters the channel from the cytosolic side of the membrane to inhibit $K^+$ ion flow through the channel. Although all four subunits carry this inactivation peptide, just one is sufficient for a $K_v$ channel to inactivate. How does the inactivation peptide interact with the channel pore? It was found that in rapidly inactivating $K_v$ channels the first ten amino acids at the N-terminal end are hydrophobic while the following ten are hydrophilic. Since the amino acids lining the inner pore and the central cavity are also hydrophobic (see above) this led to the conclusion that the hydrophobic residues apparently enter the inner pore of the channel by one of the 'side windows' these $K_v$ channel molecules possess. The N-terminus ($-NH_3^+$) locks then into the central cavity and so blocks the conduction pathway.

In addition to the N-type inactivation a secondary, much slower inactivation mechanism has been identified. This inactivation mechanism is called 'C-type inactivation' and can be induced by prolonged (i.e., seconds to minutes long) depolarizations. It has been identified in some $K_v$ but also in $Na_v$ channels. C-type inactivation persists in inactivating $K_v$ channels even when the N-terminal inactivation peptide is ablated. It involves primarily conformational changes around the selectivity filter and extracellular entrance to the channel. C-type inactivation has been found to slow down in response to rises in extracellular [$K^+$] and thereby may counteract extracellular [$K^+$] accumulation.

Inactivation of voltage-gated ion channels is – like their activation – a steeply voltage-dependent process and is sensitive to the membrane potential before and during activation, indicating that direct transitions from both the resting and the activated (open) state of a channel to the inactivated state are possible. The sigmoidal 'activation curve' shown in Fig. 7.7d is obtained by stepping from $V_{m\_rest}$ to various activating potentials. 'Inactivation curves' can be obtained by starting from various holding potentials to a fixed activation voltage. The current (and the open probability of the ion channels) will then start at maximum and decrease sigmoidally to zero. Activation and inactivation curves intersect at a certain potential. In the area under this cross-over point the voltage-dependent ion channel is open with a finite probability. Thus in this potential region tonic, so-called window currents occur. For example, the window current region for the delayed rectifier $K_v$ channel in the squid giant axon is near its resting potential while that of T-type $Ca^{2+}$ currents is around –60 mV.

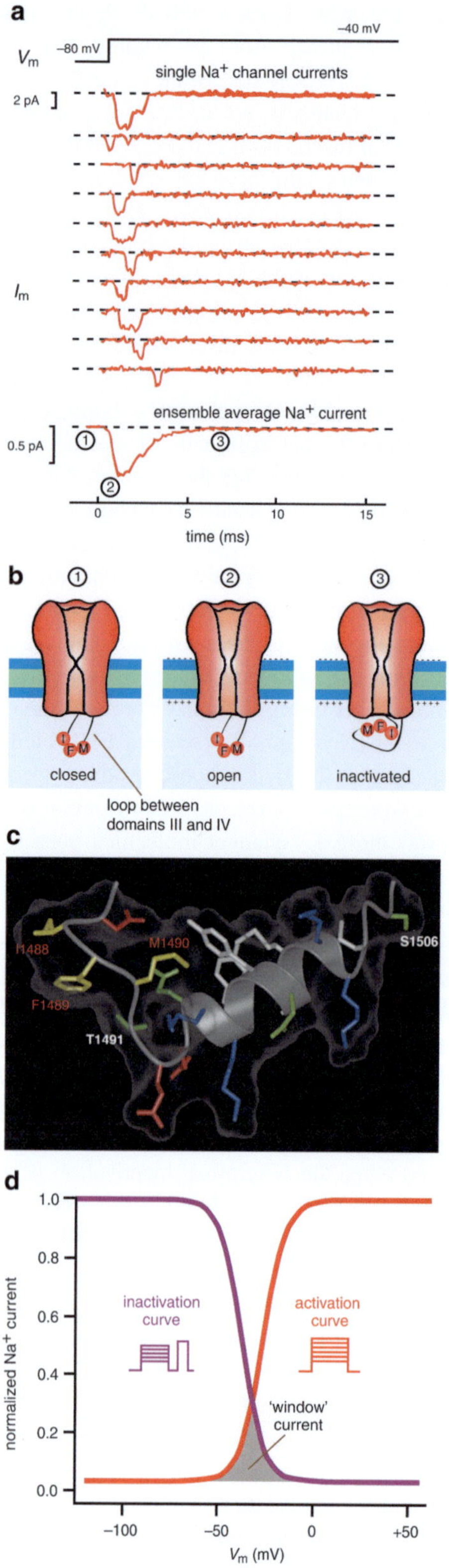


**Fig. 7.7 Mechanism of inactivation of $Na_v$ channels. (a)** Patch clamp recording of transient single $Na_v$ channel openings in response to a depolarization from –80 to –40 mV. $Na_v$ channels open only during the first few milliseconds of the depolarization and subsequently close again. The ensemble average $Na^+$ current calculated from the sum of the single channel currents shows that from rest there is a rapid upstroke that is followed by a decline to almost zero. This last process is called **inactivation** (Modified after Hille [13] with permission). These recordings also illustrate the stochastic nature of channel kinetics which shows in the variability in the time points of channel activation and inactivation. The ensemble current could also be obtained by the simultaneous recording from a large number of $Na_v$ channels, e.g., in the whole-cell mode. Compare with Fig. 7.3f. **(b)** Ball-and-chain model of $Na_v$ channel inactivation. The onset of a depolarization will eventually take the channel from the closed (*1*) to the open state (*2*). During the depolarization the polar amino acids (the ball on a chain) move into the inner vestibule of the ion channel pore and block the conduction pathway. This so-called 'inactivation particle' consists of a sequence of three amino acids, isoleucine (*I*), phenylalanine (*F*), and methionine (*M*). Now the channel is in the inactivated state (*3*). **(c)** Three-dimensional structure of the central segment of the inactivation gate. Side chains of the critical IFM motif residues are shown in *yellow*, and those of threonine1491 (*T1491*), which is important for inactivation, and serine1506 (*S1506*), which is a protein kinase C-dependent phosphorylation site, are also indicated (Adapted from Rohl et al. [27] with permission). **(d)** Activation and inactivation curves of an $Na^+$ current. The voltage pulses used to measure activation and inactivation curves are shown as insets (increasing square pulse for the activation curve; increasing square prepulse and subsequent pulse to a fixed potential for the inactivation curve). In a very narrow voltage range, the activation and inactivation relations overlap (*shaded area*). At these potentials a steady-state or 'window' current can exist. This current is small compared to the maximum $Na^+$ current in a neuron. 'Window' currents have also been found for some L- and T-type $Ca^{2+}$ currents

## 7.3 Chloride Channels

$Cl^-$ channels are found in almost all tissues, both neuronal and non-neuronal. They play a particularly important role in stabilising the resting membrane potential in skeletal muscle cells, in cell volume and pH regulation, but also in chloride transport. Functionally, $Cl^-$ channels can be broadly classified as voltage-gated, volume-regulated, second messenger-activated channels (e.g., the bestrophins and the anoctamins) and finally ligand-gated $Cl^-$ channels, the $GABA_A$ and glycine receptors, that mediate synaptic inhibition (see Chap. 8).

So far, nine mammalian voltage-gated chloride channel (ClC) genes have been identified and their gene products characterized. The ClC proteins show a ubiquitous distribution and are found not only in nerve and glia cells but also in epithelial cells. They are located both in the cell membrane and in the membrane of intracellular organelles. However, only the ClC-0, ClC-1, ClC-2, ClC-Ka, and ClC-Kb subunits function as true voltage-gated ion channels, while the ClC-3, ClC-4, and ClC-5 proteins are $Cl^-/H^+$ transporters. For two other proteins (ClC-6, ClC-7) the exact transport function has not yet been determined.

$Cl^-$ channels are dimers in which each of the two identical subunits consists of 17 intramembrane and 1 intracellular α-helices and forms its own pore. This so-called 'double-barrel' model was based on the biophysical analysis of ClC-0 channels from the electric ray *Torpedo* electroplax, which demonstrated that the channel had three conductance states, namely closed, one pore open, and both pores open (Fig. 7.8a). Subsequently, the idea that a pore is formed entirely within each subunit was supported by patch-clamp analysis of the cloned ClC-0 channel and structural analysis of prokaryotic ClCs (Fig. 7.8b, c).

Voltage-gated $Cl^-$ channels are considerably less selective than voltage-gated cation channels and permeable not only to $Cl^-$ but many other small anions, in particular $HCO_3^-$. The selectivity filter of ClC channels is somewhat shorter than that of $K_v$ channels. It contains positively charged amide NH groups that serve as surrogate water molecules for $Cl^-$ ions in the filter pore. Remember that the hydration shell for anions is oriented with the electropositive H-atoms inwards (see Sect. 7.2.3). ClC channels strongly differ from cation channels in the voltage-dependence of ion channel gating and its mechanism. Instead of an intramembraneous voltage-sensor the permeating anions themselves appear to serve as *extrinsic* voltage sensors, so gating and permeation are intertwined.

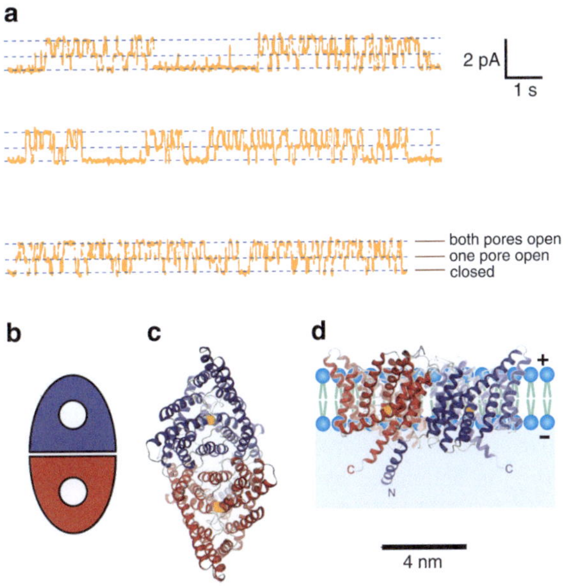


**Fig. 7.8 Chloride channels.** (**a**) Openings of a single $Cl^-$ channel from *Torpedo electroplax*. Recordings were collected at a membrane potential of −90 mV. The channel has three conductance levels, namely 0 (no conductance), 1 (one subunit conducting) and 2 (both subunits conducting). Channel openings occur in bursts (Modified after Miller and White [24] with permission). (**b**) In contrast to cationic voltage-gated ion channels where a single pore is formed by four identical or structurally similar subunits, $Cl^-$ channels from the ClC family are dimers, in which *each* subunit (*blue* and *red*) has its own pore. (**c**) Structure of a prokaryotic ClC protein from *Salmonella enteritidis*. Ribbon representation of the ClC dimer as seen from the extracellular side. Again, the two subunits are *blue* and *red*. Each of the subunits contains a pore holding a $Cl^-$ ion (*orange*). (**d**) View of the protein from within the membrane with the extracellular space above. Note that this ClC protein has lately been shown to be a transporter and not an ion channel. However, the true $Cl^-$ channels are supposed to have a similar structure (Modified after Dutzler et al. [7] with permission)

Besides ClCs, there are also voltage-sensitive, second messenger-activated $Cl^-$ channels. For example, very recently $Ca^{2+}$-activated $Cl^-$ channels, so-called 'anoctamins', have been identified. The name refers to the anion-selectivity (**AN**) and the eight (**OCT**) transmembrane segments. These channels are involved in secretion, neuronal excitability, olfactory and retinal transduction, and many more processes. The anoctamin channel family is found in all eukaryotic kingdoms and has ten mammalian members. With their eight TM segments anoctamins are profoundly distinct from any previously described ion channels. Their voltage-sensitivity depends on the intracellular $[Ca^{2+}]$.

## 7.4 Resting Potential

Following a brief introduction to principles governing the movement of ions in electrolytic solutions we describe the origin of the resting membrane potential $V_{m\_rest}$, fundamental for all electrical signaling in neurons. Most importantly, it requires differing $K^+$ concentrations at the intra- versus the extracellular compartment and a membrane that is semipermeable for $K^+$ ions. The ensuing electrochemical balance between the concentration gradient that is pre-set by the activity of various pumps and the electrical gradient resulting from charge separation at the membrane gives rise to the Nernst equilibrium potential. While $V_{m\_rest}$ is dominated by the $K^+$ equilibrium potential, other ions such as $Na^+$ and $Cl^-$ also exert some influence, together with the action of electrogenic pumps. Finally, the neuronal membrane at rest is described in terms of an equivalent electrical circuit, an important concept to understand its response to small polarizing inputs. In its simplest form this circuit consists of the membrane resistance and capacitance wired up in parallel.

### 7.4.1 Properties of Ions in Aqueous Solutions

A salt that has been dissolved in water will be mostly dissociated into anions and cations. Such a solution is *electrolytic*: A net movement of either type of charged particle into a certain direction will carry a corresponding electrical current. However, such a net movement is unlikely to happen in the absence of external influences: The principle of *electroneutrality* requires equal numbers of positive and negative charges within the bulk solution and also on a local scale; otherwise repulsive forces would disintegrate the solution. The local electroneutrality underlies the shielding of individual ions by polar water molecules that form the ion's hydration shell (see Sect. 7.2.3).

Ions in solution move constantly on random paths, as well as the water molecules surrounding them, because particles in a fluid experience *Brownian motion*. This motion also mediates *diffusion* along a concentration gradient (see Eq. 7.3/Box): if there are differing concentrations of particles in connected compartments, the second law of thermodynamics requires maximal disorder and the particles will mix evenly via diffusion.

### 7.4.2 Prerequisites for the Establishment of an Electrical Potential at a Membrane

There are two main requirements for the establishment of a potential at the cellular membrane:

1. Semipermeability of the membrane
2. Concentration gradient of the ion(s) that the membrane is permeable for

The differences in concentration are established by the activity of various pumps and transporters (see Chap. 6; Table 7.1). The semipermeability of cellular membranes originates in their structure: membranes consist of a lipid bilayer and integrated proteins. Because lipids are highly hydrophobic, ions cannot diffuse directly across the membrane. However, some of the membrane proteins function as channel pores that span the membrane. Most of these ion channels are selective (Sect. 7.2.3), allowing only for the passage of certain ions under certain conditions. *Semipermeability* means that a membrane is permeable only for a subset of the ions in the surrounding electrolyte. In neurons, ionic permeabilities depend on the membrane potential. At $V_{m\_rest}$ the cellular membrane is always permeable for $K^+$ ions via passive (also called leak or background) $K^+$ channels, while larger anions cannot pass. This conductance is mainly provided by the $K_{2P}$ channels (see Sect. 7.2.2).

Figure 7.9 illustrates how a potential is generated across a semipermeable membrane. We start with two aqueous solutions, namely a higher and a lower concentration of a potassium salt (e.g., $K^+Cl^-$, here $K^+A^-$). The respective $K^+$ concentrations are chosen such that they mimick $[K^+]$ in the intracellular and extracellular regime. Each solution is electrically neutral since it contains the same number of positive and negative charges: for every $K^+$ there is a 'partner' $A^-$. If we now pour the two solutions into one beaker (Fig. 7.9a *left*), the solutions will mix evenly within a fairly short time due to Brownian motion. Initial local concentration gradients will decline because of the diffusion of the ions, until the system equilibrates to the final mixture with a uniform ion concentration. Electroneutrality is not violated throughout the entire process.

Now if we put the solutions in the opposite compartments of a beaker separated by a semipermeable membrane that allows only $K^+$ to permeate but not $A^-$ (Fig. 7.9a *right*), the $K^+$ ions will follow their chemical

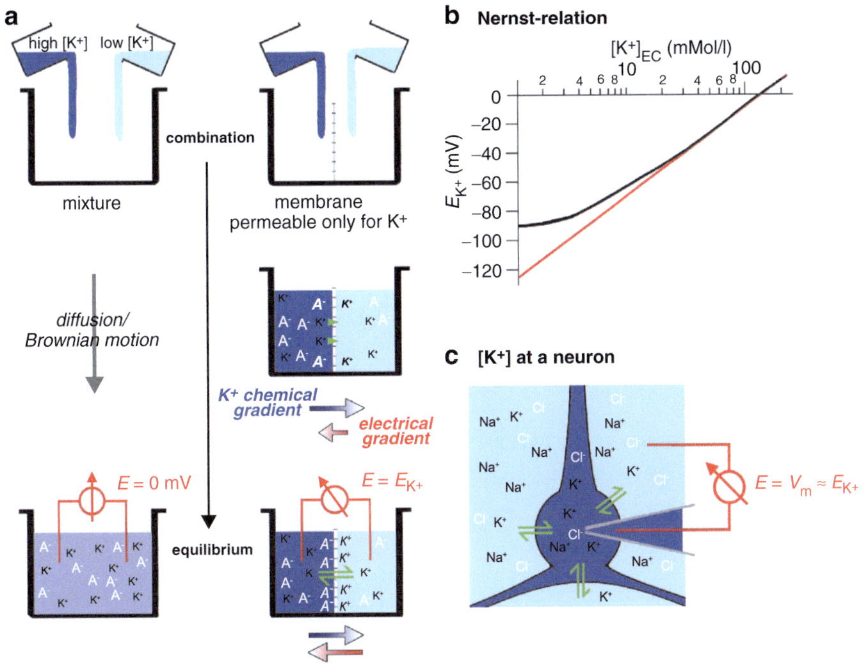


**Fig. 7.9 The origin of the resting membrane potential: the Nernst potential.** (**a**) A higher and a lower concentration of an aqueous solution of a $K^+$ salt (e.g., KCl, here KA), combined in two different manners. The respective $K^+$ concentrations mimick $[K^+]$ in the intracellular and extracellular regime. The *left column* shows the direct mixture. The two solutions mix evenly due to diffusion. At the bottom, there is equilibrium. Since for every $K^+$ there is a 'partner' $A^-$, there is no net charge difference between the two electrodes; no electrical potential can be detected. At the *right* the solutions are poured into two compartments that are separated by a membrane permeable only for $K^+$. Now the $K^+$ ions follow the chemical concentration gradient across the membrane, while their larger anionic partners $A^-$ are left behind (***separated partner ions***). This results in a charge separation at the membrane due to the accumulation of $A^-$ on the left side and $K^+$ on the right side, until the electrical gradient fully counteracts the chemical gradient in a dynamic equilibrium. Note that the bulk concentrations of $K^+$ in the two compartments are unchanged; the charge separation is restricted to the vicinity of the membrane. (**b**) The dependence of the Nernst equilibrium potential (*red line*; Eq. 7.4) on the extracellular $[K^+]_{EC}$ for a fixed intracellular $[K^+]_{IC} = 110$ mM. On this logarithmic scale $E_K$ appears linear. The actual resting potential of a neuron (*black line*) is less negative at physiological values of $[K^+]_{EC}$ due to extra membrane conductances, mostly a small $Na^+$ current. Since $Na^+$ currents are inward due to the positive $E_{Na}$, this effect shifts the actual resting potential towards more depolarized values in comparison to $E_K$. This shift can be quantified via Eq. 7.6 (see text). (**c**) The ionic composition surrounding a neuron. At rest, mainly $K^+$ ions cross the cellular membrane in dynamic equilibrium, whereas the other types of ions can barely pass. A recording pipette measures the potential gradient between the cytoplasm and the extracellular milieu. The situation is similar to the *right panel* in (a)

concentration gradient and flow towards the compartment with the lower concentration. If the particles were neutral, this flow would ultimately lead to a similar $K^+$ concentration in both compartments. But the net flow of $K^+$ across the membrane now results in a separation of charges, because for every $K^+$ that crosses the membrane an $A^-$ is left behind. Positive charges accumulate on the right side and negative charges on the left. This separation results in the build-up of an electrical gradient across the membrane (measurable as a potential with the voltmeter). In turn, it becomes more and more 'difficult' for $K^+$ ions to cross the membrane to the right: the negative charge accumulation on the left attracts the ions and thus retains them, while the positive charge accumulation on the right acts repulsively on the ions and thus also keeps them from moving to the right.

Ultimately there is a state of electrochemical balance with no more net movement of $K^+$ ions. The electrical gradient now fully counteracts the chemical gradient. The equilibrium is dynamic, i.e., $K^+$ ions still cross the membrane, but in equal numbers in every direction, so their net current is zero. The potential $E$ will settle at the Nernst potential $E_K$ (Eq. 7.4).

### 7.4.3 Electrochemical Balance at the Cellular Membrane: Nernst-Equation

Table 7.1 shows the relevant ion species concentrations and respective transport mechanisms and their $E_i$.

The Nernst equation entails that there is no net current of ion species i across the membrane if the membrane potential $V_m$ equals $E_i$. On the other hand, if $V_m$ has a value away from $E_i$, the ions i will flow to counteract this discrepancy, with a net current proportional to $(V_m - E_i)$. This concept is called the driving force, resulting in a modification of Ohm's law:

$$I_i = g_i\left(V_m - E_i\right) = \frac{V_m - E_i}{r_{mi}} \tag{7.5}$$

We have already met a similar relation regarding the current through single ion channels (Fig. 7.1, Eq. 7.1) with the reversal potential $V_{rev}$ instead of $E_i$.

**Derivation of Nernst electrochemical balance**

Thermal flow of particles along concentration gradient (CG; Fick's first law):

$$J_{CG} = -D_i \frac{dc_i}{dx} \tag{7.3}$$

Nernst-Einstein-Relation (links the diffusion of charged particles to their drift in an electric field):

$$D_i = \mu_i \frac{RT}{z_i F}$$

Electrical flow of ions along electrical gradient (EG):

$$J_{EG} = -\mu_i c_i \frac{dV}{dx}$$

Total flow:

$$J_{Total} = J_{CG} + J_{EG} = -D_i \frac{dc_i}{dx} - \mu_i c_i \frac{dV}{dx}$$

Equilibrium: no net flow of particles $J_{Total} = 0$

$$\frac{dV}{dx} = -\frac{RT}{z_i F}\frac{1}{c_i}\frac{dc_i}{dx}$$

Integration across cellular membrane:

$$V_m = E_i = \frac{RT}{z_i F}\ln\left(\frac{c_{iEC}}{c_{iIC}}\right) \tag{7.4}$$

**Nernst equilibrium potential for ion i**

Simplified version for $K^+$ at physiological temperature (37 °C):

$$E_K = 61\ \text{mV} \log\frac{\left[K^+\right]_{EC}}{\left[K^+\right]_{IC}}$$

Table of parameters and constants:

| Symbol | Name | Unit | Value |
|---|---|---|---|
| $J$ | Flow density of particles (not current!) | mol/(cm²·s) | |
| $D$ | Diffusion coefficient | cm²/s | Ions: ~$10^{-5}$ |
| $i$ | Ionic species | | |
| $c$ | Concentration of particles | M (or mol/l) | |
| $c_{EC}$ | Extracellular concentration | M | |
| $c_{IC}$ | Intracellular concentration | M | |
| $x$ | Distance | cm | |
| $\mu$ | Electrical mobility of particles in membrane | cm²/(Vs) | Relevant ions: 5–8·$10^{-4}$ |
| $R$ | Gas constant | J/(mol·K) | 8.31 |
| $T$ | Temperature | K | |
| $V$ | Potential | V | |
| $F$ | Faraday constant | C/mol | 9.65·$10^{4}$ |
| $z_i$ | Charge number of ion $i$ | | |
| $I$ | Current | A | |
| $E_i$ | Equilibrium potential for ion $i$ | V | |

**Table 7.1** Ion distribution, transporters, and Nernst equilibrium potential according to Eq. 7.4 in a skeletal muscle cell

| Ion | $c_{IC}$ (cytosol) (mM) | $c_{EC}$ (mM) | $E_i$ (mV) | Main pumps/transporters |
|---|---|---|---|---|
| $Na^+$ | 12 | 145 | +67 | $Na^+K^+$-ATPase |
| $K^+$ | 155 | 4 | −98 | $Na^+K^+$-ATPase |
| $Cl^-$ | 4 | 123 | −90 | NKCC, KCC, NCDBE |
| $Ca^{2+}$ | $1 \cdot 10^{-4}$ | 1.5 | +129 | $Na^+Ca^{2+}$-ATPase, SERCA |
| $HCO_3^-$ | 8 | 27 | | NCDBE |
| Other larger anions$^-$ | 155 | 5 | | |

Ionic distributions in the CNS are quite similar, except for the relative proportions of bicarbonate and the large anions. The sum of charges within each column – intracellular or extracellular compartment – amounts to zero, as required by the overall electroneutrality. Note that the composition of the extracellular fluid closely resembles a saline solution, dating from the evolution of cellular organisms in sea water

Equation 7.1 is the more general formulation: Since ion channels are not necessarily perfectly selective for a particular ion (e.g., many ligand-gated channels admit both $Na^+$ and $K^+$; chloride channels admit both $Cl^-$ and $HCO_3^-$), $V_{rev}$ denotes the potential with zero net current through the respective ion channel. Only for the case of a perfectly selective ion channel $E_i$ equals $V_{rev}$. Note also that Eq. 7.5 is linear only for non-voltage-dependent conductances $g_i$.

As stated above, the neuronal $V_{m\_rest}$ is largely governed by passive $K^+$ conductances and thus close to the Nernst potential $E_K = -98$ mV. If the cell is now depolarized to $V_m = -30$ mV, $K^+$ ions will flow out of the neuron to bring $V_m$ back towards $E_K$. Accordingly, if $V_m$ was −100 mV, $K^+$ ions would flow into the neuron. The initial net inward current in the latter case is smaller than the initial outward current in the former case, since the driving force ($V_m - E_K$) is much lower.

In spite of the net movement of $K^+$ ions across the membrane during the establishment of the electrochemical balance in Fig. 7.9a, the overall concentration gradient will remain the same because only a small fraction of the total number of ions participates in the charge separation at the membrane. The size of the fraction depends on the surface-to-volume ratio of the membranous structure; in a typical neuronal dendrite it is on the order of $10^{-5}$. Therefore the general principle of electroneutrality is violated only in the close vicinity of the membrane.

### 7.4.4 Beyond $E_K$: Additional Influences on the Resting Potential

The actual $V_{m,rest}$ of neurons is usually between −60 and −80 mV because of additional net depolarizing influences via small conductances for the other ion species. The resulting shift can be estimated via the summation of the I–V relationships (Eq. 7.5) for all the individual conductances present in the membrane (see also Fig. 7.10b; Sects. 7.4.5 and 7.5.1). If the total movement of charges across the membrane is in equilibrium, there is zero total current and no capacitive current. The membrane potential then equals the average of the Nernst potentials of the individual ions $i$ weighted with their conductances:

$$I_{\text{Total}} = \sum I_i = \sum \bar{g}_i (V_m - E_i) = V_m \sum \bar{g}_i - \sum \bar{g}_i E_i = 0$$

$$\Rightarrow V_m = \frac{\sum_i \bar{g}_i E_i}{\sum_i \bar{g}_i} \tag{7.6}$$

Note that in this equation the conductances comprise all available conductances for a certain ion and are therefore marked with bars (e.g., $\bar{g}_K$ contains both leak or passive conductances and voltage-gated channels: $\bar{g}_K = g_{KL} + g_K(V_m)$). Because of the equilibrium condition $V_m$ will settle at a constant value, so the individual voltage-gated conductances will assume fixed values and $V_m$ will be close to the Nernst potential of the ion with the largest conductance. For example, a situation that resembles the resting conditions such as $\bar{g}_{Na} = 0.2\,\bar{g}_K$, $\bar{g}_{Cl} = 0.1\,\bar{g}_K$, $\bar{g}_{Ca} = 0$ with their equilibrium potential taken from Table 7.1 yields $V_{m\_rest} = -72$ mV, close to $E_K$. For the origin of $\bar{g}_{Na}$ at rest see Sect. 7.5.1. Since there is no net current flow, $V_{m\_rest}$ equals the reversal potential $V_{rev}$ for the entire neuron.

Equation 7.6 also allows to estimate qualitatively $V_m$ at conditions away from rest. For example, during the upstroke of the action potential $\bar{g}_{Na}$ drastically increases whereas the voltage-gated fraction of $\bar{g}_K$ activates much more slowly (see below). So $\bar{g}_{Na} = 5\,\bar{g}_K$, $\bar{g}_{Cl} = 0.1\,\bar{g}_K$, $\bar{g}_{Ca} = 0$ yields $V_m = 37$ mV, predicting the positive overshoot that is characteristic of the $Na^+$ action potential and results from the positive value of $E_{Na}$ in conjunction with the high transient $\bar{g}_{Na}$.

As we have just seen there is also a small contribution of $g^{-}_{Cl}$ to $V_{m\_rest}$. The following example for the impact of $E_{Cl}$ on neuronal function also illustrates the role of transporters for the establishment of Nernst potentials. During the development of neurons in many species a shift in the relative expression of the ubiquitous KCC and NKCC1 cotransporters changes the relative extra- and intracellular $Cl^-$ concentrations from $E_{Cl} > V_{m\_rest}$ to $E_{Cl} < V_{m\_rest}$. So the sign of the driving force for $Cl^-$, $(V_{m\_rest} - E_{Cl})$, flips from negative to positive: in the young organism synaptic openings of $Cl^-$ channels result in depolarizing, outward $I_{Cl}$ instead of the adult hyperpolarizing, inward $I_{Cl}$. This early excitatory action of synaptically gated $Cl^-$ channels helps to promote neuronal activity, which in turn drives synaptogenesis and the formation of neuronal networks.

Finally, $V_{m\_rest}$ in a neuron is not only determined by the conductances and equilibrium potentials for $K^+$, $Na^+$, $Cl^-$, and $Ca^{2+}$ but also by the ongoing activity of the electrogenic $Na^+K^+$-ATPase. All in all, this pump removes cations from the cytoplasm (3 $Na^+$ ions out vs 2 $K^+$ ions in, see Chap. 6) and thus exerts a hyperpolarizing influence of ~−5 mV on $V_{m\_rest}$.

### 7.4.5 Passive Membrane: Equivalent Electrical Circuit – Role of Capacitance

The electrical responses of the cellular membrane can be modeled in terms of an equivalent circuit. This approach is very helpful to understand electrical signaling in neurons and can incorporate any type of additional electrical mechanism (see below for the treatment of the action potential). The so-called passive case without voltage-gated conductances suffices to explain the response of a neuron at rest to small de- or hyperpolarizing inputs. The equivalent circuit consists of a resistance and a capacitance in parallel (Fig. 7.10a).

Since ions cannot directly pass the phospholipid bilayer due to its hydrophobic interior, the ionic current depends on the presence of pores or ion channels. Each single channel open at rest is represented by a resistor $r_{SC}$, the inverse of its single channel conductance $g_{SC}$. All channels are wired up in parallel and together yield the membrane resistance $r_m$. The summed conductance of these channels is called leak conductance $g_L$:

$$g_L = \frac{1}{r_m} = \sum \frac{1}{r_{SC}} = \sum g_{SC}$$

The polar head groups of the membrane phospholipids can store charges, similar to the two plates of a plate type capacitor. The capacitance $c$ relates the amount of separated charges $q$ to the potential difference induced by them ($c = q/V$). For a plate type capacitor

$$c = \varepsilon\varepsilon_o \frac{A}{d} \tag{7.7}$$

with

| | | | |
|---|---|---|---|
| $\varepsilon$ | Polarizability | | |
| $\varepsilon_o$ | Polarizability of free space | $8.85 \cdot 10^{-12}$ | C/(Vm) |
| $A$ | area of plates | | $m^2$ |
| $d$ | distance of plates | | m |

For the lipid bilayer with $d \approx 2$ nm and $\varepsilon \approx 2$, the so-called specific capacitance $C_m$ of 1 $cm^2$ membrane area is about 1 $\mu F/cm^2$. This value is rather universal across cellular membranes. Specific units are commonly used to compare structures of different size – similarly, $R_m$ is the resistance of 1 $cm^2$ of membrane (unit $\Omega cm^2$).

How does the presence of the capacitance affect the polarization of the membrane? Since the membrane potential $V_m$ is proportional to the current across the membrane $I_m$ (Ohm's law), $V_m$ follows the same time course as $I_m$. If the cellular membrane was unable to store charges (no capacitance in the circuit, $I_C = 0$), then $V_m$ would be directly proportional to an injected current step $I_{inj}$ and follow the same rectangular shape (see Fig 7.10a, *right*). In the presence of the capacitance, the total injected current is the sum of the ionic current across the membrane $I_m$ and the capacitive current $I_C$: $I_{inj} = I = I_C + I_m$. $I_C$ represents the flow of charges onto or off the capacitor; its contribution to the total current flow is best illustrated by the discharging at the end of a depolarizing current pulse: The current injection has ended, but there is still current flowing off the capacitor until the charges accumulated during the loading at the beginning of the pulse have been unloaded. For a spherical cell, this decay of the current is exponential with the so-called membrane time constant $\tau_m = R_m C_m$. Similarly, during the onset of the injection the charging

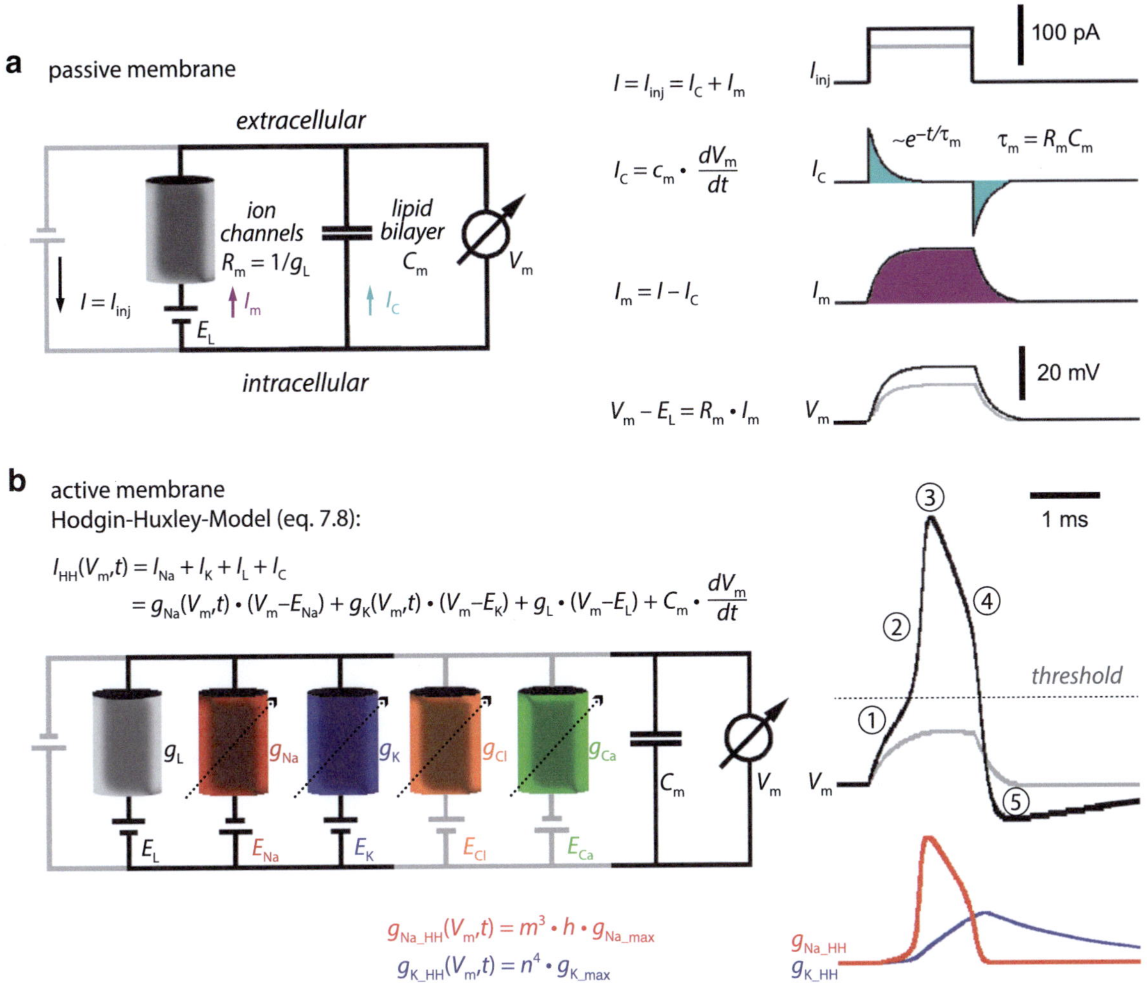


**Fig. 7.10** **Passive and active membranes.** The $Na^+$ action potential. *Left panels*: Equivalent circuits for passive and active membranes. In the passive case (panel **a**), the membrane is represented by a parallel circuit of the membrane resistance $R_m$ and capacitance $C_m$. The size of $R_m$ is determined by the density of open ion channels – the more channels, the lower the resistance (see Sect. 7.4.5). The ion channels open at rest are also called leak channels, and the inverse of $R_m$ is thus the leak conductance $g_L$. The associated battery $E_L$ represents the driving force for the ions that the leak is permeable for. $C_m$ arises from the ability of the polar head group of the bilayer phospholipids to store charges. The battery on the very left (*gray*) allows for external current injections $I_{inj}$. The voltmeter (*right*) is used to monitor the membrane potential $V_m$. In the active case (panel **b**), there are several active conductances $g_i$ whose voltage dependence is indicated by *slanted arrows*. The respective batteries represent the driving forces for the particular ions. All voltage-independent conductances open at rest fall into $g_L$ and $E_L$. The external battery on the *left* allows for current injections $I_{inj}$ to depolarize the circuit above threshold. Only the *black* parts of the circuit are part of the original HH model (Eq. 7.8). *Right panels*: Responses of the circuits to the same two depolarizing current injections (top trace, strong and weaker injection in *black* and *gray*, resp.). In the passive case (a), the total injected current $I_{inj}$ is the sum of the ionic current across the membrane $I_m$ and the capacitive current $I_C$ that represents the charges that flow onto or off the capacitor. The charging of the capacitor results in a gradual increase in the membrane current rather than an instant increase. In the case of active conductances (b), an AP is elicited once the depolarization due to the current injection surpasses the threshold (*1*) ($V_m \approx -50$ mV). At that point $g_{Na}$ increases substantially and the depolarization rises steeply during the upstroke (*2*) and reaches the overshoot (*3*). The subsequent repolarization (*4*) is mostly due to the inactivation of $Na^+$ channels and the increased conductance $g_K$. The afterpolarization in this neuron (*5*) is due to the prolonged opening of the $K_v$ channels. The very *bottom* shows the respective changes in $g_{Na}$ and $g_K$ according to the HH model. The responses shown were generated by the same HH model, except for the removal of $Na_v$ and $K_v$ conductances in the passive case

of the capacitor results in a gradual increase in the membrane current rather than an instant increase. Thus the presence of the membrane capacitance slows the time course of de- and repolarization, thereby affecting the conduction velocity of neuronal electrical signals within dendrites and axons (see Sect. 7.6).

The passive circuit can be used to model the electrical response of any type of cell (e.g., glial cells or epithelial cells). Neurons also contain active (i.e., voltage-gated) conductances that enable them to generate action potentials, thus the passive circuit is not sufficient.

Today, pharmacological tools such as the highly potent $Na_v$ channel blocker tetrodotoxin (a neurotoxin known from the order of Tetraodontiformes, that includes puffer fish) and various $K_v$ channel blockers allow a more direct observation of $g_{Na}(V_m)$ and $g_K(V_m)$. In addition, the dynamic clamp technique – a highspeed feed-back circuit that allows to manipulate $I_m$ and $V_m$ in parallel – can help to selectively investigate conductances of interest.

## 7.5 Regenerative Activity: $Na^+$ and $Ca^{2+}$ Action Potentials

The $Na^+$ action potential arises from the tight sequential activation of two voltage-gated conductances – $g_{Na}(V_m)$ and $g_K(V_m)$. This groundbreaking concept was first established in the 1930s by Hodgkin & Huxley (HH; [14, 15, 16] Nobel Prize 1963). Their success is due to the combination of clever experimentation and biophysical modelling, allowing them to predict many essential features of the action potential (AP), such as the existence of selective ion channels that underlie these conductances. The HH model has become canonical, with only a few details that are being challenged.

Many classes of neurons also feature a regenerative $Ca^{2+}$ spike, a signal arising from the activation of voltage-dependent, low-threshold $Ca^{2+}$ channels. These $Ca^{2+}$ spikes often contribute to neuronal pacemaking.

Hodgkin and Huxley recorded from the giant axon of the squid (*Loligo*). Its diameter is ~1 mm (vs 1–10 μm for most axons in vertebrate nervous systems). So HH were able to insert wire electrodes into a piece of axon from both sides and apply the voltage-clamp technique (see also Sect. 7.1.2): The voltage within the axon is controlled via one electrode and the current resulting from voltage changes is measured via the other using a feed-back circuit that injects current to keep up the preset voltage. HH dissected $g_{Na}(V_m)$ and $g_K(V_m)$ by substituting all $Na^+$ ions with membrane-impermeant choline and subtracting the currents measured under physiological and substituted conditions.

### 7.5.1 $Na^+$ Action Potential

The HH experiments and their interpretation revealed the special nature of the action potential that relies on a cooperative, avalanche-like opening of $Na_v$ channels once the membrane has been depolarized past the threshold. At threshold, the open probability $P_o$ of $Na_v$ channels increases so that progressively more channels open and as a result the potential depolarizes further – a sort of self-excitation. Next, the inactivation of $Na_v$ channels (Fig. 7.7) and the increased $P_o$ of $K_v$ channels cause a repolarization that is almost as swift as the upstroke, limiting the total duration of the AP to <1 ms. The entire process is highly nonlinear and stereotyped and cannot be reversed once the AP threshold is crossed. Therefore the AP is often called an all-or-nothing event. In addition, its amplitude does not depend on the total magnitude of the depolarizing input but always attains the same value. Thus the AP is a binary signal, unlike the postsynaptic or sensory potentials that encode the strength of the stimulus via their amplitude.

Note that the AP threshold is *not* an absolute threshold for the opening of $Na_v$ channels, rather a threshold for the opening of a sufficiently large number of $Na_v$ channels to trigger the avalanche. Due to the broad distribution of open probabilities across $V_m$ (e.g., Fig 7.1b) and thermal noise, a few $Na_v$ channels are also open at $V_{m\_rest}$. This effect is also responsible for the small $Na_v$ conductance that depolarizes the resting potential away from $E_K$ (see also Sect. 7.4.4).

Important features of the HH model are illustrated in Fig 7.10b. While Fig. 7.10a shows the passive case – i.e., an equivalent circuit for a purely passive membrane without voltage-dependent mechanisms – in active membranes there are several voltage-gated conductances $g(V_m)$ operating in parallel. The respective batteries represent the driving forces for the particular ions at the membrane potential. The external battery on the left allows for current injections $I_{inj}$ to depolarize the circuit above threshold. In a real-world neuron, these depolarizations would be generated by the summation of a sufficient amount of excitatory postsynaptic potentials or sensory potentials at the site of AP initiation, usually the axon initial segment.

The original HH model contains the following conductances: a leak conductance $g_L$, that comprises all voltage-independent conductances, no matter for which ion, and the voltage- and time-dependent $g_{Na}(V_m, t)$ and $g_K(V_m, t)$ (see Eq. 7.8 and the scheme in Fig. 7.10b without $g_{Cl}$ and $g_{Ca}$). HH described the voltage-dependence using maximal conductance values $g_{Na\_max}$ and $g_{K\_max}$ that are modulated by the voltage- and time-dependent gating of the respective channels. The gating occurs stochastically, based on probabilities for the proposed individual gating steps n, m, and h that are required to activate or to inactivate the channel (n: activation of $g_K$; m: activation of $g_{Na}$; 1 – h: inactivation of $g_{Na}$). Based on the fits of their data, HH predicted a dependence of $g_{K\_HH}(V_m, t)$ on $n^4$. The fourth power of n means that the same gating step has to occur four times in a row to fully open the channel. $g_{Na\_HH}(V_m, t)$ depends on $m^3h$, where m and h are independent processes.

The main success of the HH model includes the explanation of the time course of the AP in terms of conductance changes and the prediction of the existence of selective, voltage-gated ion channels. Modern molecular biological and structural data have revealed that indeed four identical elements, the S4 segments in each part of the tetramer, have to perform the same movement to open the $K_v$ channel (see Sect. 7.2.4, Fig. 7.3) – precisely the predicted $n^4$-dependence, whereas the situation is more complicated for $Na^+$ channel gating (see also below). Moreover, the sequence of activation and inactivation of $Na_v$ channels explains the refractory period of several milliseconds following each AP (Fig. 7.7): no more (full) action potentials can be elicited because there are not yet enough $Na^+$ channels available. Because of this refractoriness, APs cannot double back at the axon terminals or cross each other (see also Fig. 7.14). The activation-inactivation sequence also may result in repetitive AP firing at a fixed frequency, often in conjunction with the rhythmic activation of other conductances that control the subthreshold regime (see Sect. 7.5.3).

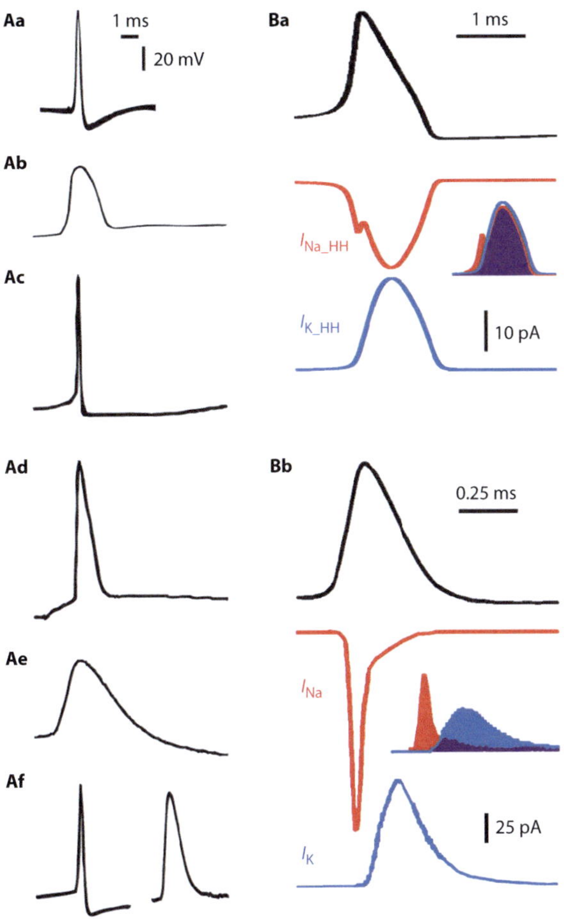


**Fig. 7.11** **$Na^+$ action potential waveforms in different species and neuronal types.** Differences in sequence of $Na^+$ current and $K^+$ current activation. (**A**) Different $Na^+$ AP waveforms. Afterpolarizations (the phase following the $Na^+$ spike and its fast repolarization) often depend on the resting membrane potential before the spike and may be quite variable. All APs have the same scale. (Aa) Squid giant axon of motoneuron (*Loligo*) by Hodgkin and Huxley ([14] with permission). (Ab) Jellyfish giant axon of motoneuron, *Aglantha* (From Greenspan [9] with permission). (Ac–f) Somatic recordings from rat and mouse neurons ((c–e) modified from Bean [4] with permission). (c) Inhibitory cerebellar Purkinje neuron. (d) Excitatory hippocampal CA1 pyramidal neuron. (e) Dopaminergic midbrain neuron. (f) *Left*: fast-spiking, inhibitory cortical neuron). *Right*: regular spiking, excitatory cortical neuron (Both from [12] with permission). (**B**) Comparison of sequential activation of $Na^+$ and $K^+$ currents in squid axon and mammalian mossy fiber. The two APs have the same voltage scale as the APs in (**A**). Along the time axis, the two APs were scaled to roughly the same half width. The *small insets* demonstrate the overlap between the currents (*white area*); *red*, $Na^+$ current; *blue*, $K^+$ current. (Ba) AP and $Na^+$ and $K^+$ currents in squid giant axon according to a full Hodgkin-Huxley model (Modified from Koch [19] with permission). (Bb) AP and currents recorded from a mammalian hippocampal mossy fiber bouton (Modified from Alle et al. [1], with permission)

Like any model, HH is based on some simplifications that may not quite hold in all types of neurons. In contrast to HH, inactivation was found experimentally to depend on activation, and modern kinetic models thus require additional $Na_v$ channel states. Moreover, there are usually many more $Na_v$ and $K_v$ channel types present in an axon instead of just one $I_{Na}$ and one $I_K$ (see Sect. 7.2.2). This multitude of possible combinations is reflected in the variable shapes of $Na^+$ APs, across both phyla and the neuronal classes within one species. A small selection is shown in Fig. 7.11A, including the classical recording by HH. While all APs show a fast upstroke due to $Na_v$ channel activation, their width differs widely. For example, action potentials in neocortical pyramidal neurons are wider (>1 ms) than in inhibitory neurons in the same brain region (<1 ms), allowing to discriminate the responses of both neuronal types in extracellular *in vivo* recordings. Moreover, the afterhyperpolarization seen in squid is by no means universal; many different types of afterhyperpolarization on fast and slow timescales are known that mostly depend on the presence of certain $K_v$ channel subtypes. Afterdepolarizations are also very common, often involving persistent $Na^+$ currents and/or $Ca^{2+}$ currents or $Ca^{2+}$-activated $K^+$currents ($K_{Ca}$, Sect. 7.2.2).

It has been noted early on that the time course of the activation of $g_K(V_m, t)$ according to HH is suboptimal with respect to energy expenditure: Because $I_{K\_HH}$ substantially overlaps with $I_{Na\_HH}$, positive charges are flowing in and out of the cell at the same time (Fig. 7.11b). All in all, four times more ions are flowing than required to simply depolarize the membrane to overshoot levels. All these extra ions need also to be pumped back via the $Na^+$-$K^+$-ATPase. Indeed, recent data indicate that the activation of $I_K$ relative to $I_{Na}$ is likely to occur more delayed in mammalian neurons and thus APs may cost far less.

### 7.5.2 $Ca^{2+}$ Action Potential or Low-Threshold Spike

An important requirement for regenerative spiking is the window current that shows between the activation and inactivation curve of the $Na_v$ channel (Fig. 7.7b). Such a window also exists for T-type $Ca^{2+}$ channels ($Ca_v$3.1,2,3, Fig. 7.3g) that are present predominantly in the dendritic membranes of many types of neurons. Substantial activation of T-type $Ca^{2+}$ channels begins well below the threshold for $Na^+$ APs (< −50 mV), boosts itself in a cooperative manner similar to $Na_v$ channels, and generates a so-called $Ca^{2+}$ or low-threshold spike. These $Ca^{2+}$ spikes often originate in dendrites and are several dozens of ms wide, appearing as characteristic humps in voltage recordings. A classical example of $Ca^{2+}$ spiking occurs in thalamic relay neurons, and some invertebrates also rely on $Ca^{2+}$ spikes alone for signaling – e.g., *Paramecium* or a special neuron in *C. elegans* (Fig. 7.12a, b).

While $Ca^{2+}$ spikes are obviously less suitable for fast signaling than $Na^+$ spikes, they expand the processing capabilities of neurons. Since substantial increases in dendritic $[Ca^{2+}]_{IC}$ can induce synaptic long-term plasiticity, $Ca^{2+}$ spikes may serve to trigger synaptic plasticity (see Chap. 26). In mammals, the dendritic origin of $Ca^{2+}$ spikes allows for specialized spatiotemporal patterns of activity within individual neurons with long dendrites such as cortical or hippocampal pyramidal cells. For example, $Ca^{2+}$ spikes may entail a somatic burst of $Na^+$ spikes (Fig. 7.12c). Bursts are generally believed to enhance the reliability of synaptic transmission and are of special importance in the thalamocortical, intracortical, and corticomotor communication but also have been observed in invertebrates such as *Aplysia*. Related phenomena are the so-called UP-states of network activity – long-lasting depolarizations that often begin with a burst and again are known from the mammalian thalamus and cortex. These levels of activity can be correlated to behavioral states and network oscillations (see below). Finally, $Ca^{2+}$ spikes are also often essential in promoting rhythmic activity of neuronal pacemakers. The proper timing of many repetitive functions in organisms is ensured by pacemaking neuronal clocks. Whereas circadian pacemakers require protein biosynthesis (see Chap. 27), pacemakers operating in the millisecond to second regime rely on the rhythmic interplay of several conductances including T-type $Ca^{2+}$ currents. Another conductance that contributes to pacemaking in some classes of neurons is the current $I_{HCN}$ (Fig. 7.4, Sect. 7.2.2). These interactions have been explored in particular in the mammalian thalamus, as shown in Fig. 7.12d, e.

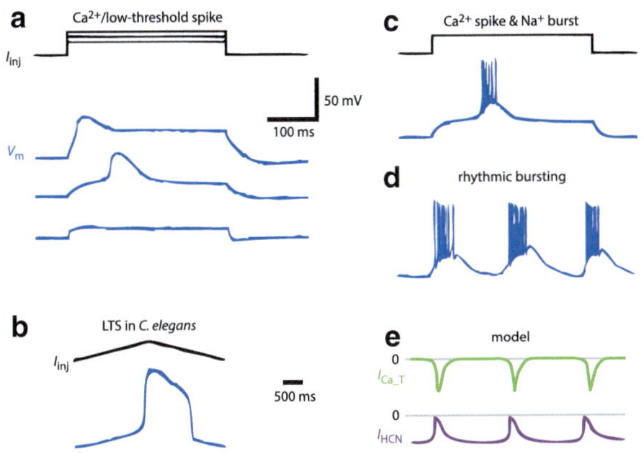


**Fig. 7.12** **$Ca^{2+}$ spikes & pacemakers.** Throughout the figure, current injections are colored in *black* and voltage recordings from neurons in *blue*. (**a**) Activation of the T-type $Ca^{2+}$ current in a mammalian thalamic relay neuron by increasing current injections. The T-type current occurs at depolarizations of ~ −50 mV and shows as a broad $Ca^{2+}$ spike (a.k.a. low-threshold spike, LTS). With increased current injection the spike is elicited earlier on and moves to the beginning of the depolarization. $Na^+$ channels and thus $g_{Na}$ are blocked with TTX (Modified from Zhan et al. [31] with permission). (**b**) $Ca^{2+}$ spike recorded from the so-called RMD neuron of *C. elegans* (Modified from Mellem et al. [23] with permission). Here the spike was activated with a current ramp instead of a step. Note the slow time scale; voltage scale as in (a). (**c**) Combination of $Ca^{2+}$ spikes and $Na^+$ spikes in a burst, i.e., a high-frequency volley of $Na^+$ APs. Burst recorded from the same neuron as in (a) before blockade of $g_{Na}$. Same scaling as in (a). (**d**) Recording of rhythmic activity from a thalamic reticular neuron *in vivo* (no external current injection!). Same scaling as in (a, c) (Modified from Mulle et al. [25] with permission). (**e**) Model of pacemaking as shown in (d) based on McCormick and Huguenard [22] with permission. Not all involved conductances are shown. $I_{HCN}$ is activated by hyperpolarization and results in a slow depolarization that ultimately leads to the next burst. The burst inactivates $I_{Ca}$ and $I_{HCN}$; repolarization is achieved via $K^+$ currents $I_K$ as in the classical HH model

## 7.6 Conduction of Electricity in Neurons

The properties of the passive propagation of electrical signals are essential for conduction in both dendrites and axons – in axons, if a depolarization is subthreshold for the opening of voltage-gated channels or if there are only few active channels present as in the internodal sections of myelinated axons. Here, we use 'neurite' to encompass axons and dendrites. The passive propagation of signals in neurites can be described in terms of conduction in a cable. Regenerative or active conduction requires a sufficient density of voltage-dependent conductances and signals large enough to activate these conductances. It may also occur in dendrites, for example in the case of dendritic $Ca^{2+}$ spikes. APs that propagate backwards into dendrites are also often supported by dendritic voltage-gated conductances. In vertebrate axons, conduction is often accelerated by means of myelination, mainly due to a decrease in membrane capacitance.

### 7.6.1 Passive Conduction

Due to the loss of charges across the leaky membrane, a local depolarization $V_0$ of a neurite will exponentially decline with distance from the generation site $x_0$, similar to the pressure in a leaky garden hose as shown in Fig. 7.13. The decay occurs in both directions along the cable and is quantified by the length constant λ that depends on the passive parameters $R_m$ (specific resistance across the membrane, see below), $R_i$ (specific resistance along the neurite) and its diameter $a$. For a cylindrical neurite:

$$\lambda = \sqrt{\frac{aR_m}{R_i}} \qquad (7.9)$$

> **Compartmental Models**
> In linear cable theory, neuronal structures are represented by a series of equivalent circuits like the one shown in Fig. 7.10 that are interconnected by axial resistances $R_i$. Any neuronal structure including the soma, branching dendrites, and axon collaterals can be represented by such a compartmental model (see Chap. 30). Linear differential equations link the voltage changes and currents between the compartments. Compartmental models allow to investigate the conduction of electrical signals in greater detail, in particular if combined with experimental techniques such as dendritic electrophysiological recordings and wide-field observation of optically visualized dendritic signals at high temporal resolution.

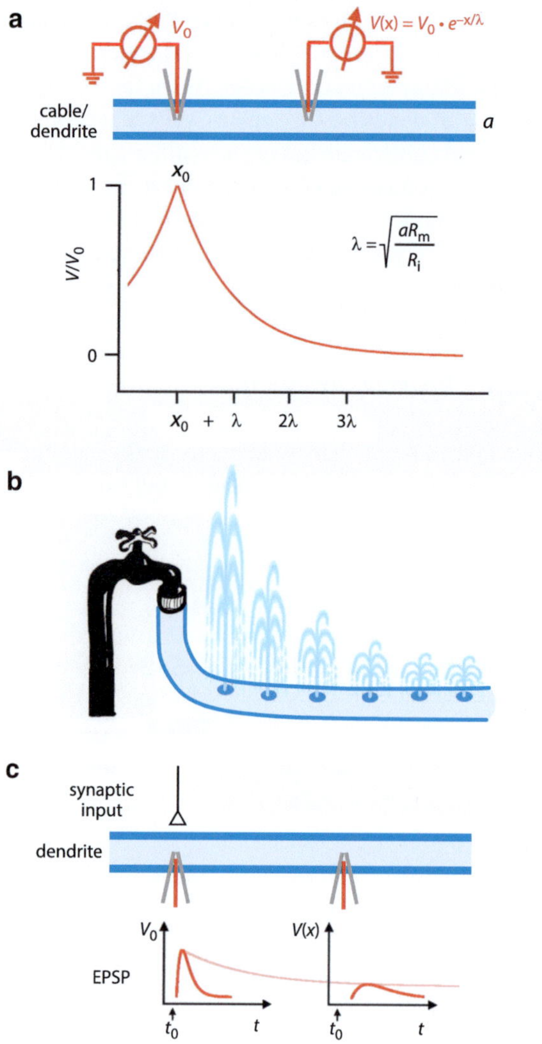


**Fig. 7.13** **Passive or electrotonical conduction.** (**a**) Cylindrical cable/dendrite of diameter $a$. Due to the loss of charges across the leaky membrane, a local depolarization $V_0$ (generated, e.g., by current injection or a synaptic input) will exponentially decline with distance $x$ from the generation site $x_0$ in both directions along the dendrite. The decay is determined by the length constant λ that depends on the passive parameters $R_m$, $R_i$, and $a$ (see text). (**b**) Analogy between loss of charges across the dendritic membrane and loss of water from a leaky garden hose. The loss causes the water pressure (corresponding to the potential $V(x)$) to decline exponentially which is reflected in the reduced maximal height of the fountainlets emerging from the hose. Similarly, the larger the hose is in diameter, the less loss of water relative to the total amount of flowing water will occur and thus more water can be piped over a certain distance. In other words, the surface-to-volume ratio becomes more favorable with increasing diameter (~$2/a$). (**c**) The attenuation of synaptic input as an example for the functional importance of electrotonic loss: a synapse is activated at the time $t_0$. The ensuing excitatory postsynaptic potential is measured by dendritic recordings, one close to the synapse and one at some distance. Relative to the potential at the synapse the more distantly measured potential is decreased in amplitude, occurs with a delay, and is broadened due to temporal filtering

Again, $R_m$ is the specific resistance of 1 cm$^2$ of membrane (unit Ωcm$^2$; range 1,000–50,000 Ωcm$^2$). The higher $R_m$, the better the neurite is isolated, so the larger λ. $R_m$ can be dynamically variable, for example due to active inputs that open ion channels and thus intermittently decrease the local membrane resistance. $R_i$ is the specific axial resistance of a cylindrical piece of neurite of 1 cm length and cross-sectional area of 1 cm$^2$ (unit Ωcm), so that $R_i/a$ is the actual resistance along 1 cm length. In contrast to $R_m$, its value is more stable and depends on the ionic composition of the cytoplasm ($R_{i\_cytoplasm} \approx 30$ Ωcm) and the presence of cellular organelles and other structures in the neurite that reduces the effective cross section for charge conduction and thereby increases $R_i$ (in mammals: $R_i \approx 200$ Ωcm). Since λ is established under conditions of steady-state, the membrane capacitance $C_m$ is not relevant here. Dendritic values of λ are on the order of 100–1,000 μm. The squid giant axon boasts a λ of 10 mm due to its large diameter, while for thin unmyelinated mammalian axons λ is about 100 μm. For the influence of myelin on λ see below.

Functionally, passive conduction results in a severe dampening of signals travelling along dendrites. In the absence of active mechanisms, the amplitude of a postsynaptic potential will decline substantially with distance from the synaptic input location. Thus, postsynaptic potentials from more remote synapses would be much less influential on synaptic integration at the soma and eventual AP generation than postsynaptic potentials from more proximal synapses, violating 'synaptic democracy' (Fig. 7.13c). This effect may be counteracted by so-called 'synaptic scaling' as described for mammalian hippocampal pyramidal cells, where more remote synapses generate larger postsynaptic potentials – or used as a computational feature to regulate the efficiency of inputs (see Chap. 30). More examples for the role of passive conduction see Figs. 7.15 and 7.16.

The conduction velocity $v_C$ of electrical signals in a passive neurite is proportional to $\lambda/\tau_m$. The smaller the current losses and the faster the charging of the membrane capacitance occurs, the faster a signal can travel (mammalian dendrites: $v_C \approx 0.1$–1 m/s).

### 7.6.2 Active Conduction and Conduction in Myelinated Axons

The active conduction of action potentials in an axon requires a sufficiently high density of

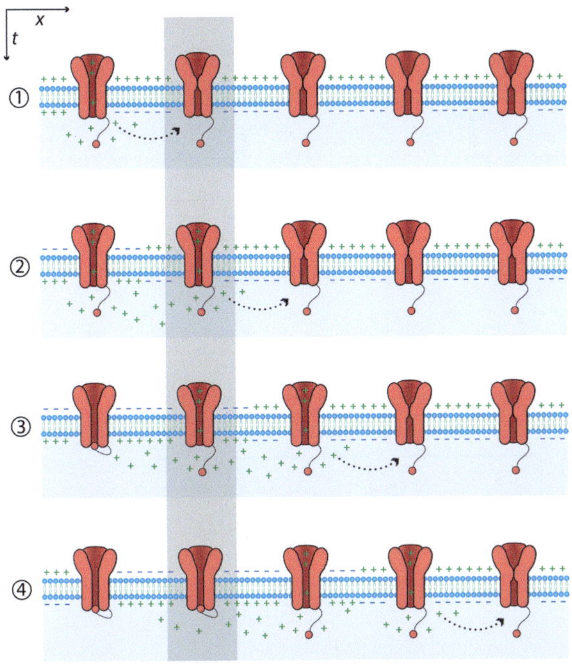


**Fig. 7.14 Active conduction in nonmyelinated axons.** Longitudinal cut through an axonal membrane. Nonmyelinated axons feature a homogenous density of $Na_v$ channels. For sake of clarity, $K^+$ channels are not shown. At the level of the second $Na_v$ channel from the left highlighted in *gray*, the numbers correspond to the stages of the action potential as shown in Fig. 7.10, *bottom right*, and also correspond to the states of the $Na_v$ channel shown in Fig. 7.7b. (*1*) The very right $Na_v$ channel in the membrane senses the depolarization associated with the upstroke of an action potential in the left adjacent membrane segment and opens. The positive charges that enter the cell via the channel spread passively, resulting in a local depolarization. (*2*) This local depolarization spreads also to the right, opening the next channel. At the same time, more charges entering via the first and second channel further depolarize the local membrane (upstroke). (*3*) The whole process moves further to the right. The additional charges entering via the second channel cause a local reversal in the membrane potential (overshoot). The first channel from the right inactivates. (*4*) Again, the whole process moves further to the right. The second channel becomes inactivated and the inward current flow stops. Repolarization will be achieved via other conductances ($g_K$, $g_{Cl}$) that are not shown here

$Na_v$ channels – first to reach the threshold for AP initiation and then to regenerate the full AP within the adjacent membrane segment. Even in the case of active conduction, note that at the microscopic level the depolarization spreads *passively* within the axon from one $Na_v$ channel to the next, opening channels in the neighboring membrane segments – more $Na^+$ ions flow in, further depolarize the membrane locally and in addition provide depolarization to the adjacent membrane segment in the direction of propagation, as shown in Fig 7.14. Thus, in the active region the 'wave front' of activity spreads farther to the left via local current loops. In the wake of the wave, $Na_v$ channels become fully inactivated and $K_v$ channels (not shown in Fig 7.14) strongly activate, cutting short the duration of local depolarizations. As already mentioned, the inactivation of $Na_v$ channels renders the excited membrane segment refractory, preventing APs from doubling back at any point.

Active conduction is not *per se* faster than passive conduction, because the gating of the involved ion channels requires time due to its stochastic nature (Sect. 7.1.1, Figs. 7.6 and 7.7a). Active $v_C$ is proportional to the temperature, because the conformational changes of the channels are strongly temperature-dependent, and also proportional to the axon diameter. Characteristically, unmyelinated axonal fibers in mammals feature a diameter of 1 μm and a fairly slow conduction velocity $v_C \approx 1$ m/s. The famous squid giant axon obtains a respectable 60 m/s due to its large diameter of up to 1 mm. The neuron that gives rise to this axon mediates an escape response and thus a high $v_C$ is essential for survival. However, this solution would not work in vertebrates – imagine the size of a spinal cord with millions of 1-mm-thick axons running in parallel. Instead, fast conduction in thin fibers is achieved by means of myelination.

The myelin sheath is formed by specialized glial cells and can be found in all vertebrates but also in some oligochaetes, shrimp, and copepods (see Chap. 9). In vertebrates, glial cells (Schwann cells in the PNS and oligodendrocytes in CNS) are tightly wrapped around the axon forming 20–300 membrane sheets, like a ribbon wound up on a reel or an insulating tape glued around a bare wire. Neighboring ribbons – the internodes – are separated by the nodes of Ranvier that expose the bare axon. Nodes are 1–2 μm wide, internodes are longer by a factor of about 1,000. The high density of $Na_v$ channels required to regenerate the AP is restricted to the node.

Myelination greatly accelerates the conduction as illustrated in Fig. 7.15a. The capacitance is reduced in comparison to the unmyelinated case because the 'plates' separating the stored charges move further apart (see Eq. 7.7). The myelin layers can be also regarded as a serial connection of the capacitors of each membrane layer, yielding the same result: the total $C_m$ is reduced and therewith the membrane time

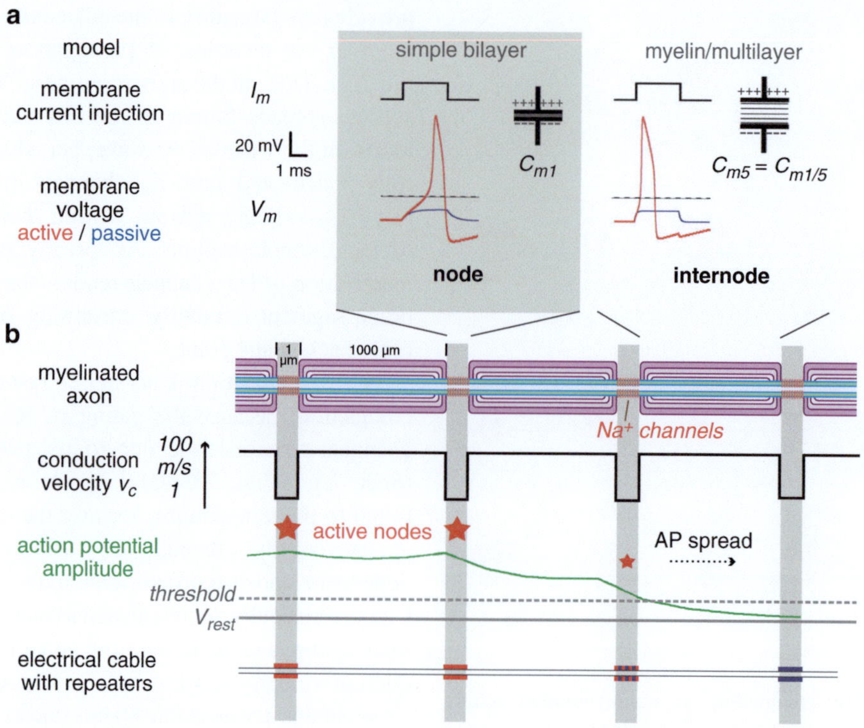


**Fig 7.15** **Conduction in myelinated axons.** (**a**) Main effect of myelination on conduction velocity: myelination reduces the membrane capacitance $C_m$ and thus speeds up action potential propagation. For sake of simplicity, this effect is shown in the absence of any other effects of myelin on the passive properties of the axonal membrane (see text). The *left side* shows the response of an unmyelinated compartment to a rectangular current pulse, in the passive case (*blue*; model contains no voltage-gated conductances, see also Fig. 7.10) and the active case (*red*; with $Na_v$ and $K_v$ channels). The *right side* shows the same current injection into the compartment with a capacitance that is reduced five times with respect to the nonmyelinated case due to four extra layers of myelin that cover the membrane. The reduction in $C_m$ allows for a much faster depolarization. (**b**) This panel illustrates the nature of saltatory propagation from one node of Ranvier to the next. As above, nodes are highlighted in *gray*. *First row*: Schematic depiction of a myelinated axon. The nodes are ~1 µm long, the internodes ~1,000 µm. $Na_v$ channels are located exclusively at the nodes and $K_v$ channels (not shown) at the nodes and their immediate vicinity. *Second row*: Discontinuous jumps in conduction velocity between nodes and internodes. Along the node, the AP propagates actively (Fig. 7.14) with a conduction velocity on the order of 1 m/s. Along the internode, the AP spreads passively (Fig. 7.13) but with increased speed because of the reduced capacitance (see (a)). Note that many nodes are excited at the same time (see Fig. 7.10). *Third row*: The amplitude of the action potential during its propagation across several nodes (*green line*; snapshot at one point in time). The forefront of the AP moves to the right in a purely passive manner. Once a node becomes depolarized beyond threshold, it begins to admit current. The ensuing additional depolarization spreads electrotonically both in the direction of propagation and backwards (see also Fig. 7.13). Thus the amplitude of the AP between two fully excited nodes is the superposition of the electrotonic contributions from these two nodes. This explains the tent-like shape of the amplitude curve (see [8]). The complete spatial amplitude curve resembles the temporal shape of the AP but mirrored horizontally: the right forefront of the AP along the spatial axis corresponds to the left initial depolarization to threshold in time, i.e., the upstroke of the AP shown in Fig. 7.10b. *Fourth row*: Since the voltage-gated channels are restricted to the immediate vicinity of the nodes, the AP is only fully regenerated at the nodes. In analogy, long electrical cables such as transatlantic telephone lines use repeaters to restore the signal to its original amplitude (every 50 km in copper cables; see *bottom row*; repeaters colored in *red* are active)

constant $\tau_m = R_m C_m$ is reduced along the internode, increasing $v_C \sim \lambda/\tau_m$. The myelinated axonal membrane can be depolarized much more quickly and thus the AP can spread faster between the nodes. Although the membrane resistance $R_m$ increases with myelin insulation, this increase does not fully counteract the effect of the reduced $C_m$ on $\tau_m$ and $v_C$. First, $R_m$ of the myelin appears to be smaller than $R_m$ of the axonal membrane because of extra leak conductances. Second, an increasing $R_m$ also benefits $v_C$ via an increase in the length constant (Eq. 7.9), so more charges are available for depolarization in comparison to the unmyelinated case. Taken together, myelination increases $v_C$ by a factor of ~10.

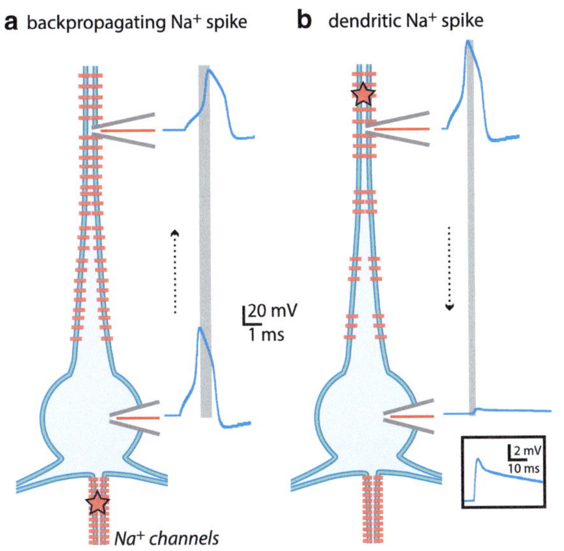


**Fig 7.16 Backpropagating Na$^+$ action potentials and dendritic Na$^+$ spikes.** (**a**) Schematic pyramidal neuron with long apical dendrite. The AP originating at the axon hillock (indicated by the red star) will propagate both forwards into the axon and backwards into the dendrite. Within the dendrite, propagation is often supported by dendritic $Na_v$ channels. Such backpropagating APs can be observed via dendritic recordings. The dendritic AP will be delayed and broadened with respect to the somatic AP because of the conduction along the dendrite. (**b**) Na$^+$ spikes may also originate in dendrites if there are regions with a sufficiently high density of $Na_v$ channels (*red star*), for example, in mammalian pyramidal neurons. Usually, the density of Na$^+$ channels declines away from the dendritic AP generation site and there is a strong drop in input resistance at the dendrite-soma junction. Thus the dendritic AP cannot sustain its full amplitude and will be registered at the soma as a so-called spikelet with a characteristic shape (shown in higher magnification in the *inset*). Here, the delay is reversed between dendritic and somatic recording

This acceleration notwithstanding the term 'saltatory conduction' is potentially misleading. The excitation is *not* hopping from node to node, since the depolarization is spreading continuously within the axon. At any point in time there are always several excited nodes (Fig. 7.15b). For example, in a fast fiber with 100 m/s conduction velocity and internodes of length 2 mm, 50 adjacent nodes will be excited simultaneously during an AP (1 ms). The term 'saltatory' rather refers to the discontinuous jumps in conduction velocity between the nodes and internode regions. In the illustrated case, the conduction time along the entire internode is just about 20 μs vs 2 μs for the much shorter node. In front of the AP upstroke the depolarization spreads passively along the axon. In the internodal regions between any pair of fully excited nodes the amplitude of the AP is barely diminished, because the local depolarization is fed by the superposition of the passively spreading action potentials generated at the adjacent nodes (Fig. 7.15b).

Another major benefit of myelination is the reduction of energy consumption per conducted action potential. Since the excitable membrane area is limited to the small nodes, much less Na$^+$ and K$^+$ ions need to be pumped to bring the concentration back to the resting values following the AP. Myelination decreases also the space requirements for fast AP propagation since unmyelinated fibers with the same conduction velocity are far larger in diameter than myelinated ones.

The length of the individual internodes is constrained by the requirement that the passively spread AP still reaches threshold upon arrival at the next node. There even is a built-in safety factor, since experimental blockade of $Na_v$ channels at one node does not prevent propagation. On the other hand, AP propagation failures have been observed across a wide range of organisms from the leech to the rat, mostly at axonal branch points and in the course of repetitive activity at moderate frequencies (from 10 Hz upwards; see [5]).

### 7.6.3 Passive and Active Conduction of Action Potentials in Dendrites

Not only do action potentials propagate actively along the axon, they also spread backwards into dendrites, a phenomenon that is called backpropagation (see [29]; Fig. 7.16a). Due to the passive spread these backpropagating APs decrease in amplitude and broaden temporally; the dendritic conduction velocity is lower than that of the myelinated axon and thus will cause a propagation delay on the order of 0.2 ms per 100 μm dendritic length. While at a first glance passive backpropagation seems like an epiphenomenon of neuronal activity it actually serves functional roles:

Backpropagating spikes can push plasticity mechanisms via membrane depolarizations that enhance certain synaptic conductances such as NMDA receptor channels (see Chaps. 8 and 26) and thus admit additional $Ca^{2+}$. Proof of the biological importance of backpropagation is the fact that the passively conducted AP is often boosted by active conductances in the dendrites. These conductances include pretty much all of the voltage-gated ion channels described in Sect. 7.2.3 – i.e., $Na_v$, $K_v$, $Ca_v$, and also HCN channels. Their overall density is lower than in the active parts of axons, and their distribution along the dendrites may be quite inhomogenous.

Finally, $Na^+$ spikes may originate not only at the axon initial segment but also in dendritic regions with a sufficiently high density of $Na_v$ channels, similar to dendritic $Ca^{2+}$ spikes. Often the density of $Na_v$ channels declines away from the dendritic AP generation site; moreover, there is a strong drop in input resistance at the dendrite-soma junction. Under these circumstances the dendritic AP cannot sustain its full amplitude and will be registered at the soma as a so-called spikelet with a characteristic shape (Fig. 7.16b). Signals with a similar appearance may also be generated by the spread of $Na^+$ APs between neurons via dendritic or axonal gap junctions (Chap. 8). Spikelets are known mainly from mammalian pyramidal neurons and GABAergic interneurons but may also occur in other vertebrate nervous systems such as in frog and turtle.

## 7.7 Summary and Outlook

Neuronal electrical activity – both in the manifestation of action potentials (or 'presynaptic potentials') and postsynaptic potentials – is based on intermittent current flow through voltage-gated ion channels and thus voltage-gated ion channels are key elements of the nervous system. Most of the toxins that organisms have evolved to attack the nervous system of their prey or predators are powerful voltage-gated ion channel inhibitors or modulators. Many of these toxins are also widely used in research rather than synthetic antagonists, e.g., the aforementioned tetrodotoxin that blocks most $Na_v$ channels at very low concentrations (see Sect. 7.5.1), or the conotoxins, peptides produced by sea snails (*Conus*), and the agatoxins from the funnel spider *Agelenopsis*, that are used to specifically block certain $Ca_v$ channel subtypes.

We have seen how the microscopic parameters that determine the conductance and the gating of single ion channels – such as protein structure, activation and inactivation mechanisms – govern macroscopic currents. Only a population of channels can sustain cooperative mechanisms, and the positive feedback between the macroscopic $Na_v$ channel-mediated current and the membrane potential lies at the core of the regenerative action potential. The action potential is rather stereotyped across brain regions and serves as the main 'currency unit' used by the nervous system to exchange information. We know now how this information can be transmitted across considerable distances almost without loss.

## References

1. Alle H, Roth A, Geiger JR (2009) Energy-efficient action potentials in hippocampal mossy fibers. Science 325: 1405–1408
2. Armstrong CM (2007) Life among the axons. Annu Rev Physiol 69:1–18 [review]
3. Armstrong CM, Bezanilla F (1973) Currents related to movement of the gating particles of the sodium channels. Nature 242:459–461
4. Bean BP (2007) The action potential in mammalian central neurons. Nat Rev Neurosci 8:451–465 [review]
5. Debanne D (2004) Information processing in the axon. Nat Rev Neurosci 5:304–316 [review]
6. Doyle DA, Morais Cabral J, Pfuetzner RA, Kuo A, Gulbis JM et al (1998) The structure of the potassium channel: molecular basis of $K^+$ conduction and selectivity. Science 280:69–77
7. Dutzler R, Campbell EB, Cadene M, Chait BT, MacKinnon R (2002) X-ray structure of a ClC chloride channel at 3.0 Å reveals the molecular basis of anion selectivity. Nature 415: 287–294
8. Fitzhugh R (1962) Computation of impulse initiation and saltatory conduction in a myelinated nerve fiber. Biophys J 2:11–21
9. Greenspan J (2007) An introduction to nervous systems. CSHL Press, Cold Spring Harbor
10. Halling DB, Aracena-Parks P, Hamilton SL (2005) Regulation of voltage-gated $Ca^{2+}$ channels by calmodulin. Sci STKE 2005, re15
11. Hamill OP, Marty A, Neher E, Sakmann B, Sigworth FJ (1981) Improved patch-clamp techniques for high-resolution current recording from cells and cell-free membrane patches. Pflugers Arch 391:85–100
12. Helmstaedter M, Sakmann B, Feldmeyer D (2009) The relation between dendritic geometry, electrical excitability, and axonal projections of L2/3 interneurons in rat barrel cortex. Cereb Cortex 19:938–950
13. Hille B (2001) Ion channels of excitable membranes, 3rd edn. Sinauer, Sunderland
14. Hodgkin AL, Huxley AF (1939) Action potentials recorded from inside a nerve fibre. Nature 144:710–711
15. Hodgkin AL, Huxley AF (1952a) The dual effect of membrane potential on sodium conductance in the giant axon of *Loligo*. J Physiol 116:497–506
16. Hodgkin AL, Huxley AF (1952b) A quantitative description of membrane current and its application to conduction and excitation in nerve. J Physiol 117:500–544

17. Jiang Y, Ruta V, Chen J, Lee A, MacKinnon R (2003) The principle of gating charge movement in a voltage-dependent $K^+$ channel. Nature 423:42–48
18. Johnston J, Griffin SJ, Baker C, Skrzypiec A, Chernova T, Forsythe ID (2008) Initial segment $K_v2.2$ channels mediate a slow delayed rectifier and maintain high frequency action potential firing in medial nucleus of the trapezoid body neurons. J Physiol 586:3493–3509
19. Koch C (2004) Biophysics of computation. Oxford University Press, New York
20. Lacinová L (2005) Voltage-dependent calcium channels. Gen Physiol Biophys, 24 Suppl 1, 1–78 [review]
21. MacKinnon R (2004) Nobel lecture. Potassium channels and the atomic basis of selective ion conduction. Biosci Rep 24:75–100
22. McCormick D, Huguenard J (1992) A model of the electrophysiological properties of thalamocortical relay neurons. J Neurophysiol 68:1384–1400
23. Mellem JE, Brockie PJ, Madsen DM, Maricq AV (2008) Action potentials contribute to neuronal signaling in *C. elegans*. Nat Neurosci 11:865–867
24. Miller C, White MM (1984) Dimeric structure of single chloride channels from *Torpedo* electroplax. Proc Natl Acad Sci USA 81:2772–2775
25. Mulle C, Madariaga A, Deschenes M (1986) Morphology and electrophysiological properties of reticularis thalami neurons in cat: in vivo study of a thalamic pacemaker. J Neurosci 6:2134–2145
26. Rodriguez-Menchaca AA, Navarro-Polanco RA, Ferrer-Villada T, Rupp J, Sachse FB et al (2008) The molecular basis of chloroquine block of the inward rectifier $K_{ir}2.1$ channel. Proc Natl Acad Sci USA 105:1364–1368
27. Rohl CA, Boeckman FA, Baker C, Scheuer T, Catterall WA, Klevit RE (1999) Solution structure of the sodium channel inactivation gate. Biochemistry 38:855–861
28. Rush AM, Dib-Hajj SD, Waxman SG (2005) Electrophysiological properties of two axonal sodium channels, Nav1.2 and Nav1.6, expressed in mouse spinal sensory neurons. J Physiol 564:803–815
29. Stuart G, Spruston N, Häusser M (eds) (2008) Dendrites. Oxford University Press, Oxford
30. Yu FH, Yarov-Yarovoy V, Gutman GA, Catterall WA (2005) Overview of molecular relationships in the voltage-gated ion channel superfamily. Pharmacol Rev 57:387–395 [review]
31. Zhan XJ, Cox CL, Rinzel J, Sherman SM (1999) Current clamp and modeling studies of low-threshold calcium spikes in cells of the cat's lateral geniculate nucleus. J Neurophysiol 81:2360–2373


## Internet


Ion channel primer. http://www2.montana.edu/cftr/IonChannelPrimers/methods_to_study_ion_channels.htm TCDB: © 2005–2013 Saier Lab

Transporter classification database TCDB. http://www.tcdb.org/tcdb/index.php?tc=1.A.1

# The Synapse

# 8

Christian Lüscher and Carl Petersen

The term synapse is derived from the Greek "syn-" (together) and "haptein" (to clasp). Synapses have fascinated neuroscientists ever since their discovery in the late nineteenth century because they are the primary locus of information exchange between nerve cells. Moreover, much experimental evidence demonstrates that the efficacy of synaptic transmission may change as a function of experience. Such synaptic plasticity is of utmost interest because it may be the cellular mechanism that underlies learning and memory. More recently, it has been proposed that malfunctioning of synapses may be core features of several brain diseases, including dementia and addiction.

This chapter will distinguish two types of synapses in the central nervous system (CNS): electrical and chemical synapses. In the former, specialized proteins physically connect two neurons, which allow electrical signals to pass directly from one cell to the next. The overwhelming majority of synapses are chemical. These are more complex structures where neurons do not directly touch, but are separated by the synaptic cleft. An action potential invading the presynaptic specialization drives calcium influx through voltage-gated calcium channels and the calcium signal in turn triggers the exocytotic release of synaptic vesicles filled with a chemical neurotransmitter. There are many different neurotransmitters, each associated with specific brain functions. In the mammalian CNS glutamate is the most important excitatory neurotransmitter and GABA is the major inhibitory neurotransmitter. This contrasts with *C. elegans*, where both excitatory and inhibitory ionotropic glutamate receptors exist. Conversely, in insects acetylcholine is the most important excitatory neurotransmitter. The released neurotransmitter rapidly crosses the synaptic cleft where it binds to postsynaptic receptors, which are either ionotropic (rapid opening of an ion channel) or metabotropic (delayed signaling via G proteins). Synaptic transmission from presynaptic to postsynaptic specializations is the most important pathway for information exchange in the brain. However, neurons can also release neurotransmitters (e.g., endocannabinoids) from their dendrites affecting presynaptic mechanisms with such a retrograde signal.

Synaptic function is highly adaptable and regulated through many different processes on different timescales. The amplitude of the postsynaptic response is strongly influenced by the recent history of action potential firing of the presynaptic neuron. Such short-term synaptic plasticity occurring on the timescale of milliseconds and seconds is primarily mediated by presynaptic changes governing the release of synaptic vesicles. Short-term plasticity can either lead to transient increases in synaptic efficacy, termed **facilitation**, or to decreases in synaptic efficacy termed **depression**.

Long-term changes in synaptic strength lasting hours, days, and perhaps entire lifetimes can also be induced by specific patterns of neuronal activity. The most intensely studied form of plasticity is hippocampal long-term potentiation (LTP) of excitatory synapses.

C. Lüscher (✉)
Department of Basic Neurosciences
and Department of Clinical Neurosciences,
University of Geneva, Geneva, Switzerland
e-mail: christian.luscher@unige.ch

C. Petersen (✉)
Brain Mind Institute, Faculty of Life Science,
Ecole Polytechnique Federale de Lausanne (EPFL),
Lausanne, Switzerland
e-mail: carl.petersen@epfl.ch

C.G. Galizia, P.-M. Lledo (eds.), *Neurosciences - From Molecule to Behavior: A University Textbook*,
DOI 10.1007/978-3-642-10769-6_8, © Springer-Verlag Berlin Heidelberg 2013

LTP is elicited when glutamate is released onto a depolarized target neuron. We will discuss several of the key molecular steps involved in controlling LTP.

Other forms of long-term synaptic plasticity rely on changes in the presynaptic release of neurotransmitters. Along with functional changes, the structure of a synapse can change in an experience and activity-dependent manner. Such structural plasticity might underlie mechanisms for long-term rewiring of synaptic circuits.

We conclude the chapter with a discussion of emerging evidence that altered synaptic function may underlie some brain diseases.

## 8.1 Phylogeny of the Synapse

The ur-synapse, which is the last common ancestor of all synapses, appeared about 1 billion years ago in the Cnidaria clade, of which jellyfish are a member. Within several hundred million years, first GABA receptors and metabotropic glutamate receptors appeared, followed by the three major ionotropic glutamate receptors. Many of the synaptic signaling proteins such as CaMKII and calcineurin and the cell adhesion molecules ephrin and caherins are much older. Invertebrate protostoma such as *Drosophila* and *Aplysia*, which are about 700 million years old, have served as model organisms for the description of many basic properties of mammalian CNS synapses. The gill-withdrawal of *Aplysia*, for example, has served to study the synaptic mechanisms underlying nonassociative learning. Vertebrate synapses, and the mammalian synapse in particular display a greater signaling complexity, in particular, because of the several genome duplications. The NMDA receptor expansion (in particular the GluN2 subunit), as well as the duplication of many proteins in the postsynaptic density (e.g., PSD-95 or SAP102) have shaped the mammalian synapse. In the *Drosophila* synapse, for example, a simple GluN2 subunit is found that through a SVL-motive binds to the PSD homolog Dlg (discs large). On the other hand, in the mouse, four isoforms of GluN (A–D) with a much longer C-terminal tail exist that allow for multiple protein–protein interactions and therefore a greater degree of NMDA receptor signaling complexity.

Here, we focus on vertebrate synapses. We refer readers interested in specific features of invertebrate synaptic transmission to the review articles provided.

## 8.2 Structure and Diversity of Synapses

Synapses are subcellular specializations whose role is a regulated signal exchange between two neurons. There are many types, but they fall into two broad categories: electrical and chemical synapses. Phylogenetically, electrical synapses were first and indeed similar structures are also prominent in many non-neuronal tissues. Electrical synapses are made at points of physical connection between two cells and they will be dealt with in more detail below (see Sect. 8.8).

At chemical synapses, on the other hand, cells do not actually touch and transmission relies on the release and diffusion of a chemical substance, the neurotransmitter. Chemical synapses represent the majority of synapses in the CNS, and are subject to many forms of use-dependent regulation, referred to as synaptic plasticity.

When examined under an electron microscope, several subcellular elements common to all synapses can be identified (Fig. 8.1). The presynaptic specialization, commonly called the **synaptic bouton** (terminal), contains the synaptic vesicles that are filled with the neurotransmitter to be released. All vesicles are of similar size (~40 nm diameter) and careful inspection reveals that some of them are actually in contact with the membrane; they are docked and ready to be released. The area containing this readily releasable pool of vesicles is termed the active zone. Mitochondria are also commonly observed in presynaptic boutons providing the energy required for synaptic signaling. On the postsynaptic side, facing the boutons, the membrane is thickened. This part is called the **postsynaptic density (PSD)** where transmitter receptors along with a large number of other proteins are tethered. Some of them serve as auxiliary subunits for receptors, others are involved in receptor signaling and scaffolding, regulating the insertion and removal of the receptors. Close to the PSD one typically observes membrane compartments that are extensions of the endoplasmic reticulum (ER). This part of the ER is called the **spine apparatus** and contributes to the maintenance of the synapse, for example, by serving as the site for local protein synthesis. Occasionally, vesicles can also be observed on the postsynaptic side. They are somewhat larger compared to their presynaptic counterparts, and their

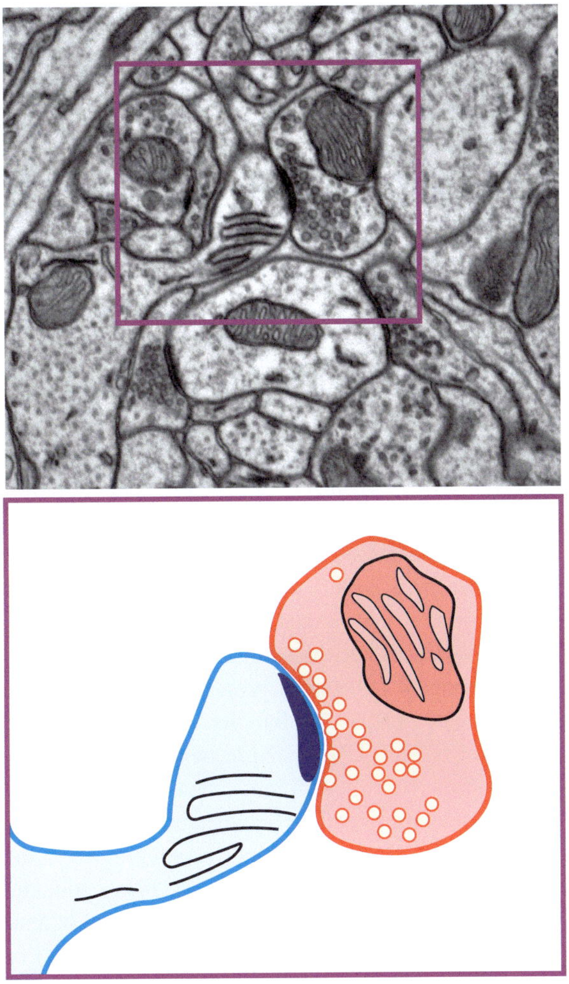

**Fig. 8.1 A synapse in the central nervous system.** This electron microscopic image shows the densely packed neuropil, and a highlighted asymmetric synapse. The schematic allows the identification of the presynaptic element with a mitochondrion and the vesicles that contain the neurotransmitter, shown in *red*. In *blue* the postsynaptic spine with the postsynaptic density apposed to the presynaptic bouton. The dense streaks are called the spine apparatus and consists of endoplasmatic reticulum. The length of the spine is approx. 1 μm (Courtesy of Graham Knott)

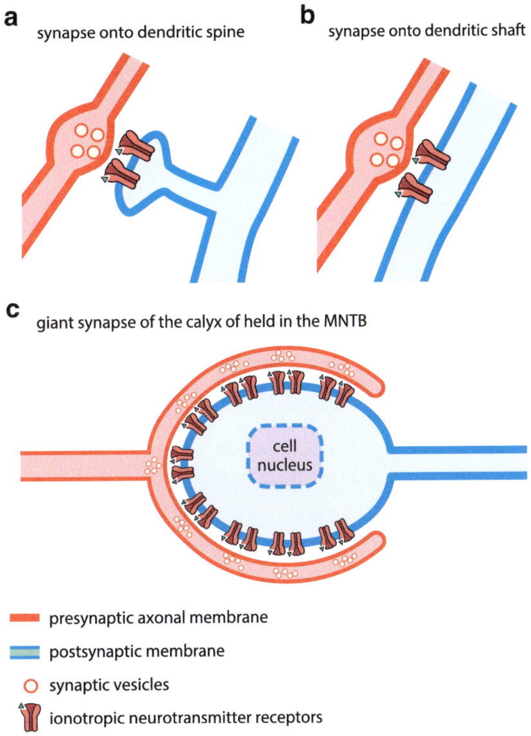


**Fig. 8.2 Synapses come in many flavors.** (**a**) Asymmetric synapses are formed onto spines, and are excitatory. (**b**) Synapses can also be formed onto the shaft of the dendrite, they can be excitatory but also inhibitory. (**c**) A special case is the giant synapse of the Calyx of Held in the medial nucleus of the trapezoid body located in the brain stem. This synapse has several release sites, also called active zones

preferential site of fusion with the postsynaptic membrane is just adjacent to the PSD. It is thought that such postsynaptic vesicles deliver new receptors or remove old receptors.

Beyond their common ultrastructural features, synapses in different parts of the brain vary considerably. The connection between CA3 and CA1 pyramidal neurons of the hippocampus is one of the best-studied synapses. CA3 neurons give off branches from the main projection of CA3 neurons called Schaffer collaterals, which are the axons that travel in the stratum radiatum. Along its course a given axon has many thickenings, which each represent a presynaptic element. These "boutons en passant" are juxtaposed with small membranous protrusion from the dendrites of CA1 neurons. These dendritic spines (Fig. 8.2) are the structural hallmark of excitatory synapses, i.e., they use glutamate as neurotransmitter. In contrast, inhibitory connections from GABAergic neurons onto the same CA1 pyramidal neurons are made directly onto the shaft of the dendrite or onto the cell body. Shaft synapses can also be excitatory, and some neurons, e.g., midbrain dopaminergic neurons, receive all their excitatory synaptic input directly onto dendritic shafts.

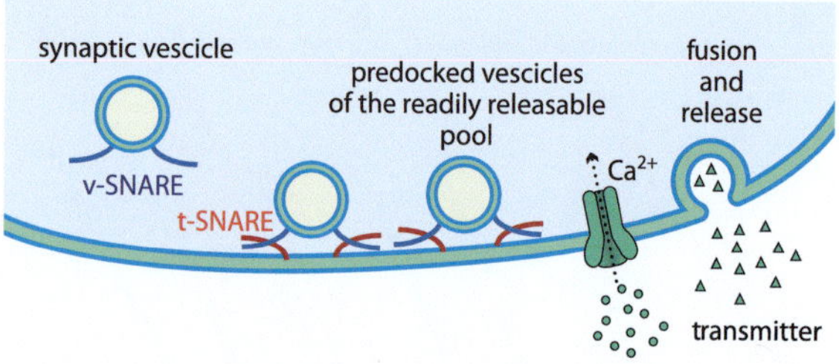


**Fig. 8.3** **Steps to release of neurotransmitter from synaptic vesicles.** Interactions between v-SNARE and t-SNARE tether the vesicle to the membrane, forming a readily releasable pool. When calcium enters the cell, the membranes fuse and the content is released. All these events happen in a fraction of a millisecond

Small synapses (like the CA3–CA1 synapses) make up the vast majority of synapses in the brain, but specific types of neurons connect with much larger synapses. One well-studied example is the "giant" synapse in the medial nucleus of the trapezoid body (MNTB), located in the brain stem. This connection is part of the auditory pathway and also called a calyx synapse because of its flower petal-like shape. In the MNTB calyx synapse, the large presynaptic terminal practically engulfs the postsynaptic neuron, allowing direct patch-clamp electrophysiological analysis of the presynaptic terminal, which is not possible at the majority of small synapses. The MNTB calyx synapse is also special as it has many release sites (i.e., several hundred active zones), offering the nervous system extremely fast and reliable signaling, which may be essential for sound localization. Other highly specialized types of synapses are found in the mossy fiber pathways of the cerebellum and the hippocampus. Both are glutamatergic, but despite their common name, they have distinct morphology and function. The cerebellar mossy fibers are axons originating from the pontine nuclei, which give off fine branches that twist through the granule cell layer and connect at slight enlargements giving a knotted appearance. Hippocampal mossy fibers travel in the stratum lucidum and display large varicosities all along their trajectory. These varicosities are very complex and contain many release sites.

Unlike the point-to-point neurotransmission of glutamate and GABA discussed so far, other substances, such as monoamines, including dopamine, adrenaline or serotonin, are often released into the extracellular space without an obvious postsynaptic target, in a process that is sometimes called volume transmission. In the striatum, for example, boutons of dopamine fibers arising from the midbrain are located at the neck and not at the base of a dendritic spine, without a distinct postsynaptic specialization on their own. Dopamine, once it is released binds to dopamine receptors and changes the functioning of the nearby glutamatergic synapses. The three major monoamine are therefore neuromodulators rather then neurotransmitters.

## 8.3 Synaptic Release

The release of neurotransmitter from the presynaptic specialization needs to occur with precise timing in the millisecond range. Some neurons fire action potentials at high frequencies and neurotransmitter release should therefore also be able to occur in rapid succession. It is then surprising that transmitter release is stochastic in nature, i.e., the invasion of an action potential will trigger release only with a certain probability, which is often quite low. For example, the mean release probability of synapses between cultured hippocampal neurons is about 0.15 (with a pronounced skew to low values). The release probability is a fundamental parameter of each CNS synapse, with different types of synapses having different release probabilities. Reliable synaptic signaling between two neurons often relies upon multiple active zones originating from the same axon, either distributed across many separate small synapses (as in the neocortex where pairs of neurons are often connected by about five synapses) or within the same large synapse (as in the MNTB calyx which has several hundred active zones). Release probability is not a fixed parameter; rather it varies during development and with activity. Activity-dependent changes in presynaptic release probability are forms of synaptic plasticity that will be discussed below (see Sects. 8.9 and 8.11). Although the molecular correlate of this stochastic property of transmitter release is not fully understood, it has been observed that the release probability correlates with the number of the vesicles predocked in the active zone (i.e., the size of the readily releasable pool). Holding vesicles close to the membrane is the task of the **SNARE proteins** (Fig. 8.3). Synaptic vesicles express the vesicle-SNARE (v-SNARE) proteins on their surface, which contain coiled-coil domains that intertwine with target-SNAREs (t-SNAREs). This core SNARE complex is formed by three α-helical proteins: synaptobrevin (v-SNARE), syntaxin (t-SNARE), and SNAP-25. Release-competent synaptic vesicles are thought to have SNARE complexes in a primed conformation, which requires little activation energy in order for membrane fusion to proceed.

When an action potential invades the presynaptic bouton, its membrane will transiently depolarize strongly and voltage-gated calcium channels will open. Calcium flows into the bouton mainly during the repolarization phase of the action potential. For a brief period, calcium concentrations can reach several micromolar in the vicinity of calcium channels localized at the active zone. The presynaptic calcium binds to the calcium-sensing protein synaptotagmin. Calcium-induced conformational changes in synaptotagmin are then relayed to the core SNARE complex, driving the fusion of the synaptic vesicle membrane with the plasma membrane of the presynaptic specialization. Several calcium ions (typically 4–5) must be simultaneously bound to synaptotagmin to induce vesicle fusion. Because there is a high degree of cooperativity between the calcium ions, the rate of exocytosis is a high power function of calcium concentration. The synaptic vesicles do not necessarily fuse completely with the presynaptic membrane, and neurotransmitter release can also occur via a fusion pore, in a so-called "kiss and run" event. Pore fusion can then be followed by full vesicle fusion and release of the entire content into the synaptic cleft. Because the release probability and the number of docked vesicles is typically low, an action potential will usually trigger the exocytosis of zero or one vesicle at a given synapse. However, multivesicular release has also been observed at a number of synapses and might be important at large and strong synapses.

Vesicles that have released their cargo are subsequently endocytosed and recycled. This mechanism involves specialized proteins such as dynamin, amphiphysin, and clathrin coating of the vesicle. In this process, the vesicles are reacidified by the V-type ATPase, causing an electrochemical gradient for protons that is necessary to refill the synaptic vesicle with neurotransmitter by special transporter proteins, such as VGLUT, VGAT, and VMAT for glutamate, GABA, and monoamine, respectively. Although a given vesicle typically only fills with a single transmitter, examples of corelease of transmitters have been reported (e.g., GABA and glycine, which are both transported by VGAT).

The unitary event of synaptic transmission is therefore the exocytosis of one synaptic vesicle, suggesting that a well-defined quantity of neurotransmitter is released that will give rise to postsynaptic potentials of a given unitary size. Such quantal transmission has been well documented at the neuromuscular junction but it has been difficult to resolve at CNS synapses. Synaptic events with well-defined quantal amplitudes may not be clearly resolved in the CNS because of small differences in the volume of individual vesicles; variability in the amount of transmitter released; the possibility of multiquantal release; and the fluctuation of the response of the postsynaptic membrane.

## 8.4 Ligand-Gated Synaptic Membrane Channels for Glutamate

Neurotransmitter diffuses across the synaptic cleft after exocytotic release of the content of the synaptic vesicle from the presynaptic specialization. The transmitter then reaches the postsynaptic membrane and binds to its respective receptors. This is a very dynamic process such that the neurotransmitter concentration remains high in the cleft for a very brief period. Ionotropic receptors are ligand-gated ion channels that respond instantaneously to the brief pulse of neurotransmitter. As a consequence, a synaptic current is evoked that typically lasts for only a few milliseconds. Here we will first discuss the ionotropic receptors for glutamate, while GABA receptors and metabotropic G protein-coupled counterparts will be dealt with in the next section.

Glutamate binds to three different receptors: the **AMPA** (α-amino-3-hydroxy-5-methyl-4-isoxazole propionic acid) the **NMDA** (*N*-methyl-D-aspartate), and the **KA** (kainic acid) **receptors**. All ionotropic glutamate receptors are permeable to $Na^+$ and $K^+$. Activation of ionotropic glutamate receptors leads to strong influx of $Na^+$ and only a small efflux of $K^+$ such that the net effect is the depolarization of the postsynaptic neuron.

The functions of **KARs** are least known, because until recently only very limited pharmacological tools were available and because only a subset of excitatory synapses express them. KARs play a role in glutamate transmission onto GABAergic neurons and in the mossy fiber projection onto CA3 pyramidal neurons. In many situations several action potentials in rapid succession are required to cause a measurable KAR-mediated response, because the currents are small and have slow kinetics.

The workhorses of glutamatergic transmission are the **AMPARs** (Fig. 8.4) encoded by the four genes *GluA1–4*. Each receptor is composed of four subunits, which can be a homomeric or heteromeric mixture of GluA1–4. Most AMPARs contain at least one subunit of

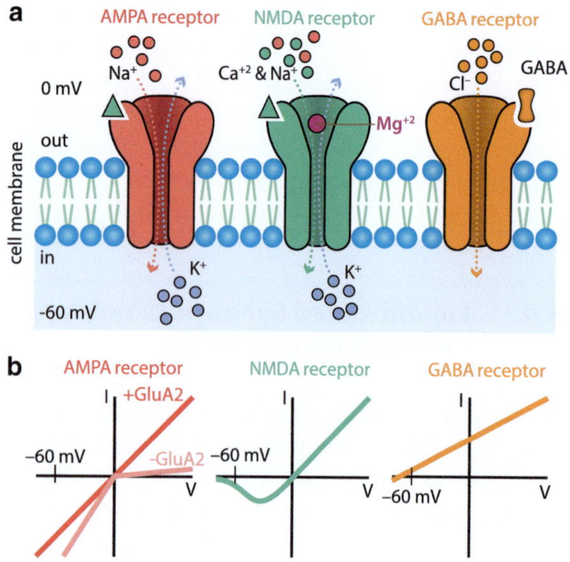


**Fig. 8.4 Major ionotropic neurotransmitter receptors.** (**a**) When glutamate binds to AMPA, many sodium ions flow into the cell while only some potassium ions leave the neuron, causing a net depolarization of the membrane. NMDA receptors are also permeable for calcium but only if the magnesium is chased off by a slight depolarization of the neuron. $GABA_A$ receptors are permeable to $Cl^-$, and therefore inhibitory in most instances. (**b**) The current–voltage (i–v) relationship provides a biophysical signature for the different receptors. AMPA receptors have a linear i–v relationship when they contain the subunit GluA2, but are inward rectifying (see text for definition) without GluA2. NMDA receptors have a complex i–v curve because Mg blocks the pore at negative potentials. For $GABA_A$ receptors the curve is linear and reverses below the resting membrane potential here indicated at −60 mV

GluA2. Following transcription, this subunit invariably undergoes RNA editing, whereby the RNA coding for a glutamine residue in the ion channel pore-forming region is exchanged for the RNA codon for arginine. As this process is an essential step in the quality control, nonedited GluA2 will be retained in the endoplasmic reticulum. In the edited form, an arginine residue located in the pore region of the channel is large enough to limit the flow of $Na^+$ and $K^+$ ions, thus preventing divalent ions from entering the cell rendering such AMPARs calcium-impermeable.

Currents mediated by AMPARs are inward at negative potentials and outward at positive potentials. In GluA2-containing AMPARs this relationship is symmetrical, i.e., the current–voltage relationship is linear, and the currents reverse at 0 mV. Receptors that lack GluA2 (e.g., GluA1 homomeric or GluA1/3 heteromeric channels) the current–voltage relationship in contrast exhibits a nonlinear voltage-dependent curve. GluA2-lacking subunits have a glutamine residue in the pore region and therefore have a high conductance for sodium. They are even permeable for calcium. Moreover, these channels are inhibited at positive potentials because endogenous polyamines access a site close to the cytoplasmic mouth of the pore. Therefore, GluA2-lacking AMPARs have an inward rectifying current–voltage relationship (i.e., they conduct current more easily into the cell than out of the cell). While most glutamatergic neurons express GluA2-containing calcium-impermeable AMPARs, many GABAergic neurons express calcium-permeable AMPARs.

In the case of **NMDARs**, the current–voltage relationship is even more complex. At resting membrane potential (that is, at negative potentials below approx. −60 mV or even −70 mV), magnesium ions get trapped in the pore of the NMDAR, blocking the passage for all other ions. Only when depolarized, magnesium is expelled from the pore, opening the pore for sodium, potassium, and calcium. With further depolarization, at potentials above 0 mV, NMDARs are maximally permeable (i.e., large outward currents are observed, Fig. 8.4). Compared to AMPARs, NMDARs have much slower kinetics. Following the binding of glutamate, NMDARs activate with peak current amplitudes long after the corresponding current in AMPAR. As a consequence, NMDARs, in response to a presynaptic release of glutamate can remain open for hundreds of milliseconds compared to a few milliseconds for the AMPARs.

Most importantly, NMDARs only conduct when two conditions coincide. Glutamate must bind and the postsynaptic neuron must depolarize. In other words, both the pre- and postsynaptic neurons need to be active in order to relieve the $Mg^{2+}$ block from NMDARs. Therefore, NMDARs play the role of molecular coincidence detectors, which is essential for several forms of synaptic plasticity as discussed below.

Ionotropic glutamate receptors interact with many proteins of the postsynaptic density (Fig. 8.6). They control and modulate several parameters of their electrical functions, and they also control their membrane insertion and removal. Interactions with proteins in the postsynaptic density thus control the number of glutamate receptors at a synapse (normally only a few dozens of receptors). A particularly important interacting protein is "stargazin", which is a member of the transmembrane AMPAR regulatory protein (TARP) family. Stargazin is more than an interacting protein, it is in fact an auxiliary subunit of most AMPARs that affects conductance, kinetics, and rectification. Because it is present in most physiological conditions, kinetics and pharmacology of neuronal AMPA receptors differ from those predicted by classical *in vitro* expression studies of AMPA receptors (where TARPs are not normally present). PICK1 is another example of a protein that regulates the number of receptors at the synapse. In contrast to the non-subunit-selective TARPs, PICK1 only interacts with GluA2. PICK1 along with other interacting proteins controls the dynamic recycling of synaptic AMPARs that occurs normally within tens of minutes. This constitutive recycling relies on mechanisms that allow rapid removal from and insertion into the postsynaptic density, such as exocytosis and endocytosis of clathrin-coated vesicles. This ensures that under baseline conditions the number of receptors remains stable, but also allows for a rapid redistribution in response to specific synaptic activity, leading to a change in the number of synaptic receptors. A transient imbalance between endo- and exocytosis is a widely used mechanism for synaptic plasticity (see below).

Ionotropic glutamate receptors have, above all, postsynaptic functions, but they can also be observed on presynaptic boutons where they may participate in the regulation of transmitter release. Presynaptic receptors may become activated by glutamate that spills over onto the terminal after massive release from the same bouton they target (homosynaptic modulation) or by glutamate released by neighboring synapses (heterosynaptic modulation). The molecular mechanism through which subsequent transmitter release is enhanced or inhibited is not fully understood and varies from synapse to synapse.

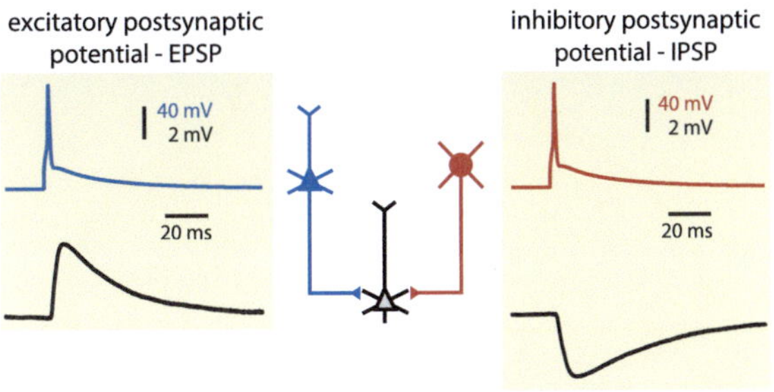


**Fig. 8.5** **Synapses are the building blocks of neural circuits.** When a glutamatergic neuron fires an action potential, the membrane of the neuron downstream shows a brief excitatory postsynaptic potential. Conversely an AP in an inhibitory neuron may release GABA evoking an inhibitory postsynaptic potential

## 8.5 Ligand-Gated Synaptic Membrane Channels for GABA

Ionotropic GABA receptors are called **$GABA_A$ receptors**. They are pentameric structures made of two α, two β, and one γ or ε subunit. There are two binding sites for the endogenous ligand GABA, which are between the α and β subunits. In addition, there are several allosteric modulation sites, the most important being activated by the benzodiazepines (e.g., Valium) located between the α and γ subunit. The activation of this site strongly enhances the single channel conductance and increases the mean open time. To date, no endogenous ligand for the benzodiazepine site has been identified. $GABA_A$ receptors are anion channels permeable for chloride. Since the intracellular chloride concentration is low (~5 mM) in adult neurons, the reversal potential for chloride is typically around −70 mV, which is below the action potential threshold. $GABA_A$ receptor activation is therefore inhibitory. Under physiological conditions, synaptically released GABA binds and rapidly opens $GABA_A$ channels giving rise to an inhibitory postsynaptic potential (IPSP, Fig. 8.5). Intracellular chloride concentration changes dramatically during development. During early postnatal development the potassium chloride cotransporter (KCC2) is expressed only at low levels, and neonatal neurons have a higher intracellular chloride concentration (reaching ~25 mM) compared to adult neurons. In young animals, when $GABA_A$ receptors open, chloride will flow out of the cell thus causing its depolarization. $GABA_A$ receptors, like glutamate receptors, are also part of a postsynaptic protein network. Its major component is gephyrin that together with collibystin is responsible for receptor clustering.

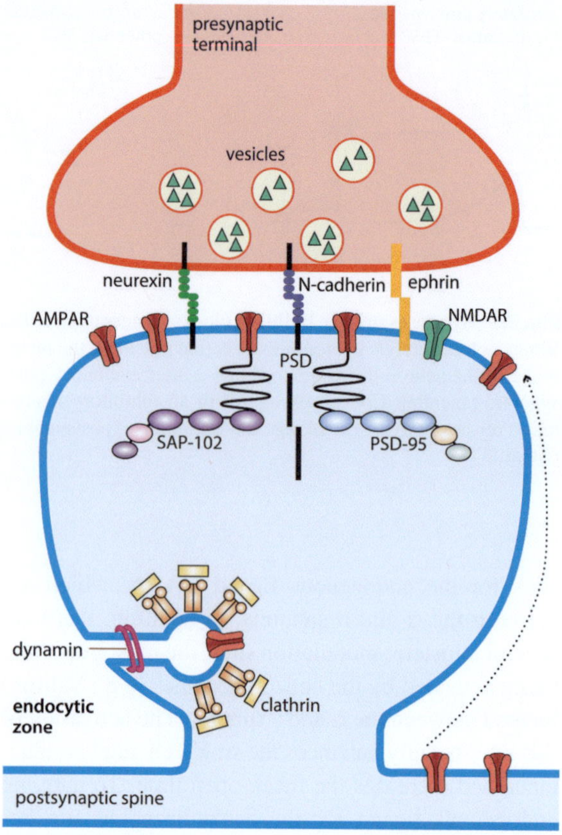


**Fig. 8.6 Scaffolding proteins control the dynamics of synaptic receptors.** Pre- and postsynaptic elements are tethered by neurexins, N-cadherins, and ephrins. In the spine, proteins such as *SAP-102* and PSD95 ensure the interactions with the cytoskeleton. New receptors are typically inserted into the membrane outside of the spine, reach the postsynaptic density through lateral movement, and are endocytosed at the neck to the spine. Such recycling of transmitter receptors can regulate the number of receptors in the synapse and thus control the strength of transmission

## 8.6 Slow Synaptic Currents Mediated by G Protein-Coupled Receptors

Glutamate and GABA also activate metabotropic receptors, which mediate their effects through G proteins and therefore have slower kinetics than ionotropic receptors. Metabotropic receptors share several structural features, such as the seven transmembrane domains and a C-terminal tail that is responsible for G protein activation. According to the G protein they activate, three major classes can be distinguished: $G_s$-, $G_{io}$-, and $G_q$-coupled receptors.

The five types of metabotropic glutamate receptors or mGluR have been divided into three groups: Group I (mGluR1 and mGluR5) couple to $G_q$ proteins, and are primarily expressed on the postsynaptic membrane, typically just outside the postsynaptic density. They mediate their effects through activation of phospholipase C (PLC) generating the second messengers diacylglycerol (which activates PKC) and $InsP_3$ (which evokes cytosolic calcium signals via release of calcium from intracellular stores). In subsequent steps, mGluR1 and mGluR5 can activate ion channels (e.g., calcium-activated potassium channels); trigger the release of endocannabinoids (see below); and control protein synthesis. Group II (mGluR2 and mGluR3) and III mGluRs (mGluR4, mGluR6-8), defined by their distinct pharmacology, inhibit adenylyl cyclase and control presynaptic release probability in many synapses.

**$GABA_B$ receptors** are metabotropic receptors that bind GABA. These receptors are encoded by two genes, which form obligatory dimers of the GB1 and GB2 subunit. GB1 contains the dominant GABA binding site, while GB2 is responsible for G protein activation. They are held together by coiled-coil domains in their C-terminal tail. $GABA_B$ receptors activate $G_{io}$, which inhibits adenylyl cyclase and calcium channels, and activates potassium channels of the Kir3 family. $GABA_B$ receptors are localized in both presynaptic and postsynaptic membranes. When located on a presynaptic bouton of either a glutamatergic or GABAergic axon, $GABA_B$ activation will decrease the release probability mainly through inhibition of calcium influx, but also through direct effects on the release machinery. Conversely, on the postsynaptic membrane the activation of Kir3 channel dominates, causing a hyperpolarization of the neuron. This hyperpolarization can be evoked pharmacologically (e.g., baclofen) or by synaptic release of GABA thus causing a slow inhibitory postsynaptic potential lasting several hundreds of milliseconds.

The three major monoamines **dopamine**, **serotonin** (5-HT), and **noradrenaline** (norepinephrine) activate metabotropic receptors (with the exception of the 5-HT3 receptor, which is a ionotropic receptor). Dopamine receptors, of which there are five, fall into two major groups. D1-like (D1 and D5) activate $G_s$ proteins while D2-like (D2–4) couple to $G_{io}$ protein. D1 and D2 receptors are expressed very densely in the striatum, where they mark the direct (D1) and indirect pathway (D2). Dopamine neurons give rise to four central projections: the nigro-striatal, the meso-cortico-limbic, the tubuloinfundibular and intrinsic olfactory. Seven serotonin receptors ($5\text{-}HT_{1-7}$) have been identified. The 5-HT3 is an ionotropic receptor that conducts $Na^+$ and $K^+$; all others are G protein coupled. 5-$HT_1$ and 5-$HT_5$ are $G_{io}$ coupled, 5-$HT_2$ coupled to $G_q$ and 5-$HT_4$, 5-$HT_6$, and-$HT_7$ activate $G_s$ proteins. Serotonergic neurons are particularly dense in the dorsal raphe, a brain-stem nucleus, which projects broadly across the brain. There are five adrenergic receptors: α1 couples to $G_q$; α2 couples to $G_i$; and β1–3 couple to $G_s$. Neurons in the locus coeruleus release noradrenaline widely across the CNS, which acts primarily on α1 and α2 receptors. Noradrenaline is thought to have a prominent role in alertness and arousal.

**Acetylcholine** is the major neurotransmitter of the neuromuscular junction. However, cholinergic neurons also exist in the CNS and there is ample evidence for nicotinic (ionotropic) and muscarinic (metabotropic) transmission. The CNS nicotinic receptors fall into two major classes: the homomeric α7 receptors and the heteromeric α4/β2 receptors. Both cause a cation flux and depolarize the target neurons, but have distinct kinetics and desensitization properties. Metabotropic muscarinic receptors couple to $G_q$ (M1, M3, and M5) or $G_i$ (M2 and M4) and are thought to participate in mediating the effects of attention upon thalamic and cortical neurons.

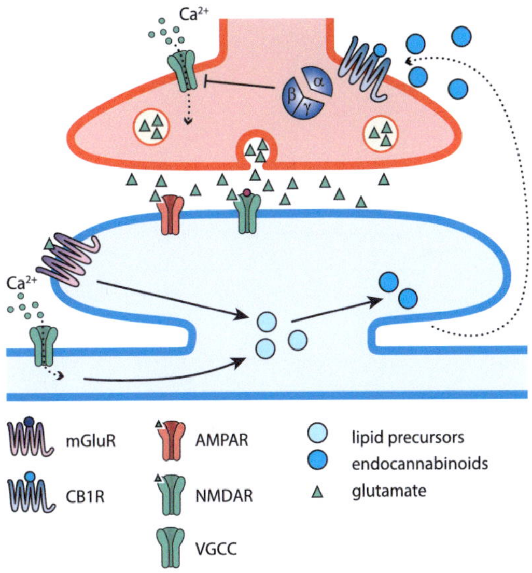


**Fig. 8.7** **Retrograde signaling by endocannabinoids.** When calcium levels rise in the spine, either because the neurons are depolarized or after the activation of metabotropic glutamate receptors (*mGluR*) and the activation of voltage-gated calcium channels (*VGCC*), lipid precursors are metabolized into endocannabinoids that easily cross the membrane and diffuse to CB1 receptors located on the presynaptic terminal. These receptors activate $G_{io}$ proteins, which inhibit calcium channels and reduce the release probability

## 8.7 Retrograde Signaling at Central Synapses

Synapses in the CNS have also been shown to signal retrogradely, that is signals that start in the postsynaptic neuron can affect the function of the presynaptic bouton (Fig. 8.7). **Endocannabinoids** (i.e., lipophilic substances that target the same receptors as tetrahydrocannabinol, the major ingredient of the plant-derived mind-altering drug marijuana) are such retrograde messengers. Anandamide and 2-arachidonoylglycerol (2-AG) are the major representatives of this family of lipophilic substances that easily cross cell membranes. Group I mGluRs and activation of diacylglycerol lipase (DAGL) synthesize 2-AG from arachidonic acid, which then freely diffuses across membranes without the need for exocytosis. When released from the postsynaptic neuron, endocannabinoids diffuse back to the presynaptic bouton where they bind to CB1 receptors. Activated CB1 receptors inhibit calcium channels through $G_{io}$ proteins thus reducing the release probability.

Presynaptic inhibition can also be elicited when presynaptic $GABA_B$, opioid, adenosine, D2Rs, or group II mGluRs are activated. The signaling pathway responsible for the reduction of the release probability is mediated by the βγ dimer of the heterotrimeric $G_{io}$ proteins and involves a reduction of presynaptic calcium current as well as a direct inhibition of the release machinery.

## 8.8 Electrical Synapses

Electrical synapses, also called **gap junctions**, involve a physical contact between two neurons, typically at the level of their dendrites (Fig. 8.8). The gap junctions are composed of channel-like protein structures allowing for the exchange of ions and other molecules. Each partner cell contributes with a transmembrane hemi-channel formed by the assembly of six connexin subunits. The pore is large (1.2 nm), which explains that small signaling molecules and metabolites can also be exchanged through gap junctions along with ions moving from one cell to the other.

Gap junctions have functional properties that contrast with chemical synapses. The electrical signal transmitted to the postsynaptic neuron always has the same sign as the source neuron. In other words, an

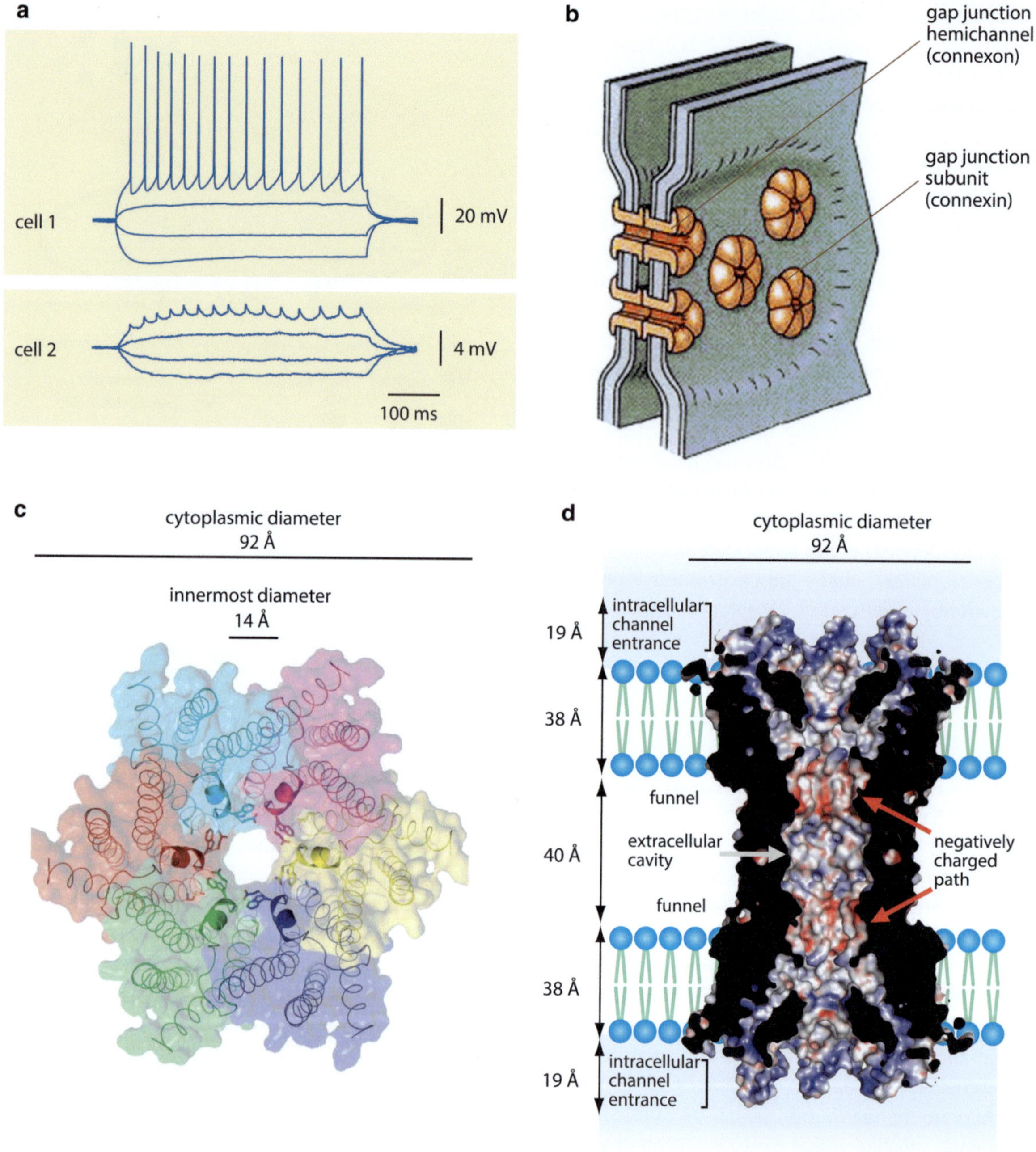


**Fig. 8.8** **Structure of electrical synapses.** (**a**) Electrical activity (voltage deflection and action potentials) in cell 1 is seen, in attenuated and filtered form, in cell 2 via gap junction coupling; (**b**) schematic drawing of a gap junction consisting of two apposed hemichannels; (**c**, **d**) molecular structure of connexin and gap junction (panel **a** is from Venance et al. [2] with permission, panel **b** modified from http://en.wikipedia.org/wiki/File:Gap_cell_junction_en.svg, panels **c**, **d** are from Nakagawa et al. [1] with permission; figure provided by V. Egger and D. Feldmeyer)

excitation of the presynaptic neurons cannot produce an inhibition of the postsynaptic cell, such as is the case for an inhibitory chemical synapse. Electrical synapses do not amplify the signal of the presynaptic neuron. In fact, the signal in the postsynaptic neuron is always smaller than that in the presynaptic neuron and high frequency signals (such as action potentials) are particularly strongly attenuated. However, the direct electrical connection of gap junctions allows for faster transmission compared to chemical synapses and gap junctions are thought to be responsible for zero-time lag synchronization of some populations of strongly gap-junction coupled neurons. For example, fast-spiking GABAergic neurons in the neocortex and thalamic reticular nucleus are strongly coupled through gap junctions, which may help generate high-frequency rhythmic behavior within the connected neuronal networks. Another example is the neurons in the locus coeruleus, the main adrenergic nucleus of the brain. The efficacy of transmission by gap junctions can be regulated by phosphorylation, allowing the degree of electrical coupling between gap junction-coupled neurons to be controlled by other signaling pathways. An example of an organism that relies on gap junctions for neuronal activity that controls motor behavior is the earthworm (*Lumbricus terrestris*). In *Drosophila*, the gap junction proteins are called innexins, as opposed to connexins and pannexins in vertebrates.

## 8.9 Short-Term Synaptic Plasticity

Many synapses change their efficacy during a train of action potentials (Fig. 8.9). Short-term synaptic facilitation is said to occur if the amplitude of postsynaptic responses increases during the train of action potentials. On the other hand, if response reduces in amplitude during the train, then this is termed short-term synaptic depression. Whether a synapse will exhibit facilitation or depression depends on several factors, including the initial release probability. For example, the low release probability of hippocampal mossy fibers makes this synapse prone to strong frequency facilitation, while GABAergic transmission in interneurons has a high probability to begin with and usually depresses after several stimuli in a train. Some synapses even show both, an initial facilitation, followed by activity-dependent depression. Short-term facilitation and depression can alter the postsynaptic response dramatically. The effects of multiple action potentials can accumulate, so that for some types of synapses, trains of presynaptic stimuli can increase the synaptic efficacy by more than an order of magnitude. The mossy fiber to CA3 synapse of the hippocampus is an example of a synapse that undergoes strong

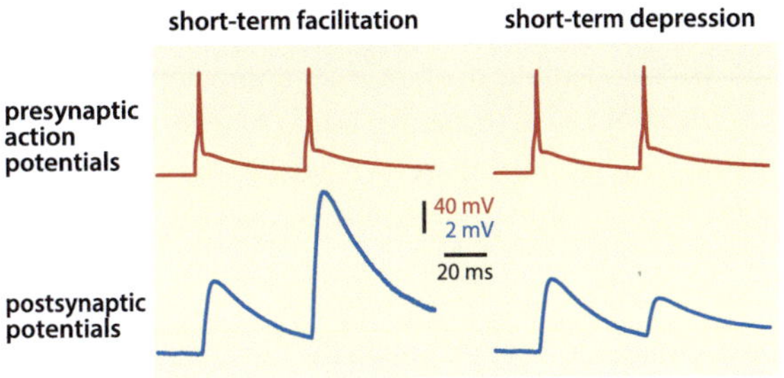


**Fig. 8.9 Short-term synaptic plasticity.** When two action potentials are elicited in rapid succession, some synapses display a paired-pulse ratio larger than one (short-term facilitation) while in others the second postsynaptic potential is smaller than the first (short-term depression)

frequency facilitation. This is linked to the low release probability at baseline. As a general rule, short-term synaptic facilitation and baseline release probability are inversely correlated.

Facilitation and depression do not lead to long-lasting changes in synaptic efficacy. They are transient changes that last on the order of 1 s, with the peak of facilitation or depression often occurring on the time-scale of tens of milliseconds. Short-term plasticity is therefore responsible for a dynamic regulation of synaptic efficacy during normal brain function, where neurons often fire action potentials in bursts or at high frequencies. Facilitating synapses will act as high-pass filters, preferentially transmitting information relating to high-frequency firing of the presynaptic neuron. Depressing synapses will act as low-pass filters, responding best to single isolated action potentials. However, it is important to note that these short-term dynamics in synaptic efficacy must be considered in the context of the temporal summation of postsynaptic potentials. When the frequency of postsynaptic potentials are high enough that the rise of the subsequent potential begins before the previous one ends there will be temporal summation. As single postsynaptic potentials are typically insufficient to generate an action potential, it is through temporal summation that several stimulations in a train allow the cell to reach the threshold. Similarly, postsynaptic potentials from different inputs (i.e., distinct presynaptic neurons) can summate to trigger action potentials. This mechanism is called spatial summation.

**Facilitation** and **depression** are usually the result of changes in the probability of neurotransmitter release. Depression is often considered to result from the partial depletion of release-competent synaptic vesicles docked at the active zone. A second invading action potential is therefore unable to release the same amount of neurotransmitter, because physically there are fewer available synaptic vesicles for exocytosis. Facilitation is thought to result primarily from an accumulation of calcium in the presynaptic bouton. The first action potential increases the cytosolic calcium concentration evoking neurotransmitter release. The calcium in the presynaptic bouton subsequently begins to return to baseline levels, mainly being extruded into the extracellular space or being pumped into intracellular organelles such as mitochondria and endoplasmic reticulum. If the second action potential arrives whilst the cytosolic calcium concentration in the synaptic bouton remains above baseline, then the peak calcium concentration evoked by the second action potential is larger than that evoked by the first action potential. Since neurotransmitter release depends steeply upon calcium concentration (see also above), even small changes in peak calcium concentration can make large differences to the release probability.

Often facilitation and depression can be observed in the same synapse under different conditions. A synapse may initially have high release probability and be depressing upon repetitive stimulation. If the release probability is reduced by presynaptic inhibition (for example, through the activation of $GABA_B$ receptors), then the response to the first action potential is reduced, but subsequent action potentials can evoke larger responses, thus converting an initially depressing synapse into a facilitating synapse.

## 8.10 Long-Term Synaptic Plasticity: Postsynaptic Forms

Long-term synaptic plasticity is a generic term that applies to a lasting experience-dependent change in the efficacy of synaptic transmission. Long-term changes in synaptic efficacy are thought to underlie learning and memory, and have therefore attracted considerable interest. There are many forms of long-term synaptic plasticity, which can be characterized by how long they last and by the molecular mechanisms that are required to elicit the change (induction) and to maintain it (expression). Here, we focus on long-term synaptic plasticity lasting more than 1 h, which can be divided according to a presynaptic or a postsynaptic locus of expression (see also Chap. 26).

A well-characterized form of postsynaptic long-term potentiation (LTP) occurs between CA3 and CA1 pyramidal neurons of the hippocampus (Fig. 8.10). Here, at these CA3–CA1 Schaffer collateral synapses, both induction and expression are postsynaptic. When glutamate is released onto a depolarized CA1 neuron, NMDA receptors become activated, evoking a rise of calcium in the spines. This initial step requires both activity of the presynaptic neuron (necessary for glutamate release) and activity of the postsynaptic neuron (necessary to relieve the NMDAR of the voltage-dependent $Mg^{2+}$ block). Such NMDAR-dependent LTP is therefore an associative form of plasticity (unlike the mossy fiber LTP that can occur independently of the activity in the postsynaptic neuron, see Sect. 8.11). *In vitro*, such concomitant activation can be achieved with different protocols. If a high-frequency stimulation protocol (e.g., 100 Hz for 1 s) is applied to Schaffer collaterals, this will sufficiently depolarize the postsynaptic neuron, such that the stimulations late in the train will release glutamate onto the neuron that is now depolarized. With other experimental approaches, associativity can be locally controlled by depolarizing the postsynaptic neuron when glutamate is released. The protocol probably closest to the physiological situation is spike timing-dependent plasticity (STDP). Here, an action potential is elicited in the postsynaptic neuron following within ~10 ms of glutamate release from the presynaptic neuron. The action potential backpropagates into the dendrites, thereby transiently relieving NMDARs of their voltage-dependent $Mg^{2+}$ block and allowing calcium influx at the activated synapses to drive plasticity. Such a sequence of presynaptic action potential followed by a postsynaptic action potential evokes LTP at many different CNS synapses.

The opening of NMDARs increases spine calcium concentration, activating CaMKII (a calcium and calmodulin sensitive kinase), which leads to the phosphorylation of AMPARs in the synapse. This

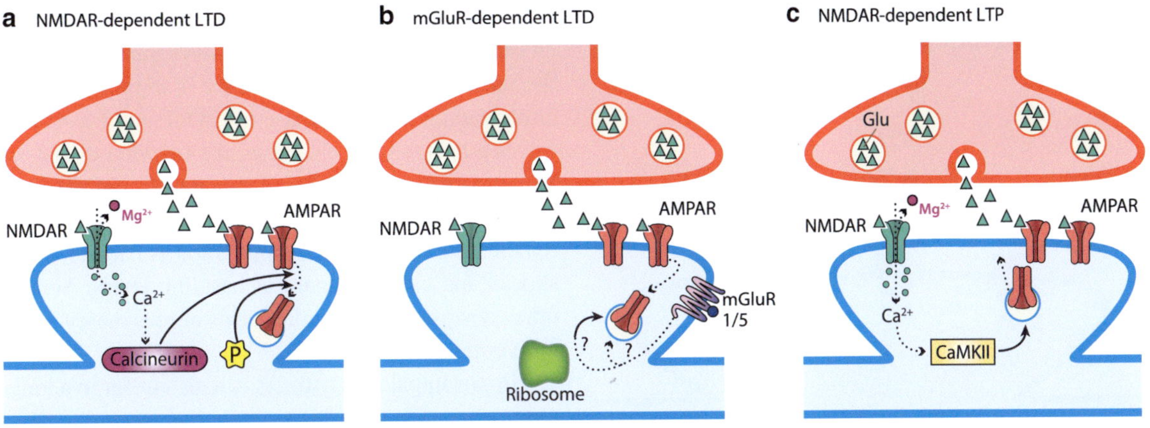


**Fig. 8.10 Postsynaptic expression mechanism of LTP and LTD.** (**a**) NMDAR-dependent LTD, postsynaptic AMPARs are internalized; (**b**) mGluR-dependent LTD, postsynaptic AMPARs are internalized; (**c**) NMDAR-dependent LTP, postsynaptic AMPARs are inserted. See text for details

leads to a rapid increase of the conductance and a larger postsynaptic potential. Subsequently, through a number of intermediate steps, CaMKII also drives the synaptic insertion of additional AMPARs. Some synapses contain only NMDARs and are therefore silent under baseline conditions. After LTP and insertion of AMPARs these silent synapses become functional. The discovery of this un-silencing of NMDA-only synapses has importantly contributed to elucidate the postsynaptic expression mechanisms of LTP. In fact, AMPARs are quite mobile and recycle even under baseline conditions within tens of minutes between the cytoplasm and the cell membrane. It is this mobile pool of AMPARs that allows for rapid but sustained changes in synaptic efficacy. The insertion and removal of receptors occurs by classical exocytosis and endocytosis via clathrin-coated vesicles at slightly perisynaptic locations, these receptors then reach the postsynaptic density by lateral redistribution. Several proteins associate with the AMPARs to regulate their mobility and biophysical properties. Of particular interest is stargazin, a member of the tarp family, which controls not only membrane insertion, but also lateral redistribution (see also above).

Interestingly, a weaker but prolonged activation of NMDARs (e.g., 1 Hz stimulation for 15 min even without depolarization of the postsynaptic neuron or STDP induced by postsynaptic action potentials *preceding* presynaptic action potentials) leads to long-term synaptic depression (LTD). This protocol leads to only modest calcium rises that activate phophatases rather than kinases leading to a dephosphorylation of AMPARs and their internalization through the pathways discussed above. How the cell detects such subtle differences in intracellular calcium remains elusive.

In the hippocampus and the cerebellum, LTD can also be induced by the activation of Group I mGluR receptors. Like in the NMDAR-dependent LTD, this form of synaptic plasticity is expressed by a removal of AMPARs. Even though it is expressed within minutes, mGluR-LTD is unique in that it requires protein synthesis. Translation of prefabricated mRNA stored in the dendrites produces "LTD proteins". In the hippocampus, one such protein is Arc that then controls the receptor redistribution necessary for the expression of LTD.

Different AMPAR subunits play distinct roles in this redistribution process. Heteromeric GluA1/GluA2 receptors are inserted in response to coordinated activity of the pre- and postsynaptic neuron. In addition, there are also forms of synaptic potentiation that are expressed by an exchange of GluA2-containing for GluA2-lacking AMPARs. Since the latter have a higher conductance, this will potentiate the synapse even if the total number remains the same. Conversely, in synapses that express GluA2-lacking receptors, mGluR-LTD is expressed by the replacement of these receptors with GluA2-containing ones. Local, dendritic protein synthesis is required, akin to mGluR-LTD in the hippocampus.

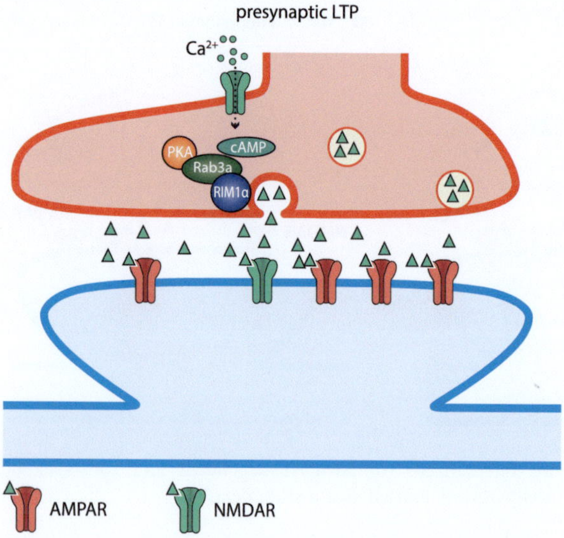


**Fig. 8.11** **Presynaptic LTP.** In this form of LTP, the presynaptic vesicle fusion machinery is changed such that the release probability increases permanently

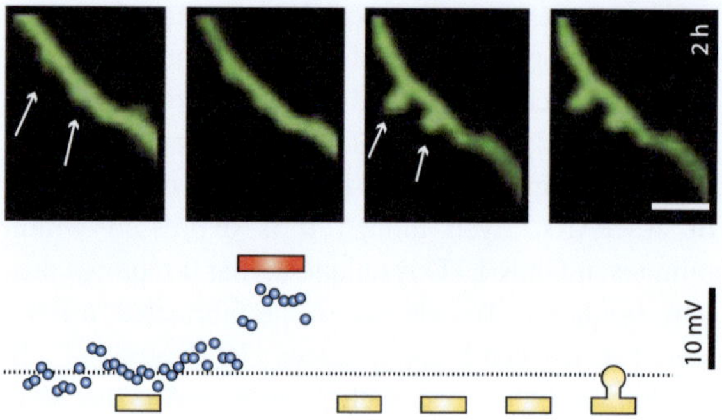


**Fig. 8.12** **Initial stages of structural plasticity.** Within less than 2 h of the potentiation of a synaptic connection in the hippocampus new spines appear (compare the *white arrows* in the first and third panel). The trace at the bottom shows LTP after stimulation (*red bar*). The shape of the icons below indicates whether new spines were detected (Taken with permission from Engert and Bonhöffer [3])

## 8.11 Long-Term Synaptic Plasticity: Presynaptic Forms

Presynaptic long-term potentiation (LTP) can be elicited in mossy fibers of the hippocampal granule cells that project onto CA3 neurons. This form of LTP is induced and expressed exclusively by the presynaptic neuron and is therefore nonassociative (Fig. 8.11). When a high-frequency train of stimuli is applied, calcium enters the presynaptic bouton and accumulates. This leads to an increase of the release probability and an immediate increase of the postsynaptic current. At the same time a signaling cascade is triggered by calcium leading to the increase of cAMP and PKA. This drives a long-lasting change in the function of at least two proteins of the release machinery: Rab3A and Rim1α. As a consequence, the release probability remains high, even when stimulation is back to single shocks and calcium again at baseline levels. Several other synapses show similar forms of presynaptic long-term potentiation.

Hippocampal mossy fibers can also undergo a long-term depression once presynaptic metabotropic glutamate receptors are activated. In this situation expression is again presynaptic by a decrease of the release probability.

A common form of LTD relies on retrograde signaling of endocannabinoids (see also above). Massive release of glutamate activates mGluR5, which triggers the release of endocannabinoids from the postsynaptic cell that by diffusion reaches that presynaptic bouton where they activate CB1 receptors. In some cells this is sufficient to induce long-lasting reductions in release probability. This form of plasticity is not limited to glutamatergic transmission but also occurs at inhibitory GABA-releasing synapses.

## 8.12 Structural Synaptic Plasticity

Changing the weight of synaptic connections is not the endpoint of synaptic plasticity. For example, structural changes are associated with NMDAR-dependent LTP *in vitro*. Within tens of minutes after a high-frequency train sufficient to potentiate existing connections in hippocampal slices, new spines appear (Fig. 8.12). When axons are labeled, new presynaptic elements are also observed, which then may form new synaptic connections onto the emerging spines at a later stage.

Over the last decade, high-resolution microscopy techniques have allowed longitudinal imaging of axonal boutons and dendritic spines over periods of many days in the cortex *in vivo*. It turns out that structural plasticity is an ongoing process in the adult brain of mammals. However, under baseline conditions, most dendritic spines and axonal boutons are stable, and only a subpopulation of 5–10 % of all synapses appears and disappears. For both spines and boutons the turnover is tightly balanced, leaving the total density of connections unchanged. Most new spines therefore rapidly disappear again, but in response to learning or changes in sensory experience, new spines may become long-lasting or even permanent. Such stabilization of selected synapses from the transient pool may underlie the experience-dependent rewiring of cortical circuits. Little is known about structural plasticity of excitatory shaft synapses or inhibitory synapses.

## 8.13 Synapses and Disease

Altered synaptic function may underlie some of the key symptoms of brain diseases. In the present chapter we will discuss the evidence for synaptic alteration in the context of three diseases: Alzheimer dementia, mental retardation, and addiction.

In **Alzheimer's disease (AD)** soluble Aβ oligomers may excessively activate mGluRs and lead to a depressed state of many synapses (a sort of mGluR-LTD), such that further plasticity is occluded (Fig. 8.13). This observation has received much attention because the extent of insoluble Aβ (i.e., plaques) only correlates poorly with the clinical manifestations. On the other hand, synaptic failure reflects the early cognitive decline at the beginning of AD quite well.

Synaptic dysfunction may also underlie mental retardation. The evidence is particularly strong for the **fragile X mental retardation**, a disease caused by an increase of triplet repeats in the DNA of *FMR1* gene, which codes for an RNA-binding protein. If more than 100 repeats are present, FMRP is no longer fully functional. Normally, the FMR1 protein exerts an inhibitory regulatory role on local dendritic protein synthesis (translational repression). Synaptic plasticities depending upon dendritic protein synthesis, such as mGluR-LTD in the hippocampus and the cerebellum, are enhanced if FMR1 function is disrupted. This leads to a downsteam retraction of spines and therefore less synaptic connections.

The currently leading hypothesis on **addiction** posits that the disease is a pathological form of learning. Addictive drugs have in common that they induce synaptic plasticity of glutamatergic and GABAergic transmission by increasing dopamine levels in the midbrain. Such drug-evoked synaptic plasticity may represent the cellular correlate underlying circuit reorganization and eventually lead to addictive behavior (Fig. 8.14). In fact, within hours after a first exposure to addictive drugs, excitatory afferents onto dopamine neurons of the ventral tegmental area (VTA) are potentiated. This drug-evoked plasticity lasts for about a week. Typically, multiple injections over a time course of several days are required to cause synaptic changes in target areas of the VTA, such as the nucleus accumbens. It is believed that this cascade of events eventually leads to a reorganization of circuits in the striatum and prefrontal cortex. These systems are involved in the control of goal-directed behavior and following chronic drug exposure, it seems that networks are privileged that

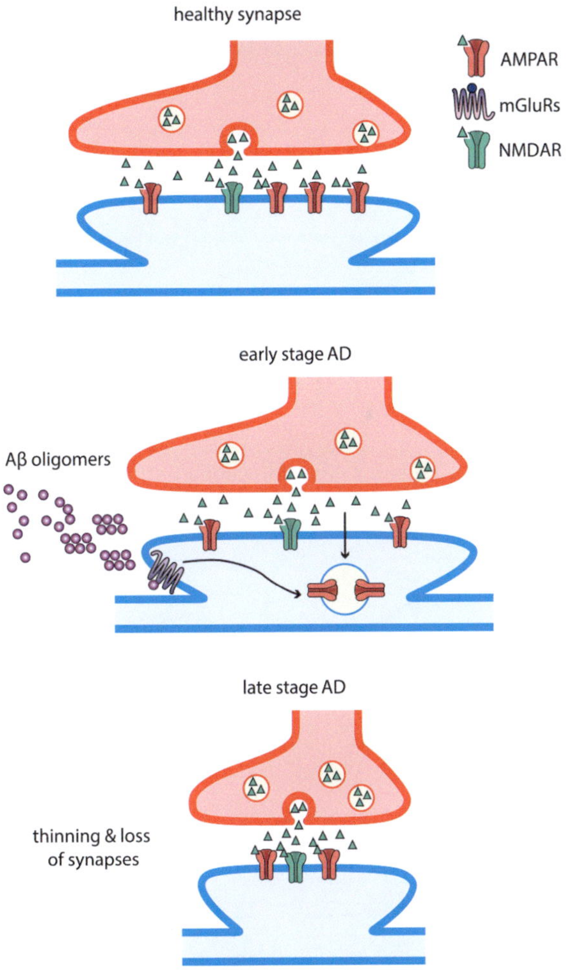


**Fig. 8.13 mGluR-hypothesis of synaptic alterations associated with Alzheimer disease.** Soluble Ab oligomers strongly activate mGluRs, which lead to the internalization of AMPA receptors and LTD. In a second step synapses become thinner and some disappear. It is thought that these changes underlie the early cognitive decline of the disease

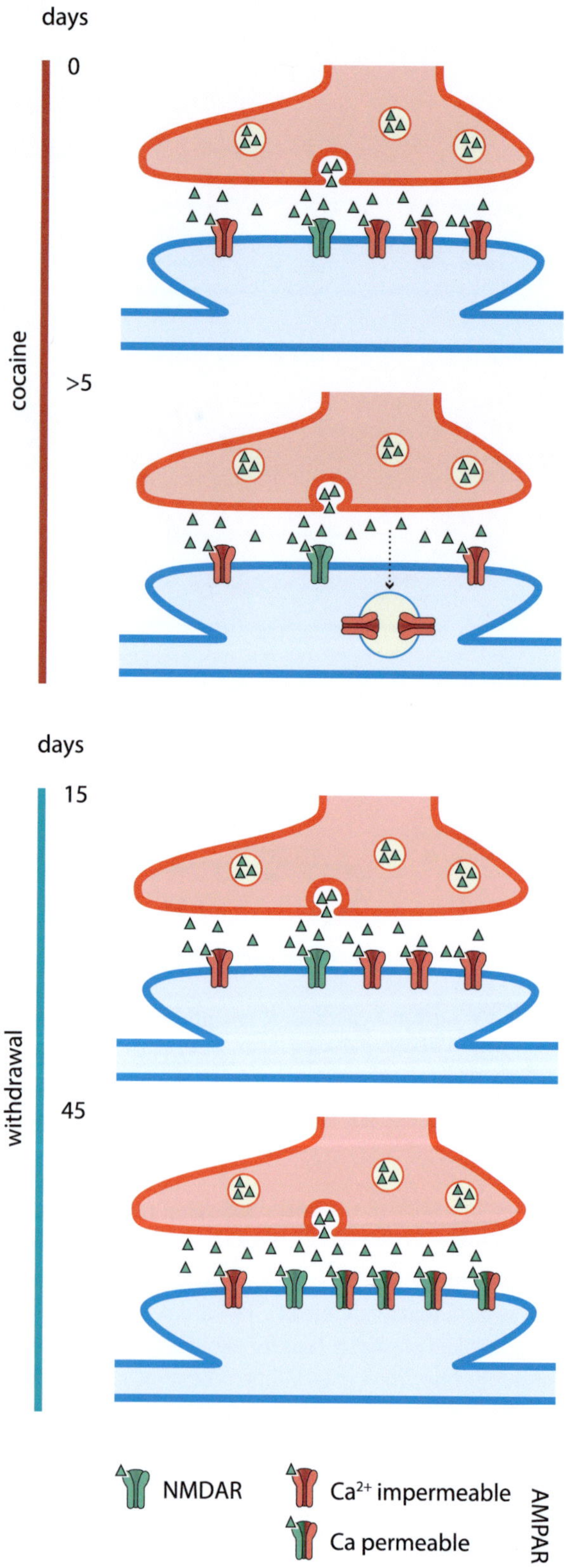


lead to automatic, compulsive decisions, such as often observed in addicts.

These three examples show that dysfunctional synaptic transmission along with pathological synaptic plasticity may represent the cellular correlates of mental diseases through network reorganization. Such a scenario may not be limited to the conditions discussed here, and efforts are underway to test whether these concepts can be expanded to depression, schizophrenia, or anxiety disorders. And although working out the details will still need much effort, the emerging concept of "synaptic disease" may apply to a multitude of mental disorders that are currently without cure.

## 8.14 Summary

Neurons in the brain interact at specialized structures called **synapses**, which primarily transmit information via chemical **neurotransmitters**. Electrical information can also be directly transmitted via **gap junctions**. At chemical synapses, the invading action potential induces calcium influx into the presynaptic bouton, driving exocytotic release of synaptic vesicles packaged with neurotransmitter. The major excitatory neurotransmitter in the CNS is glutamate, which acts upon ionotropic glutamate receptors, which are ligand-gated cation channels. The AMPA-type glutamate receptor mediates large and rapid synaptic currents. The NMDA-type glutamate receptor functions as a coincidence detector of presynaptic and postsynaptic activity, via its voltage-dependent $Mg^{2+}$ block. Strong activation of NMDA receptors drives a long-lasting enhancement in synaptic efficacy (**long-term potentiation – LTP**), mediated via calcium-dependent kinases and changes in postsynaptic glutamate receptors. Other

**Fig. 8.14 Drug-evoked synaptic plasticity in the nucleus accumbens.** Several injections of cocaine lead to a reduction of the number of AMPA receptors in excitatory synapses in the medium spiny neurons, which receive their inputs from the cortex. Within 10 days of withdrawal, the synapse subsequently becomes potentiated through the insertion of additional receptors some of which are homomeric GluA1 assemblies. Such GluA2-lacking AMPARs are calcium permeable. It is thus that these synaptic changes control the long-term relapse

forms of long-term plasticity can increase synaptic efficacy by enhancing the release of neurotransmitter, or decrease the efficacy of synaptic transmission by reducing presynaptic release or changes in postsynaptic glutamate receptors.

It is worth pointing out that synaptic function in the mammalian CNS has so far primarily been analyzed in reduced preparations, and it will be important in the future to re-examine what we have learned in intact preparations where brain function and dysfunction can be studied in the context of ongoing sensory processing, learning, and behavior.

## References

1. Nakagawa S, Maeda S, Tsukihara T (2010) Structural and functional studies of gap junction channels. Curr Opin Struct Biol 20:423–430
2. Venance L, Rozov A, Blatow M, Burnashev N, Feldmeyer D, Monyer H (2000) Connexin expression in electrically coupled postnatal rat brain neurons. Proc Natl Acad Sci USA 97:10260–10265
3. Engert F, Bonhoeffer T (1999) Dendritic spine changes associated with hippocampal long-term synaptic plasticity. Nature 399:66–70

## Additional Readings

Bennett MV, Zukin RS (2004) Electrical coupling and neuronal synchronization in the mammalian brain. Neuron 41:495–511 [Review]

Brockie PJ, Maricq AV (2006) Building a synapse: genetic analysis of glutamatergic neurotransmission. Biochem Soc Trans 34:64–67 [Review]

Brockie PJ, Maricq AV (2006) Ionotropic glutamate receptors: genetics, behavior and electrophysiology. WormBook 19:1–16 [Review]

Castillo PE, Schoch S, Schmitz F, Sudhof TC, Malenka RC (2002) RIM1alpha is required for presynaptic long-term potentiation. Nature 415:327–330

Holtmaat A, Svoboda K (2009) Experience-dependent structural synaptic plasticity in the mammalian brain. Nat Rev Neurosci 10:647–658 [Review]

Holtmaat A, Wilbrecht L, Knott GW, Welker E, Svoboda K (2006) Experience-dependent and cell-type-specific spine growth in the neocortex. Nature 441:979–983

Isaac JTR, Nicoll RA, Malenka RC (1995) Evidence for silent synapses. Implications for the expression of LTP. Neuron 15:427–434

Kaupmann K, Malitschek B, Schuler V, Heid J, Froestl W, Beck P, Mosbacher J, Bischoff S, Kulik A, Shigemoto R, Karschin A, Bettler B (1998) GABA(B)-receptor subtypes assemble into functional heteromeric complexes. Nature 396:683–687

Kessels HW, Malinow R (2009) Synaptic AMPA receptor plasticity and behavior. Neuron 61:340–350 [Review]

Lledo PM, Hjelmstad GO, Mukherji S, Soderling TR, Malenka RC, Nicoll RA (1995) Calcium/calmodulin-dependent kinase II and long-term potentiation enhance synaptic transmission by the same mechanism. Proc Natl Acad Sci USA 92:11175–11179

Lledo PM, Zhang X, Sudhof TC, Malenka RC, Nicoll RA (1998) Postsynaptic membrane fusion and long-term potentiation. Science 279:399–403

Lou X, Scheuss V, Schneggenburger R (2005) Allosteric modulation of the presynaptic $Ca^{2+}$ sensor for vesicle fusion. Nature 435:497–501

Lüscher C, Huber KM (2010) Group 1 mGluR-dependent synaptic long-term depression (mGluR-LTD): mechanisms and implications for circuitry and disease. Neuron 65:445–459 [Review]

Lüscher C, Malenka RC (2011) Drug-evoked synaptic plasticity in addiction: from molecular changes to circuit remodeling. Neuron 69:650–663 [Review]

Lüscher C, Slesinger P (2010) Emerging concepts for GIRK channels in health and disease. Nat Rev Neurosci 11:301–315 [Review]

Lüscher C, Xia H, Beattie EC, Carroll RC, von Zastrow M, Malenka RC, Nicoll RA (1999) Role of AMPA receptor cycling in synaptic transmission and plasticity. Neuron 24:649–658

Malenka RC, Kauer JA, Zucker RS, Nicoll RA (1988) Postsynaptic calcium is sufficient for potentiation of hippocampal synaptic transmission. Science 242:81–84

Marin-Burgin A, Kristan WB Jr, French KA (2008) From synapses to behavior: development of a sensory-motor circuit in the leech. Dev Neurobiol 68:779–787 [Review]

Milstein AD, Nicoll RA (2008) Regulation of AMPA receptor gating and pharmacology by TARP auxiliary subunits. Trends Pharmacol Sci 29:333–339 [Review]

Neher E (1998) Vesicle pools and $Ca^{2+}$ microdomains: new tools for understanding their roles in neurotransmitter release. Neuron 20:389–399

Neher E (2008) Details of $Ca^{2+}$ dynamics matter. J Physiol 586:2031 [Review]

Nestler EJ, Hyman SE, Malenka RC (2008) Molecular neuropharmacology – a foundation for clinical neuroscience, 2nd edn. McGraw Hill, New York [Textbook]

Purves D, Augustine GJ, Fitzpatrick D, Hall WC, LaMantia AS, McNamara JO, White LE (eds) (2011) Neuroscience, 5th edn. Palgrave Macmillan, New York [Textbook]

Reuveny E, Slesinger PA, Inglese J, Morales JM, Iniguez-Lluhi JA, Lefkowitz RJ, Bourne HR, Jan YN, Jan LY (1994) Activation of the cloned muscarinic potassium channel by G protein beta gamma subunits. Nature 370:143–146

Rohrbough J, Broadie K (2002) Electrophysiological analysis of synaptic transmission in central neurons of *Drosophila* larvae. J Neurophysiol 88:847–860 [Review]

Ryan G (2009) The origin and evolution of synapses. Nat Rev Neurosci 10:701–712 [Review]

Shi SH, Hayashi Y, Esteban JA, Malinow R (2001) Subunit-specific rules governing AMPA receptor trafficking to synapses in hippocampal pyramidal neurons. Cell 105: 331–343

Shi SH, Hayashi Y, Petralia RS, Zaman SH, Wenthold RJ, Svoboda K, Malinow R (1999) Rapid spine delivery and redistribution of AMPA receptors after synaptic NMDA receptor activation. Science 284:1811–1816

Sobolevsky AI, Rosconi MP, Gouaux E (2009) X-ray structure, symmetry and mechanism of an AMPA-subtype glutamate receptor. Nature 462:745–756

Sudhof TC, Rothman JE (2009) Membrane fusion: grappling with SNARE and SM proteins. Science 323:474–477 [Review]

Sutton RB, Fasshauer D, Jahn R, Brunger AT (1998) Crystal structure of a SNARE complex involved in synaptic exocytosis at 2.4 Å resolution. Nature 395:347–353

Watanabe S, Kirino Y, Gelperin A (2008) Neural and molecular mechanisms of microcognition in *Limax*. Learn Mem 15:633–642 [Review]

Wickman KD, Iniguez-Lluhl JA, Davenport PA, Taussig R, Krapivinsky GB, Linder ME, Gilman AG, Clapham DE (1994) Recombinant G-protein β, γ-subunits activate the muscarinic-gated atrial potassium channel. Nature 368: 255–257

Wilson RI (2011) Understanding the functional consequences of synaptic specialization: insight from the *Drosophila* antennal lobe. Curr Opin Neurobiol 21:254–260 [Review]

Wilson RI, Nicoll RA (2001) Endogenous cannabinoids mediate retrograde signalling at hippocampal synapses. Nature 410:588–592

Zhang Q, Li Y, Tsien RW (2009) The dynamic control of kiss-and-run and vesicular reuse probed with single nanoparticles. Science 323:1448–1453

# Biology and Function of Glial Cells

# 9

Magdalena Götz

The nervous system is generally composed of two cell types, neurons and glia. During the evolution of nervous systems, both become more numerous, but the glial cells even more so than neurons. Glial cells perform a complex panel of functions ranging from key roles in development to a diversity of functions in the adult nervous system.

During development glial cells help neurons to survive and guide both dendrites and axons towards their appropriate targets. Glial cells also act as neural stem and progenitor cells.

In the mature nervous system glial cells are crucial for several aspects of neuronal communication, like improving conduction velocity and synaptic transmission, controlling the extracellular milieu and optimizing it for the needs of neural networks. Some glial cells even receive synapses from neurons further highlighting that glia are an intricate part of neural networks. Thus, a complex nervous system will not function properly without glial cells.

This is also reflected when the brain is damaged, as glial cells perform the wound reaction of the central nervous system (CNS) and their reaction is critical to remove debris by **phagocytosis**, minimize the damage, restore the extracellular milieu and the **blood brain barrier (BBB),** a barrier to create a special extracellular milieu within the nervous system, especially in regard to ion and transmitter concentrations. Also after injury, glial cells act as neural stem and progenitor cells replacing neurons and glia in many species. But before dealing with all their fascinating functions, what are these glial cells that are so versatile and even outnumber the neurons in complex brains?

The first to define 'neuroglia' ('nervenkitt' or nerve cement) in the mid nineteenth century was the neuropathologist Rudolf Virchow who refers to 'substance … which lies between the proper nervous parts, holds them together and gives the whole its form in a greater or less degree' (Virchow, 1858 in [22]). Thereby all non-neuronal cell types and even the matrix, i.e., the noncellular material between the nerve cells, are included in this definition. As this definition is clearly too broad, it has been refined in 1965 [16] by explicitely excluding specific non-neuronal cells such as blood vessels, trachea, muscle fibers, glands, epithelia, and connective tissue. Thus, glia are the remaining types of non-neuronal cells typically in close vicinity of neurons [16, 20]. As most neurons generate fast $Na^+$-dependent action potentials, a **minimal definition of glial cells** is that they are **non-neuronal cells within nervous systems (except the above mentioned cell types) which normally fail to generate fast $Na^+$-dependent action potentials and are closely associated with neurons often ensheathing parts of them**, e.g., axons or synapses.

In many cases, these criteria and analogous functions across phyla allow unequivocal identification of glial cells. For example, glia enwrap axon fascicles in many species ranging from invertebrates to vertebrates. However, such densely packed wrapping of membranes around axons to improve conduction velocity can also be formed by neurons themselves (Klämbt, Chap. 1 in [20]). The close association of glia with neurons in most cases reflects a common developmental origin, i.e., from the ectoderm including neural crest in

M. Götz
Physiological Genomics Ludwig-Maximilians-Universität,
Pettenkoferstr.12
80336 München, Germany
e-mail: magdalena.goetz@lrz.uni-muenchen.de

C.G. Galizia, P.-M. Lledo (eds.), *Neurosciences - From Molecule to Behavior: A University Textbook*,
DOI 10.1007/978-3-642-10769-6_9, © Springer-Verlag Berlin Heidelberg 2013

vertebrates. This applies for most glia, but there are also subsets of glia derived from the mesoderm (see below). Likewise, cytological criteria for glial cells, such as the content of **intermediate filaments** and **glycogen granules**, have exceptions. Thus, like neurons, glia are so diverse that a common definition has to be rather broad possibly even including exceptions to the rule. For example, in many species with a neural tube, glial cells with intermediate filaments line the lumen of this tube (ependymoglia/radial glia, see below). These are then sometimes not in direct contact with neurons, but rather establish a border zone between the fluid-filled lumen and the nervous system. This is also the case for glial cells sealing off the nervous system to create the nervous system-specific extracellular milieu by forming the BBB. Thus, while a strict definition of neural glia appears somewhat confusing, these considerations reiterate a common theme in developmental and evolution biology: **specific functions are taken over by cells of various origins**. Obvious examples are cells involved in skeleton formation, often derived either from the mesoderm or the ectoderm, including neural crest. This shows that functional aspects are most important and indeed the need for the roles typically exerted by glial cells emerges with increasing complexity of nervous systems.

The importance of glial cells in nervous systems with increasing complexity is best highlighted by evolution. In multicellular organisms cells specialize towards sensory functions and 'supporting cells' often become associated with these. For example, in Hydrozoa (such as *Hydra*) sensory cells and nematocytes have distinct associated supporting cells that resemble accessory cells of complex peripheral sense organs in bilaterians. To which extent these may represent glial cells, their ancestors, or just some other supporting cells is still a matter of discussion [16, 20, 27]. In acoelomorphs, sediment-dwelling flatworm-like animals now considered as being at or near the base of present-day bilaterians, glia-like cells next to neurons have been described [16]. Interestingly, even within the same species glial cells are present in some, but not necessarily in every developmental stage. In the actively swimming larval stages of urochordates, e.g., *Ciona*, glial cells line the neural tube, while the reduced nervous system of the sessile adult forms has no obvious glia [16]. This highlights again the key aspect of glial cell function in more complex, condensed nervous systems. Accordingly, glial cells are present amongst deuterostomes in Echinodermata, Cephalochordata, and Craniata (including vertebrates) and amongst the Protostomia in Ecdysozoa, including Cycloneuralia (of which we will describe the nematode *C. elegans*) and arthropods (of which we shall describe *Drosophila melanogaster*), and Spiralia [16]. Thus, while some early supporting cells can not clearly be identified as glia, a general rule is that as soon as neurons evolved, supporting cells evolved along with them.

While diffuse nerve nets often have poorly defined supporting cells, glial cells become more clearly detectable when neurons condense into more complex nerve cords extending through the axis and brain-like structures in the head region.

Accordingly, the glia:neuron ratio is higher in neuronal networks of higher complexity with, e.g., 1:30 in the leech, 1:15 in *Drosophila*, 1:6 in the nematode *C. elegans*, 1:1 in the rat brain, to more than 2:1 in human and 3:1 in dolphin brains. Glial cells are not only more frequent in nervous systems of higher complexity, but also more diverse with many different classes of glial cells that even differ between brain regions in vertebrates and mammals. Indeed, this is only the tip of the iceberg, as new glial subtypes are continuously being identified, such as the NG2 glia or polydendrocytes (see below). A better understanding of the heterogeneity and diversity of glial cells will improve the understanding of brain function: the field of glial cell research still holds many surprises.

## 9.1 Invertebrate Glia

### 9.1.1 Glia in *Caenorhabditis elegans*

Glial cells perform important roles in nervous system development and function. They regulate dendrite and axon extension, remove the debris of cells undergoing programmed cell death, and are important in synapse formation and function.

In the nematode *C. elegans* 50 glial cells derived from neural/epithelial progenitors associated with sensory organs and 6 glial cells of mesodermal origin (GLR glia) associated with head muscle cells and motor neurons have been identified using the above criteria [27] (Fig. 9.1a). The nervous system of *C. elegans* contains 302 neurons (6× more than glia) arranged in a ventral nerve cord and a circumpharyngeal nerve ring (Fig. 9.1). Information from sensory organs is transmitted to the dense arrangement of axons and interneurons in the head, the nerve ring, where it is integrated. The output is sent to motoneurons and other efferents in the head and ventral nerve cord. The 50 glia associated with sensory organs ensheath the dendritic endings of sensory neurons [27]. Some of them, the 4 CEPsh glia, extend sheath-like processes that wrap around the nerve ring with processes that contact synapses within it (Fig. 9.1a). These glia in contact with synapses are reminiscent of astrocytes in vertebrates or the 'astrocyte-like' glia in *Drosophila* (see below). Interestingly, the processes of these glia are also organized in sharply-defined, nonoverlapping domains as described for mammalian astrocytes [27].

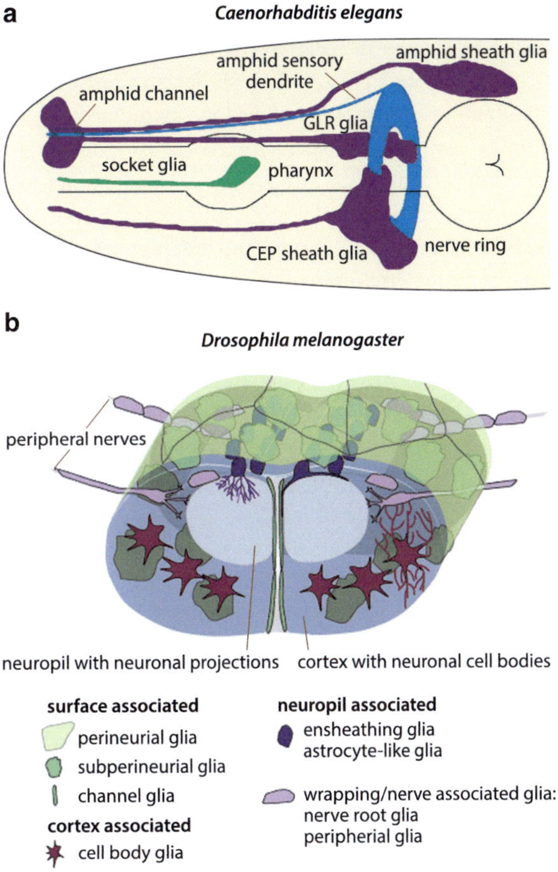


**Fig. 9.1** **Invertebrate glia.** (**a**) Examples of *C. elegans* glial cells. Cell bodies of sheath glial cells and GLR glia depicted in *purple* and socket glia in *green*. (**b**) Cell bodies of glial cells of the ventral nerve cord of *D. melanogaster*, embryonic stage 16. Three classes of glial cells can be distinguished based on their association with either surface, cortex, or neuropil

### 9.1.1.1 Roles During Development

Glial functions were first described for sensory receptive endings in the amphid, a bilateral sensory organ located in the head of the animal. This is composed of 12 sensory neurons and two specialized glial cells, the sheath and the socket cell that create a sealed environment via **junctional coupling**. If glial cells are lacking or fail to express key molecules, sensory cell development and/or function is impaired, depending on when the glial defect is first manifested. For example, early ablation of the amphid sheath glia progenitors results in failure of the amphid sensory dendrites to extend properly at the tip of the nose [2, 27]. Notably, expression of extracellular matrix molecules in glia is sufficient to allow normal anchoring of sensory neuron dendrites in *C. elegans*. These proteins resemble tectorins, which are proteins anchoring the hair cell cilia in the inner ear of vertebrates. When socket glia are ablated, sensory dendrites infiltrate a different sensory organ [27]. CEPsh glia are important for normal dendrite development, and ablation of CEPsh glia precursors results in animals either entirely lacking a nerve ring or manifesting severe axon guidance defects. At least some of these defects are due to the lack of UNC6/netrin secretion which is critical for axon guidance in the nerve ring [27]. Thus, an important developmental function of glia is to position neuronal processes, dendrites, and axons near to their appropriate targets. They utilize the same molecular cues involved in such functions in many other species.

### 9.1.1.2 Roles of Postembryonic Glia

Even when sensory neurons have formed normally, abnormal glia impairs sensory function. Postdevelopmental ablation of amphid sheath glia results in animals unable to respond to many environmental cues, including tastants, odorants, and temperature. Glia ablation results in the elimination of sensory-receptive endings of some of these neurons, suggesting important roles in maintenance of neuronal structures [2, 27]. In other neurons in which receptive-ending structure is intact, dysfunction can be traced to defects in calcium ion channel function [2]. A glial protein, FIG-1, which shares similarities with mammalian thrombospondin, an astrocyte-secreted protein required for synaptogenesis, is required for function and physical properties of sensory neurons [2]. Moreover, amphid sheath glia express epithelial $Na^+$ channels (acd-1), suggesting that they regulate ion balance in the environment of neurons [27]. Thus, **glia are required for sensory organ function.** Notably, this is a bidirectional communication and defects in sensory processes, e.g., cilia formation, subsequently causes defects in glial physiology, such as accumulation of secretory vesicles, demonstrating that communication between neurons and glia is key for sensory organ development and function.

*C. elegans* glia may also be involved in regulating synapse functions, reminiscent of such roles in *Drosophila* and vertebrates. This has been shown by an experiment where CEPsh glia was ablated after the establishment of the nerve ring. Neuronal morphology and locomotion was affected, probably by damage to the few neuronal synapses in the nerve ring which are enwrapped by CEPsh glia [27].

### 9.1.2 Glia in *Drosophila melanogaster*

Like the above described glia, also glial cells in *Drosophila melanogaster* perform key roles in development and at later more mature stages.

> During development glial cells regulate neurite guidance, perform trophic support of neurons, remove the debris of neurons subject to programmed cell death, and are involved in the formation of the blood brain barrier (BBB). At later stages, glial cells ensheath axons and function in synaptic communication, BBB formation and function and phagocytosis of damaged cells.

*D. melanogaster* has a ventral nerve cord with about 60 glial cells per segment and glia constituting about 15 % of cells in the CNS. Besides the glia in the peripheral nervous system (PNS), a significant number of glial subtypes is associated with the CNS. These comprise the surface-associated glia consisting of perineurial (mesoderm derived) and ectodermally derived subperineurial glia, forming the BBB, cortex-associated cell body glia thought to supply trophic and metabolic support for neurons, and the neuropil-associated glia comprising glia wrapping axons (Fig. 9.1b) [40].

#### 9.1.2.1 Roles During Development

While in *C. elegans* mostly sensory neurons are enwrapped by glia, many more neurons are accompanied by **wrapping glia** in *Drosophila*. During development these glial cells play important roles for guidance of both commissural and longitudinal axons tracts. Ablating midline glia has dramatic effects on commissure formation because glial cells are the source of guidance cues which attract motor neuron fibers towards the midline (secretion of netrin, see above) as well as repulsive cues (such as slit) which repel axons from the midline after they crossed it [11, 28] (see also Chap. 3).

Many glial cells migrate along the axons and ensheath them, either as groups of axons or ensheathing single axons with up to 8 glial wraps. These sheaths serve as electrical insulators and mediate the close bilateral communication between neurons and glia. Glia-neuron signaling between wrapping glia and the axon ensures survival of neurons through trophic support mechanisms. Conversely, neuron-glia communication regulates the number of glial cells surrounding a nerve fiber by axons releasing members of the epidermal growth factor family (the EGF receptor ligand spitz in *Drosophila* and neuregulins in mice).

Glia can also perform the opposite function, namely **execute the death of a neuron** or its processes. Programmed cell death is an important hallmark in *Drosophila* development, as about half of the cells in the nervous system are removed, e.g., during remodelling in pupal development. This important task is performed almost exclusively by glia engulfing debris of dead cells, as well as degrading axons or entire cells via the engulfment receptors Simu and Draper. Hence, a key function of glia is **phagocytosis**.

#### 9.1.2.2 Roles of Postembryonic Glia

Once the nervous system has been formed and sculpted, glial cells perform crucial functions also in the adult nervous system. Some roles are common in development and adulthood, such as the phagocytosis function that is continued in the adult *Drosophila* brain and critical after neuronal damage [10, 40].

Within the *Drosophila* brain, glia associates either with the cell bodies of neurons (the **cortex-associated cell body glia**) or extends processes into the neuropil (neuropil glia: **ensheathing and astrocyte-like glia**) where axons and dendrites communicate. Neuropil glia regulates synaptic functions in various aspects, such as neurotransmitter uptake and synaptic strength. Two high-affinity excitatory amino acid transporters (dEAAT) are expressed in glia – either in the CNS (EAAT1) or the PNS (EAAT2). Targeted silencing of EAAT1 leads to widespread degeneration in the adult fly CNS [28] indicating the importance of glia-specific removal of glutamate. This is reminiscent of **excitotoxicity** observed in vertebrate brain upon inefficient removal of glutamate or other transmitters. Typically this toxicity results from excessive $Ca^{2+}$ influx which triggers enzymatic cascades leading to cell death. *Drosophila* glia also expresses glutamine synthase, an essential component of the glutamate recycling pathway. Thus, glial cells remove the glutamate from the synapse after its release by the presynaptic neuron, convert glutamate to glutamine and release it to be taken back up by the neuron. They provide not only efficient removal of glutamate, but also efficient recycling (Fig. 9.2).

Glial cells also inactivate neurotransmitters such as dopamine, histamine, and serotonin by conjugating β-alanine. Dopamine and serotonin are involved in entrainment of activity rhythms. Ebony is an enzyme expressed specifically in astrocyte-like glia and its cell-autonomous diurnal regulation profoundly affects circadian locomotor behavior [18, 19]. *Ebony* mutant flies exhibit arrhythmic patterns of locomotor activity. The key role of Ebony in recycling of amine neurotransmitters suggests that this phenotype is caused by defective amine recycling by Ebony-expressing glial cells [19]. Ebony-expressing glial cells are involved in removing amine-containing transmitters from the synapses, but also in recycling by returning N-β-alanyl amine compounds which can then be processed by enzymatic cleavage in neurons for transmitter recycling [19]. This is an example of how glia-specific neurotransmitter removal and inactivation has direct effects on behavior.

Glial cells not only take up and recycle transmitters, but are active partners in synaptic communication. They release glutamate or other **gliotransmitters** in a regulated manner. Also in vertebrates, glia release transmitter at synaptic sites, either by reversal of the transporters taking up the transmitters or by de-novo synthesized transmitters which may be released from vesicles

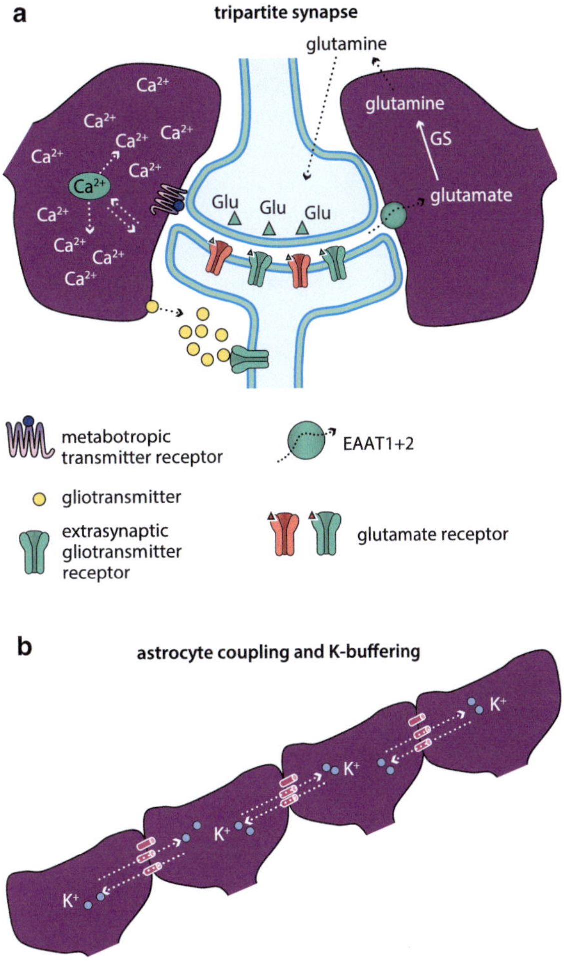


**Fig. 9.2 Tripartite synapse (a) astrocyte coupling and K-buffering (b).** Schematic drawings showing in (**a**) some of the major roles of astrocytes at the synapse; (**b**) buffering of some ions by distribution via gap junctions

reminiscent of synaptic vesicles at the presynaptic, typically neuronal side. The density of postsynaptic glutamate receptors and synaptic strength is regulated via gliotransmitter release. This key synaptic function of glia was discovered by mutation of the cystine/glutamate-type amino acid transporter encoded by the gene *genderblind*, whose mutation leads to lack of discrimination between male and female flies [13]. Thus, the ensheathing- and astrocyte-like glial cells in contact with synaptic sites in the neuropil of *Drosophila* regulate synapse function by various mechanisms and constitute the crucial third part of a **'tripartite synapse',** consisting of a presynaptic, postsynaptic, and a glial side (Fig. 9.2a). In vertebrates, this function is performed by radial glia, tanycytes, or astrocytes (see below).

Cortex-associated cell body glial cells are in close contact with neuronal somata, ensheathing dozens of neuronal cell bodies. These glial cells are also in contact with the subperineurial sheath, with the neuropil-associated glia described above, or with both. These cells therefore seem to move gas and nutrients between neurons and the hemolymph, reminiscent of a function of astrocytes in the mammalian brain. Moreover, these cortex-associated glia take up $K^+$ ions and dilute them via gap junctions amongst many coupled glial cells (Fig. 9.2b) thereby contributing to ion homeostasis within the brain.

The insect hemolymph has relatively high $K^+$ and low $Na^+$ concentrations, which is incompatible with neuronal communication mediated by action potentials (see Chap. 7). Therefore, the formation of an efficient blood nerve barrier protecting against high $K^+$ levels is essential. The blood brain barrier (BBB) is formed by a sheath of flattened, ectoderm-derived surface glia, the **subperineurial glia**. They generate an ionic seal by tight connections referred to as **'septate junctions'** composed of neurexin IV, contactin, and neuroglian. These junctions resemble the axon-glia paranodal junctions in vertebrates composed of the homologous proteins Caspr, Contactin, and Neurofascin 155 (Klämbt in [20]). In vertebrates, these junctions are established between the axon and the glial cells forming the myelin sheaths and hence critical for conduction velocity (see below). If septate junction formation fails in *Drosophila*, e.g., by mutation of a G protein-coupled receptor affecting cortical actin formation [36], the BBB is compromised and action potentials can no longer propagate, thus the animals are paralyzed. In vertebrates, the BBB is formed by tight junctions between endothelial cells. Pericytes, a neural crest-derived cell type closely associated with blood vessels in the brain, play instructive roles in vertebrates [1]. Summing up, glia in *Drosophila* and vertebrates play a similar role in the formation of junctions, but this is recruited into the distinct functional contexts.

## 9.2 Vertebrate Glia

The key functions of glia performed in invertebrates are also present in vertebrate glial cells. Wrapping of axons is further elaborated by the formation of **myelin**, either by the neural crest-derived Schwann cells in the PNS or by the neuroepithelium-derived oligodendrocytes in the CNS.

The synaptic functions of glia are numerous, including ensheathing of synapses by various glial cells (radial glia, astrocytes). Some (NG2 glia) are even postsynaptic partners of neurons. Mesodermally derived microglia specialize on phagocytotic functions and interact with the immune system, but also partake in synaptic communication and pruning. Ependymal cells line the ventricles. Radial glia, a ubiquitous type of glia in the developing brain, guide migrating neurons and act as neural stem and progenitor cells.

### 9.2.1 Vertebrate Glia in Development

The developing CNS of vertebrates contains a single, ubiquitous type of glia, the **radial glia**. Radial glia (Fig. 9.3) are named by their morphology, the long radial processes spanning the entire thickness of the neural tube, and their glial nature: they do not form action potentials and they express hallmarks of glial cells, such as glutamine synthase, glutamate transporters (GLAST and GLT-1), the glia fibrillary acidic protein (GFAP), and the content of glycogen granules, a typical feature of many glia [20, 31, 32]. Radial glial cells are amongst the few glial cells with epithelial properties (in vertebrates besides radial glia only ependymal cells). They are coupled by adherens and gap junctions, delineating the apical from the basolateral membrane domain [12] (Fig. 9.3a). Radial glial cells are derived from earlier neuroepithelial cells that gradually acquire the glial characteristics listed above around the onset of neurogenesis (Chap. 5 in [20]).

By way of their long radial morphology (in primates up to 7 mm long), radial glial cells serve as architectural pillars in the developing brain and spinal cord, which both undergo many morphological movements during development. They anchor at the apical, ventricular surface via special types of adherens junctions and at the basal surface by a tight contact to the **basement membrane** (BM) underlying the meninges via specialized enlargements, their subpial endfeet (Fig. 9.3), providing a stable reference systems for the brain. Integrins and dystroglycans anchor the radial glia endfeet to the BM. During growth of the brain, the BM is under continuous pressure and ruptures easily if molecules involved in the anchoring of radial glia endfeet at the BM are deleted [31]. Defects are even more severe if adherens junctions connecting radial glial cells at the ventricular side are disturbed, either by interfering with cadherins or the connections between the cadherin molecules and the actin cytoskeleton. The radial processes of radial glial cells further serve as guidance structure for migrating neurons (dark blue in Fig. 9.3a). Neurons often have to migrate long distances from their place of birth close to the ventricle towards the outer part of the neural tube (see also Chap. 3). Neuron-glia interactions help to guide the neurons over this distance and may also determine their exact final position, namely the point at which neurons stop to migrate.

#### 9.2.1.1 Radial Glia as Progenitors and Stem Cells

Given the importance of such structural roles, radial glia were perceived as being static and stable structures. Evidence that they themselves divide was rather ignored. Moreover, the cell biological dogma that dividing cells round up retracting their processes during M-phase further cemented the view that glial cells with such long radial processes could not divide. However, both these dogmas turned out to be wrong – live imaging in the developing CNS of zebrafish, chicken, and mouse allowed to directly observe that radial glial cells divide without retracting their long radial processes (Götz, Chap. 5 in [20]). Radial glial cells are heterogeneous, their progeny comprising neurons and glial cells which differentiate at later stages (Chap. 3; [12, 20]). Some act as multipotent **neural stem cells** generating neurons and glial cells, while others are more fate-restricted and generate only neurons or only specific types of glia. During development, radial glial cells self-renew for several rounds of cell division. In most regions of the CNS in the mammalian species examined so far their self-renewal is limited, and radial glial cells disappear at the end of development. The exceptions to this rule are some niches where radial glial cells persist as neural stem cells into adulthood and continue to generate neurons [23]. Adult neurogenesis and the persistence of radial glial cells is more widespread in nonmammalian vertebrates [5, 32].

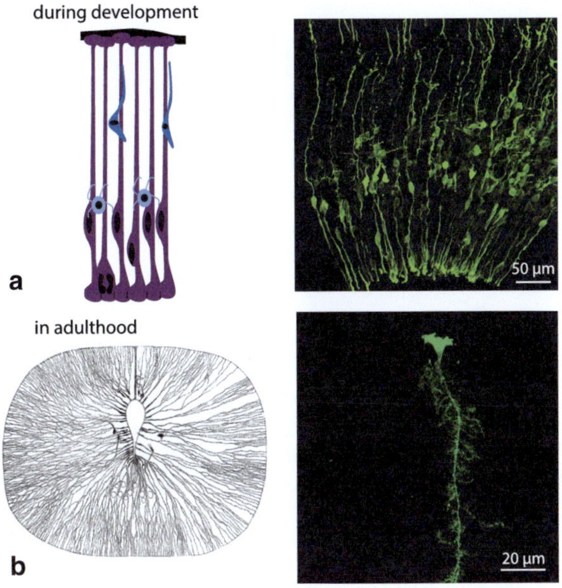


**Fig. 9.3** **Radial glia in vertebrates** during development (a) and in adulthood (b). (**a**) The *left panel* depicts a schematic drawing of radial glial cells (*purple*) and their anchoring at the ventricular side (*bottom*) and the pial surface (*top*). Migrating neurons are depicted in *dark blue*, intermediate progenitors in *light blue*. The *right panel* shows an electroporated mouse cerebral cortex during development (E14) displaying radial glial cells. Scale bar: 50 µm. (**b**) The *left panel* depicts a drawing by G. Retzius of radial glial cells/tanycytes in the spinal cord of an adult lamprey. The *right panel* depicts a GFP-labeled radial glia/tanycyte with the soma at the ventricular surface (*upper side*) and the branched radial processes extending into the adult zebrafish brain parenchyma. Scale bar: 20 µm

### 9.2.2 Vertebrate Glia in Adulthood

#### 9.2.2.1 Radial Glia in Adult Brains

In various species, radial glia disappear and scattered star-shaped astrocytes appear in brain regions with increased thickness (Fig. 9.3b, Chap. 5 in [20]). For example, scattered astrocytes appear in the expanded telencephalon of some rays and sharks, while tanycytes persist in the spinal cord. These observations lead to the suggestion that increased brain size and thickness exceeds the support capacity of the elongated radial glial fibers, but rather requires equally spaced and scattered glial cells, such as astrocytes and NG2 glia. Along with these changes, a new cell type then lines the ventricle, the multiciliated ependymal cells. In many other vertebrates, such as cartilaginous and bony fish, amphibia, reptiles, and many birds, radial glia persist into adulthood (Fig. 9.3b). These vertebrates maintain the presence of neural stem cells and progenitor cells into adulthood resulting in rather more widespread neurogenesis and the capacity of regeneration after brain injury. When radial glial cells lining the ventricle persist, no cuboid multiciliated ependymal cells (Fig. 9.5) are formed at the ventricle, and there is a lack of **star-shaped astrocytes** (Chap. 5 in [20]). Therefore,

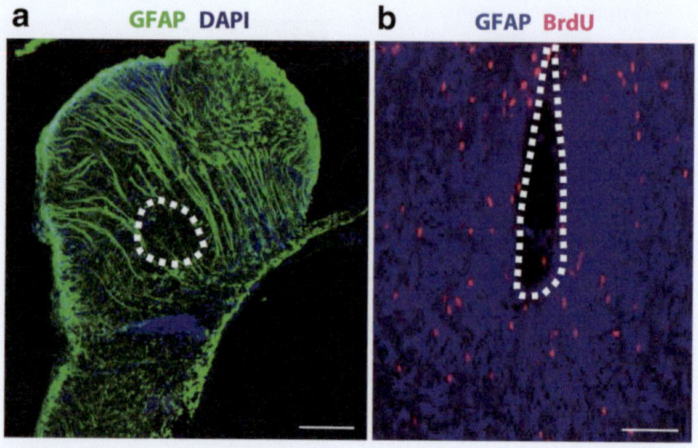


**Fig. 9.4** **Glial reaction after brain injury** in zebrafish (a) and mouse telencephalon (b). (**a**) Radial glia/tanycyte fibers depicted by GFAP (*green*) and nuclei labeled by DAPI (*blue*) 2 days after stab wound injury in the telencephalon of zebrafish. The *dashed line* indicates the injury site. Note the absence of reactive gliosis compared to (**b**). (**b**) Reactive astrogliosis after stab wound injury in the telencephalon of mice; hypertrophic astrocytes are labeled by GFAP (*blue*) and the proliferation marker BrdU (*pink*). The *dashed line* indicates the injury site. Scale bars: 80 μm (**a**), 100 μm (**b**)

radial glial cells that persist into adulthood and hence continue to line the ventricles are often referred to as **ependymoglia** or **tanycytes.**

The function of radial glial cells in an adult vertebrate brain is best understood in the zebrafish. Radial glial cells in the adult zebrafish brain continue to proliferate and give rise to new neurons that are added to the parenchyma from the ventricular site [5]. This allows the continued growth of the zebrafish brain, following the continuing growth of the entire organism including its sense organs. Despite their ongoing proliferation, radial glia in the adult zebrafish brain also possess many elaborated horizontal protrusions (Fig. 9.3b), reminiscent of the protrusions exhibited by astroglial cells enwrapping synaptic contacts between neurons in the mammalian brain. Probably, these cells also perform the typical synaptic functions of glia. They express glutamine synthase and are involved in transmitter recycling as described above.

The persistence of radial glial cells into adulthood has also important consequences after brain injury. Some radial glial processes are disrupted by brain injury in zebrafish (Fig. 9.4a), but there is no scar formation and the injury site is healed and hardly detectable after a few days [3]. In mammalian brains, however, astrocytes and NG2 glia become hypertrophic and upregulate many extracellular matrix molecules, leading to wound closure and scar formation, a process called **reactive gliosis** (Fig. 9.4b) [32]. While the deposits of extracellular matrix molecules help to shield the remaining brain parenchyma from the injury site, these deposits also contain molecules, such as chondroitin sulfate proteoglycans that inhibit axonal growth. Thus, the persistence of radial glia in the adult zebrafish brain is beneficial in injury and reminiscent of the fast wound closure without scar formation during mammalian development.

#### 9.2.2.2 Ependymal Cells and Their Functions

Ependymal cells line the ventricle, forming a barrier between the fluid in the ventricle, the cerebrospinal fluid (CSF), and the CNS by their tight junctional coupling. They also contribute to the flow of the CSF by beating of their motile cilia.

In mammalian brains, ependymal cells are cuboidal epithelial cells lining the ventricle (Fig. 9.5). Ependymal cells do not divide (in contrast to the radial glial cells described above) and are characterized by passive electrical properties similar to astrocytes but different from radial glia and NG2 glia (see below; [24]). Ependymal cells are coupled by gap junctions and other junctional complexes and hence form a **partial barrier between the CSF and the brain parenchyma,** which is less tight than the BBB [7, 20]. This barrier is open for many CSF proteins which then enter the adult brain parenchyma. Therefore the CSF composition has to be tightly regulated, which is done by the tightly coupled choroid plexus epithelium. It forms the blood CSF barrier, the interface between the blood and the CSF.

The overall flow of the CSF is achieved by active secretion from the choroid plexus cells and reabsorption at the arachnoid villi in the walls in the superior sagittal sinus and at the cribiform plate via extracranial lymphatics. However, ependymal cells contribute to the CSF flow by beating of cilia (Fig. 9.5b). The most crucial location is the cerebral aqueduct, the narrowest lumen between the communicating ventricles.

Defects in cilia movements result in stenosis at this critical passage and hydrocephalus due to accumulating CSF.

Interdispersed amongst the ependymal cells are radial glial cells/tanycytes that are particularly frequent in specific brain regions (e.g., the hypothalamus and the adult neural stem cell niches, Fig. 9.5c). These remaining radial glial cells act as **adult neural stem cells** in some regions of the adult mammalian brain, where they continue to generate neurons in the adult brain. For example, new interneurons migrate from the wall of the lateral ventricle to the olfactory bulb [23]. In addition, these radial glia-like adult neural stem cells self-renew and replenish neurogenesis when neuronal progenitors have been eliminated and also possess the potential to generate non-neuronal progeny, e.g., oligodendrocytes and astrocytes. These hallmarks, self-renewal and multipotency, define them as adult neural stem cells. Notably, also in the second neurogenic niche of the adult mammalian brain, the dentate gyrus, adult neural stem cells are radial glial cells, reminiscent of their role during development [23, 32].

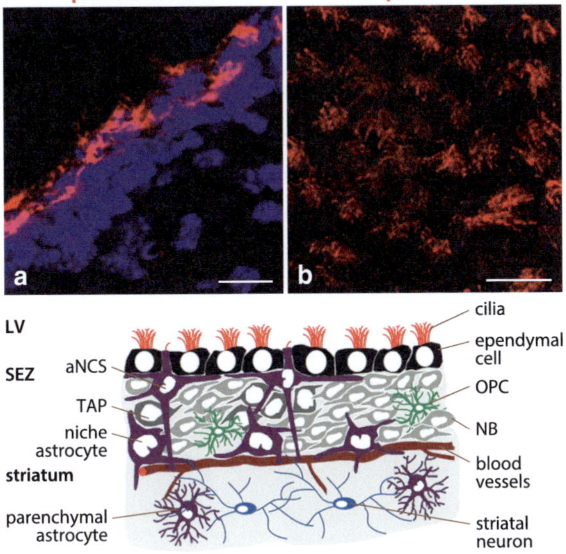


**Fig. 9.5** **Ependymal cells and adult neural stem cells.** (**a**) Fluorescent micrograph depicting ependymal cells lining the ventricle in side view. DAPI (*blue*) labels cell nuclei, prominin1 (*red*) stains cilia. Scale bar: 20 µm. (**b**) Fluorescent micrograph depicting ependymal cells in en-face view onto the ventricular surface. In both panels cilia of ependymal cells are labeled in *red* by prominin1 immunostaining. Scale bar: 20 µm. (**c**) Schematic drawing of the cellular composition of the lateral wall of the lateral ventricle in the adult mouse brain. Adult neural stem cells (*aNSCs*) exhibiting radial glia hallmarks are intercalated into the ependymal layer. Prominin1 (*red*) labels both multiple cilia of ependymal cells and primary cilia of aNSCs. *OPC* oligodendrocyte precursor cell, *TAP* transient amplifying cell, *NB* neuroblast, *LV* lateral ventricle

### 9.2.2.3 Astrocytes and Their Functions

The most frequent glial type in the adult mammalian brain parenchyma are the star-shaped astrocytes. They enwrap many synapses as part of the **tripartite synapse** (Fig. 9.2) and form connections to the blood vessels. They link neuronal activity to the supply with oxygen and nutrients by regulating blood flow.

Astrocytes have been named according to their star-shaped morphology (von Lenhossék, 1893 in [20, 22]) and comprise two major subtypes, the **protoplasmic astrocytes** forming a regular lattice extending fine nonoverlapping processes in close contact with synapses in the gray matter of the brain (Fig. 9.6a) and the **fibrous astrocytes** with thicker, stellate processes in the white matter of the brain where axon bundles project over long distances (Fig. 9.6b). In contrast to the radial glial cells persisting in adult vertebrates, both astrocytes as well as ependymal cells are postmitotic glial cells with little turnover in the healthy adult brain. They are further characterized by their passive electrical properties and large $K^+$ conductance due to inward rectifying $K^+$ channels with large conductance (see Chap. 7).

#### Functions of Astroglia at the Tripartite Synapse

Synapses are almost completely enwrapped by glial cells – in the CNS by star-shaped or radial astrocytes (such as **Bergmann glia** in the cerebellum or **Müller glia** in the retina) and in the PNS by **satellite cells** and nonmyelinating **Schwann cells**. Notably, a single astrocyte may enwrap and hence interact with thousands of synapses thereby integrating synaptic activity of many neurons (a single reconstructed Bergmann

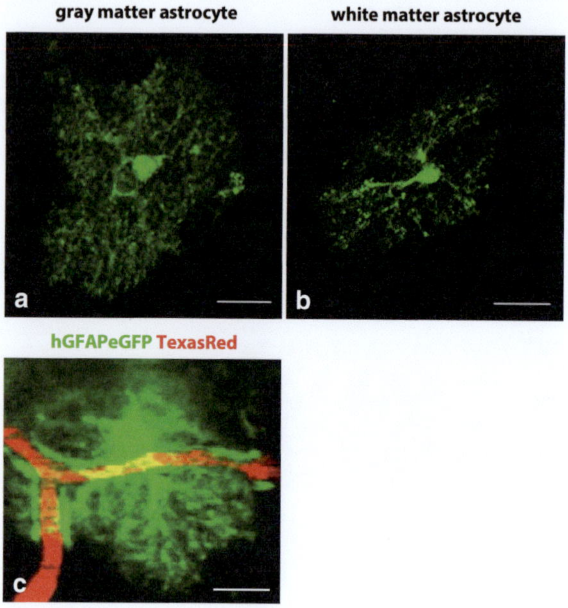


**Fig. 9.6** **Astrocytes.** Fluorescence micrographs of GFP-labeled astrocytes in the adult mouse cerebral cortex gray matter (**a**, **c**) and white matter (**b**). Note the astroglial endfeet aligned to the Texas Red labeled blood vessel in (**c**). Scale bars: 20 μm (**a**, **b**); 10 μm (**c**)

glia contacted about 6,000 synapses [20]). Astroglia are already important during synapse formation providing key signals, such as cholesterol and thrombospondins to promote both the number and the strength of synaptic connectivity. When synapses have been established, the enwrapping glial cells are responsible for removal and recycling of transmitters, sensing synaptic activity via their neurotransmitter receptors and modulating synaptic activity by gliotransmission. Genetic deletion of the AMPA glutamate receptor exclusively in astrocytes in the adult brain resulted in retraction of the fine astrocyte processes from synapses and deficits in fine motor coordination in mice [33], demonstrating the key role of astrocyte processes at the synapse and of astroglial neurotransmitter receptors as the sensing units.

Astrocytes also extend processes to the BM surrounding the blood vessels (Fig. 9.6c) and contribute to adapt the blood flow to local brain activity [17]. Local brain activity needs to be matched by increased supply of blood flow, a phenomenon called functional hyperemia or **neurovascular coupling**. Upon synaptic activity, neurotransmitters, such as glutamate, also bind to metabotropic glutamate receptors on astrocyte processes at the synaptic sites. This elicits increased intracellular $Ca^{2+}$ concentrations within astrocytes resulting in activation of cyclooxygenase-1 generating vasoactive prostaglandins which are released and cause vasodilation [30]. Thus, astrocytes are important to match synaptic activity with blood supply by neurovascular coupling.

The role of *Drosophila* neuropil glia in transmitter removal at synaptic sites is exerted by **protoplasmic astrocytes** in the mammalian brain. Astrocytes express high levels of the astrocyte-specific glutamate transporter GLAST (Slc1a3; EAAT1) and Glt-1 (Slc1a2; EAAT2). Lack of these glial transporters results in severe defects of synaptic communication and brain function. Astrocytes also possess glutamine synthase fuelling the **glutamate-glutamine shuttle** (Fig. 9.2). Astroglia **buffer $K^+$** when extracellular $K^+$ levels increase upon neuronal activity using their Kir channels and their large effective volume due to gap junctional coupling across astrocytes (Fig. 9.2b).

Besides these passive functions in synaptic communication, astrocytes also act as direct and active synaptic partners that receive synaptic signals from neurons via their transmitter receptors and signal to neurons via the release of gliotransmitters ([29, 33], Fig. 9.2a). Astrocytes possess functional purinergic as well as glutamatergic, GABAergic, cholinergic, and adrenergic receptors that affect $Ca^{2+}$ signals in and between astroglial cells as described above. Rather than by fast $Na^+$-dependent action potentials, astrocytes largely **communicate by slow $Ca^{2+}$ waves**, prompting the concept that glia provides analog information processing as opposed to the binary code employed by most neurons [22]. Increase in intracellular calcium in astrocytes also results in release of gliotransmitters, such as glutamate, D-serine or ATP. Astrocytes possess various mechanisms by which they release glutamate, including the vesicular release machinery. Selective deletion of the SNARE-based release machinery in astrocytes has profound effects on synaptic efficiency and sleep homeostasis [14]. When released, the gliotransmitters, such as glutamate or D-serine, bind to extrasynaptic receptors of neurons, including NMDA receptors. If the neuron is still depolarized, this can trigger the influx of $Ca^{2+}$ through these NMDA receptors resulting in **modulation of synaptic strength**. Thus, glia are active and direct synaptic partners shaping neuronal communication in a dynamic manner with profound behavioral relevance [29, 33] (Fig. 9.2).

Astroglial cells **react to injury** by becoming hypertrophic, increasing in number, and accumulating around the injury site: they contribute to wound healing as well as scar formation [32]. One of the first changes of reactive astrocytes appears to be the loss of polarity. Astroglial endfeet at the BM surrounding blood vessels possess enriched densities of $K^+$ channels next to water channels, the aquaporins. In this way, K-buffering is accompanied by water uptake to avoid osmotic pressure changes. Disruption of clustering of these channels is one of the first effects after injury. Reactive astrocytes appear to be impaired in K-buffering due to loss of polarity and gap-junctional coupling. Extracellular $K^+$ concentration is increased after injury. Due to downregulation of glutamine synthase mRNA and protein levels, the glutamate-glutamine shuttle is reduced with a subsequent increase in extracellular glutamate. Both these changes result in overexcitability of neurons leading to a wave of secondary neuronal death, called **excitotoxicity**. As a result, epileptic foci with abnormally high activity can form in regions of brain injury [37]. Besides these changes aggravating injury, reactive astrocytes also perform beneficial roles, such as limiting the size of injury and the degree of inflammation [32]. Some astrocytes even dedifferentiate into proliferating cells with neural stem cell hallmarks and provide a novel cell source to regenerate neurons [32]. This is particularly relevant for sites distant from the niches of adult neural stem cells (see above), which is the case for most brain regions.

### 9.2.2.4 Oligodendrocytes and Their Functions

Glial cells specialized to wrap and ensheath axonal processes are also present in vertebrates. These are the **oligodendrocytes** in the CNS (Fig. 9.7a) and **myelinating Schwann cells** in the PNS [20].

The insulation is achieved by vertebrate-specific **myelin**. Myelin is formed by the tightest possible packing of glial membranes with hardly any cytoplasm between the closely apposed membranes connected by myelin-specific proteins (Fig. 9.7b). This lipid-enriched sheath insulates the axon to such an extent that the action potentials' conduction speed can reach

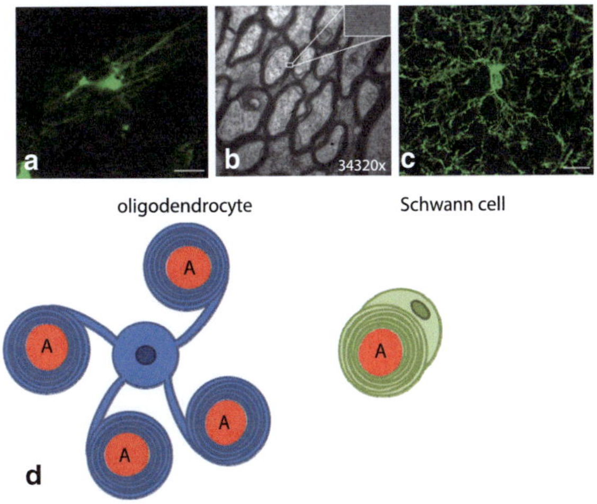


**Fig. 9.7 Oligodendrocytes and NG2 glia.** (**a**) Fluorescent micrograph of a GFP-labeled oligodendrocyte in the adult mouse corpus callosum shows the orientation of the myelin fibers parallel to the axons. (**b**) Electron microscopic picture of myelin in the adult mouse corpus callosum. (**c**) A polydendrocyte/NG2 cell labeled with NG2-immunostaining (*green*) in the adult mouse cerebral cortex gray matter with its highly ramified processes. (**d**) Schematic drawing depicting an oligodendrocyte myelinating several axons (*A* axon) in contrast to Schwann cells of the PNS myelinating only one axon. Scale bars: 20 μm

up to 120 m/s (Chap. 7). Insulation and accordingly conduction velocity depend on the number of **lamellas**, i.e., the thickness of the isolating myelin sheath, and the length of the **internodes**, i.e., the distance between **nodes of Ranvier**, a short myelin-free gap where ions can flow and action potentials are regenerated. Thus, axon potentials 'jump' in a saltatory fashion from node to node thereby greatly increasing conduction velocity (see Chap. 7). Periaxin is a protein interacting with the dystroglycan complex at the BM surrounding the nerve in the PNS and necessary for lateral growth of the myelin sheaths. In the mouse periaxin mutant, internode length is reduced to half and conduction velocity is also reduced to half of the normal value [6].

**Myelin sheath extension** is crucial because the organism may still grow after myelination occurred and hence axon length still increases, e.g., by > 4× after myelination has been completed in peripheral nerves. Thus, some molecular signals, such as those mediated by periaxin, are responsible for the lateral myelin sheath extension and internode length. Other signals, such as **neuregulin**, are key to regulate **myelin thickness**. Myelin thickness depends on the axon diameter with thicker axons being enwrapped with thicker myelin sheaths, supposedly due to higher levels of neuroregulin signaling. There is also a correlation between small and big oligodendrocytes enwrapping smaller and bigger caliber axons. **Small oligodendrocytes** form multiple processes contacting and myelinating many different axons. These axons are typically smaller with a diameter below 2 μm, such as in the optic nerve or the corpus callosum. The myelin formed by these smaller oligodendrocytes amounts to about 500 $\mu m^3$ per cell. **Large oligodendrocytes** myelinate few, but large diameter (>4 μm) axons and form much thicker myelin sheaths with about 50.000 $\mu m^3$ myelin/cell. These are amongst the first to myelinate during development, while the myelination of smaller axons occurs later in development. Thus, **thicker axons have thicker myelin** than smaller axons, a relation reflected in the constant ratio of the axon diameter divided by the diameter of its myelin sheath (g-ratio = 0.6–0.7). If neuregulin is overexpressed in neurons, their myelin sheaths become thicker, and reduced amounts of neuregulin due to haploinsufficiency (when one of the two alleles of the gene is deleted) results in reduced thickness of myelin sheaths [38]. Thus, axons signal to the ensheathing glia to regulate myelination. Conversely, myelinating oligodendrocytes or Schwann cells also signal in various ways to the neuron via their close axonal contacts. For example, oligodendrocytes modulate impulse propagation when depolarized, provide signals influencing neuronal survival, and regulate fast axonal transport. Especially the latter is affected in many neurodegenerative diseases. Myelinating cells in close contact with the neurons are important for research of neurodegenerative diseases. Glia-specific mutations are sufficient to elicit some neurodegenerative diseases, a concept referred to as **gliopathy**, i.e., diseases initiated by defects in glial cells.

The process of myelination is increasingly protracted in more complex brains. It is completed in the first postnatal weeks in mice, but continues for longer than 20 years in the human brain. This has been correlated to the prolonged phase of high neuronal plasticity, the **critical period** (see Chap. 25). Intriguingly, recent evidence suggests that myelination continues even thereafter life-long. Oligodendrocyte progenitors during development are characterized by expression of the transcription factor Olig2 which is a master regulator of their development [20]. Oligodendrocyte progenitors also express the platelet-derived growth factor receptor and the **proteoglycan NG2**. Proliferating cells with these proteins are also present in the adult rodent and human brain [8, 25, 39]. Genetic fate mapping turning on a marker gene in these progenitors in adult rodent brains revealed that their progeny differentiates into new myelinating oligodendrocytes in the white matter even at 6 or 9 months of age, when myelination has been previously thought to be completed. Thus, the continuation of neuronal plasticity into adult stages (see Chap. 25) seems to be accompanied by the continuation of myelination and a certain degree of oligodendrocyte turnover [9].

#### 9.2.2.5 NG2 Glia

NG2 glia are a relatively recently discovered type of macroglia, i.e., ectodermally derived glia [26]. Their main tasks are sensing neuronal activity and generating mature oligodendrocytes.

These cells are characterized by the **proteoglycan NG2** (chondroitin sulfate proteoglycan 4) on their cell surface and constitute around 5 % of the glia in the mature CNS [34]. They also express further genes characteristic for oligodendrocyte progenitor cells, such as the transcription factor Olig2 and the platelet-derived growth factor receptor alpha (PDGFRA). Accordingly, NG2 glial cells proliferate and generate oligodendrocytes, both at postnatal stages when myelination occurs, as well as in the adult brain [8, 34]. This lineage relation has prompted the names **polydendrocytes** (dendrocytes with many processes) or nonmyelinating oligodendrocytes [34]. Importantly, both proliferation as well as differentiation of these cells is regulated by physiological stimuli, such as motor activity, or brain injury [9, 34]. This suggests that the generation of new myelinating oligodendrocytes in the adult mouse brain is a highly regulated process.

In order to match oligodendrocyte generation to neuronal activity, the NG2 glia should be able to sense this activity. Indeed, NG2 glia, e.g., in the cerebral cortex white and gray matter, are electrically active, possess transmitter receptors, such as AMPA, NMDA, and $GABA_A$ receptors, and receive synapses [34]. Only few NG2 glia are coupled by gap junctions, but they rather serve as individual postsynaptic units with a presynaptic vesicle-containing site formed by a neuronal process and a postsynaptic density on the NG2 glia [34]. Strikingly, these neuron-glia synapses can also undergo plastic changes, such as long-term potentiation (LTP). NG2 glia are depolarized by synaptic activity. Depolarization seems to inhibit their proliferation and differentiation [9, 34] suggesting that reduced activity levels may result in release of this brake and differentiation into mature oligodendrocytes. Indeed, when NG2 cells differentiate into oligodendrocytes, this coincides with a loss of the synaptic contacts between neurons and NG2 glial cells [34] consistent with neuronal input regulating the proliferation and differentiation of NG2 glia into myelinating oligodendrocytes. This has obvious relevance for adult brain plasticity [9, 34]. Myelinated axons can not form side branches and establish new connections. However, if myelin is turned over in the adult brain this may provide an opportunity for establishing new connections, as supported by the accumulating evidence for unprecedented plasticity in the adult brain. Moreover, alterations in myelination change the conduction velocity and timing with profound implications in functional adaptations in the mature brain.

Notably, NG2 glia do not proliferate and generate mature oligodendrocytes in all brain regions [9, 34], neither in all species examined so far [3]. Accordingly, NG2 glia possess additional roles, such as formation of **perineural nets** that stabilize synaptic structures. Again these structures seem involved in regulating neuronal plasticity as digestion of the extracellular matrix-rich nets by chondroitinase in the adult cerebral cortex further enhances neuronal plasticity (see Chap. 25). Thus, both these types of macroglial cells, astrocytes and NG2 glia, are exquisite sensors of synaptic activity and partake in synaptic plasticity, but do so in different manners. Astrocytes are an active part of a tripartite synapse, while NG2 glia often are the postsynaptic compartment. Accordingly, these glial cells exert different effects – astrocytes influence the neuronal synapse by gliotransmission and transmitter recycling while NG2 glia appear so far to exert their effects largely to signal to axons and form myelinating oligodendrocytes.

#### 9.2.2.6 Microglia Functions in Vertebrates

> Microglia cells are the defense system in the adult brain, and also monitor synaptic connections.

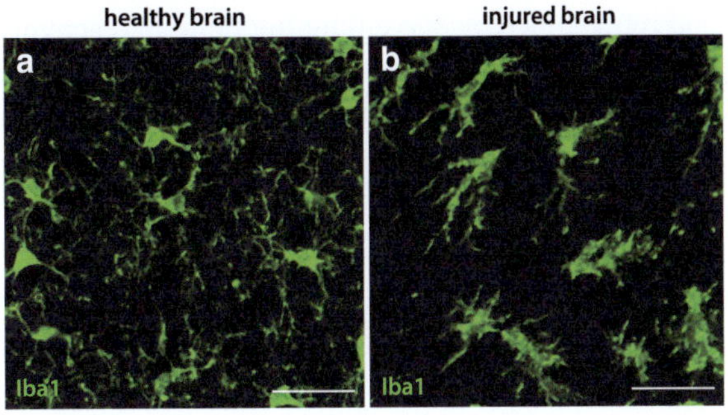


**Fig. 9.8** **Microglia.** Fluorescent micrographs of microglial cells in the adult mouse cerebral cortex labeled by *Iba1* without (**a**) and 3 days (**b**) after injury. Scale bars: 50 μm

Microglial cells (Fig. 9.8) are specialized in **phagocytosis** to engulf and remove any cellular debris and extracellular protein aggregates (including plaques of β-amyloid or prion protein). They also act as the **immune cells of the brain**, guarding against foreign particle invasion and communicating with cells of the immune system including antigen presentation [21].

##### Ontogeny of Microglia

Microglial cells are distinguished from all the above discussed macroglial cells by their mesodermal origin. Progenitors labeled by the lectin Iba1 (Fig. 9.8) enter the neuroepithelium during the first two embryonic trimesters in humans and around embryonic day 10 in rodents [4]. Early myeloid progenitors expressing the transcription factor PU.1 appear to act as the source for the early microglial progenitors entering the brain during embryogenesis. To which extent they play a role in the embryonic brain remains to be determined. A second wave of ameboid microglia arrives at early postnatal stages in the brain [15], but it is not yet known whether this is a transient population or gives rise to a second possibly distinct pool of microglia. In the adult brain, virtually no turnover of microglia occurs in the healthy state. This is in contrast to mononuclear phagocytes, such as perivascular, meningeal, and choroid plexus macrophages, which are constantly replaced by circulating blood progenitors.

##### Microglia Function in the Adult Brain

Microglial cells populate all brain regions with slight variations in density, e.g., lower in the white matter compared to gray matter regions. Microglial cells are highly active. Their processes continuously extend and retract, thereby surveilling the brain parenchyma [25]. They form transient (a few minutes) contacts with synapses which are modulated by synaptic activity [35]. Strikingly, synapses that have been contacted for longer appear to disappear, and microglia are actively involved in pruning synaptic connections during development [35]. As they also remove any cell debris or extracellular accumulations of proteins, microglial cells play important roles for the homeostatic surveillance in the normal brain and hence are best described as **'surveying'** (rather than resting) **microglia** (Fig. 9.8a) [15].

Upon minor damage, microglia orient all their processes towards the site of injury, shielding off this area [21]. A major damage signal is ATP, but microglia possess a plethora of receptors allowing their chemotactic reaction towards many damage, cytokine, and other signals [21]. Microglia remove debris (e.g., myelin components) or dead cells. They then release either inflammatory or anti-inflammatory **cytokines** [15, 21]. These cytokines exert complex effects on many other cells, including astrocytes which become hypertrophic and 'reactive' as described above and recruitment of specific immune cells, e.g., T-cells, into the brain. Strongly 'activated' microglial cells change their morphology by retracting their small processes and assuming a more ameboid state as **reactive microglia** (Fig. 9.8b). The reactive state of microglia differs depending on the cause for reaction. Microglia activated by interleukin 4 promote generation of new cells, such as oligodendrocytes and neurons in specific regions. Conversely, microglia activated by the bacterial stimulus LPS, by microbial infiltration or by foreign DNA and RNA detected via distinct Toll-like receptors that differentiate viral, bacterial, and fungal infection, do not support cell renewal, but rather result in efficient removal of the infective agents by lysis or **phagocytosis** [15].

Microglia also adopt full immune effector function by presenting antigens associated with MHC class II molecules to circulating T cells. They produce multiple inflammatory cytokines, reactive oxygen intermediates, and nitric oxide [15]. Highly phagocytotic microglia also remove neuronal processes and entire neurons and thereby aggravate some disease states, as observed in mouse models of demyelination and Parkinson's disease. Thus, similar to the dual roles of reactive astrocytes, also reactive microglia exert negative as well as positive, protective functions ameliorating the disease outcome (as, e.g., in mouse models of Alzheimer's or prion disease). Depending on the type of injury, **reactive microglia** consist of resident microglia which become activated as described above as well as of up to 50 % microglia newly recruited from a subpopulation of macrophages circulating in the blood. The latter are recruited into the brain across the BBB. To which extent these distinct populations may exert specific functions is not yet clear, but research will be crucial for progress towards promoting the beneficial and inhibiting the adverse functions of microglia after brain injury.

## 9.3 Summary and Outlook

Taken together, glial numbers and subtypes have dramatically increased during evolution of highly complex nervous systems. Widespread functions of glia across the animal kingdom include guidance and ensheathment of neurites, phagocytosis, and regulation of the extracellular milieu affecting synaptic communication in various ways.

The key role of glial cells in synaptic communication has been much extended by development of several glial subtypes, astrocytes, NG2 glia, and even microglia in the mammalian brain.

Glial cells mediate the wound reaction of the brain and include populations specialized to remove cell debris as well as cell sources to regenerate either oligodendrocytes (from NG2 glia) or neurons (from radial glia). Neural stem cells are a specialized set of vertebrate glia, typically radial glia, able to generate neurons and glial cells and the ability to repair several cell types after brain injury. In addition, a subset of reactive astrocytes can reacquire neural stem cell hallmarks after injury in the adult mouse brain. Glia clearly still holds many surprises which can now be addressed due to novel tools allowing deletion of genes of interest in a subtype-specific and inducible manner in the adult brain.

## References

1. Armulik A, Genove G, Mäe M, Nisancioglu MH, Wallgard E, Niaudet C, He L, Norlin J, Lindblom P, Strittmatter K, Johansson BR, Betsholtz C (2010) Pericytes regulate the blood–brain barrier. Nature 468:557–561
2. Bacaj T, Tevlin M, Shaham S (2008) Glia are essential for sensory organ function in *C. elegans*. Science 322:744–747

3. Baumgart EV, Barbosa J, Bally-Cuif L, Götz M, Ninkovic J (2012) Stab wound injury of the zebrafish telencephalon – a model for comparative analysis of reactive gliosis. Glia 60:343–357
4. Chan WY, Kohsaka S, Rezaie P (2007) The origin and cell lineage of microglia – new concepts. Brain Res Rev 53: 344–354
5. Chapouton P, Jagasia R, Bally-Cuif L (2007) Adult neurogenesis in non-mammalian vertebrates. Bioessays 29:745–757
6. Court FA, Sherman DL, Pratt T, Garry EM, Ribchester RR, Cottrell DF, Fleetwood-Walker SM, Brophy PJ (2004) Restricted growth of Schwann cells lacking Cajal bands slows conduction in myelinated nerves. Nature 431:191–195
7. Del Bigio MR (2010) Ependymal cells: biology and pathology. Acta Neuropathol 119:55–73
8. Dimou L, Simon C, Kirchhoff F, Takebayashi H, Götz M (2008) Progeny of Olig2-expressing progenitors in the grey and white matter of the adult mouse cerebral cortex. J Neurosci 28:10434–10442
9. Dimou L, Götz M (2012) Shaping barrels: activity moves NG2 glia. Nat Neurosci 15:1176–1178
10. Doherty J, Logan MA, Tasdemir OE, Freeman MR (2009) Ensheathing glia function as phagocytes in the adult *Drosophila* brain. J Neurosci 29:4768–4781
11. Freeman MR, Doherty J (2006) Glial cell biology in Drosophila and vertebrates. Trends Neurosci 29:82–90
12. Götz M, Huttner WB (2005) The cell biology of neurogenesis. Nat Rev Mol Cell Biol 6:777–788
13. Grosjean Y, Grillet M, Augustin H, Ferveur JF, Featherstone DE (2008) A glial amino-acid transporter controls synapse strength and courtship in *Drosophila*. Nat Neurosci 11: 54–61
14. Halassa MM, Haydon PG (2010) Integrated brain circuits: astrocytic networks modulate neuronal activity and behaviour. Annu Rev Physiol 72:335–355
15. Hanisch UK, Kettenmann H (2007) Microglia: active sensor and versatile effector cells in the normal and pathologic brain. Nat Neurosci 10:1387–1394
16. Hartline DK (2011) The evolutionary origins of glia. Glia 59:1215–1236
17. Iadecola C, Nedergaard M (2007) Glial regulation of the cerebral microvasculature. Nat Neurosci 10:1369–1376
18. Jackson FR, Haydon PG (2008) Glial cell regulation of neurotransmission and behavior in *Drosophila*. Neuron Glia Biol 4:11–17
19. Jackson FR (2011) Glial cell modulation of circadian rhythms. Glia 59:1341–1350
20. Kettenmann H, Ranson BR (2012) Neuroglia, 3rd edn. Oxford University Press, Oxford/New York
21. Kettenmann H, Hanisch UK, Noda M, Verkhratsky A (2011) Physiology of microglia. Physiol Rev 91:461–553
22. Kettenmann H, Verkratskhy A (2008) Neuroglia: the 150 years after. Trends Neurosci 31:653–659
23. Kriegstein A, Alvarez-Buylla A (2009) The glial nature of embryonic and adult neural stem cells. Annu Rev Neurosci 32:149–184
24. Liu X, Bolteus AJ, Balkin DM, Henschel O, Bordey A (2006) GFAP-expressing cells in the postnatal subventricular zone display a unique glial phenotype intermediate between radial glia and astrocytes. Glia 54:394–410
25. Nimmerjahn A, Kirchhoff F, Helmchen F (2005) Resting microglial cells are highly dynamic surveillants of brain parenchyma in vivo. Science 308:1314–1318
26. Nishiyama A, Komitova M, Suzuki R, Zhu X (2009) Polydendrocytes (NG2 cells): multifunctional cells with lineage plasticity. Nat Rev Neurosci 10:9–22
27. Oikonomu G, Shaham S (2011) The glia in *C. elegans*. Glia 59:1253–1263
28. Parker RJ, Auld VJ (2006) Roles of glia in *Drosophila* nervous system. Semin Cell Dev Biol 17:66–77
29. Perea G, Navarrete M, Araque A (2009) Tripartite synapses: astrocytes process and control synaptic information. Trends Neurosci 32:421–431
30. Petzold GC, Murphy VN (2011) Role of astrocytes in neurovascular coupling. Neuron 71:782–795
31. Pinto L, Götz M (2007) Radial glia heterogeneity – the source of diverse progeny in the CNS. Prog Neurobiol 83:2–23
32. Robel S, Berninger B, Götz M (2011) The stem cell potential from glia – lessons from reactive gliosis. Nat Rev Neurosci 12:88–104
33. Saab AS, Neumeyer A, Jahn HM, Cupido A, Simek AAM, Boele HJ, Scheller A, Le Meur K, Götz M, Monyer H, Sprengel R, Rubio ME, Deitmer JW, De Zeeuw CI, Kirchhoff F (2012) Bergmann glial AMPA receptors are required for fine motor coordination. Science 337:749–753
34. Sakry D, Karram K, Trotter J (2011) Synapses between NG2 glia and neurons. J Anat 219:2–7
35. Schafer DP, Lehrman EK, Stevens B (2013) The "quad-partite" synapse: microglia-synapse interactions in the developing and mature CNS. Glia 61:24–36
36. Schwabe T, Bainton RJ, Fetter RD, Heberlein U, Gaul U (2005) GPCR signalling is required for blood–brain barrier formation in *Drosophila*. Cell 123:133–144
37. Seifert G, Schilling K, Steinhäuser C (2006) Astrocyte dysfunction in neurological disorders: a molecular perspective. Nat Rev Neurosci 7:194–206
38. Sherman DL, Brophy PJ (2005) Mechanisms of axon ensheathment and myelin growth. Nat Rev Neurosci 6: 683–690
39. Staugaitis SM, Trapp BD (2009) NG2 positive glia in the human central nervous system. Neuron Glia Biol 29: 1–10
40. Stork T, Bernardos R, Freeman MR (2012) Analysis of glial cell development and function in *Drosophila*. Cold Spring Harb Protoc 1:1–17

# The Autonomic Nervous System

# 10

Wilfrid Jänig

## 10.1 Neural and Neuroendocrine Regulation of Visceral Body Functions: An Overview

The body's motor activity responding to internal and external demands is only possible under controlled internal conditions maintaining an optimal environment for the function of the component cells, tissues, and organs. The associated mechanisms include the control of the fluid matrix of the body; gas exchange with the environment; ingestion and digestion of nutrients; transport of gases, nutrients, and other substances throughout the body; excretion of substances; body temperature; reproduction; and defense of body tissues.

The *efferent signals* from the brain to the periphery of the body by which the control of the internal milieu is achieved are *neural* (by the autonomic nervous system [ANS]) or *hormonal* (by the neuroendocrine systems). The *afferent signals* from the periphery of the body to the brain are neural, hormonal, or physicochemical (e.g., blood glucose level, temperature, etc.). The maintenance of the physiological parameters in a narrow range by the autonomic and endocrine systems (including the respiratory system) is called *homeostasis* [23].

The brain provides "sensorimotor programs" for the coordinated regulation of the internal environment of the body's tissues and sends efferent commands to the peripheral target tissues through autonomic and endocrine pathways. Integration – within the brain, between the representations of the autonomic, endocrine, and somatomotor system and of sensory systems – is essential for generating purposeful behaviors of the organism within its environment.

The scheme in Fig. 10.1 outlines the role of the ANS within the generation of behavior of vertebrates. The behavior consists of the coordinated activation (a) of somatomotor neurons to move the body in the environment and (b) of autonomic and neuroendocrine motor neurons to prepare and adjust the internal milieu of the body enabling the body to move. Thus, behavior is dependent on the somatic, the autonomic, and the neuroendocrine motor system [58].

These motor systems are closely integrated in spinal cord, brainstem, and hypothalamus and hierarchically organized, the neuroendocrine motor system being represented at the top of this hierarchy.

The motoneuron pools (final motor pathways) extend from the midbrain to the caudal end of the spinal cord for the somatomotor system and ANS. The neuroendocrine motor neurons are located in the periventricular zone of the hypothalamus.

The activity of the motor systems generating behavior is dependent on three major classes of input: (a) the interoceptive sensory systems monitoring the mechanical, thermal, and metabolic states of the body tissues and the exteroceptive sensory systems monitoring events in the or related to the environment and leading to reflex activity *(reflex)*; (b) the cortical system generating purposeful behavior based on higher nervous system processing (*cortical*); and (c) the behavioral state system consisting of intrinsic neural systems of the brain that control arousal, attention, vigilance,

W. Jänig
Physiologisches Institut,
Christian-Albrechts-Universität,
Olshausenstr. 40, 24098 Kiel,
Germany
e-mail: w.janig@physiologie.uni-kiel.de

C.G. Galizia, P.-M. Lledo (eds.), *Neurosciences - From Molecule to Behavior: A University Textbook*,
DOI 10.1007/978-3-642-10769-6_10, © Springer-Verlag Berlin Heidelberg 2013

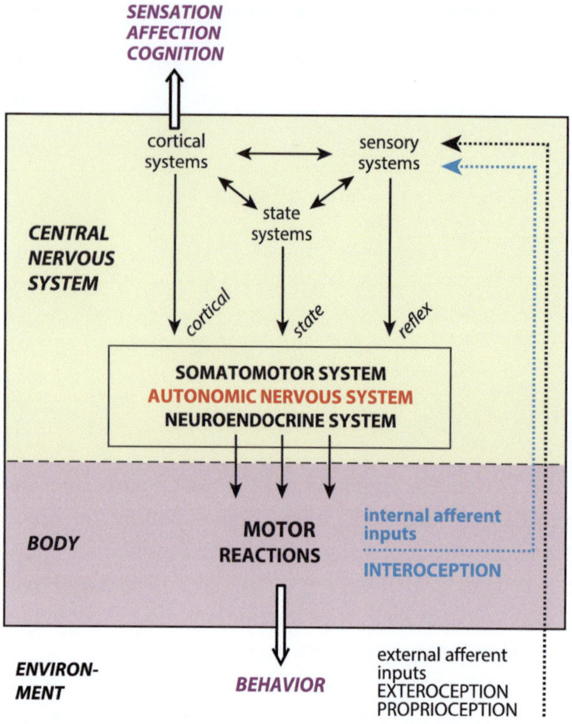


**Fig. 10.1 Functional organization of the nervous system to generate behavior in vertebrates.** The motor system, consisting of the somatomotor, the autonomic (visceromotor), and the neuroendocrine system, controls behavior. It is hierarchically organized into spinal cord, brainstem, and hypothalamus and receives three global types of synaptic input: from the sensory systems monitoring processes in the body or in the environment to all levels of the motor system generating reflex behavior (*reflex*); from the cerebral hemispheres responsible for cortical control of the behavior (*cortical*) based on neural processes related to cognitive and affective-emotional processes; from the behavioral state system controlling attention, arousal, sleep/wakefulness, circadian timing etc. (*state*). The three general input systems communicate bidirectionally with each other (*upper part* of the figure). Integral components of behavior are sensations, affective-motivational processes and cognitive processes which are dependent on cortical activity (Designed after Swanson (in Squire et al. [58]) with permission)

sleep, and wakefulness, and circadian timing of all body functions (*state*).

Thus in *vertebrates* the brain contains the "sensorimotor programs" for the coordinated regulation of the body's tissues and organs as well as skeletal muscles and the activity in the autonomic neurons is dependent on the intrinsic structure of the sensorimotor programs of the motor hierarchy and on its three global input systems. The peripheral correlates of the precise autonomic regulation are the functionally distinct motor pathways of the sympathetic and parasympathetic system and the enteric nervous system.

This chapter concentrates mainly on the peripheral ANS of mammals since almost all functional studies on the ANS have been done in these vertebrates [12, 23] (see series The Autonomic Nervous System [series editor Burnstock] [5]). Organization of the central circuits connected to the peripheral ANS will only be summarized [23, 38]. The pathophysiology of the ANS will not be covered [41, 53]. Some neurophysiological, anatomical, histochemical, and pharmacological studies have been conducted on the peripheral ANS of non-mammalian vertebrates [19, 47, 49]. These studies will be summarized, assuming that the principles of functioning of the ANS as have been worked out in mammals also hold for non-mammalian vertebrates. Neurobiological studies of the central circuits connected to the peripheral ANS are virtually absent in non-mammalian species.

In *invertebrates* regulation of the internal environment is also under neural control and coordinated with the behavioral repertoires such as feeding, reproduction, and defense. This involves the regulation of respiration, blood circulation, ingestion and digestion of nutrients, water regulation, excretion, and defense of body tissues. Most or many internal organs are innervated by neurons of the so-called **visceral nervous system**. The correlates of the regulation of these organs with behavior are represented in the cerebral and other ganglia of invertebrates.

These systems are extremely diverse and remain poorly understood in most cases. Here the text will concentrate exemplary on the neural regulation of the internal environment of mollusks, leeches, and crustaceans and the subject will be summarized including other invertebrates.

## 10.2 Organization of the Autonomic Nervous System in Mammals

### 10.2.1 Divisions of the Autonomic Nervous System

The generic term **autonomic nervous system** (ANS) and its division into the sympathetic, parasympathetic, and enteric nervous system, which was originally proposed by Langley [37], describes the efferent innervation of all tissues except striated muscle fibers and the brain and is now universally applied. The definition of the **parasympathetic** and **sympathetic nervous systems** is primarily anatomical by the outflow from the central nervous system (CNS). The parasympathetic (or craniosacral) system originates from the tectal and bulbar region of the brainstem and the sacral spinal region and the sympathetic (thoracolumbar) system from the spinal cord. These origins correspond to the somitic levels from which neural crest cells migrate to become parasympathetic neurons, sympathetic neurons, or enteric neurons.

The sympathetic and parasympathetic systems consist of two populations of neurons in series which are connected synaptically. The neurons that innervate the target organs lie entirely outside the CNS. Their cell bodies are grouped in **autonomic ganglia.** Their axons are unmyelinated (conduction velocity 0.1–1 m/s) and project from these ganglia to the target organs in peripheral nerves. These neurons are called sympathetic or parasympathetic **ganglion cells** or **postganglionic neurons.** The neurons that connect the CNS to the postganglionic neurons are the **preganglionic neurons** whose cell bodies lie in the spinal cord or brainstem. They send thinly myelinated or unmyelinated axons (conduction velocity 1–15 m/s; B fibers) from the CNS into the ganglia where they form synapses on the dendrites and somata of the postganglionic neurons (Fig. 10.2).

The **enteric nervous system** is intrinsic to the wall of the gastrointestinal tract and arises from neural crest cells of primarily vagal origin. It consists of synaptically interconnecting plexuses of neurons along its length which are responsible for the reflex activity involved in motility, secretion, and absorption during transit of food along the alimentary canal [12, 23]. This system will be described in Sect. 10.5.

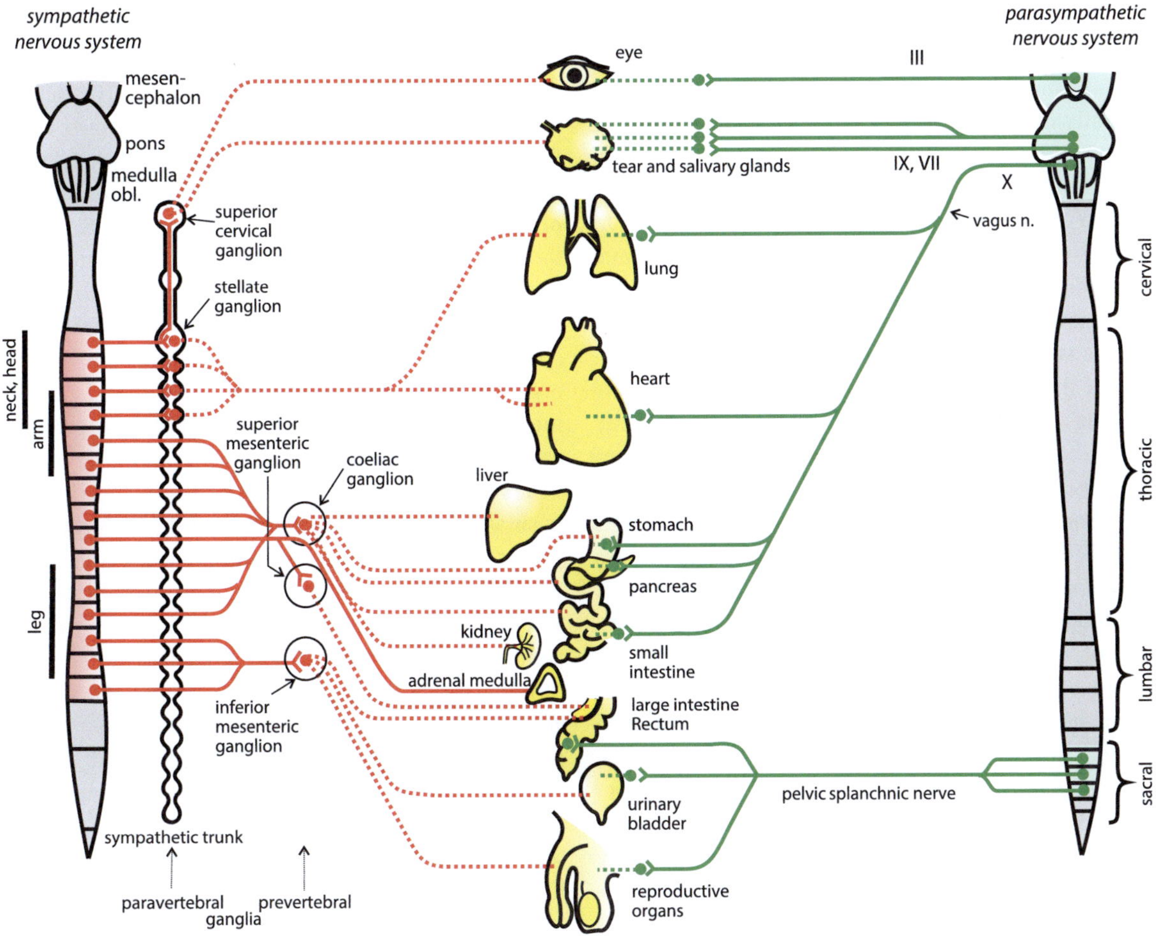


**Fig. 10.2 The peripheral autonomic nervous system.** *Bold lines*: preganglionic axons, *dotted lines*: postganglionic axons. The sympathetic innervation of blood vessels, sweat glands, and arrector pili muscles originate in all thoraco-lumbar spinal segments. The origin of the sympathetic innervation for the extremities and the head are indicated on the *left side*. (From Jänig [25] with permission)

#### 10.2.1.1 Sympathetic (Thoracolumbar) System

The cell bodies of the sympathetic preganglionic neurons lie in the spinal cord, extending from segments T1 to the rostral lumbar part. They are mainly clustered in columns of the intermediolateral zone and form a ladder-like arrangement throughout the thoracolumbar segments. The neurons receive inputs from interneurons in the dorsal horn and intermediate zone as well as direct projections from the medulla, pons, and hypothalamus and other spinal segments (propriospinal inputs).

The preganglionic neurons project ipsilaterally from the cord via the segmental ventral root, spinal nerve, and white ramus to the paravertebral chain or prevertebral ganglia (Fig. 10.2). Here they synapse with postganglionic neurons in the same and several adjacent segmental ganglia. Most of the functional (i.e., effective) synaptic connections occur segmentally. Neurons in the upper thoracic segments project rostrally to regulate targets in the head; many lumbar preganglionic neurons project caudally to regulate targets in the lower trunk or hindlimbs.

The *paravertebral ganglia* are interconnected to form a chain on either side of the vertebral column, extending from the base of the skull to the sacrum. There is usually one pair of ganglia per segment, except for the **superior cervical ganglia** (SCG) at the rostral end of the cervical sympathetic trunk and the stellate ganglia at the rostral end of the thoracic sympathetic chain, both of which represent several segmental ganglia fused together. Paravertebral ganglion cells project through the gray rami, the respective spinal nerves and the peripheral nerve trunks to the somatic tissues in all parts of the body and innervate blood vessels, pilomotor muscles, and sweat glands. Most postganglionic neurons in the SCG send their axons via the internal carotid nerve to join the nerves to target organs in the head; some postganglionic neurons in the SCG project to the upper two or three cervical segments. Postganglionic neurons in the stellate ganglion project through nerve branches to the heart and lungs and through gray rami to C4–T2 spinal nerves to supply the neck and upper limbs. Some paravertebral neurons project via the splanchnic nerves (including the pelvic nerves) to the viscera, where they mainly innervate blood vessels.

Some preganglionic neurons in the mid to lower thoracic and upper lumbar segments project across the chain into the splanchnic nerves (major, minor, and lumbar) and synapse in the *prevertebral ganglia* (celiac, superior mesenteric, and inferior mesenteric) which lie around the base of the large arterial branches from the abdominal aorta and supply the abdominal and pelvic viscera. These ganglia contain vasoconstrictor neurons and neurons that regulate both motility and secretion in the gastrointestinal tract and pelvic organs. They project in nerve bundles accompanying the vascular supply to the abdominal organs or in specialized nerves.

#### 10.2.1.2 Parasympathetic (Craniosacral) System

The cell bodies of the parasympathetic preganglionic neurons in the brainstem are grouped in distinct nuclei (e.g., dorsal motor nucleus of the vagus, nucleus ambiguus, salivary nuclei, Edinger-Westphal nucleus). They project via distinct cranial nerves to the ocular smooth muscles and glands (III, oculomotor, nerve), to the nasal and palatal glands (VII, facial and XI, glossopharyngeal nerves), and to the thoracic and abdominal viscera (X, vagus and XI nerves). In the sacral cord segments S1–S3 (S4), the preganglionic neurons lie in clusters across the intermediate zone and around the lateral part of the ventral horn, and project via the sacral ventral roots and pelvic splanchnic nerves to supply the pelvic organs. They receive inputs from local spinal interneurons as well as supraspinal and propriospinal projections [23].

*Parasympathetic* postganglionic neurons are located in small *ganglia* (often interconnected in a network or plexus) just outside, or even within the wall of the target organs. They are found in the head (ciliary ganglia to eye; pterygopalatine ganglia to lachrymal, nasal, and palatal glands; otic and submandibular ganglia to salivary glands) and near or in the wall of various effector organs (heart, airways, pancreas, gall bladder, pelvic organs). The cranial parasympathetic ganglia receiving preganglionic innervation from brainstem nuclei are generally larger aggregations of neurons than the parasympathetic ganglia of the trunk. Preganglionic neurons which project to the gastrointestinal tract synapse either with neurons that are part of the enteric nervous system (see Sect. 10.5) or with postganglionic neurons in the pelvic plexus which in turn synapse with enteric neurons of the colon/rectum.

### 10.2.2 Visceral Afferent Neurons

About 85 % of the axons in the vagus nerve and about 50 % of those in the splanchnic (spinal) nerves (greater, lesser, least, lumbar, and pelvic) are afferent. These **visceral afferents** come from sensory receptors in the internal organs. Their cell bodies lie in the ganglia of cranial nerves X and XI and in the dorsal root ganglia of the segments corresponding to the autonomic outflow (spinal visceral afferents). Most visceral afferent axons are unmyelinated, some are thinly myelinated. Spinal visceral afferents project through the *white rami* to the respective spinal segments. There is no reason to associate these visceral afferent neurons with only one of the autonomic systems [23].

Vagal and sacral visceral afferents from the lungs, cardiovascular system, gastrointestinal tract, evacuative organs, and reproductive organs project to the nucleus of the solitary tract in the brainstem or to the sacral spinal cord. Most of these afferents react to distension and contraction of the organs. Their activity encodes intraluminal pressure (e.g., arterial baroreceptor afferents, afferents from the urinary bladder) or volume (afferents from the gut, atria, and lungs). Some are chemosensitive (arising from arterial chemoreceptors in the carotid bodies, chemosensors in the gut mucosa, and osmosensors in the liver). Most vagal visceral afferents do not signal noxious events. Pelvic visceral afferents signal noxious events as well as distension and contraction; pain arises during strong contractions and distensions of the pelvic organs as well as after inflammation of the organ [23, 59].

The sensory receptors of thoracolumbar visceral afferents are situated in the serosa, the mesenteries, and the walls of some organs. Most are mechanosensitive and some are active only during tissue inflammation and ischemia. These afferent neurons are involved (a) in organ-specific spinal reflexes (e.g., cardio-cardiac, reno-renal) and (b) in nociception of all visceral organs [23].

## 10.3 Functional Autonomic Motor Pathways

### 10.3.1 Effector Responses to Activation of Autonomic Neurons

Table 10.1 describes the overall responses of the peripheral targets to activity in sympathetic and parasympathetic neurons. These responses have been defined by electrical stimulation of the respective nerves or by reflex activation:

1. Most target tissues react predominantly to only *one* of the autonomic systems.
2. A few target organs react to activity in both autonomic systems (e.g., iris, heart, urinary bladder).
3. Opposite reactions to activity in sympathetic and parasympathetic neurons are more the exception than the rule.
4. Most responses are excitatory, i.e., inhibition (e.g., relaxation of muscle, decrease of secretion) is rare.

Table 10.1 shows that the idea of antagonism between the parasympathetic and sympathetic nervous systems is largely a misconception. Where there are reciprocal effects on the target organs, it can usually be shown either that the systems work synergistically or that they exert their influence under different functional conditions.

### 10.3.2 Reflexes in Neurons of Functional Autonomic Motor Pathways

The data in Table 10.1 imply that there may exist many functionally separate autonomic pathways from spinal cord or brainstem to the autonomic target tissues. This is supported by functional neurophysiological studies *in vivo* on anesthetized animals for several peripheral autonomic pathways. Individual pre- or postganglionic neurons may exhibit ongoing activity and/or can be activated or inhibited by physiological peripheral or central afferent stimuli. The reflexes observed correspond to the effector responses that are induced by changes in activity in these neurons. The reflex patterns elicited by stimulation of various afferent input systems are characteristic for each functional autonomic pathway and are the result of integrative processes in spinal cord, brainstem, and hypothalamus. They therefore are physiological 'fingerprints' for each pathway.

Examples for the functional 'fingerprints' are shown in Fig. 10.3 for sympathetic postganglionic neurons innervating skeletal muscle or skin – cutaneous vasoconstrictor (CVC), muscle vasoconstrictor (MVC), sudomotor (SM) neurons innervating sweat glands – and in Fig. 10.4 for four functional types of sympathetic preganglionic neuron – CVC, MVC, and "inspiration"-type neuron in Fig. 10.4a;

**Table 10.1** Effects of activation of sympathetic and parasympathetic neurons on autonomic target organs

| Organ and organ system | Activation of parasympathetic nerves | Activation of sympathetic nerves | Adrenoceptor |
|---|---|---|---|
| *Cardiac muscle* | Decreased heart rate | Increased heart rate | $\beta_1$ |
| | Decreased contractility (only atria) | Increased contractility (atria, ventricles) | $\beta_1$ |
| *Blood vessels* | | | |
| Arteries | | | |
| In skin of trunk and limbs | 0 | Vasoconstriction | $\alpha_1$ |
| In skin and mucosa of face (nose, mouth, etc.) | Vasodilation | Vasoconstriction | $\alpha_1$ |
| In visceral domain | | Vasoconstriction | $\alpha_1$ |
| In skeletal muscle | 0 | Vasoconstriction | $\alpha_1$ |
| | | Vasodilation (cholinergic)[a] | |
| In heart (coronary arteries) | | Vasoconstriction | $\alpha_1$ |
| In erectile tissue (helical arteries and sinusoids in penis and clitoris) | Vasodilation | Vasoconstriction | $\alpha_1$ |
| In vagina, cervix, uterus | Vasodilation | ? (either vasodilation or vasoconstriction or both) | |
| In cranium | Vasodilation | Vasoconstriction | $\alpha_1$ |
| Veins | 0 | Vasoconstriction | $\alpha_1$ |
| *Gastrointestinal tract* | | | |
| Longitudinal and circular muscle | Increased motility | Decreased motility | $\alpha_2$ and $\beta_1$ |
| Sphincters | Relaxation | Contraction | $\alpha_1$ |
| *Capsule of spleen* | 0 | Contraction | |
| *Kidney* | | | |
| Juxtaglomerula cells | 0 | Release of renin | $\beta_1$ |
| Tubuli | 0 | Sodium resorption increased | $\alpha_1$ |
| *Urinary bladder* | | | |
| Detrusor vesicae | Contraction | Relaxation (minor) | $\beta_2$ |
| Trigone, internal urinary sphincter | 0 | Contraction | $\alpha_1$ |
| Urethra | Relaxation | Contraction | $\alpha_1$ |
| *Reproductive organs* | | | |
| Seminal vesicle, prostate | 0 | Contraction | $\alpha_1$ |
| Vas deferens | 0 | Contraction | $\alpha_1$ |
| Uterus | 0 | Contraction | $\alpha_1$ |
| | | Relaxation (depends on hormonal state) | $\beta_2$ |
| *Eye*[b] | | | |
| Dilator muscle of pupil | 0 | Contraction (mydriasis) | $\alpha_1$ |
| Sphincter muscle of pupil | Contraction (miosis) | 0 | |
| Ciliary muscle | Contraction (accommodation) | | |
| Tarsal muscle | 0 | Contraction (lifting of lid) | |
| Orbital muscle | 0 | Contraction (protrusion of eye) | |
| *Tracheo-bronchial muscles* | Contraction | Relaxation (mainly by circulating adrenaline) | $\beta_2$ |
| *Piloerector muscles* | 0 | Contraction | $\alpha_1$ |
| *Exocrine glands*[c] | | | |
| Salivary glands | Copious serous secretion | Weak serous secretion (submandibular gland) | $\alpha_1$ |
| | | Mucous secretion | |
| Lacrimal glands | Secretion | 0 | |

**Table 10.1** (continued)

| Organ and organ system | Activation of parasympathetic nerves | Activation of sympathetic nerves | Adrenoceptor |
|---|---|---|---|
| Nasopharyngeal glands | Secretion | | |
| Bronchial glands | Secretion | ? | |
| Sweat glands | 0 | Secretion (cholinergic) | |
| Digestive glands (stomach, pancreas) | Secretion | Decreased secretion or 0 (depends also on changes of blood flow) | |
| Mucosa (small, large intestine) | Secretion (mainly in large intestine) | Decreased secretion or reabsorption | |
| *Pineal gland* | 0 | Increased synthesis of melatonin | $\beta_2$ |
| *Brown adipose tissue* | 0 | Heat production | $\beta_3$ |
| *Metabolism* | | | |
| Liver | Decrease of glucose secretion | Glycogenolysis, Gluconeogenesis | $\beta_2$ |
| Fat cells | 0 | Lipolysis (free fatty acids in blood increased) | $\beta_2$ |
| β-cells in islets of pancreas | Secretion | Decreased secretion | $\alpha_2$ |
| α-cells in islets of pancreas | 0 | Secretion of glucagon | β |
| *Adrenal medulla*[d] | 0 | Secretion of adrenaline and noradrenaline | |
| *Lymphoid tissue* | 0 | Depression of immune response | $\beta_2$ |

Modified from Jänig [23] with permission
[a]Only in some species
[b]Chorioid blood vessels are also innervated by parasympathetic neurons and dilate on their activation
[c]Secretion of exocrine glands by activation of secretomotor neurons accompanied by dilatation of associated vasculature
[d]Cells in the *adrenal medulla* are ontogenetically homologous to sympathetic postganglionic neurones. When activated by their preganglionic axons, subgroups of medullary cells release either adrenaline or noradrenaline directly into the circulation. In humans, 85 % of adrenal medullary output is adrenaline. Most of the effects of circulating catecholamines are mediated through adrenaline's actions on β-adrenoceptors
0 no effect

motility-regulating (MR) neuron projecting to pelvic organs in Fig. 10.4b.

1. MVC neurons are inhibited by arterial baroreceptors and excited by arterial chemoreceptors and nociceptors (Figs. 10.3a, b and 10.4a).
2. Most CVC neurons are inhibited by stimulation of cutaneous nociceptors, spinal visceral afferents, arterial chemoreceptors, and central warm-sensitive neurons in the spinal cord and hypothalamus (Figs. 10.3 and 10.4a).
3. SM neurons are activated by stimulation of Pacinian corpuscles in skin (Fig. 10.3c) and by stimulation of nociceptors.
4. MR neurons innervating pelvic organs are excited or inhibited by stimulation of sacral afferents from the urinary bladder and hindgut and show powerful reflex activation (with afterdischarge) to mechanical shearing stimulation of the anal canal with absence of activation to mechanical stimulation of perigenital skin (Fig. 10.3b). They are not affected by arterial baroreceptor or chemoreceptor activation.
5. "Inspiration"-type neurons are active in inspiration and are powerfully activated by noxious trigeminal and other noxious stimuli (Fig. 10.4a), most of them not being under baroreceptor control.

This excursion can be generalized for the sympathetic and parasympathetic nervous system (Table 10.2): autonomic neurons show distinct reflex patterns suggesting that the messages generated in the CNS are faithfully transmitted to the autonomic effector tissues by separate parasympathetic and sympathetic pathways (Fig. 10.5). This is supported by two further observations: (a) The reflex patterns are the same in the preganglionic neurons and in the postganglionic neurons. Thus functionally similar pre- and postganglionic neurons are synaptically connected in the autonomic ganglia probably with no 'cross-talk' between different peripheral autonomic pathways (Table 10.2). (b) Various **neuropeptides** are colocalized

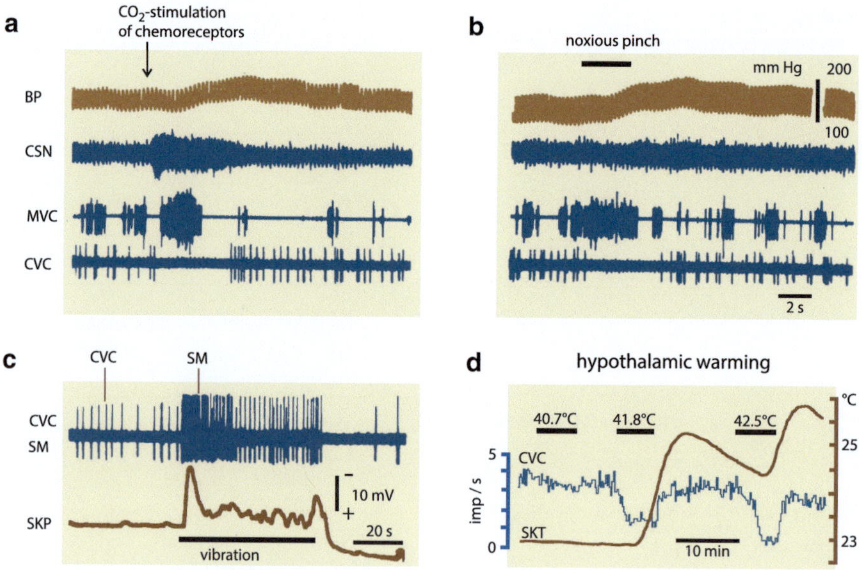


**Fig. 10.3** **Reflexes in muscle (MVC) and cutaneous (CVC) vasoconstrictor and sudomotor (SM) neurons** recorded from postganglionic axons in anesthetized cats. (**a**) Stimulation of carotid chemoreceptors by a bolus injection of $CO_2$-enriched saline into the carotid artery (at *arrow*) activated the MVC neurons (multi-axon preparation) and inhibited the CVC neuron (recorded simultaneously). Increased chemoreceptor afferent activity in the carotid sinus nerve (*CSN*) was monitored. The increase of blood pressure (*BP*) evoked by chemoreceptor stimulation led to a baroreceptor-mediated inhibition of MVC activity but not of CVC activity. (**b**) Stimulation of cutaneous nociceptors by pinching the ipsilateral hindpaw (indicated by bar) also excited the MVC neurons and inhibited the CVC neurons but did not affect blood pressure. Same experiments as in (**a**). (**c**) Simultaneous recording of a single CVC neuron (small signal) and a single SM neuron (larger signal) innervating the central paw pad and the skin potential (*SKP*) recorded from the central paw pad. Stimulation of Pacinian corpuscles by vibration excited the SM neuron (correlated with change in SKP) and inhibited the CVC neuron. (**d**) Inhibition of CVC neurons to warming of the anterior hypothalamus. Note that the increase of skin temperature (*SKT*) on the central paw pad followed the depression of CVC activity (Data for a, b and c – from Jänig and Kümmel, unpublished. d – modified from Grewe et al. [15] with permission)

with the classical transmitters acetylcholine or noradrenaline ([14], Gibbins in [43]) in the autonomic neurons with some correlation between the function of the autonomic neuron (as defined by the target tissue) and the peptide or combination of peptides (see also Chap. 11). This *'neurochemical coding'* supports the idea that the peripheral ANS consist of many functionally and anatomically separate pathways, although the functions of most peptides are unknown and although there are considerable species differences.

## 10.4 Transmission of Signals in Peripheral Autonomic Pathways

In Sects. 10.2 and 10.3 we have learnt that the peripheral ANS consists of many functionally separate pathways which are called here **final autonomic motor pathways** (Fig. 10.5). This section describes the transmission of centrally generated impulse activity in preganglionic neurons to the effector cells [52, 54] (see other textbooks of physiology and Table 10.4). The description includes the transmitters used in the peripheral ANS, transmission of impulses through autonomic ganglia (including ganglionic integration and peripheral reflex activity), and neuroeffector transmission.

### 10.4.1 Transmitter Substances in Parasympathetic and Sympathetic Neurons

#### 10.4.1.1 Adrenaline and Noradrenaline

**Synthesis, release, and inactivation of noradrenaline (NAd) and adrenaline.** The catecholamines NAd and adrenaline are synthesized in the postganglionic neurons. The amino acid tyrosine is taken up by the neurons and converted to dihydroxyphenylalanine (DOPA). Decarboxylation of DOPA leads to dopamine and hydroxylation of dopamine to NAd. The three enzymes catalyzing these reactions are tyrosine hydroxylase (TH), DOPA-decarboxylase (DDC), and dopamine-β-hydroxylase (DBH). In the adrenal medulla and in some adrenergic neurons of some non-mammalian vertebrates including birds, NAd is changed into adrenaline by methylation at the nitrogen atom by the enzyme phenylethanolamine N-methyl-transferase (PNMT; Fig. 10.6a). These catalytic steps occur either in the cytoplasm of the neurons (TH, DDC) or are associated with the synaptic vesicles (DBH, PNMT). The activity of the enzymes TH und DBH, and therefore also the synthesis of NAd and adrenaline, are dependent on the neural activity of the postganglionic neurons (neuronal induction of the enzyme activity) [46].

NAd is stored in the small and large granular vesicles of the varicosities of the adrenergic neurons (Fig. 10.6b). Activation of the postganglionic axon leads to opening of voltage-sensitive $Ca^{2+}$ channels, increase of the intracellular concentration of $Ca^{2+}$, fusion of the vesicle with the plasma membrane, and release of the vesicle content into the extracellular space (quantal release of transmitter; see Chap. 8). Individual varicosities release the transmitter of *one* vesicle with a low probability of about $p = 0.01$ per action potential at a low discharge frequency. This probability of transmitter release increases up to ten times with increasing discharge rates of the postganglionic neurons. Thus, only 100–1,000 of the 10,000 varicosities formed, e.g., by a vasoconstrictor axon in its target tissue, release one vesicle each upon one action potential [23]. The released NAd diffuses from the varicosity to the effector cell membrane and reacts with adrenoceptors. NAd disappears from its site of release in three ways (Fig. 10.6b): (a) about 60 % is actively taken up by the varicosities and recirculated into the vesicles; (b) a small part diffuses to the venous site and gets into the general circulation; (c) a small part is taken up by the effector cells. NAd and adrenaline are degraded mainly in the liver but also in the target tissues and in the varicosities. The molecules are deaminated by *monoamine oxidase (MAO)* or methylated by *catechol-O-methyltransferase (COMT)* (Fig. 10.6b).

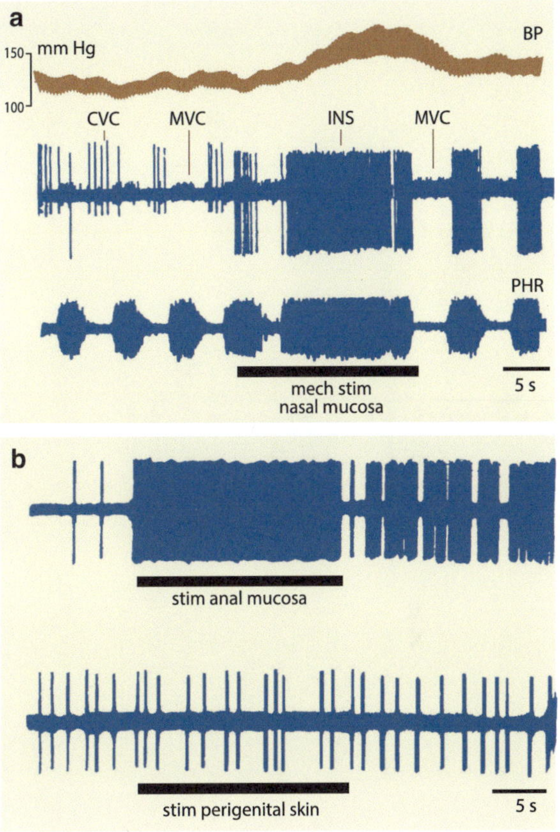


**Fig. 10.4** **(a) Reflexes in sympathetic neurons elicited by mechanical stimulation of the nasal mucosa.** Simultaneous recording of the activity in a preganglionic cutaneous vasoconstrictor (*CVC*) neuron, a preganglionic inspiratory (*INS*) neuron and a preganglionic muscle vasoconstrictor (*MVC*) neuron in an anesthetized cat. The activity was recorded from a strand of nerve fibers isolated from the cervical sympathetic trunk and from the phrenic nerve (*PHR*). Before stimulation, the CVC neuron was active in expiration, the MVC neuron in inspiration and expiration and the INS neuron was almost silent. Mechanical (probably noxious) stimulation of the nasal mucosa exciting trigeminal afferents inhibited the CVC neuron, activated the MVC neuron in inspiration as well as expiration and activated the INS neuron, but only in inspiration. Note that the reflexes in the neurons outlasted the stimulus and that the increase in blood pressure (*BP*) was correlated with the continuous *MVC* discharge. *BP* blood pressure (Modified from Boczeck-Funcke et al. [4]). **(b) Reaction of a preganglionic motility-regulating neuron projecting in a lumbar splanchnic nerve in cat** in response to mechanical stimulation of sacral afferents of the mucosa of the anal canal. Anal stimulation consisted of a light shearing stimulus applied at a moving frequency of about 0.5–1 Hz with a spatula to the anal mucosa for 20 s. Activation and afterdischarge in response to anal stimulation. No activation in response to mechanical stimulation of the perigenital skin (Modified from Bahr et al. [2] with permission)

**Table 10.2** Functional classification of autonomic neurons based on their reflex behavior *in vivo*

| Likely function | Location | Target organ | Likely target tissue | Major identifying stimulus[a] | Ongoing activity[b] |
|---|---|---|---|---|---|
| *A. Sympathetic neurons* | | | | | |
| *Vasoconstrictor muscle* post, pre | Lumbar[c] Cervical[d] | Muscle hindlimb; neck, head | Resistance vessels | Inhibition by baroreceptor stimulation | Yes |
| *Vasoconstrictor skin* post, pre | Lumbar Cervical | Skin hindlimb, head, neck | Thermoregulatory vessels | Inhibited by CNS warming | Yes |
| *Vasoconstrictor viscera* (incl. kidney) post, pre | Lumbar splanchnic | Pelvic viscera | Resistance vessels | Inhibition by baroreceptor stimulation | Yes |
| *Vasodilator muscle* post | Lumbar | Hindlimb muscle | Muscle arteries | Hypothalamic stim. | No |
| *Vasodilator skin* post | Lumbar | Hindlimb skin | Skin vasculature | ? | No |
| *Sudomotor* post, pre | Lumbar | Paw pads | Sweat glands | Vibration[e], Warming[f] | Yes, some |
| *Pilomotor* post | Lumbar | Tail, Back | Piloerector muscles | Hypothalamic stim. | No |
| *Inspiratory* pre | Cervical | Airways? | Nasal mucosal vessels | Inspiration | Yes |
| *Pupillomotor* post | Cervical | Iris | Dilator papillae muscle | Inhibition by light | Yes, some |
| *Motility regulating* | | | | | |
| *Type 1* post, pre | Lumbar splanchnic | Hindgut, urinary tract | Visceral smooth muscle, glands? | Bladder distension | Yes |
| *Type 2* post, pre | Lumbar splanchnic | Hindgut, urinary tract | Visceral smooth muscle, glands? | Inhibited by bladder distension | Yes |
| *Reproduction* | Lumbar splanchnic | Internal reproductive organs | Visceral smooth muscle, glands? | ? | No |
| *B. Parasympathetic neurons* | | | | | |
| *Pupilloconstrictor* post | Ciliary ggl | Iris | Constrictor pupillae m. | Excitation or inhibition by light | Yes |
| pre | EW lateral | Iris | Constrictor pupillae m. | Light | Yes |
| *Accomodation* pre | EW lateral | Ciliary body | Ciliary muscle | Target moving | Yes |
| *Cardiomotor* pre | NA | Heart | Pacemaker, atrial m. cells | Baroreceptor | Yes |
| *Bronchomotor* post, pre | NA, tracheobronch ggl | Trachea, bronchi | Smooth muscle cells | Tracheal mucosa, inspiration | Yes |
| *Bronchosecreto-motor* post | Tracheobronch ggl | Trachea, bronchi | Glands | Expiration | Yes |
| *Gastromotor excitatory* pre | DMNX | Stomach | Visc. smooth muscle | Inhibition by duodenal distension | Yes |
| *Gastromotor inhib.* pre | DMNX | Stomach | Visc. smooth muscle | Duodenal distension | Yes |
| *Urinary bladder* pre | Sacral SC | Urinary bladder | Visc. smooth muscle | Bladder distension | No |
| *Colon* pre | Sacral SC | Colon | Visc. smooth muscle | Colon distension | Yes |

Details about rates of ongoing activity and reflexes in pre- and postganglionic neurons see Refs. [22, 23, 27]

*Abbreviations*: *ggl* ganglion, *pre* preganglionic, *post* postganglionic, *EW* Edinger-Westphal nucleus, *DMNX* dorsal motor nucleus of the vagus, *NA* Nucleus ambiguous, *SC* spinal cord

[a]Only responses to the major (afferent) stimulus are listed. For most autonomic pathways more than one identifying afferent stimulus exists. Excitation to the stimulus unless inhibition is specified

[b]Not all neurons have spontaneous activity

[c, d]'Lumbar' refers to preganglionic and postganglionic axons in lumbar outflow; 'Cervical' refers to preganglionic axons in the cervical sympathetic trunk

[e, f]'Vibration'stimulation of Pacinian corpuscles in the cat paw; 'Warming' central warming in humans

**Adrenoceptors mediate the effects of NAd and adrenaline on the effector tissues.** Adrenoceptors are transmembrane proteins with seven helix structures in the membranes of the target cells. All adrenoceptors are G protein-coupled receptors, the coupling to the catecholamine occurring at an extracellular loop and the coupling to the G protein an intracellular loop. Based on pharmacological criteria (sensitivity to the artificial β-adrenoceptor agonists isoproterenol, NAd, and adrenaline; selectivity of adrenoceptor blockers) and molecular cloning the adrenoceptors are divided into three classes, each being further subdivided into three types [46, 53]:

$\alpha_1$-adrenoceptors ($\alpha_{1A}$, $\alpha_{1B}$ $\alpha_{1D}$) are coupled to $G_q$/$G_{11}$ proteins which activate phospholipase C to induce the hydrolysis of phosphatidylinositol-4′,5′-bisphosphate ($PIP_2$). This releases $IP_3$ followed by increased intracellular $Ca^{2+}$ and diacylglycerol (DAG). DAG and $Ca^{2+}$ activate protein kinase C.

$\alpha_2$-adrenoceptors ($\alpha_{2A}$, $\alpha_{2B}$, $\alpha_{2C}$) are coupled to $G_i$/$G_o$ proteins which inhibit adenylyl cyclase (with a decrease of cAMP) and stimulate phospholipase $A_2$.

β-adrenoceptors ($\beta_1$, $\beta_2$, $\beta_3$) are coupled to $G_s$ proteins and activate adenylyl cyclase with an increase of cAMP followed by activation of protein kinase A.

The cells in most tissues which can be influenced by adrenaline and NAd have α- as well ß-adrenoceptors in their membranes, both receptor types mediating mostly opposite effects. However, under physiological conditions the response of an organ to activation of its adrenergic innervation is dependent on which adrenoceptor dominates. The biological meaning of the adrenoceptors into $\alpha_1$, $\alpha_2$, $\beta_1$, $\beta_2$, and $\beta_3$ appears to be clear. These receptors are therefore listed in Table 10.1 (right).

**Adrenaline released by chromaffine tissues.** Chromaffine tissues consist of cells that have the same precursor cells as sympathetic postganglionic neurons. They synthesize either adrenaline or NAd, are concentrated in the **adrenal medullae**, and are synaptically innervated by sympathetic preganglionic neurons. They release either adrenaline or NAd into the circulation during excitation of these preganglionic neurons. The proportion of adrenaline and NAd varies between species, e.g., in humans 80 % adrenaline and 20 % NAd, in rabbits 100 % adrenaline and in whales 100 % NAd. Circulating adrenaline is a metabolic hormone in mammals leading to fast mobilization of glucose and free fatty acids from glycogen or fat deposits during emergency situations of the organism.

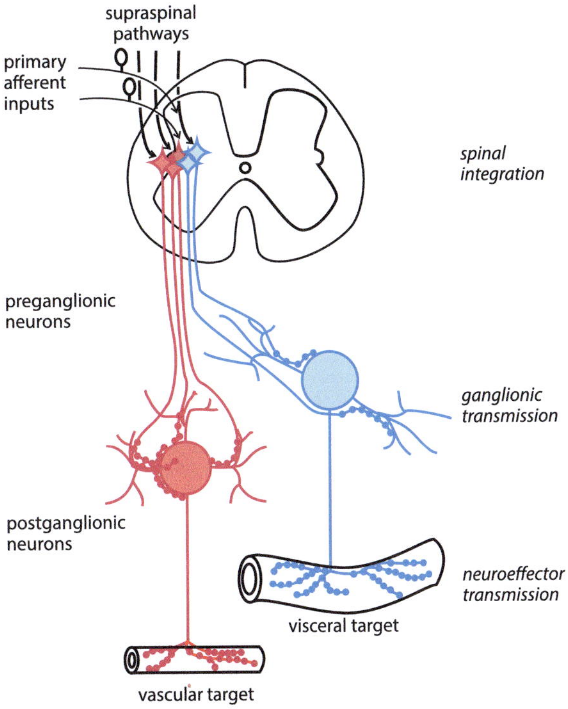


**Fig. 10.5 Organization of the spinal autonomic nervous systems into building blocks or final autonomic motor pathways.** Separate functional pathways extend from the CNS to the effector organs. Preganglionic neurons located in the spinal intermediate zone integrate signals descending from brainstem or hypothalamus and arising segmentally from primary afferent axons (via interneurons). The preganglionic neurons project to peripheral ganglia and converge onto postganglionic neurons. The postganglionic axons form multiple neuroeffector junctions with their target organs (Modified from Jänig [23] with permission)

#### 10.4.1.2 Acetylcholine (ACh)

In the nineteen-twenties, Otto Loewi demonstrated for the heart of the frog that the signaling from the vagus nerve to the heart is chemical [46]. He called the chemical substance mediating the excitation of the vagal neurons "Vagusstoff". This substance turned out to be acetylcholine (ACh). Otto Loewi's experiments were the first to demonstrate neurochemical transmission. ACh is the transmitter in all preganglionic neurons, in all parasympathetic postganglionic neurons, in most neurons of the enteric nervous system, and in a few groups of postganglionic sympathetic neurons (sudomotor neurons, muscle vasodilator neurons for some mammalian species).

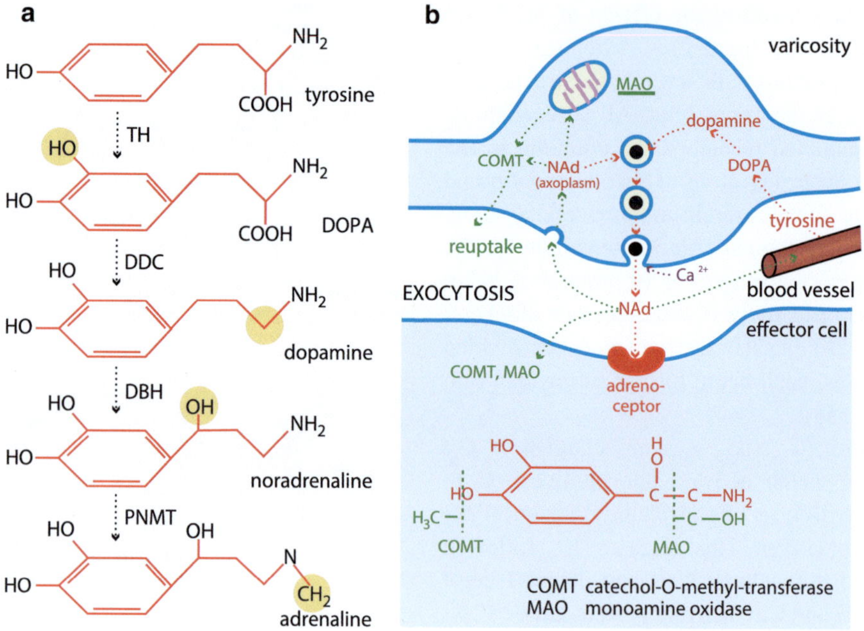


**Fig. 10.6** **Adrenalin and noradrenaline (*NAd*):** synthesis, release, inactivation. (**a**) Synthesis of NAd and adrenaline. *TH* tyrosine hydroxylase, *DDC* DOPA-decarboxylase, *DBH* dopamine-β-hydroxylase, *PNMT* phenylethanolamine *N*-methyltransferase, *DOPA* deoxyphenylalanine. (**b**) Release and inactivation of NAd. Small granular vesicles contain NAd and other substances (e.g., ATP). "Inactivation" of NAd by active reuptake into the varicosities by a membrane-associated transporter protein (*blue circles*) and uptake into storage vesicles (about 60 %), by diffusion away from the receptor sites and dilution in extracellular fluid or plasma, and by catabolism by monamine oxidase (*MAO*) and catechol-α-methyltransferase (*COMT*)

**Synthesis of ACh.** ACh is synthesized in the varicosities of axons from choline and acetyl-coenzyme A and is pumped into the vesicles by vesicular acetylcholine transporter. Release and inactivation of ACh occur in the same way as at the neuromuscular endplate (see Chap. 22). Depolarization of the nerve membrane is followed by inflow of $Ca^{2+}$ through voltage-sensitive $Ca^{2+}$ channels. This triggers the fusion of the vesicles with the cell membrane and the release of its content (quantal release of transmitter, see Chap. 8). ACh reacts with cholinergic receptors in the postsynaptic membrane and is quickly degraded by acetylcholinesterase into choline and acetate. Choline is actively taken up into the presynaptic terminal and recycled.

**ACh receptors.** Postganglionic neurons and effector cells have two types of cholinoreceptors:

Autonomic ganglia use ligand-gated **nicotinic cholinergic receptors**. They can be activated by nicotine and blocked by curare, as at the neuromuscular endplate. The muscular and the neuronal nicotinic cholinergic receptors differ slightly in their pentameric molecular structure and in their pharmacology.

All effector responses generated by parasympathetic postganglionic neurons, most motor neurons of the enteric nervous system, and some sympathetic cholinergic postganglionic neurons (see Table 10.1) are mediated by **muscarinic cholinergic receptors**. The cholinergic muscarinic effects can be mimicked by **muscarine**, the toxin of the fly agaric mushroom *Amanita muscaria*, and selectively blocked by **atropine**, an alkaloid of *Atropa belladonna*. Muscarinic cholinergic receptors are 7-transmembrane G protein-coupled receptors (7TM-GPCRs) (see Chap. 6). They are divided into five types according to their G protein coupling characteristics.

#### 10.4.1.3 Other Transmitters

**Adenosine 5′-triphosphate (ATP)**. In some postganglionic neurons or neurons of the enteric nervous system ATP is colocalized with NAd or ACh in the same vesicles. ATP is released during depolarization of the presynaptic terminals and reacts with **purinoceptors** ($P_{2x}$) in the effector cell membranes. This purinergic transmission can selectively be prevented by specific blockers.

The best known examples are the synaptic transmission from postganglionic noradrenergic neurons to the smooth muscle of arterioles and of the ductus deference.

**Neuropeptides.** Neuropeptides are colocalized in many functionally types of autonomic post- or preganglionic neurons with the classical transmitters ACh, adrenaline or NAd in the large granular vesicles. Up to four peptides or more are present in the same autonomic neurons. The peptide content in the autonomic neurons correlates to some extent with the function of the autonomic neurons, leading to the concept of neurochemical coding of the neurons. However, there are considerable species differences and differences between target organs. The function of most neuropeptides in autonomic neurons is unknown ([14], Gibbins in [43]). It is believed that they amplify or modulate the effect of the classical transmitters. Examples are:

Neuropeptide Y (NPY) in sympathetic vasoconstrictor axons potentiates the contractile effects of nerve-released NAd. In the heart, NPY released by sympathetic postganglionic axons inhibits vagal slowing generated by parasympathetic cardiomotor neurons.

Vasoactive intestinal peptide (VIP) is released from cholinergic vasodilator or secretomotor neurons. It contributes to vasodilation in, e.g., erectile tissue (where NO seems to be the primary transmitter), to vasodilation generated around sweat glands by activation of sudomotor neurons and to vasodilation generated within salivary glands by activation of secretomotor neurons.

Inhibitory motor neurons of the enteric nervous system innervating the circulatory musculature (see Figs. 10.5 and 10.14) use VIP and nitric oxide (NO) as primary inhibitory transmitter.

**Nitric oxide (NO).** NO is presumably the first example of a class of synaptic transmitters in the CNS and in the peripheral ANS which is not stored in presynaptic vesicles and does not mediate its effect on target cells via receptors. Together with the gases carbon monoxide (CO) and hydrogen sulfide ($H_2S$), NO is called a *gasotransmitter* [45]. NO is used by postganglionic (cholinergic) parasympathetic neurons to relax arteries and sinusoids of the erectile tissue, by motoneurons of the enteric nervous system to relax the circular musculature of the gut (Fig. 10.14) and probably other postganglionic neurons. The underlying mechanism is as follows (Fig. 10.7): Excitation of the neurons opens

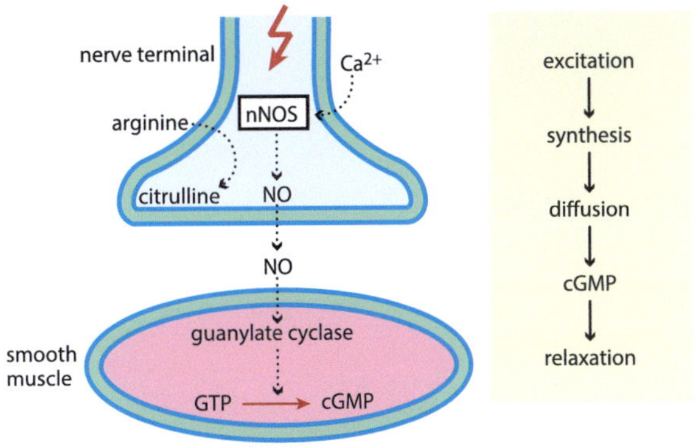


**Fig. 10.7 Synthesis and release of nitric oxide (*NO*) as transmitter** in the peripheral ANS during excitation of the neurons. Excitation leads to entry of $Ca^{2+}$ through voltage-sensitive $Ca^{2+}$-channels. Intracellular $Ca^{2+}$ regulates the expression of neural NO synthase (*nNOS*) which catalyzes the synthesis of NO from arginine. Its half time decay is about 5 s. NO acts on neighboring target cells leading, e.g., to relaxation of smooth muscle cells by activation of guanylate cyclase and increase of *cGMP* (cyclic guanosine monophosphate). *GTP* guanosine tri-phosphate

voltage-sensitive $Ca^{2+}$ channels with inflow of $Ca^{2+}$ into the presynaptic terminals. Activation of NO synthase via calmodulin leads to the synthesis of NO from arginine. NO diffuses through the cell membranes to the effector cells. Here, it activates guanylate cyclase leading to an increase of the intracellular messenger cyclic guanosine monophosphate (cGMP). cGMP generates, by decrease of the intracellular concentration of $Ca^{2+}$ and some other processes, a relaxation of the smooth musculature. All postganglionic neurons which synthesize and release NO use also other transmitters. For example, the vasodilator neurons to the erectile tissue also make use, in addition to NO, of the peptide VIP and the inhibitory enteric motor neurons to the circulatory musculature VIP and acetylcholine.

#### 10.4.1.4 Signal Transduction to and Gating of Cellular Effector Responses to Nerve-Released Acetylcholine or Noradrenaline

The principles of the signal transduction pathways between the membrane receptors for ACh and NAd and the cellular effectors are demonstrated in Fig. 10.8 for the case of effector tissues of the postganglionic autonomic neurons and for the postganglionic neurons. Column I shows the signal transduction pathways between the membrane receptors and gating molecules (in red) and column II the cellular effectors. Columns

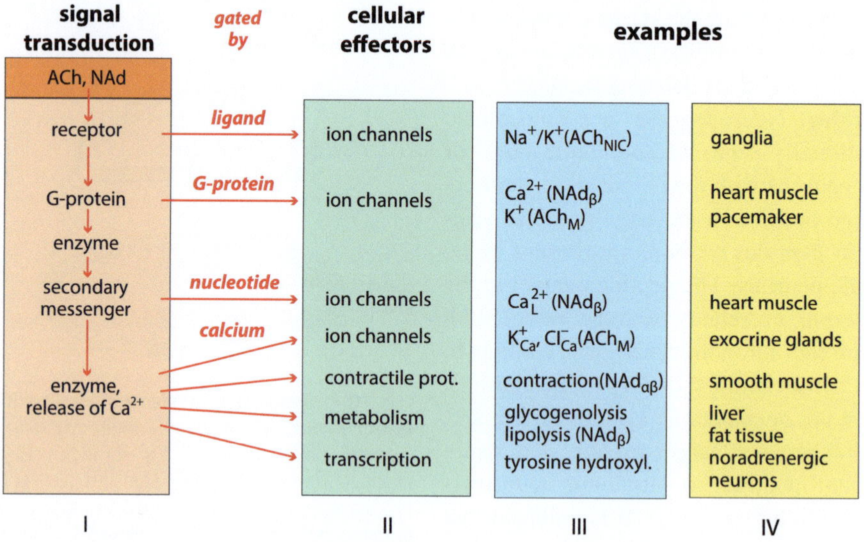


**Fig. 10.8 Relations between the adrenoceptors for noradrenaline (*NAd*) or cholinoceptors for acetylcholine (*ACh*), intracellular pathways, and cellular effectors of the autonomic target tissues.** *Red*: The cellular effectors can be directly ligand-gated, gated directly by the G protein, gated by nucleotides, or gated by intracellular $Ca^{2+}$. Examples of cellular effectors in *blue*: channels for $Na^{+}$-, $K^{+}$-, $Cl^{-}$-, or $Ca^{2+}$-ions that are opened by nicotinic (NIC) or muscarinic (M) effects of ACh ($ACh_{NIC}$, $ACH_{M}$) or by reaction of NAd with α- or β-adrenoceptors ($NAd_{\beta}$, $NAd_{\alpha}$). $Ca^{2+}{}_{L}$, L-type calcium channel; $K^{+}{}_{Ca}$, $Cl^{-}{}_{Ca}$, activation by increase of intracellular $Ca^{2+}$ (e.g., in exocrine glands). Increase of tyrosine hydroxylase activity by neuronal activity (enzyme induction by neural activity)

III and IV show examples of cellular effectors. These pathways are the basis of the target tissue responses to activation of parasympathetic or sympathetic neurons as listed in Table 10.1.

1. Fast synaptic transmission is mediated by ligand-gated ionic channels.
2. Relatively fast synaptic transmission occurs via ionic channels that are directly coupled to the G proteins or gated by nucleotides (e.g., cAMP).
3. Relatively slow effector responses are mediated in effector cells by second messengers and the enzymes involved (adenylyl cyclase, protein kinases, phospholipases) and intracellular increase of $Ca^{2+}$.
4. Finally, the intracellular second-messenger signaling pathways mediate contractions, metabolic changes (adrenoceptor-mediated glycogenolysis or lipolysis) by increase of $Ca^{2+}$ in the cytosol, and long-term changes by transcription of genetic information, e.g., increase of the activity of tyrosine hydroxylase in adrenergic neurons by excitation of the neurons or long-term changes of ionic channels (by phosphorylation).

## 10.4.2 Organization and Function of Autonomic Ganglia

### 10.4.2.1 Divergence and Convergence in Autonomic Ganglia

The major function of autonomic ganglia is to distribute the centrally generated signals by connecting each preganglionic neuron with several postganglionic neurons. The extent of divergence of individual preganglionic neurons on postganglionic neurons (see preganglionic axon *b* in Fig. 10.9a) is probably a function of the size of target organ (and therefore of body size) and of the functional type of autonomic final pathway. Thus, it varies significantly between a low degree of divergence to small targets, e.g., with a ratio of pre- to postganglionic neurons of 1:4 in the pathways to the iris and ciliary body via the ciliary ganglion, and a high degree of divergence to anatomically extensive effectors, e.g., of about 1:200 or more in vasoconstrictor pathways. The degree of divergence is not a pertinent characteristic of parasympathetic or sympathetic systems [43].

Synaptic convergence of preganglionic axons on postganglionic neurons (Fig. 10.9a) guarantees in some autonomic ganglia a high safety factor of synaptic transmission. However, in many if not most sympathetic and parasympathetic ganglia the function of convergence is unclear (see Sect. 10.4.2.2). The degree of convergence varies considerably. For example, only few preganglionic axons converge on individual parasympathetic postganglionic neurons innervating the pupil, but many (up to 20) converge on sympathetic postganglionic vasoconstrictor neurons.

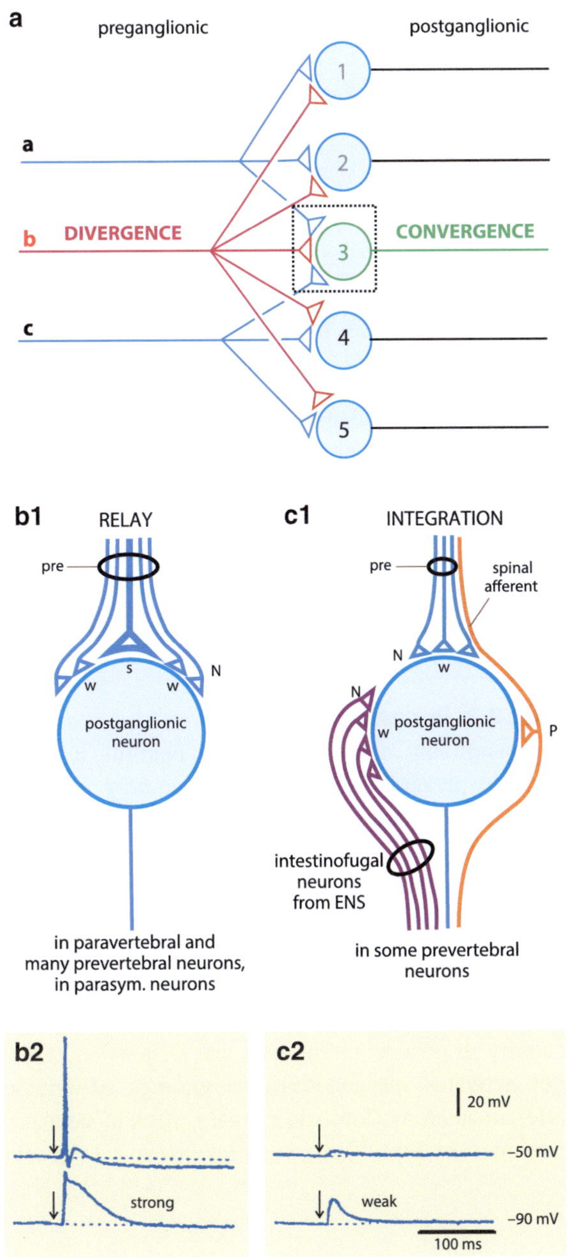


### 10.4.2.2 Dominant and Weak Synapses in Autonomic Ganglia

Transmission of impulses through paravertebral sympathetic ganglia projecting to somatic tissues, parasympathetic ganglia, and some sympathetic nonvasoconstrictor systems projecting through prevertebral ganglia to viscera occurs in a *relay-like fashion*. In these ganglia the incoming impulse activity of one or a few of the converging preganglionic axons produce large excitatory postsynaptic potentials (EPSP) of several 10 mV, generated by the release of many quanta of acetylcholine, which are always suprathreshold. These synaptic inputs have, like in the skeletal neuromuscular junction, a high safety factor and always initiate an action potential. They are called *strong synaptic inputs* (Figs. 10.9b1, b2). Thus, the impulse activity in the postganglionic neurons of these autonomic pathways

**Fig. 10.9 Transmission of impulses in autonomic ganglia. (a) Convergence and divergence in autonomic ganglia.** The degree of divergence of individual preganglionic neurons on postganglionic neurons (see preganglionic axon *b*) probably is a function of the size of target organ (and therefore body size) and of the type of autonomic final pathway (e.g., low divergence in the pupillomotor pathway and high degree of divergence in vasoconstrictor pathways). The degree of convergence (here on boxed neuron *3*) varies between different autonomic pathways. **(b, c) Two types of transmission in autonomic ganglia. (b1)** Most ganglion cells receive one or a few preganglionic inputs that, when activated, produce a suprathreshold ('strong', *S*) response. Most also receive several weak (*w*) convergent inputs that evoke only subthreshold (ineffective) synaptic potentials. Transmission occurs via the S input, which relays its signals directly to the postganglionic neuron. Summation of weak responses is rare. **(c1)** In some sympathetic prevertebral neurons, the preganglionic inputs are of the weak type but other cholinergic inputs in intestinofugal neurons that arise in the enteric nervous system (*ENS*) also converge on the cell. Activation occurs by temporal and/or spatial summation. Peptides released from spinal afferent neurons (P) may potentiate transmission. N, nicotinic cholinergic. **(b2, c2)** Synaptic potentials recorded in postganglionic neurons at resting membrane potential (upper traces) and with the membrane hyperpolarized to block action potential initiation (lower traces). Preganglionic axons were electrically stimulated (*arrows*). Stimulation of a single strong input **(b2)** generates a suprasthreshold response at resting membrane potential and a large amplitude excitatory synaptic potential at −90 mV. Stimulation of a single weak preganglionic axon **(c2)** evokes a small subthreshold excitatory synaptic potential that increases in amplitude with hyperpolarization. The differences in amplitude between strong and weak inputs reflect differences in the number of quanta of acetylcholine released from each preganglionic input. **(b, c)** Modified from [23]. After McLachlan in [23]

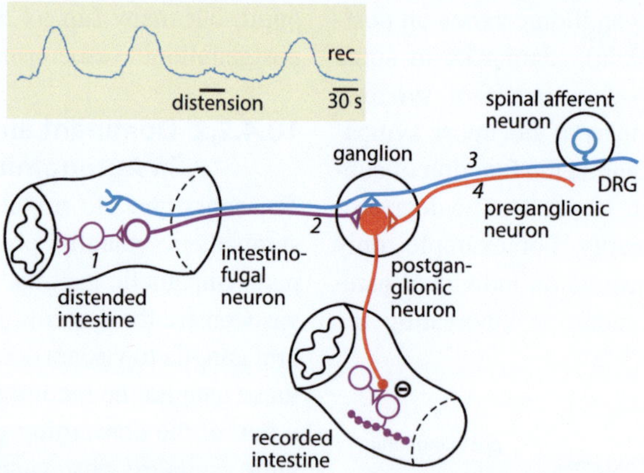


**Fig. 10.10** **Peripheral reflexes involving motility of the gastrointestinal tract** and mediated by sympathetic prevertebral ganglia. Distension of the bowel activates intrinsic primary afferent neurons of the enteric nervous system (neuron *1*) which connect with intestinofugal neurons (neuron *2*) that project to the prevertebral ganglia. Visceral primary afferent neurons (neuron *3*) with cell bodies in the dorsal root ganglia (DRG) are also activated. The intestinofugal neurons excite postganglionic motility-regulating noradrenergic neurons (*red*) projecting to another part of the bowel and lead to inhibition (–) of enteric excitatory muscle motor neurons and relaxation. Collateral branches of primary afferent neurons (neuron *3*) activated by the same stimulus release substance P around the same postganglionic neurons producing a slow depolarization. *4*, preganglionic axon. *Inset*: Recording of the intraluminal pressure in an isolated proximal segment of the colon and distension of the rectum in an anesthetized cat. The prevertebral ganglion was decentralized by section of the preganglionic axons in the lumbar splanchnic nerves. The regular contraction waves of the proximal colon are inhibited during distension of the rectum mediated by the prevertebral ganglion (From Kuntz [35]. Modified from Jänig [23] with permission)

is dependent on the activity in one or a few preganglionic axons and not on spatial and temporal summation of EPSPs. The synaptic inputs of the remaining converging preganglionic axons are weak, generating usually EPSPs of a few millivolt. The function of the weak (subthreshold) synapses is not clear.

Sympathetic neurons consist of three broad groups in prevertebral ganglia based on electrophysiological (by their resting $K^+$ channels and the voltage- and $Ca^{2+}$-dependent $K^+$ channels that control their excitability), morphological (by their size and dendritic branching), and neurochemical (by their neuropeptide content) criteria [23]. Two groups are similar to paravertebral sympathetic neurons (Fig. 10.9b), but the mode of synaptic transmission in the third group is different. Neurons of this group receive weak preganglionic synaptic inputs that do not necessarily activate them (Fig. 10.9c). However, they also receive many weak nicotinic synaptic inputs from intestino-fugal neurons that are activated by distension of the intestines. Summation of the EPSPs arising from the intestines with those from preganglionic inputs is necessary to initiate their discharge. Some of these neurons also depolarize slowly when the intestines are distended leading to activation of peptidergic visceral spinal unmyelinated afferents. These afferents form collaterals synapsing with the postganglionic neurons and release the neuropeptide Substance P so that the small nicotinic responses are brought closer to threshold for firing. The activity in these prevertebral neurons therefore depends on temporal and spatial integration of incoming excitatory signals.

#### 10.4.2.3 Peripheral Reflexes

Postganglionic nonvasoconstrictor neurons in sympathetic prevertebral ganglia are not only involved in reflexes organized at spinal and supraspinal levels but also in **peripheral reflexes** that do not involve the CNS. Neurons within the enteric nervous system (known as **intestino-fugal neurons**, see Sect. 10.5) and possibly other effector tissues have axons that project back into autonomic prevertebral ganglia through the mesenteric nerves and may initiate reflex activity in postganglionic neurons. Distension of the gut activates mechanosensitive endings of enteric afferent neurons (intrinsic primary afferent neurons, neuron *1* in Fig. 10.10; see Figs. 10.9c, 10.13 and

10.14) that activate the intestino-fugal neurons (neuron *2* in Fig. 10.10). The axons of intestino-fugal neurons from a considerable length of the intestine (in particular the proximal colon) converge on the ganglion cells. Summation of their input triggers the activation of the postganglionic neurons and leads to the release of noradrenaline and thus relaxation of other parts of the intestines by inhibition of enteric neurons generated pre- and/or postsynaptically (Fig. 10.10). These peripheral reflexes presumably contribute to the storage function of the large intestine aiding to reabsorb fluid and electrolytes. Substance P released in sympathetic prevertebral ganglia from collaterals of spinal visceral afferent neurons (neuron *3* in Fig. 10.10) activated by distension can enhance cholinergic synaptic transmission from both preganglionic axons (*4* in Fig. 10.10) and enteric neurons.

Postganglionic neurons in the sympathetic chain ganglia projecting to somatic tissues are not involved in extracentral peripheral reflex activity. Parasympathetic cardiac and pelvic ganglia may mediate peripheral reflexes [11].

### 10.4.3 Mechanisms of Neuroeffector Transmission

Autonomic effector cells are very diverse, the most common being smooth muscle cells, secretory epithelia, and cardiac muscle cells (Table 10.1). The cytoplasm of adjacent cells in these autonomic effector tissues is connected by gap junction channels that allow electrical and chemical signals to pass directly between neighboring cells. Thus, these cells form *functional syncytia*. Axons of postganglionic neurons form many branches close to the effector cells with hundreds to thousands of presynaptic varicosities. Excitation of the postganglionic neurons spreads over all branches and normally invades all varicosities. Signal transmission from postganglionic neurons to most effector cells occurs through neuroeffector synapses which are morphologically characterized by a close junction between varicosities and membranes of the effector cells, by loss of Schwann cell sheath, by fusion of the basal laminae, and by accumulation of synaptic vesicles close to the synaptic junction

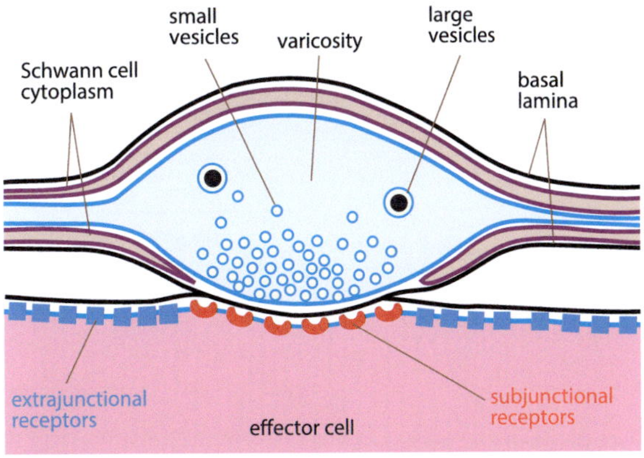


**Fig. 10.11 Simplified scheme of neuroeffector transmission to autonomic target cells.** The cell surface at which this nerve-effector communication occurs is <1 % of the total effector cell surface. Subjunctional receptors mediate the effect of transmitter released by the nerve terminals during excitation under physiological conditions. These subjunctional receptors are ligand gated or second messenger coupled to cellular effectors (e.g., ionic channels). Extrasynaptic receptors for the transmitter are either different from the subsynaptic ones and/or are coupled by different intracellular second-messenger pathways to the cellular effectors. The function of the extrajunctionally located receptors are unclear for most innervated effector organs. The neuroeffector junction is characterized by absent Schwann cell cytoplasm, a close synaptic cleft and a fused basal lamina. Small vesicles containing the transmitter are localized close to the synaptic cleft. Large vesicles are also present in many varicosities and not located close to the synaptic cleft. In these large vesicles neuropeptides are colocalized with "classical" transmitter. The physiological role of the neuropeptides is, in most cases, unknown

(Fig. 10.11). These junctions cover about 0.1–1 % of the surface of the effector cells. The release of the content of a vesicle from a sympathetic varicosity leads to a short lasting increase of the concentration of the transmitter in the junctional cleft which may reach into the millimolar range and the subsequent interaction of the transmitter with junctional receptors in the effector cells which are connected directly or indirectly, via second messenger pathways, to the cellular effectors. The mechanisms underlying the effect of neurally released transmitter acting on postjunctional receptors of autonomic effector cells, as it occurs under physiological conditions, has been studied in only a few autonomically innervated tissues (sinoatrial node of the heart, submucosal arterioles, mesenteric and tail

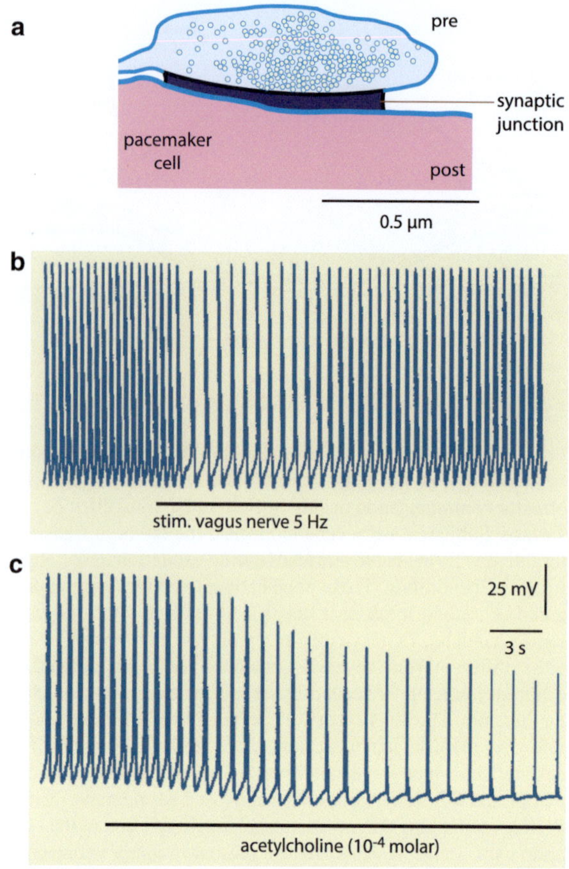


**Fig. 10.12 Neuroeffector transmission to cells of the sinoatrial (SA) node in guinea pig.** (**a**) Schematic outline of a cholinergic neuromuscular junction (*blue*) formed by a varicosity on a cell of the SA node based on serial sectioning and tracings of a cholinergic varicosity. These junctions cover less than 1 % of the cell surface of the pacemaker cells. Pre, prejunctional; post, postjunctional. (**b, c**) Intracellular record from a pacemaker cell in an *in vitro* preparation with attached vagus nerve. The cell shows pacemaker potentials with action potentials. Electrical stimulation of the vagus nerve at 5 Hz shows no hyperpolarization but a reduced slope of the pacemaker potential and a reduced frequency of action potentials. The mechanism is reduction of $Na^+$ current (due to closing of voltage-dependent $Na^+$ channels and voltage-independent background $Na^+$ channels and increase in membrane resistance). The intracellular second-messenger pathway connecting subjunctional muscarinic ACh receptors and $Na^+$ channels is unknown. Exogenous ACh hyperpolarizes the membrane and shunts the action potentials (due to opening of $K^+$ channels and decrease in membrane resistance), reducing their amplitude and duration. The intracellular second-messenger pathway connecting extrajunctional muscarinic ACh receptors and $K^+$ channels is the cAMP pathway (a – According to Klemm et al. [30] and Choate et al. [8]. b, c – Modified from Campbell et al. [6]

arteries, mesenteric vein, smooth muscle of ileum, dilator myoepithelium of eye [18, 23]). These studies clearly show that the effect of neurally released transmitter, acting on subjunctional receptors, cannot necessarily be mimicked by exogenously applied transmitter, acting on extrajunctionally located receptors (Fig. 10.11).

This is exemplarily illustrated in Fig. 10.12 for the sinoatrial node of the guinea pig. Neuronally released ACh reacts with specialized subjunctional ACh receptors and does not leave the synaptic cleft (blue in Fig. 10.12a) due to ACh esterase. This leads to inhibition of sodium channels (sodium background current, voltage-dependent sodium current), mediated by a yet unknown second messenger pathway, and slowing of pacemaker activity. The action potentials of the pacemaker cells do not change in size and duration, the membrane does not hyperpolarize, and the membrane resistance increases (due to closing of sodium channels). Exogenously applied ACh reacts with extrajunctionally located ACh receptors and causes hyperpolarization of pacemaker cells by opening of potassium channels; the membrane resistance decreases and the action potentials are reduced in size and duration (due to shunting of the membrane by opening of potassium channels) (Fig. 10.12c). In the arrested atrial preparation (action potentials of the pacemaker cells generated by inflow of $Ca^{2+}$ prevented by blocking calcium channels) barium ions (which block potassium channels) prevent the hyperpolarization generated by exogenously applied ACh but do not block the effect of vagus nerve stimulation (which is now hyperpolarizing since the open sodium channels keep the membrane potential in a depolarized state of about −35 mV). In conclusion, neurally released ACh and exogenously applied ACh cause slowing of pacemaker cells via muscarinic ACh receptors (both effects being blocked by muscarinic antagonists) by different mechanisms, involving different cellular effectors (ionic channels) and different intracellular pathways. The mechanism underlying the effect of neurally released ACh operates under physiological conditions. A similar situation occurs for NAd released by the varicosities of the sympathetic cardiomotor axons reacting with junctional $\beta_1$-adrenoceptors of pacemaker cells [18].

Arterioles and small arteries are influenced by neural release of noradrenaline and ATP from the varicosities of the vasoconstrictor axons. ATP reacts with junctional purinoceptors ($P_{2x}$) and opens ligand-gated cation channels. Noradrenaline released from the postganglionic vasoconstrictor terminals reacts largely with extrajunctionally located α-adrenoceptors leading to slow depolarization in some blood vessels. The role of both mechanisms of neuroeffector transmission varies between different types of blood vessel. Furthermore, the integration between these cellular processes during ongoing regulation of small arteries and arterioles is unclear. Circulating catecholamines cannot be involved in the regulation of the *innervated* blood vessels in the concentrations as they occur *in vivo*.

Cholinergic neurons of the enteric nervous system which innervate the longitudinal musculature of the ileum activate the smooth muscle cells via junctional muscarinic receptors [18]. The functions of neuropeptides in autonomic neuroeffector transmission, located in the large vesicles of varicosities (Fig. 10.11), have been little explored and are, in almost all cases, unknown (see Sect. 10.4.1.3).

## 10.5 The Enteric Nervous System

The **gastrointestinal tract** (GIT) serves ingestion of energy-rich compounds, water, electrolytes, and other vital substances, transport and mixing of its content, enzymatic breakdown and resorption of the various substances, evacuation of waste products, and protection of the body against toxic compounds. Various effector tissues are involved in these functions: *smooth GIT musculature*, *secretory* and *resorptive epithelia*, *endocrine cells*, *blood vessels*, and the *gut-associated lymphoid tissue*. The **enteric nervous system** (ENS) regulates and coordinates these effector systems. Most neurons of the ENS are located in the **myenteric (Auerbach) plexus** between longitudinal and circular musculature of the GIT or in the **submucosal (Meissner) plexus** *internal to the circular* musculature ([12, 23], Brookes & Costa in [5]).

The ENS contains about as many neurons as the spinal cord and is an ANS in its own right. The enteric neurons consist functionally of the following three types: (a) Afferent neurons (intrinsic primary afferent neurons). Their dendrites are projecting in the wall of the GIT and to the mucosa and are receptive for mechanical (distention, contraction, shearing) or intraluminal chemical stimuli. These neurons form synaptically connected networks. Their primary transmitters are ACh and the neuropeptide tachykinin. (b) Motoneurons, innervating circular or longitudinal musculature, secretory epithelia, endocrine cells in the mucosa, or blood vessels (see Table 10.3 for the functional characteristics and transmitters of the motoneurons). (c) Interneurons intercalated between afferent neurons and motoneurons (Fig. 10.13). These Neurons form synaptically connected networks which project either orally or anally. Their primary transmitter is ACh.

The ENS contains *sensori-motor programs* for the regulation and coordination of the effector systems, related to *transport, mixing, secretion* and *resorption* in the GIT. These neural programs are composed of several reflex circuits including intrinsic primary afferent neurons, interneurons and motoneurons. The CNS interferes with these local regulations exerted by the ENS by way of the extrinsic, parasympathetic and sympathetic, innervation. Preganglionic parasympathetic neurons either excite motoneurons of the ENS directly or via enteric interneurons or via postgan-

**Table 10.3** Functions of motoneurons of the enteric nervous system: effectors, transmitters, co-localized neuropeptides and functions

| Effector tissue | Transmitter(s) (Co-transmitter) | Neuropeptide(s) | Function |
|---|---|---|---|
| *1. Motoneurons to smooth musculature (myenteric plexus)* | | | |
| Circ. musculature | ACh (TK) | TK/ENK | contraction (asc., desc.) |
| Long. Musculature | ACh (TK) | CALRET/TK | contraction (asc., desc.) |
| Circ. Musculature | NO, VIP, ATP, PACAP | VIP/PACAP/ENK/NPY, | Relaxation (desc.) |
| *2. Motoneurons to secretary epithelia and blood vessels (submucosal plexus)* | | | |
| Mucosa | ACh | NPY/CCK/CGRP/SOM/DYN | Secretion |
| Mucosa | ACh | CALRET/DYN | Secretion/vasodilation |
| Mucosa | VIP | VIP/CART/GAL/PACAP | Secretion/vasolidation |

After Furness [12]

Transmitter: *ACh* acetylcholine, *ATP* adenosine-tri-phosphate, *NO* nitric oxide, *PACAP* pituitary adenylyl cyclase activating peptide, *TK* tachykinin, *VIP* vasoactive intestinal peptide

Co-localized neuropeptides: *CALRET* calretinin, *CART* cocaine and amphetamine-regulated transcript peptide, *CCK* cholecystokinin, *CGRP* calcitonin gene-related peptide, *ENK* enkephalin, *DYN* dynorphin, *GAL* galanin, *NPY* neuropeptide Y, *SOM* somatostatin

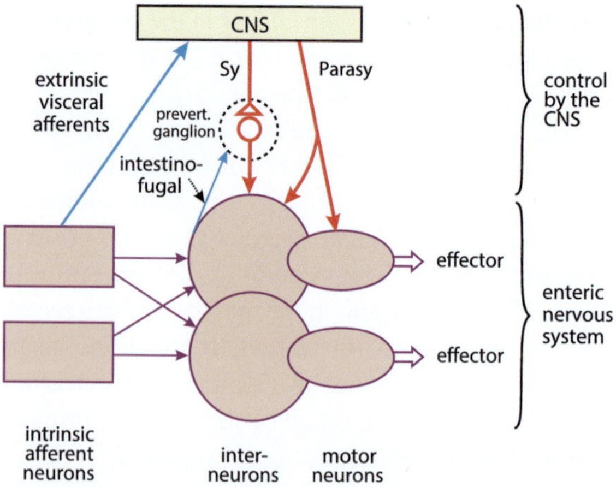


**Fig. 10.13** **Functioning of the enteric nervous system (*ENS*) and its control by the central nervous system (*CNS*).** The ENS (in *brown shading*) consists of intrinsic primary afferent neurons, interneurons and motoneurons which form various reflex circuits. Effector targets can be smooth muscle cells, interstitial cells of Cajal (ICC), secretory epithelia, hormone-secreting cells or arterioles. Each category of neuron is subdivided into several types according to its physiological response properties, morphology (location of cell body, projection of axon and dendrites), immunohistochemistry, and transmitter(s). The circuits and their reciprocal synaptic connections represent the "sensory-motor programs" of the ENS which regulate movement patterns and secretory processes. Extrinsic visceral afferent neurons (*blue*) projecting to the spinal cord or to the nucleus tractus solitarii in the lower brainstem provide detailed information about events in the gastrointestinal tract (GIT) to the CNS. The CNS acts on the ENS via sympathetic and parasympathetic pathways (*red*). It does not regulate single motor functions of the GIT but interferes with the sensory-motor programs of the ENS. It has direct control over intake of nutrients, disposal of waste products and resistance vessels. Intrinsic neurons (intestinofugal interneurons) project to prevertebral sympathetic ganglia giving afferent feedback to postganglionic secretomotor and motility-regulating neurons (Modified from Jänig [23] with permission)

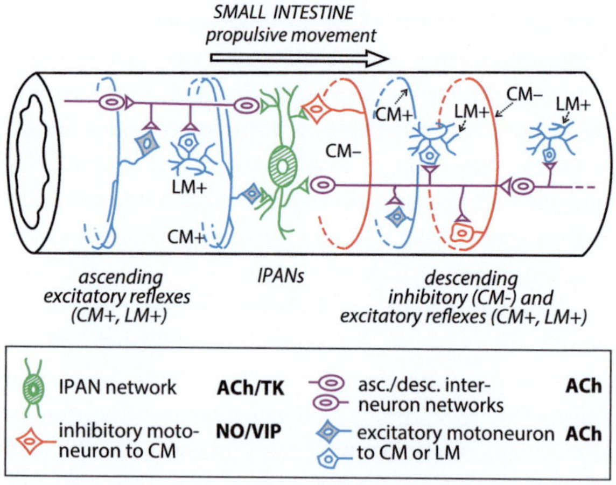


**Fig. 10.14** **Organization of reflex circuits responsible for peristalsis in the small intestine.** Intrinsic primary afferent neurons (*IPANs*), ascending interneurons and descending interneurons form synaptically connected networks. Three reflex circuits to the circular muscles (*CM*; ascending excitatory, descending inhibitory and descending excitatory) and two reflex circuits to the longitudinal muscles (*LM*; ascending excitatory and descending excitatory) are indicated. Only one ascending and one descending network of interneurons, respectively have been indicated. Each may consist of two (ascending) or three (descending) networks. The transmitters are indicated in bold. *ACh* acetylcholine, *NO* nitric oxide, *TK* tachykinin, *VIP* vasoactive intestinal peptide, + excitation, – inhibition (Modified from Furness [12] with permission)

glionic neurons. Sympathetic postganglionic non-vasoconstrictor neurons in the prevertebral ganglia innervate (pre- and postsynaptically) neurons of the ENS and inhibit them. The sphincteric musculature is directly innervated and excited by sympathetic postganglionic neurons.

The CNS is involved in the regulation and coordination of the programs of the ENS rather than regulating individual functions (Fig. 10.13). However, it keeps full control of intake and processing of food at the oral part and of the evacuation at the hindgut. Furthermore, overall blood flow through the GIT is under the control of the CNS. The CNS is continuously informed about the mechanical and chemical state of the GIT by impulse activity in *extrinsic visceral afferent neurons* projecting to the medulla oblongata (nucleus of the solitary tract) through the vagus nerves or projecting to the spinal cord through the splanchnic nerves. An extraspinal afferent feedback occurs to the prevertebral ganglia via intestino-fugal neurons (see Fig. 10.10). Finally, the CNS is informed about the state of the GIT by hormones acting on central circuits via the area postrema of the medulla oblongata or the arcuate nucleus of the hypothalamus, such as gastrin, cholecystokinin, ghrelin, peptide YY, etc.

The reflex circuits of the ENS as defined by the effector cells and the afferent neurons are coordinated in the context of the functions of the GIT. The neural mechanisms underlying **propulsive peristalsis**, i.e., the transport of the intraluminal content from oral to aboral, have been most extensively studied and shall serve here as an example for the integrative activity of the ENS (Fig. 10.14). This process is triggered by activation of intrinsic primary afferent neurons during distension of the GIT wall and shearing stimuli along the mucosa and is independent of the CNS. In the small intestine it consists orally of a contraction of the circular and longitudinal musculature and aborally of a relaxation of the circular musculature and contraction of the longitudinal muscles. Figure 10.14 depicts the synaptic connections of afferent neurons, interneurons, and motoneurons and the transmitters involved resulting in peristalsis of the small intestine: (a). Activation of *inhibitory motoneurons* to the aborally located circular musculature. The inhibitory synaptic transmitters to the musculature are nitric oxide (NO) and vasoactive intestinal peptide (VIP). (b) Activation of *cholinergic excitatory motoneurons* to the longitudinal musculature located aborally (in the colon this musculature is inhibited). (c) Activation of *cholinergic excitatory motoneurons* innervating circular or longitudinal musculature located orally.

Propulsive (oral-aboral) peristalsis and the synaptic effects of the excitatory and inhibitory motor neurons on the muscle layers in some parts of the GIT (e.g., circular muscles of small intestine, colon and stomach) are also dependent on the so-called *interstitial cells of Cajal*. These cells are of mesenchymal origin, as are smooth muscle cells. They form networks of electrically coupled pacemaker cells and are coupled electrically to the smooth muscle cells. The electrical activity of these cells is responsible for generation and propagation of the rhythmic movements of the stomach and small intestine. Furthermore, they mediate some excitatory and inhibitory synaptic activity of enteric motor neurons to the circular musculature [12, 23, 60].

## 10.6 Central Organization of the Autonomic Nervous System: A Summary

Pre- and postganglionic neurons of the peripheral autonomic pathways exhibit discharge patterns to physiological stimuli that are typical for their function in regulating particular target tissues, as described in Sect. 10.3 and Table 10.2. These discharge patterns are the functional fingerprints of the peripheral autonomic pathways and dependent on distinct neural circuits in spinal cord, brainstem, and hypothalamus connected to these peripheral autonomic pathways (Fig. 10.15). Known details of the neurobiology of these central circuits are described in the literature [23, 38]. The spinal cord contains spinal reflex circuits associated with the regulation of pelvic organs (hindgut, genital organs, urinary bladder), GIT, kidney, heart, vasculature, and some other organ systems. These spinal reflex circuits are di- or polysynaptic and integrated in the homeostatic regulations represented supraspinally in the brainstem and hypothalamus. The homeostatic regulations of the cardiovascular system (heart, arteries, veins), respiration (gas transport), and GIT are represented and integrated in the lower brainstem. The circuits in the medulla oblongata involved in these

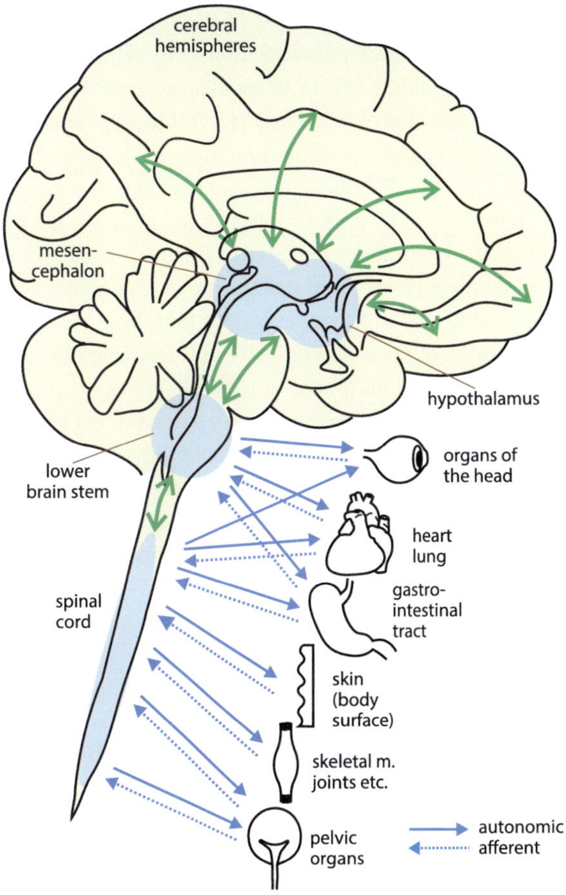


**Fig. 10.15 Reciprocal communication between brain and body tissues** by efferent autonomic pathways and afferent pathways. The global autonomic centers in spinal cord, lower and upper brain and hypothalamus are shaded in *blue*. These centers consist of the neural circuits that are at the base of the homeostatic autonomic regulations (Table 10.4) and their coordination with the regulation of the neuroendocrine systems, the somatomotor systems, and the sensory systems establishing behavior (see Fig. 10.1). The brain sends efferent commands to the peripheral target tissues through the peripheral autonomic pathways. The afferent pathways consist of groups of afferent neurons with unmyelinated or small diameter myelinated fibers. These afferent neurons monitor the mechanical, thermal, chemical, and metabolic states of the body tissues

regulations are under control of the upper brainstem, hypothalamus, and telencephalon (Fig. 10.15). They are neural building blocks of regulations represented in the supramedullary brain centers. Table 10.4 summarizes various data about the functions of the hypothalamus (and upper brainstem) in which the autonomic systems are involved (see Card et al. in [58]). These complex integrative functions include somatomotor, neuroendocrine, and autonomic components. The important point to be made here is that these complex regulations require anatomically and functionally precisely organized peripheral autonomic pathways innervating various target organs and precisely organized central autonomic circuits in the spinal cord and lower brainstem. Finally, most functions mediated by the autonomic nervous system are potentially also under the control of the telencephalon (cerebral cortex and limbic system) [23].

An additional function in which hypothalamus and mesencephalon are involved and which very much seems to depend on functioning sympathetic systems (in addition to the neuroendocrine systems) is the protection of body tissues during acute and chronic pain and stress; this includes the neuronal control of the immune system [24, 26].

## 10.7 The Peripheral Autonomic Nervous System in Non-Mammalian Vertebrates: A Comparative View

Knowledge about the ANS in non-mammalian vertebrates (Fig. 10.16) is largely restricted to the peripheral ANS and in particular to the anatomy. Neurobiological studies of central circuits associated with the peripheral autonomic systems and functional (neurophysiological) studies of peripheral autonomic neurons *in vivo* (see Figs. 10.3 and 10.4) are almost absent in non-mammalian vertebrates. However, the comparison of the anatomy of the peripheral ANS of the major vertebrate groups indicates that the principal organization of this system appears to be highly conserved in evolution over a time period of up to about 500 million years. This is further supported by studies of the principal transmitters acetylcholine, adrenaline, and noradrenaline and of the neuropeptides in the neurons of the peripheral ANS [19, 47, 49]. Table 10.5 shows the autonomic innervation of the main target tissues (cardiovascular system, gastrointestinal tract, lung, spleen, urogenital tract, eye, chromaffine tissue) for the major non-mammalian vertebrate groups of animals, their reactions to stimulation of the (cranial) parasympathetic, or of the spinal autonomic innervation and the cholinergic receptors (nicotinic, muscarinic) or adrenoceptors involved. Common ancestors of these vertebrates cover an evolutionary time of about 500–200 million years from cyclostomes to birds (Fig. 10.16):

The peripheral ANS consist of the *cranial parasympathetic system* and *spinal autonomic systems*. Spinal sympathetic and spinal parasympathetic systems cannot be distinguished in non-mammalian vertebrates.

The peripheral ANS consists of populations of cholinergic preganglionic neurons and populations of adrenergic and cholinergic postganglionic neurons which are synaptically connected. Adrenaline is the primary transmitter in some adrenergic neurons of teleosts, amphibians, and birds. The receptors of these transmitters are cholinergic nicotinic, cholinergic muscarinic, or adrenergic. Preganglionic neurons of the cranial parasympathetic system project through the cranial nerve III (oculomotor), VII (facial), IX (glossopharyngeal), or X (vagal) to the postganglionic neurons located close to the target tissues.

Preganglionic neurons of the spinal autonomic systems project through the ventral roots to the peripheral ganglia. In cyclostomes, elasmobranchs, and amphibians they may also project through the dorsal roots. However, this is very much debated for elasmobranchs and amphibians.

Sympathetic chains are present in dipnoans, teleost fishes, amphibians, reptiles, and birds. In cyclostomes these chains do not exist or consist of scattered ganglia along the aorta. In elasmobranchs they are incomplete. Figure 10.17 depicts the sympathetic paravertebral chain, the sympathetic prevertebral systems, and peripheral cranial parasympathetic system in an elasmobranch fish, in a teleost fish, and in a reptile. These groups of vertebrates are about 100 to more than 200 million years apart in their evolutionary origin.

The ENS of the GIT is present in all groups of non-mammalian vertebrates and under the control of the brain via the (cranial) parasympathetic system (stomach and proximal gut) and via the spinal autonomic systems (small intestine and hindgut).

Chromaffin tissues consist of cells that synthesize and release adrenaline or noradrenaline. With

**Table 10.4** Integrative functions of hypothalamus and mesencephalon

| Function | Behavior | Afferent feedback: neural, hormonal, cytokines | Autonomic systems | Endocrine systems, hormones |
|---|---|---|---|---|
| Thermoregulation | Thermoregulatory behavior | Peripheral thermoreceptors, central thermosens. (preopt.), cytokines (fever) | SyNS (skin [CVC,SM]), (BAT [rodents]) | TRH/Thyr (anterior pituitary) |
| Reproduction, sexual behavior | Sexual behavior and sexual orientation | Afferents from sexual organs, other sensory systems | SyNS (thor.-lumb.), PaNS (sacral) (sexual organs) | GnRH, FSH/LH (anterior pituitary) |
| Volume-, osmoregulation (fluid homeostasis) | Drinking behavior, thirst | Osmorecept. in OVLT & liver, volume rec.ri. atrium (vagal) angiotensin II via SFO | NTS, SyNS (kidney) | Vasopressin (posterior pituitary) |
| Regulation of nutrition<br>Regulation of metabolism | Nutritive behavior, hunger/satiety | Vagal afferents & hormones from GIT (CCK, ghrelin, GLP-1, insulin, glucagon), leptin from adipose tissue, nutritive signals | Enteric nervous system, PaNS (DMV), SyNS (BAT [rodents], WAT) | Insulin, glucagon, orexin, leptin |
| Temporal organisation of body functions | Sleep-waking behavior, circadian/endogenous rhythm of body functions | Afferents from retina (retino-hypothalamic tract) | SyNS, PaNS SyNS to gl. pinealis | Melatonin (gl. pinealis) |
| Body protection (acute, e.g. during pain and stress) | Defense behavior (fight, flight, quiescence) | Nociceptive afferents (from body surface, deep somatic tissues, viscera) | SyNS, PaNS (cardiovascular system [CMN,MVC,VVC etc.]) | CRH/ACTH (anterior pituitary gland), adrenaline (SA system) |
| Immune defense | Defense of toxic substances and situations (sickness behavior) | Cytokines | SyNS (to immune tissue) | CRH/ACTH (anterior pituitary gland), adrenaline (SA system) |

The specific autonomic systems which are mainly involved in these functions are shows in the shaded vertical column

*Abbreviations*: *BAT* brown adipose tissue, *CCK* cholecystokinin, *CRH/ACTH* corticotropin-RH/adrenocorticotropic hormone, *CVC* cutaneous vasoconstrictor neurons, *CMN* cardiomotor neurons (symp., parasymp.), *DMV* dorsal motor nucleus of the vagus, *FSH/LH* follicle-stimulating hormone/luteinizing hormone, *GIT* gastrointestinal tract, *GLP-1* glucagon-like peptide 1, *GnRH* gonadotropine RH, *MVC* muscle vasoconstrictor neurons, *N.* nucleus, *NTS* nucleus tractus solitarii, *OVLT* organum vasculosum laminae terminalis (osmosensors), *PAG* periaqueductal gray, *PaNS* parasympathetic nervous system, *RH* releasing hormone, *SA system* sympathoadrenal system (adrenal medulla), *SM* sudomotor neurons, *SyNS* sympathetic nervous system, *thor.-lumb.* thoraco-lumbar, *TRH/Thyr* thyreotropin RH/thyroxin, *VVC* visceral vasoconstrictor neurons, *WAT* white adipose tissue

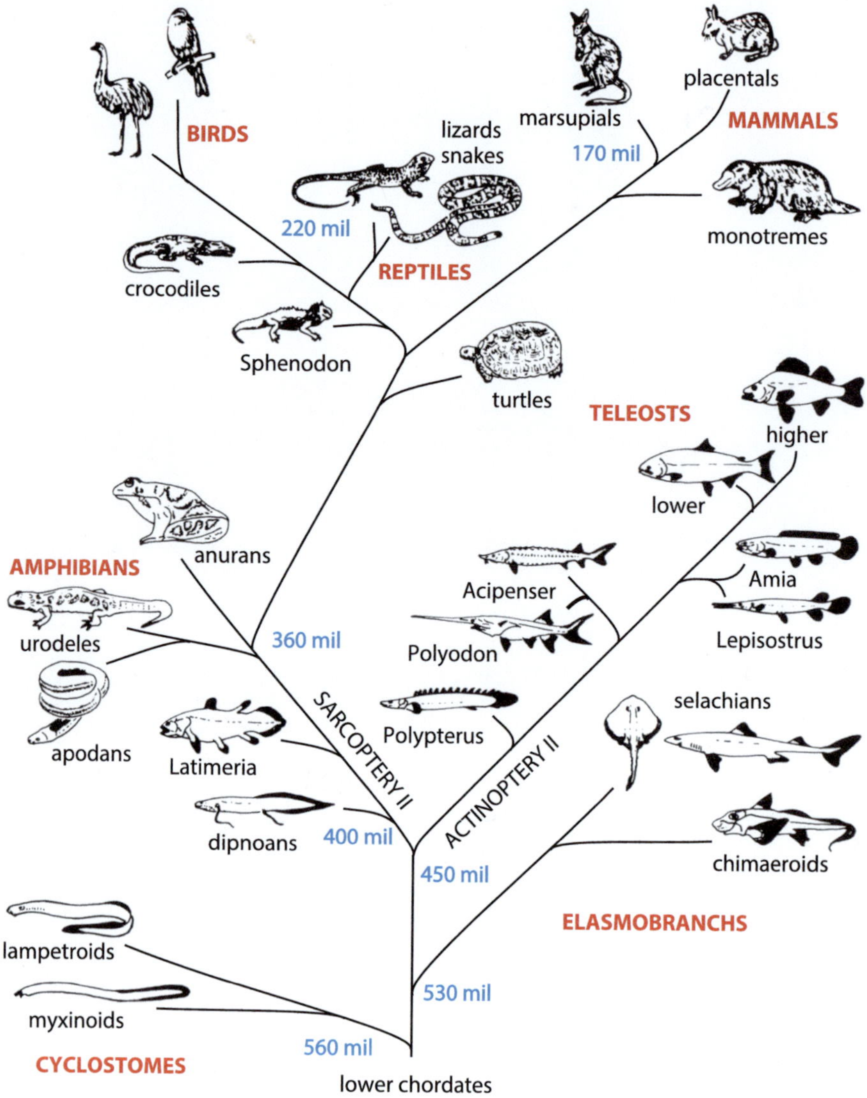


**Fig. 10.16** **Relationships among the major vertebrate groups** (in *red*): cyclostomes, elasmobranchs, teleosts, amphibians, reptiles, birds, and mammals and some of the phylogenetically interesting but smaller groups. The times in evolution are indicated in million years (*blue*). Times in evolution when the groups diverged taken from Kumar and Hedges [34] with permission. See also Table 10.5 (Modified from Nilsson [47] with permission)

the exception of the cyclostomes, these tissues are innervated by spinal preganglionic neurons. In fish, these tissues are located within the heart (in cyclostomes) and close to the atria, posterior cardinal vein, intercostal arteries, and ganglia of the paravertebral chain (in elasmobranchs, blue in Fig. 10.17a). In amphibia, reptiles, and birds this tissue is organized in adrenal glands. Adrenaline and noradrenaline released by the chromaffin tissue probably is primarily involved in cardiovascular regulation in cyclostomes, elasmobranchs, and dipnoans, but not in teleosts, amphibians, reptiles, and birds.

Overall we can conclude and hypothesize:

The general anatomical plan of the peripheral ANS (including the enteric nervous system) is strikingly similar in all vertebrate groups (except in cyclostomes). The dominance of the brain in the autonomic regulation is already present in early evolution of vertebrates and this dominance of neural control by the brain increases with evolution. The dominance of the control of autonomic functions by the brain probably is paralleled by an increasing complexity and functional differentiation of the autonomic regulations, related particularly to the adaptation of the vertebrates to terrestrial life. The increased

**Table 10.5** Peripheral autonomic systems in non-mammalian vertebrates. Anatomy, neurotransmitters, effects of cranial parasympathetic and spinal autonomic systems

| | Cyclostome | Elasmobranch | Dipnoan | Teleost | Amphibian | Reptile/Bird |
|---|---|---|---|---|---|---|
| **Time in evolution** | About 560 million years | About 530 million years | About 400 million years | About 450 million years | About 360 million years | About 310–220 million. years |
| **Cranial nerves**[a] | X (VII,IX) | III,X,(VII,IX) | X (III) | III, X | III,X (VII,IX) | III,VII,IX,X |
| **Sympathetic chain** | No, scattered ganglia along dorsal aorta | Incomplete | Poorly developed | Yes (continues into the head) | Yes | Yes, like in mammals |
| **CVS** | | | | | | |
| Heart | **P, S** Ø; CT β+ | **P** m–<br>**S** Ø; via CT β+ | **P** m–<br>**S** Ø; via CT β+ | **P** m–<br>**S** β+; via CT β+ | **P** m–, **S** β+ anurans,<br>**S** urodeles? | **P** m–, **S** β+ |
| Artery[b] | **S** Ø (H), S?(L); CT α | **S** α+, β– | **S** α+, β- | **S** α+, β– | **S** α+, β– | **S** α+, β– |
| Vein[b] | **S** Ø (H), S? (L); CT α | Ø; CT α+, β– | **S** via CT? | **S** α+ | **S** α+ | **S** α+ |
| **GIT**[c] | **ENS**[d]**(P,S)** | **ENS, P,S** | **ENS, P?,S?** | **ENS, P,S** | **ENS, P,S** | **ENS, P,S** |
| **Lung** | | | **P** m+ | | **P** ±; **S** ± (anurans)<br>**P** – (urodeles?) | **P** m+ (birds)<br>**P** m+; – (reptiles) |
| **Spleen** | Embedded in gut wall | **S α+** | Embedded in gut wall | **S** α+, m+ | **S** α+ | **S α+ (birds), S**? (reptiles) |
| **Kidney** | Ø | ?[e] | **S**? | **S** adrenergic | **S** anurans, urodeles? | ? reptiles, **S** birds |
| **Urinary bladder/ ureter** | ? | ? | ?[e] | **S** ACh, adrenergic (osmoregulation) | **S** adrenergic? ACh? (osmoregulation)[f] | **S** ACh, adrenergic[f] |
| **Gonads** | ? | **S** ♂ ♀ | ? | **S** ♂ ACh<br>**S** ♀ ACh, adrenergic | **S** ♂ ♀ adrenergic | **S** ♂ ♀? Reptiles<br>**S** ♂ ♀ adrenergic, birds |
| **Eye** | | | | | | |
| Sphincter | Ø | Ø | Not studied | **P** Ø, **S** α+[g] | **S** β– | **P** m+; also nicotinic+ |
| Dilator | **P** m+? | **P** m+ | Not studied | **P** m+[g] | Ø | **S** α+ |
| **Chromaffin tissue (CT)**[h] | Within heart, great veins; not innervated | Paravertebral ganglia, axillary bodies | Intercostal artery, atrium, posterior cardinal vein | Intercostal artery, posterior cardinal vein, atrium | Adrenal gland | Adrenal gland |

Data from: Anatomy: Nilsson in [19, 47]. Cardiovascular system: Morris & Nilsson and Sandblom & Axelsson in [19]. Chromaffin tissue: Perry & Capaldo in [19]. Eye: Neuhuber & Schrödl in [19]. Gastrointestinal tract: Holmgren & Olsson and Olsson & Holmgren in [19]. Kidney, urogenital tract, gonads: Jobling in [19]. Lung: Campbell & McLean in [49]. Spleen: Nilsson in [49]. Lungfish: Nilsson & Holmgren personal communication. Time in evolution after Kumar and Hedges [34]; all with permission

Abbreviations: *ACh* acetylcholine, *CT* chromaffin tissue, *CVS* cardiovascular system, *ENS* enteric nervous system, *GIT* gastrointestinal tract, *H* hagfish, *L* lamprey, *m* muscarinic cholinergic, *P* parasympathetic (cranial autonomic), *S* spinal autonomic, *α* α-adrenoceptor, *β* β-adrenoceptor, + activation, – inhibition/relaxation, Ø innervation absent (or not found), *?* autonomic innervation not investigated

[a]Cranial nerves with parasympathetic preganglionic fibers (III, oculomotor nerve; VII, facial nerve; IX, glossopharyngeal nerve; X vagal nerve)

[b]Blood vessels involved in regulation of the CVS;

[c]Stomach & proximal intestine vagally innervated, small intestine mainly spinally innervated

[d]No stomach (lack of acid secreting mucosa)

[e]Rectal gland secreting hypertonic NaCl under spinal autonomic control

[f]Cloacal bladder

[g]Only true for some species

[h]Chromaffin tissue (CT) innervated by spinal preganglionic neurons (except in cyclostomes)

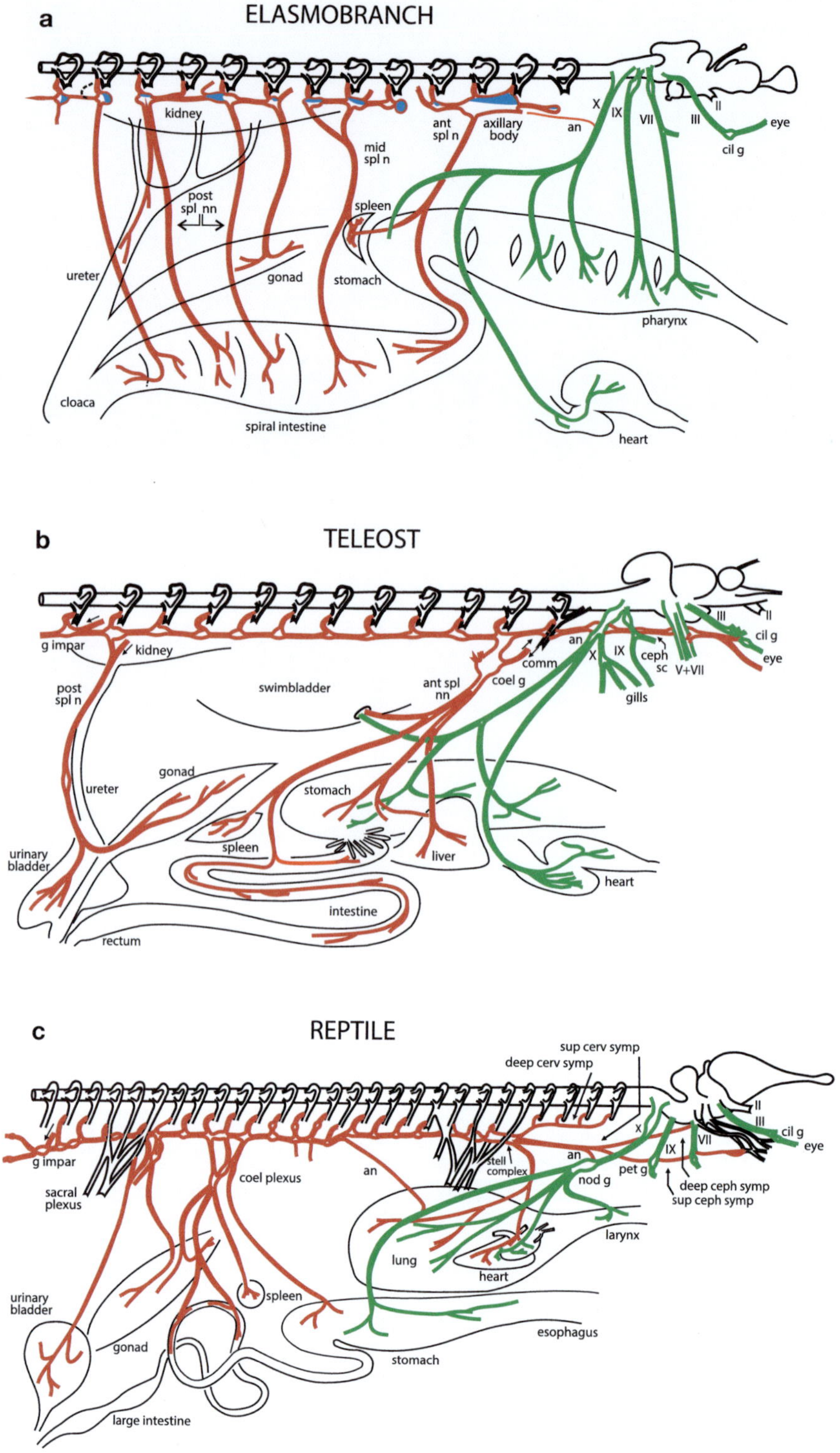


**Fig. 10.17** **Arrangement of the peripheral autonomic nervous system** of an elasmobranch (**a**), a teleost (**b**) and a reptile (**c**). Cranial parasympathetic systems in *green*; spinal autonomic systems in *red*. Note chromaffine tissue associated with paravertebral ganglia in elasmobranch (*blue*). Abbreviations: *an*, anastomosis between spinal autonomic and cranial nerves; *ant spl n*, anterior splanchnic nerve; *ceph sc*, cephalic sympathetic chain; *cil g*, ciliary ganglion; *coel g*, coeliac ganglion; *deep ceph symp*, deep cephalic sympathetic; *g imp*, ganglion impar; *mid spl n*, middle splanchnic nerve; *nod g*, nodose ganglion; *pet g*, petrous (glossopharyngeal) ganglion; *post spl nn*, posterior splanchnic nerves; *sup cerv symp*, superior cervical sympathetic; *sup cerv g*, superior cervical ganglion; *stell g*, stellate ganglion; *Roman numbers* refer to cranial nerves (Modified from Nilsson [47] and Nilsson in [19] with permission)

complexity of regulation may be reflected, e.g., in the cardiovascular regulation in the field of gravity, in the regulation of body core temperature, in the regulation of fluid homeostasis (volume- and osmoregulation), in the regulation of metabolisms (in relation to the terrestrial changes of climate), and in the regulation of body defense (see Table 10.4). The assumption of increased complexity of autonomic regulation with evolution does not imply that complex autonomic regulations have not already been developed relatively early in evolution, e.g., for the swimbladder in fish ([48, 49], Campbell and McLean in [49], Smith and Croll in [19]) or for chromatophores in fish (Grove in [49]). The molecular mechanisms used by the peripheral autonomic neurons and their target cells in the regulation of autonomic target tissues probably are very similar or identical in all vertebrate groups (transmitters and their receptors, neuropeptides, intracellular pathways) (Hoyle in [19]).

## 10.8 Regulation of Body Tissues in Invertebrates

### 10.8.1 The Visceral Nervous System in Invertebrates

Most internal organs of invertebrates have an efferent innervation. This efferent innervation, including the afferent feedback from the organs, is described under the term **visceral nervous system**. Compared to vertebrates – and in particular mammals –we know rather little about the biology of the neural regulation of internal organs in invertebrates. A peripheral ANS that compares morphologically, functionally, histochemically, and in its structure and connections with the CNS with the peripheral ANS in vertebrates (see above), does not exist in invertebrates. However, the physiological studies of the neural regulation of internal organs of invertebrates conducted so far clearly show three principles:

1. Practically all visceral organs of invertebrates are under neural control.
2. The neural control of internal organs is dominated by the central nervous system (such as the buccal and cerebral ganglia in mollusks, crustaceans, hexapods, and annelids).
3. The neural control of the internal organs is coordinated with the neural control of the somato-motor system showing that this control of internal organs is an integral component of the behavioral repertoire of invertebrates as it is in vertebrates (e.g., defense behavior, feeding behavior, reproductive behavior, egg-laying behavior, temporal organization of behavior including circadian and endogenous timing of body functions).

Any neural activity related to the regulation of internal organs has to be seen in the context of the overall behavior of the animals and this applies to invertebrates as well [28, 29].

We have some knowledge about the neural regulations of the cardiovascular system, in particular the heart, of mollusks (e.g., *Aplysia*, snails), cephalopods, annelids (e.g., leech), insects, and crustaceans, some knowledge about neural regulation of the digestive tract, in particular the foregut, in crustaceans, mollusks (*Aplysia*, snail), hydra (cnidaria), earthworm (annelid), and some insects and some knowledge about the regulation of evacuative organs [7, 16, 17, 28, 39, 44]. This knowledge is largely based on the investigation of species of invertebrates that have been established as laboratory animals in neurobiological research (mainly annelids, mollusks, and arthropods). It is sometimes restricted to the anatomy of the neural systems innervating internal organs of these representatives of invertebrates with only little physiology of the neuronal control of the internal organs.

Many visceral organs of invertebrates generate repetitive rhythmically organized movements (e.g., heartbeat, peristalsis of the digestive tract). Neural circuits of invertebrates being involved are characterized by their reciprocal, largely inhibitory, synaptic connections and generate repetitive patterns of activation. The neurons can readily be recognized and identified by their locations, synaptic and electrical connections, transmitters, and membrane properties and the networks can be studied *in vitro* independent of the target tissues. Therefore, these neural circuits together with the visceral effector systems served as model systems to study rhythms generated by neuronal networks (“**central pattern generators**”, CPG), in particular the stomatogastric ganglion in crustaceans (see Chap. 23), but also the neural coordination of heartbeat in annelids (leech, Fig. 10.18), mollusks, and crustaceans, feeding in *Aplysia* and *Limnea* (pond snail), or swimming in mollusks (*Tritonia*, *Clione*) [9, 39, 40, 50, 55].

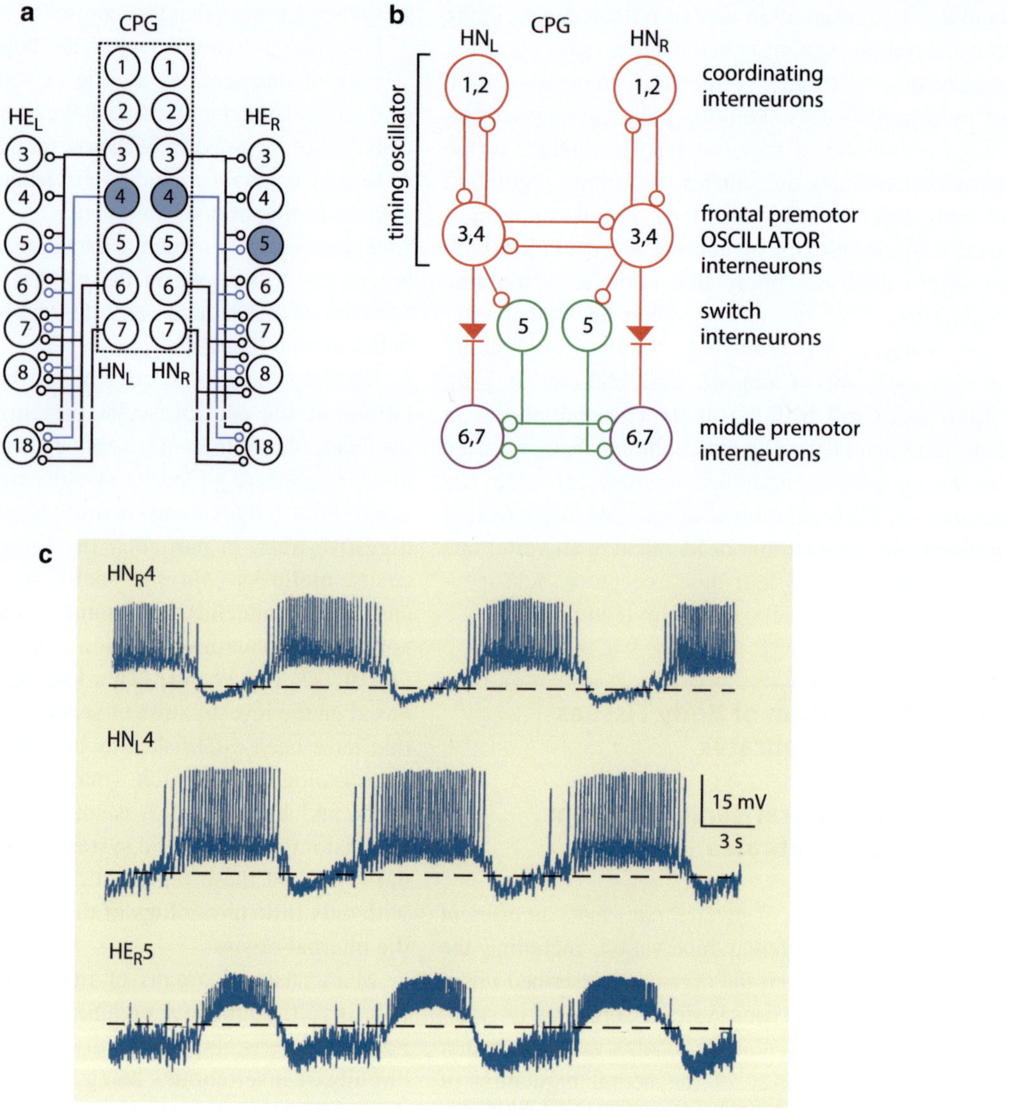


**Fig. 10.18 Neural regulation of the heart in leech.** (**a**) The heart tubes are innervated by 16 pairs of heart exciter motoneurons ($HE_L$ [*left*] and $HE_R$ [*right*]) which are located in the midbody segmental ganglia 3–18. The central pattern generator (CPG) consists of seven pairs of heart interneurons (HN) in the ganglia 1–7 [two additional pairs of heart interneurons in segments 15 and 16 are not shown]. HN3, 4, 6, and 7 interneurons form inhibitory synapses with the motoneurons HE. (**b**) Organization of the CPG. All reciprocal synaptic connections of HN1 to 4 are inhibitory. Synaptic inputs to HN5 from HN3 and 4 and from HN5 to HN6 and 7 are inhibitory. HN1 to 4 form the timing oscillator whereby HN3 and HN4 left and right form the frontal premotor oscillator between right and left. HN5 interneurons are responsible for the switching of rear-to-front peristaltic contraction and synchronous contraction between right and left heart tube. HN 3&4 and HN 6&7 are electrically coupled (indicated by a *diode sign*). (**c**) Simultaneous discharge of HN4 interneurons left and right (*upper traces*) and HE motor neuron in the right segment 5 (neurons indicated in *blue* in (a)). The dashed lines indicate a membrane potential of –50 mV. (Modified from Arbas and Calabrese [1] and Lamb and Calabrese [36] with permission)

Studies of these neuronal networks show whether repetitive movements are dependent on pacemaker neurons or on the emergent (intrinsic) properties of the networks. They may serve to understand the mechanisms underlying rhythmic movements not only in invertebrates but also in vertebrates (e.g., neural mechanisms underlying breathing or locomotion).

The remaining text will concentrate exemplarily on the leech heart system, on some aspects of the crustacean stomatogastric ganglion, on the digestive tract, and on integration of neural regulation of the cardiovascular system and respiration in *Aplysia*.

### 10.8.2 The Beating of the Heart in Leeches is Neuronally Regulated and Under the Control of the Central Nervous System

The cardiovascular systems (heart and often large arteries) in invertebrates are innervated and either closed or open. Their neural regulations are extremely diverse reaching from leeches (annelid), in which about 50 neurons are involved in the regulation of the heart tubes, and crustaceans, in which 6–16 neurons in the cardiac ganglion are involved in regulation of the heart [9], to cephalopods (e.g., *Octopus*) in which about one million neurons may be involved [16, 17].

The leech has two tubular hearts, one on each side. The myogenically generated rhythm of these heart tubes is superimposed by a rhythm generated by a neural network which is located in the cardiac ganglia and results in a typical rhythmic pattern of contraction of the heart tubes: One heart tube contracts peristaltically from rear-to-front, leading to an increase of systolic pressure while the other tube contracts almost synchronously along its length. This left-right pattern switches every 20–40 heartbeats. The heart tubes are synaptically innervated by 16 cholinergic heart exciter (HE) motoneurons on each side located in the cardiac ganglia 3–18. These motoneurons are tonically active. They are rhythmically inhibited by a network of heart interneurons (HN neurons) that form the central pattern generator (CPG) for the heart (Fig. 10.18a, b). The HN-interneurons are located in the rostral seven segmental ganglia (and in ganglia 15 and 16, not shown in Fig. 10.18a). The HN-neurons of segments 1–4 are reciprocally connected by inhibitory synapses and constitute the heart oscillator. The two HN neurons in segment 5 are switch interneurons which generate the regular switch of contraction of the heart tubes from the rear-to-front peristalsis mode into the synchronous mode (i.e., left peristalsis/right synchronous to left synchronous/right peristalsis). These HN5 interneurons are inhibited by HN interneurons 3 and 4 and inhibit HN interneurons 6 and 7 [36, 40].

About 6–16 neurons in the cardiac ganglion of crustaceans form a central pattern generator which probably is the smallest of its kind. For example, in decapod crustaceans the cardiac CPG consist of a network of four synaptically coupled interneurons which drive five motoneurons that drive the neurogenic heart. This cardiac network is under excitatory and inhibitory control of the CNS and can also be influenced by hormonal factors (biogenic amines, neuropeptides) [9, 20, 55].

### 10.8.3 Regulation of the Digestive System: The Stomatogastric Ganglion and the Enteric Nervous System

The foregut of crustaceans consists of esophagus, cardiac sac, gastric mill, and pylorus. Food taken up by the esophagus is mixed with digestive fluids in the cardiac sac, chewed by internal teeth in the gastric mill, and filtered by the pylorus before moving on to the midgut. This process is controlled by the stomatogastric nervous system. The *stomatogastric ganglion* consist of about 30 large neurons that are involved in regulation and coordination of the movements of the different parts of the foregut. Apart from four interneurons, most are motoneurons. The neurons are interconnected by inhibitory or electric synapses and form three neural networks: the cardiac sac network, the gastric mill network, and the pyloric network. These networks are responsible for the rhythmic movements of the foregut. They are coordinated with each other in a dynamic way, the type and degree of coordination being dependent on the functional state of the foregut, on the afferent feedback from the foregut, on the command signals from the cerebral ganglion, and on various hormonal signals. The coordinated rhythmic regulation of the foregut musculature is reminiscent of the rhythmic regulation of ventilation by the pontomedullary respiratory neural network in vertebrates [21, 38]. Thus, the stomatogastric nervous system regulating the crustacean foregut would functionally correspond to the neural circuits in the lower brainstem being involved in the regulation of the foregut in mammals which includes part of the nucleus of the solitary tract, the nucleus dorsalis nervi vagi (which is essentially a motor nucleus of the foregut), and the area postrema [23]. The cellular neural mechanisms underlying these rhythms in crustaceans will be described in Chap. 27 [20, 39].

Although not investigated systematically, it is likely that the digestive tract of invertebrates is innervated by an ENS with functions similar to the mammalian ENS.

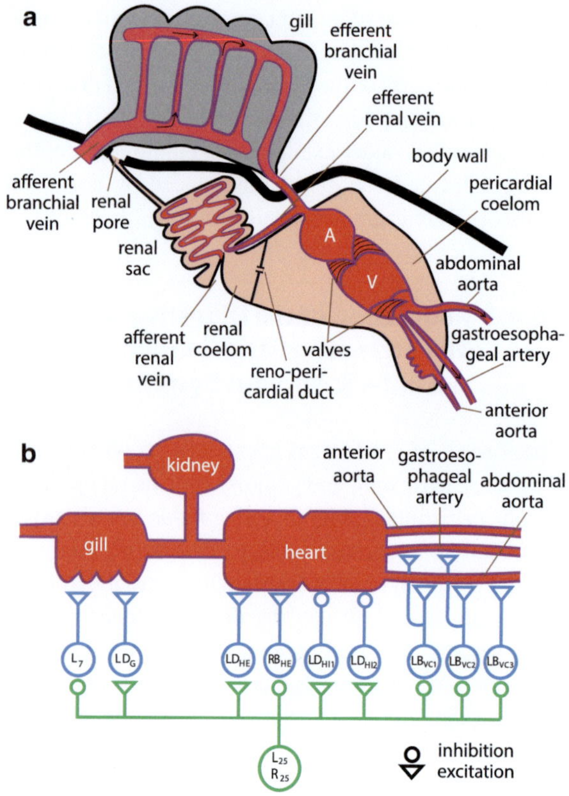


**Fig. 10.19 Respiratory and cardiovascular system of the mollusk *Aplysia*.** (**a**) Anatomical relations between cardiovascular system, gill and kidney (*renal sac*). *A* atrium, *V* ventricle. (**b**) Innervation of gill muscles, heart, and main body arteries by motoneurons (*blue*) and innervation of motoneurons by neurons L25/R25 which are neurons of the respiratory network. All neurons are located in the abdominal ganglion (Modified after Mayeri et al. [42] with permission)

It can function independently of the CNS. This is based on experimental studies of the mollusks *Aplysia* and *Limnea* (snails) [21, 51], the earthworm (annelid [3]), insects [10], and echinoderms (e.g., sea cucumber [13]).

Finally, digestive movements reminiscent of esophageal reflex, segmentation, and defecation of the GIT in mammals have been shown to exist already in *Hydra* (cnidarian). These movements are dependent on a nerve network associated with the circular and longitudinal musculature of the body column of hydra [56]. Interestingly, the nerve ring of neurons running circumferentially around the hypostome above the tentacle zone in *Hydra*, from which the putative ENS has its origin, is hypothesized to be related phylogenetically to the development of the CNS in bilaterian animals. Cnidarians diverged about 600 million years ago [33].

### 10.8.4 Integration of Visceral and Motor Functions with Behavior in Invertebrates

The neural regulation of visceral organs (cardiovascular system, digestive system, kidney) in invertebrates is closely adapted to the behavior of the animals. This adaptation is controlled by the CNS. In some cases it could be shown that visceral functions and somatomotor functions are coordinated by the activation of functionally specialized single neurons or small groups of neurons. This has been demonstrated for the gastropod *Aplysia* showing the neural mechanisms underlying coordination and integration of neural regulation of circulation and respiration (Figs. 10.19 and 10.20). Respiratory system (gill) and cardiovascular system are hydraulically coupled in many gastropods. The gas exchange between seawater and hemolymph occurs in the gill. The hemolymph is pumped by the heart through the large body arteries into body cavities. The contractions of the mantle, parapodia, and siphon push the seawater from the mantle cavity through the siphon leading to a passive inflow of seawater to the large exchange surface of the gill. The simultaneous contraction of the gill pushes the blood into the heart cavity (Fig. 10.19a). Heart and large body arteries relax to prevent that the blood is pumped into the kidney.

The heart of *Aplysia* consists of auricle and ventricle which are separated by valves. The arterial blood is pumped through the large body arteries to the body tissues (digestive organs, somatic tissues, nervous system, reproductive organs) and collects in the hemocoel. The backflow of blood into the ventricle is prevented by valves. The rhythm of the heart is generated by myogenic pacemaker cells. Heart and body arteries are innervated by motoneurons (Fig. 10.19b): (a) The *heart* is innervated by two cholinergic inhibitory cardiomotor neurons ($LD_{HI1,2}$) and two excitatory cardiomotor neurons ($LD_{HE}$, $RB_{HE}$), one of them using serotonin as transmitter. Their activation leads to a decrease of heart rate or to increase of contraction, respectively. (b) The *abdominal aorta* and *gastroesophageal artery* are innervated by three cholinergic vasoconstrictor neurons. Their activation generates a vasoconstriction – and inhibition of their activity, a vasodilatation.

The respiratory rhythm in *Aplysia* is generated by the respiratory neural network consisting of two clusters of electrically and synaptically connected neurons (L25/R25) located in the abdominal ganglion. The respiratory musculature is innervated by excitatory motoneurons that have their cell bodies in the abdominal ganglion, too (Fig. 10.19b). Most respiratory motoneurons are activated by interneurons of the respiratory network and some are inhibited [31, 57].

Neurons of the *respiratory network* (neurons L25/R25 in Fig. 10.19b) which synaptically activate respiratory motoneurons leading to contraction of the respiratory musculature (e.g., via the motoneuron $LD_G$, Fig. 10.20), form inhibitory synapses with the excitatory cardiomotor neurons $RB_{HE}$ – and the vasoconstrictor neurons LB and excitatory synapses with the inhibitory cardiomotor neurons ($LD_{HI}$) and with the excitatory cardiomotor neuron $LD_{HE}$ innervating the heart (Fig. 10.19b). Activation of this respiratory neuron network generates in this way not only a contraction of the gill musculature but also an inhibition of the heart and a dilation of the large body arteries followed by decrease of heart rate, ventricle contraction, and blood pressure (Fig. 10.20). This preprogrammed pattern of the coordination of the *neural regulation of respiration and circulation* is integrated in the regulation of several behaviors of *Aplysia*: feeding behavior, locomotion, egg-laying behavior, defense behavior during low oxygen content, and increased $CO_2$-concentration of the seawater, etc.

*Aplysia* is an herbivore and takes several hours per day to graze algae and seaweed. This process occurs by rhythmic movements of the radula generated by the buccal muscles of the head. The cycle consists of protraction and retraction of the radula, each half cycle lasting about 3–6 s. During protraction the three vasoconstrictor neurons ($LB_{VC}$ in Fig. 10.19b) that innervate the gastroesophageal and abdominal aorta are activated and the resistance in these blood vessels is increased leading to an increased blood flow through the anterior aorta to the head region. During retraction, the vasoconstrictor neurons are inhibited leading to an increase of blood flow through the viscera and decrease through the head region. In this way the cardiac output is pumped either into the head and/or into the viscera. Furthermore, heart rate and arterial pressure increase by activation of the cardiomotor neuron $RB_{HE}$ (Fig. 10.19b). The alternating distribution of blood flow and increase of heart rate and blood pressure during the feeding behavior are abolished after severance of the nerve to the heart and body arteries. This example demonstrates that the neural regulations of the buccal muscles and of the cardiovascular system are temporarily precisely adjusted by the CNS [31, 57].

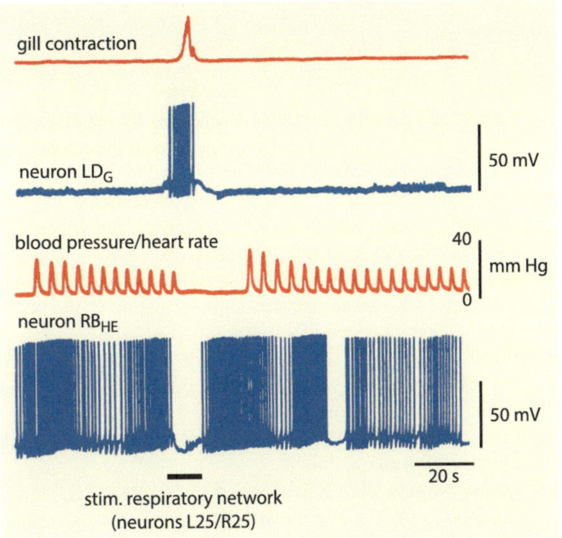


**Fig. 10.20** **Stimulation of the neurons of the respiratory network** activates the motoneurons innervating gill muscles (neuron $LD_G$) followed by gill contraction and inhibits excitatory motoneurons to the heart (neuron $RB_{HE}$) followed by decrease of heart rate and blood pressure (Modified after Koester et al. [32] with permission)

## References

1. Arbas EA, Calabrese RL (1987) Slow oscillations of membrane potential in interneurons that control heartbeat in the medicinal leech. J Neurosci 7:3953–3960
2. Bahr, Bartel B, Blumberg H, Jänig W (1986) Functional characterization of preganglionic neurons projecting in the lumbar splanchnic nerves: neurons regulating motility. J Auton Nerv Syst 15:109–130
3. Barna J, Csoknya M, Lazar Z, Bartho L, Hamori J, Elekes K (2001) Distribution and action of some putative neurotransmitters in the stomatogastric nervous system of the earthworm, *Eisenia fetida* (Oligochaeta, Annelida). J Neurocytol 30:313–325
4. Boczeck-Funcke A, Häbler HJ, Jänig W, Michaelis M (1992) Respiratory modulation of the activity in sympathetic neurones supplying muscle, skin and pelvic organs in the cat. J Physiol 449:333–361
5. Burnstock G (ed) (1992–2003) The autonomic nervous system, vol 1–13. Harwood Acad Publ, Chur; Vols 14–15, Taylor and Francis, London
6. Campbell GD, Edwards FR, Hirst GDS, O'Shea JE (1989) Effects of vagal stimulation and applied acetylcholine on

pacemaker potentials in the guinea-pig heart. J Physiol 415:57–68.
7. Chase R (2002) Behavior and its neural control in gastropod molluscs. Oxford Univ Press, Oxford/New York
8. Choate JK, Klemm M, Hirst GDS (1993) Sympathetic and parasympathetic neuromuscular junctions in the guinea-pig sino-atrial node. J Auton Nerv Syst 44:1–15
9. Cooke IM (2002) Reliable, responsive pacemaking and pattern generation with minimal cell numbers: the crustacean cardiac ganglion. Biol Bull 202:108–136
10. Copenhaver PF (2007) How to innervate a simple gut: familiar themes and unique aspects in the formation of the insect enteric nervous system. Dev Dyn 236:1841–1864
11. Edwards FR, Hirst GD, Klemm MF, Steele PA (1995) Different types of ganglion cell in the cardiac plexus of guinea-pigs. J Physiol 486:453–471
12. Furness JB (2006) The enteric nervous system. Blackwell Science, Oxford
13. Garcia-Arrarás JE, Rojas-Soto M, Jiménez LB, Diaz-Miranda L (2001) The enteric nervous system of echinoderms: unexpected complexity revealed by neurochemical analysis. J Exp Biol 204:865–873
14. Gibbins IL (2004) Peripheral autonomic pathways. In: Paxinos G, Mai JK (eds) The human nervous system, 2nd edn. Elsevier, Amsterdam/San Diego/London, pp 134–189
15. Grewe W, Jänig W, Kümmel H (1995) Effects of hypothalamic thermal stimuli on sympathetic neurones innervating skin and skeletal muscle of the cat hindlimb. J Physiol 448:139–152
16. Hill RB (ed) (1987) Cardiovascular control in mollusca (multiauthor review). Experientia 43:953–997
17. Hill RB (ed) (1992) Control of circulation in invertebrates. Experientia 48:797–858 [Multiauthor review]
18. Hirst GDS, Choate JK, Cousins HM, Edwards FR, Klemm MF (1996) Transmission by post-ganglionic axons of the autonomic nervous system: the importance of the specialized neuroeffector junction. Neuroscience 73:7–23
19. Holmgren S, Olsson C (eds) (2011) Comparative physiology of the autonomic nervous system. Auton Neurosci 165: 1–148
20. Hooper SL, DiCaprio RA (2004) Crustacean motor pattern generator networks. Neurosignals 13:50–69
21. Ito S, Kurokawa M (2007) Coordinated peripheral neuronal activities among the different regions of the digestive tract in *Aplysia*. Zoolog Sci 24:714–722
22. Jänig W (1985) Organization of the lumbar sympathetic outflow to skeletal muscle and skin of the cat hindlimb and tail. Rev Physiol Biochem Pharmacol 102:119–213
23. Jänig W (2006) The integrative action of the autonomic nervous system. Neurobiology of homeostasis. Cambridge Univ Press, Cambridge/New York
24. Jänig W (2009) Autonomic nervous system and pain. In: Basbaum AI, Bushnell MC (eds) Science of pain. Academic, San Diego, pp 193–225
25. Jänig W (2010) Vegetatives Nervensystem. In: Schmidt RF, Lang F, Heckmann M (eds) Physiologie des Menschen, 31st edn. Springer Medizin, Berlin Heidelberg New York, pp 403–434
26. Jänig W, Levine JD (2013) Autonomic-endocrine-immune responses in acute and chronic pain. In: McMahon SB, Koltzenburg M, Tracey I, Turk D (eds) Wall and Melzack's textbook of pain, 6th ed. Elsevier Churchill Livingstone, Amsterdam/Edinburgh
27. Jänig W, McLachlan EM (1987) Organization of lumbar spinal outflow to distal colon and pelvic organs. Physiol Rev 67:1332–1404
28. Kandel ER (1976) Cellular basis of behavior. Freeman, San Francisco
29. Kandel ER (1979) Behavioral biology of *Aplysia*. Freeman, San Francisco
30. Klemm M, Hirst GDS, Campbell G (1992) Structure of autonomic neuromuscular junctions in the sinus venosus of the toad. J Auton Nerv Syst 39:139–150
31. Koester J, Koch UT (1987) Neural control of the circulatory system of *Aplysia*. Experientia 43:972–980
32. Koester J, Mayeri E, Liebeswar G, Kandel ER (1994) Neural control of circulation in *Aplysia*. II. Interneurons. J Neurophysiol 37:476–496
33. Koizumi O (2007) Nerve ring of the hypostome in hydra: is it an origin of the central nervous system of bilaterian animals? Brain Behav Evol 69:151–159
34. Kumar S, Hedges SB (1998) A molecular timescale for vertebrate evolution. Nature 392:917–920
35. Kuntz A (1940) The structural organization of the inferior mesenteric ganglia. J Comp Neurol 72:371–382
36. Lamb DG, Calabrese RL (2011) Neural circuits controlling behavior and autonomic functions in medicinal leeches. Neural Syst & Circ 1:13
37. Langley JN (1921) The autonomic nervous system. Part I. W Heffer, Cambridge
38. Llewellyn-Smith IJ, Verbene AJM (eds) (2011) Central regulation of autonomic functions. Oxford Univ Press, New York
39. Marder E, Bucher D (2007) Understanding circuit dynamics using the stomatogastric nervous system of lobsters and crabs. Annu Rev Physiol 69:291–316
40. Marder E, Calabrese RL (1996) Principles of rhythmic motor pattern generation. Physiol Rev 76:687–717
41. Mathias CJ, Bannister R (eds) (2013) Autonomic failure, 5th edn. Oxford Univ Press, New York
42. Mayeri E, Koester J, Kupfermann I, Liebeswar G, Kandel ER (1974) J Neurophysiol 37:458–475
43. McLachlan EM (ed) (1995) Autonomic ganglia. In: Burnstock G (ed) The autonomic nervous system, vol 6. Harwood Academic Publ, Luxembourg
44. Miller TA (1997) Control of circulation in insects. Gen Pharmacol 29:23–38
45. Mustafa AK, Gadalla MM, Snyder SH (2009) Signaling by gasotransmitters. Sci Signal 2:re2
46. Nicholls JG, Martin AR, Fuchs PA, Brown DA, Diamond ME, Weisblat DA (2012) From neuron to brain, 5th edn. Sinauer, Sunderland
47. Nilsson S (1983) Autonomic nerve function in the vertebrates. Springer, Berlin Heidelberg New York

48. Nilsson S (2009) Nervous control of fish swimbladders. Acta Histochem 111:176–184
49. Nilsson S, Holmgren S (eds) (1994) Comparative physiology and evolution of the autonomic nervous system. In: Burnstock G (ed) The autonomic nervous system, vol 4. Harwood Acad Publ, Chur/Switzerland
50. Nusbaum MP, Beenhakker MP (2002) A small-systems approach to motor pattern generation. Nature 417:343–350
51. Okamoto T, Kurokawa M (2010) The role of the peripheral enteric nervous system in the control of gut motility in the snail *Lymnaea stagnalis*. Zoolog Sci 27:602–610
52. Randall DJ, Burrgren W, French K (2001) Eckert animal physiology, 5th edn. Palgrave Macmillan, New York
53. Robertson D, Biaggioni I, Burnstock G, Low PA, Paton JFR (eds) (2012) Primer on the autonomic nervous system, 3rd edn. Academic Press/Elsevier, Boston/Amsterdam
54. Schmidt RF, Lang F, Heckmann M (eds) (2010) Physiologie des Menschen, 31st edn. Springer Medizin, Berlin Heidelberg New York
55. Selverston AI (2010) Invertebrate central pattern generator circuits. Philos Trans R Soc Lond B Biol Sci 365:2329–2345
56. Shimizu H, Koizumi O, Fujisawa T (2004) Three digestive movements in *Hydra* regulated by the diffuse nerve net in the body column. J Comp Physiol A Neuroethol Sens Neural Behav Physiol 190:623–630
57. Skelton M, Alevizos A, Koester J (1992) Control of the cardiovascular system of *Aplysia* by identified neurons. Experientia 48:809–817
58. Squire LR, Bloom FE, Spitzer NC, du Lac S, Ghosh A, Berg D (eds) (2008) Fundamental Neuroscience, 3rd edn. Academic Press, San Diego
59. Undem B, Weinreich D (eds) (2005) Advance in vagal afferent neurobiology. CRC Press, Boca Raton
60. Ward SM, Sanders KM, Hirst GD (2004) Role of interstitial cells of Cajal in neural control of gastrointestinal smooth muscles. Neurogastroenterol Motil 16(Suppl 1): 112–117

# Neuropeptides and Peptide Hormones

# 11

Dick R. Nässel and Dan Larhammar

Several types of chemical messengers are employed by the nervous system for local or more diffuse signaling. Among these the peptides are the most diverse in structure and function. In nervous tissues they are typically produced by neurons or neurosecretory cells, and can therefore be specified as **neuropeptides** or **peptide hormones**, respectively. Additionally, many peptides are produced by endocrine cells or other cell types in different locations. In fact, the same peptide can be expressed by all these cell types in a given animal. Neuropeptides and peptide hormones are ubiquitous in the nervous and endocrine systems of all metazoans. Not only do these peptides exist in a large number of distinct molecular forms, they are also very diverse in their actions and signaling mechanisms. Thus, in a single animal species there may be more than a 100 different neuropeptides and peptide hormones (and their receptors), and each can have multiple functions. The peptides are encoded in the genome as parts of larger precursor proteins, referred to as **prepropeptides**. This direct coding means that when whole animal genomes have been sequenced the total inventory of neuropeptides and peptide hormones can be predicted. Such sequence data can also be used for analysis of neuropeptide evolution and show that some peptide sequences are well conserved across a broad range of species, whereas others display considerable variability and some are even unique to certain taxa. In this chapter we show that in some cases not only peptide sequences, but also the receptor structures and mechanisms of action and physiological functions can be conserved from invertebrates to vertebrates.

Classically the peptides were investigated in neurosecretory and endocrine systems and many important roles of peptides as hormones could be established. Peptide hormones often regulate basic mechanisms in development, growth, reproduction, and metabolism and thus feature as critical players in the homeostasis of the organism. Several of these peptidergic systems are targets of therapeutics due to their importance in, for instance, diabetes, growth, pain regulation, and mental health.

The neuropeptides are produced by interneurons, sensory neurons, and efferent neurons such as motoneurons. Their actions are spatially restricted and depend on the morphology of the neurons releasing them, and distribution of cognate peptide receptors. Thus, neuropeptide function in neuronal circuits can be either as local or more global activators or modulators depending on the extent of their neuronal arborizations. In some cases assemblies of peptidergic neurons form neuropeptide systems with specific regulatory functions. In this chapter we will use the terms **neuropeptide** and **peptidergic neuron** for cases where peptides are produced by a neuron to act on adjacent neurons or other cell types. This signaling can be at synapses or nonsynaptic by means of volume transmission (see Chap. 8). In contrast, we use **peptide hormone** and **neurosecretory cell** (or when appropriate **endocrine cell**) where peptides are released into the circulation or body cavity and act on distant target cells with appropriate receptors.

D.R. Nässel (✉)
Department of Zoology, Stockholm University,
Svante Arrhenius väg 18 B, SE-10691 Stockholm,
Sweden
e-mail: dnassel@zoologi.su.se

D. Larhammar
Department of Neuroscience, Uppsala University,
Box 593, SE-75124 Uppsala, Sweden
e-mail: dan.larhammar@neuro.uu.se

C.G. Galizia, P.-M. Lledo (eds.), *Neurosciences - From Molecule to Behavior: A University Textbook*,
DOI 10.1007/978-3-642-10769-6_11, © Springer-Verlag Berlin Heidelberg 2013

We show here that peptides have a huge spectrum of functions both in central neuronal circuits and as circulating hormones. We provide details on the organization and function of peptidergic neurons and neurosecretory cells, as well as other hormonal systems signaling with peptides, both in invertebrates and vertebrates. The hypothalamus-pituitary system and its analogs in insects are used to illustrate neuroendocrine regulatory systems utilizing peptides. Furthermore, we include some comparisons between distantly related animal groups to highlight the ancient evolutionary origins of several peptide signaling systems.

## 11.1 Neuropeptides, Peptide Hormones, and Their Receptors

The earliest neuropeptides to be sequenced and synthesized were purified from mammals, namely oxytocin and vasopressin from the posterior pituitary of cattle in the 1940s. For these achievements Vincent du Vigneaud (1901–1978) was awarded the Nobel Prize for Chemistry in 1955. With improved techniques further peptides were discovered in the brains of pig and sheep in the late 1960s. For instance, the hypothalamic peptides thyrotropin-releasing hormone (TRH) and gonadotropin-releasing hormone (GnRH) were characterized independently in the laboratories of Andrew Schally and Roger Guillemin, who both received the Nobel Prize for Medicine in 1977. In the years that followed, numerous peptides were sequenced from a broad range of animals, including invertebrates. Gradually molecular biology techniques allowed discovery of peptides deduced from DNA sequences, thereby enabling identification of peptides in an even wider range of species.

### 11.1.1 Neuropeptides Are Derived from Larger Prepropeptides, and Generally Act on GPCRs

Neuropeptides and peptide hormones are chains of amino acids (AA) that vary in length from 3 to more than 80 AA, and may act as monomers or oligomers. They are derived from precursor proteins, prepropeptides, or preprohormones, that are generated by regular protein synthesis (Fig. 11.1). Peptides are thus encoded in the genome and their biosynthesis requires gene transcription and subsequent translation by ribosomes. **Prepropeptides** typically consist of about 100–300 amino acids. After removal of the signal peptide by signal peptidase they are called **propeptides**. These are processed further by proteolytic enzymes acting at specific cleavage sites to release the biologically active **peptides** (Fig. 11.1). After further posttranslational processing steps the mature neuropeptides are stored in vesicles and ready for release (Fig. 11.2). Usually peptides are stored in **large dense core vesicles**, located away from the synaptic zone (Figs 11.2a, b), and a stronger depolarization is required for peptide release than for classical neurotransmitters (Fig. 11.2b). Large dense core vesicles have diameters of around 200 nm, as opposed to the clear vesicles of classical neurotransmitters with diameters of 35–45 nm. Prepropeptides often give rise to more than one peptide. In some cases, especially in invertebrates, multiple copies of identical peptides can be liberated; in other cases the precursor contains several copies of structurally related or unrelated peptides. After release neuropeptides and peptide hormones commonly act on G protein-coupled receptors (GPCRs) of different types. For a given peptide there may be more than one GPCR, especially in vertebrates. Some peptide hormones, like insulin-like peptides, activate receptor tyrosine kinases, others like atrial natriuretic peptide act on membrane guanylyl cyclase, and in mollusks one peptide (FMRFamide) is known to activate a ligand-gated ion channel. Activation of a GPCR triggers a cascade of events, usually involving second messengers, that via protein kinases produce various cellular responses such as alteration of ion channel and membrane properties, or long-term alterations that involve gene transcription and protein synthesis (see Chap. 6). After release the action of neuropeptides is limited by dilution over distance and by enzymatic inactivation by ubiquitous extracellular peptidases. Another way to terminate peptide action is by desensitization of the GPCRs, and upon sustained activation the receptors even internalize from their cell surface locations.

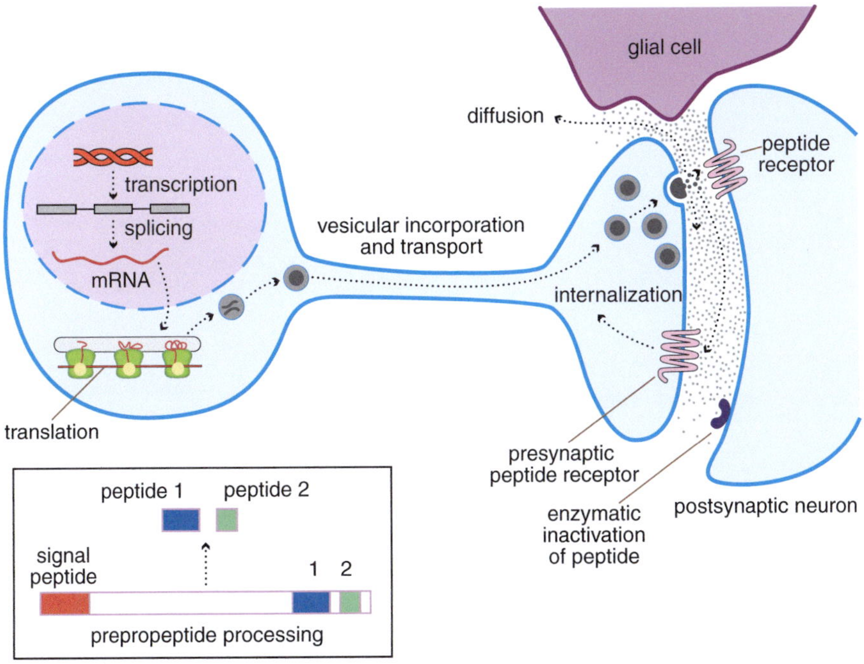


**Fig. 11.1 A peptidergic neuron.** Peptide biosynthesis starts with gene transcription and splicing, followed by mRNA translation at ribosomes and production of a prepropeptide (precursor). The precursor is processed into bioactive peptides (peptide 1 and 2) after incorporation into dense core vesicles. These are transported down the axon and the vesicles with mature peptides are stored in the axon termination (disproportionately enlarged in the figure). After release, peptides act on postsynaptic receptors (GPCRs). Peptide spillover is monitored by presynaptic receptors that can regulate release of the peptide or the neuron's classical neuro-transmitter. Enzymatic inactivation of peptide occurs in the synaptic region by means of specific membrane-bound peptidases. Another factor for diminishing peptide action is diffusion away from the receptor sites (Modified from Squire 2003 [1])

## 11.2 Morphology and Function of Neurosecretory Cells and Peptidergic Neurons

### 11.2.1 Peptides Released from Neurohemal Organs Act on Distant Targets via the Circulation

Peptidergic neurosecretory cells are organized in a rather conserved fashion from arthropods to mammals. They commonly have large cell bodies (somata), where peptide biosynthesis occurs, and axons terminating in storage and release sites with access to the circulation. Often such neurosecretory axon terminals are located in specific structures called **neurohemal organs**, where several types of neurosecretory cell axons coalesce. In invertebrates it is also common that neurosecretory axon terminations are diffusely distributed on the outer surface of nerves, or even of the brain or ventral ganglia, along the intestine, or on body wall muscle. These diffuse release sites are referred to as **neurohemal areas**. The targets of peptide hormones are reached via the circulation. Therefore, potentially, all cells in the body might respond to neurohormones, and target specificity is reached by the selective expression of hormone receptors in the target organs. At least in vertebrates the target organs of many centrally released hormones produce their own hormones that often elicit both local physiological effects and feedback on the central systems.

In many invertebrate taxa (e.g., cnidarians, flatworms, and nematodes) there is no bona fide circulatory system. Animals of these taxa have peptide-producing neurons that probably release their messengers mostly in a paracrine fashion, i.e., the peptide is released nonsynaptically to act on adjacent neighbor cells. As an example, peptide-producing neurons of the nematode *Caenorhabditis elegans* are commonly sensory-, motor-, or interneurons and part of the circuitry of the nervous system. Nevertheless, these types of neurons

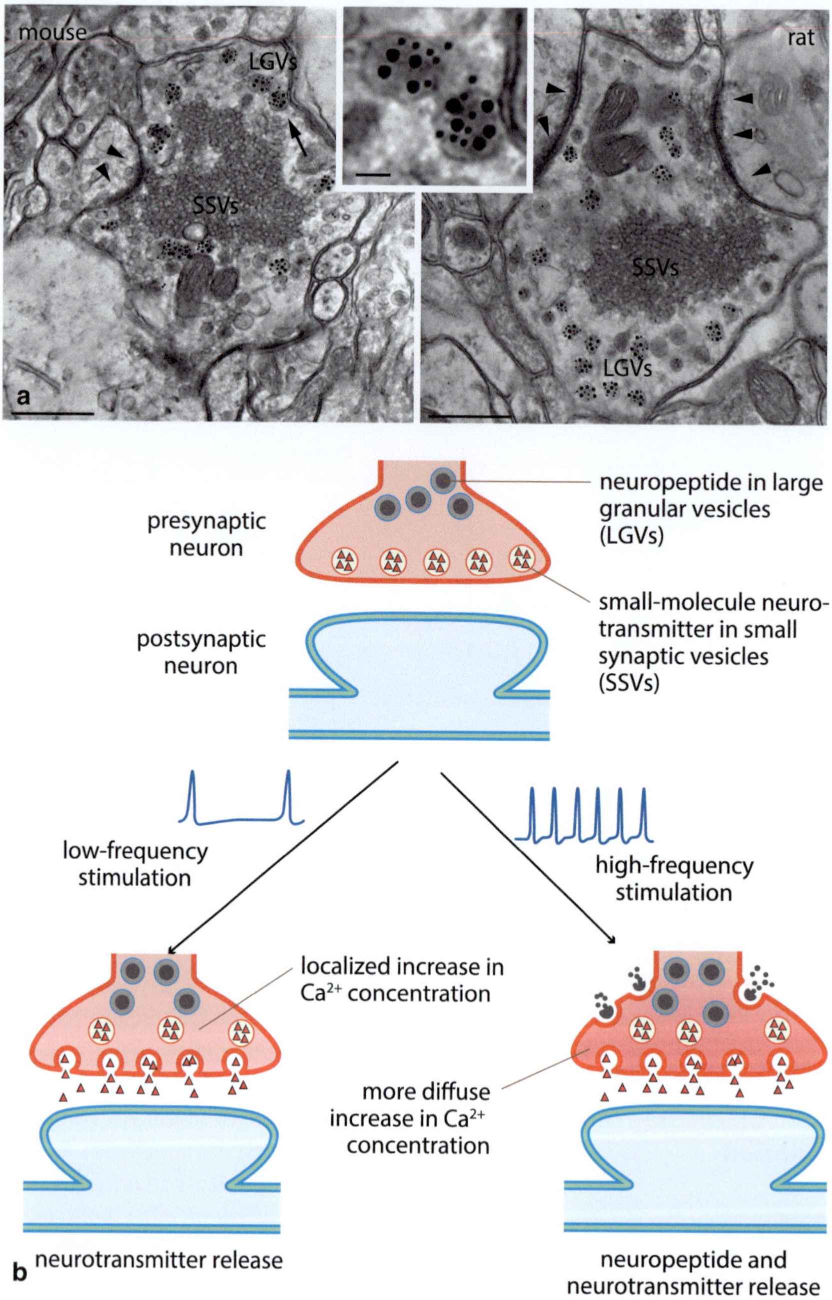


**Fig. 11.2** **Neuropeptides at the synapse.** (**a**) Electron microscopic image of neuron terminations in substantia gelatinosa in spinal cord of mouse and rat with small molecule transmitter in small clear-core vesicles (small synaptic vesicles, *SSVs*) and two different neuropeptides in dense-core vesicles (large granular vesicles, *LGVs*). The peptides, calcitonin gene-related peptide (CGRP) and substance P were detected with immunogold labeling (two sizes of gold particles, shown in *inset*) in the same vesicles. The small synaptic vesicles are stored close to the active zone (*arrowheads*), whereas the dense core vesicles are located away from this, in the perisynaptic area (*arrow*) (From Salio et al. [2] with permission from A. Merighi and Springer-Verlag). (**b**) Depolarization and release of small classical neurotransmitter and colocalized neuropeptide. A neuropeptide is stored in large dense vesicles that are localized perisynaptically and a small molecule (classical) transmitter in small synaptic vesicles in the active zone. These are affected differently by low- and high-frequency stimulation of the axon termination. The synaptic vesicles fuse with presynaptic membrane at low-frequency stimulus and localized calcium influx, whereas the peptidergic vesicles require more massive increase of calcium for release perisynaptically

may produce peptides that in other taxa function as peptide hormones, for instance, insulin-like peptides.

### 11.2.2 Peptides in the CNS Often Act Together with Other Neurotransmitters as Modulatory Cotransmitters

In the CNS there is a large variety of interneurons and sensory neurons that utilize neuropeptides for signaling or signal modulation within central circuits. These commonly have relatively small cell bodies and produce and store smaller amounts of peptides than neurosecretory cells. Target specificity is secured not only by the expression of the right GPCR in the target cell (as for neurohormones), but also by the specific branching patterns of the peptidergic neurons. The peptidergic interneurons display vast variations in morphology and their branching can be either restricted to smaller portions of the CNS or be quite extensive, suggesting more global influence. It is quite common that interneurons colocalize neuropeptides with either of a range of classical neurotransmitters, such as acetylcholine, glutamate, GABA, or monoamines. Thus, the neuropeptides may act as **cotransmitters** that modulate fast neurotransmission. Different interneurons use neuropeptides in two major forms of neuromodulation: intrinsic or extrinsic modulation. **Intrinsic modulation** is when the neuropeptide is released synaptically or nonsynaptically by a neuron within the circuit that is modulated. **Extrinsic modulation** is by neuropeptide released by neurons from outside the circuit.

## 11.3 Organization of Neuroendocrine and Peptidergic Systems in Invertebrates

### 11.3.1 From Early Surgical Experiments Demonstrating Hormones to Modern Molecular Approaches to Study Peptide Systems

The first demonstration of an invertebrate hormone produced in the nervous tissue was in experiments by Stefan Kopec (1888–1941) with the gypsy moth already in the early 1920s. In this study, removal of the brain within a certain time window in late larval development led to a failure to pupate and timely reimplantation of a brain anywhere in the organism triggered the pupariation process (i.e., the onset of prepupal development). From these results it could be concluded that the brain secretes a pupariation-inducing hormone. This hormone, later designated **prothoracicotropic hormone (PTTH)** was identified and chemically elucidated as a large dimeric peptide produced by brain neurosecretory cells more than 60 years after Kopec's study. Interestingly, it was found that the PTTH receptor is a membrane tyrosine kinase.

Neurosecretory cells were first visualized with histochemical staining techniques that did not identify the specific hormones, but led to anatomical descriptions of some major neuroendocrine systems. Mollusks, decapod crustaceans, and larger insects were commonly used for these earlier studies. The anatomical findings made it possible to perform more precise extirpation and replacement experiments where bioactivity of neurosecretory cells could be determined. Once peptide sequences were elucidated they could be synthesized for bioassays and for production of antisera to be used for immunocytochemical identification of the cells producing them. Peptide immunocytochemistry provided a great step forward in understanding the organization of neurosecretory systems and also revealed that other types of neurons or endocrine cells produce peptides.

Today there is a huge array of molecular genetic techniques enabling the scientist to identify and specifically interfere with various components of neuropeptide and peptide hormone signaling. Over the last 10 years, the entire genomes of many species have been sequenced enabling comprehensive identification of the genes encoding neuropeptide precursors and GPCRs and many other proteins involved in peptide signaling. This has been a tremendous help for studies of the evolution of neuropeptide signaling, and has also provided directions for functional studies. In addition, tools developed for proteomics to identify and quantify proteins (e.g., MALDI-TOF mass spectrometry) have allowed identification of peptides in increasingly smaller samples down to single neurons.

It appears as if neuropeptide signaling occurs in all animal phyla alongside that conducted by small molecule neurotransmitters of different types. We focus here on a few model invertebrates, although peptide signaling has been studied also in cnidarians, flatworms, and other taxa.

### 11.3.2 Peptidergic Systems in Mollusks

Recently, the genome of the limpet *Lottia gigantea* was sequenced and neuropeptide-encoding genes annotated. This small genome contains 59 genes that were predicted to encode neuropeptide precursors and 8 that can give rise to insulin-like peptides or cysteine-knot protein hormones. Many of these peptides are probably ancestrally related to ones identified in insects and other invertebrates described below. Unfortunately, *L. gigantea* has not been utilized in neurobiological studies. For the two major model mollusks, the pond snail *Lymnaea stagnalis* and the marine slug *Aplysia californica*, the identification of neuropeptide precursor genes and expressed neuropeptides is mostly based on traditional cloning and biochemical techniques. More than 80 peptides have been identified this way in *Aplysia* and about 75 in *Lymnaea*, and several additional ones were predicted in connection with the *L. gigantea* genome analysis. However, only few peptide GPCRs are known so far in mollusks. Most of our knowledge on molluskan neuroendocrinology and neuropeptide function is based on research on these two snails. The snail neuroendocrine system is easily accessible and mostly composed of large individually identifiable neurons with cell bodies in constant positions. Many well defined and easily studied physiological processes were found to be under hormonal control in snails, including growth, water and ion balance, metabolism, and reproduction, as well as a number of stereotypic behaviors such as feeding, copulation, and egg laying. Here we shall look at some neurosecretory cell systems of the central ganglia in *Lymnaea* and the abdominal ganglion of *Aplysia* that have predominated in classical studies.

In *Lymnaea* there are 18 different types of neurosecretory cells within the central ganglia as distinguishable by classical staining techniques. Some cells are pigmented and named after the pigment color. Many of the snail neurosecretory cells have two sets of release sites: neurohemal areas and directly innervated peripheral targets, such as heart, genital tract, intestine, and kidney. The neurohemal areas are distributed over the surface of different nerve roots of the ganglia (Fig. 11.3). One set of cells is termed **yellow cells**. These are located in the parietal and visceral ganglia and send processes both to neurohemal areas and to direct innervation of the kidney, urether, and some other organs (Fig. 11.3). The yellow cells produce a peptide termed

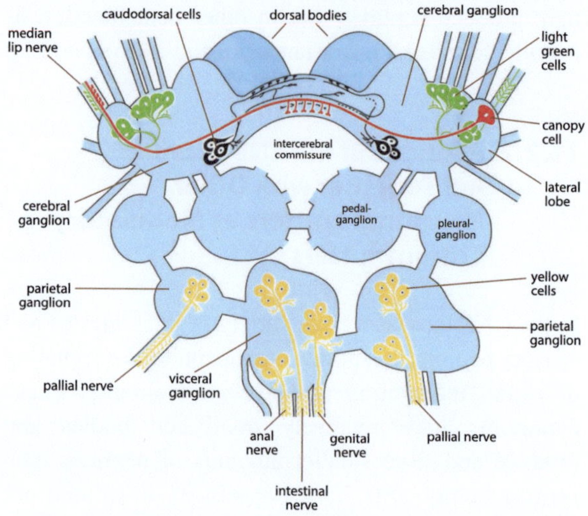


**Fig. 11.3 Molluskan neuroendocrine systems.** Neurosecretory cells in the CNS of the pond snail *Lymnaea stagnalis*. Four main types of neurosecretory cells are shown here: the caudodorsal cells (*black*), the light green cells (*green*), and the yellow cells (*yellow*), and a single so-called canopy cell (*red*) is drawn to the *right*. Note neurohemal release sites in several of the nerve roots and the neurohemal area of the intercerebral commissure. For clarity the number of cells shown here is not accurate for any of the cell types (Redrawn from Kobayashi [3] with permission)

sodium influx stimulating peptide that is important in water and ion homeostasis. The **light green cells** are located in the cerebral ganglia, and with terminations in neurohemal areas in the median lip nerves (Fig. 11.3). These cells produce insulin-like peptides and are known to be involved in growth control.

#### 11.3.2.1 Peptides Control Metabolic Physiology and Complex Behavioral Sequences, Like Egg-Laying Behavior

One distinct neurohemal organ in *Lymnaea* is the so-called **intercerebral commissure** located in the cerebral ganglion (Fig. 11.3). This neurohemal organ is innervated by the approximately 100 peptidergic **caudodorsal cells** (CDCs) that also extend axons into cerebral ganglion neuropils. The CDCs produce a peptide precursor that contains egg-laying hormone and several smaller peptides, all of which contribute to initiation and integration of processes connected to egg-laying behavior. In *Aplysia* the hormonal cascade that initiates and orchestrates egg-laying behavior has been studied in detail [4]. Egg laying lasts more than an hour and involves specific head movements, increase of respiratory pumping, and inhibition of locomotion and feeding, and eventually release of eggs. Initiation of egg laying is preceded by extended bursts of firing of two clusters of so-called **bag cells** located around nerve roots anteriorly in the abdominal ganglion. The bag cells produce a peptide precursor that contains egg-laying hormone (ELH) and several smaller peptides, such as α, β, and γ-bag cell peptides (BCPs), all of which are co-released. The bag cells are multipolar and supply branches to neurohemal areas in the abdominal nerves as well as to cell bodies of certain neurons inside the abdominal ganglion. When activated, the bag cells release massive amounts of ELH and BCPs in a coordinated fashion. The actions of these peptides are complex: ELH and all three BCPs act nonsynaptically on neurons in the abdominal ganglion and ELH acts as a hormone to induce ovulation. The ELH gene-derived peptides thus act at different levels: in the CNS they induce prolonged modulation of circuits controlling egg-laying behavior, leading to inhibition of feeding and locomotion and stimulation of respiratory pumping, whereas in the periphery the peptides act on muscles of the hermaphrodite gland and heart (see Chap. 23).

### 11.3.3 Peptidergic Systems in Crustaceans

The only report so far on a genome-wide sequencing of crustacean neuropeptide genes is from the water flea *Daphnia pulex*. According to this there are 43 neuropeptide precursor genes that encode 73 predicted peptides. Most of these are more related to neuropeptides found in insects than those found in decapod crustaceans. Expressed sequence tag data analysis combined with mass spectrometry of expressed peptides has indicated quite a large number of neuropeptides in some other crustaceans. For example, in the lobster *Homarus americanus* 84 peptides representing 15 or more precursor genes were identified. Peptide GPCRs have, however, not yet been identified in crustaceans, although four putative insulin receptors were annotated in *Daphnia*.

#### 11.3.3.1 Peptide Hormones Produced by Neurosecretory Cells in the Eyestalk Control Many Physiological Functions

In decapod crustaceans, such as crabs, crayfishes, and lobsters, the predominant cerebral neuroendocrine system is that of the eyestalks, the X-organ-sinus gland system shown in Fig. 11.4. There are, however, also thoracic and abdominal neurohemal organs or areas (Fig. 11.5): the pericardial organ (PO), postcommissural organs, as well as the surface of ganglia and several nerves. The PO is situated within the pericardial cavity on either side of the heart (Fig. 11.5). Finally, the intestine contains large numbers of peptide-producing endocrine cells.

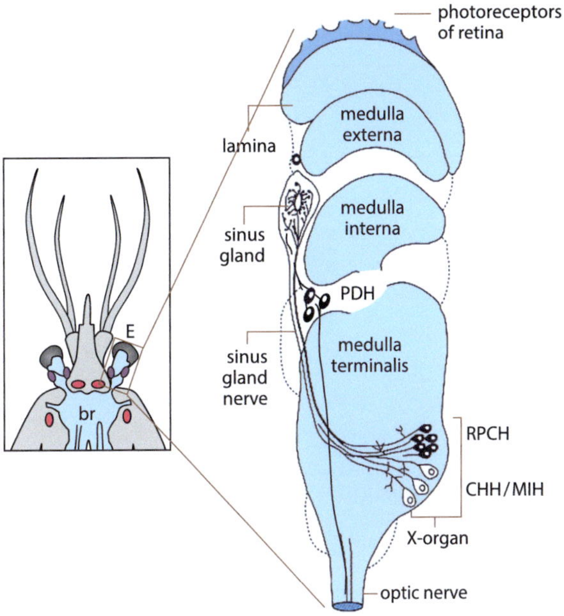


**Fig. 11.4 A crustacean neuroendocrine system.** Neuroendocrine system associated with the optic lobes of the shore crab *Carcinus maenas*: the X-organ–sinus gland. Neurosecretory cells producing crustacean hyperglycemic hormone (*CHH*), molt-inhibiting hormone (*MIH*), and red pigment-concentrating hormone (*RPCH*) are located in the X-organ and send axons via the sinus gland nerve to the sinus gland. In the center of the sinus gland there is a hemolymph lacuna (part of the ophthalmic artery). Another set of neurosecretory cells located near the medulla interna produce pigment-dispersing hormone (*PDH*). Also these have axon terminations in the sinus gland. The *inset* shows the head with position of eyestalks (*E*) and brain (*br*) (Redrawn from Hartenstein [5] with permission)

The neurosecretory cells of the X-organ are localized in the so-called medulla terminalis of the eyestalk neuropils (Fig. 11.4). Their axons terminate in the sinus gland, a neurohemal organ adjacent to the ophthalmic artery in the distal eyestalk. Different X-organ cells produce the peptide hormones red pigment-concentrating hormone (RPCH) and members of the crustacean hyperglycemic hormone/molt inhibiting hormone (CHH/MIH) family.

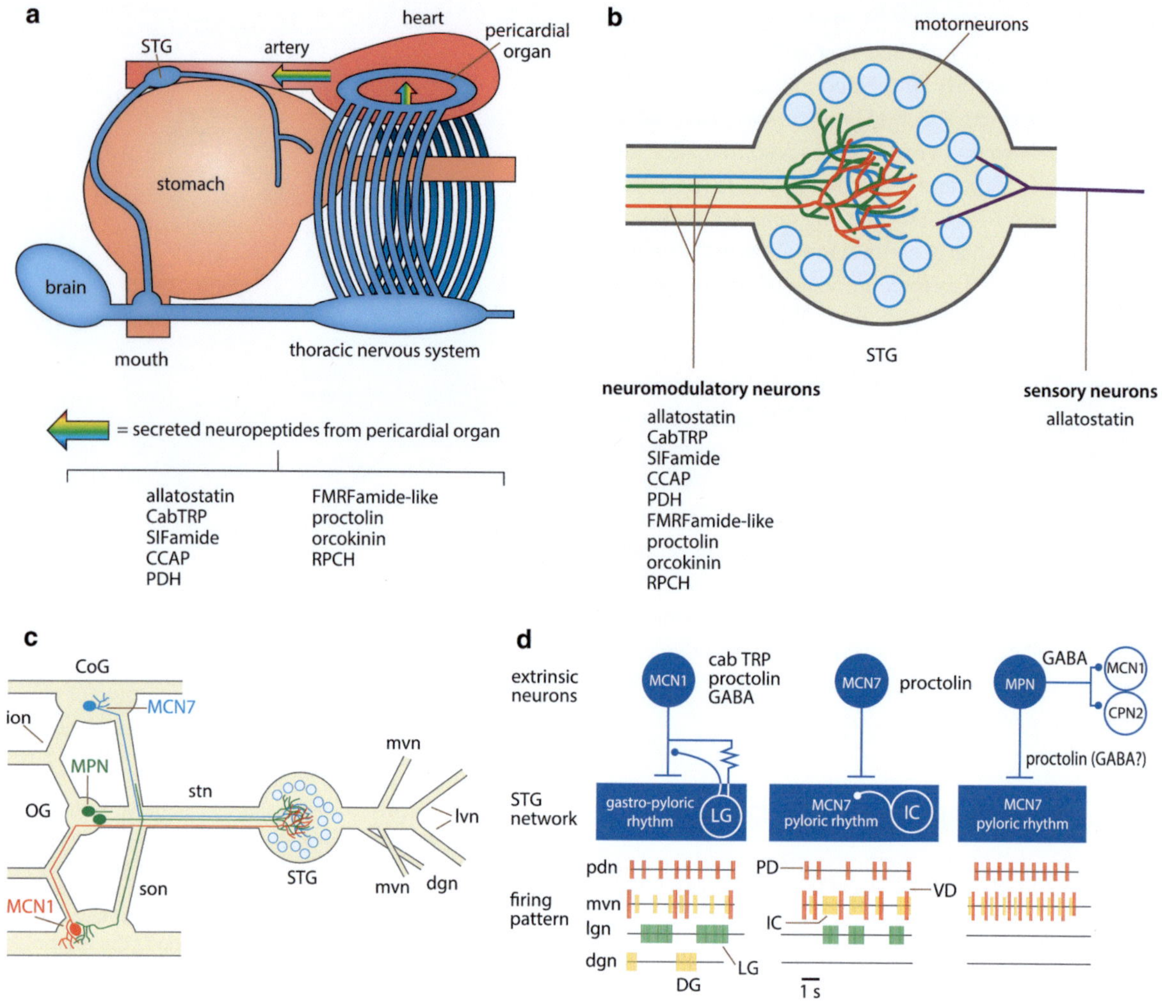


**Fig. 11.5** **Central actions of neuropeptides.** Peptides in the stomatogastric nervous systems act both as hormones and as centrally released neuromodulators. (**a**) The stomatogastric ganglion (*STG*) is situated in the dorsal artery, directly anterior to the heart. The pericardial organs are neurosecretory structures that release many amines and neuropeptides directly into the circulatory system at the level of the heart. Some of the studied neuropeptides of the system are listed. (**b**) The STG is directly modulated by terminals of descending neurons and ascending sensory neurons. These direct neural inputs release many small classical neurotransmitters and neuropeptides into the neuropil of the STG (Redrawn from Marder and Bucher [6] and Nusbaum et al. [7] with permission). (**c**) The stomatogastric nervous system, including the soma location and axon projection patterns of the three proctolin neurons that innervate the stomatogastric ganglion (*STG*). The modulatory proctolin neuron (*MPN*) (shown in *green*) occurs as a functionally equivalent pair of neurons in the esophageal ganglion (*OG*). Each commissural ganglion (*CoG*) contains a single copy of modulatory commissural neuron 1 (*MCN1*) (shown in *red*) and modulatory commissural neuron 7 (*CN7*) (shown in *blue*). Abbreviations: *dgn* dorsal gastric nerve, *ion* inferior oesophageal nerve, *lvn* lateral ventricular nerve, *mvn* medial ventricular nerve, *son* superior esophageal nerve, *stn* stomatogastric nerve (Redrawn from Nusbaum et al. [7] with permission. (**d**) Concerted actions of the neuropeptide proctolin, tachykinin-related peptide (*CabTRP*), and the inhibitory transmitter GABA in different neurons of the crab stomatogastric system. The distinct stomatogastric ganglion motor patterns elicited by the three proctolin-containing projection neurons (MCN1, MCN7, and MPN) are shown. This includes a summary of their transmitter content and additional synaptic actions by which they elicit the indicated stomatogastric ganglion (*STG*) motor patterns. These additional actions include: (1) MCN1: presynaptic inhibition of MCN1 by the LG neuron in the STG and electrical coupling to the *LG* neuron; (2) MCN7: strong excitation of the IC neuron; (3) MPN: synaptic inhibition of projection neurons in the commissural ganglia. *Lower panels*: rhythmic impulse bursts in STG neurons are represented by labeled boxes. Abbreviations: Nerves: *pdn* pyloric dilator nerve, *mvn* medial ventricular nerve, *lgn* lateral gastric nerve, *dgn* dorsal gastric nerve, Neurons: *PD* pyloric dilator neuron, *IC* inferior cardiac neuron, *VD* ventricular dilator neuron, *LG* lateral gastric neuron, *DG* dorsal gastric neuron. Legend: *t-bar* transmitter-mediated excitation, *filled circle*, transmitter mediated inhibition; resistor, electrical coupling (Modified from Nusbaum et al. [7] with permission)

Another group of peptidergic cells outside the X-organ also send axons to the sinus gland; these produce pigment-dispersing hormone (PDH). PDH and RPCH regulate pigment migrations or other light adaptational mechanisms in the compound eyes and they act antagonistically on pigment distribution in chromatophores in the epidermis. Both these peptides are also produced by neurons in other parts of the CNS and thus act in neuronal circuits. Circulating CHH has pleiotropic functions, including regulation of blood sugar levels, salt and water balance, molting, and reproduction, whereas MIH inhibits ecdysteroid hormone production by acting on the molting glands (Y-organs). Production of both these peptide hormones appears to be restricted to the X-organ cells.

#### 11.3.3.2 In the Stomatogastric System, Peptides Control and Regulate the Motor Rhythms in the Foregut

One part of the nervous system where neuropeptide modulation within neuronal circuits has been extensively studied is the stomatogastric system of decapod crustaceans. This system is composed of a small set of ganglia, including the stomatogastric ganglion (STG) depicted in Fig. 11.5. The STG contains rhythm-generating networks composed of about 30 interneurons and motoneurons that control the foregut and stomach muscles during feeding. The circuitry of the STG is functionally flexible and can produce a set of dynamic output patterns, subserving a repertoire of chewing and filtering behaviors. The initiation and modulation of motor rhythms in the STG circuits are largely regulated by neuropeptides or peptide hormones. These reach the circuits either from extrinsic or intrinsic STG neurons, or via the circulation from the pericardial organ. More than a dozen different neuropeptides have been identified in neurons innervating the STG (Fig. 11.5b). Most peptides are excitatory and initiate activity in the quiescent STG or increase the frequency of existing rhythms, some produce a fusion of two rhythms to form a novel pattern. A case of peptide cotransmission can be illustrated by a set of STG neurons expressing the neuropeptide proctolin that colocalizes with another peptide, the tachykinin CabTRP, and the inhibitory classical transmitter GABA in different patterns (Fig. 11.5c, d). Experiments showed that proctolin can produce three different actions in the rhythm-generating network depending on which transmitters are coreleased (Fig. 11.5c, d). The STG of crustaceans is the best studied invertebrate neuronal network in terms of peptide and transmitter actions in intrinsic and extrinsic neuromodulation. The stomatogastric ganglion also provides excellent insight into how the same network can generate very different activity patterns depending on its modulatory input.

### 11.3.4 Peptidergic Systems in Insects

Genome sequencing projects have provided good estimates of the total number of peptide precursors in several insects, and proteomics, by means of mass spectrometry, has identified many neuropeptides and peptide hormones expressed in nervous and neuroendocrine tissues. Thus, we know that there are 40 or more peptide precursor genes in insects and approximately 70–100 peptides can be produced [8]. The number of peptide-activated GPCRs is between 40 and 50. Many of the insect peptides have been investigated for functional roles. In general, insect neuropeptides and peptide hormones regulate many aspects of development, growth, metabolism, homeostasis, and reproduction, as well as specific behaviors. Neuropeptides also act as neuromodulators or cotransmitters in central neuronal circuits.

#### 11.3.4.1 The Corpora Cardiaca and the Corpora Allata are the Main Cerebral Neurohemal Structures in Insects

The most prominent neuroendocrine system in insects is that of the protocerebral portion of the brain, referred to as the **brain-retrocerebral complex** (Fig. 11.6a). This complex is anatomically well conserved among insects and the brain portion consists of neurosecretory cells in two bilateral groups of neurons, the median and the lateral neurosecretory cells (MNCs and LNCs, Fig. 11.6b). Both cell groups send axons posteriorly to release sites in neurohemal organs, the **corpora cardiaca** (CC) and **corpora allata** (CA), that are associated with the anterior aorta (Fig. 11.6a, b). Several different peptide hormones are produced in MNCs and LNCs, as will be detailed later. In each of the CC and CA there is also a glandular portion with endocrine cells producing hormones: the peptide adipokinetic hormone (AKH) in CC and the terpenoid juvenile hormone (JH) in CA. In both CC and CA, there are also axon terminals

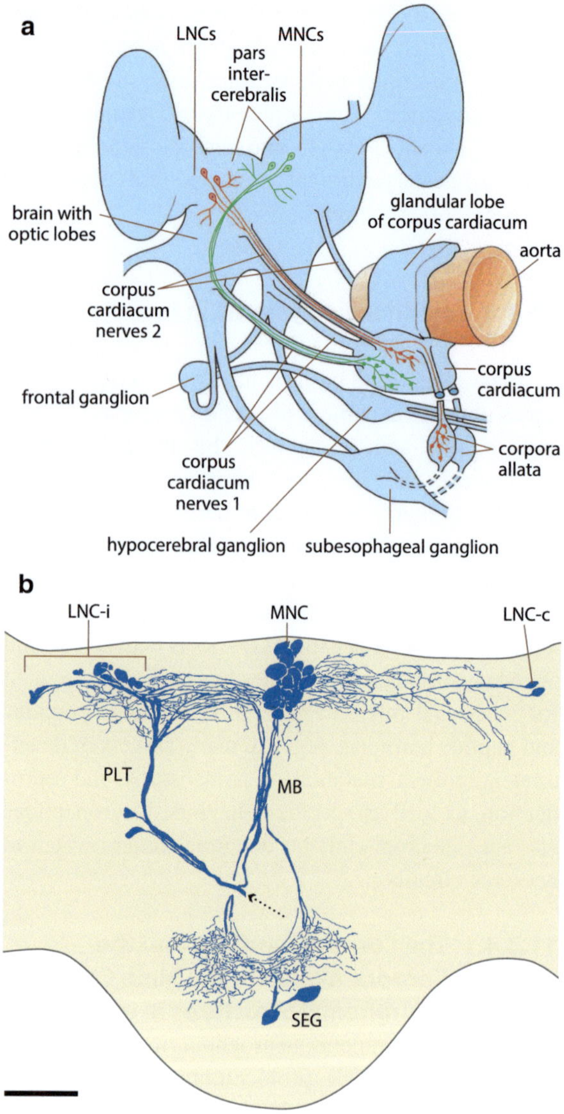


**Fig. 11.6** **(a) The neuroendocrine system of the brain-retrocerebral complex of the locust *Locusta migratoria*.** Two groups of neurosecretory cells are shown schematically in the brain, lateral (*LNCs*) and median (*MNCs*) neurosecretory cells. These cells are shown only on one side of the brain (note that the MNCs send axons across to the contralateral corpus cardiacum nerve 1). There are many more neurons in each group and the different cell types produce a total of up to ten different peptide hormones. (**b**) Neurosecretory cells of the blowfly brain revealed by cobalt backfilling of one of the nerves to corpora cardiaca (*arrow*). Two main groups of neurosecretory cells are visualized in the dorsal brain, lateral (*LNCs*) and median (*MNCs*) neurosecretory cells. The majority of the *LNCs* are ipsilateral (LMC-i) and a few (LNC-c) are found on the contralateral side. Two axon tracts from the cells join the nerve to the corpora cardiaca: the posterior lateral tract (PLT) and the median bundle (MB). Two cells can also be seen in the subesophageal ganglion (SEG) (Redrawn from Shiga et al. [9] with permission)

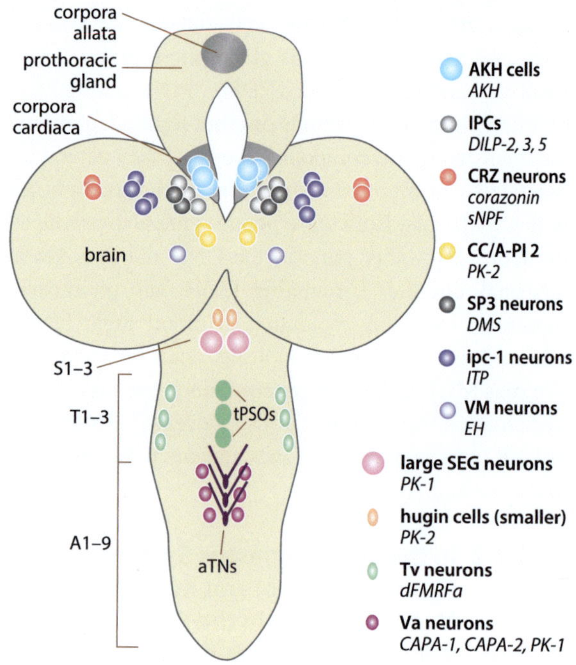


**Fig. 11.7** **Neurosecretory cells in a *Drosophila* larva.** The brain neurosecretory cells send axons to neurohemal sites in the corpus cardiacum or corpus allatum in the ring gland and in ventral nerve cord to thoracic perisympathetic organs (*tPSOs*) or abdominal transverse nerves (*aTNs*). The color-coded cells express different neuropeptides: adipokinetic hormone (*AKH*), insulin-like peptides (*DILP-2,3,5*), corazonin, short neuropeptide F (*sNPF*), pyrokinins 1 and 2 (*PK-1* and 2), myosuppressin (*DMS*), ion transport peptide (*ITP*), eclosion hormone (*EH*), FMRFamides (*dFMRFa*), and capability peptides (*CAPA-1* and 2) (Redrawn and updated from Wegener et al. [10] with permission)

from peptidergic neurons in the brain that control production or release of AKH and JH by the endocrine cells. One set of brain neurosecretory cells that release peptides into the circulation via CA controls hormone production elsewhere. This is the system initially studied by Kopec in the 1920s: a few LNCs release PTTH, which controls production of the steroid hormone ecdysone in the prothoracic glands. Thus, the brain-retrocerebral complex of insects bears some similarities to the hypothalamus-pituitary of vertebrates that will be described later: brain neurosecretory cells with terminals in a neurohemal organ that also displays intrinsic endocrine cells and where hormone production and release is under control by systems of peptidergic brain neurons producing releasing factors.

Many MNCs and LNCs produce peptide hormones (Figs. 11.6b and 11.7). Peptide products of 11 precursors have been detected in the *Drosophila* MNCs and LNCs. In adult flies three of these are insulin-like, others are diuretic or antidiuretic hormones, or peptides regulating aspects of feeding and metabolism. The larval neurosecretory cells produce the same peptide hormones, and there are a few additional ones that have developmental roles, such as, for example, PTTH and pyrokinins (Fig. 11.7).

#### 11.3.4.2 Many Peptidergic Cells are Organized Segmentally in Thorax and Abdomen, Including Some That Control Molting Behavior

There are also peptidergic neurosecretory cells in most of the thoracic and abdominal ganglia (Fig. 11.7). These cells have their release sites in various locations in the body segments. From each ganglion an unpaired median dorsal nerve bifurcates and runs laterally to release sites in the lateral heart nerves, the heart muscles, tracheal trunks, and body wall diaphragms (Fig. 11.8). Along the median and transverse nerves there are enlarged neurohemal organs, designated perisympathetic organs, that are densely supplied by peptidergic axon terminals (Fig. 11.8). In some insects, like in adults of *Drosophila* and other flies, the dorsal neurohemal organs and areas of the body ganglia have merged with the neural sheath on the dorsal ganglion surface. Finally, there are endocrine cells in the intestine that produce a number of different peptides with local paracrine or remote hormonal functions.

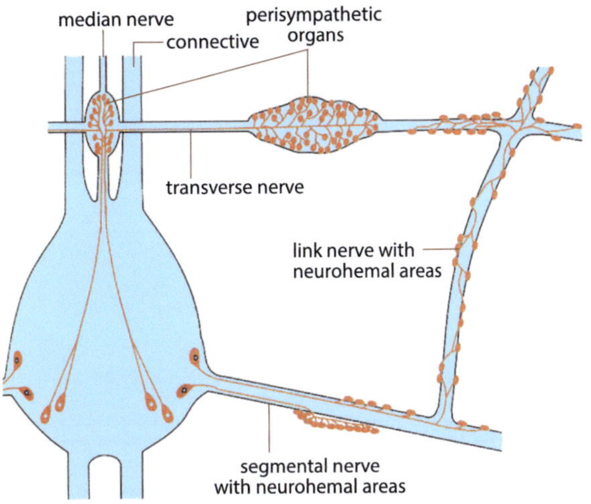


**Fig. 11.8 Neuroendocrine systems in the body of insects.** Details of neurohemal organs and areas in the abdominal neurosecretory system of an insect (generalized). This arrangement is typical for insects with unfused abdominal ganglia such as moths, locusts, and cockroaches. In some insects, like adults of dipteran flies (including *Drosophila*), the axons from neurosecretory cells all terminate in the dorsal neural sheath of the fused ganglia or in some segmental nerve roots

More recently, a system of endocrine cells was discovered attached to the main tracheal trunks of the body. These so-called **Inka cells** produce ecdysis-triggering hormones (ETH) and are part of a complex peptidergic regulatory system which also includes various neurons and regulates ecdysis motor behavior in the moth *Manduca sexta* and in *Drosophila* (Figs. 11.9 and 11.10) [11]. Ecdysis behavior occurs when the developing insect sheds its old inflexible cuticle to enable growth. In *Manduca*, a cascade of hormonal actions, starting with release of the neuropeptide corazonin, followed by ETH released from Inka cells and eclosion hormone (EH) are formed by central neurosecretory cells in a positive feedback loop. ETH triggers responses in central interneurons and motoneurons that lead to synchronized and rhythmic muscle contractions in the body wall. These contractions break the old cuticle and enable the growing larva to escape from it. In *Drosophila* a very similar cascade has been identified and the central peptidergic neurons expressing the ETH receptor were identified which provided clues to action of further peptides in the ecdysis behavior control as shown in Figs. 11.9 and 11.10. An interesting discovery was recently made in the Oriental fruitfly, *Bactrocera dorsalis*, namely that EH acts on a **receptor guanylyl cyclase** (GC) expressed on Inka cells to increase cyclic GMP levels and thereby induce massive release of ETH. This is the first demonstration of a peptide receptor of the membrane-bound GC type in invertebrates. A GC receptor for atrial natriuretic peptide and related peptides has been known for several years in mammals.

#### 11.3.4.3 In the CNS, Peptides act as Neuromodulators and Cotransmitters

Neuropeptides representing most of the identified precursors have also been localized to different types of interneurons in the insect CNS suggesting that they serve as neuromodulators or cotransmitters in central circuits. Examples of neuropeptides expressed in

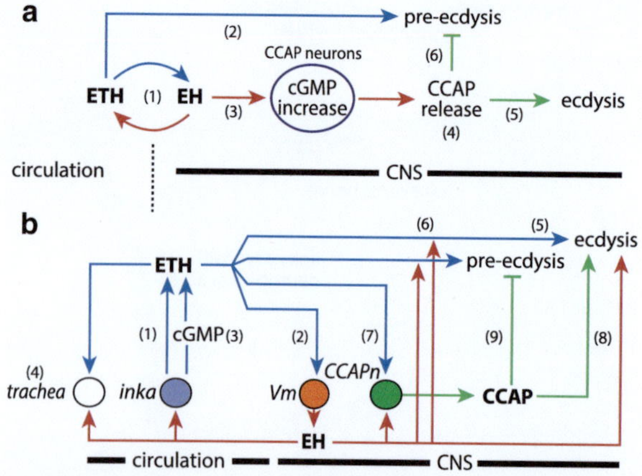


**Fig. 11.9** **Peptidergic cascade regulating ecdysis in insects.** (**a**) Model for actions and relationships among ecdysis controlling neuropeptides in the moth *Manduca sexta*. Ecdysis is triggered by two neuropeptides: eclosion hormone (*EH*) and ecdysis triggering hormone (*ETH*). At ecdysis, *EH* is released from brain neurons (*Vm*) into the CNS and into the blood, and ETH is secreted from the peritracheal Inka cells into the blood. The release of these two neuropeptides is controlled by a positive feedback, in which *EH* and ETH stimulate release of each other (*1*). ETH also acts on the CNS to trigger pre-ecdysis behavior that prepares the animal for ecdysis (*2*). *EH* released within the CNS elicits an increase in cGMP levels in neurons that produce the peptide CCAP (*3*) which induces release of CCAP (*4*). CCAP then induces ecdysis (*5*) and inhibits pre-ecdysis (*6*). (**b**) Model for ecdysis control in *Drosophila*. The endocrine cascade is initiated by an EH-independent ETH secretion from Inka cells (*1*). This release induces release of *EH* from central *Vm* neurons (*2*), which stimulates further ETH release via increases in cGMP (*3*). Increased *EH* and ETH in circulation causes air-filling of the trachea (*4*). ETH peptides also act on neurons in the CNS to turn on pre-ecdysis and ecdysis behavior (*5*). This also requires action of *EH* neurons (*Vm*) on ETH output (*6*). ETH and *EH* regulate CCAP release from central CCAP neurons (*CCAPn*) (*7*) which inhibits pre-ecdysis (*9*). ETH mutants show some ecdysis-like behaviors, which could be controlled independently by *EH* and CCAP (*8*). *Colored circles* indicate known target cells and *EH*, ETH, and CCAP actions are indicated in *red*, *blue*, and *green*, respectively (Redrawn from Ewer [11] with permission)

central interneurons of *Drosophila* are neuropeptide F (NPF), pigment dispersing factor (PDF), and tachykinin-like peptide (DTK), which have been implicated in regulation of feeding behavior and aggression (NPF), circadian clock function (PDF), and modulation of olfactory sensory input and locomotor behavior (DTK), respectively. Some further examples of peptide functions are given in Sect. 11.4.

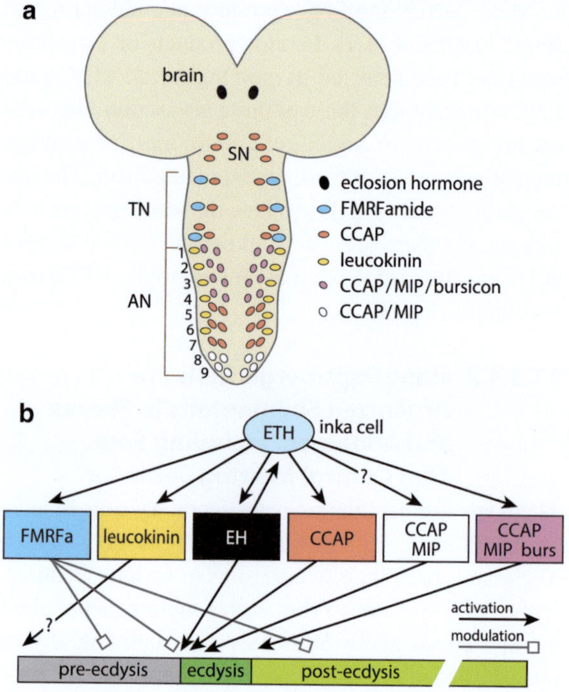


**Fig. 11.10** **Neurons and neuropeptides in ecdysis control in *Drosophila*.** (**a**) Schematic depiction of peptidergic neurons in the larval CNS expressing the ecdysis triggering hormone receptor ETHR-A. These neurons respond to ETH in a sequence and trigger ecdysis behavior as depicted in (**b**). The *color coding* shows the expression of various neuropeptides in the neurons. ETH is released from peritracheal Inka cells. The Inka cells express a membrane-bound receptor guanylyl cyclase that is activated by eclosion hormone (*EH*). The first trigger of pre-ecdysis has not been conclusively identified. *CCAP* crustacean cardioactive peptide; *MIP* myoinhibitory peptide; burs, bursicon (tanning hormone) (Redrawn from Kim et al. [12] with permission)

### 11.3.5 Peptidergic Systems in the Nematode *C. elegans*

#### 11.3.5.1 Peptides in Nematodes are Produced by Neurons, Sensory Cells, and Muscle

In *C. elegans* there is a fixed number of neurons and other cells. The nervous system of the adult hermaphrodite worm is composed of 302 neurons, classified into 118 different neuron types, connected by approximately 7,000 synapses [13]. These neurons form an anterior nerve ring, a ventral and a dorsal nerve cord, and small ganglia anteriorly and posteriorly (Fig. 11.11). No distinct neuroendocrine system has been distinguished in nematodes in the sense that there are no neurons with axons terminating in specific neurohemal organs. However, in the pharynx there are two neurons considered to be neurosecretory cells because they have varicose axons running along the border to the body cavity, the **pseudocoel**. These cells produce serotonin and peptides generated from at least two precursors (neuropeptide-like precursors, nlp 13 and 18). In *C. elegans* most neuropeptides are located in sensory, motor, and interneurons of various types, but also in other cell types. We may thus have to consider peptidergic signaling in *C. elegans* and other nematodes mostly as a form of paracrine signaling. As an example, insulin-like peptides that act in hormonal signaling to regulate growth, reproduction, and metabolism in mollusks, insects, and vertebrates exist also in this nematode. Here, many insulin-like peptides are produced primarily in sensory neurons and other neurons, suggesting paracrine action as the main signaling form.

In *C. elegans* 109 peptide precursor genes and more than 50 peptide GPCRs have been identified, some of which are ancestrally related to ones in insects and vertebrates. The peptide precursors have been divided into three major groups: FMRFamide-like (flp), insulin-like (ins), and neuropeptide-like precursors (nlp). There are 26 *flp*, 37–39 *ins*, and 45 *nlp* genes and many of these have been localized to neurons. Out of the 302 neurons, 160 express *flp* genes; some of these neurons even express multiple *flps*. Additionally, muscles in head and pharynx as well as cells of the uterus and vulva express *flps*. The *ins* genes are seen in sensory and other neurons as well as in intestine, hypodermis, pharynx, and vulva. Finally, the *nlps* are seen in different patterns in various types of neurons and in gonadal cells. Screens using RNA interference to diminish levels of peptides or their GPCRs have shown effects on various behaviors such as egg laying, locomotion, and sensory processing. Since many neuropeptides are expressed in sensory neurons it seems that peptide functions in the worm are important in mediating environmental inputs. We will discuss further the issue of peptide signaling in *C. elegans* in the next section.

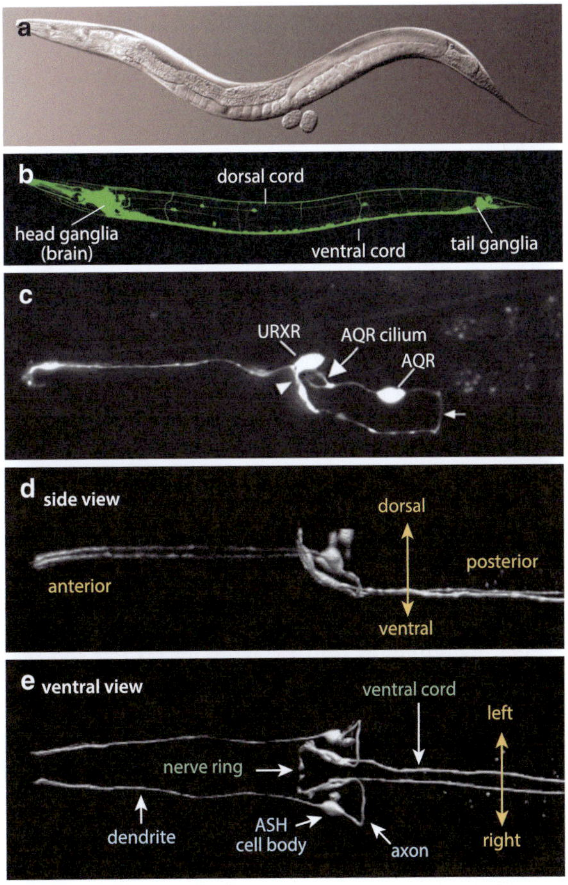


**Fig. 11.11 Peptides in the nervous system of the nematode worm *C. elegans*.** Anterior is to the left in all images. (**a**) Nomarsky contrast image of the worm (The image was kindly provided by Maria Gallegos, California State University, East Bay, CA, who owns the copyright*)* (**b**) The worm's nervous system is displayed with green fluorescent protein revealing the head and tail ganglia and cords connecting these (Image kindly provided by Dr. Harald Hutter, Simon Fraser University, Burnaby, BC, Canada, who owns the copyright) (**c**) A pair of neurons designated URX are oxygen-sensing neurons and express neuropeptide Y receptor-like receptor NPR-1. The ligands of this receptor are the peptides encoded by *flp-18 and flp-21* and the neurons are important for social feeding behavior. Also the AQR neuron expresses NPR-1. Image from Wormatlas (www.wormatlas.org), with permission from Dr. Zeynep Altun-Gultekin, Albert Einstein College of Medicine, Bronx, NY, who owns the copyright. (**d** and **e**) Side view and ventral view of a pair of peptidergic sensory neurons designated ASH. These neurons express glutamate and the neuropeptides NLP-3, NLP-15, and FLP-21 (the latter FMRFamide like), as well as the receptor NPR-1 (see Wormatlas). These neurons are involved in avoidance responses to touch of the "nose" of the worm (Image from Dr. Harald Hutter, who owns the copyright). We also thank Drs Cori Bargmann and David Hall for help with these images

## 11.4 Comparing Functional Roles of Some Neuropeptides and Peptide Hormones Across Phyla

Gene sequencing projects have provided ample evidence for a strong evolutionary conservation of amino acid sequences of a number of peptides and peptide GPCRs. Some peptide signaling systems appear to be functionally conserved among invertebrates and even extend into the vertebrates. This section will provide a few examples of peptide signaling in invertebrates that seem more or less conserved over great evolutionary distances.

### 11.4.1 Insulin-Like Peptides Are Conserved Across Phylogeny

Genes encoding insulin-like peptides (ILPs) and ILP receptors have been identified in many invertebrate species, including the mollusks *Aplysia*, *Lymnaea*, and *Lottia* as well as *C. elegans* and *Drosophila*. There are 38 insulin-like peptides (ILPs) in the worm and 8 ILPs in the fly. In *Drosophila* some of the ILPs resemble mammalian insulin, others are relaxin-like, and at least one relates to insulin-like growth factors (IGFs). In both the worm, the fly, and the snails a single tyrosine kinase type ILP receptor has been identified, designated daf-2 in *C. elegans* and dInR in *Drosophila*. The entire signal pathway in growth control, downstream of the ILP receptor, is conserved from *C. elegans* to mammals (see Fig. 11.12).

Experimentally impaired ILP signaling in *C. elegans* leads to pleiotropic effects, including increased fat storage, stress resistance, and constitutive entry into a diapause-like state called the **dauer**, as well as extended life span. Similarly, in *Drosophila* interference with insulin signaling affects growth, carbohydrate and lipid storage, stress resistance, female fertility and longevity [15]. The ILPs of mollusks are involved in regulation of growth and associated metabolic processes, in control of glycogen levels and gonadotropic activity. Thus, in general, ILP signaling in invertebrates combines some functions seen for vertebrate insulin, IGFs, and relaxin.

### 11.4.2 NPF and NPY Control Several Behaviors – Including Feeding – Across Species

Another peptide-signaling system that appears to be partly conserved throughout evolution is that of the invertebrate neuropeptide F-like (NPF-) and vertebrate NPY-like peptides. As described in Sect. 11.6, NPY is a strong stimulator of feeding in mammals. In *Drosophila* the 36-amino-acid-long NPF is expressed in about 20 brain interneurons and activates an NPF receptor (NPFR1) distantly related to the mammalian NPY receptors and the NPF receptor of *C. elegans*. Genetic interference with the expression of NPF or NPFR1 in *Drosophila* produced phenotypes with modified feeding behavior (e.g., appetite and food choice), foraging, social behavior, ethanol sensitivity, and aggressive behavior. A male-specific role in the **circadian clock** has also been discovered. Intriguingly, a developmental change in NPF signaling in late larvae of *Drosophila* coincides with the distinct behavioral switch from continuous feeding to food aversion and wandering. All the above-mentioned roles of NPF in *Drosophila* are assumed to be mediated by release from brain interneurons and not hormonal actions.

In *C. elegans* NPF signaling also regulates feeding behavior. The worm NPF receptor (Npr1) displays a natural polymorphism that positively correlates with a drastic behavioral change: the two variants of Npr1 are associated with social or solitary feeding, respectively. *Npr1* gene knockout worms display clumping, which suggests that the normal role of this receptor is to inhibit social feeding. The natural ligands for Npr1 are peptides derived from the two genes, *flp-21* and *flp-18*. These are, however, shorter than NPF and NPY and not clearly evolutionarily related (in contrast to their receptors). A peptide resembling NPF of *Drosophila* has also been identified in *Aplysia* where NPF signaling has been implicated in circuits regulating feeding behavior.

### 11.4.3 AKH/GnRH Are Similar Peptides but Have Different Functions in Different Species

An example where sequences of peptide and GPCR are conserved, but functions are divergent is a *C. elegans*

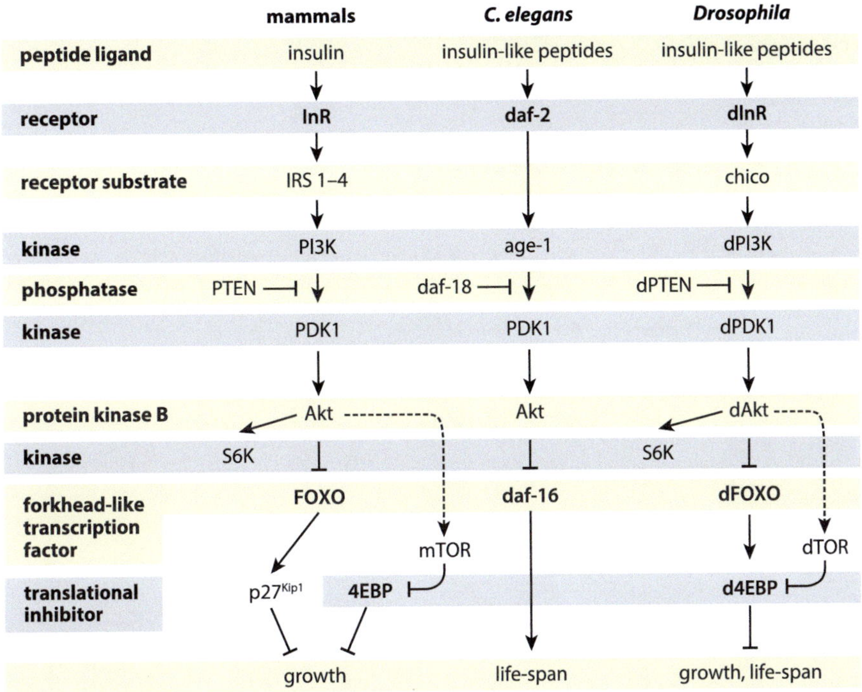


**Fig. 11.12** **Insulin signaling pathway in mammals, *C. elegans*, and *Drosophila*.** The depicted pathways are involved in regulation of growth; throughout their evolution these molecules have been conserved between species. To the *left* the type of molecule is indicated. All three pathways signal through PI3K to Akt/PKB homologs, and are negatively regulated at this step by homologs of PTEN. Abbreviations: *IRS* insulin receptor substrate, *PI3K* phosphatidylinositol 3-kinase, *PTEN* lipid phosphatase, *PDK1* phosphoinositide-dependent kinase-1, *Akt* protein kinase B; *FOXO* forkhead-like transcription factor, *S6K* ribosomal S6 kinase, *TOR* target of rapamycin, *4EBP* eukaryotic initiation factor 4-binding protein (a potent translational inhibitor and growth suppressor); $p27^{Kip1}$, a cell cycle regulator (Redrawn from Puig et al. [14] with permission)

peptide resembling insect AKH and vertebrate gonadotropin-releasing hormone (GnRH). In insects AKH is a hormone regulating carbohydrate and lipid metabolism and in vertebrates GnRH regulates release of gonadotropin from the pituitary and thus ovary maturation. In *C. elegans* the AKH-GnRH-like peptide and its receptor are important for timing of sexual maturation and ensuing egg laying, but not for lipid metabolism. Thus the functional role of the worm peptide resembles that played by the hormone GnRH in vertebrates, and not AKH of insects.

Finally, examples of a peptide and GPCR that can be found in many invertebrates and generally among vertebrates (see Fig. 11.16a), but not in *Drosophila* and several other insects are the vasopressin-oxytocin-like peptides (VLPs) and their receptors. In vertebrates the VLPs play important regulatory roles in reproduction, blood pressure, diuresis, and clock functions. In the snail *Lymnaea* conopressin is expressed in neurons controlling male sexual behavior and induces muscular contractions in the vas deferens. In snail females the peptide inhibits neurons that control sexual behavior. Similarly, annetocin induces egg-laying behavior in the earthworm. Thus, in these animals VLPs have a functional role reminiscent of oxytocin rather than of vasopressin in vertebrates. However, in those insects where VLPs have been identified, like the beetle *Triboleum*, they seem to be indirectly regulating diuresis, like vasopressins.

## 11.5 Vertebrate Neuropeptides and Peptide Hormones

Some of the major sites of synthesis of peptide hormones in the vertebrates are the anterior pituitary, the **hypothalamus** (releasing peptides via the

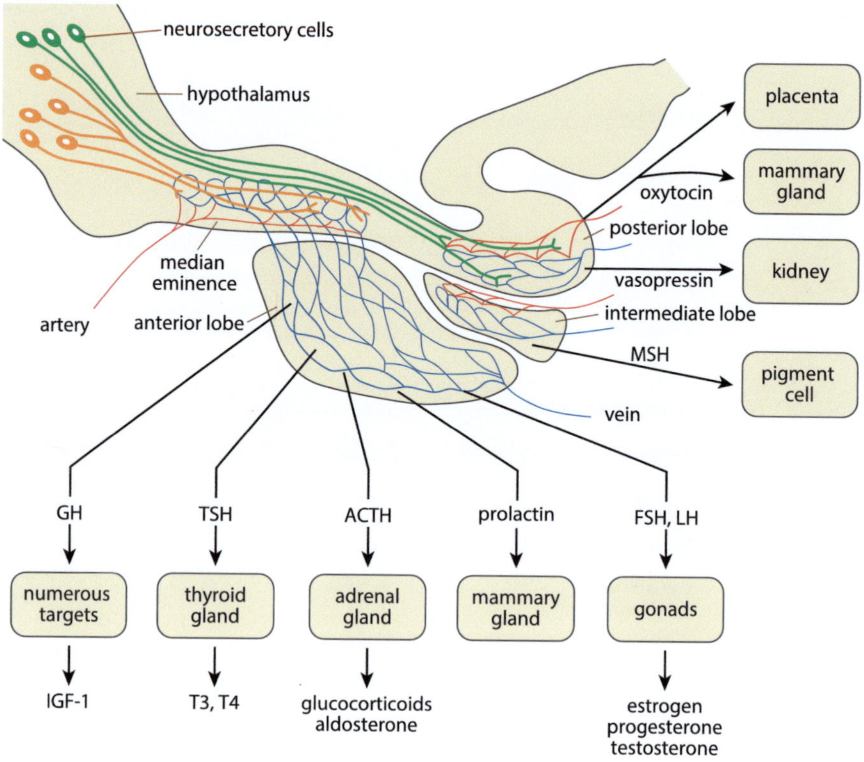


**Fig. 11.13 Hypothalamus and pituitary in mammals.** Peptides are released from hypothalamic neurons into separate capillary networks. The anterior network transports neuropeptides to the anterior lobe of the pituitary where they regulate the release of pituitary peptide hormones. The posterior network receives peptides from large hypothalamic neurons whose axons project to the posterior lobe of the pituitary. These peptides are forwarded to the general circulation that transports them to distant target organs. Some mammals also have an intermediate pituitary lobe with a separate capillary network

**posterior pituitary**), and the urophysis in bony fishes and cartilaginous fishes. In addition, a large number of neuropeptides are produced in the brain and the spinal cord as well as in peripheral neurons. Some peptides are also synthesized in various non-neuronal cell types and released as hormones. In this section we will focus on the peptides involved in the communication between the nervous system and various endocrine systems, primarily the hypothalamus-pituitary connection. Often these peptides have been referred to as neuroendocrine peptides due to their role in this communication and several of these are produced both in neurons and in endocrine or neurosecretory cells. However, we will use terms neuropeptide and peptide hormone in the following. The neuropeptides that serve functions more exclusively within the brain will be discussed below in Sect. 11.7.

### 11.5.1 Hypothalamus, Neurohypophysis, Adenohypophysis

The hypothalamus and pituitary are usually considered to comprise the most prominent **neuroendocrine** system in vertebrates. The **pituitary gland** (also called hypophysis) has even been named the "master gland" because it controls, via its various hormones, many functional systems throughout the body. Anatomically the pituitary consists of two clearly distinguishable parts with distinct developmental origins (Fig. 11.13). The anterior lobe is called the **adenohypophysis** and is formed by the dorsal part of the oral cavity, an embryonic structure named **Rathke's pouch** after the nineteenth century German embryologist Martin Rathke (1793–1860). The posterior lobe is called the **neurohypophysis** because it is an extension of the hypothalamus located in the diencephalon (the posterior part of the forebrain). Between the anterior and posterior pituitary lobes an intermediate lobe arising from Rathke's pouch is present in many groups of vertebrates. The intermediate lobe is very modest or almost nonexistent in some adult mammals but is prominent in other vertebrate classes.

#### 11.5.1.1 Hypothalamic Neuropeptides act as Releasing Factors in the Pituitary Gland

The hypothalamus and pituitary are highly vascularized and contain 2–3 capillary networks, depending on the particular class of vertebrates. Peptides are released into these networks and transported to their target receptors. In mammals an anterior artery enters at the base of the hypothalamus in a region called the median eminence. Peptides released here from hypothalamic neurons are transported by capillaries to the endocrine cells in the anterior lobe of the pituitary (Fig. 11.14a) where they bind to receptors that regulate the release of hormones. Therefore, these hypothalamic peptides are called releasing hormones/factors. The pituitary hormones are small proteins which are transported to target cells elsewhere in the body. Dorsally or rostrodorsally to the median eminence is a group of hypothalamic neurons called the arcuate nucleus which has important functions as a sensory region where the hypothalamus can detect hormones in the blood arriving from elsewhere in the body. This sensing is possible because the arcuate nucleus, like the median eminence, lacks the blood–brain barrier (BBB). Here hormones such as insulin from the pancreas, leptin from adipose tissue, and ghrelin from the stomach can influence the activity of hypothalamic centers involved in the regulation of feeding and metabolism.

One striking difference in the anterior pituitary between vertebrate classes is that teleost fishes do not depend on a capillary network to transport the hypothalamic neuropeptides to the endocrine cells. Instead, these neurons extend their axons directly all the way to the endocrine cells and release their neuropeptides (releasing factors) onto them (Fig. 11.14b). In the posterior pituitary of all species, on the other hand, peptides are released from axon terminals of large hypothalamic neurons and are transported directly to the rest of the body via the systemic blood circulation and reach target organs such as the kidneys or, in mammals, the uterus and mammary glands.

Finally, the intermediate lobe, where present, has a distinct set of endocrine cells that are innervated by hypothalamic neurons, and there is also a separate capillary network for the intermediate lobe. This lobe releases a set of related peptide hormones called **melanocyte-stimulating hormones (MSH)** that are derived from the same precursor as ACTH. This prepropeptide is called POMC, for pro-opiomelanocortin. One of the MSH peptides, alpha-MSH, corresponds to the aminoterminal 13 amino acids of ACTH and is released after cleavage by an endopeptidase present in the intermediate pituitary but not in the ACTH-producing cells of the anterior pituitary. Some adult mammals lack an intermediate lobe, including humans, whales, and elephants and several others, most of which have few or no melanocytes.

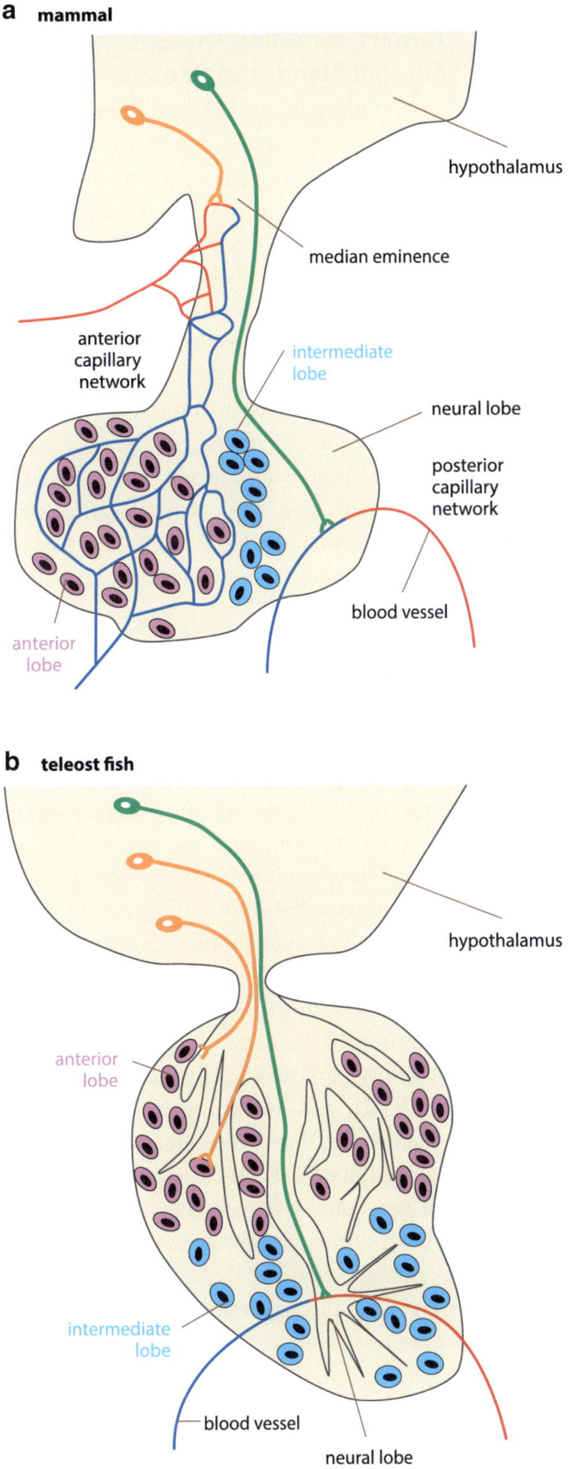


**Fig. 11.14 Comparison of hypothalamus-pituitary connections** in mammals (a) and bony fishes (b). The major difference is that the mammalian hypothalamus releases peptides into a capillary network that carries these signals to the anterial lobe, whereas in bony fishes all three lobes are innervated by hypothalamic neurons. The intermediate lobe is minor or absent in some mammals (Redrawn from Zohar et al. [16])

#### 11.5.1.2 Pituitary Hormones act on Distant Targets, Including Thyroid, Adrenal Gland, and Gonads

The release of pituitary hormones in vertebrates constitutes some of the most well-characterized systems of integration of multiple regulatory inputs as well as extensive feedback mechanisms. Each pathway of pituitary hormone regulation and action is called an "axis", such as the hypothalamus-pituitary-thyroid axis (HPT), the hypothalamus-pituitary-adrenal axis (HPA), and the hypothalamus-pituitary-gonad axis (HPG). For each of these, multiple inputs are integrated in the hypothalamus and result in modulation of the release of each of the pituitary hormones. The mechanisms have been investigated in mammals, chickens, amphibians, and bony fishes. The central features of the regulation are described below. Not all differences between vertebrate classes can be elaborated here.

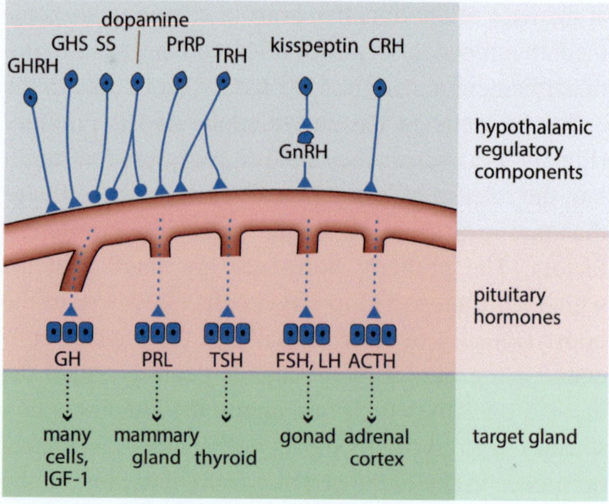


**Fig. 11.15 Regulation of hormonal factors** from the brain (*top*), offer hypothalamic regulatory components and pituitary hormones to the target glands and their hormones

### 11.5.2 Hypothalamus-Anterior Pituitary-Peripheral Organ Axes

In descriptions of the hypothalamic-pituitary-target networks, the pituitary hormones are often taken as starting point and the present description will use this approach. The **anterior pituitary** of mammals typically manufactures six peptide hormones belonging to three families (Fig. 11.15). One family consists of growth hormone (GH) and prolactin (PRL), the second family is comprised of the three glycoproteins called thyroid-stimulating hormone (TSH), follicle-stimulating hormone (FSH), and luteinizing hormone (LH), and the third family has a single member named adrenocorticotropic hormone (ACTH). A seventh hormone is present in fishes, namely somatolactin (SL) which is related to growth hormone. Some vertebrate groups, particularly among teleost fishes, have additional duplicates of some of the hormones. Functional differences between these duplicates are still incompletely known.

#### 11.5.2.1 The HPT Axis: Thyroid Hormones Control Hormones that Control Metabolic Rates

The pituitary cells that produce TSH (a heterodimer whose two polypeptides are called alpha and beta) are primarily regulated by stimulation from the hypothalamic neuropeptide TRH (TSH-releasing hormone, Fig. 11.15). Its production and release from the hypothalmic neurons is regulated by factors such as ambient temperature and metabolism. TRH is unusual in vertebrates because it is one of very few peptides that occurs in multiple identical copies in the same precursor: six in human, seven in the frog *Silurana (Xenopus) tropicalis*, and eight in goldfish. In chickens there are at least four copies with flanking consensus processing sites and a few more that may no longer be cleaved. TRH triggers the release of TSH, which binds to receptors in the thyroid gland and stimulates a variety of processes that lead to release of the thyroid hormones T3 and T4, which are iodinated derivatives of the amino acid tyrosine. These will reach almost all cells in the body and stimulate their metabolic rate by binding to nuclear receptors after uptake by transport proteins.

#### 11.5.2.2 The HPG Axis Controls the Gonads and Sexual Drive

The two relatives of TSH, namely FSH and LH, have their primary roles in reproduction. They have the same alpha subunit as TSH but have distinct beta subunits. FSH and LH are released from a population of endocrine cells in the pituitary that are stimulated by the hypothalamic peptide GnRH (gonadotropin-releasing hormone). Recently it has been found that the GnRH-producing neurons are themselves regulated by another peptide called kisspeptin which stimulates GnRH release [17]. (The name is derived from the famous so-called 'kisses' manufactured by a prominent U.S. chocolate manufacturing company located in Pennsylvania, U.S.A. where the research team was based that discovered the peptide). The hypothalamus integrates a broad range of stimuli to regulate reproduction with these neuropeptides. FSH and LH are released into the systemic circulation and bind to receptors in the gonads to regulate gametogenesis and ovulation, often in response to seasonal changes.

#### 11.5.2.3 Growth, Metabolism, and Electrolytes are Controlled by Several Systems

GH, PRL and intermediate lobe SL are produced by separate cell types in the anterior pituitary. The two first-mentioned have been investigated in great detail, whereas SL, as the most recently discovered member of the family, is still only partially known. SL is present in bony fishes (Actinopterygii) as well as in lungfish, but has been lost in the tetrapod lineage. GH release is stimulated by the peptides GHRH (growth hormone releasing hormone) and ghrelin, also called GHS (growth hormone segretagogue), while inhibition is exerted by somatostatin (SS) and a non-peptide, namely dopamine. GH acts on many cell types and has anabolic effects. Its release occurs primarily after onset of sleep in mammals and after physical activity. Many of the peripheral effects of GH are mediated by IGF-1 (insulin-like growth factor 1) that acts locally in the target tissues.

PRL cells are mainly regulated by a tonic inhibitory signal from dopamine and in some vertebrates also stimulatory signals from TRH and the neuropeptide PrRP (PRL-releasing peptide). PRL has an extremely long list of effects in various vertebrates, the most well-known of which is lactation in mammals. However, PRL existed long before this role evolved and its most prominent effects are on reproduction, growth (particularly seasonal growth), and on electrolyte balance in fishes that alternate between freshwater and seawater.

Also several other neuroendocrine systems that utilize neuropeptides and peptide hormones have been investigated in great detail. One such system with fascinating complexity concerns the appetite-regulating peptides (Fig. 11.16). Ingestion of food leads to the release of gastrointestinal peptides such as insulin, CCK (cholecystokinin), peptide YY (PYY), and pancreatic polypeptide that bind to receptors in the nervous system and reduce appetite. The peptide leptin, now often described as a "long-term reporter" of fat stores in adipocytes, also reduces appetite. Absence of leptin in mice or humans results in lack of appetite inhibition, resulting in overeating and obesity. Insulin, leptin, and PYY bind to receptors on neurons in the basal part of the hypothalamus called the arcuate nucleus, resulting in inhibition of

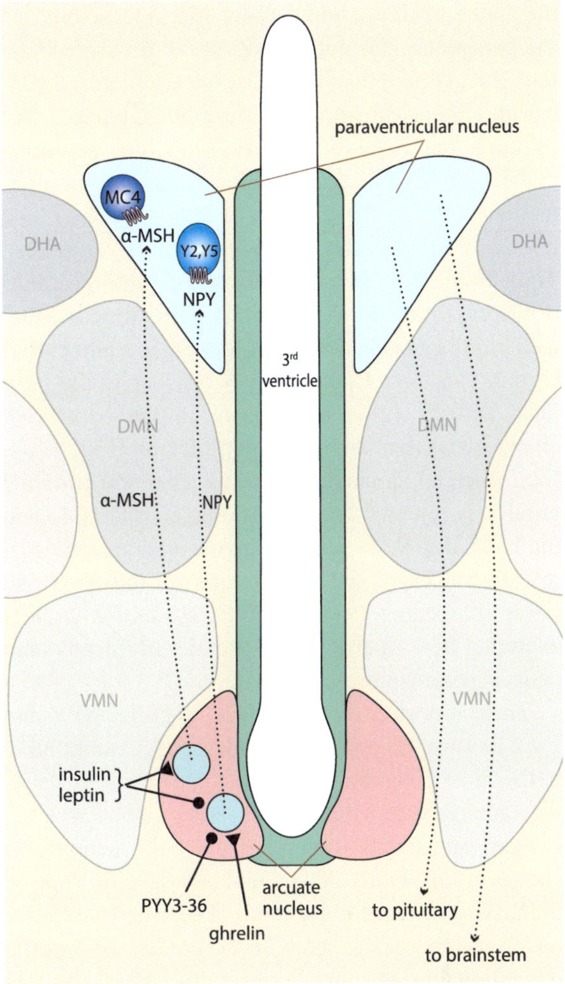


**Fig. 11.16 Schematic outline of neuronal pathways that regulate hunger and satiety in a coronal section of the rat hypothalamus.** The neurons in the arcuate nucleus can respond to hormones in the blood. MC4 is a receptor for α-MSH. Y1 and Y5 are NPY-receptor subtypes. Ascending pathways are shown in the *left part* of the figure and descending pathways in the *right part* for clarity only, in reality there is bilateral symmetry

the appetite-stimulating neurons that release NPY in the paraventricular nucleus (PVN) of the hypothalamus. In addition, insulin and leptin stimulate a different population of neurons in the arcuate nucleus that produce the peptide α-MSH from the precursor POMC (the same peptide that is produced in the intermediate lobe of the pituitary in other species), and release α-MSH in the PVN where it induces satiety. Thus, food intake is reduced by the combined effects of increased release of the satiety-generating α-MSH and diminished release of the hunger-generating NPY. Output pathways from the PVN neurons to the pituitary and the brain stem result in the behavioral changes that terminate feeding. In the opposite situation, when the stomach is empty, the peptide hormone ghrelin is released from gastric endocrine cells and binds to the NPY neurons in the arcuate nucleus, resulting in increased NPY release in the PVN and onset of feeding behavior. This system is a clear example how "peripheral" peptide hormones and neuropeptide pathways interact.

Thus, it is clear that the related peptides NPY and PYY have opposing roles in appetite regulation: NPY is the most powerful endogenous stimulator of feeding, acting within the hypothalamus, whereas PYY reduces appetite by inhibiting the NPY neurons as shown in Fig. 11.16. Also a tetrapod-specific copy of PYY, the pancreatic polypeptide (PP), is released after meals and reduces appetite. In the context of appetite regulation the three peptides exert their effects through distinct receptors as shown in the figure. NPY has its receptors in the paraventricular nucleus of the hypothalamus. PYY as an endocrine signal binds to receptors on neurons in the arcuate nucleus in a more open region of the blood–brain barrier. The site of action of PP is not yet entirely clear, but its receptors may reside on the vagal nerve, bringing the signal to the brain stem for further relay to the hypothalamus.

#### 11.5.2.4 The HPA Axis Controls Stress Responses, with a Network of Positive and Negative Feedback Loops

ACTH release is stimulated by the neuropeptide CRH, corticotropin-releasing hormone (or CR*F*, for factor). ACTH release is stimulated by stressful stimuli, both metabolic stress and in mammals more psychological types of stress (see Chap. 24). ACTH binds primarily to receptors in the adrenal cortex where it stimulates production of steroids such as cortisol in mammals which leads to increased blood glucose and increased blood pressure, both of which may be part of a stress reaction.

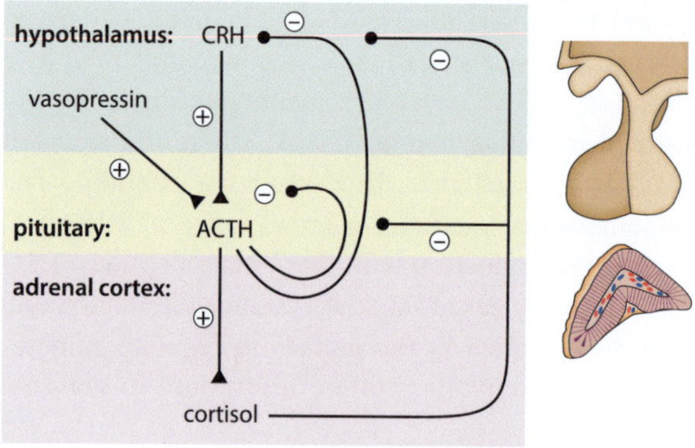


**Fig. 11.17** **Hypothalamus-pituitary-adrenal axis.** *Plus signs* mean stimulation, *minus signs* mean inhibition

The HPA axis (Fig. 11.17) is centered around the pituitary peptide hormone ACTH excised from the POMC precursor. The biosynthesis and release of ACTH is primarily regulated by the hypothalamic neuropeptide CRH. ACTH acts on its receptor MC2 (melanocortin receptor 2) on adrenal cortex cells that produce steroid hormones. One of these, cortisol, exerts feedback regulation on multiple steps in the axis: it inhibits transcription of prepro-CRH as well as the release of CRH and it inhibits the transcription of the CRH receptor on the ACTH-producing pituitary cells as well as the transcription of POMC and the release of ACTH. In addition, ACTH has been reported to have an inhibiting effect on CRH release from the hypothalamic neurons as well as on its own release from the pituitary. One of the effects of ACTH on the adrenal gland is to stimulate the release of aldosterone, a mineralocorticoid that increases sodium retention in the kidneys and thereby increases blood pressure. The release of ACTH is stimulated by vasopressin. Thereby vasopressin stimulates retention of both salt and water because its direct effect on the kidneys is to increase water retention.

### 11.5.3 Posterior Pituitary Neurohormones

In mammals, the **posterior pituitary lobe** releases two peptide hormones from axons extending from

**a**

**vertebrate oxytocin and vasopressin**

| | |
|---|---|
| oxytocin most mammals | CYIQNCPLGamide |
| mesotocin birds, amphibians | -------I-amide |
| isotocin bony fishes | ---S---I-amide |
| oxytocin elephant shark | ---------amide |
| vasopressin most mammals | --F----R-amide |
| vasotocin most non-mammals | -------R-amide |
| cephalotocin octopus | --FR----I-amide |
| octopressin octopus | -FWTS--I-amide |
| annetocin earthworm | -FVR---T-amide |
| lys-conopressin pond snail | -F-R---K-amide |
| inotocin locust | -L-T---R-amide |

octopus is *Octopus vulgaris*
earthworm is *Eisenia foetida*
pond snail is *Lymnaea stagnalis*
locust is *Locusta migratoria*

**b**

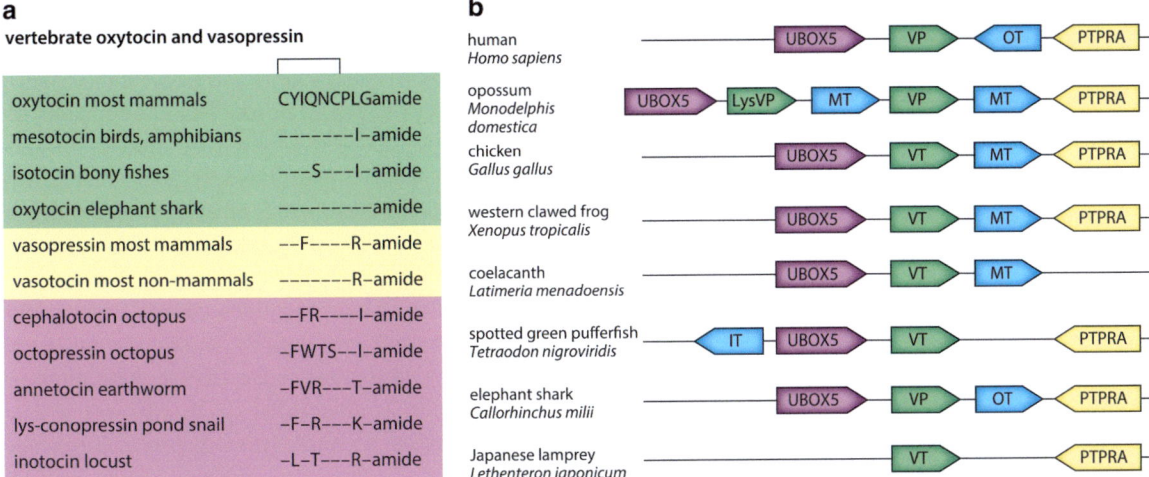

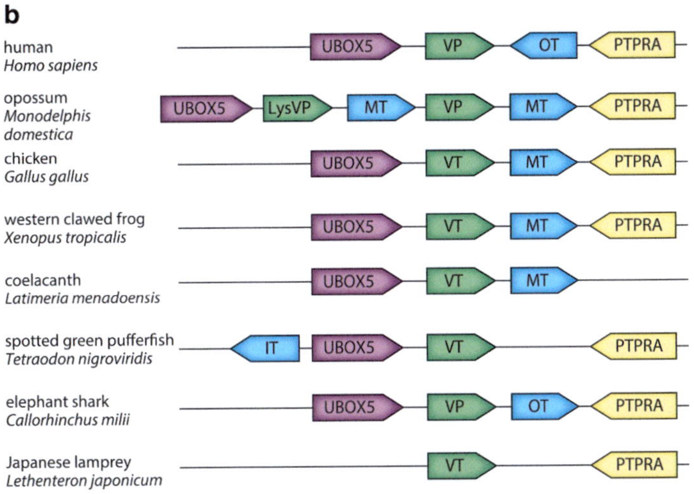


**Fig. 11.18 Oxytocin and vasopressin sequence comparison.** (**a**) Amino acids are shown with single-letter code. The top sequence is shown as master sequence. *Dashes* mean identity to the master sequence. Oxytocin orthologs are shown with *gray* background and vasopressin orthologs with *yellow* background. Invertebrate sequences are shown with *pink* background (*Octopus vulgaris, Eisenia foetida, Lymnaea stagnalis, Locusta migratoria*). All peptides are cyclized by a disulfide bond between the two cysteines and they all have a carboxy-terminal amide group. (**b**) Schematic outline of chromosomal region harboring the oxytocin (*OT*)/mesotocin (*MT*)/isotocin (*IT*) (*blue*) and the vasopressin (*VP*)/vasotocin (*VT*) (*green*) genes in various gnathostomes. LysVP is vasopressin with a lysine residue instead of arginine (see Fig. 11.18a). Adjacent genes: *Ubox5* stands for "U box domain containing 5" and *Ptpra* is "protein tyrosine phosphatase, receptor type, A". As the lamprey has a single peptide gene, it is likely that the gene duplication that generated oxytocin and vasopression from a common ancestral gene took place in a gnathostome ancestor after the cyclostomes had branched off

large ("magnocellular") hypothalamic neurons – the closely related nonapeptides oxytocin and vasopressin. **Vasopressin** acts as a peptide hormone to increase blood pressure and decrease urinary volume. **Oxytocin** in its hormonal role in mammals stimulates the milk letdown reflex (milk ejection) as well as uterus contractions during parturition. The corresponding two hormones are found in all gnathostomes (jawed fishes), although they have been named differently in various animal groups as will be described below. Here, oxytocin has other roles in reproduction and water-salt balance. Notably, both oxytocin and vasopressin also have prominent roles as neuropeptides within the brain, influencing for instance, social behaviors.

These peptides were two of the earliest peptides to be characterized in multiple vertebrate species, and when different peptide sequences were reported the peptides were given new designations, which unfortunately resulted in a confusing plethora of names. Only recently has gene characterization confirmed that oxytocin is the ortholog (species homolog) of mesotocin (birds, amphibians) and isotocin (bony fishes), whereas vasopressin is the ortholog of non-mammalian vasotocin (Fig. 11.18a). A local gene duplication gave rise to this ancestral pair before the origin of gnathostomes (Fig. 11.18b). Other peptide families have been described by using information about chromosomal location of the genes for assignment of orthology versus paralogy (sequence comparisons have been ambiguous).

### 11.5.4 The Urophysis

A population of large neurons was discovered in the caudal part of the spinal cord in skates by the American zoologist Ulric Dahlgren (1870–1946) in 1914. These cells were also identified in teleost fishes and were named **Dahlgren cells** (Fig. 11.19). They were found

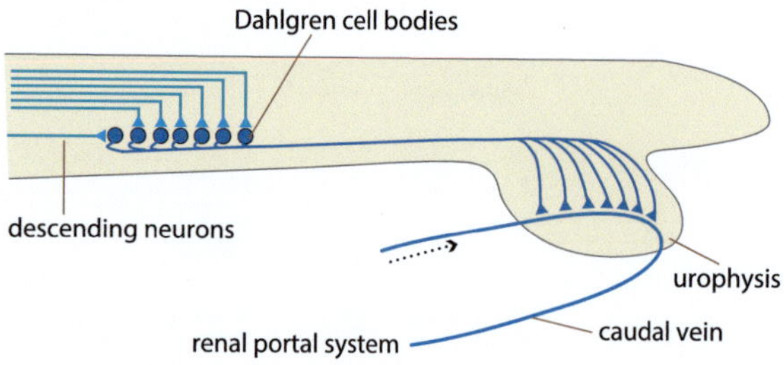


**Fig. 11.19** **Organization of Dahlgren cells in the teleost fish urophysis** in the caudal spinal cord

to extend their axons to the posterior tip of the spinal cord where they form a neurohemal organ called the **urophysis** or the caudal neurosecretory system. The presence of the urophysis in both cartilaginous fishes, actinopterygian fishes, and the sarcoptergyian lungfishes shows that it was present in the gnathostome ancestor and was subsequently lost in tetrapods. Two peptides have been discovered in the urophysis and were named urotensin I and II. Both of these are also expressed elsewhere in the nervous system. They are structurally unrelated: urotensin I also exists in mammals where it is called urocortin and belongs to the CRH family (corticotropin-releasing hormone). Urotensin II, on the other hand, is a distant relative of somatostatin. In teleost fishes both urotensin I and urotensin II are involved in osmoregulation.

## 11.6 Neuropeptides as (Co)Transmitters in the Brain of Vertebrates

Many of the peptides mentioned in the preceding sections also have roles as neuropeptides entirely within the brain. To name a few, these include oxytocin and vasopressin that were originally discovered in the posterior pituitary but are now known to have prominent effects on behavior, CCK that was first found in the gastrointestinal tract, and substance P found in multiple tissues. In the vertebrates the number of peptides participating in neuronal signaling is in the order of 100, accompanied by a roughly equal number of GPCRs that mediate their actions.

The nerve terminals that release neuropeptides from large dense-core vesicles also frequently release classical neurotransmitters from small synaptic vesicles (see Fig. 11.2). The first case of such coexistance was reported in 1977 by Tomas Hökfelt and colleagues [18] at the Karolinska Institute in Stockholm and concerned somatostatin colocalization with noradrenalin (or rather the enzyme that makes noradrenaline, i.e., dopamine-β-hydroxylase). This was followed by findings of substance P coexistance with 5-hydroxytryptamine, CCK with dopamine, and numerous other examples with one or more neuropeptides coexisting with classical neurotransmitters.

The number of neuropeptides and receptors seems to have increased considerably in the early stages of vertebrate evolution due to the two genome duplications (tetraploidizations) that took place in the gnathostome ancestor. Peptide genes that were duplicated as a result of these tetraploidizations include NPY, somatostatin, tachykinins, opioid peptides, relaxins, and other insulin-like peptides. Several peptide families expanded further in the third tetraploidization that took place in the bony fish lineage before the radiation of teleost fishes. In addition, many local peptide gene duplications have occurred in various vertebrate lineages. Although many new peptides have been discovered, several more probably remain to be identified because there are many GPCRs whose ligands are still unknown (so-called orphan receptors) but are expected to be peptides. Many of the vertebrate neuropeptides have homologs in invertebrates, but some may be unique to the vertebrates (see above).

Prominent peptidergic systems in the brains of mammals include oxytocin and vasopressin, somatostatin, CRH, and others. Two of the more widespread systems in mammals are those involving CCK and somatostatin. The octapeptide CCK was initially discovered as a peptide hormone produced by endocrine cells in the duodenum, released after meals and contributing to satiety as well as gastrointestinal functions (see above). Later it was found to be widely distributed in the brain in mammals, including the cerebral cortex and particularly in interneurons where it is thought to have modulatory roles. One effect of CCK is increased anxiety. Somatostatin consists of 14 amino acids and was initially identified as a hypothalamic peptide inhibiting the release of growth hormone (somatotropin) from the pituitary. It has subsequently been found in many parts of the brain and like CCK it is present in many cortical interneurons, as well as in the gastrointestinal tract in both endocrine cells and neurons. Somatostatin in the cerebral cortex is involved in inhibitory modulation of, for instance, somatosensory information.

NPY (neuropeptide Y) is also widespread in interneurons of the brain cortex. The NPY system consists of 3–4 related peptides all of which are 36 amino acids in length. One of these is almost exclusively neuronal in mammals, namely NPY itself. PYY (peptide YY), on the other hand, is almost exclusively produced in endocrine cells in mammals, but is expressed in the brain in teleost fishes as well as lampreys. This illustrates the close interrelationship of peptides in the neuronal and endocrine systems. The roles that NPY and PYY play in satiation control are explained above. Teleost fishes have duplicates of both NPY and PYY, but it remains to be explored what these duplications mean regarding functional diversification or specialization.

A family of vertebrate opioid peptides is expressed in the brain, including endorphin, dynorphin, enkephalins, and orphanin, arising from four large precursors encoded by four separate genes. This family, too, expanded at the dawn of vertebrate evolution along with its four receptors. The opioid peptides are important signals to reduce pain and many pharmaceuticals that relieve pain are agonists on opioid receptors (see Chap. 21). They also stimulate reward mechanisms and are thought to be involved in motivation. In humans they play key roles in the development of certain types of drug dependence and substances of abuse such as morphine and heroin.

### 11.6.1 Circadian Rhythms Are Controlled by Peptides

Several peptides influence the circadian rhythm in vertebrates. The two related peptides VIP (vasoactive intestinal peptide) and PACAP (pituitary adenylate cyclase-activating polypeptide) are released from different neurons in the suprachiasmatic nucleus of the hypothalamus where they act on GPCRs. PACAP binds to two related receptors whereas VIP binds to only one of them. The receptors have slightly different roles in the circadian regulation. Other peptides that affect the circadian rhythm include NPY and vasopressin.

A somewhat unusual neuropeptide system is formed by the peptides orexin A and B, also called hypocretins, that arise from the same precursor. Each peptide is approximately 30 amino acids long and they are produced by a rather small population of neurons in one nucleus of the hypothalamus. These neurons have quite widespread projections to be a neuropeptide system and reach many remote targets via long axons (Fig. 11.20) [19]. One important role of these neurons in mammals is to maintain wakefulness by acting on two GPCRs. In mammals, this alerting system can fail for a number of reasons resulting in dysregulation of sleep-wakefulness, more specifically loss of wakefulness and immediate entrance into so-called REM sleep, rapid eye movement sleep, i.e., the type of sleep associated with dreaming in mammals. The sleep disorder is called **narcolepsy** and has been described not only in humans but also several other mammalian species, particularly dogs. Human narcolepsy is usually due to loss of the neurons that produce the orexins, perhaps due to autoimmunity, but one case is known where early onset narcolepsy is due to a deleterious mutation in the gene encoding prepro-orexin. In dogs with an hereditary

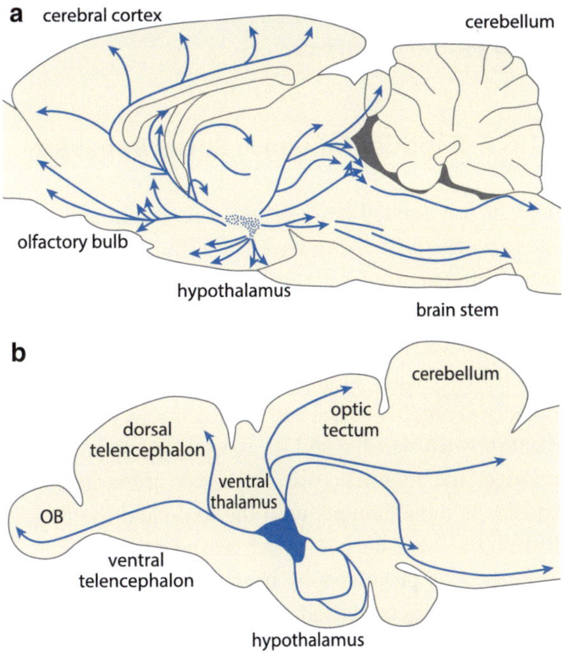


**Fig. 11.20 Orexin distribution in the rat (a) and zebrafish brain (b), shown in parasagittal sections.** Note that the brains are not drawn in the same scale. The rather limited population of hypothalamic cells that produce the two orexins have unusually wide axonal projections for a peptide system (a, Redrawn from Kilduff and Peyron [19] with permission. b, Redrawn from Panula [20] with permission)

type of narcolepsy, deleterious mutations are present in one of the two orexin receptors.

### 11.6.2 Peptides Influence Social Behavior

The closely related neuropeptides oxytocin and vasopressin described above have been found to have profound effects on social behavior in several mammals. Two species of North American voles have become famous because their reproductive and social behaviors differ greatly although they are closely related, namely the highly social and monogamous prairie vole *Microtus ochrogaster* and its asocial and promiscuous relative, the montane vole *Microtus montanus*. One species is monogamous with the male and the female forming a life-long pair, whereas the other species is described as promiscuous because male and female only meet for copulation whereupon the male leaves the female to rear the young. Many observations point to prominent roles for the oxytocin-vasopressin system in the regulation of these behaviors, such as clearly different receptor distribution in the brains of the two species. Inactivation of the oxytocin gene in mice results in specific blocking of reproductive memory. Recent studies have reported that subsets of humans with disturbed social behavior, such as autism, have genetic differences in the genes encoding these peptides or their receptors. In line with this, it is believed that oxytocin is essential for social cohesion, allowing people to form productive and meaningful relationships, in particular by mediating the development of trust.

## 11.7 Summary

Neuropeptides and peptide hormones play critical roles in development, growth, and reproduction and they are also involved in the regulation of most aspects of daily life of an animal. In mammals peptidergic regulatory systems are complex and often include multiple neuropeptides and peptide hormones that act at different levels in so-called neuroendocrine axes, that also involve feedback regulation from the distal cells in the pathway. Three prominent axes are distinguished in vertebrates: the hypothalamus-pituitary combined with either the thyroid gland, the adrenal glands, or the gonads. Invertebrate neuroendocrine systems are based on similar principles. Functional overlap between components is present in many neuroendocrine systems, particularly in the vertebrates (due to gene duplicates), but defects in single genes of peptides or their receptors can nevertheless produce serious alterations in physiology and behavior. Thus, research in peptide signaling is important for future development of pharmaceutical therapies as well as pest control agents. Examples were given to illustrate how neuropeptides can be involved in the regulation of anxiety, pain, appetite, reward systems, clock functions, and sleep.

Peptide functions in invertebrates are less known than in vertebrates. One reason is that the invertebrates display tremendous diversity in organization of nervous and neuroendocrine systems, as well as considerable differences in physiology and behavior. At the molecular level some ancestral evolutionary relationships are obvious because the amino acid sequences of many peptides and peptide receptors display striking similarities between *Drosophila, C. elegans*, and vertebrates. Nevertheless, it has been more rare to find convincing examples of evolutionary conservation of the functional aspects of peptide signaling. One reason for this is that most peptides appear to be pleiotropic in their functions even in invertebrates. Therefore it is not clear what the original or even main function of a peptide is and more studies are required in invertebrates to resolve this before meaningful comparisons can be made. Probably the best example of peptide signaling that is conserved both at the molecular and functional levels is provided by the insulin-like peptides and their tyrosine kinase receptors. The original role of insulins may be in the regulation of growth since this is a common feature in all studied organisms.

One of the most striking conclusions emerging from studies of biology concerns the enormous diversity of paths taken in evolution, yet often using the same or similar components in a variety of different ways (a process coined exaptation by Stephen Jay Gould to refer to a feature that performs a given function but that was not produced by natural selection for its current use). Thus, we should view the diversity in organization of neuroendocrine and peptidergic systems, especially among invertebrates, as a fascinating challenge rather than a nuisance. Analysis of the complexity of neuropeptide signaling systems will certainly benefit from a broad comparative approach involving studies of many different types of animals.

## References


1. Squire LR (ed) (2003) Fundamental Neuroscience. Academic Press, Amsterdam
2. Salio C, Lossi L, Ferrini F, Merighi A (2006) Neuropeptides as synaptic transmitters. Cell Tissue Res 326:583–598
3. Kobayashi H (ed) (1987) Atlas of endocrine organs, vertebrates and invertebrates. Kodansha, Tokyo
4. Brown RO, Pulst SM, Mayeri E (1989) Neuroendocrine bag cells of *Aplysia* are activated by bag cell peptide-containing neurons in the pleural ganglion. J Neurophysiol 61:1142–1152
5. Hartenstein V (2006) The neuroendocrine system of invertebrates: a developmental and evolutionary perspective. J Endocrinol 190:555–570
6. Marder E, Bucher D (2007) Understanding circuit dynamics using the stomatogastric nervous system of lobsters and crabs. Annu Rev Physiol 69:291–316
7. Nusbaum MP, Blitz DM, Swensen AM, Wood D, Marder E (2001) The roles of co-transmission in neural network modulation. Trends Neurosci 24:146–154
8. Roller L, Yamanaka N, Watanabe K, Daubnerova I, Zitnan D et al. (2008) The unique evolution of neuropeptide genes in the silkworm *Bombyx mori*. Insect Biochem Mol Biol 38:1147–1157
9. Shiga S, Toyoda I, Numata H (2000) Neurons projecting to the retrocerebral complex of the adult blow fly, *Protophormia terraenovae*. Cell Tissue Res 299:427–439
10. Wegener C, Reinl T, Jänsch L, Predel R (2006) Direct mass spectrometric peptide profiling and fragmentation of larval peptide hormone release sites in *Drosophila melanogaster* reveals tagma-specific peptide expression and differential processing. J Neurochem 96:1362–1374
11. Ewer J (2005) Behavioral actions of neuropeptides in invertebrates: insights from *Drosophila*. Horm Behav 48:418–429
12. Kim YJ, Zitnan D, Galizia CG, Cho KH, Adams ME (2006) A command chemical triggers an innate behavior by sequential activation of multiple peptidergic ensembles. Curr Biol 16:1395–1407
13. Hall DH, Altun ZF (2008) *C. elegans* Atlas. Cold Spring Harbor Laboratory Press, Cold Spring Harbor, p 348
14. Puig O, Marr MT, Ruhf ML, Tjian R (2003) Control of cell number by *Drosophila* FOXO: downstream and feedback regulation of the insulin receptor pathway. Genes Dev 17:2006–2020
15. Géminard G, Arquier N, Layalle S, Bourouis M, Slaidina M et al. (2006) Control of metabolism and growth through insulin-like peptides in *Drosophila*. Diabetes 55:S5–S8
16. Zohar Y, Muñoz-Cueto JA, Elizur A, Kah O (2009) Neuroendocrinology of reproduction in teleost fish. Gen Comp Endocrinol 165:438–455.
17. Roseweir AK, Millar RP (2009) The role of kisspeptin in the control of gonadotrophin secretion. Hum Reprod Update 15: 203–212
18. Hokfelt T, Elfvin LG, Elde R, Schultzberg M, Goldstein M et al. (1977) Occurrence of somatostatin-like immunoreactivity in some peripheral sympathetic noradrenergic neurons. Proc Natl Acad Sci USA 74:3587–3591
19. Kilduff TS, Peyron C (2000) The hypocretin/orexin ligand-receptor system: implications for sleep and sleep disorders. Trends Neurosci 23:359–365
20. Panula P (2010) Hypocretin/orexin in fish physiology with emphasis on zebrafish. Acta Physiol (Oxf) 198:381–386


## Further Readings


Anctil M (2009) Chemical transmission in the sea anemone *Nematostella vectensis*: A genomic perspective. Comp Biochem Physiol Part D Genomics Proteomics 4:268–289

Cerdá-Reverter JM, Larhammar D (2000) Neuropeptide Y family of peptides: structure, anatomical expression, function, and molecular evolution. Biochem Cell Biol 78: 371–392

Dircksen H, Neupert S, Predel R, Verleyen P, Huybrechts J, Strauss J, Hauser F, Stafflinger E, Schneider M, Pauwels K, Schoofs L, Grimmelikhuijzen CJ (2011) Genomics, transcriptomics, and peptidomics of *Daphnia pulex* neuropeptides and protein hormones. J Proteome Res 10: 4478–4504

Grönke S, Clarke DF, Broughton S, Andrews TD, Partridge L (2010) Molecular evolution and functional characterization of *Drosophila* insulin-like peptides. PLoS Genet 6(2): e1000857

Gwee PC, Tay BH, Brenner S, Venkatesh B (2009) Characterization of the neurohypophysial hormone gene loci in elephant shark and the Japanese lamprey: origin of the vertebrate neurohypophysial hormone genes. BMC Evol Biol 9:47

Hartenstein V (2006) The neuroendocrine system of invertebrates: a developmental and evolutionary perspective. J Endocrinol 190:555–570

Husson SJ, Mertens I, Janssen T, Lindemans M, Schoofs L (2007) Neuropeptidergic signaling in the nematode *Caenorhabditis elegans*. Prog Neurobiol 82:33–55

Kastin AJ (ed) (2006) The handbook of biologically active peptides. Elsevier, Amsterdam

Kits KS, Boer HH, Joose J (1991) Molluscan neurobiology. North-Holland, Amsterdam

Kobayashi H (ed) (1987) Atlas of endocrine organs, vertebrates and invertebrates. Kodansha, Tokyo

Lopez-Bermejo A, Buckway CK, Rosenfeld RG (2000) Genetic defects of the growth hormone-insulin-like growth factor axis. Trends Endocrinol Metab 11:39–49

Nässel DR, Winther ÅM (2010) *Drosophila* neuropeptides in regulation of physiology and behavior. Prog Neurobiol 92: 42–104

Papadimitriou A, Priftis KN (2009) Regulation of the hypothalamic-pituitary-adrenal axis. Neuroimmunomodulation 16:265–271

Strand FL (1999) Neuropeptides: regulators of physiological processes. MIT Press, Cambridge

Taghert PH, Nitabach MN (2012) Peptide neuromodulation in invertebrate model systems. Neuron 76:82–97

Veenstra JA (2010) Neurohormones and neuropeptides encoded by the genome of *Lottia gigantea*, with reference to other mollusks and insects. Gen Comp Endocrinol 167:86–103

Winter MJ, Ashworth A, Bond H, Brierley MJ, McCrohan CR, Balment RJ (2000) The caudal neurosecretory system: control and function of a novel neuroendocrine system in fish. Biochem Cell Biol 78:193–203

Wynne K, Stanley S, McGowan B, Bloom S (2005) Appetite control. J Endocrinol 184:291–318

# The Biological Function of Sensory Systems

# 12

Rainer Mausfeld

Sensory systems provide the organism with information on functionally relevant aspects of its physical and biological environment.

Sensory systems serve to link the organism to functionally relevant aspects of the physical environment. A mobile organism requires diverse information from the biological and physical environment and about its internal state for orientation and movement in space and in order to regulate and control its body and behavior. In the course of evolution, with the increased complexity of tasks serving towards orientation and behavioral control there was a growing need to interrelate the diverse sensory channels and also to integrate information about the internal state of the body. This sensory integration required a regulation of awareness to develop which would be able to filter the external signals according to internal motivational and emotional states. As sensory systems are central for behavioral control, the neuronal architecture of sensory systems is intricately interwoven with the motor system. Furthermore, evolutionary considerations suggest that fundamental features of perception formed the basis for more abstract cognitive achievements and that the underlying general principles are thus also reflected in the organization of cognitive processes.

For humans, the sensory integratory achievements pertain not only to haptic, visual, auditive, olfactory, and gustatory perception, but also to the perception of the body and its parts [6, 27] and the relative position of these parts in relation to each other (**proprioception**) and to their environment, the perception of the viscera (**entero-** or **visceroception**), the perception of pain, the perception of physiognomy and body movements and the thus communicated affective expressions and signals, as well as the perception of speech, events, or time.

Biological species may differently exploit and utilize the physical energies impinging on the organism and organize these energies in the form of **sensory modalities**. By far the largest part of the impinging spatiotemporal energy pattern is not processed for biological purposes. Only a highly restricted range of this energy pattern is used for the biological function of coupling the organism to its environment (e.g., humans neither can perceive the plain of polarized light nor the direction of the magnetic field). The physical energy is transduced into neural codes in such a way that the particular physical origin of the resulting code is unidentifiable (e.g., light perception at the eye may originate from optical, mechanical, or electrical stimuli). This already demonstrates that the function of perception is not to recognize the nature of "physical reality". Accordingly, the idea that the world, as it really is, independently of an observer, is mirrored in perception, i.e., naïve realism, is utterly inappropriate. Sensory systems (and more generally: cognitive systems) have rather attained, in the course of evolution, a highly specific internal structure, by which these systems as a whole transform certain structural aspects of physical energy into "information" that codes functionally relevant aspects of its physical and biological environment. Particularly, sensory systems do *not* serve to instruct the organism about the nature of the physical *input*. The function of sensory systems rather is determined by the entire functional organization of the organism.

R. Mausfeld
Institute of Psychology, University of Kiel,
D-24098 Kiel, Germany
e-mail: mausfeld@psychologie.uni-kiel.de

C.G. Galizia, P.-M. Lledo (eds.), *Neurosciences - From Molecule to Behavior: A University Textbook*,
DOI 10.1007/978-3-642-10769-6_12, © Springer-Verlag Berlin Heidelberg 2013

The study of perception and the sensory systems was initially motivated by epistemological questions regarding the possibility, in face of our biological constitution, of achieving reliable and true knowledge of the "external world". The pre-Socratics were the first to derive the idea of the "fallibility" of the senses from the observation that different senses can lead us to different beliefs about the world (think of a rod half-dipped in water). Since then, the question as to the nature of perception pervades the entire history of epistemology, and figures prominently in controversies between rationalism and empiricism, and in philosophical attempts at understanding the nature of consciousness. The study of perception, in which epistemology, psychology, ethology, neurophysiology, and "artificial intelligence" are closely interwoven, forms the basis of cognitive science.

In antiquity the doctrine of the five senses (vision, audition, haptic perception, olfaction, gustation) had already assigned certain classes of perception (**sensory modalities**) to the sensory organs. In the course of these conceptions, the sensory modalities have been considered as natural units of analysis that can be investigated largely in isolation. As sensory physiology developed, the idea of sensory organs was replaced by more specific ideas referring to sensory types. Correspondingly, thermoreceptors, mechanoreceptors, photoreceptors, chemoreceptors etc. became to be differentiated. Such a classification turned out to be immensely fruitful for sensory physiology. At the same time, however, it may impede a deeper theoretical understanding of the nature of perception. Corresponding elementaristic conceptions bear the danger of neglecting the intricate interplay of the different modalities within the functional architecture of the organism. In humans, for instance, visual experiences are organized in terms of an "external world" only due to the interaction of visual, haptic, and kinetic experiences; simple visual impressions lacking kinetic sensations strongly impair perceptual achievements. It is a core feature of the perceptual system that its structure is determined in an essential manner by transmodal representations of the physico-biological environment, and is thus not reducible to the structure of isolated input channels.

## 12.1 Key Terminology and Key Issues of General Sensory Physiology

A **stimulus** (or **sensory input**) can be regarded as a physical (or chemical) energy pattern that impinges on a primary receptor and is transduced into a neural signal. Perceptual psychology also differentiates between **proximal stimuli** which refer to the above kind of sensory input, and **distal stimuli** which refer to those objects in the physical environment that are causally responsible for the energy pattern impinging on the receptors. For instance, the surface of an illuminated object can be regarded as a distal stimulus for color perception leading to a proximal stimulus of a physico-geometrical light pattern impinging on the retina.

What constitutes a stimulus for a particular perceptual achievement depends on the nature of the specific mechanisms on which a particular achievement is based and hence cannot be determined independently.

So-called **adequate stimuli** are those kinds of energy that fit the specific properties and demands of a sensory receptor. For instance, electromagnetic waves of 380–700 nm or $0.43–0.79 \times 10^{15}$ Hz are an adequate stimulus for human photoreceptors; all other electromagnetic waves, as those in the range of $10^{13}$ Hz (infrared) or 50 Hz (AC) do not serve as an adequate stimulus for human photoreceptors, nor do mechanical stimuli.

Within individual sensory modalities, one can distinguish different perceptual **qualities**. In the case of vision, for instance, these qualities comprise color, brightness, depth, motion, or form. Perceptual experiences are furthermore distinguished in terms of intensity, duration, and location.

> The quality of perception is determined not by the physical nature of the stimulus, but by the kind of activated neural substrate.

According to the **Law of Specific Sensory Energies** by Johannes Müller (1801–1858), perception is not determined by the nature of the stimulus, but rather by the "terminal organ" of the central nervous system. Different kinds of stimuli thus can lead to the same kind of perceptual quality, while the same kind of physical stimulus may lead to different perceptual qualities. For instance, a perception of light may arise from an optical, a mechanic, or electrical stimulus of the eye; on the other hand, sunlight can elicit retinal receptors for light perception but also can lead to thermal perception by stimulating thermoreceptors of the skin. An essential aspect of sensory and motor systems is that the "information" generated by a neural signal is mainly determined by the neural pathway by which the signal is transmitted ("*labeled line*") and by the activated neural structures.

### 12.1.1 Sensory Systems Can Be Investigated on Different Levels of Analysis

The scientific inquiry into sensory systems involves a variety of disciplines, ranging from neurophysiology, neuroethology, psychophysics to perceptual psychology, and is also linked to technological fields, such as computational perception and robotics.

Neurophysiology and molecular biology investigate the mechanisms of sensory transduction and of neural coding at different stages of sensory pathways. Neuroethology focuses on the biological processes and mechanisms by which complex behavioral accomplishments of an organism are controlled by sensory information. Psychophysics and perceptual psychology investigate complex perceptual achievements as well as phenomenal aspects of perception; they pursue a level of analysis that largely abstracts away from the specific (and often unknown) biophysical and neural processes by which these achievements are neurally implemented. Each of these domains constitutes a comparably independent level of analysis and theory development (analogous to, for instance, chemistry and biology with respect to physics).

The field of mechanic sensing and computational perception, in contrast, does not aim at the understanding of a natural biological system but rather attempts to derive abstract computational principles from the functional requirements and properties of natural sensory systems in order to take advantage of this knowledge for technological purposes.

## 12.2 Neurophysics and Neurobiology of Sensory Systems

> Sensory systems share common features of the encoding of neural information and principles of neural organization.

The similarity of tasks of sensory systems within the entire functional organization of the organism is mirrored in its neural architecture. Accordingly, different sensory systems share common fundamental features. These commonalities pertain both to the principles of neural coding and to principles of the functional architecture of the sensory system and subsequent systems.

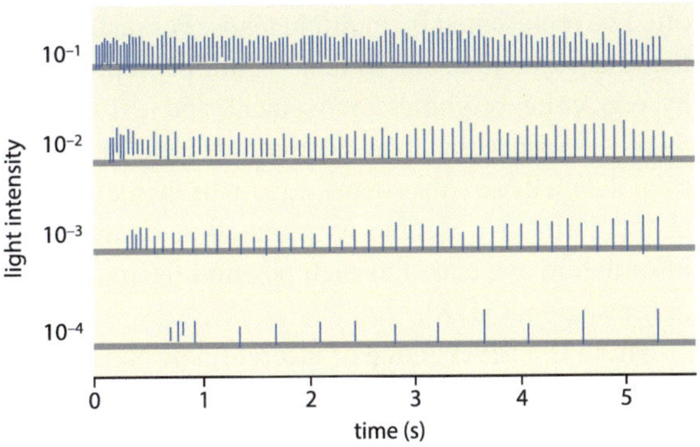


**Fig. 12.1** **Action potentials elicited by light stimuli in a receptor of the compound eye of *Limulus*** (horseshoe crab) are directly dependent on light intensity. For strong light intensities (*upper row*) there are about 30 impulses per second. Reducing the light intensity to one ten-thousandth of the initial value lowers the impulse frequency to two or three impulses per second (after Miller et al. [32] with permission)

Sensory systems are structured such that certain cells, serving as sensory cells or **receptor cells**, convert specific kinds of physical energies into an electrochemical form. This first step is referred to as **transduction**. Different kinds of physical energies thus are converted into one and the same kind of energy – electrochemical energy. This is the first prerequisite for organisms to utilize the various kinds of energies encountered in the physico-biological environment for the purpose of its own internal data processing. Thus, transduction uses the stimulus energy to produce an electrotonic **receptor potential** or **generator potential**. The next step, which can occur in the receptor cell, or one or two synapses further down, is the conversion of this electrotonic (analog) signal into a pattern of **action potentials**, i.e., a digital signal. This step is called **transformation**. These action potentials are then transmitted to higher processing centers via a small number of subsequent neurons. In vertebrates, signals arriving from primary and secondary afferent neurons are relayed in the specific contralateral sensory nuclei of the thalamus to downstream neurons. The sensory pathways of receptor cells of the same kind are organized independently and linked to certain primary sensory areas in the cerebral cortex. In most cases these pathways are **somatotopic** (except those of the gustatory and olfactory pathways), i.e., they retain spatial relations at the receptor level (and thus also the spatial neighborhood relationship of the sensory input). This results in a representation in form of so-called **sensory maps**. A significant feature of the sensory architecture is that one and the same receptive area can often be represented by multiple maps. For instance, in the visual system a sensory area is multiply represented by, e.g., color, brightness, movement, and texture; also there is evidence that there are multiple representations even *within* these subsystems, e.g., in the visual system in the form of a double representation of the color information associated to each position (as in the case of transparency) [28].

There is a **divergence** of afferent neurons to various central neurons in different central areas. Conversely, central neurons can receive inputs from various afferent neurons (**convergence**). Despite this divergence and convergence, cortical neurons have only a very limited interconnectivity, as compared to the potentially possible range of interaction, which reflects the somatotopic organization as much as the **modularity** of the cortical architecture. An essential feature of sensory systems is that afferent sensory information is to a great part modulated via corticofugal connections. For instance, in the visual system of humans the connections between *lateral geniculate nucleus* and the primary visual cortex is mostly corticofugal; also, sensory information in retinotopically organized regions are modulated by corticofugal connections from non-retinotopically organized regions.

Different kinds of receptor cells transform relevant physical characteristics of stimuli into a neural signal.

Sensory receptor cells are adapted in a highly specific manner to particular functions and biologically relevant physical properties of the environment. Accordingly, evolution has brought forth a great variety of receptor cell types. Their high degree of selectivity for a particular kind of stimulus (**stimulus specificity**) pertains to the type of physical energy (e.g., mechanical, electromagnetic) as well as to particular energy characteristics (e.g., mechanoreceptors with selective sensitivity for pressure, touch, or vibrational stimuli, or photoreceptors with maximal sensitivity in different spectral ranges). With respect to the functional specialization of the receptor array, we often find a differentiation into a center and a periphery; the sensory center has a higher sensitivity and resolution (typical example: the fovea), is connected to the accordingly specialized higher processing centers and can be attentionally focused onto relevant aspects by an active motor control.

In the process of transduction, the stimuli impinging on the receptor molecules trigger processes which ultimately either open or close ion channels in receptor cell membranes (see the respective chapter for each sensory system). The membrane's resting potential is altered by an increase or decrease of the ionic current; the resulting potential change is referred to as a **receptor potential**: a local depolarization of the cell membrane of the receptors (or, as in the case of vertebrate photoreceptors, a hyperpolarization). The stimulus thus does not simply cause the receptor potential in a passive way, but rather modulates molecular-biological processes that occur within the membrane and within intracellular structures of the receptor cells. As one single activated receptor molecule can effect an entire biochemical reaction cascade, where the signal may become substantially amplified by second messengers, the potential change will have a several times higher energy as compared to the eliciting stimulus. Physical laws certainly limit the maximally attainable sensitivity and reliability in the process of transducing the sensory input into a neural signal. Surprisingly, receptor cells and neural transduction do in fact reach close to the theoretical limits of sensitivity (e.g., some photoreceptors are sensitive to single photons – the sensory information is only limited by the quantum fluctuation of the input and the thermal stability of the photopigments).

Neural coding encodes the relevant sensory information by a temporal pattern of action potentials.

Receptor potentials trigger action potentials in a neuron. Transformations convert the amplitude of the receptor potential into the interval length between consecutive action potentials. The result is a frequency-modulated sequence of impulses of action potentials (*spikes*). Because of the all-or-none character of action potentials, such a sequence of action potentials can be symbolically represented by a sequence of zeros and ones. This series of impulses is conveyed via the axon to subsequent neurons. The medium of neural encoding thus alternates from predominantly chemical (at the synapse) to electric (for instance, in the axon).

The dynamic range that a neuron is capable of encoding into action potentials is highly restricted. It is difficult to judge what relevance this has for the dynamic range of neural codes, as the exact temporal features are yet unknown. But it seems that the crucial variable is the number of impulses within a short period of time. For instance, if the time interval relevant for coding would be 100 ms, the maximal number of impulses (for a refractory period of 1 ms) would be about 50, and the dynamic range (measured in spike numbers) would be equivalent to an information content of about 6 bits – a time interval of 10 ms would correspond to approximately 2 bits. By comparing this with the continuously variable receptor potential (which for, e.g., the quantum absorption rates of photoreceptors is in the range of $10^{12}$ or 40 bits) one can clearly see the discrepancy between the wide input range and the narrow dynamic range of nerve fibers.

The sequence or *train* of action potentials encodes the relevant sensory information, in particular intensity and duration, and forwards this information to the central areas (Fig. 12.1). Time and duration can, within certain limits, have an equivalent effect on the coding in action potentials (Fig. 12.2).

## 12.3 Sensory Information Can Be Coded in Frequency Codes or in Temporal Codes

Which characteristics of an impulse series are encoding the relevant sensory information? This can be the median of impulses within a time interval or the interval between consecutive impulse series, or the time difference between the arrival of impulses in two or more fibers. In any case, it is essential to know the time interval that is relevant for coding, i.e., which duration of a sequence of action potentials is equivalent to a particular abstract symbol of the neural code. A relevant coding interval of 100 ms at maximally 50 impulses would result in a binary neural "alphabet" of $2^{50}$ symbols, i.e., a single symbol would in the ideal case be able to encode an information content of 50 bits. This would require the exact temporal position of an action potential within a corresponding sequence representing a symbol to be highly reliable. Because of restrictions as given by the existing noise [16] within the neural system, by limitations in the postsynaptic decoding, etc., a neural code can only be based on a much smaller set of symbols.

A drastic reduction could be achieved if the relevant coding quantity were based on the **number of action potentials** within a certain time interval, and if the temporal pattern within this interval did not carry any relevant information. Coding schemes of this kind are called **frequency codes** or **rate codes** [4, 36]. In such a

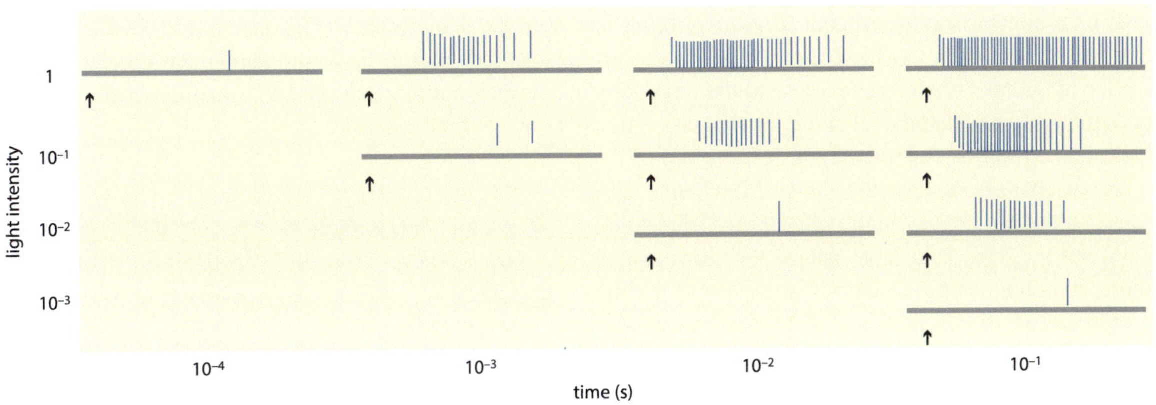


**Fig. 12.2** **The reaction of a receptor of the compound eye of *Limulus*** upon a light stimulus is determined by the sum total of the energy impinging per time interval, regardless of the distribution of this energy in time. A brief light stimulus (*above left*) elicits an approximately equally strong response as a light stimulus whose intensity is 1000th, but whose duration is 1000 times longer than the first stimulus

coding scheme all relevant stimulus parameters are encoded in the (average) number of action potentials within a coding interval, and there would be no correlation between stimulus parameters and higher-order statistics for the action potentials within this interval. A special case of a frequency code is a code where the action potentials within different subintervals of the coding interval are integrated with different weights in the calculation of the mean or in frequency estimation.

The concept of a frequency code dates back to the earliest work on neural coding. Various forms of evidence have meanwhile confirmed this concept. Hence, the assumption of frequency codes allows a first rough idea of the nature of neural coding. It has been shown, however, that the temporal structure within a sequence of action potentials is also exploited by subsequent neurons. Coding schemes that are based not only on the number of action potentials within a certain coding interval but also on the kind of temporal distribution within that interval are referred to as **temporal codes** or **timing codes**. Although temporal codes and frequency codes are coding schemes with completely different characteristics, this distinction loses significance when it comes to very short coding intervals. This raises the question as to the role of very small coding intervals, and hence of codes that are based on a very small number of action potentials. In fact, there is evidence that perceptual performances can rest on a very small number of action potentials (often only 1–2) and that, from the point of view of information theory, individual action potentials can convey a high amount of information [4]. These include **latency codes** which are dependent on a comparison of different neurons in order to extract the latency. It has been shown that the performance of neurons often reaches close to the physically possible limit, and that the reliability of signals is often close to the limits set by noise in the sensory input.

If one changes the level of analysis from single neurons to more complex systems, which range from small neuronal aggregates to large neural networks, much more complex coding schemes can be found for the representation of information by neural codes – in contrast to the situation for merely single neurons. In more recent time mounting evidence has shown that sensory information of various kinds is encoded at the level of small cell aggregates. For large cell aggregates there is evidence for coding schemes that can encode information that goes beyond that encodable by individual neurons, e.g., by combinatorial codes (see Chap. 13).

A general difficulty associated with an information theoretical perspective on neural coding [7, 10, 13] is due to the fact that the functional tasks of the perceptual system cannot simply be identified with a reconstruction of the sensory input. According to Shannon's conception of information in information theory, a signal carries information about a source if we can predict the state of the source from the signal. A source is characterized by the property that it has a number of alternative states that can be realized on a particular occasion. A signal carries information about the source if its state is correlated with the state of the source; a signal that is a better predictor of the state of the source carries, according to this conception, more information about a source than a signal that is a worse predictor. However, sensory organs are not to be understood as measuring devices that inform the organism about physical characteristics of an input [29]. To determine which aspects of the sensory input are of relevance to the organism, and thus can be regarded as potential information, requires taking into account the entire neural architecture into which the receptors are embedded, and the functional tasks associated with them. Only on this basis it is possible to determine whether variations in an input signal are exploited by subsequent cells as information. Changes in the physical input signal that are not represented by codes of subsequent neural units cannot be regarded as information with regard to the neural processes underlying perceptual tasks and achievements. Hence, neurophysiology (and more generally biology, or cognitive science) requires notions of information that are much richer than the one used in mathematical information theory. However, there is as yet no agreement how to formulate such notions that refer to functional and semantic aspects of information.

> Sensory information processing comprises a great variety of sensory mechanisms, such as adaptation, redundancy reduction, or contrast amplification.

Sensory systems comprise a number of mechanisms that can be regarded as (abstract or idealized) computational building blocks of perceptual processing. These include a variety of mechanisms, not discussed in this chapter, such as mechanisms for improved signal-to-noise ratio, contrast coding and contrast amplification, spatial and temporal filtering, spatial and temporal adaptation, mechanisms for a decorrelation of redundant input channels, as well as for more complex achievements [15, 21].

### 12.3.1 Sensory Adaptation

Sensory systems are able to adapt to the physical range of particular input values – this is referred to as sensory adaptation [3, 5, 25, 38]. A simple example is the adaptation of the receptor potential to temporally constant stimuli (Fig. 12.3).

The visual system, for instance, functions in an intensity range (quanta/time) ranging from only a few light quanta up to $10^{14}$ quanta, i.e., over 14 logarithmic scale units; the auditory system can function for signals in an intensity range of 12 log units. The question then arises as to how the perceivable physical range is transposed into the dynamic range of the neural code which is by several powers of ten smaller. In the light of the smaller dynamic range of nerve fibers, the high sensitivity of the sensory organs over such a large input range can only be accomplished by a multitude of biochemical and neural adaptation mechanisms.

For instance, one distinguishes in the visual system:

- mechanisms of input control (e.g., a change of pupil diameter allowing an adaptation of about 1:10; an input control of the photoreceptors that leads to a lower number of absorbed quanta, possibly as a result of the bleaching of photosensitive molecules; the change of integration time; changes of membrane conductivity; or inhibition of individual enzymes that are part of the reaction cascade);
- mechanisms that are based on a response compression of neural transduction functions, as typically described by saturation functions of the relationship between input and response, and
- mechanisms of output control that rescale the primary sensory code in relation to temporal or spatial conditions (e.g., mechanisms based on a Weber law response behavior that guarantee a constant contrast sensitivity).

While adaptation allows a sensory system to function over a wide range of stimulus intensities, it has to be clearly differentiated from habituation. **Habituation** is a form of non-associative learning allowing an animal to ignore specific repetitive stimuli (see Chap. 26).

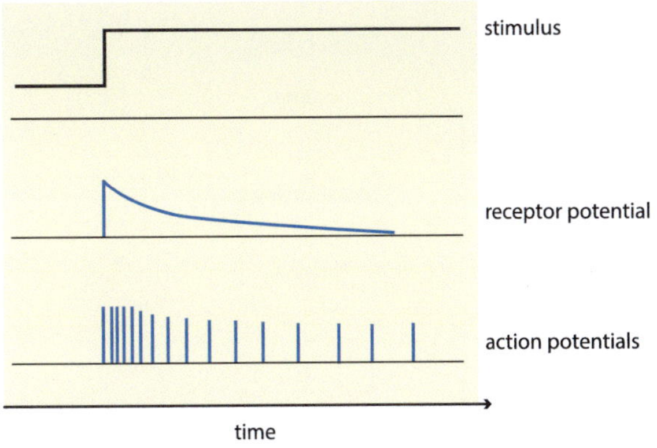


**Fig. 12.3 In the simplest case of sensory adaptation,** the receptor potential resulting from a stimulus of temporally constant intensity will become weaker in a time-course that is characteristic for a sensor type. Sensor types with fast adaptation predominantly encode dynamic stimulus features, an example being the mechanoreceptive Pacini corpuscles

Habituation can be distinguished from adaptation in that it is immediately reversible by a dishabituating stimulus, while adaptation reverts following its own time course.

The essential principle behind sensory adaptation is the ability to remove (spatial or temporal) steady components from the signal and thus to achieve a renormalization of the signal. This guarantees that the complete output range can be retained for encoding. By these means the sensory system is able to adjust its sensitivity over a large range of physical intensities, thus retaining its high capacity of discrimination for differences of sensory inputs, i.e., for **contrasts**. The price paid for retaining discriminatory sensitivity is a certain loss of information regarding *absolute* intensity, which in most cases is biologically irrelevant. Adaptations can also refer to more complex image statistics of the input signal rather than to its mean value [35].

### 12.3.2 Redundancy Reduction and Contrast Amplification

Due to the limited capacity of neural information transmission it is possible to increase the efficiency of transmission by removing such signal components that contain no or a minor amount of information, i.e., redundant information (unless the signal is corrupted by high levels of noise). Information is redundant when it can be predicted from a spatial or temporal neighborhood in the sensory input. Removing this "irrelevant"

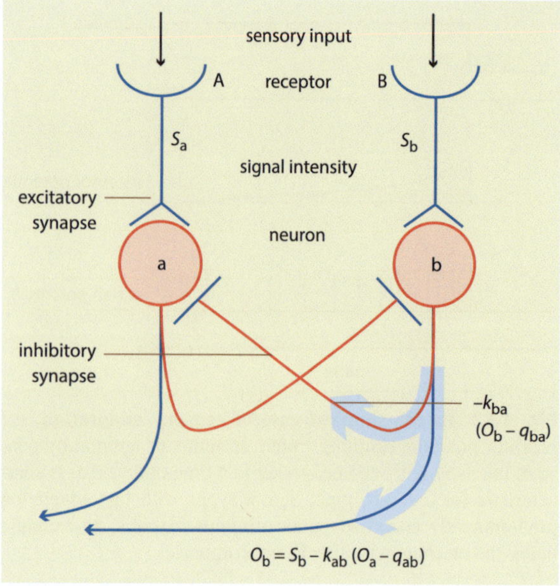


**Fig. 12.4** **For this simple example of a feedback lateral inhibition,** the output frequency $O_b$ of the neuron *b* is a function of signal intensity $S_b$ transmitted by the receptor and also of the inhibitory input $O_a$ determined by the output frequency of the neuron *a*. $O_b$ is determined for a set inhibitory threshold $q_{ab}$ and a set coupling coefficient $k_{ab}$ as: $O_b = S_b - k_{ab}\,(O_a - q_{ab})$

information allows the signal amplitude to be lowered without losing information. However, as the presence of redundancy protects the system against noise, a reduction of redundancy will improve the transmission of information only if the latter remains constant under the specific noise conditions [7].

Redundancy reduction can be realized by a multitude of neural schemes. A simple case is lateral inhibition (Fig. 12.4) as occurring in the compound eye of *Limulus*.

Instead of transmitting the mean value of the local sensory input over several nerve fibers, it is only the difference from the mean that is transmitted. Thus, the uniform surfaces are not encoded, but the edges of a visual scene are.

A neural code that is designed to encode the sensory inputs by a small number of cell types so that the redundancy and the dimensionality of sensory representation are reduced is called **compact coding**. This kind of encoding allows the relevant information to be transmitted by a comparably small population of cells [8].

Another principle aside from redundancy reduction is represented by **sparse coding** [26]. Here the relevant features of the sensory input are represented by a comparably large number of cells, where, however, the number of cells reacting to a specific input is minimized. In such a code, all cells have an equal response probability across potential inputs but have a low response probability for any single sensory input. Sparse coding does not reduce the dimensionality but rather transforms the redundancy of the input into the redundancy of the firing pattern of cells. While in the case of compact coding, different stimuli or perceptual objects are represented by different activating conditions of the same fraction of cells, in sparse coding these are in each case represented by a specifically assigned fraction of cells. In sparse coding, different objects are represented on the basis of which *kinds of cells* are activated, and not on the basis of the intensity at which these cells are activated. This type of coding is particularly advantageous for recognition of specific patterns [18].

## 12.4 Gestalt-Filters for Complex Environmental Features Can Be Realized by Suitable Neural Networks

A characteristic feature of neural architecture is that particular arrays of sensory receptors provide an input for a subsequent neuron. Such arrays of receptors are referred to as the **receptive fields** of neurons, which in sensory systems correspond to particular fields of sensory input. The spatial organization of a receptive field mirrors the kind of circuitry by which a neuron is connected with the sensory periphery. Receptive fields can be of various size and spatial structure. In the simplest case, a receptive field can be a summation array of receptors from which a neuron receives an excitatory input. Many neurons have receptive fields with an antagonistic concentric infield-surround structure, through which the neuron integrates excitatory afferents from one field (e.g., the infield) with inhibitory afferents of the other field. By suitable neural interconnections, receptive fields of higher-level neurons can take on very specific forms of spatial organization (Fig. 12.5).

Neurons with a suitable structure of receptive fields can in turn be considered as basic elements of neural circuitry whose higher-level neurons act as **Gestalt filters**, i.e., they react to specific properties of the input that are invariant under certain classes of transformations. For instance, the response behavior of a neuron can exhibit the characteristics, as in frogs, that it reacts only to small stimuli bearing a negative contrast to its environment and moving with a particular speed (thereby serving as a Gestalt filter for a typical prey animal, namely a fly). Through an increasingly higher specificity neurons can become sensitive to complex features of the environment, which allows sensory mechanisms to be placed in the service of specific behavioral action patterns [39].

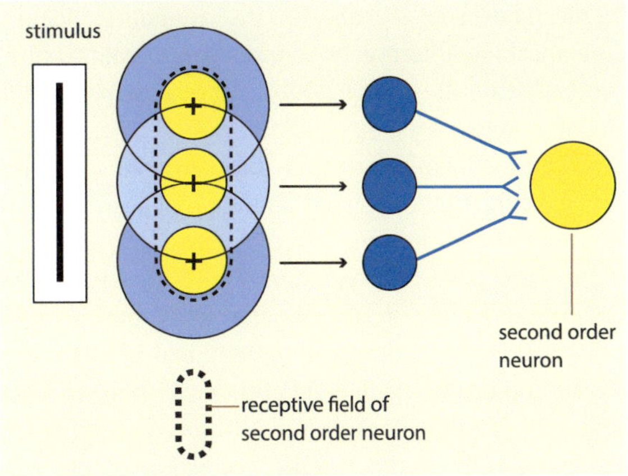


**Fig. 12.5 Circuitry by which simple circular receptive fields** of several neurons are interconnected in order to yield an oblong slit-shaped receptive field of a subsequent neuron (given suitable weights of excitatory and inhibitory areas)

## 12.5 Behavioral Biology of Sensory Systems: Neuroethology and Comparative Sensory Physiology

Empirical findings from neuroethology and comparative sensory physiology show that sensory systems are not only highly adapted to specific physico-biological features of the environment, but are also highly adapted to the behavioral repertoire with which the organism is endowed.

The range of a comparative approach in the study of sensory systems can be illustrated by the investigations of, e.g., complex pattern recognition achievements in pigeons [23], mechanisms of form and distance recognition by electroception in fish [11, 22], color vision in bees and its relation to spectral remission properties of flowers [12], navigation of insects [20], regulation of behavior by pheromones [2, 37], properties of the unusual visual systems of mantis shrimp [13] or jumping spiders [19], as well as the investigations of echolocation of bats [24, 34].

The results of neuroethology and comparative sensory physiology regarding the high specificity of sensory systems again illustrate that these systems are not designed to correctly identify or recover physical features of the external world, but rather to serve specific biological achievements of coupling the organism to its environment. For this purpose, an organism can also take advantage of correlations between input features and environmental aspects that are, from a physical perspective, merely contingent and happen to occur in a certain ecological context only. For instance, magnetotactic bacteria take advantage of a correlation between a magnetic sensory signal and a property of their environment that is physically contingent and holds only in the northern hemisphere; they orient themselves along the geomagnetic field in direction of water with low oxygen content [9].

In neuroethology, the **systems character** of the sensory-motoric architecture is placed in the foreground because behavioral regulations of an organism are usually based on several input systems and hence can only be understood by taking into account the internal **integration** of many sensory input channels.

For instance, in humans the motor control is based on at least four sensory systems: the visual system, the somatosensory system, the vestibular system, and the auditory system. The visual and auditory systems serve, among others, for planning and anticipatory functions, while the somatosensory and vestibular systems supply information on changes of states that have already occurred.

## 12.6 Psychophysics and Perceptual Psychology

Psychophysics and perceptual psychology are devoted to the question of how the relevant perceptual aspects mediated by the sensory input are internally represented. Classical psychophysics has primarily dealt with attempts to construct, for elementary physical magnitudes, so-called **subjective scales** by which subjective relations, such as equality, similarity, distances, or the detectability of differences between elementary physical stimuli, are internally represented. Additionally, psychophysics focuses on investigations of **modality-independent** sensory mechanisms, such as mechanisms of signal-noise discrimination, adaptation, masking, filtering, or contrast amplification, similarity generalization, probability learning, and perceptual learning in general.

Computationally oriented psychophysics, on the other hand, is task-oriented and approaches the study of internal information processing backwards, as it were, from a specific perceptual goal that has to be achieved on the basis of the available sensory information. Accordingly computational psychophysics attempts to determine the abstract processing steps that lead from the sensory input to the corresponding perceptual achievement. Take, for instance, the perceptual achievement that, under appropriate ecological conditions, the perceived color of an object does not change under variations of the color of the illumination ("color constancy"). In the local retinal input, the color signal pertaining to the objects spectral surface characteristics, and the color signal pertaining to the spectral energy function of the illumination are completely confounded. From a computational point of view, the question then arises by which kinds of computational processes these two components can be disentangled in order to yield an illumination-invariant color designator for the object.

"Absolute threshold" and "difference threshold" are the basic concepts of psychophysics.

The **absolute threshold** is defined as the smallest physical value of a stimulus that is perceptually detectable. The **difference threshold** is the smallest physical difference between two stimuli by which the two stimuli can be perceptually distinguished. Stimuli below threshold are referred to as subthreshold.

With respect to the notion of a threshold for conscious perception, the question arises whether stimuli that are not consciously perceived can affect human behavior (**subliminal perception**). As sensory processes themselves (as all neural processes) are not accessible to consciousness, sensory inputs can be chosen (e.g., by using tachistoscopic presentation) that do not yield a conscious perception, but still are able to elicit processes which are internally effective in another way and thereby affect processes underlying perceptual and motoric achievements. This can be illustrated by the finding that sensory stimuli can elicit internal cortical reactions without the person actually consciously being aware of the stimuli. For instance, subliminal stimuli (e.g., in the case of olfaction: pheromones) can regulate internal "attention filters" and thus modulate a sensitivity within one or in a different modality.

The detection and discrimination performance of a sensory domain are determined by its corresponding absolute and difference thresholds. However, these threshold values cannot be determined in a simply way because they vary according to internal states (e.g., random fluctuations of neural states, adaptation, motivational components), external factors (e.g., context conditions), and the particular response criterion chosen for detectability.

The Weber law describes the sensitivity and discriminatory power of sensory systems.

Weber's Law states that (within an average intensity range of physical inputs) the discriminatory behavior is characterized by a **linear relationship** between the respective initial stimulus $S$ and the difference threshold or just noticeable difference $\Delta S$:

$$S = c\,\Delta S,$$

where $c$ is a positive constant. The value of the stimulus increment $\Delta S$ that is just detectable increases as a linear function of the reference stimulus. Each sensory modality has its specific sensitivity and discriminatory power, and hence is characterized by its specific value of the Weber constant $c$.

For instance, under certain context conditions, brightness discrimination will have a Weber constant of about 0.02, for the discrimination of weight ranging around 0.3 kg the Weber constant is about 0.04, for volume discrimination of a sound of 1,000 Hz the Weber constant is about 0.10, and the vibrational discrimination of skin has a Weber constant of about 0.20.

The crucial empirical content of Weber's Law is in the assumption of a linear relationship between $S$ and $\Delta S$ and thus of an internal **contrast coding**. Hence, the relevant sensory information is given by the ratio of two values, whereas absolute values are no longer identifiable. The numeric (dimensionless) value of the Weber constant $c$ itself can vary with contextual aspects. Furthermore, its value depends on the physical units chosen for a description of the stimulus (e.g., whether sound is characterized by sound intensity or by sound pressure) [33]. Also, Weber constants vary with judgmental methods and other parameters of the experimental situation. Because of these dependencies, Weber constants are of rather limited use for a sensitivity comparison of different modalities.

As the discriminatory performance and the sensitivity of a biological sensory system depend on temporal fluctuations (noise), the threshold $\Delta S$ cannot be determined, even under fixed boundary conditions, in a deterministic way. Rather, it must be determined by suitable statistical procedures. Psychophysics has developed a rich set of tools for this purpose, among which signal detection theory and procedures based on **psychometric functions** (Fig. 12.6) play a prominent role [17]. The psychometric function gives for each physical stimulus in the neighborhood of a given standard stimulus (or, as a special case, the threshold) the probability by which this stimulus is judged as "larger" than the standard stimulus.

The psychometric function thus represents a local operation characteristic that describes the discriminatory behavior of a sensory system in the neighborhood of a given reference stimulus. The steeper the slope of the psychometric function is, the better is the discriminatory performance in the respective range. Weber's Law then states that the discriminatory performance decreases with an increase of the value of the physical reference stimulus. This means that the slope of psychometric function decreases when one moves from a lower value of the physical stimulus continuum to a higher one.

Weber's Law refers to the discriminatory behavior only, i.e., the sensitivity of a sensory system, and makes no reference (beyond contrast coding) to the way in which the physical input is subjectively represented. This question of the metrics of subjective scales is addressed by the psychophysical scaling, which originated in the work of Fechner [1].

> The Fechner's Law is based on Weber's Law and expresses a definition of the notion of sensation.

Fechner's Law claims that the subjective value or sensation $P_S$ elicited by a physical stimulus $S$ is given by the logarithm of the ratio of a stimulus to the stimulus threshold $S_0$:

$$P_S = k \log (S / S_0)$$

It was developed by Theodor Fechner (1801–1887) in a (philosophically motivated) effort to prove the measurability of the mental and to determine the quantitative relation between physical stimuli and the resulting sensation. Fechner wanted to show that sensations can be measured, on the basis of an additive cumulation of sensitivity values [14], in the same strict sense as physical quantities. Fechner was inspired to consider a logarithmic relationship between stimulus and perception by D. Bernoulli's investigations on utility, by which Bernoulli had concluded that the slope of the utility function is inversely proportional to the initial value.

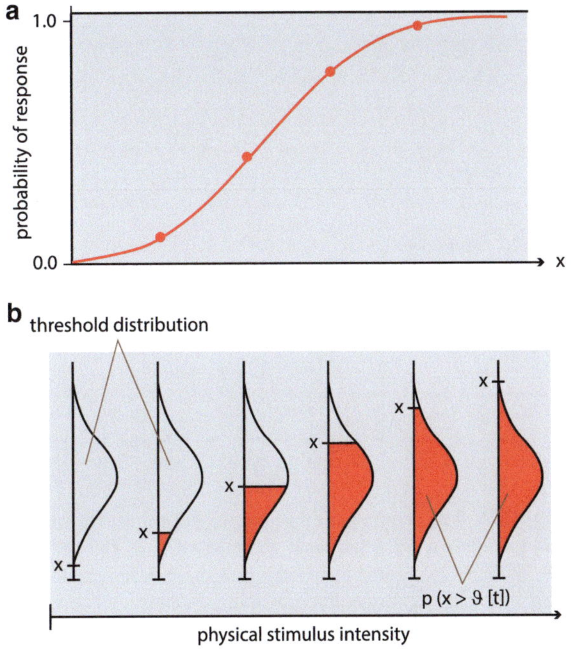


**Fig. 12.6** **(a) A psychometric function in the neighborhood of the absolute threshold.** For each physical stimulus value (abscissa) it gives the probability with which this stimulus value can be detected. (**b**) The psychometric function can be regarded as the result of a mechanism that assigns to a threshold whose value $\vartheta(t)$ at time $t$ can be understood as a normally distributed random variable, the probability of detection. The larger the physical stimulus value, the larger the probability $p(x > \vartheta(t))$ that it is located above the threshold. The psychometric function then takes the form of a cumulative normal distribution

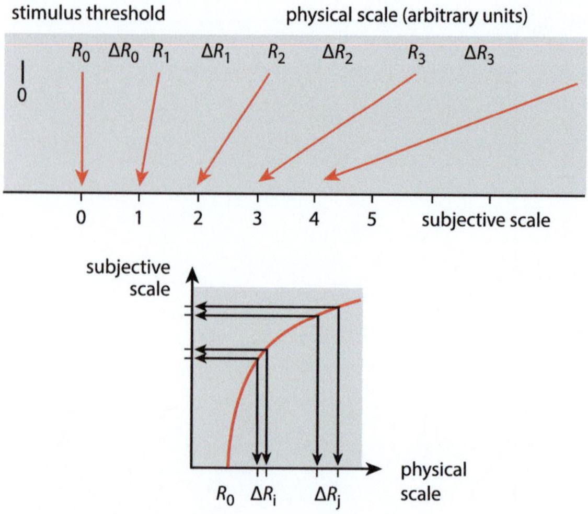


**Fig. 12.7** **The locations of a sequence difference thresholds $\Delta R$ are shown on a physical scale,** starting at the absolute threshold $R_0$. According to Weber's Law these difference thresholds increase proportionally with value of the reference stimulus. According to Fechner's assumption that the difference threshold $\Delta R$ defines the **unit** of the subjective scale, every $\Delta R_i$ yields an increase of the subjective value by *one* unit, hence giving rise to a logarithmic relationship between physical stimulus intensity and the thus defined sensation

Fechner's law is derived on the basis of two components: an empirical regularity and a speculative assumption. The empirical regularity is Weber's Law. Fechner's additional assumption states that just noticeable differences $\Delta S$, though being physically different, are subjectively equal and thus can be regarded as unit of measurement for the construction of a sensation scale. According to Fechner's assumption, just noticeable (or more generally: equally perceived) differences correspond to equal distances on the subjective scale (Fig. 12.7). In other words, Fechner assumed that equal ratios of physical stimuli **are mapped to equal differences on a subjective scale.** This is precisely achieved by a logarithmic function.

The qualitative type of the function between stimulus and sensation as assumed by Fechner mirrors a kind of saturation in the operation characteristics of sensory systems. This saturation behavior is closely related to mechanisms of sensory adaptation, by which the biological relevant physical range is mapped into a much smaller dynamic range of internal codes.

In contradistinction to Fechner, several of his contemporary authors (a.o., Brentano, Merkel, Plateau) proposed not to assign the equal stimulus ratios to equal perceptual differences, but rather to equal ratios of subjective values. This assumption leads to a power function between stimulus and sensation. S.S. Stevens carried out a variety of investigations on the basis of this assumption, so that this relationship between stimulus and sensation is often referred to as **Stevens' power function**. Stevens' Law claims that the subjective value or sensation $P_S$ elicited by a physical stimulus $S$ is given by power function of $S$, with an exponent $n$ varying with the kind of sensory input:

$$P_s = k\,S^n$$

Because the graph of a power function for exponents smaller than 1 is qualitatively very similar to that of a logarithmic function, the appropriateness of the two competing assumptions is difficult to evaluate empirically.

Psychophysics has developed a multitude of methods and judgmental procedures [1, 15] that allow the construction of corresponding (one- or multidimensional) scales for the subjective representation of physical quantities.

The interpretation of the value of the exponent of a Stevens' power function as a characteristic of a sensory modality is subject to similar reservation as the value of the Weber constant, because the value of the exponent depends on the physical units employed (in the case of loudness, the exponent doubles when measuring the physical stimulus in units of sound pressure instead of sound energy).

> Perceptual psychology focuses on perceptual phenomena and achievements that go beyond single sensory modalities and pertain to the internal conceptual organization of the 'perceptual world'.

In dealing with more complex perceptual achievements, perceptual psychology acknowledges that such achievements cannot simply be explained on the basis of sensory codes. One rather has to assume, in line with empirical evidence, that the perceptual system is biologically endowed with a rich structure that cannot be derived, by whatever mechanisms of inductive inference, from the elementary sensory predicates delivered by the senses. In more complex organisms, the sensory input serves as a kind of sign or trigger for

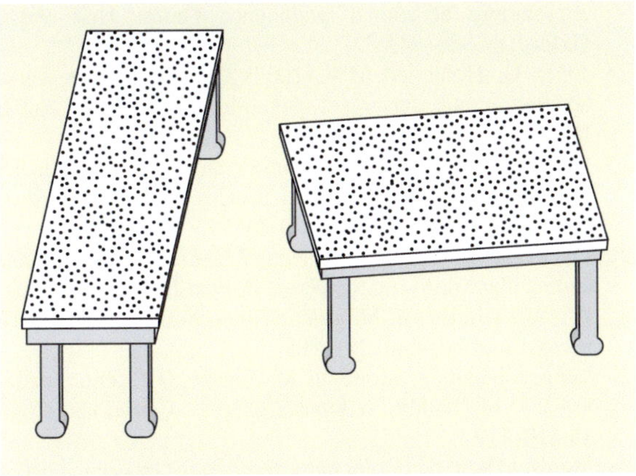

**Fig. 12.8** **In this Shepard-demonstration, the shape and size of the dotted parallelograms,** as simple two-dimensional drawings, are to be compared. They evidently appear to be quite different in form and size. However, if one measures the length and width of these parallelograms, it turns out that they are exactly *identical*. Because the visual system seems to have internalized rules of projective geometry and accordingly to interpret the sensory input in the context of these internalized rules, interpretations are favored where all angles are right angles in three-dimensional space

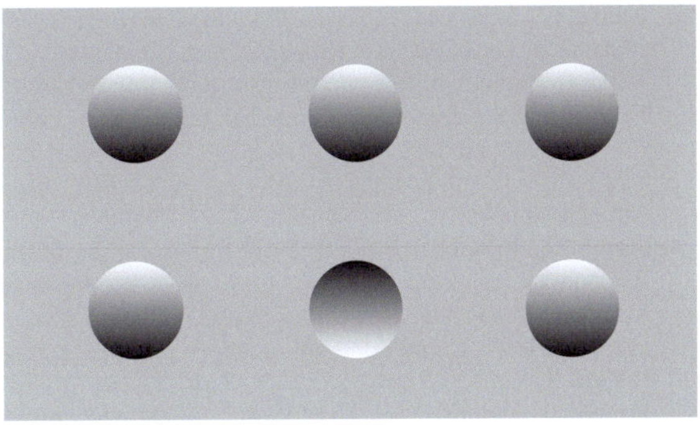

**Fig. 12.9** **In this demonstration the visual input consists of a two-dimensional physico-geometrical pattern of circles,** each with an intensity or brightness gradient. The perceptual impression, however, goes far beyond the information available in the input. What we actually see are three-dimensional indentations or elevations, i.e., convex and concave structures, each with a shading or lightning gradient. By concentrating on any single circle and turning the page upside-down, an indentation will become an elevation and vice versa. The perceptual impression, i.e., the interpretation of the input provided by the visual system is thus changed as if the visual system only allowed the one interpretation by which the light is shining from above

the activation of biologically given conceptual forms, which determine the data format of the computational processes involved [30, 31]. A central achievement of more complex perceptual systems is their ability to detach their modes of operation from the sensory input channels and to achieve **transmodal representations** of the environment that by far exceed the aspects that can be captured by the sensory organs.

The perceptual system would not be able to achieve its complex tasks by means of the available sensory information alone due to the fact that each sensory input is consistent with a large number of possible external scenes. In order to take advantage of the sensory input, the perceptual system must be equipped with some form of in-built knowledge about relevant aspects and regularities of its physico-biological environment. Arguably, it has internalized, in the course of evolution, relevant regularities of the environment as well as of the potential relations between organism and environment that remained stable over very long periods of time (e.g., that surfaces are steady and objects mostly solid, that the spectral energy distribution of solar light is "smooth", that light usually shines from above, and that vision is governed by the laws of projective geometry). With these **phylogenetically internalized regularities** the perceptual system possesses "*a-priori* knowledge", as it were, about the environment that can be exploited for establishing an appropriate relation between sensory codes and the categories and concepts underlying the organization of its perceptual world. Figures 12.8 and 12.9 provide two examples that illustrate how internalized regularities constrain visual information processing.

A distinctive design principle of more complex perceptual systems is their functional compartmentalization and their ability to seal themselves largely off from motivational and emotional states of the organism, and in humans also from conscious volitional interventions. As compared to mental processes of thinking, perceptual processes are rapid, relatively rigid and stereotypic, and barely accessible by conscious control. Only by this kind of design principle can perceptual systems guarantee a quick and stable link to biologically relevant aspects of the physical environment.

## References


1. Baird JC, Noma E (1978) Fundamentals of scaling and psychophysics. Wiley, New York [Textbook]
2. Dusenbery DB (1996) Life at small scale: the behavior of microbes. WH Freeman, New York [Textbook]

3. Kandel ER, Schwartz JH, Jessell TM (eds) (2000) Principles of neural science. McGraw-Hill, New York [Textbook]
4. Rieke F, Warland D, deRuyter van Steveninck R, Bialek W (1997) Spikes: exploring the neural code. MIT Press, Cambridge [Textbook]
5. Smith CUM (2009) Biology of sensory systems. Wiley, New York [Textbook]
6. Armel KC, Ramachandran VS (2003) Projecting sensations to external objects: evidence from skin conductance response. Proc R Soc Lond B Biol Sci B270:1499–1506
7. Atick JJ (1992) Could information theory provide an ecological theory of sensory processing? In: Bialek W (ed) Princeton lectures on biophysics. World Scientific, Singapore, pp 223–289
8. Barlow HB, Foldiak P (1989) Adaptation and decorrelation in the cortex. In: Durbin R, Miall C, Mitchison G (eds) The computing neuron. Addison-Wesley, Reading, pp 54–72
9. Blakemore RP, Frankel RB (1981) Magnetic navigation in bacteria. Sci Am 245:42–49
10. Borst A, Theunissen FE (1999) Information theory and neural coding. Nat Neurosci 2:947–957
11. Bullock TH, Hopkins CD, Popper AN, Fay RR (eds) (2005) Electroreception. Springer, Berlin Heidelberg New York
12. Chitta L, Menzel R (1992) The evolutionary adaptation of flower colours and the insect pollinators' colour vision. J Comp Physiol A 171:171–181
13. Cronin TW, Marshall NJ, Land MF (1994) The unique visual system of the mantis shrimp. Am Sci 82:356–365
14. Dzhafarov EN, Colonius H (2011) The fechnerian idea. Am J Psychol 124:127–140
15. Fabian AS, Wasserman EA (2010) Comparative vision science: seeing eye to eye? Comp Cogn Behav Rev 5:148–154
16. Faisal AA, Selen LPJ, Wolpert DM (2008) Noise in the nervous system. Nat Rev Neurosci 9:292–303
17. Falmagne JC (1986) Psychophysical measurement and theory. In: Boff KR, Kaufman L, Thomas JP (eds) Handbook of perception and human performance, vol I, Sensory processes and perception. Wiley, New York
18. Field DJ (1994) What is the goal of sensory coding? Neural Comput 6:559–601
19. Forster L (1982) Vision and prey-catching strategies in jumping spiders. Am Sci 70:165–175
20. Gallistel CR (1998) Symbolic processes in the brain: the case of insect navigation. In: Scarborough D, Sternberg S (eds) An invitation to cognitive science, vol 4, Methods models and conceptual issues. MIT Press, Cambridge, pp 1–51
21. Gollisch T, Meister M (2010) Eye smarter than scientists believed: neural computations in circuits of the retina. Neuron 65:150–164
22. Hara TJ, Zielinski B (eds) (2006) Fish physiology: sensory systems neuroscience, vol 25. Academic, New York
23. Huber L, Aust U (2006) A modified feature theory as an account of pigeon visual categorization. In: Wasserman EA, Zentall TR (eds) Comparative cognition: experimental explorations of animal intelligence. Oxford Univ Press, Oxford, pp 325–342
24. Jones G, Holderied MW (2007) Bat echolocation calls: adaptation and convergent evolution. Proc R Soc Lond B Biol Sci 274:905–912
25. Lazova MD, Ahmed T, Bellomo D, Stocker R, Shimizu TS (2011) Response rescaling in bacterial chemotaxis. Proc Natl Acad Sci USA 108:13870–13875
26. Lee H, Battle A, Raina R, Ng AY (2007) Efficient sparse coding algorithms. In: Schölkopf B, Platt J, Hoffman T (eds) Advances in neural information processing systems. MIT Press, Cambridge, pp 801–808
27. Lenggenhager B, Mouthon M, Blanke O (2009) Spatial aspects of bodily selfconsciousness. Conscious Cogn 18:110–117
28. Mausfeld R (1998) Color perception: from Grassmann codes to a dual code for object and illumination colors. In: Backhaus W, Kliegl R, Werner J (eds) Color vision. De Gruyter, Berlin/New York, pp 219–250
29. Mausfeld R (2002) The physicalistic trap in perception. In: Heyer D, Mausfeld R (eds) Perception and the physical world. Wiley, Chichester, pp 75–112
30. Mausfeld R (2010) The perception of material qualities and the internal semantics of the perceptual system. In: Albertazzi L, van Tonder G, Vishwanath D (eds) Perception beyond inference. The information content of visual processes. MIT Press, Cambridge, pp 159–200
31. Mausfeld R (2011) Intrinsic multiperspectivity. Conceptual forms and the functional architecture of the perceptual system. In: Welsch W, Singer WJ, Wunder A (eds) Interdisciplinary anthropology: continuing evolution of Man. Springer, Berlin Heidelberg New York, pp 19–54
32. Miller WH, Ratliff F, Hartline HK (1961) How cells receive stimuli. Sci Am 205:223–238
33. Narens L, Mausfeld R (1992) On the relationship of the psychological and the physical in psychophysics. Psychol Rev 99:467–479
34. Popper AN, Fay RR (eds) (1995) Hearing by bats: Springer handbook of auditory research, vol 5. Springer, Berlin Heidelberg New York
35. Rieke F, Rudd ME (2009) The challenges natural images pose for visual adaptation. Neuron 64:605–616
36. Theunissen F, Miller JP (1995) Temporal encoding in nervous systems: a rigorous definition. J Comput Neurosci 2:149–162
37. Touhara K, Vosshall LB (2009) Sensing odorants and pheromones with chemosensory receptors. Annu Rev Physiol 71:307–332
38. Walraven J, Enroth-Cugell C, Hood DC, MacLeod DIA, Schnapf JL (1990) The control of visual sensitivity: receptoral and postreceptoral processes. In: Spillmann L, Werner JS (eds) Visual perception. The neurophysiological foundations. Academic, San Diego, pp 53–101
39. Wehner R (1987) Matched filters – neural models of the external world. J Comp Physiol A 161:511–531

# 13 Olfaction

C. Giovanni Galizia and Pierre-Marie Lledo

## 13.1 The Theory of Olfactory Coding

Problems relating to smell and taste can have a big impact on our lives. Because these senses contribute substantially to our enjoyment of life, our desire to eat, and to be social, smelling and tasting disorders can lead to depression and other mental disorders related to feeling disconnected with the external world. When smell and taste are impaired, we eat poorly, socialize less, and as a result, feel worse. Many older people experience this problem. But smell and taste also warn us about dangers, such as fire, poisonous fumes, and spoiled food. Certain jobs require that these senses be accurate – chefs, winemakers, and firefighters rely on taste and smell. Loss or reduction of the sense of smell (respectively **anosmia** or **hyposmia**) may be due to damage to the olfactory mucosa (e.g., in smoking) or to the olfactory bulbs or tracts after a brain trauma. Central nervous system (CNS) disorders (e.g., some types of epileptic seizures) can cause **parosmia** (disturbed sense of smell). Like olfaction, the sense of taste is important in regulating appetite and to some degree, dietary intake. Loss or reduction of the ability to taste is termed **ageusia** or **hypogeusia.** More than 200,000 people visit a doctor with smell and taste disorders every year in the United States.

Strikingly, olfactory and gustatory systems are maintained by rejuvenating their properties throughout life. Olfactory and taste cells are one of the few cell types of the nervous system to be continuously replaced when the sensory organs become old or damaged. This phenomenon, called adult **neurogenesis**, might help in finding ways to use this potential to replace other damaged nerve cells of the CNS.

Although our sense of smell is our most primal, it is also very complex. Our noses are extremely sensitive for some substances (e.g., thiols, which are added to gas to render it detectable), and anosmic to others (e.g., CO or $CO_2$). To identify the smell of a rose, the brain simultaneously analyzes over 300 odor molecules. An olfactory perception also depends on internal brain states: The aroma of a baking apple pie sends a pleasant message when someone is hungry and an opposite message of disgust when that same person has just finished a six-course meal! Sigmund Freud, addressing the members of the Vienna Psychoanalytical Society, said about olfaction: "*the organic sublimation of the sense of smell is a factor of civilization*". Animals need the sense of smell to survive. Although a blind rat might survive, a rat without its sense of smell cannot mate or find food: it will die as a pup, incapable of finding the mother's nipple. Many species find their mates using olfactory cues (pheromones and other odorant molecules), and most odors are, in fact, organic substances, i.e., produced by other living species of the planet.

### 13.1.1 What Are Odors? The Nature of the Stimulus

Odorants are small chemicals released by a substrate and transported in the medium (generally air or water) that can be perceived by an animal. They range from

C.G. Galizia (✉)
Department of Neurobiology, University of Konstanz, Universitätsstr. 10, D-78457 Konstanz, Germany
e-mail: galizia@uni-konstanz.de

P.-M. Lledo
Laboratoire Perception et Mémoire, Pasteur Institute and CNRS, 25 rue du Dr. Roux, 75724 Paris Cedex, France

C.G. Galizia, P.-M. Lledo (eds.), *Neurosciences - From Molecule to Behavior: A University Textbook*,
DOI 10.1007/978-3-642-10769-6_13, © Springer-Verlag Berlin Heidelberg 2013

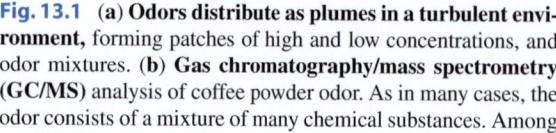

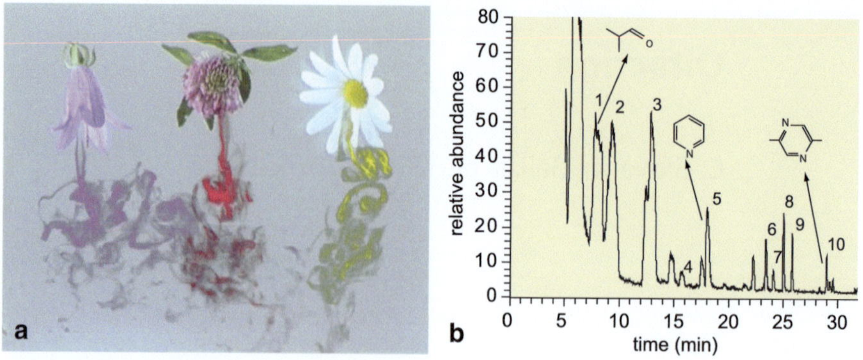


**Fig. 13.1** (a) **Odors distribute as plumes in a turbulent environment,** forming patches of high and low concentrations, and odor mixtures. (**b**) **Gas chromatography/mass spectrometry (GC/MS)** analysis of coffee powder odor. As in many cases, the odor consists of a mixture of many chemical substances. Among the most prominent are the following peaks: (*1*) 2-methylpropanal; (*2*) 2-butanone; (*3*) 2-methylpentanal; (*4*) 3-hydroxy-2-butanone; (*5*) pyridine; (*6*) 4-methylpyrimidine; (*7*) 3-furaldehyde; (*8*) 2-furanmethanol; (*9*) 1,2-ethanediol diacetate; (*10*) 2,5-dimethylpyrazine (b – Courtesy of David Gustav, Konstanz)

small molecules such as gases ($CO_2$ or $H_2S$), simple (aldehydes, esters, nucleotides), to complex organic molecules or even peptides. Most odorous substances have a molecular weight of less than 350 Da, and are generally organic and (for terrestrial animals) hydrophobic. In fact, because the required physico-chemical properties are not so constrained, the vast majority of molecules give rise to olfactory perception and so are used to communicate among individuals. When an odor is produced by a conspecific for communication that benefits both individuals, it is called a **pheromone**. The first chemically identified sex pheromone was bombykol, when Alfred Butenandt isolated 1 mg from 500,000 female abdominal glands of the moth *Bombyx mori* [1]. But many substances are released and communicate between species, with a benefit to either the emitter only (**allomone**) or to the recipient only (**kairomone**). For example, damaged plants release wound substances that attract herbivores, to the sole advantage of the herbivore (though, in some cases, damage induced by an herbivore induces the release of substances by the plant to attract predators and parasites that attack herbivores: so-called tritrophic interactions). Many odors, such as pheromones, have innate meaning – others are learned. A human newborn baby, as most mammals, has the innate behavior to search the mother's nipple. It will immediately learn that particular odor, and use it for individual recognition. A powerful, inborn capacity to quickly learn a complex odor (a typical case for imprinting, see Chap. 25) allows for creating a personal chemical communication channel between the mother and the baby, using a learned odor cue.

Odorants are transported either by diffusion (in which case their concentration diminishes rapidly over distance) or by convection, together with the moving substrate (Fig. 13.1). In the latter case, turbulences lead to generally chaotic sequences of high and low concentration, and odor concentration cannot be used to tell the distance to the odor source. Different animals smell different substances. For example, humans are anosmic to CO and to $CO_2$, but many animals smell these substances. Most perceptually perceived odors are not single substances: coffee, or in fact most food odors, consist of a mixture of hundreds of substances, and the flavor industry is highly involved in finding which of the components are more relevant than others. Often, changing the relative concentration of a few components can have massive effects on the perceived odor. Odors, together with taste (in a narrow sense), thermosensation, and mechanosensation make up our "taste" experience of food (see Chap. 14). In this chapter, we will use "odorant" to refer to the stimulus, and "odor" for the percept created by the brain. In other words, while the wine glass only contains odorants, the exquisite odor of your favorite Burgundy inside is in the *brain* and not in the glass!

Arguably, chemosensation may be the oldest sense in biology, certainly preceding animals and neural systems in evolution. Bacteria sample their environment for chemicals, and swim towards attractive chemicals (odors), and away from aversive substances (**chemotaxis**). Taste and smell are both chemosensory systems: we use "taste" for those instances where direct contact is made with the source of the chemical being sensed, and "olfaction" for **distance chemosensation**. Similarly, unicellular prokaryotes have elaborated capacities to sample the environment. For example, the ciliate *Paramecium* possesses dedicated ion channels that sense $Na^{+}$ and $Mg^{2+}$ concentration, which induce characteristic avoidance behavior. Several prokaryotes also use organic substances for orientation, *Paramecium* uses folic acid to track bacteria. Interestingly, our own neutrophile granulocytes also use chemotaxis to find bacteria; the signal is not folic acid but formylmethionyl-leucyl-phenylalanine. It is fascinating to see that despite this old evolutionary history, some fundamental aspects of olfaction in neurobiology have independently evolved several times – for example, the families of olfactory receptor genes do not all derive from the primordial eukaryotes (unlike, for example, some $Ca^{2+}$ channels), but several such families exist that have evolved independently. At the same time, we find a large similarity in neural coding across species, for example when comparing the mammalian olfactory bulb with the insect antennal lobe, even though they have arisen independently in evolution. This convergence argues for some basic properties that all neural networks need to express in order to compute olfactory signals.

The very first process that occurs in olfaction consists of the interaction between odorant and its cognate receptor. Far from being understood, what is the physical energy that is active here? The lock-and-key principle assumes that the receptor has a pocket that geometrically corresponds to a negative image of the odorant molecule, and the better the fit the higher the response (Fig. 13.2). The odotope assumption states that not the entire odorant molecule needs to bind, but just a particular part of it (say, an aldehyde functional group). Also, the interaction might be purely geometrical, it might involve electric charges (with polar odorants), or even vibrational or resonant interactions. All of these mechanisms, and more, are likely to occur in the one or the other receptor, and no receptor is likely to use all of them.

## 13.1.2 Odor Information: Identity

Which are the characteristic aspects and special traits of an odor? First: identity (this is a rose, or chocolate). Identity is not trivial, because, as seen above, in most cases complex mixtures of many substances create an odor. Concentration is another important characteristic. Consider, for example, the simple question "how many odors can we smell?" This question has at least three answers: how many chemical substances elicit an olfactory experience either conscious or unconscious (i.e., we have receptors that react to them; this question is only useful for pure substances), how many odors can we distinguish (i.e., when presented in short succession we are able to tell that they differ; this question is also relevant to mixtures), how many odors can we remember and name (or associate with a particular event, when presented at different times). We do not know the answer to any of the three questions, neither for humans nor for any nonhuman species, but we do know that for many species the number is certainly higher than several thousand to any of the three.

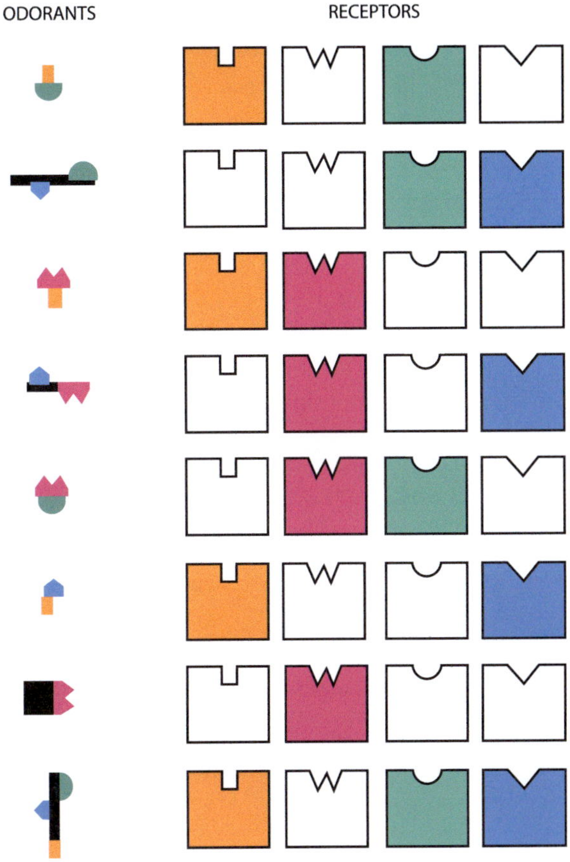


**Fig. 13.2 The odotope model states that each olfactory receptor recognizes a particular feature of an odorant molecule** (e.g., a functional group). Note, however, that interaction with a receptor is never limited to the odotope alone – the rest of the molecule does also influence the binding properties (Adapted from Malnic et al. [2], with permission from Elsevier, Copyright (1999))

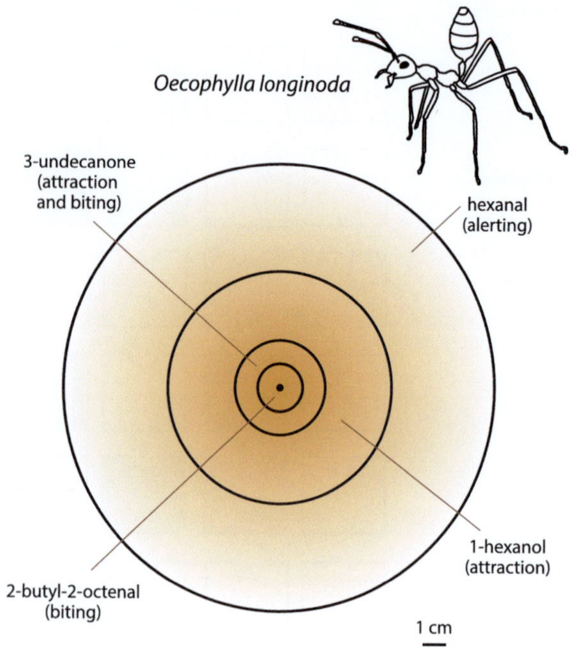


**Fig. 13.3 When an *Oecophylla longinoda* ant releases alarm pheromone,** the different components diffuse to different distances, so that a receiving ant will know the distance from the releasing ant, and responds with an adequate behavior (Adapted from Hölldobler and Wilson [3] with permission)

### 13.1.3 Concentration, Mixtures, and Blends

Olfactory systems need to extract not only the identity of an odor (for example, the odor of a predator, such as a mouse smelling a cat) but also its concentration. In some instances, concentration is crucial for the elicited behavior. In most cases, however, the difficulty in coding lies not in extracting concentration information, but rather in extracting **concentration invariance**, i.e., in recognizing the identity of an odor irrespective of its concentration.

Relative concentration is equally important. Many pheromones are mixtures of two or more substances in a specific ratio. The sexual pheromone of the moth *Heliothis virescens* consists of (*Z*)-11-hexadecenal (main component) and (*Z*)-9-hexadecenal at the ratio 60:1. This pheromone is released into the air, and moths use the relative ratio of these two main components to distinguish it from the pheromone of other moth species that use the same substances, but in a different ratio. Moth sexual pheromone transport is dominated by convection. Ants, however, live close to the substrate, where air movement is negligible. Here, odor movement is dominated by diffusion. Interestingly, some ant species use this property for their alarm pheromone: *Oecophylla longinoda* releases a mixture of 3-butyl-2-octenal, 3-undecanone, 1-hexanal, and 1-hexanol when attacked (Fig. 13.3). Because of volatility differences, some will diffuse faster. Ants close to the attacked individual will smell all substances; they walk to the release site, join the attacked ant and bite. Ants farther away only smell hexanol – thus they know that a sister is being attacked, but some distance away. The behavioral effect of the pheromone here is to be alert – but not to approach the site of disturbance.

Mixtures and blends also need to be distinguished. In a blend, many odorants together create a single percept (such as in coffee). In a mixture, it is possible to extract the information of the component odorants (e.g., peanut butter and jelly sandwich), and these odorants may themselves be blends. One of the capacities of olfactory systems is to extract components from some mixtures, and to create blends from other mixtures. The former is called **analytical coding**, because the components are analyzed, the latter **synthetic coding**, because a synthesis of the blends is made.

## 13.2 Olfaction in Insects

Insect olfactory systems vary enormously. Some species have highly specialized olfactory systems, possibly limited to few odorants with innate significance, such as the carrot psyllid with only 50 receptor cells of possibly only four types [4]. **Oligolectic pollinators** (some moth or bee species that live on nectar of very few flower species) find their preferred flower based on a few **key compounds**. Some jewel beetles (e.g., *Melanophila acuminata*) fly long distances to find burning forests, where they deposit their eggs in burnt trunks. They are attracted by the characteristic smell of guaiacol derivatives, together with infrared radiation that they sense with pit organs in their thorax. The highest sensitivity described so far is that of male moths following the sexual pheromone trail of female conspecifics, often at a distance of several miles. Indeed, the receptors of the moth *Bombyx mori* are capable of responding to single molecules of the pheromone bombykol! (But note that while a single molecule can excite a single receptor cell, this is not sufficient to elicit behavior. Also, *Bombyx* uses the pheromone to know *when* to fly, but uses wind direction to know *where* to fly, a technique that allows for target finding from farther away than a concentration descent tactic). At the other end of the spectrum, some insect species have very generalistic olfactory systems. Honeybees visit whatever flower will deliver nectar: their olfactory system can learn hundreds of odors and distinguish them [5, 6].

## 13.2.1 Olfactory Receptors and Receptor Cells in Insects

### 13.2.1.1 ORs: Odorant Receptors

Olfactory receptor genes in insects code for olfactory receptor proteins. These are 7-TM molecules somewhat similar to G protein-coupled receptors (GPCRs) with the main difference that they are inverted within the membrane: the N-terminal is cytosolic, the C-terminal is extracellular. Species differ in the number of *OR* genes: approx. 62 in *Drosophila,* 163 in honeybees, and over 400 in some ant species, as examples. Each receptor cell expresses one *OR* gene, and in addition the Orco **gene**. This gene belongs to the same family, is also 7-TM, and functions as a coreceptor to the ligand-specific *OR* gene. It is the only gene in the family that is highly conserved across insect species. Its function is not entirely clear yet: it does not bind to odorants, but it is necessary for the translocation of the ORs to the cell membrane, and it is necessary for transduction. It might be a nonselective cation-channel, or in other ways involved in the primary transduction cascade. Thus, the unique OR confers the specific response profile to a receptor cell (e.g., best response to odorant A, weak responses to odorants B, C, no response to odorant D), while Orco acts as **coreceptor**. More receptors in a species thus indicate that more odorants can be perceived, though the relationship is not directly linear: a small number of receptors with overlapping response profiles can be as potent as a larger number with less overlap (think about how many colors we can see with three photoreceptor types). The transduction cascade used by the OR-Orco heteromer is still under debate: currents indicating the involvement of the direct activation of an ion channel have been reported, but the role of a $G_s$-protein cascade and cAMP has also been shown. For pheromone-sensitive cells, $IP_3$ might also be involved, suggesting a PLC/$IP_3$/DAG/$Ca^{2+}$ cascade at least in some insect olfactory receptor cells. Olfactory transduction in insects remains an open field for research (Fig. 13.4).

The *OR* gene family constitutes a single subtree of the gustatory receptors in insects. The $CO_2$ receptor cell in *Drosophila* (and other dipterans) is a special case: the heteromer consists of dGR21a and dGR63a, with no need of Orco. While $CO_2$ is clearly an odor for a fly, and not a taste, the receptors are within the gustatory clade of the gene tree (hence GR), i.e., the $CO_2$ receptor is more related to other gustatory receptors than to the olfactory receptors (Fig. 13.4).

This $CO_2$ receptor has phasic response properties. Honeybees and ants also have good $CO_2$-sensing capacity, but their genome lacks an ortolog for the *Drosophila* $CO_2$ receptor, indicating that $CO_2$ sensitivity might have evolved at least twice. Indeed, the ecological necessity of $CO_2$ sensing is different in social insects, because they monitor the levels of $CO_2$ in their hives in order to control it (a process related to homeostatic control), and therefore need a receptor with tonic response properties.

### 13.2.1.2 IRs: Ionotropic Odor Receptors

A second family of olfactory receptor genes has been called *IR* (from ionotropic receptor). These receptors are related to ionotropic glutamate receptors (iGluRs), but have lost the binding site for glutamate. *Drosophila* has 11 neuron types expressing IRs in 4 coeloconic sensillum types (Fig. 13.4). Each receptor cell expresses at least a pair of IRs, generally one specific IR and either one or both of IR8a and IR25a, suggesting that they work as receptor complexes, most likely as odorant-activated ion channels [10]. ORs and IRs are not coexpressed. IR response profiles to odorants are similarly broad as ORs, and some odorants elicit activity both in IRs and in ORs, though for other odorants responses have only been found in the one or the other family. Typical ligands for IRs include amines and acids, and – unlike ORs – their diversity across species is low. IR receptors have been found in all protostomian species looked at so far, and may indeed be their primordial olfactory receptors, predating insect ORs, and predating the terrestrial lifestyle.

### 13.2.1.3 Other Important Molecules

In addition to IRs and ORs, several other protein families play important roles in olfaction. These include **olfactory binding proteins** (**OBP**) and

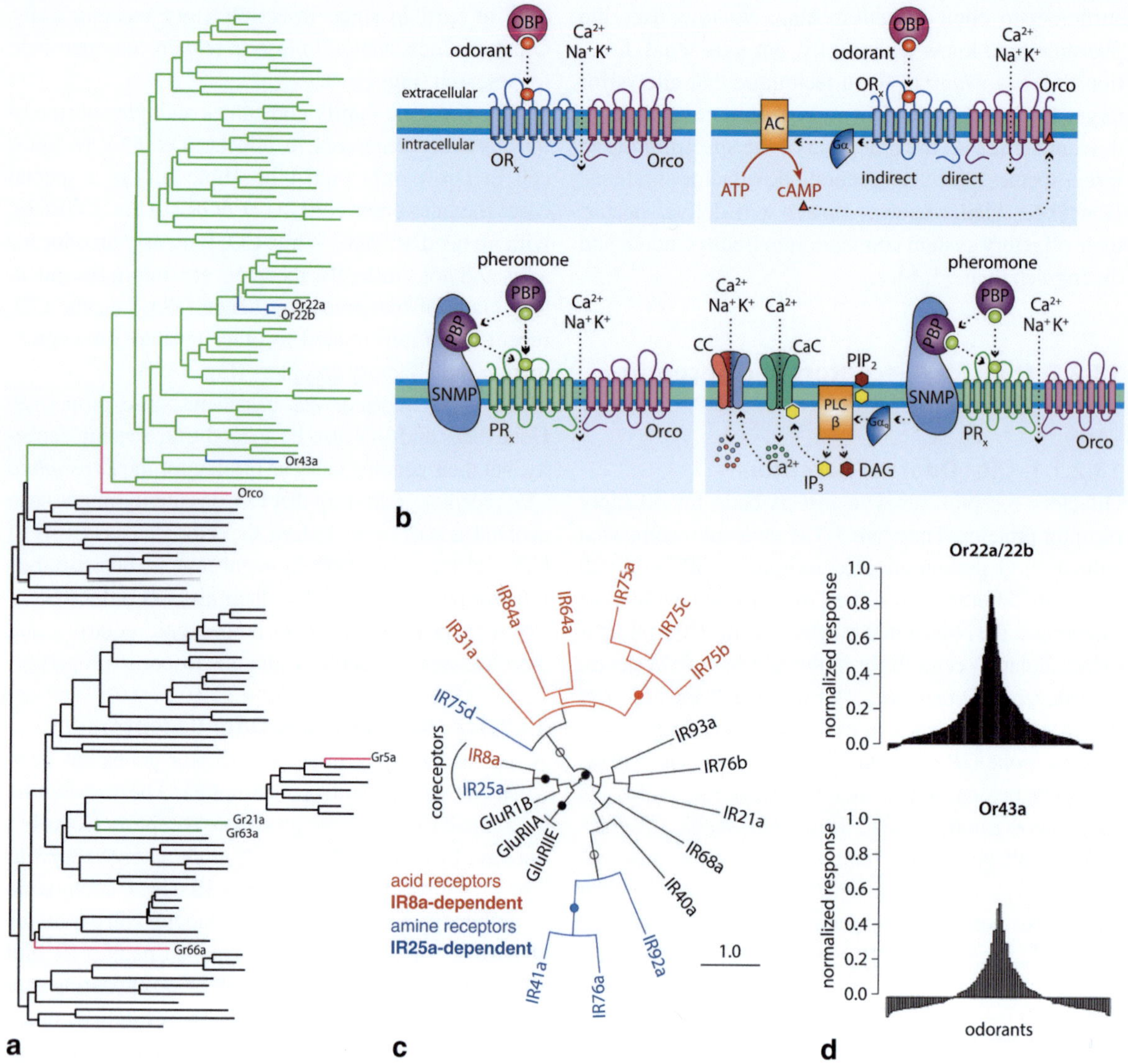


**Fig. 13.4** **Olfactory receptors in insects.** (**a**) Phylogenetic tree of insect gustatory (*black*) and olfactory (*green*) receptors. Sensory cells generally express a receptor and a coreceptor. Olfactory cells express Orco as coreceptor (*pink*), gustatory cells either Gr5a or Gr66a (*pink*). Special cases include Gr21a co-expressed with Gr63a in olfactory receptors sensitive to $CO_2$. Note that olfactory receptors are a subgroup within the gustatory tree. (**b**) The transduction used by olfactory receptors remains unclear: OR and Orco might form an ion channel complex (*upper left*), or combine ion channel and metabotropic signaling via cAMP (*upper right*). In particular for pheromone-sensitive receptors, the involvement of pheromone-binding proteins (PBP) and other membrane proteins (SNMP) has been shown (*lower left*), including metabotropic pathways that activate other ion channels (*lower right*). Note that ORs are different from 7-TM receptors in that the N-terminus is cytosolic. (**c**) Another family of insect olfactory receptors is given by IRs, which are closely related to glutamate-sensitive ion channels (GluR). (**d**) Examples of odor-response profiles for the two receptors Or22a/22b and Or43a (shown in *blue* in *a*) in *Drosophila*. Many different odorants were tested, and their response plotted with the strongest response in the center, and decreasing response strengths to the sides. Or22a has a fairly broad response profile (responds to many odors), Or43a gives inhibitory responses to many odorants (a – Adapted from Robertson et al. [7], National Academy of Sciences, USA, Copyright (2003), b – Adapted from Sachse and Krieger [8] with permission, c – Adapted from Silbering et al. [9] with permission, d – From http://neuro.uni-konstanz.de/DoOR)

**sensory neuron membrane proteins (SNMP).** OBPs are produced by the accessory cells in the sensillum and released into the lymph at high concentrations (10–20 mM). They are globular proteins (12–16 kDa) with three characteristic disulfide bridges across six cysteines. They probably serve several functions which are still debated, but helping the transport of lipophilic odorants across the aqueous

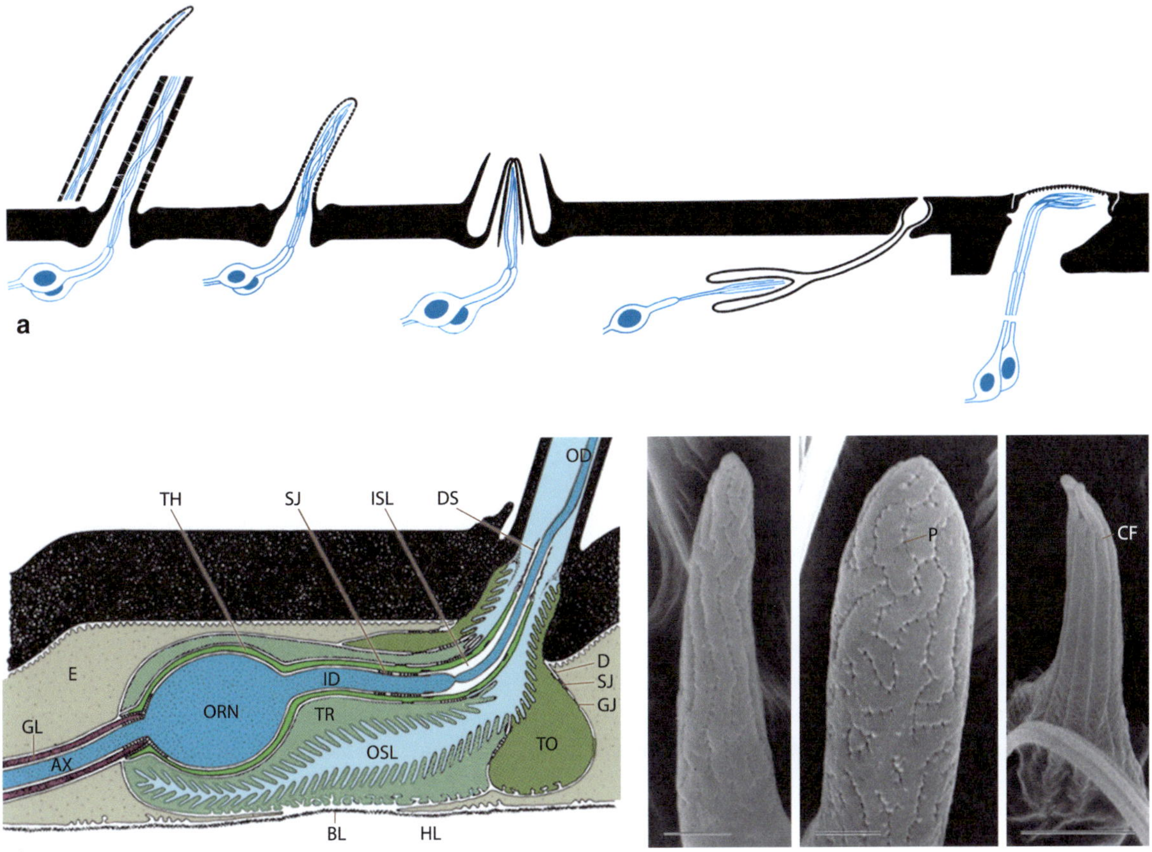


**Fig. 13.5 Olfactory sensilla in insects.** (**a**) Schematic of some sensilla forms, from *left* to *right*: sensillum trichodeum (note that the image is cut – these are long, hairlike sensilla), s. basiconicum, s. coeloconicum, s. placodeum, s. ampullaceum. (**b**) Schematic of an adult sensillum, simplified. *AX* axon, *BL* basal lamina, *D* desmosomes, *DS* dendrite sheath, *E* epidermis, *GJ* gap junctions, *GL* glia, *HL* hemolymph space, *ID* inner dendritic segment, *ISL* inner sensillum symph space, *OD* outer dendritic segment (= ciliary dendrite), *ORN* olfactory receptor neuron, *OSL* outer sensillum lymph space, *SJ* septate junction, *TH* thecogen cell, *TO* tormogen cell, *TR* trichogen cell. (**c**) Scanning electron micrographs of (from *left* to *right*) thin s. basiconicum, large s. basiconicum, and s. coeloconicum. Note the pores (*P*) in the cuticle that allow odorant molecules to enter the sensillum, and the cuticular fingers (*CF*) in the s. coeloconicum (a – Adapted from Keil [12] with permission. b – Adapted from Keil and Steinbrecht [13] with permission from Elsevier, Copyright (1987). c – Adapted from Shanbhag et al. [14] with permission from Elsevier, Copyright (1999))

lymph is uncontroversial [11]. A special family of OBPs is represented by the **PBPs** (**pheromone binding proteins**). In *Drosophila*, the PBP *lush* is necessary for the detection of the sexual pheromone 11-*cis*-vaccenyl acetate by dOr67d. SNMPs are located in the receptor cell membrane, and might belong to the receptor complex, participating in olfactory transduction (at least for pheromones). SNMPs belong to the CD36 protein family, which have two transmembrane domains, and large extracellular binding domains. These proteins typically bind and transport cholesterin, fatty acids, and other hydrophobic molecules.

### 13.2.2 The Olfactory System

#### 13.2.2.1 Olfactory Sensilla

The olfactory organ of an insect is the olfactory sensillum. These are cuticular structures, generally on the antennae (and, in some species, on the maxillary palps). They have many different forms: some are long hairs (**sensilla trichoidea**), some are short hairs, some pegs (**sensilla basiconica**), some are flat pore-plates (**sensilla placodea**), some form invaginations (**sensilla ampullacea**) (Fig. 13.5). Different species have different complements of these sensilla. The functional role of the sensillar form is yet unclear, with only few tentative suggestions: invaginated sensilla may act as a low-pass filter, reducing the effect of fast odor fluctuations (such as for $CO_2$ receptors in ants, which are housed in sensilla ampullacea), and long hair may increase the sieve effect of sensilla, acting like a basket capturing as many molecules as possible (such as pheromone receptors in moths). The basic structure of a sensillum is comparable across all types: cuticular pores in the wall of the sensillum allow odorant molecules to enter the sensillum. The sensillum itself is filled with the aqueous sensillar lymph. Thus, the odorant – which in most cases is a lipophilic small molecule – needs to be dissolved in the watery lymph, possibly with the help of OBPs. The odorant then reaches the membrane of the receptor cell, and interacts with the olfactory receptor complex, initiating the transduction cascade. Some sensilla house a single receptor cell, others house several dozen receptor cells. Many sensilla types are not purely olfactory, but also contain mechanoreceptors or receptors for other modalities (e.g., heat or humidity – modalities that sometimes are treated in the brain in conjunction with odorant information). In addition to the receptor cells, sensilla contain three types of accessory cells: the **thecogen**, the **tormogen**, and the **trichogen supporting cell**. These cells produce the sensillar lymph including OBPs, and create an electrical insulation against the rest of the body.

#### 13.2.2.2 Olfactory Receptor Cells

Insect olfactory receptor cells are **primary sensory cells** that generate their own action potential. Dendrites protrude into the sensillar lymph, with the receptor proteins expressed on the dendritic membrane. Binding to an odorant leads to the transduction cascade being activated. The dendrite is divided into proximal, ciliary and distal segments. The distal segment is bathed in tightly controlled and electrically separate lymph, while the somata and axons are in a different ionic and electric environment. These differences are used for generating the receptor potential. Action potentials are generated at the soma, and transmitted via the axon to the antennal lobe in the insect brain.

#### 13.2.2.3 Antennal Lobe

The axons of the receptor cell enter the brain in a structure called the **antennal lobe** (**AL**). The AL is situated in the **deutocerebrum**, i.e., the neuromere ancestrally associated with the body segment of the antennae (Fig. 13.6). The AL is functionally comparable to the vertebrate olfactory bulb (see below). Generally, all axons that express the same receptor gene coalesce into a single glomerulus, forming a spatial separation corresponding to the different receptor types on the antenna. How many are there? In a fruit fly, roughly 1,200 receptor cell axons converge onto about 60 glomeruli in the antennal lobe in honeybees, about 60,000 receptor cells on each antenna converge onto 160 glomeruli in the male moth *Manduca sexta* (which is particularly sensitive to the female pheromone), 360,000 receptor cells converge onto 60 glomeruli, while the female has the same number of glomeruli, but only 300,000 receptor cells (the 1:1 relationship of receptor genes and glomeruli does not hold in all species: the coleopteran *Tribolium* has many more genes than glomeruli). The antennal lobe itself is a complex neural network. The receptor cell axons make synaptic contact on **projection neurons** – these are neurons that exit the antennal lobe toward higher brain areas. Most projection neurons are uniglomerular, i.e., they branch in a single glomerulus in the antennal lobe. Each glomerulus has 3–5 such projection neurons. In addition, multiglomerular projection neurons branch in many glomeruli, and send their axons to the **protocerebrum**. However, the antennal lobe is not only a feed-forward relay station. The receptor neuron axon terminals also receive synaptic input from **inhibitory local neurons** in the antennal lobes, i.e., neurons that branch between glomeruli, but do not exit the lobe. And they make synapses onto these and other local neurons. The local neurons in turn make synapses onto other local neurons and onto projection neurons. Most local neurons are inhibitory, and use GABA as transmitter, but not all: some use acetylcholine as transmitter, which is excitatory in the insect brain, or other transmitters – modulatory, inhibitory, and excitatory. The picture is even more complicated given that many synapses are dyadic, i.e., a single presynaptic site has two postsynaptic partners, triadic (three partners), or reciprocal: two or three cells have a contact with pre- and postsynaptic structures. In addition, gap junctions form electrical synapses within the antennal lobe! No wonder that the functional network of the antennal lobe is far from understood. The most important role that has been described for the antennal lobe network is **contrast enhancement** (e.g., lateral inhibitory connections reduce activity in some weak glomeruli, making the response pattern more unambiguous), temporal coding (e.g., imposing a temporal structure onto projection neuron firing patterns that might be used by the next brain areas), temporal

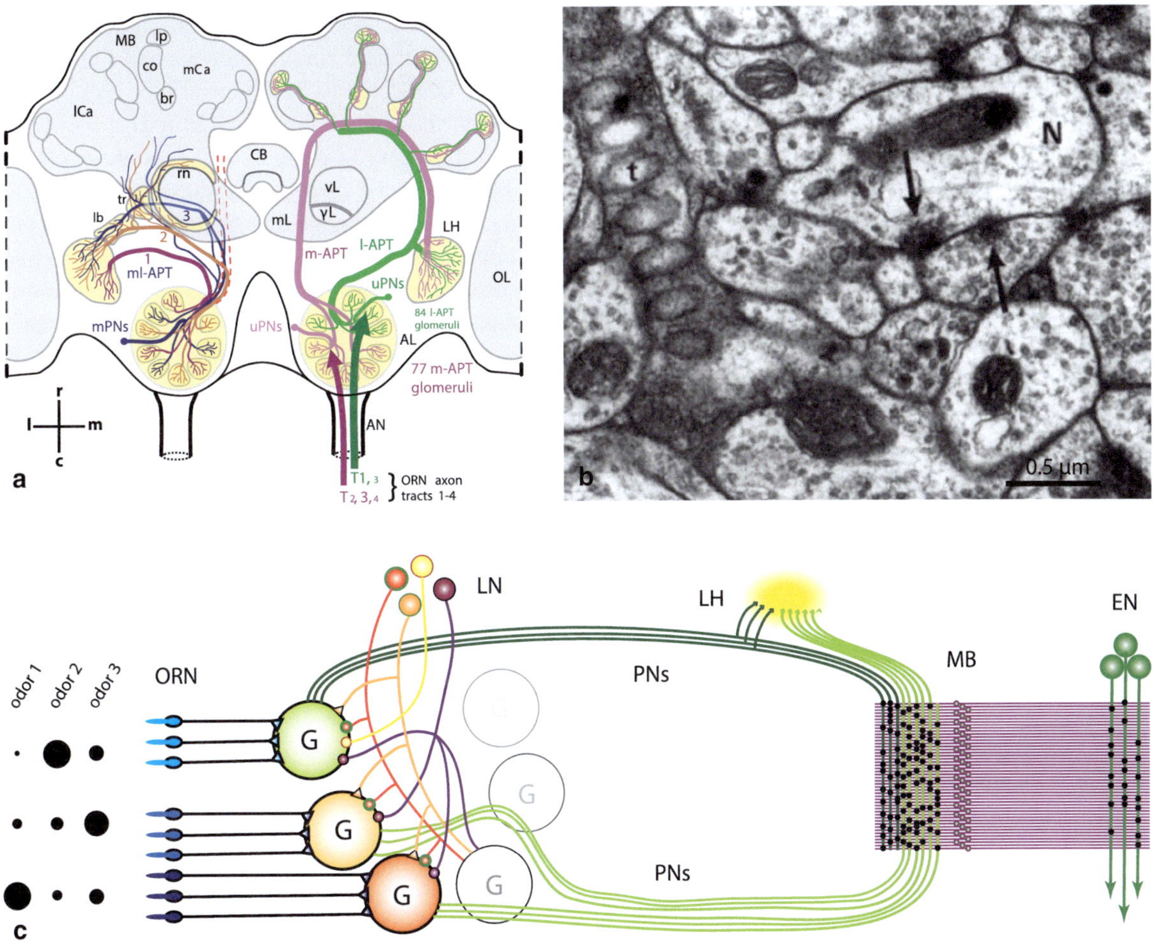


**Fig. 13.6 Olfactory processing in the insect brain.** (**a**) Schematic view of a honeybee brain. On the *left*, multiglomerular projection neurons leave the antennal lobe (*AL*) along three tracts 1,2,3 and innervate brain structures around the α-lobe and the lateral horn (*LH*). On the *right*, uniglomerular projection neurons leave the AL along the two tracts l-APT and m-APT (lateral and median antenno-protocerebral tract) towards the mushroom bodies (*MB*) and the LH. *OL* optic lobes, *lCa* lateral calyx, *mCa* medial calyx, *CB* central body, *vL* vertical lobe, *γL* gamma lobe. (**b**) Electromicrograph from the AL. Note the reciprocal synapses (*arrows*). *N* neurites, *t* trachea. (**c**) Olfactory coding in the bee brain, as a simplified sequential process. From *left* to *right*: each odor elicits a combinatorial pattern of activity in families of receptor neurons (*ORN*). Each family converges onto a single glomerulus (*G*). Glomeruli are interconnected by a diverse family of local neurons (*LN*). This information is transported to higher-order brain structures by projection neurons (*PNs*), including the lateral horn (*LH*) and the mushroom bodies (*MB*). Here, a high number of Kenyon cells generate a new combinatorial code that is read by extrinsic neurons (*EN*) (a – Obtained from Galizia and Rössler [6] with permission. b – From Gascuel and Masson [15] with permission from Elsevier, Copyright (1991))

filtering (e.g., extracting short activity bouts as they would occur when an animal flies through an odor plume), and learning (e.g., enhancing patterns that correspond to a learned, and thus relevant, odorant).

#### 13.2.2.4 Mushroom Bodies and Lateral Protocerebrum

The axons of antennal lobe projection neurons project to the protocerebrum (the most anterior neuromere in the insect brain), and here to the **mushroom bodies** and the **lateral protocerebrum**. The mushroom bodies have been studied in detail, for two reasons: first, early on already they were shown to be important for learning and memory (e.g., by cooling them during a learning experiment, and showing that learning was impaired). Second, they have a highly structured and geometrically regular architecture: their intrinsic cells, the **Kenyon cells** (named after Frederick Courtland Kenyon, 1867–1941, who first described them), form dense bundles of very fine axons that bifurcate to form the mushroom body peduncle and lobes. The cell bodies are arranged in a circular manner, forming a cup-like structure (called the **calyx**), and the axon bundles are like stalks of a mushroom – hence the name. In most insects, Kenyon cells are the largest cell population in the brain: 180,000 in each hemilobe of the bee correspond to more than a third of all bee brain cells. In most insect species, mushroom bodies get olfactory and visual input, allowing multimodal processing and memory (e.g., learning that an odor in a particular visual context has a different meaning from the same odor in a different visual context).

The lateral protocerebrum does not have the geometrical beauty of the mushroom bodies. Maybe this is part of the reason why studying it has been more difficult. Nevertheless, it is an important part of the brain olfactory circuitry: premotor neurons receive input in the lateral protocerebrum, and thus input from the antennal lobe to premotor neurons in the lateral protocerebrum is the fastest connection from a stimulus to a behavioral response. Indeed, many innate odor-responses are thought to take this route.

#### 13.2.2.5 Overall Architecture

The olfactory system is characterized by a sequence of convergence/divergence steps. For example, many thousand receptor cells enter the antennal lobe, but only few hundred projection neurons exit it. One of the main tasks of the antennal lobe is to be a **relay station**, and to increase **signal-to-noise** ratio. Assume that 300 (a typical number) receptor cells converge onto 3 projection neurons – this is a ratio of 1:100, which increases the signal-to-noise (for stochastic noise) by 1:10 (square root of 100)! In addition, the antennal lobe does preprocessing and filtering, also based on previous experience (i.e., learning). The next step is one of divergence: a few hundred projection neurons synapse onto thousands of Kenyon cells in the mushroom bodies (approx. 800 onto 180,000 in the bee, and 180 onto 2,500 in the adult *Drosophila*). These Kenyon cells can extract complex activity patterns across glomeruli, and because there are so many of them, many different patterns can be extracted, explaining the very high number of odors that can be distinguished. The Kenyon cells in turn converge onto very few output cells (in the range of 10–100), corresponding to behavioral responses (e.g., aversion/escape, or attraction/approach, or food/eat). In computational terms, this sequence of divergence and convergence has been likened to a **support vector machine**. Assume a simple integrate and fire neuron would be connected to the antennal lobe output neurons, with the task to fire if and only if a particular pattern of activity is present. This neuron would be capable to extract a very specific subset of combinatorial patterns (mathematically, all linear dissections of the multidimensional space spanned by glomerular dimensions, see Chap. 30). By expanding the number of inputs including Kenyon cells into the circuit, non-linear dissections are possible, giving the system more power to extract components in mixtures, to detect odors in varying environments, to be concentration insensitive or concentration selective, depending on the situation.

### 13.2.3 Physiology of the Olfactory System

#### 13.2.3.1 Physiology of Receptor Neurons

The characteristic of olfactory receptor neurons is their odorant-response profiles: a receptor neuron will respond to some odorants, and not to others. Their large spectrum of odor sensitivity supports a fuzzy concentration coding. Responses are graded: some odorants elicit strong responses at a low stimulus concentration already, others need higher concentrations to elicit responses. Remember that odorant concentration is also an ecological parameter: the pheromone plume of a female moth far away upwind has a strong ecological significance at very low concentration, while the pine-tree odor of an entire forest is released with many more molecules into the air. Odorant-response profiles do not only differ in **sensitivity** (the lowest odorant-concentration that elicits a response), but also in **efficacy** (the maximum response at high concentration). Some odorants elicit long-lasting responses even if only a short puff is given, others elicit short responses for short stimuli, and long responses for long stimuli, and others again elicit short responses irrespective of stimulus length (phasic responses). Furthermore, not all responses are positive: many receptor cells are continuously active, and react to some odorants with an activity decrease (inhibitory responses). “Invisible” inhibitory responses also exist: receptor cells without continuous (“spontaneous”) activity can still have inhibitory ligands, but these become apparent only when an excitatory ligand is given at the same time or slightly before: here, we speak of mixture interactions at the level of receptor neurons. The ecological relevance of phasic and/or tonic responses becomes clear when considering $CO_2$ as an odor: the female mosquito *Aedes aegypti* responds to fluctuating $CO_2$ stimuli (phasic receptor responses), which amplifies the response to nonfluctuating lactic acid and ammonia: signals for a breathing human that can be bitten for a blood meal. Ants and bees, however, respond to $CO_2$ with tonic responses, because they control $CO_2$ concentrations in their hives (in a different way, but functionally akin to how our body controls $CO_2$ homeostasis in the blood).

An important physiological property of all receptor cells is **adaptation** to ongoing stimulation. Adaptation allows shifting the dose-response curve to the current situation. For example, if we enter a smelly room, after a while we get used to the odor because our receptors have adapted to it, and will then be able to smell fluctuations at that concentration range. Unlike other sensory systems, however, the molecular cascades involved in insect olfactory receptor neuron adaptation are not known yet.

#### 13.2.3.2 Combinatorial Patterns and Labeled Lines

Because each receptor neuron responds to many odorants, and because most odorants activate several receptors, each olfactory stimulus elicits a **combinatorial pattern** of activity across receptor neurons, which translates into a combinatorial pattern of active glomeruli in the antennal lobe, and across Kenyon cells in the mushroom bodies. Combinatorial patterns greatly enhance the coding capacity of a system. Assume you have 50 information channels (the magnitude of receptor cell types in *Drosophila*). With each receptor responding to a single substance, you could code for 50 odorants, and for all possible combinations (mixtures) of them. Your analytic capacity for these 50 substances would be very high, but you would be anosmic to everything else. Such a system is called a "**labeled line**" system: there would be 50 labeled lines for 50 odorants. No insect olfactory system works like this, but subsystems do: for example, the sexual pheromone subsystem in moths is a labeled line system. In the noctuid moth *Heliothis virescens*, four receptor types respond to pheromone components, and project to four glomeruli that together are called the **macroglomerular complex**. Together, they form a perfect analytical device for conspecific and similar pheromones: the conspecific pheromone consists of two substances activating two of the four glomeruli: ((*Z*)-11-hexadecenal activates the cumulus and (*Z*)-9-hexadecenal the dorsomedial compartment). A closely related species produces similar pheromones, but including (*Z*)-11-hexadecen-1-ol. Thus, activity in the corresponding ventrolateral glomerulus signals that the pheromone mixture comes from a nonrelated female, and the behavioral response is inhibited (Fig. 13.7). Thus, combinatorial patterns across labeled line glomeruli help the male moth to find their females even in an environment where closely related species have similar pheromones.

Most systems are purely combinatorial, without a labeled line component. Here, 50 glomeruli with overlapping and diverse odor-response profiles allow for many combinatorial response patterns. Assume, as a simplifying example, that every odorant would activate three glomeruli, and that responses would be yes/no, and not graded. In this situation, 50 glomeruli could code for 117,600 odorants! Given that responses are graded, and that more than three glomeruli are activated by most odorants, the real number is much higher; however, many odors are coded by multiple patterns (think, for example, about different concentrations of the same odor), and noise reduces the system's resolution, reducing the number again. Because the combinatorial pattern is spatially visible as a pattern of active glomeruli in the antennal lobe, sometimes it is referred to as "**spatial pattern**". However, to really be spatial, the particular position of a glomerulus needs to have a functional reason (e.g., as for the tonotopical arrangement of cells in the auditory system, see Chap. 17). In most cases, in the olfactory system, no functional reason is known for the spatial arrangement of glomeruli, so that the term "**identity code**" is more appropriate [18].

#### 13.2.3.3 Temporal Patterns

Receptor neurons do not only respond with different intensity (i.e., spike frequency) to odors, but also with different temporal properties. We have already seen different properties at the tail end of the stimulus (tonic vs. phasic neurons). In addition, receptor neurons differ in their temporal resolution, i.e., whether they are able to follow fast pulses. Furthermore, response onset time differs. Thus, by measuring response onset in receptor neurons, it is often possible to extract sufficient information to identify the odorant that was presented.

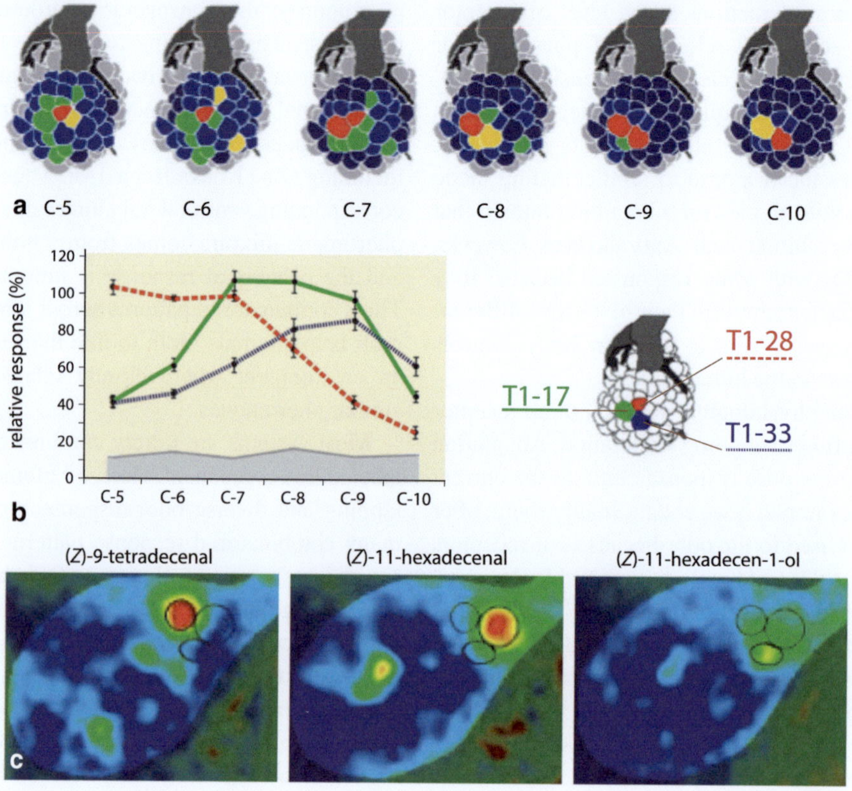


**Fig. 13.7 Combinatorial odor coding in the insect antennal lobe.** (**a**) Glomerular response patterns in the honeybee antennal lobe to aliphatic alcohols, C-5 (1-pentanol) to C-10 (1-decanol). From strong to no response: *red*, *yellow*, *green*, *blue*. (**b**) Responses of the three glomeruli T1-17 (*green*), T1-28 (*red*), and T1-33 (*blue*) to these odors. Note that no glomerulus alone is sufficient to decode the odor information. Also note the gradual shift from T1-28 being dominant (C-5, pentanol) to T1-33 being dominant (C-10, decanol). (**c**) Responses in the antennal lobe of a male moth *Heliothis virescens* to the pheromone components of a conspecific female ((*Z*)-9-tetradecenal and (*Z*)-11-hexadecenal, the two main components of the pheromone), and to a pheromone component of another species ((*Z*)-11-hexadecen-1-ol). The proximity of these glomeruli allows for a rapid discrimination of conspecific or alien pheromone (a and b – Adapted from Sachse et al. [16] with permission. c – From Galizia et al. [17] with permission)

Does the brain use delay information for coding? What we know is that it might, and in theoretical models using this information accelerates odor detection. In addition, the neural network in the antennal lobe introduces temporal fingerprints to the across-glomerular activity patterns in projection neurons. Among the strongest such temporal fingerprints are oscillations: odors induce oscillations (between 20 and 50 Hz) in the antennal lobe-mushroom body network (a property also found in the olfactory bulb of mammals). Oscillations in turn offer at least two more possibilities for coding: spikes across different neurons have a higher probability of being synchronous (and thus of producing a stronger input to a postsynaptic neuron), and the relative position of a spike with respect to oscillation phase can be used to code information (i.e., temporal coding). Whether and how this temporal component across projection neurons is used remains an open field of research.

#### 13.2.3.4 Learning and Memory

Many odors have innate meanings, in particular in insects. Many other odors acquire their meaning due to learning events, by being associated to pleasant (food, shelter) or unpleasant (heat, predation) situations. The mechanisms underlying learning are treated in Chap. 26. For the purposes of this chapter, it is important to realize that odor-learning is spatially distributed. Take, for example, the situation of a honeybee that learns to associate a particular flower odor with a nectar reward. In this situation, a template ("flower odor") needs to be created, and discriminated from other odors ("the wrong flower"). However, the same flower species might grow on different soil, with a slightly different odor, so a certain degree of **generalization** is necessary (see Chap. 28). Thus, in addition to the Pavlovian association (odor-reward association, in its short-term, middle-term, and long-term form, localized in several brain areas, including antennal lobe and mushroom bodies), the olfactory system also generates a sensory memory that allows for better **discrimination** and sufficient generalization. This latter memory is likely localized in the antennal lobe, by modifying synaptic connectivity of local neurons.

## 13.3 Olfaction in Crustaceans

### 13.3.1 The Olfactory Environment of Crustaceans

Most crustaceans live in water. Relevant odors are either prey- or predator-derived substances, or signaling chemicals such as pheromones. For example, the cuticle of adult barnacles contains alpha$_2$-macroglobulins. Larval barnacles use this odor to choose appropriate settlement sites – and predatory snails use the same odor to find their prey. Being waterborne, fluid dynamics is important to understand the turbulent and/or laminar nature of stimuli. Small crustaceans float in the water, and thus live in an environment dominated by diffusion. Large crustaceans live in an environment dominated by turbulent odor plumes. Terrestrial crustaceans use both airborne and waterborne odors: their antennules have evolved for volatiles, while mouthparts and legs are specialized for water-soluble substances.

### 13.3.2 The Crustacean Olfactory System

Crustaceans – being arthropods – share many aspects with insects (which is not surprising, given that they are a paraphyletic group, with hexapods being one branch within the pancrustacean clade). However, they also differ in several aspects, many of which related to their aquatic lifestyle. A fundamental similarity is the segmental structure of the body: each segment has appendages, and afferent axons from the appendages innervate the corresponding ganglion and motor control for these appendages comes from that ganglion. The supraesophageal ganglion consists of three fused ganglia, the appendages being the eyes, the antennules, and the antenna. The subesophageal ganglion consists of six fused ganglia innervating the mouthparts. All appendages in crustaceans have some chemosensory capacity, with sensilla generally being sensitive to chemosensory stimuli and also mechanosensory stimulation (bimodal sensilla, with a wide variety of sensillar morphologies). These "distributed" chemosensilla are also used to locate food, with narrowly tuned olfactory receptors, innervating the respective ganglion.

The only appendages with sensilla that are purely olfactory are the antennules, innervating the deutocerebrum. These sensilla are called **aesthetasc sensilla**, generally arranged in dense tufts at the distal end of the lateral flagellum of the antennule (Fig. 13.8). Not all crustaceans have aesthetasc sensilla, and not all references to aesthetasc may relate to homologous structures across clades. The olfactory neurons are bipolar, with two cilia that arise from the ciliary rootlet. The dendritic cilia bifurcate, in some species only few times, in some many times with up to 30 branches innervating the sensillum to the tip. A single aesthetasc sensillum may contain as few as a single neuron in some species (brachiopods), and up to several hundred in other species (palinurids and brachyurans). The number of aesthetasc sensilla generally increases with each molt, with distal segments of the antennules being lost, and new aesthetascs being added at the proximal end. For long-lived animals such as decapods this increase in sensillar number can be very large, and also implies reorganization in the CNS, where the axons of new ORNs have to find the right glomerulus and integrate into the circuitry. In species such as copepods with sexual dimorphism that use chemical communication, reproductive males (terminal phase) increase the number of aesthetascs enormously with their final mold.

Unlike insects, no olfactory binding proteins have been found in crustaceans, but molecules related to odorant-inactivation are abundant, including ectonucleotidases and amino acid transporters. A typical behavior of a "smelling" crustacean is the antennule flicker, a behavior that might serve several purposes: it might "wash" the sensilla from old fluid, it increases the speed of molecules hitting the sensilla, and it creates a synchrony in afferent signals. Thus, antennule flickering is as integral to crustacean olfaction as sniffing is to mammals or eye movement to visual perception.

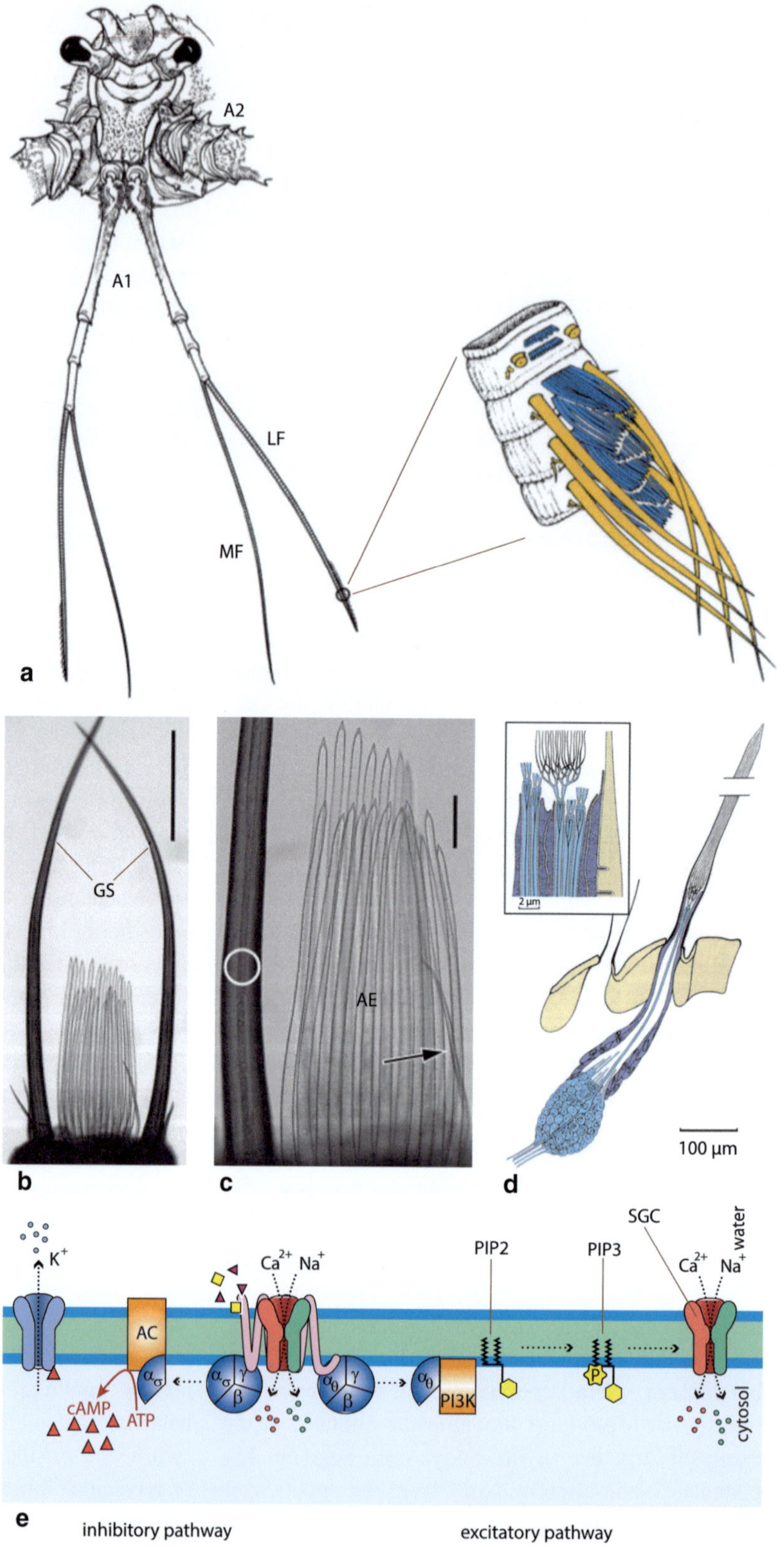


**Fig. 13.8 Olfaction in crustaceans.** (**a**) Morphology of the lateral antennular flagellum of the Caribbean spiny lobster, *Panulirus argus*. A1, A2 first and second antennae, LF, MF lateral and medial flagellum, *blue*: aesthetasc sensilla, *yellow*: guard setas. (**b**, **c**) High power images of aesthetasc sensilla (*AE*) and guard setas (*GS*). (**d**) Simplified longitudinal section through one hair, with only one of up to 370 dendrites shown. Note the many sensory cell somata innervating a single hair (*blue* at *bottom*). Note the fine branching of the outer dendritic section of the dendrites (*inset*). (**e**) Diagram of a working model for olfactory transduction in lobster olfactory receptor neurons (ORNs). Different ligands (*triangles*, *squares*) activate a heteromeric ionotropic receptor (IR) homologous to insect olfactory IRs that also couples to two different intracellular signaling pathways in a ligand-selective manner. Excitatory ligands for the IR activate phosphoinositide signaling (only *PI3K*-mediated signaling shown) through a GTP-binding protein that, once activated, targets a sodium-gated nonselective cation channel (*SGC*) and increases the output of the ORN. Permeant sodium and calcium ions recurrently activate the SGC and amplify the output. Inhibitory ligands for the IR activate cyclic nucleotide signaling through a distinct GTP-binding protein that targets a cyclic nucleotide-gated potassium channel and decreases the output of the ORN. Both pathways working in concert potentially shape the dynamics as well as the intensity of the output (a – Adapted from Schmidt et al. [19] with permission and Grünert and Ache [20] with permission. b, c – from Schmidt and Mellon [21] with permission. d – Courtesy of Barry Ache)

Olfactory receptor genes have been poorly characterized to date. Those found, e.g., in the genome of *Daphnia*, belong to the IR family, the ionotropic chemosensory receptors (see Sect. 13.2.1.2). Crustacean receptors respond to a variety of signals, including amino acids, nucleotides, or other organic material that is present in water. The transduction cascade is also still unclear, but what is known is that in addition to the ionotropic action of IR receptors, several metabotropic transduction cascades coexist in single receptor neurons, with some being inhibitory (mediated by cAMP activating an outward $K^+$ current via a CNG channel, and responding to some ligands) and some excitatory (mediated by $IP_3$ activating a cation current, and responding to other ligands). In addition, ORNs are modulated by centrifugal modulation. Thus, mixture interactions and complex odor-processing is already present at the very periphery.

### 13.3.3 The Crustacean Central Olfactory System

The coding logic in non-insect crustaceans appears to be similar to that in insects: ORNs innervate glomeruli in the paired olfactory lobes of the deutocerebrum. Glomeruli are interconnected by local neurons, and projection neurons send that information to higher-order brain centers. Crustacean glomeruli have a laminar substructure: ORNs innervate the cap, while subcap and base are differentially innervated by distinct subpopulations of LNs and PNs. Neurogenesis in crustaceans is not limited to ORNs: LNs and PNs also form during their entire lifetime, adding complexity to the circuit. Odor coding is likely to follow a combinatorial logic, much like what is known from insects and mammals, with each odorant eliciting a specific pattern of activated and inhibited glomeruli, possibly with complex temporal structures, partly induced by flickering behavior, partly by the neural networks in the olfactory lobe.

Another prominent structure in the deutocerebrum of many crustaceans is the (paired) accessory lobe, which does not receive direct sensory input, but rather input from the olfactory lobe and from other modalities. Structurally, the accessory lobe is formed by many microglomeruli, complex substructuring, and multimodal input, suggesting a higher-order multimodal function for this structure.

## 13.4 Olfaction in Nematodes

### 13.4.1 Olfactory Receptors in Nematodes

The olfactory receptor gene family in nematodes is a large family consisting of more than 1,000 predicted genes, coding for GPCRs that are localized in the olfactory receptor cell dendrites. These genes are unrelated to the mammalian olfactory GPCRs. *C. elegans* has 32 olfactory receptor neurons [22]. Thus, the number of receptor genes and hence proteins is much larger than the number of receptors: in this species, every receptor cell expresses many receptor genes, a situation very different from insects or vertebrates. At the same time, intracellular computation is more complex, and the expression of the receptors themselves is also controlled by food and pheromones. Not all receptors in a given cell use the same transduction cascade, and the way a receptor cell responds to olfactory stimulation also depends on centrifugal modulation by peptides, so that the same cell will generate quite different responses depending on the circumstances. The dominant transduction cascade involves the $G_i$-related proteins ODR-3 and GPA-13, activating a guanylate cyclase (*daf-11* and *odr-1* genes), followed by a cGMP-triggered cation channel (*tax-4* and *tax-2* genes). Activation of a phosphodiesterase reduces cGMP and thus closes the cation channel. Several other pathways participate in olfactory transduction. In addition to GPCRs, other receptors contribute to chemosensation. Receptor cells are ciliated, and are primary sensory cells, generating their own action potentials.

### 13.4.2 Olfactory Coding in Nematodes

Given the small number of cells involved in the nematode system, it is not a surprise that much of olfactory coding that is done in neural networks in more complex nervous systems is resolved with intracellular pathways in the nematode [22]. The receptor cell itself responds to many odors (also due to the many receptors it expresses), and by itself has a behavioral meaning. For example, in *C. elegans*, the AWA cell (there is one on each side of the body) has an attractive effect: when firing action potentials, the animal will move forward. The AWB cell has a repellent effect: when firing, the animal will retract. AWA expresses, among others, the ODR-10 receptor, a receptor with diacetyl

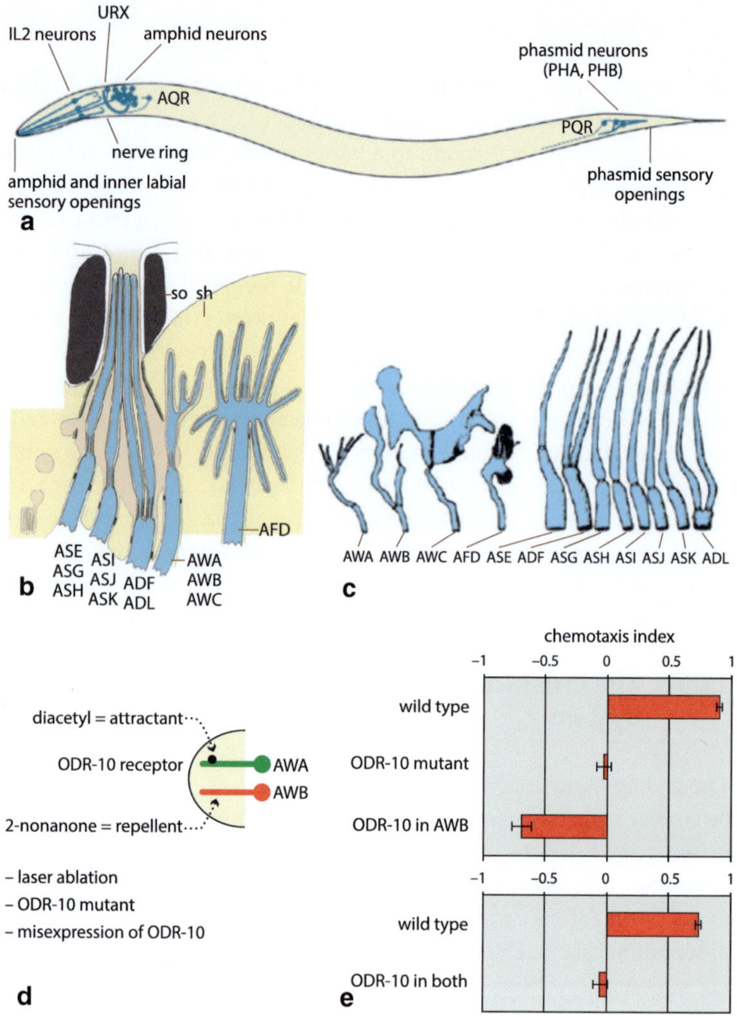


**Fig. 13.9 Olfaction in nematodes.** (**a**) Ciliated chemosensory neurons are found at the rostral end (amphid neurons, each of the two amphids contains 12 associated chemosensory or thermosensory neurons), in the IL2 neurons in the labial organs, in two URX, one AQR, and one PQR neuron with endings within the animal and at the caudal end (each of the two phasmids contains two chemosensory neurons, PHA and PHB). (**b**) Detailed structure of an amphid sensory opening. Note the ciliated endings of the neurons. AWA, AWB, AWC, and AFD are not exposed through the amphid pore. so socket, sh sheath. (**c**) Structure of amphid cilia. (**d**) The neuron AWA responds to the attractive odor diacetyl via the receptor ODR-10, AWB to the repellent odor 2-nonanone. (**e**) Behavioral experiment: wild type is attracted by diacetyl, if ODR-10 is genetically ablated the animal does not respond to diacetyl, and when ODR-10 is expressed only in AWB neurons diacetyl becomes repulsive. With ODR-10 in both AWB and AWA the animal is indifferent to diacetyl (a, c – Adapted from Bargmann [22] with permission. b – Adapted from Perkins et al. [23] with permission from Elsevier, Copyright (1986). e – From Troemel et al. [24] with permission from Elsevier, Copyright (1997))

as potent ligand. Diacetyl is highly attractive to nematodes, but when in transgenic animals *odr-10* expression is removed from AWA, and instead *odr-10* is expressed in AWB cells, the animal will loath diacetyl and avoid this odor (Fig. 13.9). Thus, the meaning of an odor is not encoded in a receptor, nor in the second messenger cascade associated with it, but rather in the identity of the receptor cell expressing it. Why then, does every receptor cell express many receptors? For one, this is the simplest way to code for several attractive or repulsive substances with a limited set of cells. Furthermore, second messenger networks allow for some odor-mixture coding capacity. Many more details in nematode olfaction remain to be explored.

## 13.5 Olfaction in Other Invertebrates

In gastropods, behavioral observations have revealed the importance of olfaction in mediating food finding, conspecific attraction, behavioral avoidance reaction, and homing. The neural substrate for their sense of smell is characterized by an extraordinarily large proportion of the number of neurons relative to the rest of the nervous system, and by the fact that many of them are unusually small in diameter with small projections. Morphological studies have revealed that there exist multiple serial and parallel pathways connecting the olfactory organ (Fig. 13.10), located at the tip of the tentacle, with integrative centers in the central nervous system [25].

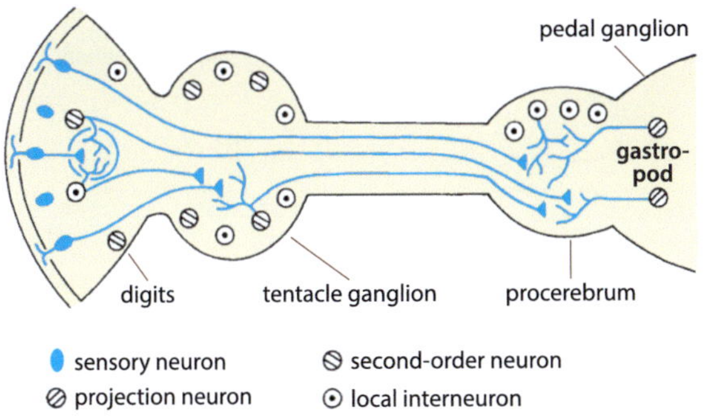


**Fig. 13.10** **Olfactory processing in mollusks.** Sensory neurons are in the tentacle digits. They project their axons to glomeruli within the digits, but also to the tentacle ganglion and to the protocerebrum. Local neurons, projection neurons, and second-order neurons are found in all of these structures, indicating a much more distributed processing network as compared to insects or mammals (Adapted from Chase and Tolloczko [25] with permission)

## 13.6 Olfaction in Vertebrates

The olfactory system in vertebrates regulates a wide range of multiple and integrative functions such as physiological regulation, emotional responses (e.g., anxiety, fear, pleasure), reproductive functions (e.g., sexual and maternal behaviors), and social behaviors (**social chemosignals** are involved in the recognition of conspecifics, family, clan, or outsiders, for example). Birds use odors to navigate, by creating cognitive spatial maps of the odors associated with places that they have been to.

### 13.6.1 The Peripheral Olfactory System

#### 13.6.1.1 The Main Olfactory Epithelium

In mammals, the initial event of odor detection takes place at a peripheral olfactory system, the olfactory epithelium of the nasal cavity, which is located at the posterior end of the nose. The presence of turbinates in the main olfactory epithelium dramatically increases the surface area of the sensory organ and generates air turbulences. The surface area of the olfactory mucosa differs widely among species. It has been used to categorize animals with high olfactory sensitivity due to a large olfactory epithelial surface ("macrosmatic" group) and low sensitivity due to a small epithelial surface ("microsmatic" group). Humans and other primates have been considered microsmats with a concomitant increased emphasis on visual cues. The olfactory mucosal surface in humans is about 300 $mm^2$, which is only 3 % of the total surface area of the nasal cavity. Mice and dogs are much closer to rats (50–60 % of the total surface area of the nasal cavity) than humans or monkeys in respect to the relative amount of the olfactory mucosa within their nasal passages (Fig. 13.11). In rabbits, it is up to 80 %.

The olfactory neuroepithelium is made up of three major cell types: sensory neurons, supporting sustentacular cells, and several types of basal cells including the olfactory stem cells. Sensory neurons are unusual in that they are short-lived cells that exist for only 30–60 days. Once mature, the sensory bipolar neurons extend a single dendrite to the neuroepithelial surface from the apical pole. Numerous **cilia** protrude from this dendrite and extensively invade the mucus lining of the nasal cavity. Odor molecules that dissolve in the nasal mucus bind to olfactory receptors on the cilia of olfactory sensory neurons. Therefore, the first step takes place at the interaction between odorant molecules and their respective receptors in sensory neuron dendrites. From its basal pole, the sensory neuron sends a single axon through the **basal lamina** and **cribiform plate** (of the ethmoid bone) to terminate in the **olfactory bulb**, the first central relay. The unmyelinated sensory neuron axons merge into densely packed fascicles to form the olfactory nerve, which transmits the electrical signals to the olfactory bulb. Throughout the olfactory pathway, a unique population of glia, the **ensheathing cells**, form the bundles of axons that make up the olfactory nerve. The gathering of sensory axons to bundles may lead to ephaptic transmission that allows synchronizing action potentials in neighboring fibers, with submillisecond precision, by an extracellular electrical field.

Axons of sensory neurons in the main olfactory epithelium project to the main olfactory bulb, while axons of sensory neurons in the vomeronasal organ (see below) project to the accessory olfactory bulb. The olfactory epithelium is also involved in the first processing step. ORNs express a range of receptors for neurotransmitters and other signaling molecules, e.g., cannabinoid receptors, cholinergic receptors, purinergic receptors, and many nonolfactory G proteins, suggesting strong modulation already at the input site. For example, activating purinergic receptors decreases odor sensitivity, activating cannabinoid receptors

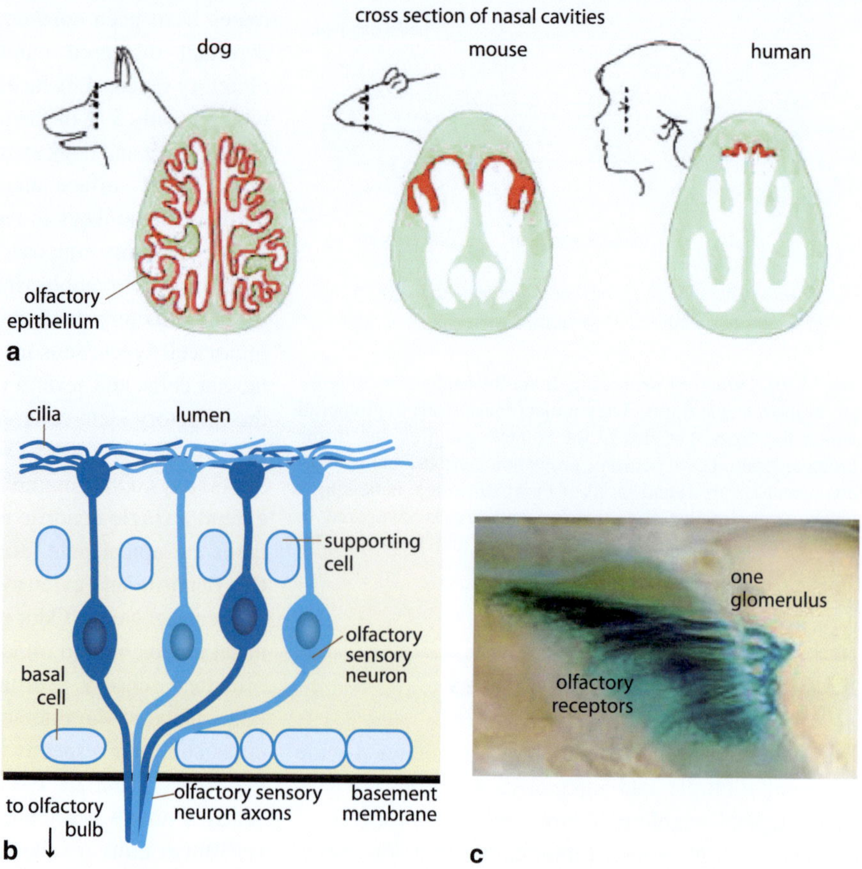


**Fig. 13.11** **Anatomical organization of the mammalian sensory organ.** (**a**) The olfactory epithelium (or mucosa) located in the nasal cavity covers the entire roof and sides in dogs, the roof in rodents, and only a portion of the roof in human beings. (**b**) The mucosa is a pseudostratified epithelium, containing three different cell types from *bottom* to the *top*: (1) basal cells (stem cells), (2) olfactory sensory neurons (bipolar nerve cells), and (3) supporting cells. Sensory neurons project their axons to the olfactory bulb. (**c**) Olfactory sensory neurons that express the same odorant receptor gene (here shown in *blue*) project their axons to either of two glomeruli in the olfactory bulb (only one is depicted in this picture) (a – Taken from Axel [26] with permission from the Nobel Foundation. c – From Mombaerts et al. [27] with permission from Elsevier, Copyright (1996))

enhances odor sensitivity. The cannabinoid system might be related to satiation: hungry animals release 2-arachidonoylglycerol into the mucosa, reducing odor detection threshold and helping the animal find food sources. Cholinergic receptors are related to the autonomic innervation of the epithelium by sympathetic and parasympathetic fibers.

The olfactory organs of fishes are diversely developed. At one extreme they are well developed (macrosmatic) such as in sharks and eels, and at the other they are poorly developed (microsmatic) such as in pike and stickleback. The nasal cavity is lined with the olfactory epithelium, which is raised from the floor of the organ into a series of lamellas to make a rosette. The arrangement, shape, and degree of development of the lamallas in the rosette vary considerably from species to species. Several studies in fishes have shown that the sensory epithelium is not distributed uniformly over the surface of the olfactory lamellas. This suggests that there is no simple relation that links the surface area of the olfactory epithelium to the odor sensitivity. Like other vertebrates, the olfactory epithelium of fish consists of three cell types including the receptor cells, supporting cells, and basal cells. The olfactory receptor cell, which is a bipolar primary sensory cell, sends a slender cylindrical dendrite toward the surface of the epithelium and is directly connected with the olfactory bulb by its axon. The dendrite terminates in an olfactory knob encompassing a variable number of cilia. Like in insects and vertebrates, the information from the receptor cell is conveyed into the olfactory bulb, where signals are processed and integrated.

#### 13.6.1.2 Vomeronasal and Other Systems

Vertebrates have several olfactory systems in parallel (Fig. 13.12). The main olfactory system, the **accessory olfactory system**, the Grüneberg organ, the septal organ, the gustatory system, and the so-called **common chemical sense** mostly carried by **trigeminal sensory neurons**, belong to chemosensation but all differ with respect to receptor molecules,

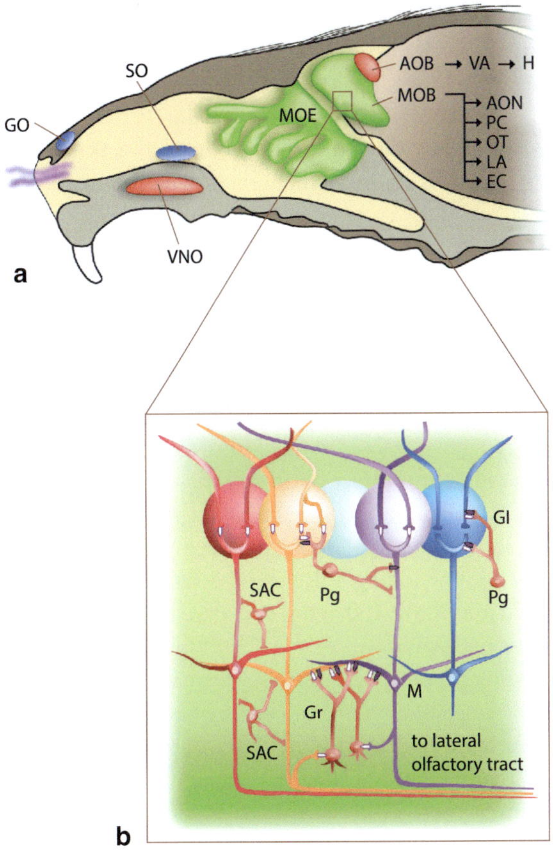


**Fig. 13.12 Schematic organization between all chemical sensory organs and the first central relay.** (**a**) General layout of the nasal chemoreceptive areas. *VNO* vomeronasal organ, *SO* septal organ, *GO* Grüneberg organ, *MOE* main olfactory epithelium. The diagram shows also two central areas: the main olfactory bulb (*MOB*) and the accessory olfactory bulb (*AOB*). The latter conveys olfactory information first into the vomeronasal amygdala (*VA*) and then to the hypothalamus (*H*). Output projections of the MOB target the primary olfactory cortex that include the anterior olfactory nucleus (*AON*), the piriform cortex (*PC*), the olfactory tubercle (*OT*), the lateral part of the cortical amygdala (*LA*), and the entorhinal cortex (*EC*). (**b**) Basic circuitry of the main olfactory bulb. Olfactory sensory neurons that express the same odorant receptor gene project their axons to the same glomerulus in the main olfactory bulb. Four populations of sensory neurons, each expressing a different odorant receptor gene, are depicted by different colors. Their axons converge on specific glomeruli (*Gl*), where they synapse with the dendrite of local interneurons (periglomerular neurons, *Pg*) and second-order neurons (mitral cells, *M*). The lateral dendrites of mitral cells contact the apical dendrites of granule inhibitory cells (*Gr*). Short axon cells (*SAC*) provide inhibition onto mitral cells and granule cells (Modified from Lledo et al. [28] with permission)

receptor cells, and wiring of the receptor cells with the CNS.

The evolution of the accessory olfactory system (the **vomeronasal system**) has long been seen as an adaptation to terrestrial life. Today, accumulated evidence rather contests this assumption. The evolution of a vomeronasal system in aquatic species might rather provide a selective advantage for terrestrial life, and consequently it could have been retained in many species of terrestrial vertebrates. Old World primates, apes, and humans might not have retained a functional vomeronasal system. Alternatively, species without a distinct vomeronasal system may still have an accessory olfactory system intermingled within the main system. Thus, it is yet possible that the accessory system did not "arise" at some point of the vertebrate evolution, but rather it just became anatomically separated from the main system.

The typical tongue flick of snakes is related to the vomeronasal system: the tongue picks up odorant molecules from the surroundings, and is then inserted into two pits in the roof of the palate (hence the need for a split tongue). Here, the vomeronasal system analyzes the chemical composition of the odor on the tongue looking for an exquisite prey.

The **septal organ** – a little-explored organ since its first identification 65 years ago – expresses the receptor SR1, which has an unusually broad response spectrum. This sensorial organ resembles the main olfactory epithelium in many aspects. It is a distinct chemosensory organ located in the mammalian nose (Fig. 13.12), and is essentially constituted of small islands of olfactory neuroepithelium located bilaterally at the ventral base of the nasal septum. It projects to the main olfactory bulb, too. The special properties and function of the septal organ may result from its position within the nasal cavity, the specific olfactory receptor genes it expresses, and/or the specific second-order projection pattern to higher brain centers.

Olfactory sensory neurons that are located in a discrete pocket of the rostral nasal septum, are referred to as the septal organ of Grüneberg (also called **Grüneberg ganglion**). These sensory neurons found in the region of the septum are located in the submucosa, in small grape-like clusters, rather than in a pseudostratified neuroepithelium, as seen in both the olfactory and vomeronasal neuroepithelia. This organ is present at birth and maintained during adult life. Despite their unusual location, axons projecting from the septal organ of Grüneberg neurons fasciculate into several discrete bundles and terminate in a subset of main olfactory bulb glomeruli. These glomeruli most likely represent a subset of atypical glomeruli that are spatially restricted to the caudal main olfactory bulb. The unique rostral position of the septal organ of Grüneberg suggests that this organ may be functionally specialized for the early detection of biologically relevant odorants. The glomerular targets of the septal organ of Grüneberg are structures previously proposed to be associated with suckling behavior. This suggests that this peculiar olfactory neuronal population plays a sensory role, possibly linked to chemoperception, although other functions such as response to temperature or tactile stimuli cannot be yet ruled out.

Which substances act as pheromones in vertebrates? In mice, an important signal consists in small MHC peptides. These peptides are found in urine, and derive from the major histocompatibility complex of the immune system (MHC), which is highly diverse across individuals. Mice use these peptides to smell genetic relatedness. Typically, $V_2R$ receptors (see below) are used to sense MHC peptides. Formyl peptides are another family of odorants used to assess the health state of conspecifics, since they are bacterial degradation products. The cognate FRP receptors have been identified in rodents, so far. Fish use prostaglandins and steroids as pheromones.

#### 13.6.1.3 Trigeminal System

The trigeminal nerve is the principal sensory innervation of the head. It supplies most of the face to near the vertex of the skull. The trigeminal nerve is the somatic afferent nerve dedicated to the mucosa of the nose, mouth, and eyes. It is a sense of alert which prompts to react when the mucosa is endangered. As such, the nasal trigeminal chemoreception conveys important information about the physical and chemical qualities of inspired air (e.g., irritants). For instance, "nasal pungency" refers to the nasal trigeminal impact of inhaled air pollutants as well as spicy foods and selected chemical compounds. Diverse sensations as cooling, numbness, tingling, itching, burning, and stinging are all conveyed by the trigeminal system. Most of the time, information conveyed by the trigeminal nerve is not differentiated in psychophysical testing to the olfactory information. Also, because the trigeminal system concerns collectively nasal, oral, and facial chemoreception, it is generally qualified as a "common sense".

### 13.6.2 Olfactory Receptors and Receptor Cells

At least five receptor gene families are currently known in vertebrates, though more may be waiting for discovery (Fig. 13.13). All of them show evidence of positive selection and dynamic evolution, with high species specificity of the sequences. These families are the "odorant receptors" (ORs, the first family to be found and probably the oldest together with $V_1R$), two families of vomeronasal receptors ($V_1R$ and $V_2R$, which are called ORA and OlfC in fish, respectively), trace amine-associated receptors (TAARs), and formylpeptide receptors (FPRs, derived from the immune system and coopted to olfaction in rodents). All of these are GPCRs. Gene family sizes vary for receptor type and species from none to thousands: for example, humans have no functioning $V_2R$ genes but a few pseudogenes, while mice have 100 genes and more than 200 pseudogenes in this family.

Vomeronasal receptors are activated by sex- and species-specific chemical cues such as those found in the urine or saliva. They allow recognition of predators, competitors or potential mates. $V_1Rs$ respond to multiple cues of different sources (predators, species, sex) while $V_2Rs$ are activated by a small number of cues of particular type. Because $V_1Rs$ and $V_2Rs$ activate distinct neural pathways both in the vomeronasal organ and in the brain, this may reflect a difference in the type of information that they encode. In the mouse olfactory epithelium, the function of TAARs has been discovered recently. This receptor type is also present in humans, mice, and fish. Like their counterpart odorant receptors, TAARs are expressed in unique subsets of neurons dispersed in the epithelium. This new family of odorant receptors is engaged in social cues that trigger innate responses. Some TAARs recognize volatile amines found in urine: one detects a compound linked to stress, whereas the other two detect compounds enriched in male versus female urine – one of which is reportedly a pheromone. Remarkably, the evolutionary conservation of the TAAR family suggests a chemosensory function distinct from odorant receptors, but associated with the detection of social cues.

Rodents have about 1,000 odorant receptor genes. This constitutes the largest gene family so far in the mammalian genome (about 1 % of the total genome), perhaps in any genome. These genes have a compact

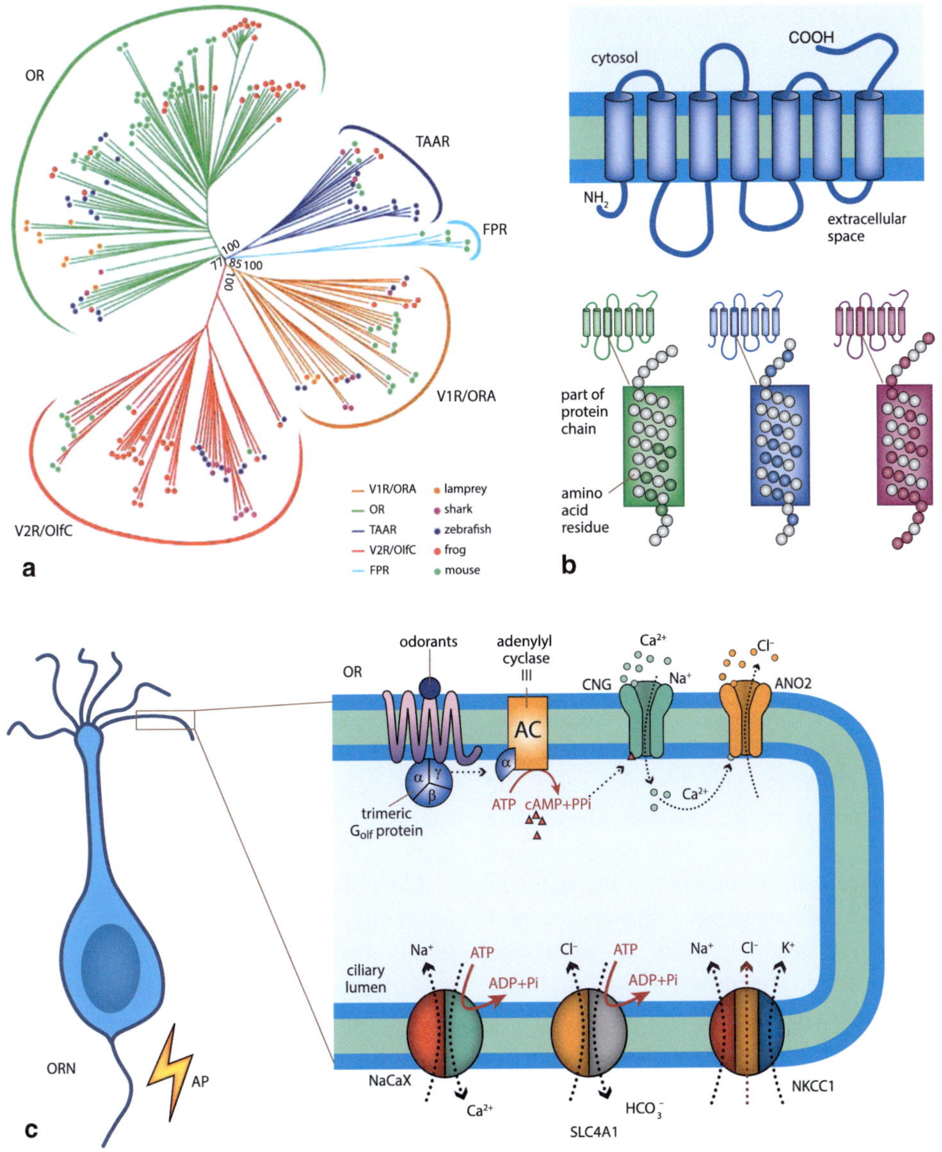


**Fig. 13.13** **Vertebrate olfactory receptors: from genes to sensory neurons.** (**a**) Phylogenetic tree of five olfactory receptor gene families coding for the expression of odorant receptor (*OR*), the trace amine-associated receptor (*TAAR*), the formyl peptide receptor (*FPR*), and the vomeronasal receptors (V1R and V2Rs). (**b**) Molecular structure of a human olfactory receptor protein. The amino acid chain passes through the cell membrane seven times, with the N-terminal starting extracellularly while the C-terminal ends inside the cell. (**c**) Activation of odorant receptor (*OR*) initiates a voltage depolarization that generates action potentials (*AP*) in the olfactory receptor neurons (*ORN*). *Inset*: Binding of odorant compounds to an odorant receptor (*OR*) initiates a transduction cascade involving a G protein ($G_{olf}$) and activation of adenylate cyclase 3 (*ACIII*), which in turn produces the second messenger cyclic AMP from ATP. cAMP binds to a cyclic nucleotide-gated (CNG, showed in *green*) channel and results in the influx of cations ($Na^+$ and $Ca^{2+}$), which depolarize the cell membrane. $Ca^{2+}$ can also activate a $Ca^{2+}$-dependent $Cl^-$ channel (ANO2). Olfactory receptor neurons maintain a high intracellular $Cl^-$ concentration because they express the NKCC1 $Cl^-$ pump and the SLC4A1 $Cl^-$ exchanger. The NaCaX exchanger guaranties $Ca^{2+}$ extrusion from inside the cell (a, c – Adapted from Manzini and Korsching [29] with permission. b – Adapted from Axel [26] with permission from the Nobel Foundation)

gene structure and are scattered throughout the genome in clusters of various sizes. This extremely large repertoire of odorant receptors is undergoing rapid evolution, with at least 20 % of the genes lost to frameshift mutations, deletions, and point mutations that are the hallmarks of **pseudogenes**. Facing a changing environment, this characteristic may reflect the pressure exerted on a gene family to diversify and generate large numbers of new receptors that might confer new selective advantages. Interestingly, approximately 50 % of human odorant receptor genes carry one or more coding region disruptions and are therefore considered pseudogenes (leaving about 350 functional genes for our species). There has been a decrease in the size of the intact odorant receptor repertoire in apes relative to other mammals, with a further deterioration of this repertoire in humans. Since such decline occurred concomitant with the evolution of full trichromatic vision in two separate primate lineages, it is possible that the weakening of olfaction resulted from the evolution of full color vision in our primate ancestors. However, several specific features such as the structure of the nasal cavity, retronasal smell (i.e., food-derived odors that enter the nasal cavity from the palate; they are perceptually different from the same substance when given orthonasally, i.e., from the nose), olfactory brain areas, and language call for reassessing the status of the sense of smell in human beings. Olfactory receptor genes form also the largest category of genes with monoallelic expression. This principle was originally demonstrated by single-cell rtPCR on pools of olfactory sensory neurons using limiting dilution and polymorphic alleles, and led to the 'one receptor–one neuron' hypothesis. However, an alternative model proposing that a single neuron expresses during differentiation zero, one, or a few odorant receptor genes has challenged this view. According to this hypothesis, termed "oligogenic expression", the developmental phase of oligogenic expression is followed by positive and negative selection for cells with only one expressed receptor. $V_2R$ genes are not monogenic: each receptor cell expresses a pair of receptors, often with a conserved partner. Possibly, these receptors work in pairs forming a receptor complex. Little is known about expression rules for the other receptor families.

Odorant receptors can bind a number of volatile compounds with rather moderate affinity despite the overall high sensitivity of the system. Since each individual receptor is substantially cross-reactive for different ligands, the receptor repertoire evolves according to the concentration and the mixture of odorants. The transduction of ORs uses the heterotrimeric G-protein subunit $G_{olf}$, type III adenylyl cyclase (ACIII), and a CNG ion channel to mediate odor detection. Inflowing calcium ions open calcium-activated chloride channels, which due to the specific chloride resting potential in the ORN/mucosa system leads to $Cl^-$ outflow, and thus further depolarization. In addition to this major pathway, several other second messenger cascades (e.g., $Ca^{2+}$, $IP_3$, or cyclic GMP) that are activated upon odorant detection are thought to regulate secondary events such as odorant adaptation, for instance. Some receptor neurons use other transduction cascades, including phospholipase C (PLC), cGMP-dependent CNG ion channels, receptor guanylate cyclases (GC-D), and TRP channels (TRPC6 and TRPM5). For example, $V_1R$ use $G_i$ proteins, and transduce via PLC; $V_2R$ use $G_o$ proteins and also transduce via PLC. PLC activates $IP_3$, DAG, and arachidonic acid, which open TRPC2 channels. However, the receptor gene does not have a binding instructive role: TAAR is transduced via $G_i$ and $G_o$ in the Grüneberg organ, but via $G_{olf}$ in the main olfactory epithelium, suggesting that the transduction cascade is more related to the neuron than to the receptor itself.

Electrophysiological studies indicate that odorant sensitivity and the odorant-induced current are uniformly distributed along the cilia, suggesting that all the components of the immediate responses to odorants are localized to the cilia.

Several processes contribute to receptor **adaptation** in order to reduce odorant-responses for long stimuli. Increasing intracellular calcium linked to transduction binds calmodulin. $Ca^{2+}$-bound calmodulin has three actions: it reduces the cAMP sensitivity of CNG channels, closing them and thus terminating the odor response, it activates a phosphodiesterase which hydrolyzes cAMP, and it inhibits ACIII directly. In addition, arrestins are activated that in turn deactivate the olfactory receptors and mediate their internalization, thus reducing their number in the membrane.

In mammals, olfactory receptor cells in the main olfactory epithelium, the septal organ, and the Grüneberg organ are ciliated. In the vomeronasal system, ORNs have **microvilli**. Fish have ciliated and microvillous receptor cells that are all located in a single olfactory epithelium, and in addition a cell type that has cilia and microvilli: the crypt cells. Ciliated cells use $G_{olf}$ and cAMP-gated CNG channels; microvillous cells use $G_o$ and TRPC2 channels. The situation for crypt cells is not yet clear, with $G_i$, $G_o$, and $G_q$ present in different species. The situation in amphibians is particularly interesting: an airborne odor system expresses genes related to terrestrial animals, while the waterborne "nose" expresses fish-related genes.

### 13.6.3 Olfactory Bulb

Upon reaching the olfactory bulb, each sensory neuron axon enters a specific glomerulus and arborizes to form approximately 15 synapses with target dendrites. Axonal projections from the sensory neurons to the olfactory bulb form reproducible patterns of glomeruli in two widely separated regions of each bulb, creating two mirror-symmetric maps of odorant receptor projections. Importantly, it has been shown that odorant receptor identity in epithelial neurons determines not only glomerular convergence and function, but also organizes the neural bulbar circuitry.

There is no strict spatial relationship between the arrangement of excitatory projections of the olfactory sensory neurons in the olfactory bulb and the regions of the mucosa from which they originate (Fig. 13.14). This feature contrasts with the spatial organization of other sensory systems where afferent inputs are organized in a rather precise topographical mode. Similarly, as described below, much evidence indicates that bulbar outputs do not have point-to-point topographical projections to their target structures, which are characteristic of other sensory systems. In mammals, the convergence ratio of sensory neurons-to-olfactory bulb output neurons is very large – about 1,000:1. A bulbar output neuron thus forms its responses to odors from very large numbers of converging inputs, ensuring that postsynaptic averaging increases signal-to-noise ratios. Target finding of receptor neuron axons uses several mechanisms: large-scale arrangement is accomplished by two chemotopic gradients, and small-scale refinement is activity-dependent. The olfactory receptor itself is involved in target finding.

Several observations indicate that descending forebrain axons from various areas can selectively modulate olfactory bulb odorant-evoked responses. These data clearly show, at the very least, that olfactory processing does not only involve simple feed-forward

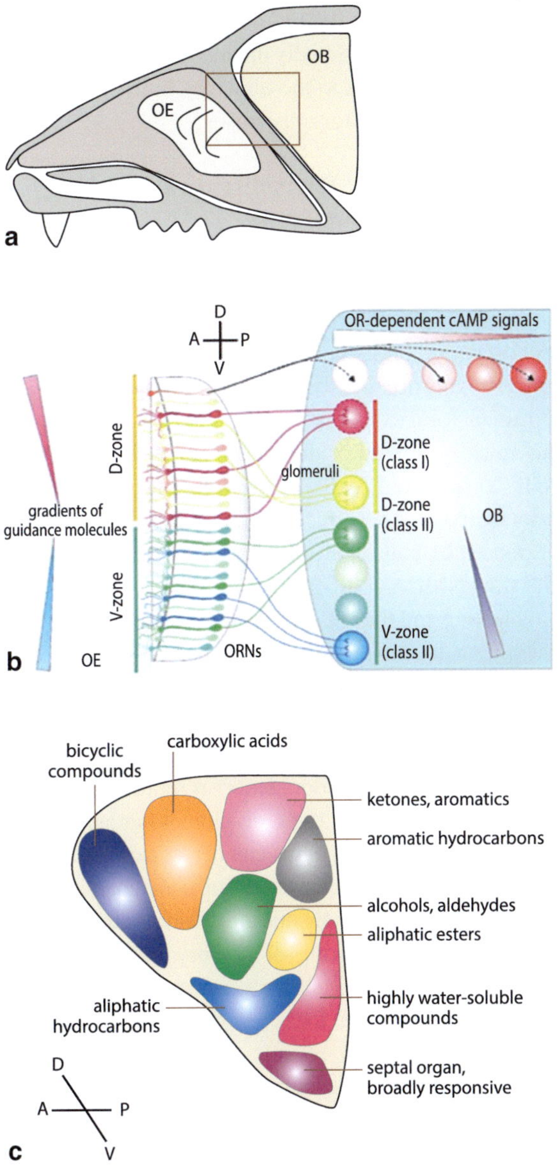


**Fig. 13.14 Olfactory map formation in the mouse.** (**a**) Olfactory receptor neurons located in the olfactory epithelium (*OE*) send axons to the olfactory bulb (*OB*). Square delineates panel B. (**b**) Axonal segregation and olfactory map formation. Olfactory receptor neurons (*ORNs*) axons are guided to approximate destinations in the OB by a combination of dorsal-ventral (*D-V*) patterning and anterior-posterior (*A-P*) patterning. D-V projection occurs based on anatomical locations of ORNs in the olfactory epithelium (*OE*). A-P projection is regulated by OR-derived cAMP signals. The map is further refined in an activity-dependent manner during the early neonatal period. (**c**) Molecular feature clusters of glomeruli. The spatial arrangement of molecular feature clusters in a dorsal view of the mouse olfactory bulb (Adapted from Manzini and Korsching [29] with permission)

pathways. Rather, in real-world situations where information has to be continually updated, forebrain circuits and their projections to the olfactory bulb circuit continuously reset olfactory responses.

#### 13.6.3.1 Synaptic Transmission at the First Processing Stage

Unlike in other sensory neurons, axonal termini of olfactory sensory neurons synapse directly onto second-order neurons within the forebrain. This makes the olfactory bulb the major site of integration for olfactory information. In the glomeruli, olfactory nerve terminals form excitatory, glutamatergic synapses with the apical dendrites of the bulbar output cells (**mitral cells** and **tufted cells**) and with **periglomerular interneurons**. The olfactory nerve-evoked excitatory responses of both neuron types comprise fast AMPA and slow NMDA components. The latter is particularly long lasting and thus plays an important role in the bulbar output by maintaining a pattern of sustained discharge of output neurons.

Remember that a given glomerulus can respond to multiple odorants, and a given odorant activates multiple glomeruli. As a result, odor identity is represented rather combinatorially by patterns of glomerular activation. Because the olfactory system can detect extremely faint signals from the outside world, some mechanisms are necessary to reliably transmit information contained in the odorant-evoked firing of sensory neurons to the brain. In other sensory systems, some devices such as the synaptic ribbons in the retina or the cochlea indeed enable sustained and reliable synaptic transmission. Such a presynaptic specialization is not present in the terminals of olfactory sensory neurons, but a single axon makes multiple synapses. In addition, analyses of olfactory nerve-evoked excitatory responses in olfactory bulb slices have shown a marked paired-pulse depression, supporting that glutamate release from sensory neuron terminals is high under normal conditions.

Since Ramón y Cajal's (1852–1934) pioneering studies, it is known that the main output neurons (or relay neurons) of the bulb consist of tufted and **mitral cells**. Somata of the latter are located in a single lamina, the mitral cell layer (Fig. 13.15). Their primary (or apical) dendrite, extending vertically from its soma, contacts one glomerulus (in rodents; some species have multiglomerular mitral cells), where massive interactions with bulbar interneurons (the **periglomerular cells**) and olfactory nerve terminals occur. Most of the local interneurons have dendrites restricted to one glomerulus and impinge onto olfactory nerve terminals or primary dendrites. In contrast to the primary dendrite, mitral cell secondary (or basal) dendrites radiate horizontally, up to 1,000 μm, to span almost entirely the olfactory bulb. In the external plexiform layer, they interact with inhibitory axonless interneurons, named **granule cells** – the most numerous cellular populations within the bulb. Thus, firing activity of output neurons is controlled both by sensory excitatory inputs and by the intrabulbar circuit, which mainly includes two distinct connections: between primary dendrites and periglomerular cells; and between secondary dendrites and granule cells. The main difference between periglomerular and granule cells is that the former mediate mostly interactions between cells affiliated with the same glomerulus while granule

**Fig. 13.15 The olfactory bulb synaptic organization.** (**a**) Layers and cell types of the olfactory bulb, olfactory nerve, and olfactory epithelium (*top right*). Glomeruli are round neuropil structures (80 μm in diameter in rodents) made of interneuron dendrites (*PGC* periglomerular cells, *SAC* short-axon cells), the apical dendrites of the mitral and tufted cells (*TC*), as well as olfactory receptor neuron endings. SACs connect neighboring glomeruli and sometimes have very long axons – contrary to their historical name. Each glomerulus receives converging axonal inputs from olfactory receptor neurons that form the olfactory nerve. Olfactory receptor neurons marked in the same *color* have one receptor type and project to the same glomerulus. Within each glomerulus, olfactory axons make excitatory synaptic terminals on the primary dendrites of mitral cells and tufted cells, the two types of projection neurons in the OB. Each glomerulus together with ORN axon inputs and associated mitral and tufted cells form an OR channel. Tufted cells project lateral dendrites in the superficial half of the external plexiform layer, whereas mitral cells extend long lateral dendrites to the deeper half of the external plexiform layer. Mitral cells and tufted cells project axons to the olfactory cortex and axons from different parts of the brain (i.e., cortex, hippocampus, etc.), project to the olfactory bulb as well. (**b**) Functional architecture of the dendrodendritic synapse (*gray dotted* region in *a*). Lateral dendrites of the mitral cells shown in *a* (*blue* and *red*), and granule cell dendrites (*yellow*). *Arrows* indicate the signal flow: when an action potential backpropagates to the lateral dendrite, it triggers glutamate release onto granule cell spines (step 1) that leads to GABA release onto the given activated mitral/tufted cell (step 2). This forms the recurrent inhibition. If the granule cell spines activation is strong enough, it might trigger GABA release onto a neighbor mitral/tufted cell, thus supporting lateral inhibition (Adapted from Kuner and Schaefer [30] with permission)

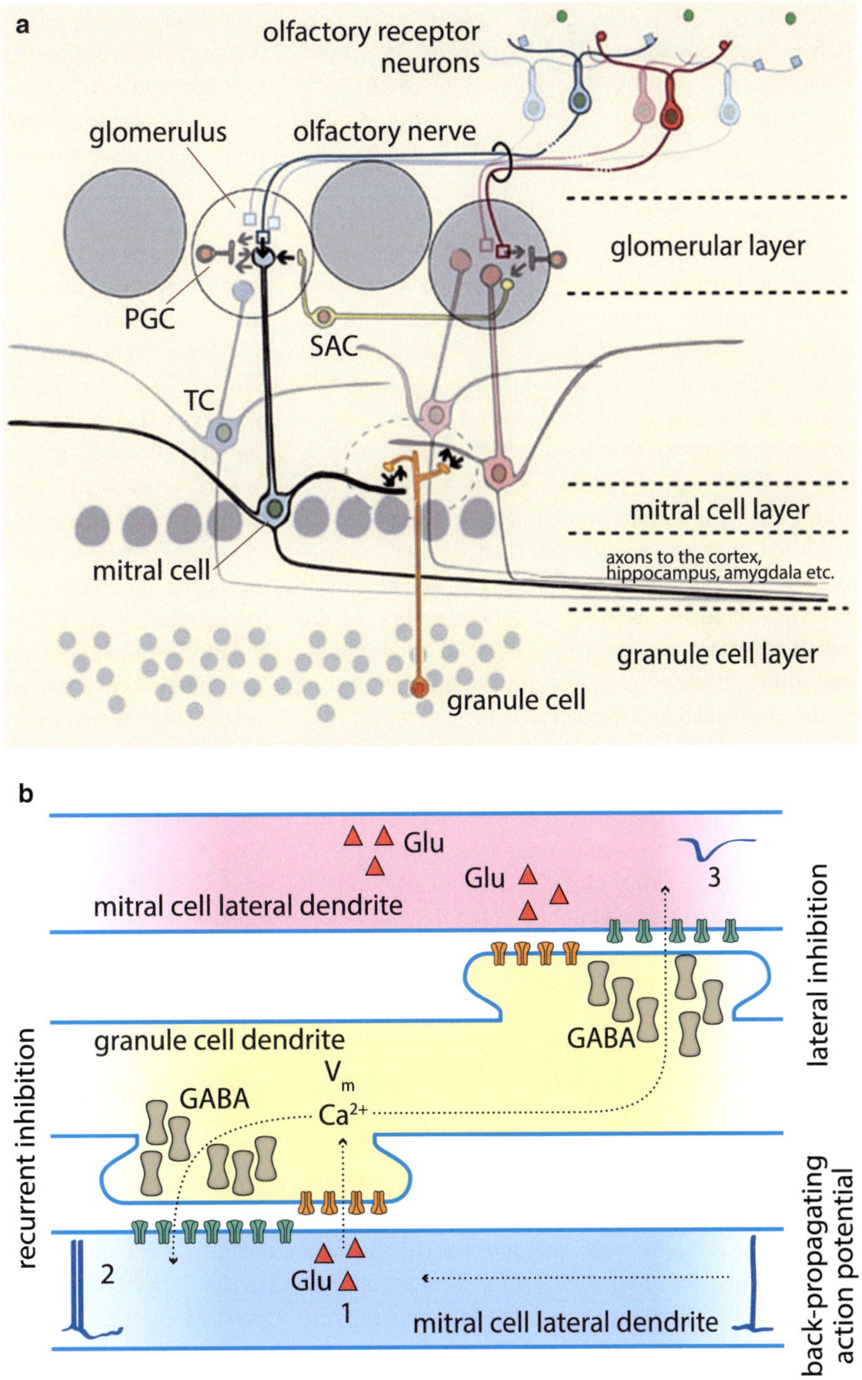
a
olfactory receptor neurons
glomerulus
olfactory nerve
glomerular layer
PGC
SAC
TC
mitral cell layer
axons to the cortex, hippocampus, amygdala etc.
mitral cell
granule cell layer
granule cell
b
Glu
Glu
3
mitral cell lateral dendrite
lateral inhibition
granule cell dendrite
GABA
recurrent inhibition
V
m
GABA
Ca2+
2
Glu
1
mitral cell lateral dendrite
back-propagating action potential

cells mostly mediate interactions between output neurons projecting to many different glomeruli. The functional consequences of this synaptic organization will be described below.

The synaptic mechanisms that play a key role in the circuits of the olfactory bulb have two unusual features. First, many bulbar neurons communicate via reciprocal dendrodendritic synapses. The reciprocal circuit provides inhibition that forms the basis for a reliable, spatially localized, recurrent inhibition. A mitral cell's synaptic depolarization driven by the long-lasting excitatory input from the sensory neurons, triggers glutamate release by dendrites and thus depolarizes interneuron dendrites and spines. This, in turn, elicits the release of GABA directly back onto the mitral cell itself (recurrent inhibition), as well as to the dendrites of other mitral cells ("lateral" inhibition). Second, several bulbar neuronal types are known to modulate their own activity via autoreceptors for the transmitters that they themselves release. In addition, transmitter release might occur through an action potential-independent manner in some cell types.

Since secondary dendrites have large projection fields and extensive reciprocal connections with interneurons, each bulbar local neuron may contact the dendrites of numerous output neurons. This suggests not only that dendrodendritic interactions provide a fast and graded feedback inhibition, but they also offer a unique mechanism for lateral processing between output neurons that innervate different glomeruli. Thus, the propagation of action potentials into the lateral dendrites, and the possible spread of excitation through granule cell dendrites, contributes to a "spatial" contrast mechanism that sharpens the tuning of output neuron odorant receptive fields. Furthermore, the so-called **short axon cells**, located near glomeruli, send interglomerular axons over long distances (the name is a clear misnomer) to excite inhibitory periglomerular neurons. This interglomerular center-surround inhibitory network, along with the mitral-granule-mitral inhibitory circuit described above, forms a serial, two-stage inhibitory circuit. Therefore, information is transmitted not only vertically across the glomerular relay between sensory neurons and output neurons, but also horizontally through local interneuron connections that are activated in odor-specific patterns (Fig. 13.16). Such a model based on lateral connection, originally introduced into neuroscience to explain visual contrast enhancement in the retina, has been mathematically extensively characterized. Both anatomical and functional analyses support the existence of lateral inhibitory mechanisms, in the olfactory bulb, through which activity in few stimulated output neurons may lead to suppression of other neurons innervating distinct glomeruli. For instance, examination of the responses of individual bulbar neurons to inhalation of aliphatic aldehydes reveals that many individual cells are excited by one subset of these odorants, inhibited by another subset, and unaffected by yet a third subset.

Alternatively, or in addition, the long-range projections of secondary dendrites could support a novel perspective that has emerged recently and views bulbar microcircuits as a nonlinear dynamical system. According to this view, the olfactory bulb transforms stationary input patterns into time-varying output patterns, moving along input-specific trajectories in coding space. In this framework, the main function of bulbar microcircuits would be to enable odor-specific dynamics that can decorrelate input patterns. Such a decorrelation function would distribute clustered input patterns more evenly in coding space, thus optimizing the use of the coding space for discrimination and other olfactory tasks. In other words, since secondary dendrites project to long distances, the olfactory bulb networks aim to reformat combinatorial representations so as to facilitate their readout by downstream olfactory centers. This model is consistent with experiments showing that GABAergic reciprocal inhibition contributes to synchronizing the output neuron activity mainly through granule cell activity.

Taking into account these data, it is obvious that the balance between excitation and inhibition in the olfactory bulb, and thus the interplay between local interneurons and output neurons, provides a combinatorial device to the representation of olfactory information. From this framework, two combinatorial encoders that take part in information processing within the bulbar network could be distinguished. The first one consists of the olfactory receptor repertoire expressed by the sensory neuron ensemble that transduces receptor activation patterns into glomerular odor maps throughout a highly reliable synaptic transmission (*stimulus space*). The secondary encoder lies in the intricate interneuron network that extracts higher-order features from the odor images to convert them as timing relationships across the firing output neuron ensemble (*representation space*). It has become apparent that the coding logic in insects and in mammals is very similar,

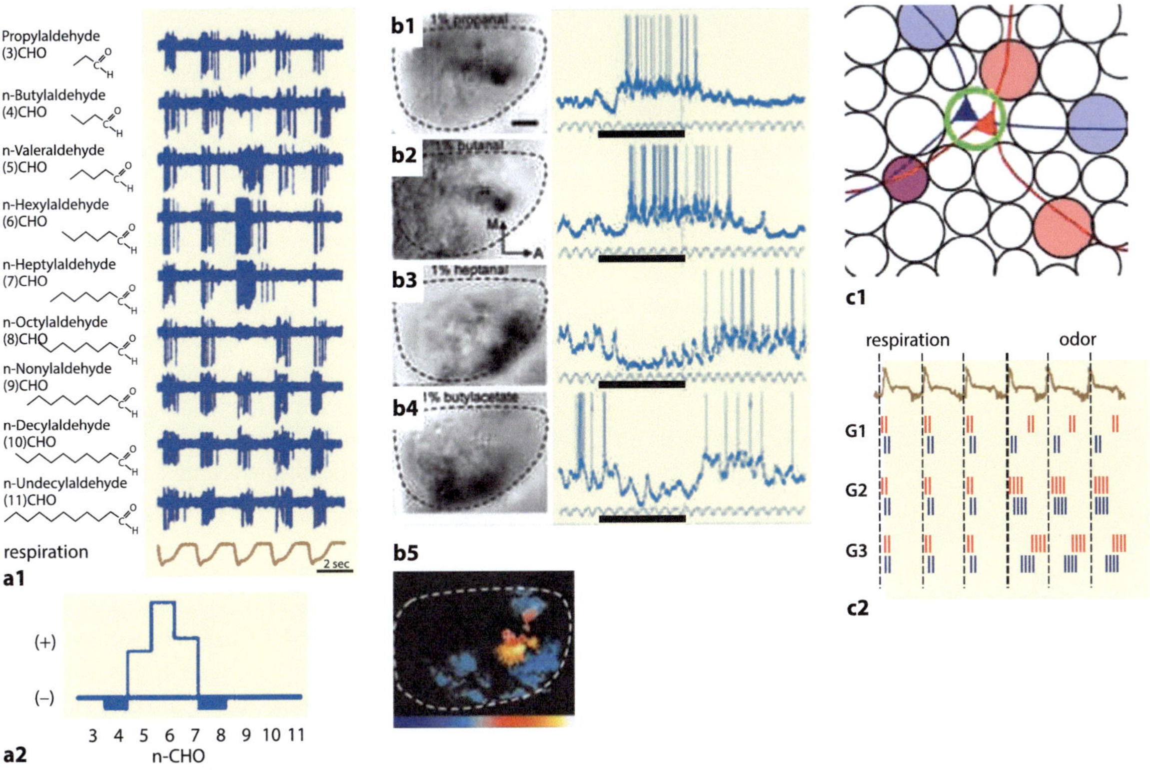


**Fig. 13.16** (**a1**) ***In vivo*** **extracellular recordings of action potential propagation of mitral cells.** The cell shown reacts most strongly to odorant stimulations (*bars*) with aldehydes comprising 5–7 carbon atoms. Similar odorants with a chain length of 4 or 8 carbons inhibit spontaneous cell activity. (**a2**) The generation rate as a function of the olfactory stimulus. (**b1–b4**) Spatial activity pattern induced by four different odorants (imaging using intrinsic signals from the dorsal side of the olfactory bulb in an anesthetized mouse). *Right*, electrical activity in a mitral cell (identical cells in 1–4). (**b5**) A summary of b1–b4 with regions which preferentially evoked inhibitory (*red*, *yellow*, see also, e.g., b1, b2) or excitatory responses in the mitral cells measured (*blue*, e.g., b3, b4). (**c1**) Two "sister" mitral cells whose primary dendrites receive input signals from the same glomerulus (*green circle*). Their olfactory response varies by virtue of the fact that both mitral cells are affected by different glomeruli (mediated via other mitral cells and granule cells) via their lateral dendrites. (**c2**) This leads to different olfactory responses in two "sister" mitral cells (*red* and *blue*). G1, G2, and G3 are three different odor stimuli. Differences are particularly visible in the action potential generation phase relative to the respiration cycle (a – Adapted from Yokoi et al. [31] with permission National Academy of Sciences, U.S.A., Copyright (1995). b – Adapted from Luo and Katz [32] with permission from Elsevier, Copyright (2001). c – Adapted from Dhawale et al. [33] with permission from Macmillan Publishers Ltd.: [Nature Neuroscience], Copyright (2010))

despite their convergent evolution. However, the circuitries have one major anatomical difference: while lateral connectivity in mammals occurs in the plexiform layers beneath the glomeruli, all synaptic contacts in insects are within the glomeruli.

### 13.6.4 Brain Circuitry

Odor information received by the olfactory bulb is first processed and refined prior to being transmitted to downstream centers (Fig. 13.17). The final processing occurs in higher-order brain structures comprising the primary and accessory olfactory cortex. The axons of mitral and tufted cells project in the olfactory tract to higher-order brain structures without contacting the thalamus. These higher centers include the anterior olfactory nucleus, which connects the two olfactory bulbs through a portion of the anterior commissure, the olfactory tubercle, the piriform cortex (considered to be the primary olfactory cortex), the cortical nucleus of the amygdala, and the entorhinal area. The olfactory

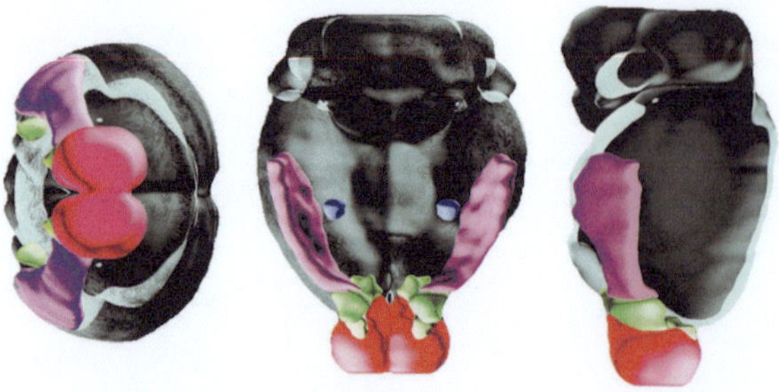

**Fig. 13.17 Rostral (*top*), dorsal (*middle*), and sagittal (*bottom*) views of the rodent brain.** *Colored areas* indicate the major olfactory brain centers. *Red*, the main olfactory bulb; *green*, the accessory olfactory bulb, the anterior olfactory nucleus; *purple*, the piriform cortex, and *blue*, the nucleus of the lateral olfactory tract (data modified from the Allen Mouse Brain Atlas, Allen Institute for Brain Sciences, http://mouse.brain-map.org)

tubercle projects to the medial dorsal nucleus of the thalamus, which in turn projects to the orbitofrontal cortex, the region of cortex thought to be involved in the conscious perception of smell. Thus, olfactory information is also relayed through the thalamus to the neocortex, but unlike the other sensory systems where the main route to the cortex is the thalamus, in the olfactory system this is a multi-synapse additional route.

Olfactory information must be relayed from convergent synapses in the olfactory bulb to higher brain centers, where it is decoded to yield a coherent odor image. Experiments in mice that traced the olfactory circuitry of olfactory sensory neurons expressing a given odorant receptor suggest a distributed olfactory code in the olfactory cortex. Genetically encoded transneuronal tracers in the mouse olfactory system have shown that output neurons which receive input from a single glomerulus project to defined regions of the piriform cortex that are more extensive than the glomerular segregation [34]. Overlap in the projection patterns of different glomeruli affords the opportunity for integration of olfactory information at higher olfactory centers. In both the olfactory bulb and higher centers, odor information seems to be encoded by activity across the entire neuronal network, similar to the antennal lobe/mushroom body projection in insects. This divergent connectivity may enable animals to still discriminate odors even after ablation of a large fraction of the olfactory bulb.

Despite the recent progress, considerable functional analysis has yet to be performed before we will completely understand how odorant molecules are represented in higher olfactory centers. The nature of the olfactory stimulus itself seems crucial for olfactory processing in higher centers. For instance, the lateralization of odor information processing in the human brain is thought to depend on odor identity. Most odorants simultaneously activate the olfactory system, which generally projects ipsilaterally, and the trigeminal system, which crosses to the contralateral side. Thus, hemispheric asymmetry for olfactory function strongly depends on quality of the nasal stimulus, more specifically its olfactory or trigeminal stimulating properties. Also, based on functional magnetic resonance imaging data, some researchers hypothesized two separate systems mediating positive and negative emotions. The right hemisphere might be involved in encoding negative affects while the left one in encoding positive affects.

Even in humans, during the first hours of life in the open air, the newborn child behaves like a "macrosmatic" animal. While adult humans are considered to be microsmatic, olfaction does remain the sense that opens the most direct route to the affective system. Part of this is likely due to the direct connection from the bulb to the cortico-medial nucleus of the amygdala. Recent results suggest that there is a hedonic map of the sense of smell in brain regions such as the orbitofrontal cortex. These results have implications for understanding the psychiatric and related problems that follow damage to these brain areas. It is remarkable that amongst all the senses, olfaction possesses a direct link with the limbic system that was taken to be the 'nose-brain' (the actual meaning of rhinencephalon). Today, it is clear that the primary olfactory cortex projects to the entorhinal area, which in turn projects to the **hippocampus**. Thus, we see reintroduced, after years of fervent affirmation followed by years of fervent denial, the idea that the hippocampus receives olfactory inputs.

Noninvasive functional imaging studies of the human olfactory system revealed that the sense of smell is organized similarly to other sensory modalities which all depend on the genesis of sensory maps, and that the specific psychological characteristics of olfaction should be attributed to an early involvement of the limbic system rather than a conceptually different mode of processing. Taking into account the high connectivity of limbic structures and the fact that activation of the amygdala immediately induces emotions and facilitates the coding of memories, one should not

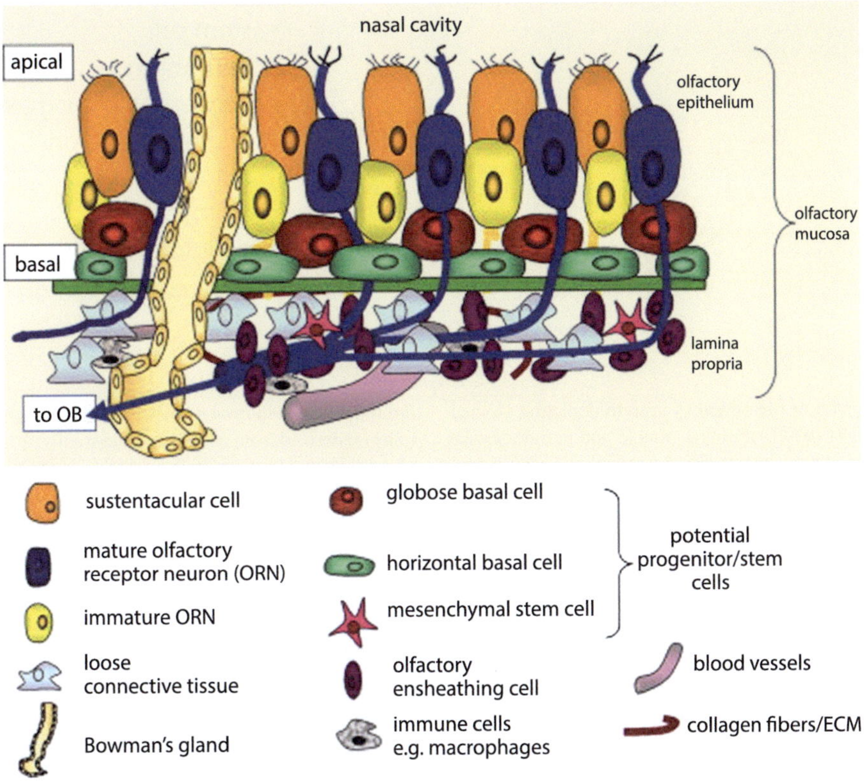


**Fig. 13.18** **The cellular composition of the olfactory mucosa.** The olfactory mucosa is comprised of the olfactory epithelium (*OE*) and the lamina propria (*LP*). The OE contains olfactory receptor neurons (in *blue*) that detect odorants. These neurons project cilia into the nasal cavity and extend axons into the LP where they form bundles (fila olfactoria or fascicles) with axons from other ORNs before crossing the cribriform plate to reach the olfactory bulb (*OB*). Within the OE are also sustentacular cells (in *orange*, non-neuronal supporting cells), Bowman's glands (in *brown*) and ducts and putative progenitor/stem cells the globose basal cells (in *purple*, *GBCs*), and the horizontal basal cells (in *green*, *HBCs*). The LP consists of loose connective tissue together with collagen fibers, extracellular matrix, and olfactory ensheathing cells (*OECs*) that enwrap bundles of ORN. Many cell types make up connective tissue, including fibroblasts, macrophages, pericytes, endothelial cells, and smooth muscle cells from the lining of blood vessels, in addition to Schwann cells that ensheath the nerves and blood vessels. It is also believed that mesenchymal-like stem cells reside in the LP (From Lindsay et al. [35] with permission)

be so surprised to uncover the special relationship that links olfaction with emotions and memory.

### 13.6.5 Neurogenesis

There are at least two germinative zones present in the adult olfactory system. The first one is located in the sensory organ where cell renewal persists throughout life to replace olfactory sensory neurons (Fig. 13.18). The second area resides near the ventricle of the forebrain. The former site of neurogenesis is made possible by the presence of globose and horizontal basal cells found deep in the olfactory epithelium, near the basal lamina that separates the epithelium from the underlying lamina propria. The mature sensory neurons have only a limited life span of about 90 days, but if mice are reared in a laminar flow hood to prevent rhinitis, sensory neurons can survive as long as 12 months, close to the life span of the animals. Thus, the turnover of sensory neurons, and by extension, the rate of neurogenesis in the olfactory epithelium, is normally regulated by environmental factors provided by a direct contact with the external world. But olfactory epithelium neurogenesis is also enhanced by ablation of the main olfactory bulb. Therefore, peripheral and central signals act in concert to regulate mitotic rates. This plasticity, coupled with the fact that there are a limited number of cell classes in the olfactory epithelium, makes this area attractive for studying mechanisms that control

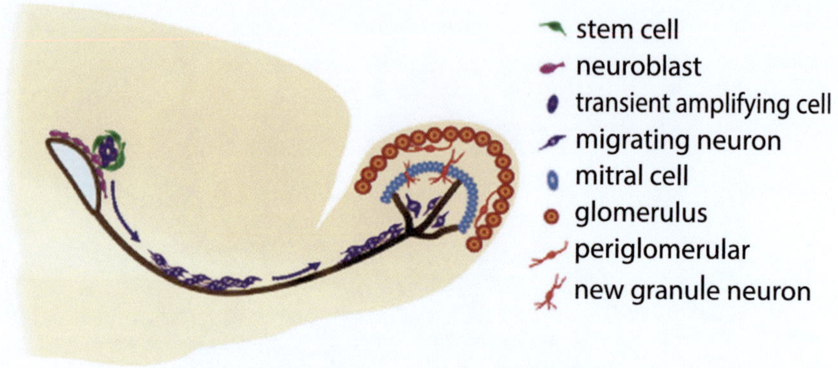


**Fig. 13.19** **Diagram of the olfactory system** depicting glomeruli (*red*), mitral cells (*pink*), newborn granule and periglomerular cells (*red*), and the regions involved in the process of adult neurogenesis: the subventricular zone and the rostral migratory stream (both delineated in *brown*). Also shown are migrating neuroblasts (*purple*) derived from adult neural stem cells (*green*) and migrating *en route* to the olfactory bulb

the rate of neuronal formation of neurons. Subgroups of stem cells can be made to differentiate into mature olfactory receptor cells by stressing them mechanically or biochemically. In addition, differentiated neurons send back regulatory signals to inform progenitor cells about the number of new neurons that need to be produced to maintain cell population equilibrium.

What could be the functional significance of this neurogenesis? Mature sensory neurons that have been damaged by exposure or by pathogens, and immature ones that cannot find adequate synaptic targets in the olfactory bulb, are two obvious candidates to support the existence of this neurogenesis. Once mature, sensory neurons must extend along a long route to the correct glomerulus. As odor quality remains constant throughout life, the glomerular array must be constant to a certain degree. Apoptotic cell death has been observed in cells representing all stages of regeneration, implying apoptotic regulation of neuron numbers at all levels of the neuronal lineage. Currently, there is a considerable interest in chemical factors that inhibit or promote neurogenesis or differentiation or actively produce apoptotic cascades.

New local interneurons of the olfactory bulb also are continuously produced throughout life (Fig. 13.19). Neuroblasts fated for the olfactory bulb are produced in the walls of the lateral ventricles (the so-called **subventricular zone**). They migrate tangentially toward the core of the olfactory bulb, forming the **rostral migratory stream**. Upon reaching the olfactory bulb, adult-born neuroblasts migrate radially within the lamina of the olfactory bulb to reach their final position and start to differentiate. The newly generated neurons form the two main classes of OB inhibitory interneurons of the olfactory bulb: granule cells and periglomerular cells. Surprisingly, after 1 month of their birth, almost one half of the newcomers have been eliminated while the other half are durably integrated into the olfactory bulb. Intense ongoing studies try to decipher the meaning of this never-ending neuronal turnover.

## 13.7 Summary

Animals discriminate and recognize chemical signals in their environment, which provide them with essential information that profoundly influences their behavior. Among others, gathering information about the environment and communicating with other organisms through chemical cues is an essential process for the survival of all multicellular systems.

**Chemosensation** is a fundamental process shared by most animals. This very conserved process across species originates from the molecule detection that dates back to prokaryotes and has evolved into four distinct modalities in most vertebrates. The main olfactory system, the accessory olfactory system, the gustatory system, and the so-called "common chemical sense" carried mostly by trigeminal sensory neurons, all differ with respect to receptor molecules, receptor cells, and wiring of the receptor cells.

As our knowledge about the neurobiology of olfaction is growing, it is becoming evident that the main olfactory systems of animals in disparate phyla have many striking features in common. For instance, vertebrate and insect olfactory systems display common organizational and functional characteristics. Importantly, most systems have a strong convergence of receptor cells onto olfactory glomeruli, and a later divergence onto second-order neurons. They code odors as spatiotemporal dynamic and combinatorial activity patterns across these glomeruli. However, this solution is not universal: nematodes, for example, have evolved a different coding strategy. Whatever the processing architecture, the initial common event, shared by all olfactory systems, requires the interaction of odorant molecules with specific receptors expressed on the cilia or microvilli of sensory olfactory neurons before conveying information to central structures.

## References


1. Butenandt A, Groschel U, Karlson P, Zillig W (1959) Über *N*-Acetyl-Tyramin, seine Isolierung aus *Bombyx*-Puppen und seine chemischen und biologischen Eigenschaften (*N*-acetyl tyramine, its isolation from *Bombyx* cocoons and its chemical and biological properties). Arch Biochem Biophys 83:76–83
2. Malnic B, Hirono J, Sato T, Buck LB (1999) Combinatorial receptor codes for odors. Cell 96:713–723
3. Hölldobler B, Wilson EO (1990) The ants. Springer, Berlin/Heidelberg/New York
4. Kristoffersen L, Larsson MC, Anderbrant O (2008) Functional characteristics of a tiny but specialized olfactory system: olfactory receptor neurons of carrot psyllids (Homoptera: Triozidae). Chem Senses 33:759–769
5. Galizia CG (2008) Insect olfaction. In: Firestein S, Beauchamp GK (eds) The senses: a comprehensive reference, vol 4, Olfaction & taste. Academic, San Diego, pp 725–770
6. Galizia CG, Rössler W (2010) Parallel olfactory systems in insects: anatomy and function. Annu Rev Entomol 55:399–420
7. Robertson HM, Warr CG, Carlson JR (2003) Molecular evolution of the insect chemoreceptor gene superfamily in *Drosophila melanogaster*. Proc Natl Acad Sci USA 100: 14537–14542
8. Sachse S, Krieger J (2011) Olfaction in insects. e-Neuroforum 2:49–60
9. Silbering AF, Rytz R, Grosjean Y, Abuin L, Ramdya P, Jefferis GSXE, Benton R (2011) Complementary function and integrated wiring of the evolutionarily distinct *Drosophila* olfactory subsystems. J Neurosci 31:13357–13375
10. Benton R, Vannice KS, Gomez-Diaz C, Vosshall LB (2009) Variant ionotropic glutamate receptors as chemosensory receptors in *Drosophila*. Cell 136:149–162
11. Kaissling KE (2001) Olfactory perireceptor and receptor events in moths: a kinetic model. Chem Senses 26:125–150
12. Keil TA (1999) Morphology and development of the peripheral olfactory organs. In: Hansson B (ed) Insect olfaction. Springer, Berlin/Heidelberg/New York, pp 6–47
13. Keil TA, Steinbrecht RA (1987) Diffusion barriers in silkmoth sensory epithelia: application of lanthanum tracer to olfactory sensilla of *Antheraea polyphemus* and *Bombyx mori*. Tissue Cell 19:119–134
14. Shanbhag SR, Müller B, Steinbrecht RA (1999) Atlas of olfactory organs of *Drosophila melanogaster.* 1. Types, external organization, innervation and distribution of olfactory sensilla. Int J Insect Morphol Embryol 28:377–397
15. Gascuel J, Masson C (1991) A quantitative ultrastructural study of the honeybee antennal lobe. Tissue Cell 23:341–355
16. Sachse S, Rappert A, Galizia CG (1999) The spatial representation of chemical structures in the antennal lobe of honeybees: steps towards the olfactory code. Eur J Neurosci 11:3970–3982
17. Galizia CG, Sachse S, Mustaparta H (2000) Calcium responses to pheromones and plant odours in the antennal lobe of the male and female moth *Heliothis virescens*. J Comp Physiol A 186:1049–1063
18. Laurent G (1999) A systems perspective on early olfactory coding. Science 286:723–728
19. Schmidt M, Van Ekeris L, Ache BW (1992) Antennular projections to the midbrain of the spiny lobster. I. Sensory innervation of the lateral and medial antennular neuropils. J Comp Neurol 318:277–290
20. Grünert U, Ache BW (1988) Ultrastructure of the aesthetasc (olfactory) sensilla of the spiny lobster, Panulirus argus. Cell Tissue Res 251:95–103
21. Schmidt M, Mellon D (2011) Neuronal processing of chemical information in crustaceans. In: Breithaupt T, Thiel M (eds) Chemical communication in crustaceans. Springer, Berlin/Heidelberg/New York, pp 123–147
22. Bargmann CI (2006) Chemosensation in *C. elegans*. In: The *C. elegans* Research Community (ed) WormBook, doi: 10.1895/wormbook.1.123.1, http://www.wormbook.org
23. Perkins LA, Hedgecock EM, Thomson JN, Culotti JG (1986) Mutant sensory cilia in the nematode *Caenorhabditis elegans*. Dev Biol 117:456–487
24. Troemel ER, Kimmel BE, Bargmann CI (1997) Reprogramming chemotaxis responses: sensory neurons define olfactory preferences in *C. elegans*. Cell 91:161–169
25. Chase R, Tolloczko B (1993) Tracing neural pathways in snail olfaction: from the tip of the tentacles to the brain and beyond. Microsc Res Tech 24:214–230
26. Axel R (2004) Nobel Lecture. 8 Dec 2004 (Nobelprize.org)
27. Mombaerts P et al. (1996) Visualizing an olfactory sensory map. Cell 87:675–686
28. Lledo PM, Gheusi G, Vincent JD (2005) Information processing in the mammalian olfactory system. Physiol Rev 85:281–317
29. Manzini I, Korsching S (2011) The peripheral olfactory system of vertebrates: molecular, structural and functional basics of the sense of smell. e-Neuroforum 2:68–77
30. Kuner T, Schaefer A (2011) Molecules, cells and networks involved in processing olfactory stimuli in the mouse olfactory bulb. e-Neuroforum 2:61–67

31. Yokoi M, Mori K, Nakanishi S (1995) Refinement of odor molecule tuning by dendrodendritic synaptic inhibition in the olfactory bulb. Proc Natl Acad Sci USA 92:3371–3375
32. Luo M, Katz LC (2001) Response correlation maps of neurons in the mammalian olfactory bulb. Neuron 32: 1165–1179
33. Dhawale AK, Hagiwara A, Bhalla US, Murthy VN, Albeanu DF (2010) Non-redundant odor coding by sister mitral cells revealed by light addressable glomeruli in the mouse. Nat Neurosci 13:1404–1412
34. Stettler DD, Axel R (2009) Representations of odor in the piriform cortex. Neuron 63:854–864
35. Lindsay SL, Riddell JS, Barnett SC (2010) Olfactory mucosa for transplant-mediated repair: a complex tissue for a complex injury? Glia 58:125–134

# Taste

# 14

Wolfgang Meyerhof

The sense of taste evaluates the quality of food in the mouth. For flavor formation gustatory, olfactory, and trigeminal cues have to be integrated.

All animal species require organic nutrients. Food delivers the energy and essential building blocks necessary to maintain their metabolism. In this respect, the sense of taste is a crucial part of a control loop that supplies energy and acts to replenish elements, ions, and molecules, including electrolytes, acids, and nitrogen, that are lost by excretion – thereby maintaining body homeostasis. On the other hand, ingestion of food contaminated with hazardous substances could be life-threatening, impair health, and diminish the probability of successful reproduction. Thus, every time before food is ingested, animals have to elucidate if a piece of food contains enough calories and if it is sufficiently free of toxins.

The sense of taste can be considered an instrument evaluating the quality of food [4]. It sorts calories from potentially harmful substances and governs the decision to ingest or reject. Interestingly, throughout the animal kingdom from humans to worms the same types of chemicals are often preferred, such as certain sugars or amino acids which contain calories, or rejected, such as the bitter compounds quinine and denatonium benzoate. Bitter substances are generally avoided because they are often, though not always, toxic. Moreover, at least in higher vertebrates, taste can provide positive hedonic tone or pleasure that provides a reward beyond the caloric content of the ingested food and compensate animals for the effort spent on the acquisition of food. The evaluation of food is the only function that taste fulfils, in contrast to smell (which is linked to multiple functions as described in Chap. 13).

Unlike smell, taste acts on site [7]. In order to perceive taste, the food has to be in contact with the sensing organs. Several sensory modalities mediate the 'taste' of food in our mouth. In order to be palatable, food must have appropriate temperature, because above or below certain thresholds temperature elicits tissue damage. In addition, the consistency or texture must allow us to masticate the food in order to swallow it safely. Moreover, the diet should largely lack chemical irritants. Chemesthesis, the taste of chemicals, recognizes and causes avoidance of irritant chemicals such as strong acids or salts, capsaicin from hot chilli pepper, mustard oil from cruciferous vegetables, or camphor from laurel (*Laurus nobilis*).

All of these sensations are transmitted by the trigeminal nerve and not the gustatory nerves. Trigeminal afferent fibers carry information to the trigeminal complex in the brainstem. Brainstem processing of somatosensory stimulation is spatially segregated from gustatory processing. At higher brain areas, however, trigeminal cues contribute to flavor formation. Finally, the sense of smell has a decisive role in our taste experience.

Not only the smell of food from a distance is important, but also odors released from the food within the oral cavity. Each time we swallow, volatile compounds liberated from the food in the mouth during chewing

W. Meyerhof
German Institute of Human Nutrition Potsdam-Rehbruecke, Department Molecular Genetics,
Arthur-Scheunert-Allee 114-116, 14558 Nuthetal, Germany
e-mail: meyerhof@dife.de

C.G. Galizia, P.-M. Lledo (eds.), *Neurosciences - From Molecule to Behavior: A University Textbook*,
DOI 10.1007/978-3-642-10769-6_14, 

reach the olfactory sensory epithelium via the oral-nasal connection eliciting odor responses transmitted by the olfactory nerve. These olfactory responses contribute to the generation of aromas that, combined with the gustatory and oral somatosensory responses, yield flavors. For food recognition the aroma of a food is a crucial parameter. Recognition scores drop heavily if test persons use nose-clips disabling their sense of smell. What remains is the gustatory response or taste response in the restricted sense of the word. Gustation refers to the on-site detection of water-soluble molecules by specialized chemosensory cells and the transmission of this information by gustatory nerves to the brain as well as its representation in the gustatory brain areas which trigger the behavioral output [7].

In the following we will use the terms gustation and taste synonymously. This chapter focuses on the gustatory mechanisms solely, olfaction and trigeminal mechanisms will not be covered even though they greatly contribute to the perception of food.

## 14.1 Taste Qualities and Taste Molecules

Sweet, umami, and salty tastes are appetitive and drive food ingestion. Together with the repulsive tastes bitter and sour, they form the five basic tastes in humans.

Unlike smell, taste has little discriminative power. Humans distinguish only five basic tastes, also called taste qualities or modalities: sweet, bitter, umami (the taste of protein in meat, fish, cheese, or savory food in general), sour, and salty. It remains to be seen if other oral sensations including metallic, electrical, fatty, or watery will be assigned an own primary taste quality in the future. At present evidence in support of this is insufficient. We also do not know for sure how animals perceive chemicals that we describe as sweet, bitter, sour, or salty. In general, different animals detect the same chemicals as humans do. Moreover, recordings from the gustatory nerves after stimulation of the tongue with tastants and behavioral experiments suggest that numerous animals across phyla avoid substances that taste bitter or intensely sour and are attracted by compounds that taste salty, sweet, or umami – strongly suggesting that many species may have similar taste categories. However, some animals

**Table 14.1** Chemical diversity of taste-active molecules (in humans)

| | |
|---|---|
| Sweet | Monosaccharides, disaccharides, oligosaccharides, glycine, L-alanine, polyols, nitroanilines, sulfonyl amides, arylureas, benzophenones, oximes, benzimidazoles, hydrazides, sulfamates, dipeptides, dihydrochalcones, pyrimidines, D-amino acids, anilides, halogenated sugars, guanidoacetic acids, lead and beryllium salts, monoterpenes, sesquiterpenes, diterpenes, triterpenes, steroidal saponins, phenolics; sweet proteins brazzein, thaumatin, monellin |
| Umami | Proteinogenic L-amino acids, L-ibotenic acid, L-tricholomic acid; enhancers ribonucleotides, ribonucleotide derivatives, alapyridaine, *N*-gluconyl ethanolamine phosphate, (*S*)-monelin |
| Salty | NaCl, NaBr, $NaHCO_3$, $Na_2SO_4$, Na-glutamate, Na-acetate, LiCl, KCl, $KNO_3$, KI, $NH_4Cl$, $CaCl_2$, $MgSO_4$ |
| Bitter | Hydroxyl fatty acids, peptides, amino acids, amines, azacycloalkanes, *N*-heterocyclic compounds, amids, ureas, thioureas, carbamides, esters, lactones, carbonyl compounds, phenols, crown ethers, alkaloids, metal ions, glycosides, diterpenes, nortriterpenes, sesquiterpene lactones, secoiridoids, ribonucleosides, sugar derivatives |
| Sour | Hydrochloric acid, sulfuric acid, phosphoric acid, acetic acid, lactic acid, malic acid, tartaric acid, fumaric acid, citric acid, benzoic acid, adipic acid |

such as pig, cattle, or cat may have additional or other taste categories. For example, cats appear to have a taste for pyrophosphate and none for sweet, whereas for pigs and even more so for calves it is uncertain what primary taste categories they have and to what extent they generalize tastants.

Each of the common five basic tastes has a special function in the evaluation of food [13]. Sweet taste is considered to be a detector for calories in form of carbohydrates and, indeed, sugars released from high-molecular-weight precursors by salivary amylase are potent stimulators of sweet taste. However, not only sugars elicit sweet taste. Numerous natural and synthetic substances do so ranging from proteins to metal ions (Table 14.1). Intriguingly, only minor chemical modifications can convert bitter into sweet and vice versa. For example, L-tryptophane is bitter, D-tryptophane sweet. Similarly, naringin is bitter, whereas its dihydrochalcone is sweet. Thus, various members of the same chemical class can be sweet or bitter. Even a single substance can have a sweet and a

**Table 14.2** Threshold values for some common taste stimuli (in humans)

| Quality | Substance | Threshold (M) | Function |
|---|---|---|---|
| Sweet | Sucrose | 0.01 | Calorie detector |
| | Glucose | 0.08 | |
| | Na-saccharin | 0.000023 | |
| Umami | Na-glutamate | 0.01 | Calorie detector |
| Salty | NaCl | 0.01 | Electrolyte detector |
| Bitter | Quinine sulfate | 0.000008 | Warning against potential toxins |
| | Aristolochic acid | 0.0000014 | |
| | Denatonium benzoate | 0.000000035 | |
| | Na-saccharin | 0.009 | |
| Sour | HCl | 0.0037 | Warning against potential toxins |
| | Acetic acid | 0.0012 | |
| | Citric acid | 0.0030 | |

bitter taste such as saccharin (Tables 14.1 and 14.2). Sweet taste is strongly appetitive and drives ingestion of food. Similarly, umami taste detects calories. It is stimulated by many proteinogenic amino acids (Table 14.1) and enhanced by the simultaneous presence of ribonucleotides, both compounds being abundantly present in meat. Like sweet, umami is appetitive and promotes the intake of calories in form of protein and, at the same time, delivers essential amino acids and nitrogen. Salty, is the third appetitive taste modality. It is mediated, in particular, by table salt, NaCl, but also by other salts (Table 14.1). All animals continuously loose NaCl and other electrolytes with their excretions and must compensate the loss through intake. In nature, NaCl usually occurs not in pure form but is mixed with other minerals. Thus NaCl intake is accompanied by uptake of other minerals as well. In this sense, salty taste can be considered as a detector for electrolytes and an important regulator in electrolyte homeostasis. For appetitive chemicals recognition thresholds are ~10 mM, i.e., the lowest concentration of a compound that can be identified (Table 14.2), and preferences usually increase with higher concentrations. This ensures that animals avoid wasting energy on the acquisition of food sources with poor caloric or electrolyte contents. However, high-potency sweeteners such as saccharin used as sugar substitutes usually have much lower thresholds than sugars (Table 14.2).

In marked contrast, bitter taste is repulsive inducing strong rejection. Even intensely sweet compounds associated with bitter off-tastes are increasingly rejected as concentration increases. Numerous toxic chemicals are bitter (Table 14.1), even though a clear correlation of bitterness and toxicity has not been observed. Truly, we know of toxins such as α-amanitin or tetrodotoxin that are devoid of bitterness in the poisonous concentration range accounting for numerous deaths each season after consumption of death cap mushrooms (*Amanita phalloides*) or spoiled puffer fish (Tetraodontinae). On the other hand, certain bitter compounds have been used in human medication for thousands of years such as the antipyretic and analgesic glycoside salicin of willow (*Salix*) bark. Most natural bitter chemicals are from plants which use the secondary metabolites to protect themselves against infections, oxidative stress, and pests. By ingesting such compounds we may take on some of these 'beneficial' effects resulting in appreciation of bitter food and beverages. Examples are certain types of vegetables and fruits, chocolate, coffee, or digestives. However, bitter chemicals also occur in the animal kingdom, originate during food processing and aging, or are generated by chemical synthesis. Even though there are no valid counts, tens of thousands of bitter compounds may exist as illustrated by the fact that some 10,000 plant species produce glycosides, many of which are bitter. Bitter substances also show extreme chemical diversity (Table 14.1). Like bitterness, also moderate or strong sourness is repulsive. It is elicited by inorganic and organic acids (Table 14.1) and may prevent animals from feeding on unhealthy unripe fruits or spoiled food which is rich in organic acids due to bacterial contamination. To effectively avoid potential chemical noxa the thresholds for such stimuli are well below those of the appetitive stimuli (Table 14.2).

## 14.2 The Mammalian Gustatory System

### 14.2.1 Functional Morphology of the Peripheral Taste System

#### 14.2.1.1 Taste Papillae and Taste Buds

Vertebrate taste receptor cells assemble to principle taste organs, called taste buds, which on the tongue are organized in taste papillae.

Taste molecules are sensed by special small organs called taste buds (Fig. 14.1). They are present in the oral epithelium. Only on the tongue do taste buds occur in

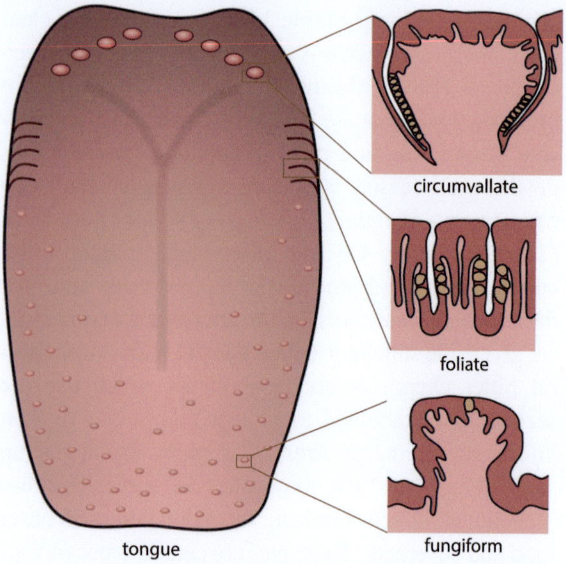


**Fig. 14.1 The human tongue and the distribution of the three types of taste papillae.** The middle surface of the tongue appears to be partly devoid of taste buds. This region is least taste-sensitive. The enlargements show the structure of the various papillae and the location of taste buds (*light brown dots*) lining the trenches of the foliate and circumvallate papillae or present at the surface of the fungiform papillae. Please note that rodent fungiform papillae contain only one taste bud. These species also have only one central vallate papilla

the taste papillae. We distinguish three chemosensory papilla types, (1) the fungiform papillae that are distributed over the anterior tongue surface, (2) the foliate papillae on the posterior tongue edges, and (3) the circumvallate papillae on the posterior tongue (Fig. 14.1). The small and thin filiform papillae do not engage in chemosensation, they respond to tactile stimuli [4].

Humans may have on average 300 fungiform papillae with 1,100 taste buds, 9 circumvallate papillae with 2,200 taste buds, and 2 foliate papillae with 5 fissures that harbor together 1,300 taste buds [4]. In contrast, rodents possess only one papilla on the posterior tongue referred to as vallate papilla due to the fact that the trench-like epithelial invagination is not fully circular assuming a v- or u-shaped circumference. Taste buds occur also frequently in the soft palate, and are found in the pharyngeal, laryngeal, and epiglottal epithelium. Rodents differ in this respect from humans as they possess taste buds at the oral exit of the nasal-incisor duct connecting the nasal and oral cavities. The ducts of minor salivary glands are connected with the grooves of the vallate and foliate papillae. The secretions of these glands form a perireceptor milieu. Its composition is complex and not fully characterized. It changes over time and depending on the physiological condition and very likely modulates recognition of taste molecules by vertebrate taste receptor cells (TRCs) of the posterior papillae in various ways.

Tastants of all qualities interact with the apical membrane of specialized chemosensory cells in the oral epithelium. Unlike olfactory sensor cells, which are neurons, TRCs are specialized cells of epithelial origin and are sometimes referred to as secondary sensory cells. They assemble with other cells to small principle taste organs with onion-shaped appearance, the taste buds (Fig. 14.2). A taste bud may contain up to ~100 cells and is embedded in the oral epithelium. The bud is shielded from the environment through epithelial tight junctions. Only the apical tips of some elongated taste bud cells have contact with the environment through the taste pore, a depression in the epithelium devoid of tight junctions (Figs. 14.2 and 14.3a). Interactions of taste molecules with the TRCs are confined to these sites. Taste buds are innervated by primary afferent nerve fibers that are excited by neurotransmitters released from taste bud cells following taste stimulation. ATP plays a major role as neurotransmitter in taste but other transmitter substances appear to contribute as well.

#### 14.2.1.2 Taste Bud Cell Types and Functions

> Taste buds may contain up to five different cell types: supportive cells, receptor cells, presynaptic cells, basal cells, and perigemmal cells.

Historically, taste buds have been considered unidirectional sensors receiving taste stimuli at the apex and releasing transmitter substances at the basolateral side. However, this view has changed and we now know that these small taste organs also process taste information [14]. Various common transmitter substances (norepinephrine, serotonin, γ-amino butyrate) and neuropeptides (neuropeptide Y, cholecystokinin, glucagon-like peptide-1) as well as their receptors are present in taste bud cells and appear to modulate taste bud function in autocrine and paracrine fashion. Moreover, the taste buds also respond to circulating hormones involved in energy balance, such as leptin released from adipose tissue and hypothalamic endocannabinoids [11]. These endocrine control loops adapt taste sensitivities to the metabolic needs of the organism. In the past, taste bud cells have been characterized based on morphological properties seen in light or electron micrographs only. Recently, however, specific functions have been assigned to the morphologically distinct cell types. Table 14.3 and Fig. 14.3 summarize some important characteristics of the taste bud cell types.

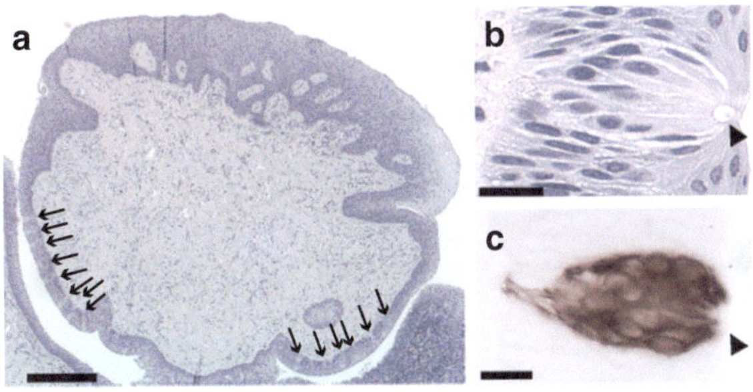


**Fig. 14.2 Human circumvallate papilla and taste buds.** (**a**) Micrograph of a hemalm-stained cross-section through a circumvallate papilla. Taste buds are difficult to identify but are visible as small light interruptions of the epithelial cell layer (*arrows*). (**b**) High magnification of a selected taste bud shown in (a). The onion-shaped structure of the taste buds and the taste pore (*arrowhead*) are clearly visible. (**c**) A circumvallate taste bud visualized by indirect immunohistochemistry using an anti-cytokeratin 20-antiserum. Single taste bud cells are visible. Also the elongated shapes of several taste cells are to be seen. Scale bars, 200 μm (a), 20 μm (b, c)

Type I cells, also referred to as supportive cells, are not electrically excitable. They are characterized by cytoplasmic cell extensions that wrap around adjacent taste bud cells and by potassium conductance that clears potassium from the extracellular milieu (Table 14.3 and Fig. 14.3). Moreover, they also express enzymes or transporters removing transmitter substances from the extracellular environment, thereby promoting signal termination. These properties resemble those of glial cells in the nervous system and are consistent with the supportive function of these cells. Recently, circumstantial evidence implicated type I cells in the transduction of salty taste.

Type II or receptor cells appear to be the principle sensors for bitter, sweet, and umami stimuli. They express the G protein-coupled receptors that recognize these taste molecules and the relevant intracellular signaling proteins (Table 14.3 and Fig. 14.3). Individual receptor cells are only dedicated to the detection of stimuli of a single taste quality, i.e., either bitter, sweet, or umami – each cell expressing either sweet or bitter or umami taste receptors. Thus, the mutually exclusive expression of the taste receptors matches well their response profiles determined in physiological experiments. This is important because the findings indicate that each taste quality appears to be represented by an own, separate population of sensor cells in the peripheral taste organs. Type II cells are excitable and fire

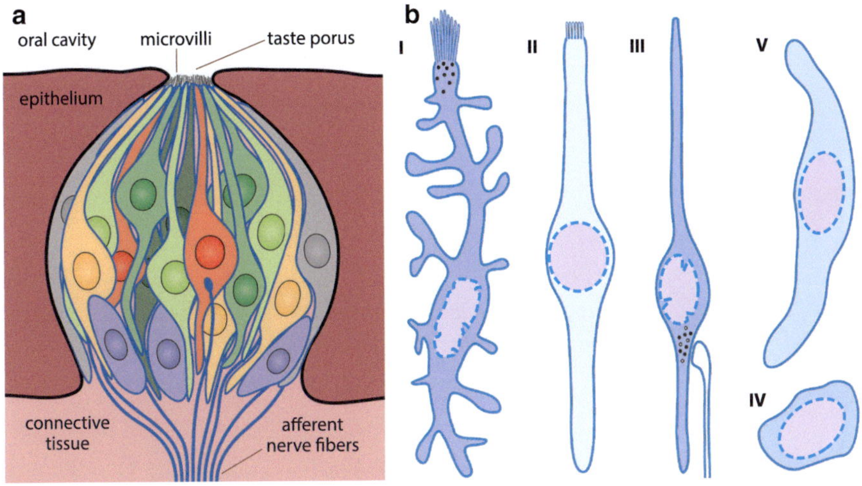


**Fig. 14.3 Taste bud and taste bud cell types.** (**a**) Scheme of a taste bud. The various cell types are displayed in different colors, *gray*, perigemmal cells; *blue*, basal cells; *yellow*, type I cells; *orange*, type III cells; *green*, type II cells. The various green tones indicate the various types of sweet, bitter, or umami receptor cells. Type III cells are the only taste cells that engage in synaptic contacts (indicated by the knob-like structure) with the afferent nerves. Only the elongated type I, II, or III cells send their apical aspects to the taste pore. (**b**) Schematic representation of taste bud cell types. For details see main text and refer to Table 14.3

**Table 14.3** Properties of taste bud cells (in mammals)

| Type I cells or supportive cells | Type II cells or receptor cells | Type III or presynaptic cells | Type IV cells or basal or precursor cells | Type V cells or marginal or perigemmal cells |
|---|---|---|---|---|
| Most numerous, 50–60 % | 15–30 % | Small proportion, 5–15 % | | |
| Slender spanning the entire length of the taste bud | Wider than Type I cells, most extend through the entire taste bud, some sit in the upper third without basal processes | Slender | Present in basolateral part, small, irregular in shape, do not extend processes to pore region.<br>Bundles of intermediate filaments | Present in the periphery of taste buds, if they extend into the taste bud they resemble type I cells, do not extent processes to pore region |
| Electron dark | Electron lucent, light | Moderately electron lucent | | |
| Dense core vesicles in the apical cytoplasm | | | | |
| Irregular cytoplasmic processes | | | | |
| Elongate electron-dense nuclei with small invaginations | Clear, large and round, lack invaginations | Elongate nucleus with large invaginations | | |
| Several long and slender microvilli, extending into pore | Several short microvilli of uniform length giving them brush-like appearance<br>Extending into the pore | A single large blunt microvillus extending into the pore | | |
| No synapses | No synapses but subsurface cisternae | Synapses with afferents | No synapses | No synapses |
| Ecto-ATPase, glutamate-aspartate transporter | Bitter, sweet, or umami taste receptors, α-gustducin, phospholipase C-β2, TRPM5, Gγ13, type III inositol trisphophate receptor, pannexin 1 | Denso-cored and clear vesicles at presynaptic sites<br>SNAP25, NCAM, serotonin, glutamic acid decarboxylase, chromogranin A, voltage-gated calcium channels, PKD2L1, aromatic amino acid decarboxylase | Sonic hedgehog | |

action potentials even though they lack voltage-gated calcium channels. Taste transduction releases calcium from intracellular stores which opens transient receptor potential M5 (TRPM5) channels and the resulting depolarizing cation fluxes opens pannexin or connexin hemichannels to extrude the neurotransmitter substance, ATP [9]. The nonsynaptic transmitter release is in line with the lack of synaptic contacts between type II cells and the afferents, even though afferent fibers are in close proximity to type II cells. The released ATP excites afferent fibers through binding to ionotropic purinergic receptors of the $P2X_2/P2X_3$ type and propagates gustatory information to the brain (please see below and Fig. 14.5). ATP appears to be the crucial transmitter in taste since mice which lack these channel proteins lost their responses to stimuli of all five basic tastes [6]. ATP also stimulates type II cells in form of a positive feed forward loop to boost its own release [14].

Type III cells are also called presynaptic cells. They express various neuronal marker proteins (Table 14.3 and Fig. 14.3) and are excitable, firing action potentials. They are the only cell type that forms synapses with the afferent fibers and thus express synaptic proteins. Moreover, integrity of synapses is required for gustatory function of type III cells. Type III cells appear to be the principle sour sensors. Ablation of these cells in genetically manipulated mice selectively abolished sour taste leaving detection of stimuli of the other basic tastes intact. In accordance with their role as sour sensors, acidic stimuli elicit calcium influx into type III cells leading to serotonin release. However, neither the role of serotonin in taste transduction nor the molecular identity of the sour receptor is established yet. Type III cells also detect gaseous $CO_2$ and carbonation. Carbonic anhydrase 4, an enzyme attached to the cell surface by a glycosylphosphatidylinositol anchor, functions as the principal $CO_2$ sensor. Enzymatic activity generates $HCO_3^-$ and protons which are thought to stimulate the cellular sour sensing protein.

Taste cells are short-lived with a half-life of less than 10 days [4]. Therefore, they must continuously be regenerated to replace the aged taste cells. Type IV or basal cells function as precursor cells. Phenotypically, they resemble the cells of the stratified squamous epithelium having a spherical shape (Table 14.3 and Fig. 14.3). They have no elongations and do not contact the taste pore. They do, however, express the transcription factor sonic hedgehog that has a role in taste cell development.

Finally, type V cells or perigemmal cells are found in layers in the periphery of taste buds and form a network of keratin bundles. Their shape is similar to that of type IV cells. If they extend into the taste buds they assume an elongated shape resembling type I cells. However, unlike type I cells they do not contact the taste pore (Table 14.3 and Fig. 14.3).

The fact that different cell types are responsible for sensing the different basic tastes raises the question whether all taste buds are functionally equivalent and, if not, is there a chemotopic map on the tongue? Taste buds contain different proportions of sour, sweet, umami, or bitter sensor cells (sensor cells for salty have not been sufficiently characterized). In rodents, umami cells appear to be more frequent in fungiform taste buds of the anterior tongue than in vallate and foliate taste buds of the anterior tongue. The opposite applies to the sweet sensing cells. They are more frequent in foliate and vallate than in fungiform taste buds [8]. In humans, this is different and sweet sensing cells are frequent on the anterior tongue surface, while bitter cells are more frequent in the posterior papillae but rare in taste buds of the anterior tongue. Despite the different distribution of the chemosensors, sensory experiments in humans revealed that all taste qualities can be sensed at the entire tongue surface, even though the middle of the tongue appears less sensitive in accordance with a lower taste bud density there (Fig. 14.1). Still, in accordance with the different densities of chemosensor cells, sensitivity for bitterness appears to be somewhat greater on the posterior tongue, for sweet the opposite has been observed in humans [4].

### 14.2.2 Taste Receptors and Signal Transduction

#### 14.2.2.1 GPCR-Mediated Transduction of Umami-, Sweet-, and Bitter Taste

> In vertebrates the taste receptors for sweet, umami, and bitter belong to the superfamily of GPCRs and are encoded by genes of the Tas1r and Tas2r families.

Three genes in the vertebrate genomes encode the so-called taste 1 receptors (Tas1r) [8]. They belong to the class C heptahelical G protein-coupled receptors

(GPCR) characterized by large amino-terminal domains (ATD) that contain venus flytrap binding motifs (Fig. 14.4a). A cysteine-rich domain (CRD) connects the ATD with the heptahelical domain (HD). The encoded polypeptides form obligatory dimers to form functional taste receptors. The rodent Tas1r1-Tas1r3 dimer is activated by several L-amino acids, yet the human counterpart is specific for L-glutamate, a hallmark of human umami taste. Together with the fact that receptor responses can be largely enhanced by ribonucleotides, such as GMP, IMP, or AMP, this indicates that the Tas1r1-Tas1r3 dimer is a umami receptor. Both compounds, L-glutamate and the ribonucleotide, bind to the ATD within the venus fly trap motif. The Tas1r2-Tas1r3 heteromer forms the sweet taste receptor [18].

Numerous sweet tasting compounds of different chemical classes activate Tas1r2-Tas1r3 suggesting that it is a general sweet taste receptor. Binding sites for various sweeteners have been identified [13]. They occur in both ATDs as well as in the CRD and HD of Tas1r3 (Fig. 14.4b). The HD binding sites interact with the sweeteners cyclamate and the sweet taste blocker lactisole, indicating that it functions as an allosteric modulatory site. G protein activation is likely mediated by the specific Tas1r1 or Tas1r2 subunits, not by the common Tas1r3 subunit. Numerous mammalian species possess 3 *Tas1r* genes suggesting that they have a sense for sweet and amino acid stimuli. Unlike other carnivores, cats, however, are indifferent to sweets because they lack a functional *Tas1r2* gene and are therefore unable to assemble a functional sweet taste receptor. In chickens the *Tas1r2* gene is also missing and in frogs no *Tas1r* gene has been found at all. Tas1rs are present in fish where they form amino acid receptors suggesting the ancestral Tas1rs to be amino acid receptors. Only later in evolution the Tas1r2 subunit has evolved from an amino acid receptor to a sweet receptor in accordance with the observation that the sweet amino acids glycine and alanine are strong activators of Tas1r2-Tas1r3. The evolutionary comparisons therefore clearly indicate that vertebrates can differ tremendously in their ability to taste the calorie-rich sugars and amino acids.

Bitter compounds are detected by an own family of GPCRs encoded by the taste 2 receptor (*Tas2rs*) genes [18]. These receptors have very short amino- and carboxy-termini and are not related to any of the other GPCR families, i.e., they lack sequence relationship and the typical signatures of other GPCRs (Fig. 14.4c). Vertebrate genomes vary largely in the number of *Tas2r* genes ranging from 3 in chicken to 49 in frogs. Humans possess 25 genes found in extended clusters on few chromosomal loci. The *Tas2r* gene family is characterized by rapid diversification showing lineage-specific amplifications and birth and death of genes suggesting that the *Tas2r* gene repertoire in the various species adapted to the different nutritional needs (Fig. 14.4d). Numerous bitter compounds have been identified for most human Tas2rs but only for very few Tas2rs of other species. Tas2rs are generally sensitive to several bitter substances with some receptors responding more broadly and others more restricted to bitter tasting chemicals. Three of the human receptors, however, recognize 50 % of the bitter substances tested. Thus, the ability to detect thousands of compounds as bitter appears to be due to the property of TAS2rs to accommodate numerous bitter tastants in their binding pockets. *Tas2r* genes in humans are rich in genetic polymorphisms that are responsible for the observed inter-individual perceptual differences in the population [3].

Sweet, umami, and bitter receptors are linked to similar if not identical signal transduction cascades (Fig. 14.5) [18]. Receptor stimulation results in activation of a heterotrimeric G protein consisting of $G\gamma_{13}$ and $G\beta_1$ or $G\beta_3$. The α subunit could be α-gustducin or a few other candidates [12]. Following exchange of GDP for GTP, the β/γ dimer stimulates PLC-β2 to hydrolyze the membrane lipid phosphatidyl inositol bisphosphate into inositol trisphosphate ($IP_3$) and diacylglycerol (DAG). Inositol trisphosphate binds to its type III receptor to release $Ca^{2+}$ from intracellular stores in the smooth endoplasmic reticulum. $Ca^{2+}$ opens TRPM5 and the resultant cation flux depolarizes the cell and opens hemichannels to extrude ATP [9]. α-gustducin decreases cAMP levels by a still unknown mechanism. Low cyclic AMP levels and low activity of cAMP-dependent protein kinase (PKA) are required for adequate $Ca^{2+}$ release through $IP_3$ receptor III. Therefore, α-gustducin sensitizes taste receptor cells for an adequate response. Thus, whereas sweet, bitter, and umami transduction employ different receptors in segregated cells, the intracellular signaling cascades are pretty much similar [19].

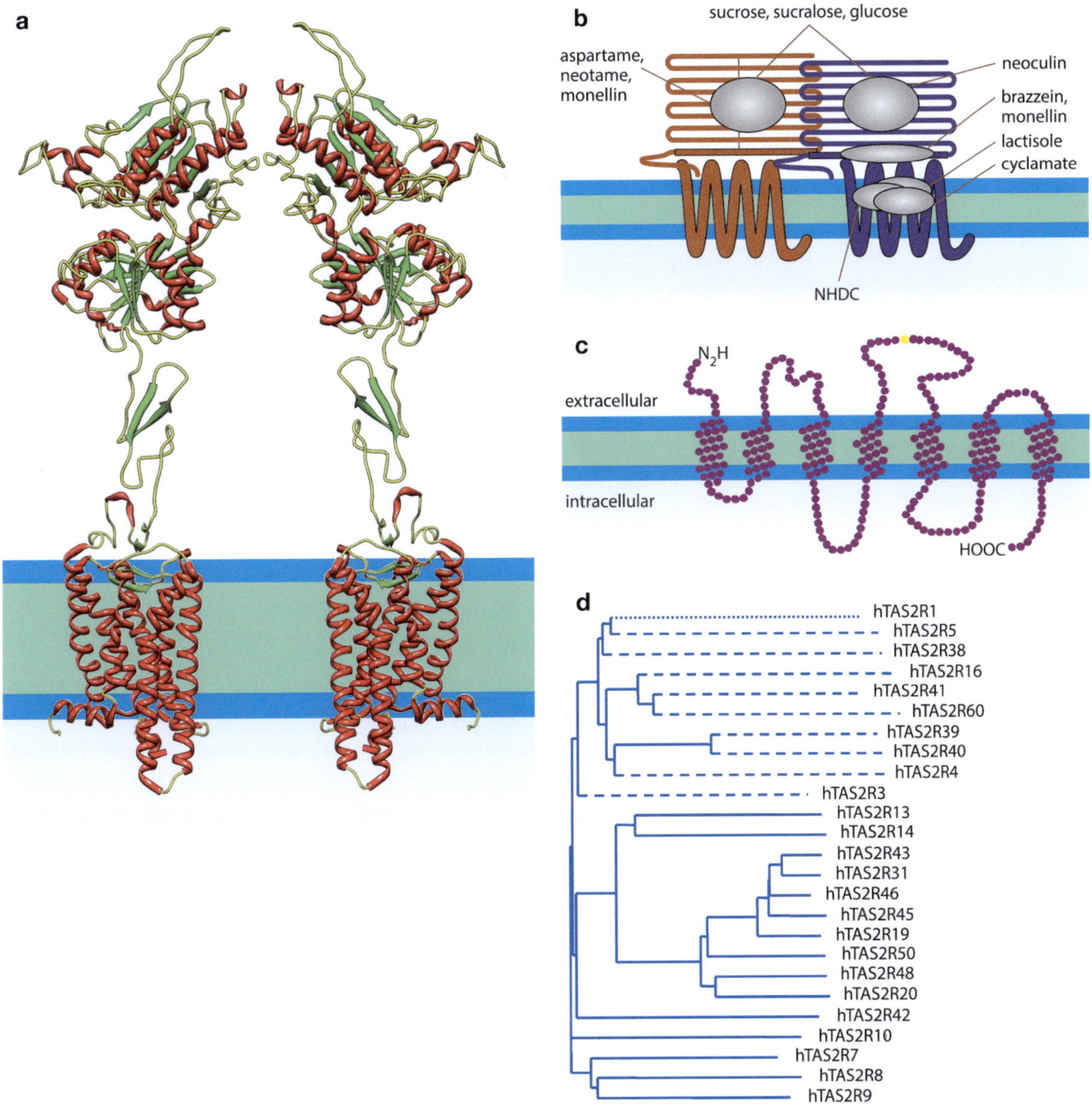


**Fig. 14.4 GPCRs as taste receptors.** (**a**) Schematic general structure of the sweet taste T1R2/T1R3 receptor. The homology model of the extracellular domain of T1R2 and T1R3 is based on the crystal structure of the extracellular domain of metabotropic glutamate receptor 3 [20] which belongs like the Tas1rs to the so-called class C GPCRs. The venus flytrap modules (*VFTM*) of both subunits with their two lobes and hinge regions as well as the cysteine-rich domains (*CRD*) connecting the *VFTM* to the transmembrane segments are visible. The 3D-structure of the transmembrane domains has been derived from the crystal structure of rhodopsin [21]. *Green*, β-structures; *red*, α-helices. The figure was prepared by using UCSF Chimera software [22] and kindly provided by Loic Briand (Dijon) (**b**) Scheme of the TAS1R2-TAS1R3 sweet receptor dimer with its identified binding sites in the amino-terminal domains of TAS1R2 and TAS1R3, the cysteine-rich domain of TAS1R3 and the transmembrane region of TAS1R3. TAS1R2 is shown in *brown*, TAS1R3 in *blue*. *NHDC*, neohesperidin dihydrochalcone. (**c**) A prototypical TAS2R. Please note that it contains very short amino- and carboxy-termini. The amino acid depicted in *yellow* marks a canonical glycosylation site present in all TAS2Rs that has been show to be indispensable for receptor biosynthesis. (**d**) The family of human *Tas2rs* depicted in a dendrogram. The length of the lines is a measure of sequence relationship. Please note the lineage specific amplification of some *Tas2rs*. The *Tas2r* genes occur in extended clusters on chromosome 12 (*solid lines*) and 7 (*broken lines*). One gene is found on chromosome 5 (*dotted line*)

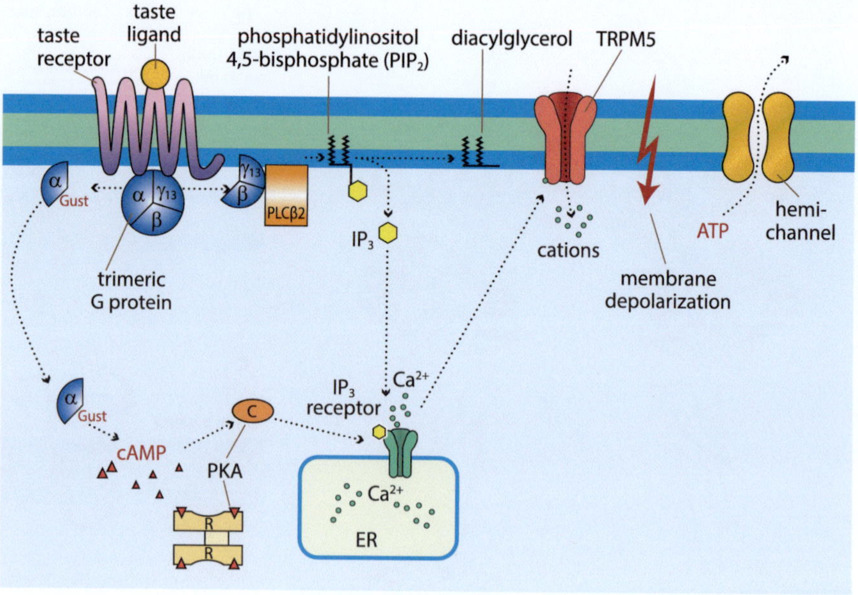


**Fig. 14.5** **Schematic outline of sweet, bitter, and umami taste signal transduction.** For details, see main text

### 14.2.2.2 Salty and Sour Transduction

> Salty taste in rodents is transmitted at least via two different pathways, one for appetitive concentrations of NaCl and one for several metal ions and ammonium. The 'receptors' for sour transduction are still unknown. Protons as well as the protonated acids appear to act as sour stimuli.

Salty taste uses at least two different transduction mechanisms [7]. One is specific for low appetitive concentrations of NaCl. This pathway can be blocked by amiloride in rodents and involves the major amiloride target, the α-subunit of the epithelial sodium channel ENaC. However, the exact channel composition has not been uncovered. Apparently, the electrochemical gradient drives sodium ions through the open ENaC pore when oral sodium concentrations rise above taste threshold. Details of the mechanisms of transmitter release remain unknown as are the exact channel composition and the taste bud cell type expressing ENaC. Although salt taste in humans is amiloride insensitive, ENaC subunits have been observed in human taste buds suggesting that it also plays a role in human salty transduction. The other salt transduction pathway mediates responses to various metal ions and ammonium. It can be pharmacologically separated from the ENaC currents by the drug cetylpyridinium chloride. However, the molecular identity has not been elucidated yet.

Astonishingly, although the cells mediating sour taste are known, sour transduction remains elusive and the receptor(s) have not been identified. Even the proximate sour stimulus is still under debate. Probably, both protons and the undissociated acid activate sour transduction explaining that weak organic acids may be more potent sour stimuli than strong inorganic acids. The undissociated organic acids could pass the plasma membrane by simple diffusion, dissociate in the cell and acidify the cytosol. A number of candidates have been proposed to mediate the action of extracellular protons, including potassium channels, activation of acid-stimulated ion-channels (ASIC) or hyperpolarization-and-cyclic nucleotide-gated-channels (HCN). Two-pore-forming potassium channels (KCNK) have been proposed as intracellular pH sensors. Nevertheless, acidic stimuli cause calcium influx and transmitter release from type III sour taste cells.

### 14.2.3 Taste Transmission to and Representation in the Brain

Taste buds are innervated by branches of three different cranial nerves: the facial nerve, the glossopharyngeal nerve, and the vagus nerve.

Fibers of three cranial nerves innervate taste buds (Fig. 14.6) [4, 7]. Those of the anterior tongue and palate are innervated by the chorda tympani and greater superficial branches, respectively, of cranial nerve VII, the facial nerve. Taste buds of the posterior tongue are contacted by fibers of the lingual-tonsillar branch of the glossopharyngeal nerve, the IX. cranial nerve. Taste buds in the larynx and pharynx are supplied by fibers of the superior laryngeal and pharyngeal branches of the X. cranial nerve, the vagus nerve. The somata of these nerves are located in the peripheral geniculate, petrosal, and nodose ganglia. Recordings classified single fibers of the peripheral nerves into distinct groups, although variations have been seen across fiber types and animal species [16, 17]. In rodents S-, N-, Q-fibers have been characterized that responded fairly specifically to sweeteners, sodium salts, and bitter chemicals, whereas the A- and H-fibers are less specific being sensitive to salts and acids, and H-fibers additionally to the bitter compound quinine. Thus, the sensitivity of peripheral nerve fibers corresponds reasonably well with that of taste receptor cells and the taste categories defined by behavioral experiments.

Fibers of the three gustatory nerves transmit taste information to the first-order neurons of the gustatory nucleus in the brainstem. There, taste information is integrated with other information relating to feeding such as stomach and gut tension as well as energy status.

Ganglionic neurons send neurites to the medulla where they contact 1st order central neurons in a small region in the rostral central division of the solitary tract nucleus (NTS), also called gustatory nucleus [7]. Primary afferent fibers of the VII. cranial nerve are located most rostral and are followed caudally by those of IX. and X. cranial nerves. Thus, the different regional chemical sensitivities of the oral cavity segregate to some extent in the central nervous system and may be used for different purposes as seen in fish. These vertebrates use the VII. cranial nerve for food seeking and the IX. cranial nerve for ingestion. In mice, each palatal or fungiform taste bud is contacted by only 3–5 geniculate neurons and each geniculate neuron sends fibers only to a single taste bud. In marked contrast, the geniculate neurons contact a much larger number of first-order neurons in the gustatory nucleus leading to multiple representations of a single taste bud in the NTS. Caudally to the sites of gustatory input the NTS also receives numerous inputs from the viscera conveying information about stomach and gut tension and their energy status. This information is integrated with gustatory information already at the brainstem level to regulate ingestive behavior. From the NTS, gustatory information is transmitted through the reticular formation to various brainstem nuclei that control tongue and jaw movements and flow of saliva. Thus, gustatory information is used in local sensory-motor reflex arches to control chewing, salivation, and swallowing of palatable food or gaping to eject food contaminated with harmful compounds [17]. Decerebrated rats and human newborns with severe cerebral malformations still show gaping or normal gusto-facial reflexes in response to aversive or appetitive taste stimuli supporting gustatory control of ingestive behavior at the brainstem level. Compared to ganglionic neurons, NTS neurons display increased breadth of tuning usually responding to stimuli of different taste modalities. They are classified as sucrose-, sodium-, HCl (acid)-, or quinine (bitter)-best cells based on their most effective stimulus. Lesions of the NTS impair the animal's ability of taste discrimination.

From the gustatory nucleus, taste information is conveyed via a thalamic relay station to the primary gustatory area in the cerebral cortex. From there taste information is transmitted to the secondary taste area in the orbitofrontal cortex to form flavors.

In primates the lemniscal pathway connects NTS neurons to the parvocellular division of the ventral

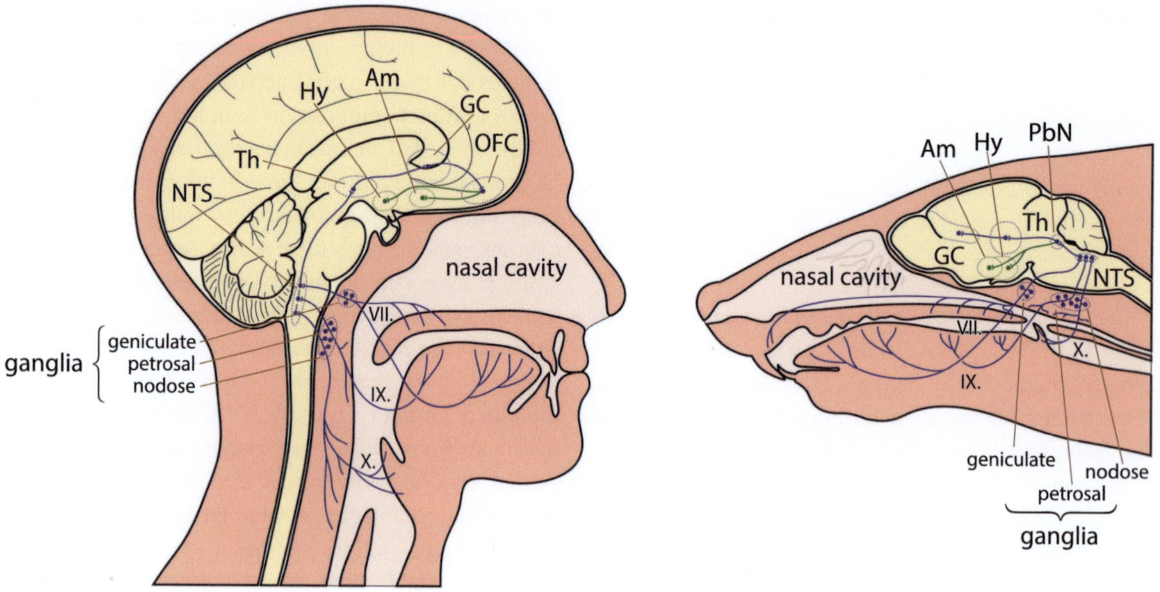


**Fig. 14.6** Transmission pathways of gustatory information in rodents and humans. Please note that in rodents there is an obligatory relay in the pontine parabrachial nucleus (*PbN*) which does not exist in humans and some other primates. Pathways shown in green differ between humans or old world monkeys and rodents. *VII.–X.*, cranial nerves; *NTS* nucleus of the solitary tract, *Th* Thalamus, *GC* gustatory cortex, *OFC* orbitofrontal cortex, *Am* amygdala, *Hy* hypothalamus, *PbN* parabrachial nucleus

posteromedial thalamic nucleus which in turn contacts gustatory neurons in the cerebral cortex. In rodents gustatory information is relayed obligatorily in the parabrachial nucleus (PBN) where gustatory information bifurcates into the thalamic-cortical axis and a ventral forebrain axis connecting to various limbic regions including the lateral hypothalamus, the central nucleus of the amygdala, the bed nucleus of the stria terminalis, the ventral pallidum and the substantia innominata [4, 7]. In primates, gustatory information reaches the limbic structures only indirectly by descending pathways from the cerebral cortical regions. Whereas the thalamo-cortical route is associated with stimulus identification, the limbic route has been related to ingestive motivation involving hedonic tone, palatability, and reward. The PBN not only processes gustatory information, yet taste and visceral information converges on single PBN neurons suggesting that it is also a major site of integrating physiological signals with taste. This role is confirmed by the observation that generation and maintenance of conditioned taste aversion, a situation caused by pairing a taste stimulus with gastrointestinal malaise, require an intact PBN. Animals with damaged PBN were unable to associate the stimulus with the malaise.

Thalamic gustatory neurons are also relatively broadly tuned to stimuli of more than one quality – many of them are multimodal responding to tactile and thermal stimuli as well. Thalamic lesions did not largely impair the ability of rodents to perceive and respond to taste stimuli but were deficient in gustatory memory. From the thalamus, taste information reaches the primary taste or gustatory area in the cerebral cortex including regions in the frontal operculum and the adjacent insula. Activities of the relatively small portion of taste neurons in these areas are thought to represent taste qualities. Imaging methods of the human and mouse brain revealed indeed somewhat segregated though overlapping regions of activity in response to various taste stimuli. Neurons in these areas respond to other stimuli associated with food including sight of food, tactile or thermal stimulation of the mouth, as well as tongue and jaw movements. From the primary taste area gustatory information is further conveyed to the secondary taste area in the orbitofrontal cortex. In this region complex representations are constructed as indicated by the multimodal responsiveness of neurons also to visual and olfactory cues. Stimulus-specific satiety appears to be generated in the orbitofrontal cortex. The secondary taste area sends gustatory information to the lateral hypothalamus and the amygdala to be used in energy homeostasis and hedonic value.

### 14.2.4 Neural Coding

Two different major theories try to explain how taste qualities are encoded, the "labeled line theory" and the "across fiber pattern theory".

The question of how taste qualities are encoded by the neural system has been under debate for several decades [16]. Two types of spatial quality coding have been proposed. The **labeled line hypothesis** purports that taste qualities are encoded by dedicated separate populations of neurons. The gustatory neurons of a given population are not tuned across modalities. Activity in this population signals the presence of a stimulus belonging to that quality. If the neuronal population is silent, a stimulus is absent. The **across fiber pattern theory** proposes that taste quality is represented by patterns of activity across the entire afferent gustatory neuronal population to which both, excited and nonexcited cells contribute. Experimental support is available for both theories. Investigations using engineered mice are largely supporting a labeled line type of coding in the periphery. Experiments with genetically engineered mice in which the taste receptors or sensor cells of each taste quality have been individually ablated clearly demonstrated that the animals lost their nerve and behavioral responses to only that taste, leaving intact all other qualities. More impressively, mice which have been manipulated to express a human taste receptor in their bitter cells acquired aversion to the receptors cognate bitter compound to which the wild-type littermates were indifferent. The same receptor expressed in the mice's sweet cells led to strong attraction. The data not only show segregated transmittance channel for each quality in the periphery, they also demonstrate that the taste cells are hard-wired to neuronal networks driving stereotyped behavioral outputs, i.e., aversion or attraction. They also indicate that the sensor cells are characterized by the taste receptor molecules they express but that the receptors do not define the identity of the sensors (for further reading please refer to [18] and references therein). Peripheral labeled lines are also supported by the mutually exclusive expression of taste receptor types of the sensor cell populations. Finally, it is consistent with the narrow tuning breadth of single peripheral nerve fibers. However, it is difficult to stipulate a labeled line type of coding with the breadth of tuning of central gustatory neurons which showed in numerous studies in various vertebrate species that they do respond to stimuli of more than one quality. Probably, it is too naïve to believe that either of the two theories applies in its most extreme outlines. Moreover, evidence is available suggesting that also the timing of neuronal activity can be used for coding. Researchers recorded a temporal pattern of spikes from a rat while it was exposed to oral stimulation with the bitter compound quinine. When they stimulated the NTS of a rat, while it was drinking water, with this pattern of spikes, the animal responded to the water as if it were the quinine solution. Future work, involving combinations of innovative physiological, anatomical, behavioral, and molecular biological methods are required to solve this important question.

## 14.3 The Insect Gustatory System

The taste-sensing organs of insects are called sensilla and pegs. They are distributed over a great part of the body surface.

Like vertebrates, insects use the sense of taste to evaluate the quality of their food [5]. Amazingly, despite the great evolutionary distance of invertebrates to vertebrates, they are similarly attracted by sugars and avoid substances that we classify as bitter. Unlike mammals, most insects are small compared to their food and have whole-body contact with it. Accordingly, the chemosensing organs for nonvolatiles are not only found in or around the mouthparts. Taste organs are also present at the insect's tarsi (the feet), the wing margins, the thorax, antennae, and the ovipositor (Fig. 14.7). The sensing organs are hair-like structures called bristles or **sensilla** and conical structures referred to as **pegs** (visible in Fig. 14.7a on the outer and inner labellar lobes). Their number and location differs across insect species and gender. The fruit fly *Drosophila melanogaster* possesses 200–300 sensilla and pegs. The taste sensilla are innervated by the single

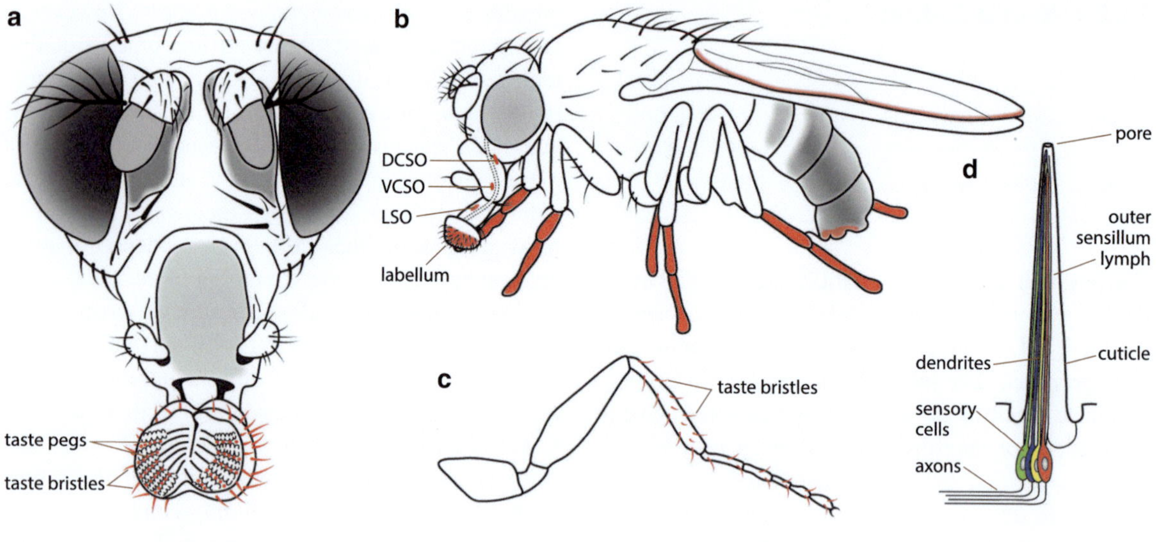


**Fig. 14.7** The gustatory system of *Drosophila*. (**a**) Fly head with proboscis. The scheme shows the labellar lobes open displaying taste sensilla on the outer and taste pegs at the inner ridges of the labellar lobes. (**b**) Fly with sites that are taste sensitive shown in *red*, i.e., proboscis, tarsae, wing margins, and ovipositor (in females). (**c**) Taste bristles on the tarsae. (**d**) A lymph-filled taste sensillum associated with various types of sensory neurons as indicated by different colors that extend their dendrites throughout the sensillum up to the pore. The axons target different regions of the fly central nervous system. For details, please see main text. *DCSO* dorsal cibarial sense organ, *VCSO* ventral cibarial sense organ, *LSO* labral sense organ

unbranched dendrites of two to six primary sensory neurons which extend to near the tip of the sensillum, whereas the somata are found at its base (Fig. 14.7d). Certain types of sensilla contain mechanosensors in addition to gustatory neurons. Their axons reach directly the central nervous system. Thus, unlike vertebrates, insects use neurons as taste sensors, not specialized epithelial cells. Taste transduction initiates when a tastant reaches the tip of a sensillum and diffuses through the receptor lymph filling the interior of the sensillum. Subsequent interaction of the taste molecule with receptors and/or ion channels of the dendritic aspects changes membrane conductance and induces depolarization.

Like mammals, insects link taste information to stereotyped behavioral outputs [18]. In response to initial contact of the tarsi sensilla (Fig. 14.7c) with attractive taste molecules, flies extend their proboscis. This results in direct contact of the tastants with the labellar sensilla and induces opening of the fly's labellar lobes (Fig. 14.7a). This in turn exposes the taste pegs that reside in the ridges of the opened labellar lobes to initiate ingestion of the food (Fig. 14.7a). Finally, the pharyngeal taste organs monitor the quality of food during ingestion (Fig. 14.7b). Repellents, in contrast, suppress the proboscis extension reflex and lead to proboscis retraction.

In *Drosophila*, the taste sensilla of the labellum are distinguished by their length. Long and short type sensilla contain four gustatory neurons classified as S-, W-, L1-, or L2-cells responding to sugars, water, and low or high concentrations of salts, respectively. Bitter compounds activate L2-type cells and inhibit W- and S-neurons. Intermediate sensilla are associated with two types of gustatory neurons responding to attractants such as sugars and low salt concentrations or to repellents such as bitter substances and high salt concentrations. Long sensilla respond best to sugars, short sensilla better to sucrose than to other sugars, and sensilla types also vary in their responses to bitter compounds.

> The insect gustatory receptors are members of the Gr family. Receptors for attractive and repulsive chemicals are segregated into different subsets of receptor cells.

The neurons associated with taste sensilla and pegs express members of the Gr family of gustatory receptors which in *Drosophila* comprises 68 members with generally little sequence conservation to detect chemical cues [15]. Most members of this family function in contact chemosensation detecting sweet or bitter compounds or nonvolatile pheromones, even though few may have distinct roles such as sensing $CO_2$ in olfactory receptor neurons. Multiple GRs appear to coexist in gustatory neurons. Although the precise subsets of receptors present in different gustatory neurons have not been fully elucidated yet, some general principles became evident. The receptor molecules Gr5a and Gr66a required for recognition of the sugar trehalose or the bitter compound caffeine are expressed in entirely distinct populations of cells [10]. Moreover, other receptors that detect aversive compounds appear to be coexpressed with Gr66a. Similarly, Gr5a is coexpressed with several other Grs that have been implicated in the attraction to numerous sugars. This segregation of receptors for attractive and repulsive chemicals into different subsets of taste receptor cells closely resembles the situation in the mammalian gustatory system. All three sites, tarsi, labellum, and pharynx deliver inputs to the suboesophageal ganglion where taste information is used to organize ingestive behavior. Interestingly, the axons of Gr5a and Gr66a mediating attraction or repulsion terminate in spatially segregated areas of the ganglion suggesting that central processing of the two types of stimuli remains separated.

## 14.4 The Chemosensory System of *Caenorhabditis elegans*

> *C. elegans* possesses three types of chemosensory sensilla in the amphid, inner labial, and phasmid organs.

Because of its simple organization, limited cell number, and ease of manipulation worms have become an attractive model to link chemosensory input to neural processing and behavioral output [1]. The worm *C. elegans* lives in the soil where it moves to search ambient environmental conditions, access food, escape noxa, identify mating partners, and find favorable sites for egg laying. In the absence of vision and audition the worm mainly uses thermal, mechanical, and particularly chemical signals to navigate. For detecting food or signs of food as well as for avoiding noxa the worms use both, volatile and soluble cues. Sometimes this is referred to as olfaction and taste. The chemical cues induce positive and negative chemotaxis as well as the avoidance reflex, i.e., the inversion of the direction of movement in response to an abruptly encountered repellent. Moreover, chemosensation is also used by the worm to regulate physiological processes including entry into and exit from an alternative larval stage called 'dauer', lifespan, fat storage, and body size.

*C. elegans* possesses three types of chemosensory sensilla in the amphid, inner labial, and phasmid organs. The amphids are bilaterally symmetric and, like the six labial sensilla located at the sides and around the mouth. Each amphid contains 12 different types of neurons which extend through sensory openings of the cuticle to expose their cilia to the environment. One type is tuned to thermal, not chemical, cues. The cilia are embedded by glial-like cells called sheath and socket cells. The somata of the amphid neurons lie in the head between the two pharyngeal bulbs. The axons are connected and contact other neurons. The six labial sensilla are each associated with the endings of two sensory neurons of which only one pair reaches the outside of the animal. The phasmids which are also bilaterally symmetric and equipped with two sensory neurons reside in the worm's tail. By laser microbeam-guided cell killing different stimuli and various output behaviors have been assigned to the chemosensory neurons (Table 14.4). The table shows that the *C. elegans* chemosensory system detects each substance by more than one neuron and each neuron recognizes more than a single compound [2]. However, sensitivity to repellents and attractants appear to segregate to different sets of neurons with the exception of ASK and ASE cells. Although the phasmid neurons PHA and PHB sense repellents, they are not mediating avoidance, but rather they counteract the avoidance response of the amphid neurons. This polarized organization of repellent detectors enables the worm to perform the very rapid avoidance reflexes.

> Worms find their way through chemotaxis critically involving chemosensory information transmitted by the ASE neuron pair.

The ASE neuron pair is the most relevant gustatory neuron since its ablation affects sensitivity to all known

**Table 14.4** Properties of *C. elegans'* sensory chemosensory neurons

| Neuron[a] | Response | Stimuli |
|---|---|---|
| ASE | Water-soluble chemotaxis, egg laying | Food, lysine, cAMP, $Na^+$, $Cl^-$ biotin, serotonin |
| AWC | Volatile chemotaxis, lifespan, navigation | Benzaldehyde, butanone, thiazole, isoamyl alcohol, pentanedione |
| AWA | Volatile chemotaxis | Diacetyl, pyrazine, thiazole |
| AWB | Volatile avoidance | Octanol, nonanone |
| ASH | Polymodal nociception, osmotic and nose touch avoidance, chemical avoidance, social feeding | Nose touch, $Cu^{2+}$, $Cd^{2+}$, quinine, octanol, $H^+$, high ionic strength, sodium dodecylsulfate, octanone, nonanone |
| ASI | Dauer formation, navigation | Food, pheromone, lysine, cAMP, $Na^+$, $Cl^-$, biotin |
| ADF | Dauer formation | cAMP, $Na^+$, $Cl^-$, biotin |
| ASG | Dauer formation, lifespan | Lysine, cAMP, $Na^+$, $Cl^-$, biotin |
| ASJ | Dauer formation and recovery, lifespan | Food, pheromone |
| ASK | Chemotaxis, lifespan, navigation, egg laying | Food, lysine |
| ADL | Social feeding | Octanone, nonanone |
| URX, AQR PQR | Aerotaxis, social feeding | Oxygen |
| PHA | Antagonistic avoidance | Sodium dodecylsulfate |
| PHB | Antagonistic avoidance | Sodium dodecylsulfate |

[a]A neurons are located in the amphids, P neurons in the phasmids

classes of gustatory cues including amino acids, salts, and metabolites, while simultaneous ablation of all other amphid and phasmid neurons spares chemotaxis. Ablation of the ASE neurons also affects migration in chemical gradients. *C. elegans* does not respond to the absolute level of attractants. Instead it measures the concentration change over time using an unknown molecular or neural mechanism. A worm travels up a gradient of attractant in long phases of even movement rarely interrupted by pirouettes to change direction. Pirouettes are more frequent when traveling down the gradient. In this way the worm moves more in the right direction than in the wrong direction. Thus, the animal uses short-term memory of chemical concentrations to maintain its direction of migration or randomly move into a different one.

The chemoreceptors in *C. elegans* also belong to the superfamily of GPCRs but lack vertebrate orthologs.

The genes that encode chemoreceptors are amazingly numerous in the worm. Chemosensory GPCR genes amount to ~1,400 and additional ~400 pseudogenes exist [1]. The gene sequences are highly divergent and do not resemble any of the mammalian chemosensor gene families. The number of genes exceeds those in any known vertebrate species reflecting the worm's strong dependence on chemical signals. This high number also indicates that the very few chemosensory neuron types in *C. elegans* express multiple chemoreceptor genes. Thus, the precise combination of receptors and their agonists determines the broad response profile of the worm's chemosensory cells enabling the animal to cope with the complex mixture of chemicals simultaneously present in its environment. Like the mammalian genes, the *C. elegans* chemoreceptor genes are often found in clusters indicating rapid diversification through duplication and loss of genes. Odr-10 was the first chemosensory receptor for which a ligand, i.e., diacetyl has been identified. Loss of function mutations fail to sense the attractive substance and manipulated animals expressing Odr-10 in AWA neurons acquire avoidance to it, impressively illustrating the crucial role of this type of chemosensor molecules in the worm. The *C. elegans* genome also contains a number of genes encoding downstream signaling molecules, including but not limited to G protein subunits, cGMP-gated ion channels, guanylate cyclases, transient receptor potential channels, GPCR kinases, protein phosphatases and kinases, cGMP-dependent kinase, and types of stimulus-specific molecules that lack mammalian counterparts.

## 14.5 Role of Taste in Feeding and Nutrition

Taste is linked to various important physiological systems.

We have seen above that the primary role of the sense of taste is to evaluate the quality of food. The food should be rich in calories and low in toxin content. Within seconds an animal must decide to ingest or reject the food from the oral cavity. Erroneous acceptance of food with a high toxin load could cause intoxication leading to immediate death or long-lasting malaise whereas rejection of calorie-rich food and a low toxin load would result in malnutrition making the animal vulnerable to disease and handicap it in competition for food and reproduction. It is evident that powerful and well-controlled interactions with multiple other physiological systems ensure that taste sensation appropriately fulfils its complex task [17]. Such mechanisms include reflex arches at the brainstem level that regulate unconscious rejection, swallowing, vomiting, and salivation. Another example is the integration of taste with the needs of energy homeostasis in the ventral forebrain.

Tastes are paired with postingestive consequences for taste preference and aversion learning.

Perhaps the most important phenomenon underlying appropriate nutrition is the formation of taste preferences and aversions [7]. We could also refer to flavor preferences and aversions, because somatosensory and olfactory cues make large contributions. Only some taste preferences appear to be innate, such as the liking of sugars and the dislike of bitter substances. Most others have to be learned through the formation of a taste recognition memory which allows the animal to classify the taste of previously encountered food into 'safe' and 'poisonous'. Young animals and humans are neophobic towards novel food. When potential food or tastant is encountered first time, consumption is low minimizing the risk of fatal effects should the food be toxic. The physiological consequences of the ingested material determine the animals' future behavior. If the ingested material lacks toxins and contains calories such that the animal doesn't get sick but satiated, its neophobia vanishes and it will increase consumption on future encounters of that particular food. Had the opposite happened and the animal survived, aversion towards that food becomes manifest and the animal rejects that kind of food. Thus, the animals associate the flavor of a food with its postingestive consequences and by doing so they learn to prefer certain flavors and avoid others. In this way, they condition themselves to reject dangerous food. This taste recognition memory is remarkably stable and can last for a lifetime. This associative learning is not confined to mammals or vertebrates, it is observed in worms and insects and even in more primitive insect larvae. Conditioning of taste preferences or aversions has been frequently used to experimentally monitor animals' ingestive behavior. In these experiments a conditioned stimulus (a flavor) is paired with an unconditioned stimulus (for example, a foot shock or an injection of lithium chloride into the peritoneum of a mouse to induce visceral malaise). Using this paradigm much has been learned about the time profile of taste preference/aversion, the behavior towards new or familiar flavors, as well as taste categories in animals.

Neither have all brain areas been identified that engage in the formation of a taste recognition memory nor have all transmitter systems of this process been uncovered. However, different taste neuronal networks are crucial for taste memory. For example, the gustatory part of the NTS is required for the attractive or aversive innate but not learned responses to tastants. The insula, however, is critical for safe taste memory formation but not for taste responsiveness. Release of the neurotransmitter acetylcholine and activation of muscarinic receptors in the insular cortex accompany the presentation of a taste stimulus. Repetitive stimulations lead to reduced release eventually declining to the level of water suggesting that the muscarinic pathway signals the novelty of a stimulus during the initial stages of taste memory formation and adapts if the stimulus is familiarized. Thus, in the absence of malaise muscarinic signaling is critically involved in safe memory formation. This trace, however, can be modified for establishing aversive taste memory formation. In this case the experienced taste has to be associated with the visceral painful consequences of the unconditioned stimulus. This association requires the activation of glutamatergic pathways from the viscera through the vagus nerve to and in the amygdala. For the establishment of long-term taste memory *de novo* protein synthesis and posttranslational modification of neural proteins are crucial in the insula and amygdala.

This type of memory divided in two branches, safe and aversive, has also been observed in invertebrates. *Drosophila* flies learn to associate electric shock punishments or sugar rewards with an odor. Both responses require cAMP signaling in mushroom body neurons. However, whereas association of odors with attractive cues generally requires octopamine, association to aversive stimuli generally involves dopamine (see Chap. 26).

## References

1. Bargmann CI (2006) Chemosensation in *C. elegans*. In: Wormbook. The *C. elegans* Research Community. doi: 10.1895/wormbook.1.123.1. http://www.wormbook.org [Review]
2. Bargmann CI, Horvitz HR (1991) Chemosensory neurons with overlapping functions direct chemotaxis to multiple chemicals in *C. elegans*. Neuron 7:729–742
3. Bufe B, Breslin PA, Kuhn C, Reed DR, Tharp CD, Slack JP, Kim UK, Drayna D, Meyerhof W (2005) The molecular basis of individual differences in phenylthiocarbamide and propylthiouracil bitterness perception. Curr Biol 15:322–327
4. Doty RL (ed) (2003) Handbook of olfaction and gustation. Marcel Dekker, New York
5. Ebbs ML, Amrein H (2007) Taste and pheromone perception in the fruit fly *Drosophila melanogaster*. Pflugers Arch 454:735–747 [Review]
6. Finger TE, Danilova V, Barrows J, Bartel DL, Vigers AJ, Stone L, Hellekant G, Kinnamon SC (2005) ATP signaling is crucial for communication from taste buds to gustatory nerves. Science 310:1495–1499
7. Firestein S, Beauchamp GK (eds) (2008) The senses. Elsevier, Amsterdam
8. Hoon MA, Adler E, Lindemeier J, Battey JF, Ryba NJ, Zuker CS (1999) Putative mammalian taste receptors: a class of taste-specific gpcrs with distinct topographic selectivity. Cell 96:541–551
9. Huang YJ, Maruyama Y, Dvoryanchikov G, Pereira E, Chaudhari N, Roper SD (2007) The role of pannexin 1 hemichannels in ATP release and cell-cell communication in mouse taste buds. Proc Natl Acad Sci USA 104: 6436–6441
10. Ishimoto H, Matsumoto A, Tanimura T (2000) Molecular identification of a taste receptor gene for trehalose in *Drosophila*. Science 289:116–119
11. Kawai K, Sugimoto K, Nakashima K, Miura H, Ninomiya Y (2000) Leptin as a modulator of sweet taste sensitivities in mice. Proc Natl Acad Sci USA 97:11044–11049
12. McLaughlin SK, McKinnon PJ, Margolskee RF (1992) Gustducin is a taste-cell-specific G protein closely related to the transducins. Nature 357:563–569
13. Meyerhof W, Korsching S (eds) (2009) Chemosensory systems in mammals, fishes, and insects. In: Richter D, Tiedge H (series eds) Results and problems of cell differentiation. Springer, Heidelberg
14. Roper SD (2007) Signal transduction and information processing in mammalian taste buds. Pflugers Arch 454:759–776 [Review]
15. Scott K, Brady RJ, Cravchik A, Morozov P, Rzhetsky A, Zuker C, Axel R (2001) A chemosensory gene family encoding candidate gustatory and olfactory receptors in *Drosophila*. Cell 104:661–673
16. Simon SA, de Araujo IE, Gutierrez R, Nicolelis MA (2006) The neural mechanisms of gustation: a distributed processing code. Nat Rev Neurosci 7:890–901 [Review]
17. Spector AC, Travers SP (2005) The representation of taste quality in the mammalian nervous system. Behav Cogn Neurosci Rev 4:143–191 [Review]
18. Yarmolinsky DA, Zuker CS, Ryba NJ (2009) Common sense about taste: from mammals to insects. Cell 139:234–244 [Review]
19. Zhang Y, Hoon MA, Chandrashekar J, Mueller KL, Cook B, Wu D, Zuker CS, Ryba NJ (2003) Coding of sweet, bitter, and umami tastes. Different receptor cells sharing similar signaling pathways. Cell 112:293–301
20. Muto T et al. (2007) Structures of the extracellular regions of the groupII/III metabotropic glutamate receptors. Proc Natl Acad Sci USA 104:3759–3764
21. Palczewski K et al. (2000) Crystal structure of rhodopsin: a G protein-coupled receptor. Science 289:739–745
22. Pettersen EF et al. (2004) UCSF Chimera – a visualization system for exploratory research and analysis. J Comput Chem 25:1605–1612

# Thermosensation

# 15

Carlos Belmonte Martinez and Elvira de la Peña García

## 15.1 Influence of Temperature on Living Organisms

All biophysical and biochemical processes in living creatures depend critically on temperature and operate optimally within a narrow temperature range. However, organisms are subject to a continuous exchange of heat energy with their environment. For instance, a living organism exposed to the Sun on a beach will be heated by energy radiated from the Sun above (radiation) and energy conducted by the sand below (conduction). Winds from the sea will also help to transfer energy because the kinetic energy of the molecules of gas or liquid in contact with the body (air, in our example) increases by heat transfer, reducing its density. The less dense region will rise, to be substituted by cooler, denser air. This creates a fluid movement called convection currents.

Variation of environmental temperature values in the habitats of animals is considerable. Temperature at the Earth's surface may range from −65 to +70 °C while surface water temperatures vary from −2 °C up to 100 °C in geothermally heated waters. The average climate is additionally subject to seasonal changes with frequent shifts in air temperature, precipitation in the form of rain or snow, and solar radiation.

Living organisms are thus continuously challenged to actively regulate their body temperature. Accordingly, during evolution they developed highly sensitive thermosensing mechanisms that detect temperature changes and provide critical information for internal temperature regulation. The strategies adopted for temperature sensing and regulation vary among species, as the result of the evolutionary adaptation to their specific environments. In pluricellular animals, thermosensor molecules are incorporated into specialized thermoreceptor cells that supply information on both environmental and internal temperature values, and serve ultimately to regulate metabolic and behavioral responses aimed at keeping a relatively constant body temperature in spite of external thermal oscillations.

**Box 15.1: Temperature**

Temperature is a physical property of matter that can be defined in terms of the kinetic energy content of the atoms and molecules comprising the body (internal energy of the body), or in terms of the thermal equilibrium between two systems in thermal contact. If they are not in equilibrium, energy (heat) is transferred from the body with the higher thermal energy to the other with lower thermal energy. Temperature measurements are made quantitative by introduction of a temperature scale. The Centigrade scale uses the melting point of pure ice as 0 °C and the boiling point of pure water at 1 atm pressure as 100 °C. In the Kelvin scale, these two points correspond, respectively, to 273 and 373 K; the absolute zero temperature (0 K) corresponds to an absence of atomic or molecular movement.

C.B. Martinez (✉) • E. de la Peña García
Instituto de Neurociencias,
Universidad Miguel Hernández-CSIC, Avda. Ramón y Cajal s/n,
Sant Joan d´Alacant, 03550 Alicante,
España
e-mail: carlos.belmonte@umh.es; elvirap@umh.es

C.G. Galizia, P.-M. Lledo (eds.), *Neurosciences - From Molecule to Behavior: A University Textbook*,
DOI 10.1007/978-3-642-10769-6_15, © Springer-Verlag Berlin Heidelberg 2013

Molecular sensors are specific for transduction of those environmental energy changes that are particularly relevant for survival. Most forms of energy representing sensory stimuli (light, mechanical forces, chemical compounds) act transiently on living organisms. In contrast, thermal energy affects continuously every molecule. Therefore, in the case of thermosensing, a temporary interaction between heat energy and its specific sensor is not possible because all molecules are already influenced to a variable degree by temperature and the condition of no stimulus is only obtained at absolute zero values. Thermodetection is thus based upon discrimination of temperature gradients. Thermal sensors in living organisms change their state according to variations in absolute temperature values and produce a response proportional to the gradient between different temperature levels. The capability of thermosensors to modify their state within a larger or smaller temperature range determines their sensitivity. A useful way to express the temperature dependence of a process is through the $Q_{10}$ temperature coefficient that measures the rate of change of a biological or chemical system as a consequence of increasing the temperature by 10 °C. In the case of thermosensors, their $Q_{10}$ will be given by the slope of the stimulus–response curve within this 10 °C range. $Q_{10}$ ranges between 1 and 3 in most biological processes but exceeds 10 in specific thermosensors.

## 15.2 Temperature Sensing

Sensing of environment's temperature is already present in aneural organisms like bacteria, ciliates, and cellular slime molds. Bacteria preferentially growing at one particular temperature accumulate at that temperature when they are exposed to a thermal gradient. A similar effect is observed in ciliates – they exhibit thermotaxis, i.e., migration guided by differences in temperature that can be positive or negative (swimming towards the optimal temperature or away from potentially harmful thermal values). This is due to change in the pattern and frequency of ciliary or flagellar beating that cause a change in navigation direction. In *Paramecium*, reversal of the direction of ciliary beating (ciliary reversal) occurs when the cell swings away as part of the avoidance response to potentially harmful temperatures detected through thermosensitive molecules.

Pluricellular animals are equipped with specialized thermoreceptor cells with a remarkable capacity to detect environmental temperature changes. These organisms have additionally developed different strategies to use such thermal information for body temperature regulation. **Ectothermic** (or poikilothermic) organisms, that include, for instance, arthropods, nematodes, and reptiles, lack the capacity to regulate the endogenous generation of heat (**thermogenesis**). They usually inhabit thermally heterogeneous environments and are forced to exploit that ambient heterogeneity to maintain body temperature within a relatively narrow range despite massive spatial and temporal changes in ambient conditions. For instance, they bask in sunlight to regulate their temperature (**heliotherms**), or press their bodies against warm substrates to facilitate heat flow (**thigmotherms**), or rely on large body mass to maintain thermal constancy (**gigantotherms**). **Endothermic** (or homeothermic) animals, such as mammals and birds use heat-producing mechanisms (shivering and nonshivering thermogenesis) and adaptive behaviors (search of favorable places, adjustments of the exposed body surface) to maintain a steady core temperature. To avoid mismatches between energy supply and demand, many species of endotherms use daily torpor and/or hibernation (heterothermy) in extreme environments.

The ratio of the temperature gradient to the heat flux is known as the volume conductivity. The efficacy of the thermoregulating mechanisms depends on various factors. Living tissues are in general poor conductors and animals often have developed insulators such as fat, blubber, feathers, and fur to reduce heat loss. The body surface area is another important factor in determining heat flux because heat loss is proportional to the body surface area. Surface generally increases with body mass, raised to an exponent close to 0.75. A number of key ecological traits depend on body mass and temperature, because the rate at which organisms process energy and materials determines how fast they grow and reproduce, how long they live, and how rapidly they move through space.

## 15.3 Thermotransduction Molecules

Several classes of thermosensitive molecules have been identified as thermal sensors. In most cases these are ion channel proteins whose gating by temperature initiates the cascade of events that lead ultimately to metabolic and/or behavioral responses to external temperature changes.

All ion channels are thermosensitive but some of them have been evolutionarily developed as temperature

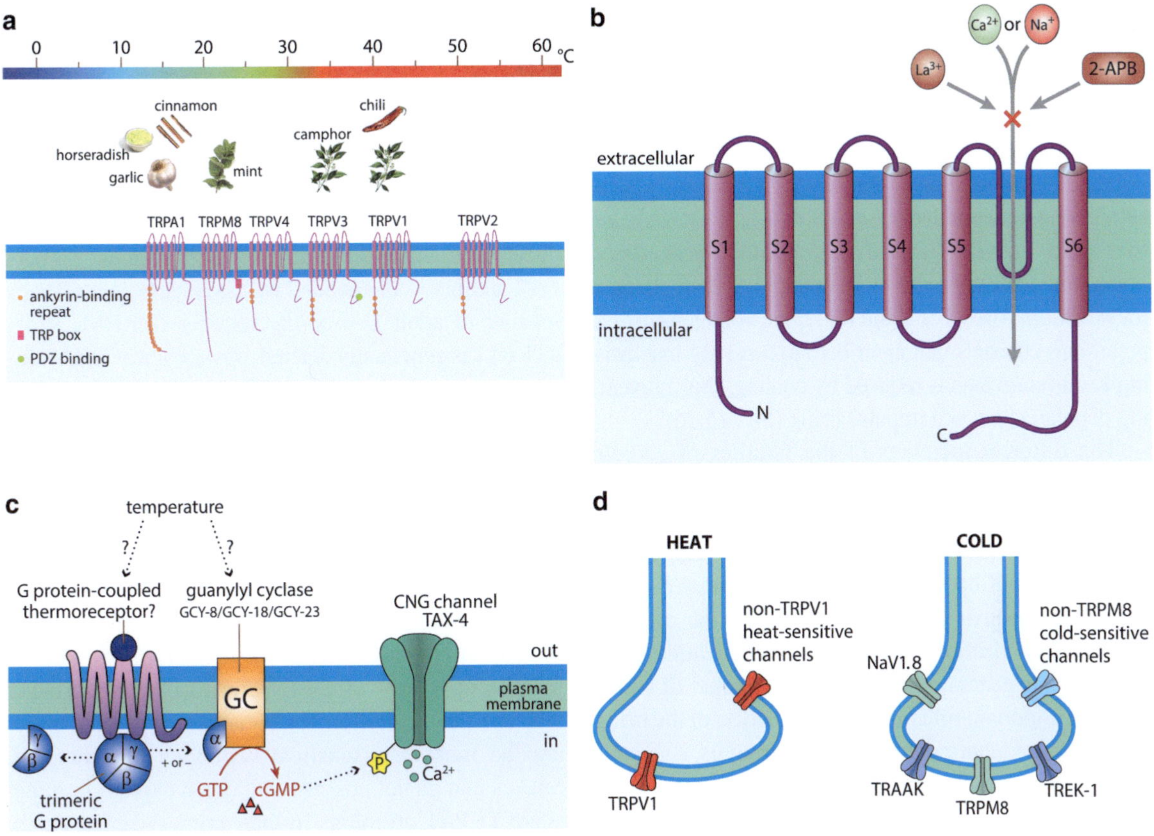


**Fig. 15.1** (**a**) **TRP channels belong to the large superfamily of cation channels.** They differ in ion selectivity, modes of activation, and physiological functions, and only some, as those shown in the figure, are temperature-sensing (From Dhaka et al. [7]. With permission). (**b**) TRP channels possess six transmembrane-spanning segments *S1–S6* forming a transmembrane domain with a pore loop inserted between S5 and S6 (From Clapham et al. [5] with permission from Macmillan Publishers Ltd. [Nature Reviews Neuroscience], Copyright (2001)). (**c**) In *C. elegans* temperature is indirectly sensed through cyclic nucleotide gated (CNG) channels. Temperature possibly acts on a thermoreceptor molecule, leading to changes in intracellular cGMP concentration via the function of three guanylyl cyclases; (From Inada et al. [14]. With permission). (**d**) Two models of sensory nerve terminals that express several transduction molecules. Heat-sensitive nociceptors express TRPV1 and possibly other thermosensitive channels. The majority of cold-sensitive afferents express TRPM8 and a host of potassium channels (such as TRAAK and TREK-1) and other ion channels that modulate their excitability (From Basbaum et al. [2] with permission from Elsevier, Copyright (2009))

detectors. This is the case of thermosensitive transient-receptor potential (TRP) channels, which are members of the superfamily of **TRP ion channels** displaying a high sensitivity to temperature within a defined range. There are three distinct families of thermoresponsive TRP channels: **TRPV**, **TRPA**, and **TRPM** (Fig. 15.1a). TRP channels contain six transmembrane domains with large cytosolic carboxyl and amino termini, a pore loop region, and in some cases, several ankyrin repeat domains in the amino terminus (Fig. 15.1b, see also Chap. 7). Four subunits co-assemble to form a functional ion channel. Thermosensitive TRP channels are directly gated by changes in temperature and behave as nonselective cationic channels, i.e., when opened they allow the flow of $Na^+$, $K^+$, and $Ca^{2+}$ ions through the cell membrane following their electrochemical gradient. Gating temperatures are different for the various types of TRP channels, covering the range of thermal oscillations at which living organisms are usually exposed. Moreover, the properties of temperature-sensitive TRP channels are under strong evolutionary pressure to conform to a physiologically relevant temperature range.

In addition to TRPs, other temperature-sensitive ion channels participate in thermal transduction. These include temperature-sensitive, cyclic nucleotide-gated (CNG) ion channels, like cGMP-gated channels of thermosensory neurons in the nematode *C. elegans* (Fig. 15.1c). Other ion channel classes belonging to the large family of potassium channels exhibit remarkable temperature dependence and contribute to shape the final thermal sensitivity of specific thermoreceptor neurons in vertebrates. Leak or background potassium channels are opened at temperatures around 35 °C and tend to hyperpolarize neurons, closing rapidly and reversibly with small temperature reductions: TREK-1 and TRAAK are two pore domain ($K_{2P}$) thermosensitive background potassium channels with a functional role in thermal transduction. Also, Kv1.1 and Kv1.2 are voltage-sensitive potassium channels that contribute to a slowly inactivating K current which is reduced by cooling, thus preventing depolarization and impulse firing (Fig. 15.1d).

The different members of the families of sodium and calcium voltage-sensitive ion channels involved in the regulation of neuronal excitability have variable temperature-dependence and participate in shaping the firing pattern of nerve impulses by peripheral endings of thermosensitive neurons. This is also the case for the hyperpolarization-activated cyclic-nucleotide-gated channels that generate an ionic current named **Ih** which plays an important role in the determination of the firing pattern of cold thermoreceptors in mammals.

In summary, an ample group of ion channels and other membrane molecules characterized as 'thermosensitive' contributes variably to build the temperature transduction mechanisms evolutionarily developed by living organisms of the different phyla for thermal adaptation to the environment. In the following, we present a few specific examples across the animal kingdom.

In **ciliates**, lipids surrounding the calcium channels in the ciliary membrane act as thermotransducers. Thermal gradients acting on them cause a change in calcium conductance, an increase in intraciliary $Ca^{2+}$ levels and thereby a change in the internal calcium concentration and membrane depolarization. The existence of two sensors, one regulating positive thermotaxis and the other regulating negative thermotaxis has been proposed.

In the **nematode** *C. elegans*, temperature modulates the intracellular calcium concentration of the thermosensitive AFD neurons via a cGMP-dependent signaling cascade and a cGMP-gated channel. The precise mechanism that activates guanylyl cyclase activity is still incompletely known (Fig. 15.1c). This indirect thermosensory mechanism, similar to the nonlinear amplification process found for vision and olfaction allows that temperature oscillations as small as 0.1 °C generate either cooling- or warming-activated thermoreceptor currents, thus providing *C. elegans* with a thermosensitivity similar to snakes and several orders of magnitude higher than the one provided by mammalian TRP channels.

In *Drosophila,* adult animals possess distinct sensors for innocuous cool and warm temperature detection. Brv1, a member of the TRPP (polycistin) subfamily of TRP ion channels appears to be the peripheral cold transducer, in conjunction with two other channel proteins brv2 and brv3, all encoded by the *brivido* genes. Warm sensing in adult *Drosophila* requires dTRPA1. When dTRPA1 is genetically deleted, physiological and behavioral responses to warmth are disrupted, eliminating heat responses in brain thermosensory neurons and causing flies to accumulate in warmer than normal regions on a thermal gradient. Noxious heat over 38 °C is transduced by two separate members of the TRPA subfamily of ion channels, named Painless and Pirexia, located in peripheral sensory nerves and central neurons.

Pit-bearing **snakes** (vipers, pythons, and boas) use the cation channel TRPA1 for temperature detection. TRPA1 orthologs of these infrared-detecting snakes that use heat radiation for localization of their preys are the most heat-sensitive vertebrate ion channels. Snakes that do not possess a pit organ express less sensitive TRPA1 orthologs in their primary sensory neurons for thermal sensitivity of the skin surface.

In **birds** and **mammals** transduction of innocuous and noxious thermal stimuli are mediated by distinct classes of TRP channels. Innocuous cold detection is mediated by TRPM8 channels that are vigorously activated by temperatures below 26 °C and by menthol and other cooling agents. As mentioned above, other thermosensitive ion channels contribute to specific cold thermosensitivity. These include background $K_{2P}$ channels of the TREK family, in particular TREK-1 that are partially open at resting temperatures and close with cooling, thus depolarizing the neuron. In addition, specific cold-sensitive neurons lack or express poorly the thermosensitive Kv1 potassium channel, that carries a slowly inactivating current ($IK_D$) thus acting in most sensory neurons as a break against cold-induced depolarization. The expression level of these channels determines the variable cold sensitivity found among different cold thermoreceptor afferent fibers of mammals. Several TRP candidates have been identified as transduction channels for warmth. TRPV4 responds to moderate temperature increases and is located in nerve endings and also in some of their surrounding cells, such as the keratinocytes. TRPV3, another TRP channel sensitive to warm temperatures is expressed only by keratinocytes and may also contribute to detection of moderate temperature increases.

In addition, in mammals distinct TRP channels are involved in the detection of cold and heat at noxious levels. TRPA1 appears to be important for transduction of noxious cold. Additionally, the sodium channel Nav1.8 plays a critical role in noxious cold detection because unlike other classes of voltage-activated sodium channels it is not inactivated by low temperatures, thus allowing the generation of propagated nerve impulses at nociceptive terminals during strong cold.

Noxious heat transduction in birds and mammals involves the polymodal channel TRPV1, which is activated by temperatures over 42–43 °C. In mammals, TRPV1 is also activated by capsaicin, while the homolog channel of birds is capsaicin-insensitive. Thermosensory cells of warm-blooded vertebrates also express TRPV2, a channel with a high thermal threshold (over 52 °C). TRPV1 is responsible for heat activation in a fraction of heat nociceptors (type II), while TRPV2 channels possibly mediate responses to heat of another subclass (type I) of heat nociceptor fibers (see below). Recently, anoctamin 1 (ANO1), a calcium-activated chloride channel that is activated by temperatures over 44 °C, has been proposed as another heat transducer channel in mammalian nociceptor neurons.

Selective elimination of the different molecular candidates for thermotransduction using gene deletion techniques evidenced that thermal sensitivity in the intact animal ultimately depends on multiple thermosensitive ion channels whose interplay determines the final level of sensitivity of thermosensory endings as well as their final impulse firing response to temperature changes. Moreover, many of the transduction channels involved in specific thermotransduction appear to be modulated by endogenous mediators that participate in processes such as inflammation, fever, etc. This may be the basis of the changes in thermal sensitivity and thermoregulatory responses often observed in mammals, including humans under many pathological conditions.

Figure. 15.2 summarizes the sensitivity and contribution of the different mammalian TRP channels to temperature detection by the various types of mammalian thermosensory neurons. It must be emphasized that thermosensitive channels are not always highly selective for temperature, and are often activated also by other modalities of stimuli like mechanical forces (e.g., TRPA1), osmolarity (TRPV4), exogenous irritants, or endogenous inflammatory mediators (TRPV1 and TRPA1).

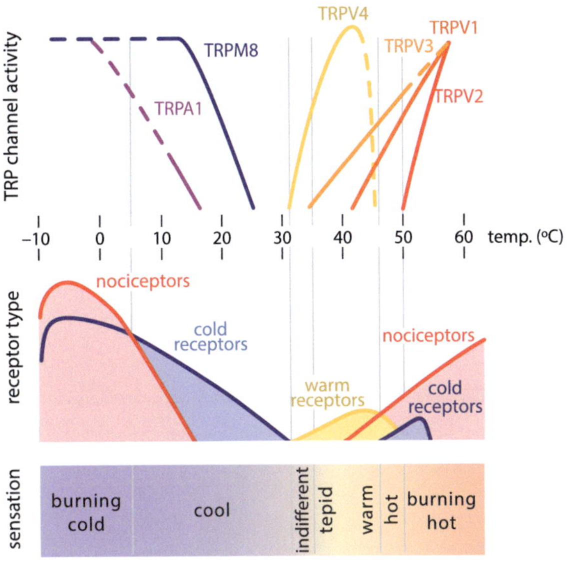


**Fig. 15.2 Hypothetical correspondence between activation of TRP channels, body surface temperature and evoked sensations.** *Upper part*: Schematic representation of the thermal activation profile of various TRP channels when expressed in recombinant systems. All of them have been located in sensory neurons and/or skin cells (Adapted from Patapoutian et al. [17] with permission from Macmillan Publishers Ltd. [Nature Reviews Neuroscience], Copyright (2003)). *Middle part*: Schematic representation of impulse activity in various cutaneous sensory receptors during application to their receptive fields of temperatures indicated in the thermal scale. *Lower part*: Quality of sensations evoked in humans by application to the skin of different temperature values (From Belmonte and Viana [3]. With permission)

## 15.4 Thermoreceptor Cells

Cells expressing thermal transduction channels possess peripheral branches that end on the body surface, directly exposed to environmental temperature oscillations. They distinguish from other modalities of sensory receptors by their high sensitivity to small temperature variations and the capacity to encode in their impulse firing frequency both the static values and the dynamic changes in temperature of the tissues where they are located as well as by their relative insensitivity to other stimulating forces. Nonetheless, the functional and morphological characteristics of thermoreceptor cells and of their surrounding accessory structures vary widely along the animal evolutionary tree, reflecting the adaptative strategies developed by the different species for survival in their specific environmental conditions.

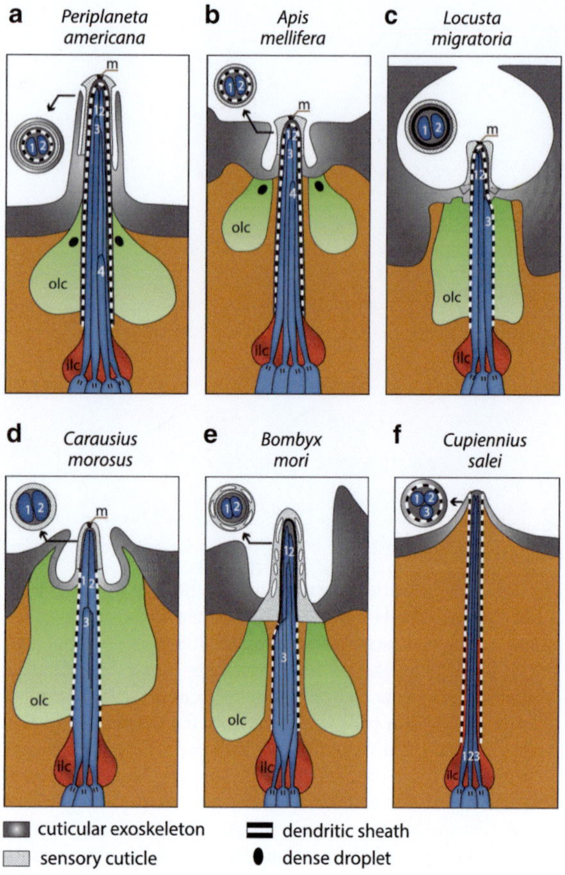


**Fig. 15.3 Diagrams of cuticular processes, cellular associations, and lymph cavities in hygro-thermosensitive sensilla** of various insects (**a–e**) and a spider (**f**). The molting pore (*m*) appears filled with dense material. *Ilc* inner receptor lymph cavity, *olc* outer receptor lymph cavity 1–4, dendrites of sensory cells (in *blue*) (From Tichi and Loftus [21] and Barth and Schmid [1] with permission)

## 15.5 Temperature Detection in Ectotherms

In their outer covering (cuticle) arthropods possess sensory organs called sensilla, protruding from the cuticle or lying within it, some of which are specifically sensitive to temperature and often also to humidity changes. The sensilla are present in the antennae, appearing as a small cuticular peg that is enclosed by a cuticular wall. The dendrites of 2–3 thermosensitive cells extend into the lumen of the peg. Sensilla from different subphyla within arthropods exhibit marked structural differences (Fig. 15.3).

In insects they generally have an unperforated cuticular wall, while in spiders, sensilla have an opening at their tip. Thermosensory dendrites are of variable size and may remain unbranched, reaching the tip of the peg or ramifying at the base depending on the species. Sensory neurons from thermal sensilla in insects respond with a discharge of nerve impulses to temperature decreases. All cells in the sensillum are also sensitive to decreases in vapor pressure. This is partly because evaporation causes cooling of the sensillum. When either vapor pressure or temperature are kept constant, some cells respond to increases in relative humidity at fixed temperature and are classified accordingly as moist cells. Others respond to cold and to vapor pressure changes, being defined as cold cells. Thus, neurons innervating the sensillum act as thermo- and hygroreceptive elements, with a striking specialization in the detection of cold stimuli, dryness, or wetness of the cuticular surface.

In the fly *Drosophila melanogaster*, peripheral thermoreceptor neurons located in the antenna use TRPP channels as moderate cold transducers. Peripheral warm thermoreceptor neurons have been identified nearby the cold ones, but their transducer channel is still unknown (Fig. 15.4). In addition, in the anterior part of the brain of *Drosophila,* central thermosensory neurons named AC neurons that respond directly to temperature increases above preferred temperatures have been identified. These central warm thermosensory neurons use dTRPA1 as thermotransducer channels and possibly contribute to monitoring external temperature changes working as 'discomfort receptors' and, together with the peripheral neurons innervating the antenna allow the fly to avoid all but the most optimal temperatures. In general, differences in the anatomy of thermosensory sensilla and in the sensitivity of their thermoreceptor cells explain the variable working range of thermoregulatory processes found among species in arthropods.

Peripheral thermosensor cells are also present in nematodes. The nematode *C. elegans* possesses a pair of bilaterally symmetric, bipolar sensory neurons (named AFD neurons) that terminate in modified ciliated endings which are located in the amphid sensilla, a sensory organ placed in the worm's nose. AFD neurons display bidirectional temperature sensing capabilities with high sensitivity (Fig 15.5a). Another thermosensory neuron named AWC has been identified in *C. elegans*. In addition, two PVD neurons project a single axon branching highly into dendritic processes that envelop the animal. These endings behave as

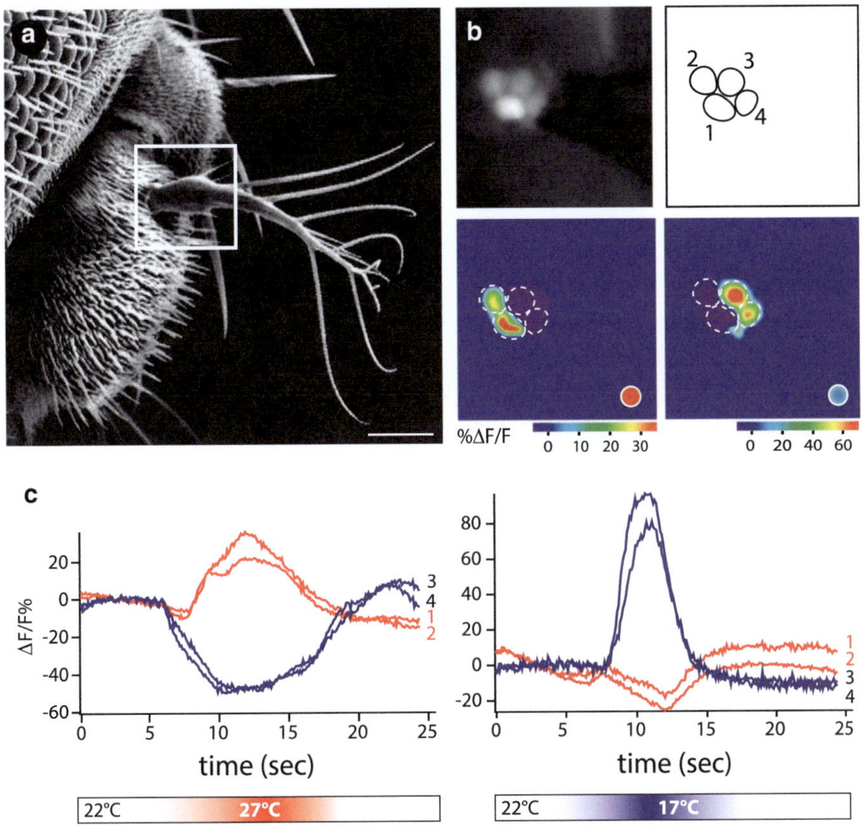


**Fig. 15.4 Hot and cold temperature receptors in the *Drosophila* antenna.** (**a**) Scanning electron micrograph of the *Drosophila* antenna. The arista (*white box*) houses six neurons, four of which are visible on the focal plane shown in (b), scale bar: 20 μm. (**b**) Basal fluorescence and maximal response images of four neurons expressing the calcium-sensitive protein G-CaMP under the control of the pan-neuronal promotor *elav-Gal4*. Functional imaging reveals that these cells respond to either hot (cells *1* and *2*) or cold (cells *3* and *4*) thermal stimuli; stimuli are a change from 22 to 27 °C (hot stimulus, *red dot*) or to 17 °C (cold stimulus, *blue dot*). (**c**) Response profile of the two hot-neurons (cells *1* and *2* in panel b) and the two cold-sensing neurons (cells *3* and *4* in panel b) to a temperature change of 5 °C; *red traces* denote responses of hot cells, and *blue traces* depict the cold cells. Note that cold-sensing neurons display a decrease in intracellular calcium in response to hot stimuli, and the hot-sensing neurons display a decrease in intracellular calcium in response to warming (a–c: From Gallio et al. [8] with permission from Elsevier, Copyright (2011))

polymodal nociceptive neurons detecting extreme cold and noxious heat as well as strong mechanical forces. TRPA1 and DEG/ENaC channels have been suggested respectively as transducers for such noxious thermal and mechanical stimuli. It is worth noting how in the nervous system of *C. elegans*, equipped with only 302 neurons in total, the complex navigational behavior of unrestrained worms is attributable to the sophisticated properties of very few individual neurons.

In ectotherm vertebrates (amphibians, reptiles, fishes), thermosensor cells have functional similarity with those found in mammals and birds (see below). They are used for thermoregulation but also for other behavioral purposes. For instance, the pit organs present in vipers, pythons, and boas are thermosensory devices that allow these snakes to detect infrared radiation with a high sensitivity and directionality. Thermally sensitive facial pits are usually located midway between the eye and nostril on either side of the head (Figs. 15.5d, e). These organs possess a rich vasculature and a dense sensory innervation made by peripheral branches of trigeminal ganglion neurons whose sensory terminals display a regular background firing at 24–25 °C that increases markedly with small temperature elevations. The main role ascribed to pit organs in snakes is to detect infrared radiation emitted by warm-blooded prey. Sensory information triggers the motor response aimed at catching the prey. Snakes

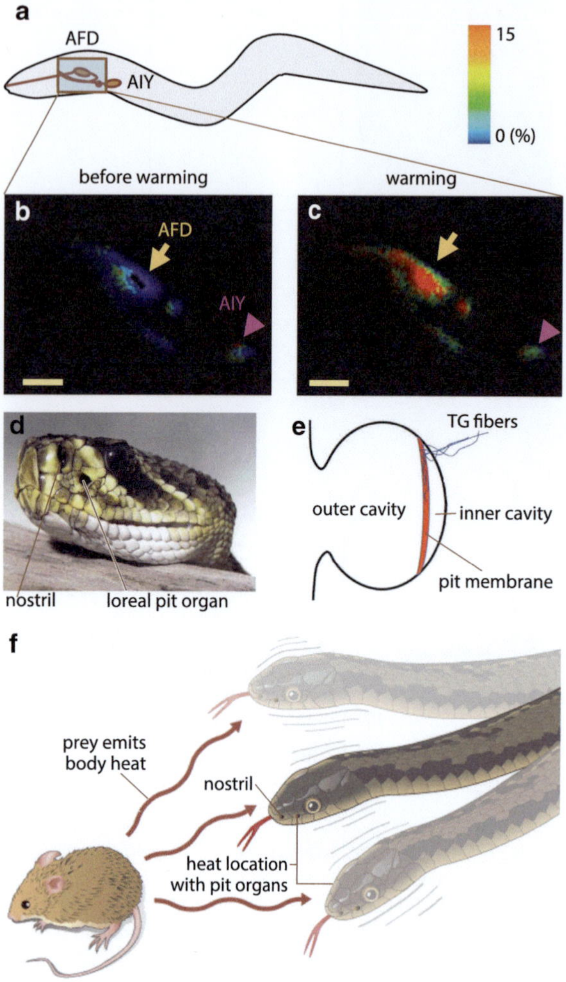


**Fig. 15.5** (**a**) **Schematic diagram showing the location in the head of *C. elegans* of the thermosensitive neuron *AFD* and the interneuron *AIY*.** (**b**, **c**) *In vivo* intracellular calcium imaging measured as changes in the YFP/CFP fluorescence ratio of the fluorescent cameleon protein before and during warm stimulation. *Arrow* and *arrowhead* indicate AFD cell body and AIY neurite, respectively and corresponding pseudocolor images depicting the fluorescence ratio. Scale bar: 10 μm (a–c: From Kuhara et al. [15] with permission from Macmillan Publishers Ltd.: [Nature Communications], Copyright (2011)). (**d**) Rattlesnake head showing location of nostril and loreal pit organ (*black* and *red* arrows, respectively) d: from Wikimedia Commons. (**e**) Schematic diagram of pit organ structure showing innervations by trigeminal ganglion (*TG*) fibers of the pit membrane suspended within a hollow cavity (e: From Gracheva et al. [9] with permission from Macmillan Publishers Ltd.: [Nature], Copyright (2010)). (**f**) Heat emitted by the prey's body as radiation and convection serves to direct the attack of the rattlesnake (f: From http://michaelfriel.net/heat_detecting_organs.html, used with permission from Microsoft)

use this information also to locate thermal radiation emitted by natural surfaces in their environment serving as a reliable and efficient mean to distinguish among microsites of varying thermal quality.

## 15.6 Temperature Detection in Endotherms

In birds and mammals there are specific thermoreceptor neurons whose transducing terminals are located in the skin and mucosa of the body surface and serve to detect environmental temperature changes and to regulate internal body temperature. Based upon the modality of sensation that they evoke in humans, which depends on their dynamic working range and central connections, thermoreceptors in mammals and birds have been generally classified as **cold receptors** that respond to reductions of temperature of the body surface, and **warm receptors** that are activated when the skin temperature rises. High or low temperatures approaching injurious levels are detected by a separate population of receptor neurons, the **nociceptors**, evoking sensations of pain (see Chap. 21). Neurons with a remarkable sensitivity to temperature are also found within the central nervous system of endotherms. They supplement and reinforce the thermoregulatory functions of peripheral thermoreceptors but do not evoke thermal sensations.

The cell bodies of peripheral cold and warm thermosensory neurons are located in cephalic and spinal ganglia. They are pseudomonopolar neurons giving a thin myelinated or unmyelinated neurite (conduction velocity: 10–0.5 m/s) that soon divides into a central branch that enters the spinal cord or brain stem and another branch that travels to the periphery (Fig. 15.6a), where they ramify sparsely and end as bare dendritic terminals extended in a small area of the skin or mucosa (Figs. 15.6b and 15.7d, e).

Cold receptor terminals in the skin of mammals lie in or just below the epidermis and warm receptors more in the upper and middle layers of the corium. Cold receptors have thin myelinated or unmyelinated nerve axons, while warm receptor axons are in most cases unmyelinated. The parent axon of specific thermoreceptor neurons branches sparsely and innervates one or few spots in the skin (Fig. 15.6). This pattern is generally valid for birds and mammals. However, the density of innervation by peripheral thermoreceptors is variable among different areas of the body and also among different species. In general, cold thermoreceptors are more abundant than warm thermoreceptors and accumulate principally in the head area, in particular in the beak skin in birds, and around nose and mouth in mammals. In humans, the density and distribution of cold and warm receptors was estimated more than a century ago through the identification of discrete areas of 1–2 mm² in the skin surface in which thermal or electrical stimulation evoked sensations of cool or warmth. A comparison of the densities of cold and warm points on the human skin has confirmed that there are clearly more of the former than the later. The hand's palm surfaces have 1–5 cold points per cm², but only 0.4 warm points per cm². As expected from sensory fiber distribution, the greatest density of cold points is in the most temperature-sensitive region of the skin, the face.

The functional characteristics of specific cold and warm thermoreceptors of ecto- or endotherm vertebrates are quite similar. They usually exhibit an ongoing discharge of nerve impulses when the skin temperature is maintained constant (**tonic response**). This activity rate rises (or falls) during a change in skin temperature (**phasic response**) (Fig. 15.7). None of them is normally activated by mechanical forces.

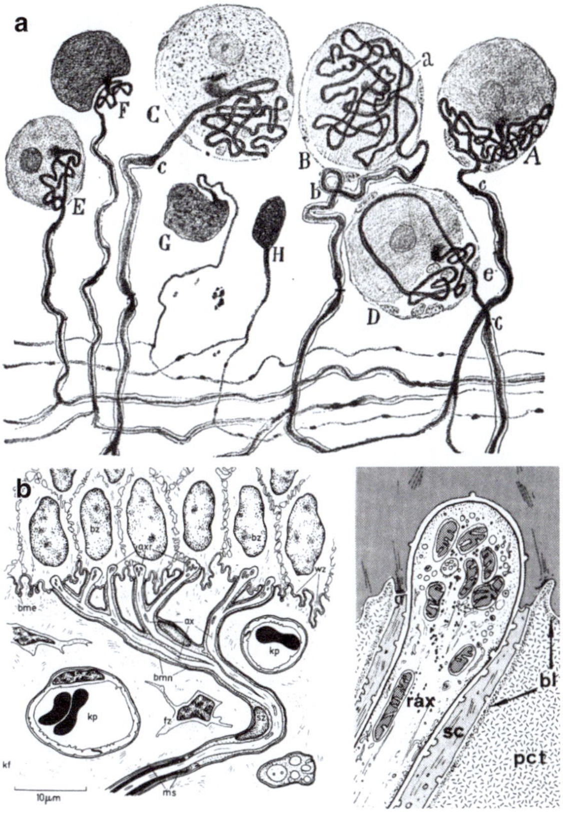


**Fig. 15.6** (**a**) **Primary sensory neurons of the cat's trigeminal ganglion** showing differences in body size and axonal myelination between neurons (From Ramón y Cajal [19]). 'A–G' 'a–c' letters indicate neurons with different axonal tangle's shapes. (**b**) Schematic representation of a cold thermoreceptor fiber of the cat's nose branching below the epithelium cell layer (*left*) and detail of the relationship between a cold nerve terminal and the epithelium cell (*right*). *rax* receptor axon, *sc* Schwann cell, *bl* basal lamina, *pct* papillary connective tissue, *ms* Myelin sheat of afferent nerve fiber, *sz* non-myelinated Schwann cell, *ax* axon, *bme* basement membrane of epidemis, *bmn* basement membrane of nerve terminals, *axr* receptive endings with mitochondria, *bz* basal epidermal cells, *wz* root feet of basal cells, *kp* capillaries, *fz* fibrocyte, *kf* collagen fibrils of stratum papillare (From Hensel et al. [12]. With permission)

### 15.6.1 Cold Thermoreceptors

Impulse frequency of the tonic response of cold fibers depends of static temperature values (Fig. 15.7a). When temperature is plotted against firing frequency, a characteristic stimulus–response function is obtained which is bell-shaped, with maximal activity at a temperature that varies widely depending on the species, and lower activity below or above this value (Fig. 15.7b).

Cold thermoreceptors respond vigorously often in bursts when the skin is actively cooled; conversely, when the skin is warmed, impulse activity is inhibited. However, heating over 40 °C activates again about half of the cold thermoreceptor fibers. This firing response is called 'the paradoxical response' and may be responsible for the sensation of cold that is sometimes experienced in hot water (Fig. 15.7c). Firing frequency of the phasic response to cooling increases rapidly during temperature reduction to a value proportional to the rate of temperature change and then decreases to its tonic discharge rate. When the skin is rewarmed to the original temperature there is a transient cessation of activity before the tonic discharge is once again attained.

Menthol, a compound present in mint essential oils and used in a wide range of cosmetic and alimentary products, evokes a coolness sensation in humans when applied on the skin or mucosal membranes. This is the consequence of the potent effect of these compounds

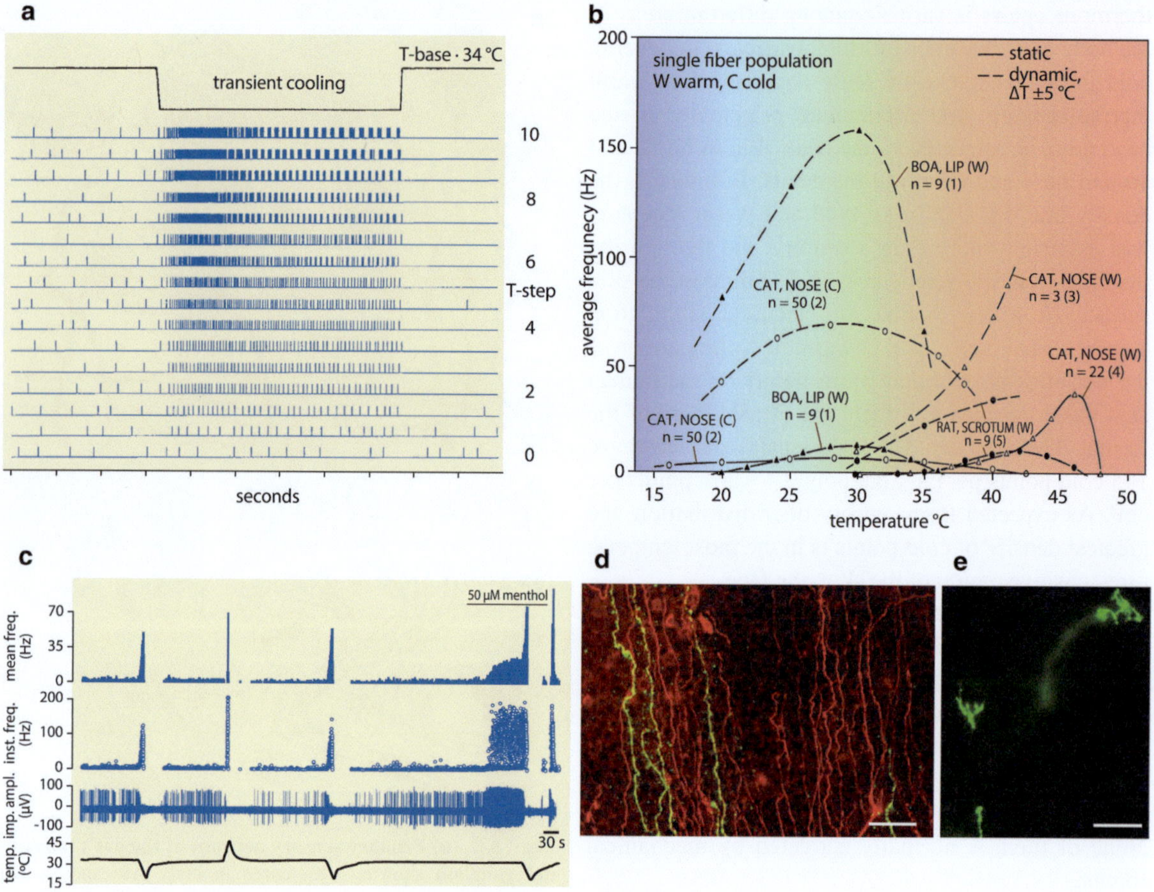


**Fig. 15.7** (**a**) **Nerve impulse response of a cold thermosensitive fiber of the monkey's hand to cooling steps of increasing magnitude** between 34 (basal temperature) and 24 °C. Step size is indicated in the right. The duration of the cooling pulse is represented in the upper trace. The lowest record shows the static discharge at 34 °C. Notice that the initial high frequency impulse discharge occurring at the onset of the pulse (dynamic response) decreases gradually to reach a lower impulse frequency value (static response) (From Darian-Smith et al. [6]. With permission). (**b**) Intensity-response curves of cold and warm thermoreceptor fibers of different vertebrates. Curves were constructed plotting the impulse frequency value at different temperatures. Mean impulse frequency values of the dynamic response (dynamic) and of the static response (static) obtained in cold and warm thermoreceptor fibers of different animal species have been represented; (From Hensel [11]. With permission). (**c**) Example of nerve impulse activity in a cold-sensitive nerve terminal of the mouse cornea to illustrate the ongoing activity at basal temperature of 34 °C, the response to cooling pulses, to heat (paradoxical response) and to the TRPM8 agonist menthol. *Traces* from *top to bottom*: mean firing rate (impulses per second); instantaneous frequency (Hz); nerve terminal impulses and temperature of the bathing solution. (**d**, **e**) Whole-mount double immunofluorescence staining of corneal sensory nerve fibers of a genetically manipulated mouse whose TRPM8-containing (cold-sensitive) nerve fibers express a yellow-fluorescent protein and are stained in *yellow*, whereas the remaining sensory fibers are stained in *red*. d, sensory axons running between basal cells of the corneal epithelium. e, cold nerve terminals branching between superficial corneal epithelium cells. Scale bars in d–e, 40 μm (c–e: From Parra et al. [16]. With permission)

on cold thermoreceptor activity. Menthol enhances the stationary discharge rate of cold thermoreceptors; also, under the effects of this compound the amplitude of the dynamic impulse response to sudden cooling increases markedly. Thus, these 'cooling' compounds sensitize cold thermoreceptors and evoke intense cooling sensations (Fig. 15.7c).

### 15.6.2 Warm Thermoreceptors

Warm fibers give tonic responses at static temperatures of 30 °C or more and silence upon cooling. As for cold fibers, the function of the tonic discharge rate versus steady-state stimulus temperature follows a bell-shaped curve, with maximum discharge at 39–42 °C that decreases markedly at temperatures over 45 °C (see Fig. 15.7b). Warm fibers respond intensely to dynamic heat ramps with an accelerating discharge that reaches a peak and then decreases to adapt to the frequency typical for the steady-state temperature. In contrast with cold receptor fibers, they are insensitive to menthol.

### 15.6.3 Thermosensitive Nociceptors

#### 15.6.3.1 'Cold' Nociceptors

A subpopulation of peripheral Aδ nociceptors is responsible for the immediate sensation of pricking pain experienced by humans when skin temperature decreases below values of 20–10 °C depending on the location of the stimulus, area of exposure, etc. Sensation of burning pain follows, reflecting the additional activation of a fraction of C polymodal nociceptors. The impulse firing frequency rise induced by strong cooling in nociceptor fibers is modest (Fig. 15.8a), suggesting that the augmented pain intensity evoked when cooling becomes stronger and faster is the consequence of the recruitment of a progressively larger number of nociceptor fibers, located in the epidermis, but possibly also around vein vessels running deeper within the skin.

#### 15.6.3.2 'Heat' Nociceptors

Aδ and C nociceptors also sense cutaneous heating. A large proportion of them already start to fire when surface temperature rises over 38–39 °C. Thin myelinated (Aδ) nociceptors are responsible of the first, sharp sensation of pain experienced by humans and presumably other vertebrates when the skin surface is heated over 42 °C (Fig. 15.8b). There are two subtypes of Aδ heat-sensitive cutaneous nociceptors. Type I are present in the glabrous and hairy skin, have a high threshold (over 53 °C), exhibit a delayed, long-lasting response to heating pulses, and become sensitized with repeated stimulation (Fig. 15.8c). Type II heat nociceptors, innervate the hairy skin and have a lower heat threshold (over 48 °C) and vigorous, immediate firing upon heat stimulation (Fig. 15.8d). The different firing patterns of both classes of heat nociceptors can explain the variable quality of the heat pain sensations experienced during heating of the areas of the skin that they innervate.

In addition, activation by heat of C-polymodal nociceptors, that distribute extensively into the epidermis and dermis including venous vessels evokes a slower sensation of dull, aching pain. Heat-sensitive C nociceptors also exhibit an ample range of heat thresholds (between 37 and 49 °C in primates) and variable impulse firing pattern in response to heat as well as different capability to develop sensitization and fatigue with repeated heating. Accordingly, heat stimuli of larger magnitude determine the recruitment of a growing number of nociceptors firing at higher frequency thereby causing increasingly intense pain sensations.

### 15.6.4 Other Thermosensory Nerve Fibers

Some low-threshold mechanoreceptors of the skin and muscles exhibit an additional responsiveness to cooling. The functional meaning of this property is obscure, but has been suggested to be responsible of the subjective feeling that cold objects are heavier (Weber paradox).

Additionally, sensory fibers responding to cold and warm have been also described in nerves innervating some viscera (respiratory pathways, digestive tract, urinary bladder). Likewise, many visceral polymodal nociceptors respond to cold and heat. They are involved in autonomic regulations (flow in respiratory pathways, local blood flow, secretory processes) as well as in nonconscious thermoregulatory adjustments.

#### 15.6.4.1 Central Thermoreceptor Neurons

As mentioned above, central neurons directly stimulated by temperature have been found in the antennal lobe of insects. In endotherms, specific thermosensory neurons located within the brain (central thermoreceptors) are in charge of core temperature detection. Most of these neurons are warm-sensitive, i.e., they increase their impulse activity when brain temperature rises. Warm-sensitive neurons of the preoptic anterior hypothalamus (POA) display spontaneous membrane depolarization and impulse firing whose rate is modulated by temperature (Fig. 15.9a–c). Other central neurons increase their activity with decreases in brain temperature. However, the cold sensitivity of most of them seems to be due to cessation of an inhibitory synaptic input from nearby warm-sensitive neurons. Thermosensitive neurons of

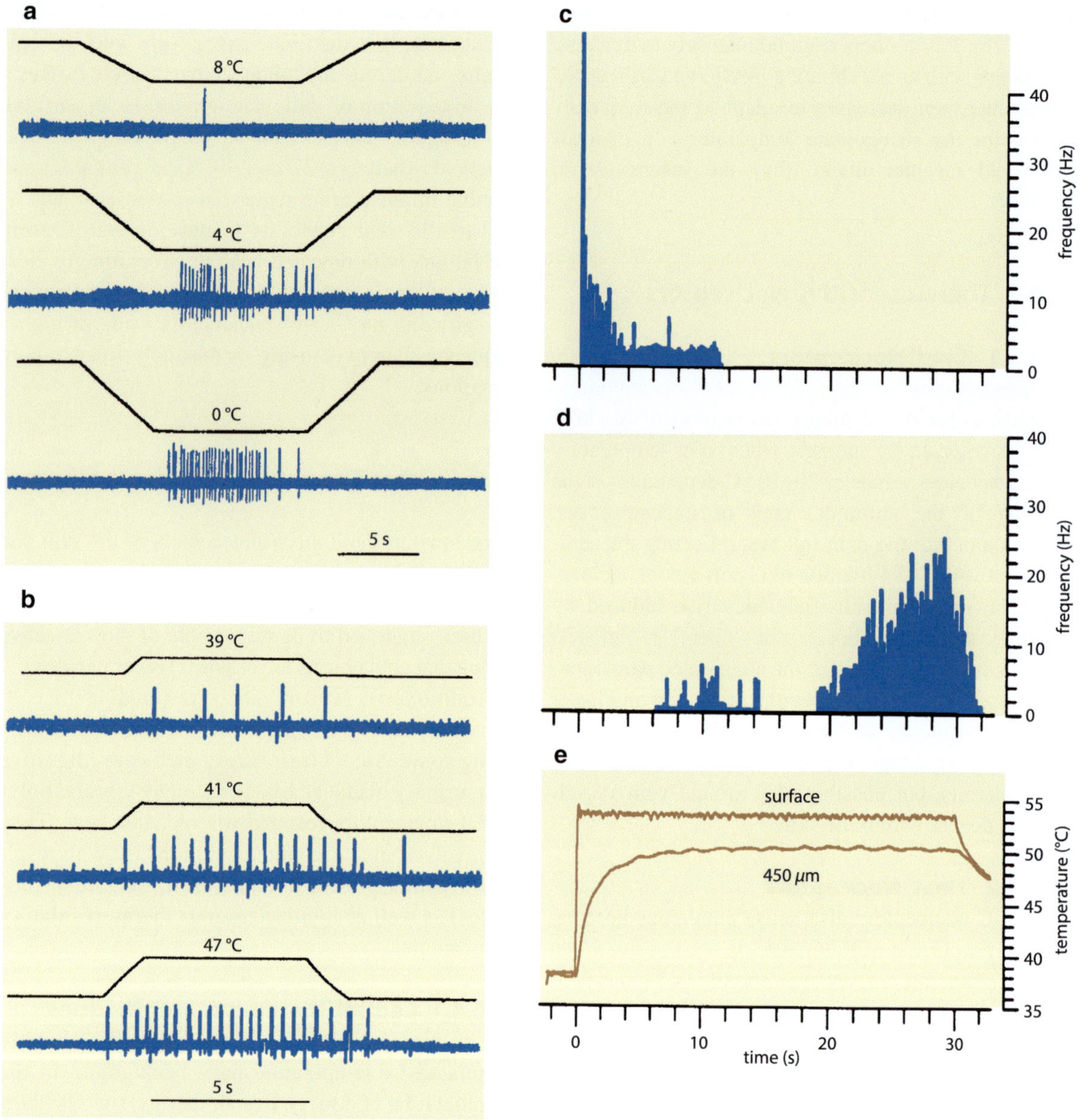


**Fig. 15.8** **Thermal sensitivity of cutaneous nociceptors.** (**a**) Nerve impulse response evoked in an Aδ nociceptor fiber by cooling of the skin surface down to 8, 4, and 0 °C. Analog traces of stimulus temperature are provided above each impulse discharge trace. (**b**) Impulse response evoked in an Aδ nociceptor fiber (type 2) by heating of the skin surface from 35 to 39, 41 and 47 °C (a–b: From Cain et al. [4]. With permission). (**c**) Firing frequency plot of the response to a 30 s duration radiant heat pulse in a type II mechano-heat nociceptor (thermal threshold: 44 °C, latency of the peak response: 0.1 s). (**d**) Response to the same stimulus of a type I mechano-heat nociceptor (thermal threshold > 53 °C, latency of the peak response: 29 s). (**e**) Temperature change at the skin surface and at a 450 μ depth induced by the radiant heat pulse (c–e: From Treede et al. [22] with permission from Wiley)

the POA are critical for triggering autonomic thermoeffector responses in mammals. Local increases of temperature evoke heat loss responses such as panting, sweating, and vasodilatation, whereas cooling elicits heat gain responses that include shivering, vasoconstriction, piloerection, decreased respiratory rate, and increased food intake. Thermoregulatory responses can be also evoked by thermal stimulation of other neuronal groups in the brain stem and the spinal cord.

## 15.7 Behavioral Responses to Thermosensory Information

The information provided by peripheral thermoreceptors is used by animals to regulate their body temperature through the development of adaptative behaviors organized at higher levels of the central nervous system.

In insects, thermal information provided by peripheral thermoreceptors to central brain neurons located in the antennal lobe, serves them to select preferred environment temperatures. Thermal information is also used by certain species like the mosquito to localize their prey: a combination of humidity and warmth appears to be the best attractant to stimulate landing of a female mosquito on the skin. In *Drosophila*, projections from cold- and hot-sensing neurons converge onto a few glomeruli lying in the **antennal lobe**. There, projections of cold and warm neurons are clearly segregated. Central thermoreceptor neurons are also located within this area. Although thermal behavior of insects has been investigated mainly in terms of positive or negative thermotaxis, thermal sensory capacities are certainly more elaborated. For example, ants can be trained to distinguish temperatures as small as 0.1 °C in a classical conditioning paradigm, using their directional sense of thermosensation for this purpose.

Central processing of thermal information in nematodes like *C. elegans* is critical because in their natural soil environment, temperature can vary by tens of degrees. Worms use this thermal information for the search of their optimal temperature (thermotaxis) and for isothermal tracking of such temperature. Neuronal mechanisms and circuits involved in this behavior are now known in some detail (see Fig. 15.10): Temperature information provided by AFD and AWC sensory neurons is processed at three interneurons (a) the AIY interneuron-mediated pathway drives warm-seeking (**thermophilic**) movement, (b) the AIZ interneuron-mediated pathway drives cold-seeking (**cryophilic**) movement, and (c) the counterbalancing regulation between the AIY and AIZ activities through the RIA-integrating interneuron is essential for the execution of such motor outputs (Fig. 15.10a). It has been suggested that large calcium concentration changes in AFD in response to thermal stimuli activate both stimulatory and inhibitory neural signaling to AIY, then inducing a relatively weak increment in the calcium concentration in AIY (Fig. 15.10b1). Small calcium concentration changes in AFD induce a more active state of stimulatory neural signaling to AIY, leading to a relatively large increment in calcium concentration in this neuron, important for thermophilic movement (Fig. 15.10b2). A complete or virtually complete loss of sensory signal eliminates both stimulatory and inhibitory neural signaling, in which an AIY-independent cryophilic driving signal generates cryophilic movement (Fig. 15.10b3). Altogether, temperature information provided by AFD and AWC thermosensory neurons as well as from other chemosensory neurons exhibiting thermal sensitivity

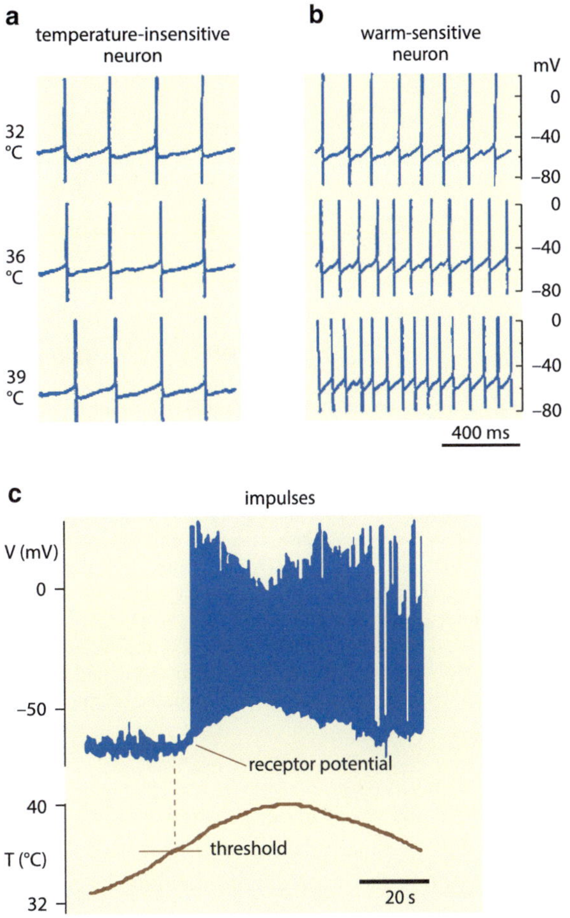


**Fig. 15.9 Central thermosensitive neurons.** Effect of three different temperature values on the firing rate of a temperature-insensitive neuron (**a**), and a warm-sensitive neuron (**b**) recorded intracellularly in a rat hypothalamic tissue slice; (From Griffin et al. [10] with permission from Wiley). (**c**) Whole-cell current-clamp recording of intracellular membrane potential of a heating-activated neuron in a rat anterior hypothalamus slice. When temperature (*T*) increased above threshold T (*bar*), a heating-induced depolarization (receptor potential) develops, generating a repetitive discharge of impulses (From Hori et al. [13] with permission from Elsevier, Copyright (1999))

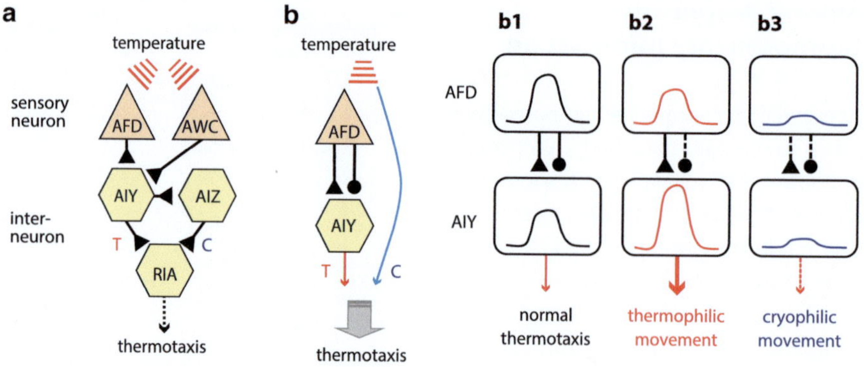


**Fig. 15.10** (**a**) **Proposed model of neural circuit for thermotaxis in *C. elegans*.** Temperature is sensed by AFD and AWC sensory neurons. The activated AIY interneuron accelerates the movement to a temperature higher than the cultivation temperature (thermophilic drive; *T*), the activated AIZ accelerates the movement to a temperature lower than the cultivation temperature (cryophilic drive; *C*), and RIA interneuron integrates thermal signaling of 'T' and 'C'. (**b**) Wiring diagram of stimulatory and inhibitory neural signaling in the AFD–AIY circuit. Triangular and round synapses between AFD and AIY indicate stimulatory or inhibitory neural signaling, respectively. *Red 'T'* and *blue 'C'* indicate thermophilic and cryophilic driving signal, respectively. *Blue line* indicates AIZ-mediated cryophilic driving pathway as described in (a). Neural calculation in AFD and AIY is essential for thermophilic movement. A strong thermosensory signal activates both stimulatory and inhibitory neural signaling to AIY, inducing a weak activation of AIY (**b1**): A weak sensory signal induces relatively more active state of stimulatory signaling, leading to a strong activation of AIY important for thermophilic drive (**b2**): A lack of sensory signal eliminates AIY activity, then AIY-independent cryophilic driving signal generates cryophilic movement (**b3**): *Dotted connections* indicate weak signals (From Kuhara et al. [15]. With permission)

(ASI) allows the worm to detect whether it is heading up or down a thermal gradient, such that it could bias its reorientation rate accordingly, defining its navigation strategy.

In mammals, the discriminative ability for skin temperature changes has been explored using behavioral responses to assess temperature preferences. Humans and monkeys exhibit remarkable similarities in discrimination performance across a wide range of noxious and innocuous temperatures. In other mammalian species, discrimination capacities vary widely. This reflects the different behavioral demands to the thermoregulatory neural mechanisms generated by the thermal characteristics of the natural habitat of each species. In humans, thermal sensations are interpreted at cognitive levels. How such information is perceived by nonhuman mammals remains a difficult question to answer (see Chap. 28).

Thermal pathways in vertebrates begin at the peripheral endings of thermosensory primary sensory neurons of dorsal root and cephalic ganglia, whose central branch projects centrally into the superficial spinal and brain stem gray matter. Spinal neurons located in **lamina I of the dorsal horn** or the equivalent area in the brainstem, receive direct input from Aδ and C thermoreceptor and nociceptor fibers that connect with three major classes of modality-selective neurons: nociceptive-specific **fusiform cells**, thermoreceptive-specific **pyramidal cells**, and polymodal nociceptive **multipolar cells**. In deeper laminas of the dorsal horn, modality-ambiguous 'wide dynamic range' (WDR) nociceptive neurons also respond to innocuous and noxious cold and heat and encode intensity over a wide range of temperatures. Lamina I neurons project through the **lateral spinothalamic tract** to nociceptive- and thermoreceptive-specific relay nucleus in the **posterior thalamus** and to other sites in the brain that together provide input to the cortical regions that are activated by thermal and nociceptive stimuli (Fig. 15.11a). Thermoreceptive-specific lamina I neurons also connect with neurons of the mesencephalic lateral parabrachial nucleus. These convey thermal sensory information of the body surface mainly to the **median preoptic nucleus** which is part of the thermoregulatory center in the preoptic area of the hypothalamus, where body temperature control mechanisms are integrated (Fig. 15.11b).

In humans, harmless temperature changes on the skin and exposed mucosa evoke a particular conscious sensation modality named sense of temperature or

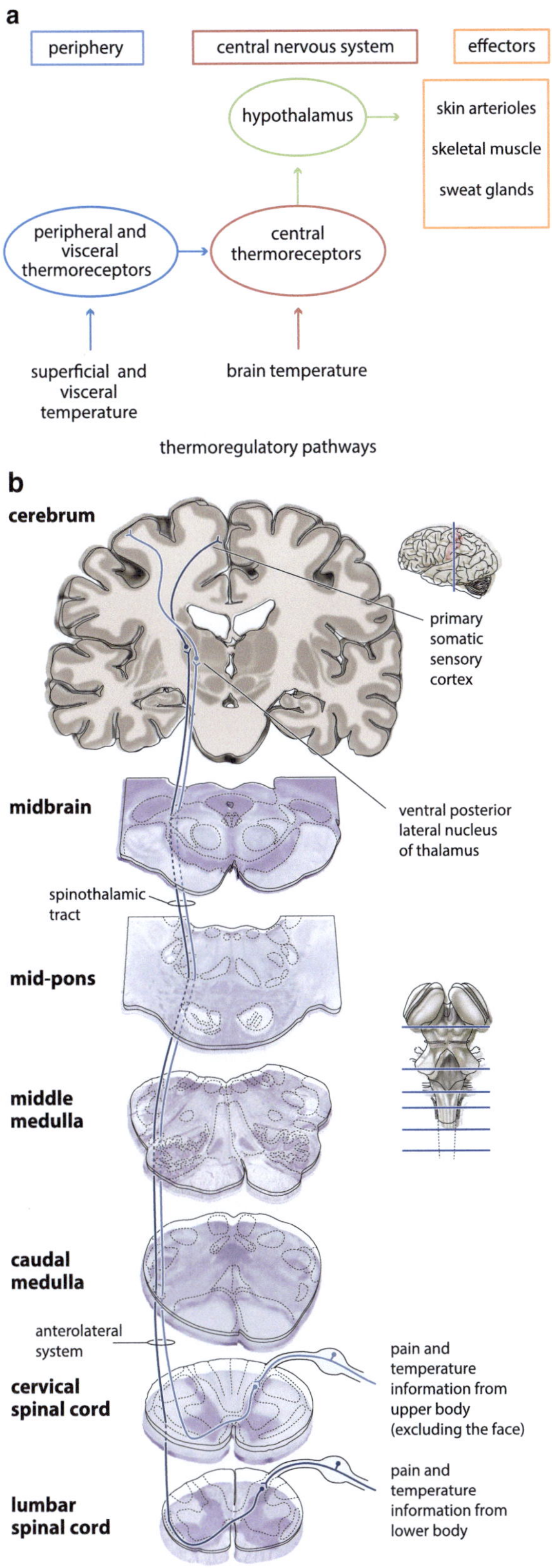


**Fig. 15.11** (**a**) **Schematic representation of peripheral and central thermoregulatory mechanisms** in mammals. (**b**) Neural pathways for transmission of the sensory information provided by peripheral thermoreceptors to the cerebral cortex, evoking conscious temperature sensations (From Purves [18]. With permission)

**thermoperception**, with two qualities: cold and warm. The quality of the experienced thermal sensation changes when the stimulating temperature is varying and attenuates with maintained temperatures. For instance, when entering into a pool with water at about 35 °C on a warm day, the first feeling is that the water is warm. Contrarily, entering into water around 28 °C evokes a first experience of cold. In both cases, after some time this sensation fades away, even if the water temperature is kept constant. The range of temperatures at which there is no thermal sensation and there is complete fading of the transient sensations of cold or warmth evoked by moderate thermal changes, is named 'neutral or comfort zone' (Fig. 15.12). In this range of temperatures there is an essentially complete **adaptation** of temperature sensation to the new skin temperature. In terms of peripheral information, it can be concluded that the overall sensory input from cold and warm thermoreceptors during static temperature values within the comfort zone is not strong enough to evoke a sustained thermal sensation.

Above or below the neutral zone, permanent sensations of heat or cold are produced even when the skin temperature is kept constant for a long time (Fig. 15.12). In experiments with naked humans in a climate-controlled room, the lower and upper temperature limits of the neutral zone are 33 and 35 °C, respectively. When smaller areas of skin are studied, the zone expands, being 30 and 36 °C respectively for a 15-$cm^2$ skin surface, an indication of central summation of the impulses coming from peripheral thermoreceptors.

The maintained sensations of warmth that persist at constant skin temperatures above 36 °C are more intense the higher the temperature of the skin. At temperatures of more than 43–44 °C the sensation of warmth gives way to a painful heat sensation (heat pain). Similarly, at temperatures below 30 °C the maintained cold sensation increases in intensity, the colder the skin. Actual cold pain sets in at skin temperatures of 17 °C and lower, but at skin temperatures as high as 25 °C the sensation of cold has an unpleasant component, especially when large areas of skin are affected.

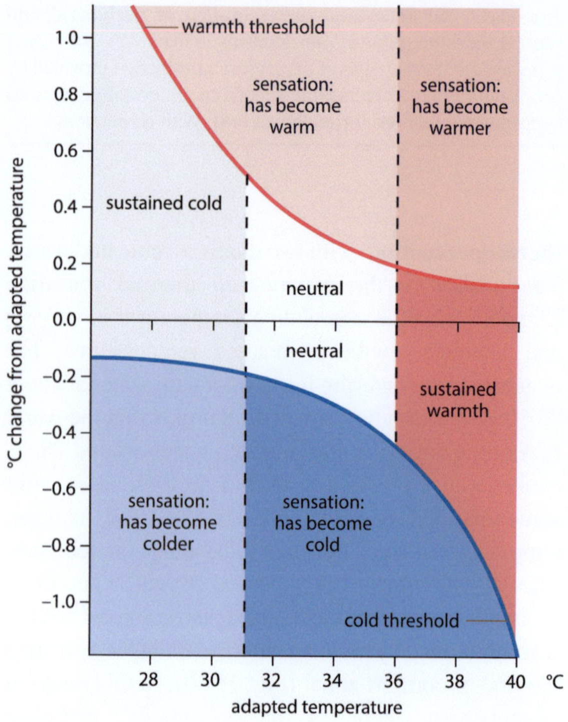


**Fig. 15.12 Dependence of threshold for sensation of warmth and cold** upon the initial skin temperature in humans. When the skin is adapted to the temperature value shown on the abscissa, the skin temperature has to be changed by the number of degrees centigrade (°C) shown on the ordinate in order to elicit a conscious sensation of cold or warmth. When the initial temperature of the skin is low, for example 28 °C, a large temperature elevation is required to elicit a warmth sensation while a cold sensation is readily evoked with a small temperature reduction. Conversely, smaller temperature elevations are required to feel warmer if the skin was adapted within the persisting warm skin temperature values. Thus, a cool skin has to be cooled further by less than 0.2 °C in order for the maintained cold sensation to give way to the sensation "colder". The same skin must be warmed by almost 1 °C before a sensation of warmth occurs. However, if the initial temperature is 38 °C a slight warming (<0.2 °C) elicits the sensation "warmer", whereas the skin must be cooled by around 0.8 °C if a "cooler" sensation is to be experienced (From Schmidt [20]. With permission)

Unpleasantness of cold is probably associated with the progressive recruitment of the population of cold thermoreceptors that exhibit a high threshold while overt pain evoked by cold or heat requires the activation of polymodal nociceptors. Adaptation after a stepwise change of skin temperature within the neutral zone takes minutes before temperature becomes neutral. Outside the neutral zone, adaption after a new skin temperature has been established is incomplete and reproducible estimation of skin temperature is possible.

With fast rates of temperature change (exceeding 0.1 °C/s) warmth and cold thresholds do not change, but they shift rapidly with slower temperature changes. Likewise, thresholds to cooling or warming depend on the size of the stimulated area of the skin. Also, the intensity of the sensation evoked by a given suprathreshold temperature change becomes larger when the surface of the stimulated area increases. These phenomena are the consequence of spatial and temporal summation at the central nervous system of the information provided by peripheral thermoreceptors. The ability of human subjects to detect skin cooling or warming in the innocuous range is an order of magnitude better than their ability to discriminate low and high temperatures in the noxious ranges. This supports the interpretation that cutaneous thermal information from thermoreceptors and nociceptors is differentially processed within the central nervous system.

## References


1. Barth FG, Schmid A (eds) (2001) Ecology of sensing. Springer, Berlin/Heidelberg/New York
2. Basbaum AI, Bautista DM, Scherrer G, Julius D (2009) Cellular and molecular mechanisms of pain. Cell 139:267–284
3. Belmonte C, Viana F (2008) Molecular and cellular limits to somatosensory specificity. Mol Pain 4:14–17
4. Cain DM, Khasabov SG, Simone DA (2001) Response properties of mechanoreceptors and nociceptors in mouse glabrous skin: an in vivo study. J Neurophysiol 85:1561–1574
5. Clapham DE, Runnels LW, Strübing C (2001) The TRP ion channel family. Nat Rev Neurosci 2:387–396
6. Darian-Smith I, Johnson KO, Dykes R (1973) "Cold" fiber population innervating palmar and digital skin of the monkey: responses to cooling pulses. J Neurophysiol 36:325–346
7. Dhaka A, Viswanath V, Patapoutian A (2006) TRP ion channels and temperature sensation. Annu Rev Neurosci 29:135–161
8. Gallio M, Ofstad TA, Macpherson LJ, Wang JW, Zuker CS (2011) The coding of temperature in the *Drosophila* brain. Cell 144:614–624
9. Gracheva EO, Ingolia NT, Kelly YM, Cordero-Morales JF, Hollopeter G, Chesler AT, Sánchez EE, Perez JC, Weissman JS, Julius D (2010) Molecular basis of infrared detection by snakes. Nature 464:1006–1011
10. Griffin JD, Kaple ML, Chow AR and Boulant JA (1996) Cellular mechanisms for neuronal thermosensitivity in the rat hypothalamus. J Physiol 492:231–242
11. Hensel H (1975) Static and dynamic activity of warm receptors in *Boa constrictor*. Pflugers Arch 353:191–199
12. Hensel H, Andres KH, von Duering M (1974) Structure and function of cold receptors. Pflugers Arch 352:1–10

13. Hori A, Minato K, Kobayashi S (1999) Warming-activated channels of warm-sensitive neurons in rat hypothalamic slices. Neurosci Letters 275:93–96
14. Inada H, Ito H, Satterlee J, Sengupta P, Matsumoto K, Mori I (2006) Identification of guanylyl cyclases that function in thermosensory neurons of *Caenorhabditis elegans*. Genetics 172:2239–2252
15. Kuhara A, Ohnishi N, Shimowada T, Mori M (2011) Neural coding in a single sensory neuron controlling opposite seeking behaviors in *Caenorhabditis elegans*. Nat Commun 2:355
16. Parra A, Madrid R, Echevarria D, del Olmo S, Morenilla-Palao C, Acosta MC, Gallar J, Dhaka A, Viana F, Belmonte C (2010) Ocular surface wetness is regulated by TRPM8-dependent cold thermoreceptors of the cornea. Nat Med 16:1396–1399
17. Patapoutian A, Peier AM, Story GM, Viswanath V (2003) ThermoTRP channels and beyond: mechanisms of temperature sensation. Nat Rev Neurosci 4:529–539
18. Purves D et al. (2012) Pain. In: Purves D et al. (ed) Neuroscience, 5th edn. Sinauer, Sunderland, MA, USA, pp 209–227
19. Ramón y Cajal S (1899) Histología del Sistema Nervioso del Hombre y de los Vertebrados, vol 1. Imprenta y Libreria de Nicolas Mora, Madrid, p 358
20. Schmidt RF (1978) Fundamentals of sensory physiology. Springer, Berlin/Heidelberg/New York
21. Tichy H, Loftus R (1996) Hygroreceptors in insects and a spider: Humidity transduction models. Naturwissenschaften 83:255–263
22. Treede RD, Meyer RA, Raja SN, Campbell JN (1995) Evidence for two different heat transduction mechanisms in nociceptive primary afferents innervating monkey skin. J Physiol 483:747–758

# Mechanosensation

# 16

Jörg T. Albert and Martin C. Göpfert

**Mechanosensation**, the ability to detect – and respond to – mechanical stimulus force, is a basic property shared by virtually all organisms and cells: tension forces acting on cells, for example, can influence cell shape by acting through integrin receptors, and mechanosensitive ion channels mediate volume changes in many pro- and eukaryotic cells. Dedicated mechanosensory (or mechanoreceptor) cells and organs are found in metazoans where they serve the detection of, e.g., medium flows, body movements, gravity, touch, sound, and noxious mechanical stimuli such as pinching of the skin.

Force is a vector quantity that, according to **Newton's second law**, equals mass ($m$) times acceleration ($a$), $\vec{F} = m\vec{a}$. The unit of force is Newton (N=kg m/s$^2$), with 1 N being the force required to accelerate a mass of 1 kg by 1 m/s$^2$. Forces acting on elastic structures such as springs cause displacements that, provided the structure behaves linearly, obey **Hook's law**, $F=kX$. Here, $k$ is the stiffness, or spring constant, of the system, and $X$ denotes displacement. The inverse of stiffness, $\frac{1}{k}$, is compliance, so the larger the compliance, the larger will be the displacement caused by a given stimulus force. Some mechanosensory systems are exquisitely sensitive to displacements, responding to nanometer-range (1 nm=$10^{-9}$ m) movements whose amplitudes hardly exceed the **Brownian motion** imposed by **thermal noise**. The energy of thermal noise is 1 $k_BT$, where $k_B$ is the Boltzmann constant ($1.38 \cdot 10^{-23}$ J/K) and $T$ is the absolute temperature (in Kelvin). At room temperature, this thermal energy is approximately $4 \cdot 10^{-21}$ J or 4 zJ, which is circa 100 times less than the energy of a single green photon (ca. 400 zJ). Mechanosensation can thus operate at lower stimulus energies than vision, simply because at physiological temperatures thermal energy is smaller than the energy of photons.

In this chapter, we will explore how organisms detect and decode mechanical stimulus force. Starting with osmotic pressure regulation in bacterial cells and gravity responses of ciliates, we will discuss how invertebrate and vertebrate mechanosensory cells transduce mechanical stimulus forces into electrical signals that are forwarded to mechanosensory pathways in the central nervous system (CNS). We shall see that despite the ubiquity of mechanosensation, surprisingly little is known about the molecular machineries that mediate the sensory detection of stimulus force. Using vertebrate inner-ear hair cells as an example, we will discuss the physical properties of such machineries. We will then provide an overview over the current state of the dissection of these machineries in genetic model organisms such as nematodes, flies, and mice. Prime systems that are widely used for studying mechanosensory mechanisms will be introduced, ranging from single cells to elaborate mechanosensory organs that can comprise sophisticated accessory structures and may harbor thousands of mechanosensory cells. Physiological response characteristics of mechanosensory cells will be discussed using the somatosensory system as example, and

J.T. Albert (✉)
Ear Institute, University College London, 332 Gray's Inn Road, London, WC1X 8EE, UK
e-mail: joerg.albert@ucl.ac.uk

M.C. Göpfert (✉)
Department of Cellular Neurobiology, University of Göttingen, Julia-Lermontowa-Weg 3, 37077 Göttingen, Germany
e-mail: mgoepfe@gwdg.de

C.G. Galizia, P.-M. Lledo (eds.), *Neurosciences - From Molecule to Behavior: A University Textbook*,
DOI 10.1007/978-3-642-10769-6_16, © Springer-Verlag Berlin Heidelberg 2013

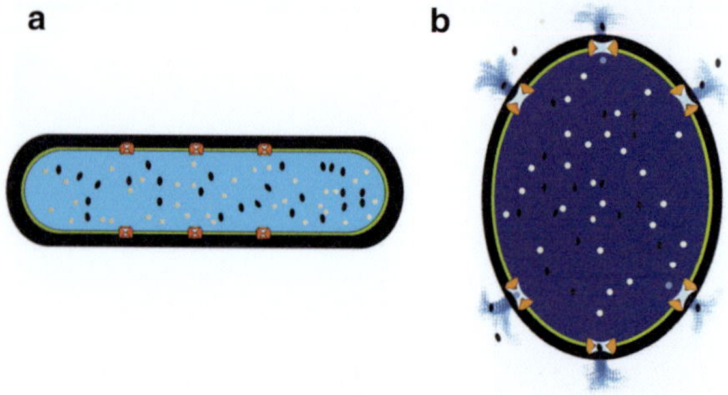


**Fig. 16.1** **The osmotic shock response of *E. coli*.** (**a**) Under physiological conditions (e.g., inside the intestines of a warm-blooded animal) *E. coli* bacteria display the characteristic rod-like shape. (**b**) When exposed to hypoosmotic conditions, however, water enters the cell and the bacteria start swelling. Mechanosensitive ion channels in the bacterial membrane open in response to the increased membrane tension and allow for the permeation of both solutes and solvent through the channel pores which acts to reduce the life-threatening membrane tension

central processing mechanisms will be illustrated for the whisker system that, feeding into the barrel cortex, provides an attractive system for analyzing how mechanosensory information is processed within the brain. A particularly sensitive form of mechanosensation, hearing, is the focus of a separate chapter (Chap. 17). To gain first ideas about what mechanosensation and mechanosensory systems are about, we now will leave the nervous system for a moment and look at organisms that respond to forces in a quite elegant manner though having neither neurons nor a brain.

## 16.1 A Mechanosensory System in a Nutshell: The Osmotic Shock Response of *E. Coli*

The probably most comprehensive description of an entire mechanosensory system can be given for a unicellular organism, the bacterium *Escherichia coli*. As a ubiquitous inhabitant of the lower intestine of warm-blooded animals, the life cycle of *E. coli* involves the regular excretion and uptake by its host. Outside the rather constant environment of the host's body, *E. coli* is confronted with the forces of nature: when diluted by a rain shower, the **osmolarity** of the external medium drops, causing large amounts of water to enter the bacterial cell (Fig. 16.1). To prevent bursting, the cell must sense – and counteract – the life-threatening increase in membrane tension. Both functions, the sensing and the release of membrane tension, are mediated by **mechanosensitive, stretch-activated ion channels** in the membrane of the bacterial cell: The channels **MscL** (mechanosensitive channel of large conductance) and **MscS** ('S' for small conductance) form two pores in the lipid bilayer that open when the tension of the membrane increases, allowing water and small intracellular solutes to leave the cell; for review, see [1]. In a nutshell, this molecular pressure-control system illustrates fundamental aspects of mechanosensory system function: firstly, external force acting on the cell membrane can open certain ion channels directly, without intermittent second-messenger cascades. And secondly, by altering the open probability of these **force-gated ion channels**, the mechanical stimulus force is transduced into a cellular signal, i.e., a particle current that flows through the channels' pores. This cellular signal can directly or indirectly trigger responses – in *E. coli*, water flowing through the channels directly decreases the membrane tension, reducing the force that acts on the channels and causing the channels to reclose. Hence, the stretch-activated MscL and MscS channels provide **cell volume regulation** by monitoring and modulating membrane tension, serving as sensors and actuators in a **negative feedback loop**.

## 16.2 Mechanoelectrical Transduction – Ciliates and the Advent of Ion Selectivity and Specificity

MscL and MscS are largely **nonselective ion channels**: They are permeable for most solutes and only large macromolecules are retained. An example for a unicellular organism that is endowed with **ion-selective** mechanosensitive channels is the ciliate *Paramecium*, which feeds on aerobic bacteria and relies on **gravity sensing** to find the upper oxygenated water layers where the bacteria live. When *Paramecium* is turned upwards, the load of the cytoplasm exerts force on the posterior half of the cell, which harbors $K^+$-selective stretch-activated channels that transduce the force into an electrical current: $K^+$ efflux through the open channels hyperpolarizes the cell, resulting in enhanced ciliary forward beats that propel the cell further up. The anterior half of *Paramecium* contains $Ca^{2+}$-selective stretch-activated channels that open when the cell is turned downwards. $Ca^{2+}$ influx through these channels depolarizes the cell, reversing the direction of the ciliary beats and propelling the cell backwards up [2]. Promoting movements towards the water surface, the asymmetrically arranged $K^+$- and $Ca^{2+}$-selective mechanosensitive channels make the cells swim against the gravity vector, a behavior that is known as **negative gravitaxis** (or antigravitaxis). It should be noted that also the asymmetric cell shape of *Paramecium* fosters upwards movements, and that the respective contributions of cell shape and mechanosensitive channels to gravitactic behavior are still a matter of debate [3].

## 16.3 Mechanoelectrical Transduction in Sensory Cells – Mechanisms and Genes

In line with the evolution of nervous systems, metazoan animals have dedicated certain cells to the detection of stimulus forces, whereby the forces can principally originate within the body (**enteroceptors**) or externally in the outside world (**exteroceptors**). **Mechanosensory** (or **mechanoreceptor**) **cells** often bear microvilli or cilia that serve as mechanosensory organelles that couple stimulus forces to – and also house – the molecular apparatus for **mechanoelectrical transduction**. This apparatus converts the forces into electrical currents, giving rise to graded receptor potentials that are subsequently encoded into action potentials. In **primary mechanosensory cells** (= **mechanosensory** or **mechanoreceptor neurons**) that bear axons and generate action potentials, this encoding can take place in the mechanosensory cell proper. In **secondary mechanosensory cells** that do not generate spikes, encoding happens in downstream neurons that make afferent synaptic contacts with the sensory cells.

The ability of mechanosensory cells to transduce stimulus forces into electrical signals relies on **mechanoelectrical transduction (MET) channels**. Like the bacterial channels MscL and MscS, MET channels can be mechanosensitive in that they are directly gated by stimulus force (for reviews, see [4–7]). Alternatively, MET channels can act downstream of mechanoreceptor proteins and be activated indirectly via second-messenger cascades. Such an **indirect channel gating** has, for example, been proposed for the **transient receptor potential (TRP) family** member **TRPC6**, a candidate MET channel that occurs in endothelial cells and the smooth muscle layers of our arteries [8]. Heterologously expressed TRPC6 becomes mechanosensitive in the presence of the angiotensin receptor **AT1R**. This induction of TRPC6 mechanosensitivity is angiotensin-independent, suggesting that AT1R may respond to mechanical stimuli with ligand-independent conformational changes and act as a mechanoreceptor protein. AT1R itself is a **G protein-coupled receptor** that seems to signal to TRPC6 via the second messenger **diacylglycerol (DAG)**.

Evidence that MET channels can also directly be activated by the stimulus force comes from the observation that many mechanosensory cells respond to mechanical stimulation within less than one millisecond, which is too fast to allow for intermittent **second messenger cascades**; for a review, see [9]. Such **direct channel gating** requires that stimulus forces are mechanically coupled to the channels, either via elastic tethers and/or via the plasma membrane itself (see Fig. 16.2 for an overview of gating paradigms). A direct MET channel activation via tethers is best documented for vertebrate hair cells, which will be used in the following to exemplify how such tether activation of MET channels works.

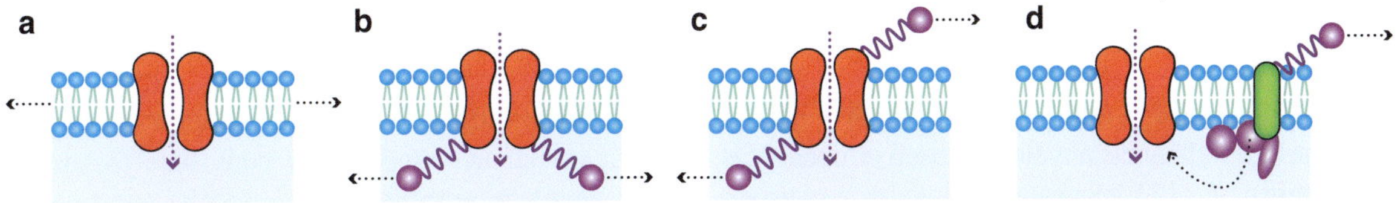


**Fig. 16.2 Paradigms of mechanical channel gating.** (**a**) Direct gating via membrane stretch. (**b**) Direct gating via the pull of intracellular tethers. (**c**) Direct gating via the pull of intra- and extracellular tethers. (**d**) Indirect gating involving a mechanoreceptor protein and a second messenger cascades (Adapted from Christensen and Corey [9]. With permission)

### 16.3.1 Hair Cell Mechanotransduction – Gating MET Channels with Gating Springs

Our ability to detect sound and acceleration relies on sensory **hair cells** in our inner ear. These cells are named after their apical microvilli (the **stereocilia)** that, being arranged in rows of increasing height, form a staircase-like **hair bundle** (Fig. 16.3). The structural asymmetry of the hair bundle is reflected by function: deflecting the hair bundle towards the tallest stereocilia depolarizes the cell, whereas the cell hyperpolarizes when the bundle is deflected in the opposite direction.

The tips of the stereocilia bear cation-selective MET channels that, according to the **gating spring model** of **hair cell transduction**, are associated with elastic tethers that act as gating springs; for reviews, see [4, 12]. Deflecting the hair bundle changes the tension of these **gating springs** and, as a consequence, alters the open probability of the MET channels. The **open probability** of the channels can be written as

$$p_o(X) = \frac{1}{1+e^{-\frac{z(X-X_0)}{K_B T}}}$$

where $K_B$ is the Boltzmann constant and $T$ is the absolute temperature, $z$ the change in force in a single gating spring as the channel opens, $X$ the displacement of the hair bundle from its resting position, and $X_0$ the hair bundle position at which the open probability of the MET channels is 0.5.

At rest, the gating springs exert a certain tension on the MET channels so that some of the channels are already open. Deflecting the hair bundle towards the tallest stereocilium will stretch the gating springs further, thereby opening additional MET channels and depolarizing the cell. The force $F$ required to hold the hair bundle in a certain position is given by

$$F = K_\infty X - p_o(X)Nz + F_0$$

Here, $K_\infty$ is the dynamic stiffness of the hair bundle if all the MET channels are either closed or open, $N$ is the number of MET channels, and $F_0$ is a constant offset force. The negative term $-p_o(X)Nz$ links the force required to deflect the bundle to the open probability of the channels such that increasing the open probability makes the force drop: as the channels open, the gating springs relax, making the hair bundle more compliant and allowing it to move further in the forcing direction. This effect is known as **nonlinear gating compliance**: the dynamic stiffness of the hair bundle $\frac{dF}{dX}$ is constant if the hair bundle is strongly deflected towards the tallest stereocilium such that all MET channels are open ($p_0 = 1$) or in the opposite direction such that all channels are closed ($p_0 = 0$). In between, the stiffness is diminished by MET channel gating, reaching a minimum at $p_0 = 0.5$ where the bundle's mechanical sensitivity peaks. To maintain this sensitivity, the MET channels must adapt when the hair bundle is kept deflected such that the hair bundle's new position coincides with its sensitivity peak. This mechanical adaptation is accomplished with the help of **adaptation motors**, which are serially arranged with the springs and channels and whose movements partially restore the initial channel open probability by readjusting the gating spring tension: If the hair bundle is strongly deflected towards the tallest stereocilium so that the gating springs are stretched and all channels are open, the initial gating spring stiffness can be restored if the motors slide down the stereocilium. Conversely, if the hair bundle is strongly deflected towards the shortest stereocilium so that the gating springs are compressed and all channels are closed, the initial gating spring tension can be restored if the motors crawl up the stereocilium.

The gating spring model provides a detailed quantitative description of mechanotransduction, providing estimates of, e.g., the numbers of MET channels (only ca. 50–150 per hair cell), their single channel gating energies (hardly exceeding thermal noise), and the stiffness of the gating springs (ca. 500–700 μN/m). The association of MET channels and adaptation motors that is proposed by the gating spring model also provides a mechanism for **active mechanical amplification** that can make hair bundles to oscillate spontaneously in the absence of external stimuli. These **active hair bundle movements** are thought to promote (non-mammalian tetrapods) or contribute to (mammals) **cochlear amplification**, a positive mechanical feedback mechanism which accounts for the exquisite sensitivity of vertebrate ears (see Chap. 17).

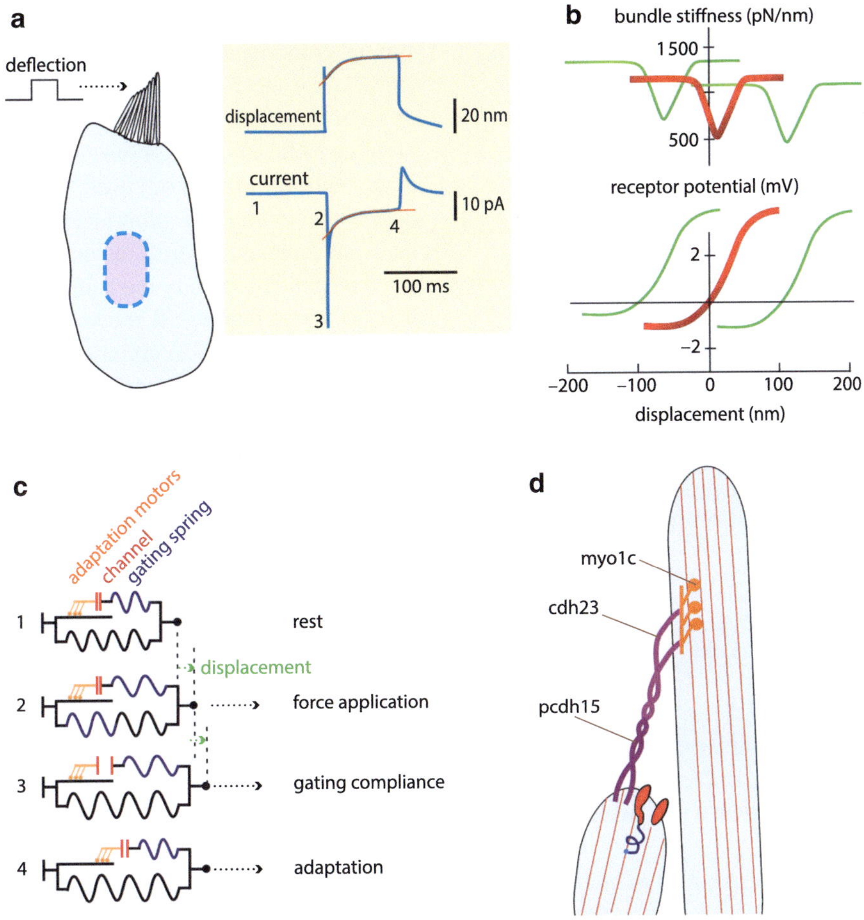


**Fig. 16.3** **Transduction in hair cells.** (**a**) Step deflection of the sensory hair bundle (*left*) evokes mechanical (*top*) and electrical (*bottom*) responses. The current response sharply peaks upon the deflection and then adapts, first rapidly and then more slowly. The exponential describing the latter slow adaptation is shown in *red*. The peak of the current coincides with a twitch in the receiver's displacement response (*top*), and the slow adaptation is reflected by a creep with the same time constant (*red line*, same as in bottom panel). Numbers refer to panel c. (**b**) *Top*, *red line*: Nonlinear compliance obtained by plotting the stiffness of the hair bundle at the displacement twitch in a against the amplitude of hair bundle deflection. *Green line*: adaptive shifts of the nonlinear compliance if the hair bundle's resting position is offset in negative (towards the left in *a*) and positive (towards the right in a) direction. *Bottom*: Associated receptor potentials, documenting that the nonlinear compliance covers the range of displacements at which transduction channels gate. (**c**) Gating spring model. The model assumes that the transduction channels are serially arranged with adaptation motors and gating springs and that they operate in parallel to a linear spring that represents the linear elasticity of the stereociliar hair bundle. Numbers, referring to time points as indicated in panel a: *1*: arrangement at rest; *2*: upon force application, gating springs and linear springs are extended, whereby they behave like Hookean springs; *3*: nonlinear gating compliance, whereby channel opening reduces the apparent stiffness of the hair bundle, making it move further in the forcing direction; *4*: adaptation, whereby the movements of the adaptation motors restore the initial open probability of the channels by adjusting the gating spring tension. (**d**) Sketch of two adjacent stereocilia depicting their actin cytoskeleton (*red lines*), the adaptation motors (myosin 1c), the tip link formed by cadherin 23 and protocadherin 15, and the presumptive positions of the transduction channel and the gating spring (Adopted from Bechstedt and Howard [10] (a, b), Nadrowski and Göpfert [11] (c), and Christensen and Corey [9] (d))

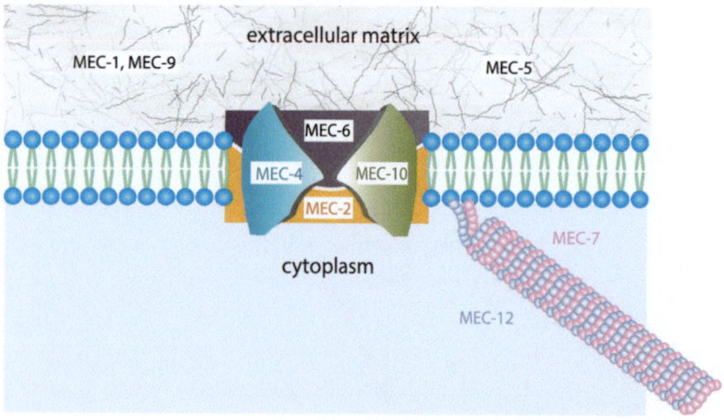


**Fig. 16.4** ***C. elegans*** **mechanotransduction complex.** Combined genetic and functional studies have identified the molecular complex that mediates the response to gentle touch in *C. elegans*. Two members of the degenerin (DEG) family of ion channels, MEC-4 and MEC-10, jointly form the mechanotransducer channel pore. The force sensitivity of this channel is thought to be modified by two other proteins which interact with the channel's lipid environment: the stomatin-like protein MEC-2 and the paraoxonase-like protein MEC-6. Both on the cytoplasmic side (microtubules MEC-7, MEC-10) and in the extracellular space (MEC-1, MEC-5, and MEC-9) further proteins contribute to set the channel's mechanical environment and gating properties

### 16.3.2 *C. elegans* Mechanosensation and the Genetic Dissection of Mechanosensory Transduction

Studies on hair cells have provided a detailed quantitative picture of how MET channels function and shown that the MET channels of hair cells are associated with adaptation motors and gating springs. **Myosin 1c** has been identified as the presumptive adaptation motor (for a review, see [13]), but the molecular identities of the MET channels proper and their gating springs are unclear. Tip links, fine filaments that interconnect adjacent stereocilia at their tips, have long been surmised to act as gating springs, yet recent findings have challenged this notion: the tip links are made of **cadherin 23** and **protocadherin 15** filaments that are joined in the middle, and are too stiff for acting as gating springs; for a review, see [14] (Fig. 16.3). Using these tip link proteins as starting points, various hair bundle proteins have been identified, yet the identities of the MET channels and the gating springs remain elusive.

An organism that has allowed to pin down the molecular identities of MET channels and associated proteins is the free-living roundworm *Caenorhabditis elegans*. The ~1-mm-long, transparent animals feed on bacteria in rotting organic material and come along in two sexual configurations: hermaphrodites and males. The nervous system of hermaphrodites is particularly simple, comprising only 302 neurons, 10 % of which are primary mechanosensory cells. The sensory dendrites of these mechanosensory neurons can be ciliated or extend into long, microtubule-filled processes that innervate the worm's body wall. Some of the mechanosensitive cells serve proprioception by monitoring body movements, whereas others detect touch. In *C. elegans*, gently touching the animal at its front makes it back away, but it will move forward again when touched at its rear. The simplicity and robustness of this **touch-avoidance behavior**, which is mediated by only six touch-sensitive neurons in the worm's body wall, have been used as an experimental paradigm for the world's first screens for genes involved in mechanotransduction [15–17]. These screens have led to the identification of a set of new proteins (collectively termed **MECs** and numbered through from 1 to 18) whose disruption causes mechanosensory abnormalities, leaving the worms insensitive to gentle touch (Fig. 16.4). Two of these proteins, **MEC-4** and **MEC-10**, form ion channels of the **degenerin (DEG)** family, which is named after the swelling (and subsequent cell degeneration) that follows gain-of-function mutations in certain family members. Interestingly, *C. elegans* DEG channels turned out to bear close structural and functional similarities to epithelial $Na^+$ channels (ENaC) of vertebrates. Reflecting this resemblance, the family is commonly referred to as **DEG/ENaC** family of ion channels. But it was not only ion channels that surfaced during the investigations. Arranging all MECs, according to their sequence similarities with known proteins, into distinct functional classes soon led researchers to postulate that mechanotransduction rather than being the exclusive task of a single transducer channel was more likely to originate from the concerted activity of a larger complex of mechanosensory proteins that form a **mechanotransduction apparatus**. In the case of *C. elegans* touch sensation, the core of this channel complex was found to consist of four proteins, which (1) are coexpressed in touch neurons *in vivo*, (2) can immunoprecipitate each other *in vitro*, and (3) form voltage-independent, amiloride-sensitive $Na^+$ channels when coexpressed in *Xenopus* oocytes [18, 19]. These four comprise the two DEG/ENaC channels MEC-4 and MEC-10, which are thought to form the actual channel pore [20], and in addition the **stomatin-like protein MEC-2** and the **paraoxonase-like protein MEC-6**. Both MEC-2 and

MEC-6 interact with cholesterol; they are expected to affect the lipid environment of the MEC-4/MEC-10 channel and, as a result, its position and integration into the membrane. This finding is particularly interesting as it has been calculated that the intrinsic force profile of cellular membranes depends on membrane composition and that variations thereof can strongly influence the functions of proteins embedded in the membrane by changing the lateral pressure acting on them [21] (Fig. 16.4). It has been proposed that these intramembranous forces could provide, or at least contribute to, the forces that gate mechanotransducer channels [6]. Consistent with a role of the bilayer in mechanotransducer activation, the coexpression of MEC-2 and MEC-6 has been shown to synergistically increase the currents flowing through an activated form of the MEC-4/MEC-10 channel by up to 200-fold [18, 19]. The current view is that, by creating a specific membrane environment, MEC-2 and MEC-6 shape the forces that act on the MEC-4/MEC-10 channels and, by doing so, help to specify the transducers' sensitivity to external force [22]. In the natural context, these external forces take their origin in the worm's cuticular body wall and are then transmitted to the membranes of microtubule-filled processes of the touch-sensitive neurons via an **electron-dense extracellular matrix (ECM)** called the **mantle** [16]. Given this line of force transmission, it is not surprising that two other groups of *mec* genes turned out to code for ECM proteins (**MEC-1, MEC-5, and MEC-9**) and microtubules (**MEC-7, MEC-12**). Especially the discovery of the touch neuron-specific, 15-protofilament microtubules MEC-7 and MEC-12 inspired a spring-operated model of MEC-4/MEC-10 activation resembling the gating-spring mechanism proposed for mechanotransduction by hair cells (see above): here, the transducers would be coupled to the body wall by the ECM on the extracellular side and tethered to the cytoskeleton by the 15-protofilament microtubules on the intracellular side. Together, ECM and microtubules would thus provide both the elasticity and the mechanical support that are required for a gating-spring-like transducer activation. However, this simple proposal has meanwhile been challenged by a number of experimental findings. First, neither MEC-7 nor MEC-12, are required for the gating of the MEC-4/MEC-10 channel [20, 23]: Mechanotransducer currents, though greatly diminished, persist in MEC-7 and MEC-12 loss-of-function mutants. Second, immunoelectron microscopy failed to detect direct physical interactions between MEC-4, MEC-5, and microtubules [24]. Even if the reality of transducer gating now seems more complex (and more resilient) than was surmised by the initial models, it is nonetheless clear that both ECM proteins like MEC-5 as well as the unconventional 15-protofilament microtubules play a pivotal role in touch-evoked mechanotransduction in *C. elegans*, as abolishing the function of either of them virtually abolishes the worm's sensitivity to gentle touch.

### 16.3.3 Mechanosensation in *Drosophila* – Mechanosensory Organs and TRPs

While the analysis of touch avoidance in *C. elegans* has led to the identification of DEG/ENaCs as mechanotransducers, the study of mechanosensation in the fruit fly *Drosophila melanogaster*, has drawn attention to **transient receptor potential (TRP)** channels (for TRPs, see [25]). In *Drosophila*, DEG/ENaCs are implicated in the response to noxious mechanical stimuli, whereas sensitive mechanosensation seems mediated by TRPs. Before discussing the role of TRPs in fly mechanosensation, we will briefly introduce the sensory repertoire insects use for detecting force (see Fig. 16.5 for an overview).

The peripheral nervous system of insects contains a vast amount of mechanosensory neurons that can be categorized into two main types. Type 1 mechanosensory neurons are monodendritic, bearing a single ciliated dendrite. Type 2 neurons are also known as multidendritic (md) neurons because they have many nonciliated dendrites. Type I neurons associate with accessory cells, with whom they form complex **mechanosensory organs**. **Type II neurons**, by contrast, do not associate with accessory cells; their dendritic projections broadly innervate the fly's joint membranes, muscles and the larval body wall, and the neurons are implicated in noci- and proprioception. The mechanosensory organs that are innervated by type I neurons can be subdivided into **external sensory (es)** organs and **chordotonal (ch)** organs. es organs are associated with external cuticular structures such as touch-sensitive hairs (**mechanosensory bristle organs**) or domes (**campaniform sensilla**) that can be deflected by touch or wind (hairs) or deformed by cuticular stretch (domes). ch organs, in turn, are internal stretch-receptors that span between

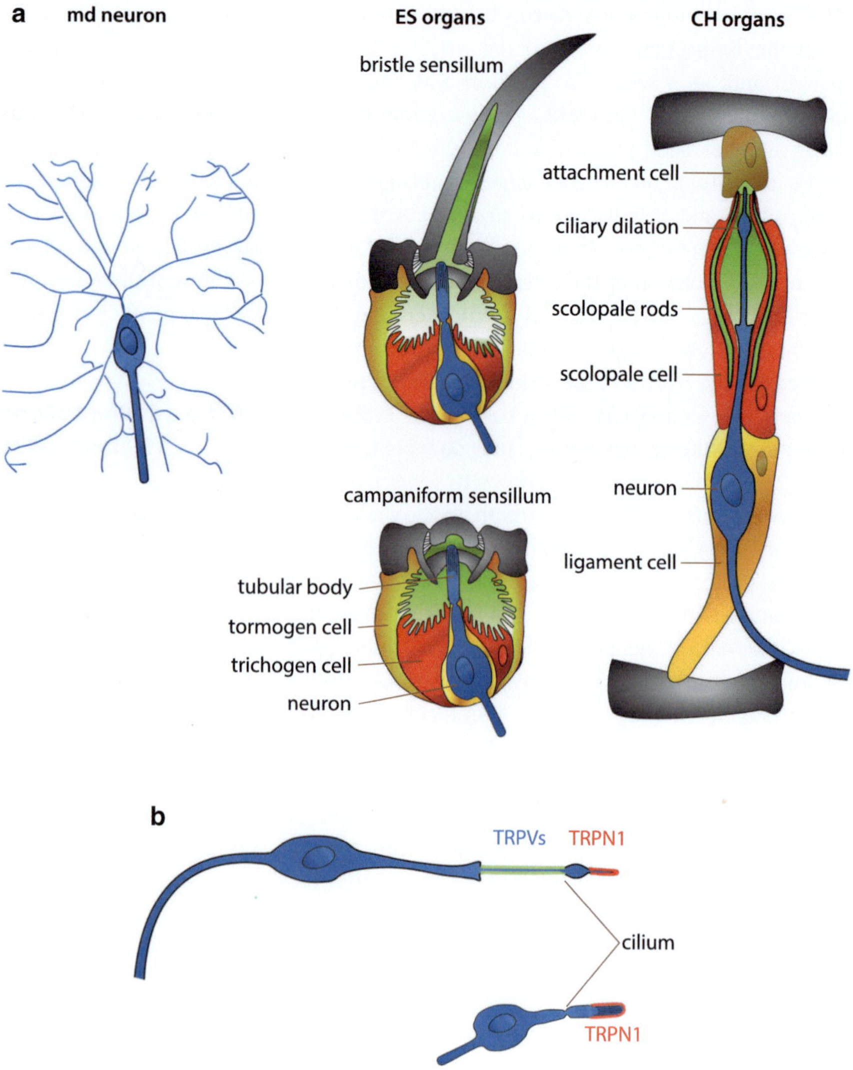


**Fig. 16.5** **Insect mechanosensory neurons and organs.** (**a**) Neuron and organ types. (**b**) Dendritic localization of TRP channels in mechanosensory neurons of chordotonal (*top*) and external sensory (*bottom*) organs

different parts of the body and monitor relative movements in between. The organs primarily serve proprioception and can also be used for detecting airborne (tympanal organs, Johnston's organ; see Chap. 17) and substrate (e.g., subgenual organs) vibrations. Anatomically, ch organs are composed of one to thousands of chordotonal sensilla (= scolopidia) that share developmental and functional parallels with es organs:

1. They arise from a series of asymmetric, *Notch*-dependent mitoses that lead from a single **sensory organ precursor (SOP) cell** to the mechanosensory neurons (only one in bristle and campaniform sensilla and up to three in chordotonal sensilla) plus three to four supporting cells.
2. The distal parts of the neurons' ciliated dendrites, which are the presumptive site of mechanotransduction, are tightly sealed against their environment by a cellular barrier formed of tight-junction-linked epithelial and supporting cells. The narrow cavity that is thereby created around the distal dendrites is filled with a **receptor lymph** which differs from the canonical extracellular condition in that it is high in $K^+$ and low in $Na^+$. Electrogenic transport by associated supporting cells holds the receptor lymph at high positive potentials of +20 to +80 mV (the transepithelial potential, TEP) with respect to the surrounding extracellular medium, providing a strong electrochemical driving force for currents through the transduction channels.

Various genes that are required for the function of es and ch organs have been identified by screening mutagenized *Drosophila* larvae for impaired touch-responses [26]. In several of these touch-insensitive mutants, deflecting the bristles of adult flies failed to elicit TEP voltage responses, and the mutations were named accordingly no-mechanoreceptor potential (*nomp*). The first nomp mutations that were molecularly characterized affected the *nompC* gene, which encodes a TRP family channel [27]. This channel, **TRPN1**, has been lost in the course of vertebrate evolution, but can be found in the genomes of, e.g., nematodes, insects, frogs, and fish. Loss of TRPN1 abolishes adapting transepithelial currents in *Drosophila* mechanosensory bristles, reducing the amplitude of the current to ca. 10 % of that in controls. A gain-of-function allele of the *nompC* gene left the current amplitude unaffected but increased the speed of adaptation. This suggests that TRPN1 forms, or is a subunit of, an adapting MET channel in the fly's bristle neurons, a notion that is corroborated by the localization of TRPN1 in the **tubular body** of es neurons – a distal dendritic swelling that, harboring a dense array of parallel microtubules, presumably is the site of mechanotransduction.

TRPN1 is also expressed in hair cells of fish and frogs, yet its importance for transduction is uncertain: in zebrafish hair cells, TRPN1 seems required for mechanosensation, yet in frog hair cells it localizes to the tip of the **kinocilium** that, unlike the actin-based stereocilia, is a true cilium with a microtubule-based axoneme and is dispensable for transduction. A similar ciliary localization of TRPN1 has also been reported for *Drosophila* ch neurons, where the channel's distribution is confined to the ciliary tips. In the fly's antennal chordotonal organ, Johnston's organ, this TRPN1 localization can be seen in virtually all the 500 mechanosensory neurons, yet only about half of the neurons functionally require TRPN1: those neurons that need this channel are **auditory neurons** that the fly uses for hearing courtship songs. The other neurons are specialized for detecting wind and gravity, which they still can in the absence of TRPN1. Auditory and gravity/wind-sensitive neurons in the fly's Johnston's organ notably differ in function as only the former cells are motile and actively amplify the vibrations they transduce. Like the active hair bundle motility of vertebrate hair cells, this active **neuronal motility** can be explained by the interplay between MET channels and associated adaptation motors, which in the fly's auditory neurons might be axonemal dyneins. Loss of TRPN1 abolishes this transducer-based mechanical amplification, supporting the idea that the fly's transduction channel for hearing might well be this TRP.

TRPN1, apart from being a candidate MET channel, has also been speculated to serve as gating spring. Like other TRPs, TRPN1 bears **ankyrin** repeats at its N-terminal end, yet the number of these repeats is exceptionally large, forming a helix with one turn. The stiffness of a helical spring depends on the number of its turns: Increasing (or decreasing) this number by adding (or removing) ankyrins should soften (or harden) the spring, providing a direct test whether TRPN1 ankyrins act as gating spring. Unfortunately, the subcellular localization of TRPN1 is disturbed when its ankyrin spring is shortened, leaving it open whether TRPN1 ankyrins serve as gating spring. That TRPN1 can serve as the pore-forming subunit of native MET channels, however, has been demonstrated for certain *C. elegans* mechanosensory neurons, illustrating that MET channels can be TRPs.

Apart from TRPN1, various other TRPs have been implicated in mechanosensation, including, for example, vanilloid (**TRPV**) subfamily members such as *Drosophila* Inactive (Iav) and Nanchung (Nan). Together, these two TRPV proteins seem to form a heteromeric Nan-Iav channel that localizes to the basal region of the cilium of ch neurons, proximally to TRPN1. Consistent with this localization, Nan-Iav operates downstream of TRPN1 in fly auditory neurons and negatively regulates the active motility of these cells: Loss of Nan-Iav makes the fly's auditory neurons hypermotile, causing excessive mechanical feedback and continuous feedback oscillations in the ear.

The fly's ability to respond to harmful mechanical stimuli is mediated by certain md neurons. The mechanosensory function of these neurons reportedly requires the DEG/ENaC channel pickpocket along with the *Drosophila* Piezo channels. Piezos are a newly defined, evolutionarily conserved family of ion channels that are unusual in that they bear 24 and 36 predicted transmembrane domains. When heterologously expressed, Piezos induce large mechanosensitive currents. Two different Piezo channels, Piezo1 and Piezo2, are found in vertebrates, and one of these channels, Piezo2, is expressed in – and required for the adapting mechanosensitive currents of – many of the dorsal root ganglion (DRG) neurons that innervate the skin. In the following section we will use this somatosensory system to exemplify the structural and functional diversities that mechanoreceptors can assume.

## 16.4 The Mammalian Somatosensory System

The mammalian skin harbors a set of morphologically, and functionally, distinct types of mechanoreceptors which have historically been assigned to various overlapping classes according to (1) their anatomy (e.g., large or small soma, specialized or unspecialized terminal structures, myelinated or unmyelinated axons), (2) their response properties (e.g., slowly or rapidly adapting, high or low threshold, or (3) their axonal conduction velocities. The arguably most important descriptor of somatosensory cells, however, is their response behavior or, more specifically, their *stimulus thresholds*, *receptive fields*, and *adaptation properties*. In sensory biology, the term stimulus threshold is used to describe the *stimulus magnitude* (e.g., skin indentation depth) that is necessary to make a sensory cell respond (e.g., by firing an action potential). A sensory cell's receptive field is the *spatial region* (e.g., skin area) within which a given stimulus will evoke a response. The term adaptation commonly refers to two interlinked phenomena: the cessation of a sensory response despite the continued presence of the stimulus and the, full or partial, restoration of a receptor's pre-stimulus sensitivity.

### 16.4.1 The Structural Basis of Somatosensation – Receptors, Innervation, and Afferent Responses

The cellular architecture of all types of mammalian somatosensory neurons shares two common features: their cell bodies cluster to form ganglia in the central nervous system (*trigeminal ganglion* or *dorsal root ganglia*) and they send out axonal projections (*sensory afferents*) into the periphery that transmit action potentials from the periphery to the brain [5, 28–30]. The below section gives an overview of the main types of cutaneous mechanoreceptors (see Fig. 16.6 for a summary) and a brief introduction into the functional logic of their operation. In a crude first distinction, all cutaneous mechanoreceptors can be classified as either mechanoreceptive nociceptors or mechanoreceptors *sensu stricto*.

**Mechanoreceptive nociceptors.** Nociceptors hold an exceptional position among the various senses. In accordance with their primary task of detecting stimuli that are potentially harmful to an animal's body, many nociceptors are polymodal: Just as physical damage can be caused by heat, chemical reagents, or mechanical impact, nociceptors often display a combined thermo-, chemo-, and mechanosensitivity. The mechanoreceptive nociceptors of the mammalian skin are formed by a class of slowly adapting (SA) nerve fibers with high thresholds (HT) and low conduction velocities (Aδ and C-fibers). The peripheral endings of these fibers terminate freely in the subepidermal corium and do not show any overt specializations (see also Chap. 21).

**Merkel cell-neurite complexes (syn. Merkel disk receptors, SA1 afferents).** The main cellular substrate of fine tactile discrimination and texture perception in mammals is the touch-receptive complex formed by the endings of myelinated, slow adapting (SA1), low-threshold nerve fibers (Aδ type), and specialized epidermal cells, the so-called *Merkel cells*. Although the specific mechanosensory roles of Merkel cells have remained unclear, evidence suggests that they are essential for the response to light touch. Merkel cell complexes are widely distributed across the mammalian body and can be found in the superficial layers of both glabrous and hairy skin. In the human fingertips Merkel receptors are clustered beneath the dermal ridges that create the fingerprints; here they reach densities of about 100/cm$^2$ enabling human beings to resolve minute textural details of only ~0.5 mm spatial dimension with their fingertips (as exploited in tactile alphabets such as the Braille system).

**Ruffini corpuscles (SA2 afferents).** Ruffini corpuscles are spindle-shaped structures that surround the endings of myelinated, low-threshold mechanosensitive nerve fibers (Aδ type) in the deep skin; their response to a step-like mechanical stimulus is of the slowly adapting pattern (SA2). Ruffini corpuscles occur in both the hairy and the glabrous skin; they mediate the perceptions of objects moving relative to the skin as well as those of hand shape and finger position and thereby contribute to *kinesthetic control*.

**Meissner corpuscles (RA afferents).** The advent of bipedalism, habitual or facultative, was one of the key events in the evolution of primates. Freeing one pair of limbs from the task of locomotion made it available for adaptive modifications which enabled novel types of behavior such as object manipulation or the use of tools. When handling objects, however, we must be able to sense and control the forces we exert on them. The type of mechanoreceptor largely responsible for this task is associated with myelinated, rapidly adapting (RA), low-threshold nerve fibers. The (unmyelinated) terminals of these afferents are embedded in a capsule of connective tissue where they form a characteristic complex with a set of flattened support cells arranged in horizontal lamellae. These terminal structures are called Meissner corpuscles. Just as the Merkel SA1 afferents, Meissner corpuscles are located in surface-near areas of the skin where they can reach densities of up to 150/cm$^2$ (human fingertip). In contrast to the SA1 afferents, however, Meissner-RA afferents are insensitive to static skin indentation and respond exclusively to the dynamic parts of skin deformation, to which they are about four times more sensitive than Merkel receptors. These response properties enable the RA afferents to detect the small (and sudden) forces between an object and the hand that is holding it. Meissner corpuscles are thus the cellular basis of grip control.

**Pacinian corpuscles.** Another type of myelinated, rapidly adapting, low-threshold nerve fibers terminates in oval, lamellar-shaped structures formed by alternating layers of connective tissue, gelatinous material and fibroblasts: the so-called Pacinian corpuscles. Pacinian corpuscles measure ~1 mm in size and are located in deeper layers of the skin; in their center is the inner bulb, a fluid-filled cavity harboring the single, unmyelinated ending of a mechanosensory afferent. Pacinian corpuscles are vibration detectors. In this context, their deep location in the skin together with the multilayered, gelatinous wrapping build an effective high-pass filter system that renders the receptors more than 100-times more sensitive to vibration frequencies of 100 Hz than to vibrations at 1 Hz. At their best frequencies (100–250 Hz), Pacinian receptors display an exquisite mechanical sensitivity with the most sensitive corpuscles responding to skin vibrations of only 10 nm amplitude! Vibrations propagating from their source through a substrate are seismic signatures of distant events. Pacinian corpuscles enable us to read, and decode, such signatures.

**Hair follicle afferents.** Another mechanoreceptor subtype is associated with dermal follicles, i.e., the miniature organs that produce the hairs of the mammalian skin. Just as the afferents innervating Meissner and Pacinian corpuscles, hair follicle afferents represent myelinated, rapidly adapting, low-threshold nerve fibers. They are maximally activated by movements of the associated hair shafts (e.g., small sinus or down hairs) but usually possess receptive fields large enough to allow for activation by motions of juxtaposed hairs

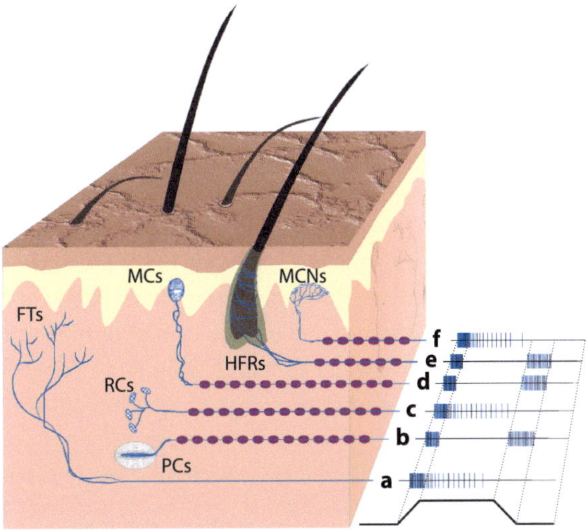


**Fig. 16.6** **Mammalian cutaneous mechanoreceptors.** The mammalian skin harbors a variety of mechanosensory cells which differ in their morphologies, response behaviors and location within the skin. (**a**) Free terminals (*FTs*) of unmyelinated C-fibers innervate the upper layers of the skin; they represent low-threshold (*LT*), slowly adapting (*SA*) mechanoreceptors which, e.g., contribute to the detection of social body contact in humans. (**b**) Pacinian corpuscles (*PCs*) occur in deeper skin layers; they are innervated by myelinated, rapidly adapting (*RA*), LT fibers. PCs act as vibration sensors and, e.g., enable our ability to remotely assess a material's mechanical properties while 'scanning' it with a hand-held probe. (**c**) Ruffini corpuscles (*RCs*) are innervated by myelinated, LT, SA fibers; they are restricted to deeper layers of the skin and contribute to the monitoring of hand shape and finger position. (**d**) Rapidly adapting Meissner corpuscles (*MCs*) associate with myelinated, low-threshold nerve fibers that are particularly sensitive to skin motion that occurs when handling objects ('slipping'). MCs are found in the upper layers of the skin and mediate human grip control. (**e**) Hair-follicle receptors (*HFRs*) associate with myelinated LT, RA fibers; they are maximally activated by deflections of the associated hair shaft and constitute the primary cellular substrate for the detection of light touch in mammals (D-hairs). (**f**) Merkel cell-neurite complexes (*MCNs*) are found in high densities in upper skin layers; they represent ultrasensitive mechanosensors (*LT*, *SA*) which mediate the tactile analysis of substrate texture. MCNs underlie the human ability to read tactile alphabets. On the right-hand side, schematic activity patterns are shown for each unit's response to a ramp-like skin indentation

or nearby skin areas as well. One subtype of hair follicle afferent, the D-hair receptors, is considered to be among the most sensitive mechanoreceptors in the skin. Using a von Frey test, which measures the response to forces applied to the skin (typically involving a calibrated nylon filament pressed against the skin until it buckles) it was found that D-hair receptors have response thresholds that are about ten times lower than those of other low-threshold receptors. Generally, hair follicle afferents are the major cellular substrate for detecting light touch in mammals. Given the fact that the anatomical design of these receptors closely resembles that of insect touch-receptive bristles (see above), it seems worth noting that a small insect crawling over our skin would most likely to be detected by a hair follicle receptor.

### 16.4.2 The Functional Logic of Somatosensation: Thresholds, Receptive Fields, and Adaptation Properties

From a somatosensory point of view, the cultural evolution of human beings could be recounted as the evolution of the hand. When anonymous, prehistoric artists started carving small figurines, such as the famous Venus of Willendorf out of pieces of limestone, their success crucially relied on feedback signaling from the four types of mechanoreceptors in the glabrous skin of the human hand. Using the example of these four mechanosensitive afferents, i.e., Merkel cells (SA1), Meissner corpuscles (RA), Pacinian corpuscles (PC), and Ruffini corpuscles (SA2), we will here briefly illustrate how their specific response properties enable them to serve distinct, but complementary, tactile functions.

As described above, somatosensory cells can be, and have been, categorized on the basis of their stimulus thresholds, receptive fields, and adaptation properties.

In response to static skin indentations, both RA and PC fibers fire short bursts of action potentials that remain restricted to the transients at the onset, and offset, of the stimuli. SA1 and SA2 afferents, in contrast, respond with prolonged (only slowly decaying) bursts; also, they do not respond to the stimulus offset, at all.

SA1 and RA afferents innervate mechanoreceptors of the *surface-near skin* (Merkel cells and Meissner corpuscles, respectively); both share *small receptive fields* and a *high receptor density*. Taken together, these features provide SA1 and RA afferents with a high spatial resolution and low skin indentation thresholds which are the basis for their involvement in the local probing of, and interaction with, tools and objects (e.g., texture analysis or grip control).

SA2 and PC afferents, in turn, innervate mechanoreceptors of the *deeper skin* (Ruffini and Pacinian corpuscles, respectively); both share *large receptive fields* and *low receptor densities*. These properties stress their more global tactile roles (e.g., monitoring hand shape or analyzing vibratory cues).

The anonymous creator(s) of the Venus of Willendorf, for example, will have had to keep firm grip (mediated by RA afferents) on their burin, which they did not only use as an engraving tool to carefully scratch off layer by layer of limestone but also as a tactile prosthesis to remotely probe the nature and quality of the stone material (mediated by PC afferents). A permanent control of finger position of both hands, the one that was holding the burin as well as the one that was holding the Venus, was required (and partly mediated by SA2 afferents). A haptic appreciation of the completed sculpture, finally, would have been impossible without signaling from low-threshold, high-resolution mechanoreceptors (such as SA1 afferents).

## 16.5 Central Processing of Mechanosensory Information – Converting Mechanosensory Information into Behavioral Responses

Mechanosensation enables organisms to respond to stimulus forces. In many cases, the conversion of stimulus forces into responses is direct, as, for example, in bacteria where membrane tension is monitored and regulated by MscL. MET channels and associated adaptation motors likewise monitor and actuate movements in hair cells and fly auditory neurons, where they serve as sensors and actuators in amplificatory feedback loops.

Apart from such cellular responses, mechanosensory input can drive animal behavior, and metazoans have dedicated considerable parts of their CNS to mechanosensory processing tasks. Until now, we have

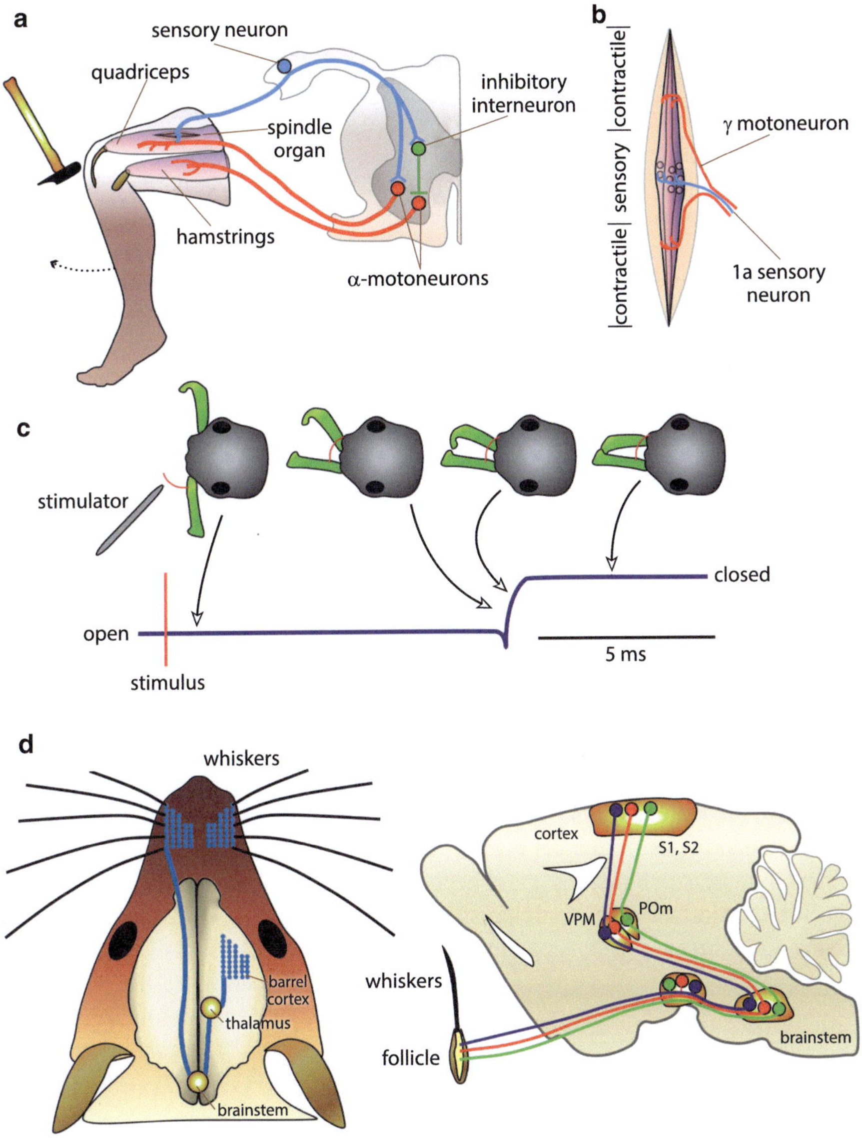


**Fig. 16.7 Mechanosensory circuits and mechanically evoked behaviors.** (**a**) Patellar reflex arc. (**b**) Muscle spindle organ depicting its sensory and motor zones. The motor zones are innervated by motoneurons and are contractile, allowing the spindle organ to modulate its stiffness and to mechanically adapt. (**c**) Mandible snap of a Dacetini ant. The snap commences ca. 10 ms after the trigger hair was deflected and the snap itself takes about 1 ms. (**d**) Whisker system in rodents (*left*: top view, *right*: side view). The vibrissae form a two-dimensional grid on each side of the snout. First-order neurons that innervate the whisker follicles synapse in the brainstem, and the axons of second-order neurons cross the midline and travel to the thalamic somatosensory nuclei (VPM, POm) whose neurons target the barrels in the primary somatosensory cortex (Adapted from Gronenberg et al. [31] (c) and Diamond et al. [32] (d))

dealt with the peripheral processing of mechanosensory information by mechanosensory organs and cells. We now will have a look at central mechanosensory pathways and discuss some of the basic mechanisms how mechanosensory information is centrally processed. Depending on the complexity of the processing task, central mechanosensory circuits can be quite simple, sometimes involving only one synapse and no interneurons, or more elaborate, including parallel pathways and multiple processing centers with reciprocal information flow in between. We will discuss such simple and complex mechanosensory circuits, with examples including mechanosensory reflexes and the whisker system that many mammals use to explore the world.

### 16.5.1 Mechanosensory Reflexes – Kicking Legs and Snapping Jaws

The classic example for a **monosynaptic** mechanosensory circuit is the **patellar reflex** that ensues from hits on our knee: the hit stretches the patellar tendon, thus activating a muscle spindle organ in the quadriceps muscle of the leg. Muscle spindles are mechanosensory organs that are located within muscles and consist of contractile parts and a central noncontractile sensory region in between. Stretching this sensory region activates a **type 1a afferent fiber** that signals back to the spinal cord. There, the fiber synapses onto an **alpha motoneuron**, whose activity makes the quadriceps contract, resulting in a kick of the leg (Figs. 16.7a, b).

Mechanosensory reflex arcs can be remarkably fast, allowing for fast responses with only millisecond delays. An example for such ultrafast reflex is the **trap-jaw reflex** that ants of the tribe Dacetini use for catching swift prey and to defend themselves [31]. The reflex is triggered by the deflection of a trigger hair on the ants' labrum, which activates a sensory neuron (Fig. 16.7c). This neuron synapses onto a motoneuron in the subesophageal ganglion, whose activity triggers a catapult contraction of the jaw muscles so that the mandibles snap. The snap itself lasts only about one millisecond, and it takes place only some 10 milliseconds after the hair was deflected, which is about the time it takes an insect photoreceptor cell to electrically respond to light. Several mechanisms account for this speedy behavior: the involved axons are thick, the synapse is electrical, and the energy generated by the muscle power is stored in elastic strain energy that, once the motoneuron fires, is suddenly released. When comparing the latency of the reflex with that of the phototransduction cascade it also becomes obvious that such fast responses are only possible if transduction is direct because there is no time for second messenger cascade.

### 16.5.2 Probing the World by Touch – The Whisker System

Pinnipeds monitor water flow with facial **vibrissae** on their snouts and seals use these whiskers to trace the hydrodynamic trails of fish on which the prey. Also nocturnal mammals such as mice and rats rhythmically move their whiskers to actively extract information about spatial object properties such as texture, size, and shape. Whiskers are arranged in a stereotyped manner and differ with respect to their length and their resonant frequency (for reviews, see [32, 33]) (Fig. 16.7). Proximally, each whisker is anchored in the skin by a **follicle**. This follicle contains muscles that move the whisker and is afferently innervated by some 200 **trigeminal neurons**. The peripheral endings of these afferents are mechanosensitive and generate action potentials that they propagate past their somata in the **trigeminal ganglia** to the **trigeminal nuclei** in the **brainstem**. Here, the trigeminal neurons synapse onto second-order neurons that project in parallel pathways to the **thalamus** and then continue to the **barrel field** in the **somatosensory cortex**. Here, each whisker is mapped to one barrel: neurons of one barrel have **receptive fields** that extend beyond one whisker, yet they are most responsive if the appropriate whisker is moved.

At least three parallel afferent pathways that arise in the trigeminal nuclei (TN) can be distinguished, a lemniscal, an extralemniscal, and a paraleminiscal one. The three pathways differ in their circuitry and are thought to have different functions [32]: Detailed whisking and touch information, for example, may be processed by the **leminiscal pathway**, which arises from second-order neurons in the TN. The respective TN neurons are somatotopically arranged into **barrelettes** and send axons to **barreloids** of the ventral **posterior medial nucleus (VPMdm)** of the thalamus whose neurons project into the somatosensory cortex. Information about the timing of vibrissal contact, in turn, may be processed via the extralemniscal pathway, which starts with second-order neurons that are also arranged in barrelettes and project to the somatosensory cortex via the **ventrolateral domain of the VPM (VPMvl)**. The paralemniscal pathway finally has been proposed to convey information about whisking kinematics. TN neurons of this pathway are not arranged in a spatially specific manner, and they target somatosensory and motor cortices via the posterior nucleus (POm) [32].

Neurons of the **barrel cortex** and the **motor cortex** target each other, **forming cortico-cortical feedback loops**. This feedback between sensory and motor centers is crucial for guiding active movements of the whiskers, which, much like the MsCL and MsCs channels of *E. coli*, thus act as sensors and actuators in a feedback loop.

## References


1. Kung C, Martinac B, Sukharev S (2010) Mechanosensitive channels in microbes. Annu Rev Microbiol 64:313–329
2. Machemer H, Bräucker R (1992) Gravireception and graviresponses in ciliates. Acta Protozool 31:185–214
3. Roberts AM (2010) The mechanics of gravitaxis in *Paramecium*. J Exp Biol 213:4158–4162

4. Sukharev S, Corey DP (2004) Mechanosensitive channels: multiplicity of families and gating paradigms. Sci STKE 3:re4
5. Lumpkin EA, Marshall KL, Nelson AM (2010) The cell biology of touch. J Cell Biol 191:237–248
6. Kung C (2005) A possible unifying principle for mechanosensation. Nature 436:647–654
7. Chalfie M (2009) Neurosensory mechanotransduction. Nat Rev Mol Cell Biol 10:44–52
8. Mederos y Schnitzler M, Storch U, Meibers S, Nurwakagari P, Breit A, Essin K, Gollasch M, Gudermann T (2008) Gq-coupled receptors as mechanosensors mediating myogenic vasoconstriction. EMBO J 27:3092–3103
9. Christensen AP, Corey DP (2007) TRP channels in mechanosensation: direct or indirect activation? Nat Rev Neurosci 8:510–521
10. Bechstedt S, Howard J (2007) Models of hair cell mechanotransduction. Curr Top Membr 59:399–424
11. Nadrowski B, Göpfert MC (2009) Modeling auditory transducer dynamics. Curr Opin Otolaryngol Head Neck Surg 17:400–406
12. Hudspeth AJ, Choe Y, Mehta AD, Martin P (2000) Putting ion channels to work: mechanoelectrical transduction, adaptation, and amplification by hair cells. Proc Natl Acad Sci USA 97:11765–11772
13. Gillespie PG, Cyr JL (2004) Myosin-1c, the hair cell's adaptation motor. Annu Rev Physiol 66:521–545
14. Peng AW, Salles FT, Pan B, Ricci AJ (2011) Integrating the biophysical and molecular mechanisms of auditory hair cell mechanotransduction. Nat Commun 1:523
15. Chalfie M, Au M (1989) Genetic control of differentiation of the *Caenorhabditis elegans* touch receptor neurons. Science 243:1027–1033
16. Chalfie M, Sulston J (1981) Developmental genetics of the mechanosensory neurons of *Caenorhabditis elegans*. Dev Biol 82:358–370
17. Sulston J, Dew M, Brenner S (1975) Dopaminergic neurons in the nematode *Caenorhabditis elegans*. J Comp Neurol 163:215–226
18. Chelur DS, Ernstrom GG, Goodman MB, Yao CA, Chen L, O'Hagan R, Chalfie M (2002) The mechanosensory protein MEC-6 is a subunit of the *C. elegans* touch-cell degenerin channel. Nature 420:669–673
19. Goodman MB, Ernstrom GG, Chelur DS, O'Hagan R, Yao CA, Chalfie M (2002) MEC-2 regulates *C. elegans* DEG/ENaC channels needed for mechanosensation. Nature 415:1039–1042
20. O'Hagan R, Chalfie M, Goodman MB (2005) The MEC-4 DEG/ENaC channel of *Caenorhabditis elegans* touch receptor neurons transduces mechanical signals. Nat Neurosci 8:43–50
21. Cantor RS (1997) Lateral pressures in cell membranes: a mechanism for modulation of protein function. J Phys Chem B 101:1723–1725
22. Brown AL, Liao ZW, Goodman MB (2008) MEC-2 and MEC-6 in the *Caenorhabditis elegans* sensory mechanotransduction complex: auxiliary subunits that enable channel activity. J Gen Physiol 131:605–616
23. Bounoutas A, O'Hagan R, Chalfie M (2009) The multipurpose 15-protofilament microtubules in *C. elegans* have specific roles in mechanosensation. Curr Biol 19: 1362–1367
24. Cueva JG, Mulholland A, Goodman MB (2007) Nanoscale organization of the MEC-4 DEG/ENaC sensory mechanotransduction channel in *Caenorhabditis elegans* touch receptor neurons. J Neurosci 27:14089–14098
25. Venkatachalam K, Montell C (2007) TRP channels. Annu Rev Biochem 76:387–417
26. Kernan M, Cowan D, Zuker C (1994) Genetic dissection of mechanosensory transduction: mechanoreception-defective mutations of *Drosophila*. Neuron 12:1195–1206
27. Walker RG, Willingham AT, Zuker CS (2000) A *Drosophila* mechanosensory transduction channel. Science 287: 2229–2234
28. Delmas P, Hao JZ, Rodat-Despoix L (2011) Molecular mechanisms of mechanotransduction in mammalian sensory neurons. Nat Rev Neurosci 12:139–153
29. Johnson KO (2001) The roles and functions of cutaneous mechanoreceptors. Curr Opin Neurobiol 11:455–461
30. Lumpkin EA, Caterina MJ (2007) Mechanisms of sensory transduction in the skin. Nature 445:858–865
31. Gronenberg W, Tautz J, Hölldobler B (1993) Fast trap jaws and giant neurons in the ant *Odontomachus*. Science 262:561–563
32. Diamond ME, von Heimendahl M, Knutsen PM, Kleinfeld D, Ahissar E (2008) 'Where' and 'what' in the whisker sensorimotor system. Nat Rev Neurosci 9:601–612
33. Petersen CC (2007) The functional organization of the barrel cortex. Neuron 56:339–355

# 17 Auditory Systems

Günter Ehret and Martin C. Göpfert

Sound waves are the adequate physical stimulus for hearing organs of vertebrates and insects. Sound waves may originate from abiotic events, for example, from running water, breaking waves at beaches, movement of bushes and trees in the wind, howling storms, or thunder. Most interesting for animals, however, are sounds generated by other moving animals – rustling noises may signal the presence of prey or predator – or by special sound-producing organs of insects and vertebrates which use sounds in communication. The morphology of hearing organs is adapted to the physical properties of the sounds to be perceived. Central to all hearing organs are mechanoreceptors (see Chap. 16). In the present chapter, we deal with mechanoreceptors serving in hearing organs as the sensitive elements through which sound-induced motion is translated into activation of the central auditory systems of animals, thus providing information about the presence of sound waves in surrounding water or air. Sound waves traveling in and being picked up from solids are not considered here (see Chap. 16).

We will see that hearing organs of vertebrates are rather homogeneous in morphology and function because hearing via the natural routes is always coupled with functions of the inner ear. In contrast, hearing organs of insects are morphologically diverse and located at various places of the body indicating different evolutionary roots. In any case, it must be emphasized that the presence and activation of hearing organs is not sufficient for sensing sounds, not even necessary in the case of electrical stimulation of the central auditory system via modern hearing aids. It is the processing in the central nervous system that leads to the perception of sounds. Vertebrates (humans included) "hear" with their brains.

Several examples will show that sound perception can happen at different levels of complexity of neural processing. Reflexes such as the startle response to a loud sound or the orienting response to a soft sound are at the base level of processing. Instinctive responses to sounds require motivation to respond and often discrimination of the acoustic quality of sounds. Acoustic quality refers to many parameters such as intensity, spectral content, and temporal structure of a sound, and to the audibility of a sound in masking background noise, i.e., to the signal-to-noise ratio. Instinctive responses occur, for example, in sound communication, predator avoidance, and predator location of insects. Finally, sound recognition requires learning of sound patterns in order to use certain sounds for orientation and/or communication. This happens in many birds and mammals. Necessary for pattern recognition is again the ability to discriminate the acoustic qualities of sounds which is, besides sound source localization, the main purpose of auditory systems.

## 17.1 The Physics of the Stimulus

Oscillating bodies and membranes (e.g., vocal chords, air bladders, elytra, membranes of loudspeakers) are the sources of sound waves when they push and pull attached

G. Ehret (✉)
Institute of Neurobiology, University of Ulm,
Albert-Einstein-Allee 11, D-89081 Ulm, Germany
e-mail: guenter.ehret@uni-ulm.de

M.C. Göpfert
Department of Cellular Neurobiology,
Schwann-Schleiden-Centre for Molecular Cell Biology,
University of Göttingen, Julia-Lermontowa-Weg 3,
D-37077 Göttingen, Germany
e-mail: mgoepfe@gwdg.de

C.G. Galizia, P.-M. Lledo (eds.), *Neurosciences - From Molecule to Behavior: A University Textbook*,
DOI 10.1007/978-3-642-10769-6_17, © Springer-Verlag Berlin Heidelberg 2013

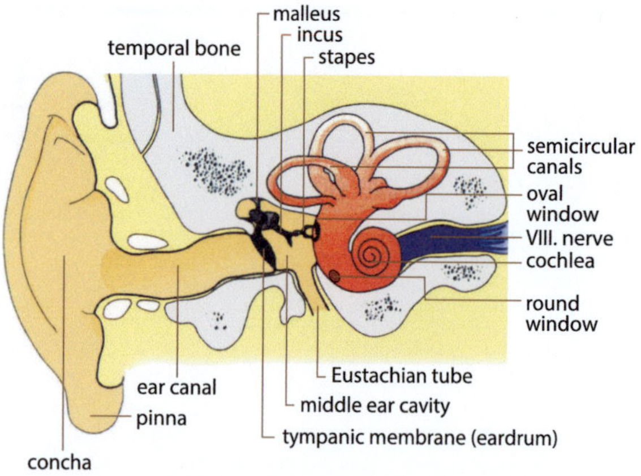


**Fig. 17.1** **The human ear** as an example of a mammalian ear with outer ear (pinna and concha, ear canal, tympanum = eardrum), middle ear with the three ossicles, and the inner ear with the cochlea as the actual hearing organ

air or water molecules, and the oscillation of these molecules spreads out in the medium. The movement of the molecules back and forth in the axis of the direction of the propagation of the sound wave creates local compressions and dilutions of the molecules. The distances between these compressions or dilutions along the axis of sound propagation equal the wavelength ($\lambda$) of the sound, the rate of oscillations per second (Hz) expresses the frequency ($f$) of a tone. The product of $f$ and $\lambda$ is the velocity $c$ of the sound wave ($c = f\ \lambda$ [m/s]) which amounts to about 330 m/s in air (at sea level) and to about 1,500 m/s in water. This means that tones of a given frequency have about five times the wavelength in water compared to propagating in air. Sounds in water travel much longer distances before they are attenuated below the hearing threshold. For example, whales can hear songs of other whales over many hundreds of kilometers depending on the noise conditions in the sea.

Taking young humans as reference, frequencies are classified as belonging to the audible range (20–20,000 Hz), to infrasound ($f < 20$ Hz), or ultrasound ($f > 20{,}000$ Hz). Because of the high frequencies of oscillations (short wavelengths) ultrasounds lose energy very rapidly when traveling through air. Thus, ultrasound can be used only for short-range orientation (e.g., echolocating bats) or communication (e.g., social calls of bats and rodents). Infrasound, however, can travel hundreds of kilometers with little attenuation so that animals can use infrasound for long-distance communication (e.g., elephants) or for locating sources of infrasound (breaking waves at beaches, wind blowing through mountain valleys, thunderstorms) for navigation (e.g., pigeons).

The pressure difference between the compression and dilution maxima of molecules in propagating sounds equals the sound pressure amplitude ($p_A$) expressed in Pascals (Pa = N/m$^2$). Microphones measure the sound pressure ($p$) which is the effective sound pressure amplitude ($p = 0.5\ p_A/\sqrt{2}$)). The working ranges of most of the hearing organs of animals comprise many orders of magnitude between the lowest just perceptible pressure ($p$ at the absolute hearing threshold) and the highest pressure beyond which the hearing organs are irreversibly damaged, in humans $10^{13}$ in sound intensity ($I$ [Watt/m$^2$]) or $3.2 \times 10^6$ in sound pressure ($I \sim p^2$). These extraordinarily large ranges and the logarithmic translation of sound pressure or intensity into perception (see Fechner's rule, Chap. 12) led to the introduction of the **sound pressure level** (**SPL**), a dimensionless quantity which is expressed in units of **Decibels** (**dB**). For measurements in air the SPL = 20 log $p/p_0$ [dB] with $p_0 = 20$ µPa (reference sound pressure at the average human hearing threshold at 1 kHz frequency). These definitions adjust the working range of the human ear from 0 to 130 dB at 1 kHz frequency. At 0 dB SPL, the movement amplitude of air molecules is in the order of $10^{-12}$ m which is about 100 times less than the diameter of a hydrogen atom. Animals like cats being even more sensitive than humans, may reach −18 dB at their absolute hearing threshold [9]. This incredible sensitivity to sound pressure from movement amplitudes of molecules is the maximum to be biologically meaningful because the thermal noise in air masks the perception of even fainter sounds.

Sound pressure is the sound parameter appropriate for stimulating animals having tympanic membranes (eardrums) or comparable structures such as air bladders. The velocity of the molecules or particle movements in sound waves is the appropriate stimulus for animals having antennae or hair sensilla on their body (insects) or hair cells with a long kinocilium, often covered by membranes or otoliths in organs of their inner ear (vertebrates). Literature about the physics of sounds can be found in many general textbooks about sound and hearing (e.g., [3, 30, 48]).

## 17.2 Mammals

### 17.2.1 Peripheral Auditory System

Peripheral auditory systems of mammals (Fig. 17.1) divide into three parts, (i) the **outer ear** and the **middle**

**ear** capturing sound to transfer it to the inner ear, (ii) the **cochlea** which houses the sound sensitive epithelia of the inner ear, and (iii) the **auditory nerve** transmitting the encoded sound information from the cochlea to the brain.

#### 17.2.1.1 Outer Ear, Middle Ear

Most mammals are land vertebrates, and so for them sound in air has to be transferred into the fluid spaces of the cochlea. Since the impedance of air against the propagation of sound waves is about 3,600 times smaller than the impedance of water, about 99 % of the sound energy would be reflected at the entrance to the cochlea without special structures of the outer and middle ear amplifying the sound pressure and, thus, matching the impedances of the media [3, 21]. The outer ear and the middle ear provide two steps of sound frequency-dependent pressure amplification. The outer ear includes an outstanding, often movable part, the pinna, a folded cartilaginous part, the concha, and the external ear canal, the meatus, ending at the eardrum (tympanic membrane). By resonances and filtering, this construction selectively amplifies certain sound frequencies, about 2–4 and 11–13 kHz in human adults (Fig. 17.2) [3, 37], thus improving the auditory sensitivity at these frequencies up to about 20 dB depending on the angle of incidence of the sound in the horizontal plane (azimuth angle, Fig. 17.2). The spectral shape of the amplification depends also on the vertical angle of incidence of the sound providing cues for the estimation of the elevation of the sound source relative to the ear. Human babies and young mammals learn to use the direction-dependent frequency transfer function of their outer ears in order to locate sound sources in azimuth and elevation and to discriminate between front and back locations.

The cavities of the middle ear and the mouth are connected with the Eustachian tube through which static pressures between the cavities can be equalized. A chain of three bones, the **middle ear ossicles** – **malleus (hammer), incus (anvil), and stapes (stirrup)** – connects the eardrum with the membrane overlying the oval window which is the entrance to the fluid spaces of the inner ear (Fig. 17.1). This middle ear construction serves as a sound pressure amplifier via three mechanisms (Fig. 17.3a; [3, 21, 37]):

1. The area of the tympanic membrane ($A_T$) is always larger than the area of the stapes footplate ($A_S$) providing amplification according to the area ratio ($A_T/A_S$) which in humans amounts to 0.55 cm$^2$/0.032 cm$^2 \approx 17$.

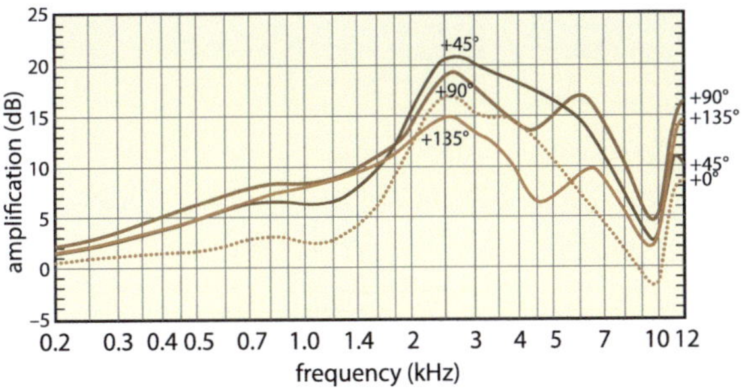


**Fig. 17.2** **Transfer functions of the human outer ear.** The relative increase of the sound pressure level at the eardrum relative to outside the ear is shown for tones arriving at the ear from several horizontal directions (angles of incidence). Angle- and frequency-dependent amplification up to about 20 dB is evident (Modified from Buser and Imbert [3] with permission)

2. The ossicles function as a lever system with the lever arm of the malleus ($l_M$) being always larger than the lever arm of the incus ($l_I$). In humans the level ratio ($l_M/l_I$) is about 1.3.
3. The malleus attaches asymmetrically at the tympanic membrane which provides another lever amplifying the sound pressure by a factor of 1.4 in humans.

The whole amplification of the pressure at the stapes footplate ($p_S$) from the pressure at the tympanic membrane ($p_T$) is equal to the product of all three factors which for humans amounts to $p_S = p_T \times 17 \times 1.3 \times 1.4 = 31\ p_T$. Other mammals have amplification factors between about 10 and 90 leading to pressure gains between about 20 and 39 dB. The largest pressure gain of 39 dB is known from the cat providing this mammal with the most sensitive ear of all animals tested so far.

In the frequency domain, **the middle ear response is that of a bandpass filter** with a broad, species-specific passing range and more or less steep slopes of attenuation at the low- and high-frequency sides (Fig. 17.3b) [37]. The cutoff frequency at the low-frequency side is determined mainly by the mechanical properties and sizes of the eardrum and the oval window. The larger the eardrum and the more elastic the stapes is coupled to the oval window, the more effectively can these membranes follow the long wavelengths of low-frequency sounds. Thus, elephants with their large eardrums can hear infrasound down to about 10 Hz. The cutoff frequency at the high-frequency side is determined mainly by the inertia of the ossicles and energy losses by friction in and bending of the ossicular chain. Small (light) and stiff ossicles such as those of bats and rodents favor transmission of ultrasounds.

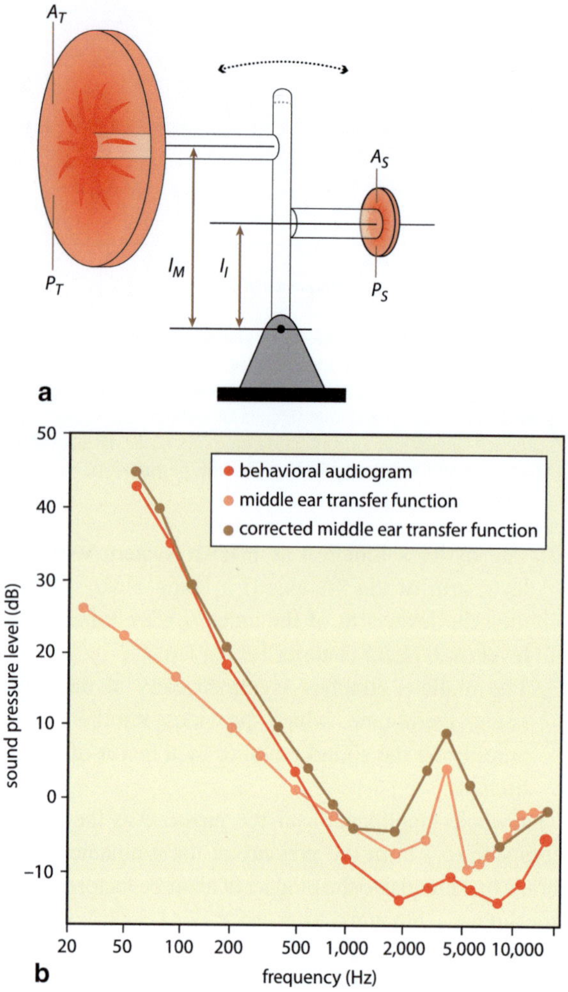


**Fig. 17.3** (**a**) **Functional principle of sound transmission through the mammalian middle ear.** The sound pressure at the stapes footplate ($p_S$) is amplified relative to the sound pressure at the tympanic membrane ($p_T$) by (1) the area ratio $A_T : A_S$, (2) the ratio of the lever arms of malleus and incus ($l_M : l_I$), and (3) by the flexure of the area of the tympanic membrane ($A_T$). (**b**) Comparison of the cat audiogram with its middle ear transfer function. This function (1/amplitude of stapes movement) is plotted on an arbitrary scale. The larger the amplitude is, the better is the sound transfer and the more sensitive is the ear. In the corrected transfer function, the highpass filter of the helicotrema at the cochlear apex is considered. The shape of the audiogram is very similar to the shape of the corrected middle ear transfer function (Compare Buser and Imbert [3] and Pickles [30])

Two muscles, the tensor tympani attached to the malleus, and the stapedius muscle influence the sound transmission through the mammalian middle ear. A reflexive contraction of the muscles (**middle ear reflex**) to loud sounds increases the stiffness of the middle ear and attenuates the passage of sounds, especially those of low frequencies (in humans below about 2 kHz). The effect is (*i*) a protection of the inner ear against damage by loud sounds except when the sound onset is instantaneous as in gun shots, (*ii*) a damping of self-produced sounds transmitted by bone conduction to the middle ear so that listening to others is less disturbed by simultaneously produced own sounds, and (*iii*) an attenuation of low-frequency sounds that can mask the perception of higher frequencies in the inner ear important for the discrimination of communication sounds [3, 37].

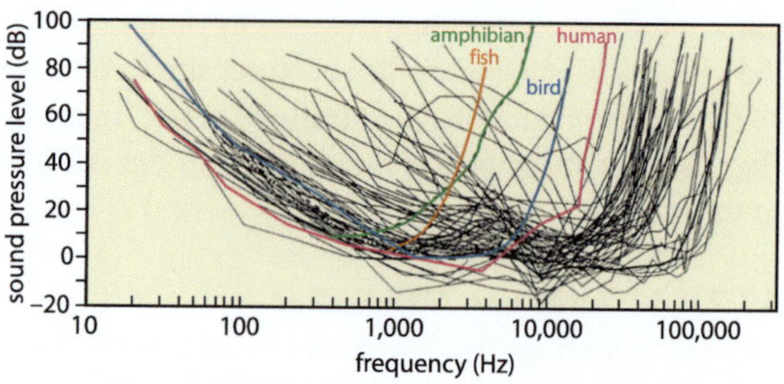


**Fig. 17.4** **Audiograms of 64 mammalian species obtained from behavioral tests.** The human audiogram is shown in *red*. The average audiogram for birds (*blue*) and the average high-frequency parts of the amphibian (*green*) and fish (*orange*) audiograms are also shown for comparison (Modified from Fay [9] with permission). On average, high-frequency hearing is improved and extended by about six octaves during vertebrate evolution from fish to mammals

In summary, the outer and middle ears of mammals largely determine the shape of their species-specific frequency response curves and the absolute sensitivity for hearing tones, both expressed in the **absolute auditory threshold curve, the audiogram**. In Fig. 17.4, audiograms for 60 terrestrial mammalian species are shown, all measured in behavioral hearing tests [9]. It is evident that humans are rather non-specialized low-frequency listeners.

#### 17.2.1.2 Cochlea

**Fluid spaces and membranes**. The cochlea (Fig. 17.1) equals an elongated and coiled basilar papilla of the inner ear of birds and reptiles [3, 22, 30]. It consists of three main fluid spaces and several membranes (Fig. 17.5).The scala vestibuli, starting at the oval window and running up to the helicotrema at the apex of the cochlea, and the scala tympani, running from the helicotrema to the round window, are filled with $Na^+$-rich perilymph, resembling extracellular fluids. The scala media is separated from the scala vestibuli by the Reissner membrane and from the scala tympani by the basilar membrane. It contains intracellular-like $K^+$-rich endolymph which is provided

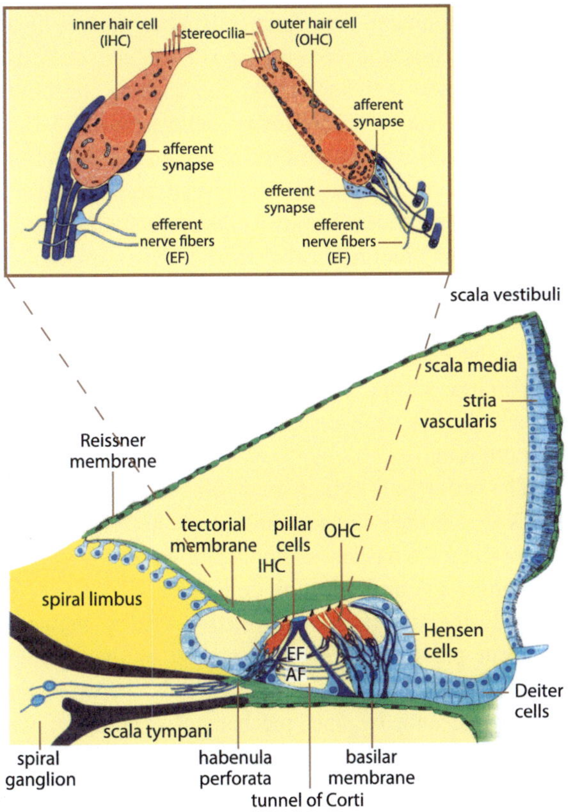


**Fig. 17.5 Radial section through the scala media of the mammalian cochlea.** Reissner's membrane separates the scala media from the scala vestibuli. The organ of Corti with the hair cells and supporting cells sits on the basilar membrane which separates the scala media from the scala tympani. The afferent and efferent innervation of an inner and an outer hair cell is enlarged. *AF* afferent nerve fibers, *EF* efferent nerve fibers (Modified from Smith [42] with permission)

by the stria vascularis at the lateral cochlear wall. **The actual hearing organ is the organ of Corti**, sitting on the basilar membrane in the scala media (Fig. 17.5). The organ of Corti is covered with the tectorial membrane. The tips of the longest stereocilia of the outer hair cells are inserted in the tectorial membrane, the tips of the stereocilia of the inner hair cells are only loosely attached.

**Basilar membrane displacement and tonotopy**. A movement of the stapes in response to, for example, the sine wave of a tone leads to a push-and-pull movement of the perilymph in the scala vestibuli inducing the same movement to the very flexible Reissner membrane and the endolymph of the scala media, i.e., a movement towards and away of the basilar membrane. The basilar membrane is stiff (thick and narrow) at the base and flexible (thin and wide) at the apex of the cochlea. Therefore, it will move at the base only in response to high-frequency (high-energy) moves of the endolymph. According to the stiffness gradient of the basilar membrane, fluid movements induced by tones of different frequencies will induce movements of the basilar membrane the farther at the apex the lower the frequencies are. Thus, sounds with a complex frequency spectrum such as vowels of human speech, induce waves of fluid movements that travel along the basilar membrane (**traveling waves**; [29, 34]) up to those locations of stiffness allowing a maximum displacement of the membrane to matching frequencies in the sound (Fig. 17.6a, b). At the displacement maximum, there is a maximum movement of the fluid in the scala tympani towards the flexible round window, which moves towards the middle ear cavity where the energy from the initial stapes movement is lost. This loss of energy leads to the rapid

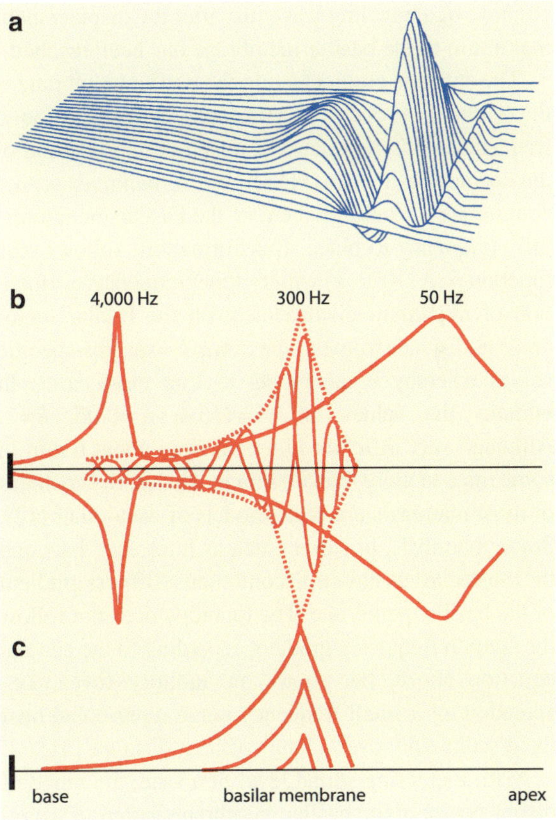


**Fig. 17.6** (**a**) **Instantaneous picture of the deflection of the mammalian basilar membrane** by a travelling wave initiated by a pure tone. (**b**) Travelling waves and their envelopes due to a high, medium, and low frequency tone. (**c**) Rectified envelope of a travelling wave in response to a medium tone frequency at three sound pressure levels. Loud tones lead to an overproportional extension of the basilar membrane deflection towards the cochlear base

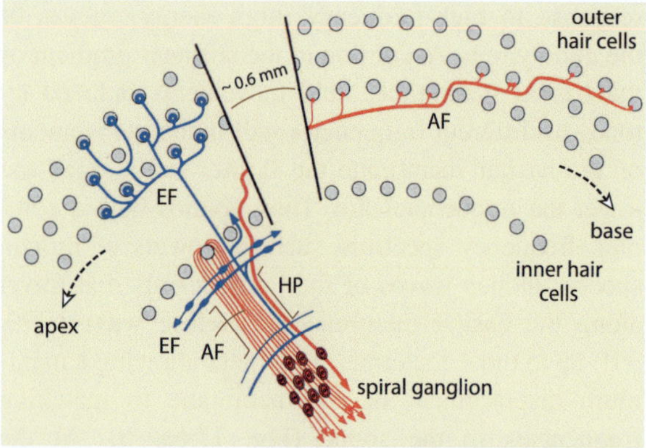


**Fig. 17.7 Scheme of the innervation pattern of inner and outer hair cells in the mammalian cochlea.** *AF* afferent fibers, *EF* efferent fibers, *HP* habenula perforata (the holes in the bone, where the fibers enter the organ of Corti; see Fig. 17.5) (Modified from Spoendlin [43] with permission)

collapse of a traveling wave just after the displacement maximum of the basilar membrane has been reached.

The stiffness gradient along the basilar membrane is the origin of the **cochlear tonotopy**, giving each sound frequency a different place of maximum stimulation of the cochlear hair cells (Fig. 17.6). In mammals with a continuous stiffness gradient of the basilar membrane, this frequency-to-place transformation follows the function $f = A\,(10^{ax} - k)$ with $f$ = tone frequency; $x$ = location of maximum displacement on the basilar membrane as distance from the apex; $A, a, k$ = species-specific values whereby $k$ can be set to 1 in most cases. In humans the values are $A = 165.4$, $a = 0.06$, $k = 1$. Although very different in length from about 6 mm in some mice to more than 60 mm in whales, the cochleas of most mammals are scale models of each other [12]. Some mammals, however, such as horseshoe bats and the mustache bat, have a discontinuous stiffness gradient of the basilar membrane. The tonotopy does not follow the smooth frequency gradient according to the general equation above, but shows an auditory fovea, i.e., spreading of a small frequency range represented near the discontinuity over a long cochlear distance [31].

With increasing sound level of a tone, the range of displacement of the basilar membrane increases asymmetrically around its maximum with a much larger involvement of basal parts when sounds are loud (Fig. 17.6c; [34]). The physiological effect is that loud sounds of **any** frequency stimulate the hair cells in the high-frequency range of the tonotopy so that these hair cells suffer more from a continuous stimulation in a loud environment than hair cells near the cochlear apex. The consequence is damage of hair cells near the cochlear base with increasing age causing the high-frequency hearing loss of many senior people.

**Hair cells – mechanoelectrical transduction** [1, 29]. There are two populations of secondary sensory cells (without own axon) in the mammalian organ of Corti, the inner hair cells (IHCs) in one row (about 4,000 in humans), and the outer hair cells (OHCs) in three to five rows (together about 12,000 in humans) along the cochlea (Figs. 17.5 and 17.7). The apical membrane of the hair cells has 50–100 elongated microvilli, the actin-containing stereocilia are about 2–5 μm long.

The **outer hair cells** regulate the sensitivity of the IHCs to the shearing motions of the tectorial membrane when the basilar membrane moves up and down. Upward movements cause deflection of stereocilia towards the tallest ones, a relative sliding between adjacent pairs of stereocilia, tension of the tiplinks (catherin/protocadherin protein threads) between the stereocilia, and, by that, mechanical opening of ion channels for cations (compare Fig. 16.3, Chap. 16). A gradient of about 150 mV of electrical potential between scala media (+80 mV) and the interior (−70 mV) of the OHCs leads to an influx mainly of $K^+$ from the $K^+$-rich scala media into the cells and, thus, to pulses of depolarization of the OHCs in the rhythm of the stereocilia movement (**mechanoelectrical transduction**). The depolarization pulses shorten the OHCs, repolarization and hyperpolarization lengthen them. This electromotility is based on the protein 'prestin' densely packed along the lateral walls of the OHCs (see Chap. 16). Since the OHCs are coupled with the tectorial membrane via their stereocilia, and are part of the organ of Corti which is attached to the basilar membrane, the electromotility of the OHCs amplifies the displacement amplitude of the basilar membrane and changes the width of the fluid-filled space between the upper hair cell surface and the tectorial membrane: shortening of the OHCs pulls the tectorial membrane closer to the hair cell surface. This affects also the IHCs because their stereocilia get in closer contact with the tectorial membrane and/or are more susceptible to the fluid motions between hair cell surface and tectorial membrane. The net effect is an increase in IHC sensitivity to basilar membrane motion in the order to 40–60 dB. This boosting of IHC sensitivity by OHC action becomes most evident when mammals with increasing age suffer from a considerable hearing loss in the high-frequency range due to progressive loss of OHCs from the basal end of the cochlea.

The OHC motility is also origin of the amazing phenomenon of '**otoacoustic emissions**', which are tones produced in the cochlea. Spontaneous local OHC movements induce traveling waves of the basilar membrane feeding back to the middle and outer ear. Thus, otoacoustic emissions occur as displacements of the cochlear round and oval windows or of the tympanum or even as tones in the ear canal. Measurements of the presence and amplitude of otoacoustic emissions are, therefore, an important noninvasive diagnostic tool for assessing functional OHCs in the cochlea [29]. Otoacoustic emissions and amplification also occur in non-mammalian ears that lack OHCs. In these animals, amplification arises from active hair bundle motions that are powered by the transduction apparatus [22, 34].

**Inner hair cells** have the same mechanoelectrical transduction through the deflection of the stereocilia as the OHCs. However, IHCs are not motile so that the depolarization by $K^+$ influx has only chemical and electrical consequences in the cells.

**Hair cells – electrical potentials, synapses, and innervation** [3, 22, 30, 43]. The mechanoelectrical transduction mainly of the OHCs leads to an alternating potential called **cochlear microphonic potential** which is a receptor potential that closely reproduces the waveform of the sound input. Synchronous cochlear microphonics from many hair cells can be recorded with electrodes outside the cochlea even from the scalp. Another type of receptor potential of the hair cells is the **summating potential** reflecting a slow depolarization of the resting potential due to cation influx that is not immediately compensated by a respective cation efflux. The receptor potentials cause a $Ca^{2+}$ influx leading to the release of the neurotransmitter glutamate from synaptic vesicles (see Chap. 16). The presynaptically released glutamate binds mainly to postsynaptic AMPA receptors to open $Na^+$ channels for ultimately generating action potentials in the bipolar cells of the cochlear spiral ganglion. The axons of the spiral ganglion cells join together as cochlear nerve fibers in the auditory nerve and run to the first auditory center of the brain, the cochlear nucleus.

The IHCs are innervated by at least 90 % of the afferent cochlear nerve fibers (about 60,000 in humans). Mostly, one afferent fiber contacts only one IHC, and one IHC is innervated by 10–30 fibers (divergent innervation pattern; Fig. 17.7). Thus, almost all auditory information is sent to the brain via the IHCs. They provide the necessary and sufficient cochlear output via the cochlear nerve fibers for many perceptual abilities including sound localization and pattern analysis and discrimination. The function of the few afferent fibers from the OHCs is not known.

OHCs, especially those responding best to low- and mid-frequencies of a species' hearing range, may receive strong innervation by efferent auditory nerve fibers sending information from auditory brain centers back to the cochlea (Figs. 17.5, 17.7, and 17.11). Efferent activity hyperpolarizes OHCs causing reduction of the depolarization by $K^+$ influx through the deflected stereocilia. Thus, the electromotility and the amplification effect of the OHCs on the mechanical stimulation of the IHCs are reduced. Ultimately, efferent activity reduces the cochlear sensitivity to tones by about 10–20 dB. Since efferent activity can be focused on small clusters of OHCs representing a certain frequency of the cochlear tonotopy, the perception of frequency components in a sound important for communication can be enhanced by efferent suppression of processing unimportant but masking frequency components. By this mechanism of **peripheral contrast enhancement**, the listening brain can modulate its own perception according to the knowledge of what sound spectra are expected.

IHCs do not receive direct efferent innervation. Efferent fibers contact afferent fibers at or near their synapse with the IHCs (Figs. 17.5 and 17.7) suggesting a modulation of the afferent activity of auditory nerve fibers. This modulation may rather relate to long-term adjustments in sensitivity and temporal precision of coding sounds than to short-term control as is the case with the efferent fibers to the OHCs.

#### 17.2.1.3 Coding of Sounds in the Auditory Nerve

The auditory nerve of vertebrates is that part of the VIIIth brain nerve which innervates the hearing organs in the inner ear, here the mammalian cochlea. The bipolar neurons of the cochlear spiral ganglion innervate the hair cells with the distal part of their axons and send the information with the proximal part to the cochlear nucleus (Figs. 17.5, 17.7, and 17.10).

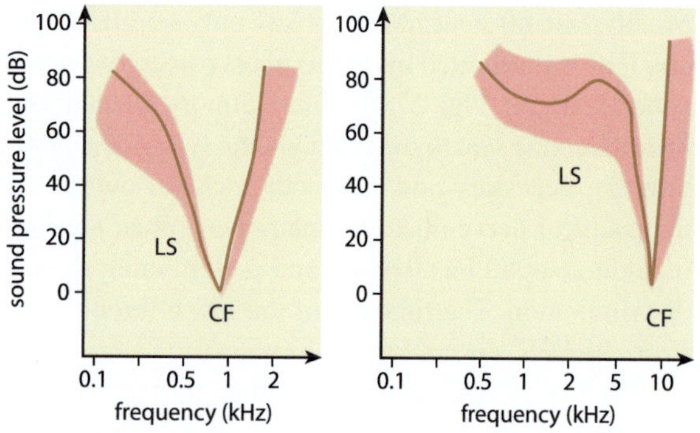


**Fig. 17.8** **Examples of frequency tuning curves of auditory nerve fibers** having a medium (*left*) or a high (*right*) characteristic frequency (*CF*). Areas of lateral suppression (*LS*) are indicated in *red*

**Coding of sound frequency** [3, 30]. There are two codes for sound frequencies in the auditory nerve fibers: the **place code based on the cochlear tonotopy** is functional for all frequencies that can stimulate the cochlea, the **phase code based on time-locking of action potentials** to the sound waveform is functional only up to about 4 kHz.

The afferent fibers from the hair cells code the local displacement pattern of the basilar membrane to traveling waves by the position of their innervation (Figs. 17.6 and 17.7). The frequency tuning of auditory nerve fibers expressed by the excitatory tuning curve (Fig. 17.8) is derived from the envelope function of the traveling wave (Fig. 17.6c). The excitatory tuning curve borders the frequency range (receptive field) within which the fiber discharges in response to tones above the spontaneous activity. Thus, the excitatory tuning curve describes the frequency-dependence of the neuronal response threshold. The lowest threshold is at the fibers characteristic frequency (CF) which equals the fibers best frequency (BF) where it responds with the highest rate at any sound level. The slopes of the excitatory tuning curve are flanked by areas of **lateral suppression** (Fig. 17.8). Tones in these areas partly or totally suppress the response of the fiber to tones in the receptive field. Different from lateral inhibition caused by neural interaction in the retina (see Chap. 18), lateral suppression in auditory nerve fibers is caused by mechanical nonlinearities in the cochlea which also give rise to the formation of distortion tones (e.g., $2f_1 - f_2$ and $f_2 - f_1$; in both cases $f_2 > f_1$) when two tones simultaneously stimulate the ear [29, 34].

The locking of action potentials in auditory nerve fibers to the phase of a low-frequency tone or the phase of amplitude modulation of a high-frequency tone or a frequency complex is a neural code for tone frequency or rate of amplitude modulation. The spike pattern is decoded in higher auditory brain centers and leads to the perception of **pitch** and to the ability to perceive and discriminate **musical intervals**. Because action

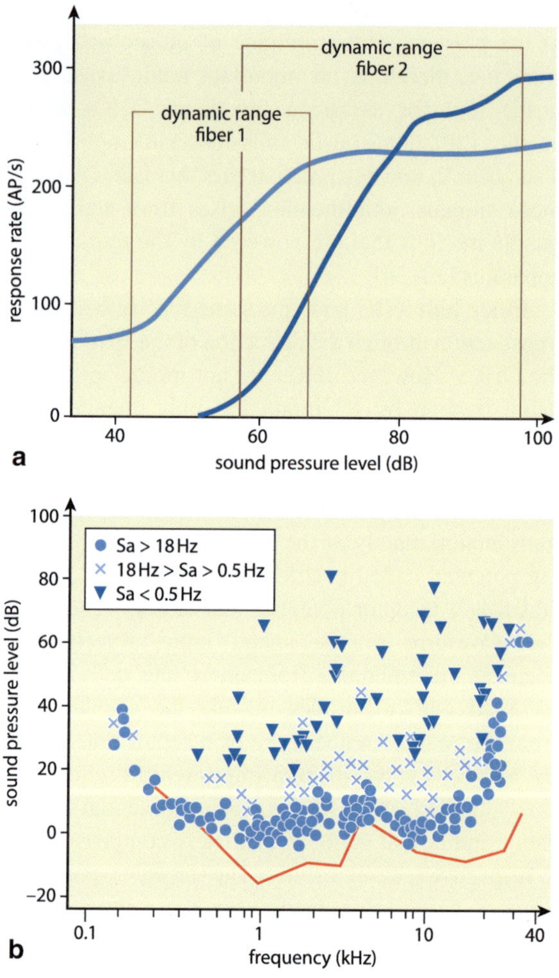


**Fig. 17.9** (**a**) **Two examples of rate-level functions of auditory nerve fibers.** Fiber 1 has a high spontaneous activity (about 80 spikes/s), a low tone-response threshold (near 40 dB), and a small dynamic range (about 25 dB) of response rate (action potentials/s, AP/s) increase with increase of the sound pressure level. Fiber 2 has very low spontaneous activity, a higher threshold (almost 60 dB), and a larger dynamic range (about 40 dB). (**b**) Behavioral audiogram of the cat and absolute response thresholds (the thresholds at the CF) of cat auditory nerve fibers with high, medium, and low spontaneous activity (Sa) (From Liberman and Pickles [19, 30] with permission)

potentials in auditory nerve fibers of mammals occur with a statistical variation of 165 μs, the phase code for frequency is limited to 1/165 μs = 6 kHz in theory. The 0.5–1 ms duration of action potentials and their refractory period allow phase coding by a single fiber only up to about 800 Hz. Several fibers can cooperate, however, to code pitch and musical intervals in the time domain up to about 4 kHz by locking their spikes to different cycles of the sound stimulus so that every cycle is labeled by at least one spike (volley principle of coding).

**Coding in the time domain.** Auditory nerve fibers respond to sound onsets with a phasic discharge and continue to respond at a lower tonic level as long as the sound lasts (phasic-tonic response; Fig. 17.12). Thus, the onsets and durations of sounds in a series (e.g., syllables in human speech) are expressed in the temporal response patterns of the fibers. This coding of sound series or rhythms by onset responses is possible up to intersound intervals of 20–30 ms. If the intervals are shorter, i.e., the repetition rate of the sounds is higher than about 30–50 Hz, the sound series is coded as an amplitude-modulated sound with a given modulation frequency (see above) and is perceived as a continuous sound with a certain pitch.

**Coding of sound intensity** [3, 19, 30]. Auditory nerve fibers differ in their response properties depending on the place where they innervate a given inner hair cell. Fibers innervating at the side of the tunnel of Corti (Fig. 17.5) have high spontaneous activity, low response thresholds and small dynamic ranges (about 10–20 dB) of spike rate increase with increasing tone level (Fig. 17.9a, fiber 1). Fibers innervating at the side of the spiral limbus (Fig. 17.5) have low spontaneous activity, higher response thresholds, and larger dynamic ranges (about 30–70 dB; Fig. 17.9a, fiber 2). Thus, fibers of the same CF, coding for the same sound frequency, can cover together a threshold range of about 60 dB (Fig. 17.9b) and a dynamic range of

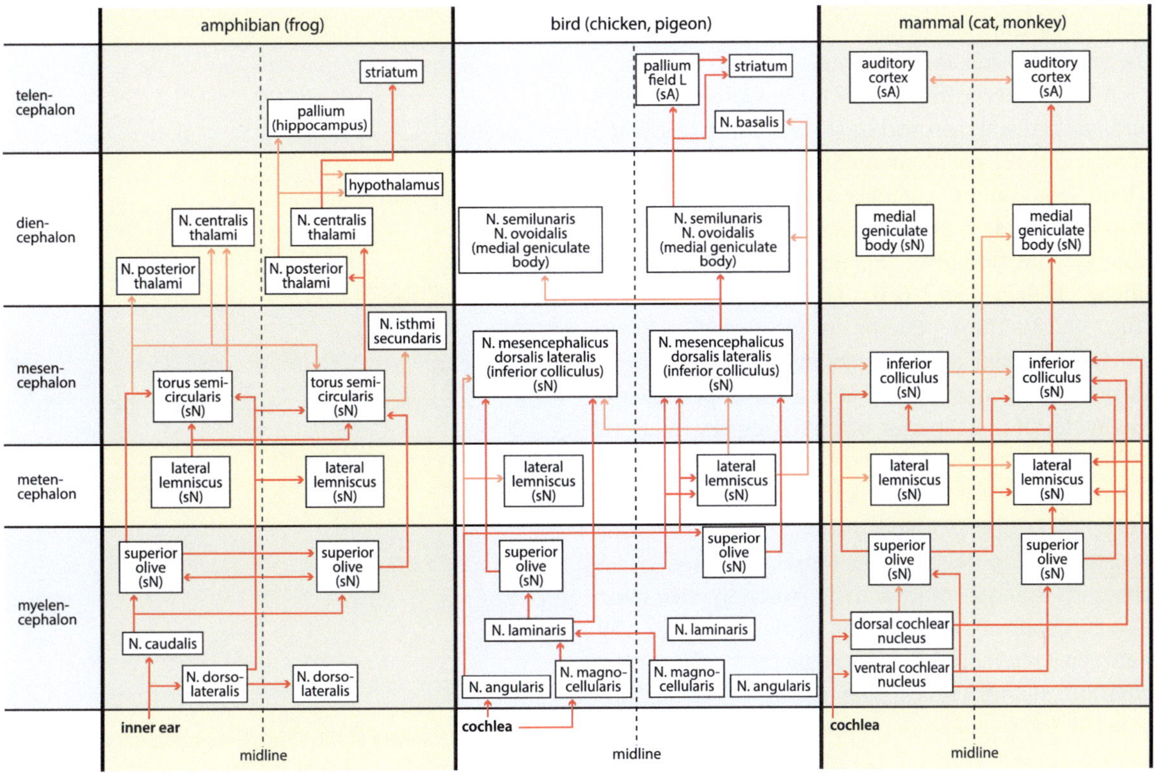


**Fig. 17.10 Diagrams of the ascending (afferent) auditory pathways of an amphibian, bird, and mammal.** The diagrams show the most important connections in only one half of the brain with origin in the left inner ear (cochlea). The main pathways cross to the other side, so that the projections starting in the left inner ear mainly end in the right-side midbrain and further in the right telencephalon. *Heavy lines* strong projections, *thin lines* weaker projections, *sA* several areas, *sN* several nuclei, *N* nucleus

about 70 dB. By adding the ranges of response threshold and rate increase, the fibers from a given cochlear location can encode a 130 dB range of sound intensity just via their total number of spikes. Therefore, it is assumed that the intensity of a tone perceived over the dynamic range of hearing, which is about 130 dB in the most sensitive frequency range of humans, is encoded at the level of the auditory nerve by the population of active nerve fibers, originating at the cochlear place of tone frequency representation, and their cumulated spike rate.

## 17.2.2 Central Auditory System

### 17.2.2.1 General Anatomy – Auditory Pathways

The central auditory system consists of **ascending** (**afferent**) and **descending** (**efferent**) projections with several centers of processing at each level of the brain (Figs. 17.10 and 17.11) [20, 38]. This parallel and hierarchical processing is unique among vertebrate sensory systems and certainly different from those of the visual and somatic sensory systems.

Auditory nerve fibers enter the medulla oblongata and, in mammals, project to several cell types in the complex of the **cochlear nucleus** (Fig. 17.12; [35]). There, the rather homogenous phasic-tonic tone responses, the V-shaped frequency tuning curves, and sigmoid rate-level functions of auditory nerve fibers change to cell type-specific patterns (Fig. 17.12). These patterns are forwarded to different target nuclei in the ascending auditory system. Primary patterns reach the next level of processing, the nuclei of the superior olivary complex (medulla oblongata) where the inputs from both ears interact to extract the information for sound localization (see Sect. 17.2.2.5). Derived patterns occur as tone responses of purely phasic (onset) activity, or as choppers (phasic-tonic activity whereby the tonic part is chopped in bursts) or pausers (a short gap between a phasic and the following tonic activity). In addition, tuning curves with inhibitory side bands (caused by synaptic inhibition), and peaked rate-level functions (Fig. 17.12) are observed. Neurons with such derived tone response characteristics mainly provide excitatory input to the nuclei of the lateral lemniscus (pons of the metencephalon) and the inferior colliculus of the midbrain. Ascending output from the nuclei of the superior olivary complex (excitatory and inhibitory) and the nuclei of the lateral lemniscus (mainly inhibitory) also reaches the inferior colliculus. That is, ascending information from lower brainstem nuclei, mainly from the contralateral side (**auditory chiasm**; Fig. 17.10) converges in the main (central) nucleus of the inferior colliculus of each brain hemisphere. From there, both excitatory (glutamatergic) and inhibitory (GABAergic) neurons project to the nuclei of the medial geniculate body (thalamus of the diencephalon) from where auditory information is distributed to many brain areas. In mammals, primary targets are the auditory areas of the cerebral cortex (Figs. 17.10, 17.13, and 17.16a). Other targets are the nuclei of the striatum, the hippocampus (part of the pallium), and nuclei of the basal telencephalon such as the amygdala and the preoptic area. These

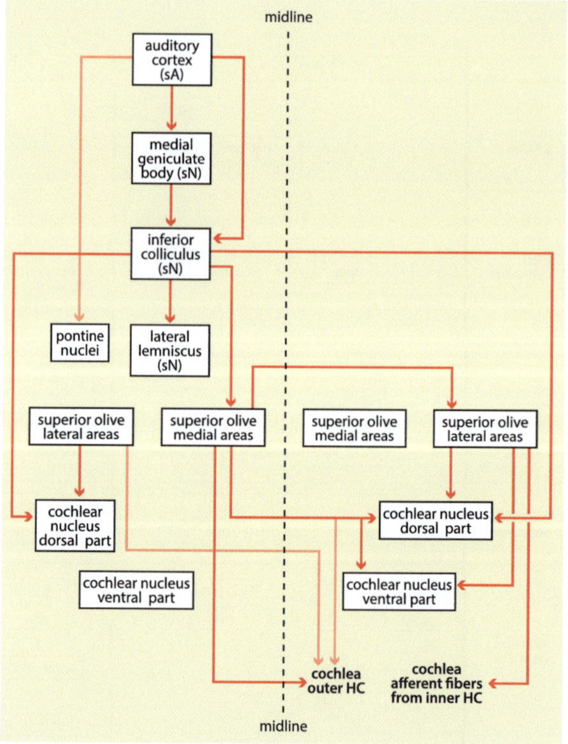


**Fig. 17.11** **Diagram of the main descending (efferent) auditory pathways of a mammal** starting at the left-side auditory cortex and ending finally in the right-side cochlea at the outer hair cells (HC) or at the afferent fibers from the inner HC. As for the ascending pathways (Fig. 17.10), the main efferent pathways cross to the other side below the midbrain level. *sA* several areas, *sN* several nuclei, *N* nucleus

projections provide direct auditory input to motor coordination, especially for vocalizations (striatum), to associative learning (pallium), and to the coordination of instincts (basal telencephalon, limbic system).

Figure 17.10 shows clearly that the ascending auditory pathways in the brainstem form a system of **parallel/hierarchical processing**, indicating a division of labor between the auditory centers in the analysis of sound parameters which is followed by an **integration of information in the inferior colliculus**. The integrated information is then passed on to the medial geniculate body and the auditory cortex, which have strong reciprocal connections and serve as a unit to allow auditory learning and complex behavioral control through sounds.

Descending auditory pathways have been studied mainly in mammals. They start at the auditory cortex and end contralaterally (chiasm at the level of the brainstem) at the cochlear outer hair cells or at the afferent fibers from the inner hair cells (Fig. 17.11). They connect in a parallel and hierarchical way the same auditory centers as the ascending pathways. Thus, there are many functional loops operating by positive or negative feedback – depending on excitatory and/or inhibitory interactions – between ascending and descending pathways. The knowledge about the importance of certain environmental sounds or the expectation of communication sounds or vocalizations of a certain speaker in a given listening situation leads to activities of the auditory cortex that decrease the response thresholds, increase the frequency selectivity, and enhance the processing of the pitch for the expected sounds in lower auditory centers [46]. Altogether, **the efferent system improves the contrast** between important and unimportant (masking) background sounds.

#### 17.2.2.2 Coding of Sound Frequency

Common to most subcortical nuclei of the ascending pathways and to primary (core) fields of the auditory cortex is a reproduction of the cochlear tonotopy (Figs. 17.13 and 17.16a). Thus, a **place-code** for frequency is realized at all levels of the auditory system

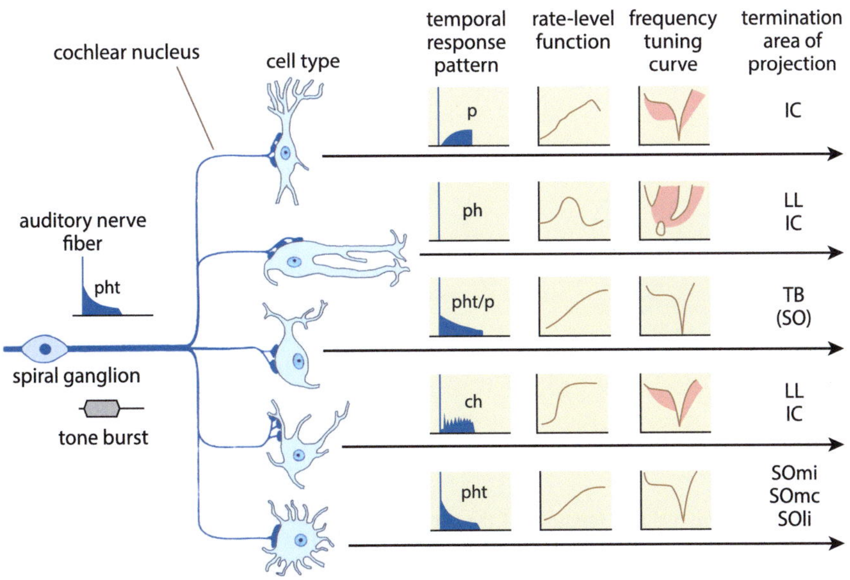


**Fig. 17.12 Diagram of the divergent innervation of five different cell types in the cochlear nucleus by an auditory nerve fiber.** The cell types are named (from above) fusiform, octopus, globular, multipolar, and spherical (bushy). The phasic-tonic (*pht*) temporal response of the auditory nerve fiber to a tone burst may change to different cell type-specific patterns such as pauser (*p*), phasic or "on" (*ph*), and chopper (*ch*) or remain unchanged (in the spherical cells). The rate-level functions of the cell types may have a sigmoid shape or show a *peak* (best level) at a certain level. The frequency tuning curves also differ in shape and may include inhibitory areas, i.e., sounds in these areas inhibit the response to sounds in the excitatory area bordered by the tuning curve. The axons of the cell types in the cochlear nucleus terminate in different areas of the auditory pathway: *IC* inferior colliculus, *LL* lateral lemniscus, *TB* trapezoid body of the superior olivary complex (*SO*), *SOmi*, *SOmc*, *SOli* medial superior olive ipsilaterally, or contralaterally, or lateral superior olive ipsilaterally (respectively) (Compare Romand and Avan [35] and Rouiller [38])

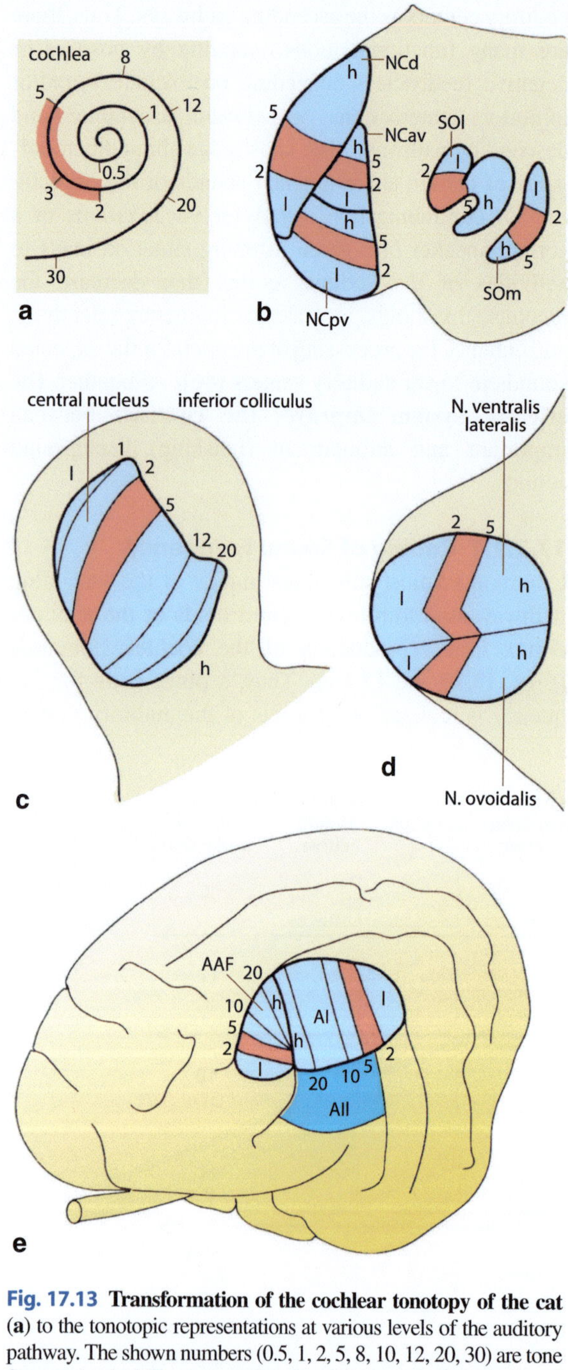


**Fig. 17.13** **Transformation of the cochlear tonotopy of the cat** (**a**) to the tonotopic representations at various levels of the auditory pathway. The shown numbers (0.5, 1, 2, 5, 8, 10, 12, 20, 30) are tone frequencies in kHz representing the characteristic frequencies of neurons at the indicated places of the tonotopic map. (**b**) *NCd* nucleus cochlearis dorsalis, *NCav*, *NCpv* nucleus cochlearis anteroventralis and posteroventralis, *SOl* lateral superior olive, *SOm* medial superior olive. (**c**) Central nucleus of the inferior colliculus. (**d**) Nucleus ventralis lateralis and ovoidalis of the medial geniculate body. (**e**) The primary (*AI*), secondary (*AII*), and anterior (*AAF*) fields of the auditory cortex. (**b**) and (**d**) show a transverse, (**c**) a sagittal section, (**e**) is a lateral view on the neocortex. *N* nucleus, *h* high frequencies, *l* low frequencies (Compare Romand and Avan [35] and Rouiller [38])

by tonotopy [38] and sharp frequency tuning of the neurons, evident by narrow frequency tuning curves (Figs. 17.8 and 17.12). Depending on the ecology and the importance of frequency ranges for social communication, certain frequency ranges of the cochlear tonotopy can be enlarged in their representation in brain areas. This is the case, for example, for frequencies in the ultrasonic auditory fovea of some echolocating bats (e.g., horseshoe bats and the mustached bat [31]).

It is important to note that the one-dimensional representation of frequency in the cochlea is transformed to a two-dimensional one (parallel isofrequency stripes) in the auditory cortical fields and to a three-dimensional one (a pile of isofrequency sheets or frequency-band laminas) in the auditory nuclei (Figs. 17.13 and 17.14a). This inflation of a frequency point in the cochlea to a strip or even a sheet provides neural tissue for the **spatial representation in maps** of neuronal response parameters (e.g., tone response threshold, sharpness of frequency tuning, latency, best modulation frequency, best azimuth angle) related to coding of further sound parameters (e.g., intensity, frequency bandwidth, pitch, direction) besides frequency, as shown for the inferior colliculus in Fig. 17.14 [7].

In the central nucleus of the inferior colliculus, neurons with their CFs form **two tonotopic gradients** [40]. The first is a steep one, running from low frequencies dorsolaterally to high frequencies ventromedially. This is shown by the central frequencies of the sheets of the frequency-band laminas in Fig. 17.14a. The second is a shallow one running on a frequency-band lamina from low frequencies medially to high frequencies laterally as indicated for the 20 kHz sheet (Fig. 17.14a) by a gradient from black (medial) to white (lateral). The two tonotopic gradients add together to represent the whole frequency range in a smooth increase of neuronal CFs over all frequency-band laminas.

Coding of frequency via coupling of action potentials to the phase of a tone or of an amplitude-modulated sound producing a pitch percept is observed up to the level of the inferior colliculus. There, this **phase-code is reorganized in a place-code for pitch** with average neurons responding best to low pitches located caudomedially and those responding best to high pitches located rostrolaterally [32]. This is shown by the color-gradient from blue-green to yellow-green on the 10-kHz frequency-band lamina (Fig. 17.14a).

A place-code for pitch may continue to the primary auditory cortex.

#### 17.2.2.3 Coding in the Time Domain

Coding in the time domain at higher centers of the auditory pathways (inferior colliculus and beyond) refers to sound series or rhythms with intersound intervals longer than about 20 ms, i.e., repetition rates of less than about 50 Hz [48]. Sound series with repetition rates higher than 50 Hz are perceived as continuous sounds with a certain pitch (see Sect. 17.2.1.3). Repetition rates between about 0.5 and 10 Hz (intervals between about 100 and 2,000 ms duration between sound elements) are most important to be perceived precisely, because communication sounds of animals, including rhythm of speech syllables of humans, are in this range [48].

The repetition rate of syllables or the rate of amplitude modulation in a sound sequence depends both on the duration of the intervals between the sound elements and on the durations of the sound elements. Hence, neural coding in the time domain comprises duration coding of presence and absence (or low-intensity) times of sounds in a series and coding the regularity of change. Coding in the time domain requires neurons that either respond with short latencies and high temporal precision to the onsets and offsets of sounds, i.e., to rapid increases and decreases of amplitudes or to the onsets or offsets of sounds and, in addition, have a best response to a certain sound duration and/or intersound interval. All such kinds of neurons have been found in the superior olivary complex, the inferior colliculus, and auditory cortex. This ensures coding of slow amplitude modulations and rhythms for perception.

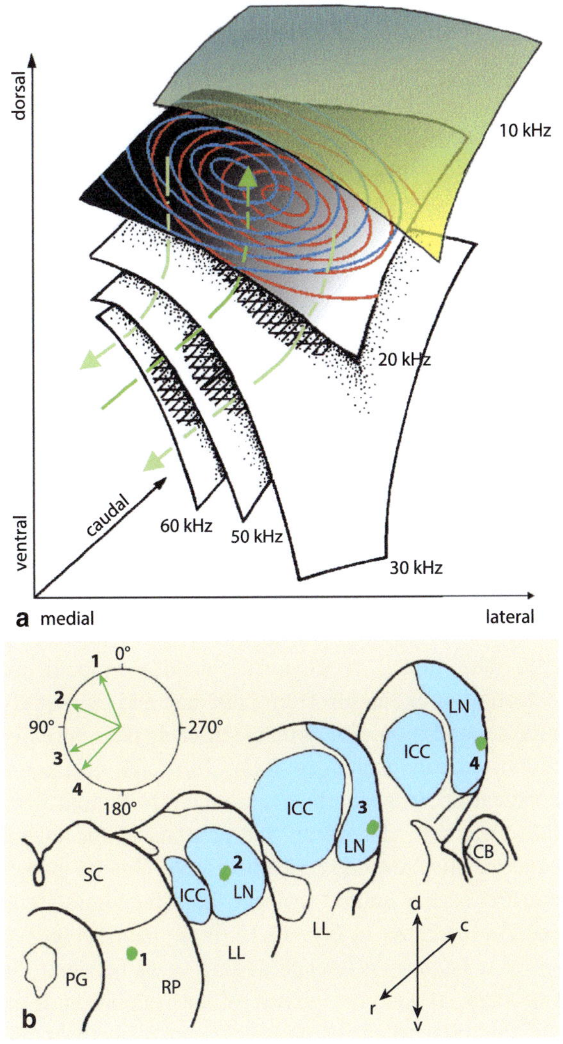


**Fig. 17.14** (**a**) **Three-dimensional sketch of functional maps** (orderly representations of neuronal response characteristics) in the central nucleus of the inferior colliculus (*ICC*). Several frequency-band laminas are shown representing neurons with characteristic frequencies (*CFs*) around 10, 20, 30, 50, and 60 kHz, respectively, in the ICC of the house mouse as an example. Within a frequency-band lamina, the CFs increase from medial to lateral as shown by the gradient from *black* to *white* on the 20-kHz lamina. On average, the neurons with the lowest tone-response thresholds are located in the center of a lamina (*hatched areas*) and neurons with increasingly higher thresholds in *circles* around the center (*red circles* on 20-kHz lamina). Also, neurons with sharp frequency tuning (narrow frequency tuning curves) tend to be located in the center of a lamina while neurons with broader tuning are located more peripherally (*blue circles* on 20-kHz lamina). Neurons in the center of a frequency-band lamina prefer rapid downward frequency changes (sweeps) in sounds while neurons located more medially or laterally prefer upward frequency sweeps (*green arrows*). The gradient from *blue-green* to *yellow-green* on the 10-kHz lamina indicates neuronal preferences for low-pitched (caudomedially) to high-pitched (rostrolaterally) tones, whereby the pitch is due to rapid amplitude modulations. (**b**) The lateral nucleus (*LN*) of the IC contains a map of the azimuth angle of a sound source in the hemifield contralateral to the IC. Neurons in the rostral LN prefer sound source locations right in front of the listener (0°), neurons in the caudal LN prefer locations behind the listener. *CB* cerebellum, *ICC* central nucleus of the IC, *LL* lateral lemniscus, *PG* periaqueductal gray, *RP* rostral pole of IC, *SC* superior colliculus, *c* caudal, *d* dorsal, *r* rostral, *v* ventral (Ehret [7] with permission; compare also Rees and Langer [32] and Schreiner and Langer [40])

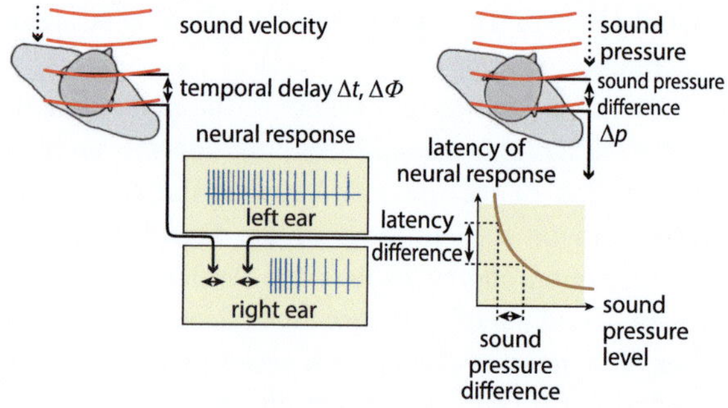


**Fig. 17.15 A listener is hit by sound waves from a source to the left of the body midline.** This situation creates differences between the two ears which are time delays ($\Delta t$), phase differences ($\Delta \Phi$) (*left part*), and, if the wavelengths in the sound are smaller than the head, also pressure differences ($\Delta p$) (*right part*). The function of the relationship between sound pressure level and latency of a neural response shows that the $\Delta p$ (sound at the right ear is less loud than at the left ear) leads to a certain latency difference that adds to the $\Delta t$ of sound arrival time at the right ear which is hit by the sound wave later than the left ear. Thus, both $\Delta t$ ($\Delta \Phi$) and $\Delta p$ sum up to code a binaural latency difference from the activation of the auditory nerve fibers of both sides

#### 17.2.2.4 Coding of Sound Intensity

The code for sound intensity in the central auditory system is not yet fully understood. Neurons of similar CFs in a given center of the auditory pathways differ in their response thresholds to tones and noises by about 30–50 dB with the most sensitive neurons located in the center of a frequency-band lamina in the inferior colliculus [7] or in the center of an isofrequency strip in the primary auditory cortex (Figs. 17.13 and 17.14a). Thus, the SPL of sounds from the threshold of perception to that of a low voice can be represented in threshold maps in the auditory system. The representation or coding of louder sounds is unclear. Presently, mainly two hypotheses are discussed: (*i*) neurons with rate-level functions covering a large dynamic range as described for part of the auditory nerve fibers add activity on top of the threshold range. Such a code may be present in the inferior colliculus and auditory cortex. (*ii*) Neurons with nonlinear (peaked) rate-level functions (see Fig. 17.12, first two neurons from the top) code the sound level by the location of the peak on the intensity scale mapped on a spatial dimension in a given auditory brain center. Such a code has been found as a circular intensity map (map of sound pressure level) in an enlarged area of the mustache bat's primary auditory cortex processing the main echolocation frequency near 60 kHz [45].

Since many neurons in higher auditory centers have rate-level functions with peaks at various SPLs, the total average spike rate in such a center does not vary much for sounds from a low voice (40 dB SPL) to sounds of a pneumatic hammer (110 dB SPL). Amazingly, loud sounds may not "flood" higher auditory centers such as the inferior colliculus with activity. Inhibitory interactions between neurons are responsible for peaked rate-level functions that keep the average activity levels constantly rather low.

#### 17.2.2.5 Coding of Sound Direction

**Coding of angles in the horizontal plane (azimuth).** Sound waves from sources displaced horizontally from the head midline lead to three types of disparities when arriving at the two ears (Fig. 17.15), namely in $\Delta t$ (arrival time), $\Delta\Phi$ (phase of frequency components and/or modulation frequencies in ongoing sounds), and $\Delta p$ (sound pressure). Differences in sound pressure occur only if the wavelengths of the sounds are smaller than the diameter of the head (in humans $\lambda < \sim 20$ cm, i.e., $f > 1{,}650$ Hz) so that the head becomes an obstacle to sound propagation and creates a sound shadow. For longer wavelengths (lower frequencies), the sound is diffracted around the head without detectable intensity differences. High-frequency hearing in mammals with small heads such as bats and rodents may have evolved in order to make $\Delta p$ available for sound localization when $\Delta t$ is very small.

The coding of binaural information for sound localization starts at the level of the superior olivary complex (Fig. 17.10) and involves complex excitatory and inhibitory interactions of the outputs of the cochlear nuclei of both sides in a number of superior olivary nuclei [13]. Roughly, the **medial superior olive** receives monosynaptic excitatory input and disynaptic inhibitory input from both sides and calculates $\Delta t$ and $\Delta\Phi$. The neural mechanisms include coincidence detection of excitation from both ears and temporally precise inhibition. The **lateral superior olive** receives monosynaptic excitatory input from the ipsilateral cochlear nucleus and disynaptic inhibitory input from the other side. Therefore, the neurons of the lateral superior olive respond best when sounds hit the ipsilateral ear earlier and/or stronger than the contralateral ear. As indicated in Fig. 17.15, these neurons recode $\Delta p$ as a latency difference adding to $\Delta t$ between the two ears so that the arrival of the inhibition from the contralateral ear is the later the larger both $\Delta t$ and

$\Delta p$ are. In humans, the smallest detectable differences in $\Delta p$ are about 1 dB, in $\Delta t$ about 10 µs, corresponding to an angle of deviation from the midline of about 2°.

The inferior colliculus of one side receives the information about binaural disparities reflecting the location of a sound source in the horizontal plane mainly of the contralateral side via the projections from the medial and lateral superior olivary nuclei (Fig. 17.10). In the lateral nucleus of the inferior colliculus, contralateral azimuth angles of 0–180° are mapped along the rostrocaudal axis [7] (Fig. 17.14b). That is, neurons in the lateral inferior colliculus have spatially tuned receptive fields with the best response shifting from rostrally to caudally located neurons when the sound source moves horizontally from the frontal midline position through the contralateral space to the caudal midline (Fig. 17.14b). This shows that the complex processing of binaural disparities in the lower brainstem ends in two **maps of horizontal space** (one for each side) in the auditory midbrain of mammals.

**Coding of angles in the vertical plane (elevation).** Mammals use differences in the sound spectrum arriving at the two ears for localizing in the vertical plane. How this information is extracted and processed in the auditory system is not yet understood. The formation of a map of auditory space containing both azimuth and elevation angles in the **superior colliculus of the midbrain** indicates that the auditory cues for the evaluation of elevation have to be learned during the individual development in a process of calibrating them through visual input.

#### 17.2.2.6 Coding of Complex Sounds Including Speech, and Auditory Cortical Function

Most natural sounds and, especially, communication sounds of mammals characterizing a behavioral context and/or the individual voice of a sender are complex in many respects. They consist of (*a*) constant frequency components varying in number (e.g., number of harmonics) and relative intensity (e.g., predominant harmonics or formants) over certain spectral bandwidths, (*b*) modulations of frequency components in frequency and intensity over time with varying depths, duty cycles, and speeds, (*c*) varying intensities and frequency bandwidths of noise (e.g., roughness or harshness of a voice), (*d*) varying durations of sound components and intersound intervals leading to varying time structures (e.g., repetition rates and rhythms) in a sound stream; and they come from geocentrically or egocentrically fixed or variable or moving locations. All the coding explained under Sects. 17.2.2.2, 17.2.2.3, 17.2.2.4, 17.2.2.5 and 17.2.2.6 predicts that the whole spectrum of potential variability of complex sounds can principally be analyzed and differences between sounds be detected. We have little knowledge, however, about the pathways, places, and neural mechanisms that provide the "result" of complex sound analysis, i.e., the basis for the perception of sounds as acoustic objects having certain meanings.

The starting point for coding of acoustic objects, which includes a resynthesis of a sound or a sound series from its analyzed parameters, is the **combination sensitivity** of the neurons arising from their locations in an auditory brain center. This location defines their participation in maps of neuronal response characteristics (Fig. 17.14a). For example, neurons in the center of a frequency-band lamina in the inferior colliculus respond, on average, very well to single frequencies or downward frequency sweeps of low intensity. Neurons located more laterally respond very well and temporally precise to a noisy loud sound, the frequency components of which may be upwards modulated. This combination sensitivity is continued in the superposition of patchy distributions of neuronal response characteristics in the primary auditory cortex [16]. **Combination sensitivity of neurons creates local hot spots of activation** in the inferior colliculus and primary auditory cortex due to hearing sounds of certain parameter combinations. How this combination sensitivity continues in nonprimary auditory cortex and finally leads to the identification of auditory objects is not understood yet. Cortical gamma-band oscillations may be involved in detecting synchronous and temporally coincident hot spot activations to be bound together for perception [16].

Neuronal combination sensitivity is the basis for at least five maps in specialized higher-order **auditory cortical areas of the mustache bat** representing acoustic parameter combinations from echolocation calls and the perceived echoes [45]. In two areas, the time delay between call emission and echo arrival is mapped for delays from about 0.4 to 18 ms representing a distance range of 7–310 cm between the bat and another object, e.g., a pursued prey. In one area, the Doppler shift in frequency between call and echo is measured producing a map of the relative speed between bat and object. In another area, the echo

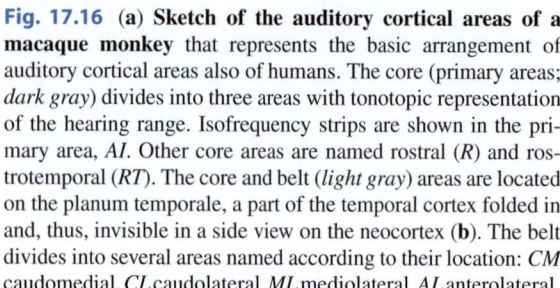

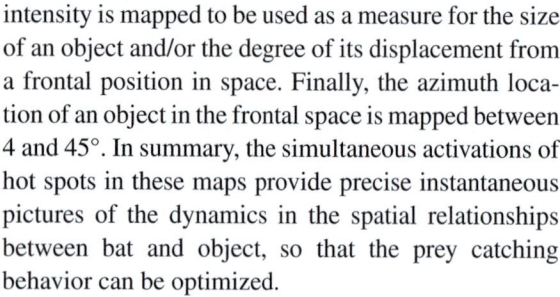

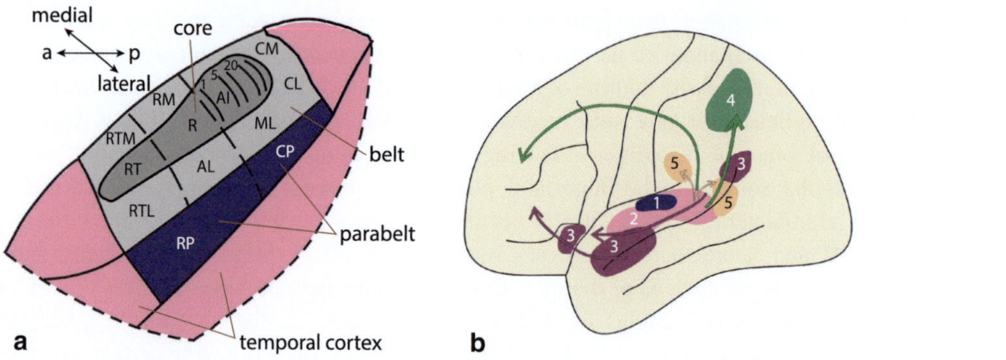


**Fig. 17.16** (**a**) **Sketch of the auditory cortical areas of a macaque monkey** that represents the basic arrangement of auditory cortical areas also of humans. The core (primary areas; *dark gray*) divides into three areas with tonotopic representation of the hearing range. Isofrequency strips are shown in the primary area, *AI*. Other core areas are named rostral (*R*) and rostrotemporal (*RT*). The core and belt (*light gray*) areas are located on the planum temporale, a part of the temporal cortex folded in and, thus, invisible in a side view on the neocortex (**b**). The belt divides into several areas named according to their location: *CM* caudomedial, *CL* caudolateral, *ML* mediolateral, *AL* anterolateral, *RTL* lateral rostrotemporal, *RTM* medial rostrotemporal, *RM* rostromedial. The auditory cortex continues lateral of the belt in the rostral (*RP*) and caudal (*CP*) parabelt (*dark blue* in **a** and **b**) and in other temporal cortex (*pink* in **a** and **b**). (**b**) Lateral view on the monkey (human) neocortex showing the information flow in the "what" pathway from the auditory to the frontal cortex (*1*, *2*, *3*, frontal), in the "where" pathway from the auditory to the parietal (*4*) and frontal cortex (*1*, *2*, *4*, frontal), and in a pathway related to speech processing (*1*, *2*, *5*) in humans (see related text). *a* anterior, *p* posterior (All modified from Kaas and Scott [15, 41] with permission)

intensity is mapped to be used as a measure for the size of an object and/or the degree of its displacement from a frontal position in space. Finally, the azimuth location of an object in the frontal space is mapped between 4 and 45°. In summary, the simultaneous activations of hot spots in these maps provide precise instantaneous pictures of the dynamics in the spatial relationships between bat and object, so that the prey catching behavior can be optimized.

The highly specialized mustache bat case is the only one known so far for neuronal combination sensitivity being transformed in auditory cortical maps from which sound perception and behavior can directly be derived. An example of a more common case of auditory cortical representation for the coding of complex sounds is the **primate (human) auditory cortex** as shown in Fig. 17.16 [15, 41]. Neuronal combination sensitivity on the background of tonotopic maps is found in the **core areas** of the auditory cortex. In primates, these are named primary (AI), rostral (R), and rostrotemporal (RT) areas (Fig. 17.16a), in other mammals often AI and anterior auditory field (AAF; see cat, Fig. 17.13). The core is surrounded by various **belt areas** (often without tonotopy), a **parabelt**, and further temporal cortex that responds to sounds. Neurons in the belt areas are especially responsive to species-specific and other complex sounds and may evaluate the meaning of species-specific sounds in relation to the behavioral context. Parabelt neurons of the left hemisphere in humans are sensitive to the phonological structure of a learned language and neurons of the nearby temporal cortex respond to various phonetic cues and syntax in speech [41]. Additional areas of the left hemisphere are sensitive to intelligible speech, others are related to verbal working memory or articulation of speech (Fig. 17.16b; [41]). It is important to note that processing in all auditory cortical areas from core to parabelt and beyond is highly plastic to be shaped and modified by **learning** (e.g., language) and **experience** [39, 50]. Thus, the functional organization of auditory cortical areas is not only the result of adaptation through evolution to the species-specific ecological and communication-related requirements but also the result of individual experience.

Common to mammals seems to be a division of functional pathways from the core auditory cortex [41]: anterior and ventral auditory areas and pathways process information for **acoustic object identification**, i.e., "what is heard", while posterior and dorsal areas and pathways relate to **auditory spatial perception and tasks**, i.e., "where is the sound source" requiring responses. A third system may be special for humans leading from heard speech to a reuse for articulation or an update of verbal working memory (Fig. 17.16b).

Mammals may also have in common a **left-hemisphere dominance** for processing and perception of the semantic content of communication sounds as mentioned above for speech [8]. This left-hemisphere dominance seems to be part of a more general division of functions between the left and right hemispheres of the brain in vertebrates [8]. Time-critical processing, perception, and categorization of stimuli, and paying attention to such stimuli, which all are important features in sound communication, are special domains of the left brain.

## 17.3 Birds, Reptiles, and Amphibians

### 17.3.1 Peripheral Auditory System

The following mammalian specializations **are not found** in amphibians, reptiles, and birds:

(a) **Structures of the outer ear.** In general, amphibians, reptiles, and birds have no pinna, concha, and, if at all, only a short ear canal. Birds may form an ear canal with specialized feathers. Barn owls are a very interesting case because they have ear canals with external openings at different heights at the right and left sides of their head [17]. This is the base for an auditory space map in their midbrain (see below).

(b) **Three middle ear ossicles as derivatives of the primary jaw bones** [4, 6, 21]. Amphibians, reptiles, and birds still use the quadrate (malleus) and articular (incus) for mastication, which have evolved to the hammer and anvil, respectively, in the mammalian middle ear. The mammalian stapes is present as columella in amphibians, birds, and reptiles to which another bone derived from the scalp, the plectrum in amphibians or the extracolumella in reptiles and birds, is added (Fig. 17.17a, b) to provide a lever system for sound amplification. Because of the mass of the plectrum or lacking stiffness of the extracolumella, this lever is ineffective in many species for transmitting high frequencies, so that most amphibians (anurans) have an upper frequency limit of hearing between about 2 and 5 kHz, reptiles and birds in the range of 1 kHz (turtles), 4–6 kHz (most reptiles), up to 12 kHz (birds, especially barn owls), compare [9] and Fig. 17.4. Recent studies have shown, however, that with specialized middle and inner ears some frogs may hear up to about 40 kHz and pygopod geckos to about 16 kHz. That is, specializations mainly of the middle ear can shift the general high-frequency limits of hearing as shown in Fig. 17.4 towards and into the ultrasonic range.

(c) **Mechanical separation of air-filled middle ears** [21]. In amphibians and reptiles, the Eustachian tubes are wide and open to the mouth cavity so that the air spaces of the right- and left-side middle ears are coupled. In birds, the middle ear spaces are coupled with the complex air spaces in the bones around the brain capsule. Thus, stimulation of one tympanic membrane by external sounds leads to the stimulation of the internal side of the other eardrum through the internal air spaces. In frogs, sounds picked up by the lungs and transferred via an open throat to the mouth cavity may also reach the internal sides of the eardrums. This coupling of both eardrums (with additional input from the lungs in frogs) makes them sensitive to phase differences between sound waves reaching them from inside and outside [28]. The eardrums function, at least in certain frequency ranges, as sound pressure-difference or pressure-gradient receivers, not just as sound pressure receivers as in mammals. As shown for insects (see Sect. 17.5.1), this has important consequences for the perception of the sound direction in the horizontal plane because the instantaneous amplitude of the motion of both eardrums and, therefore, the magnitude of the cochlear (inner ear in frogs) stimulation depends on the wavelength of the sound and the location of the sound source relative to both eardrums. Different from mammals, each amphibian, reptile, and bird left and right ear has an inherent sensitivity for sound direction, which is transmitted by the timing and the rate of action potentials in the auditory nerve fibers to the brain. This may help amphibians, reptiles, and birds with small heads to localize sound sources, especially those emitting low-frequency sounds.

(d) **Inner and outer hair cells in a coiled cochlea.** The basilar papilla (cochlea) is not coiled and varies in length between about 0.3 and 3 mm in reptiles and 3–10 mm in birds [21, 22]. Most amphibians have two hearing organs, the amphibian papilla (a specialty of amphibians) for relatively low-frequency hearing, and the basilar papilla, that seems to have evolved in sarcopterygian fish [10], for high-frequency hearing (Fig. 17.17a). The structure of the basilar papilla in reptiles is a playground of evolutionary tendencies [21, 22], i.e., it is very diverse with regard to the arrangement of the hair cells in a single or several clusters and various rows, the presence of a tectorial structure and the type of its connection to the hair cells. The arrangement of

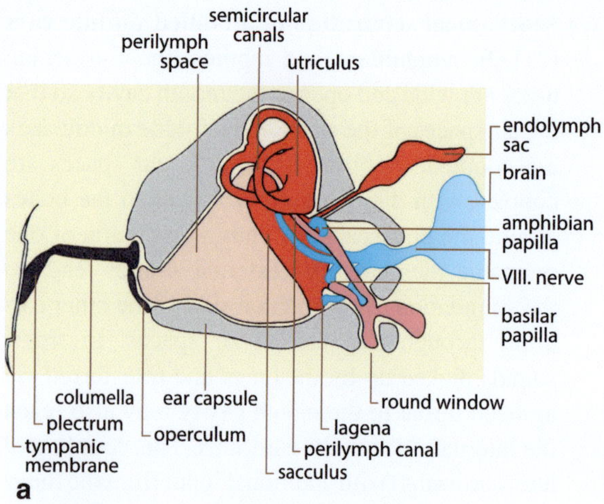


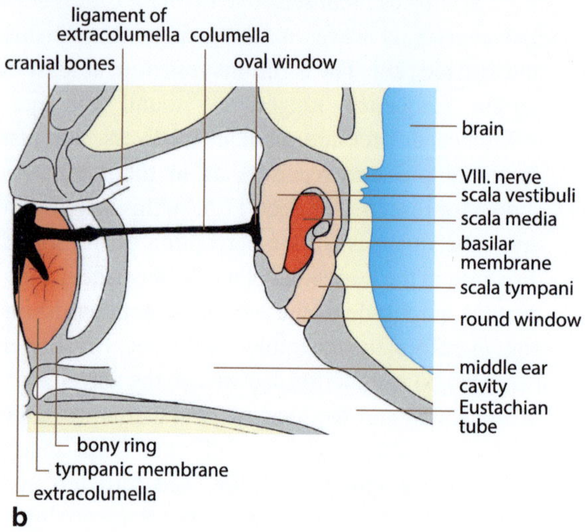


**Fig. 17.17 Diagram of the middle and inner ears** of an amphibian (**a**) and a reptile/bird (**b**) (Modified from Capranica and Manley [4, 21] with permission)

the tonotopic order can run from the base (high-) to the apex (low frequencies) or vice versa or can have two gradients from low frequencies in the middle to high frequencies at both the apical and basal ends. The basilar papilla in birds has always the same tonotopic gradient as in mammals. Tonotopy in the amphibian papilla of frogs and the basilar papilla of reptiles and birds results from mechanical tuning, plus electrical tuning of the hair cells (see Chap. 16), and a local amplification of the shearing motion of the tectorial structure by active motions of the kinocilium (they have a kinocilium besides many stereocilia) probably also of the stereocilia. The bird basilar papilla contains two types of hair cells – tall and short hair cells – whereby only the tall hair cells have afferent innervations [21, 22]. Thus, short hair cells in birds have a similar function as the OHCs of mammals, i.e., to mechanically amplify the stimulation of the tall hair cells, from which the information is transferred to the brain. Specialized hair cells in the apical part of the pigeon basilar papilla are important for infrasound sensitivity down to less than 1 Hz so that pigeons can detect sources of infrasound such as coastlines, valleys, thunderstorms and use them for navigation. In summary, the outer, middle, and inner ears of amphibians, reptiles, and birds differ from those of mammals in many details with regard to mechanics of sound transfer into the inner ear, hair cell stimulation and function. In addition, auditory nerve fibers transmit information related to the location of a sound source. The other sound parameters (frequency, rhythm, intensity) are encoded by the auditory nerve fibers in the same way as in mammals. Only phase-locking of spikes to the sound waveform is limited to about 1–2 kHz in amphibians and reptiles.

### 17.3.2 Central Auditory System

The ascending central auditory pathways of amphibians (anurans) and birds (also valid for most reptiles) are shown in Fig. 17.10 [5, 25]. Although centers of processing from the myelencephalon to the telencephalon may have different names, principle features such as **auditory chiasm**, **parallel/hierarchical** organization, and the **midbrain as an integrative center** are very similar to mammals. Main differences to mammals occur (*a*) in the myelencephalon for binaural interactions, which are found already at the cochlear nucleus level (dorsolateral nucleus) in amphibians and involve a special nucleus, the nucleus laminaris, in birds, and (*b*) in the auditory telencephalic centers that are little developed in amphibians and appear as a highly differentiated pallium in birds, which takes the place of the auditory cortex in mammals.

Especially interesting is the function of the nucleus laminaris of birds as **coincidence detector for interaural time differences** (Fig. 17.18a; [18]). Neurons in this nucleus receive excitatory input from both sides and respond maximally when these inputs are active at the very same time. Disparities in arrival time ($\Delta t$) and/or phase ($\Delta\Phi$) of sounds to the two ears by sources located aside the head midline lead to coincident activation according to the map shown in Fig. 17.18a. The topography runs in the nucleus of the barn owl in isodelay laminas from dorsolateral to ventromedial representing disparities from about 100 µs contralateral ear leading to 20 µs ipsilateral ear leading. The delay map is superimposed on the orthogonally running tonotopic gradient represented by neuronal CFs, which means that the delays are calculated for the whole hearing range. With a head diameter of about 5 cm in the barn owl, a delay of 100 and 20 µs equal angles ($\alpha$) of sound incidence of about 90° and 8°, respectively ($\sin \alpha = \Delta tc/d$; $c$ = sound velocity, $d$ = head diameter). Taking the maps in the nuclei of both sides together, the **azimuth angle of a sound source** relative to the owl is transformed to a **place code** of neural activation already at the level of the superior olive.

The neural topography for the azimuth angle is extended to a **three-dimensional neural map of auditory space** in the lateral nucleus of the barn owl inferior colliculus (Fig. 17.18b; [17]). Neurons at a given position in the nucleus are optimally activated by sound coming from a certain location in space that is of relevance for the owl, especially for prey catching. Azimuth angles are mapped on the anterior-posterior axis, elevation angles on the dorsoventral axis of the nucleus. Thus, a space from about 60° contralateral to 15° ipsilateral and about 30° above to almost 90° below the owl is mapped in the inferior colliculus of both sides of the brain. Since the frontal space is mapped twice, prey (e.g., moving mice) in this area is located with very high precision.

## 17.4 Fish

### 17.4.1 Peripheral Auditory System

The peripheral auditory system of fish differs in many respects from those of land vertebrates [25, 47, 49]:

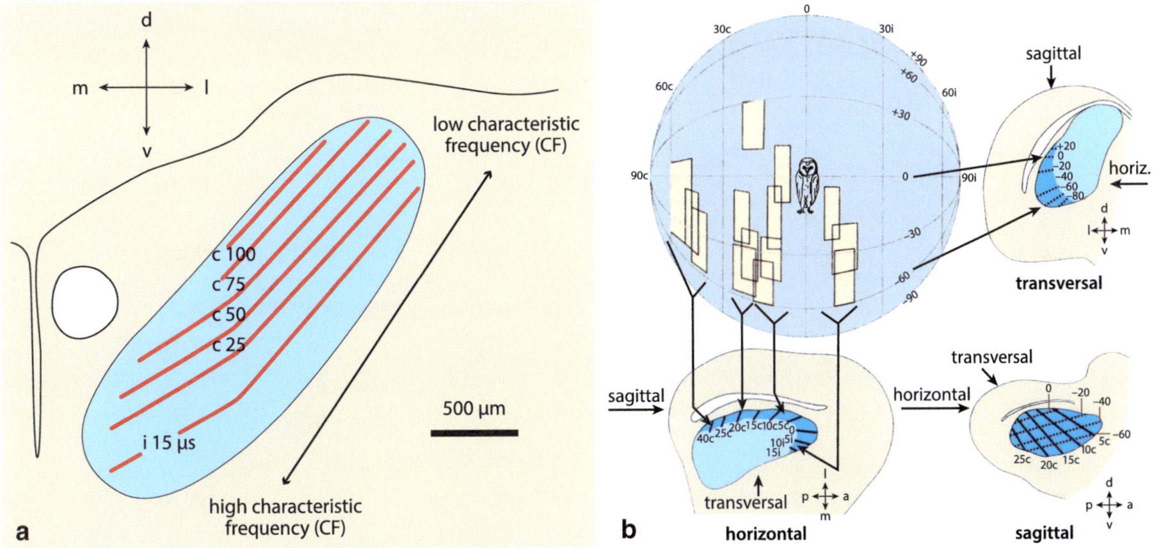


**Fig. 17.18** (**a**) **Frontal section through the brainstem with the nucleus laminaris of the barn owl.** This nucleus receives binaural input (see Fig. 17.10). The *lines* indicate the positions of neurons which respond best when the sound hits the contralateral (*c*) ear 25–100 µs or the ipsilateral (*i*) ear 15 µs earlier than the other ear. Altogether, these neurons represent a map of coincidence detectors coding the azimuth angle of a sound source by evaluation of the time delay ($\Delta t$) of sound incidence between both ears. The gradient of characteristic frequencies (low to high CFs) of the neurons runs perpendicular to the map of time delay (*azimuth*) representation, so that azimuth angles can be coded in the same way across the whole hearing range of the owl (Modified from Konishi et al. [18]). *m* medial, *l* lateral, *d* dorsal, *v* ventral. (**b**) Distribution of neurons in the right-side inferior colliculus of the barn owl according to their spatial receptive fields (response areas) that are shown on the sphere around the owl. Thus, locations in space contralateral (*c*) and ipsilateral (*i*) to the owl's IC are orderly represented (mapped) by the locations of neurons responding best to sound from a certain point in space. The space maps in the colliculus are shown in three sectioning planes: horizontal, transvers (frontal), and sagittal. In the sagittal and transvers sections, angles with negative signs relate to locations *below*, those with positive signs to locations *above* the owl. *a* anterior, *p* posterior, *l* lateral, *m* medial, *d* dorsal, *v* ventral (Modified from Knudsen and Konishi [17] with permission)

(a) **Cartilaginous and most bony fishes have no outer and middle ears**. In fish, no impedance-matching device (middle ear) is necessary when sounds in water stimulate hair cells in fluid-filled inner ears. Since these fish lack tympanic membranes or similar structures sensitive to sound pressure, the hair cells are stimulated by the velocity of the sound waves and not by the sound pressure.

(b) **Fish with a swimbladder are sound pressure sensitive.** Bony fishes such as carps (Cypriniformes), catfish (Siluriformes), herrings (Clupeiformes), Mormyriformes, and Beryciformes (including Holocentridae) have a swimbladder (or air bladder) which can serve as a "sound bladder" when picking up sound pressure waves from the water. These bladders have various extensions to or even into the

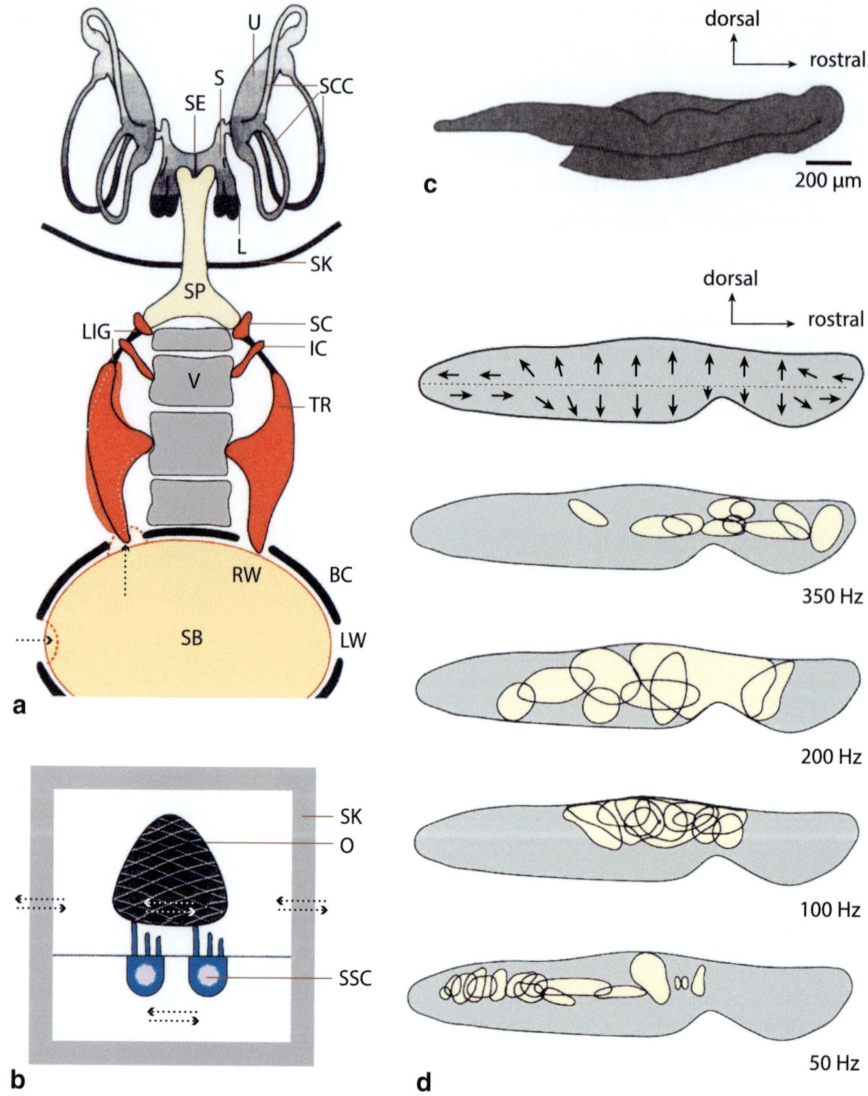


**Fig. 17.19** (**a**) **Sound transmission via the Weberian ossicles** (*TR* tripus, *IC* intercalarium, *SC* scaphium) from the air ('sound') bladder (*SB*) to the inner ear of ostariophysan fish. The SB is located in a bony capsule (*BC*). Sound pressure waves enter the capsule through the lateral windows (*LW*) and stimulate movements of the SB at the rostral windows (*RW*) in which, on both sides, the TR is inserted. The vibration of the TR travels via the IC and SC to the sinus perilymphaticus (*SP*; filled with perilymph and reaching into the skull (*SK*)), and finally to the sinus endolymphaticus (*SE*; filled with endolymph) in the inner ear. The SE connects to the possible hearing organs, the sacculus (*S*), the lagena (*L*), and the utriculus (*U*). *LIG* ligament, *SCC* semicircular canals, *SK* skull, *V* vertebra. (**b**) Displacement of sensory cells (*SSC*) of an otolithic organ in a fish without air bladder. In the near-field of a sound source the skull vibrates relative to the inertial mass of the otolith (*O*) leading to a displacement of the kinocilium and the stereocilia of the sensory cells. (**c**) Otolith from the sacculus of a minnow. (**d**) Orientation of the cilia of the sensory cells in the sacculus of a cod. The main areas of stimulation by tones of different frequencies are also shown (All modified from Tavolga et al. and Ziswiler [47, 54] with permission)

skull so that the vibrations can be transferred via bone conduction or directly into the inner ear. Several of the ostariophysan fish have one to three so-called Weberian ossicles, a special sound-conducting apparatus connecting the swimbladder with the inner ear (Fig. 17.19a). Thus, vibrations of the swimbladder are picked up on both sides by the tripus attached to the swimbladder through its bony capsule, and finally transmitted as pressure waves into the peri- and endolymph sacks of the inner ear. Because there is only a single perilymph sack entering the skull, the pressure waves to the inner ear are without directional information. However, fish with such "sound bladders" and Weberian ossicles are more sensitive to sounds and can hear higher frequencies (up to about 5 kHz) compared to other fish (about 1 kHz).

(c) **Fish have several hearing organs**. All three sensory organs with otoliths in the inner ear – sacculus, utriclus, and lagena – and the macula neclecta (without otolith) can function as hearing organs besides their tasks as organs of equilibrium and position control. The sacculus has auditory function most probably in all fish, the other organs contribute in a species-specific way. The crossopterygian fish *Latimeria* may be the only living species with a basilar papilla, which becomes the main auditory organ in land vertebrates [10]. Figure 17.19b explains the principle function of otolith organs in hearing. In the near field of a sound source the skull vibrates with the sound frequency relative to the inertial mass of the otolith. This relative movement by the sound velocity leads to displacements of the kinocilium and the stereocilia of the hair cells in the rhythm of the sound waveform (e.g., sine wave for low-frequency tones). Depending on the mass to be moved and the shape of the otolith, only relatively low frequencies ($f < 1$ kHz) lead to displacements of the cilia and to auditory stimulation. The shape of the otolith together with the orientation of the hair cells and their attachment to the otolith are responsible for local stimulation maxima on the sensory epithelium as a function of the stimulation frequency (Fig. 17.19c, d). This shows that one principle of frequency analysis in auditory organs of vertebrates is already realized in a rather crude way in fish, namely the **frequency-place-transformation leading to a tonotopy** on the sensory epithelium. The main mechanism of frequency coding in auditory nerve fibers of fish, however, is the phase-locking of spikes to the sound waveform, which is functional up to about 1–2 kHz.

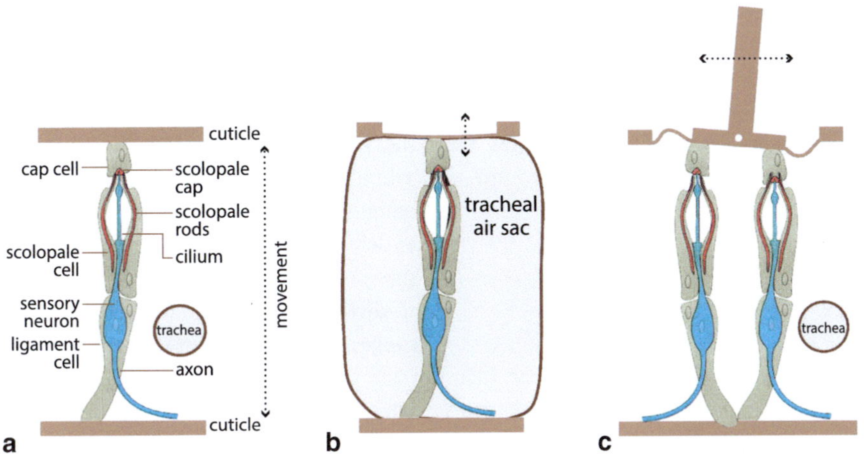


**Fig. 17.20 Proprioceptive chordotonal organ and derived tympanal and antennal hearing organs.** (**a**) Proprioceptive chordotonal organs span between different regions of the cuticular exoskeleton and monitor relative movements. The organs can comprise one or more chordotonal sensilla that each consist of one to three mechanosensory neurons and three supporting cells (cap cell, scolopale cell, ligament cell). (**b**) Tympanal hearing organs are obtained if the cuticule the organ connects to is thinned to a sound-receiving tympanal membrane and backed by a tracheal air sac so that it vibrates in response to sound. (**c**) In antennal hearing organs, the antenna's distal part acts as the sound receiver that directly follows the air particle displacements in the sound field. Vibrations of this receiver are picked up by Johnston's organ, a chordotonal organ in the second antennal segment. The chordotonal sensilla of Johnston's organ usually perpendicularly connect from different sides to the antennal receiver and are alternately activated as the antenna moves back and forth

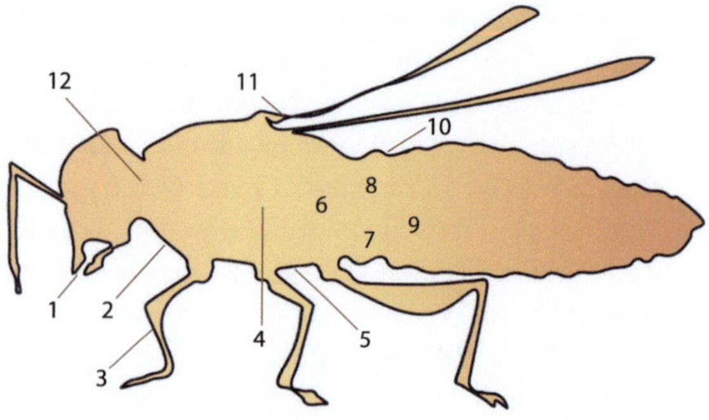


**Fig. 17.21 Sketch of a generalized insect depicting the different body parts on which tympanal ears can be found in certain insect taxa.** *1* mouthparts (some sphingid moths), *2* prosternum on the neck's ventral side (some tachinid flies), *3* tibia of foreleg (crickets, bushcrickets), *4* mesothorax (some corixid water bugs), *5* grove between metathoracic legs (some praying mantids), *6* posterior metathorax (noctuid moths), *7* ventral (pyralid moths) and anterior regions (geometrid moths, drepanid moths) of the first abdominal segment, *8* laterally on the first abdominal segment (grasshoppers), *9* second abdominal segment (cicadas, some uraniid butterflies), *10* dorsal surface of first abdominal segment (tiger beetles), *11* wing base (lacewings), *12* dorsally on prothorax (some scarab beetles)

### 17.4.2 Central Auditory System

Since the hearing organs of fish are also intimately related to the functions of equilibrium and position control, it is difficult to separate these functions in the brain centers for processing information from the inner ear. In addition, central projections of the lateral line system partly overlap areas receiving projections from the inner ear. Therefore, the central auditory system of fish deviates from the diagrams shown in Fig. 17.10 and possible homologies are difficult to be assessed. Reviews can be found in [25, 47, 49].

## 17.5 Insects

Apart from vertebrates, various insects can hear. Insects use hearing for acoustic communication or to detect the ultrasonic echolocation signals emitted by hunting bats [44]. Some parasitoid flies also employ hearing to locate their hosts, singing crickets [33] and some predatory bushcrickets listen to, and acoustically mimic, the songs of sexually receptive females of certain cicada species to misguide and devour their males [23]. The ability to hear has evolved many times independently within different insect taxa, which all transformed proprioceptive stretch receptor organs into hearing organs [53] (Fig. 17.20). Because of these multiple evolutionary origins, insect hearing organs vastly differ in their anatomies and functional properties, often reflecting adaptations for specific tasks. This diversity, which also extends to the central auditory pathways, may seem bewildering, yet it provides an enticing field for studying evolutionary solutions and innovations of how acoustic signals are processed.

### 17.5.1 Peripheral Auditory System

Some insects detect sounds with mechanosensory bristles (Chap. 16), yet the more sophisticated hearing organs that can be found in insects seem all derived from **proprioceptive chordotonal organs** (Fig. 17.20a). These internal stretch receptors are serially arranged along the insect body and, spanning between different regions of the cuticular exoskeleton, monitor body movements. An auditory function is obtained if the cuticle in one of these regions is made sufficiently compliant to act as a sound receiver that vibrates in response to sound (Fig. 17.20b, c).
With respect to the physical component of the sound wave that sets this receiver into vibration, sound **particle velocity-sensitive ears** and **sound pressure--sensitive ears** can be distinguished. Particle velocity-sensitive ears are known from honeybees, mosquitoes, and certain drosophilid flies, which hear with their antennae [27] (Fig. 17.20c). In these insects, the mobile distal part of the antenna directly follows the particle movement in the sound field, and these vibrations are monitored by Johnston's organ, the antennal chordotonal organ. Antennal hearing is used to detect the wing-beat sounds of conspecifics; because the sound particle velocity rapidly drops with distance, it allows for intimate sound communication at close range. Pressure-sensitive ears can occur on various parts of the insect body (Fig. 17.21), depending on which of the serially arranged chordotonal organs has been modified for the detection of sound. Modifications include the thinning of one of the cuticular regions the chordotonal organ connects to, so that, when backed with an air sac, a pressure-sensitive tympanal membrane is obtained [14] (Fig. 17.20b). In some cases, sound also enters through the tracheal system and acts on both the membrane's in- and outside, turning the tympanum into a **pressure-difference receiver** measuring

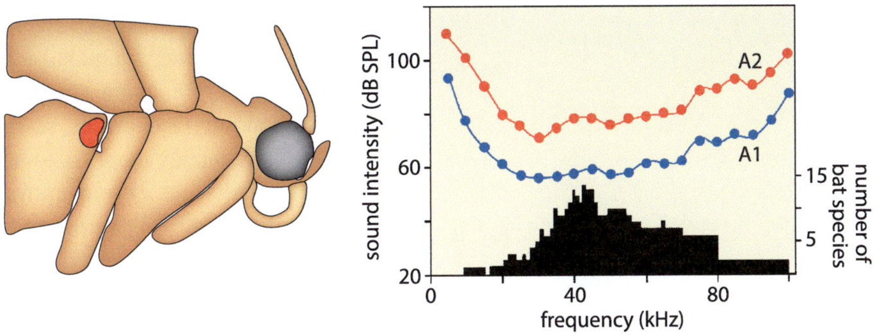


**Fig. 17.22 Ultrasound hearing in noctuid moths.** *Left*: Sketch of a moth highlighting the position of the tympanum (*red*). *Right*: Frequency tuning of the ear's two sensory cells and corresponding frequency distribution of the echolocation signals of sympatric bat species. The two sensory cells named *A1* and *A2* display a similar frequency tuning, yet the *A1* cell is more sensitive (Right graph modified from Fullard [11] with permission)

cycle-by-cycle the pressure differences resulting from the different phases of a sound wave having traveled slightly different distances when arriving at the tympanum's outer and inner sides [44]. As explained for amphibia, reptiles, and birds, the mechanical coupling of tympana in a pressure-difference receiver system provides excellent sound localization abilities for these small animals. Tympanal ears allow for hearing at long distances in the kHz range of frequencies. They are found in, e.g., cicadas, grasshoppers, and crickets that communicate with loud sounds and in nocturnal insects such as moths, mantids, and lacewings that are faced with predation by bats.

Insect sound receivers are mechanically tuned to relevant sound frequencies and their vibrations are directly coupled to chordotonal sensory neurons (Fig. 17.20). They transduce the vibrations into electrical signals and encode them in action potentials forwarded to the CNS. Sound-induced vibration with amplitudes of less than a nanometer can suffice to activate the neurons [52]. For antennal ears, it has also been shown that the neurons can be motile and actively augment the receiver's vibrations on a cycle-by-cycle basis, thus increasing the ear's mechanical susceptibility for sound. The number of neurons per ear varies greatly between insect taxa, ranging from ca. 16,000 in male mosquitoes down to only a single neuron in certain moths. In the mosquitoes, the neurons are radially arranged around the antennal sound receiver, which is used like a joystick to pin down the direction of sound sources, i.e., females flying nearby. This is possible because sound particle velocity is a vectorial quantity so that the receiver vibrates in the direction in which the sound propagates. Moth ears perform less sophisticated tasks, and a single neuron suffices to determine whether a bat is around or not. In some moth ears, a second, less sensitive neuron has been added (Fig. 17.22). If only the more sensitive neuron fires, that bat must be far, and a turning behavior is triggered that brings away the moth from the bat. If the bat should nonetheless come sufficiently close so that its echolocation signals also activate the less sensitive neuron, the moth immediately stops flying and drops down to the ground.

Because sound pressure is a scalar quantity, turning away from the bat requires the use of binaural cues. The most prominent insects using binaural cues are the small parasitoid flies that home in on singing crickets in the dark. Because the two ears that sit on their neck are directly located beside each other, there is little room for using binaural cues (Fig. 17.23). Measured directly in front of the ears with the loudspeaker placed at an angle of 90° from the animal's midline, interaural intensity differences are virtually zero and interaural time differences are only 1.5 μs, which is smaller than the jitter of the action potentials generated by the receptor cells (ca. 70 μs). The flies nonetheless show exquisite directional hearing, and they manage by using a trick: instead of keeping them separately, they have mechanically coupled the eardrums of their two ears so that they rock like a seesaw over a small ridge. When the sound wave hits the ipsilateral tympanum, it immediately starts to move, making the contralateral tympanum move in the opposite direction (Fig. 17.24). Once initiated, this oscillation continues as long as the sound stimulus lasts. This seesaw mechanisms increases interaural time difference by a factor of about 1,100 at the level of the tympana, which is then further increased by a factor of about 5.5

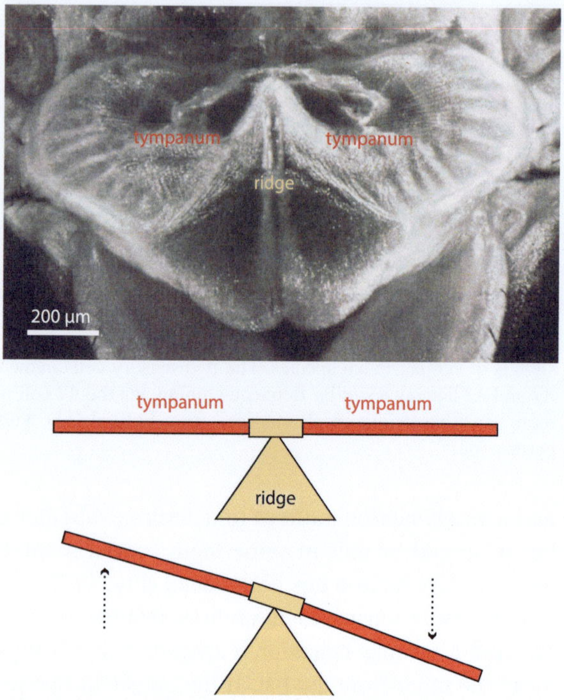


**Fig. 17.23** **Mechanically coupled tympana on the neck of a parasitoid fly.** *Top*: Electron micrograph of the two tympana, which are mechanically coupled via a cuticular ridge (Reprinted with permission from Miles et al. [26]). *Bottom*: Sound-induced deflection of the tympana upon sound stimulation from the *right*. Downwards movement of the ipsilateral tympanum makes the contralateral one move up, a seesaw mechanism that amplifies binaural cues, allowing the flies to acoustically pinpoint their singing cricket hosts at night (Modified from Robert and Göpfert [33] with permission)

at the level of the first-order interneuron onto which the receptors converge. Thus, an interaural time difference of only 50 ns resulting from a sound source displaced only 2° from the fly's midline [24] is enlarged to a 300 μs difference which is close to the threshold of perception.

Like vertebrate hearing organs, the tympanal ears of many insects also decompose sound into frequency components. Bushcrickets, for example, have ears on their front legs that comprise a linear arrangement of some 60 receptor neurons, the **crista acoustica** (Fig. 17.24). Each neuron has a different best frequency, and this best frequency gradually increases from the crista's proximal to its distal end. In grasshoppers, the ears sit laterally on the abdomen, and their auditory receptor organ is called **Müller's organ**. This organ comprises several groups of receptor cells that differ in their frequency characteristics and with respect to their tympanal attachment sites. At these sites, the tympanum is mechanically tuned to the corresponding frequencies, so that its mechanics provides a frequency map. Like the basilar membrane in our ears, the locust tympanum also propagates travelling waves that, depending on the sound frequency, funnel sound energy to the respective receptor cells [51].

### 17.5.2 Central Auditory System

Insect auditory receptors are primary sensory cells that send axons to the central nervous system. The axons usually arborize in the ganglia of the body segment that carries the ear. The receptors of the antennal Johnston's organ synapse in the antennal mechanosensory motor center in the deuterocerebrum of the brain. Tympanal receptors synapse in the ventral thoracic or abdominal ganglia, unless, as is the case in some sphingid moths, they are located on the head. A considerable amount of stimulus processing takes place already in the ventral ganglia [44]. For example, local interneurons with an Ω-shaped morphology (omega neurons of crickets) enhance directional cues by inhibiting the contralateral side. Movements of the animals towards or away from the sound source (positive and negative phonotaxis) critically depend on the action of these interneurons. In some insects, thoracic ganglia are tonotopically organized with regard to the projection areas of the peripheral sensory cells so that the peripheral tonotopy is reproduced there [36].

From the thoracic or abdominal ganglia ascending interneurons forward the preprocessed information to the brain. There, auditory neurons seem rather scattered in insects with tympanal ears, with some preference for the lateral protocerebrum. Female bushcrickets and crickets listening to singing males at night might be attacked by echolocating bats. So they have to discriminate between their species-specific communication sounds and echolocation signals. Based on frequency differences, this discrimination is done in a categorical way in the ventral ganglia leading to either approach or avoidance behavior. The respective information is also forwarded to the brain via parallel ascending pathways [44]. Descending interneurons may prime the ventral ganglia with information from

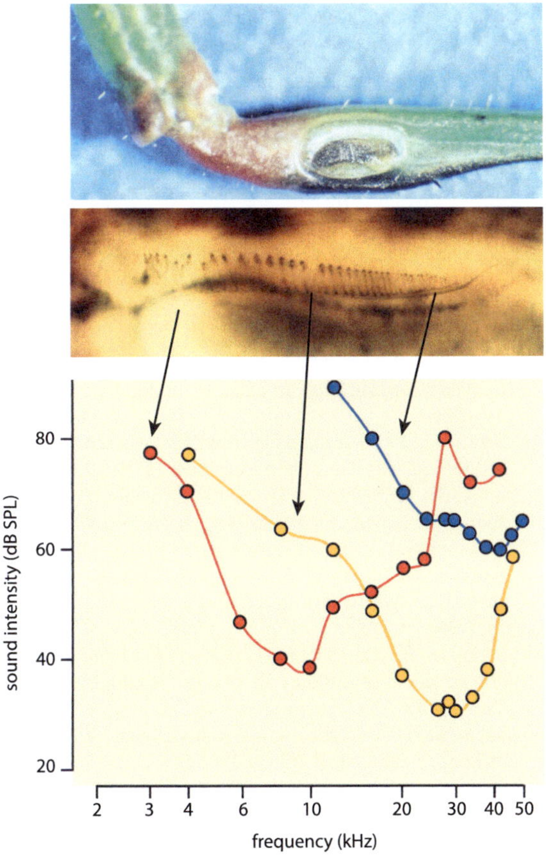


**Fig. 17.24** **Tonotopic arrangement of auditory receptors in a bushcricket ear.** *Top*: Position of the tympanum on the tibia of the foreleg. *Middle*: Close-up of the crista acoustica, which comprises ca. 60 receptor cells that gradually decrease in size. *Bottom*: Corresponding frequency tuning of the receptors, illustrated for three cells at different sites

the brain. Apparently, only few synapses are involved in sound processing and response generation. Bushcrickets may start an avoidance response within less than 20 ms after sound onset. More complex behaviors such as phonotaxic approaches of a singing male by a female involve more complex neuronal circuits and require more processing time.

## 17.6 Other Invertebrates

Detection of far-field sound or pressure waves through specialized structures such as mechanosensory bristles or tympanic organs is not known for nonarthropod invertebrates (see discussion and review [2]).

## References

1. Ashmore J (2008) Cochlear outer hair cell motility. Physiol Rev 88:173–210
2. Budelmann BU (1992) Hearing in nonarthopod invertebrates. In: Webster DB, Fay RR, Popper AN (eds) The evolutionary biology of hearing. Springer, Berlin Heidelberg New York, pp 141–155
3. Buser P, Imbert M (1992) Audition. MIT Press, Cambridge
4. Capranica RR (1976) Morphology and physiology of the auditory system. In: Llinás R, Precht W (eds) Frog neurobiology. Springer, Berlin Heidelberg New York, pp 551–575
5. Carr CE, Code RA (2000) The central auditory system of reptiles and birds. In: Dooling RJ, Fay RR (eds) Comparative hearing: birds and reptiles. Springer, Berlin Heidelberg New York, pp 197–248
6. Clack JA, Allin E (2004) The evolution of single- and multiple-ossicle ears in fishes and tetrapodes. In: Manley GA, Fay RR (eds) The evolution of the vertebrate auditory system. Springer, Berlin Heidelberg New York, pp 128–163
7. Ehret G (1997) The auditory midbrain, a "shunting yard" of acoustical information processing. In: Ehret G, Romand R (eds) The central auditory system. Oxford Univ Press, New York, pp 259–316
8. Ehret G (2006) Hemisphere dominance of brain function – which functions are lateralized and why? In: van Hemmen JL, Sejnowski TJ (eds) 23 problems in systems neuroscience. Oxford Univ Press, New York, pp 44–61
9. Fay RR (1988) Hearing in vertebrates: a psychophysics databook. Hill-Fay Assoc, Winnetka
10. Fritzsch B (1992) The water-to-land transition: evolution of the tetrapod basilar papilla, middle ear, and auditory nuclei. In: Webster DB, Fay RR, Popper AN (eds) The evolutionary biology of hearing. Springer, Berlin Heidelberg New York, pp 351–375
11. Fullard JH (2006) The evolution of hearing in moths: the ears of *Oenosandra boisduvalii* (Noctuoidea: Oenosandridae). Aust J Zool 54:51–56
12. Greenwood DD (1990) A cochlear frequency-position function for several species – 29 years later. J Acoust Soc Am 87:2592–2605
13. Grothe B, Pecka M, McAlpine D (2010) Mechanisms of sound localization in mammals. Physiol Rev 90:983–1012
14. Hoy RR, Robert D (1996) Tympanal hearing in insects. Annu Rev Entomol 41:433–450
15. Kaas JH (2011) The evolution of auditory cortex: the core areas. In: Winer JA, Schreiner CE (eds) The auditory cortex. Springer, Berlin Heidelberg New York, pp 407–427
16. Kanwal JS, Ehret G (2011) Communication sounds and their cortical representation. In: Winer JA, Schreiner CE (eds) The auditory cortex. Springer, Berlin Heidelberg New York, pp 343–367
17. Knudsen EI, Konishi M (1978) Space and frequency are represented separately in the auditory midbrain of the owl. J Neurophysiol 41:870–884
18. Konishi M, Takahashi TT, Wagner H, Sullivan WE, Carr CE (1988) Neurophysiological and anatomical substrates for sound localization in the owl. In: Edelman GM, Gall WE, Cowan WM (eds) Auditory function. The neurobiological bases of hearing. Wiley, New York, pp 721–745

19. Liberman MC (1978) Auditory-nerve response from cats raised in a low-noise chamber. J Acoust Soc Am 63:442–455
20. Malmierca MS, Ryugo DK (2011) Descending connections of auditory cortex to the midbrain and brain stem. In: Winer JA, Schreiner CE (eds) The auditory cortex. Springer, Berlin Heidelberg New York, pp 189–208
21. Manley GA (1990) Peripheral hearing mechanisms in reptiles and birds. Springer, Berlin Heidelberg New York
22. Manley GA (2000) Cochlear mechanisms from a phylogenetic viewpoint. Proc Natl Acad Sci U S A 97:11736–11743
23. Marshall DC, Hill KB (2009) Versatile aggressive mimicry of cicadas by an Australian predatory katydid. PLoS One 4:e4185
24. Mason AC, Oshinsky ML, Hoy RR (2001) Hyperacute directional hearing in a microscale auditory system. Nature 410:686–690
25. McCormick CA (1992) Evolution of central auditory pathways in anamniotes. In: Webster DB, Fay RR, Popper AN (eds) The evolutionary biology of hearing. Springer, Berlin Heidelberg New York, pp 323–350
26. Miles RN, Robert D, Hoy RR (1995) Mechanically coupled ears for directional hearing in the parasitoid fly *Ormia ochracea*. J Acoust Soc Am 98:3059–3070
27. Nadrowski B, Effertz T, Senthilan PR, Göpfert MC (2011) Antennal hearing in insects – new findings, new questions. Hear Res 273:7–13
28. Narins PM, Ehret G, Tautz J (1988) Accessory pathway for sound transfer in a neotropical frog. Proc Natl Acad Sci USA 85:1508–1512
29. Patuzzi R (1996) Cochlear micromechanics and macromechanics. In: Dallos P, Fay RR (eds) The cochlea. Springer, Berlin Heidelberg New York, pp 186–257
30. Pickles JO (2008) An introduction to the physiology of hearing, 3rd edn. Emerald, Bingley
31. Pollak GD (1992) Adaptations of basic structures and mechanisms in the cochlea and central auditory pathway of the mustache bat. In: Webster DB, Fay RR, Popper AN (eds) The evolutionary biology of hearing. Springer, Berlin Heidelberg New York, pp 751–778
32. Rees A, Langer G (2005) Temporal coding in the auditory midbrain. In: Winer JA, Schreiner CE (eds) The inferior colliculus. Springer, Berlin Heidelberg New York, pp 346–376
33. Robert D, Göpfert MC (2002) Novel schemes for hearing and acoustic orientation in insects. Curr Opin Neurobiol 12:715–720
34. Robles L, Ruggero MA (2001) Mechanics of the mammalian cochlea. Physiol Rev 81:1305–1352
35. Romand R, Avan P (1997) Anatomical and functional aspects of the cochlear nucleus. In: Ehret G, Romand R (eds) The central auditory system. Oxford Univ Press, New York, pp 97–191
36. Römer H (1983) Tonotopic organization of the auditory neuropile in the bushcricket *Tettigonia viridissima*. Nature 306:60–62
37. Rosowski JJ (1994) Outer and middle ears. In: Fay RR (ed) Comparative hearing: mammals. Springer, Berlin Heidelberg New York, pp 172–247
38. Rouiller EM (1997) Functional organization of the auditory pathways. In: Ehret G, Romand R (eds) The central auditory system. Oxford Univ Press, New York, pp 3–96
39. Scheich H, Ohl FW (2011) A semantic concept of auditory cortex function and learning. In: Winer JA, Schreiner CE (eds) The auditory cortex. Springer, Berlin Heidelberg New York, pp 369–387
40. Schreiner CE, Langer G (1997) Laminar fine structure of frequency organization in auditory midbrain. Nature 388: 383–386
41. Scott SK (2005) Auditory processing – speech, space and auditory objects. Curr Opin Neurobiol 15:197–201
42. Smith CA (1981) Recent advances in structural correlates of auditory receptors. Prog Sens Physiol 2:135–187
43. Spoendlin H (1970) Structural basis of peripheral frequency analysis. In: Plomb R, Smoorenburg GF (eds) Frequency analysis and periodicity detection in hearing. Sijhoff, Leiden, pp 2–36
44. Stumpner A, von Helversen D (2001) Evolution and function of auditory systems in insects. Naturwissenschaften 88:159–170
45. Suga N (1990) Cortical computational maps for auditory imaging. Neural Netw 3:3–21
46. Suga N, Gao E, Zhang Y, Ma X, Olsen JF (2000) The corticofugal system for hearing: recent progress. Proc Natl Acad Sci USA 97:11807–11814
47. Tavolga WN, Popper AN, Fay RR (eds) (1981) Hearing and sound communication in fishes. Springer, Berlin Heidelberg New York
48. Terhardt E (1998) Akustische Kommunikation. Springer, Berlin Heidelberg New York
49. Webb JF, Fay RR, Popper AN (eds) (2008) Fish bioacoustics. Springer, Berlin Heidelberg New York
50. Weinberger NM (2011) Reconceptualizing the primary auditory cortex: learning, memory and specific plasticity. In: Winer JA, Schreiner CE (eds) The auditory cortex. Springer, Berlin Heidelberg New York, pp 465–491
51. Windmill JF, Göpfert MC, Robert D (2005) Tympanal travelling waves in migratory locusts. J Exp Biol 208:157–168
52. Windmill JF, Fullard JH, Robert D (2007) Mechanics of a 'simple' ear: tympanal vibrations in noctuid moths. J Exp Biol 210:2637–2648
53. Yager DD (1999) Structure, development, and evolution of insect auditory systems. Microsc Res Tech 47: 380–400
54. Ziswiler V (1976) Die Wirbeltiere, vol 1. Thieme, Stuttgart

# Vision

# 18

Jutta Kretzberg and Udo Ernst

Vision is for humans usually the most present sense. Our eyes gather light reflection from surrounding objects, triggering neuronal activity in photoreceptors. After extracting important aspects of the light stimulation, the retina sends sequences of action potentials to the brain. Large parts of our brain are devoted to processing of visual information, providing the basis for our behavioral decisions relying on our percept of visual stimuli. Even though the visual system is not as prominent for many animals and some animals are even more specialized to vision than humans, the main characteristics of their visual systems are very similar across species. Even vertebrates and invertebrates share many of the principles of visual information processing.

In this chapter, two visual systems are explained in some detail. Primate (including human) vision is shown as an example for mammalian vertebrates (Sect. 18.2). Some specific differences to other vertebrates are given in Sect. 18.3. Insect vision is explained based on the well-studied visual system of the fly (Sect. 18.4), and some characteristics of other invertebrates are mentioned in Sect. 18.5. However, space limitation only allows the presentation of a small selection of the huge amount of specializations developed by different vertebrate and invertebrate species to adapt to the particular conditions of their sensory environments. To point out the computational aspects of visual information processing, seven boxes introducing important methods and concepts from computational neuroscience supplement this chapter.

J. Kretzberg (✉)
University of Oldenburg, Carl-von-Ossietzky-Str. 9-11, 26129 Oldenburg, Germany
e-mail: jutta.kretzberg@uni-oldenburg.de

U. Ernst
Institute for Theoretical Physics, University of Bremen, Hochschulring 18, 28359 Bremen, Germany
e-mail: udo@neuro.uni-bremen.de

## 18.1 The Physics of the Stimulus

Light is electromagnetic radiation. Wavelength, direction, polarization, and intensity of light are relevant properties for visual perception.

Visible light is electromagnetic radiation, which can be absorbed by photopigments. Light exhibits both particle-like and wave-like properties. Both aspects are connected by the formula $E_\gamma = h\nu = hc/\lambda$ with $E_\gamma$ the energy of a single photon of wavelength $\lambda$ and $h$ being the Planck constant. The wavelength $\lambda$ and the frequency $\nu$ of the electromagnetic wave are connected by $c = \lambda\nu$, with $c$ being the velocity of light in vacuum. Both, the particle and wave models of light are important for visual perception: Photons are needed as particles to be absorbed in phototransduction (see Sect. 18.2.2). On the other hand, light is only visible in a certain range of wavelengths, and polarization vision can only be explained using the wave model. In this section, we will briefly introduce wavelength, direction, polarization, and intensity of light, which are relevant properties for visual perception, and give a short summary of biologically relevant light sources.

**Wavelength** – The basis for color vision is that light waves have specific wavelengths. For humans,

C.G. Galizia, P.-M. Lledo (eds.), *Neurosciences - From Molecule to Behavior: A University Textbook*,
DOI 10.1007/978-3-642-10769-6_18, © Springer-Verlag Berlin Heidelberg 2013

visible light ranges from approximately 400 nm (violet) to approximately 700 nm (red), while many other animals are able to perceive also ultraviolet (down to ~300 nm) and near infrared (up to ~800 nm) light, depending on the visual pigments expressed by their photoreceptors (see Box 18.1) and the transmission of the lens. Electromagnetic radiation with shorter wavelength is called ultraviolet light, X-rays, and gamma rays. Infrared light, microwaves, and radio waves have longer wavelengths. The restriction of visible light to this particular range of wavelengths can be explained by two physical factors: Only a small fraction of the sunlight reaching the earth's surface has wavelengths shorter than 300 nm, because short wavelengths are effectively filtered out by ozone molecules in the atmosphere. On the long wavelength side, low photon energy limits receptor efficiency. Since photon energy is inversely proportional to the wavelength, the energy difference between photons of long-wavelength light (>700 nm) and thermal quanta is small. Receptors for long wavelengths are prone to thermal noise when activated by heat rather than light.

**Direction** – The basis for spatial vision is that light rays are directional. As long as they travel through the same optic medium (e.g., air or water) they have a constant speed and direction. However, when a light beam reaches the boundary between two optic media (e.g., air and eye tissue) in an oblique angle, the light beam is split. A fraction of the light is reflected by the interface. The remainder is refracted; it travels through the medium with a change in speed and direction. Since the refractive index of optic media depends on the wavelength of light, prisms split white light into light beams with separate wavelengths. This effect called dispersion is also the reason for chromatic aberration at lenses, causing, e.g., the lens of the human eye to refract blue light more than red light. Moreover, when light shines through openings (like a pupil) in a nontransparent surface, diffraction causes bending of the light waves. In particular, when the opening is smaller than the wavelength, interference of the waves spreading out past the opening can cause complex wave (interference) patterns.

**Polarization** – The human eye is not very sensitive to the polarization of light. However, in many animals, particularly invertebrates, vision of polarized light is used to determine the position of the sun and thereby plays an important role for navigation, because it helps to determine the position of the sun. For a light ray traveling in free space, the electric field vector oscillates perpendicular to the direction of travel. When this field vector is oriented in a single plane, the light beam is called linearly polarized.

In nature, there is no direct source of polarized light. However, scattering of sunlight in the atmosphere causes a three-dimensional polarization pattern that indicates the position of the sun even when clouds obstruct the direct view. Moreover, reflection from shiny surfaces like water, vegetation, or some animals can cause polarization of light.

**Light intensity** – Differences in light intensity provide the basis for image-forming vision. Light intensity is defined as the power of electromagnetic radiation per unit area, measured in watts per square meter ($W/m^2$). It needs to be kept in mind that light consists of individual photons, impinging stochastically on a certain area in space.

For light of a specific wavelength, light intensity can be measured as photons per square meter per second. However, since photon energy varies with wavelength, the exact spectral composition must be considered when measuring the light intensity of mixtures of different wavelengths – like, for instance, "white" light. To express the brightness perception of a human observer, the number of photons for each wavelength must be weighted with the wavelength-specific photopic (high light) or scotopic (low light) sensitivity of the human eye, leading to the unit candela per square meter ($cd/m^2$). Ambient moonlight has ~0.1 $cd/m^2$, a typical computer display emits 100 $cd/m^2$, the sun at noon $\sim 1.6 \times 10^9$ $cd/m^2$.

It should be noted that only a small fraction of photons reaching the cornea are absorbed in photoreceptors. In human eyes more than 90 % of light in the visible spectrum are absorbed or diffracted by other eye tissue, or pass the photoreceptors.

**Light sources** – The most important light source on earth is sunlight, which is reflected by objects. However, each body at a given temperature also emits a characteristic spectrum of radiation by itself. Most relatively cool objects like animals emit mainly infrared wavelengths, hot objects like gas flames or heated metal also emit light in the visible range. Some chemicals produce visible light by **chemoluminescence**. Organisms ranging from dinoflagellate algae, bacteria, and fungi to fireflies and anglerfishes use this effect to actively produce light to attract prey or mates or to repel potential predators. This ability, called **bioluminescence**, is much more common in marine life, in particular in deep-sea organisms, than in terrestrial organisms.

## 18.2 Mammals: Vision in Primates

In this chapter, we will present the visual system of mammals, using the example of the primate (and human) visual system. Differences in the visual systems of other vertebrates [33], including other mammals are presented in Sect. 18.3.

### 18.2.1 Sensory Organs and Peripheral Mechanisms

#### 18.2.1.1 Anatomy of the Eye

> The vertebrate eye optimizes the transfer of light from the outside environment to the retina, where it triggers neuronal activity.

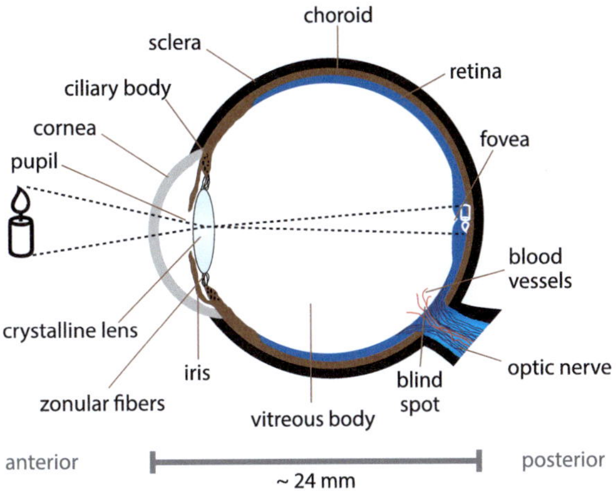


**Fig. 18.1** **Sketch of the anatomy of the human eye** (vertical sagittal section). Visual objects are projected *upside-down* onto the retina, the innermost layer of the eyeball. See text for explanation (Thanks to Sirko Straube for his help with the artwork)

The eye (see Fig. 18.1) consists of three layers of tissue, and is filled with a transparent, viscous fluid, the vitreous body. The external layer forms the supporting wall of the eyeball. It consists of the white-colored sclera, which can be seen as the white part of the human eye, and the transparent cornea. The cornea is the first and most powerful lens in the eye. Together with the crystalline lens it produces a sharp image on the photoreceptor level of the retina. In the anterior part of the eye, the intermediate layer contains the iris and the ciliary body. The iris is a circular muscle that controls the size of the pupil and therefore the amount of light that enters the eye. This beautifully pigmented muscle gives us our eye's colors, depending on the amount of eumelanin (appearing brown) and pheomelanin (appearing blue-green). The ciliary muscles change the shape of the transparent crystalline lens for accommodation, the adjustment of the eye's optical power to objects at various distances. In most mammals, the relaxed eye produces a sharp image of distant objects. To focus on close objects, the animal accommodates by contracting the ciliary muscles and loosening the zonular fibers to give the crystalline lens a rounder shape and smaller radius. In the posterior part of the eye, the intermediate layer is called choroid. It consists of connective tissue and blood vessels providing the retina with oxygen and nourishment. The ~0.2–0.5-mm-thin internal layer is the sensory part of the eye, the retina. Since the retina derives from the neural tube during embryonic development (see Chap. 3), it is a part of the central nervous system. It transforms the incoming light into electrical signals, processes the visual information, and sends it via the optic nerve to the brain. The cornea and crystalline lens focus the incoming light rays upon the retina. In the human eye, the fovea is the central point for the image focus, the visual axis. The foveal pit (in humans ~1.5 mm in diameter) allows seeing the finest details with highest resolution. In the blind spot the axons of the retinal ganglion cells (see Sect. 18.2.3.2) leave the eye to approach the brain. In this region, no photoreceptors are present and no light can be detected.

#### 18.2.1.2 Eye Movements

> Eyes move all the time with alternating periods of fast saccades shifting the visual attention and slower movements stabilizing gaze during self-motion.

The eyeballs are held in position by several muscles and ligaments. Three pairs of muscles are inserted into the sclera and attached to the skull. They rotate the eyeball and allow the image to be focused on the fovea of the central retina even during head and body movements. Generally, during natural viewing behavior, sequences of fixations, holding the image of the surrounding stable on the retina, are interwoven with saccades, fast movements of the eyes, rapidly shifting the direction of gaze. An important function of saccades is to scan for interesting objects on which the visual attention will be focused (see Sect. 18.2.5). However, during the fast saccadic eye movements visual perception is largely suppressed, segmenting the visual input into discontinuous samples (in humans on average two per second). In addition to saccades, primates (but probably not many other vertebrates) voluntarily use smooth pursuit, stabilizing the picture of a moving object on the retina by following the object with the gaze. In addition to the voluntary eye movements, the eyes are frequently moved involuntarily and usually independent in both eyes by microsaccades (involuntary saccades with a small amplitude), drift (slow movements between microsaccades), and tremor (fast, wave-like movements with tiny amplitudes). During self-motion, the retinal picture of the environment is stabilized between saccades by **oculomotor reflexes**. For instance, the **vestibular-ocular reflex** is induced by head movements and causes smooth eye movements in the opposite direction to stabilize gaze.

### 18.2.2 Signal Transduction in Photoreceptors

When a retinal molecule absorbs a photon, the vertebrate phototransduction cascade leads to hyperpolarization of the photoreceptor membrane and to decreased transmitter release.

The first step of vision is phototransduction, the conversion of a physical light stimulus into electrical signals triggered by photosensitive visual pigments, which are contained in the outer segments of vertebrate photoreceptors (see Fig. 18.2). Visual pigments consist of two parts: the light-sensitive chromophore retinal (a derivative of vitamin A) and opsin (a protein consisting of ~380 amino acids, which is not light-sensitive by itself). Opsins differ between photoreceptor types (e.g., rhodopsin in rods) and determine their spectral sensitivity (see Box 18.1).

The absorption of a photon by the retinal molecule leads to a conformational change from 11-*cis*-retinal to all-*trans*-retinal. Since this isoform of retinal does not fit into the opsin-binding pocket, opsin changes to the semistable conformation metarhodopsin II (see Fig. 18.3). This molecule catalyzes the activation of hundreds of G-protein molecules by exchanging a tightly bound GDP (guanosine diphosphate) molecule with a GTP (guanosine triphosphate) molecule before it splits into opsin and all-*trans* retinal. In vertebrate rods, all-*trans* retinal is transported to the pigment epithelium where it is re-isomerized and sent back to the photoreceptor. Cones transport retinal to Müller glia cells (see Sect. 18.2.3.2) for re-isomerization.

After the step of G-protein activation, the transduction cascades of vertebrates and invertebrates differ considerably (see Fig. 18.3 and Sect. 18.4.1.3) [12]. In vertebrates, each of the regulatory G-protein transducins activate a phosphodiesterase (PDE) molecule, which causes hydrolysis of up to 1,000 molecules of cGMP (cyclic guanosine 3′–5′ monophosphate) per second in the cytoplasm. In darkness, cGMP-gated ion channels are constantly open, allowing $Na^+$ and $Ca^{2+}$ to enter the outer segment. This steady inward current (called **dark current**) depolarizes the membrane and causes continuous glutamate release. When phototransduction leads to a decrease in cGMP concentration, cGMP-gated ion channels close, reducing the cation influx. Hence, the membrane hyperpolarizes and the glutamate release decreases in response to light.

Adaptation tunes the sensitivity of the visual system to the current range of ambient light intensities. The most important mechanisms of light/dark adaptation are located in the photoreceptors.

When we enter a dark room coming from bright sunlight, we are not able to see details for several minutes. Even though the human visual system is able to react to more than ten orders of magnitude of light intensities, at any given instance of time the dynamic

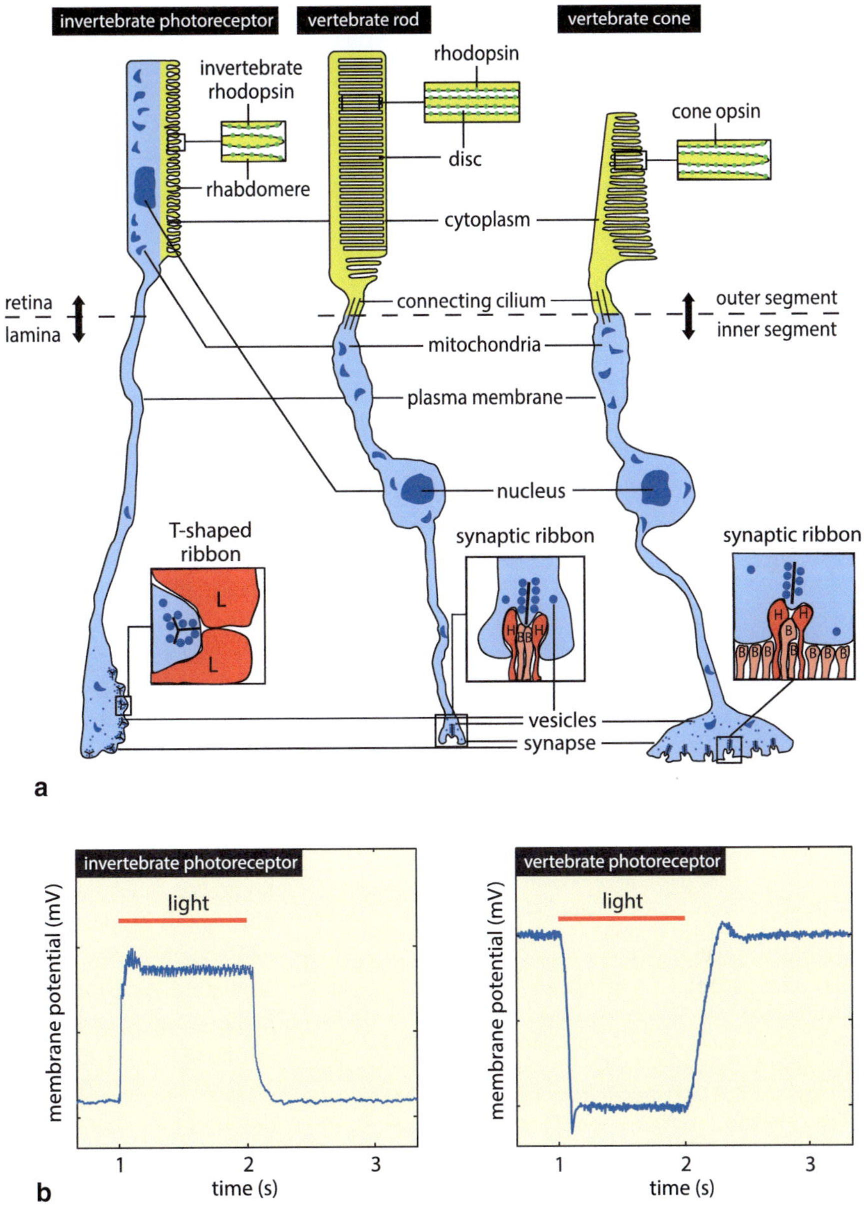


**Fig. 18.2** **Photoreceptors.** (**a**) Sketch of insect microvillar photoreceptor (*left*), and vertebrate ciliar receptors rod (*middle*) and cone (*right*). Specialized photosensitive membrane regions with embedded opsins (see *upper insets*) are shown in *yellow*. *Lower insets* show highly specialized ribbon synapses transferring signals to postsynaptic neurons, e.g., to lamina cells (*L*) in flies and to horizontal cells (*H*) and bipolar cells (*B*) in vertebrates. (**b**) Sketch of electrical responses of invertebrate (*left*) and vertebrate (*right*) photoreceptors. Invertebrate photoreceptors respond to light with depolarization, vertebrate photoreceptors with hyperpolarization (Thanks to Sirko Straube for his help with the artwork)

range of the photoreceptors is much more limited. Several mechanisms are employed to adapt the sensitivity of the visual system to the current range of ambient light intensities. While the pupillary light reflex controlling the amount of light reaching the retina accounts for only one order of magnitude of light intensity, adaptation of human photoreceptor responses covers up to nine orders of magnitudes of light intensities. An important mechanism is response compression due to a reduced gain of the photoreceptor response. When high light intensities cause strong hyperpolarization, the dynamic range to react to even brighter light is reduced, because hyperpolarization cannot exceed a maximum of –70 mV given by the $K^+$ equilibrium potential (see Chap. 7). The second important factor of photoreceptor adaptation is the conformational change (called bleaching, because rhodopsin loses its color) from 11-*cis*-retinal to all-*trans*-retinal. When a large percentage of photopigments is bleached by high light intensities, they first need to be

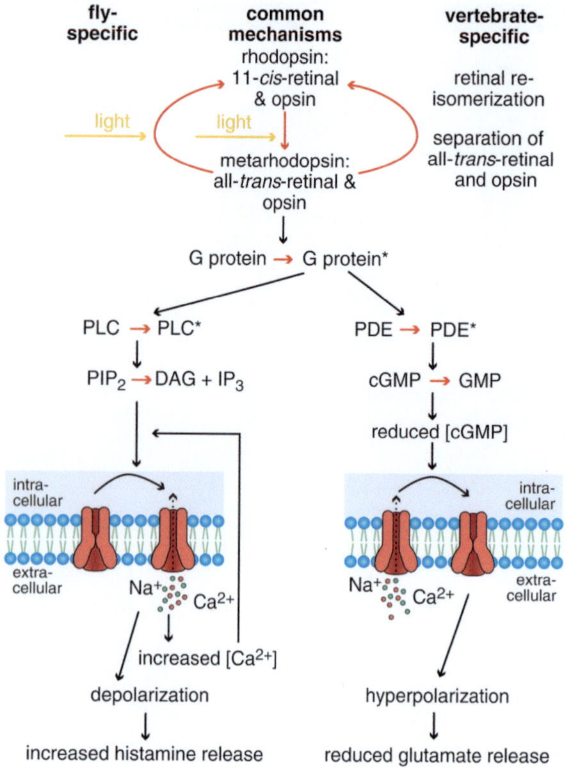


**Fig. 18.3 Main steps of phototransduction cascade in flies and vertebrates.** While the first steps of the transduction cascade are shared among vertebrates and invertebrates, the cascade following the activation of a G protein differs considerably and leads in invertebrates to light-induced membrane depolarization and increased transmitter release, but in vertebrates to hyperpolarization and reduced transmitter release (see text in Sects. 18.2.2 and 18.4.1.3 for details). *Arrows*: *black*: chemical reactions and conformational changes, *red*: causal steps, e.g., catalysis, *yellow*: light triggers photoreaction, *dotted*: ion influx into cell. Abbreviations: *PLC* phospholipase C, $PIP_2$ phosphatidyl inositol, *DAG* diacylglycerol, $IP_3$ inositol-3-phosphate, *PDE* phosphodiesterase, *cGMP* cyclic guanosine-3′-5′-monophosphate, * denotes activated molecules, [ ] denotes intracellular concentration

re-isomerized by the pigment epithelium before they are available for phototransduction again, leading to reduced sensitivity. Thirdly, several steps of the phototransduction cascade depend on the level of intracellular calcium, which reduces in the light due to the closing of cGMP-gated ion channels. The retinal network further enhances light adaptation, e.g., feedback from horizontal cells to photoreceptors depends on light intensity (see Sect. 18.2.3.2).

## 18.2.3 Primate Retina

### 18.2.3.1 Anatomy

> The vertebrate retina consists of seven major cell classes and is structured into clearly separated layers.

Santiago Ramón y Cajal (1852–1934) was the first person to stain retina sections with the Golgi method [34]. His beautiful drawings revealed an extremely clear structure that inspired him to postulate two major concepts of neuroscience, for which he received the Nobel Price in 1906: Firstly, nervous tissue consists of separated neurons; secondly, neurons transmit information from their dendrites to their axons and towards other neurons. He found that the vertebrate retina is arranged in clearly visible retinal layers (see Fig. 18.4) consisting of only six major classes of nerve cells (rod and cone photoreceptors, horizontal cells, bipolar cells, amacrine cells, ganglion cells) and two types of glia cells (Müller cells and astrocytes). The three nuclear layers of the retina contain the cell bodies. Photoreceptor somata are located in the outer nuclear layer, the inner nuclear layer contains the somata of horizontal, bipolar, and amacrine cells, the ganglion cell somata form the ganglion cell layer. In the well-structured plexiform layers, the different cell types make synaptic contacts. In the outer plexiform layer, output synapses of photoreceptors contact horizontal and bipolar cells. Bipolar, amacrine, and ganglion cells stratify in the inner plexiform layer, consisting of several sublayers (see Fig. 18.4). The axons of all ganglion cells are directed towards the blind spot of the retina, where they leave the eye and draw towards the brain.

When looking at the anatomy of the retina in more detail, however, a more complicated picture arises. Currently, approximately 80 cell types are known in the mammalian retina [28], with anatomical and physiological cell properties correlating well in most cases. This diversity allows the specialization of certain cell types to visual features like light intensity, color, and movement, which are processed in parallel pathways [45].

### 18.2.3.2 Retinal Cell Types

> Cone photoreceptors react to bright light of specific wavelengths, rod photoreceptors are specialized to low light levels. Together, both photoreceptor types cover a large range of light intensities of at least ten orders of magnitude.

**Cones** provide the basis for fast, high-acuity vision and color perception under daylight conditions. Cones are able to adapt to a wide range (nine orders of magnitude) of light intensities. In humans, cone-dominated photopic (high light) conditions range from 1 to $10^6$ cd/m$^2$. Rods and cones are both active in the mesopic range from $10^{-2}$ to 1 cd/m$^2$.

Cones possess visual pigment molecules that are embedded in infolded sections of the cell membrane in the outer segments (see Fig. 18.2a). In primates, cones express three different opsins with different spectral absorption maxima (see Box 18.1), leading to three types of cones: S-cones (short wavelengths), M-cones (medium), and L-cones (long). Even though most of the primate retina is clearly rod-dominated (human: 120,000,000 rods compared to 6,400,000 cones), the fovea contains exclusively cones.

In response to light, the cone membrane potential hyperpolarizes (see Fig. 18.2b), reducing the continuous release of the transmitter glutamate. The cone pedicle (see inset in Fig. 18.2a) is probably the most complex synapse in the nervous system, allowing to rapidly react to graded changes of the membrane potential with up- or downregulation of the continuous transmitter release. This **ribbon synapse** contains ~30 presynaptic **ribbons**, specialized membrane structures tethering 100 or more synaptic vesicles. Peripheral cones make up to 500 contacts with over 100 bipolar and horizontal cells, while cones located in the fovea have substantially fewer postsynaptic targets. Moreover, cone pedicles are coupled electrically to neighboring cones and rods via gap junctions (see Chap. 8). Inhibitory feedback from horizontal cells depolarizes the membrane of the cone pedicle when neighboring cones respond to light, resulting in a concentric **receptive field** (see Box 18.2).

**Rods** are specialized to low-light vision, covering the purely rod-dominated scotopic range from $10^{-5}$ to $10^{-2}$ cd/m$^2$, and the mesopic range of $10^{-2}$ to 1 cd/m$^2$. When higher light intensities are present, rod responses

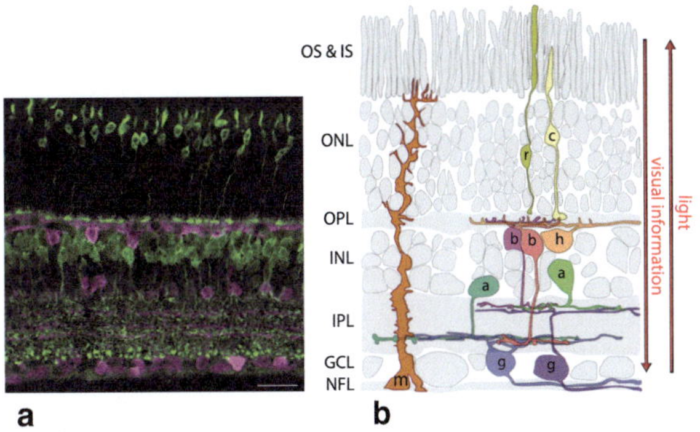


**Fig. 18.4** **Anatomy of vertebrate retina.** (**a**) Double staining of mouse retina. Cone photoreceptors and bipolar cells are stained *green*, horizontal cells, amacrine cells, and ganglion cells appear *purple*. Scale = 25 µm (Picture kindly provided by Silke Haverkamp, 2009, unpublished data). (**b**) Drawing of a retinal cross-section derived from two-photon laser scanning fluorescence microscope analysis of mouse retina to illustrate layering (*NFL* nerve fiber layer, *GCL* ganglion cell layer, *IPL* inner plexiform layer, *INL* inner nuclear layer, *OPL* outer plexiform layer, *ONL* outer nuclear layer, *OS&IS* outer and inner segments of photoreceptors) and main cell classes (*g* ganglion cells, *a* amacrine cells, *h* horizontal cell, *b* bipolar cells, *r* rod photoreceptor, *c* cone photoreceptor, *m* Müller cell) (Kindly provided by Thomas Euler, modified from [11] with permission)

saturate to maximum hyperpolarization and transmitter release stops. Rods respond much slower to light than cones and also adapt their sensitivity more slowly. However, after 30 min of dark adaptation rods are sensitive enough to detect single light quanta.

Rods have larger outer segments than cones (see Fig. 18.2a). Each rod contains stacks of several hundred isolated membrane disks packed with rhodopsin molecules (50 million rhodopsin molecules in a human rod). These disks are constantly renewed by adding disks at the base of the outer segment and displacing old disks for phagocytosis by the pigment epithelium.

The rod synapses, called spherules, also continuously release glutamate. They contain two presynaptic ribbons and make contact to two rod bipolar cells and two horizontal cells.

> Bipolar cells establish the separation into ON and OFF visual pathways and transmit photoreceptor signals directly to the ganglion cells.

**Bipolar cells** transfer signals vertically from the photoreceptor input layer to the ganglion cell output layer via excitatory synapses. They can be classified into two main types which establish the separation into ON and OFF visual pathways: OFF-center cone bipolar cells express the ionotropic glutamate receptors AMPA and kainate and hyperpolarize when presynaptic photoreceptors hyperpolarize in response to light. In contrast, ON-center cone bipolar cells respond with depolarization to light, because their metabotropic glutamate receptor mGluR6 inverts the sign of the synaptic potentials. ON- and OFF-center bipolar cells establish the separation into ON and OFF visual pathways, which are characteristic for the entire visual system of vertebrates.

Depending on their anatomy and physiology, human bipolar cells can be further classified into ten types receiving input from 1 to 20 cones and one type (ON-center rod bipolar) contacting 15–50 rods. Bipolar cells have circular receptive fields, consisting of center and surround, providing the basis for the ganglion cell receptive field structure (see Box 18.2).

Horizontal cells and amacrine cells provide lateral interaction of light responses.

**Box 18.1 Color Vision**

The perception of luminance differences is sufficient for most behavioral tasks. However, color vision introduces an additional perceptional dimension. Some objects can better be detected and discriminated based on the wavelength composition of their light reflection. For example, a red fruit between green leaves can be hard to detect solely based on luminance, but clearly stands out for observers who are able to perceive color differences.

Color perception is based on the responses of photoreceptors which respond specifically in a certain range of light wavelengths. The visual pigment based on the opsin molecule expressed by a photoreceptor determines the spectral range the cell is sensitive to. In the figure, the sensitivity curves show the relative probability that a photon of a given wavelength elicits phototransduction in the photoreceptor. When many photons are present, the sensitivity curves also determine the amplitude of the photoreceptor response. The *top row* of the figure shows typical types of sensitivity curves: Most vertebrate phyla are tetrachromats. They express four opsins in different types of cones, usually covering the spectral range from ~350 to ~650 nm. Many insects are trichromats (expressing three opsins for color vision) with stronger sensitivity in the short wavelengths range (UV light, 300–400 nm). Mammals have lost two of the four vertebrate opsins in evolution. Hence, they are usually dichromats with only one type of green/red-sensitive cones and a lower number (~10 %) of blue-cones. Only few mammals, like Old-World primates, have regained a third opsin for trichromatic color vision during evolution, leading to the unusually similar absorption maxima of L- (long wavelength) and M- (medium wavelength) cones (in human 564 and 533 nm) in addition to the S- (short wavelength) cone (absorption maximum 437 nm). The panels in the *bottom row* of the figure show three specific examples of spectral sensitivity of different photoreceptors involved in color vision (*black lines*, in primate cones), together with the spectral sensitivity of a wavelength-unspecific photoreceptor type (*dashed lines*, in primate rod).

The single photoreceptor is color-blind. For instance, a medium response amplitude of a cone could be elicited either by a relatively low intensity of the optimal wavelength or by a high intensity of a suboptimal wavelength. For the discrimination of colors, it is necessary to combine the responses of photoreceptors with overlapping spectral sensitivity curves. Their relative responses are evaluated in subsequent steps of visual information processing. Both in vertebrates and invertebrates, postsynaptic neurons have color-opponent receptive fields responding in their center to one wavelength with increased and to another with decreased neuronal

**Horizontal cell** dendrites contact several cone pedicles in the outer plexiform layer. Since horizontal cells of the same type are electrically coupled via gap junctions, their network establishes large-field integration of photoreceptor signals. In humans, three types of horizontal cells were found. Horizontal cells contact rods and cones, receiving synaptic input and providing electrical and chemical feedback inhibition to photoreceptors. When a photoreceptor hyperpolarizes in response to light, the horizontal cell also hyperpolarizes. The negative feedback onto neighboring photoreceptors depolarizes the photoreceptor. This adjustment of the photoreceptor response gain leads to contrast enhancement and shapes the center-surround receptive field of the postsynaptic bipolar cells. In addition, direct feedforward input from horizontal cells to bipolar cells was shown in some vertebrates. Depending on the light level, the electric coupling strength of the horizontal cell network, as well as the lateral inhibition of photoreceptors change dynamically. Therefore, horizontal cause variable sizes of the receptive field surrounds in bipolar and ganglion cells depending on the light level.

In the inner plexiform layer, **amacrine cells** provide an abundance of lateral and feedback interactions between bipolar cells and ganglion cells. Coarsely, amacrine cells are classified into narrow-field, small-field, and wide-field cells, corresponding to their dendritic field diameters. The number of different

**Box 18.1 Color Vision** *(continued)*

activity, as can be seen, e.g., in primate retinal ganglion cells (see Box 18.2 and Sect. 18.2.3.2). Dichromats are not able to discriminate many colors, because many different mixtures of wavelengths lead to the same activation pattern of the two types of photoreceptors.

Color perception, at least in humans and primates, is not unambiguously determined by cone activation. Several high-level computations introduce context information and prior experience into color sensation. They lead, e.g., to the phenomenon of color constancy, the ability to perceive the color of an object as constant under varying illumination, even though the actual composition of the reflected wavelengths changes drastically. (Figure after Dowling [8] and modified from Kelber in [21, 22] after [26, 31]).

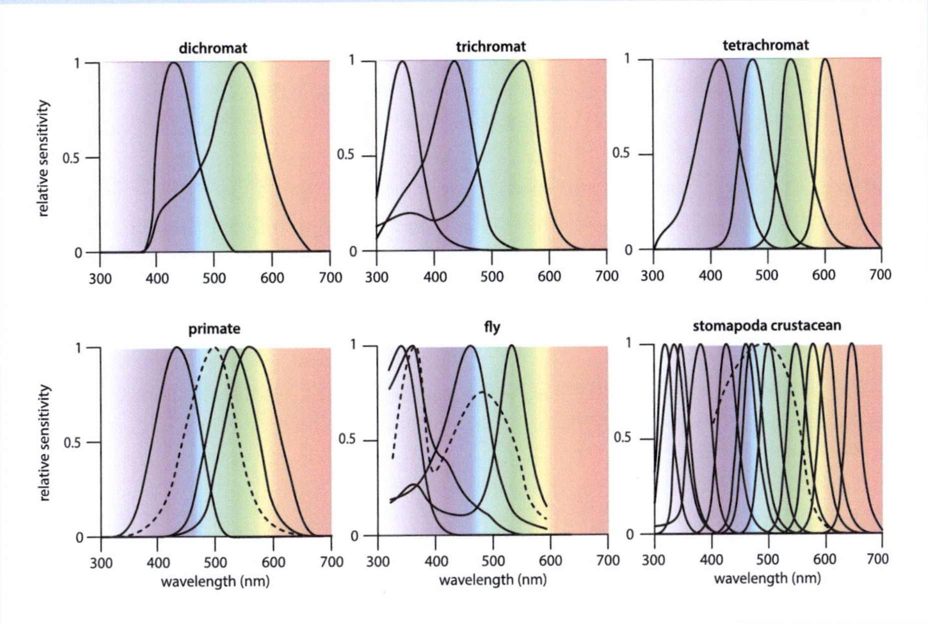

amacrine cell types is still a matter of debate, with estimations ranging from 20 to 50.

Amacrine cells integrate and modulate signals representing light stimuli. Most amacrine cells are connected via an extensive serial inhibitory network. Amacrine cells can have complex receptive field structures and nonlinear response properties. All three coarse types of amacrine cells concertedly shape the synaptic transmission from bipolar cells to ganglion cells by providing inhibitory glycinergic or GABAergic feedback to bipolar synaptic terminals. Thereby, they drastically influence the temporal structure of ganglion cell responses, providing the basis for complex visual functions like detection of wide-field and object motion, as well as local edges and looming stimuli.

Examples for well-studied amacrine cells are AII (a key player in the rod pathway; see 18.2.3.3) and the starburst amacrine cell (involved in processing of motion information) (see Box 18.7).

**Box 18.2 Receptive Fields**

The receptive field of a neuron in the visual system is the visual space, in which light influences the responses of the neuron under study. In the year 1938, Hartline (1903–1983) defined the receptive field of a retinal ganglion cell: "Responses can be obtained in a given optic nerve fiber only upon illumination of a certain restricted region of the retina, termed the receptive field of the fiber" [16]. In 1953, using small light spots for stimulation, Kuffler (1913–1980) found that the receptive fields of retinal ganglion cells usually have a center-surround organization [25], depicted in panel (**a**) of the figure. *Light gray* areas show ON-responsive regions of the receptive field, and *dark gray* areas OFF-responsive regions. On-center retinal ganglion cells increase the number of spikes (symbolized by *black dots*) when a small light spot (shown in *yellow*) shines on the receptive field center (panel a *top row left*) and decrease the spike frequency when the suppressive surround is activated by light (panel a, *bottom row left*). Therefore, the optimal stimulus to elicit strong responses of an ON-center retinal ganglion cell is a small bright spot on dark background (e.g., a star). OFF-center cells show the opposite response characteristics (panel a, *right*). Additionally, many retinal ganglion cells show an antagonistic receptive field structure for light of different wavelength, e.g., increasing activity when red light shines on the receptive field center and green light on the surround. (The optimal stimulus for such a cell could be a red fruit between green leaves).

The concentric receptive field structure originates at the level of cone responses. Horizontal cells provide inhibitory feedback, causing the cone membrane to depolarize when neighboring cones are hyperpolarized by light. In consequence, each individual cone has a center-surround receptive field. Depending on the membrane potential at the cone pedicle the amount of transmitter release can be regulated gradually up and down. Based on this mechanism, the receptive field structure is passed to the postsynaptic bipolar cells and then to the downstream ganglion cells, both of which use additional mechanisms to further shape their receptive fields (see Sect. 18.2.3.2 and 18.2.3.3). Along the visual pathway, receptive fields increase in size and complexity (compare Sects. 18.2.3, 18.2.4, and 18.2.5). While responses of thalamic neurons strongly resemble the small circular center-surround receptive fields of ganglion cells, the receptive fields of cortical neurons are much more diverse. In 1962 Hubel (*1926) and Wiesel (*1924) introduced the terms **simple cells** and **complex cells** [18]. Simple cells possess an elongated receptive field that is separated into ON- and OFF-regions, and therefore respond optimally to a bar of light with a specific orientation. In contrast, complex cell receptive fields do not have separate subregions. They respond optimally to movement in specific directions, but not necessarily at specific positions.

For a complete picture, the spatial receptive field structure should be complemented with the temporal aspects of stimulus changes, leading to spatiotemporal receptive fields. Temporal aspects of receptive fields are particularly important for cells responding specifically to moving stimuli, e.g., direction-selective cells (see Box 18.7), but also reveal additional information about cells with classical receptive fields.

Panel (**b**) shows the spatiotemporal receptive field of a macaque parasol ON-RGC. The retina was stimulated with a sequence of random black and white

In the primate retina, ganglion cells with different response characteristics establish the magnocellular, parvocellular, and koniocellular pathways for achromatic, red/green, and blue/yellow vision.

**Ganglion cells** are responsible for the retinal output by firing the action potentials, which travel along the optic nerve to the brain.

Resulting from bipolar cell inputs, most ganglion cells possess circular antagonistic receptive fields (see Box 18.2) and are either excited by stimuli that are brighter than the background (ON-center ganglion cells) or by stimuli darker than the background (OFF-center ganglion cells). These major physiological types can also be distinguished morphologically, because OFF-center cells stratify in the outer sublamina of the inner plexiform layer, where they receive input from OFF-bipolar cells, while the inner

**Box 18.2 Receptive Fields** (*continued*)

checkerboard patterns (*left*). Spikes elicited by this stimulus sequence (shown as *dots* below the stimulus sequence) are used to calculate the spike triggered average stimulus (see Box 18.4) for each preceding stimulus frame, showing how the spatial properties of the receptive field change in time (*middle*). For each pixel the time course of the optimal stimulus contrast can be calculated as it is shown here for the receptive field center (*right*). The optimal stimulus to trigger a spike in this cell is a dark spot in the center of the receptive field ~100 ms before spike occurrence that changes into a bright spot with a dark annulus ~50 ms before spike occurrence (corresponding to the classical picture of an ON-center cell). (Panel b kindly provided by Martin Greschner and EJ Chichilnisky, "unpublished ").

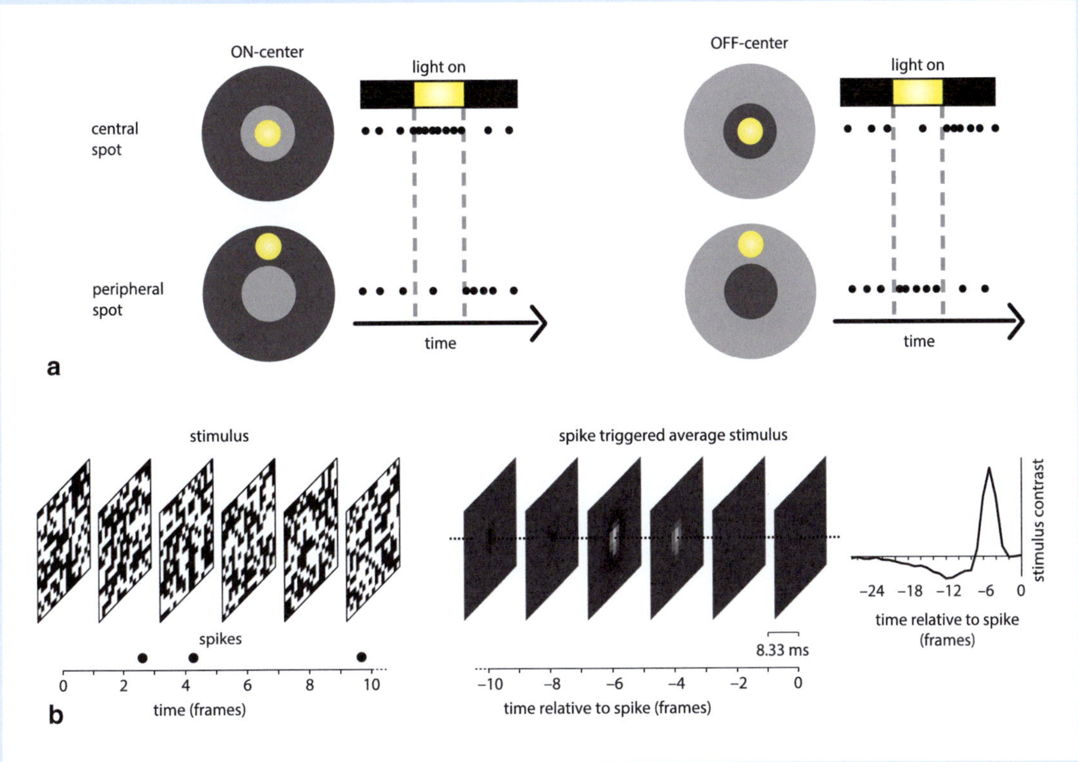

sublamina provides ON input. The morphological dendritic field size corresponds to the physiological receptive field center size. In total, there are 10–15 different morphological ganglion cell types in any mammalian retina, presumably corresponding to different physiological types and representing separate information channels to the brain [45]. Here, we concentrate on the main ganglion cell types of the primate retina providing the basis for the different pathways of visual information processing in the brain (compare Fig. 18.5):

**Parasol ganglion cells** in the primate retina receive mixed input from rods and cones and establish the magnocellular (M) pathway (see Sect. 18.2.4) for achromatic vision. These large cells probably provide signals to initiate visual attention, because their vigorous transient ON- or OFF-center responses to movement and to rapid changes in light intensity are transmitted to the brain very fast by their thick axons. Parasol ganglion cell axons contact several other brain regions (e.g., superior colliculus and pretectum) in addition to magnocellular (M) layers of the lateral geniculate nucleus (LGN) (see Sect. 18.2.4).

**Midget ganglion cells** are the most common RGC type in primates (~80 %). They receive their inputs from midget bipolar cells and transfer their outputs to the parvocellular (P) layers of the LGN (see Sect. 18.2.4). Midget cells respond in a sustained way to small, high-contrast, stationary stimuli. Their circular receptive field are red/green antagonistic and can have an ON or an OFF center. It remains controversial whether the parvocellular pathway is influenced by rod responses.

**Small bistratified ganglion cells** form the morphological basis of the primate koniocellular (K) blue-yellow pathway (see Sect. 18.2.4). Blue light signals are transferred usually from a single S-cone via an S-cone bipolar cell to the blue-ON-center of the receptive field. The inhibitory surround combines inputs from L- and M-cone signals, resulting in a high sensitivity for yellow light. The opposite type of receptive field structure is found only occasionally.

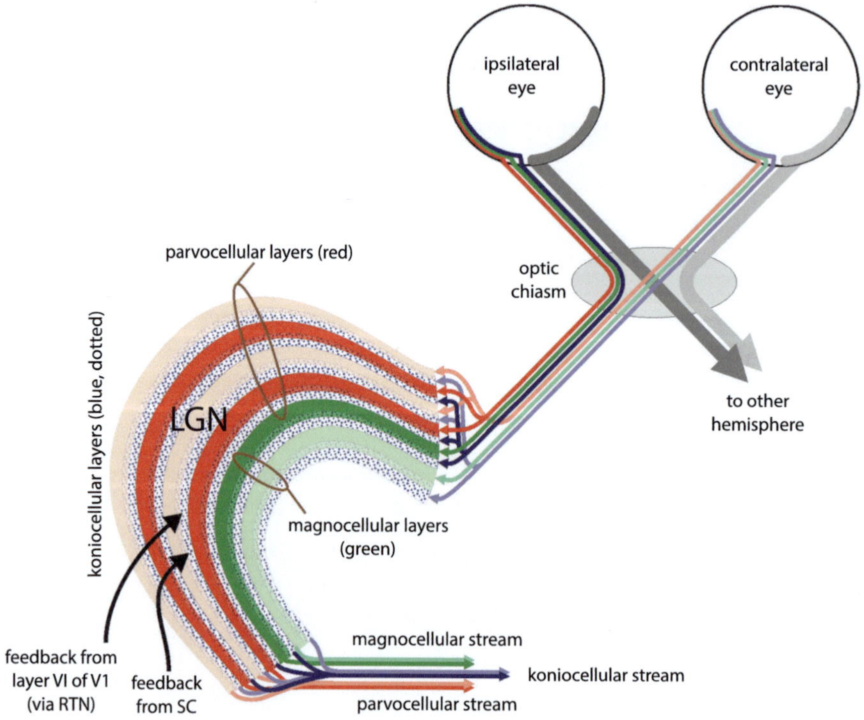


**Fig. 18.5 Sketch of the lateral geniculate nucleus** (*LGN*) in the left brain hemisphere, which receives information from the right visual hemifield. This information is collected in the left halves of the retinas in the ipsi- and contralateral eye, and rerouted via the optic chiasm. Streams from different ganglion cell types (*red*: midget, *green*: parasol, *blue*: bistratified) and different eyes (*light* and *dark* coloring) remain segregated in the six LGN main layers, and are subsequently relayed to visual cortex. Feedback projections from SC and from the visual cortex via the reticular nucleus (RTN) strongly modulate information processing in the LGN. In the right hemisphere, a similar structure processes information from the left visual hemifield (not shown)

**Intrinsically photosensitive retinal ganglion cells** occur only rarely (primates ~0.2 %). The putative photopigment melanopsin makes them directly sensitive to light, and they also respond to conventional photoreceptor input. These ganglion cells are not used for image-forming vision, but project predominantly to the suprachiasmatic nucleus, entraining the circadian rhythm, and to the pretectum, regulating pupil constriction.

**Müller cells** are glia cells spanning the entire retina. They play important roles by ensheathing the neuronal processes, maintaining the homeostasis of the retinal extracellular milieu, recycling neurotransmitters, and re-isomerizing cone retinal. Moreover, they are discussed to act as optical fibers to guide the incoming light through the retina to the photoreceptors. In the axon layer of the retinal ganglion cells, **astrocytes** are additionally involved in the control of the homeostasis.

The **retinal pigment epithelium** is located between the blood vessels of the choroid and the outer segments of the photoreceptors. The heavily pigmented cells absorb scattered light and form the blood/retina barrier with a highly selective transport between blood and retina. The cells re-isomerize rod retinal and phagocytose discarded pieces of photoreceptor outer segments that were damaged by photo-oxidation.

#### 18.2.3.3 Retinal Pathways

> In primates the cone pathway provides most of the visual input to the brain. The rod pathway is the high-sensitivity system of vision used in low light conditions.

When **cones** respond to light stimuli, ON- and OFF-cone bipolar cells receive synaptic input, causing their membrane potentials to de- or hyperpolarize. They transmit visual information to all kinds of ganglion cells described above. In addition, lateral interactions by horizontal cells and amacrine cells strongly influence the spatial and temporal structure of RGC receptive fields (see Box 18.2). Depending on the mixture of inputs from different cones, RGC responses are more or less color specific.

Specifically in the fovea of primates, the receptive field center of each midget ganglion cell receives input from a single midget bipolar cell, which contacts only a single pedicle of an L- or M-cone. This private line from the photoreceptor to the brain provides excellent visual acuity. In peripheral regions of the retina, midget ganglion cells have much larger dendritic fields and integrate synaptic inputs from several cones.

**Rods** transmit their signals to specific rod bipolar cells, all of which are ON cells. Each rod bipolar cell integrates inputs from several rod spherules. However, rod bipolar cells do not contact ganglion cells directly, but synapse with a specialized amacrine cell type called AII. These amacrine cells integrate inputs from several rod bipolar cells and transfer their signals via gap junctions to the axon terminals of ON-cone bipolar cells and via inhibitory synapses onto OFF-cone bipolar cells. Recent findings propose an additional route of rod information processing, showing that some OFF-cone bipolar cells contact rod spherules directly [45]. There is good evidence that this high sensitivity system developed later in evolution, because it reuses parts of the cone pathway.

#### 18.2.3.4 Principles of Retinal Information Processing

All visual information is transmitted via the optic nerve to the brain. However, the number of photoreceptors (~120 million in the human retina) vastly exceeds the number of ganglion cells (~1 million). The retinal network needs to efficiently encode visual information by the RGC population to minimize the negative effects of this information bottleneck. Key to achieving this goal are the center-surround receptive fields, which suppress responses to spatially or temporally homogeneous parts in the retinal image, and enhance responses to features like edges, corners, and other inhomogeneities in the image content.

> Adaptation adjusts the sensitivity of the retinal network to the surrounding light intensity. Spatiotemporal receptive field properties show adaption also to higher stimulus statistics like contrast, pattern, and motion of the stimulus.

The brain does not recover the original visual image, but mainly its behaviorally relevant aspects. Therefore, the retina reduces information by reporting only the latest news: RGCs respond to stimulus changes rather than to constant features. This strategy is similar to data compression methods for movies on a DVD, which stores only the temporal changes between subsequent frames, thus greatly reducing data size.

In addition to the photoreceptor adaptation, several mechanisms of network adaptation shift the dynamic range of most RGC responses in a way that stimulus contrast (the relative ratio of light intensities) is encoded more or less independently from the absolute level of light intensity. Moreover, RGCs responses were shown to adapt to higher stimulus statistics like contrast, pattern, and motion of the stimulus. Such temporal receptive field structures are visible in the spike-triggered average of stimulus properties (see Boxes 18.2 and 18.4). Adaptation affects the visual perception so strongly that constant stimuli can only be perceived when eye movements lead to continuously changing retinal stimulation. When eye movements are blocked pharmacologically, perception fades away.

The retina processes several properties of visual stimuli in parallel and sends this information to the brain via several information channels. Ganglion cells, the output elements of the retina, are specialized to certain features of light stimuli (e.g., contrast, wavelength, size, movement). Each of the different ganglion cell types of the mammalian retina covers the entire retina with their dendritic trees. While there is only marginal overlap of receptive fields between ganglion cells of the same type, each point in space is represented by several parallel channels of information processing represented by the different ganglion cell types.

Visual information reaches the brain encoded in a distributed way by the combination of several RGC

**Box 18.3 LNP Models**

LNP models are a mathematically simple, but very successful tool to understand the emergence of linear receptive field properties in cortex; see [1]. The acronym LNP describes the components of such a model: a **L**inear summation stage, a point **N**onlinearity, and finally the conversion of the output firing rate into a **P**oissonian spike train. LNP models can characterize response properties of neurons in the retina and in early visual areas (LGN and primary visual cortex) extremely well, but are less useful for understanding receptive fields with stronger nonlinearities.

The input to a LNP model is a time-varying visual stimulus $S(x, y, t)$, for example, a movie displayed on a computer screen. In such a case S would be the luminance of a pixel at spatial coordinates $(x, y)$ at time $t$ (one frame $S(x, y, t - \tau_0)$ is shown in *black and white*). The initial stage of a LNP model computes the product between the pixel values S and a weighting function $w(x, y, \tau)$ (the 'response kernel'), and sums the results over space and time (*green arrows*, graphs and formulas). $\tau$ denotes the time the response of the model is delayed relative to the onset of a stimulus, which is in the order of 10–100 ms for neurons in the retina or in the primary visual areas (compare Box 18.2).

$$I(t) = \sum_x \sum_y \sum_\tau S(x, y, t-\tau)\omega(x, y, \tau)$$

The total input $I(t)$ is then passed through a point nonlinearity g giving the output firing rate $f = g(I)$ (*blue* graph and *arrow*). This firing rate is converted into a spike train by a Poissonian point process. A Poisson process is a random process where the probability for an event (here, the emission of a spike) does not depend on previous events. The probability $P(n = k \mid \Delta t, f)$ to observe $k$ spikes in a time interval $\Delta t$, given a firing rate $f$, is

$$P(n = k \mid \Delta t, f) = \frac{(f \Delta t)^k}{k!} \exp(-f \Delta t)$$

The LNP model has a simple biophysical interpretation: a neuron sums over all its synaptic inputs which carry time-delayed and weighted information about the stimulus. Then it converts the total synaptic current into a (stochastic) output spike train, whose mean rate is determined by the cell's gain function. The dependence on $\tau$ is used (1) to model filter processes (for example, in synaptic transmission or dendritic conductance) which spread an instantaneous change in the stimulus over time, and (2) to model the delays caused by previous processing stages.

The response kernel $w$ can be estimated by computing the spike-triggered average from spike trains obtained as responses to random noise stimuli (see Box 18.4). Once $w$ has been determined, LNP

responses of the same and different types. In these responses two types of temporal dynamics interact: The visual environment changes dynamically over time and the cells respond with a temporal sequence of action potentials which can be highly complex even in response to very simple stimuli. While classical studies usually investigated the spike rate of single cells, modern approaches focus on the fine temporal structure of the responses and on interactions between cell responses. For example, it was proposed that the relative timing of first spikes occurring in a retinal ganglion cell population could encode stimulus changes faster than spike rates. Moreover, one needs to keep in mind that responses of the same cell to repeated presentations of the same stimulus vary considerably (see Box 18.5 for an example). In spite of this variability, the nervous system needs to estimate the current stimulus situation based on the neuronal responses, without being able to average over several stimulus presentations. Therefore, a good way to investigate neural coding is to compare stimulus reconstruction based on different coding assumptions (see Box 18.5).

### 18.2.4 Primary Central Processing

#### 18.2.4.1 Lateral Geniculate Nucleus (LGN)

**Anatomy**. The lateral geniculate nuclei (Fig. 18.5), which are part of the **thalamus**, collect sensory information from a particular region of the visual field from both eyes, and relay these signals to the

**Box 18.3 LNP Models** (*continued*)

models explain a great proportion of the response properties of retinal or cortical neurons.

An important extension of the model was motivated by the observation that predicted responses do not generalize well to different contrast regimes in the visual input. The general idea of this extension is to feed back the output of a LNP model (*red arrows*), or the output of a group of LNP models, as a subtractive (via inhibition) or divisive signal (via shunting inhibition) to the summation stage. This mechanism implements adaptation of the model to the dynamical range of the stimulus: if the model's output is too high, the inputs are attenuated which in turn reduces the output to a more moderate level. If the output is too small, the inputs are enhanced, which in turn increases the output and brings the model back into a suitable working range.

Although we assumed a discrete space and time $\Delta t$ as is practical for application to real, sampled data, all equations can be written as integrals, being more appropriate for analytical calculations (for more information, see [1]). Both notations are used in the literature, but describe the same ideas and concepts. Combining different LNP models in multiple stages leads to GLMs (generalized linear models), which are currently subject of intense research.

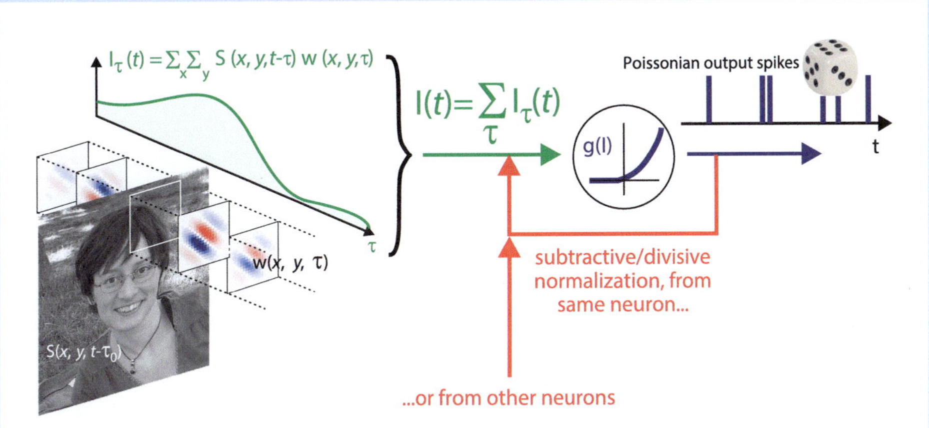

**primary visual cortex** (V1, Fig. 18.5) in the same (ipsilateral) brain hemisphere. Hereby the LGN in the left hemisphere receives only signals from the right (contralateral) visual hemifield, and the LGN in the right hemisphere processes solely signals from the left visual hemifield. For this purpose, midway between retina and LGN in the optical chiasm, fibers from the left and right visual hemifields of the retina are separated and rerouted to the right and left brain hemispheres, respectively. The LGN is divided into six main layers containing large cell bodies easily seen in Nissl stainings, separated by intermediate layers consisting of neuropil (a mixture of axons, dendrites, and extensions of glial cells) and smaller cell bodies. This structure preserves the segregation between input streams from different eyes and different types of retinal ganglion cells. While layers 2, 3, and 5 process information from the ipsilateral eye, layers 1, 4, and 6 receive input from the contralateral eye. Assignment of the LGN layers to the

**Box 18.4 Reverse Correlation**

How does one find the (linear) receptive field of a neuron?

The most popular method is to compute the 'spike-triggered average' or 'reverse correlation' [37]. Here, a sequence of stimuli in which one or several features are varied randomly is presented, and a neuron's response $r(t)$ to this stimulation is recorded (a). Commonly used features for the visual system are, for example, the orientation of a grating, the pixel color, or the pixel luminance values at positions $(x, y)$ in the visual field. For correlated feature sequences like natural images or movies, one has to compute and to invert the stimulus covariance matrix for removing spatial and/or temporal covariations between the features.

We will consider variations at the level of single image pixels in the following paragraphs. An estimate of the linear receptive field of a neuron is obtained from stimulus $S$ and response $r$ by computing the reverse correlation

$$\hat{w}(x,y,\tau) \propto \frac{1}{T} \sum_{x',y',\tau'} C^{-1}(x,y,\tau;x',y',\tau') \times \sum_{t'} S(x',y',t'-\tau')\, r(t')$$

$C^{-1}$ is the inverse of the covariance matrix C of the stimulus $S$, and $T$ the stimulus sequence length. With $E[\ldots]$ denoting the average over the ensemble of all possible stimuli S, C is given by the equation

$$C(x,y,t;x',y',t') = \mathrm{E}[(S(x,y,t)-\mathrm{E}[S(x,y,t)]) \times (S(x',y',t')-\mathrm{E}[S(x',y',t')])]$$

For sparse and checkerboard noise, spatiotemporal variations of any two pixels in the stimuli are uncorrelated. Thus $C^{-1}$ becomes the identity matrix, for which the first equation simplifies to

$$\hat{w}(x,y,\tau) \propto \frac{1}{T} \sum_{t} S(x,y,t-\tau)\, r(t)$$

An example for this procedure is shown in panel (a) (*blue curve* and expression). If the neuron's response is a spike train with $N$ spikes observed at times $t_1, t_2, \ldots, t_N$, this expression further simplifies to a sum over all stimuli preceding the spikes by a time $\tau$, which is termed the spike-triggered average (*red curve* and expression in panel (a)):

$$\hat{w}(x,y,\tau) \propto \frac{1}{N} \sum_{k} S(x,y,t_k-\tau)$$

If a neuron can be described by an LNP model (see Box 18.3), the estimate $\hat{w}$ is proportional to $w$, which together with $g$ fully specifies the cell's receptive field properties.

Receptive fields can have space-time separable or inseparable kernels $w$.

Panel (b) shows an example for a space-time separable kernel, which can be written as the product of two functions $w_x(x)$ and $w_t(t)$ depending only on space and on time, respectively. *Blue* regions correspond to OFF-zones, where a darker stimulus leads to a higher firing rate, while *red* regions correspond to ON-zones, where a brighter stimulus leads to higher firing rates.

Panel (c) shows an inseparable response kernel, which is sensitive to the direction of movement of a

different pathways from retina to thalamus is as follows:

(a) Layers 1 and 2 receive input from parasol ganglion cells, forming the **magnocellular-(M-) pathway**.
(b) Layers 3–6 receive input from midget ganglion cells, forming the **parvocellular-(P-) pathway**.
(c) All intermediate layers ventrally to the main layers, which extend into the magno- and parvocellular layers, receive input from small and large bistratified (i.e., having a two-layered structure) ganglion cells, forming the **koniocellular-(K-) pathway**.

Single LGN cells receive afferents from multiple, neighboring ganglion cells. A particular feature of the pathway from the retina via LGN to the visual cortex is that the **retinotopy** is preserved, which means that neighboring fibers or neurons carry information from neighboring locations on the retina. In addition, information from different eyes becomes represented in adjacent layers of the LGN. This convergence is a prerequisite for the detection of differences in

**Box 18.4 Reverse Correlation** (*continued*)

grating. The slope of the ON- and OFF-zones determines the preferred speed of a grating, which maximally excites the cell, whereas the spacing of the zones corresponds to the preferred spatial frequency of the neuron.

For neurons responding nonlinearly to a visual stimulus, $w$ can maximally quantify the linear part of their RFs. In extreme cases, as, e.g., for complex cells in V1 which respond preferentially to an oriented grating irrespectively of its spatial phase, reverse correlation may yield an average $w$ which is zero. It should be noted that for such nonlinear cells, a better description of their response properties may either be obtained by parameterizing their RF by a different variable or by approximating their RFs by a Volterra series expansion. Estimation of these receptive fields is done by computing the so-called Wiener kernels of second and higher orders (Wiener filters). Heuristically speaking, these kernels quantify how the response of a neuron to a combination of two or more activated pixels in a stimulus deviates from the sum of the responses to each of the pixels activated individually. In practice, estimating these kernels requires a large amount of data and is not possible in many experimental settings.

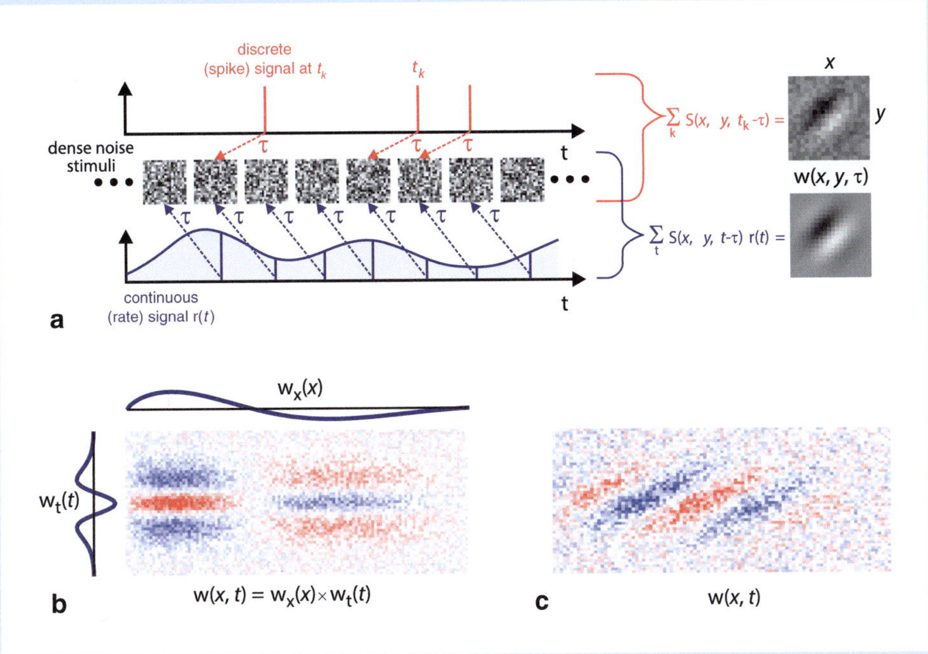

horizontal angular position ('disparities') between the visual images seen by the left and right eye, which are used in visual cortex to estimate depth for stereo vision. Information processing in LGN is not purely feed-forward, as massive feedback projections from layer VI of V1 and **superior colliculus** (**SC**) innervate this structure. These mainly inhibitory connections are routed via the thalamic reticular nucleus (TRN).

**Box 18.5 Stimulus Reconstruction**

What is the neural code used by the brain, and how are different stimuli or computations being represented in the different visual areas?

These questions can be investigated by trying to reconstruct, or to decode, stimuli from neural activities. Decoding is not only important for investigating information processing strategies in the brain, but also useful for applications: current research projects use suitable algorithms to construct brain-computer interfaces (BCIs). An application of BCIs is to use them as neural prostheses for helping patients that are partially or fully paralyzed. For example, an intended movement could be decoded from brain activity, which would then be executed by an external device like a robot arm or a wheelchair.

For understanding how decoding works, let us first consider the encoding process (**a**, *top*). In a very simplified way, encoding can be described as a transformation $T$ from a stimulus $S$ into a neural response $r$. One example for such an encoding scheme is the LNP model (see Box 18.3). Encoding is stochastic, as the exact outputs $r$ to identical stimuli $S$ can vary considerably. The source of this stochasticity is both due to noise in the transmission of information by neurons and synapses, and due to hidden variables, e.g., ongoing activity in the brain or the attentional state, which cannot be observed or controlled in the experiment. A mathematical description of the transformation $T$ and the noise model $\eta$ is also called a 'generative model' (i.e., an exact description of the probability distribution for neural responses $r$, given a particular stimulus $S$).

Decoding algorithms use the observation of neural activity $r$ to seek for the most likely stimulus $\hat{S}$ which has caused this response (**a**, *bottom*). Hereby the decoder has to 'invert' the transform $T$ under consideration of the nature of the source of noise. This process is often termed 'inference'.

A variety of decoding algorithms has been used on brain data, but there are only few main classes whose areas of application depend both on the amount of knowledge about the encoding process, and on the number of observed variables in $r$. We mention but a few:

(i) the nearest-neighbor classifier works as a lookup table on sample data. The closest match of a new observation $r$ to previous observations registered in a dictionary is computed, and the corresponding stimulus $\hat{S}$ taken as the estimate.
(ii) Bayesian estimation uses Bayes' formula to compute the conditional probability for a stimulus $S$, given the observation of a response $r$; by multiplying a prior distribution $p(S)$ with the likelihood to observe $r$, given a stimulus $S$. Bayesian estimation can be performed iteratively on successive observations r.
(iii) Support vector machines (SVMs) are nonlinear classifiers which achieve high classification performances even on noisy data. An advantage of SVMs is the availability of ready-to-use implementations, which require little knowledge about the particular mathematics behind the algorithms.

In general, when nothing about the encoding $T$ or the noise process $\eta$ is known, estimation is more difficult and the correct decoding strategy must be extracted from the data itself. Conversely, the more knowledge we have about the encoding, the more precise decoding will be, sometimes even allowing us to compute the optimal decoder explicitly.

Particular attention should be given to an appropriate preprocessing of the data: for practical purposes one should boil down a high-dimensional dataset to a manageable number of observables before applying any framework for decoding. From the observation of action potentials in some time period, one could alternatively take the mean rate, or the length of the first interspike-interval, or all the points in time where spikes have been observed, as the observation $r$. In general, there will be a

**Receptive fields and response properties**. For localized stimuli, the firing rate of an LGN cell is proportional to the logarithm of the physical stimulus contrast. The spatial shapes of the RFs are similar to the ON-OFF, center-surround structures found in retinal ganglion cells (Box 18.2 and Sect. 18.2.3). In addition, a considerable fraction of LGN cells responds maximally for a particular stimulus orientation

**Box 18.5 Stimulus Reconstruction** (*continued*)

tradeoff between loss of information and reduction in complexity of the estimation problem. Estimation performance will strongly depend on this decision, but it might also give us hints about the coding mechanisms used by the brain (in the above example, a rate code, a code considering only the first interspike interval, or a full spike-timing code, respectively).

Panels (b) and (c) show an experiment and stimulus reconstruction to analyze motion processing in the turtle retina. To stimulate the retina, a dot pattern was moved horizontally with one of nine different velocities, which changed abruptly every 500 ms (**b**, *bottom graph*). The two upper graphs of (**b**) display simultaneously recorded responses of two cells (*purple*: DS cell preferring movement to the left, shown as negative velocities, *mauve*: DS cell preferring movement to the right, shown as positive velocities). Each *dash* represents the occurrence of an action potential. Responses to seven subsequent, identical stimulus presentations are shown underneath each other, indicating the variability of neuronal responses. (**c**) Reconstruction of the motion stimulus based on the responses of the individual cells (*purple*, *mauve*) and of the combined responses of the two cells (*purple-mauve-dashed*) in comparison to the actual stimulus that elicited the responses (*light gray*). Reconstruction was done with metric-based clustering using spike count tuning functions derived from many (560) presentations of the constant velocity stimuli. The combined responses of two cells which are specialized to represent different aspects of motion stimuli already allow a very exact reconstruction of direction and speed of the motion stimulus. (Data for panels (b) and (c) kindly provided by León M. Juarez Paz, unpublished).

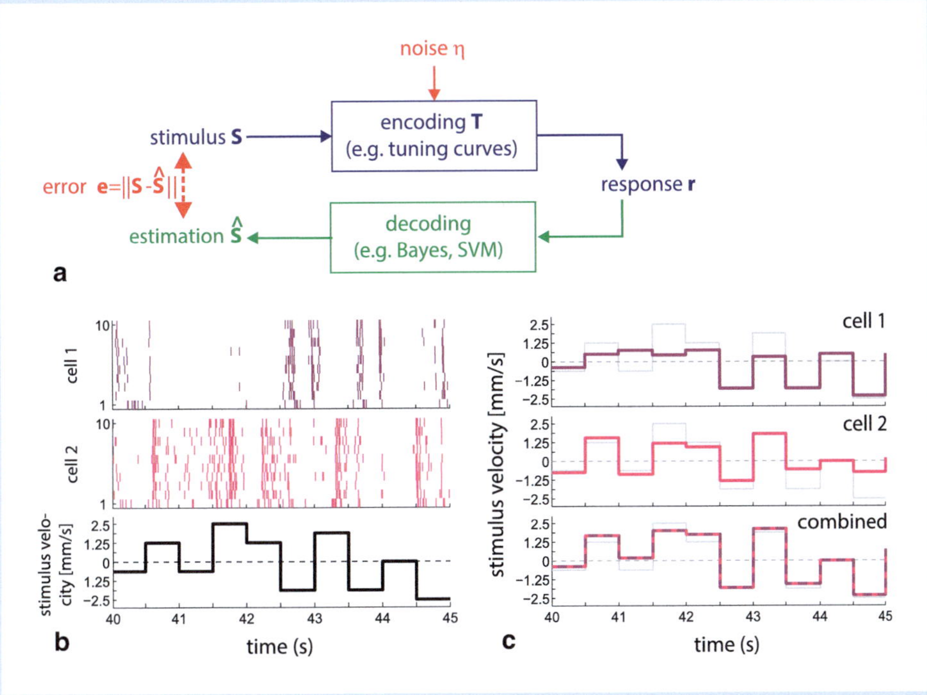

(orientation preference) and direction (direction preference). Direction selectivity may either be inherited from the ganglion cells, or generated within the LGN by similar mechanisms as in the retina (see Box 18.7). In general, these preferences are weak and elongations of nonconcentric RFs are small. In primary visual cortex, the selectivity of neurons for these stimulus features is greatly enhanced.

The spatial extent, temporal shape, and color tuning of LGN receptive fields closely resemble those of the ganglion cells in the corresponding pathway:

(a) **M-stream** (input from parasol ganglion cells): large receptive fields (RFs), insensitivity to color, and a preference for high temporal frequencies which leads to a high luminance contrast sensitivity for moving stimuli.
(b) **P-stream** (input from midget ganglion cells): red-green, color-opponent RFs of small size, with a preference for low temporal frequencies and maximum luminance contrast sensitivity for static stimuli.
(c) **K-stream** (input from bistratified ganglion cells): a blue-yellow, color-opponent RF. Sensitivities for spatial and temporal frequencies are reported to vary greatly between cells in the same animal.

> The LGN is more than just a relay station for transmitting visual input from the retina to the cortex; signals are gated by feedback from higher areas, and receptive fields become more diverse than in the retina.

**Principles of information processing**. Relaying visual information in parallel streams from the retina to the visual cortex is only part of the LGN's functional role [29]: Local connections within LGN mediate **contrast gain control**, which adjusts the sensitivity of a neuron for input changes to the actual mean input. Furthermore, feedback projections are known to strongly modulate LGN activity. Putative functional goals of feedback are, e.g., suppression of visual input as related to the attentional state, implementation of gain control to enhance the dynamic range of neuronal responses, or sharpening of the spatial tuning curves of the LGN neurons. In consequence, visual computation in the LGN and in all subsequent visual areas can rarely be regarded as a purely feed-forward process. Instead neuronal responses are caused by both, the actual visual stimulus and the activation of other visual areas.

#### 18.2.4.2 Primary Visual Cortex

**Anatomy**. The major target for the output from the LGN is the primary visual cortex (named V1 in monkeys and humans and area A17 in cats – we will use the abbreviation V1 in the following paragraphs). Like other areas in the neocortex, the visual cortex is composed of morphologically different layers (Fig. 18.6), numbered in ascending order, starting from the cortical surface. Connectivity within V1 is dense, and since Ramón y Cajal uncovered in the nineteenth century the first constituents of the cortical tissue by applying the newly developed Golgi staining technique, many scientists worked to disentangle the functional parts of this complex network. V1 connections can be divided into three different classes:

(a) **Vertical fibers** link different cortical layers. Ascending connections project to layers situated more closely to the cortical surface, while descending connections project away from the cortical surface. Some major pathways established by these fibers are indicated by purple arrows in Fig. 18.6.
(b) **Horizontal axons** link cells within the same cortical layers, parallel to the surface of cortex. Their spatial extension is several times larger than the range of the local connections (up to 3–4 mm from soma), and they link preferentially neurons with similar response properties (more details below in "Receptive Fields"). In Fig. 18.6, these connections are symbolized by orange arrows.
(c) **Local connections** are dense, extend over a much smaller range than vertical and horizontal fibers, and can be found in all layers with the least number of neurons in layer I. They connect neurons within but not across layers, and are not shown in Fig. 18.6.

Some cortical layers have been further divided into sublayers, partly because of their different morphology, connectivity, and partly because they are innervated by different retinothalamic afferent pathways. In Fig. 18.6, these afferents are indicated by black arrows. Most LGN afferents terminate in layer IV, from which activity is relayed to other V1 layers. The M-pathway innervates sublayer IVc$\alpha$, while the

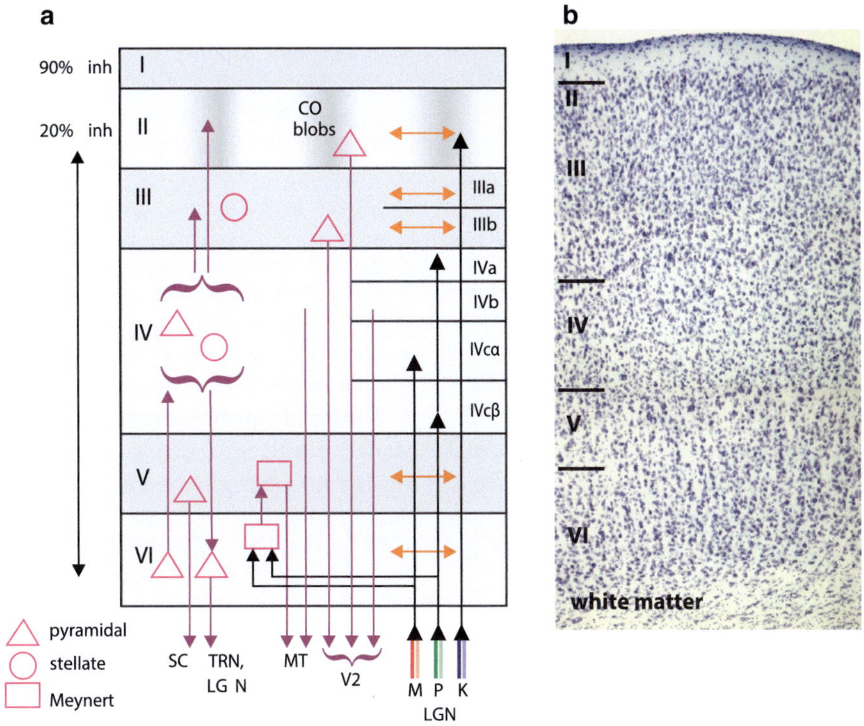


**Fig. 18.6** **Anatomy of the primary visual cortex.** (**a**) Sketch of major connections, afferents, efferents, and cell types in primary visual cortex. Inputs from LGN are shown as *black arrows*. *Mauve* symbols depict some of the major cell populations found in the different layers, together with selected ascending and descending fibers shown as *purple arrows*. Horizontal fibers are indicated by the *orange arrows*. In layer II, CO blobs are indicated by *darker shading*. Typical proportions for excitatory/inhibitory cell types in primates and cats are shown on the left. (**b**) Nissl staining of the primary visual cortex in the cat, vertical size about 2 mm. Layer I is almost free of cell bodies. The chief constituents are apical dendrites of layer II and layer V pyramidal cells, together with axons of intra- and subcortical origin. Layer II is a zone of small pyramidal cells with typically bifurcating apical dendrites. Layer III comprises medium to large pyramidal cells, many of which contribute to local as well as long-range lateral connections. They also represent output to other cortical areas ipsilaterally and to the contralateral visual cortex. Layer IV is specific for spiny stellate and star pyramidal cells. Layer V consists of small and large pyramidal cells, some of the largest pyramids (Meynert cells) are found here. These pyramids represent output to a number of subcortical structures (e.g., superior colliculus, pulvinar). Layer VI: A mixed population of pyramidal cells (at least three types). Chief targets include, respectively, layer IV and the thalamus, claustrum, and layer V/VI. Inhibitory GABAergic neurons are found in all layers including layer I. They represent a morphologically versatile population (This staining is courtesy of Zoltán Kisvárday and Kitti Csorba)

P-pathway projects onto sublayers IVcβ and IVa. An exception is a large part of the afferents from the koniocellular pathway which terminate in localized subregions of layer II named **cytochrome oxidase (CO) blobs**. This term is derived from the corresponding staining method revealing parts of neuronal tissue with high metabolic activity. In case of the CO blobs this enhanced activity is most probably directly related to the additional koniocellular input, and not caused by anatomical differences within layer II itself.

Distinct pathways from the retina/LGN terminate in different layers of V1, suggesting that the cortical layers are specialized in processing different aspects of visual information.

V1 output is relayed to higher cortical areas like secondary visual cortex (V2) and middle temporal area (MT, named PMLS in cats), and back to subcortical areas like SC and thalamus. Some major projection pathways are symbolized in Fig. 18.6 by the purple arrows protruding from the cortical layers. Note that Fig. 18.6 shows only the major, but not all pathways within V1. Even from this simplified scheme, it is obvious that V1 contains multiple feedback and recurrent loops such that understanding response properties of V1 neurons becomes more difficult than in LGN or in the retina.

In V1, there are about 80 % excitatory neurons and 20 % inhibitory neurons. These numbers do not necessarily reflect their effective contribution to the cortical dynamics. Inhibitory **postsynaptic potentials** (**PSPs**) can be larger and more sustained than excitatory PSPs, thus being very efficient in suppressing neural responses. Inhibitory neurons form only local connections, while excitatory interactions extend over longer ranges and even project to different cortical layers and areas. Some inhibitory neurons are known to make synapses near the soma of excitatory neurons, thus being able to shut off the more distal input from whole dendritic branches.

Different neurons can be distinguished by their morphology. Prominent examples for excitatory neurons are **pyramidal** and **star-shaped** (**stellate**) **cells** with spiny dendrites. Most of the inhibitory cells have dendrites without protusions (spines) and are termed **smooth cells**. Other types are described in more detail in the literature and comprise, e.g., **chandelier**, **neurogliaform**, **double bouquet**, and **basket cells**. From this variety, Fig. 18.6 symbolically shows only the most common types of pyramidal and stellate cells and the so-called **Meynert cells**. Meynert cells are special because they have asymmetric dendrites which form an anatomical substrate for establishing directional tuning of a neural response (see Box 18.7).

**Receptive fields** (**RF**). The responses of primary cortical cells encompass a variety of dynamical properties; prototypical pyramidal cells have medium firing rates of up to 100 Hz which adapt to lower rates when stimulated with a sustained input current. Inhibitory neurons [27] spike very fast and in a sustained manner, sometimes reaching frequencies of several 100 Hz. In the following, we will focus on the response properties of excitatory neurons unless stated otherwise.

RFs of V1 cells may be characterized according to different criteria. One distinguishes between **simple cells** and **complex cells**, between the classical- and nonclassical RF properties of a neuron, and one describes the response of V1 cells in terms of the preferred features in a visual stimulus that elicit a maximum activation.

> Most V1 neurons are highly selective for the orientation of a bar in their receptive field. They can serve as edge detectors for supporting visual scene segmentation.

**Tuning properties of simple and complex cells**. Many V1 cells are tuned to the orientation of a moving bar or grating (Fig. 18.7b), with the maximum activity observed when the stimulus is shown at the cell's preferred orientation. Static bars or gratings may also evoke large transient responses when the stimulus is switched on, but sustained responses are often much smaller than with moving stimuli. The spatial RF profile of orientation-selective cells is well characterized by a plane sine wave with a Gaussian envelope (**Gabor function**), which are therefore often used in experiments as stimuli to maximally drive a V1 neuron. While simple cells are activated only by a specific phase of such a grating, complex cells respond to gratings regardless of their phase (for a discussion of putative models and mechanisms of orientation selectivity, see Box 18.6). In addition to **orientation tuning** (OT), many V1 neurons in layer IVb also prefer movement into a specific direction. While **direction tuning** (**DT**) was also found in the retina of several vertebrate species, the underlying cellular and network mechanisms seem to differ (see Box 18.7).

Besides OT, V1 cells may also be tuned to the spatial and temporal frequency of a moving grating, to color (see Fig. 18.7a), to **binocular disparity**, and to other stimulus features. For understanding information processing in the cortex, one has to bear in mind that most neurons are tuned to a combination of many different features, although experiments often target only one particular feature of a neuron's RF.

The distribution of RF properties in V1 is linked to the anatomical architecture of the corresponding cortical layers. For example, cells in layer IVc$\alpha$ (terminals of the M-pathway) have larger receptive fields than cells in layer IVc$\beta$ (terminals of the P-pathway). Cells in CO blobs in layer II receive direct input from the

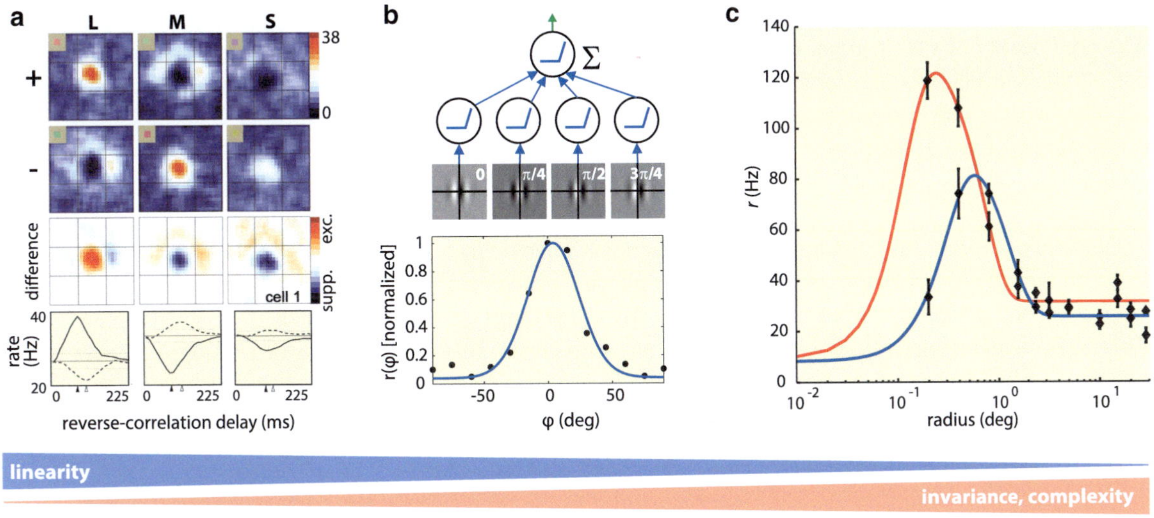


**Fig. 18.7 From linear to nonlinear response properties of cortical neurons.** (**a**) Organization of cone inputs to a double-opponent cell (*red*-ON/*green*-OFF) in alert macaque V1. The 3×3 *top panels* show spatial RF maps obtained from reverse correlation using sparse noise stimuli for isolating the contributions of different cones (which are sensitive for light of *L* long, *M* medium, and *S* short wavelengths). The *bottom* graphs show the time courses of the spike-triggered average traces for the RF differences displayed in the third row (*solid* center quadrant, *dashed* surround quadrants). The RF center was excited by an increase in L-cone activity (L+) or a decrease in *M* or *S* activity (M−, S−) (*top row*), and suppressed by a decrease in *L* (L−) or an increase in *M* or *S* activity (M+, S+) (*second row*); the RF surround gave the opposite pattern of chromatic tuning. The *inset* indicates the color of each stimulus, although the actual stimuli were presented on a computer monitor and carefully color calibrated (Data provided by Bevil Conway, see also [6] with permission). (**b**) *Top*, model of a complex cell in V1: to generate phase-invariant orientation-tuned responses, rectified outputs of simple cells (symbolized by the threshold-linear gain functions inside the neurons; see also Box 18.3) with linear RFs of different preferred phases are summed by the complex cell. *Bottom*, orientation tuning curve of a neuron recorded in V1 of alert macaque monkey (original data in *black*, von-Mises curve fit in *blue*; data provided by Simon Neitzel). (**c**) Nonclassical receptive field properties of a V1 neuron: The graph shows the responses of one cell to circular gratings of increasing diameter (horizontal axis), for two different contrasts (original data in *black*, curve fits in *blue* and *red*, for two different stimulus contrasts). The cell's firing rate first increases with stimulus size, reaches a maximum, and then becomes suppressed by a further increase in the grating's diameter. The different positions of the response maxima demonstrate that the preferred stimulus size changes with contrast (Data kindly provided by S. Shushruth and A. Angelucci, see also [40]). From (**a–c**), nonlinearity and complexity of receptive fields increase and/or responses become more invariant

K-pathway, and indirect input from the P-pathway over layer IVcβ and hence are strongly selective for color [41].

**Nonclassical receptive fields.** By definition (see Box 18.2), a neuron does not respond to localized stimuli presented outside its 'classical' RF. However, if a stimulus inside the classical RF is combined with a stimulus outside the classical RF ('surround stimulus', often called flanker), one observes a strong modulation of the responses of many cortical neurons by the surrounding stimulus. In consequence, response to center plus flanker is different from the sum of responses to center and flanker presented alone. These nonlinear response properties are termed the nonclassical RF (ncRF) of a neuron [39].

> Mediated by recurrent feedback from distant neurons in the same or higher visual areas, many neurons in visual cortex integrate more global, contextual information of a local stimulus in a highly nonlinear way.

ncRFs come in a variety of 'flavors' and there is yet no commonly accepted classification scheme for these effects. For example, with respect to orientation tuning some cells display enhanced firing rates for orthogonal configurations of center and surround stimuli. If center and flanking stimulus have the same orientation, firing rates of some neurons are suppressed at high contrasts but enhanced at low contrasts of the center stimulus. The nonlinearity of the neuronal responses is the reason why it is difficult to map nonclassical RFs systematically, because one cannot predict the response to a novel stimulus from the responses to the single parts or components of this stimulus (see Fig. 18.7c for an example).

Nonclassical RFs may emerge from interactions between cells within primary visual cortex as well as from feedback projections from higher cortical areas. This was demonstrated by cooling experiments in which V2 was inactivated, removing the suppressive influence of certain stimulus configurations on the firing rates of V1 neurons. A surprising fact is that feedback from higher areas can be much faster (V1-V2-V1 within a few milliseconds) than horizontal interactions within V1. The reason for this difference in speed is that axons establishing long-ranging horizontal connections are not myelinated (see Chap. 7) and thus have slow action potential propagation speeds.

**Cortical maps**. Similar to the LGN, cortical receptive fields of V1 are organized in a retinotopic manner, which means that the spatial receptive fields of neighboring neurons are adjacent or overlapping in the visual field (preservation of topology). However, the transformation of visual into cortical space is often distorted and not linear like the mapping of a camera image onto a computer screen. This distortion greatly

**Box 18.6 Orientation Preference**

Over the past 50 years, the emergence of orientation selectivity in cortical cells has been in the focus of many experimental and computational studies. The classical, first model [18] by Hubel and Wiesel (1962) proposed a feed-forward scheme visualized in the yellow center box: LGN cells, whose receptive fields are aligned in visual space, project to V1 neurons, which sum these afferents and fire spikes when a threshold (shown as a *gray dotted* line in the synaptic input graph) is exceeded. Stimuli matching the RF alignment and polarity (phase) lead to a high response, while orthogonal stimuli elicit small or no responses. Soon it became obvious that there are differences between model and reality: When stimulus contrast in the model is increased, the synaptic input for different stimulus orientations is scaled by a corresponding factor. This is depicted by the *blue curves* in the center diagram which represent the synaptic input in dependence of the difference $\Delta\Phi$ between stimulus orientation and the cell's preferred orientation (*dashed* – low contrast; *dashed-dotted* – medium contrast; *solid* – high contrast). With a fixed threshold, the width of the orientation tuning curve increases with increasing contrast (shown above as the cortical output by the *green curves*). In experiments, however, the width of the tuning curves turned out to be contrast-invariant and orientation tuning was much sharper than predicted from the model. These problems started a discussion about alternative mechanisms.

Ben-Yishai and collaborators proposed an extension to this framework, which is known under the term ring model. In the ring model [3], neurons in a cortical hypercolumn (which contains cells with all orientation preferences between 0 and 180°) are recurrently coupled (network scheme in *left panel, red arrows*) such that neurons with similar orientation preferences excite each other, whereas neurons with different orientation preferences inhibit each other. The resulting coupling function for the center cell in the sketched network is shown in the *lower graph* of the *left panel*. If the coupling is strong, recurrent interactions dominate over the feed-forward input and amplify orientation selectivity. In consequence, the shape of the tuning curve is almost completely determined by the shape of intracortical interactions, which deliver this dominant feedback, and stimulus contrast only scales this curve in a multiplicative manner. Hence tuning width becomes invariant to contrast.

Further experimental evidence raised doubt whether recurrent interactions could be as strong as required by the ring model; also, investigation of synaptic inputs to orientation-selective cells revealed a more complex pattern than Hubel and Wiesel proposed originally. Feed-forward alternatives to the ring model explain contrast-invariance of orientation tuning by extensions to the Hubel-and-Wiesel model: an inhibitory input from LGN, which is not tuned to orientation. This input serves

varies over species and may have both anatomical and functional reasons: An anatomical constraint is that visual space has to be morphed to match the anatomical shape of cortical areas. Functionally, it may be beneficial to enhance the cortical representation of important stimulus aspects. One example is the cortical magnification factor in humans: Near the fovea, one square degree of visual space is mapped onto 25 $mm^2$ in cortex. At 10° of eccentricity, the cortical representation of the same area in visual space is represented on only about 5 $mm^2$. This supports the idea that more cortical resources are allocated towards the processing of the part of a stimulus which is currently fixated, and is consistent with the increase in midget ganglion cell density in the retina near the fovea. Cortical maps are also found for other sensory modalities, and show a similar overrepresentation of behaviorally important parts of the stimulus space, for example in the somatosensory system.

In many species, smoothly changing maps of neural response properties across the cortical surface demonstrate that neighboring neurons often have similar RFs. More distant neurons with similar RFs tend to be connected by horizontal axons.

**Box 18.6 Orientation Preference** (*continued*)
to implement a form of contrast gain control, by subtracting an offset from the synaptic input which increases with contrast (see resulting *blue curves* in the *right panel*; the receptive field of this additional input is shown as the big *red circle*. Elements of the Hubel-and-Wiesel model are shown in *light gray*).

Current evidence seems to favor a combination of all three models: the Hubel-and-Wiesel scheme for having an initial bias for the preferred orientation, and the interplay between antagonistic afferent inputs, normalization, and recurrent feedback of intermediate strength for achieving contrast-invariant orientation tuning.

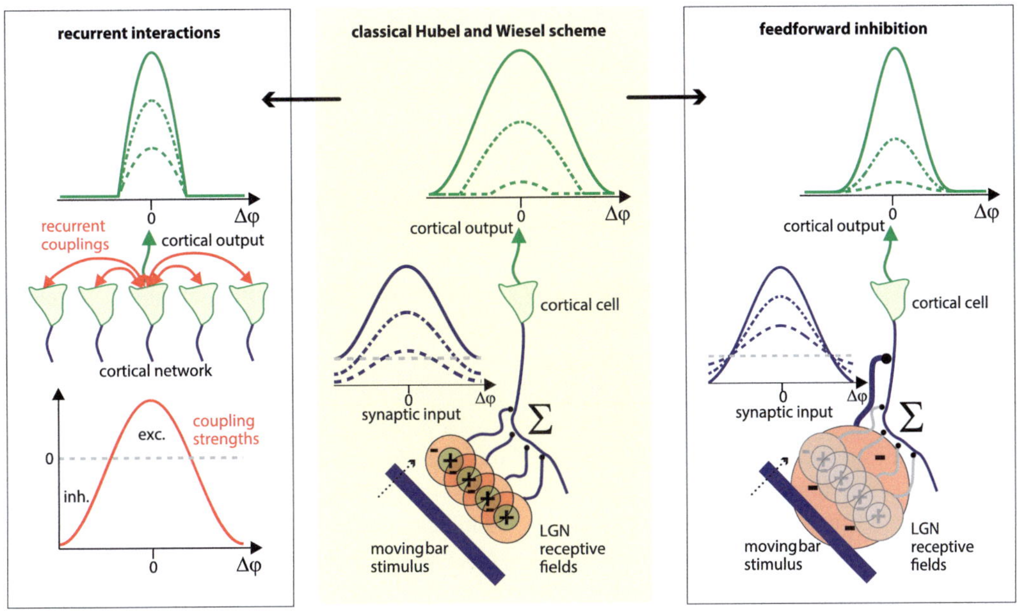

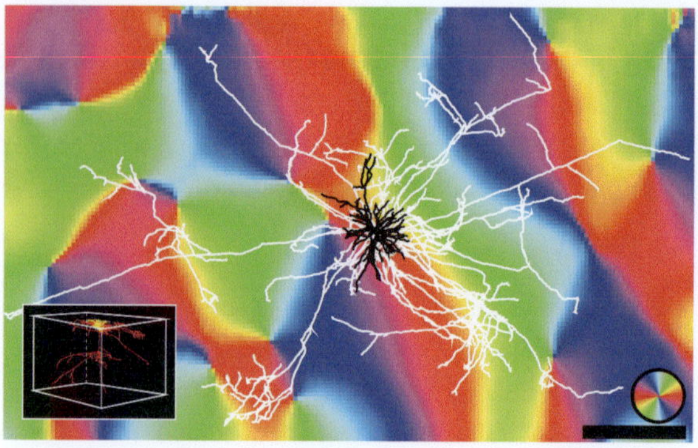

**Fig. 18.8 Orientation preference map and layout of horizontal connections of a layer II pyramidal cell in cat primary visual cortex.** Preferred stimulus orientation is coded according to the color palette in the *lower right corner*. The *black bar* corresponds to 500 μm in cortex. The map has been obtained from intrinsic signal optical imaging, using full-field square wave gratings at four orientations as stimuli. From the topology of the orientation map it is obvious that hypercolumns are rather abstract functional modules that are not defined in strict anatomical or physiological terms. Their size is related to the average distance between two regions with similar orientation preferences. The dendrites of the cell are shown in *black*, and its axon in *white*. The *lower left inset* displays the same pyramidal cell from a side view (This figure is courtesy of Zoltán Kisvárday and Kitti Csorba)

In primate visual cortex, maps for several stimulus features other than location in visual space (retinotopy) have been found. Examples are **orientation maps**, **direction selectivity maps**, **spatial frequency maps**, and **ocular dominance (OD-) maps**. OD maps quantify whether, for a particular stimulus, neurons are more strongly driven by the left or by the right eye. Cortical maps are measured *in vivo* by using optical imaging or two-photon microscopy or post mortem by using staining methods. Neighboring neurons have similar orientation preference (OP) except for singular points in the map called 'pinwheels' where OP changes abruptly (Fig. 18.8). Regions where all orientations are represented roughly once have been termed hypercolumns. There is a strong link between the electrophysiological response of individual neurons and activation measured with optical imaging methods. However, investigating some properties of OP maps like single neuron tuning near orientation pinwheel centers yields different results when using different measurement techniques.

The interrelationship between different cortical maps is still under investigation. There is evidence that at locations where the spatial layout of one map is changing abruptly, maps for other stimulus features exhibit only smooth changes. Such a principle in map organization would guarantee that the coverage of visual space with detectors of different features (or feature combinations) is as uniformly as possible. 'Holes' in the neural representation, as e.g., locations in visual space where horizontal bars could not be detected, are hereby avoided.

The origin of feature maps and of their particular structure is still discussed [19]. There is both evidence for a genetic origin and a strong influence of visual experience. Genetic factors were revealed by studies showing that in siblings of the same litter the shape of cortical areas and inhomogeneities in the maps are more correlated than in young animals from different parentages. Other studies demonstrated stimulus-dependent plasticity of maps in adult animals by showing that microstimulation of neurons caused a change in map structure around the stimulation site. One hypothesis is that genetic factors determine the initial layout of maps, which are then refined to achieve an ordered structure, which in addition may reflect the particular statistics of the visual environment the individual was subjected to during the adaptation process. This view is supported by evidence from **computational neuroscience**, where self-organizing map models like the Kohonen map (named after the Finnish scientist Teuvo Kohonen) have been successful in reproducing many cortical map structures and their interrelationships [7].

Even less clear than their origin is the functional relevance of maps. If visual information processing requires more interactions between neurons having similar RFs than between neurons with different RFs, an ordered map would minimize wiring length. However, some mammals as, e.g., squirrels or mice do not have orientation maps, yet some of them are animals which strongly rely on vision to survive. Therefore maps could also be an epiphenomenon of developmental and adaptive processes.

Some cortical maps are linked to the pattern of anatomical connections between neurons [15]. For example, the ocular dominance map in layer IVc reflects the pattern of afferent connections from ipsilateral and contralateral LGN terminating here. Another example is **long-ranging horizontal axons** in layers II and V which predominantly connect cells with similar orientation preference (Fig. 18.8). These axons may contribute to specific computations in early vision, for example by enhancing the neural responses to colinearly aligned edge elements for supporting contour integration. Although these horizontal interactions are excitatory, they can also exert an inhibitory effect when they target inhibitory interneurons. This mechanism termed disynaptic inhibition could be used by the brain to suppress responses to large fields of elements with identical orientations. Functionally, this effect can support the detection of inhomogeneities in visual scenes like object borders by attenuating neuronal responses to homogeneous textures.

#### 18.2.4.3 Principles of Information Processing

The primary visual cortex performs computations that extend beyond contrast enhancement and gain control, which we already described for neurons in retina and LGN. One major part of these computations appears to be devoted to the extraction of localized image features, where the most prominent example is edge orientation. At the same time, primary visual cortex is the first stage in the visual system where the strict segregation between the different pathways from the retina over the LGN is lifted. Hence, a further principle of information processing is the combination and integration of spatially and temporally separated, local features. This principle of visual computation is iterated in subsequent visual areas, such as V2, V3, and V4, for building more and more complex representations of a scene. Christoph von der Malsburg and Wolf Singer proposed **synchronization** between cortical neurons as a mechanism to integrate or to 'bind' these local features. Population coding based on modulated firing rates is an complementary idea for implementing feature integration.

Functionally, extraction and combination of image features are believed to be the fundamental steps necessary for scene segmentation and object recognition. Of all areas in the visual cortex, responses in primary visual cortex are still closely linked to the actual stimulus, and less influenced by other behavioral variables like attention. Hence it is assumed that computations in this area reflect rather general operations on visual images which are useful for any kind of behavioral task, while higher areas are more specialized and dynamically reconfigured in relation to other cortical states and the current needs.

> V1 neurons seem to extract elementary features or feature combinations, which are perceptually salient and highly informative elements for scene segmentation like edges or corners of visual objects.

Psychophysical experiments have revealed a set of principles according to which the visual system combines different image features. Classical examples for these principles are the so-called '**Gestalt laws**', which were first formulated by Max Wertheimer (1880–1943) and his colleagues during the 1920s. The anatomical structure of primary visual cortex (and higher visual areas), like the functional specificity of long-ranging horizontal axons, seem to support some of these principles. Further principles become visible in nonclassical RF properties, like the enhancement of contrasts and borders (in different features like luminance, orientation, and color), or the detection of elementary feature combinations like corners or curve segments which are more complex than single, oriented bars.

Still the functional role of primary visual cortex is subject of intense research. While **LNP models** (see Box 18.3) can explain RGC and LGN responses to a large extent, similar models for cortical neurons can explain only about 30–40 % of response variability in the best cases [5]. In addition, LNP models constructed from artificial stimuli do not adequately predict responses to natural stimuli. Furthermore, experiments using optical imaging for recording spatiotemporal activity in cortex have shown that neural responses strongly depend on the 'spontaneous' **on-going activity** observed prior to stimulus onset [43]. Taken together, this evidence demonstrates that cortical neurons are substantially influenced by recurrent, contextual interactions and top-down feedback. Hence V1 cells might perform much more complex operations than previously thought, which poses an interesting challenge for future investigations.

### 18.2.5 Higher Central Processing

Area V2 (or in cats A18) is the immediate successor of the primary visual cortex, from which it receives the majority of its inputs. Its response properties and

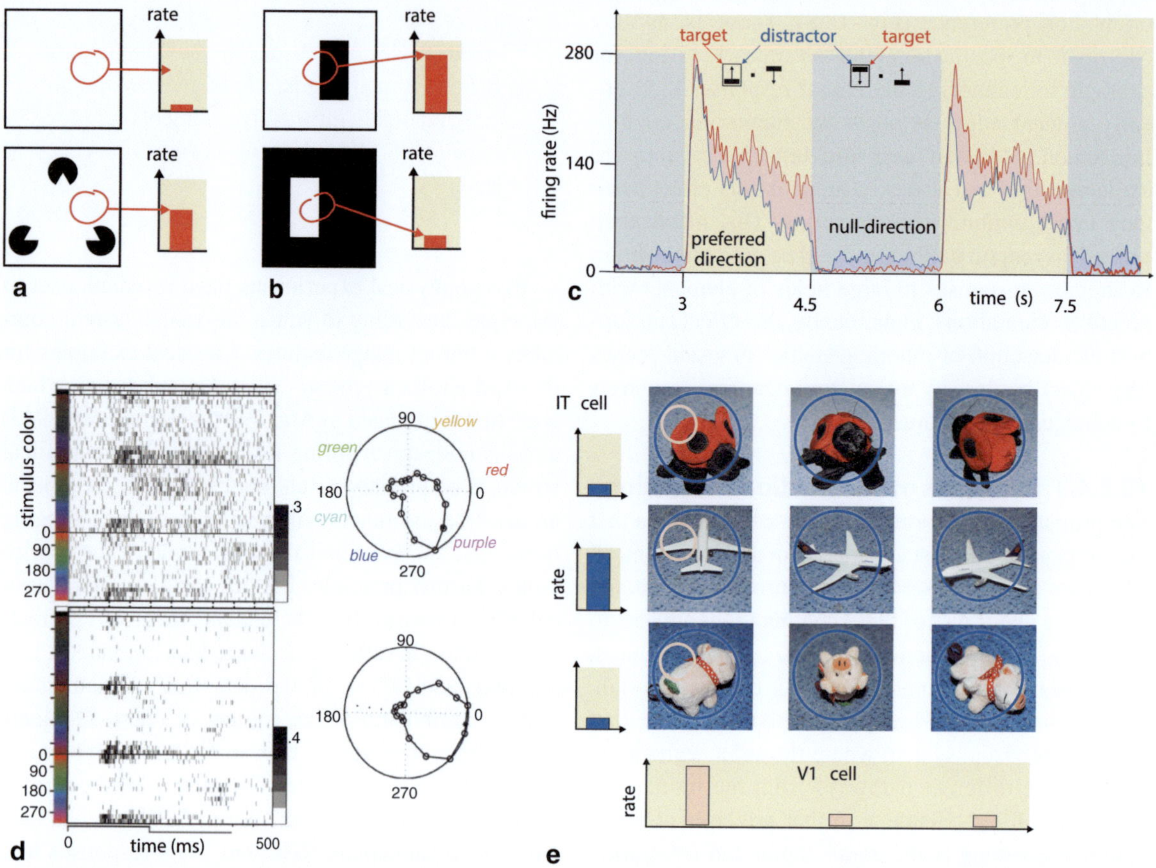


**Fig. 18.9 Response properties of cells in V2 and higher cortical areas.** (**a**) Sketch of an orientation-selective V2 cell with very low spontaneous activity in absence of stimuli (*top*), but responding to an illusory edge in a Kanisza triangle as strongly as to a real edge (*bottom*). The percept of human observers is a triangle whose corners occlude the *black circles*. (**b**) Sketch of a border-ownership cell, which responds only to the edge if it is part of a figure to the right of the edge. (**c**) Response of an MT cell (dorsal stream) in awake, behaving macaque monkey to bars moving in its preferred and in the opposite (null) direction (shaded in *gray*). The *red curve* displays the response when the stimulus within the RF is the behaviorally relevant target, the *blue curve* shows the response to the same stimulus when it is a distractor. Clearly the firing rate in all movement phases is strongly influenced by the task ('attentional modulation') (Data kindly provided by Detlef Wegener, see also [46]). (**d**) Color-tuning of two glob cells recorded in alert macaque PIT (ventral stream). The cell in the *top panel* is tuned to *purple*, while the one in the *bottom panel* is tuned to *red* (*right panels*). *Left panels* show poststimulus time histograms to an optimally shaped bar of various colors (see color scale to the left of the graphs). Stimulus on- and offset is indicated by the step function at the *bottom*, and the *gray* scale to the right denotes mean number of spikes per stimulus repeat per bin (Data provided by Bevil Conway). (**e**) Schematic response of a V1 complex cell (*mauve*) and an IT cell (*blue*, ventral stream) to different 'natural' stimuli. While the V1 cell strongly responds to an oblique edge within its receptive field (*mauve circle*), regardless of the object in the scene, the IT cell responds strongest to different views of the same object, in this case the toy airplane

anatomical organization bears many similarities to V1, although RFs tend to be larger in size and more complex in shape. V2 is the first area in which neurons respond to visual illusions, as e.g., to the physically nonexistent, but perceptually present contours in a Kanisza triangle (Fig. 18.9a). Responses also depend more strongly on the global context of a local image patch: Some neurons are activated by oriented edges only if these are part of a figure to the left of these edges ('border-ownership neurons', Fig. 18.9b). It seems that the V2 circuitry supports basic computational principles necessary for performing **figure-ground segregation** and object recognition in visual scenes. For example, there is strong evidence that contour integration is predominantly performed in V2.

From V1 and V2, visual information is propagated in parallel to several higher visual areas, which are strongly interconnected by feedforward and feedback

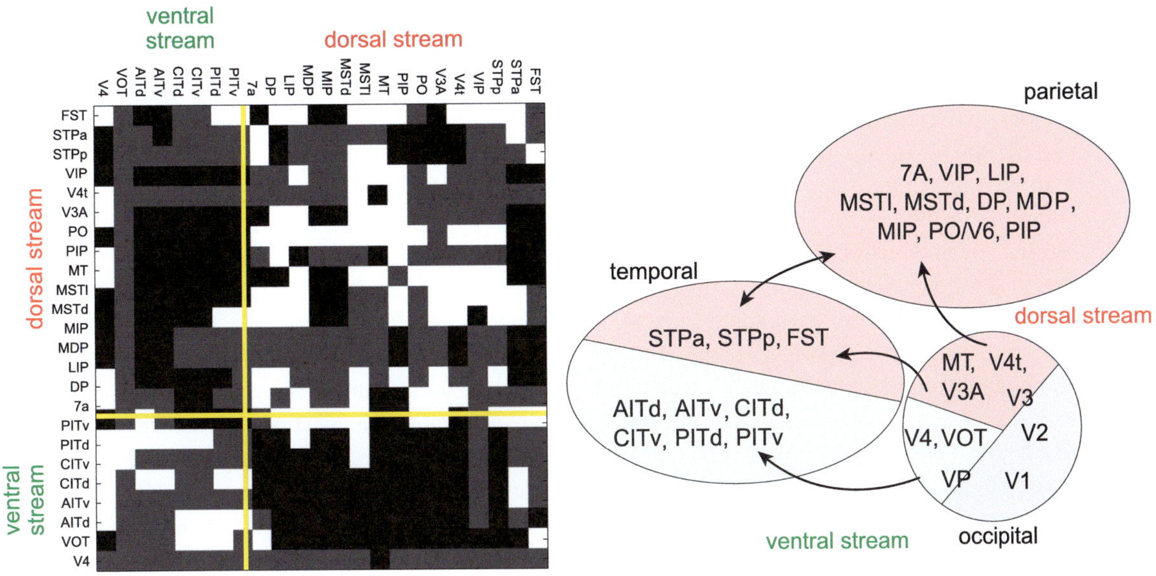


**Fig. 18.10 Separation of visual information processing into two major streams in higher cortical areas.** *Left*, connectivity matrix between visual areas, ordered with respect to the dorsal and ventral pathways (Data provided by Claus Hilgetag, modified from the original data from Felleman and Van Essen [13] with permission). *White boxes* show existing connections, *black boxes* absent connections. *Dark gray* boxes are connections for which there is currently no consistent information available. Areas within the same pathway are preferentially connected (diagonal submatrices), areas between pathways are less strongly connected. The diagram to the *right* summarizes the locations and most important areas of the dorsal and ventral pathway. Note that these diagrams emphasize anatomical, but not necessarily functional connectivity

connections. Pioneering work in structuring these visual cortical networks was done in David van Essen's lab [13]. A rule of thumb is that spatial receptive fields become larger the more 'distant', in anatomical terms, a visual area is from the retinal input. Also, the 'higher' a cortical area, the more are neuronal responses influenced by other cortical variables like, e.g., attention (Fig. 18.9c), the behavioral task, or the emotional state. In addition, neural activity tends to be sparser, i.e., the mean activation level under natural stimulation decreases because only few of the neurons in a specific area are active at any given time. This trend is compatible with the observation that in higher areas receptive fields get more and more specialized for complex feature combinations (Fig. 18.9e). It is not always possible to find homologs of a certain higher visual area across all mammals.

RFs of neurons in higher visual areas become larger, but also more specialized and more nonlinear. In parallel, one observes an increasing invariance of the neural response against local changes in the retinal image.

A computational method to analyze the structure of the networks engaged in higher central processing is to perform a cluster analysis on the connectivity matrix between the visual areas. A cluster analysis seeks to identify different groups of brain areas that are as sparsely connected as possible among the groups, and as densely connected as possible within the groups. Such quantitative and unbiased analyses have confirmed the existence of two major streams for visual processing, which were suggested from earlier studies focusing on the specific response properties of different visual areas. Named after their relative location within the brain, these pathways are termed the **dorsal stream** and the **ventral stream** (Fig. 18.10).

The dorsal and ventral stream have often been called the 'What'- and 'Where'-pathways, respectively suggesting that one pathway identifies the objects being present in a visual scene, while the other pathway analyzes where these objects are located and in which directions they are moving. This interpretation suggests that certain aspects of the full visual information are available only to one of the pathways, but not to the other. However, there is a huge amount of crosstalk between these streams. Refined experiments have demonstrated that properties like shape, motion, and disparity may be processed in both streams, but to varying extents depending on the current behavioral task. Current consensus is that the dorsal and ventral streams rather serve different behavioral goals, hereby having access to all aspects of a visual scene [24] rather than relying on separated 'what' and 'where' cues. Examples for these goals are object recognition for the ventral stream versus tasks of locomotion, pursuit, or tracking for the dorsal stream. For instance, the behavioral task to track a specific person on a bustling crossroads (multisensory integration) requires information both about motion and about object identity, which must be combined appropriately. An alternative idea is that both paths essentially process and represent similar information, but in different reference systems: allocentric (w.r.t. an external reference frame) in the ventral path, and egocentric (self-centered) in the dorsal path. As the computational functions of higher visual areas are still barely understood, it will be interesting to observe how these current views will be refined in future.

#### 18.2.5.1 Dorsal Stream

**Anatomy, receptive fields, and principles of information processing**. Emerging in the occipital region from V1 and V2, and passing through areas like MT, V3(A), STP, and FST, the (monkey) dorsal stream projects to the parietal regions and includes areas VIP, LIP, MST, PO, and others (in humans, areas in the dorsal stream have a different naming scheme, see more specialized literature for details about these areas and their naming schemes). In area MT (PMLS in cats), neurons are highly direction-selective, responding strongly to gratings, bars, or random dot patterns which move into the preferred directions of the cells (Fig. 18.9c). Most neurons are also selective for disparity. MT cells inherit their basic selectivity for direction and disparity from their afferents from layer IVb in V1 and from the thick strips in V2, which are known to host dense clusters of direction- and disparity-selective neurons. These tuning properties become enhanced within MT.

Areas MST, FST, and STP are involved in the tasks of locomotion, pursuit, or tracking. For example, many MST neurons respond selectively to **optic flow** and even to movement of the animal in darkness. Some particularly important functions of these higher areas are target selection for eye and hand movement in conjunction with visuospatial attention, visual control of actions directed to objects, and navigation. A further division of the dorsal stream into a pathway for online control of actions, and a pathway for understanding the actions of others, has been suggested.

#### 18.2.5.2 Ventral Stream

**Anatomy, receptive fields, and principles of information processing**. The ventral stream originates at area V4 in the occipital region (or even earlier in the P-pathway in V1), and further projects to different areas in temporal regions, like the inferotemporal cortex (area IT in monkeys and complex LO in humans) with its many subdivisions. Neurons in area V4 (in humans, a putative homolog is area LOS/LOC) are especially selective for form and color (Fig. 18.9d). Similar to the emergence of disparity and motion selectivity in MT, the preference for color emerges from the existence of anatomical connections from layers II/III of V1 and from the pale and thin strips of V2, which comprise neurons selective for color or color contrasts. Selectivity for form includes a preference for specific contour configurations and for specific shapes or parts of complex forms within the spatial extent of V4 neuron's receptive fields.

Ascending higher in the ventral stream, area PIT is activated by even more complex stimulus configurations. Responses of neurons in area AIT are often invariant to the orientation and/or location of an object. Computing such **invariant representations** is a very important step in object recognition. For example, the retinal image of a cup of tea can completely change when this object is seen from different viewing angles and under different lighting conditions. However, a neural representation of the cup should be identical in all these cases, because the object itself did not change (see an example in Fig. 18.9e). AIT is the first visual area in which invariance to spatial transformations like object translations or rotations has been observed. Prominent examples are IT 'face neurons' which only fire when a stimulus contains a face.

A further important aspect of computation in the ventral stream is the processing of relationships of different objects to each other.

#### 18.2.5.3 Attentional Modulation and Other Modulatory Influences on Neural Responses

A remarkable feature of information processing in higher visual areas (and to a lesser extent also in LGN and the primary visual area) is the modulation of neuronal responses by factors other than the current stimulus. These factors are related to activities in other (nonvisual) brain areas, which reflect the current mental or physical 'state' of the brain.

> Information processing in the visual system is constantly functionally reconfigured to adapt computational strategies to the current needs or behavioral task.

A prominent example is **attention**, which can be described as a behavioral condition where a subject directs its neuronal resources to a specific aspect of a stimulus. The term attention subsumes a great variety of different behavioral conditions and observed phenomena, possibly involving many distinct networks and pathways. One distinguishes spatial attention (i.e., attention to a specific location in visual space) from 'object-based' and 'feature-based' attention (i.e., attention to a specific content of a visual stimulus).

Attentional modulation in areas V4 and MT can be substantial: As a rule of thumb, activities of neuronst responding to an attended stimulus are enhanced, while activities of neurons responding to a nonattended part of the stimulus either remain unchanged (typically at locations far from the focus of attention), or are substantially suppressed (typically at locations near the focus of attention). For an example, see Fig. 18.9c.

Functionally, attention is regarded as a selection mechanism, which can either substantially reduce the amount of information to be processed for a given task, or which can enhance the processing of aspects of a stimulus which are currently behaviorally relevant [20]. The neural mechanisms of how attention acts on neural representations are still unclear. Candidates are **neuromodulators**, which could change the characteristics of synaptic transmission, or neural activity provided by other neurons: additional excitatory synaptic input may serve to increase the firing rate of neurons processing important parts of a stimulus, while **shunting inhibition** could be used to suppress neurons processing irrelevant aspects of a stimulus. These neurons could be located within the same area, but also in different visual or nonvisual areas [35] like the parietal or prefrontal cortex (**working memory**).

#### 18.2.5.4 Connections to Other Brain Areas

> Understanding the visual system is not possible in isolation: Vision controls movements and behavior, vision changes through action, vision relies on memory and experience, and vision integrates other sensory evidence to improve the percept.

The visual system is linked to many other brain areas. One aspect of these interactions is to transfer results from visual computations to areas serving other functional roles. Conversely, information from different sensory modalities, or from brain areas involved in higher cognitive functions, is conveyed to visual cortex and modulates the processing of visual stimuli.

From all areas receiving information from the visual cortex, the **superior colliculus** (**SC**) plays a central role [23]. SC is part of the tectum mesencephali and located in the midbrain, superior to the brainstem and inferior to the thalamus. The SC comprises several layers, in which different sensory modalities are represented in spatial maps. Visual information in the upper layer is combined with auditory and somatosensory information in other layers, forming a multimodal representation in the lower layers (multisensory integration, see [42]). This representation is used in directing movements of the eyes, ears, and head to actively select behaviorally relevant aspects of the sensory environment. In primates, the visual field representation is lateralized as in the LGN, which is different to other mammals.

Interestingly, the **tectum** and **nucleus of the optic tract** (**NOT**) also receive direct input from the retina, which allows for a fast processing of visual information, e.g., in the context of visual reflexes for initiating express saccades (tectum), or for smooth saccadic pursuit (NOT).

Areas in the ventral stream of the visual cortex project to the **perirhinal cortex**, a part of the medial temporal lobe which is important for memory. The perirhinal cortex plays a major role in the recognition and identification of objects, and forms direct connections to the **hippocampus** whose functional role is linked to memory consolidation and spatial orientation. The perirhinal cortex also receives afferents from other sensory modalities, which are integrated with its input from the ventral stream.

Higher areas in the dorsal stream as the posterior parietal cortex (in monkeys: VIP, LIP, 7a, PIP) project to areas involved in motor control, like the premotor cortex. This pathway serves to select, to plan, and to initiate actions based on visual computations. A particularly interesting feature of the premotor cortex is the existence of **mirror neurons**, which fire both, in response to an action being performed, and in response to the observation of that same action being performed by a different individual. One part of the premotor cortex is the frontal eye field (FEF). Together with the supplementary eye fields (SEF) and the intraparietal sulcus (IPS) it forms a system for controlling and directing saccadic eye movements. A major input to FEF comes from SC.

Visual processing itself is modulated by other processes throughout all visual areas. An example for a very early modulation is through the existence of connections from the auditory cortex to primary visual cortex in monkey. Another example are afferents from dorsolateral prefrontal cortex which become activated in delayed-match-to-sample tasks, where an observer has to match the actual visual input to a memorized template. Currently, it is not known whether the functional role of these interactions is linked to short-term memory or to attentional selection.

### 18.2.6 Vision and Behavior

Since primates extremely rely on their visual system to gather information about their environment, it is quite obvious to us that behavior is strongly influenced by vision: The visual system developed to identify food sources, mates, and predators from greater distances than, e.g., using smell and touch. Moreover, many animals use visual signals like body posture or facial expression for communication. In addition, the circadian rhythm (sleep-wake cycle) is strongly influenced by the perception of light and darkness. In mammals, the **nucleus suprachiasmaticus** (SCN), located directly above the optic chiasm, is involved in adjusting the internal clock in response to changing levels of light in the environment.

One should keep in mind that it is not the physical light stimulus per se that determines the behavioral reactions of an animal or human, but the percept of it – the subjectively perceived appearance. The link between stimulus and percept is investigated in psychophysical experiments, which query behavioral response in response to well-defined stimulus parameters and therefore help to identify behaviorally important stimulus features. Recognizing objects [32], for example, requires invariance of the percept with respect to different positions, orientations, and lighting conditions, even though these features lead to considerably different patterns of photoreceptor activation. In this context, visual illusions are particularly useful for discovering principles of information processing [9].

> Visual illusions are not failures of our visual system, but attempts at making sense of incomplete or noisy sensory information.

For example, two image patches of exactly the same luminance may be perceived as a 'dark' and a 'bright' patch, depending on whether these patches are located in the shadow of other objects in a scene or not (Adelson's checker-shadow illusion). This example gives us some hints about how the visual system uses contextual clues to estimate 'real' object color when the latter is illuminated. Such a perspective on the visual system refers to the ideas of Hermann von Helmholtz (1821–1894), as exemplified in his Treatise on Physiological Optics (1856–1867) [17]. He considered perception as an inference process on sensory information, which constructs a meaningful, internal model of the outside world. Hereby vision uses prior information about the statistics of our environment to interpret a particular scene. In accordance with this concept, there is evidence that receptive fields and anatomical structures in the visual system reflect the statistics of the visual input, hereby providing the neural substrate for such priors.

A good example for the interplay of vision and behavior is the processing of optic flow during self-motion. On the one hand, the (visual) perception of objects determines the movements of an animal or human: After turning the eyes with a saccade towards a potentially "interesting" object (i.e., an object with high conspicuity or salience, see Fig. 18.11), we might move towards the object and reach out for it. On the other hand, movement of eyes, head, and body shift the picture of the environment on the retina, thus influencing the visual perception of objects. Hence, vision is a recurrent process, which interacts with the environment in a closed-loop situation.

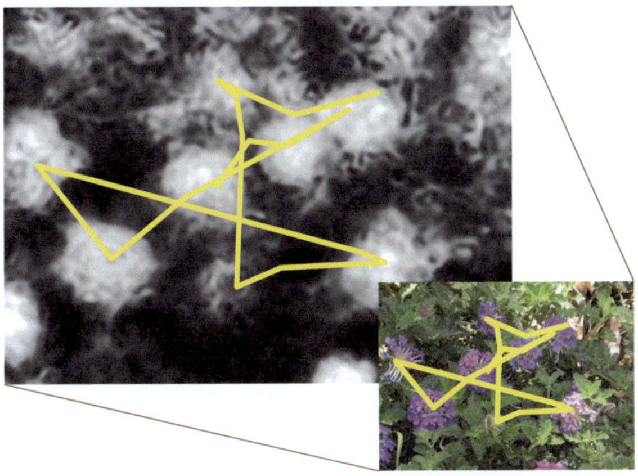

**Fig. 18.11** **Example for saccadic eye movements:** recent advances in understanding fundamental computations in the visual systems allow to predict the saliencies of image parts with high accuracy: the result of such a prediction from a computer model is visualized in a *gray-scale* representation, with brighter regions indicating image areas of high perceptual saliency (i.e., the flowers). In experiments, the original image (*bottom right*) was shown to human observers and their saccadic eye movements were recorded. The sample eye trajectory (*yellow lines*) demonstrates that saccade targets are located in regions with a high predicted saliency (Data and image kindly provided by Peter König)

Realizing or inferring the aspects of a visual scene that might be perceived by other individuals is one of the most complex computations in the visual system and very important for social behavior.

Predicting what others can see is mainly a domain of primate vision (apart from the exceptional example of the western scrub jay, see Sect. 18.3.2). For example in humans, children learn to follow the gaze of their parents to discover objects of interest (see Chap. 29).

## 18.3 Specific Differences in Other Vertebrates

Most of the principles of visual information processing are common to all vertebrates. Non-mammalian vertebrates are in general better equipped for color perception than mammals.

The visual systems of vertebrates in general share most of the principles of information processing with the primate visual system described in detail in Sect. 18.2. However, comparing the visual system of different animals reveals that many features of their visual systems (e.g., photoreceptor absorption spectra, light adaptation, dynamics of motion vision) are specifically adapted to their habitats and their behavioral tasks.

All vertebrates share in particular phototransduction mechanisms and the structure of the retina. All main cell types and the major retinal pathways are found in all studied vertebrate species. However, major differences exist, e.g., in color vision, which is based on different numbers of cone types (see Box 18.1). In contrast to the trichromatic primates and usually dichromatic mammals, most non-mammalian vertebrates possess at least four types of cones. Therefore, birds, reptiles, amphibians, and fish are generally better equipped for **color perception** in particular in the ultraviolet light range. Compared to mammals, color perception of non-mammalian vertebrates is further enhanced by retinal network properties, e.g., by wavelength-specific responses of horizontal cells.

A major difference in central processing of visual information is that the retinal input to the **optic tectum** is the standard visual pathway in non-mammalian vertebrates, while connections from the retina to the superior colliculus (the corresponding structure in mammals) plays only a minor role for visual tasks performed by mammals. The optic tectum of non-mammalian vertebrates shows some similarity to the LGN of mammals, in particular a strong retinotopy and a clearly layered structure. However, tectal processing of visual information is less well-studied than the mammalian LGN and visual cortex.

Behaviorally, many vertebrates rely strongly on visual perception, in particular color vision. For example, in addition to body language, reptiles and fish use also a color language, in which the skin rapidly changes color depending on the motivational status of the animal (see, e.g., Fig. 18.12).

### 18.3.1 Non-primate Mammals

Even though the visual systems of all mammals are very similar, individual species developed several specializations to their habitats and life styles. For example, the percentages of rods and cones vary

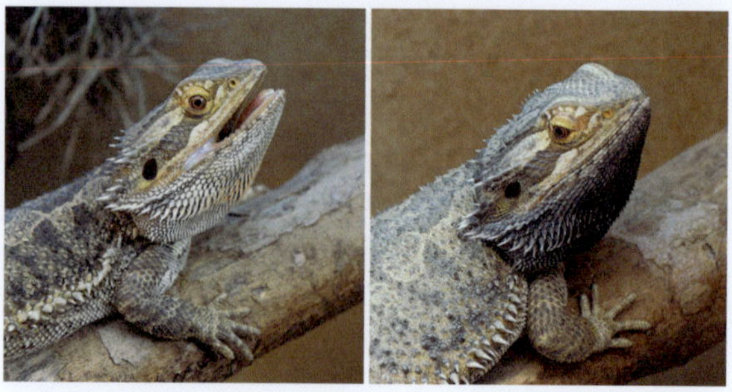

**Fig. 18.12** **Color language.** Many non-mammalian vertebrates communicate visually by changing color such as the Australian bearded dragon shown here. The same male during rest (*left*) and during courtship (*right*), trying to impress females and potential rivals with his *black* "beard" and *yellow* back. These reptiles can change color within few seconds

greatly between animal species. While nocturnal animals have a clearly rod-dominated retina (~95–99.5 % rods, e.g., mouse 97 %), the percentages of rods and cones vary considerably in retinas of diurnal animals, e.g., humans and rhesus monkeys possess ~95 % rods, pigs 80–90 %, tree shrews only ~5 %. Many mammals, in particular nocturnal carnivores, further improve the percentage of photons triggering phototransduction in low-light conditions with a tapetum lucidum. This light-reflecting layer of tissue is part of the choroid in the back of the eyes and causes, e.g., the pupils of cats to glow in the dark when a light is shone into the eye.

Another example for a habitat-specific anatomy is the shape of the region of highest visual precision. While the fovea of primates is round, cats and dogs have an elongated central band called visual streak, which might have developed to scan the horizon for prey or predators.

Being among mammals the genetically best controllable model organism, mice are used frequently in visual research, even though the visual abilities of these nocturnal animals are rather limited. Therefore, diurnal animals like rabbits, guinea pigs, and squirrels also receive some scientific attention, in particular in retina research. Since Hubel and Wiesel used cats for their groundbreaking work in the 1960s (see Box 18.2 and e.g. [18]), they are standard animals to study the central visual system of non-primate mammals. The anatomy and physiology of their central visual system was found to be very similar as in primates, even though a different nomenclature is used.

### 18.3.2 Birds

Birds have a highly developed visual system, providing excellent spatial and temporal acuity as well as color vision.

Birds in general rely very strongly on their visual sense. They have developed several anatomical and physiological specializations to achieve excellent spatial and temporal acuity, on the one hand, and brilliant color vision, on the other. Among vertebrates, birds have the largest eyes relative to their body size. In their eyes, birds have developed a special comb-like pigmented structure of blood vessels, the pecten oculi. This structure helps to improve eyesight by keeping the blood vessels away from the retina. The photoreceptor density of birds is – depending on species – two to five times higher than in humans. While most birds can see best monocularly to the side, some vision specialists like birds of prey have evolved forward-facing eyes with a wide field of binocular vision, and some have two foveas (one looking forward for binocular vision and one sideways) to further improve their eyesight.

As an adaptation to the rapid optic flow (the flow of images on the retina) during flight, the temporal processing of visual stimuli in birds is very fast. Their **flicker fusion frequency**, the number of flashes per second for which the animal starts to perceive a continuous light beam instead of flashes, is with ~100 Hz approximately twice as high as that of humans. So birds would perceive a movie on a television screen as a sequence of individual static pictures. The visual system of birds might have evolved to additionally be the basis for magnetic field perception (see Chap. 20).

Like other animals, birds try to keep the picture of objects steady on their retina during self-movement. However, since their eye and head geometry leaves less space for eye movements, they move their entire head rather than their eyes for compensation movements, explaining the peculiar head movements of many birds during walking.

Birds easily learn to discriminate stimuli of different colors and many species rely on color information, e.g., for sexual selection. Bird color vision uses four types of cones, one of which is tuned to light in the UV range. The spectral tuning is additionally sharpened by **oil droplets** in the photoreceptors, which cut off light of short wavelengths. The best-studied birds in visual neuroscience are pigeons and chicken. However, an extraordinary example for the interaction of vision and behavior was found in the western scrub jay. This North American bird hides its food in numerous places scattered over wide areas in montane terrain, and is capable to remember and to find the majority of these caches in wintertime. Apart from an excellent memory, this behavior requires a highly sophisticated visual system, which can imagine how a specific location will look like when it is covered in snow. Moreover, these birds are the only known example of non-primate animals who can also imagine what others might see: when they are observed by 'fellow' scrub jays during foraging, they often only pretend to hide their food, but retain it in their beak to later hide it elsewhere.

### 18.3.3 Reptiles

The visual system of most reptiles – maybe except for the freshwater turtle – is less well studied than vision in mammals and birds. Like birds, most reptiles have one type of rod and four types of cones with oil droplets to sharpen their spectral sensitivity. The ability to discriminate colors plays a role in the social behavior of many reptiles, e.g., for color language during courtship behavior (see Fig. 18.12). In reptiles, major aspects of motion vision take place in the retina. In freshwater turtles it was shown that a high percentage of retinal ganglion cells are **direction-selective**, reacting specifically to motion in one direction (see Box 18.7).

In contrast to birds and amphibians, many reptiles use eye movements in behavioral contexts, e.g., freshwater turtles adjusting their visual streak (elongated band of highest visual acuity) to the water surface and chameleons scanning a full 360° arc of vision with separately moving eyes.

Snakes have developed a particular organ to perceive radiation in the infrared spectrum, the **pit organ**. This organ is used mainly to detect warm-blooded prey based on the temperature difference to the background (see Chap. 15). It is located near the mouth of the snakes and anatomically strongly resembles an eye. Neuronal information coming from eyes and pit organs are merged in the snakes' tectum.

### 18.3.4 Amphibians

Since amphibians provide very stable experimental conditions, they play a certain role in retina research, in particular the tiger salamander, the clawed frog *Xenopus* (mainly for developmental studies), *Rana* frogs, and mudpuppies. Behaviorally, a striking feature of the amphibian visual system is the absence of eye movements, making, e.g., toads blind to nonmoving scenes, as long as they do not move their head or body. On the cellular side, some specific features were found in the retina, e.g., *Rana* frogs and *Xenopus* possess two types of rods.

### 18.3.5 Fish

In visual neuroscience research, zebra fish are standard organisms to study, e.g., color vision and optomotoric responses due to the genetic possibilities they offer. Carp, gold fish, and catfish are also used frequently. However, visual systems of different fish species vary greatly. Fish eyes have evolved a range of different specializations to adapt to their specific habitats, ranging from colorful coral reefs to complete darkness of the deep oceans.

For example, in most fish the crystalline lens is the only source of refraction, because the refractive index of the cornea is very similar to that of water, making it a purely protective tissue. However, some fish that spend most of the time at the water surface use specific adaptations of their cornea. Most strikingly, the eyes and also the visual brain regions of four-eyed fish are divided into two sets, one for seeing under water, and one for searching the air and water surface for insects.

Deep-sea fish have evolved several adaptations to deal with the very low light conditions in their habitat. Anatomical adaptations include extremely large eyes, secondary retinas, stacking of several layers of photoreceptors, photopigments with absorption maxima in the blue range (470–480 nm), reflecting layers of tissue in the back of the eyes, heating of the eyes, and bioluminescence of different body parts to increase the available number of photons.

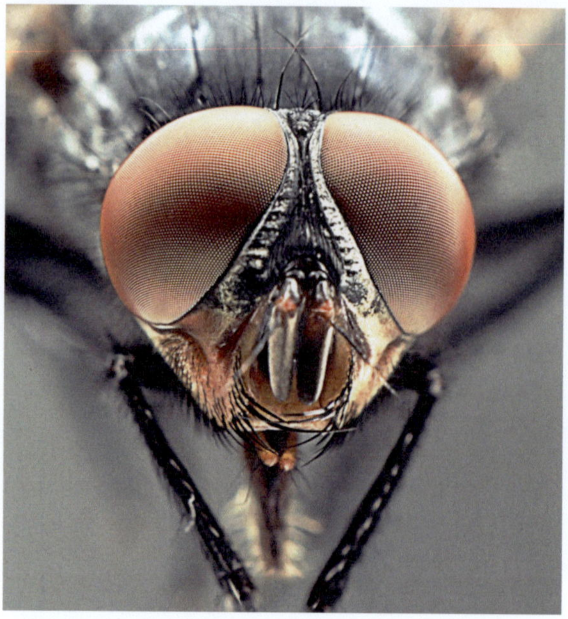

**Fig. 18.13** **Fly eyes.** Head of the fly *Calliphora vicina* with the two huge facet eyes consisting of several thousands of ommatidia, and the three small ocelli on the top of the head (Picture kindly provided by Hein Leertower)

## 18.4 Insects

The visual systems of different invertebrates can differ considerably, depending on their lifestyles, ranging from almost blind nematodes and deep-sea crabs to color-specialists like butterflies and mantis shrimp, motion-specialists like dragonflies, or acuity-specialists like jumping spiders and octopuses. Accordingly, the types of eyes and the subsequent processing of visual information differ enormously between different invertebrates, and many species have developed a highly specialized anatomy for specific visual tasks. We will focus on the highly developed and well-studied visual system of the fly (reviews e.g. [4, 44]), and only mention some specific differences in other invertebrates.

Cajal discovered a century ago striking similarities of neural circuits underlying vision in vertebrates and flies. Many studies confirmed that the visual systems of vertebrates and insects share several structural, functional, and developmental features, e.g., the arrangement of cells in parallel layers, the parallel processing of different visual features like color and motion, and some developmental strategies and molecules. Some basic principles, however, like the anatomy of the eye (compare Figs. 18.1, 18.13, and 18.14), the last steps of phototransduction (Fig. 18.3), and central processing of visual information are fundamentally different in vertebrates and flies.

### 18.4.1 Sensory Organs, Receptors and Transduction

#### 18.4.1.1 Fly Eyes

Insects have two types of eyes. In flies, the major compound eyes consist of hundreds or thousands of facets containing eight photoreceptors. Ocelli are simpler secondary eyes.

The major eyes of insects, called **compound eyes**, consist of hundreds or thousands of individual photoreceptor units with separate lenses, called **ommatidia** or facets (Fig. 18.13). The visual acuity of animals with compound eyes is generally poor compared to animals with lens eyes. For example, each eye of the fruit fly covers almost 180° of visual space with only 700 ommatidia, resulting in an approximately 500-fold lower resolving power compared to the human visual system. In fly eyes, each ommatidium contains eight photoreceptors, two of which (R7, R8) are specialized for color perception. The other six (R1–R6) provide the basis for motion vision.

The three **ocelli** (Latin “small eyes”) on the dorsal surface of the fly head (Fig. 18.13) consist of a single lens and a photoreceptor layer. They give input to a separated visual pathway. Since several ocelli photoreceptors converge to one second-order neuron, ocelli are presumably not used for the detection of fine structural details of objects but for large-scale changes of the light intensity such as for localizing the horizon.

#### 18.4.1.2 Receptors

Insect photoreceptors depolarize, while vertebrate photoreceptors hyperpolarize in response to light.

In contrast to the photoreceptors of vertebrates, in which the photoactive region evolved from a cilium, insect photoreceptors (see Fig. 18.2a) possess a photoactive structure, called **rhabdomere**, consisting of numerous microvillar membrane foldings. Rhabdomeres are densely packed with light-sensitive molecules and together they act as light-guide channel. Their strongly compartmentalized structure assures short diffusion distances and hence rapid response kinetics. This is one of the reasons why fly photoreceptors respond faster and can adapt more quickly to a larger range of light intensities than vertebrate rods and cones. Having only one major type of photoreceptor, the same cell is sensitive enough to detect individual photons and to provide the basis for extremely fast motion vision. However, this all-round performance of fly photoreceptors comes at the price of high energy consumption [12].

While vertebrate and invertebrate photoreceptors share several features like their gross anatomy and the complex multiple contact ribbon synapses (see Fig. 18.2a), there are major differences between their electrical responses: In contrast to the hyperpolarizing responses and reduced glutamate release of vertebrate rods and cones, insect photoreceptors depolarize in response to light and increase histamine release (see Fig. 18.2b and Sect. 18.2.2).

In flies, which are specialized to motion vision, six (R1–R6) of the eight photoreceptors contained in each ommatidium (see Fig. 18.14) express the same opsin with a broad absorption spectrum. They provide achromatic input to the motion processing system. Studies with genetically modified *Drosophila* revealed that optomotor responses (body movements used in visual course control to stabilize the retinal image) rely exclusively on these photoreceptors. Chromatic information is processed in a separate pathway, based on the responses of photoreceptors R7 and R8. In each ommatidium, photoreceptor R7 contains one of two opsins sensitive to UV, and R8 an opsin sensitive either to blue or to green light.

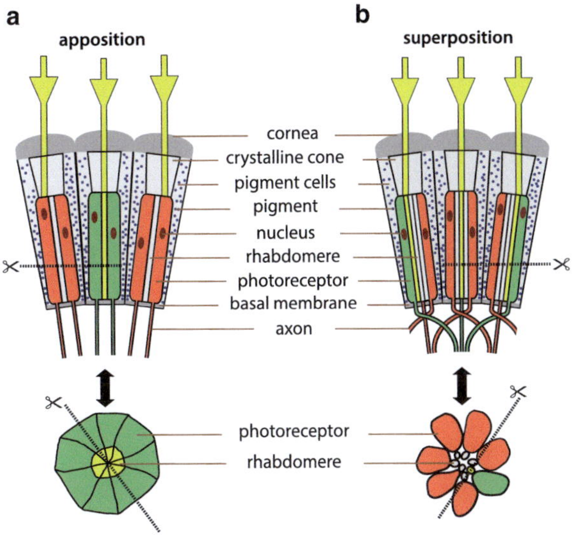


**Fig. 18.14** **Sketch of apposition and neural superposition eyes** (shown in two dissection planes). Rhabdomeres are shown in *yellow* when activated and in *gray* when inactivated, activated somata and axons are shown in *green*, inactivated in *orange*. (**a**) In apposition eyes, e.g., of honeybees the rhabdomeres of all photoreceptors are fused, leading to combined activation of all photoreceptors of one ommatidium (see Sect. 18.5.1). (**b**) In neural superposition eyes of flies, activation is restricted to individual photoreceptors. Signals of photoreceptors from different ommatidia are combined postsynaptically (see Sect. 18.4.2.2 for details). Photoreceptor R8 is not shown, because it is located underneath R7 and both photoreceptors form the central rhabodomere of the ommatidium (Thanks to Sirko Straube for his help with the artwork)

#### 18.4.1.3 Phototransduction

The first steps of phototransduction, namely absorption of a photon, the isomerization of retinal, the conformational change from rhodopsin to metarhodopsin and the activation of a G protein are common to vertebrate and invertebrate photreceptors (see Sect. 18.2.2 and Fig. 18.3) [12]. However, in contrast to vertebrate metarhodopsin, which splits into opsin and retinal, the insect **metarhodopsin** is stable and stays safely positioned in the microvillar membrane. By absorbing another photon (usually of a different wavelength) insect metarhodopsin directly reverts back into its resting state. Since retinal does not need to diffuse out of the cell and back, this mechanism increases the speed of fly photoreceptor responses compared to rods and cones. Moreover, it provides a great advantage for adaptation, because the more photons are present, the faster can insect rhodopsin be activated again.

The transduction steps following the activation of a G protein differ considerably between vertebrate and insect systems (see Fig. 18.3). The activated G protein activates phospholipase C (PLC), which cleaves phosphatidyl inositol ($PIP_2$) into diacylglycerol (DAG) and

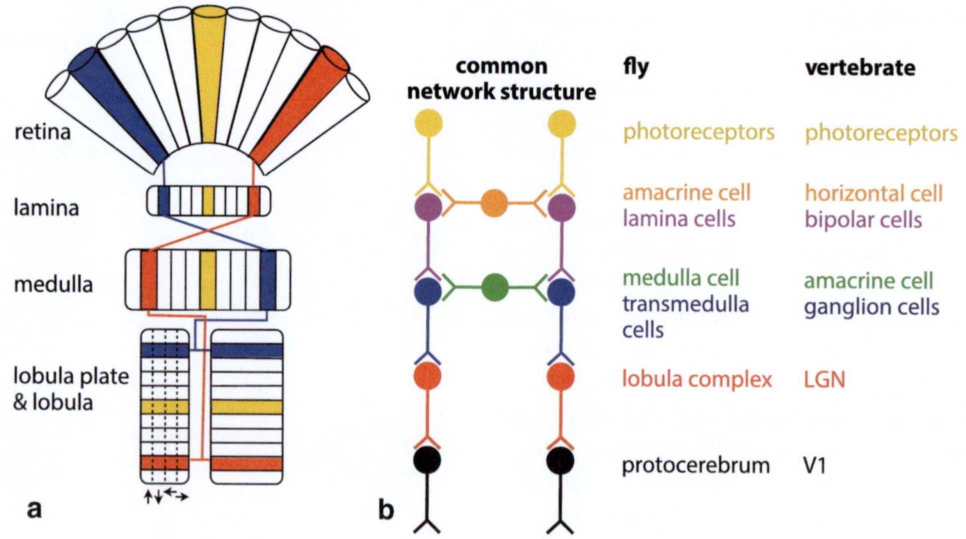


**Fig. 18.15** **Sketches of the anatomy and basic network structure of the fly visual system.** (**a**) Sketch of the anatomical structure of the fly visual system, consisting of the retina in the compound eye and three consecutive optic lobes. Retinotopic organization is depicted by three colored ommatidia and corresponding columns in the optic lobes. In the lobula plate, borders of perpendicular layers are shown with *dashed lines*, *arrows* below indicate the preferred directions of motion stimuli for tangential cells in the respective layers. (**b**) Sketch of the common anatomical network structure of the visual systems of flies and vertebrates, showing corresponding cell types and brain regions. It has to be noted that functionally the fly lobula complex resembles the primary visual cortex in vertebrates, rather than the LGN (**b** – after Sanes and Zipursky [38] with permission)

inositol triphosphate ($IP_3$). The exact opening mechanism of the light-gated TRP channels is still unresolved for fly photoreceptors. However, it is known that opening of the first TRP channel leads to $Ca^{2+}$ and $Na^+$ influx. Due to the tiny volume of the microvillar cell structure, this increase in microvillar $Ca^{2+}$ concentration is sufficient to trigger a positive feedback loop, leading to rapid opening of additional TRP channels. $Na^+$ and $Ca^{2+}$ depolarize the membrane, causing increased synaptic release of the transmitter histamine.

## 18.4.2 Central Processing

### 18.4.2.1 Anatomy

> The visual system of the fly consists of retinotopically arranged optic lobes.

The fly retina consists solely of photoreceptors, arranged in ommatidia. The subsequent processing takes place in optic lobes (Fig. 18.15) of the protocerebrum (see Chap. 2), which are organized in columns forming a retinotopic map (Fig. 18.15a). The first optic lobe following the retina is the **lamina**, which is organized in repetitive cartridges. Each photoreceptor R1–R6 provides input to four types of lamina cells in a single cartridge. The axons of photoreceptors R7 and R8, as well as the axons of lamina cells, arborize in distinct layers of the **medulla**, where they contact different interneurons and transmedullary neurons. The ten layers of the medulla, which are perpendicular to the retinotopic cartridges, can be attributed to separate processing of information, such as motion and color. The medulla contains a large number of different cell types, most of which are not functionally characterized yet. Many of them extend across several columns. The most central region of visual information processing in flies is the **lobula complex**, consisting of the **lobula** and the **lobula plate** (the lobula plate is a fly-specific structure among insects). Both lobula and lobula plate are innervated by axons from neurons in specific layers of the medulla. Many of the numerous different types of neurons in the lobula plate are involved in the processing of motion information, in particular 60 different individually characterized **tangential cells**. Their large dendrites integrate information from hundreds of photoreceptors, which is still present in a retinotopic way, even though the lobula complex is not organized in separated columns. The dendrites of tangential cells form four layers in the lobula plate, corresponding to their preferred directions of motion stimuli (Fig. 18.15a).

#### 18.4.2.2 Function: Processing of Stimulus Motion and Color

The fly visual system is specialized for motion processing based on neuronal superposition eyes, local motion detection, and large-field integration of motion information. Color information is transmitted via a separate pathway.

Like in the vertebrate visual system, information about stimulus color and about stimulus motion is transmitted via separate pathways [44]. The visual motion pathway of the fly consists of photoreceptors R1–R6 and cells in the three optic lobes lamina, medulla, and lobula plate. The receptive fields of neurons involved in motion vision increase in size and complexity in each step of this pathway.

In the first step, lamina cells pool responses of photoreceptors with the same optical axis. Each light beam activates only a single photoreceptor in each ommatidium, because the photoreceptors have slightly different optical axes (Fig. 18.14b). Therefore, the optical axis of each photoreceptor R1–R6 is shared by photoreceptors in six neighboring ommatidia. These six photoreceptors, which are activated by parallel light beams, send their axons to the same postsynaptic targets in the lamina. This convergence of photoreceptor signals, called **neural superposition**, results in a better signal-to-noise ratio.

In the next step, local motion computation presumably takes place in the medulla; however, the functions of this neuropil are not understood well yet. Finally, wide-field direction-selective tangential cells in the lobula plate integrate local motion information. Their complex spatiotemporal receptive fields are matched to optic flow patterns occurring in specific behavioral situations, e.g., turns around a distinct body axis during flight. To achieve this goal, many tangential cells integrate inputs from both eyes and are interconnected with each other. For the mechanisms underlying fly visual motion detection and direction selectivity, see Box 18.7 [4, 10].

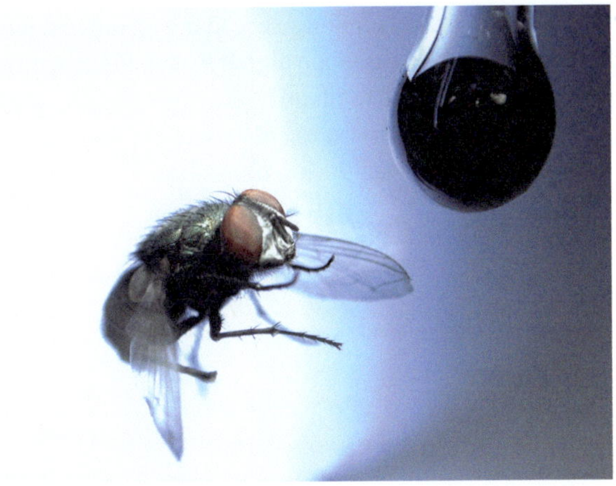

**Fig. 18.16** **Blowfly approaching a dummy fly.** Male blowflies follow flying females with virtuosic flight maneuvers and catch them in the air. The male fly shown here follows a little moving glass globe (dummy fly) and stretches out its legs to catch it (Picture kindly provided by Norbert Boeddeker)

Extraction of color information starts in the medulla. In flies, specific cells of the medulla receive direct input from photoreceptors R7 and R8 and also input from one type of lamina cells, which transfers information from photoreceptors R1–R6. They could act as color-opponent cells (see Box 18.1), comparing the spectrally broadly tuned input from the lamina cell with the wavelength-specific direct input from photoreceptors R7 or R8. Other insects probably use the signals of receptors with different spectral sensitivities for direct comparison. In bees and butterflies, relying behaviorally much more on color vision than flies, color-opponent cells were found in the medulla, the lobula, and the central brain.

### 18.4.3 Vision and Behavior

Insects show an abundance of impressive visually guided behaviors, some of which show principles of visual information processing even clearer than in vertebrates. Flies perform highly acrobatic flight maneuvers, in particular male flies chasing females and catching them in the air (Fig. 18.16). For this behavior, the eyes of male flies have specific zones with increased photoreceptor density and a specific pathway of information processing, which are absent in females. During cruising flight, the fly's course is a sequence of relatively straight translation movements and extremely fast body turns, resembling saccadic eye movements in vertebrates, but at a much higher speed. Presumably, this strategy is used to identify the three-dimensional structure of their environment based on the motion information perceived during translation when nearby objects produce faster optic flow than the background.

## 18.5 Specific Differences in Other Invertebrates

Not all invertebrates use microvillar photoreceptors with a phototransduction cascade based on $Ca^{2+}$ and PLC like insects and other arthropods. In annelids and many mollusks **microvillar photoreceptors** with

**Box 18.7 Direction Selectivity (DS)**

In behavioral situations, the visual input is full of motion: External objects are moving and a behaving animal rarely sits completely motionless, but usually induces optic flow, movement of the retinal picture by self-motion of eyes, head, or body. It is important to estimate self-motion and object motion correctly. Optic flow induced by self-motion also provides cues about the three-dimensional structure of the environment.

From a theoretical point of view, the prerequisites to detect motion are two inputs, which interact asymmetrically. In 1961, Reichardt (1924–1992) developed a phenomenological model for motion detection [36] to explain insect **optomotor responses**, compensatory self-movement in response to large-field stimulus motion: The correlation-type motion detector (see panel **a**) explains the origin of direction selectivity by a multiplicative interaction (M) of two photoreceptor inputs to an integrating postsynaptic target and the temporal delay ($\tau$) of one of these signals. Subtraction of two mirror-symmetrical subunits, summation (SUM) of a positive (+) and a negative (−) signal, enables the detector to respond in a fully direction-selective way, i.e., to produce a positive output signal during motion in one direction and a negative output during motion in the opposite direction.

In the fly visual system (panel **b**) (see also Sect. 18.4.2), the subtraction of local movement inputs with opposite preferred direction is thought to be implemented by pairs of excitatory cholinergic and inhibitory GABAergic synapses located next to each other on the dendrites of direction-selective tangential cells in the lobula plate. The responses of each local input to stimulus patterns moving with a constant velocity are temporally modulated, because responses are only induced when brightness changes locally in their very small receptive field. Excitation and inhibition are both activated during motion, but to different extents depending on the velocity. The integration of many phase-shifted local input pairs causes stable widefield responses, reflecting the time course of visual motion.

In many vertebrates, e.g., rabbits and turtles, direction-selective responses can be found in specific retinal ganglion cells (DS-RGC). In 1965, Barlow and Levick proposed a computational model for DS-RGC responses [2], in which inhibition induced by movement in the antipreferred direction is the most important factor (panel **c**). In contrast to the multiplication implemented in the model applying to the fly visual system, the nonlinearity in this model consists of a logical AND-operation, receiving one input is temporally delayed ($\tau$: temporal delay) and negated ($\neg$: negation), leading to a veto-gate ("AND-NOT").

The key to understanding the origin of retinal direction-selectivity is the spatial distribution (see panel **d**) of excitatory and inhibitory synapses on the dendrite of the direction-selective ganglion cell (DS-RGC, *blue*). Inhibitory synapses of starburst amacrine cells (AC, *green*) are located more proximal than the excitatory bipolar cell synapses (BC, *mauve*). During stimulus motion in preferred direction, photoreceptor PR1 (*yellow*) is activated first and excites the ganglion cell body via an OFF-bipolar cell, causing spike responses before a small inhibition from a starburst amacrine cell arrives. In antipreferred

cGMP-based phototransduction like in vertebrates were found additionally to the **ciliar photoreceptors**. Jellyfish (cnidarians) display exclusively ciliar photoreceptors, suggesting that this type used by vertebrates is evolutionarily older than the fast and highly dynamic photoreceptors of insects [12].

Most arthropods use complex eyes and optic lobes in the central nervous system for image-forming vision. However, other invertebrates, in particular mollusks have evolved several different eye designs and also differ in their central processing of visual information. Many lower invertebrates, e.g., many annelids do not have image-forming vision, but possess light-sensing organs on different body parts, allowing at least to discriminate light and darkness.

**Box 18.7 Direction Selectivity (DS)** *(continued)*

direction, however, inhibition from the more proximal amacrine cell synapse arrives at the cell body prior to the excitation caused by bipolar cells. The inhibitory inputs prevent spike responses in the ganglion cell by shunting the excitatory signals traveling along the ganglion cell dendrite. The cells providing the asymmetric inhibitory input are starburst amacrine cells, which were shown to be direction-selective themselves. Their dendrites act as isolated, separate units of computation, in which motion away from the amacrine cell soma causes larger calcium transients than motion towards the soma. However, the underlying details of retinal direction selectivity are even more complicated (e.g., because many DS-RGCs are of ON/OFF type, responding to objects both brighter and darker than the background, and because starburst amacrine cells also give input to bipolar and amacrine cells) and still not completely understood.

In contrast to other vertebrates, no DS-RGCs were found in the primate retina. In primates, V1 is the first brain region in the motion vision pathway showing directional selectivity. Directional selectivity of V1 neurons is again based on the asymmetry of dendritic integration. In this case the asymmetric dendrites of Meynert cells in layer IVc of V1 provide the anatomical substrate for implementing directional-selective neural responses. In addition to the inhibition-based mechanism described for retinal ganglion cells, Meynert cells implement an excitation-based mechanism that relies on the exact timing of inputs. For this mechanism, it is assumed that the RF consists of two excitatory subregions and that one of the signals arrives temporally delayed at the Meynert cell dendrite. If a bar moves in the preferred direction of such a neuron, fast and slow inputs arrive in phase and add up, allowing the neuron to cross a firing threshold and become active. In nonpreferred direction, inputs are out of phase and do not cross threshold. (Source: (b) after [10], (d) simplified after Oesch et al. [30] all with permission).

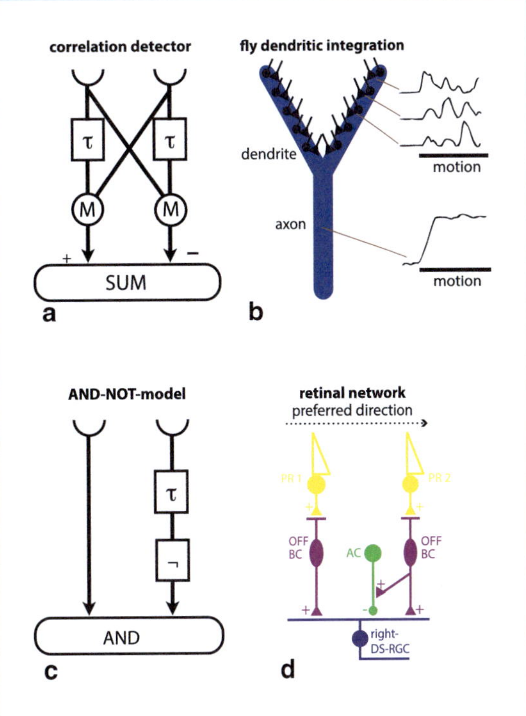

### 18.5.1 Other Insects

The eye anatomy of insects reflects their lifestyles: Diurnal color-specialists like bees have apposition eyes, nocturnal insects like moths optical superposition eyes, and motion vision specialists like flies neuronal superposition eyes.

There are two main types of eye anatomy in insects, superposition and apposition eyes:

The principle of **superposition** is that signals of photoreceptors located in different ommatidia are combined according to their optical axes. While flies use neural superposition (see Sect. 18.4.2.2, and Fig. 18.14b) optimized for high-speed motion vision, nocturnal insects like moths have developed optical superposition eyes for optimal quantum efficiency under low-light conditions. Their eyes possess a clear zone between crystalline lenses and photoreceptors. In darkness, a shading pigment retracts between the lenses and the clear zone directs light beams to photoreceptors in several ommatidia, increasing the number of photons triggering phototransduction.

In **apposition eyes** (Fig. 18.14a), the rhabdomeres of all photoreceptors in one ommatidium are fused, while ommatidia are optically isolated from each other. Each individual ommatidium produces an isolated image, which is transmitted isolated from the signals coming from the other ommatidia, until these images are combined in the brain. Apposition eyes, which are typical for diurnal insects (e.g., honeybees, butterflies, and grasshoppers) are advantageous for color vision. The fused rhabdomeres sharpen spectral sensitivities by transmitting monochromatic light to the photoreceptor expressing the best-matching opsin. The color vision of most insects with apposition eyes is probably much better than that of most mammals. Already in 1914, von Frisch (1886–1982) showed that honeybees trained to feed from sugar solution presented on a blue or yellow card could discriminate these colors from 30 different shades of gray, demonstrating color vision in insects [14]. Butterflies are color vision specialists, probably not only because color helps to identify flowers during food foraging, but also for sexual selection. Many butterfly species show a sexual dimorphism in wing colors in the UV range, invisible to humans.

In addition to light intensity and wavelength, many insects are able to perceive the pattern of **light polarization** (see Sect. 18.1) that is characteristic for the position of the sun. In photoreceptors, which are sensitive to polarized light, the elongated chromophore molecules are placed along the direction of the microvilli, which are arranged in parallel to each other. Absorption is most effective when the direction of light polarization equals the direction of the molecules. Desert ants use their **sun compass** based on the specialized polarization-sensitive dorsal rim region of their eyes to find their way straight back to the nest after a meandering foraging tour. Bees communicate the position of food sources relative to the sun position by their waggle dance.

### 18.5.2 Other Arthropods

The visual systems of crustaceans resemble insect visual systems. Spiders use lens eyes evolved from ocelli for image-forming vision.

The visual systems of crustaceans are very similar to the insect visual system. Their compound eyes are usually of the apposition type (Fig. 18.14a). Except for deep-sea crustaceans, many species have a highly developed system for color vision. Having up to 12 different receptor types, stomatopod crustaceans hold the world record in opsin expression. However, these receptors are limited to a narrow band of ommatidia, with which the animals scan their visual environment for color cues.

Among the Chelicerata, the horseshoe crabs (e.g., *Limulus*) have two compound eyes of the apposition type and two ocelli. In contrast, spiders usually have four pairs of **lens eyes**, the location of which varies between spider species. The two main eyes at the front of the spider's head are ocelli, which in contrast to insect ocelli are specialized to image-forming vision. The secondary eyes were probably derived from compound eyes, but do not show separate facets. Jumping spiders are specialized on the visual task of hunting, showing similarities to the visual systems of predatory mammals. The two principal forward-facing eyes function like a mammalian fovea, providing high spatial acuity (ten times better than the best compound eye of dragonflies) and color vision with four receptor types. The six secondary eyes are color-blind and specialized to detect movement at the periphery with wide fields of view, similar to the peripheral retina in vertebrates.

### 18.5.3 Nematodes and Mollusks

Nematodes live in a dark environment and do not have specialized light-sensing organs. However, they are able to avoid strong light. In their nervous system, four pairs of light-sensitive neurons were found which depolarize in response to light. These responses are based on cGMP-sensitive cyclic nucleotide-gated (CNG) channels, resembling the vertebrate photoreceptor membrane.

While most mollusks have only limited visual abilities, cephalopods have highly developed visual systems.

Mollusks have evolved a broad spectrum of different eye designs. For example, most clams have quite poor visual abilities. However, the giant clam *Tridacna* has trichromatic vision and high enough resolution to detect approaching fish with hundreds of small pinhole eyes. Cephalopods (octopuses, squids, cuttlefishes, and nautiluses) all have highly developed vision, which in most species is based on a lens eye similar to the vertebrate eye. However, most of these animals only have one spectral type of photoreceptor and therefore are unable to discriminate colors. It remains a mystery how they are nevertheless able to change their body colors to match their environment. An exception to this rule is the firefly squid. It has a two-colored bioluminescence and a narrow region of specialized photoreceptors, which differ peculiarly not in their opsins but in three types of chromophores.

## 18.6 Summary

The sense of vision is a very prominent source of information for behavioral decisions of most organisms ranging from mollusks to man. Many of the fundamental principles of visual systems are evolutionarily very well preserved.

The first step of vision is always phototransduction in specialized nerve cells. Photoreceptors of all animals possess either a microvillar structure and phototransduction based on phospholipase C and $Ca^{2+}$ dynamics like in insects, or they use the evolutionarily older ciliary type employing a cGMP-based photoreceptor cascade like vertebrates and simple invertebrates.

Based on photoreceptor responses, the visual systems of all animals have the common task to extract behaviorally relevant information, which is performed by a number of common mechanisms. Photoreceptor responses converge onto a smaller number of specialized cells with receptive fields, increasing in size and spatiotemporal complexity in each step of processing. Information about different features of the visual environment, e.g., color and motion, are propagated through separate, parallel channels. Interestingly, the visual system of insects and vertebrates resemble each other even in their coarse anatomical structure consisting of at least five feed-forward connections and two levels of lateral interactions. Moreover, central visual areas are in both cases organized in a retinotopic and a perpendicular columnar structure.

Color perception relies on responses of photoreceptors, which are sensitive to light of different ranges of wavelength. Color perception requires the combination of responses of different photoreceptor types in subsequent steps of visual information processing, i.e., color-antagonistic receptive fields of postsynaptic neurons. In most animals, the photoreceptor spectral sensitivities are determined by three or four different opsins. However, most mammals express only two opsins, and other animals (e.g., octopus) are completely color-blind, because they possess only one type of photopigment. On the other hand, color specialists like butterflies or stomatopod crustaceans have developed as many as 12 opsins, presumably leading to the perception of fine nuances of color differences. It should be kept in mind that most animals are sensitive to a broader spectrum of wavelength than we are, because humans do not express an opsin sensitive in the range of UV light.

Another common principle is that motion vision is processed separately from color information in different pathways. The basic principle of motion detection is an asymmetric interaction of information about light intensities at different positions. Depending on species, this computational step is performed in the retina like in turtles, or in central parts of the nervous system like in primates and also in insects.

However, despite of generally similar mechanisms and principles found in vertebrates and invertebrates, a striking feature of visual systems is the diversity between species. Individual species have developed a huge number of specializations to account for their specific visual environments and visually guided tasks.

## References


1. Dayan R, Abbott LF (2001) Theoretical neuroscience: computational and mathematical modeling of neural systems. MIT Press, Cambridge [Textbook]
2. Barlow HB, Levick WR (1965) The mechanism of directionally selective units in rabbit's retina. J Physiol (London) 178:477–504
3. Ben-Yishai R, Lev Bar-Or R, Sompolinsky H (1995) Theory of orientation tuning in visual cortex. Proc Natl Acad Sci USA 92:3844–3848
4. Borst A (2009) *Drosophila*'s view on insect vision. Curr Biol 19:R36–R47
5. Carandini M, Demb JB, Mante V, Tolhurst DJ, Dan Y, Olshausen BA, Gallant JL, Rust NC (2005) Do we know what the early visual system does? J Neurosci 25:10577–10597
6. Conway BR, Livingstone MS (2006) Spatial and temporal properties of cone signals in alert macaque primary visual cortex. J Neurosci 26:10826–10846
7. Das A (2005) Cortical maps: Where theory meets experiments. Neuron 47:168–171
8. Dowling JE (1987) The retina: an approachable part of the brain. Harvard University Press, Cambridge [Textbook]
9. Eagleman DM (2001) Visual illusions and neurobiology. Nat Rev Neurosci 2:920–926
10. Egelhaaf M (2006) The neural computation of visual motion information. In: Warrant E, Nilsson DE (eds) Invertebrate Vision. Cambridge Univ Press [textbook]
11. Euler T, Hausselt SE, Margolis DJ, Breuninger T, Castell X, Detwiler PB, Denk W (2009) Eyecup scope – optical recordings of light stimulus-evoked fluorescence signals in the retina. Pflugers Arch 457:1393–1414
12. Fain GL, Hardie R, Laughlin SB (2010) Phototransduction and the evolution of photoreceptors. Curr Biol 20:R114–R124
13. Felleman DJ, Van Essen DC (1991) Distributed hierarchical processing in the primate cerebral cortex. Cereb Cortex 1:1–47
14. von Frisch K (1914) Der Farbensinn und Formensinn der Biene. Zool Jahrb Allg Jena 35:1–182
15. Funke K, Kisvarday Z, Volgushev M, Wörgötter F (2002) Integrating anatomy and physiology of the primary visual pathway: from LGN to cortex. In: Van Hemmen L (ed) Models of neural networks IV. Springer, Berlin/Heidelberg/New York, pp 97–171
16. Hartline HK (1938) The response of single optic nerve fibers of the vertebrate eye to illumination of the retina. Am J Physiol 121:400–415
17. von Helmholtz H (1867) Handbuch der physiologischen Optik von Helmholtz. Verlag L. Voss, Leipzig. Download: vlp.mpiwg-berlin.mpg.de/library/data/lit39509?
18. Hubel DH, Wiesel TN (1962) Receptive fields, binocular interaction and functional architecture in the cat's visual cortex. J Physiol (London) 160:106–154
19. Huberman AD, Feller MB, Chapman B (2008) Mechanisms underlying development of visual maps and receptive fields. Annu Rev Neurosci 31:479–509
20. Itti L, Koch C (2001) Computational modelling of visual attention. Nat Rev Neurosci 2:194–203
21. Kelber A, Vorobyev M, Osorio D (2003) Animal colour vision – behavioural tests and physiological concepts. Biol Rev 78:81–118
22. Kelber A (2006) Invertebrate colour vision In: Warrant E, Nilsson DE (eds) Invertebrate Vision. Cambridge Univ Press, Cambridge/New York [Textbook]
23. King AJ (2004) The superior colliculus. Curr Biol 14: R335–R338
24. Konen CS, Kastner S (2008) Two hierarchically organized neural systems for object information in human visual cortex. Nat Neurosci 11:224–231
25. Kuffler SW (1953) Discharge patterns and functional organization of mammalian retina. J Neurophysiol 16:37–68
26. Marshal J, Oberwinkler J (1999) Ultraviolet vision: The colourful world of the mantis shrimp. Nature 401:873–874
27. Markram H, Toledo-Rodriguez M, Wang Y, Gupta A, Silberberg G, Wu CZ (2004) Interneurons of the neocortical inhibitory system. Nat Rev Neurosci 5:793–807
28. Masland RH (2001) The fundamental plan of the retina. Nat Neurosci 4:877–886
29. Nassi JJ, Callaway EM (2009) Parallel processing strategies of the primate visual system. Nat Rev Neurosci 10: 360–372
30. Oesch N, Euler T, Taylor WR (2005) Direction-selective dendritic action potentials in rabbit retina. Neuron 47:739–750
31. Osorio D, Vorobyev M (1997) Sepia tones, stomatopod signals and the uses of colour. Trends Ecol Evol 12: 167–168
32. Palmeri TJ, Gauthier I (2004) Visual object understanding. Nat Rev Neurosci 5:291–303
33. Passingham R (2009) How good is the macaque monkey model of the human brain? Curr Opin Neurobiol 19:6–11
34. Ramón y Cajal S (1893) La retine des vertebres. Cellule 9:121–255. English translation in: Thorpe SA, Glickstein M (1972) The structure of the retina. Thomas, Springfield
35. Raz A, Buhle J (2006) Typologies of attentional networks. Nat Rev Neurosci 7:367–379
36. Reichardt W (1961) Autocorrelation, a principle for the evaluation of sensory information by the central nervous system. In: Rosenblith W (ed) Sensory communication. MIT Press/Wiley, Cambridge/New York, pp 303–317
37. Ringach D, Shapley R (2004) Reverse correlation in neurophysiology. Cogn Sci 28:147–166
38. Sanes JR, Zipursky SL (2010) Design principles of insect and vertebrate visual systems. Neuron 66:15–36
39. Series P, Lorenceau J, Fregnac Y (2003) The "silent" surround of V1 receptive fields: theory and experiments. J Physiol Paris 97:453–474

40. Shushruth S, Ichida JM, Levitt JB, Angelucci A (2009) Comparison of spatial summation properties of neurons in macaque V1 and V2. J Neurophysiol 102:2069–2083
41. Solomon SG, Lennie P (2007) The machinery of colour vision. Nat Rev Neurosci 8:276–286
42. Stein BE, Stanford TR (2008) Multisensory integration: current issues from the perspective of the single neuron. Nat Rev Neurosci 9:255–266
43. Tsodyks M, Kenet T, Grinvald A, Arieli A (1999) Linking spontaneous activity of single cortical neurons and the underlying functional architecture. Science 286:1943–1946
44. Warrant E, Nilsson DE (eds) (2006) Invertebrate vision. Cambridge University Press, Cambridge/New York [Textbook]
45. Wässle H (2004) Parallel processing in the mammalian retina. Nat Rev Neurosci 5:747–757
46. Wegener D, Freiwald WA, Kreiter AK (2004) The influence of sustained attention on stimulus selectivity in macaque visual area MT. J Neurosci 24:6106–6114

## Internet Resources

webvision.med.utah.edu/book/ (regularly updated online book covering the visual system and related topics)

www.scholarpedia.org/article/Insect_motion_vision (fundamental information about motion processing)

www.scholarpedia.org/article/Retina (fundamental information about the retina)

brainmaps.org (anatomical connections in visual cortex and other brain areas)

www.csie.ntu.edu.tw/~cjlin/libsvm (software for applying support vector machines in classifying experimental data)

bluebrain.epfl.ch (the Blue Brain project, a large-scale simulation of a cortical column)

www.michaelbach.de/ot (for many fine examples, and some explanations, of visual illusions)

# Electroreception 19

Gerhard von der Emde

Electroreception by animals living in aquatic environments is a widespread phenomenon found in many vertebrates. With **ampullary electroreceptor organs** or **trigeminal electroreceptor** structures, these animals can detect even extremely weak electric sources in their surroundings, a process called **passive electrolocation**.

**Weakly electric fish** have specialized electric organs derived from muscle or nervous tissues, which emit high-frequency electric signals. These signals are detected by cutaneous arrays of **tuberous electroreceptor organs** for the purpose of **active electrolocation** and **electro-communication** in dark and/or turbid habitats. During active electrolocation, nearby objects locally alter the self-produced electrical signals and thus project **electric images** onto the fish's body surface. By scanning these images with their electroreceptors, weakly electric fish can **detect and localize objects** and determine their electrical and spatial properties. Two separate **electrical foveae** allow a detailed three-dimensional analysis of the surroundings. Living objects with complex electric impedances are recognized through a process called **electrical color perception**. In addition, electric fish have a true sense of **depth perception**.

Weakly electric fish have separate electroreceptor organs for measuring the **timing** (**phase**) and the **amplitude** of electric signals. Throughout the brain, there exist separate pathways for processing time and amplitude information. In African weakly electric fish, an electric organ **corollary discharge** signal associated with each electric motor command is relayed to electrosensory structures in the brain. In these fishes, the electroreceptors use a **latency code** to convey stimulus amplitude and waveform information to the brain, where it is transformed into a mixed latency and rate code through interactions with the corollary discharge. Many electrosensory brain nuclei are structures with complex neural circuitries, which often contain several parallel **somatotopic maps** of the electrosensory body surface, in which the foveae are overrepresented. **Recurrent feedback loops** from higher to lower centers allow for memories and motivational features to influence neural processing, e.g., through **associative depression** and **potentiation**. Spike time-dependent **neural plasticity** at several synapses create **negative images** of the recent sensory past and thus allow novel or unexpected features in the sensory input to stand out (**novelty detection**).

## 19.1 Electroreception in Nature

Since humans have no special receptor structures for electricity, it was long unknown that some animals can perceive naturally occurring weak electrical signals. Interestingly, however, the anatomical structures now known to be electroreceptor organs have been known at least since the seventeenth century and described in 1678 by the Italian physician Stefano Lorenzini, but their correct function was not recognized. Only in the early 1960s, it was shown that the **ampullae of Lorenzini** respond to extremely weak electric fields and that the fish use them for their perception. Because electroreception in animals needs a conductive

G. von der Emde
University of Bonn, Institute for Zoology, Neuroethology/Sensory Ecology, Endenicher Allee 11-13, 53115 Bonn, Germany
e-mail: vonderemde@uni-bonn.de

C.G. Galizia, P.-M. Lledo (eds.), *Neurosciences - From Molecule to Behavior: A University Textbook*,
DOI 10.1007/978-3-642-10769-6_19, 

medium, it is usually associated with aquatic or semiaquatic organisms. Many marine and freshwater fishes, with the important exception of most (but not all) teleosts, are electroreceptive [1–3]. Besides fishes, only a few vertebrates and maybe some invertebrates possess this sense, i.e., several aquatic urodele amphibians, the three members of the Monotremata (the platypus, *Ornithorhynchus anatinus*), the long-nosed and the short-nosed echidna (*Zaglossus bruijni* and *Tachyglossus aculeatus*) [4–6], and the Guiana dolphin (*Sotalia guianensis*) [7]. All these animals probably detect electric fields for the main purpose of prey detection, i.e., they find and identify benthic prey animals by the electric fields they emit. Some invertebrates such as cockroaches (*Periplaneta americana*) [8] or the nematode *C. elegans* [9] also respond to electric voltages, however the detection thresholds are so high that these animals probably do not use electroreception in a natural context.

## 19.2 Electroreception in Mammals

Electroreception in monotremes (egg-laying mammals) was discovered in 1986, when it was shown that the platypus could find and attack batteries that were invisibly placed underwater [4]. Weak voltage pulses applied across the bill evoke potentials from the somatosensory cortex of the brain. The behavioral and electrophysiological sensitivity was estimated to be about 50 μV $cm^{-1}$ (but see below), which would enable the animals to locate moving prey by the electrical activity associated with its muscle contractions. In the echidna, electroreception was demonstrated behaviorally and electrophysiologically [10]. This is the first known example of electroreception in a terrestrial animal, which probably uses this ability to find living food in moist soil, e.g., to detect the queens, larvae, and eggs in an ants nest. In contrast to fishes (see below), electroreception in the momotremes is associated with the trigeminal system, with the 5th cranial nerve transmitting electrosensory information from the beak of the animal to its brain. This and other lines of evidence suggest that electroreception in monotremes has evolved independently.

Recently, another aquatic mammal, the Guiana dolphin (*Sotalia guianensis*), was found to be electroreceptive [7]. Remains of vibrissal follicles, which are crypts at the rostrum, act as passive electroreceptors in these animals and are innervated by the trigeminal nerve, like in the monotremes. Their behaviorally determined sensitivity is with about 5 μV $cm^{-1}$ in a similar range as that of the platypus. Guiana dolphins probably use passive electrolocation for close-distance prey detection in murky waters.

## 19.3 Passive Electrolocation in Fishes

In fishes, electroreception is found among almost all nonteleost taxa (Fig. 19.1). The common ancestor of Teleostei and Holostei might have lost the ability to detect weak electric currents for reasons which are not yet understood [1]. During the evolution of teleosts, electroreception was independently reinvented two or maybe even four times. It is present in four teleost groups, two of which are the weakly electric fish from Africa (Mormyriformes) and South America (Gymnotiformes), and the other two electroreceptive teleost groups are the catfishes (Siluriformes) and the Xenomystinae.

Detecting electrical signals emitted from the environment and estimating the position of their source is called **passive electrolocation** [3, 11, 12]. In the aquatic environment, weak electric fields of both abiotic and biotic origin can be found. Most abiotic fields are DC or low frequency and are caused by geochemical and seismic processes, by lightning activity and magnetic storms that charge up the atmosphere, and in seawater, also by the flow of water through the earth's magnetic field. Electric signals of biotic origin are caused by muscle contraction of animals or by biochemical processes (ion flow) across thin epithelia such as gills. These signals surround the animals with very weak electric fields, which are in turn detectable by animals with electroreceptors. In contrast, weakly electric fish actively produce and detect high-frequency electric signals for the purpose of active electrolocation and for electrocommunication.

### 19.3.1 Ampullary Electroreceptors are Used for Passive Electrolocation

All electroreceptive fishes possess so-called ampullary electroreceptor organs which respond to low frequencies between 0.05 and 50 Hz and, if relative movement is considered, also to DC electric fields of

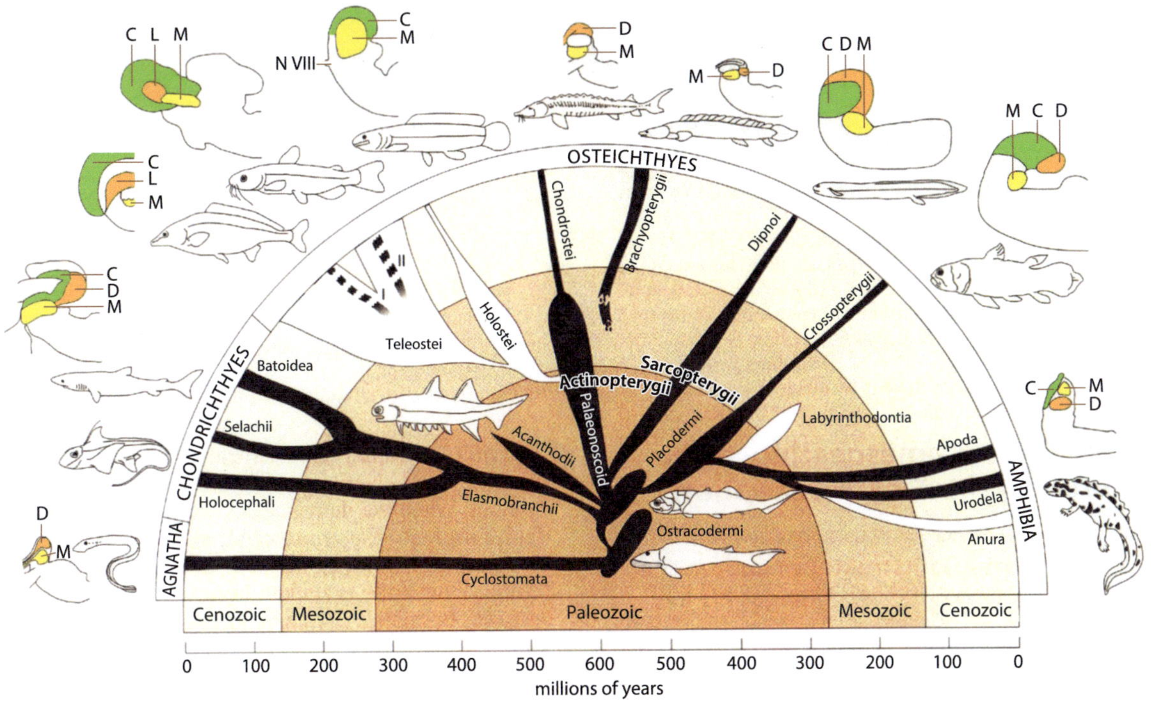


**Fig. 19.1 The evolution of electroreception in fish and amphibians within the past 600 million years.** The *black branches* represent groups that are electroreceptive. Above each *black branch*, a drawing of one representative of this electroreceptive group is shown. In addition, schematized transversal sections through one hemisphere of the hindbrain show the projections areas of the electroreceptive afferent fibers. Cerebellar (*C*) structures are marked in *green*, the electroreceptive dorsal nucleus (*D*) in *orange*, the mechanoreceptive medial nucleus (*M*) in *yellow*. In the teleosts, the electroreceptive lateral line lobe (*L*) is shown in *orange*. The two *black and white branches* of the Teleostei represent the Xenomystinae (passive electroreception) and the African Mormyriformes (active and passive electroreception) which belong to the Osteoglossomorpha (I), and the Siluriformes (passive electroreception) and the South American Gymnotiformes (active and passive electroreception) which belong to the Ostariophysi (II). These groups reinvented electroreception after it had been lost during the evolution of the Holostei (This diagram was kindly provided by Carl D. Hopkins and modified by data from B. Fritzsch and M. Wullimann)

low amplitudes. These receptor organs are used for passive electrolocation [11]. Ampullary electroreceptors were first found in elasmobranch fish, i.e., in sharks and rays, where they are called ampullae of Lorenzini. Their sensitivity to weak electric fields can be striking: Skates, for example, respond reliably to voltage gradients of less than 0.01 $\mu V\ cm^{-1}$ [13], which corresponds to a voltage of 1 V in 1,000 km. Individual receptor cells are less sensitive; e.g., in skates their firing frequency can be changed by a stimulus no smaller than 2 $\mu V\ cm^{-1}$. The much higher behavioral sensitivity of the animal might be explained by averaging over many receptor cells situated in several ampullae and by central nervous processes [11, 14].

Passive electrolocation with ampullary receptor organs is used by electroreceptive fish to locate prey or potential mates for reproduction. With their remarkable sensitivity, elasmobranchs, like some sharks and rays, can detect prey fish buried in the ground by sensing their weak electric fields (a fraction of 1 $mV\ cm^{-1}$). It was shown that in experimental tanks they dug up buried electrodes mimicking electric potentials from prey [13]. Male round stingrays (*Urolophus*) use ampullary receptors to find females, which produce low-frequency electric signals during ventilation of their gills.

Also weakly electric fish possess ampullary receptor organs that are distributed over their entire body surface (Fig. 19.2), and like other electroreceptive fish they use them for passive electrolocation, which helps, for example, to find prey items. In addition to prey sensing, wave-type gymnotiforms (see below) use ampullary organs for sensing brief discharge interruptions of conspecifics during electrocommunication. Usually wave-type EOD will not stimulate ampullary electroreceptors because they lack

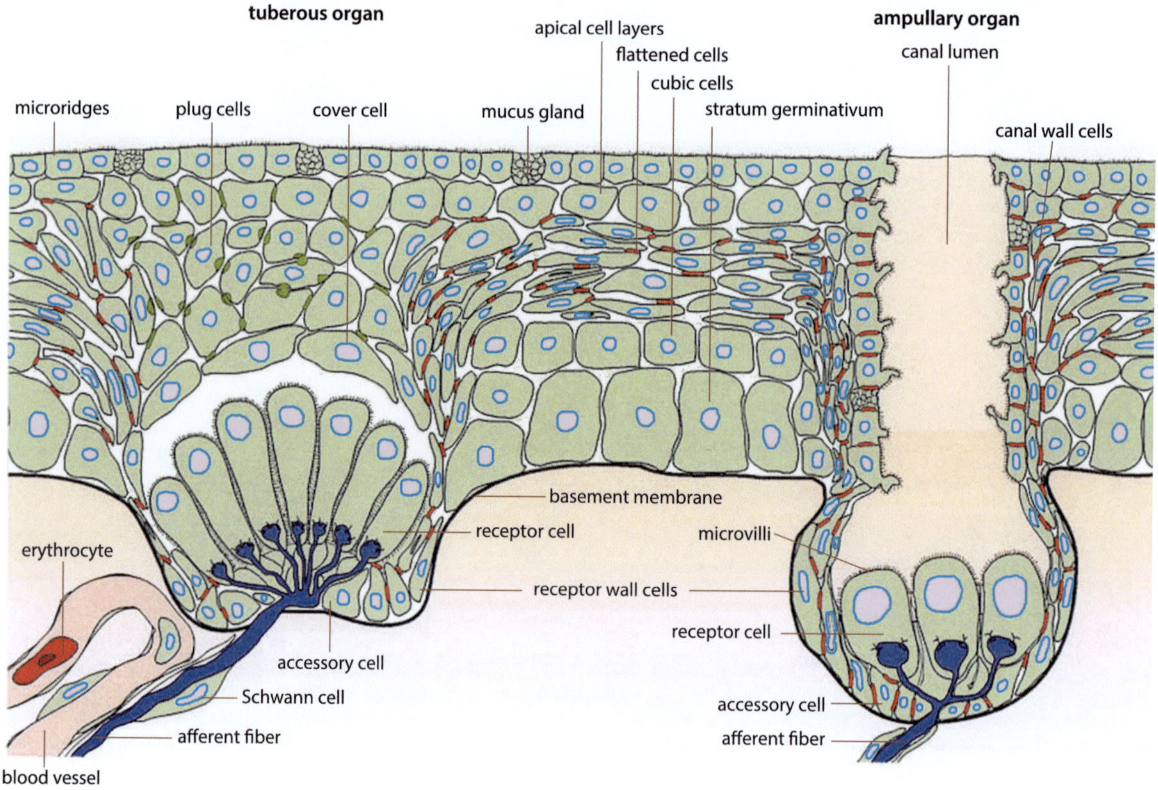


**Fig. 19.2** **Schematic drawing of a skin section with electroreceptor organs of the South American gymnotiform *Eigenmannia***, with a tuberous organ on the *left* and an ampullary organ on the *right*. Because of tight junctions (shown as *red dots*) between the cells, the epidermis has a high electrical resistance. Tight junctions also prevent electrical current flow around the receptor organs. The receptor cells have chemical synapses with afferent nerve fibers (*blue*). The ampullary organs open to the water through a canal which is filled with a conductive gel. Tuberous organs are covered with a layer of loose cells connected by desmosomes (*green dots*), which allow electricity to reach the receptor cells below. A receptor cell has a width of about 10 μm (This drawing was kindly provided by H.A. Vischer)

energy in the low frequency range. However, during discharge cessations, a remaining head-negative offset strongly stimulates them.

The primary afferents innervating the ampullary system convey their information in a somatotopic manner to a subdivision of the electrosensory lateral line lobe (ELL) of the brain (Fig. 19.11). Ampullary receptor afferents in the mormyrid *Gnathonemus petersii*, for example, were found to have their lowest threshold of 40 $\mu V\ cm^{-1}$ at low frequencies of 1–10 Hz and appear to be tuned to a mix of amplitude and slope of the input signals (fractional order filtering). The integration of simultaneously recorded afferents results in a strongly enhanced signal-to-noise ratio and increased mutual information rates. The neuronal integration of inputs from receptors from the left and the right side of the fish, which experience opposite polarities of a stimulus, were shown to theoretically enhance encoding of such stimuli, including an increase in bandwidth. Covariance and coherence analysis showed that spiking of ampullary afferents is sufficiently explained by the spike-triggered averages, i.e., electroreceptors respond to a single feature of an electrical stimulus [14].

## 19.3.2 Trigeminal Electroreceptors in Monotremes

In contrast to all electroreceptive fishes whose electroreceptors belong to the octavo-lateralis system and are innervated by the lateral line nerve (cranial nerve VIII), the three extant monotreme species have electroreceptors which are innervated by the trigeminal nerve (cranial nerve V). The platypus has 40,000 electroreceptors arranged in a characteristic stripe pattern

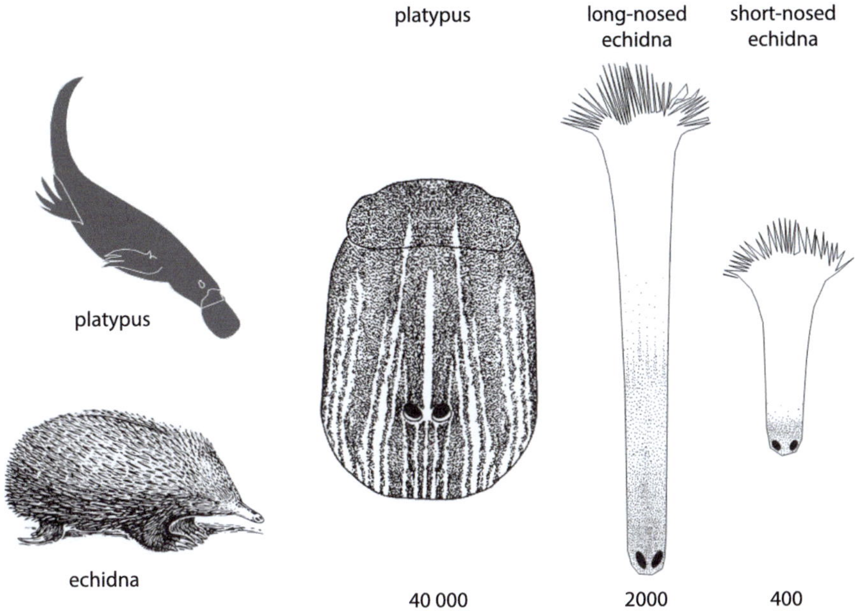


**Fig. 19.3 Distribution of mucous electroreceptors** shown as *black dots* on the bills of the platypus (*Ornithorhynchus anatinus*) and the long-nosed (*Zaglossus brujnii*) and short-nosed (*Tachyglossus aculeatus*) echidna viewed from the dorsal aspect. The number of electroreceptors per animal are given below (Modified after Pettigrew [6] with permission)

only on its bill [15]. Similar in echidnas, which have only 2,000 (long-nosed echidna) or 400 (short-nosed echidna) electroreceptors, they are located at the anterior parts of the snout [6] (Fig. 19.3). The electroreceptors appear as pits formed from secretory ducts of serous or mucous glands, each surrounded by a petal-like arrangement of epithelium that opens when the snout is immersed in water [16]. The trigeminal innervations is very complex with up to 16 different afferent nerve fibers forming specialized nerve endings, which are loaded with mitochondria. In contrast to ampullary electroreceptors of fishes, there are no electroreceptor cells and the nerve endings are depolarized directly by an outside negative voltage. Electroreceptor afferents show high spontaneous activity and respond to low-frequency alternating currents, with cathodal (negative) voltages exciting and anodal (positive) voltages inhibiting the afferent fibers. The platypus can detect electric stimuli as low as 20 $\mu$V cm$^{-1}$, [16] which is considerably lower than the threshold of single electroreceptor afferents (2 mV cm$^{-1}$) [10].

Platypus electroreceptors are closely associated with so-called push-rod mechanoreceptors of the bill, which are also innervated by the trigeminal nerve, and possibly respond to vibrations caused by prey organisms. The central projection zones of both mechano- and electroreceptor afferents are the somatosensory regions in the cerebral cortex. Interestingly, the region of the cortex receiving electrosensory information was found to overlap completely with cortical areas receiving tactile information. This shows that in monotremes, also in the brain the electrosensory system is part of the somatosensory system, which contrasts to the situation in all electroreceptive fishes, where the ampullary system is part of the octavolateralis system.

The platypus and the echidnas employ electroreception in prey search during foraging. The platypus is a very efficient forager and can catch half of its body weight of live prey (mainly shrimps) in one night. When hunting these animals close their eyes and ears – they only use their “bill” as a sense organ to locate their prey when digging amongst stones and detritus on the ground of a stream. They scan the ground by quickly moving their snout from left to right and respond to the tail flip of a shrimp, which evokes a combined mechanosensory and an electric stimulus, with a quick, reflex-like head saccade always in the correct direction towards the prey. It was suggested that the platypus can also judge the distance of its prey by an interaction of the mechanosensitive and electrosensitive receptors from the animal’s snout in the somatosensory cortex. Bimodal cells in this cortical region might be preferentially responsive to a given delay between mechano- and electroreceptive inputs, which corresponds to a certain distance of the prey because of the different propagation velocities of the two modalities [6].

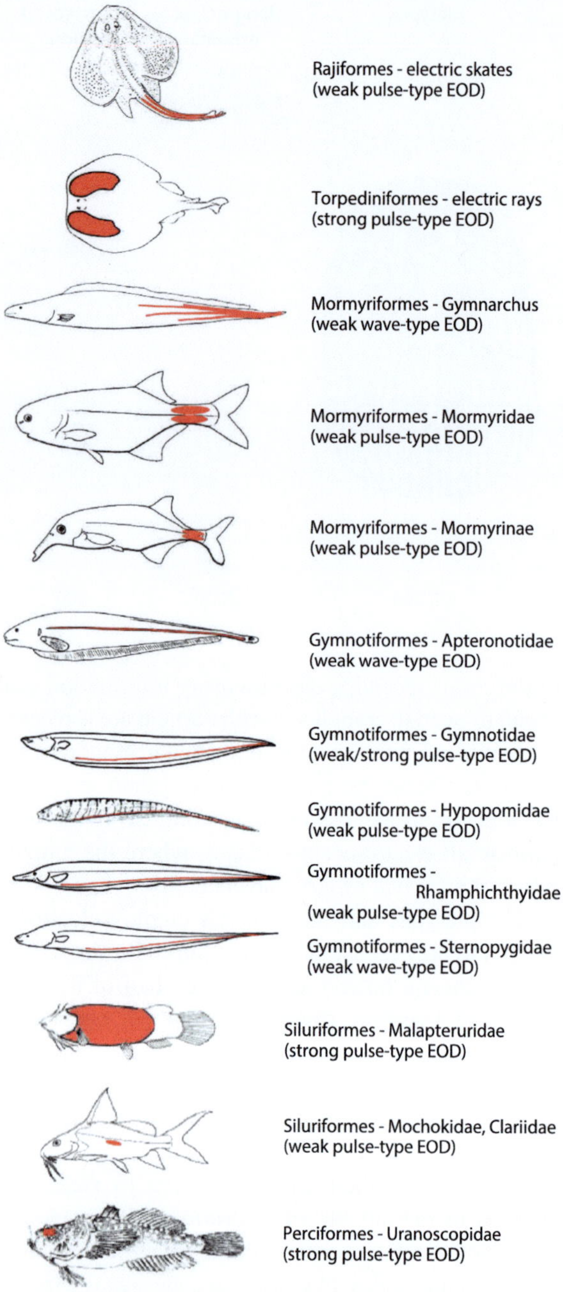


**Fig. 19.4 Location and characteristics of electric organs in different systematic groups of strongly and weakly electric fish.** The locations of the electric organ(s) are shown in *red* inside the animal's body in a representative species (Modified after Hopkins [3] with permission)

## 19.4 Strongly Electric Fish

People have been aware of the biogenic electric signals from strongly electric fish like the electric rays (e.g., *Torpedo torpedo*), the electric catfish (*Malapterus electricus*), or the electric eel (*Electrophorus electricus*), at least since ancient Egyptian times. This is because the electric signals they produce are so strong (more than 600 V in the case of the freshwater electric eel, or up to 20 A in the case of the marine electric torpedo ray [17]) that getting in close contact with a strongly electric fish can be very painful or even dangerous. These powerful discharges are used for two purposes: First, they are used as a defense against predators, which are deterred from harming even smaller strongly electric prey. For example, when touching a small electric catfish of only a few centimeters with your hand, the painful sensation their electric discharge will cause resembles the sting of a wasp. Second, they are used as hunting weapons during foraging. Strong electric discharges can stun or even kill smaller prey animals nearby, which then are easily captured by the electric predator.

## 19.5 Weakly Electric Fish

### 19.5.1 Electric Organs

The existence of weak electric signals in the aquatic environment, some of which are produced by fishes, was not known until H. W. Lissmann of the University of Cambridge recorded weak biogenic signals from the tail of a *Gymnarchus niloticus* in 1951 and subsequently demonstrated the electric sense and its use for active electrolocation in a series of experiments with K. E. Machin [18]. Later, the second general function of electric signals produced by weakly electric fish, i.e., electrocommunication, was discovered. Members of the Gymnotiformes and Mormyriformes generate only weak discharges of not more than 10 V, which is sufficient for short-range communication and for active electrolocation, but is much too weak for prey capture or defense. In fact, the discharges of most weakly electric fish cannot be felt, even if we hold the animals in our hands.

African (Mormyriformes) and South American (Gymnotiformes) weakly electric fish (Figs. 19.4 and 19.5) use specialized electric organs to produce their high-frequency electric signals (i.e., with significant energy up to about 5 kHz or more), which are called electric organ discharges (EOD, Fig. 19.5). During evolution, electric organs developed out of electrically excitable tissues, i.e., either from muscles

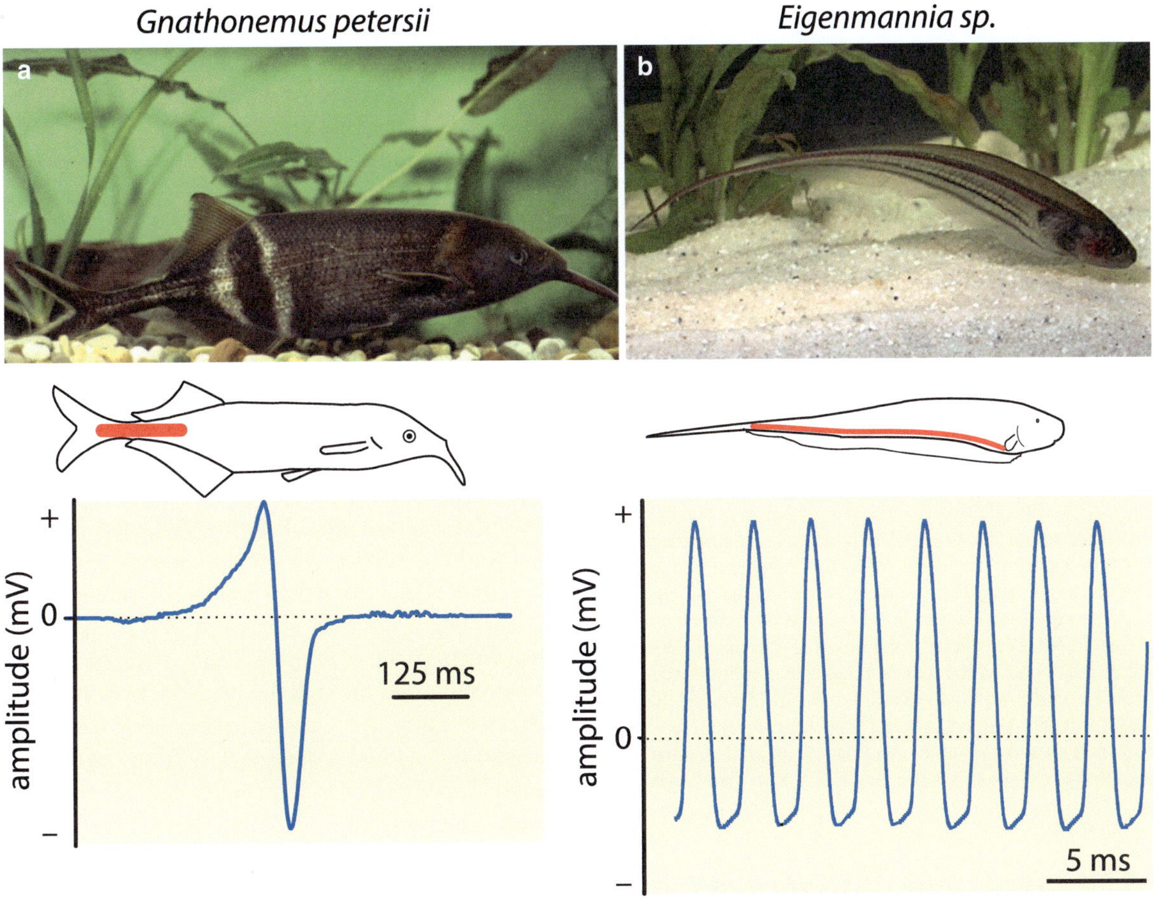


**Fig. 19.5** **The African mormyrid *Gnathonemus petersii* (a), and the South American gymnotiform *Eigenmannia* sp. (b).** In the contour drawings below the photographs, the locations of the electric organs within the bodies are marked in *red*. Below, the electric organ discharge waveforms of *G. petersii* (pulse-type EOD, *left*) and *Eigenmannia* (wave-type EOD, *right*) are shown. Note the different time scales (The photograph of *Eigenmannia* was kindly provided by T. Moritz)

(myogenic organs found in all Mormyriformes and most Gymnotiformes) or nervous tissues (neurogenic organs found in one family of Gymnotiformes, the Apteronotidae; Fig. 19.4). The cellular elements of electric organs, the electrocytes, have a special flat morphology and are arranged in a certain regular way (like coins in a rouleaux) in order to optimally sum up the electric currents. Weak or strong electric discharges are produced by the synchronized action potentials from hundreds or thousands of electrocytes acting in series and in parallel. Myogenic organs are composed of a serial array of electrogenic syncitia derived from the fusion of muscular precursor cells. Neurogenic organs generate electric fields by the coordinated activation of several hundred enlarged axons of electromotor neurons. These former axons do not end with a synapse and have unique adaptations in their geometry and also in their nodes of Ranvier to maximize current summation [19].

In most Gymnotiformes, electric organs extend almost along the entire length of the fish, except for the head (Figs. 19.4 and 19.5). In many species of this group, several electric organs are found within the fish's body and the discharges of each of them sum up in a complex way to produce the EOD of the fish. As a result, electroreceptor organs at different body regions can experience EODs of different waveforms and amplitudes. In contrast, Mormyriformes have only a single, relatively short electric organ in their caudal peduncle (Figs. 19.4 and 19.5). As a consequence, the electric field produced is an homogeneous, asymmetric dipole field. The entire trunk of the animal rostral of the electric organ forms one pole and the tip of the tail the other pole (Fig. 19.6).

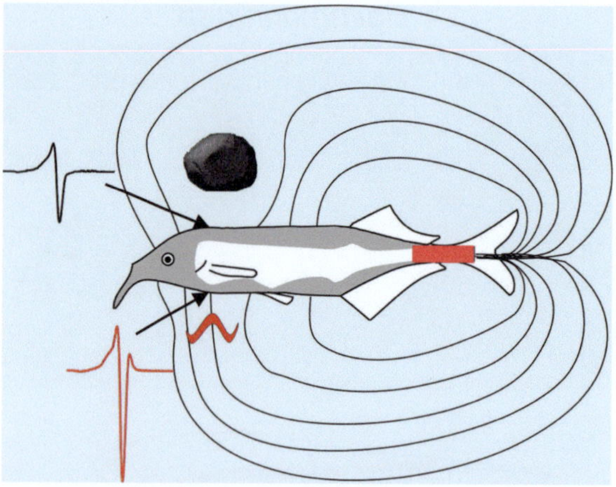

**Fig. 19.6 The electric field produced by *G. petersii.*** Field lines in the vertical plane are shown as *black lines*. The field lines are distorted by a rock (*above*, only resistive electric properties) and a living prey item (*worm below fish*, capacitive and resistive electric properties). The local EODs stimulating the electroreceptors within the electric images of the two objects are also shown in *black* for the rock and in *red* for the worm. While the local EOD opposite the rock is reduced in amplitude, the worm causes a slight amplitude increase and a distortion of the EOD waveform. The skin areas containing electroreceptor organs are shown in *gray*; the location of the electric organ within the fish is depicted in *red* (From von der Emde [20] with permission)

## 19.5.2 Electric Organ Discharges of Weakly Electric Fish

There are two main types of EODs: (i) brief, pulse-type signals and (ii) continuous wave-type discharges (Figs. 19.4 and 19.5). Pulse-type EODs have a duration that is much shorter than the interpulse intervals, which means they can be shorter than 200 μs in some mormyrids, while other species generate EODs of several milliseconds. In the case of mormyrids, interpulse intervals of single individuals are highly variable and depend on the behavioral context. Most pulse-type EODs have extremely constant waveforms, which depends on the species, the sex, and the hormonal state of the sender animal. Since the animals can not vary the EOD waveform on a short-term basis, they have to rely on other means such as modulating the interpulse intervals to change the information content during electro-communication.

Many Gymnotiformes and one single mormyriform species (*G. niloticus*) produce wave-type signals that more or less resemble distorted sine waves (Figs. 19.4 and 19.5). These signals are produced almost continuously, except during certain behavioral episodes such as courtship or aggression. Unlike electric fishes with pulse-type discharges, those that produce wave discharges do so at extremely regular rates, unless induced to change emission frequency because of a particular social situation. The quasi sine-wave output is at a stable frequency and can range from a few tens of Hertz to more than 2 kHz in some Apteronotidae.

Weakly electric fish are usually active at night and in the absence of light use their EODs for active electrolocation and electrocommunication. An advantage of the use of electric signals for these tasks in contrast to, say, acoustic or visual signals, is that the waveform is only slightly distorted by the environment. Whereas acoustic signals are often distorted by the medium and objects within it in various often frequency-dependent ways (reflection, refraction, scattering, attenuation), electric signals are only attenuated (but not in a frequency-dependent way) and their waveforms pass almost unaffected through the medium, even if this is turbid and noisy. As a consequence, the shape of the received signals varies only slightly from the emitted signals. Weakly electric fish exploit this fact by using temporal and waveform cues during both electrocommunication and active electrolocation.

## 19.5.3 Active Electrolocation

Passive electrolocation requires the presence of environmental electrical fields in order to detect objects or other structures. This disadvantage is overcome in active electrolocation, where the animal itself is the source of the electrical energy. The self-produced electrical signals build up an electrical field around the animal, which is perceived by an array of cutaneous electroreceptor organs that are distributed over the body surface of the fish (Fig. 19.6). Objects are detected because they interact with the electrical field and modulate the EOD amplitude and phase, which are detected by the animal's electroreceptor organs [21].

### 19.5.3.1 Electric Images

During active electrolocation, local modulation of the EOD is the carrier of environmental information for the animal. The modulations of the local EOD amplitude are caused by objects close to the animals' skin, which differ in impedance from the surrounding water.

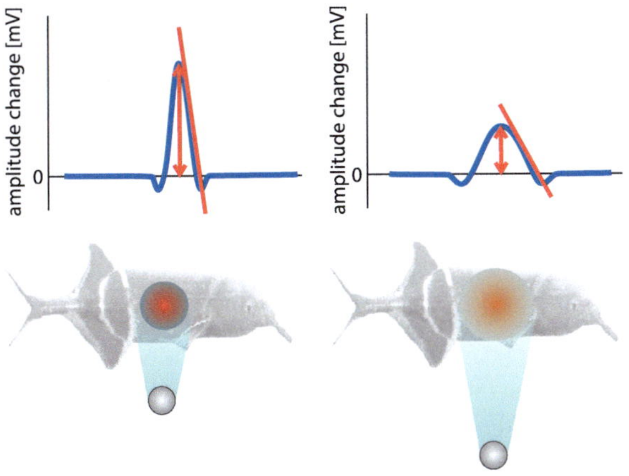


**Fig. 19.7** **The electric images of two metal spheres in *G. petersii*.** The two-dimensional image depends on the resistance and the distance of an object. The lower the resistance of the object, the higher is the signal amplitude in the center of the image (*red double arrows*). With increasing distance (*right*) the overall amplitude decreases, the image increases in width, and the slope of the image becomes shallower (*red lines*). In the *lower parts* of the figure, amplitude increases on the fish's skin are shown in *red*, amplitude decreases in *blue*. The *upper parts* show a one-dimensional horizontal measurement through the center of the electric images. The change in amplitude caused by the presence of the object is plotted versus the location of the measuring electrode at the fish's body (Figure kindly provided by S. Schwarz)

Capacitive properties of nearby objects can additionally alter the waveform of the local EOD [21]. An "*electric image*" (Figs. 19.6 and 19.7) originating from the objects is thus projected onto the animal's electroreceptive skin, which typically has a center-surround (or Mexican hat) spatial profile [22, 23]. It is important to note here that the Mexican hat profile arises because of electric field physics and not through neural processing. For example, a good conductor projects an image with a center region where the local EOD amplitude is increased, surrounded by a rim area where the amplitude decreases compared to the amplitude in absence of an object. Images of nonconductors are of opposite arrangement: local EOD amplitude decreases centrally and slightly increases in the surrounding rim area.

Electric images are always blurred, or "out of focus", since no focusing mechanisms are in place comparable to the lens of an eye. It follows that when projecting 3-D objects on the 2-D sensory surface, there is no one-to-one relationship between spatial object properties and image shape. Hence, the fish has to use complex decoding algorithms in order to extract object information, such as size, shape, and geometrical properties. In addition, electric images depend on object distance and location along the fish's body, the fish's body proportions, bending movements of the fish's body, the presence of additional objects, the background and many other factors. To make matters even worse, electric images of two objects will fuse in a nonlinear way depending on the distance between them [24].

### 19.5.4 Perception of Objects by Weakly Electric Fish

#### 19.5.4.1 Electrical Color Perception

During active electrolocation weakly electric fish can not only detect the presence of an object but in addition can recognize several object properties. Already Lissmann and Machin showed in 1958 that these fish can determine the electrical resistance of an object by measuring the amplitude changes of the local EOD caused by the object [18]. Living objects, such as water plants, other fishes, or insect larvae, which make up a major portion of the fish's diet, have complex impedances with considerable capacitive components. When a living object comes close to an electric fish, it causes not only amplitude changes but, in addition, distorts the local EOD waveform. By recording from the mormyromast electroreceptor organs (see below) and by conducting behavioral experiments, it was shown that mormyrid electric fish can perceive these waveform distortions [25]. Since amplitude and waveform modulations can be measured independently by the fish, they can quantitatively determine the capacitive and resistive components of an object's complex electrical impedance and thus discriminate unequivocally between living and nonliving objects. Since this ability resembles color vision, during which two or more parameters, e.g., hue and lightness of a visual stimulus, are perceived separately, it was called **electrical color perception** of the electric sense.

#### 19.5.4.2 Distance Determination

One of the best studied electric fish in terms of object perception and the corresponding brain circuits is the mormyrid *Gnathonemus petersii*. This species, and probably other mormyrids as well, can localize objects in three-dimensional space during active electrolocation [23]. If an object moves away from an electric fish three things happen: the size of the electric image gets larger, the amplitude in the image's center decreases, and the image gets more fuzzy, i.e., the rise (or fall) of the amplitude towards the center is less steep (Fig. 19.7). The latter parameter, called the *slope* of the image, together with the amplitude change in the center turned out to be crucial for distance measurements by *G. petersii*: the animals measure the *slope/amplitude ratio* of the electric image, which depends only on object distance and not on size or other object properties [23]. Distance determination is independent of the size or electrical properties of an object and *G. petersii* thus has a true *sense of depth perception*.

#### 19.5.4.3 Shape Detection

Even though the electric sense lacks focusing mechanisms, weakly electric fish also can perceive an object's three-dimensional shape. In training experiments *G. petersii* quickly learns to recognize an object of a certain shape and to discriminate it from objects of other shapes [26]. Shape recognition persists even when the object is rotated in space, indicating a *viewpoint independent recognition* of objects. In additional experiments, *G. petersii* demonstrated *size constancy* during object recognition [21]. For analyzing the shape or size of an unknown object, fish have to perform probing motor acts, i.e., they have to swim around the object scanning it with their sensory surface from several viewpoints. This is in contrast to distance measurements, which can be achieved instantly and do not require scanning movements.

As mentioned above, the electric images of objects located close to each other near the fish will fuse in a nonlinear way leading to complex electric images. In spite of this effect, *G. petersii* is able to perceive the shape of an object even when it is positioned right in front of a large background [21]. When trained to discriminate between two objects placed next to each other and a solid object of the same length, *G. petersii* was able to detect gaps as small as 1–2 mm between the two objects.

### 19.5.5 Tuberous Electroreceptor Organs of Weakly Electric Fish

#### 19.5.5.1 Sensory Coding of Amplitude and Timing

In addition to ampullary receptor organs, weakly electric fish from Africa and South America have a second class of epidermal electroreceptor organs, which are used to encode the amplitudes and phases (timing) of the fishes' self-produced EODs. In contrast to ampullary organs, these so-called 'tuberous receptor organs' are not connected to the water with an open canal and respond best to the dominant frequencies of the fish's own EOD (Fig. 19.2). In gymnotiform wave-type species, two types of tuberous organs were found: **T-units** (T for time) fire a single action potential to each cycle of the signal, which is coupled to the zero crossing of the signal and thus codes for the EOD's phase. **P-units** (P for probability), on the other hand, do not respond to every cycle of the EOD but instead the probability of their firing increases with an increase in EOD amplitude. T- and P-units innervate separate central nervous networks in the brain and are the beginning of two separate pathways for phase- and amplitude information [27].

The principle of *separate pathways for time (phase) and amplitude information* beginning with two different populations of receptor organs can be found in all groups of weakly electric fish. In pulse-type gymnotiforms, for example, **M-units** for timing and **B-units** for amplitude information were found [3]. Like all mormyrid fish, also *G. petersii* possess two types of tuberous electroreceptors, which are called **mormyromasts** (the name is based on the fact that these organ types only occur in members of the family Mormyridae) and **knollenorgans** (named by the zoologist Walter Stendell because of their appearance after the German word 'Knolle' for tuber). Knollenorgans are time-coders and are exclusively used to detect the EODs of other mormyrids during electro-communication [3]. Mormyromasts, on the other hand, are responsive to the animal's own EOD and modulations of the EOD by the external environment and are thus used for active electrolocation [28].

#### 19.5.5.2 Mormyromast Electroreceptor Organs Code for the Amplitude and Waveform of the EOD

Each mormyromast houses two types of receptor cells that are tuned to different aspects of the signal carrier, i.e., one channel for amplitude and one for waveform coding. Each mormyromast consists of two chambers containing several receptor cells. A-cells are found at the basal part of the outer chamber, while B-cells are located inside of an inner chamber (Fig. 19.8). Both are innervated by separate nerve fibers, which project to the electrosensory lateral line lobe (ELL) in the brain, where type A and B afferents terminate in separate areas each forming a *somatotopical map* (Fig. 19.8) [29].

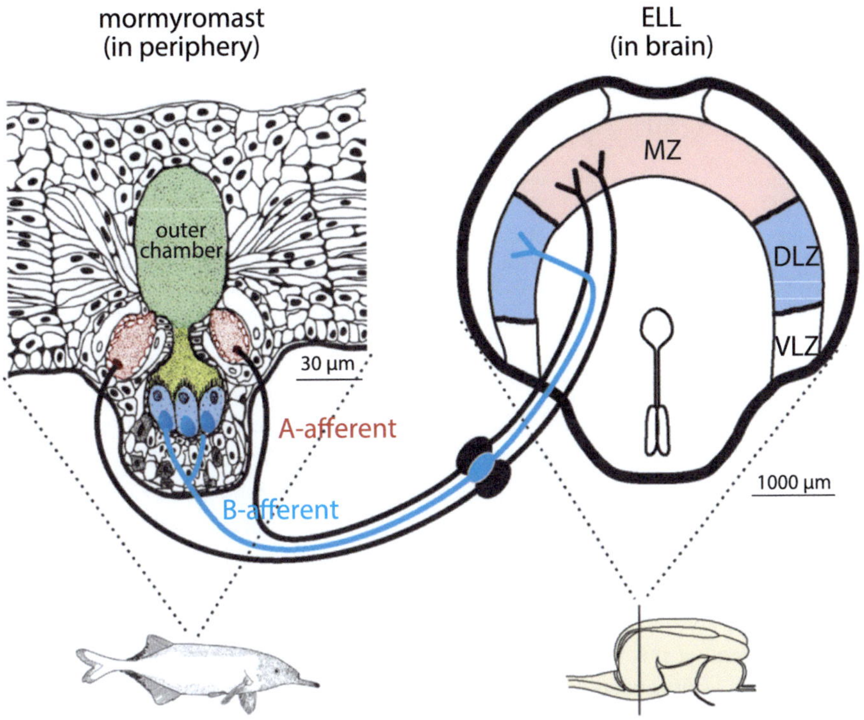


**Fig. 19.8** **Schematic drawing of a single mormyromast electroreceptor organ** in the skin from the back of the fish (*left*) and its projection into the electrosensory lateral line lobe (*ELL*) in the brainstem (schematic drawing of a horizontal section *on the right*) of *G. petersii*. Electroreceptor cells – *A* (*red*) and *B* cells (*blue*) – of the mormyromasts are innervated by electrosensory lateral line nerve afferents, which project into separate somatotopic maps in two zones of the *ELL*. *A* cells project into the medial zone (MZ, *red*), while *B* cells project into the dorsolateral zone (*DLZ*, *blue*) (The drawing of the mormyromast organ was kindly provided by M. Amey-Özel)

Physiologically, both types of fiber are silent in the absence of electrosensory stimulation and both types respond to a brief EOD-like pulse of current with one or more spikes. Stimulation near threshold evokes a single spike at a latency of about 10 ms. As stimulus intensity is increased, the latency of the first spike shows a smooth decrease to a minimum of about 2 ms, and additional spikes are added. The smooth decrease in latency of the first spike suggested to Szabo and Hagiwara [30] that mormyromast afferents may use *latency as a code for stimulus intensity*, and several physiological and behavioral observations support this hypothesis [31].

The most important difference between afferent fibers from A and B sensory receptors is the sensitivity of only the B fibers to distortions of the EOD waveform, such as those which are caused by capacitive objects [25]. Fibers from type B cells are exquisitely sensitive to such distortions, whereas those from type A cells are not. Both fibers are similarly sensitive to changes in EOD amplitude. These findings suggest that the fish sense the capacitive properties of objects, independently of the resistive properties, by centrally comparing the EOD responses of A and B fibers. If the responses of both fiber types change in the same direction and by a similar amount in the presence of a new object, then the object is resistive. If, on the other hand, the responses of B fibers are more strongly affected than those of A fibers, then the object has capacitive properties.

### 19.5.6 Electroreceptive Foveae in Weakly Electric Fish

In the human eye, the fovea plays an important role for scene recognition and object analysis. Behavioral, anatomical, and physiological results reveal several similarities between certain electroreceptive skin regions of both gymnotiform and mormyriform fishes and the fovea of the eye of several vertebrates [21, 24]. In *G. petersii*, two separate foveae have been described, one located at the movable chin appendix, the so-called Schnauzenorgan, and another one between the mouth and the nares, the so-called nasal region (Fig. 19.9). Both match the requirements of a fovea, which are: (i) a higher density of electroreceptors compared to the rest of the sensory epithelium; (ii) an overrepresentation of these areas in the brain, i.e., there are more central neurons devoted to the processing of a single receptor element at the fovea compared to a receptor in the periphery; (iii) structural/morphological and physiological specializations of the receptors and accompanying structures within the foveae including

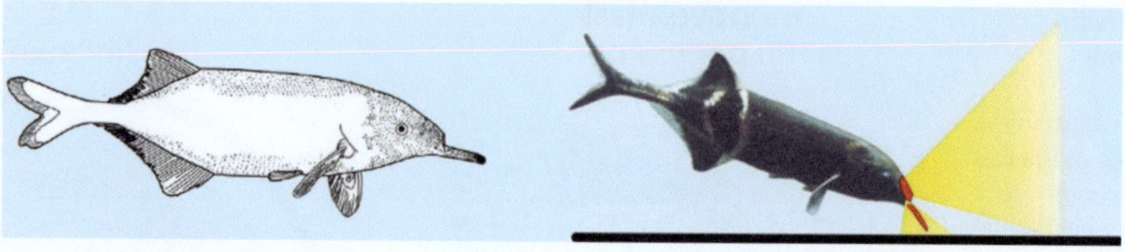

**Fig. 19.9** *Left*: *G. petersii* with the pores of the mormyromast electroreceptor organs shown as *small dots*. The density of receptor organs is highest at the tip of the Schnauzenorgan (chin) and just above the mouth (nasal region) (kindly provided by Dr. M. Hollmann). *Right*: Typical posture of a foraging *G. petersii* with the two electroreceptive foveae marked in *red*. The *yellow* areas highlight the direction of electroreceptive input. The Schnauzenorgan fovea scans the ground, while the nasal fovea "looks" ahead to detect obstacles

prereceptor mechanisms; and finally, (iv) behavioral adaptations, i.e., scanning movements, for focusing a stimulus onto the foveae and scanning the object for detailed analysis [21, 29].

The two foveae of *Gnathonemus* serve separate functions: the **nasal region** is a long-range guidance system that is used to detect obstacles or other objects in front of and at the side of the animal, e.g., during foraging. The **schnauzenorgan**, on the other hand, is a short-range movable (prey-) detection system that is used to find and identify prey on the ground or inspect details of objects. Both systems work simultaneously and ensure an optimal sensory inspection of the nocturnal environment of the fish. A similar foveal system has been found in gymnotiform pulse fishes. These have a fovea and a parafovea above and below the mouth, which like in *G. petersii* are characterized by high receptor densities and physiological specializations [24].

## 19.5.7 Processing of Electrosensory Information in the Brain

### 19.5.7.1 The Electrosensory Lateral Line Lobe (ELL)

Afferent fibers from electroreceptor organs terminate in the first station of the electrosensory pathways of the brain (Fig. 19.10), the **electrosensory lateral line lobe** (ELL), a cerebellum-like cortical structure in the medulla. While the brain circuits of mormyriforms and gymnotiforms differ in several respects, there exist many similarities [3]. Here, the ELL circuitry of the African mormyrid *G. petersii* is described exemplarily.

Fibers from A-type sensory cells terminate in the medial zone (MZ), fibers from B-type sensory cells terminate in the dorsolateral zone (DLZ), and fibers from ampullary receptors terminate in the ventrolateral zone (VLZ) (Figs. 19.8 and 19.11). In each of these three zones, the electroreceptive surface is somatotopically mapped, with the foveal regions at the head of the animal being overrepresented [29, 35]. The afferent fibers from mormyromasts terminate with mixed chemical-electrical synapses on small granular cells in the deeper layers of the ELL (Fig. 19.10). Axons of these granular cells, some of which are excitatory while others are inhibitory, terminate in turn on the basilar dendrites of larger cells located more superficially. The apical dendrites of these larger cells extend throughout the molecular layer of the ELL where they are contacted by parallel fibers that originate from an external granular cell mass known as the **eminentia granularis posterior** (EGp). The larger cells include two types of efferent cells, large ganglion (LG) and large fusiform (LF) cells, which relay electrosensory information to higher levels of the system, as well as two types of Purkinje-like inhibitory interneurons known as medium ganglion cells (MG1 and MG2). MG1 and MG2 cells show mutual inhibition, while, in addition, MG1 cells inhibit LF cells and MG2 cells inhibit LG cells [36]. LG cells are inhibited by electrosensory stimuli in the center of their receptive fields, whereas LF cells are excited by such stimuli (Fig. 19.10) [37].

These anatomical and physiological findings suggest a functional circuit within the ELL: LF cells are efferent 'on units' conveying increases in transcutaneous voltage to higher brain centers, while LG cells are efferent 'off units' conveying decreases in transcutaneous voltage. Similar to contrast-enhancing lateral inhibition in the visual system, the hypothesized mutual inhibition between MG1 and MG2 cells and their separate inhibition of the efferent cells could mediate contrast enhancement and other functions of the otherwise blurred electrosensory input [32, 36].

#### 19.5.7.2 The Electric Organ Corollary Discharge

In addition to the input from electrosensory afferent fibers from the periphery, the ELL and all other electroreceptive regions of the brain receive a so-called electric organ *corollary discharge* (CD) signal associated with each EOD motor command [34]. The CD informs the sensory structures of the brain that an EOD has been initiated and prepares these structures for the processing of sensory information that will soon arrive because of the stimulation of the electroreceptor organs by the EOD. In the brains of many other animals, vertebrates and invertebrates (e.g., in [38]), principally similar corollaries of motor commands have been described and the interaction between motor-associated and sensory information appears to be a general feature of many sensory systems.

Interaction between EOD-evoked reafferent input and the CD begins at the very first stage of central processing in the ELL. Recordings of synaptic potentials show that the CD evokes a prominent EPSP in granular cells that arrives at the same time as EOD evoked electroreceptor input. It serves two different functions: The first is to selectively enhance transmission of the electrosensory input from the fish's own mormyromasts, evoked by the fish's own EODs, which is the only input that can be used for active electrolocation. Input from other voltage sources, such as the EODs of nearby electric fishes, are noise and will not be processed in the context of active electrolocation. This gating of mormyromast input by the CD is in contrast to the suppression of the input from the other two classes of electrosensory fibers, the Knollenorgans and the ampullary receptor organs, which cannot be used for active electrolocation.

The second function of the CD-driven EPSP in granular cells is to provide a timing signal for deriving stimulus intensity from the latency code of the mormyromasts. The sum of CD-evoked and afferent-evoked EPSPs in granular cells grows larger as the latency of the afferent EPSP gets smaller, reaching a maximum when the peaks of the two EPSPs coincide at the minimal afferent latency. Thus, depolarization and activation of the granular cell by EOD-evoked electroreceptor responses depend on the latency of the spikes in the afferent fibers. Because of this, the granular cells have the unique ability to transform the latency-encoded stimulus amplitude conveyed by the afferents to a rate-code in the ELL [32].

#### 19.5.7.3 Neural Plasticity in the ELL

Functionally, neural plasticity in electric fishes enhances the detection of weak signals embedded in noise by the subtraction of predictions based on the recent sensory past from the current sensory input. This is equivalent to forward models where future states of the sensory input are predicted based on associations between recent input and motor commands. Plasticity in the electrosensory system has been studied in the ELL of both mormyrids and gymnotiforms [3]. In mormyrids it is caused by interactions between the CD, the electroreceptor input, and recurrent input from higher brain centers (Figs. 19.10 and 19.11).

CD effects on higher-order cells in the ELL are prominent, and many of these effects are plastic in that they depend on the sensory input to the cell that has occurred in the recent past [39]. Pairing the CD with a sensory stimulus leads to a marked change in the CD effect, which opposes the effect of the sensory stimulus. If a sensory stimulus excites the cell, then the CD causes an increased inhibition. If a sensory stimulus inhibits the cell, then the CD change is one of increased excitation. Thus, in the normal life of the animal, the plasticity results in a CD-evoked *negative image* of the pattern of sensory responses that followed the EOD in the recent past. This allows for novel or unexpected features in the sensory input to stand out more clearly. In the context of active electrolocation, CD plasticity and the accompanying subtraction of a previous pattern of sensory responses generates a more pronounced electric image of a suddenly appearing object. The plasticity of CD responses in the ELL appears to be due to **anti-Hebbian synaptic plasticity** at synapses between CD conveying parallel fibers and the apical dendrites of MG, LF, and LG cells [39, 40].

The neurons of EGp that give rise to parallel fibers receive not only CD-driven input but also signals from other sources, including proprioceptive signals conveying information about body and fin positions, and electrosensory signals descending from higher levels of the electrosensory system beyond ELL (Figs. 19.10 and 19.11). The generation of negative images based on past associations with proprioceptive

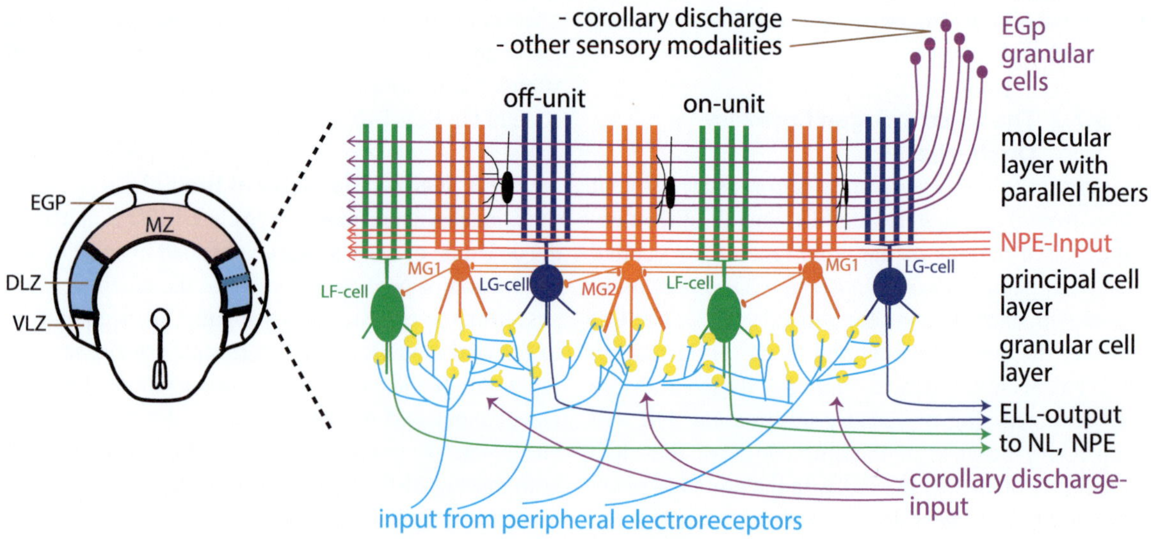


**Fig. 19.10 Schematic drawing of a small segment of the ELL cortex** (see *dark blue* rectangle in the DLZ in the horizontal section of the ELL on the *left*) with its different cell layers. The ELL receives input from mormyromast electroreceptors through afferent fibers (*blue*), which synapse on granular cells (*yellow*). Granular cells contact the principal cells that have long apical dendrites in the molecular layer, which is composed of parallel fibers (*purple*) from the eminentia granularis posterior (*EPp*). Large fusiform cells (*LF*, *green*) are efferent neurons, which together with MG interneurons (*orange*) form on-units, while off-units are composed of large ganglion cells (LG, *dark blue*). Efferent cells project to the nucleus lateralis (*NL*) of the torus semicircularis and to the preeminential nucleus (*NPE*). For brain areas see Fig. 19.11. Medial ganglionic 1-cells (MG1, *orange*) and MG2 cells (*orange*) are inhibitory interneurons of the principal cell layer. Electric organ corollary discharge input reaches ELL ventrally from the juxtalobar nucleus and in the molecular layer through parallel fibers from the EGp. Recurrent input from the preeminential nucleus is provided by parallel fibers in the deep molecular layer. *Small black cells* in the molecular layer are inhibitory interneurons (For details and further information see [33], for cell types [34], and for efferent and afferent connections including corollary discharge input [32])

or descending electrosensory signals is used to cancel out unwanted effects of, for example, body movements. When the fish bends its body, the electric organ in the tail is moved either closer to or further away from electroreceptors on one side of the body. The changes in electroreceptor input induced by such bending convey no useful information and can be predicted based on proprioceptive activity. Subtracting these bending-induced changes in the ELL allows other unpredictable, information-bearing sensory inputs to stand out more clearly. Similarly, previous electrosensory input, as signaled by descending signals from higher centers, will often permit the prediction of current electrosensory input, and subtraction of such predictions will again allow for information-bearing new sensory input to stand out more clearly. These plasticity effects have been observed in the ELL of both African mormyriforms and South American gymnotiforms, and they therefore are interesting examples of convergent evolution of neural mechanisms.

Key to this spike-timing dependent plasticity in ELL cells is the occurrence of broad spikes (comparable to complex spikes in the cerebellum) generated in the proximal apical dendrites of MG cells about 50 ms after the parallel-fiber input. These broad spikes invade the apical dendrites and unblock NMDA receptors leading to a site-specific decrease in the synaptic strength (depression) of parallel fibers to MG-cell synapses. Correlations between broad spikes and parallel-fiber input at times outside of the time window of 50 ms result in unspecific potentiation, and *in vivo* data show that associative depression as well as associative potentiation can occur [41].

#### 19.5.7.4 Higher Stations of the Electrosensory Pathway

The ELL is only the first of several structures in the mormyrid brain that process electrosensory information (Figs. 19.8, 19.10, and 19.11). At the top of the hierarchy is the **valvula cerebelli** which is also known as the mormyrid gigantocerebellum because it is

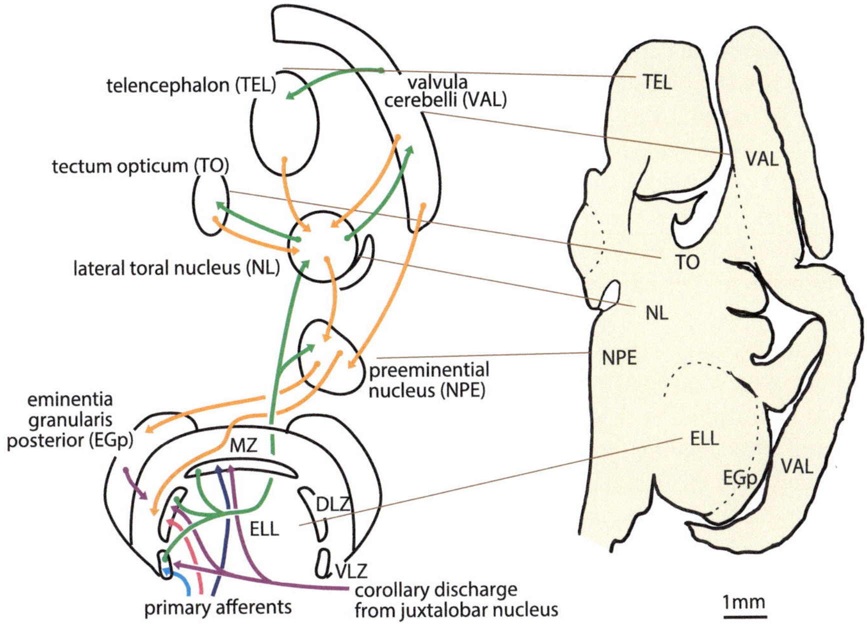


**Fig. 19.11** Electroreceptive brain circuits of the mormyrid *Gnathonemus petersii*. The ascending pathway begins with the primary afferent fibers of the A and B cells of the mormyromast electroreceptor organs (*blue* and *red*), which terminate in the medial zone (*MZ*) and the dorsolateral zone (*DLZ*) of the *ELL*. Ampullary afferents (*turquoise*) terminate in the ventrolateral zone (*VLZ*) of the *ELL*. The ascending pathway is shown in *green*. There exist several recurrent feedback pathways (*yellow*) that project from higher brain centers back to earlier stations of the pathway. The electric organ corollary discharge input is shown in *magenta*. On the *right*, a longitudinal section of the *G. petersii* brain is shown. The location of the same nuclei of the electroreceptive pathway as shown on the *left* are connected by *dotted lines* (See von der Emde and Engelmann [32] for further information)

extraordinarily large and covers all the rest of the mormyrid brain [42]. One third to one half of the valvula surface is devoted to the electrosensory system. The valvula receives from and projects back to the lateral nucleus of the torus semicircularis (NL) and to the preeminential nucleus (NPE) and also projects to the telencephalon (Fig. 19.11). The lateral nucleus receives from the ELL and projects back to the preeminential nucleus as well as to other structures. The preeminential nucleus projects back to the ELL both directly to the deep molecular layer and indirectly via EGp and the parallel fibers (Fig. 19.11) [32, 34, 43].

#### 19.5.7.5 Recurrent Pathways in the Brain

These anatomical findings demonstrate the prominent presence of feedback from higher to lower stages of the mormyrid electrosensory system used for active electrolocation. This feedback allows for the results of higher-level processing, for memories that may be stored at the higher levels or of motivational inputs to be conveyed to lower levels for interaction with ascending information from the periphery. Extensive feedback from higher to lower centers is also a major feature of many other sensory systems, including the mammalian visual and auditory systems [39, 44], but the concrete roles of such feedback are not always completely understood.

Histological studies have shown that the efferent fibers of all three ELL zones project to the nucleus lateralis (NL) of the torus semicircularis in the midbrain. The NL is one of seven subnuclei of the torus, four of which are part of the electrosensory system, and two, including the lateral nucleus, are concerned with active and passive electrolocation. The NL receives input from the ELL and projects back to the preeminential nucleus as well as to other structures [34]. Anatomical studies suggest that in the NL the information from the two mormyromast zones and the ampullary zone of the ELL converge at single locations. Studies of the corollary discharge pathway in the brain suggest that the NL is not directly connected to this pathway but instead might receive CD information indirectly from other sources (Fig. 19.11).

Neurons in the NL respond to electrosensory stimulation and different recording sites within the NL showed different receptive areas on the body of the fish. Electrophysiological results revealed that in contrast to the ELL there is no topographic representation of the body surface in the NL. Very restricted focal injections into certain areas of the NL (Mohr and von der Emde, unpublished observation) showed that the NL is heavily interconnected. Thus, the original topography arriving from the ELL is masked, resulting in neurons which respond to several body sites and require a very special set of input features. For example, some areas in the NL are responsive to certain waveform distortions of the local EOD signal independently of amplitude information. Such responsiveness can only be achieved by comparing inputs from neurons of the medial and from the dorsolateral zones of the ELL. The NL probably contains neurons which extract specific complex object features, e.g., capacitive properties, which are specific for living organisms in the fish's environment.

So far, only little is known about the physiology of most of the higher-order electrosensory structures beyond the ELL and NL. Even though electrosensory responses have been recorded in the preeminential nucleus, the lateral nucleus, parts of the valvula cerebelli, and in the telencephalon [45], the precise functional roles of these nuclei during active electrolocation still has to be uncovered by future research.

## References


1. Bullock TH, Northcutt RG, Bodznick DA (1982) Evolution of electroreception. Trends Neurosci 5:50–53
2. Albert J, Crampton G (2006) Electroreception and electrogenesis. In: Evans D, Claiborne J (eds) The physiology of fishes. CRC Press, Boca Raton, pp 431–472 [review]
3. Hopkins CD (2009) Electrical perception and communication. In: Squire L (ed) Encyclopedia of neuroscience, vol 3. Academic, Oxford, pp 813–831 [review]
4. Scheich H, Langner G, Tidemann C, Coles RB, Guppy A (1986) Electroreception and electrolocation in platypus. Nature 319:401–402
5. Proske U, Gregory JE, Iggo A (1998) Sensory receptors in monotremes. Philos Trans R Soc Lond B Biol Sci 353: 1187–1198
6. Pettigrew JD (1999) Electroreception in monotremes. J Exp Biol 202:1447–1454
7. Czech-Damal NU, Liebschner A, Miersch L, Klauer G, Hanke FD, Marshall C, Dehnhardt G, Hanke W (2011) Electroreception in the Guiana dolphin (*Sotalia guianensis*). Proc Biol Sci 279:663–668
8. Jackson CW, Hunt E, Sharkh S, Newland PL (2011) Static electric fields modify the locomotory behaviour of cockroaches. J Exp Biol 214:2020–2026
9. Gabel C, Gabel H, Pavlichin D, Kao A, Clark D, Samuel A (2007) Neural circuits mediate elecrosensory behaviour in *Caenorhabditis elegans*. J Neurosci 27:7586–7596
10. Gregory JE, Iggo A, McIntyre AK, Proske U (1988) Receptors in the bill of the platypus. J Physiol 400: 349–366
11. Bodznick D, Montgomery JC (2005) The physiology of low-frequency electrosensory systems. In: Bullock TH et al (eds) Electroreception. Springer, Berlin/Heidelberg/New York, pp 132–153
12. Wilkens L, Hofmann M (2005) Behaviour of animals with passive, low-frequency electrosensory systems. In: Bullock TH et al (eds) Electroreception. Springer, Berlin/Heidelberg/New York, pp 229–263
13. Kalmijn AJ (1987) Detection of weak electric fields. In: Atema J et al (eds) Social communication in aquatic environments. Springer, Berlin/Heidelberg/New York, pp 151–186
14. Engelmann J, Gertz S, Goulet J, Schuh A, von der Emde G (2010) Coding of stimuli by ampullary afferents in *Gnathonemus petersii*. J Neurophysiol 104:1955–1968
15. Anders KH, von During M (1984) The platypus bill. A structural and functional model of a pattern-like arrangement of cutaneous sensory receptors. In: Hamann W, Iggo A (eds) Sensory receptor mechanisms. World Scientific Publishing, Singapore, pp 81–89
16. Manger PR, Keast JR, Pettigrew JD, Troutt L (1998) Distribution and putative function of autonomic nerve fibres in the bill skin of the platypus (*Ornithorhynchus anatinus*). Phil Trans R Soc Lond 353:1159–1170
17. Bennett MV, Wurzel M, Grundfest H (1961) The electrophysiology of electric organs of marine electric fishes: I. Properties of electroplaques of *Torpedo nobiliana*. J Gen Physiol 44:757–804
18. Lissmann HW, Machin KE (1958) The mechanism of object location in *Gymnarchus niloticus* and similar fish. J Exp Biol 35:451–486
19. Caputi AA, Aguilera PA, Pereira AC (2011) Active electric imaging: body-object interplay and object's "electric texture". PLoS One 6:e22793 [review]
20. von der Emde G (2006) Non-visual environmental imaging and object detection through active electrolocation in weakly electric fish. J Comp Physiol A 192:601–612
21. von der Emde G, Behr K, Bouton B, Engelmann J, Fetz S, Folde C (2010) Three-dimensional scene perception during active electrolocation in a weakly electric pulse fish. Front Behav Neurosci 4:26
22. Caputi AA, Budelli R, Grant K, Bell CC (1998) The electric image in weakly electric fish: physical images of resistive objects in *Gnathonemus petersii*. J Exp Biol 201:2115–2128
23. von der Emde G, Schwarz S, Gomez L, Budelli R, Grant K (1998) Electric fish measure distance in the dark. Nature 395:890–894
24. Caputi AA, Budelli R (2006) Peripheral electrosensory imaging by weakly electric fish. J Comp Physiol A 192: 587–600
25. von der Emde G, Bleckmann H (1992) Differential responses of two types of electroreceptive afferents to

signal distortions may permit capacitance measurement in a weakly electric fish, *Gnathonemus petersii*. J Comp Physiol A 171:683–694
26. von der Emde G, Fetz S (2007) Distance, shape and more: recognition of object features during active electrolocation in a weakly electric fish. J Exp Biol 210:3082–3095
27. Heiligenberg W (1991) Neural nets in electric fish. MIT Press, Cambridge/London
28. Bell CC (1990) Mormyromast electroreceptor organs and their afferent fibers in mormyrid fish. III. Physiological differences between two morphological types of fibers. J Neurophysiol 63:319–332
29. Bacelo J, Engelmann J, Hollmann M, von der Emde G, Grant K (2008) Functional foveae in an electrosensory system. J Comp Neurol 511:342–359
30. Szabo T, Hagiwara S (1967) A latency change mechanism involved in sensory coding of electric fish (mormyrids). Physiol Behav 2:331–335
31. Hall C, Bell C, Zelick R (1995) Behavioral evidence of a latency code for stimulus intensity in mormyrid electric fish. J Comp Physiol A 177:29–39
32. von der Emde G, Engelmann J (2011) Active electrolocation. In: Farrell A (ed) Encyclopedia of fish physiology: from genome to environment, vol 1. Academic, San Diego, pp 375–386 [review]
33. Meek J, Grant K, Bell C (1999) Structural organization of the mormyrid electrosensory lateral line lobe. J Exp Biol 202:1291–1300
34. Bell CC (1986) Electroreception in mormyrid fish. Central physiology. In: Bullock TH, Heiligenberg W (eds) Electroreception. Wiley, New York, pp 423–451
35. Pusch R, von der Emde G, Hollmann M, Bacelo J, Nöbel S, Grant K, Engelmann J (2008) Active sensing in a mormyrid fish: electric images and peripheral modifications of the signal carrier give evidence of dual foveation. J Exp Biol 211:921–934
36. Han VZ, Bell CC, Grant K, Sugawara Y (1999) Mormyrid electrosensory lobe in vitro: morphology of cells and circuits. J Comp Neurol 404:359–374
37. Mohr C, Roberts PD, Bell CC (2003) The mormyromast region of the mormyrid electrosensory lobe. I. Responses to corollary discharge and electrosensory stimuli. J Neurophysiol 90:1193–1210
38. Crapse T, Sommer M (2008) Corollary discharge circuits in the primate brain. Curr Opin Neurobiol 18:552–557
39. Bell CC, Han V, Sawtell NB (2008) Cerebellum-like structures and their implications for cerebellar function. Annu Rev Neurosci 31:1–24
40. Bastian J, Chacron MJ, Maler L (2004) Plastic and nonplastic pyramidal cells perform unique roles in a network capable of adaptive redundancy reduction. Neuron 41: 767–779
41. Sawtell NB, Williams A, Bell CC (2007) Central control of dendritic spikes shapes the responses of Purkinje-like cells through spike timing-dependent synaptic plasticity. J Neurosci 27:1552–1565
42. Nieuwenhuys R, Nicholson C (1969) Aspects of the histology of the cerebellum of mormyrid fishes. In: Llinas RR (ed) Neurobiology of cerebellar evolution and development. American Medical Association Institute for Biomedical Research, Chicago, pp 135–169
43. Finger TE, Bell CC, Russell CJ (1981) Electrosensory pathways to the valvula cerebelli in mormyrid fish. Exp Brain Res 42:22–33
44. van Essen DC, Gallant JL (1994) Neural mechanisms of form and motion processing in the primate visual system. Neuron 13:1–10
45. Prechtl JC, von der Emde G, Wolfart J, Karamürsel S, Akoev GN, Andrianov YN, Bullock TH (1998) Sensory processing in the pallium of a teleost fish, *Gnathonemus petersii*. J Neurosci 18:7381–7393

## Videos

Nice videos about communication with electric signals (plus acoustic communication in weakly electric fishes): http://www.nbb.cornell.edu/neurobio/hopkins/video.htm

Electric field and potential animations: http://www2.fiu.edu/~efish/visitors/electric_field_animations.htm; http://alumnus.caltech.edu/~rasnow/qtmov.html

Introduction to weakly electric fish by Carl Hopkins: http://www.youtube.com/watch?v=f49ln-s99aw&noredirect=1

Electrical field and electric images during electrolocation in wave-type fish and prey capture movies: http://nelson.beckman.illinois.edu/movies.html

# The Magnetic Senses

# 20

Henrik Mouritsen

The Earth's magnetic field potentially provides information which can help animals to navigate over both short and long distances. Magnetic information can be useful to determine position (i.e., as part of a map sense) and for determining a favorable direction of movement (i.e., as part of a compass sense). An amazing variety of organisms has been shown to use the geomagnetic field to gather spatial information. Most research has focussed on compass orientation in migratory birds, map-based navigation in homing pigeons and sea turtles, and various magnetic behaviors in amphibians, but there is growing evidence that many other organisms including some mammals can sense magnetic information. Sensing magnetic fields as weak as that of the Earth is not easy using only biological materials. Three basic mechanisms can be considered: iron mineral-based magnetoreception, radical-pair-based magnetoreception, and induction in highly sensitive electric sensors. In recent years, strong evidence has been accumulated that both, iron mineral-based magnetoreception and radical-pair-based magnetoreception mechanisms exist in nature, and some animals seem to possess both types of magnetic senses. In both of these senses, plausible candidate molecules have been identified and a few brain areas involved in processing magnetic information have been identified. Despite substantial progress, we are still far away from understanding the detailed function of any magnetic sense. Many possibilities for groundbreaking research still await the scientific community in the field of magnetoreception and -perception.

H. Mouritsen
Institut für Biologie und Umweltwissenschaften and Research Center for Neurosensory Sciences,
University of Oldenburg, 26111 Oldenburg, Germany
e-mail: henrik.mouritsen@uni-oldenburg.de

## 20.1 Magnetic Fields

Moving electric charges produce magnetic fields. These moving electric charges are usually electrons. On the microscopic scale, electron [and nuclear] spins can generate magnetic fields. On the macroscopic scale, when current runs through a wire, a magnetic field, B, is generated around the wire. The magnetic field, B, is measured as magnetic flux density using the unit Tesla [T]:

$$1T = \frac{1V \times s}{m^2} = \frac{1N \times s}{C \times m} = 10,000\ G$$

[V = Volt, s = second, m = meter, N = Newton, C = Coulomb, G = Gauss]. The direction of the magnetic field around the wire can be determined by the "right-hand rule": if you grasp around the wire with your right hand so that your thumb is pointing in the direction of the current, the magnetic field around the wire runs in the direction in which your fingers are pointing. The magnetic field decreases with the distance as you move away from the wire. If you create a coil of wire, the magnetic field created is much stronger inside the coil than on the outside of the coil. Therefore, various coil constructions are typically used to produce artificial magnetic fields (for details see below). Magnetic fields can be described by three-dimensional vectors. If one adds an artificially created field to an existing field (such as that of the Earth), the resultant field is calculated by simple vector addition of the two fields. Some materials, defined as ferromagnetic, can be permanently magnetized by a magnetic field, and this magnetization remains after the magnetizing field has been removed.

C.G. Galizia, P.-M. Lledo (eds.), *Neurosciences - From Molecule to Behavior: A University Textbook*,
DOI 10.1007/978-3-642-10769-6_20, 

Magnetic fields are generated by moving electric charges and are measured in Tesla. Right-hand rule can be applied to determine the direction of the magnetic field generated by a wire.

## 20.2 The Earth's Magnetic Field

The Earth generates its own geomagnetic field, which is mostly caused by electric currents in the liquid outer core of the Earth (the "reverse dynamo effect", whereby electric currents create the magnetic field). The magnetic field, which can be measured at the Earth's surface, is similar to the magnetic field one would expect if a large dipole magnet were placed in the center of the Earth (see Fig. 20.1). The Earth's magnetic field currently has a magnetic field south pole near the Earth's geographic North Pole. This magnetic pole is referred to as "magnetic north" or "magnetic north pole" in biology. Thus, in biology and in the remainder of this text, the term "magnetic north" refers to the magnetic pole closest to the geographic North Pole, not the physical magnetic north pole. Equivalently, the magnetic field N pole near the Earth's geographic South Pole is referred to as "magnetic south" or the "magnetic south pole" in biology. Magnetic field lines leave the southern magnetic pole and re-enter at the northern magnetic pole. The polarity of the magnetic field lines always points towards the northern magnetic pole and they can therefore provide a highly reliable directional reference, which can be used as the basis for a magnetic compass.

The magnetic north pole is currently located in Northern Canada and the magnetic south pole is currently located south of Australia. Consequently, the geographical and magnetic poles do not coincide (see Fig. 20.1). The difference in direction towards geographic and magnetic North is called the declination angle or just the declination. The declination angle is mostly small near the equator, but near the magnetic poles, declination can pose a serious challenge for navigators using a magnetic compass.

Two additional characteristics of the geomagnetic field can potentially be used for navigation. Magnetic intensity ranges from ca. 30,000 nT (nT = nanoTesla $= 10^{-9}$ T $= 10^{-5}$ G) near the magnetic equator to ca. 60,000 nT at the magnetic poles (0.6 G; but in the field of animal navigation, magnetic fields are usually indicated in nT even for quite strong fields). Magnetic inclination, the angle between the magnetic field lines and the Earth's surface changes gradually from −90° at the magnetic south pole (the field lines come straight out of the Earth) to zero (field lines parallel to the Earth's surface) degrees at the magnetic equator to +90° at the magnetic north pole (the field lines run straight into the Earth; see Fig. 20.1).

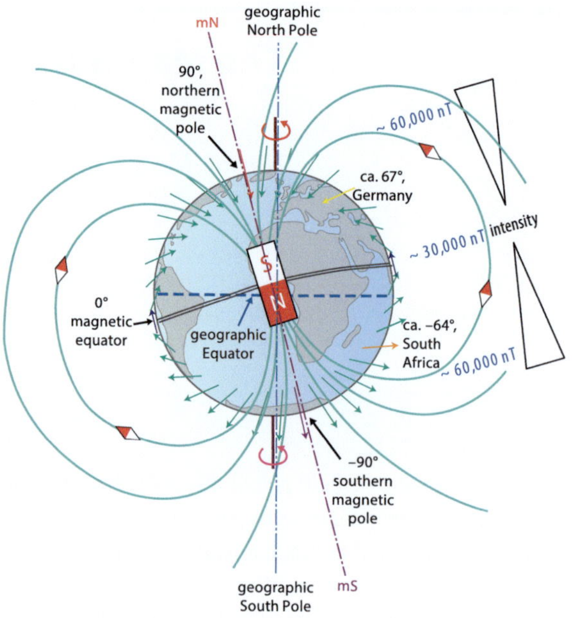


**Fig. 20.1 The Earth's magnetic field (geomagnetic field).** Notice that the magnetic southern and northern poles and the magnetic equator do not coincide with the geographical Poles and the geographic Equator. Also notice that the magnetic field lines intersect the Earth's surface at different angles depending on the magnetic latitude (*blue-green lines* and *vectors*). The intersection angle is called the magnetic inclination. Magnetic inclination is, for instance, +90° at the magnetic north pole (*red vector*), ca. +67° at the latitude of Germany (*yellow vector*), 0° at the magnetic equator (*dark blue vectors*), ~−64° at the latitude of South Africa (*orange vector*), and −90° at the magnetic south pole (*magenta vector*) (Adapted after Wiltschko and Wiltschko [37] with permission). The magnetic intensity varies from ca. 60,000 nT near the magnetic poles to ca. 30,000 nT along the magnetic equator

Declination is the angular deviation between magnetic and geographic North. Magnetic field intensity and the inclination angle changes mainly on the North–South axis. Daily variability in the geomagnetic field limits the expected accuracy of any magnetic map sense.

Since the circumference of the Earth is ca. 40,000 km, the average change in total magnetic intensity is only about 3 nT/km and the variation in magnetic inclination is only about 0.009°/km along the North–South axis. When we consider these numbers together with the fact that rather stochastic natural variations in the geomagnetic field in the order of 30–100 nT in more or less random directions occur every day (during magnetic storms generated primarily by the Sun, the variability can reach 1,000 nT), it becomes clear why any magnetic field-based map sense is unlikely to have a precision of a few kilometers. On a larger scale, both magnetic inclination and magnetic intensity can in theory be useful for determining one's position on Earth. Since magnetic inclination and intensity changes predominantly from North to South and much less from East to West on most parts of the Earth, it seems easier for animals to determine latitude than longitude from magnetic field information. However, if animals are able to measure declination accurately, the declination angle could provide useful East–west information about location in many parts of the Earth.

## 20.3 How Can We Study the Influence of the Earth's Magnetic Field on Animal Behavior?

Anyone wanting to study the use of magnetic cues in behavior must first identify a behavior, usually an orientation task, which is suspected to be influenced by magnetic cues. Thereafter, one must bring the behavior under experimental control. If the behavior of interest is spatially restricted, one can bring the needed equipment into the field and make field studies (this approach has been used highly successfully by, e.g., Rüdiger Wehner and colleagues when they studied polarized light-based orientation in the desert ant *Cataglyphis fortis*) (see Chap. 18). However, in many cases navigation takes place over too large a spatial scale to make pure field studies practicable. In these cases, the first major challenge is to design an experimental paradigm, which successfully brings the behavior under study into the lab, where the information available to the animals can be carefully controlled. For night-migrating songbirds, orientation cages such as the Emlen funnel (see Fig. 20.3a) are used; for migratory insects, flight simulators have been built [1]; and for sea turtle hatchlings, a small swimming pool, special swimsuits, and tracking arms are used [2]. Logistical challenges and/or the lack of a robust, lab-based behavioral paradigm are often the main reasons why it has not been properly tested whether migratory animals such as whales, wildebeests, and many species of migratory insects use magnetic cues for orientation.

> To study orientation, it is mostly necessary to first bring the behavior into the lab where potential cues can be carefully controlled. Various coil systems are used to experimentally change magnetic fields. To enable proper control experiments, the coils must be double-wound.

Once a behavioral paradigm has been established where the animals orient spontaneously in a consistent, biologically sensible, preferred direction, coils of electric wire are used to alter the magnetic field within the space where the behavior of the animals is being tested.

The most commonly used coil design is the Helmholtz coil (a pair of parallel coils placed one radius from each other; Fig. 20.2b). In a pair of Helmholtz coils, the magnetic field is very homogeneous (<1 % heterogeneity) within a central space of ca. 60 % of the radius of the coils. The magnetic field generated in the center of the Helmholtz coils is $B = (0.9 \times 10^{-6}\ \mathrm{T\ m/A} \times n \times I)/R$, where $R$ is the radius of the coils (measured in meter), $n$ is the number of turns in each coil, $I$ is the current flowing through the coils (measured in Ampere), and T is the unit Tesla. One pair of Helmholtz coils can alter the magnetic field only along the axis running through the center points of the two coils. To make any desired three-dimensional magnetic field, one needs three pairs of Helmholtz coils oriented perpendicular to each other. A three-axial Fluxgate magnetometer is needed to properly measure the generated fields.

The central homogeneous space can be increased to ca. 110 % of the radius of the coils by using more elaborate coil designs such as the Merritt-4-coil system (Fig. 20.2a). In future studies one would also expect the coils to be double-wound, i.e., to have two independent sets of windings equal in number within any given coil. Thereby, an experiment can be performed where the same amount of current runs through the coils whether the magnetic field is being changed or

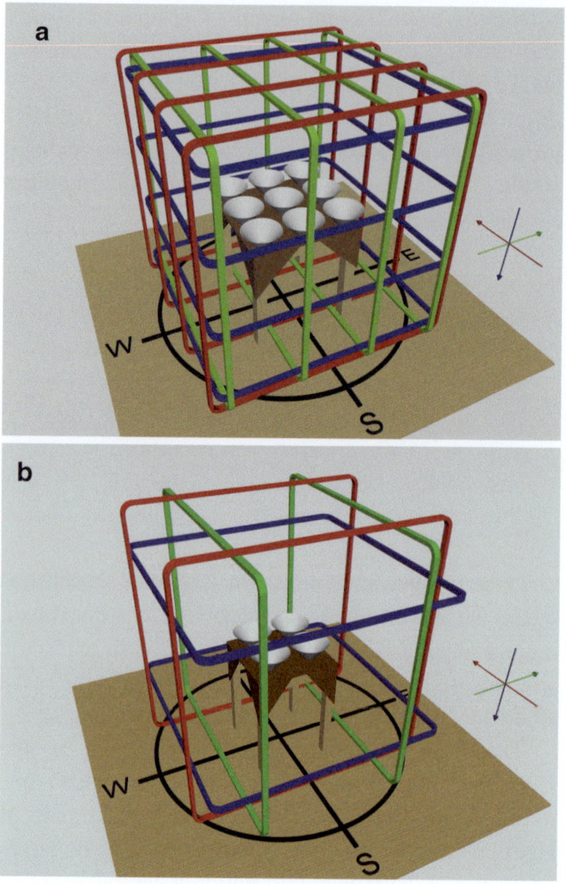


**Fig. 20.2** **Coil systems used to manipulate the magnetic field in animal experiments.** (**a**) One of the most elaborate 3-axis systems regularly applied in animal orientation experiments: The "Merritt-4-coil" system is built of aluminum profiles and copper wire [23]. When current-stabilized DC current runs through the *blue*, *red*, and *green* coils, the magnetic field vector is changed along the up-down, North–South, and East–West axis, respectively. The wooden table ensures that the orientation funnels are placed in the center of the coil system, where the homogeneity is highest. The exact distances between coils must be +0.5055 × d, +0.1281 × d, −0.1281 × d, −0.5055 × d from the center for the Merritt-4-coil system (d is the diameter of the coils), and the proportion numbers for the number of windings in the four coils per axis must be 26:11:11:26 [3]. If constructed accurately, Merritt-4-coil systems generate fields with <1 % heterogeneity within a space of ca. ±0.55 d × ±0.55 d × ±0.55 d, that is ca. 110 × 110 × 110 cm for a 200-cm-diameter coil. (**b**) For the often used Helmholtz coil system, the equivalent numbers are two coils per axis with equal number of windings (1:1) placed +0.25 d and +0.25 d from the center. Helmholtz coils generate fields with <1 % heterogeneity within a space of ca. ±0.33 d × ±0.33 d × ±0.33 d [3]

not. In the control condition, the current runs antiparallel through the two sets of windings and no change of the magnetic field occurs. Double-wound coils allow for truly double-blind experiments [3, 4].

### 20.3.1 Use of the Earth's Magnetic Field in Behavior

The body of literature reporting that various animals can sense and use magnetic field information is growing steadily [5]. Most commonly, animals can use the reference direction provided by the geomagnetic field as a compass for spatial orientation. Most studies on magnetically guided behaviors have been performed in birds, and we will therefore start by looking at this group of animals.

#### 20.3.1.1 Birds Use Geomagnetic Field Information for Several Purposes

Birds possess a magnetic compass sense. The magnetic compass behavior of migratory birds was first documented in the mid-1960s by Friedrich W. Merkel and Wolfgang Wiltschko. This and most subsequent studies made use of the migratory restlessness (Zugunruhe) phenomenon: birds primarily jump/flutter in the direction they want to fly to when confined to a round, so-called orientation cage (Fig. 20.3a). In spring on the Northern Hemisphere, the birds orient their activity towards northerly directions and in autumn they orient in southerly directions, and when birds are released, they actually, on average, fly in the direction in which they jumped/fluttered in the orientation cage [6].

To demonstrate magnetically guided compass behavior, one must document that a group of birds tested in orientation cages orient in their species-specific migratory direction in the unchanged geomagnetic field and that they change their mean orientation accordingly when the magnetic field direction is being changed (for an example see Fig. 20.7f–h).

The magnetic compass of birds has been shown to be an "inclination compass" [7] (see Fig. 20.3b). Also, the magnetic compass sense seems to have a rather narrow functional window with respect to magnetic intensity, but this window can be expanded greatly after a few hours of adaptation to a changed field intensity [5]. It had been reported that the magnetic compass of birds should only be located in the right eye, but this has turned out not to be true (see, e.g., review [8]).

> Birds have a magnetic inclination compass. They can most likely extract positional information from the geomagnetic field, but whether they use it for a detailed magnetic map is heavily debated.

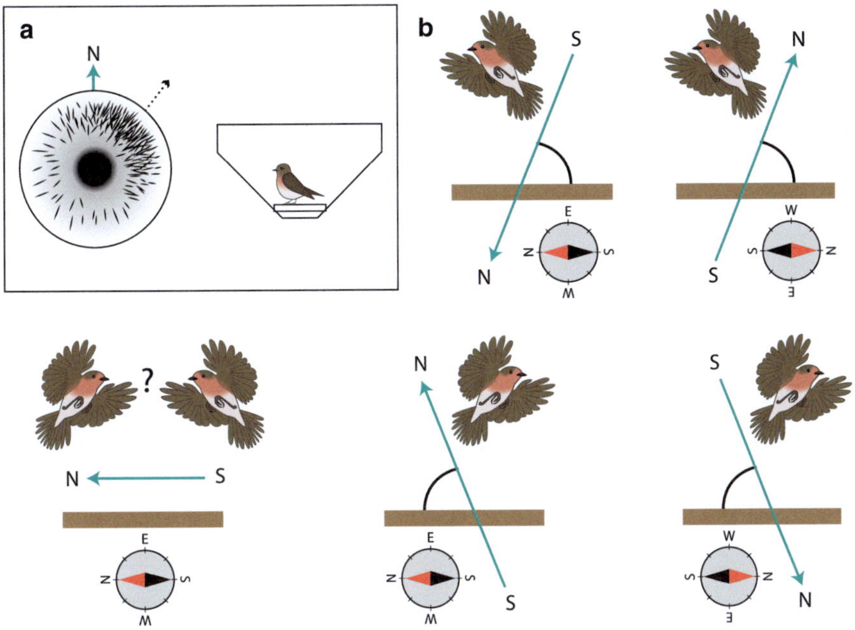


**Fig. 20.3** (**a**) **The Emlen funnel** [38] – the most commonly used orientation cage. The mean jumping direction of the birds are recorded on scratch-sensitive paper lining the inclined wall of the funnel. (**b**) Classical experiments [7] have shown that birds have an inclination compass, which means that the birds measure the angle between the magnetic field lines and the Earth's surface or gravity. Thereby, the birds distinguish between poleward and equatorward, but not between North and South as a polarity compass would do. If birds had used a polarity compass they would orient in the direction indicated by the red end of the inserted technical compass. Birds are disoriented in a horizontal magnetic field like the one occurring at the magnetic equator. The flight direction of the inserted bird indicates the springtime mean direction chosen by all bird species tested so far in the given magnetic field. The *blue-green arrows* indicate the direction of the magnetic field lines, *brown bar* the Earth's surface, *N* magnetic North, *S* magnetic South

The use of a magnetically based map has been reported in various bird species, but its existence, and in particular its spatial accuracy is still heavily debated. The views among researchers studying pigeon homing range from a magnetic map with a precision of a few kilometers being an established fact to a magnetic map sense being an evergreen phantom [5, 9]! Whether a detailed magnetic map exists or not, it remains a challenge to understand how a magnetic field-based map should be able to function on a scale relevant to pigeon homing. Because pigeons with opaque lenses preventing them from detecting any local visual landmarks returned to within <5 km of their loft, the relevant scale seems to be a few kilometers. A magnetic map precision of a few kilometers would require that the birds could sense variations in the Earth's magnetic field with an accuracy of about 10 nT or the inclination angle to a few hundredths of a degree in a background field of about 50,000 nT on top of which rather stochastic, naturally occurring variations of at least 30–100 nT in random directions occur every day. It is easier to imagine that a magnetic map or signpost sense could function on a much larger spatial scale, and there exist intriguing data suggesting that some songbirds can use magnetic cues as an approximate geographic signpost, which for example tells the birds when to increase their fat reserves before crossing the Sahara Desert [10].

#### 20.3.1.2 Specific Bacteria and Algae Align to Magnetic Fields

Even though bacteria are unicellular organisms without a nervous system, some have magnetosensors, and studying them could be instructive in the context of magnetic senses of higher animals. Several types of bacteria and algae synthesize and embed chains of magnetite ($Fe_3O_4$) crystals, forming so-called magnetosomes (Fig. 20.4a) in their cells. Thereby, the microorganisms spontaneously move in the direction of the polarity of the magnetic field vector. In the Northern Hemisphere, where the inclination is positive and the polarity therefore points earthward, the polarity of the

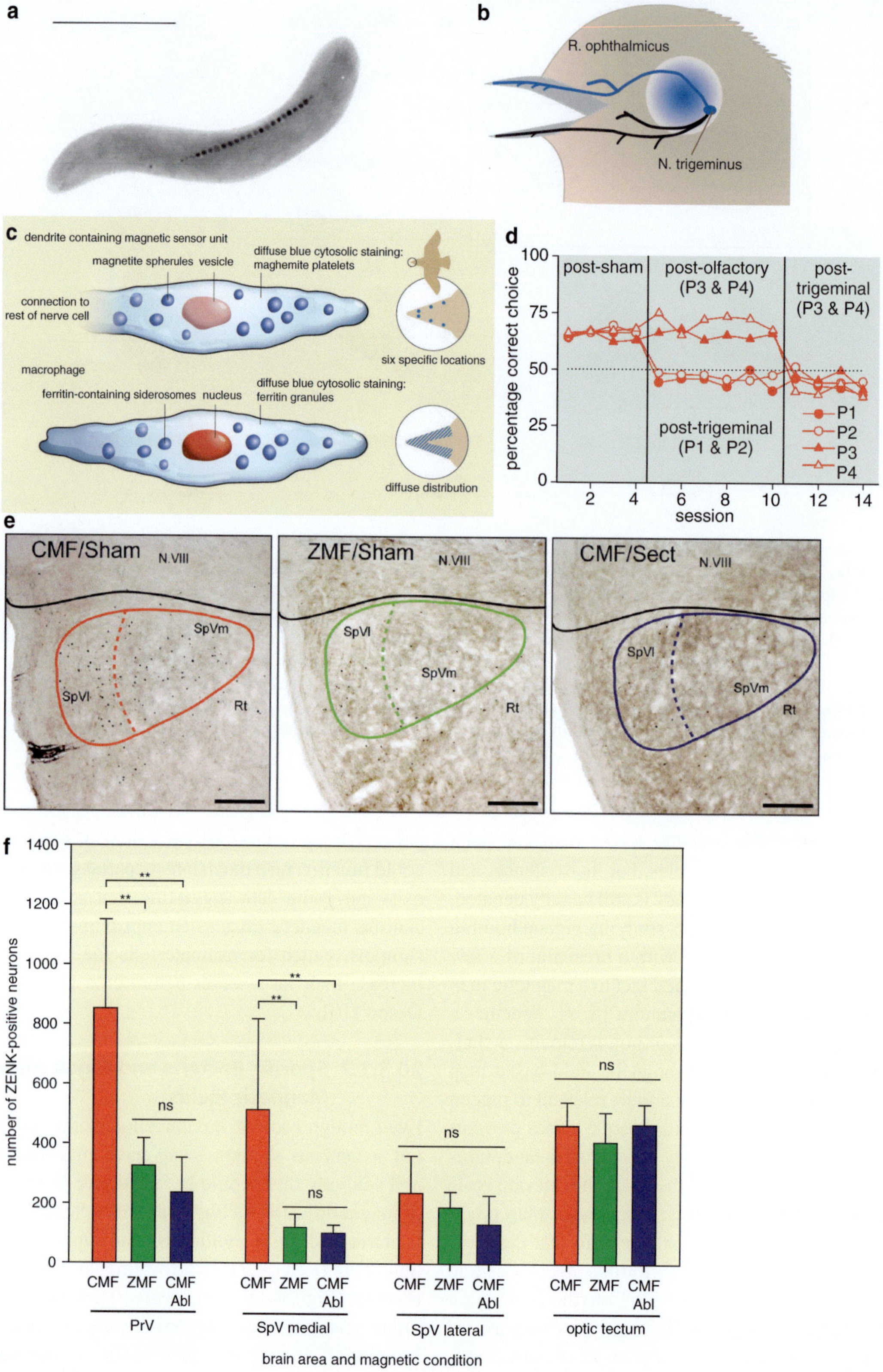
a
b
R. ophthalmicus
N. trigeminus
c
dendrite containing magnetic sensor unit
magnetite spherules vesicle
diffuse blue cytosolic staining: maghemite platelets
connection to rest of nerve cell
six specific locations
macrophage
ferritin-containing siderosomes nucleus
diffuse blue cytosolic staining: ferritin granules
diffuse distribution
d
percentage correct choice
post-sham
post-olfactory (P3 & P4)
post-trigeminal (P3 & P4)
post-trigeminal (P1 & P2)
P1
P2
P3
P4
session
e
CMF/Sham
ZMF/Sham
CMF/Sect
N.VIII
SpVm
SpVl
Rt
f
number of ZENK-positive neurons
ns
**
CMF ZMF CMF Abl
PrV
SpV medial
SpV lateral
optic tectum
brain area and magnetic condition

magnetite chains makes the magnetotactic microorganisms "North seeking". In the Southern Hemisphere, where the inclination is negative and the field lines point upwards, the magnetosomes have the opposite polarity and the microorganisms are "South seeking" [11]. If bacteria from the Southern Hemisphere are taken to the Northern Hemisphere and vice versa, they move in the wrong direction. The magnetosomes enable the light, anaerobic, magnetotactic microorganisms to move downwards towards the sediment away from the oxygen-rich open water. Building of the magnetite chains is an active process subject to natural selection, but once the magnetosomes are in place, the microorganisms passively align themselves along the magnetic field vector, making magnetic alignment a passive process which does not involve a nervous system or any other active processes.

> Some bacteria and algae contain chains of magnetite, so-called magnetosomes, which passively aligns them to the magnetic field lines.

#### 20.3.1.3 Several Arthropods Align or Can Actively Sense the Geomagnetic Field

The first magnetically guided animal behavior documented in the literature was magnetic alignment behavior in various insect species. The most prominent example is the "magnetic termites" of Australia (*Amitermes meridionalis*), which build impressive mounds predominantly oriented along the North–South magnetic axis. Later, various other behaviors of insects including honeybees (*Apis mellifera*) have been suggested to be affected by magnetic field changes. However, at present, it has not been convincingly shown whether the observed responses are based on simple passive alignment effects due to iron minerals found in many parts of the insects' bodies, or whether the behaviors observed are due to an active magnetosensory process.

Crustaceans can also detect the magnetic field and use it for compass orientation and for navigation as part of a position-determining system. The most studied species includes spiny lobsters [2] and various species of sandhoppers.

#### 20.3.1.4 Magnetic Sensitivity in Mollusks

Behavioral and electrophysiological evidence from the marine gastropod *Tritonia diomedea* suggest that mollusks may be able to detect the geomagnetic field and putatively use it for orientation [12].

> Many reports of magnetic senses exist for invertebrates, but the understanding of these systems is still very limited.

#### 20.3.1.5 Several Amphibians and Reptiles Have Both Magnetic Compasses and Magnetic "Maps"

Hatchling loggerhead sea turtles are born with innate information resulting in adaptive directional responses to specific combinations of magnetic inclination and

**Fig. 20.4 Examples of evidence supporting an the iron mineral-based and/or trigeminal-nerve-related magnetic sensing hypothesis.** (**a**) Transmission electron micrograph of the magnetotactic bacteria, *Magnetospirillum magnetotacticum*, showing the magnetosome chain inside the cell. Scale bar: 1 μm (Reproduced with permission from the photographer Richard B. Frankel). (**b**) Schematic drawing of the trigeminal nerve branches within the head of a European robin. The ophthalmic branch of the trigeminal nerve is indicated in *blue*. (**c**) Schematic drawing of iron mineral-containing structures in birds' upper beak illustrating the opposing interpretations of Fleissner et al. [18] and Treiber et al. [19] below (Fig. 20.4c is reproduced with permission from Nature (Macmillan Publishers Ltd): Mouritsen 2012, Nature 484: 320–321, © 2012 [20]). (**d**) Pigeons conditioned to magnetic stimuli could perform the task after sham sections of the trigeminal nerve and after real sections of the olfactory nerve, but lost their magnetic discrimination abilities when the opthalmic branch of the trigeminal nerve was cut (Fig. 20.4d is adapted by permission from Macmillan Publishers Ltd: Mora et al. 2004, Nature 432: 508–511, © 2004 [21]). (**e–f**) A changing magnetic field (*CMF*) leads to increased expression of the neuronal activity dependent gene *ZENK* in nuclei of neurons (*black dots*) in the hindbrain regions *SpV* (shown in (**e**)) and *PrV*, which receives the primary input from the trigeminal nerve. Much fewer activated neurons are observed when the magnetic field stimulus is absent (*ZMF* zero magnetic field) and when the ophthalmic branch of the trigeminal nerve was cut (Abl. = ablated = Sect. = Sectioned = severed nerve). No magnetic activation is seen in control regions (Figs. 20.4e–f are reproduced from Heyers et al. [22] with permission)

magnetic intensity (which seem to act as "signposts") that they could encounter within the North Atlantic Gyre system, where they spend the first years of their lives. The innate magnetic signpost sense of hatchling sea turtles is based on sensing geomagnetic inclination and intensity [2]. Whether it is the same signposts, or a more detailed map developed through experience, which enables adult loggerhead sea turtles to return to their preferred feeding grounds later in life, is not completely clear.

Newts and salamanders can extract both magnetic compass direction and magnetic "map" information (location) from the geomagnetic field [13]. In newts, the magnetic compass response seems to be light-dependent, whereas the "map" response seems to be light-independent. What is truly amazing about the magnetic "map" of newts is that it seems to work over distances of less than a kilometer. Consequently, the precision of their magnetic sense(s), their sense of gravity, and their knowledge about the irregular daily variations of the magnetic field would have to be incredibly accurate (see considerations about positional accuracy in the section on the Earth's magnetic field).

#### 20.3.1.6 Many Fish Can Sense Magnetic and Electric Fields

Many fish species can sense electric fields (see Chap. 19) and possibly also magnetic fields. There is good behavioral, histological, and physiological evidence that trout can detect magnetic field parameters [14]. Magnetic cues have been suggested to be involved in the long-distance oceanic part of the homing behavior of salmonid fishes, even though it has been shown that salmon primarily use odor cues to locate their natal stream from shorter distances. It has also been reported that zebra fishes (*Danio rerio*) can be conditioned to respond to magnetic stimuli.

#### 20.3.1.7 Some Mammals Seem to Possess Magnetic Senses, But Human Do Not

Magnetic sensing abilities have also been suggested for several mammalian species. The strongest evidence comes from subterranean mole rats of the family Bathyergidae, which, in a cage, will build their nest in a preferred magnetic compass direction [15]. Recent studies reported that ungulates (deer and cows) orient themselves in a preferred direction relative to the geomagnetic field, but other researchers have had difficulty reproducing these findings. Interestingly, this alignment has been reported to be disturbed by the presence of nearby power lines. Even though several claims have been made that humans should have a magnetic sense, which can be used to point towards home from an unknown location, repeated attempts to reproduce these findings have all failed. Various secondary effects of magnetic fields on human health have also been reported, but so far, clear studies performed double blindly and producing irrefutable evidence is lacking.

> Sea turtles, fish, mole rats, newts, and salamanders have all been shown to change behavior after magnetic cues were manipulated.

## 20.4 Interactions with Other Cues

In most orientation-related contexts, magnetic cues interact with other sources of similar and/or conflicting information. For instance, night-migrating songbirds do not only have a magnetic compass, they also have a Sun compass and a star compass. The birds only need information from any one of these compasses to orient in the appropriate direction. If the three compasses provide conflicting information, it is not consistent which compass the birds prefer. The preference is probably dependent on the ecological context and on details of the experimental setup and the conditions under which the birds were housed and tested, and it is likely that some sort of calibrations are taking place in nature. An experiment radio-tracking truly free-flying birds over hundreds of kilometers suggested that two species of North American songbirds used the magnetic compass as their primary compass in midair during spring migration, but that they calibrate this compass based on celestial cues during the sunset period [16]. Other evidence suggests that polarized light cues may be crucial for this calibration, but, so far, it is not understood how a bird's eyes can detect polarized light.

Navigation does not need to rely on magnetism exclusively: visual and olfactory cues also provide important sensory information that can allow animals to find a target over long distances.

In pigeon homing or any other map-related task, the interactions between different cues seem to be even more complicated. Homing pigeons have been shown to use olfactory cues, visual landmarks of various kinds, outward journey information, and maybe magnetic cues to estimate their position relative to home when released from a previously unknown location. The relative importance of these map cues is hotly debated with many apparently contradictory results occurring in the literature. One reason may very well be that in one location, one type of cue may be particularly reliable, whereas another cue is more reliable in a different location. Therefore, the animals might predominantly rely on different cues in different locations.

## 20.5 How Can Animals Possibly Sense the Geomagnetic Field?

It is challenging to detect the weak geomagnetic field with biological materials. Considering the anatomical constraints and known structures found within small animals, careful models of putative sensory mechanisms often find it hard to explain how a 50,000 nT magnetic field can result in reliable signals in the presence of thermal fluctuations (kT) and other sources of noise [4, 17]. In fact, any biological mechanism that can, in principle, allow detection of 50,000 nT fields is noteworthy. Only three basic mechanisms are currently considered to be physically viable: (i) induction in highly sensitive electric sensors, (ii) iron mineral-based magnetoreception, and (iii) radical-pair-based magnetoreception.

Only three different physical mechanisms are thought to be potentially able to sense Earth-strength magnetic fields using biological material.

### 20.5.1 The Induction Hypothesis

Electromagnetic induction is the production of voltage across an electric conductor situated in a changing magnetic field or a conductor moving through a stationary magnetic field. Thus, in practical terms: if one has an electric wire and one moves this through a magnetic field, a current will be generated in the wire. If this wire is ring- or coil-shaped, directional sensitivity can be achieved. In biological tissues, one would need conductive, liquid-filled, ring-like structures of sufficient size and diameter to generate measurable electrical signals, which can be picked up by an electrically sensitive receptor cell.

For electromagnetic induction, **Lorenzini ampullae** are a concrete realization of an electrically sensitive cell operating in saltwater fish (see Chap. 19). Their function uses the fact that saltwater is electrically conductive, and aquatic animals could potentially use induction to sense the geomagnetic field. Electrophysiological recordings have suggested that Lorenzini ampullae can detect movements through the geomagnetic field. However, no strong evidence currently exists that fish actually use their electric sense to deduce information from the geomagnetic field.

In land-based animals, it is difficult to imagine how induction could be used to sense the geomagnetic field since air has low conductivity and the needed structures therefore would have to be realized inside the animals themselves. In fact, biophysical considerations effectively eliminate induction as a potential source of magnetodetection in terrestrial animals: the required physiological structures filled with conductive liquid would be large and easily detectable, but no such structures have been reported. Thus, for terrestrial animals another mechanism must be responsible for magnetoreception.

Induction could enable saltwater fish to sense the geomagnetic field, whereas induction is unlikely to be an option for terrestrial animals.

### 20.5.2 The Iron Mineral-Based Hypothesis

When human beings want to use the direction of the geomagnetic field for orientation, we use a technical compass based on a needle made of magnetized iron or another magnetic compound that moves in the horizontal plane. Therefore, the first suggestion almost any human thinks of when one asks them how animals may detect the geomagnetic field is: "Maybe they have little compass needles in their head". Not surprisingly, this suggestion was also the first suggestion scientists came up with. A chain of single-domain magnetite ($Fe_3O_4$) crystals or other very similar iron oxides can most easily realize a small compass needle-like structure inside an animal, but other arrangements of iron mineral crystals could also work as a magnetic field detector. The iron mineral crystals are expected to transduce the magnetic signal by opening or closing pressure-sensitive ion channels.

#### 20.5.2.1 Experimental Evidence Supporting the Iron Mineral-Based Hypothesis

Many studies have documented the presence of magnetite or some other kind of iron mineral crystals in almost any animal, where researchers have seriously looked for such crystals (e.g., in *C. elegans*, mollusks, insects, crustaceans, and various vertebrates). However, the mere existence of iron mineral crystals or even magnetite does not represent significant evidence by itself that such structures have any relevance to magnetoreception. Iron is an important element required for proper function of most organisms. Consequently, iron homeostasis is important, and iron mineral deposits may just be a way for an organism to get rid of excess iron. Therefore, only if iron mineral structures are found at consistent, specific locations and are associated with the nervous system, the iron mineral structures qualify as serious magnetosensory candidate structures. The existence of magnetite crystal chains, which lead to a magnetically oriented swimming behavior in so-called magnetotactic bacteria [11], unequivocally proves that living cells can in principle synthesize magnetite that will align with the geomagnetic field (Fig. 20.4a). However, the magnetite crystals in these bacteria are not part of an active sensory system, they only lead to passive alignment.

> Iron mineral crystals are present in many species, but whether they act as active magnetic sensors remains unclear.

The most promising, active, iron mineral-based magnetoreceptor candidate structures are those reported from the olfactory epithelium of fish [14]. Elaborate iron mineral-based structures thought to be magnetoreceptors [18] have also been reported in the upper beak of birds (Fig. 20.4b–c). However, recent findings suggest that these structures are macrophages involved in iron homeostasis [19] (see Fig. 20.4c). Thus, at present, no convincingly documented, iron mineral-based, magnetoreceptive candidate structures are known from birds [20].

Conditioning of animals to magnetic stimuli has proven to be very difficult, and independent replication is rare. Michael Walker and colleagues [14, 21] reported that homing pigeons and various species of fish can be conditioned to respond to strong magnetic fields. In pigeons, the conditioned response to a very strong magnetic field (around two times the strength of the geomagnetic field) required intact trigeminal nerves (Fig. 20.4d). Consequently, some animals can, in principle, detect strong magnetic field changes via the ophthalmic branch of the trigeminal nerve. However, in order to use geomagnetic information for a map, animals must be sensitive to changes in the geomagnetic field, which are 3–5 orders of magnitude smaller than the anomalies used in the successful conditioning experiments.

> The ophthalmic branch of the trigeminal nerve seems to carry magnetic information in birds and fish.

The ophthalmic branch of the trigeminal nerve terminates in the principal (PrV) and spinal tract (SpV) nuclei of the trigeminal brainstem complex. Recent evidence [22] shows that subpopulations of neurons in both PrV and SpV in European robins (*Erithacus rubecula*), a night-migrating songbird, are activated by changing magnetic field stimuli, but not by a zero magnetic field. Furthermore, the activation in the changing magnetic field disappears when the ophthalmic branch of the trigeminal nerve is severed (Fig. 20.4e–f). These findings suggest that the ophthalmic branches of the trigeminal nerves carry magnetic information in birds. However, both the sensory origin (most likely iron mineral-based) and the biological significance of the trigeminally mediated magnetic information is unclear at present [8]. Information from the ophthalmic branch of the trigeminal nerve is neither required nor sufficient for magnetic compass orientation in several night-migrating songbird species (e.g. [23]; Fig. 20.7b–d). Homing experiments with pigeons have shown that pigeons tested around Pisa (Italy) need intact olfactory nerves but not intact trigeminal nerves to home [24]. The most likely function of the trigeminal nerve-related magnetic sense is to detect changes in magnetic field strength and/or magnetic inclination, which can be used to determine approximate position. It has also been suggested that the avian lagena (a part of the birds' vestibular system) plays a role in magnetodetection [25].

In birds, the brain regions PrV and SpV seem to process trigeminally mediated magnetic input. Magnetic compass orientation in European robins and pigeon homing works without trigeminal input. The avian lagena may also play an important role in magnetic sensing.

It is currently unclear what mechanism underlies magnetoreception in insects. Most evidence supports an iron mineral-based mechanism but recent studies on *Drosophila* have indicated that a light-dependent mechanism may also exist in insects [26]. The magnetic senses of mole rats and fish are most likely iron mineral-based polarity compasses [14, 27].

### 20.5.3 The Light-Dependent Hypothesis

The magnetic compass behavior of salamanders [13] and birds [28] is dependent on the wavelengths of light being available during behavioral tests. Already in the late 1970s, theoretical considerations led Klaus Schulten to suggest that chemical reactions in photosensitive molecules could form the basis of a magnetic compass sense.

The principles of the suggested light-dependent magnetic sensing mechanism are illustrated in Fig. 20.5. A light-sensitive molecule (D) absorbs light and uses the light energy to transfer an electron to an acceptor (A). Thereby, a radical pair is produced (Fig. 20.5a). If this radical pair is long-lived (>1 μs), it can, depending on the spin of the electrons, exist in one of two states, a singlet state (spins antiparallel) or a triplet state (spins parallel). It is known from

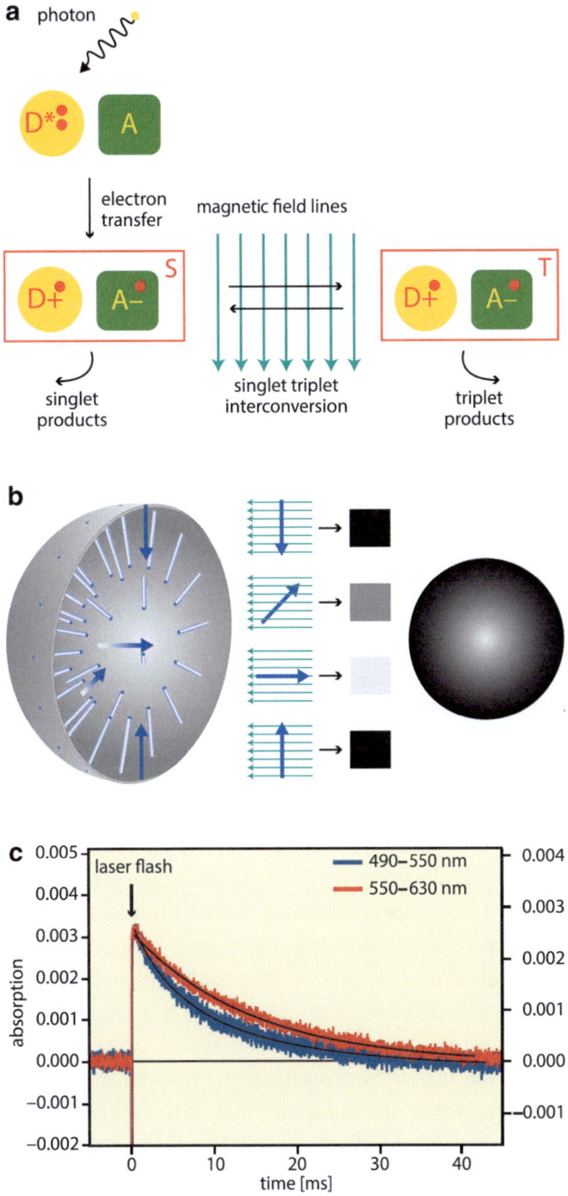


**Fig. 20.5 The principles of the suggested light-dependent, radical-pair-based, magnetic sensing mechanism.** (**a**) A simplified version of the primary reaction as suggested by Ritz et al. [29]. *D* donor, *A* acceptor, *S* singlet state, *T* triplet state. (**b**) Illustration how this mechanism could putatively lead to the formation of a virtual visual image. *Left*: three-dimensional illustration of the half-sphere of an eyeball. *White-blue pins* simulate a putative cryptochrome orientation, where all molecules point towards the center of the eyeball. *Blue-green arrows* indicate the direction of the magnetic field lines. *Blue arrows* indicate example orientations of cryptochrome molecules. (**c**) Transient absorption spectroscopy on isolated garden warbler cryptochromes show that long-lived radical pairs with a half-life of about 10 ms are produced in response to a short flash of light (Modified after Liedvogel et al. [36])

chemistry that singlet and triplet states have different chemical properties and thus often result in different chemical end products. Earth-strength magnetic fields can theoretically affect this statistical equilibrium, and thereby modulate a presently unknown biochemical pathway; Fig. 20.5a; [29, 30].

> The interaction between long-lived radical pairs and the geomagnetic field could lead to light-dependent magnetic sensing.

Which molecule can be responsible for light-dependent magnetoreception? Opsins cannot function as radical-pair-based magnetoreceptors, because opsins use the light energy to change a chemical bond, not to transfer an electron. The only currently known photoreceptor molecules found in vertebrates which can use light energy to form long-lived radical pairs (Fig. 20.5c) are the **cryptochromes**. Some cryptochromes are known to be involved in circadian clocks (see Chap. 27). However, in many animals, including migratory birds, more cryptochromes than the ones thought to be involved in the clock occur, so it is easy to imagine that they can play a role in other biochemical processes. Cryptochromes are related to the DNA repair enzymes called photolyases and consist of a photolyase homology region and a C-terminal end, which varies greatly between different cryptochromes. The C-terminal is thought to be involved in binding cryptochromes to currently unknown interaction partners. Cryptochromes non-covalently bind the cofactor flavin. The light-induced electron transfer is thought to take place between the flavin and three tryptophane residues within the cryptochrome protein. Cryptochromes are predominantly found within photoreceptor cells and ganglion cells in the eyes of birds, and cryptochromes are currently the only seriously considered candidate molecules for radical-pair-based magnetoreception in animals [31].

How can we imagine that a bird using a light-induced, radical-pair mechanism would detect the magnetic field? It is possible that a virtual visual image would literally enable an animal to "see" the direction of the magnetic field lines [17, 29]. If one makes the simple assumption that the sensory molecules are oriented perpendicularly to the eyeball, the half ball shape of the retina would mean that molecules oriented in all axial directions would occur (Fig. 20.5b). If a bird looks in the direction of the magnetic field lines, then in the line of sight, the retinal molecules would be parallel to the magnetic field and this could lead to a light pixel. At the edge of the eye, the molecules would be perpendicular to the magnetic field and this could lead to darker pixels. In between, the molecules would be oriented at different angles relative to the magnetic field, and various shades of gray pixels could appear. Altogether, this could lead to a virtual image looking somewhat like the one shown to the right in Fig. 20.5b. This pattern is for principle illustrative purposes only. We have much too little information available at present to know how an actual magnetically modulated light pattern seen by a bird would look like.

> Chryptochromes could function as magnetoreceptors in the eye, allowing birds to "see" the magnetic field.

Following the suggestion of Klaus Schulten it was shown that the compass orientation behavior of newts [13] and night-migrating songbirds [28] is influenced by the color (i.e., wavelengths) of the light available in the room where the orientation tests are performed. This wavelength dependence is difficult to explain if the eyes and/or pineal organ are not somehow involved in the magnetic compass. In birds, the pineal organ is not needed for magnetic compass orientation, whereas photoreceptor molecules in the pineal organ seem to be essential for magnetic compass orientation in newts [32].

Thorsten Ritz and colleagues have reported that oscillating magnetic fields in the low MHz range as weak as 15 nT (ca. 0.03 % of the geomagnetic field strength) disrupt the magnetic compass orientation capabilities of birds. Such effects are difficult to understand from a theoretical perspective, but if they can be independently confirmed, they are likely to be diagnostic for the involvement of a radical pair mechanism in the magnetic compass sense [33].

On the molecular level, it has been shown that putatively magnetosensitive cryptochrome molecules exist in the retina of many vertebrates

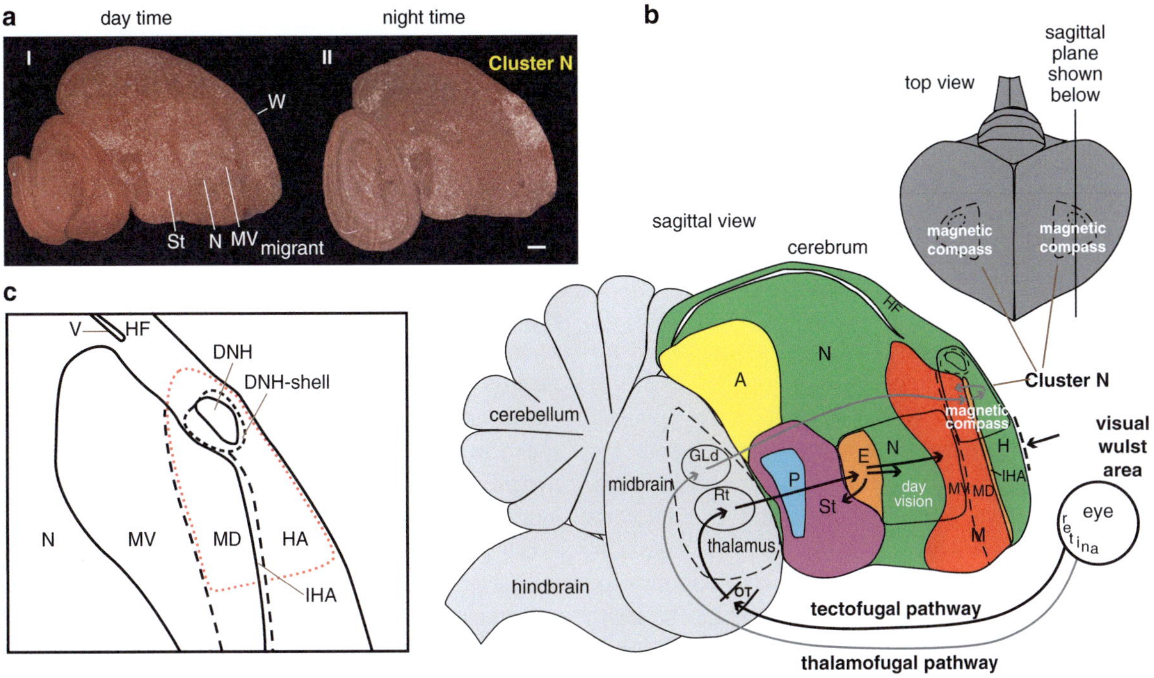


**Fig. 20.6** **Cluster N**. (**a**) Cluster N is the most active brain area (*white signal* on the brain slices indicates activity) when night-migrating birds perform magnetic sensing and/or compass orientation at night (*right brain* section) but not during the day (*left brain* section), and it is required for magnetic compass orientation (see Fig. 20.7). (**b**) Cluster N is a small part of the visual Wulst and receives input from the eyes via the thalamofugal visual pathway. Top view of the brain in *gray* indicates the medial-lateral and the frontal-caudal extent of Cluster N and the DNH and DNH-shell. (**c**) Cluster N is a functional unit consisting of a part of the hyperpallium, a part of the dorsal mesopallium, and a nucleus embedded within the hyperpallium and named DNH with a shell of cells around the DNH (**a**–**c** are modified after Mouritsen et al. [34] with permission). Anatomy: *A* arcopallium, *P* pallidum, *E* entopallium, *St* striatum, *N* nidopallium, *M* mesopallium, *MD* mesopallium dorsale, *MV* mesopallium ventrale, *H* hyperpallium, *v* ventricle, *OT* optic tectum, *HF* hippocampal formation, *IHA* HI: interstitial region of the hyperpallium intercalatum, *DNH* dorsal nucleus of the hyperpallium, *DNH-shell* shell around the DNH, *W* visual Wulst, *GLd* Lateral geniculate nucleus, dorsal part, *Rt* nucleus rotundus. Scale bar in (**a**) : 0.5 mm

including migratory birds [31]. Furthermore, cryptochromes from migratory garden warblers (*Sylvia borin*) have been shown to form long-lived radical pairs upon light excitation (Fig. 20.5c), and effects of Earth-strength magnetic fields on a radical pair reaction in an artificially produced molecule have supported the theoretical feasibility of the suggested mechanism (reviewed in [8]). Finally, behavioral evidence from genetically modified *Drosophila* has also supported the involvement of cryptochrome in magnetic sensing [26].

> Disorientation in weak oscillating magnetic fields has been suggested to be a diagnostic test for a radical-pair-based magnetoreception mechanism.

On the neuroanatomical level, a region named Cluster N (Fig. 20.6) is by far the most active forebrain region when night-migrating birds perform magnetic compass orientation and this activation disappears when the birds' eyes are covered [34]. Cluster N is the lateral-most part of the visual Wulst in European robins. Cluster N receives its neuronal input from the eyes via the thalamofugal visual pathway [35]. Could Cluster N be a processing center of light-dependent magnetic compass information?

> The brain region "Cluster N" seems to process magnetic compass information in night-migratory songbirds, since it is needed for magnetic compass orientation but not for star compass or sun compass orientation. More than one magnetoreception mechanism could easily exist side-by-side.

Double-blind experiments with European robins have shown that birds with bilateral Cluster N lesions were unable to orient using their magnetic compass (Fig. 20.7). In contrast, sham Cluster N lesions or bilateral sections of the ophthalmic branch of the trigeminal nerves did not influence the robins' ability to use their magnetic compass for orientation (Fig. 20.7b–h). Cluster N lesions only affect the magnetic compass, since Cluster N lesioned robins orient well using their sun and star compasses (Fig. 20.7j–l). These data (a) show that Cluster N is required for magnetic compass orientation in this species; (b) indicate that Cluster N may be specifically involved in processing magnetic compass information; (c) strongly suggest that a vision-mediated mechanism underlies the magnetic compass in this migratory songbird; (d) indicate that input from the lagena is not sufficient for magnetic compass orientation in robins; and (e) show that the proposed magnetic input to the brain transmitted via the trigeminal nerve is neither necessary nor sufficient for magnetic compass orientation of European robins tested in an orientation cage [23]. The exact role of Cluster N within the magnetic compass information processing circuit has not been determined, but the existing results raise the distinct possibility that this small part of the visual system enables birds to "see" magnetic compass information.

Do these results exclude the possibility that iron mineral-based and/or trigeminally mediated and/or lagena-mediated magnetoreception exists? Absolutely not! In birds, iron mineral-based magnetoreception may very well exist, and magnetic field dependent neuronal activation in trigeminorecipient regions has been documented (see above). Trigeminally mediated magnetoreception just does not seem to be the primary mechanism for the birds' magnetic compass, but they could be a primary source for magnetic positional information. In other animals such as mole rats and fish, most available evidence supports an iron mineral-based magnetic compass and/or map sense. It is likely that light-mediated, radical-pair-based magnetoreception and iron mineral-based magnetoreception mechanisms exist side by side in several animal species and that they may provide the animals with different types of magnetic information.

## 20.6 Irreproducible Results and the Urgent Need for Independent Replication

Magnetic sense research is strongly influenced by a number of claims, which nobody has ever replicated. This is particularly true for electrophysiological evidence, but there are also many other examples of contradicting or irreproducible results in the literature including all claims that humans have a magnetic sense. These problems with reproducibility do not necessarily mean that the original claims were wrong. However, it means that any result in magnetoreception – and in any other field for that matter – should be treated with caution until a given finding has been independently replicated.

This lack of reproducibility, which has appeared quite a few times in magnetic sense-related research, is unfortunately accompanied by an almost complete lack of double-blind procedures. Considering this history and the fact that humans have no intuitive feel for magnetic stimuli (and therefore are less likely to detect even obvious artifacts), double-blind procedures should become the standard. Studies representing the first independent, double-blind, replication of key findings in magnetoreception should therefore be considered almost as important as the original finding.

> Double-blind, independent replication is crucial before any finding should be accepted as an established fact.

## 20.7 Where Do We Go from Here?

Even though the magnetic senses are still not completely understood, many studies from different fields support both the iron mineral-based and the light-dependent magnetoreception hypotheses. However, fundamental questions remain in all relevant fields.

For instance, functional understanding of any particular iron mineral-based structure proven to be involved in an active sensory system is lacking.

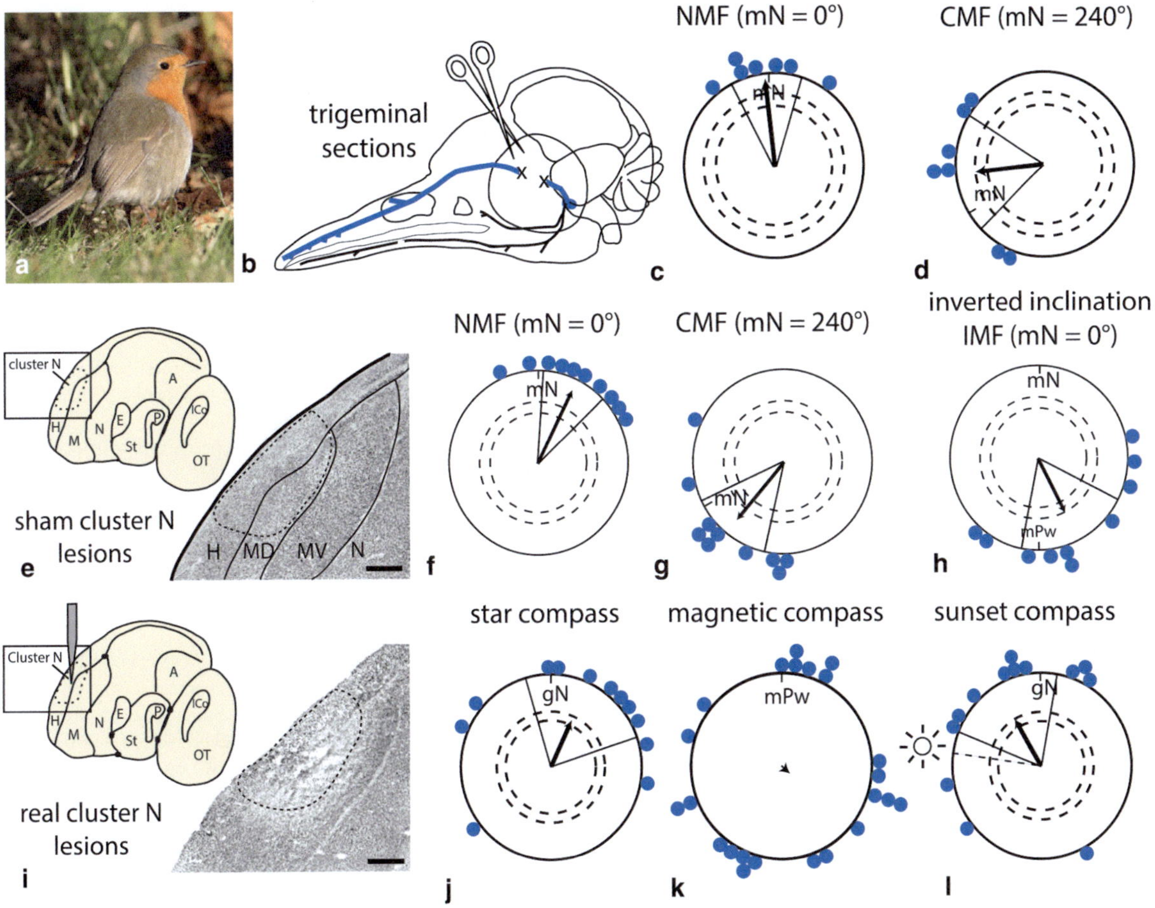


**Fig. 20.7 The brain region Cluster N is necessary for magnetic compass orientation behavior, but not for star and sun compass orientation**. The trigeminal nerve is neither necessary nor sufficient for magnetic compass orientation in European robins. (**a**) The European robin (photo © Henrik Mouritsen). (**b–d**) Bilateral sectioning of the ophthalmic branch of the trigeminal nerve (**b**) does not affect the birds' magnetic compass orientation capabilities (**c–d**; *mN* magnetic north). (**e–h**) Birds subjected to bilateral sham lesions of the forebrain region Cluster N (**e**) could perform excellent magnetic compass orientation behavior (**f–g**) using a magnetic inclination compass (**h**; *mPw* magnetic poleward). (**i–l**) Bilateral chemical lesions of Cluster N (**i**) destroyed the magnetic compass capabilities of the birds (**k**), whereas star compass orientation in a planetarium (**j**) and sun compass orientation outdoors with view of the setting Sun (**l**) was unaffected by Cluster N lesions (Fig. 20.7b–l are slightly modified by permission from Macmillan Publishers Ltd: Zapka et al. Nature 461:1274–1277, © 2009 [23]). In the circular diagrams (**c–d**, **f–h**, **j–l**), each *dot* at the *circle* periphery indicates the mean orientation of one individual bird based on ca. ten tests in the given magnetic condition. *Arrows* indicate the group mean vectors. *Inner* and *outer dashed circles* indicate the radius of the group mean vector needed for significance according to the Rayleigh Test ($p<0.05$ and $p<0.01$, respectively). *Lines* flanking the group mean vector indicate the 95 % confidence intervals for the group mean direction. (**e**) and (**i**) show schematic drawings and an example part of a brain section sagittally cut through the center of Cluster N and stained with a neuronal marker. Scale bar 500 μm. Rostral is left, caudal is right. Note that the tissue where Cluster N should have been in the lesioned bird (**i**) is destroyed compared to a brain slice from a sham lesioned control (**e**). Anatomy: *A* arcopallium, *E* entopallium, *H* hyperpallium, *ICo* intercollicular complex, *M* mesopallium, *MD* mesopallium dorsale, *MV* mesopallium ventrale, *N* nidopallium, *OT* optic tectum, *P* pallidum, *St* striatum

Likewise, we yet have to understand biophysically how nature designed radical-pair receptors so that they can be sensitive to Earth-strength magnetic fields at physiological temperatures, a feat that has been approximated, but not yet fully accomplished in man-made radical-pair reactions [30]. Furthermore, studies at the protein level suggest that cryptochromes have properties optimal for magnetic sensing, such as formation of long-lived radical pairs. But, we yet have to demonstrate Earth-strength magnetic field effects on cryptochromes, both at the protein level and *in vivo*.

On the neuroanatomical level, we have just begun to explore the brain circuits processing magnetic information, but we are still far from understanding how an animal gets from the detection of magnetic information to a directional choice, which is made, based on integration of information from multiple sensory systems. Also, so far, none of the claimed responses of single neurons to magnetic stimuli have been independently replicated. Even at the behavioral level, where most studies about magnetic senses have been published, a clear separation of experimental parameters has proven difficult, and many behaviors appear to be multimodal, or at least modulated by other modalities, such as vision and olfaction.

In conclusion, magnetoreception is an important part of life for a wide variety of animals and there are still many opportunities to perform new, ground-breaking research which is needed to conclusively elucidate the molecules, cells, and neural processes underlying any kind of magnetoreception.

## References


1. Mouritsen H, Frost BJ (2002) Virtual migration in tethered flying monarch butterflies reveals their orientation mechanisms. Proc Natl Acad Sci USA 99:10162–10166
2. Lohmann KJ, Lohmann CMF, Putman NF (2007) Magnetic maps in animals: nature's GPS. J Exp Biol 210:3697–3705
3. Kirschvink JL (1992) Uniform magnetic fields and double-wrapped coil systems: improved techniques for the design of bioelectromagnetic experiments. Bioelectromagnetics 13: 401–411
4. Kirschvink JL, Winklhofer M, Walker MM (2010) Biophysics of magnetic orientation: strengthening the interface between theory and experimental design. J R Soc Interface 7:S179–S191
5. Wiltschko R, Wiltschko W (1995) Magnetic orientation in animals. Springer, Berlin/Heidelberg/New York
6. Mouritsen H (1998) Redstarts, *Phoenicurus phoenicurus*, can orient in a true-zero magnetic field. Anim Behav 55: 1311–1324
7. Wiltschko W, Wiltschko R (1972) Magnetic compass of European robins. Science 176:62–64
8. Mouritsen H, Hore P (2012) The magnetic retina: light-dependent and trigeminal magnetoreception in migratory birds. Curr Opin Neurobiol 22:343–352
9. Wallraff HG (2001) Navigation by homing pigeons: updated perspective. Ethol Ecol Evol 13:1–48
10. Fransson T, Jakobsson S, Johansson P, Kullberg C, Lind J, Vallin A (2001) Bird migration – magnetic cues trigger extensive refuelling. Nature 414:35–36
11. Frankel RB, Blakemore RP (1989) Magnetite and magnetotaxis in microorganisms. Bioelectromagnetics 10:223–237
12. Lohmann KJ, Willows O, Pinter RB (1991) An identifiable molluskan neuron responds to changes in earth-strength magnetic fields. J Exp Biol 161:1–24
13. Phillips JB, Borland SC (1992) Behavioural evidence use of a light-dependent magnetoreception mechanism by a vertebrate. Nature 359:142–144
14. Walker MM, Diebel CE, Haugh CV, Pankhurst PM, Montgomery JC, Green CR (1997) The vertebrate magnetic sense. Nature 390:371–376
15. Thalau P, Ritz T, Burda H, Wegner RE, Wiltschko R (2006) The magnetic compass mechanisms of birds and rodents are based on different physical principles. J R Soc Interface 3:583–587
16. Cochran WW, Mouritsen H, Wikelski M (2004) Migrating songbirds recalibrate their magnetic compass daily from twilight cues. Science 304:405–408
17. Ritz T, Ahmad M, Mouritsen H, Wiltschko R, Wiltschko W (2010) Photoreceptor-based magnetoreception: optimal design of receptor molecules, cells, and neuronal processing. J R Soc Interface 7:S135–S146
18. Fleissner G, Holtkamp-Rotzler E, Hanzlik M, Winklhofer M, Fleissner G, Petersen N, Wiltschko W (2003) Ultrastructural analysis of a putative magnetoreceptor in the beak of homing pigeons. J Comp Neurol 458:350–360
19. Treiber CD, Salzer MC, Riegler J, Edelman N, Sugar C et al. (2012) Clusters of iron-rich cells in the upper beak of pigeons are macrophages not magnetosensitive neurons. Nature 484:367–370
20. Mouritsen H (2012) Sensory biology: Search for the compass needles. Nature 484:320–321
21. Mora CV, Davison M, Wild JM, Walker MM (2004) Magnetoreception and its trigeminal mediation in the homing pigeon. Nature 432:508–511
22. Heyers D, Zapka M, Hoffmeister M, Wild JM, Mouritsen H (2010) Magnetic field changes activate the trigeminal brainstem complex in a migratory bird. Proc Natl Acad Sci USA 107:9394–9399
23. Zapka M, Heyers D, Hein CM, Engels S, Schneider NL, Hans J, Weiler S, Dreyer D, Kishkinev D, Wild JM, Mouritsen H (2009) Visual but not trigeminal mediation of magnetic compass information in a migratory bird. Nature 461:1274–1277
24. Gagliardo A, Ioalé P, Savini M, Wild M (2009) Navigational abilities of adult and experienced homing pigeons deprived of olfactory or trigeminally mediated magnetic information. J Exp Biol 212:3119–3124

25. Wu LQ, Dickman JD (2012) Neural correlates of a magnetic sense. Science 336:1054–1057
26. Gegear RJ, Casselman A, Waddell S, Reppert SM (2008) Cryptochrome mediates light-dependent magnetosensitivity in *Drosophila*. Nature 454:1014–1018
27. Nemec P, Altmann J, Marhold S, Burda H, Oelschlager HHA (2001) Neuroanatomy of magnetoreception: the superior colliculus involved in magnetic orientation in a mammal. Science 294:366–368
28. Wiltschko W, Munro U, Ford H, Wiltschko R (1993) Red light disrupts magnetic orientation in migratory birds. Nature 364:525–527
29. Ritz T, Adem S, Schulten K (2000) A model for photoreceptor-based magnetoreception in birds. Biophys J 78: 707–718
30. Maeda K, Henbest KB, Cintolesi F, Kuprov I, Rodgers CT, Liddell PA, Gust D, Timmel CR, Hore PJ (2008) Chemical compass model of avian magnetoreception. Nature 453: 387–390
31. Liedvogel M, Mouritsen H (2010) Cryptochromes – a potential magnetoreceptor: what do we know and what do we want to know? J R Soc Interface 7:S147–S162
32. Phillips JB, Deutschlander ME, Freake MJ, Borland SC (2001) The role of extraocular photoreceptors in newt magnetic compass orientation: parallels between light-dependent magnetoreception and polarized light detection in vertebrates. J Exp Biol 204:2543–2552
33. Ritz T, Wiltschko R, Hore PJ, Rodgers CT, Stapput K, Thalau P, Timmel CR, Wiltschko W (2009) Magnetic compass of birds is based on a molecule with optimal directional sensitivity. Biophys J 96:3451–3457
34. Mouritsen H, Feenders G, Liedvogel M, Wada K, Jarvis ED (2005) Night-vision brain area in migratory songbirds. Proc Natl Acad Sci U S A 102:8339–8344
35. Heyers D, Manns M, Luksch H, Güntürkün O, Mouritsen H (2007) A visual pathway links brain structures active during magnetic compass orientation in migratory birds. PLoS One 2:e937
36. Liedvogel M, Maeda K, Henbest K, Schleicher E, Simon T, Hore PJ, Timmel CR, Mouritsen H (2007) Chemical magnetoreception: bird cryptochrome 1a is excited by blue light and forms long-lived radical-pairs. PLoS One 2:e1106
37. Wiltschko W, Wiltschko R (1996) Magnetic orientation in birds. J Exp Biol 199:29–38
38. Emlen ST, Emlen JT (1966) A technique for recording migratory orientation of captive birds. Auk 83:361–367

# Pain and Nociception

# 21

Maria P. Abbracchio and Angelo M. Reggiani

Pain is a fundamental self-defense mechanism for warning against damage so that action can be taken to avoid or minimize injury. According to the International Association for the Study of Pain (IASP) pain is defined as "An unpleasant sensory and emotional experience associated with actual or potential tissue damage or described in terms of such damage".

This definition contains two crucially distinct words: *sensory* and *emotional*. The sensory experience – also called **nociception** – pertains to our "sense of pain", one of the bodily senses (somesthetic senses, including pain, heat, touch). These senses have evolved early in evolution, and can be found across all animal species. The emotional experience – **pain** in the strict sense of the word – pertains to cognitive processing of (existing or imaginary) nociceptive stimuli. In this chapter, we will use "nociception" and "pain" specifically for the sensory and emotional aspects, respectively. The presence of pain across the animal kingdom cannot be always known for certain.

As depicted in Fig. 21.1, in humans, the overall pain response results from the integration of initial damage and several additional emotional and psychological factors. The key concept is that only the brain recognizes and gives "meaning" to the pain message, while the mind evaluates how "bad" the pain is. This means that individual psychological and emotional states can have a profound impact on the overall pain sensation.

Therefore, in addition to humans, the only animals capable of feeling pain are those that can feel emotions such as fear, anxiety, distress, and terror. In this respect, by interpreting the physical and behavioral reactions to noxious stimuli, it can be inferred that all vertebrates show the typical signs of pain.

While invertebrates show behavior that is related to avoiding nociception (e.g., withdrawal reflexes), they show few, if any, of the behaviors that we would recognize as evidence of emotional pain. Although it is impossible to know the subjective experience of other animals, the balance of the evidence suggests that most invertebrates do not feel pain. Indeed, the legislation for the protection of animals in research has "avoiding unnecessary pain" as the main criterion, and as a

**Box 21.1: Definitions**

**Analgesia** – absence of pain in response to stimulation that would normally be painful

**Anesthesia** – absence of all sensory modalities

**Allodynia** – pain due to a stimulus that does not normally provoke pain

**Hyperalgesia** – an increased response to a stimulus that is normally painful

**Dysesthesia** – an unpleasant abnormal sensation, whether spontaneous or evoked

M.P. Abbracchio (✉)
Department of Pharmacological and Biomolecular Sciences, Laboratory of Molecular and Cellular Pharmacology of Purinergic Transmission,
University of Milan, Milan, Italy
e-mail: mariapia.abbracchio@unimi.it

A.M. Reggiani
Department of Drug Discovery and Development, Italian Institute of Technology,
Genova, Italy

C.G. Galizia, P.-M. Lledo (eds.), *Neurosciences - From Molecule to Behavior: A University Textbook*,
DOI 10.1007/978-3-642-10769-6_21, © Springer-Verlag Berlin Heidelberg 2013

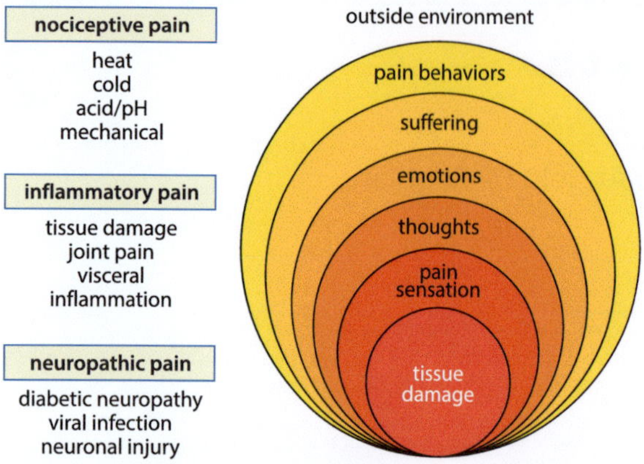


**Fig. 21.1 Multiple components in pain pathophysiology.** According to the IASP (International Association Study of Pain), pain is "an unpleasant sensory and emotional experience associated with actual or potential tissue damage, or described in terms of such damage". There is a strong psychological component in pain, since emotions such as depression and anxiety worsen while hopefulness and optimism alleviate pain sensation. Cultural influences are also important in determining whether a person is to be more stoic or more dramatic in showing pain to others. The newest theories of pain can now explain, on a physiological level, how and why people experience pain differently

consequence research on invertebrates is less restricted than research on vertebrates.

## 21.1 The Physics of the Stimulus

Nociception is always triggered by an initial injury which can be due to external causes or secondary to ongoing disease.

Most common external causes are stimuli that indicate **damage** or potential damage, such as trauma, fractures, cuts, bruises, but also strong pressure, heat and excessive cold, and **viral infections** (e.g., herpetic infection). Most common internal causes are **pressure**, such as that exerted by a cancer pressing on the tissues around it, or on a nerve, or **musculo-skeletal pain**, such as that caused by osteoarthritis, rheumatoid arthritis, low back pain, and osteoporosis. Furthermore, **nerve pain** includes diabetic neuropathy, peripheral blood vessel disorders, and stroke.

This list shows already several distinct initial stimuli: pressure (excessively high), temperature, tissue damage, or internal bodily signals. For each of these, dedicated receptors have evolved.

## 21.2 Pain in Humans

Pain in mammals can be defined as (1) **acute**, intense but short-lived, and (2) **chronic**, mild or intense (severe) and longer lasting (Table 21.1).

Acute pain is not of real medical interest. When the initial damage disappears, also pain disappears. Currently there are rather good drugs to control acute pain.

Chronic pain is the clinical manifestation of relevant pathological changes in both the CNS and PNS. It usually lingers on after the disappearance of the initial damage or even occurs in absence of injury (e.g., *phantom limb syndrome*). Few safe and effective drugs are available, thus search for novel medicines is a priority.

Pain transmission is a complex process, involving different anatomical structures at different levels (peripherally and in the CNS) and chronic pain is the consequence of a profound disruption of some key gating mechanisms.

Pain (Fig. 21.1) is classified into:

- **Nociceptive pain** (somatic and visceral pain) is the normal physiological response regardless of the injury's nature. It occurs when injury is established and disappears when injury is healed. Acute pain is mainly of nociceptive type, chronic pain may have nociceptive components.
- **Inflammatory pain** due to **peripheral tissue injury** is triggered by activation of nociceptive afferents by inflamed tissue and often, but not always, is associated to nociceptive pain. Rather frequently, it becomes chronic.
- **Neuropathic pain** (**neuronal tissue injury**) due to a metabolic disruption (diabetes), a viral infection (herpes), a traumatic event (lesion, amputation). It is only chronic and can have both inflammatory and nociceptive components.

In addition to a physical event, pain sensation is greatly influenced by psychological, emotional, and cultural background (Fig. 21.1), such as:

1. *Cultural beliefs*. In some societies, people learn to "bear the pain" without complaining; they are aware of the physical sensation of pain, but not particularly bothered by it.
2. *The personality, coping style, beliefs, and past experience of a patient*. If a medical procedure caused pain, when a patient needs to go through that procedure again, the fear of "potential" pain may cause to tense muscles involuntarily, thus setting the

**Table 21.1** Different types of pain in humans and pathophysiological correlates (From IASP Taxonomy, see www.iasp-pain.org)

| Type of pain | Pathophysiological correlate | | |
|---|---|---|---|
| | Nociceptive | Inflammatory | Neuropathic |
| Type of chronic pain | | | |
| Osteoarthritis | x | x | x |
| Rheumatoid arthritis | x | x | x |
| Chronic low back pain | x | x | x |
| Diabetic neuropathy | | | x |
| Postherpetic neuralgia | | | x |
| HIV-related neuropathic pain | | | x |
| Cancer pain | | x | x |
| Fibromyalgia | | x | x |
| Type of acute pain | | | |
| Postoperative pain | x | x | |
| Migraine | | x | x |

stage for opening the pain gate (see also below). Conversely, if patients are convinced that a certain medical protocol/treatment that they have already received relieves pain, they will experience analgesia even when receiving a "**placebo**" treatment. For example, for postoperative pain following tooth extraction, saline injection while telling the patient that it was a potent analgesic was as potent as a 6–8 mg dose of morphine. This is believed to be due to the activation of the descending inhibitory pathway (see Sects. 21.2.1.4 and 21.2.2.1).

3. *Stress*. Ordinary stresses of life, such as problems at work or in relationships, decrease the ability to withstand discomfort and pain.

A particular type of pain is the emotional (psychological) pain. This pain is not associated to a physical injury but rather can be described as heartache that results from a painful experience, such as the loss of a loved one. It can stem from depression, anxiety, disappointment, fear or guilt, and tends to worsen when you replay and relive painful, traumatic events that occurred in the past. Emotional pain can become crippling when it affects your mood, relationships, personal and professional life, and it occupies your mind constantly.

Several lines of evidence suggest a common neural link between physical and emotional pain. The anterior cingulate cortex (ACC) is involved in the affectively distressing components of both physical and emotional pain. Pain and depression are closely related. Depression can cause pain – and pain can cause depression. Sometimes pain and depression create a vicious cycle in which pain worsens symptoms of depression, and then the resulting depression worsens feelings of pain.

Chronic pain (physical and emotional) is strongly linked to depression. Common neural circuits have been described and also common neurotransmitter signaling pathways (noradrenaline and serotonin). Antidepressants are among the most effective drugs for chronic pain, e.g., duloxetine and venlafaxine, which are serotonin-norepinephrine reuptake inhibitors (SNRI), thus lengthening the action of serotonin.

## 21.2.1 Anatomy of Pain

### 21.2.1.1 Peripheral Mechanisms: Sensory Receptors

Pain generally starts with a physical event: a cut, tear, bruise, or burn. The primary neuronal relays receiving the noxious stimuli are the peripheral nociceptors.

Nociceptors are specialized receptive molecules located on the nerve ending of the primary sensory neurons innervating skin, muscle, or viscera, and are responsible to pick up the mechanical (e.g., pressure), chemical, electrical, or thermal (i.e., hot-cold sensitive) stimulations. Some nociceptors do not respond to the stimulus itself, but to a signal released by other cells upon damage. For example, strong UV light (sunburn) or mechanical trauma induce cell damage, and the resulting pain is elicited by substances released from these damaged cells. In particular, these include local changes in pH, or the release of amino acids and ATP. Several cells, in particular immune cells, release dedicated signaling molecules, such as bradykinin, prostaglandins, and, again, ATP (see below).

Nociceptor cells express the appropriate membrane receptive receptors, of which several are known or under study, and often belong to the following families:

- Transient receptor potential channels (**TRP channels**) are sodium channels that respond specifically to heat, acid, and vanilloid molecules such as capsaicin (the pungent active substance in chili peppers in the genus *Capsicum* L.). Mice lacking TRPV1 do respond less to these stimuli. The fly mutant *painless* is deficient in a TRPA family member. Most of these channels are not exclusively nociceptors, but are also used for innocuous stimuli. Of interest, these channels undergo strong desensitization with total loss of response when exposed to very high agonist concentrations. A typical example of this effect is the loss of tongue sensitivity upon ingestion of high quantities of capsaicin, which produces an initial very strong pain reaction followed by insensitivity. Physiologically, this effect suggests that nociceptors can adapt to persistence of stimuli; pharmacologically, it is currently attempted to develop capsaicin derivatives devoid of the pungent early reaction to be developed as analgesic agents.
- **DEG/ENaCs** (degenerin/epithelial $Na^+$ channels) are also sodium channels, which are sensitive to mechanical stimuli, or in some cases to acid stimulation.
- **NaV1.7**, **NaV1.8**, and **NaV1.9** channels are voltage-gated sodium channels important in the generation of action potentials in many nociceptive neurons. Mutations in NaV1.7 leading to excessive channel activity may result in primary erythromelalgia, an autosomal dominantly inherited disorder characterized by symmetrical burning pain and reddening of extremities, and elevation of skin temperature. Conversely, a nonsense mutation causing loss of function of NaV1.7, that has been detected in three consaguineous Pakistan families, is associated to incapability to experience pain.
- The **$P2X_{2/3}$** sodium channels are opened by ATP (called purinergic receptors). These receptors are phylogenetically ancient, since, in ciliates (*Paramecium* and *Tetrahymena*), they have been reported to mediate defensive/avoiding reactions. These responses can be considered as a prototype of nociception in higher animals, being a method of perceiving nearby cell lysis, which results in massive release of nucleotides, including ATP. Together with other evidence, this suggests that ATP (and other nucleotides) has universal roles as "danger" signals when perceived by purinergic receptors to initiate avoidance responses as a primitive form of nociception. In a similar way to TRP channels, also P2X receptors can undergo strong desensitization and loss of function upon exposure to high agonist concentrations.
- The bradykinin **$B_1$** and **$B_2$ receptors** are G protein-coupled receptors (GPCRs) coupled to both adenylyl cyclase inhibition, phospholipase C, and phospholipase A2 stimulation.
- The **PGE2 receptors** for prostaglandin E2 are also GPCRs coupled to the stimulatory $G_s$ protein, adenylyl cyclase activation, and cAMP formation, with consequent PKA activation and phosphorylation of TRP channels, which keeps these channels open.
- Nerve growth factor (**NGF**) **receptors** are tyrosine kinase receptors which also phosphorylate TRP channels, thus keeping them open.
- **Cytokines** (**interleukins**, **interferons**) are a diverse family of polypeptide regulators (proteins, peptides, or glycoproteins) that are secreted by specific cells of the immune system and are critical to the development and functioning of both the innate and adaptive immune response, although they are not limited to the immune system. Cytokines circulate in nanomolar to picomolar concentrations but can increase up to 1,000-fold during trauma or infection. Cytokine receptors can be subdivided into the following categories:
  - type I cytokine receptors are interleukin receptors
  - type II cytokine receptors are receptors mainly for interferons
  - immunoglobulin (Ig) receptor superfamily are ubiquitously present throughout several cells and tissues of the vertebrate body
  - tumor necrosis factor receptor family, whose members share a cysteine-rich common extracellular binding domain, and includes several other non-cytokine ligands like CD40, CD27, and CD30, besides the ligands on which the family is named (TNF)
  - chemokine receptors, two of which act as binding proteins for HIV (CXCR4 and CCR5). They are GPCRs
  - TGF beta receptors
- The **Mas-related receptors** are a large family of GPCRs including the so-called sensory neurons specific receptors (**SNSR**s). Despite the fact that several of these receptors have been implicated in nociception, their endogenous ligands are still largely unknown.

All these receptors induce nociception (Fig. 21.3); nociceptor cells can, however, concomitantly express

**Table 21.2** Classification of fibers in peripheral nerves

| Fiber group | Myelination | Diameter (μm) | Conduction velocity (m/s) | Sensory stimulus conveyed |
|---|---|---|---|---|
| Aβ | Myelinated | Large diameter (5–15) | Fast conduction (30–70) | Mechanoreceptors activated by non-noxious mechanical stimuli (touch) |
| Aδ | Myelinated | Medium diameter (1–4) | Intermediate conduction velocity (12–30) | Nociceptors activated by acute "sharp" pain stimuli, thermal perception |
| C | Unmyelinated | Small diameter (0.5–1.5) | Slow conduction (0.5–2) | Nociceptors recruited in chronic pain (responsible for secondary pain, normally burning, aching pain) |

Abnormal pain sensation may be transmitted along Aβ, Aδ, and C fibers

antinociceptive receptors that attenuate the depolarizing effect of the receptors listed above. These include:

- The G protein-coupled inwardly-rectifying potassium channels (**GIRKs**) are potassium ion channels opened by ligand-stimulated GPCRs. A variety of GPCRs coupled to inhibitory $G_i$ proteins activate GIRKs, some of which are involved in analgesia. Activation of GIRKs leads to massive entry of potassium ions into cells and the resulting hyperpolarization counteracts the depolarization induced by nociceptors. In this respect, the most known GPCRs inducing this effect are: **μ**, **∂**, and **k opioid receptors** responding to both morphine (the analgesic agent in opium, the drug derived from *Papaver somniferum*), and to endogenous morphine-like molecules like enkephalins, and adenosine **A1 receptors** responding to the analgesic agent adenosine, which is produced by the enzymatic degradation of ATP (a main algesic signal) to regulate, via a feed-back mechanism, the extent of the ATP response.
- The cannabinoid **CB1** and **CB2 receptors** are $G_i$-coupled GPCRs activated by pharmacological agents contained in the hemp plant (*Cannabis sativa*), as well as by endogenous cannabinoid-like molecules such as anandamide. It was initially believed that these receptors function as GIRKs; more recent data suggest that, besides utilizing this mechanism, these receptors are also able to influence calcium channels, protein kinases A and C, and mitogen-activated protein kinases (MAPK).
- The purinergic **P2Y receptors** are a large family of GPCRs responding to both adenine (ATP, ADP) and uracil (UPT, UDP, UDP-glucose) nucleotides. While P2X receptors mediate nociception, some P2Y receptor subtypes are analgesic. However, the exact role of these receptors is still under study, since some of them seem to contribute to nociception via an amplification of the effects mediated by P2X channels and other algesic receptors.

In conclusion, the generation and intensity of nociceptive signals in peripheral nerve terminals will depend on the **balance** between depolarizing activatory signals mediated by algesic agents and hyperpolarizing inhibitory signals mediated by antinociceptive agents.

Nociceptive signals are relayed via three different types of fibers: Aβ, Aδ, and C. These differ in diameter, myelination, conduction velocity, and type of stimulus reported (see Table 21.2).

#### 21.2.1.2 Spinal Cord Mechanisms: Signal Amplification and Gating

Sensory signals generated by nociceptors in the periphery travel to the spinal cord through nerves whose cell bodies are located in the dorsal root ganglia. These "primary" sensory neurons indeed send their sensitive dendrites to the skin or internal organs, and project their axons to the spinal cord, where they synapse onto neurons within the dorsal horn (Fig. 21.2). Afferent fibers synapse on spinal cord neurons ("second-order neurons") both in the segment they entered and on neurons 1–2 segments above and below the segment of entry (signal amplification). These multiple connections widen the alerting signal and may explain why it's sometimes difficult to determine the exact location of pain, especially internal pain. Neurotransmitters released by sensory fibers on spinal cord neurons include ATP, substance P, glutamate and possibly GABA, which activate their specific membrane receptors on spinal cord neurons.

However, whether or not these signals will generate a response and stimulate impulse generation from the spinal cord to the brain (see Sect. 21.2.1.3) will depend on two factors: (1), the state of activation of inhibitory signals coming from local opioid interneurons, and (2),

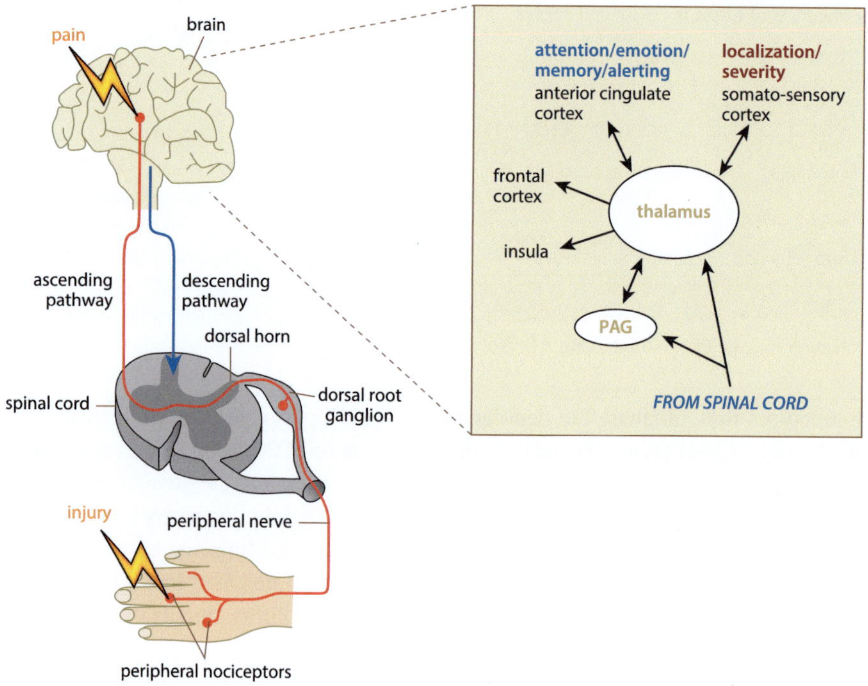


**Fig. 21.2 Pain transmission from the periphery – ascending and descending pathways.** Pain impulses arising in the periphery (e.g., an injured muscle) travel to the dorsal horn of the spinal cord through sensory nerves whose cell bodies (first-order neurons) are located in dorsal root ganglions. Three types of sensory fibers (Aβ, Aδ, and C fibers) are recognized (see also Table 21.2). In the spinal cord dorsal horn, signals are gated and transferred to second-order neurons and then to the brain via the ascending pathway. The firing of second-order neurons is also modulated by the descending pathway (see text for more detail). The first brain gating mechanism is in the periaqueductal gray matter (*PAG*). From PAG, signals are projected to the thalamus; then some projections go to the somatosensory cortex (recording of pain localization and severity) while other projections go to the cingulate and frontal cortex and to the insula (emotional response)

signals coming to these second-order neurons through a descending pathway (Fig. 21.2), a mostly inhibitory feed-back system connecting higher brain's pain areas to the spinal cord. Concerning (*i*), the firing of spinal cord second-order neurons is under the control of small local interneurons located mainly in the substantia gelatinosa, SG (Fig. 21.4), that release inhibitory aminoacids such as glycine or GABA (acting via hyperpolarizing glycine or GABA-A chloride channels), or small opioid peptides like enkephalins. As already described for peripheral sensory terminals (Sect. 21.2.1.1), these morphine-like molecules activate $G_i$-coupled GPCRs that induce the influx of potassium ions, leading to hyperpolarization. Concerning (*ii*), the descending pathway is a polineuronal pathway originating in the sensory cortex, with several stations at various telencephalic and mesencephalic areas, that importantly modulates the extent of signals going through the spinal cord (see also Figs. 21.3 and 21.4). Globally, these inhibitory systems constitute the spinal cord "gate" that has to be overcome to generate pain signals travelling to higher brain areas via the ascending pathway. Thanks to the existence of the spinal gate, innocuous mechanosensory signals are relayed to the brain through the spinal cord but they are not perceived as pain (Figs. 21.2 and 21.4).

The spinal cord is also the CNS area where one of the most primitive and fundamental pain responses (the nociceptor or flexor withdrawal reflex, also known as **withdrawal reflex**) is localized. This is a fast response intended to protect the body from potentially damaging stimuli. A classical example is when we touch something hot and immediately withdraw our hand from the hot object without thinking about it. The withdrawal reflex is based on the fact the afferent sensory fibers entering the spinal cord also synapse with ipsilateral motor neurons that exit the anterior horn of the spinal cord and work to pull the injured hand away from danger. At the same time, sensory fibers also synapse with contralateral motor neurons that, in turn, stabilize the uninjured side of the body to allow a highly

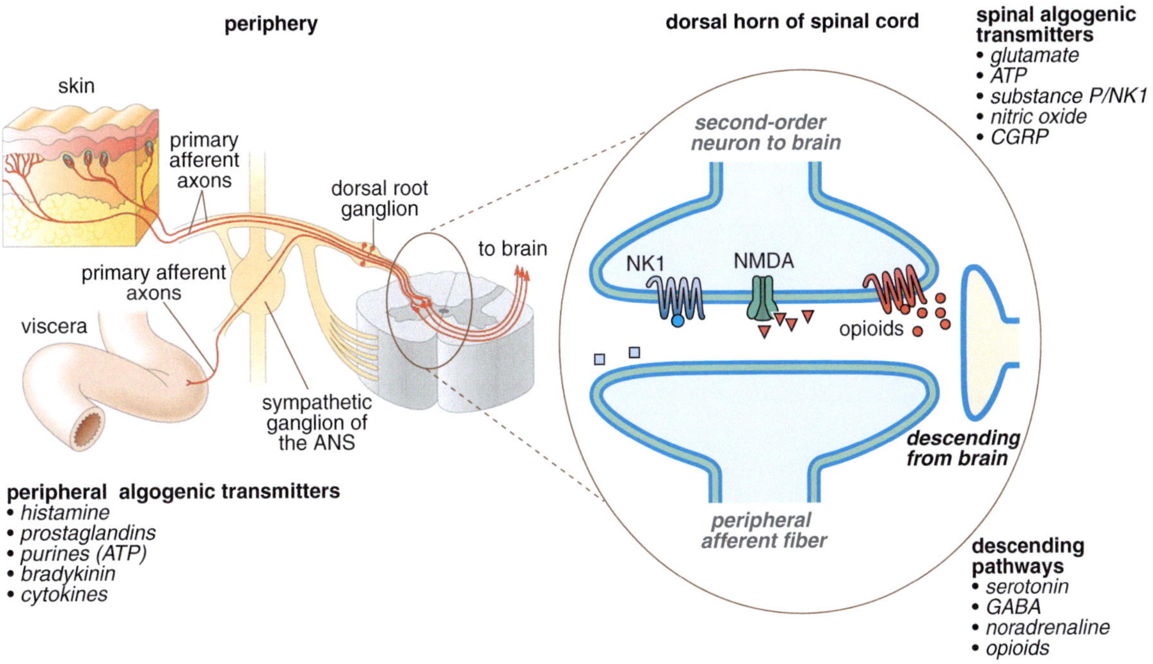


**Fig. 21.3 Molecular pathways in the periphery and in spinal cord dorsal horn.** Release of peripheral algogenic neurotransmitters by injured or perturbed tissues triggers signals traveling to the spinal cord, where they are transferred to second-order neurons by release of specific spinal neurotransmitters (ascending pathways). In the periphery, generation of signals is also modulated by local inhibitory modulators (not shown here, see text). The activity of the spinal synapse is under the control of local inhibitory interneurons (not shown here, see text) and of descending input from the brainstem. This descending system modulates (mainly inhibits) the transmission of nerve impulses via release of other specific neurotransmitters, such as serotonin, noradrenaline, GABA, or opioids. Some pharmacological agents (e.g., antidepressants) induce analgesia by potentiating these descending pathways

integrated motor response. Nevertheless, the same information is also relayed to central brain areas via ascending fibers, so that the brain can evaluate the danger message and modulate the reflex in the spinal cord via descending routes (Fig. 21.2).

#### 21.2.1.3 Central Mechanisms: Pain Recognition

From the spinal cord, pain signals are projected up to brain via the so-called ascending pathways (second-order neurons). The old spino-reticulo-diencephalic pathway mainly ends in the reticular system of the brainstem, but also sends fibers to the medial nuclei of the thalamus. Connections between the reticular system and the hypothalamus (and thalamus) are probably important – these may explain autonomic components of the pain response. The evolutionarily more recent spinothalamic tract ends in to the ventrobasal part of the lateral thalamus. Connections go from here to the sensory cortex (postcentral gyrus), which allows to localize somatic pain. This tract is NOT the main pain pathway, because lesions along this pathway do not cancel out the sensation of pain, but rather may cause severe pain (the "thalamic syndrome" – possibly due to damage to inhibitory pathways).

#### 21.2.1.4 The Descending Pathway: Signal Modulation

Descending pathways allow the brain to modulate nociceptive sensory information at the level of its first synaptic contact in the spinal cord. Most of these neurons are inhibitory, and chemical transmitters released in the spinal cord include serotonin, GABA, opioids, and noradrenaline. Antidepressant drugs that inhibit noradrenaline and serotonin reuptake at presynaptic terminals and thus increase the synaptic concentrations of these biogenic amines exert potent analgesic effects by acting on this system (Fig. 21.3).

Descending inhibitory pathways are of great research interest. Strong support to the existence of descending inhibitory paths in mammals comes from *in vivo* electrophysiological studies showing that the firing of dorsal horn neurons in response to noxious skin heating can be inhibited by stimulation of higher

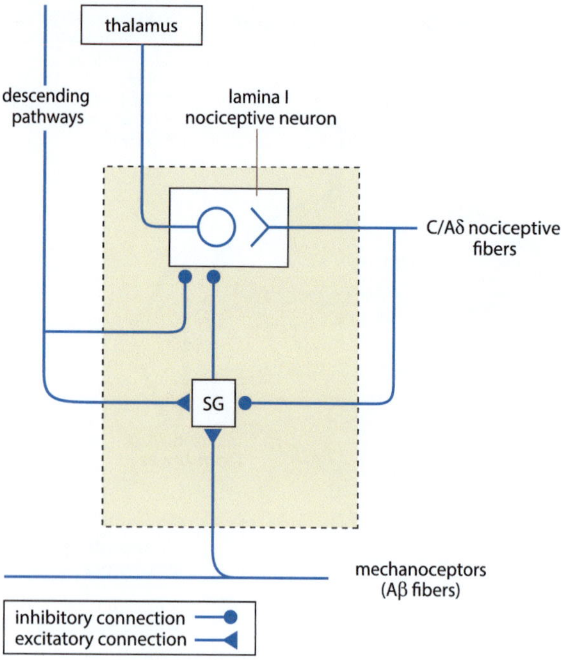


**Fig. 21.4** **The gate control theory of pain.** The activation of spinal cord lamina I nociceptive neurons by peripheral signals coming through C/Aδ nociceptive fibers is controlled by local SG inhibitory interneurons as well as by descending pathways. See text for more detail

brain centers such as the periaqueductal gray (PAG) and the lateral reticular formation (LRF) in the midbrain. In addition, inhibition of the spinal cord neurons can also be achieved by electrical stimulation of other regions of the brain, such as the raphe nuclei and the locus coeruleus (where the cell bodies of the serotonergic and noradrenergic neurons are located), and various regions of the medullary reticular formation, as well as sites in the hypothalamus, septum, orbital cortex, and sensorimotor cortex.

At present, it is not completely clear to what extent these different descending systems cooperate and interact, what their normal physiological functions are, and how they can be activated.

## 21.2.2 Peripheral and Central Processing of the Pain Signal

### 21.2.2.1 The Gate Control Theory

In more recent years, to explain why normal somatosensory inputs do not generate pain and how thoughts and emotions can influence pain perception. Ronald Melzack and Patrick Wall proposed that a gating mechanism exists within the dorsal horn of the spinal cord. This is known as the **gate control theory of pain** (Fig. 21.4), and can be briefly summarized in the following steps:

1. When no peripheral input comes in, inhibitory spinal cord interneurons, i.e., opioidergic neurons located in the SG, glycinergic and GABAergic interneurons, prevent lamina I nociceptive projecting neurons of the ascending pathways from sending signals to the brain (gate is closed).
2. Normal somatosensory input happens when there is more large-fiber stimulation (or only large-fiber stimulation, see Aβ fibers in Fig. 21.4). Both inhibitory neurons and projecting neurons are stimulated, but inhibitory neurons prevent projecting neurons from sending signals to the brain (gate is closed).
3. Pain happens when there is more small-fiber stimulation or only small-fiber stimulation (see C/Aδ nociceptive fibers in Fig. 21.4). When entering the spinal cord, these fibers not only activate lamina I nociceptive neurons but also send collaterals to the SG that inactivates inhibitory interneurons. Projecting neurons can now send signals to the brain, informing of pain (gate is open).

Descending pathways from the brain directly inhibit projecting neurons and also send collaterals to the SG that activate inhibitory interneurons, thus diminishing pain perception (gate is closed).

### 21.2.2.2 The Immune Modulatory Response

The immune system is actively involved in controlling pain. In particular, immune cells release many factors that contribute in maintaining chronic pain.

Differences exist depending on the type of chronic pain (inflammatory vs. neuropathic) and its location (CNS vs PNS), but the basic mechanism always involves an increased excitability or sensitization of neuronal cells caused by the release of endogenous mediators from immunocompetent cells. An example of the vicious circle generated by inflammation at the sites of pain signaling and involving a cross-talk between neuronal and immunocompetent cells is shown in Fig. 21.5. In the CNS, microglial cells represent the primarily involved immunocompetent cells; in the PNS (i.e., ganglia and peripheral sensory fibers at the site of damage) macrophages are the main responsible cells. T cells and blood-born circulating neutrophils recruit chemokines and cytokines in the PNS, lead to a breakdown of the blood–brain barrier and

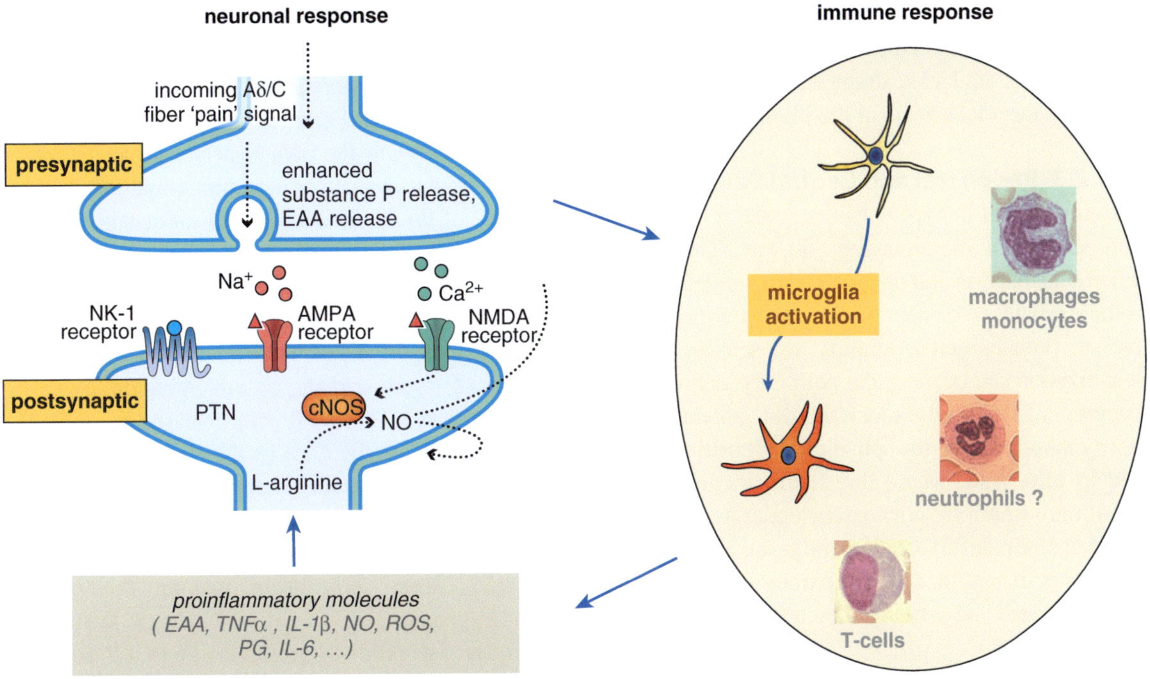


**Fig. 21.5** **Activation of immune and glial cell systems.** The neuronal response associated to pain stimulus is often accompanied by inflammation generated by excessive release of excitatory amino acids (*EAA*), peptides like substance P, and others that recruit immunocompetent cells (microglia in the CNS, macrophages, T cells and even circulating neutrophils in the PNS) at the sites of pain signaling. Under some circumstances, breakdown of the blood–brain barrier also enables circulating cells to penetrate into the CNS. These cells release pro-inflammatory molecules such as more EAA, cytokines (TNFα, IL-1β, and IL-6), or other mediators like NO, reactive oxygen species (*ROS*), and prostaglandins (*PG*) that all contribute to enhancing postsynaptic responsiveness and establishing a vicious circle that potentiates pain signaling

thus enter the brain and spinal cord, resulting in chronic inflammatory pain.

In inflammatory pain, the sequential response of immune cells is rather well known, and has led to new therapeutic hypotheses. Various immune mediators are released by activated immune cells, such as tumor necrosis factor-alpha (TNFα), interleukin-1 (IL-1), interleukin-6 (IL-6), nitric oxide (NO), ATP, bradykinin, nerve growth factor (NGF), prostaglandins, and protons, which exert their algesic effects on either peripheral sensory fibers or on PNS and CNS neuronal synapses. These agents may either act directly by activating specific receptors located on these targets or indirectly through the release of other inflammatory mediators (Fig. 21.5).

In neuropathic pain, the immune response is less characterized. The primary afferent neuron terminals conveying the pain signal are flanked by microglial cells that survey the environment in the spinal cord. In neuropathic pain states, the microglia is activated, probably by the release of transmitters or modulators from primary afferents. The activated microglia release several pro-inflammatory cytokines, chemokines, and other agents that modulate pain processing by affecting either presynaptic release of neurotransmitters and/or postsynaptic excitability.

Injuries and diseases that directly affect the CNS can also induce a strong immune reaction. Direct injury of the spinal cord leads to local breakdown of the blood–brain barrier, the release of many intracellular constituents, such as ATP, and the production of reactive oxygen species (ROS). As in other tissues, injury of the spinal cord leads to local inflammatory responses. Although direct evidence is still lacking, similar factors and mechanisms might contribute to the abnormal pain state that occurs in spinal cord injury. In response to injury, neutrophils, monocytes/macrophages, and lymphocytes are recruited to the injury site, and microglia are activated at and beyond this site.

The challenge given by these findings is to exploit this new knowledge on the immune system and pain to develop novel analgesic strategies without interfering with the beneficial effects of the immune response.

### 21.2.2.3 Peripheral and Central Sensitization

Chronic pain is a hypersensitive state reflecting the disregulation of the signaling pathway. This state is called sensitization and means an exaggerated increase in the excitability of neurons, so they are more sensitive to nonalgogenic stimuli or sensory inputs. Sensitization can occur at the periphery, acting on the receptor cells themselves, or centrally, generally at the first synaptic relay in the spinal cord. Sensitization can lead to **allodynia**, where thresholds are lowered, so that stimuli which are not noxious under normal conditions become painful, or to hyperalgesia, where responsiveness is increased, so that noxious stimuli produce an exaggerated and prolonged pain (Fig. 21.6).

#### Peripheral Sensitization

Peripheral sensitization is a reduction in threshold and an increase in responsiveness of the peripheral ends of nociceptors. This concerns the high-threshold peripheral sensory neurons that transfer input from peripheral targets (skin, muscle, joints, and the viscera) through peripheral nerves to the central nervous system (spinal cord and brainstem). Peripheral sensitization contributes to the pain hypersensitivity found at the site of tissue damage and inflammation. A good example of this is the change in heat sensitivity after sunburn, when a normally warm stimulus, such as a shower, feels burning hot in the sunburned areas.

Peripheral sensitization is due to the action of inflammatory chemicals or mediators released around the site of tissue damage or inflammation. Some of these, such as ATP, are directly released by damaged cells and activate the ends of the peripheral nociceptors. Other chemical mediators are produced by activated inflammatory cells, such as neutrophils or microglia/macrophages (see also Sect. 21.2.2.2). When activated, these cells begin making the enzyme cyclooxygenase-2 (COX-2), which leads to the production and secretion of prostaglandin PGE2, acting as a sensitizer. Increased sensitivity in the receptor cells can be both posttranslational or via altered gene expression. Activation of kinases takes minutes, changes in protein levels a day or so.

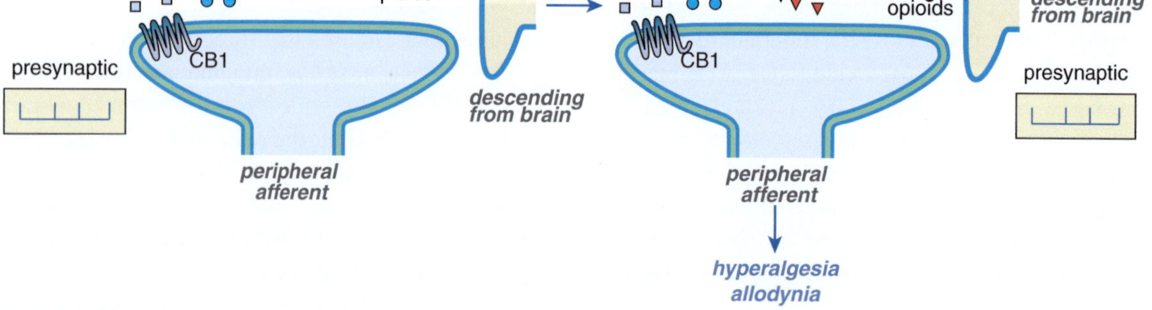

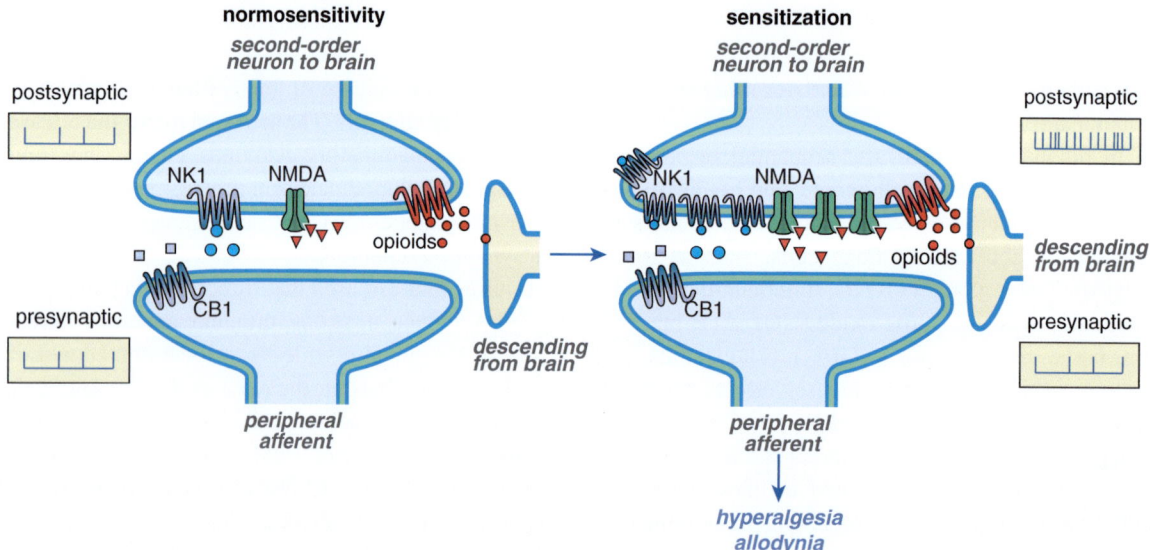

**Fig. 21.6** **Mechanisms of peripheral and central sensitization.** Continued stimulation by peripheral afferent nerves leads to distinct biochemical and physiological changes including increased postsynaptic receptor density or establishment of new nerve connections. The figure shows sensitization mechanisms within the dorsal horn of the spinal cord as an example, but similar events may occur in the periphery and at other pain signaling stations. Postsynaptic responsiveness to what would normally be perceived as mildly painful stimulus is therefore increased. **Hyperalgesia** is a heightened or exaggerated response to a painful stimulation. **Allodynia** is a painful response to a nonpainful stimulus

### Central Sensitization

Central sensitization is a very important mechanism in pain development and can be described as an increase in the excitability of spinal cord neurons, so that normal inputs begin to produce abnormal responses. Central sensitization is always triggered by repeated stimulation of dorsal horn neurons from periphery and is often the consequence of peripheral sensitization. However, more recent evidence suggests that also dorsal root sensory neurons can become a source of abnormal impulse generation and the contribution can last even when peripheral stimulation is over. At the cellular level, central sensitization can be described as an increased strength of synaptic connections at the spinal level (Fig. 21.6).

Two phases have been identified:

1. an immediate but relatively transient phase (wind up)
2. a longer-lasting phase but with a slower onset.

The wind-up phase is triggered by abnormal activation of unmyelinated C fibers and reflects initial changes in synaptic connections within the spinal cord. During wind-up, signal molecules are released, including the excitatory amino acid synaptic transmitter glutamate, neuropeptides (substance P and calcitonin gene-related peptide, CGRP) and synaptic modulators such as brain-derived neurotrophic factor (BDNF). These transmitters/modulators activate intracellular signaling pathways that lead to the phosphorylation of membrane receptors and channels, particularly the NMDA and AMPA receptors for the glutamate neurotransmitter. These posttranslational changes lower the threshold and opening characteristics of these channels, thereby increasing the excitability of the neurons (hyperalgesia).

The later transcription-dependent phase of central sensitization is mediated by increased levels of protein production. The net effect of these changes is that normally subliminal inputs begin to activate neurons and pain sensibility is drastically altered (hyperalgesia). Among the proteins mediating this effect are dynorphin, an endogenous opioid that increases neuronal excitability, and COX-2, the enzyme that produces PGE2 (see also above). As well as being involved in peripheral sensitization, prostaglandins also affect central neurons, contributing to central sensitization. Indeed, the analgesic action of aspirin-like drugs may derive more from their central than peripheral actions on COX-2.

In addition to the above mechanisms, several studies indicate that the development of new synaptic connections between afferent and spinal cord neurons can also take place. A typical consequence of this synaptic plasticity is that threshold sensory fibers activated by innocuous stimuli (e.g., a very light touch of the skin) begin to activate spinal cord neurons that normally only respond to noxious stimuli. Thus, an input that would normally evoke an innocuous sensation now produces pain (allodynia).

Central sensitization can also occur after surgery, contributing to pain on movement or touch, in migraine attacks where brushing hair is often painful, and in some patients with nerve damage where even blowing on the skin produces excruciating burning pain. More recently, it has been suggested that diseases such as fibromyalgia (a condition associated with a tender painful muscular pain) or irritable bowel syndrome may be manifestations not of peripheral pathology but of altered function of the nervous system (functional pain).

## 21.3 Pain in Animals Other than Mammals

All animals have the nociceptive sense, but – as stated above – physical hurt or discomfort caused by injury or disease (nociception) is different from emotional suffering. All vertebrates possess the primitive areas of the brain to process nociceptive information, namely the medulla, thalamus, and limbic system. However, the key area for emotional pain perception is the cortex. For this reason, many think that only animals with a large cortex, such as primates and humans can feel emotional pain. However, research has provided evidence that dogs and cats can show signs of emotional pain and display behaviors associated with depression during painful experience, i.e., lack of motivation, lethargy, anorexia, and unresponsiveness to other animals.

### 21.3.1 Birds and Reptiles

Cutaneous receptors responding to noxious stimulation have been identified in birds and have been characterized physiologically in the chicken. Following cutaneous nociceptive stimulation, chicken show cardiovascular and characteristic behavioral changes consistent with those seen in mammals and indicative of pain perception. Following major burn trauma, there is evidence for a pain-free period lasting several hours followed by pain-related behavior with anatomical and physiological characteristics similar to long-term chronic pain in mammals.

Although nociceptors have not been specifically studied in reptiles, these animals demonstrate typical responses to noxious stimuli such as flinching, muscle contractions, aversive movements away from the unpleasant stimulus and attempts to bite.

### 21.3.2 Fish

Fish have recently been shown to possess sensory neurons that are sensitive to damaging stimuli and are physiologically identical to human nociceptors. Fish show several responses to a painful event: they adopt guarding behaviors, become unresponsive to external stimuli, and their respiration increases. These responses disappear when the fish are given opioids such as morphine – evidence that they share, at least in part, some of the mechanisms at the basis of pain responses in mammals. Unlike mammals, Aδ are the dominant nociceptive fibers in fish, and not C fibers.

### 21.3.3 Nematodes

Exposure to noxious temperature induces a withdrawal reflex in the nematode *Caenorhabditis elegans*. This thermal avoidance response was demonstrated to differ significantly from the thermotaxis behavior that is based on the perception of physiological temperature, since it involves different neurons and is influenced by mutations in distinct genes. As in mammals, the strength of the thermal avoidance response has been shown to be increased by application of capsaicin, which acts on the TRPV1 channel. Thermal avoidance was strongly reduced in mutants affecting the neural transmission modulated by glutamate and neuropeptides, as well as in mutants affecting the structure and function of sensory neurons.

### 21.3.4 Arthropods

Insects share most of the nociceptor proteins with mammals, notably the TRP superfamily, and the DEG/ENaC family. For example, the fly *painless* mutant lacks a TRPA receptor, and leads to flies that do not respond to noxious mechanical or heat stimuli. It should be noted, though, that the presence of some of these receptors does not mean that nociceptors cover the same range of noxious stimuli as in mammals. Furthermore, other receptors that are selective to the mammalian pain system are not present in insects (including NaV1.8, NaV1.9, MRG, SNSR). Similarly, the opiate system and the corresponding melanocortin receptor family is not found in arthropods. There are, however, modulatory networks that downregulate nociceptive information and some of these use peptides (specifically FMRFamide) which are related to mammalian peptides of the same family.

*Drosophila* has become a major model system to study learning using aversive classical conditioning (see Chap. 26). It is likely that the noxious stimulus used, generally an electric shock, causes the activation of appropriate nociceptors. Aversive classical conditioning works well also with a noxious heat shock. Nevertheless, no behavioral or other evidence exists to date that insects or other arthropods might sense (emotional) pain. For example, an insect will withdraw a leg from a heated plate, but will not show any changed behavior if a leg is cut.

### 21.3.5 Mollusks

*Aplysia* is another important experimental model to unveil some of the cellular mechanisms at the basis of learning and sensitization (see Chap. 26). Also in *Aplysia*, elemental learning was studied using nociceptive neurons: habituation and sensitization in particular were studied here first, and later shown to occur and use the same synaptic machinery in mammals, and display many functional similarities to the alterations associated with the clinical problem of chronic pain. Presumably, this kind of adaptation represents a means to ensure that the intensity of defensive responses matches the threat posed by a noxious stimulus. Sensitization ensures that even a small stimulus – after sensitization – will induce an adaptive behavioral response. However, sensitization in *Aplysia* does not lead to ongoing activity even in the absence of the stimulus. This latter property, however, is typical of chronic pain, and likely to be necessary for emotional pain sensation.

If pain is a property that needs complex brains to evolve, some believe that cephalopods (octopuses, cuttlefish, and squid) may experience pain. Indeed, legislation has imposed more stringent limits on research with octopus based on their complex brains. However, no good evidence is available even for which nociceptors cephalopods express, and similarly no confirmation for pain in octopuses can be found in the literature.

In summarizing what is known about pain in animals other than mammals it has become clear that they possess the biological systems that are necessary to perceive and react to sensory pain and injury, and at least some of the brain structures that process pain in mammals. Therefore, some of the ethical considerations normally afforded only to mammals regarding pain should also be extended to other animal species. In this respect, a particularly important issue concerns animal experimentation. Pain has to be produced experimentally in animals to better understand the biological mechanisms of pain and to develop new methods of pain therapy (see Sect. 21.4). The ensuing moral dilemma requires that the scientists involved have a high level of responsibility for the animal. A set of ethical guidelines has been developed by the International Association for the Study of Pain in order to minimize pain and suffering in such studies.

## 21.4 Pain Therapy in Humans: Current and Future

### 21.4.1 Current Pain Drugs

Chronic pain is an area of unmet medical need. Both prescription and nonprescription drugs are used to treat chronic pain but, unfortunately, all these medicines have relevant side effects.

The most common problem is the side effect profile that is not always justified (or acceptable) by the efficacy profile of the drug. In some other cases, efficacy of available drugs is not sufficient (e.g., in neuropathic pain). Finally, it is relatively frequent that it may take several weeks before medicines start to show appreciable efficacy, which raises the need for individual titration.

A detailed discussion of the therapeutic approach to pain is beyond the scope of this chapter. Nonetheless, we include a brief summary of the drug classes currently used to treat chronic pain:

- Pain relievers (analgesics), such as acetaminophen, or nonsteroidal anti-inflammatory drugs (NSAIDs), such as aspirin, paracetamol, or ibuprofen, which may be prescribed for mild to moderate pain and to reduce inflammation. Due to their mechanism of action on peripheral sensitization (inhibition of COX-2, see Sect. 21.2), they are very efficacious in inflammatory pain.
- Antidepressants, such as tricyclic antidepressants (e.g., amitriptyline), may be used to treat chronic pain, although not all antidepressants are effective at reducing pain. Duloxetine, a serotonin/noradrenalin reuptake inhibitor, is another type of antidepressant that is approved by the U.S. Food and Drug Administration (FDA) to treat pain from peripheral neuropathy. These agents work by activating the descending inhibitory system (see Sect. 21.2).
- Corticosteroids, such as prednisone, which induce apoptosis of immune cells and inhibit the synthesis of molecules involved in inflammation and pain (see Sect. 21.2).
- Anticonvulsants are currently the gold standards in neuropathic pain. They may work by either increasing the effects of GABA (a major inhibitory neurotransmitter at the spinal level; by increasing the GABAergic transmission, a profound analgesic effect can be achieved), or by blocking sodium

**Table 21.3** Efficacy of key drug classes in pain pathophysiology

| | Efficacy on specific pathophysiological correlates | | |
|---|---|---|---|
| Drug class | Nociceptive | Inflammatory | Neuropathic |
| Traditional NSAIDs | x | x | |
| Selective COX-2 inhibitors | x | x | |
| Opioids | x | x | |
| Antiepileptics (gabapentin, pregabalin) | | | x |
| Antidepressants (duloxetine) | | | x |
| Local anesthetics | x | x | x |
| Serotonergic agents and 5HT1 receptor agonists | x | | |

channels, which, in turn, inhibits transmission of pain from the periphery to the CNS. Examples are:
   - gabapentin for postherpetic neuralgia (nerve pain from shingles)
   - pregabalin for postherpetic neuralgia and diabetic neuropathy (nerve pain from diabetes)
   - carbamazepine, oxcarbazepine, and lamotrigine which help controlling the episodes of facial pain in trigeminal neuralgia. If carbamazepine is taken daily, patients should be checked regularly to be sure they don't develop serious side effects (such as an allergic reaction or liver problems)
- Topical analgesics, such as lidocaine or prilocaine as cream or patch, can numb the skin and reduce pain. These drugs act as local membrane stabilizers by blocking sodium channels, thus abolishing nerve depolarization and transmission of action potentials from skin sensory fibers to the spinal cord.
- Cooling spray. This involves using a cooling spray directly on the skin, resulting in blockade of sensitive fibers.
- Opioid analgesics, which may relieve moderate to severe pain. Examples of opioids include morphine and oxycodone. In several cases, they are utilized in combination with analgesics (e.g., hydrocodone with acetaminophen, or acetaminophen with codeine), or with anticonvulsants. These agents are only moderately effective in neuropathic pain, however.

Other nonpharmacological therapies that may be used to treat chronic pain include:

- *Nerve block injections*. An anesthetic is injected into the affected nerve to relieve pain. The anesthetic may relieve pain for several days, but the pain often returns. Although nerve blocks do not normally cure chronic pain, they may allow physically blocked patients to begin physical therapy and improve their range of motion.
- *Epidural steroid injections* (injecting steroids around the spine). Although these injections have been used for many years and may provide relief for low back or neck pain caused by disc disease or pinched nerves, they may not work for everyone.
- *Trigger point injections*. These may relieve pain by injecting a local anesthetic into trigger points (or specific tender areas) linked to chronic facial pain or fibromyalgia. For many people, nerve blocks or other injections can relieve chronic pain for good. But it is not completely clear how this type of treatment works. These injections do not relieve chronic pain in everyone.

A summary of the efficacy of key drug classes and specific pathophysiology correlates is reported in Table 21.3.

### 21.4.2 Trends and Issues in Current Pain Research

At the end, we would like to refer to several recent approaches in pain research:

1. Increasing knowledge of the molecular consequences of nerve injury and the availability of genome databases has greatly increased the range of neuronal targets of potential interest for the pharmacological management of chronic pain. Neuronal sensitization

**Table 21.4** Most widely used animal models in pain research and their pathophysiological correlates

| Animal model | Pathophysiological correlate: Nociceptive | Inflammatory | Neuropathic |
|---|---|---|---|
| Spinal nerve ligation[a] | | x | x |
| Chronic constriction injury[b] | | x | x |
| Streptozotocin induced diabetic rats[c] | | | x |
| Chemotherapy induced pain[d] | | | x |
| Skin wound healing | | x | |
| Freund's adjuvant (FREUND's); carrageenan (CARRA), monoiodoacetate-induced osteoarthritis (MIA) | | x | |
| Hot plate, tail flick, plantar test | x | | |

[a]Kim and Chung [1]
[b]Bennet and Xie [2]
[c]Malcangio and Tomlinson [3]
[d]Xiao and Bennett [4]

and the associated alterations in gene expression, protein modification, and neuronal excitability is currently a prominent research area to achieve an efficacious neuropathic pain management.

2. Protection and amplification of inhibitory (descending) systems is gaining growing interest, again due to the well-known efficacy of such approaches. For example, opioids are effective in this way and they still remain the unsurpassed gold standard in inflammatory pain therapy. The challenge in this field is to be able to identify new targets with similar efficacy to opioids but with a less severe side effect profile.
3. The cross talk between neuronal and immune system is an emerging area of great interest. It is believed that neuronal sensitization can be efficiently controlled through the immune system. Basic research in the field is ongoing to identify the best therapeutic opportunities.
4. Great emphasis is now placed on the development of predictive animal models. As pointed out in Table 21.4, the most common animal models have a reasonable pathophysiological correlate in humans. However, this is still an area of active research, since good animal models are thought to represent a key step for a successful research strategy.

## References


1. Kim SH, Chung JM (1992) An experimental model for peripheral neuropathy produced by segmental spinal nerve ligation in the rat. Pain 50:355–363
2. Bennet GJ, Xie YK (1988) A peripheral mononeuropathy in rat that produces disorders of pain sensation like those seen in man. Pain 33:87–107
3. Malcangio M, Tomlinson DR (1998) A pharmacologic analysis of mechanical hyperalgesia in streptozotocin/diabetic rats. Pain 76:151–157
4. Xiao WH, Bennett GJ (2008) Chemotherapy-evoked neuropathic pain: abnormal spontaneous discharge in A-fiber and C-fiber primary afferent neurons and its suppression by acetyl-l-carnitine. Pain 135:262–270


## Further Readings


Burnstock G, Verkhratsky A (2009) Evolutionary origins of the purinergic signalling system. Acta Physiologica 195:415–447

Costigan M, Scholz J, Woolf CJ (2009) Neuropathic pain: a maladaptive response of the nervous system to damage. Annu Rev Neurosci 32:1–32

Hunt SP, Mantyh PW (2001) The molecular dynamics of pain control. Nat Rev Neurosci 2:83–91

Julius D, Basbaum AI (2001) Molecular mechanisms of nociception. Nature 413:203–210

Kidd BL, Urban LA (2001) Mechanisms of inflammatory pain. Br J Anaesth 87:3–11

Latremoliere A, Woolf CJ (2009) Central sensitization: a generator of pain hypersensitivity by central neural plasticity. J Pain 10:895–926

Max MB, Stewart WF (2008) The molecular epidemiology of pain: a new discipline for drug discovery. Nat Rev Drug Discov 7:647–658

Woolf CJ, Salter MW (2000) Neuronal plasticity: increasing the gain in pain. Science 288:1765–1769

Zimmerman M (1986) Ethical considerations in relation to pain in animal experimentation. Acta Physiol Scand Suppl 554:221–233

# Muscles and Motility

# 22

Ingo Morano

Molecular motors drive all types of motility of living organisms both on the microscopic and macroscopic level, including chromosome segregation, cell division, intracellular transport of organelles, as well as the movement of cells and cilia, and in fact the movement of entire organisms. Hence, "every motion has its motor" [1] appears as the common strategy. Myosins (with 35 classes) use actin filaments and kinesin, dynein, and dynamin use microtubules as tracks (see Chap. 6). DNA and RNA polymerases as well as helicases move on nucleic acids. High-frequency electromotility of the outer hair cells of the cochlea is powered by the anion transporter prestin. The $F_1F_0$ATP synthase in mitochondria is a rotary motor.

Motility is unique to animal life and may have evolved in several steps. In a first step, "**amoeboid motile systems**" are based on aggregations of actin filaments forming pseudopodia located in the direction of locomotion at the leading edge and myosin-driven contraction at the rear end of a migrating cell. Amoeboid movement of many different cell types is involved in embryogenesis, inflammation, immune defense, wound healing, angiogenesis, and formation of tumor metastases. In a second step, specialized cellular compartments become motile. **Cilia** (cellular membrane protrusions of 0.2 μm diameter and up to 10 μm length, stabilized by microtubuli) are the locomotive structures of *ciliates* and also are used to drive sperm cells. The phylum *Cnidaria* with only two cell layers (diploblastic animals) develop **epitheliomuscular cells**. Certain parts of the epitheliomuscular cells form the epithelium while the cell bases extend to form muscle fibers. Finally, in a third step multicellular organisms form specialized cells with predominantly contractile function (**muscle cells**).

Vertebrate muscle tissues can be divided into smooth and cross-striated, i.e., skeletal and cardiac muscle. Invertebrates have, in addition obliquely striated muscle tissues. Roughly 40 % of the human body mass is skeletal muscle, another 10 % cardiac and smooth muscle. Humans have more than 600 skeletal muscles which fine-tune a variety of different motor actions. Those motor actions allow fine motor skills (like speech or had-finger control), expression of specific emotions by mimic muscles, reflex responses (like withdrawal reflexes, coughing, sneezing, or swallowing), rhythmic body movements (like walking, running, or swimming), eye movements (like saccades), and postures. The human heart contracts about 86,000 times and pumps roughly 3,500 l of blood through the body per day yielding about 100 kJ of work. Smooth muscle organs regulate blood pressure and organ perfusion, they are involved in digestion, defecation, micturition, and parturition.

## 22.1 Control of Muscle Function in the Vertebrate Body

In 1751, the Berlin Academy of Sciences announced an award for the best answer to the question "... whether the communication between brain and muscles via the nerves is caused by a liquid matter which distends the muscle?" Two applicants submitted a new electrical

I. Morano
Max-Delbrück-Center for Molecular Medicine, Department of Molecular Muscle Physiology and University Medicine Charité Berlin, Berlin, Germany
e-mail: imorano@mdc-berlin.de

C.G. Galizia, P.-M. Lledo (eds.), *Neurosciences - From Molecule to Behavior: A University Textbook*,
DOI 10.1007/978-3-642-10769-6_22, © Springer-Verlag Berlin Heidelberg 2013

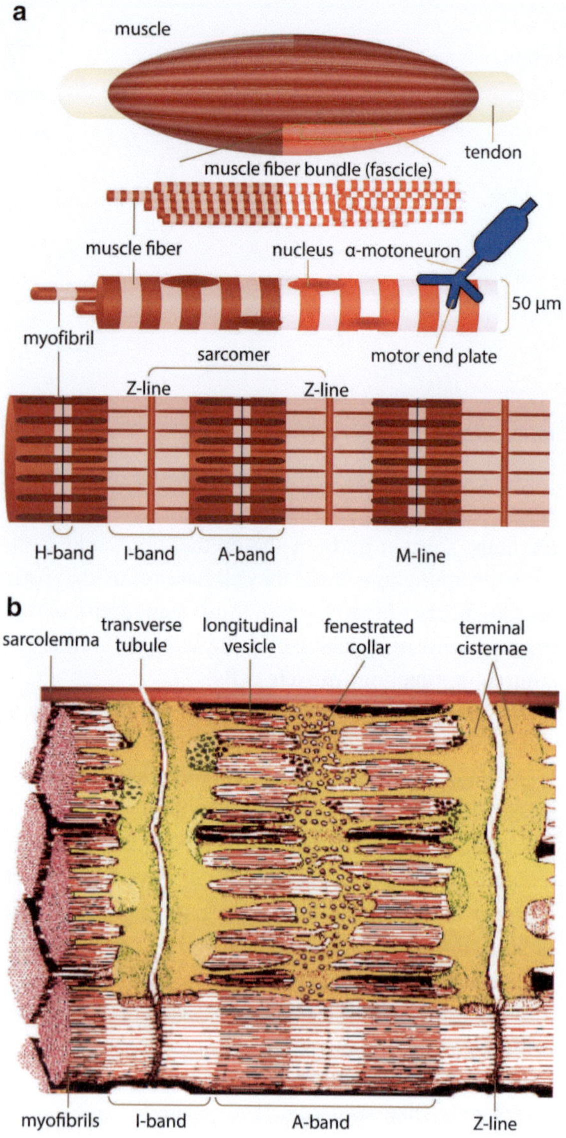


**Fig. 22.1** (**a**) **Structural organization and motor innervation of skeletal muscle.** (**b**) Structural organization of a skeletal muscle fiber with sarcolemma, sarcoplasmic reticulum, and myofibrils (Modified from Peachey [3], with permission)

hypothesis – which was refused, however, in favor of the traditional assumption regarding a "liquid matter" (cited from Bois-Reymond [2]). This idea goes back to the Ancient Greek school of Alexandria. According to Erasistratos (304–250 BC) air from the lungs is drawn into the heart during diastole, where a particular *pneuma zootiké* is formed and pumped to all parts of the body through the arteries. The pneuma zootiké is transformed in the brain into *pneuma psychiké* which is distributed to the muscles via the cranial and the spinal nerves. If the muscle is filled with *pneuma*, it expands and contracts. Galen (129–201 AD) and Descartes (1596–1659) still believed in the nerve pneuma model. In his work *De motu animalum*, Borelli (1608–1679) described an experiment in which muscles are incised longitudinally and immersed in water. Hence, bubbles of air should be released from the wound and rise in the water. Since this does not appear at all, muscles moved by a *succus nerveus* instead of a volatile spirit. Around 1663, Swammerdam observed that the volume of a contracted muscle remained constant.

In 1773, Leopoldo Caldani finally elaborated the basic concept of the contemporary view of muscle activation. He showed that a muscle contracted upon direct electrical stimulation or electrical stimulation of its innervating nerve. Luigi Galvani and Alexander von Humboldt in the late eighteenth century and Emil du Bois-Reymond and Hermann von Helmholtz in the early nineteenth century elaborated the basic principles of nerve and muscle excitability.

In vertebrates, single (**mononeuronal**) myelinated nerve fibers, referred to as **α-motoneurons**, originate from motor nuclei in the brain stem or the anterior horn of the spinal cord and innervate **skeletal muscle fibers** for voluntary motion (Fig. 22.1a). Within the muscle, axons branch to form single synaptic (**monoterminal** innervation) contacts (**motor endplates**) with distinct amounts of muscle fibers, forming **motor units**. Motor-unit innervation ratios vary from 20 to 2,000 between different human muscles [38]. Similar to nerve fibers, those muscle types respond in an **all-or-none** fashion with an action potential and a single contraction upon motoneuron stimulation (**twitch fibers**). Muscle fibers of **arthropods** are innervated by more than one motoneuron (**polyneuronal** innervation) which may be excitatory or inhibitory (cf. Sect. 22.9). **Tonic fibers** (e.g., the Rectus abdominis of the frog), however, are innervated by motoneurons which form more than one motor endplate (**multiterminal** innervation) and do not respond in an all-or-none fashion. The contractile response of those muscles to motoneuron stimulation is local and graded: high stimulation frequencies cause greater depolarizations of the muscle membranes which results in stronger contractions [25].

Motoneurons at their motor endplates release **acetylcholine**, which activates skeletal muscle contraction (cf. 22.4). Some trophic factors are also released at the motor endplate (e.g., agrin, calcitonin gene-related peptide, neuregulin) which help to form the neuromuscular junctions. There are at least two different types of α-motoneurons with distinct cellular and functional properties, namely "S" (slow) and "F" (fast) α-motoneurons. The **S motoneuron** has a small cell body, thin diameter axon, fires with low action potential frequencies due to a prolonged after-hyperpolarization (maximally 20 Hz), and innervates a lower number of muscle fibers. **F motoneurons** have large cell bodies, high (up to 120 Hz) firing frequencies, and innervate a larger number of muscle fibers. Recruitment of motor units is asynchronous and ordered by the "**size principle**": small motoneurons are the first to begin firing, while larger motoneurons are activated later during stronger excitatory input [4]. Activity of the α-motoneurons in the spinal cord depends on descending innervation from the central nervous system, propriospinal inputs as well as from sensory inputs from muscles.

Skeletal muscles contain sensory organs which help to regulate both muscle length and force generation, i.e., **muscle spindles** and **golgi tendon organs**, respectively via spinal reflex mechanisms. The muscle spindles contain specialized muscle fibers (**intrafusal fibers**) with contractile ends but noncontractile central regions, which run in parallel to the working skeletal muscle fibers (**extrafusal fibers**). The noncontractile central part of intrafusal fibers are innervated by sensory nerve ends from group II and Ia nerve fibers (flower spray or annulospiral endings), while its contractile peripheral regions are innervated by **γ-motoneurons**. Intrafusal fibers respond to γ-motoneuron stimulation like tonic fibers (**nuclear bag fibers**) or like twitch fibers (**nuclear chain fibers**). Golgi tendon organs are sensors located in series with the extrafusal fibers at the junction between muscle fibers and tendon. Each Golgi tendon organ is innervated by sensory nerve endings from group Ib nerve fibers.

Adult skeletal muscle fibers are associated with adult stem cells, called "**satellite cells**". They are able to differentiate and fuse to form new skeletal muscle fibers. Satellite cells are involved in the normal growth of muscle, as well as regeneration following injury.

## 22.2 Structure of the Vertebrate Skeletal Muscles

According to their appearance in the polarization microscope, vertebrate skeletal muscles (Fig. 22.1a) are referred to as **cross-striated muscle**.

A layer of connective tissue, the **epimysium** ensheaths the entire skeletal muscle, which itself is made of **fascicles**, i.e., **muscle fiber bundles** surrounded by connective tissue, the **perimysium**. The fascicles consist of single **muscle fibers** (at about 50 μm diameter) which may extend the entire length of the muscle surrounded by the **endomysium**. Endomysium also contains blood vessels, nerves, and lymphatics.

A muscle fiber consists of a **sarcolemma** with regular invaginations, called **T-tubules** (Fig. 22.1b). The well-developed **sarcoplasmic reticulum** (SR) consists of longitudinal vesicles and terminal cisternae storing large amounts (mM concentrations) of $Ca^{2+}$ buffered by specific $Ca^{2+}$-binding proteins in the SR, mainly **calsequestrin**. Terminal cisternae of the SR communicate with the T-tubules, thus forming "**triades**", i.e., one T-tubule forming contacts two terminal cisternae. **Myofibrils** are tubular cross-striated contractile structures with around 1 μm diameter that extend along the whole length of the muscle fiber [3]. During embryogenesis, **myoblasts** (muscle progenitor cells) fuse to form muscle fibers while their nuclei remain preserved. Muscle fibers therefore contain hundreds of nuclei located at the subsarcolemmal space (Fig. 22.1a).

The myofibrils reveal regions with high birefringence (anisotropic **A-bands**), and regions with lower birefringence, i.e., isotropic **I-bands** around the **Z-lines**, and **M-lines** in the middle of the A-bands with **H-bands** around the M-lines. The distance between Z-lines is called a **sarcomere** with a length of 2–2.4 μm in the resting muscle (Fig. 22.1a).

Sarcomeres mainly consist of longitudinally arranged **thin** (**actin**) **filaments**, **thick** (**myosin**) **filaments** and filaments made of **titin** (1 μm length) (Figs. 22.1a and 22.2). The actin filaments anchors into the Z-lines, the myosin filaments are centered within the sarcomeres interconnected by the central M-line. Titin (also called connectin) is a giant elastic protein with spring-like elements in the I-band region, roughly 3,900 kDa, which directly connects the Z-line with the M-line of the sarcomere (Fig. 22.2). The actin filament is mainly associated with filamentous **tropomyosin** as well as **troponin** complexes. The myosin filament is mainly associated with myosin-binding protein **C** (**C-protein**) at regular intervals, and **myosin light chains**.

The Z-line with its characteristic zig-zag structure is basically composed of the globular **α-actinin**, but contains many more structural and regulatory proteins

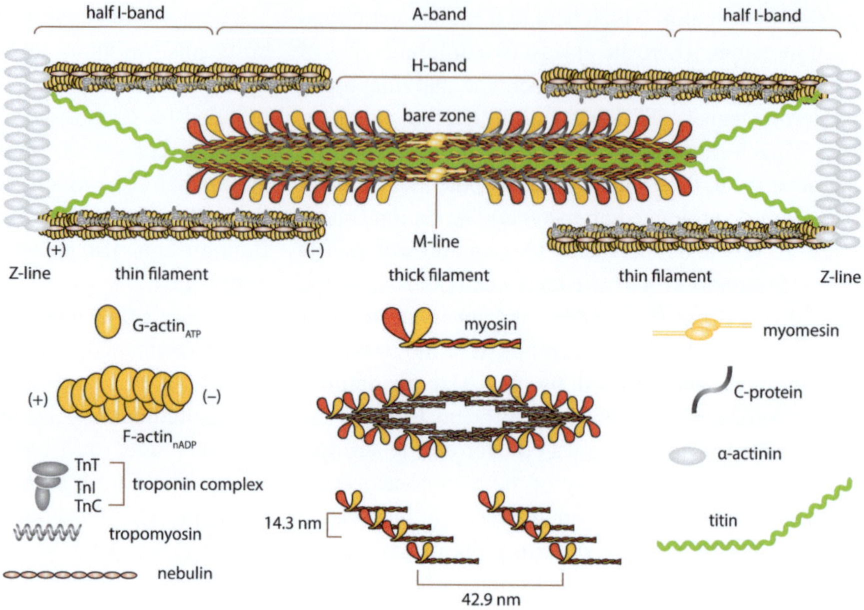


**Fig. 22.2** **Structure of a longitudinally arranged sarcomere (*upper part*).** The lower part shows details of some proteins of the thin filament (*left*), thick filament (*middle*), Z-line, and M-line (*right*). Not all sarcomeric proteins could be represented in that scheme. See text for more details (D. Fürst, with permission)

(e.g., CapZ, desmin, myotilin, telethonin). The structure maintaining the thick filament and titin arrays is the M-line (Figs. 22.1a, 22.2). The main proteins forming the M-line are muscle-specific **creatine kinase**, **M-protein**, and **myomesin**. The giant (800 kDa) protein **obscurin** tether the M-lines of the sarcomeres with the sarcoplasmic reticulum. **Costameres** are protein assemblies that align the Z-lines of subsarcolemmal myofibrils with the **dystrophin-glycoprotein complex** (DGC) at the sarcolemma via cytoskeletal actin filaments. Thus, costameres couple contraction of myofibrils with the sarcolemma in striated muscle. The DGC consists of a cytoplasmic **syntrophin** complex, a transmembrane **sarcoglycan** complex, and a transmembrane/extracellular **dystroglycan** complex, which binds to the extracellular matrix proteins **laminin** (**agrin** at the neuromuscular junction). The cytoplasmic rod-shaped 427 kDa **dystrophin** protein (**utrophin** at the neuromuscular junction) and the 700 kDa phosphoprotein **ahnak** anchors the cytoskeletal actin filaments to the DGC complex (Fig. 22.8a). Ahnak is anchored via **dysferlin** (237 kDa) to the sarcolemma.

### 22.2.1 Myosin

Myosins constitute a superfamily of molecular motors [5] which consist of heavy chain and light chain subunits (see Chap. 6). According to the sequence of their motor domain, myosins were recently classified into 35 classes [6]. Those myosins which form dimers and assemble into thick filaments are called the conventional or **type II myosins**, for the first time roughly prepared and named by Kühne in 1863 [7].

Type II myosins (around 35 % of total skeletal muscle protein) are the motor proteins of all muscle types. They form a hexamer composed of two **heavy chains** (**MYH**). Each MYH associates with two different **light chains** (**MLC**), named **essential** (**ELC**) and **regulatory** (**RLC**).

The MYH (e.g., chicken skeletal MYH = 1,938 AA, ~200 kDa) can be proteolytically cleaved into a C-terminal 150 nm α-helical C-terminal **rod domain** and a pear-shaped ~20 nm N-terminal **head domain with associated MLC**, called the **subfragment 1** or **S-1** (Figs. 22.2 and 22.3a). The coiled-coil rod domains associate to form **end-polar** (striated muscles) or **side-polar** (smooth muscle) myosin filaments. Each end-polar myosin filament (~1.6 μm length, 100 nm diameter; **thick filament**) contains about 300 myosin molecules. The myosin heads of end-polar thick filament project in a helical array with a periodicity of 42.9 nm and an axial interval of 14.3 nm – except for a 0.15 μm bare zone, where there are only overlapping myosin rods (Figs. 22.2, 22.6). The myosin heads bind as "**cross-bridges**" (**XBs**) to specific sites on the actin filament for contraction generation.

Proteolytic cleavage of the myosin rod domain yields the **subfragment 2** (**S-2**) and **light meromyosin** (LMM; Fig. 22.3a). They contain multiple heptad-repeat

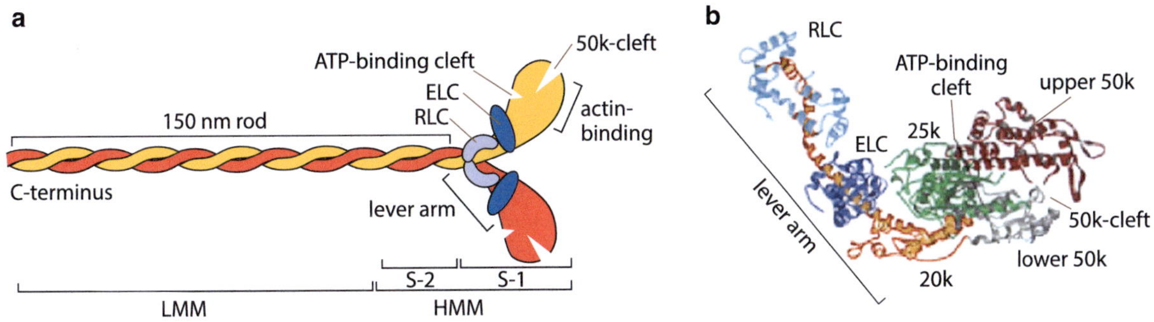


**Fig. 22.3** (**a**) **Scheme of the structure of a myosin II molecule.** It consists of two heavy chains which form a dimer through a coiled-coil of their C-terminal rod domains and a globular N-terminus which contain the catalytic center (ATP-binding cleft) and the actin-binding sites with the 50 K-cleft. *S-1* myosin subfragment 1, *S-2* myosin subfragment 2, *LMM* light meromyosin, *HMM* heavy meromyosin, *ELC* essential myosin light chain, *RLC* regulatory myosin light chain. (**b**) **3-D structure of myosin S-1** with (from N- to C-terminus) 25 K (*green*), 50 K (*red*: upper domain, *gray*: lower domain), and 20 K (*yellow*) subdomains. *ELC* essential myosin light chain (*magenta*), *RLC* regulatory myosin light chain (*blue*). *S* subdomain (From Geeves and Holmes [8], with permission)

sequences which allow dimerization to a coiled-coil-superstructure. S-1 and S-2 together is named **heavy meromyosin** (**HMM**).

Limited proteolysis of S-1 finally produces (from N- to C-terminus) a **25 K**, a **50 K** (with a large cleft and the actin-binding sites), and a **20 K** domain with the converter domain and the α-helical **lever arm** (Fig. 22.3b). The ATPase binding site forms a cleft at the 25 K/50 K interface. The lever arm contains two IQ motifs in tandem, namely IQ1 for ELC binding and IQ2 for RLC binding (Fig. 22.3b) [9]. The full-length ELC is designated as alkali 1 (A1), while alternatively spliced ELC with deleted N-terminal 46 AA are called alkali 2 (A2). The antenna-like N-terminus of A1 can bind to the thin filament (not shown on Fig. 22.3b), an interaction which regulates chemomechanical energy conversion of actomyosin.

### 22.2.2 Actin and Troponin

**Actin** (around 19 % of total skeletal muscle protein) is a globular protein (**G-actin**) with 5.5 nm diameter and 42 kDa. G-actin can be divided into four subdomains with a central ATP bound. G-actin molecules polymerize under physiologic conditions into a helical filamentous structure (**F-actin**) with 1 μm length and 50 nm diameter (**thin filament**; Fig. 22.2), each actin with a central ADP bound. In particular subdomain one which contains both the N- and C-termini of actin, is located at the periphery of the thin filament and available for **myosin interactions**. 13 actin molecules arranged on six left-handed turns repeating every 36 nm form the thin filament [8]. F-actin appears in the electron microscope as two twisting strands of globular subunits.

F-actin is anchored into the Z-line with its **plus end**, i.e., the filament site of preferential growth, and binds to the capping protein **tropomodulin** at its minus end (Fig. 22.2). Myosin S-1 forms **arrowhead** complexes with actin filaments which point to the **minus end**, i.e., towards the center of the sarcomere (hence type II myosin is called a **minus end-directed motor**). The final thin filament contains about 200 actin monomers associated with troponin-tropomyosin complexes (ca. 30 tropomyosin molecules and ca. 30 troponin complexes) located in regular spacings (in a 38 nm repeat) along the thin filament (Fig. 22.2).

**Tropomyosin** (Tm) consists of two α-helical polypeptide chains, each ca. 35 kDa, forming a ~42-nm coiled-coil. Tm molecules aggregate end-to-end to form continuous strands running along the actin filament. Interactions between tropomyosin and actin are electrostatic (Fig. 22.2).

The **troponin** (Tn) complex consists of three components, Tn I, TnT, and TnC with a total molecular weight of ~80 kDa. Tn complexes locate regularly to F-actin every 35 nm, probably at the overlap between neighboring tropomyosin molecules.

**TnT** (~36 kDa) is a rod-like protein (18.5 nm long) which binds with its C-terminal half to both TnI and TnC, and with its N-terminal half to both Tm and actin. **TnI** (~24 kDa) binds to both TnC and actin. Cardiac TnI has an N-terminal ~30 AA extension with several phosphorylation sites. **TnC** (~18 kDa) is the $Ca^{2+}$ sensor of the contractile apparatus. It has a dumbbell-like shape with an N- and a C-terminal lobe. Binding of $Ca^{2+}$ to

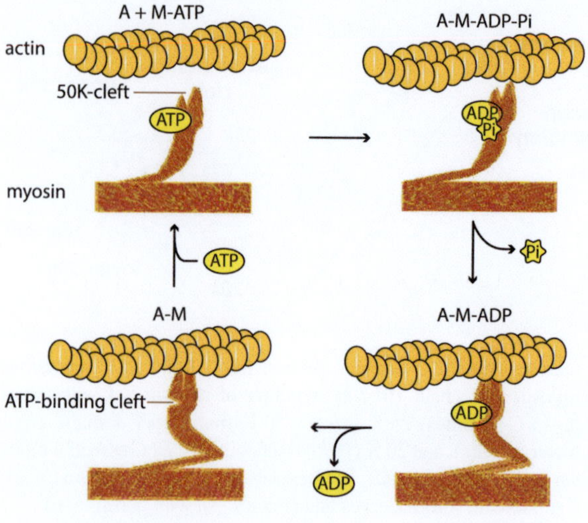


**Fig. 22.4** **Simplified scheme of the chemomechanical coupling.** A-M: the myosin cross-bridge is bound in a nucleotide-free force-generating state to actin in rigor position; lever arm "down", 50 K-cleft closed, ATP-binding cleft open. A + M-ATP: ATP binds to myosin, which then dissociates quickly from actin finally moving the lever-arm into its "up"-position, 50 K-cleft open, ATP-binding cleft closed (recovery stroke). A-M-ADP-$P_i$: myosin quickly hydrolyzes ATP, and rebinds to actin in a non-force-generating pre-power stroke state (lever arm up, 50 K-cleft open, ATP-binding cleft closed); A-M-ADP: upon release of $P_i$, myosin is shifting into its force-generating state (power stroke), with the lever arm "down", 50 K-cleft closed, ATP-binding cleft open. A-M: Release of ADP forms the rigor state; binding of ATP to the active site of the myosin cross-bridge then causes again the rapid dissociation of the A-M into the A + M-ATP state. *A* actin, *M* myosin, $P_i$ inorganic phosphate, *ATP* adenosine triphosphate, *ADP* adenosine diphosphate

TnC at the low-affinity sites at the N-lobe (two sites in skeletal, one site in cardiac TnC) represent the physiological trigger for contraction regulation. The C-lobe of TnC (with two high-affinity $Ca^{2+}$/$Mg^{2+}$ binding sites) interacts with TnI and TnT.

In addition to the regulatory Tn-Tm complexes, F-actin associates with **nebulin** filaments (skeletal muscle ~700 kDa) or its smaller cardiac homolog **nebulette** (~100 kDa). Nebulin binds to actin and Tn-Tm complexes, is anchored with the C-terminus in the Z-disk, and binds with its N-terminus to tropomodulin. Nebulin and nebulette determine the thin filament length, but may also modulate actomyosin interactions.

## 22.3 Energetics of Muscle Contraction

Muscles show characteristic passive as well as active properties. Passive elasticity of a muscle fiber, i.e., tension generated upon stretch during rest, mainly depends on the properties of the elastic titin filaments. Furthermore, compression of titin spring elements during sarcomere shortening may provide the restoring force which sets the sarcomere length to resting levels when activation ceases.

An activated muscle contracts, i.e., generates force and shortens with a load. Force generation without shortening is called **isometric**, shortening at constant load/force generation is called **isotonic**.

The energy for muscle contraction comes from the hydrolysis of ATP. The ATP concentration in living muscles is surprisingly constant (around 5 mM) and remains high even during muscle fatigue. Cleaved ATP can quickly be resynthesized by transfer of $P_i$ from phosphocreatine to ADP through creatine kinase (**Lohmann reaction**). **Adenylate kinase** (or myokinase) produces ATP by conversion of two ADP molecules to one ATP and one AMP, which is finally removed by **AMP deaminase** to form inosine monophosphate and $NH_4$. ATP can be synthesized from pyruvate (which is formed from glucose through glycolysis) in two different pathways, namely conversion of pyruvate by an **anaeorobic** metabolic process to lactate (which very quickly leaves the muscle fiber via lactate transporter), or **aerobic** conversion of pyruvate to acetyl CoA and subsequent oxidation to $CO_2$ and $H_2O$ in the mitochondria (**Krebs cycle**, **electron transport chain**), which produces 12 times more ATP than the anaerobic pathway.

Mechanochemical energy transformation is performed by cyclic interaction of the **cross-bridges** (**XBs**) with actin, in which the XBs bind independently, generate force upon a conformational change (the power stroke), and detach from the thick filaments (the **cross-bridge theory**) [10]. According to the model of Lymn and Taylor [11] (Fig. 22.4), ATP binds to the ATP-binding cleft of the myosin XB (M), which causes the rapid detachment of the XB from the actin (A) filament (A + M-ATP). ATP is rapidly hydrolyzed forming the M-ADP-$P_i$ state (**pre-power-stroke state**). The M-ADP-$P_i$ state can only weakly attach to actin through ionic interactions (A-M-ADP-$P_i$). Releasing $P_i$ from A-M-ADP-$P_i$ then forms a strong stereospecific actin-binding state and the power stroke of the XB (**force-generating state**; A-M-ADP). The power stroke is executed by movement of the myosin lever arm by ~10 nm. Detachment of the force-generating XB from the actin filament is achieved by the release of ADP, binding of ATP, and repriming of the lever arm of the XB into its non-force generating position (recovery stroke). In the absence of ATP, the nucleotide-free XBs remain strongly bound to the actin filament, the **rigor mortis state** (A-M) (Figs. 22.4 and 22.5).

Recent X-ray crystallographic analysis of myosin S-1 fixed at different states revealed the domain movements. In the pre-power stroke state, the cleft in the 50 K domain is open, the ATP-binding site is closed, while the lever arm is in the "up" position. In the strongly bound, force-generating state, the 50 K-cleft is closed, the ATP-binding pocket is open, and the lever arm is in the "down" position (the **swinging lever arm model**, Fig. 22.5) [8]. Those crystal structures also showed that a small conformational change in the ATP-binding cleft of the XB (movement of 0.5 nm) is finally amplified into a large movement of the lever arm of around 10 nm.

### 22.3.1 Isometric Contraction

Isometric force ($F_0$) during steady-state contraction depends on the number of cross-bridges (XBs) in the force-generating state and the force a XB can excert ($F'$, ~6 pN). Since XBs independently interact with actin during muscle activation, only a fraction "$n$" of the total amount of cycling cross-bridges ($n_{tot}$) generate force at any one time. "$n$" depends on the XB kinetics, i.e., the rates of transitions from force- into the non-force-generating XB states (detachment rate; $g_{app}$) and the rate of transitions from non-force into force-generating XB states (attachment rate; $f_{app}$), namely: $n = f_{app}/(f_{app} + g_{app})$ [10, 12]. Thus, $F_0$ equals $F' \cdot n_{tot} \cdot [f_{app}/(f_{app} + g_{app})]$.

There is a bell-shape dependency of force generation from the muscle/sarcomere length [13] (Fig. 22.6). Optimal force of skeletal muscle is elicited at resting sarcomere length, i.e., between 2 and 2.25 μm, while smaller or larger sarcomere lengths gradually decline force generation. The level of thick and thin filament overlap and the resulting different amounts of force-generating XBs provide the structural basis for this observation.

### 22.3.2 Isotonic Contraction

Muscle shortens by the relative sliding of thick and thin filaments while the distance between Z-lines (the sarcomere length) decrease (the **sliding filament model**; [14]). The velocity of shortening depends on the load in a characteristic hyperbolic manner [15], i.e., the common experience that a low load can be

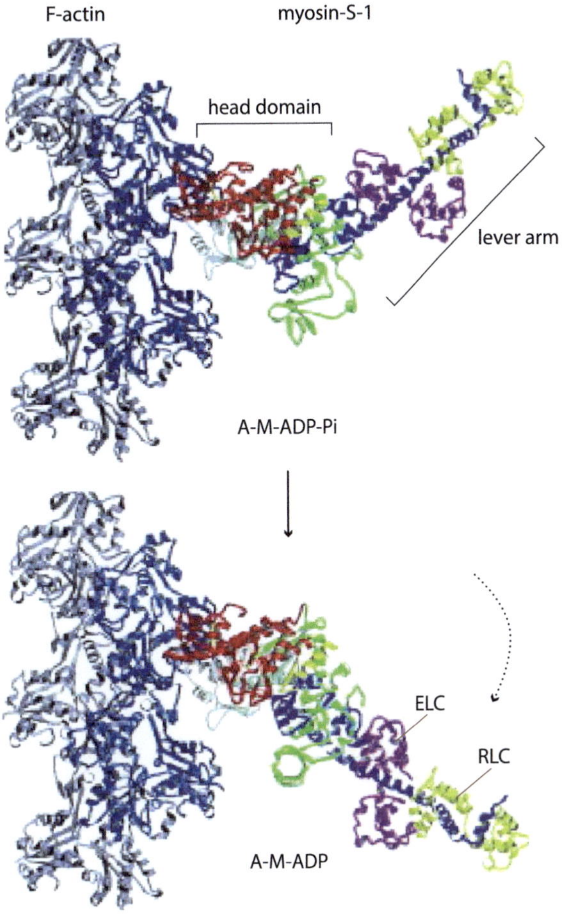


**Fig. 22.5 3D structure of actomyosin S-1 complexes modelled from crystallographic data.** Five actin molecules forming a small actin filament (F-actin) and one myosin subfragment 1 (S-1) cross-bridge are shown. *Top*: AM-ADP-$P_i$: myosin bound to actin in a non-force-generating pre-power stroke state (lever arm up); *Bottom*: AM-ADP upon release of $P_i$ myosin is shifting into its force-generating state (power stroke), with the lever arm down. Color code: *magenta*: essential light chain (ELC); *yellow*: regulatory myosin light chain (RLC); *green*: 25 K domain; *red* and *gray*: 50 K domain; *blue*: 20 K domain (Modified from Geeves and Holmes [8], with permission)

lifted faster than a high load (Fig. 22.7a). Power (and efficiency curves, if heat measurements are available) can be calculated from the force–velocity relationship, showing a characteristic bell-shaped function with maximal efficiency at around one third of maximal shortening velocity, and zero efficiency at maximal isometric force generation or maximal shortening velocity (Fig. 22.7b). Hence, we have to switch gears upon up-hill (or down-hill) cycling, since the slower (or faster) leg movements would shift muscle efficiency into nonoptimal ranges.

If an isometrically contracting muscle is quickly (less than 1 ms) allowed to shorten ("quick release"), there is

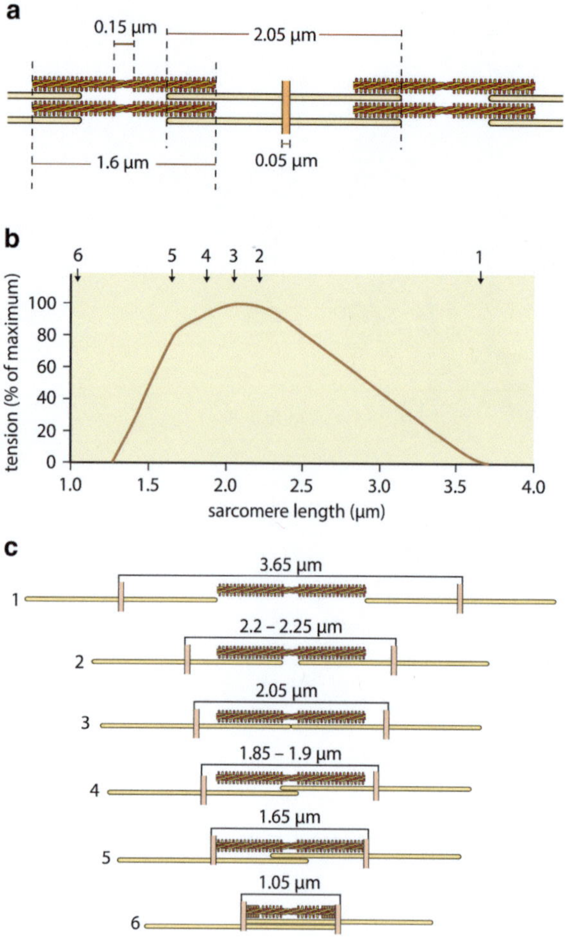


**Fig. 22.6** **Length-isometric tension relationship of skeletal muscle.** (**a**) Schematic diagram of filaments, indicating dimensions of the structural components of the sarcomere. 1.6 μm: length of thick filaments; 2.05 μm: length of two thin filaments; 0.15 μm: length of the bare region; 0.005 μm: thickness of Z-line; (**b**) Length-normalized isometric tension ratio of a single skeletal muscle fiber. The *arrows* along the top are placed opposite the striation spacings at which the critical stages of overlap of filaments occur; (**c**) sarcomere configuration at these six different sarcomere lengths (From Gordon et al. [13], with permission)

an instantaneous decrease of tension due to the discharge of the elastic element of the force-generating cross-bridges (XBs). A quick release of roughly 1 % of muscle length at optimal overlap, which equals a step size of around 10 nm per half-sarcomere, causes a complete drop to zero force. Larger shortening distances then compress elastic spring components of the XBs, generating force into the reverse direction. The fraction of compressed ("negative") XBs increases, that of pulling ("positive"), i.e., force-generating XBs decreases with increasing shortening velocities. To allow fast shortening, detachment rate of negative XBs ("$g_2$") – and consequently ADP release rate and ATP consumption – should be higher during shortening than during isometric contraction. As a consequence, maximal shortening velocity ($V_{max}$) of a muscle depends on "$g_2$". In fact, the different $V_{max}$ values of fast and slow muscle types are caused by distinct gene expression of myosin isoenzymes with higher and lower ADP-release rates, respectively. Attenuated force-bearing capacity and increased ATP consumption during muscle shortening could explain both the hyperbolic force–velocity relationship and the increased heat released during shortening (**Fenn effect**).

Hermann v. Helmholtz (1848) was the first to directly show that muscle heat rises upon electrical stimulation of frog skeletal muscles. Hence, activated muscle produces heat and work. The work that can be done by an XB is 22 kJ/mol. Since the free energy change of ATP cleavage is about 50 kJ/mol, **mechanical efficiency** of an XB is roughly 50 %. Skeletal muscle contraction revealed peak mechanical efficiencies (% mechanical power out of the total energy liberated) of around 30 %, as determined by simultaneous measurements of force, shortening velocity, and heat release [16] (Fig. 22.7b). The lower efficiency of the whole muscle compared to the XBs is due to additional ATP consumption besides the actomyosin interaction, mainly $Ca^{2+}$ sequestration into the SR by ATP-driven SERCA, sarcolemmal $Ca^{2+}$ pumps, and $Na^+/K^+$-ATPase.

## 22.4 Excitation-Contraction Coupling

More than 70 % of the resting membrane potential of skeletal muscle is due to inward chloride ion conductance generated by voltage-gated ClC-1 channels. In addition, potassium outward current by voltage-gated $K^+$ channels contribute to resting potential. **Acetylcholine** (**ACh**) released from the α-motoneuron binds to **nicotinic ACh receptors** which is formed from five subunits, two α-subunits, one ß-, one γ-, and one δ-subunit (**nAChR**; ACh-gated nonspecific cation channels) in the subsynaptic sarcolemma of the motor endplate, thus increasing the open-probability of its channel domain (Fig. 22.8a). ACh in the synaptic cleft is rapidly degraded by **ACh-esterase** into acetic acid and choline, which prevents sustained depolarization of the subsynaptic sarcolemma. Depolarization of the subsynaptic sarcolemma to threshold level activates voltage-gated $Na^+$ channels eliciting an action potential which spreads all over the sarcolemma and its T-tubular system with a conduction velocity of around 3–5 m/s. Specific voltage-gated **L-type calcium channels (dihydropyridine receptors, DHPR)** locate mainly in the T-tubular membrane of the sarcolemma. DHPRs comprise a voltage-sensing and pore-forming $\alpha_{1s}$ subunit and accessory $\alpha 2/\delta$, ß, and $\gamma$ subunits. DHPRs interact with **$Ca^{2+}$ release channels (ryanodine receptors; RyR)** which are inserted into the terminal cisternae of the SR **(junctional SR)**. **Junctophilin** tethers the gap between the T-tubule and terminal cisternae [35]. Three RyR isoforms are expressed – RyR1 in skeletal muscle, RyR2 in cardiac muscle, and all three in smooth muscle. The RyR is a homotetramer and forms a macromolecular complex with multiple interacting partners, including anchoring proteins, protein kinases and phosphatases, and accessory regulatory proteins (triadin, junctin, calsequestrin). In skeletal muscle, four DHPR molecules assemble (**tetrad**), each DHPR molecule localizing to one subunit of a RyR1 (**$Ca^{2+}$ release unit** or **couplon**). DHPRs in skeletal muscle mainly function as voltage sensors, which change their conformation in response to an action potential. As a consequence of direct interactions, RyR1s open and large amounts of $Ca^{2+}$ are released from the SR into the myoplasm, increasing the free $Ca^{2+}$ from around $10^{-7}$ M at resting state to above $10^{-5}$ M at maximal activation. Fluxes of calcium ions from the SR during release and reuptake are counterbalanced by high chloride and potassium conductance of the SR membrane, in particular TRIC channels [35].

Biochemical and structural studies led to the introduction of the **steric block model** [17, 18]. In the absence of $Ca^{2+}$ the Tn-Tm complex blocks crossbridge (XB) binding to the thin filament while allowing some weak (electrostatic) XB interactions (off-state). Activating $Ca^{2+}$ levels bind to the regulatory sites of TnC at its N-terminal lobe. This strengthens the TnC-TnI and TnC-TnT interactions, thereby weakening the binding of TnI to actin. The subsequent changes of conformation and location of the Tn-Tm complex on the thin filament relieves myosin binding sites on actin and turns the thin filament into the on-state, which allows strong stereospecific, force-generating interactions with the XBs (Fig. 22.8b).

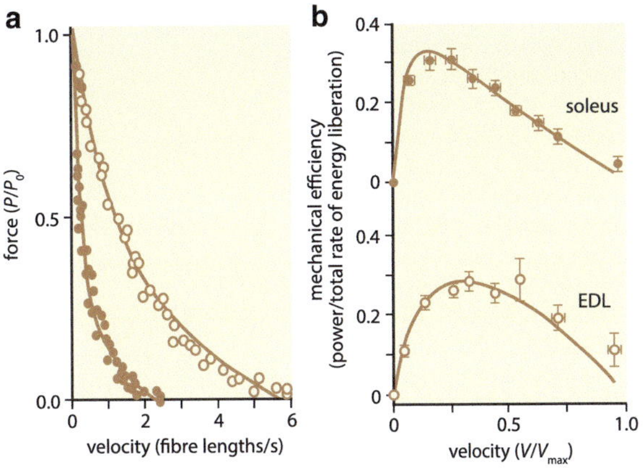


**Fig. 22.7** (**a**) **The ratios between normalized force ($P/P_0$) and shortening velocity** (given in fiber lengths/s) of mouse soleus (*filled circles*) and mouse extensor digitorum longus (EDL) (*open circles*). (**b**) **Mechanical efficiency (power/total rate of energy liberation) as a function of normalized shortening velocity** ($V/V_{max}$) of soleus (*filled circles, top*) and EDL (*open circles, bottom*): $V$ shortening velocity, $V_{max}$ maximal shortening velocity (velocity at zero force), $P$ force, $P_0$ maximal force (From Barclay et al. 1993 [16], with permission)

Repolarization of the sarcolemma upon $Na^+$ channel inactivation, increased $K^+$ outward conductance and $Cl^-$ inward current inactivates the couplons. Eliminating activating $Ca^{2+}$ concentrations from the myoplasm elicits relaxation of skeletal muscle fibers. In particular the **sarcoplasmic-endoplasmatic reticulum $Ca^{2+}$ ATPase (SERCA)** – a $Ca^{2+}$-regulated ATP-dependent $Ca^{2+}$ pump – sequesters myoplasmic $Ca^{2+}$ into the SR. In addition, a **$Na^+/Ca^{2+}$ exchanger (NCX)** in the sarcolemma eliminates activating $Ca^{2+}$ from the myoplasm into the extracellular space (Fig. 22.4).

## 22.5 Skeletal Muscle Fiber Types and Regulation of Force Generation

During development, differentiation of skeletal muscle fibers is controlled by the particular type of innervating $\alpha$-motoneuron: S motoneurons generate slow-twitch (**ST** or **type I**) skeletal muscle fibers, while F motoneuron innervation leads to fast-twitch (**FT** or **type II**)

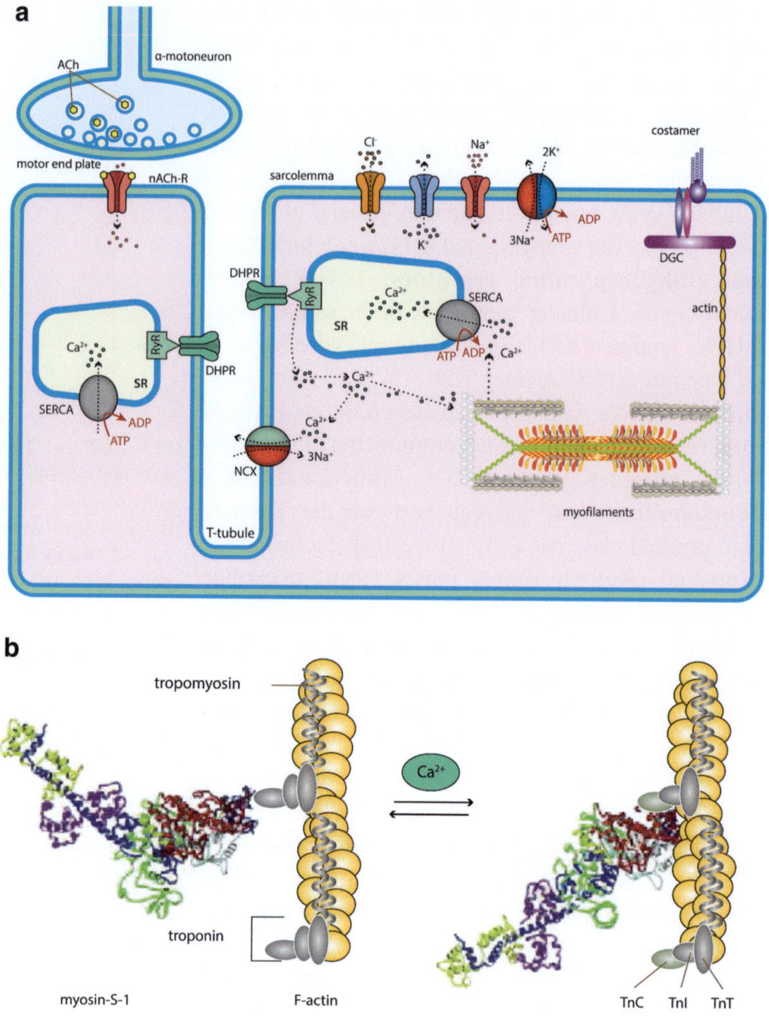


**Fig. 22.8** (**a**) **Excitation-contraction coupling and $Ca^{2+}$ transport of skeletal muscle.** *ACh* acetylcholine, *nACh-R* nicotinic acetylcholine receptor, *DHPR* dihydropyridine receptor, *NCX* potassium-calcium exchanger, *RyR* ryanodine receptor, *SERCA* sarcoplasmic endoplasmic reticulum calcium ATPase, *DGC* dystrophin-dystroglycan complex; see text for more details. (**b**) The steric block-model. *Left*: resting $Ca^{2+}$ concentrations with the actin filament (F-actin) in the "off" state, i.e., no force-generating actomyosin interactions (only one myosin S-1 is shown) are allowed. The troponin-tropomyosin complex is masking the myosin-binding sites on F-actin. *Right*: activating $Ca^{2+}$ (*green circles*) bind to troponin C of the troponin complex. Both conformations and location of the troponin-tropomyosin complexes at the thin filament change (*arrow*) thus pushing the actin filament into the "on-state", i.e., force-generating actomyosin interactions are allowed

fibers. The distinct firing pattern of the F and S motoneurons cause different spatial and temporal changes of intracellular free $Ca^{2+}$, which may be transduced into distinct changes of gene transcription. Thus, the level of calcineurin activation as well as the activation of the different multifunctional $Ca^{2+}$ calmodulin-dependent kinases are dependent on the temporal mode of muscle activation [36].

**ST fibers** have a long isometric twitch time due to a long time-to-peak contraction and slow relaxation (slow-twitch fibers). ST fibers operate mainly through an aerobic energy metabolism having abundant mitochondria, large activities of enzymes for the aerobic energy production, and contain much of the oxygen receptor myoglobin, which gives this muscle type the red color. **FT fibers** have a short isometric twitch time due to fast time-to-peak contraction and relaxation times. FT fibers can be subdivided into FT-IIa and FT-IIb. FT-IIa has moderate numbers of mitochondria and uses both the aerobic and anaerobic ATP production pathways. FT-IIb fibers use anaerobic metabolic pathways to provide ATP. Many more differences exist between ST and FT fibers, e.g., volume of the T-tubular system and sarcoplasmic reticulum, maximal rate of $Ca^{2+}$ uptake and release, etc. [25].

The mechanical efficiencies of an ST muscle (soleus) and a FT muscle (extensor digitorum longus) is shown in Fig. 22.7b. For a slow movement, say 0.5

muscle lengths per second (ML/s), the mechanical efficiency of the FT muscle would just be around 5 %. However, for the ST muscle it would be six times higher, namely 30 %. Therefore, activation of distinct fiber types enables a variety of motor actions to be performed at maximal mechanical efficiency. In fact, ST fibers are mainly involved in exercise of low intensity, while at higher work intensities mainly FT fibers are active [19]. Exclusion of ST-motor units and preferential activation of FT-motor units during high intensity exercise cannot readily be related to the above-mentioned Henneman's "size principle" of motor unit selection [4].

Force generation of a skeletal muscle is regulated by two basic principles, namely **recruiting** different amounts of motor units and modulation of the firing **frequency** of the α-motoneurons. Thus, firing of the α-motoneuron with low frequency elicits only single **twitch contractions** while stepwise increasing firing frequencies summates to unfused, and then fused **tetanic contractions** with much larger (up to threefold) contraction amplitudes compared to single twitch contractions (Fig. 22.9). High tetanic force is elicited by higher myoplasmic calcium concentration compared with twitch contractions (Fig. 22.9). The **fusion frequency**, i.e., the firing frequency required to reach a fused tetanic contraction is lower for ST muscle fibers (around 20 Hz) then for FT fibers (around 60 Hz), matching the basic physiological properties of the innervating α-motoneuron types. Except at the very beginning of a maximal contraction, when F motor units fire at 120 Hz, the mean maximum rate of discharge declines rapidly to around 30 Hz without loss of force generation of the motor unit [38, 39]. Force maintenance despite decline in firing frequency was called "**muscular wisdom**" [40], a phenomenon which could delay the onset of peripheral muscle fatigue (cf. 22.6). During motor tasks that require a slow increase in force, F-motor units start with 8 Hz when first recruited and then increase their firing rate up to ca. 30 Hz [38].

Fusion of single twitches to tetanic contractions are based on the different durations of action potential (5 ms) and twitch times (around 30 ms for FT fibers, 90 ms for ST fibers). According to the faster twitch time, fusion frequency of FT fibers is higher compared to ST fibers.

A twitch contraction elicited immediately after tetanic stimulation is higher compared to the twitch elicited before tetanic contraction (**posttetanic potentiation, PTT**) (Fig. 22.9). This cannot be explained by a higher myoplasmic $Ca^{2+}$ activation level (Fig. 22.9). Rather, a higher phosphorylation level of the RLC of myosin, which increased the attachment rate constant "$f_{app}$" of the XBs and $Ca^{2+}$ sensitivity of the myofilaments, represents the main mechanisms of PTT.

Contraction of single twitches is not dependent on extracellular $Ca^{2+}$. However, tetanic contractions triggered by high action potential frequencies are dependent on extracellular $Ca^{2+}$ [21].

## 22.6 Skeletal Muscle Fatigue

Skeletal muscle **fatigue**, i.e., the reversible decline in muscle performance associated with muscle activity, may arise at many points during voluntary contractions, starting from the cerebral cortex, descending excitation to motoneurons (i.e., **central fatigue**) as well as in the muscle fibers (i.e., **peripheral fatigue**).

Experimentally, peripheral muscle fatigue is induced by eliciting repeated short tetani of, say 40 Hz for 300 ms every second for some minutes. In those experiments, ST muscles generate almost normal contractions, while tetanic force, in particular of FT-IIb fibers, declines rapidly [24]. Physiologically, the onset of peripheral muscle fatigue is retarded by using the muscle wisdom effect, i.e., reduction of firing frequency of the innervating motoneuron without loss of contractile force (cf. 22.5). The main cause of peripheral fatigue of FT fibers is a decline of the amount of $Ca^{2+}$ released from the RyR1 during fatigue which causes incomplete activation of the contractile apparatus. An important factor in muscle fatigue is the reduced ability of the T-system to perform action potentials as a result of excitation-induced extracellular **$K^+$ accumulation**. $K^+$ depolarization induced by extracellular $K^+$ accumulation inactivates $Na^+$ channels, hampers action potential generation, thus attenuating activation of the $Ca^{2+}$ release units (couplons) which reduces $Ca^{2+}$ release from the RyR1. Historically, accumulation of lactic acid in muscle and subsequent acidification has been considered a major factor associated with fatigue. In

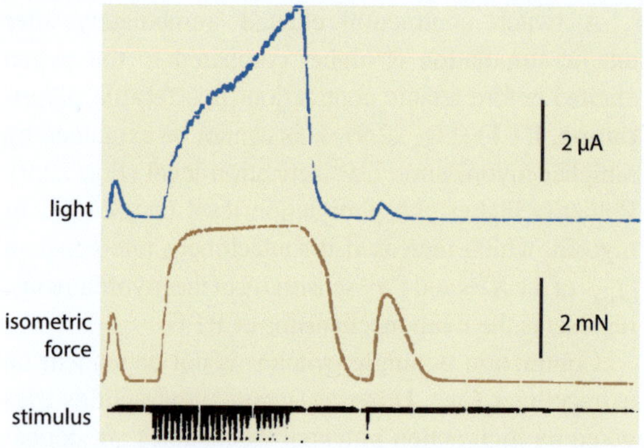


**Fig. 22.9** **Simultaneous recordings of intracellular $Ca^{2+}$** (*top*; aequorine luminescence light) and isometric force (*bottom*) of a single skeletal muscle fiber at different electrical stimulation frequencies (*stimulus*) causing single twitches (1 Hz, left and right signals) or tetanic contractions (50 Hz stimulation; signals in the middle of the record). Please note the much higher $Ca^{2+}$ signal and force generation during tetanic stimulation compared to single twitch contractions (From Blinks et al. [20], with permission)

vitro, low pH reduces $Ca^{2+}$ affinity of troponin C, which attenuates the "on" state of the thin filament and actomyosin interactions. However, recent experiments showed very small myoplasmic acidification during fatigue. Furthermore, decreasing intracellular pH caused relatively minor reduction of tetanic force at physiological temperatures (above 30 °C). In contrast, acidification could be beneficial because it counteracts the loss of muscle excitability seen in $K^+$-depolarized muscle [22]. Another candidate, **inorganic phosphate** (1–5 mM at rest) which accumulates during fatigue (30–40 mM) inhibits the force-generating $P_i$ release step from the A-M-ADP-$P_i$ state. However, this effect on XB action is weak at physiological temperatures. Augmented production of **reactive oxygen species** (ROS) and ammonia ($NH_4$) as well as glycogen depletion during contraction could also play a role in muscle fatigue [23, 24].

Central fatigue can be elicited by exercise-induced reduced central nervous system drive to muscle. Activation of muscle nociceptors, e.g., by acidification, $K^+$ accumulation, hypoxia, or heat production not only depresses mood and motivation but also inhibits α-motoneuron output via inhibitory interneurons in the spinal cord. Fatigue upon prolonged exercise was associated with increased serotonin and decreased dopamine in the brain. Considering the important functions of both transmitters in the brain, this may reflect reduced motivation, arousal, and muscular coordination during fatigue. In fact, high dopamine in the brain upon amphetamine abuse reduces fatigue [23].

## 22.7 The Musculature of the Mammalian Heart

The mammalian heart is a four-chamber pump located between the lungs behind the breastbone. The heart lies within the pericardial sack. The muscle (**myocardium**) is delimited to the outside by the **epicardium**, and by the **endocardium** on all internal surfaces. The myocardium consists of sheets of contractile cells, the **cardiomyocytes** which are interconnected by electron-dense specialized cell–cell junctions, the **intercalated disks**, which provide mechanical stability and electrical communication between cells. The resting cardiomyocyte is much stiffer than the skeletal muscle fiber, due to the presence of cardiac-specific isoforms of **titin** with distinct elastic properties.

A heartbeat, or cardiac cycle, consists of a contraction (**systole**) and a relaxation (**diastole**) generating blood ejection and blood filling, respectively. Activation of the heart by depolarization and generation of action potentials normally starts in specialized **pacemaker cells** of the **sinoatrial node** (**SA**). SA cells slowly depolarize by an inward $Na^+$ current through cAMP-regulated **HCN-channels** (**funny channels**), which open at resting membrane potentials to the threshold potential, which thereafter activate voltage-dependent **T-type $Ca^{2+}$ channels** and then **L-type $Ca^{2+}$ channels** thus eliciting an action potential. Electrical impulses then propagate between cardiomyocytes by a special class of ion channels, **connexones**, each made up of six connexin molecules (**gap junctions** or **nexus**) of the intercalated disks (see Chap. 8). The heart therefore represents a functional **syncytium**. Cardiomyocytes remain terminally differentiated. Multipotent cardiac stem cells may exist able to regenerate the myocardium during life. Cardiomyocytes are cross-striated, longitudinally arranged cells (around 150 μm long and 20 μm wide) with one or two nuclei, contractile myofibrils, which closely resemble those of skeletal muscle, and a well-developed sarcoplasmic reticulum which communicates with T-tubuli to form **diads** (a single T-tubule associated with one terminal cisterna; cf. triads, Sect. 22.2). In contrast to skeletal muscle fibers, L-type $Ca^{2+}$ channels do not directly bind to the $Ca^{2+}$-release channels (ryanodine receptors; RyR2) of the terminal cisternae of the sarcoplasmic reticulum. Between DHPR and RyR2 is a small space called the "**dyadic cleft**" of 0.05–0.2 μm. Hence, cardiomyocytes use another excitation-contraction coupling mechanism than the skeletal muscle fibers.

### 22.7.1 Excitation-Contraction Coupling of Cardiomyocytes

Compared to nerve fibers or skeletal muscle fibers, the action potential of a cardiomyocyte is long, roughly 200–300 ms, almost the duration of a twitch contraction. Hence, the cardiomyocyte cannot readily be tetanized.

Action potentials of cardiomyocytes of the myocardium start with a rapid depolarization caused by gating of the voltage-operated $Na^+$ channels and a subsequent characteristic long-lasting plateau phase due to a slow $Ca^{2+}$ inward current through the $\alpha_{1c}$ subunit of voltage-operated L-type $Ca^{2+}$ channels (DHPR). Roughly 20–40 DHPR are clustered, forming functional **couplons** (c.f. Sect. 22.3). The small $Ca^{2+}$ inward current during the action potential passes the dyadic cleft and triggers the release of large amounts of $Ca^{2+}$ from the SR through RyR2 (**$Ca^{2+}$-induced $Ca^{2+}$ release**; **CICR**). CICR through RyR occurs upon binding of low $Ca^{2+}$ concentration (pCa 6.25) to high-affinity sites, inactivation occurs by high $Ca^{2+}$ (pCa 5.5) through binding of $Ca^{2+}$ to the low-affinity site of RyR2. Inactivation of $Ca^{2+}$ release is also facilitated by calmodulin, which increases RyR2 close time. Phosphorylation of RyR2s by $Ca^{2+}$/calmodulin-dependent protein kinase II (CaMKII) and by cAMP-dependent protein kinase (PKA), i.e., upon sympathetic stimulation, increases $Ca^{2+}$ release from RyR2.

During systole, free cytosolic $Ca^{2+}$ rises to about 600 nM which only half-maximally activates the myofilaments generating half-maximal force (for ventricular pressure) and shortening (for blood ejection). Thus, there is a large **contractile reserve** which can be recruited by increasing SR $Ca^{2+}$ release, e.g., via sympathetic stimulation (cf. below). Activation of the myofilaments of the heart by $Ca^{2+}$ via binding to troponin C, conformational changes of the regulatory troponin-tropomyosin system, chemomechanical energy transformation by the ATP-driven interaction between actin filaments and XBs, and the sliding filament mechanism are similar to that of the skeletal muscle. During diastole, intracellular $Ca^{2+}$ is reduced to resting levels mainly by SERCA2, a sarcolemmal $Na^+/Ca^{2+}$ exchanger (NCX), and a **sarcolemmal $Ca^{2+}$ pump** (**PMCA**). Decreased free cytoplasmic $Ca^{2+}$ inactivates the myofilaments, the heart relaxes.

The contractile reserve of the myocardium may be recruited by its sympathetic innervation which originates from stellate ganglia and cardiac plexus, innervated mainly from the fourth and fifth thoracic segments of the spinal cord (see Chap. 10). Binding of norepinephrine to its seven-membrane-spanning Gs-coupled $ß_1$-adrenergic receptor activates adenylyl cyclase which rises the myoplasmic level of the second messenger cyclic adenosine monophosphate (cAMP). cAMP activates **cAMP-dependent protein kinase** (**PKA**) which phosphorylates DHPR and RyR, thus increasing their open-probabilities and consequently increasing myoplasmic $Ca^{2+}$ [37]. Higher myoplasmic $Ca^{2+}$ levels increase the contractile force of the cardiomyocyte (**positive inotropy**). Increased cAMP levels activate the funny channels of pacemaker cells, thus increasing diastolic depolarization velocity to threshold potential causing an increased frequency of action potentials (**positive chronotropy**). Higher frequencies can raise myoplasmic free $Ca^{2+}$ and myocardial contraction (**frequency inotropy**). Activated PKA phosphorylates gap junctions, which increases their open probabilities thus improving action potential propagation (**positive dromotropy**). PKA-dependent phosphorylation inactivates **phospholamban**, a potent SERCA inhibitor, and a cardiac-specific N-terminal extension of TnI. This accelerates $Ca^{2+}$ sequestration from the myoplasm into the sarcoplasmic reticulum and $Ca^{2+}$ dissociation from TnC, respectively, accelerating the relaxation rate (**positive lusitropy**).

The contractile reserve may be increased by raising the parasympathetic vagal tone. Vagal nerve fibers originate in the dorsal efferent nuclei of the medulla oblongata, innervating mainly the sinoatrial node cells. Acetylcholine binds to muscarinic, $G_i$-coupled GPCRs of SA cells thus decreasing myoplasmic cAMP levels via inhibition of adenylyl cyclase activity. Hence, funny channel activity is reduced, thus decreasing diastolic depolarization velocity which results in a decreased action potential frequency (negative chronotropy) and lower myoplasmic $Ca^{2+}$ activation levels of myocardial cells (negative inotropy).

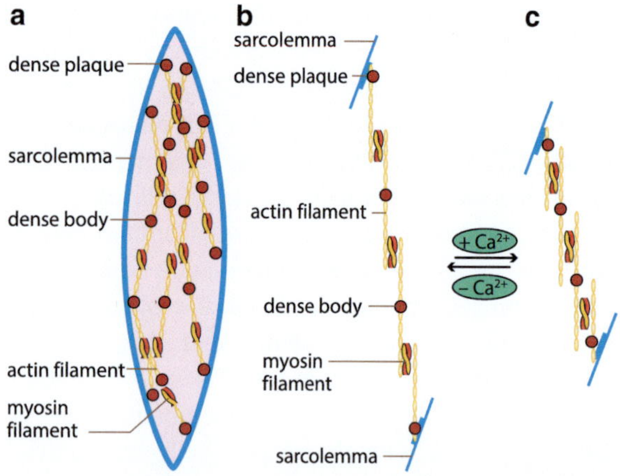


**Fig. 22.10** (**a**) **Scheme of a smooth muscle cell with contractile structures,** (**b**) contractile structures in the relaxed, and (**c**) in the contracted state (Modified from Rüegg 1992 [25], with permission)

## 22.8 The Mammalian Smooth Musculature

Nonstriated or smooth muscle cells are spindle-shaped, 5–50 μm wide and 50–500 μm long. They form the muscular part in the wall of hollow organs, e.g., intestine, urinary bladder, uterus, and blood vessels. Smooth muscle cells show no cross-striations since the thin and thick filaments of their myofibrils are not arranged in a comparable ordered manner like cross-striated muscle types (Fig. 22.10). Actin filaments are fixed by **dense bodies**, a Z-band analog within the cells and anchor into the sarcolemma by **dense** (**adhesion**) **plaques**. The sarcolemma forms small (100 nm) flask-shaped invaginations, called **caveolae** which have important functions in signal transduction.

Smooth muscle drives a variety of involuntary body functions, e.g., micturition, intestinal propulsion, myometrium contraction during child birth, or the maintenance of blood pressure. It was early recognized by Emil Bozler [26] that two classes, "**single-unit**" (or "**unitary**") and "**multi-unit**", smooth muscle can be defined on the basis of their membrane properties and innervation. The single-unit smooth muscle organ often has specialized pacemaker cells (e.g., interstitial cells of Cajal in the gastrointestinal tract) where contractions are spontaneously and rhythmically generated (**myogenic regulation**) and gap junctions which electrically connect the smooth muscle cells. Therefore, the single unit type is syncytial and its contractile state is modulated by sympathetic and parasympathetic tone (e.g., intestine, urinary bladder, uterus). The multi-unit smooth muscle types are innervated by sympathetic and parasympathetic nerve fibers (e.g., lungs, arteries, erectile tissues of hair follicles, ciliary muscle, iris of the eye, reproductive tracts) and respond independently from each other upon nerve stimulation (**neurogenic regulation**).

Contraction of smooth muscle generally operates by a $Ca^{2+}$-dependent sliding filament mechanism (Fig. 22.10), but shows distinct activation properties if compared to striated muscles. Thus, smooth muscle activation is elicited by a thick-filament activation mechanism rather than by the troponin/tropomyosin-bound thin-filament activation (steric block) mechanism of striated muscle. Shortening velocity of smooth muscle cells is slow, roughly one-tenth of the maximal shortening velocity of striated muscle. Smooth muscle cells contain only one-third of the myosin concentration found in striated muscle, but they generate similar forces.

Three Type II myosin heavy chain (MYH) genes are expressed in smooth muscle cells (SM, NMA, NMB) forming a large number of different isoenzymes by alternative splicing mechanisms. All myosin II isoforms expressed in smooth muscle cells bind ADP much stronger than striated muscle myosins. Therefore, the detachment rate $g_{app}$ of myosin II isoforms in smooth muscles is lower than that of myosin II isoforms expressed in striated muscle. Hence, smooth muscle myosins remain a longer fraction of their cross-bridge cycle time in the force-generating state (the **duty cycle** is high), while the fraction of actively cycling cross-bridges in the force-generating state increases (c.f. Sect. 22.4), explaining the high force generation with little myosin expression. Another consequence of high ADP affinity and small $g_{app}$ of smooth muscle myosins is that maximal shortening velocity ($V_{max}$) becomes very low.

During sustained activation smooth muscle generates a state of high-efficiency contraction referred to as the "**latch-state**" [27]. Intracellular $Ca^{2+}$, RLC phosphorylation, ATP consumption, and maximal shortening velocity decrease while force is maintained. A number of mechanisms including the formation of slowly cycling dephosphorylated cross-bridges, ADP affinity of smooth muscle myosin, as well as activation of NM-MYH [28] are discussed as putative mechanisms.

### 22.8.1 Excitation-Contraction Coupling of Mammalian Smooth Muscle

Action potentials in single-unit muscle occur as **spike potentials** (10–50 ms duration) or **action potentials with plateau** (up to 1,000 ms duration). The **slow-wave potentials** generated by pacemaker cells are rhythmically generated wave-like subthreshold membrane depolarizations (**pacemaker waves**) and not action potentials. However, if slow-wave potentials reach the threshold level (from −60 to −35 mV), action potentials and contractions are elicited. In particular $Ca^{2+}$ currents are responsible for smooth muscle action potential generation. Single contractions of smooth muscle may be very long, i.e., 1,000–3,000 ms, hence smooth muscle cells may become tetanized at low fusion frequencies. Neurogenic activation of single-unit smooth muscle mainly causes graded depolarizations rather than action potentials.

Smooth muscle cells may be activated by **electromechanical** and/or **pharmacomechanical** coupling, which access distinct intracellular pathways for raising free myoplasmic $Ca^{2+}$.

During **electromechanical coupling**, entry of extracellular $Ca^{2+}$ during slow-wave or action potentials through the L-type $Ca^{2+}$ channels directly activate the contractile machinery and elicit $Ca^{2+}$ release from the sarcoplasmic reticulum by CICR mechanism which opens the RyRs (mainly RyR2).

Although a rise of free $Ca^{2+}$ triggers smooth muscle contraction, the $Ca^{2+}$ receptor is calmodulin rather than troponin C (Fig. 22.11). The $Ca^{2+}$ calmodulin complex activates **myosin light chain kinase** (**MLCK**), which specifically phosphorylates Ser19 of the **regulatory myosin light chain** (**RLC**) with 20 kDa (**MLC20**) of the myosin II isoforms expressed in smooth muscle cells. The two heads of smooth muscle myosin with unphosphorylated MLC20 revealed an asymmetrical interaction in which the ATPase and actin-binding domains of one head are blocked [29] (Fig. 22.11), showing a very low ATPase activity which is hardly activated by actin, and adopted a compact, soluble 10 S form (S = Svedberg units) ("**tonomyosin**"). Phosphorylation of $Ser^{19}$ of the MLC20 confers the smooth muscle myosin into an open less soluble 6 S conformation with high ATPase activity which assemble into thick filaments and generates contraction. In contrast to striated muscle, which form end-polar thick filaments, myosins in smooth muscle cells assemble into **side-polar thick filaments**. Dephosphorylation of MLC20 by a specific **MLC20-phosphatase** (**MLCP**) enforces relaxation of smooth muscle cells. MLCP is a PP1-type phosphatase composed of three subunits: the 38 kDa catalytic subunit (PP1c), the 110 kDa myosin phosphatase target subunit (MYPT1) which plays an important role in targeting MLCP to myosin filaments, and the 20 kDa (M20) subunit [30].

**Pharmacomechanical coupling** occurs upon neurohumoral stimulation and comprises distinct mechanisms that regulate myoplasmic $Ca^{2+}$ of smooth muscle independently of the membrane potential. A major mechanism is elicited upon binding of acetylcholine (ACh) to its **muscarinic receptor** (**mAChR**; **M3 type**) at the sarcolemma of single-unit smooth muscle cells that activate heterotrimeric $G_{q/11}$ proteins which in turn stimulate **phospholipase C-ß** activity and phosphatidylinositol turnover. This finally leads to the formation of second messengers (see Chap. 6). $IP_3$ triggers $Ca^{2+}$ release from the sarcoplasmic reticulum by binding to **$IP_3$ receptors** (**$IP_3R$**), i.e. calcium release channels like RyR. DAG activates **PKC**, which phosphorylates a variety of proteins involved in signal transduction, contraction, and membrane polarization. For example, PKC phosphorylates **CPI-17**, a 17 kDa phosphatase inhibitor, which inhibits the catalytic subunit of MLCP. Inhibition of MLCP activity leads to an increase of the fraction of force-generating myosin with phosphorylated MLC20 and therefore increases of force generation. Since inhibition of MLCP activity increases force without increasing $Ca^{2+}$ activation levels, it is designated as **$Ca^{2+}$-sensitizing mechanism**. Inhibition of MLCP can also be achieved through activation of **Rho-kinase** by the small monomeric G protein Rho-A. Active Rho-kinase inhibits MLCP activity by phosphorylation of its regulatory subunit [31].

## 22.9 Muscles in Invertebrates

Muscles of invertebrates have long and thick myosin filaments which can be up to 25 μm long and 2,000 Å thick. This increased size is due to the presence of **paramyosin**, a 130-nm-long α-helical coiled-coil dimer of ~220 kDa, which forms the inner core of thick filaments with myosin II arranged at its surface. Thin filaments can reach 11 μm in length. Striated muscle types of invertebrates have highly variable sarcomere lengths with up to 14 μm in some crustacean muscles.

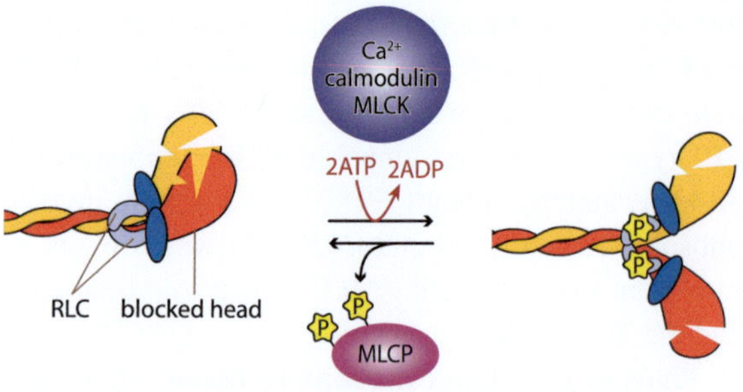


**Fig. 22.11** **Smooth muscle contraction regulation.** *Left*: unphosphorylated smooth muscle myosin is in an inactive, folded conformation with a blocked head (motor domain). Activation of smooth muscle cells increases myoplasmic free $Ca^{2+}$ ions which bind to calmodulin, a complex which activates MLCK. Activated MLCK phosphorylated the RLC which unfolds the myosin heads, increases myosin ATPase activity and allows actomyosin interaction. *MLCK* myosin light chain kinase, *RLC* regulatory myosin light chain with 20 kDa, *P* inorganic phosphate, *MLCP* myosin light chain phosphatase

Invertebrate actomyosin is regulated either as thin filaments (troponin-tropomyosin) or as thick filaments (primarily via direct $Ca^{2+}$ binding to myosin) [25, 32].

**Mollusks** have mainly smooth muscles besides obliquely striated and cross-striated muscles. The two shells of the bivalves are closed rapidly and can be kept closed for a long period by the adductor muscle which consists of two main parts: a translucent part, which is obliquely striated (in most bivalves) or a cross-striated muscle (in scallops), and an opaque part, which is smooth muscle. The striated adductor muscle can shorten rapidly, resulting in quick closure of the shells. The opaque smooth muscle part of the adductor (and the anterior byssus retractor muscle of *Mytilus edulis*) shorten slowly but can then maintain high passive tension, i.e., with negligible ATP consumption, after their active contraction (**catch state**). These muscles are controlled by two kinds of nerve fibers, namely excitatory which release acetylcholine and induce catch contraction, and inhibitory which release 5-hydroxytryptamine (5-HT) to interrupt the catch for rapid relaxation. Acetylcholine stimulation rises intracellular $Ca^{2+}$ which elicits contraction by direct binding to myosin essential light chains. The catch state is initiated by $Ca^{2+}$-stimulated dephosphorylation of the titin-like protein **twitchin** (~530 kDa), a thick filament protein composed of 24 immunoglobulin-like and 15 fibronectin type III-like domains [33, 34]. During catch, the membrane potential and the cytosolic free $Ca^{2+}$ are at resting levels. The catch state is relieved by 5-HT. 5-HT stimulation increases the intracellular cAMP concentration that results in an activation of the cAMP-dependent protein kinase which phosphorylates twitchin, thus terminating catch. However, it remains to be elucidated whether the catch state can be explained by the twitchin mechanism alone since twitchin is 15 times less abundant than myosin.

**Arthropods** have mainly striated muscle types. Non-flight muscles have thin myofibrils, I-bands of varying width, depending on the function of the muscle, and elaborate sarcoplasmic reticulum. The regulatory proteins of flight muscles of insects are located on the thin filament and consist of troponin and tropomyosin. Tropomyosin molecules move upon $Ca^{2+}$ activation. Insects with lower wingbeat frequencies have **synchronous flight muscle** while insects with high wingbeat frequencies have **asynchronous flight muscle** with large myofibrils and scarce sarcoplasmic reticulum. The A-band of asynchronous flight muscles is very close to the length of the sarcomere (almost no I-band). Unlike contractions of synchronous muscles, asynchronous flight muscles are not activated and deactivated neurogenically. Rather, they are driven myogenically, i.e., the muscles are stimulated by periodic mechanical stretches (**stretch activation**). Thus, the motor neurons of the asynchronous flight muscles in *Drosophila* fire at ~5 Hz, well below the contraction frequency of ~240 Hz. Nonetheless, $Ca^{2+}$ has a permissive role of maintaining the muscle fibers in a stretch-activatable state [25].

There are two isoforms of TnC in asynchronous flight muscles. **F1 TnC** is responsible for activating the muscle following a stretch, whereas **F2 TnC** produces a sustained contraction which depends on the concentration of $Ca^{2+}$ in the fiber. F1 TnC has an N-lobe that is completely insensitive to $Ca^{2+}$, whereas the N-lobe of F2 TnC contains one $Ca^{2+}$-binding site. Most flying insects use **indirect flight muscles** which are attached to the cuticle of the thorax and only connected indirectly to the wings (the wing muscles of mayflies and odonates insert directly at the wing bases). The high passive stiffness of indirect flight muscles can be explained by short connecting filaments (**C filaments**) which consist of the proteins **projectin** and **kettin** that anchor the thick filaments to the Z-disk in the sarcomere.

The two sets of indirect flight muscles, the dorsal ventral muscles (DVM) and the dorsal longitudinal muscles (DLM), are primed with intracellular $Ca^{2+}$ released from calcium stores by nervous stimulation. Just before flight, contraction of the tergal depressor of the trochanter lengthens the DLM, which responds to stretch activation. This triggers a deformation of the thorax and stretches the DVMs, which subsequently contract. Alternate stretch activation of opposing sets of indirect flight muscles produces oscillations in the thorax, which moves the wings at the resonant frequency of the flight system [25].

## References


1. Gelfand VI, Scholey JM (1992) Every motion has its motor. Nature 359:480–482
2. Du Bois-Reymond E (1912) Akademische Ansprache in der Leibnitz-Sitzung der Akademie der Wissenschaften am 3. Juli 1851. In: Du Bois-Reymond E (ed) Reden von Emil Du Bois-Reymond. Verlag von Veit & Comp, Leipzig
3. Peachey LD (1965) The sarcoplasmic reticulum and transverse tubules of the frog's Sartorius. J Cell Biol 25:209–231
4. Henneman E, Somjen G, Carpenter DO (1965) Functional significance of cell size in spinal motoneurons. J Neurophysiol 28:560–580
5. Coluccio LM (ed) (2008) Myosins. A superfamily of molecular motors. Springer, Heidelberg/Berlin/New York
6. Odronitz F, Kollmar M (2007) Drawing the tree of eukaryotic life based on the analysis of 2,269 manually annotated myosins from 328 species. Genome Biol 8:R196
7. Kühne W (1864) Über das Protoplasma und die Contractilität. Verlag von Wilhelm Engelmann, Leipzig
8. Geeves MA, Holmes KC (2005) The molecular mechanism of muscle contraction. Adv Protein Chem 71:161–193
9. Rayment I, Rypniewski WR, Schmidt-Bäse K, Smith R, Tomchick DR, Benning MM, Winkelmann DA, Wesenberg G, Holden HM (1993) Three-dimensional structure of myosin subfragment-1: a molecular motor. Science 261:50–58
10. Huxley AF (1957) Muscle structure and theories of contraction. Prog Biophys Biophys Chem 7:255–318
11. Lymn RW, Taylor EW (1971) Mechanism of adenosine triphosphate hydrolysis by actomyosin. Biochemistry 10:4617–4624
12. Brenner B (1988) Effects of $Ca^{2+}$ on cross-bridge turnover kinetics in skinned single rabbit psoas fibers: implications for regulation of muscle contraction. Proc Natl Acad Sci USA 85:3265–3269
13. Gordon AM, Huxley AF, Julian FJ (1966) The variation in isometric tension with sarcomere length in vertebrate muscle fibers. J Physiol 184:170–192
14. Huxley HE, Hanson J (1953) Changes in the cross-striations of muscle during contraction and stretch and their structural interpretation. Nature 173:973–976
15. Hill AV (1938) The heat of shortening and the dynamic constants of muscle. Proc R Soc Lond B Biol Sci 126:136–195
16. Barclay CJ, Constable JK, Gibbs CL (1993) Energetics of fast- and slow-twitch muscles of the mouse. J Physiol 472:61–80
17. Huxley HE (1973) Structural changes of the actin- and myosin-containing filaments during contraction. In: The mechanism of muscle contraction. Cold Spring Harb Symp on Quant Biol 37:361–376
18. Haselgrove JC (1973) Evidence for a conformational change in the actin-containing filaments of vertebrate striated muscle. In: The mechanism of muscle contraction. Cold Spring Harb Symp Quant Biol 37:341–353
19. Essen B (1977) Intramuscular substrate utilization during prolonged exercise. Ann N Y Acad Sci 301:30–44
20. Blinks JR, Rüdel R, Taylor SR (1978) Calcium transients in isolated amphibian skeletal muscle fibres: detection with aequorin. J Physiol 277:291–323
21. Sculptoreanu A, Scheuer T, Catterall WA (1993) Voltage-dependent potentiation of L-type $Ca^{2+}$ channels due to phosphorylation by cAMP-dependent protein kinase. Nature 364:240–243
22. Pedersen TH, Nielsen OB, Lamb GD, Stephensen DG (2004) Intracellular acidosis enhances the excitability of working muscle. Science 305:1144–1147
23. Kirkendall DT (2000) Fatigue from voluntary motor activity. In: Garrett WE, Kirkendall DT (eds) Exercise and sport science. Lippincott Williams & Wilkins, Philadelphia
24. Allen DG, Lamb GD, Westerblad H (2008) Skeletal muscle fatigue: cellular mechanisms. Physiol Rev 88:287–332
25. Rüegg JC (1992) Calcium in muscle contraction. Springer, Berlin/Heidelberg/New York
26. Bozler E (1941) Action potentials and conduction of excitation in muscle. Biol Symp 3:95–109
27. Dillon PF, Aksoy MO, Driska SP, Murphy RA (1981) Myosin phosphorylation and the cross-bridge cycle in arterial smooth muscle. Science 211:495–497
28. Arner A, Löfgren M, Morano I (2003) Smooth, slow and smart muscle motors. J Muscle Res Cell Motil 24:165–173
29. Wendt T, Taylor D, Trybus KM, Taylor K (2003) Three-dimensional image reconstruction of dephosphorylated smooth muscle heavy meromyosin reveals asymmetry in the interaction between myosin heads and placement of subfragment 2. Proc Natl Acad Sci USA 98:4361–4366
30. Somlyo AP, Somlyo AV (2003) $Ca^{2+}$ sensitivity of smooth muscle and nonmuscle myosin II: modulated by G proteins, kinases, and myosin phosphatase. Physiol Rev 834:1325–1358
31. Somlyo AP, Somlyo AV (2000) Signal transduction by G-proteins, rho-kinase and protein phosphatase to smooth muscle and non-muscle myosin II. J Physiol 522:177–185
32. Hooper SL, Hobbs KH, Thuma JB (2008) Invertebrate muscles: thin and thick filament structure; molecular basis of contraction and its regulation, catch and asynchronous muscle. Prog Neurobiol 86:72–127
33. Funabara D, Hamamoto C, Yamamoto K, Inoue A, Ueda M, Osawa R, Kanoh S, Hartshorne DJ, Suzuki S, Watabe S (2007) Unphosphorylated twitchin forms a complex with actin and myosin that may contribute to tension maintenance in catch. J Exp Biol 210:4399–4410
34. Galler S (2008) Molecular basis of the catch state in molluscan smooth muscles: a catch challenge. J Muscle Res Cell Motil 29:73–99
35. Zhao X, Yamazaki D, Kakizawa S, Pan Z, Takeshima H, Ma J (2011) Molecular architecture of $Ca^{2+}$ signaling control in muscle and heart cells. Channels 5:391–396

36. Buoanno A, Fields RD (1999) Gene regulation by patterned electrical activity during neural and skeletal muscle development. Curr Opin Neurobiol 9:110–120
37. Harvey RD, Hell JW (2013) Cav1.2 signaling complex in the heart. J Mol Cell Cardiol 58:143–152
38. Buchthal F, Schmalbruch H (1980) Motor unit of mammalian muscle. Physiol Rev 60:90–142
39. Bellemare F, Woods J, Johansson R, Bigland-Ritchie B (1983) Motor-unit discharge rates in maximal voluntary contractions of three human muscles. J Neurophys 50:1380–1392
40. Gandevia SC (2001) Spinal and supraspinal factors in human muscle fatigue. Physiol Rev 81:1725–1789

# Motor Control

# 23

Hans-Joachim Pflüger and Keith Sillar

## 23.1 Introduction

Survival depends upon efficient motor control behaviors which enable organisms to navigate through their environment, search for food, find a mate and avoid predation. In this chapter we provide an overview of motor systems that generate movements in a range of different animals. In vertebrates and invertebrates alike motor systems are similar: central networks drive motoneurons to fire which in turn elicit contractions in skeletal muscles to articulate the joints of the limbs or segments of the body. Biomechanical constraints influence behavior, and once behavior is produced sensory information is generated that feeds back and in turn modulates motor control. The components of motor systems, including motoneurons, muscles with their electrical and mechanical properties as well as sensory feedback are present across different phyla but the ways in which they are assembled during development have been sculpted into different species-specific solutions by the forces of natural selection.

The behavior of an animal is usually observed by its movements, which may be the result of an "inner state" or the motivation of an individual animal and caused by complex network activities in the brain and modified by neurohormones or neuromodulators, or it may be the result of a reaction to an internal (proprioceptive) or external (exteroceptive) stimulus, for example, a reflex reaction. Thus, movement is the common output in response to both sensory feedback mechanisms and central nervous mechanisms that control the inner state. Movement ranges from very simple motor behaviors such as resistance reflexes to the most complex motor behaviors such as shedding old cuticle (molting), building nests, navigating across difficult terrain, or even creating tools or performing speech. Most motor behaviors result from so-called "central motor programs" which are generated by ensembles of neurons that form networks and that produce at least some parts of the respective motor activity. For example, a central pattern generator (CPG) produces a basic alternating rhythm in the absence of any sensory feedback. Therefore, both sensory feedback mechanisms and central nervous mechanisms that regulate the inner state interact to generate movement.

Comparative modern genetics plausibly suggests that all bilaterally symmetrical organisms are related and derive from an "ur-bilaterian" ancestor [52, 60]. Similar suggestions come from the study of nervous systems where genes with a high degree of homology are responsible for the assembly of the central nervous system (CNS), including the brain in both vertebrates and invertebrates; see Chap. 2 [60]. Since all neurons rely on very similar cellular and molecular mechanisms of electrical excitability, such ideas also apply to neuronal networks. Therefore, a comparative approach that studies the neuronal mechanisms that underlie movements will reveal basic principles and, at the same time, point to special adaptations which are the result of evolutionary forces and the constraints of particular ecological niches. On the planet today there are about $50 \times 10^6$ animal species of which only 5 % are

H.-J. Pflüger (✉)
Institute of Biology, Neurobiology, Freie Universität Berlin, Königin-Luise-Str. 28-30, 14195 Berlin, Germany
e-mail: pflueger@neurobiologie.fu-berlin.de

K. Sillar
School of Psychology and Neuroscience, University of St Andrews, Carnegie Wing, Bute Building, St Andrews, KY16 9TS Scotland, UK
e-mail: kts1@st-andrews.ac.uk

C.G. Galizia, P.-M. Lledo (eds.), *Neurosciences - From Molecule to Behavior: A University Textbook*,
DOI 10.1007/978-3-642-10769-6_23, © Springer-Verlag Berlin Heidelberg 2013

vertebrates and of which less than 1 % are mammals. Thus, 95 % of all animal species are invertebrates, 70 % of which are insects. The evolutionary success of some invertebrates, in particular arthropods, may be due to their enormous adaptability to ecological constraints, and the basic mechanisms in the CNS of these animals already allow complex forms of learning and memory formation. Sometimes invertebrates such as annelids, mollusks, crustaceans, and insects are regarded as "model organisms" although one has to bear in mind that all "model systems" were shaped by evolution and species-specific constraints; no single system can be regarded as a representative for the whole range of different organisms. Some animal species are regarded as "genetic model systems", because their genome has been fully sequenced and the tools of modern genetics can be applied to produce mutant and transgenic animals. In this increasing list of animal species for which the genome is sequenced, a few stand out because they are regarded the most influential and they have also played an important role in understanding motor control: the nematode *Caenorhabditis elegans*, the fruit fly *Drosophila melanogaster*, the zebrafish *Danio rerio,* and the mouse *Mus musculus*.

For physiological studies, large invertebrates proved to be excellent experimental "model" animals mainly because they have comparatively large neurons that are readily accessible for study; see also [40]. Indeed, many of the fundamental physiological mechanisms in the nervous system were first studied in invertebrates: for example, (1) the giant axon of the squid served to unravel the ionic mechanisms underlying the action potential (see Chap. 7) as well as fundamental properties of synaptic transmission; (2) the principles of peripheral inhibition were first studied in crustaceans, where also the first electrical synapses were identified [18]; (3) the cellular mechanisms of various kinds of learning and memory formation were first studied in the sea slug *Aplysia californica* (see Chap. 26); and, in addition, (4) the roles of biogenic amines and peptides were first identified in invertebrates (see Chap. 11). The advantage of working with invertebrates is their relatively smaller number of neurons (an insect brain has between 500,000 and $10^6$ neurons compared to $10^{11}$ or $10^{12}$ of the human brain), and the possibility to uniquely identify many of the neurons. This "identified neuron concept" makes it possible to study a given neuron in the nervous system of many individuals and due to the constancy of location of their cell bodies it can be found and recorded reliably, facilitating an understanding of its role in behavior. Motor networks in particular can be analyzed at a cellular level, as rhythmic activity can be induced either in semi-intact (with reduced sensory feedback) or in completely isolated (without sensory feedback) preparations. In addition, particular neurons that persist from larvae to adults in holometabolous insects such as the tobacco hawkmoth (*Manduca sexta*), or the fruit fly (*Drosophila melanogaster*), can be identified and their profound changes during metamorphosis can be studied at the level of an individual neuron [34]. Another advantage is that the invertebrate CNS needs to be sufficiently sophisticated to generate and control their rich behavioral repertoire. Similar arguments also apply to "lower" vertebrates and, thus, animals such as the lamprey, zebrafish, and amphibians, such as the clawed frog (*Xenopus laevis*), are widely used in studies of motor control. The most frequent "higher" vertebrate species in motor control are mice, rats, cats, and monkeys. The terms "lower" and "higher" vertebrate are in common usage, but this is unfortunate phraseology since it erroneously conveys a sense of "inferior" and "superior", which is not justified.

## 23.2 The Neuromuscular Basics

### 23.2.1 Biomechanical Constraints

Muscles act upon skeletal components to generate movement. In general, each joint of a limb or each segment of the body is able to move in two opposing directions under the control of a pair of functionally antagonistic skeletal muscles, such as the elevator and depressor muscles of the wings, extensor and flexor muscles of the limbs, or segmented myotomal muscles on the left and right sides of a fish. During rhythmic locomotion the muscles of a pair contract in alternation, one muscle generates thrust against the environment during the power stroke of behavior while its antagonist generates the minimum possible back thrust during the return stroke. During walking behavior these two phases of movement are also referred to as the stance (limb on the substrate generating thrust) and swing (limb above the substrate and returning to the starting position) phases.

All muscles exert forces on the skeletal components such as exoskeletons of insects and crustaceans and internal bones of vertebrates. If, for example, a joint is moved around the pivot, the muscles produce a corresponding torque which is dependent on the force multiplied by leverage. Many different forces act on a moving joint: (1) external forces such as friction and viscosity of the media (ground, air, and water), inertia of movements or gravity; and (2) internal forces such as friction of joints and connective tissues and inertia produced by translational and rotational forces. All of these forces together with the forces generated by programming muscular activity through central circuits of the brain will generate a dynamic system that has to be calculated in the appropriate time, "online" so to speak, in order to produce a smooth and appropriate movement. This is the formidable task of all nervous systems in motor control.

The anatomical structure of the striated muscles of many invertebrates and vertebrates creates specific physical properties. The connective tissues and membranes surrounding a muscle as well as the apodemes or tendons define some of the elastic properties. Therefore, even if the sarcomeres, the basic units of striated muscles, contract, a joint may not move at all as first the forces will stretch the apodeme (or tendon) and only increase muscle tension (see Chap. 22). In addition, the geometrical arrangement of muscle fibers is important: a **parallel alignment** of fibers summates

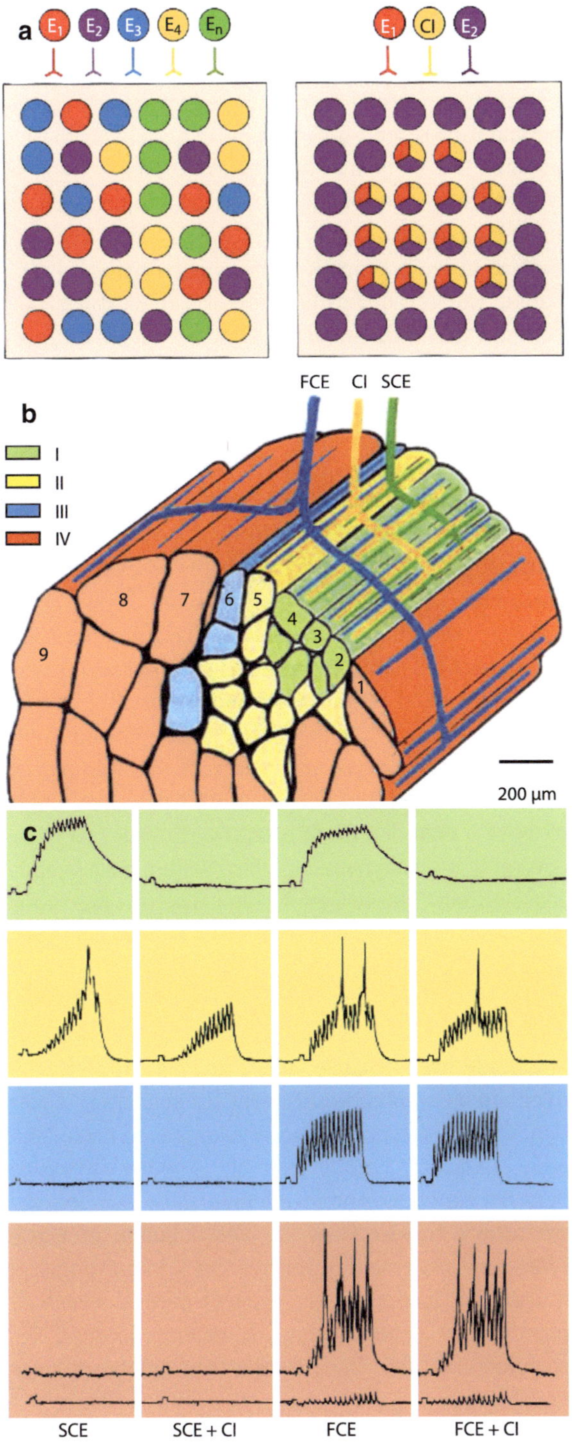


**Fig. 23.1** **Muscle fiber composition.** (**a**) Composition of skeletal muscles by different motor units in a vertebrate (*left*), and an invertebrate (*right*). A typical muscle of a mammalian (*left*), which is exclusively supplied by many exclusively excitatory (*E*) motoneurons and possesses as many motor units (see color code of motoneurons and fibers). Motor units do not overlap and each adult muscle fiber is only innervated by one motoneuron, and fibers belonging to one motor unit have similar properties. The motor units are irregularly distributed over the muscle cross section. A typical skeletal muscle of a crustacean (*right*) as an example of a polyneuronally innervated invertebrate muscle receives different types of innervation; two excitatory motoneurons (E) and one inhibitory neuron which also innervates other muscles and, thus, is called common inhibitor (CI). Individual muscle fibers can be innervated by all motor and inhibitory neurons, or by less. Fibers receiving similar innervation are often arranged in groups and the properties of fibers within one motor unit can be heterogeneous. Individual motoneurons cause very different muscular contractions that can range from slow to fast (Courtesy of W. Rathmayer). (**b**) Part of a typical crustacean muscle (proximal part of closer muscle in a walking leg) receiving three different types of innervation and composed of different types of muscle fibers (I–IV). (**c**) Differences in postsynaptic responses (EJPs) of the different fiber types following identical stimuli to the slow closer excitor motoneuron (SCE) or to the fast excitor motoneuron (FCE) or to the CI simultaneously. The inhibitory effect is complete in type-I fibers, weak in type-II fibers, and not present in type-III and -IV fibers as they do not have CI innervation. Type-I fibers cannot generate action potentials. The EJPs of the type-IV fibers are small. Only by high discharge rates of the FCE can muscle potentials be generated due to facilitation and, thus, recruit this fiber type for contractions. The calibration pulse at the beginning of each trace is 2 mV for 10 ms (Courtesy of W. Rathmayer)

the forces of the individual fibers and, thus, the larger the muscle volume the bigger the muscular force; a **serial arrangement**, in contrast, leads to summation of the shortening of each individual sarcomere and, thus, the longer a muscle the faster its contraction velocity. Similar rules apply to where exactly on the joint a muscle inserts as this will determine the leverage. Therefore, the physical (visco-elastic) properties in combination with the neural activation determine the force and how the movement is exerted. Even the geometrical arrangement of fibers on the tendon or apodeme contributes to force development.

## 23.3 Comparison Between Invertebrate and Vertebrate Skeletal Muscles

A comparison between a typical skeletal muscle in the leg of an insect or crab and in the leg of a vertebrate shows many similar properties like striation as a result of very regularly structured sarcomeres, and the biochemical machinery for the production of force. Differences between invertebrate and vertebrate skeletal muscles exist regarding the innervations patterns. A typical vertebrate skeletal muscle consists of many muscle fibers which are innervated by numerous motoneurons. The extensor digitorum longus muscle (EDL) of the rat, for example, consists of around 3,000 fibers which are supplied by approximately 100 motoneurons. In vertebrate muscle the size of an individual motor unit (i.e., all muscle fibers innervated by one motoneuron) can vary enormously. For example, in an eye muscle (rectus oculus lateralis) one motoneuron innervates 13 muscle fibers, and in the case of the human biceps brachii one motoneuron innervates about 750 muscle fibers. There are also large differences in the innervation of homologous muscles in different animals: the tibialis anterior is comprised of 56,000 fibers in a cat and 270,000 fibers in humans, supplied by approximately 250 and 445 motoneurons, respectively. These adult conditions are only gradually achieved during postembryonic development. A single muscle fiber is initially innervated by more than one motoneuron. The adult conditions are only reached through an activity-dependent competitive process during ontogeny.

The situation in arthropods is completely different, although the same definition for motor unit applies to invertebrates (Fig. 23.1). In general only 2–4 excitatory motoneurons innervate one muscle (some exceptional muscles are supplied by up to one dozen motoneurons and other muscles exist that are only supplied by a single excitatory motoneuron). Variations occur between the different arthropod groups. Spider muscles, for example, are supplied by a larger number of motoneurons, perhaps a dozen or more, and such **polyneuronal innervation** also exists for nematodes, annelids, and mollusks. Molluskan muscles, in general, have a large number of motor units and may be comprised of several tens of motoneurons. Two features only found in invertebrate muscle and not existing in vertebrate muscle are innervation by inhibitory motoneurons and by neuromodulatory neurons. Therefore, the number of motor units in invertebrate muscles is always small, and each muscle fiber, in general, receives polyneuronal innervation, and thus one particular muscle fiber can be part of more than one motor unit. However, the range of mechanical performance of an invertebrate muscle is very similar to that of a vertebrate muscle, and those specialized for particular behavioral performances, like jumping, easily out-power vertebrate muscles if scaled appropriately to the size of the animal.

Whereas vertebrate motoneurons possess one neuromuscular contact on each fiber, the so-called **motor endplate**, invertebrate motoneurons form many synaptic contacts with the muscle fiber (multiterminal innervation). The neuromuscular synapses are spaced between 10 and 20 μm apart and are usually distributed over the length of the muscle fiber, although some spatial concentrations are observed, for example on muscles of *Drosophila melanogaster* larvae. A single axon of a motoneuron in a crustacean can thus make several hundred synaptic contacts with a postsynaptic muscle fiber. The reason for this type of innervation may be that in contrast to vertebrate muscle fibers, which can produce action potentials, the postsynaptic junctional potentials in many invertebrate muscle fibers are only passively conducted with a decrease in amplitude over distance (see Chap. 7). Therefore, to maintain an amplitude sufficient for generating a particular force output, the motor axon has to make synaptic contacts at regular intervals. Another difference between vertebrate and invertebrate muscles is that the **excitatory transmitters** are acetylcholine in vertebrates and glutamate in invertebrates. Interestingly, it is normally the opposite for the transmitters of sensory neurons and of most excitatory neurons in the CNS: acetylcholine for invertebrates and glutamate for vertebrates.

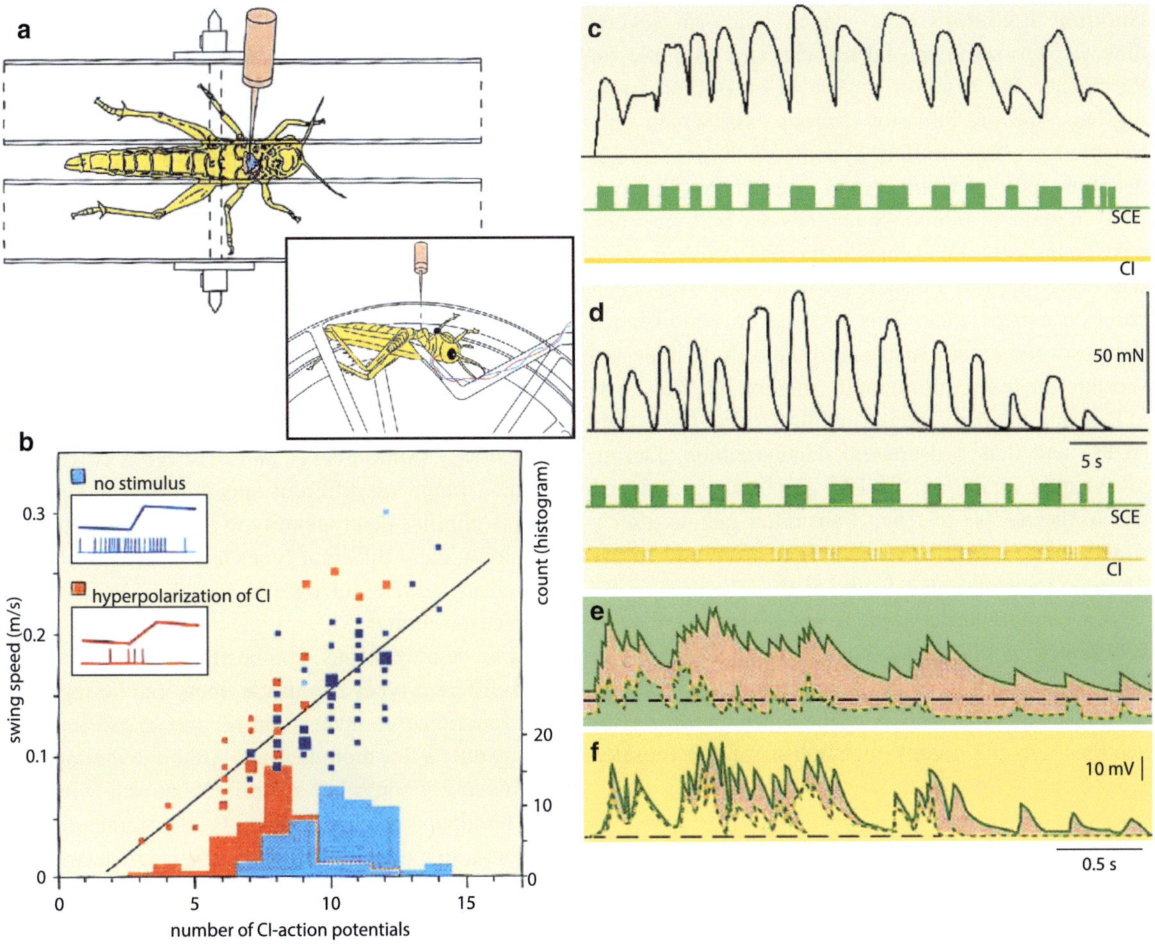


**Fig. 23.2 Effects of the common inhibitor.** (**a**) The activity of a CI neuron of a locust walking upside-down on a treadmill is recorded intracellularly and via current injection into the CI soma can be manipulated. Leg movements are registered optically (see also a side view in the *insert*). (**b**) Histogram of CI discharge in relation to the protraction speed (swing speed) of the leg. CI hyperpolarization reduces the number of action potentials generated and slows down the protraction speed. *Inserts*: registration of leg movement and CI action potentials. *Red*: effects after CI manipulation (Modified from Wolf [68] with permission). (**c** and **d**) Force development of a closer muscle in a crab walking leg due to stimulation of the slow excitor (SCE) alone (**c**) or together with the CI (**d**). The stimulation patterns were gained from a walking crab and correspond to naturally occurring activity patterns (14 steps). Without inhibition the muscle shows strong properties of hysteresis during relaxation after each contraction and, thus, builds up tonic tension which opposes a complete relaxation of the dactyl to zero values (*black base line*). (**e** and **f**) Two intracellular recordings of EJPs from a tonic (**e**) and a phasic (**f**) muscle fiber during one step without (upper trace, *green*) and with (lower trace, *yellow stippled*) simultaneous CI stimulation. The inhibitory effect (*red*) is strongest in the tonic fiber. *Dashed line* refers to the resting potential of the muscle fiber (Modified from Ballantyne and Rathmayer [1] with permission)

### 23.3.1 Invertebrate Muscles Receive Inhibitory and Neuromodulatory Innervation

All efferent signals to a vertebrate muscle are excitatory and the number of motoneurons that are activated during a particular movement is the result of complex excitatory and inhibitory interactions at the level of the CNS, for example the spinal cord. In many invertebrates, including almost all arthropods, and also nematodes, annelids, and mollusks, an additional peripheral inhibitory system exists with important roles during the various behavioral tasks in which muscles have to function (Fig. 23.2). The working range of arthropod muscles, from very slow to very fast contractions, is aided by the action of **inhibitory neurons**. For fast-moving insects and crustaceans their action is essential. Most inhibitory neurons are

**common inhibitors** (CIs) which innervate several muscles. For some special muscles, for example for the opener and extensor muscle of crustacean walking legs, specific inhibitor neurons exist which only innervate one muscle [66]. Specific inhibitors serve to decouple muscles from a common excitatory innervation. In general, inhibitory neurons act both pre- and postsynaptically. Postsynaptically, any depolarization will be reduced by an increase in a GABA-mediated chloride conductance. This increase in conductance decreases the input resistance of the muscle fiber and shortens the time constant of the membrane. This leads to less summation of **excitatory junctional potentials** (EJPs) and thus a decreased depolarization. During presynaptic inhibition the EJP amplitude is reduced due to the release of fewer transmitter quanta after a single action potential in the motoneuron, and sometimes no transmitter is released at all.

A second important system, not so far described in vertebrates, is that invertebrate muscles also receive innervation from specific neuromodulatory neurons which can alter the efficacy of the neuromuscular connections and the energy metabolism of the muscles [16, 45].

### 23.3.2 The Fiber Compositions of Invertebrate Muscles

In general invertebrate muscles are heterogeneous, comprising very different types of muscle fibers. This, in combination with the specific innervation patterns, allows for the enormous range of mechanical responses. Although many different fiber types also exist in vertebrates, a particular muscle is usually formed by very similar fiber types of similar physiological properties. The fibers of one motor unit are more or less homogeneous with respect to their electrophysiological, mechanical, and biochemical properties. Invertebrate muscles, in particular those of arthropods, possess functionally different fiber types, and even the fibers of one motor unit are heterogeneous.

The closer muscle in a crab leg has been examined in detail in this respect. This muscle has four fiber types (types I–IV). The other leg muscles like extensor and flexor are very similar in their fiber compositions. All 300–400 fibers of the closer muscle are innervated by just one **fast closer excitor** (FCE) motoneuron and, therefore, these fibers are components of this individual motor unit. Approximately 50 % of these fibers receive additional innervation by a second excitatory motoneuron, the slow closer excitor (SCE), and by an inhibitory motoneuron which also supplies other muscles and, therefore, is called the **common inhibitor (CI)**. Despite being innervated by the same motoneuron, the quantal content and the facilitation properties of their synapses is very different, and so is their ability to generate action potentials (types II–IV) or only passively conducted potentials (type I). They are also distinguished by the isoform of myofibrillar ATPase and its total activity [38, 39]. This is the reason why type-I fibers only contract very slowly (tonic fibers), and fiber types II, III, and IV at a range of different speeds (different phasic fibers). Similar heterogeneity in fiber composition of physiologically different types has been described in numerous muscles in insects, crustaceans, and also other invertebrates.

This heterogeneous composition of motor units from different types of muscle fibers can be regarded as a functional compensation for the sparse innervation by only a few motoneurons. Although the capacity for fine motor control of a vertebrate muscle with several hundreds or even thousands of motoneurons is immense, that of invertebrate muscle is not necessarily less refined. In particular, even muscles innervated by a single motoneuron (like the opener or extensor in a crustacean walking leg) are able to perform in postural control tasks and in walking at different speeds under different load conditions.

### 23.3.3 Invertebrate Muscles Are Innervated by Different Types of Motoneurons

The efficacy of neuromuscular transmission varies between individual motoneurons, and between the synapses of an individual motoneuron made with different fiber types. Motoneurons of the fast and slow type, such as those found in insects, crustaceans, and other invertebrates, generate very different contractions in the muscle fibers so that a functional division or partitioning of the working range is observed. A single action potential in a **fast motoneuron** causes a powerful twitch that may also comprise over 50 % of the maximal force output of the muscle and, thus, smooth tetanic muscle contractions are achieved rapidly and at relatively low firing frequencies. In contrast, a single **slow motoneuron** generates a slow, noticeable twitch in the muscle only at a particular firing frequency with a single action potential usually producing no noticeable contraction. These motoneurons can generate tetanic contractions in the muscle only with slow speed and only at very high firing frequencies of 300 Hz and above. The prefix "fast" and "slow" motoneuron indicates the speed of the mechanical response of the innervated muscle, that is, the contraction caused by this motoneuron, and has nothing to do with the conduction velocities of their action potentials. However, fast motoneurons often have large diameter axons and, therefore, fast conduction velocities. "Fast" and "slow" indicate only the extreme ends of a functional continuum, and, thus, many motoneurons have intermediate properties. One mechanism for this is the different amount of transmitter released. Slow motoneurons, in general, release only a few transmitter quanta following a single action potential, so that the EJP is of only small amplitude. Higher firing frequencies, in combination with summation and facilitation of EJPs lead to suprathreshold depolarization which, in turn, can generate a contraction. Fast motoneurons have larger quantal amplitudes of their transmitter release and the resulting EJPs are of much larger amplitudes. One of these EJPs may be sufficient to cross the contraction threshold and cause a considerable twitch of the muscle fiber.

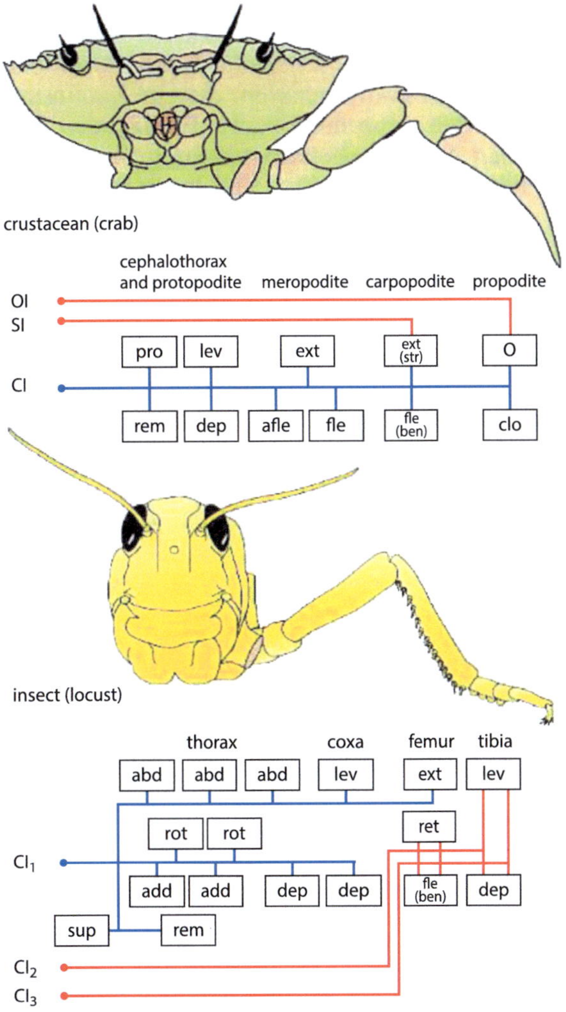


**Fig. 23.3 Comparison of CI organization in crustaceans and insects.** The CI neurons in a crab and in a locust innervate similar target muscles in legs and thorax, and reveal a great deal of evolutionary conservation. Abbreviations refer to individual muscles. OI and SI are two specific inhibitors in the crab that only supply one muscle (Courtesy of W. Rathmayer). *Abd* Abdominal, *Add* adductor, *Afle* anterior flexor, *CI* common inhibitor ($CI_{1-3}$), *Clo* closer, *Dep* depressor, *Ext* extensor, *Ext (str)* extensor (*stretcher*), *Fle (ben)* flexor (*bender*), *Lev* levator, *O* opener, *OI* opener inhibitor, *Pro* protractor, *Rem* remotor, *Ret* retractor, *Rot* rotator, *SI* stretcher inhibitor, *Sup* supressor

### 23.3.4 Different Motoneurons Activate Different Fiber Populations and Fiber Types

An additional differentiation is achieved by different **muscle fiber types**. This is important for the working range of muscles whose fibers are innervated by an individual motoneuron and which receive identical impulse patterns. Muscles of the walking legs in crustaceans are well studied in this respect, and like insects, they predominantly use slow motoneurons for walking. In addition, slow motoneurons are involved in posture and maintaining muscle tone including providing stiffness to joints. In both crustaceans and insects fast motoneurons may only be active briefly at the beginning of muscular activity, and seem to be involved in special, more rapid tasks such as escape, jumping, or defensive kicking.

### 23.3.5 Activation of Tonic Muscle Fibers

Slow motoneurons innervate tonic and phasic muscle fibers. If recorded from tonic muscle fibers the EJPs of slow motoneurons have larger amplitudes than those recorded in the phasic muscle fibers. Due to the higher membrane resistance the time constant of the decay of the EJP is 5- to 30-fold longer than in the phasic fibers. This increases the summation of EJPs in these tonic fibers and, thus, the total depolarization. Even if the slow motoneuron has only low firing rates, the tonic muscle fibers depolarize sufficiently and contract. The velocity of contraction is slow due to the low activity of the myofibrillar ATPase. At higher firing frequencies of the slow motoneuron the depolarization of tonic fibers does not increase much because pre- and postsynaptic effects counteract this. The tonic fibers show hardly any facilitation and the electrical properties of the muscle membrane of the tonic fibers prevent a further increase in depolarization at a certain amplitude (inward rectification), and, in addition, active electrical responses may also occur in these fibers.

### 23.3.6 Activation of Phasic Muscle Fibers

In phasic fibers the EJPs caused by a slow motoneuron are small and due to the short time constant of the membrane they do not summate at low firing frequencies. Therefore, in contrast to the tonic muscle fibers, the phasic muscle fibers do not show contractions at low firing frequencies of the slow motoneuron. These fibers show summation, facilitation, and thus contractions only at higher motoneuronal firing frequencies, and are recruited only then for the overall muscle contraction. Further increases in the firing frequencies of the slow motoneuron lead to increased depolarization of these muscle fibers, and action potentials may even be generated. Due to the high activity of myofibrillar ATPase in these phasic muscle fibers, the contraction speed is fast and, therefore, a high firing frequency of the slow motoneuron leads to fast contractions in these fibers. If, in addition to the slow motoneuron, the fast motoneuron, which innervates all phasic muscle fibers, is recruited, the depolarization of these doubly innervated muscle fibers is potentiated and the contraction force and speed is increased enormously. If other phasic muscle fibers that are only innervated by a fast motoneuron are recruited the contraction is even stronger and faster. Thus, whereas in vertebrates such enormous capabilities of gradual and fast contractions are achieved by the number of recruited (activated) motoneurons, in invertebrates it is a combination of motoneuron and muscle fiber properties.

### 23.3.7 Inhibitory Innervation Affects Mostly Tonic Muscle Fibers

The efficacy of the inhibitory action strongly depends on the type of muscle fiber. The most effective inhibition by the aforementioned CI neuron exists for tonic fibers. This has an important functional consequence. As tonic and phasic muscle fibers are often coactivated due to common excitatory innervation, the full contribution of the tonic fibers to the contraction properties results in a slow development of a contraction, and also a slow relaxation. This prevents the fast execution of muscle contractions, as required for rapid alternating movements during fast walking or running. Therefore, all tonic fibers are selectively and very efficiently inhibited and functionally removed from the system. This allows the whole muscle to contract and relax much faster. As the proportion of tonic fibers for the whole diameter of the muscle (which determines the force output) is only small, this has only a small impact on the total force output. Therefore, these CI neurons remove the contribution of all tonic fibers during fast movements.

Recordings from freely moving crabs and insects have revealed the necessity of CI action for the rhythmical alterations of muscle contractions and relaxations during fast walking movements (Fig. 23.3). The CI neuron is usually activated at the beginning of a walking sequence. In contrast to the rhythmically alternating excitatory motoneurons, however, the CI neuron discharges at a more tonic basal frequency for as long as the animal walks. If the discharge frequency of the CI neuron is experimentally decreased by injecting a hyperpolarizing current into its soma, the speed of the swing phase (protraction) is also decreased. This is due to a decreased efficacy of inhibition of tonic fibers and, thus, a noticeable contribution of this muscle fiber type that slows movements down [1, 68].

### 23.3.8 Neuromodulatory Control of Muscle Contractions

In arthropods, in particular in insects, many if not all skeletal and visceral muscles receive additional innervation from **neuromodulatory neurons** which often innervate whole sets of muscles with both pre- and postsynaptic functions. In insects these neurons belong to the class of dorsal or ventral unpaired median (DUM or VUM) neurons which have dorsal or ventral, median cell bodies, are unpaired and therefore possess bilaterally symmetrical axons and release the biogenic amine **octopamine**. If they are active simultaneously with the excitatory motoneurons, the resulting muscle contractions are altered. For example, the twitch amplitude and the relaxation rate are increased [16]. This is due to pre- and postsynaptic actions of octopamine. In addition, these neurons influence the energy metabolism of

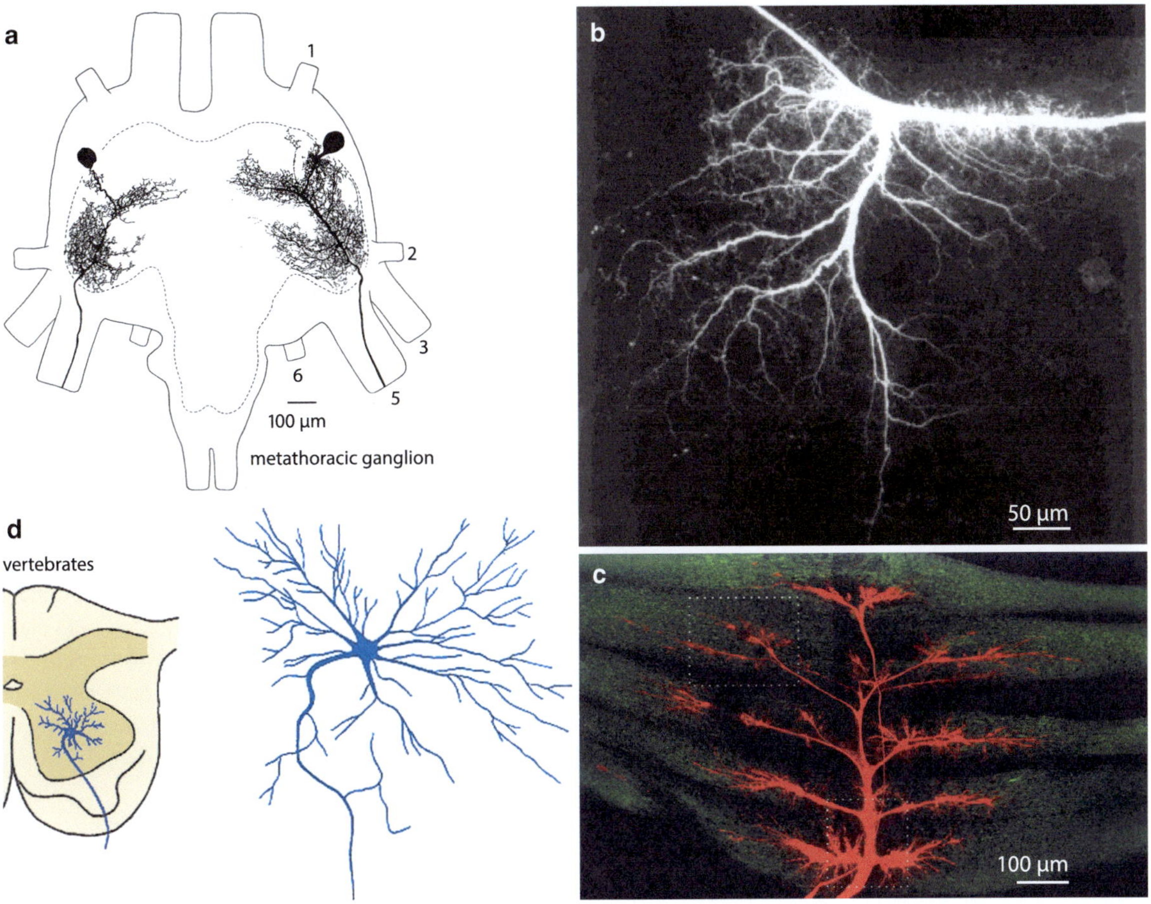


**Fig. 23.4** **Motoneurons.** (**a**) Two motoneurons in a locust metathoracic ganglion with its dendritic tree (From Watkins et al. [64] with permission). (**b**) A dendritic tree from a motoneuron of an adult tobacco hawkmoth (From Duch et al. [12] with permission). (**c**) The axonal arborizations (axonal tree) of several motoneurons each innervating a single muscle fiber of a large flight muscle in the tobacco hawkmoth, *Manduca sexta*; bar 50 µm (From Duch et al. [12] with permission). (**d**) Typical motoneuron of a vertebrate spinal cord (Modified from Reichert [51] with permission)

a muscle fiber by activating glycolysis and thus production of ATP in the muscle. Interestingly, in muscles that rely on the oxidation of lipids because the oxidation of sugars would be energetically far too costly, such as in locusts for example, these neurons are switched off during motor behavior but are active at rest [50]. Here they serve to accumulate metabolic signaling substances that will boost glycolysis if the animal is suddenly in need of a great amount of energy, such as that necessary for the take-off of a flying locust [45]. Additional experiments in which these neurons were recorded during various motor behaviors suggest that they serve to prepare the whole skeletal muscular system for dynamic action rather than postural tone. This is supported by the fact that octopamine has opposite effects on visceral muscles, where all myogenic contractions of guts and oviducts are inhibited. Therefore, in some respects, the octopaminergic neurons may be the arthropod's equivalent of the sympathetic nervous system of vertebrates. In a few insects some muscles are supplied by neurons that release allatostatin, which is one of the few inhibitory peptides.

## 23.4 Common Principles in the Generation and Control of Motor Patterns in Vertebrates and Invertebrates

Motoneurons (Fig. 23.4), according to Sir Charles Sherrington (1857–1952), represent the final common pathway for behavior. **Motoneurons** receive their inputs from either descending interneurons, local interneurons that convey preprocessed sensory information to them, or directly from primary sensory cells (invertebrates) or afferent neurons (vertebrates). The latter case would correspond to a monosynaptic reflex circuit. A spinal motoneuron of a mammal may thus receive up to 2,000 input synapses on the soma and up to 8,000 input synapses on the dendrites. Although invertebrate motoneurons have a slightly different structure, there is no fundamental difference with input synapses primarily targeting the dendritic branches as well. In vertebrates and invertebrates alike these synapses are not randomly distributed but formed according to some specific rules and, most likely, also to activity-dependent processes which ensure developmental flexibility and plasticity.

The forces that a vertebrate muscle produces due to the activity of its α-motoneurons can be increased by each individual motor unit being activated at increasing frequencies, and by additional recruitment of motor units that may actually be slightly asynchronous within a given burst, thus ensuring smooth contractions. Muscle contractions can vary from a single twitch to incomplete and complete tetanic contractions. During slow vertebrate locomotion, muscle motoneurons with small somata and axon diameters are recruited first causing low contraction velocities and in general innervate muscles that are oxidative and fatigue resistant. Fast locomotion is achieved by recruiting motoneurons with larger somata and axon diameters which innervate muscles that are glycolytic and fast fatiguing. In most fish the oxidative "red" muscle fibers form a relatively thin, peripheral band along the flank, while the bulk of the fish's trunk is composed of fast, glycolytic "white" muscle fibers that are only activated during rapid manoeuvres such as escape. The way in which the different muscle types are deployed during fish swimming is covered in more detail below.

The orderly recruitment of motoneurons according to size is known as the **size principle** (see Chap. 22). In mammals recruitment of motoneurons may be task-specific and more complex. The topographic arrangement of motoneurons in the spinal cord favors task-specific coordination of pools of motoneurons by presynaptic interneurons. A special feature of vertebrate motoneurons is the recurrent inhibition by so-called **Renshaw cells**. These receive their excitatory inputs from axon collaterals near the axon hillock of the motoneuron and themselves form inhibitory output synapses with the same motoneuron or with adjacent motoneurons. This recurrent or feedback inhibition is an interesting autoregulatory mechanism. The stronger a motoneuron is active the more recurrent inhibition it will receive, thus ensuring that the upper limit or "saturation" of firing is prevented. By modulating the Renshaw cell, for example through descending pathways, motoneuron firing can be finely regulated. Such mechanisms may play a role in movements that have to be regulated very finely and precisely such as those of the primate hand.

### 23.4.1 Proprioceptive Control of Muscle Contractions

Muscle contractions are under extensive proprioceptive control, in particular in vertebrates. Muscle length is controlled by the sensory fibers of the **muscle spindle** (predominantly Ia fibers, and also group II fibers) that surround several **intrafusal muscle fibers** of the spindle with different elastic properties. In mammals the intrafusal muscle fibers are innervated at both ends by a special class of motoneurons, the **γ-motoneurons** which are much smaller in soma and axon diameter than the α-motoneurons. Thus, through the activity of γ-motoneurons, the intrafusal muscle fibers contract and can activate the spindle afferents in a very precise way. This is important, for example, when both α- and γ-motoneurons are activated: first, the muscle shortens due to the activity of the rapidly firing α-motoneurons, and second, due to the delayed contractions of the intrafusal muscle fibers caused by γ-motoneuron firing the sensory afferents are maintained within a working range even in a very shortened muscle maintaining its sensitivity to any subtle length changes. In general, the afferents of the muscle spindle, the Ia fibers, are activated by passive stretch of the muscle, and in turn activate the α-motoneurons of their own ("homonymous") muscle. This opposes the passively induced movement and brings the joint back to its set point. The connections between the Ia fibers and the respective α-motoneurons are monosynaptic. In addition, the Ia fibers, by exciting interneurons in the spinal cord which inhibit the α-motoneurons of the antagonistic muscle, help in bringing the joint back to its set point.

To avoid the contracting muscle breaking the tendon (or apodeme) a second system controls muscle tension. The Ib fibers or **Golgi tendon organs** surround apodemal fibers and measure the tension exerted by the contracting (extrafusal) muscle fibers on the tendon. If this tension becomes too high, the α-motoneurons of

**New Techniques and the Advantage of Studying Particular Animals in Motor Control**

An obvious advantage is that the nervous systems of "simpler" animals, i.e., invertebrates and "lower" vertebrates, consist of few neurons, and many of their larger and more prominent neurons can be identified as individuals from one animal to the next due to the constancy of their location and structure within the CNS [6, 25]. This also applies to certain neurons in vertebrates such as fish and amphibians. Such identified neurons can be found in every individual animal in the same location within the CNS; they exhibit the same morphology and possess the same physiological properties. As many organisms possess a segmental organization, similar and serially homologous neurons are reiterated throughout the rostrocaudal axis of the nervous system. Usually they exhibit similar anatomy and often similar physiological features, although the functional tasks can vary depending on the particular part of the nervous system in which they are located. Nevertheless, they can be regarded as serial homologs. In closely related species homologous neurons with similar morphology and location can be identified across major phylogenetic boundaries even if the behavior they participate in may have been substantially modified in the course of evolution. One example is the Mauthner neuron in fish which is also identifiable in amphibian tadpoles (for a brief overview, see [56]). This large reticulospinal neuron located on each side of the medulla mediates a rapid "C-start" escape behavior in most teleosts when the animal is startled, but in some species it is thought to be wired into a different neuromuscular system, such as the fin adductor muscles that propel flying fish into the air.

Large identifiable neurons allow for intracellular recordings and subsequent dye filling, thus, correlations between structure and function can be made. In addition, new imaging techniques, for example by introducing voltage- or calcium-sensitive dyes into neurons or even whole neuron ensembles can visualize the activity of entire neuronal networks. This is even possible with electrophysiological recordings where multiunit recordings from brain areas can reveal the activity of more than one neuron. In addition, the techniques of electromyography, extracellular nerve recordings, and intracellular neuron recordings offer powerful methods for neurobiologists to study the neuronal networks underlying movement generation and control. A major advantage of genetically tractable systems is that transgenic animals can be made in which particular neurons can be visualized by the insertion of fluorescent markers, for example GFP (green fluorescent protein), or that particular sets of neurons can be switched on or off, for example by the insertion of heat-shock proteins or channelrhodopsins, or that whole sets of neurons can be genetically ablated. If such experiments are performed with component neurons of motor networks, their function and contribution to the motor pattern can be studied with unsurpassed precision. The extent to which the wide range of available techniques can be applied to different model systems is variable, but, in combination, major advances have been made in understanding the rich diversity of movements in the animal kingdom.

its own muscle are inhibited via inhibitory interneurons in the spinal cord and, in addition, via excitatory interneurons the α-motoneurons of the antagonistic muscle are activated. This reflex ensures then that the potentially damaging contraction is terminated.

In invertebrates muscle forces are dependent upon which type of motoneuron is activated. A single action potential of the fast motoneuron will produce a fast and powerful twitch, whereas the slow motoneuron will generate gradual and slowly developing muscle forces depending on its firing frequency. Slow motoneurons will achieve maximal muscular force by their highest firing frequencies (in locusts up to 250 Hz). As for vertebrate muscle, contractions range from single twitches to incomplete and complete tetanus. Although invertebrates do not have a similar system of muscle spindles and tendon-organs, in some cases multipolar sensory cells have been found associated with muscles or tendons and are believed to fulfil similar functions. In some special muscles such as the ovipositor in locusts, muscle tension may be monitored by as many as 200 receptor cells. In the jumping leg of a locust, for example, some multipolar cells or "joint organs" limit the amount of extension of the joint.

## 23.5 Sensory Feedback Together with Central Pattern Generators Produce Appropriate Motor Behavior

Around the turn of the twentieth century, in the early days of modern neurobiology, a fierce debate began on the question of whether motor behaviors are predominantly the result of reflexes that are linked in a "chain-like fashion", as proposed by Sherrington, or whether networks or circuits within the CNS are responsible for generating rhythmical activity patterns in the absence of any sensory feedback, as proposed by Graham Brown in 1911 (cited in [63]). As one might predict, the current view is that for the production of an appropriate motor behavior both the activity of central rhythm-generating networks as well as feedback control mechanisms are necessary. For different motor behaviors the central or peripheral contribution may vary. As a general rule, the more rhythmical and stereotyped a motor behavior is, the more dominant the centrally generated rhythm will be, while the more adaptive the motor behavior is to fast changing environmental conditions, the more dominant feedback control will be. Thus, in such a continuum of motor acts stereotyped motor behaviors like heartbeat, respiration, feeding, chewing, and swallowing, may have a strong central component similar to rhythmical movements such as swimming, flying, jumping, stridulating, copulating, or egg laying, but in more variable and adaptive motor behaviors such as walking over rough terrain or climbing, components of peripheral feedback loops play more important roles. Such variable contributions of either peripheral feedback loops or central circuits may even apply to one set of motor behaviors, with slow walking involving a greater contribution of peripheral control loops and fast running being more centrally programmed.

For many motor behaviors the alternating bursts of activity between functional antagonists, for example between depressors and elevators during flight, are generated by a network of neurons that has been termed a **central pattern generator** (CPG). The definition of a CPG is that in the complete absence of sensory feedback a self-sustaining and alternating rhythm can be generated by the CNS alone. If in this situation the phase relationships between firing of the agonist and antagonist systems remain the same regardless of frequency, and if similar sets of neurons are activated in the isolated situation compared to the intact animal, one speaks of a "fictive" motor behavior because the motoneuron output pattern resembles features of the intact behavior. A CPG consists of neurons, either interneurons or motoneurons or a mixture of the two, and a given neuron is considered part of the CPG if manipulations in the timing of its firing can reset the timing of the whole rhythm. In general, the frequency of the motor rhythm produced by the CPG in the isolated CNS is slower than the normal behavior in the intact animal. The extent to which CPG frequency is reduced in the absence of movement correlates with the degree to which the behavior requires sensory feedback to adapt it to environmental change. For example, in walking and flight systems there is often a dramatic reduction in locomotor frequency after isolation, while in swimming systems like *Xenopus* frog tadpoles where there are no known proprioceptors there is little change in locomotor output frequency once movement is prevented. Exceptions occur: an example is the centrally generated rhythm of rocking in stick insects which, with almost all sensory feedback absent, is faster than in the intact animal. One important question is how a single anatomically wired network is capable of producing more than one functional output rhythm. We shall see later that neurohormones and neuromodulators play a crucial role in increasing the operational range of CPGs.

### 23.5.1 The Stomatogastric System of Crustaceans as a Central Pattern Generator

The **stomatogastric ganglion** (STG) of crustaceans (Fig. 23.5) controls muscles which are involved in moving the gut and the different parts of the stomach. It is attached to the CNS via the unpaired **esophageal ganglion** (OG) and the paired **commissural ganglia** (CG), and is located within the large head artery. It produces four rhythms: (1) an **esophageal rhythm**, (2) a **cardiac sac rhythm**, (3) a **gastric mill rhythm** which moves three internal grinding teeth of the gastric region, and (4) a **pyloric rhythm** (for a review see [42]). In addition, if the food is inappropriate, crustaceans can also "vomit" using a fifth rhythm. These rhythms are generated by only 30 neurons and, in the case of the STG, all but two are motoneurons. If the STG is isolated together with the OG and CG, these rhythms can be produced spontaneously *in vitro* for several hours.

This preparation has several advantages: for instance, the somata within the STG can be visualized under appropriate illumination allowing intracellular

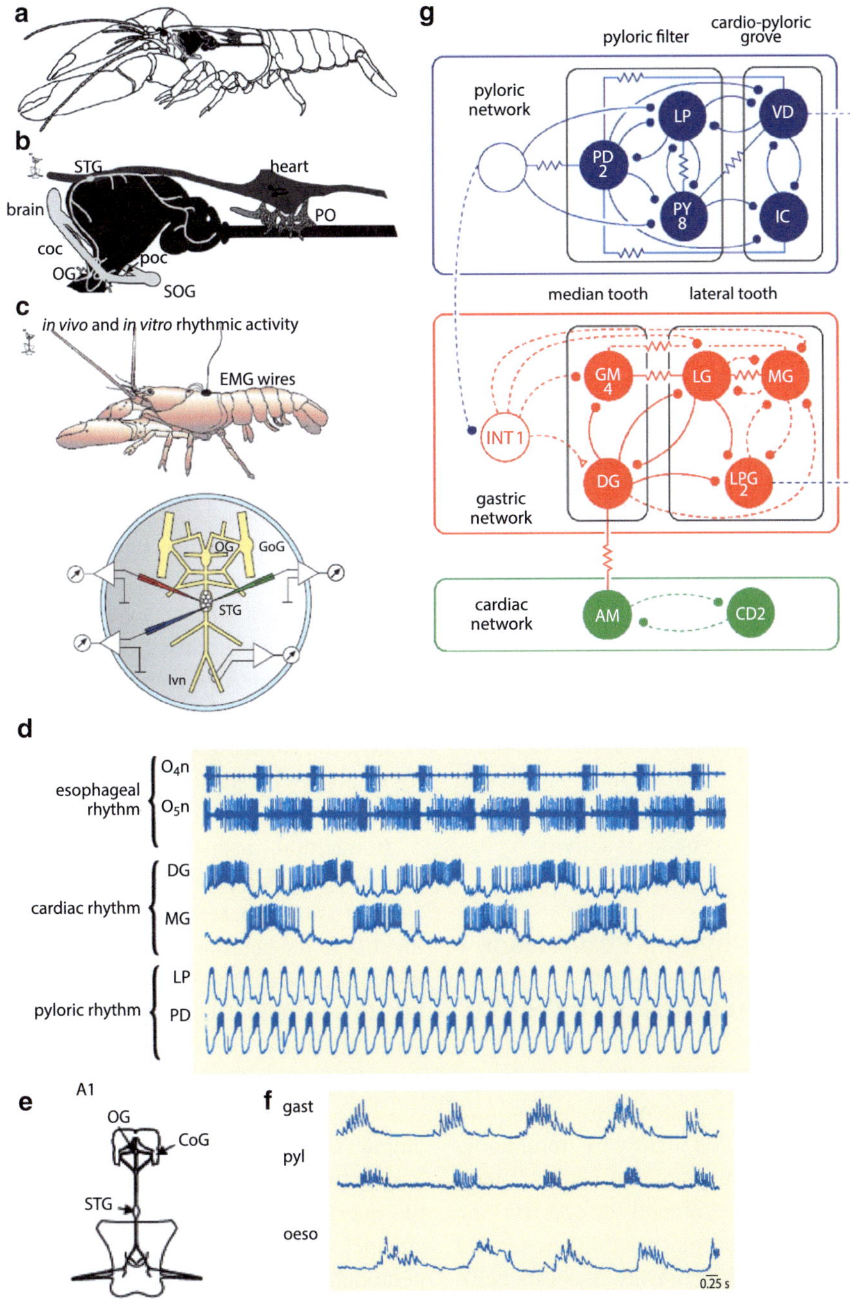


**Fig. 23.5 Stomatogastric system of crustaceans.** (**a**) Location of the stomatogastric system in a crayfish and (**b**) enlargement showing the arrangement of the CNS and the stomatogastric ganglion, the esophagus and stomach, and the heart (From Skiebe [59] with permission). (**c**) Recording the activity of muscles *in vivo* (*upper picture*) or from an isolated *in vitro* preparation of the stomatogastric nervous system in a petri dish (*lower picture*) showing three intracellular electrodes and one extracellular electrode (From Marder and Bucher [41] with permission). (**d**) The three major rhythms of the stomatogastric nervous system as recorded either extracellularly from nerves or intracellularly from motoneurons: esophageal rhythm (oeso), gastric mill rhythm (gast), and pyloric rhythm (pyl) (From Casanovas and Meyrand [7] with permission). (**e**, **f**) Outline of the embryonic preparation of the stomatogastric ganglion [13], and intracellular recordings from embryonic muscles (*first trace*, gastric muscle; *middle trace*, pyloric muscle; *lower trace*, esophageal muscle) reveal a common embryonic rhythm from which the adult rhythms are shaped by neuromodulatory input [7]. (**g**) The synaptic connections within the different subnetworks of the stomatogastric system. All synaptic connections are either made by electrical synapses (gap junctions) or chemical inhibitory synapses. The labeling in *filled circles* refers to motoneurons, in *open circles* to interneurons (Courtesy of H.G. Heinzel, University of Bonn). Abbreviations: *AB* anterior burster, *AM* anterior median neuron, *CD2* cardiac sac dilator neuron 2, *CoG* commissural ganglion, *DG* dorsal gastric neuron, *GM* gastric mill neurons, *IC* inferior cardiac neuron, *INT1* interneuron 1, *MG* median gastric neuron, *LG* lateral gastric neuron, *LP* lateral pyloric neuron, *LPG* lateral posterior gastric neuron, *lvn* lateral ventricular nerve, *OG* esophageal ganglion, *o4n* constrictor esophageal nerve, *o5n* dilator esophageal nerve, *PD* pyloric dilator neuron, *PY* pyloric neuron, *SOG* subesophageal ganglion, *STG* stomatogastric ganglion, *stn* stomatogastric nerve, *VD* ventricular dilator neuron

recordings from several of these somata simultaneously and hence the study of synaptic connections between neurons. As a result, it is known that all connections between neurons are either electrical (**gap junctions**) or chemical **inhibitory synapses**. Each of the principal rhythms is generated by a subset of neurons and their respective connections. The contribution of each neuron to the rhythm can be studied in this system by various methods: the neurons can be completely isolated from the rest of the network by pharmacological methods and/or by selective laser ablation of neurons; and the properties and activity of a particular neuron can be investigated using the "dynamic clamp", a method developed from the voltage-clamp technique in which computer-simulated currents can be inserted into the recorded neuron and then manipulated experimentally. In either case, if manipulations of a specific neuron or connection lead to alterations of the monitored rhythm, the contribution of an individual neuron to this rhythm can be ascertained. Thus, it was shown that some neurons of the STG display endogenous bursting properties. These neurons possess a particular set of ion channels that generate oscillations of the membrane potential resulting in bursts of spikes followed by periods of inhibition which gradually decrease prior to the next burst of activity.

### 23.5.2 The Different Rhythms Result from Many Neurohormones and Neuromodulators Acting on the STG

An important feature of the STG is that a large number of neuromodulators and neurohormones (see Chap. 11) have been identified mainly by immunocytochemical methods. It transpires that each of these chemical substances has profound effects on the rhythm and, in fact, the different rhythms generated by the STG are the result of the prevailing neurochemical "consommé" acting on all neurons of the network which possess receptor proteins for the respective chemical substances. For example, if **dopamine** is added to the STG, particular neurons of the pyloric rhythm including the anterior burster (AB) neuron fire more strongly whereas others fire more weakly or not at all. Alterations of the rhythms such as speeding or slowing the frequency, or slightly changing phase relationships are also the result of different neuromodulators or neurohormones which act on particular ion channels and thus change the electrical properties of neurons. As the STG is located within the head artery, all neurohormones circulating in the blood can potentially act on neurons or targets that possess the respective hormone receptors or that can easily cross the glial barrier ("blood-brain barrier"). Sources for the neuromodulators are neurons in the OG or CG which project axons via the stomatogastric nerve and, thus, deliver their transmitters directly to the STG. Other inputs to the STG come from sensory neurons that are associated with the muscles or apodemes. Whether patterns shown in isolation correspond to those of the intact animal and, thus, have biological relevance, is only known for some of the rhythms. In these cases the rhythm was recorded with extracellular electrodes from nerves while the animal was feeding, and thus a correlation between the patterns and the behavior could be made. An interesting approach was to visualize with an endoscope movements of the teeth of the gastric mill and then to describe how these movements change under the influence of different modulators.

In addition, in this system it has been possible to isolate particular neurons and by molecular methods such as single-neuron PCR, determine the molecular identity of the neuron, for example in terms of its complement of various ion channels [2]. An interesting, but still critically debated finding is that a "**homeostasis**" apparently exists with respect to the electrical properties of a neuron. Sets of different ion channels can be up- or downregulated depending on the total amount or density of particular ion channels which can vary between individual identified neurons. Therefore, the question was asked: how many variations in ion channel activity are required to produce a particular neuronal response? From such theoretical considerations it was postulated that a given neuron may have a surprisingly large number of options with regard to how to generate a particular physiological response via the expression of different combinations of ion channels in its membrane. The stomatogastric system has also made important contributions to our knowledge on the development of motor patterns.

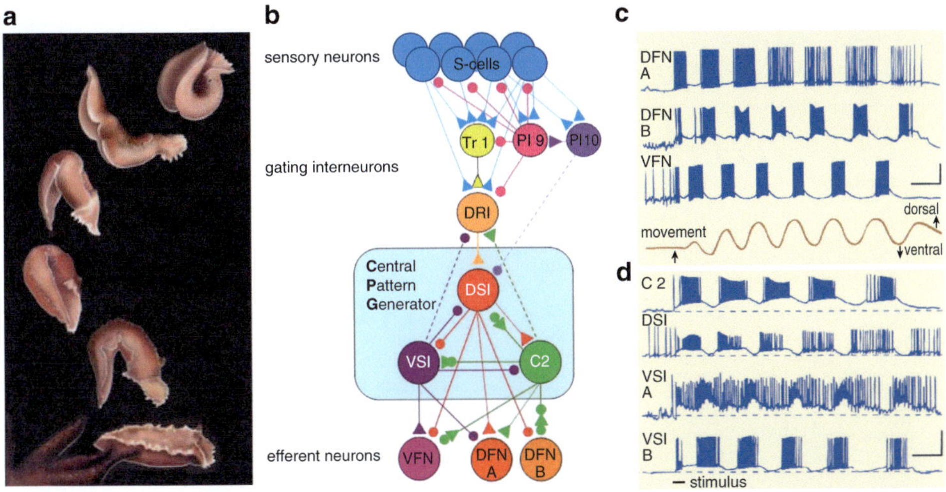


**Fig. 23.6 Swimming in the sea slug *Tritonia*.** (**a**) If an individual encounters a predator such as the tube feet of a sea star (star fish), it escapes by a series of ventral and dorsal flexion movements of the whole body that take it out of the reach of the predator (Drawing courtesy of Melisa Beveridge). (**b**) The swimming network consists of sensory neurons (S) which are chemo- and mechanosensitive and via glutamatergic synapses excite neurons (Tr1 and DRI) that trigger and gate the swimming pattern. The most powerful is DRI which can be considered a command neuron and makes excitatory synapses to the DSI neurons which are part of CPG. Via inhibitory P-neurons (P9 and P10) initiation of an escape swimming cycle can be prevented. The CPG itself consists of bilaterally represented types of neurons such as the cerebral neuron 2 (C2), three dorsal swim interneurons (DSI) and two ventral swim interneurons (VSI). The synaptic connections are either excitatory (*triangles*), inhibitory (*circles*), or more complex (*triangle and circle*). If DRI is activated by sensory input, it in turn activates the DSI neurons which activate the C2 neurons that feed back to DRI (positive feedback) and also eventually leads to activation of the VSI neurons which due to their inhibitory connections briefly interrupt swimming. The DSI neurons use serotonin (5-HT) as a transmitter and in addition to the point-to-point synaptic action also release it as an intrinsic modulator that causes longer lasting plasticity in the network. The *Tritonia* swimming network is an example of a network oscillator in which the synaptic connections are determining the network properties rather than the specific electrical properties of individual neurons (from http://www.scholarpedia.org/article/File:Tritoniaswim.jpg, home page of Paul Katz, Atlanta, USA). (**c**) Rhythmical discharges of motoneurons during swimming which is initiated by sensory stimulation (*arrow*). Activity of dorsal flexor neurons (DFNa and b) is alternating with activity in the ventral flexor neuron (VFN), Calibration: 40 mV, 5 s. (**d**) Intracellular recordings of interneurons of the CPG in the cerebral ganglion. The rhythm is initiated by electrical stimulation of a sensory nerve indicated by a *black bar* (lower trace). The DSI and C2 neurons fire more or less synchronously alternating with the VSI neurons. The *dashed lines* refer to the membrane potentials and show that neuron C2 is tonically depolarized during a swim cycle. Calibration: 50 mV (25 mV for VSIa), 5 s (Modified from Getting [19] with permission)

### 23.5.3 *Tritonia* Swimming

If the sea slug *Tritonia* wants to escape from a predator such as a sea star, it produces a series of 2–20 alternating dorsal and ventral bending movements that lift it from the substrate and propel it away from the predator (Fig. 23.6). This simple form of swimming movement is generated by muscles that are innervated by motoneurons located in the pedal ganglion, the dorsal and ventral flexion neurons (DFN and VFN), which fire rhythmically and in antiphase. These motoneurons receive monosynaptic inputs from three groups of interneurons located in the cerebral ganglion which form the CPG and produce the basic alternating rhythm: cerebral interneuron 2 (C2), dorsal swim interneurons (DSIs), and the ventral swim interneuron (VSI). Another important interneuron is the dorsal ramp interneuron (DRI) that is activated by sensory neurons through Tr1, a trigger neuron. DRI then activates the DSIs which are serotonergic and which in turn excite C2 and modulate its properties so that DRI firing is enhanced in a positive feedback loop [28]. DSIs and C2 are inhibited by VSI which interrupts the cycle. Although the *Tritonia* swimming system is amongst the simplest, it already incorporates many of the design features present in more complex motor control systems: **sensory trigger neurons, command neurons**, a CPG, feedback loops, and intrinsic neuromodulation.

### 23.5.4 Leech Swimming

Leeches start swimming by first flattening their body due to contractions of the dorsal and ventral musculature and then rhythmically bending their body in dorsal and ventral directions due to alternating contractions of the dorsal and ventral longitudinal muscles [17] (Fig. 23.7). This wave of contractions starts anteriorly and progresses posteriorly, and the forces exerted on the water provide the forward thrust. Swimming in the leech can be initiated by activation of sensory cells, for example the T ("touch") or P ("pain") cells which are repeated in each ganglion (serial homology), or by other mechanoreceptive cells of the leech skin which respond to water movements. These sensory cells connect to command neurons which in turn activate the swimming CPGs which are networks of neurons located in each of the 21 segmental ganglia of the ventral nerve cord. Each CPG consists of at least five pairs of individually identified interneurons (cells 27, 28, 33, 123, 208). Each

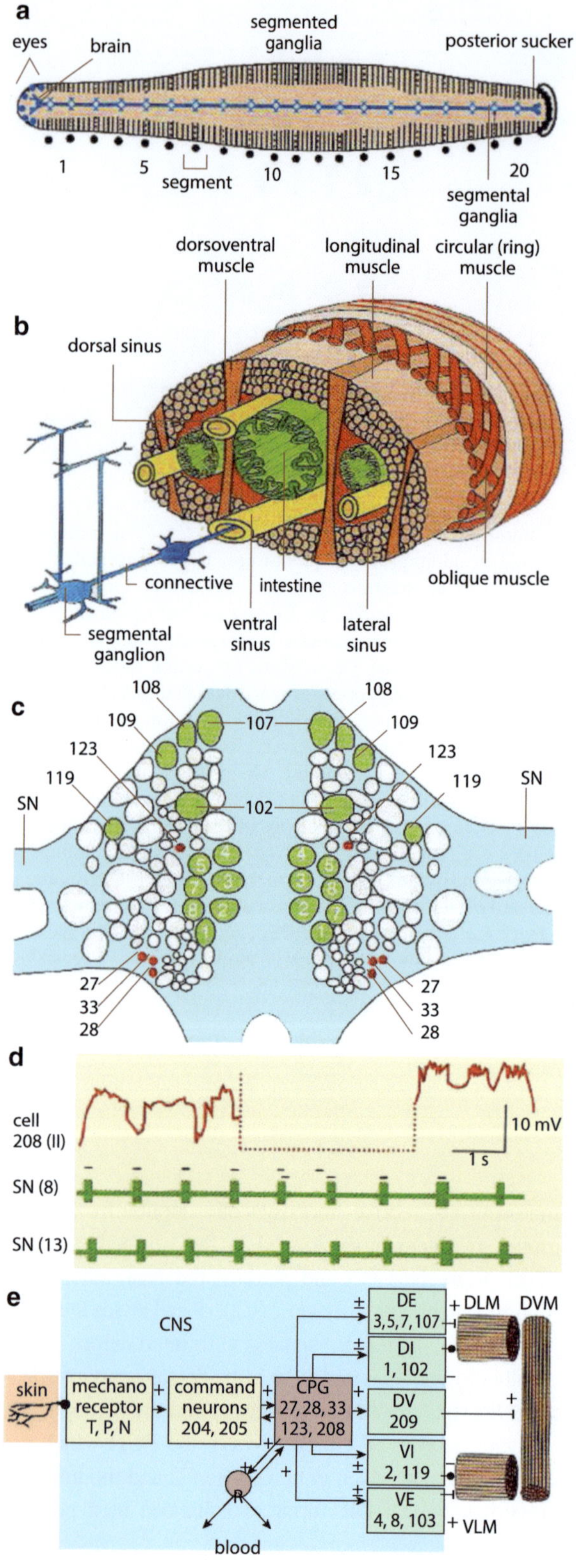


Fig. 23.7 **Swimming in the medicinal leech.** (**a**) Animal with its ventral chain of ganglia (ventral nerve cord) and peripheral nerves in each segment. (**b**) Schematic drawing of one body segment showing the different layers of muscles. (**c**) Dorsal view of a segmental ganglion showing location of identified neurons such as cell bodies of motoneurons labeled 1–8, and in *red* four pairs of interneurons of the CPG. (**d**) Activity of cell 208 of the CPG recorded intracellularly, and of motoneurons recorded extracellularly from peripheral (segmental) nerves (SN) of body segments 8 and 13. During the artifact shown in the trace of cell 208, this cell is hyperpolarized by current injection and this leads to a slight delay and slower motoneuron rhythm (see indicating *bars* above the bursts of activity). (**e**) Schematic drawing of motor control pathways (numbers refer to identified neurons, see **c**). DV, DE, VE, DI, and VI refer to motoneurons for the dorsoventral muscle (DVM) and excitors or inhibitors for the dorsal- (DLM) or ventral- (VLM) longitudinal muscle. R refers to the Retzius cell which by releasing serotonin modulates the CPG (Modified from Stent and Kristan (1981) with permission, in Kristan [32] and Müller et al. [48])

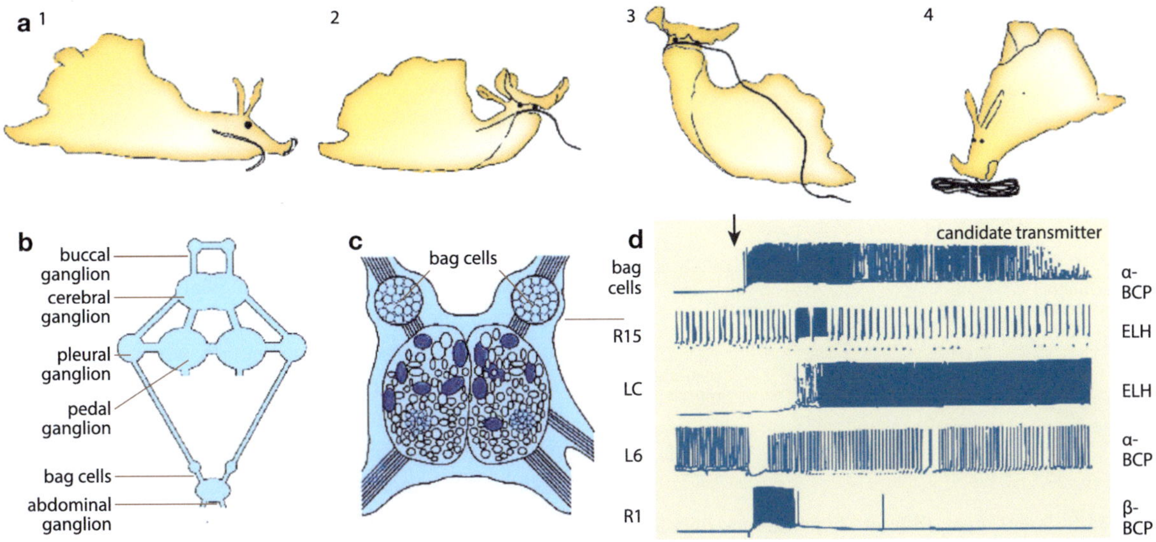


**Fig. 23.8** **Egg laying in *Aplysia*.** (**a**) Behavioral sequence: (*1*) the string of eggs is released from the gonadal opening, (*2*) and is captured by the mouth, and (*3*) further pulled out by swaying head movements, and (*4*) finally attached to objects by head movements. (**b**) Schematic drawing of the nervous system of *Aplysia*. (**c**) Ventral view of an abdominal ganglion showing bag cells (BC) and a number of identified neurons (in *blue*) which are targets for the BC peptide. (**d**) Short electrical stimulation (*arrow*) of BC leads to long-lasting activity such as two short high-frequency bursts in cell R15, an activation of cells LC (a motoneuron) and R1, and an inhibition of cell L6. The presumed transmitters are mentioned on the right margin (*α- and β-BCP* bag cell peptide, *ELH* egg-laying hormone)

interneuronal network generates a rhythmically alternating pattern of activation of the respective motoneurons within its own ganglion. Once such a CPG has been activated, it is able to maintain its activity without further sensory feedback. A swimming cycle starts with activation of the dorsal longitudinal muscles in anterior body segments by synchronous discharges of the four excitatory motoneurons in the respective ganglia. Subsequently the muscles are inhibited by two inhibitory neurons. This sequence of excitation/inhibition of the dorsal longitudinal muscles followed by excitation/inhibition of the ventral muscles now progresses in a coordinated fashion from anterior to posterior. Each cell of the CPG does not contribute equally to the pattern. For example, if the onset of the depolarization in cell 208 is experimentally delayed by hyperpolarization, the activity of the motoneurons will also be delayed. Therefore, cell 208 is able to reset the rhythm.

The morphology of many of the neurons of a CPG is such that they possess axons which extend over more than one ganglion to influence the CPG of the other ganglia and thus to coordinate the rhythm along the body of the leech. The CPGs are activated by special interneurons that act as command neurons (cells 61, 204, 205). If only one of the command neurons is stimulated, a swimming pattern in all ganglia will eventually be established, but with deficits in synchronization. One transmitter/modulator that plays an important role in inducing swimming is serotonin (or 5-hydroxytryptamine, 5-HT) which is released by several neurons in a ganglion, including the large **Retzius cells**, into the hemolymph. Serotonin acts on particular neurons of the CPG which possess 5-HT receptors. Application of serotonin induces swimming and if, for example, dopamine is applied to a leech ganglion, a neuronal network is activated that induces crawling [46]. This again shows the importance of neuromodulators for the activation of different networks.

### 23.5.5 Egg Laying and Feeding in Mollusks

**Neuropeptides** initiate egg-laying behavior in many mollusks such as *Aplysia californica* and *Lymnaea stagnalis* (Fig. 23.8). At the beginning of ovulation an *Aplysia* will cease to move and to feed. The fertilized

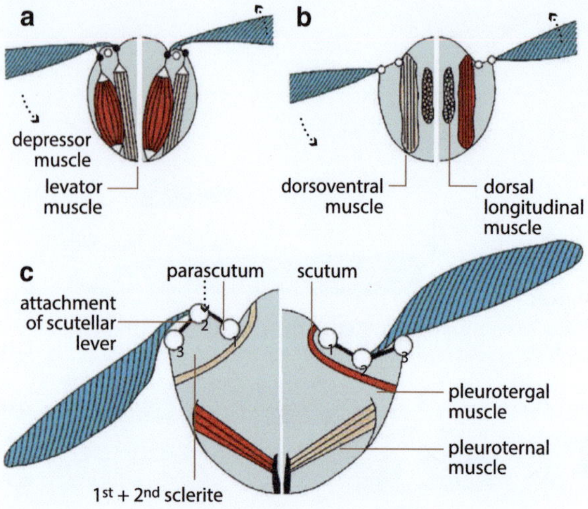


**Fig. 23.9 Insect flight: different flight motors.** In each drawing the conditions for wing depression (*down stroke*) and wing elevation (*up stroke*) are shown (*arrows*). Active (contracting) muscles are drawn in *dark red*. Direct (**a**) and indirect (**b**) muscles are drawn and either insert directly on the wing hinge (**a**) or move the entire thoracic cavity (**b**). (**c**) The indirect "click mechanisms" of dipteran flight (DLM and DVM not shown). Special "tension muscles" provide tension to the intermediary joint at points 1–3. The thorax is connected to point 2 vertically to the paper plane via a scutellar lever and transmits the forces of the thoracic cavity, and this way point 2 is bistable and oscillates between two positions (Modified from Nachtigall [49] with permission)

eggs are laid in strings from the genital pore, grasped by the mouth's upper lip, covered by sticky mucus and finally attached and distributed over the substrate and objects by oscillatory head movements. The firing patterns of the motoneurons supplying the corresponding muscles are rhythmic and more or less stereotyped. They are initiated by a peptide, the egg-laying hormone (ELH), which is synthesized and released from a cluster of cells near an abdominal ganglion, the "bag" cells (see Chap. 11). The bag cells only become active during egg laying and generate long-lasting bursts of action potentials. As is common for many other peptidergic systems, the first gene product is a "prohormone" or a "pre-ELH" from which the active hormone, ELH, is enzymatically generated, as for some other peptides. These peptides are released into the hemolymph when the bag cells themselves are activated by neurons descending from the cerebral and pleural ganglia. Their release into the hemolymph ensures that the corresponding peptides act on many organs: ELH itself influences the gonad (the ovotestis) and aids ovulation indirectly stimulating the follicle muscles. Within the buccal and abdominal ganglia ELH directly affects neurons: for example, the neurons involved in generating the motor patterns for feeding within the buccal ganglion are inhibited, whereas within the abdominal ganglia populations of neurons are either inhibited or activated. One of the peptides produced by the bag cells (the δ-BC peptide) stimulates the albumen gland which is important for producing the string of eggs. In addition, sensory feedback from the genital tracts is necessary for an appropriate egg-laying pattern. Peptidergic signaling occurs in many other behaviors (for example, feeding or molting) and many neurons may contain more than one peptide, which when tested individually may cause slightly different patterns. It is widely believed that peptides either released as cotransmitters from the same neurons or from separate neuron populations generate plasticity and induce flexibility in neuronal circuits. However, in many neuronal circuits the exact behavioral function is not known yet.

### 23.5.6 Insect Flight

During evolution the ability to actively fly has only been mastered by birds, bats, and insects, and may be one of the fundamental reasons for the evolutionary success of insects. Winged insects have been known since the Carboniferous (approx. 350 mya): some recent species of dragonflies still appear remarkably similar to those ancient forms. The evolutionary origins of insect flight are proposed to be an adaptation of segmentally paired gill-like appendages that originally functioned as ventilatory or swimming structures in aquatic forms. Supportive evidence comes from two sources: modern-day stoneflies use their wings not for flight but to skim across the water suggesting that aquatic ancestors may have used "protowings" as a novel nonaerial form of locomotion with flapping flight appearing subsequently [43]; interneurons that are part of the locust flight CPG appear as serially repeated homologs in the caudal thoracic ganglion which derives from the embryonic fusion of abdominal segments, suggesting that the flight system incorporates motor control components of ancestral abdominal structures [14].

### 23.5.7 The Flight Muscles

The forces for the wing beat are generated by contractions of powerful antagonistic muscles which reside within the thorax and produce the necessary lift and thrust (Fig. 23.9). Three general construction principles

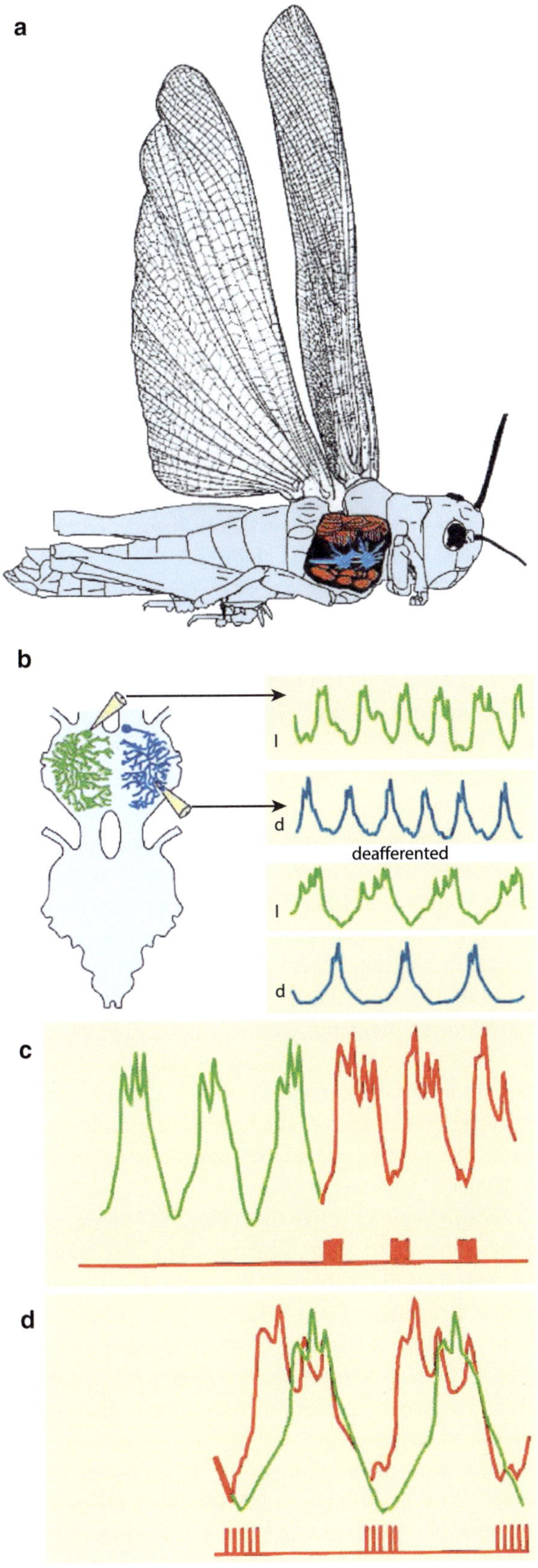


exist: (1) two sets of antagonistic flight muscles, for example wing elevators and depressors, are directly connected to the base of the wing (**direct flight muscles**). This organization is typical for insects with a low wing beat frequency such as dragonflies; (2) two antagonistic sets of muscles, for example the dorsal longitudinal muscles (DLM) as wing depressors and the dorsal ventral muscles (DVM) as wing elevators, are not directly connected to the wing hinge but indirectly via the thoracic walls (**indirect flight muscles**). This is typical for insects with fast wing beats such as beetles, bees, wasps, and flies. Mechanical elasticity of the cuticular exoskeleton of the thorax may play an additional role for the wing beats, as shown in flies; (3) the third type combines features of the direct and indirect flight musculature and is typical for butterflies and locusts. In addition to the powerful depressor and elevator muscles, some specialized and much weaker muscles that also insert on the wing hinge act as steering muscles. Due to these muscles the angle of twist of the wing can be altered (**pronation** or **supination**) and thus, lift and thrust can be altered as the angle of attack of the air flow across the wing is changed. Therefore these muscles are important for steering and in the control of yaw, roll, and pitch movements of the flying insect.

### 23.5.8 Neurogenic and Myogenic Flight

The aforementioned different structural types of flight muscles imply differences in how they are activated by motoneurons during flight (Figs. 23.10 and 23.11). For direct flight muscles or when an insect possesses a mixture of direct and indirect flight muscles, but

**Fig. 23.10** **Neural control of locust flight.** (**a**) Schematic view of a locust in which the right pair of wings has been removed to allow a view of the meso- and metathoracic ganglion of the thorax and the various wing muscles (outer row depressors, inner row elevators). (**b**) Intracellular recordings from an identified elevator motoneuron (l, traces 1 and 3) and from a depressor motoneuron (d, traces 2 and 4) during stationary flight with reduced sensory feedback intact (trace 1 and 2) and after deafferentation (traces 3 and 4). (**c** and **d**) Activity of an elevator motoneuron in an animal with removed tegulae and after electrical stimulation (stimuli indicated in second trace) of the tegula nerve (*red*). The superimposed traces in (**d**) show that the elevator motoneuron is activated earlier by stimulation of the tegula nerve (Modified from Wolf and Pearson [69] with permission)

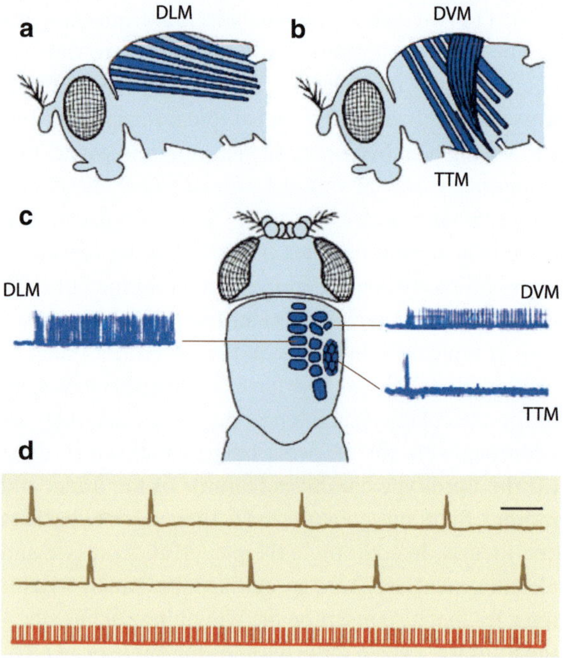


**Fig. 23.11** **Myogenic flight of dipterans.** (**a** and **b**) Position of the indirect, force-generating wing muscles in *Drosophila*: dorsolongitudinal muscle (DLM), dorsoventral muscle (DVM), and the tergotrochanteral muscle (TTM) which is the starter muscle. (**c**) Intracellular recordings from muscle fibers of TTM, DLM, and DVM at the start of a fictive flight sequence. In TTM only one action potential occurs at the beginning of flight. DLM and DVM produce action potentials throughout the entire flight sequence. (**d**) Action potentials of the dorsoventral muscles of the left (*upper trace*) and right side (*middle trace*). In the *lower trace* each wing beat is monitored optically, and each impulse corresponds to one wing beat. The wing beat frequency is much higher than the frequency of action potentials to the wing muscles

also when flying at a low frequency, each action potential of the motoneuron causes a muscle potential and hence a noticeable contraction. As most motoneurons of flight muscles belong to the fast type, one action potential causes a powerful twitch/contraction. The frequency of action potentials in the motoneuron determines the flight frequency and, therefore, these muscles are **synchronous flight** muscles and the flight motor is "neurogenic", i.e., generated by neurons. The indirect flight muscles of the fast-flying insects work differently: they develop peak forces at much higher frequencies than the frequencies of the innervating motoneurons. These muscles are out of synchrony with the discharge frequencies of the motoneurons (**asynchronous flight** muscles) and, therefore, the flight motor is called "myogenic" (as determined by the muscle itself). Due to this mechanism, these insects can produce wing beat frequencies from several hundred to over 1000 Hz. Movements of the thoracic walls cause the flight muscles to be stretched, which causes a powerful contraction of the passively stretched muscles (stretch-activated musculature; see Chap. 22), which in turn causes stretch of a different part of the thoracic wall that now activates the antagonistic muscles. These sequences of mechanical stretches determine the flight frequency. The discharges of the motoneurons are only used for maintenance of the system during flight.

### 23.5.9 Flight Control in Insects

A well-examined system is that of locust flight which combines direct and indirect flight muscles (Fig. 23.10). Usually flight starts with a jump, and by placing the fore- and middle legs at a particular angle the locust can influence its jumping trajectory. As soon as the animal has lost tarsal ground contact, and hairs on the head and frontal thoracic segment are deflected by wind stimuli and activate descending projection neurons, the wings are opened and the CPG for flight, the flight motor, which is comprised of interneurons in the meso- and metathoracic ganglion, is activated. In total approximately 40 groups of direct and indirect flight muscles which are controlled by about 100 motoneurons are involved in moving the wings. In addition, many more muscles, for example those of the legs, the abdomen, and body wall are active during flight and, as far as it is known, twitch at the flight frequency. During flight, elevator and depressor motoneurons produce 1–2 action potentials per wing beat and cause the respective muscle to contract. The hindwing precedes the forewing by about 5–10 ms.

Fast-flying insects like flies (Fig. 23.11) always start with a jump and a kind of catapult start. Two fast-conducting giant fibers descend from the brain (see **command neurons**). These fibers receive inputs from wind-sensitive head hairs of the body and from the eyes. In the mesothoracic ganglion they are connected via an interposed interneuron to the motoneurons of the DLM and monosynaptically to the motoneurons of a pair of tergotrochanteral muscles (TTMs). In *Drosophila* the giant fibers form electrical synapses with the TTM motoneurons. The latencies between the sensory stimuli and the corresponding muscle contractions are very short: only 0.8 ms to the TTM, 1.25 ms to the DLM, and between 1.9 and 3.3 ms to the DVM in *Drosophila*. An attractive hypothesis is that due to a contraction of the TTM, the tergum, the dorsal part, is pulled down which in turn may stretch-activate the DLM, that shortly after will receive additional input from the giant fibers. Thus, due to action of the TTM, the middle leg is pushed to the ground and catapults the fly into the air and, in addition, the TTM contraction also helps in activating the flight oscillator. Flight maintenance in the fly is achieved by the stretch activation of each antagonistic muscle determining the flight frequency and the less frequent neuronal activity of the motoneurons. In addition, the mechanical properties of the exoskeleton and the structure of the wing hinge contribute to the flight mechanism. Flight is terminated when the motoneurons stop discharging, for example following tarsal contact, and then the wings are refolded again.

#### 23.5.9.1 A CPG Underlies the Alternating Flight Pattern

The study of insect flight was long dominated by the finding in locusts that in the absence of sensory input (deafferented preparations) the remaining central neuronal structures are still able of producing a rhythmically alternating neuronal pattern [67]. Although the frequency was about half of that in the intact animal (10 Hz vs. 20 Hz), the phase relationships between antagonistic motoneurons were largely maintained. It is now clear that a basic alternating neuronal pattern which resembles a flight-like pattern (**fictive flight**) can be elicited in a completely isolated ventral nerve cord. For a long time it was believed that octopamine is the responsible transmitter for flight initiation, but increasing evidence suggests instead that it is acetylcholine with octopamine playing an important modulatory role. Octopamine clearly facilitates activation of the CPG.

#### 23.5.9.2 Sensory Control of Flight

It is now clear that all aspects of flight are under strong sensory control, e.g., from small hairs on the wings or hair fields on the body signaling air flow and turbulences, from chordotonal organs signaling movements between cuticular structures, from compound eyes monitoring visual flow and detecting polarized light, and from the ocelli detecting horizon cues. Proprioceptive control is used for the correct execution of the motor pattern, and the control by exteroceptors is used for steering and course control. In order to find out about the contribution of each type of sensory receptor some of them can be recorded extracellularly in tethered animals flying in a wind tunnel. Although this situation is still far from that experienced by a freely flying insect, it nevertheless gives important insights. For example, it was shown that a stretch receptor that determines the upper turning point of the wing, monosynaptically excites depressor motoneurons and via inhibitory interneurons inhibits elevator motoneurons, whereas the tegula controls elevation. Advanced or delayed stimulation of the respective sense organs influences the wing beat and its trajectory and, thus, the contribution of a particular sense organ can be studied.

#### 23.5.9.3 Motor Learning

Insects manage to fly under different load conditions, for example when the female carries eggs, or even when parts of the wings are missing. Thus, the insect must learn to apply a different motor pattern on each side. Approximately 40 muscle groups move the locust wings and it turns out that very precise and small temporal differences in the millisecond range between synergistic muscles are maintained and cause particular wing movement trajectories. Experimentally induced time changes by electrically stimulating a muscle lead to a disturbed flight. However, the locust is able to compensate for this by maintaining a pattern that cancels out the disturbance and now learns to fly with a slightly changed motor pattern on each side. Therefore, it seems that a locust by slightly varying the timing of flight muscle contractions is able to find out about the effects of this variation via its head sense organs, for example if a roll, pitch, or yaw deviation is induced. If, however, such a variation stabilizes flight, the locust is able to learn and maintain this new pattern within a few wing beats. This is an example of motor learning in an insect [47].

### 23.5.10 Postural Control and Walking in Arthropods: Insects and Crustaceans

Terrestrial locomotion, like walking on a horizontal surface or moving over an uneven terrain or climbing on trees and within leaves, involves a complex set of motor tasks that nervous systems face under these variable

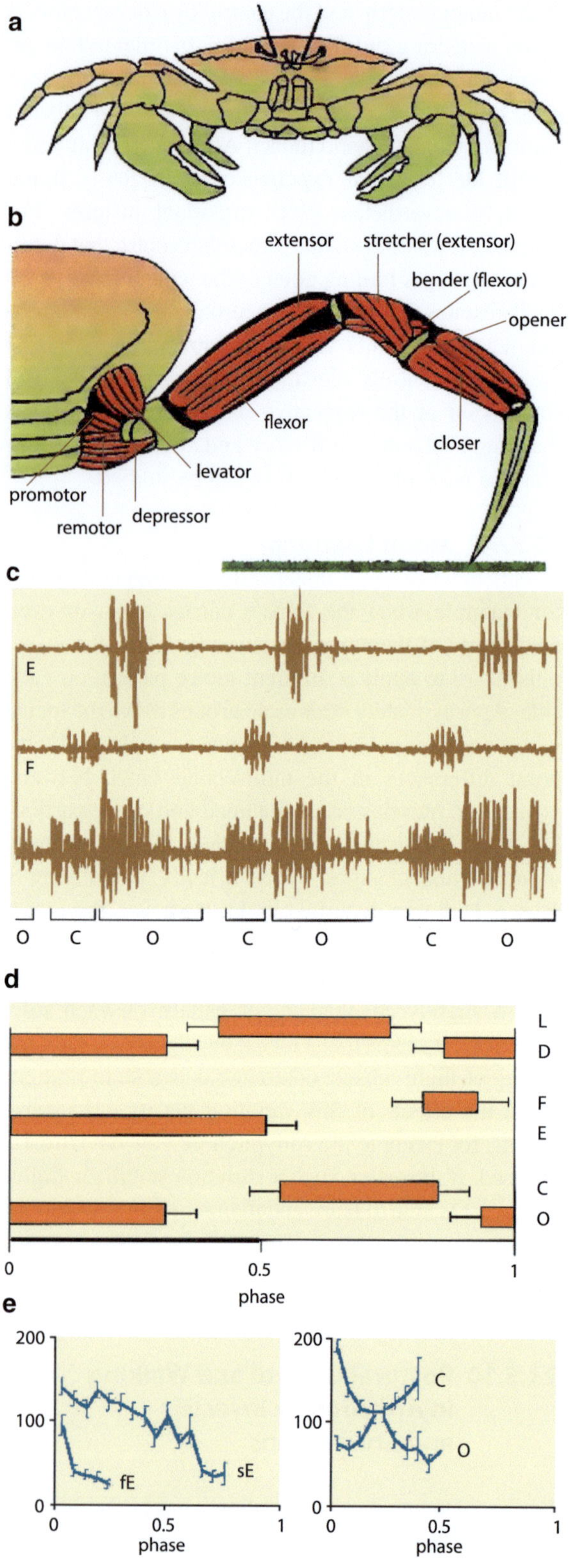


conditions and demands (Figs. 23.12 and 23.13). For the correct execution of such motor behaviors the contribution of sensory receptors is most important. Arthropods possess an exoskeleton that is well endowed with sensory receptors and they display a rich behavioral repertoire. Thus, different crustacean species such as crabs (Fig. 23.12), crayfish, or lobsters, and different insect species such as cockroaches, locusts, or stick insects are amongst the best-studied animals with regard to sensorimotor integration. The study of these animals is of considerable importance in the field of robotics because the design of new generations of multilegged automatons is based on arthropod animals. From observing the coordination of the six legs in a stick insect and monitoring the reactions to specific perturbations of the leg trajectories Cruse [10] formulated a set of simple rules (the "**Cruse controller**") which, if applied to a robot, enables simple horizontal walking. CPGs are also known to play a role in walking. In stick insects each joint is served by its own CPG. During leg movements these CPGs have to be coordinated by sensory input and mechanical coupling through the skeleton as well as by descending inputs from higher motor control centers of the subesophageal ganglion and brain.

Different types of mechanoreceptors signal various parameters such as the exact 3D position or the precise movements of an appendage, including velocity and acceleration (see Chap. 16). Among the proprioceptors that signal the animal's own movement the most

**Fig. 23.12 Arthropod walking.** (**a** and **b**) The shore crab *Carcinus maenas* in a frontal view showing appendages (**a**) and a schematic drawing of relevant muscles in a walking leg (b). (**c**) Simultaneous electromyogram recordings from the extensor (E), flexor (F), opener (O), and closer (C) muscle during pushing steps in freely sidewards walking crabs. (**d**) Score of muscular activities showing the duration of the stance phase (*bars* with standard deviation) of respective muscles and the phase relationships. (**e**) Mean discharge frequencies (ordinate) of four motoneurons, the slow and fast excitor to the extensor muscle, sE and fE, as well as the slow excitor of the closer (C) and opener (O) during pushing steps in sideward walking. During the stance phase the interburst frequency of action potentials is high for extensor motoneurons and the excitor of the opener in the beginning but then decreases. The opposite is true for the bursts of the slow excitor of the closer muscle. The discharge properties for homologous muscles of the trailing (pulling) leg are different. There, the interburst frequency remains constant during the entire burst (From Clarac et al. [8] with permission)

important are the campaniform sensilla that measure strain of the cuticle and, thus, act as load receptors. Multipolar sensory cells signal the tension of strands, apodemes, or special receptor muscles (stretch receptors). Up to several hundreds of receptor cells, so-called **scolopidial cells**, together with their supporting and accessory cells, form **chordotonal organs** which signal the movements of appendages. Exteroceptors signal changes in the environment. **Trichoid sensilla**, or "hairs", signal touch, and very flexible filiform sensilla are extremely sensitive to air vibrations or wind stimuli. The ones found on the cerci of crickets or cockroaches are so sensitive that they respond to the wind generated by the protruding tongue of a toad and often lead to successful escapes. The **subgenual organs** are highly sensitive to substrate vibrations and the **tympanal organs** to airborne vibrations. Both belong to the class of scolopidial organs. Most receptor systems are located in the head region, including the antennae with their important mechanoreceptive and olfactory functions, and the compound eyes and ocelli which provide information on different visual cues including polarized light.

In order to perform appropriate reactions to sensory stimuli, for example to precisely localize touch on the body surface, it is necessary that the body surface is mapped onto structures within the CNS, in the case of arthropods, the sensory neuropiles. A topographic organization has been described for extero- and proprioceptors, and a somatotopy already exists for the segmental ganglia of an insect ventral nerve cord. Usually many receptors of one or more modalities converge on first-order interneurons which, thus, possess complex receptive fields. The formidable task that the insect nervous system has to perform is that more than 10,000 sensory receptors of a typical thoracic segment with legs and wings map onto approximately 100 motoneurons. This shows the enormous amount of convergence that takes place. With respect to an individual sensory receptor, divergence to a number of first-order interneurons has also been described. Pathways of postural and movement control have been described in detail in the locust [5].

Coordinating several legs, as in a crustacean or in an insect (Fig. 23.13a, b), requires that information from one segment also reaches the adjacent segments. This is achieved either by the primary afferents themselves that project axons to and make synaptic connections within more than one ganglion, or via intersegmental interneurons that usually convey information from one ganglion to adjacent ganglia. The majority of sensory axons project and terminate in one ganglion, where synapses are made with either motoneurons or local interneurons that are confined in their branching pattern to the same ganglion. Although monosynaptic connections between sensory receptors and motoneurons occur, they do not seem to be the predominant connection. An exception is when, for example, a fast and powerful input from the wing stretch receptor activates a flight motoneuron. Most connections seem to be made with local interneurons of which two types exist (Fig. 23.13g). The local spiking interneurons ("**local spikers**") are first-order interneurons onto which many sensory neurons converge (for review see [5]). These interneurons connect the areas of the mostly ventral sensory neuropiles to the more dorsal neuropiles where the dendrites of motoneurons and premotor interneurons are localized. One large population of local spiking interneurons uses the transmitter GABA and is inhibitory. A separate population of local spiking interneurons is excitatory. The second type of local interneurons is the local nonspiking interneurons ("**local nonspikers**") which never produce action potentials. Instead they continuously release transmitter, the amount of which varies with the membrane potential. These interneurons are premotor interneurons as changes in their membrane potential recruit whole sets of motoneurons and in addition also connect with other local nonspiking interneurons. They are either excitatory or inhibitory. The actions of local nonspiking interneurons are important as they seem to organize the ensemble of interneurons and motoneurons that generate a meaningful and coordinated movement of a leg. Due to their graded release of transmitter and, thus, to the tonic influence on the membrane potentials of postsynaptic target neurons, they are involved in modulating, enhancing, or shunting other inputs to target neurons. Thus, the local nonspikers are involved in the modulation of muscular force and contraction velocity.

The signals from proprioceptors such as chordotonal organs are similarly processed. In postural control, the chordotonal organs are the sensors for the control loops of joints. The control loop of the femorotibial ("knee") joint in locusts and stick insects is amongst the best-studied control loops [4].

In posture each joint has a particular, energetically stable set point. Any passive deviation from this set point induces a movement that opposes this passive deviation. For example, if the tibia is passively flexed, the femoral chordotonal organ is stretched and its sensory receptors in turn activate the motoneurons of the extensor tibiae muscle and inhibit the motoneurons of the flexor tibiae muscle. As a result the tibia is actively extended until the set point of the femorotibial joint is reached again. Reflexes of this kind are called **resistance reflexes** (as they resist a passive movement).

If the tactile hairs on the dorsal side of an insect leg are stimulated, for example by a small brush, the leg is flexed, whereas stimulating the ventral hairs leads to leg extension. In both cases the leg is moved away from the stimulus so these reflexes are called **avoidance reflexes**. Such reflexes play an important role in maintaining posture. But how do these reflexes work during active movements when they might oppose the intended action? In active movements some of the reflexes are modified and may even reverse their sign. The cyclic leg movements during walking are characterized by the following two phases: the **stance phase** (retraction) is terminated when the posterior extreme position is reached and then during the **swing phase** (protraction) the leg swings forward until at the anterior extreme position the leg touches the ground and a new stance phase is initiated. Sensory signals are very important for precisely determining the points of transition between these phases and any sensory manipulation will either advance or delay initiation of the next phase. Very important for active movements are load sensors which signal the forces acting on the cuticle during a

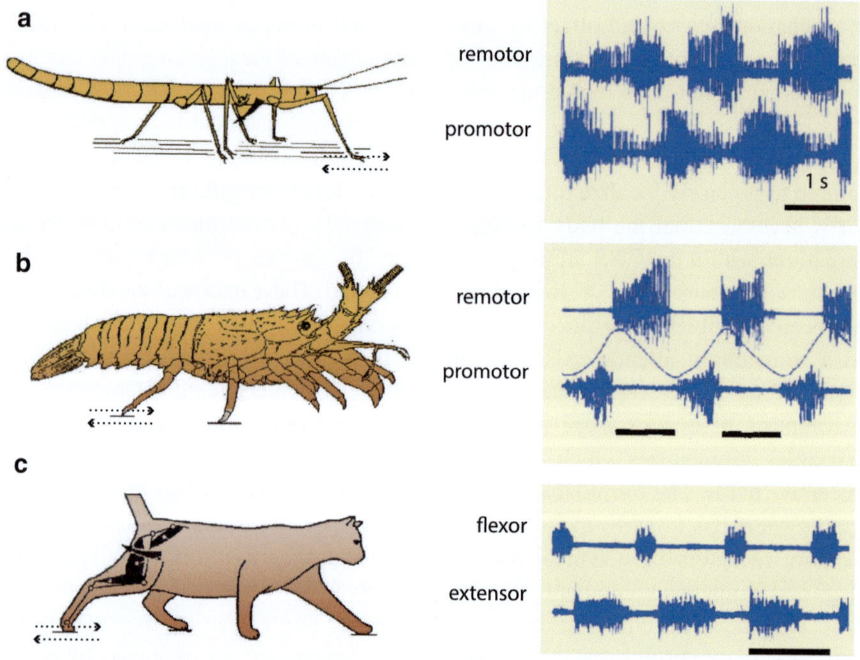


**Fig. 23.13** **Walking**. (**a–c**) Movements of individual legs of animals that are widely used in studies of locomotion [stick insect (**a**), lobster (**b**), cat (**c**)], and electromygrams of the respective muscles closely resemble each other (Courtesy of Ansgar Büschges, University of Cologne). (**d–f**) Electromygram recordings from two different muscles in a freely moving locust in different locomotory behaviors: horizontal walking (**d**), climbing (**e**), and walking upside-down (**f**), and on the right, scores of muscular contractions during the different locomotor situations in relation to the swing and stance phase (Adapted from Duch and Pflüger [11] with permission). (**g**) Schematic drawing of local sensory processing in a locust metathoracic ganglion for exteroceptors and proprioceptors of a hind leg. Main connections between major neuron classes are show. The *lines* indicate the strength of the connections, a *dashed line* refers to presumed but not yet clearly identified connections. + are excitatory connections, – are inhibitory connections. *AM* neurons of the anterior-median group, *INS* intersegmental interneurons with axons to adjacent ganglia, *MN* motoneurons which innervate antagonistic muscles such as flexor and extensor, for example, *NSLIN* nonspiking local interneurons, *SLIN* spiking local interneurons, *VML* neurons of the ventral midline group (From Burrows [5] with permission)

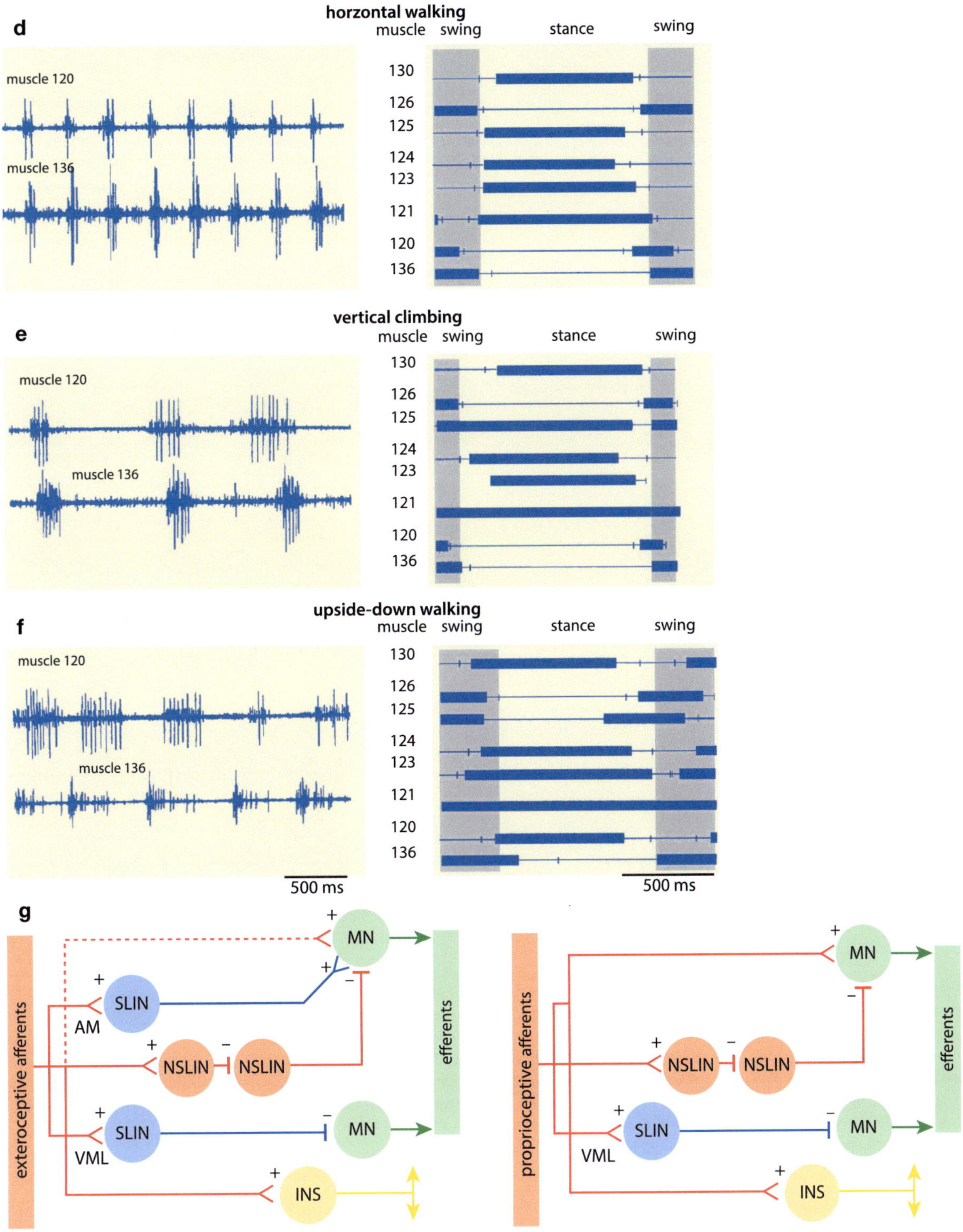


**Fig. 23.13** (continued)

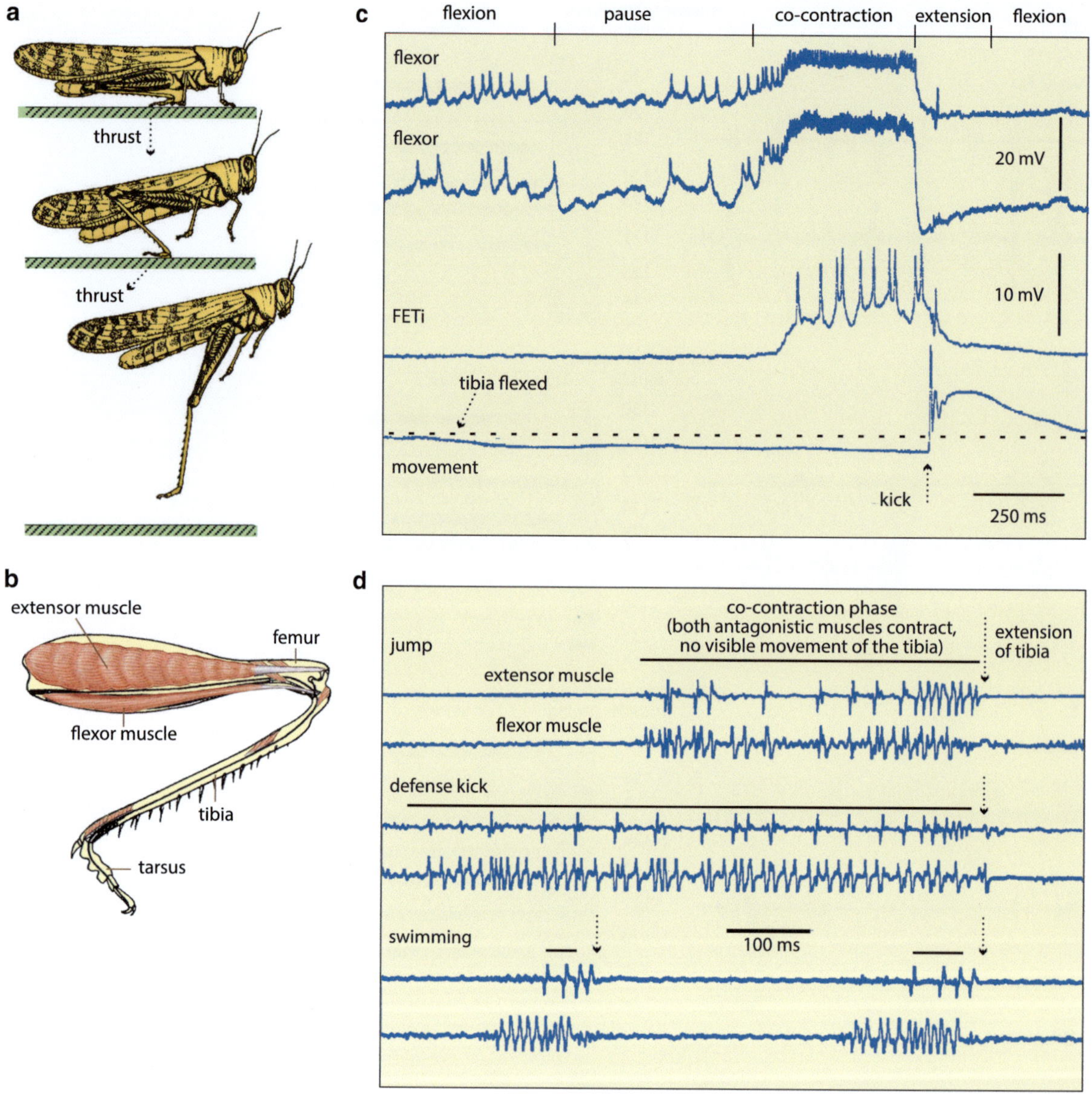


**Fig. 23.14** **Locust jump**. (**a**) A locust jump showing the hind leg providing the thrust against ground and lift off. (**b**) The muscles in a hind leg of a locust are highly adapted to their tasks. The most powerful muscle is the extensor tibiae muscle, whereas the flexor tibiae muscle is much weaker. In addition, part of the retractor unguis which moves the claw is also located in the femur. (**c**) Intracellular recordings from two flexor motoneurons (*first* and *second trace*) and the fast extensor tibiae motoneuron (FETi, *third trace*), and movement of the tibia (*fourth trace*). The different phases of the special jump motor program are indicated on the top. *Stippled arrows* in **c** and **d** point to the rapid tibial extension.(**d**) Electromyograms from freely moving locusts in different motor behaviors (defense kick, jump, and swimming) which all show the special feature of cocontraction between the two antagonistic muscles, flexor and extensor (From Burrows [5] with permission)

movement. These sensors are cutaneous sensilla, such as the campaniform sensilla, and the information they provide is important for controlling the strength of a muscular contraction, for example how many motor units are recruited. To ensure that reflexes will only exert their influence at the appropriate movement phase and time (phase-dependent reflex), the sensory signals are modulated by copies of the efferent pattern of the CPG (**efferent copy**). At all other times, the signals are prevented from reaching the postsynaptic neurons by presynaptic inhibition. In addition, many terminals of exteroceptors of arthropods receive presynaptic inhibition mediated by their immediate sensory neighbors in a kind of automatic gain control mechanism. It is a general problem for animals to differentiate between sensory information deriving from self-generated movements and external stimuli. In the auditory pathway of crickets a corollary discharge neuron inhibits its auditory interneurons while the cricket is producing its own songs (see Chap. 17).

### 23.5.11 The Locust Jump

The hindlegs of grasshoppers and locusts are specialized for jumping (Fig. 23.14). If locusts use their hindlegs for walking the two principle muscles, the flexor and extensor tibiae, strictly alternate with their activity. The extensor muscle is only innervated by four neurons, a **fast extensor tibiae motoneuron** (FETi), a **slow extensor tibiae motoneuron** (SETi), a **common inhibitory neuron** (CI), and an **octopaminergic neuromodulatory neuron** (DUMETi). The activity of some of these neurons can be recorded by electromyograms or intracellularly from a dissected and semi-intact locust. For walking the SETi motoneuron is dominant. Only during fast movements a few action potentials of FETi may occur. How about jumping or strong defensive kicks that locusts use to rid themselves of disturbing objects? These two behaviors require a totally different motor pattern that is comprised of different distinct movement phases: an initial flexion of the tibia is followed by a cocontraction period of both the flexor and extensor muscle in which the tibia does not move but force is generated isometrically and stored in a distortion of the whole femorotibial joint, the "knee-joint" of the locust [22]. This period can last from 100 to several hundreds of milliseconds. Only when the flexor muscle stops contracting due to a sudden inhibition of its many (>9) motoneurons, and the extensor muscle continues to contract, then the force is suddenly released and catapults the locust into the air. Besides the neuronal pattern, several mechanical properties aid the jump mechanism: first, the leverage is such that at very flexed joint angles the much weaker flexor has a mechanical advantage over the contracting extensor muscle, as its tendon hooks over a "lump" [23] and thus is able to keep the tibia flexed despite vigorous contractions of the extensor muscle. Disengagement can only be achieved if the flexor tibia muscle relaxes and the joint is slowly extended again. Second, and in addition to the special structure of the apodeme, the joint itself has interesting structural properties and also may contain a protein, **resilin**, that is found in many other insects where forces are stored, for example in dragonfly wings. Thus, if the release of the tibia has been triggered by inhibition of the flexor muscle, the tibia is extended slowly until at a certain angle the distortion of the joint is suddenly reversed and the stored forces released. As one can imagine the femorotibial joint bears many different sense organs that monitor all aspects of this particular behavior [23]. In fact, cuticular campaniform sensilla may be able to exactly monitor the isometrically generated forces, and special multipolar sensilla of the joint may serve as safety sensors as they only discharge at very extended joint angles and, thus, may prevent breaking of the joint. Despite the enormous precision of a jump, the two hindlegs are controlled separately by presynaptic neurons and do not seem to be coupled other than mechanically via the exoskeleton.

### 23.5.12 CPGs in Vertebrates

As for invertebrates, most rhythmic motor behaviors in vertebrates rely upon the coordination and contractions of skeletal muscles arranged as functionally antagonistic pairs which operate the different joints of the limbs or skeleton. There is also strong evidence for the existence of CPGs to control these rhythmic motor behaviors of vertebrates. In particular, central networks are capable of generating the rhythmic motor commands that support both respiratory and locomotory rhythms. Furthermore, there is evidence that the respiratory and locomotory CPGs are functionally linked at the level of the CNS, presumably to ensure that increases in locomotory behavior lead to a commensurate increase in the respiratory drive to provide the required extra oxygen. On a simple level, the CPGs for breathing and locomotion are similarly organized in terms of synaptic interactions: in each there is a core network of interneurons that utilize glutamate as the excitatory transmitter to drive the motoneurons. This group of excitatory interneurons also make connections amongst themselves, connections which contribute to the maintenance of the rhythm. Inhibition in these networks is primarily mediated by interneurons that release glycine.

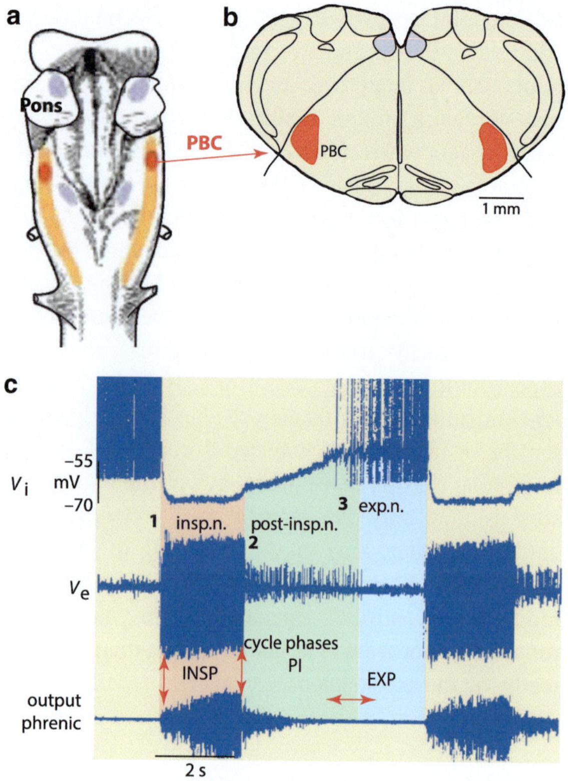


**Fig. 23.15** **Respiratory CPG**. The respiratory CPG in mammals is located in the medulla of the brainstem (**a**), just caudal to the pons. Although there are several dorsal and ventral respiratory neuron groups (DRG, VRG) involved in generating the respiratory rhythm, the pre-Bötzinger complex in the VRG is particularly important (PBC, *red label*). In slice preparations (**b**), the PBC is able to generate a respiratory rhythm in which inspiratory and expiratory neurons fire bursts of action potentials in alternation. In (**c**), from an intact nervous system, the *top trace* is an intracellular recording from an expiratory neuron. The *middle trace* is an extracellular recording of inspiratory-related neuron activity which includes inspiratory neurons and postinspiratory (PI) neurons that fire before the next expiratory bursts. The *bottom trace* represents network output recorded from the phrenic nerve (Adapted from Richter and Spyer [53] with permission)

### 23.5.13 Breathing

Breathing represents a special case in motor control because unlike locomotion, which is usually expressed "on demand", the organism is obliged to generate a continuous respiratory rhythm throughout its life in order to survive. For air-breathing mammals respiration primarily involves active inspiration, produced by rhythmic contractions of the diaphragm, and under low oxygen demand the expiration phase of breathing is largely passive. An important component of the CPG controlling inspiration is located in a highly localized, bilaterally symmetrical region of the ventrolateral medulla of the brainstem, called the **pre-Bötzinger complex** (Fig. 23.15). Even comparatively thin slices of the isolated brainstem that contain the pre-Bötzinger complex are able to generate a respiratory rhythm. Respiratory bursting in these isolated preparations persists even when synaptic inhibition mediated by glycine is blocked pharmacologically using strychnine, suggesting that interneurons in the pre-Bötzinger complex may possess intrinsic oscillatory membrane properties. Recent evidence suggests a neighboring region of the medulla, incorporating the retrotrapezoid nucleus/parafacial group contains a separate CPG for expiration that is normally silent, but which is recruited under conditions that simulate heavy exercise, a situation when active expiration is needed to ensure that oxygen supply is maintained.

Breathing is subject to modulation from a variety of sources, perhaps most significantly from the brainstem **raphe nuclei** which contain neurons that release serotonin. The effects of serotonin are to facilitate the respiratory rhythm. The conditions under which the raphe system is brought into play are not fully understood, but it is likely that this modulatory system responds to increased concentrations of $CO_2$ in the blood and, in keeping with this, the processes of serotonergic neurons in the brainstem seem to faithfully track the vasculature, as if these neurons "sniff" $CO_2$. Thus these serotonergic raphe neurons may provide the critical link between oxygen debt and enhanced respiratory output [54]. Sudden infant death syndrome (SIDS) may in part be linked to developmental deficits in the efficacy of this link.

### 23.5.14 Axially-Based Swimming in Fish and Tadpoles

The simplest form of locomotion in vertebrates involves the side-to-side oscillations of the muscular tail during axially-based swimming locomotion, for example in fish and amphibian tadpoles (Figs. 23.16 and 23.17). This type of movement is also displayed by primitive vertebrates like lampreys and hagfish, and chordates like *Amphioxus*, and even the tadpole phase of tunicates such as *Cione*. Therefore, axial swimming is probably the ancestral form of locomotion in all chordates. In each case the movement is generated by two components that unite to generate thrust. The left-right alternation of **myotomal muscle** contraction

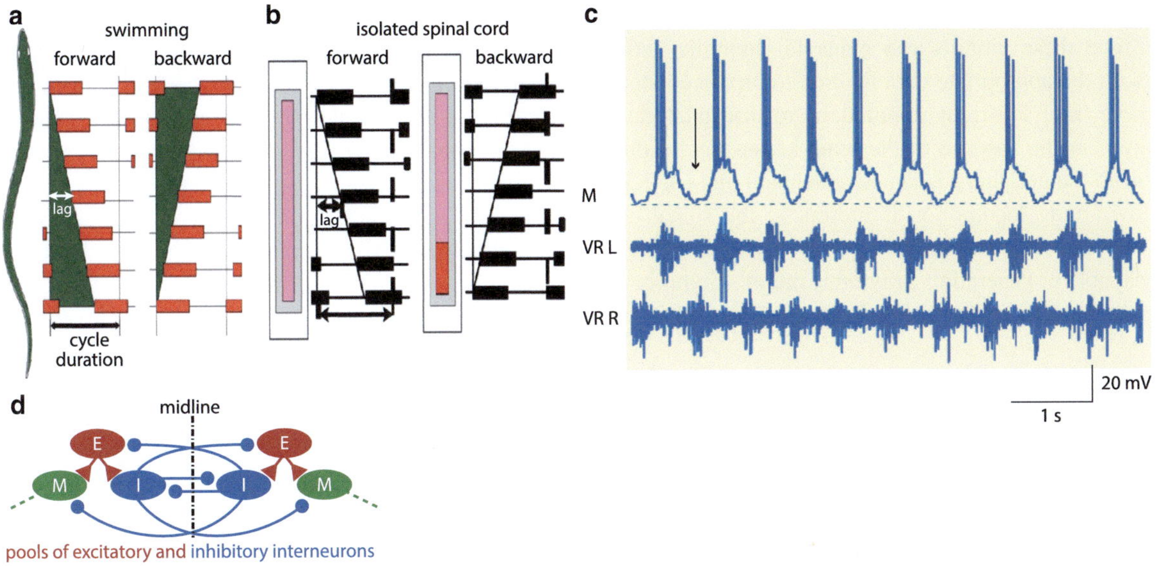


**Fig. 23.16 Lamprey swimming CPG.** The lamprey, a primitive agnathan vertebrate, has been extensively used as a model for the neural control of adult vertebrate locomotion. Lampreys are anguilliform swimmers (**a**) using oscillations of the body to generate forwards propulsion with a head to tail lag (*green*) in muscle contractions (*red*), or more rarely backward swimming with a reversed tail to head lag. The isolated spinal cord (**b**, *pink*) will generate "fictive" locomotion (**b**, *black bars* represent motoneuron bursts), with similar phase lags, when perfused with the glutamate receptor agonist, NMDA. Higher activation of caudal cord (*red* in schematic) produces fictive backward swimming. During CPG activity (**c**) the left and right ventral motor roots alternate (VR L, VR R). (**c**) Motoneurons (M) fire bursts of action potentials followed by periods of rhythmic mid-cycle inhibition (e.g., at *arrow*). The basic network that generates swimming (**d**) includes glutamatergic excitatory interneurons (E, *red*) providing ipsilateral drive and glycinergic inhibitory interneurons (I, *blue*) providing reciprocal inhibition that couples the two sides of the cord in alternation. *Dashed line* in (**d**) represents spinal cord midline (Adapted from Grillner [20] with permission. Data in (**c**) kindly provided by Abdel El Manira)

produces a local bend of the body near the front end which is then propelled rostrocaudally along the length of the body, first along one side and then along the other. It is the back-propagating region of body curvature that produces reactive thrust against the water column to produce forward propulsion.

In fish it has been demonstrated that the peripheral muscles are heterogeneous in their contractile properties possessing at least two different muscle fiber types with most teleosts possessing three (called red, white, and pink). The bulk of the fish's trunk comprises fast twitch white muscle fibers which have low concentrations of the oxygen-carrying pigment myoglobin, few capillaries, and few mitochondria. These fibers are generally larger than other fiber types and they rely on anaerobic glycogenolysis for their energy supply. They are readily fatigueable and generally only recruited during the fastest of swimming movements such as during escape or predation. At the lateral-medial border of the trunk of the fish is a thin band of red muscle fibers with high concentrations of myoglobin, numerous capillaries, and rich in mitochondria. The red muscle is fatigue-resistant and generally used during low frequency cruise swimming. Teleosts possess a third interpolated band of pink muscle fibers with intermediate properties. In these fish red muscle alone is active at low swimming speeds but as the speed of swimming is increased first the pink and then the white muscle fibers are recruited [26]. The red and white muscle fibers are innervated by motoneurons of different soma diameters and their orderly recruitment adheres to Henneman's size principle with smaller motoneurons innervating red muscle active during low swimming speeds and larger motoneurons innervating white muscle only recruited during rapid swimming movements. It seems wasteful and cumbersome to dedicate the majority of the fish's body to white muscle that is only used during escape behavior, but this underlines the survival value of escape. In addition, the heavy weight of the muscle is less of a hindrance to a neutrally buoyant aquatic organism compared to an animal that uses aerial locomotion, for example.

The underlying neuronal networks have been studied in detail in adult lampreys [20], developing

*Xenopus* frog tadpoles [55] and zebrafish larvae [44]. Since these circuits can generate rhythmic activity with the appropriate coordination to produce locomotion (i.e., left-right alternation, rostrocaudal delay) even in the absence of the brain or sensory feedback from the periphery it can be concluded that a CPG for swimming locomotion is present in the spinal cord. The circuit is organized as two "half-centers", each capable of generating a rhythmic output. Commissural interneurons in the spinal cord are involved in coupling the two sides in alternation due to reciprocal glycinergic inhibition. The excitation is generated by interneurons with processes that are restricted to one side of the nervous system. In tadpoles the excitatory interneurons extend into the caudal hindbrain [55]. The excitation is primarily mediated by the amino acid glutamate acting at two different types of receptor on motoneurons; AMPA receptors generate a fast excitation while NMDA receptors generate a somewhat slower response that in some cases can sum from one cycle to the next to produce a sustained depolarization during swimming activity. The excitatory interneurons have been shown to excite each other and so they form a positive feedback network in which reverberating excitation is responsible for maintaining the rhythm once initiated. The basic swimming circuit appears to be very similar across the different vertebrate groups, and also at different stages of development, suggesting that much of it has been conserved during evolution.

Swimming vertebrates must adapt their movements in response to afferent inputs. Exteroceptive, unpredictable sensory stimuli produce modifications to swimming activity that are strongly dependent on when they occur in relation to the phase of the movement cycle. This phase dependence is due to interneurons that perform a gating role to ensure that the reflex response reinforces movement in one phase but which is gated out to avoid conflicting movements in the other phase. Proprioceptive inputs are able to entrain the locomotor CPG output generated by the lamprey spinal cord [21]. Indeed sinusoidal movements of the isolated spinal cord can still entrain fictive swimming

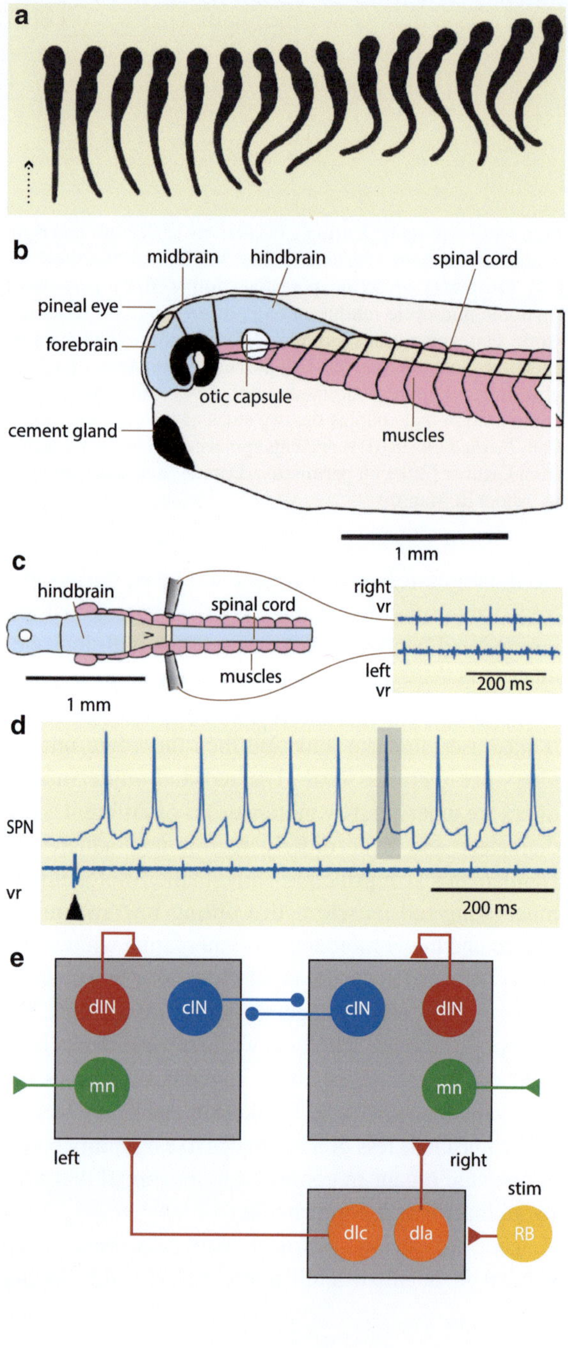


**Fig. 23.17 Tadpole swimming.** Around the time of hatching *Xenopus* embryos can already generate coordinated swimming movements (**a**), similar to the adult lamprey (see Fig. 23.16). (**a**) The first cycle (~50 ms) of a longer swim episode. Swimming is generated by neural circuitry in the CNS shown in schematic form in (**b**) which generates rhythmic activity in spinal motoneurons that in turn activate segmented myotomal swimming muscles. Fictive swimming in immobilized animals (**c**) can be recorded from ventral roots (vr) with electrodes positioned in clefts between segmented muscle blocks; activity alternates between left and right sides. (**d**) Spinal neurons (SPN) fire an action potential in each cycle (e.g., *shaded area*) and receive inhibition when the opposite side (vr) is active. (**e**) *Large gray boxes* show basic spinal CPG circuit that generates swimming activity in motoneurons (mn). dINs are ipsilateral glutamatergic excitatory interneurons, cINs are commissural glycinergic interneurons. RB, dlc, and dla are sensory pathway neurons that trigger swimming in response to stimulation (e.g., at *arrow head* in (**d**)). Note similarity between tadpole and lamprey (Fig. 23.16) CPGs (Adapted from [3, 25], all with permission)

suggesting that the proprioceptors must be intraspinal neurons. In fact the proprioceptors are called **edge cells**, which are located close to the lateral margins of the dorsoventrally flattened spinal cord and they possess end feet at the very edge of the lateral margins of the cord (Fig. 23.18). Edge cells respond to mechanical stretch of the cord as would occur during real swimming. They also make appropriate synaptic connections in the spinal cord that explain their entraining function by exciting neurons on the same side but inhibiting neurons on the opposite side. Computer simulation studies in which the effect of removing edge cells from a network driving simulated swimming suggest that the role of the edge cells is to assist navigation in conditions with turbulent flow. When the edge cells are removed the simulated lamprey is unable to cross a region of turbulent water flow.

## 23.5.15 Postural Control and Walking in Vertebrates

Tetrapod locomotion is displayed by members of several vertebrate groups including amphibians and mammals, with most known about the latter. During a cycle of walking each of the four limbs moves to and fro, switching between the stance phase and the swing phase (Fig. 23.19). The two phases appear to be subject to different control principles because the stance phase is variable in duration, matching variations in the cycle frequency, but the swing phase is generally more constant in duration. This difference can be attributed to the fact that during stance phase the limb plays the dual role of generating thrust against the environment and also bearing the load of the body. Thus, when load increases, the relative duration of the stance phase also increases. The four limbs are precisely coordinated during locomotion in order that the weight is evenly borne across the body. As a result, quadrupedal vertebrates adopt a gait during walking in which the front left and right rear limbs move forward approximately in sequence, alternating with the front right and rear left, and so on.

To accommodate different walking speeds many motor systems have evolved multiple gaits in which the interlimb coordination changes according to the speed of locomotion (Fig. 23.19b). The need for gait changes relates to a requirement to distribute the load evenly over the body. In quadrupeds, for example,

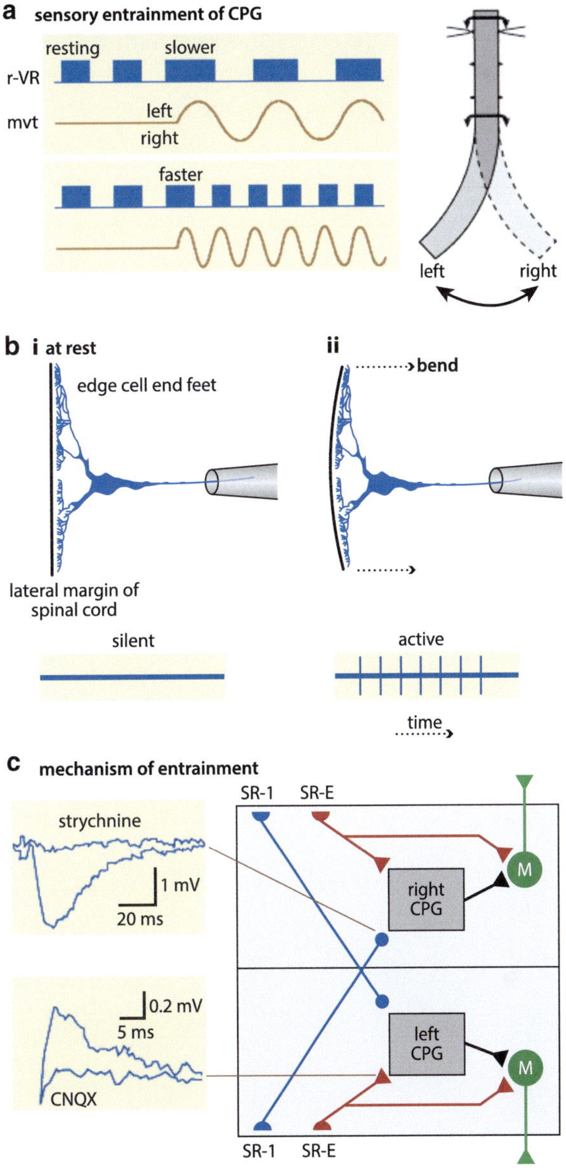


**Fig. 23.18 Entrainment in lamprey.** Sensory feedback during swimming entrains the CPG output in the lamprey. Entrainment, either faster or slower than the resting rhythm, occurs when the isolated spinal cord is moved sinusoidally (**a**) suggesting that the cord contains proprioceptive neurons that respond to local bending. The spinal cord of the lamprey is dorsoventrally flattened. (**b**) Schematic showing lateral margins of the cord contain the sensory end feet of edge cells (**bi**) that respond to mechanical stretch of the cord (**bii**). (**c**) Edge cell stretch receptors are either inhibitory (SR-I; effects blocked by strychnine) or excitatory (SR-E; effects blocked by CNQX) and make appropriate contact with the left and right sides to provide positive feedback excitation to the side of the body that is bending ((**a**, **c**) Adapted from Grillner et al. [21] with permission)

there are normally four gaits, and as the speed of locomotion increases the animal switches from walking to trotting or pacing and finally to galloping. In each gait the swing phase duration, when the limb is in the air, is relatively constant but the stance phase decreases as speed increases. Also, the coordination between the limbs is modified so that a pair of limbs alternates across the body during walking but becomes nearly synchronous during galloping. In well-trained Icelandic horses there is even a fifth gait called the tölt in which alternation is maintained and the duration of stance and swing are approximately even; the horse's back appears to be stable and the horse and rider "glide" over the ground. The neural basis of the switch between gaits is only partly understood. The source of the descending drive for locomotion includes the **mesencephalic locomotor region** (MLR) of the brainstem. When the MLR is stimulated, locomotion is initiated and when the intensity of MLR stimulation is increased this can trigger switches in the locomotor gait. Gait changes are a feature of all types of limbed locomotion: bipeds (e.g., birds and humans), hexapods (insects), and decapods (crustaceans) utilize a variable number of limbs and gaits to move over the terrain. In fish, gait changes are less common because these animals have buoyancy and so the need to distribute body load is effectively eliminated. Trout, however, display a "Karman" gait when holding position behind objects in a flowing water column [37]. This energy-saving strategy harnesses the kinetic energy of the water as the fish "bounces" off eddies and responds by reducing muscle contractions in caudal segments.

The two phases of walking (Fig. 23.19) are controlled in part by sensory feedback from proprioceptors such as 1a muscle spindle afferents of the hip flexor muscle that monitor extent of hip extension and 1b Golgi tendon organ afferents of the hip extensor that monitor limb loading. The former contribute to the timing of the transitions between phases via negative feedback,

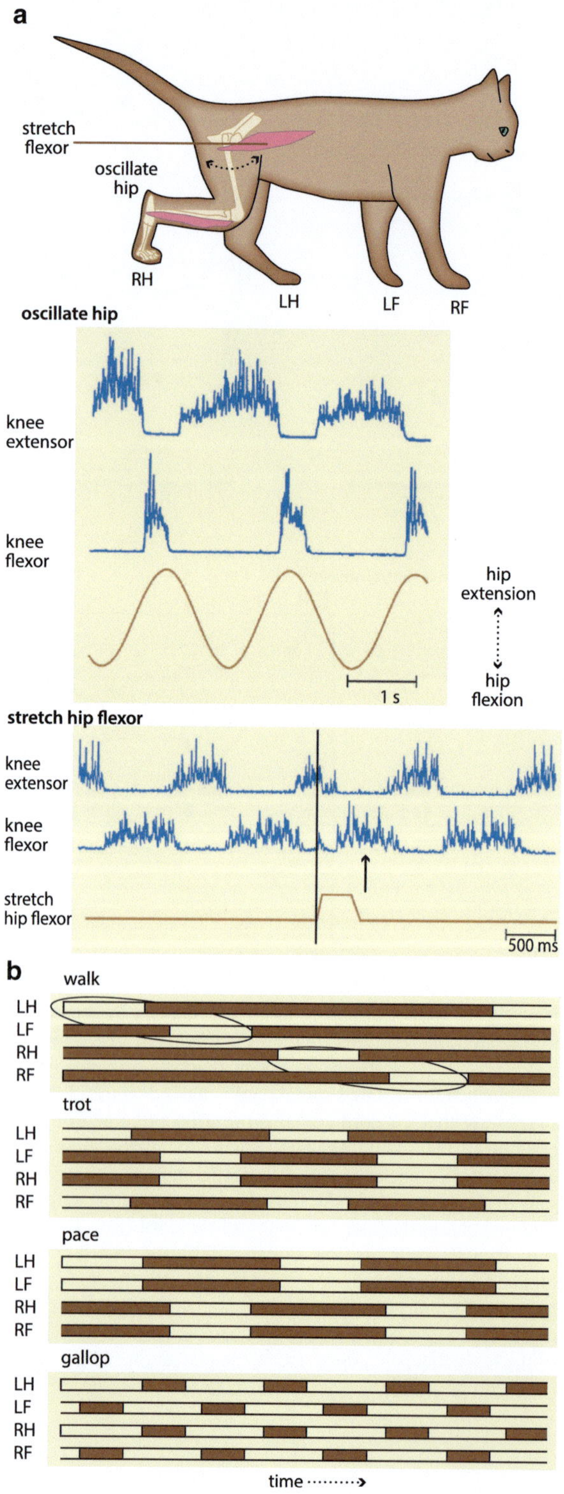


**Fig. 23.19** **Gaits.** Quadrupedal mammals like cats possess spinal locomotor CPGs (as in lamprey and *Xenopus* tadpoles) that generate coordinated extension/flexion movements of the four limbs. (**a**) The CPG output, here monitored as knee flexor and extensor activity, can be entrained by sinusoidal oscillations of the hip showing that proprioceptive feedback is important in timing the transitions between stance (extensor) and swing (flexor) phases of the movement cycle. Stretch of the hip flexor muscle (as occurs during the extension phase) can prematurely trigger a flexion phase showing the important role played by 1a spindle afferents in timing transitions from stance to swing. (**b**) During walking (e.g., on a treadmill) the four limbs are coordinated in a precise sequence, but as the speed of locomotion increases the coordination changes into the different gaits of trotting, pacing, and galloping. The *brown bars* denote the stance phase when the limbs are on the ground; *white bars* denote the swing phase when the limb is in the air. Note that duration of the swing phase is relatively constant irrespective of speed or gait; ((**a**) Adapted from Hiebert et al. [24] and Kriellaars et al. [31], all with permission)

while the latter are responsible for providing positive feedback support and hence controlling the duration of the stance phase. Thus proprioceptive reflexes are important in timing the onset and regulating the duration of the two phases of the movement cycle.

The vertebrate CNS contains CPGs for locomotion, as outlined above and described in invertebrates, but how are these walking CPGs organized? One influential idea, the "unit burst generator" hypothesis proposed by Sten Grillner, suggests that there are separate CPGs generating rhythmic commands to the antagonistic muscle groups of each limb. This organization provides for the flexible linkages that must exist to allow the different gaits to emerge and operate efficiently during changes in locomotor speed. While primarily located in the spinal cord, these CPGs have also been shown to extend into the hindbrain. As in the swimming system, glutamate seems to be the principle fast-acting neurotransmitter responsible for exciting the motoneurons during locomotor rhythm generation and the rhythm persists even after blocking fast inhibition with strychnine. This suggests that the interneurons involved may possess intrinsic oscillatory membrane properties, as in the respiratory system. These oscillations in membrane potential may be conditional upon activation of the NMDA-type glutamate receptor which leads to the activation of a sequence of different membrane currents mediating alternate depolarization towards and hyperpolarization away from action potential threshold. There is also evidence that the excitatory glutamatergic interneurons make synaptic connections with each other so that once locomotion is initiated rhythm generation becomes self-sustaining through positive feedback excitation within the network.

## 23.6 Decision Making for a Particular Motor Pattern

In a constantly changing environment animals are frequently faced with the challenge of deciding which is the most appropriate motor behavior to perform and this decision is potentially a life-or-death one. The selection of a particular motor pattern from a wide range of options (including deciding to stay still) is therefore not a trivial one. In many cases the organism has no time to deliberate, especially when faced with a threatening stimulus and, once a decision has been taken to move, there is an increased risk of becoming visible to a predator. What strategies are employed to assist in the decision-making process?

### 23.6.1 Command Neurons

A command neuron in the strictest sense is a neuron which is both sufficient *and* necessary to initiate a particular behavior. Such command neurons have been described for initiating singing (stridulation) in crickets, swimming in leeches and sea slugs, and in various escape behaviors of insects and crustaceans. Although for some of these behaviors it is still debated whether, in addition to the command neurons, other interneurons are involved, the command neuron concept has been unequivocally demonstrated for some fast escape behaviors. Arguably the best example is the rapid tailflip escape response of macruran crustaceans like lobsters and crayfish.

Crayfish escape from dangerous situations by producing powerful abdominal flexion and extension movements (tailflips) that generate a backward thrust to evade dangerous stimuli (Fig. 23.20). Depending on where on the body an abrupt stimulus impinges, the crayfish produces either a simple tailflip thrusting it backwards (in response to rostral/anterior stimuli), or a more complex tailflip that first produces a somersault and then a fast backward movement (in response to caudal/posterior stimuli). These two behaviors are mediated by two pairs of giant interneurons with very large diameter axons (therefore often called giant fibers): two **medial giant interneurons** (MGIs) and two **lateral giant interneurons** (LGIs). The LGIs are technically a ladder-like set of bilaterally paired cells repeated in each abdominal segment, but because there is extensive and powerful electrical coupling between them they function as a single unit. All the giants extend through the whole ventral nerve cord but the two MGIs are only activated by stimuli to the rostral areas and cause bending (flexion) at all abdominal segments thrusting the animal backwards. In contrast, the LGIs are activated by stimuli to caudal areas and elicit flexion in only the more rostral abdominal segments which causes the animal to pitch upward and rotate its tail end forwards (somersault) and as a result of this movement the animal is propelled away from the stimulus.

Due to it being an escape reflex the circuitry is relatively simple: primary afferents (sensory neurons) and sensory interneurons converge onto each giant interneuron which forms powerful electrical synapses (gap

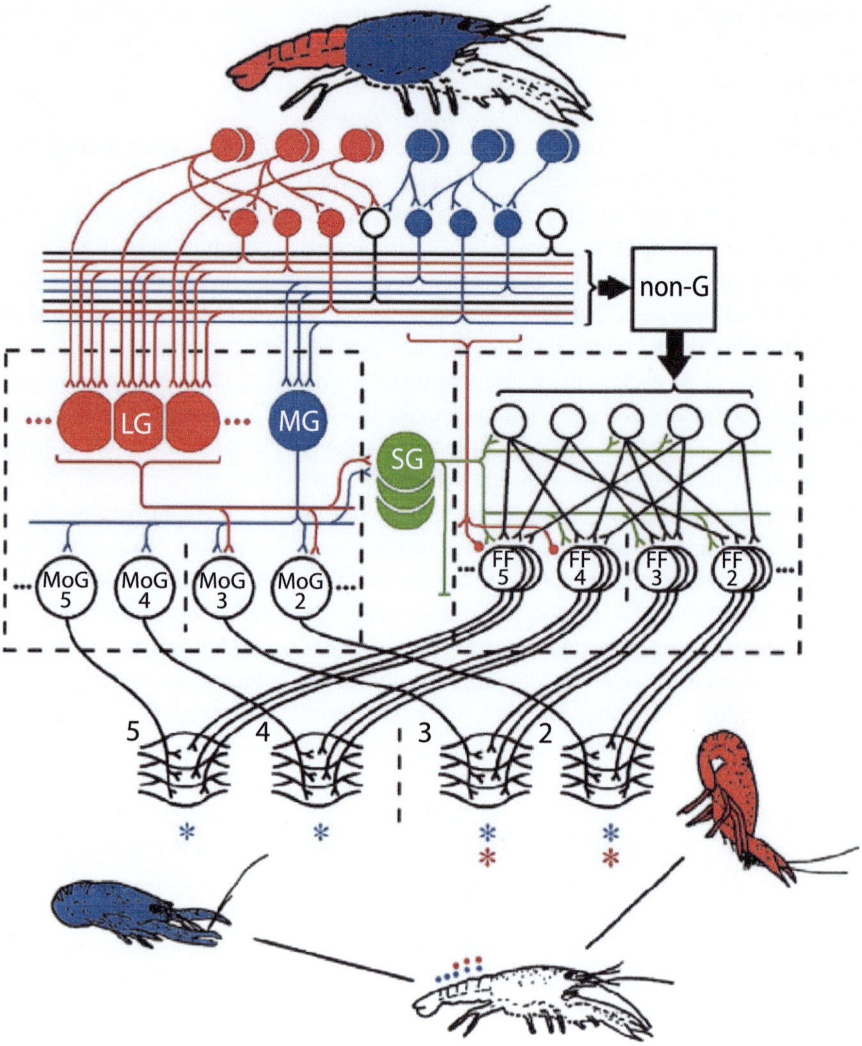


**Fig. 23.20 Crayfish tailflip circuit and swimming behavior.** Mechanical stimulation of the abdominal (*red*) or thoracic (*blue*) area leads to either a medial giant-axon (MG)-mediated response, which is a powerful retreat, or to a lateral giant-axon (LG)-mediated response, which is a somersault followed by a powerful retreat. The giant fibers connect with giant motoneurons (MoGs) that cause powerful contractions of the respective muscles. In addition, the circuitry for non-giant responses (non-G) is also shown on the *right*. Unspecified circuitry (non-G box) is involved in activating non-giant interneurons (*open circles*) that descend in the ventral nerve cord and excite a separate population of fast flexor motoneurons (FF). The giant-fiber system has access to this non-giant system by the segmental giant interneurons (SG). Thus, the crayfish can generate powerful escape movements in an "all-or-none" manner, and swimming movements that can vary from very fast to slow and allow enormous response flexibility (From Edwards et al. [15] with permission)

junctions) with giant motoneurons (**motor giants**, MoGs), which in turn innervate phasic flexor muscles in the respective abdominal segments. These kinds of escape responses are rather stereotyped, and one action potential in the giant interneuron can generate a complete tailflip. In contrast, if threats develop more gradually, different and less rapid escape responses are performed which do not involve the activity of the giant interneurons and which also activate a different set of flexor motoneurons, the fast flexor motoneurons (FFs). These responses are more variable compared to the tailflips generated by the giant interneurons, and produce fast swimming movements. Via a single pair of segmental neurons, the segmental giants (SGs), the giant interneurons, MGIs and LGIs, recruit many elements of the non-giant network. Indeed the activation of the giant fiber-mediated circuit also activates in parallel, but with a delay, the non-giant network so that a series of rhythmic swimming tailflips follows the initial escape flexion. A reciprocal contribution of the non-giant system to the giant neuron escape system does not exist. It turns out that if the giant interneurons are ablated, this stereotyped rapid escape movement can no longer be performed. This is well demonstrated by the time scales: the abrupt sensory stimulus mediated by the giant interneurons initiates muscle potentials with a delay of 5–10 ms, whereas the responses mediated by the non-giant interneurons take 50–500 ms. In this system some important basic features of fast-reacting escape systems can be learned:

1. The giant interneurons and rectifying electrical synapses work as coincidence detectors. The large size

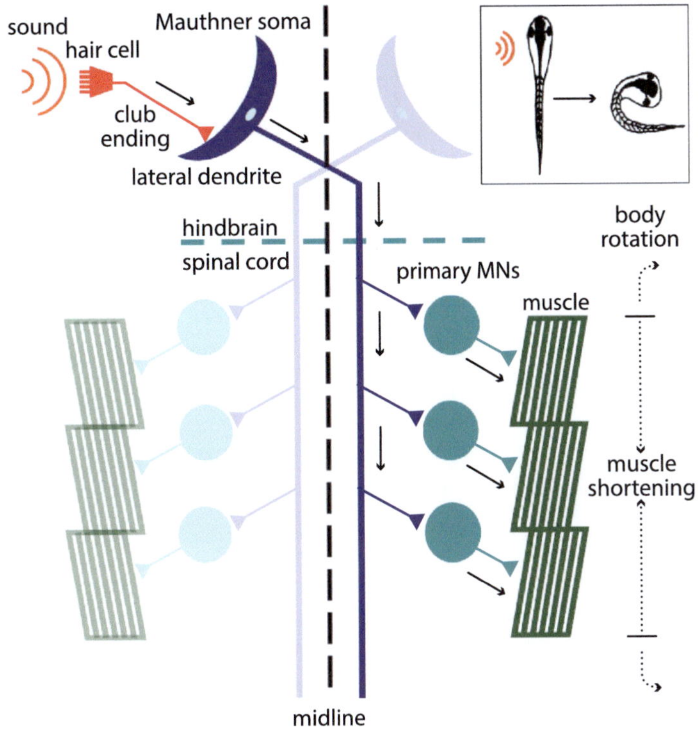


**Fig. 23.21** **Mauthner circuit for fish escape responses.** Schematic representation of the Mauthner cell circuitry of the goldfish. A simulated response to sound on the *left* is shown with *arrows* indicating the progress of activity from hair cell afferent to contralateral primary motoneurons (primary MNs) and muscles on the *right* via the Mauthner cell. Afferents excite the Mauthner cell on its lateral dendrite at specialized synapses called club endings. Equivalent Mauthner circuitry on the other side is shown *shaded*. A single Mauthner cell action potential causes near synchronous contraction and shortening of contralateral trunk muscles leading to rapid head rotation. *Inset* shows resulting bend of body into C shape to orient the fish away from the stimulus (From Sillar [56] with permission)

and low input resistance of the giant interneurons means that each sensory input causes only a small change in voltage at the spike-initiating site. The giant interneurons also have quite negative resting membrane potentials and so a large input is needed to reach spike threshold. The electrical synapses also favor coincidence detection because asynchronous EPSPs are shunted by the gap junctions that are already open as a result of preceding inputs. In combination these biophysical properties ensure that the giants are activated only if a sufficient number of coincident sensory inputs occur, as for example during the attack of a predator. At the same time the system ensures that superfluous inputs cannot trigger escape behavior which would be, on the one hand, energetically expensive and, on the other, likely to expose the animal to predators.

2. The synapse between the **giant interneurons** and the MoGs is a **rectifying electrical synapse**, so that it only functions in the direction of giant interneurons to MoGs, but not *vice versa*. In addition, nonrectifying electrical synapses contribute to the reaction speed as they introduce only a minimal synaptic delay compared to chemical synapses.
3. Moreover, as a rule, the larger the axon diameter, the faster the conduction velocity. This principle is also nicely demonstrated by the squid giant axon which controls the escape movements of cephalopods by rapidly contracting the mantle, and thus producing a backward thrust. Functionally it can be regarded as a single giant fiber. In contrast to crustaceans and insects, this giant fiber of cephalopods is not the axon of a single individual neuron but derived from fusion of the axons of many interneurons running in parallel. In other invertebrates such as annelids, the giant fibers already show some additional glial wrapping, perhaps indicating a convergent evolutionary trend towards the myelination of the vertebrates (see Chap. 9). An interneuron with a particularly large axon diameter in fish is the so-called Mauthner cell. The large myelinated axon of the Mauthner cell activates contralateral motoneurons innervating the fast-twitch white muscle fibers to initiate C-start behavior in which the body arcs into a C shape to bend the body away from a startle stimulus before the fish darts off in the opposite direction (Fig. 23.21).

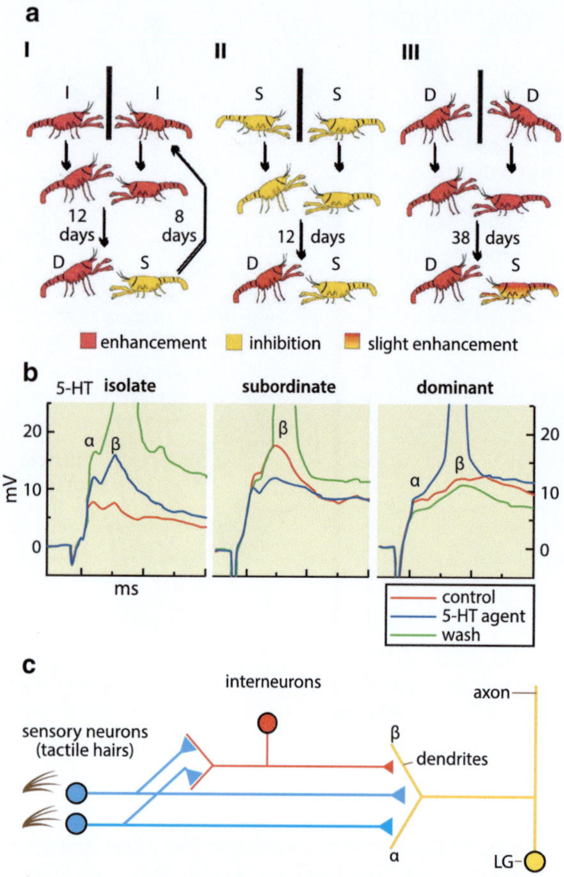


**Fig. 23.22** **Crayfish social status and circuit modulation.** (**a**) The effect of social status on modulation of 5-HT (serotonin)-mediated LG excitability. Three types of animals were paired and fought for a new dominance hierarchy (dominant-subordinate relationship). Then they spent some days together, and again at the indicated days, they were tested for the effect of 5-HT on LG (lateral giant interneuron) excitability. Abbreviations: *I* social isolate (neutral posture), *S* social subordinate (supine posture), *D* social dominant, upright posture. Experiments A: I-I pairs, when D and S were reisolated for 5 days again, the effects of 5-HT were restored to those of I animals. B: S-S pairs and C: D-D pairs. The *color code* indicates the effect of 5-HT on LG excitability. Although it appears counterintuitive that the escape response (tailflip) is reduced in subordinates, the non-giant responses are in contrast enhanced in these animals, again pointing to greater flexibility in non-giant responses. (**b**) Different effects of 50 μM 5-HT application to an LG response in isolated, subordinate, and dominant animals (After Yeh et al. [70] with permission). α and β indicate postsynaptic potentials recorded from LG and caused by either sensory receptors (tactile hairs) or groups of interposed interneurons (see **c**). (**c**) Schematic wiring diagram of the input synapses to LG that are sensitive to 5-HT modulation (After Edwards et al. [15] with permission)

4. The execution of a powerful tailflip in crayfish has to ensure that particular reflexes that usually control abdominal movements are briefly inhibited. For example, resistance reflexes normally involved in postural control to oppose any deviations from a particular set point, such as a particular angle between two abdominal segments, have to be inhibited. Sensors for such a resistance reflex are the extensor-muscle stretch receptors and they have to be prevented from functioning during the tailflip. Therefore, two kinds of inhibition can occur, postsynaptic inhibition of the sensory neuron itself, or presynaptic inhibition of the axon terminals of the sensory neuron preventing or decreasing the release of transmitter. This will ensure that the sensory receptors will be fully functional even after a powerful tailflip that usually would cause them to fatigue. Presynaptic inhibition itself is a widespread phenomenon and ensures that sensory signals only feed into a neuronal circuit when it is appropriate to use this information.
5. Autoregulatory mechanisms limit the amount of neuronal activity. Each giant interneuron also drives inhibition of both the giant interneurons and the MoGs, and by this "recurrent" and "feed forward" inhibition the neuronal response to strong sensory stimuli is limited to one or few action potentials in the giant interneurons and one action potential in the MoGs.

It is important to note that despite the fact that the tailflip leads to a powerful escape response, behavioral variability is nevertheless retained, and the same stimulus that leads to a tailflip on one occasion may be ignored on other occasions due to influences from the more rostral ganglia and from the brain. For example, a tonic inhibitory drive from anterior ganglia, acting at the level of the GIs, prevents escape during feeding, during defence against attacks, or during agonistic interactions with other crayfish. In addition, tailflips are prone to habituation and as in the sea slug *Aplysia* this is the result of decreased amounts of transmitter, acetylcholine in this case, released from the mechanosensory primary afferent terminals. Interestingly, however, the social status of an individual or the levels of aggression of a crayfish have profound influences on the excitability of the tailflip system (Fig. 23.22). Crayfish rapidly establish social dominance hierarchies with dominant individuals prepared to fight and subordinates more prone to retreat. The neuromodulator that is involved in this fighting modulation, and acts on the synapses between the sensory receptors and the command interneuron, is serotonin (5-HT). In a socially subordinate crayfish, such as the loser of a fight, 5-HT decreases the different postsynaptic potentials in LGIs whereas in isolates that have never experienced any fight before or in dominant crayfish these potentials are increased. The social status may cause different compliments of serotonin receptors in the LGIs which in turn may be the cause of these different responses [70]. The apparent paradox that subordinate crayfish are less likely to escape by a GI tailflip than dominants is solved by the fact that the aforementioned fast and much more flexible swimming responses via the non-giant system are now enhanced, for example to retreat backwards from the dominant. Therefore, it can be speculated that escape by a subordinate animal may require much more "delicate" maneuvering than the fast and powerful tailflip generated by the giant system in the escape of a dominant animal that may encounter an increased risk of predation. Interestingly, some of the effects of social experiences can be reversed, for example, when a subordinate is paired with a new partner over which it is dominant. This shows that even in a crayfish the social status of an animal can have profound effects on the functional synaptic circuitry of the brain.

### 23.6.2 Motor Program Selection

During vertebrate evolution there has been a delegation of function during which the spinal cord has been assigned the executive role of producing a basic rhythm suitable to drive locomotion. Sensory feedback and descending commands are responsible for sculpting a behaviorally relevant sequence from the resulting movements that is adaptable in the face of unexpected changes that occur while the movement is taking place. With few exceptions, locomotion is initiated and terminated during specific tasks – but how is the decision for a particular task taken? Selection of an appropriate motor behavior from the often large available repertoire is an important task and the final decision to make one movement rather than another can occur at multiple levels in the motor system from higher brain motor centers to the spinal circuitry. There is a consensus in both invertebrate and vertebrate systems that the chain of commands involves a hierarchy of executive instructions that descend from the brain to subordinate motor control centers. In the case of primates, the initial programming of a movement sequence involves areas of premotor cortex which then instruct the motor system to execute a particular movement. In turn, the motor cortex accesses the spinal circuitry via direct corticospinal pathways; this pathway is fast and axons of

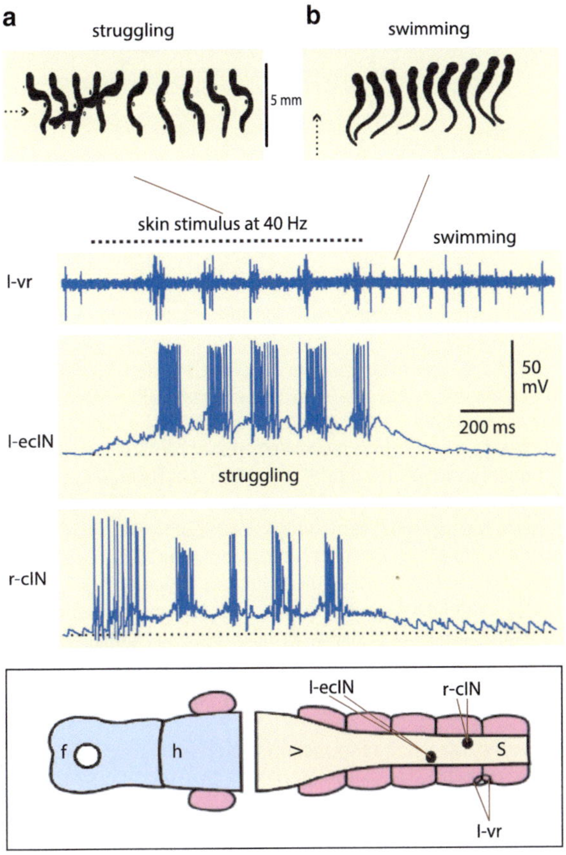


**Fig. 23.23 Tadpole motor program switching.** *Xenopus* tadpoles generate struggling movements (**a**) in responses to a continuous tactile stimulus (e.g., if grasped) but when released will generate propulsive swimming (**b**) (From Boothby and Roberts [3] and Kahn and Roberts [27], all with permission). Fictive correlates of struggling and swimming can be triggered in immobilized animals. Struggling occurs if the skin is stimulated repetitively at 40 Hz. Unlike swimming activity in which brief ventral root bursts propagate from head to tail (Fig. 23.17), struggling propagates from tail to head and involves more intense bursts of motor activity. Many spinal neurons are active in both behaviors but some are recruited during the struggling pattern (e.g., commissural inhibitory interneuron (cIN)). Excitatory commissural interneurons (ecIN) are only active during struggling. *Inset* (*boxed*) shows schematic dorsal view of CNS and recording positions; note struggling persists when brain is detached. *F* forebrain, *h* hindbrain, *s* spinal cord (Adapted from Li et al. [36], with permission)

the corticospinal tract are large and amongst the most rapidly conducting in the animal kingdom.

Motor pattern selection does not necessarily require cognitive processing and may depend solely on responses to particular sensory signals. *Xenopus* tadpoles, for example, have a restricted behavioral repertoire in which swimming or struggling are the two main options (Fig. 23.23). Unlike swimming, struggling is a forceful,

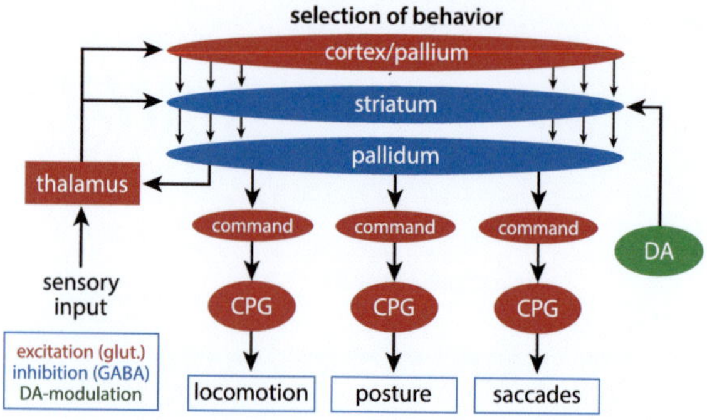


**Fig. 23.24** **The basal ganglia control CPG activity.** The cortex in mammals (the "pallium" in nonmammalian vertebrates) has an excitatory (*red*) output to the striatum, the input layer of the basal ganglia, which is also activated from thalamus directly. Striatum consists of inhibitory (*blue*) GABAergic neurons. Striatal neurons inhibit the pallidum, which also contains GABAergic neurons that tonically inhibit networks involved in control of posture, saccades, and locomotion. CPG activity is only released following disinhibition. Dopaminergic (DA, *green*) modulatory input to striatum controls the excitability of striatal neurons. The basal ganglia play an important role selecting when different motor behaviors are expressed (From Grillner [20] with permission)

but nonpropulsive response in which activity propagates from tail to head rather than head to tail, as in swimming. The two behaviors can be elicited by tactile stimulation of the skin but the response evoked depends on the timing of the stimulus (reviewed in [55]). A single touch leads to swimming while a train of stimuli, such as would occur if the tadpole were grasped by a predator, triggers struggling in an attempt to wriggle free. Thus, the firing pattern in cutaneous sensory receptors determines the selection of an appropriate response.

### 23.6.3 The Subesophageal Ganglion and Central Complex of the Insect Brain Are Important Motor Centers

In insects and other invertebrates the basic networks equivalent to those of the spinal cord are located in segmental ganglia of the ventral nerve cord. For regulating and modulating these segmental networks the subesophageal ganglion is necessary, as it is for coordination of the limbs. If the subesophageal ganglion of an insect is removed from the rest of the ventral nerve cord, the animal no longer is able to walk. If, however, the brain (the supraesophageal ganglion) alone is removed, the animals walk almost uninhibitedly, showing that for execution of a coordinated movement the subesophageal ganglion is sufficient. Much less is known about decision making and planning of movements in invertebrates in the brain (supraesophageal ganglion). One brain center that most likely plays a role in movement control and also analysis of sensory cues, for example polarized light, is the **central complex** with its different subcompartments. Mutant *Drosophila* fruit flies that lack or have disturbed central complexes show severe locomotory deficits [62].

### 23.6.4 Motor Centers in the Vertebrate Brain

There are more indirect routes between cortex and spinal cord such as through various nuclei of the brainstem like the **red nucleus** and the **vestibular nucleus**. These relay stations allow for the integration of critical corollary inputs that place the motor commands into a context that reflects the prevailing "state" of the organism. Also important in the precise control of the ongoing movement is the **cerebellum** which receives ascending signals from the spinal cord and brainstem via the **inferior olive** regarding the sensory feedback resulting from movement. It is in the cerebellum that "re-afference" error signals from climbing fibers are compared with a copy of the original command (called an "efference copy"). Mismatches between what was intended and what actually happened equate to an error signal which the cerebellum can use to correct for via its output to the various brainstem nuclei that project to the spinal cord. The capacity of the cerebellum to negate motor errors is central to its important role in motor learning such as when we learn to ride a bicycle or juggle. The **basal ganglia** are also important in various aspects of motor control (Fig. 23.24), though not in such an overt or well understood manner as other structures like the motor cortex or cerebellum. Abnormal signaling in the basal ganglia, such as the deficits that occur in dopamine signaling in the substantia nigra during **Parkinson's disease**, leads to deterioration in the quality of movements or in the ability of the premotor cortex to plan motor behavior.

### 23.6.5 The Genetic Control of Motor Behavior

Early in vertebrate embryogenesis the fates of neuronal precursors destined to form the sensory and motor control networks of the spinal cord are determined by a ubiquitous molecular code which is initially established by a dorsoventral gradient in two morphogens, **sonic**

**hedgehog (SHH)** and **bone morphogenic protein (BMP)**. The former is produced by the ventral floor plate and the latter is produced by the dorsal aspect of the spinal cord. The concentrations of SHH and BMP at a particular dorsoventral location in the cord trigger the expression of factors that regulate gene transcription. In the ventral spinal cord, where the circuitry required for locomotor rhythm generation will be located, these zones include a motoneuron-generating domain (pMN) and four additional progenitor domains (P0–P3) that generate the interneurons responsible for coordinating motoneuron activity. The different progenitor domains "sense" the relative concentrations of BMP and SHH – high BMB and low SHH signal a ventral position and so on. Thus, it is from within these discrete dorsoventral zones that differentiated neuron classes emerge sharing common anatomical features and transmitter phenotypes. Along the rostrocaudal axis the differentiation of limb-specific neuronal components is set up by a second transcriptional code which relies upon the position-dependent activation of several ***HOX* genes** aligned along the rostrocaudal axis. *HOX* gene transcription is regulated by other morphogens such as fibroblast growth factor (FGF) which is responsible for ensuring the proper assembly of circuits controlling limb motoneurons at two particular points along this axis of the nervous system.

A particularly interesting genetic approach is the identification of mutants that lack particular parts of behavior. This has been beautifully demonstrated in the fruit fly *Drosophila*, where deficits in some aspects of walking performance could be associated with alterations in the neuronal architecture of the central complex or the subesophageal ganglion. Thus, important locomotory centers can be identified. In addition, the ability to produce knockout mutants, to produce transgenic animals in which particular neurons, ion channels, receptor molecules, or intracellular signaling cascades can be silenced or overexpressed, or switched on or off at a particular time, is opening up completely new and exciting experimental possibilities.

### 23.6.6 Ontogeny of Vertebrate Motor Behavior

The neuronal networks controlling movements are extremely complex, often comprising large ensembles of neurons arranged in a series of interconnected groups and distributed throughout the CNS. It is a remarkable feat of development that this complexity is achieved at all, let alone in such a similar manner in almost every individual in each species. Many thousands or even millions of neurons must differentiate into functionally distinct groups, each having a different complement of transmitters, receptors, and ion channels. Subsequently, the members of these groups must make and receive a myriad of precise synaptic interconnections with appropriate targets. What are the mechanisms involved in the assembly of motor control circuitry? Research on a small number of amenable model systems strongly suggests that amongst the vertebrates there are several common underlying principles of motor network assembly, especially at early stages. For example, it appears that in animals as diverse as zebrafish, frogs, chickens, and mammals, the basic circuitry necessary to generate rhythmic motoneuron discharges during locomotor rhythm generation is assembled very early in development, during embryogenesis before hatching or birth. This means that circuit assembly precedes the behavior that the circuit will eventually control. Evidence from real-time ultrasonography suggests that the rhythm-generating networks for locomotion are present in humans from as early as 11 weeks into gestation. The fetus is able to produce cyclical extension/flexion movements of the legs to move around the womb and if a newborn is held upright and tilted forwards it will produce walking movements. So it seems that the locomotor CPGs are constructed early in development, before locomotion is even possible, but then undergo a period of postembryonic maturation.

### 23.6.7 Maturation of Vertebrate Motor Networks

The output generated by these immature networks often lacks the precision and versatility of the activity generated by the same networks later in development. For example, in embryonic rodents and chicks the flexors and extensors of a given limb are often coactive initially, but then switch to alternation during the course of ontogeny (Fig. 23.25). A likely reason for this is that embryonic neurons tend to have a high intracellular chloride concentration. This means that when inhibitory amino acid transmitters like GABA or glycine bind to their respective postsynaptic receptors an efflux rather than an influx of chloride ions occurs. This mediates depolarizing responses that are initially excitatory until a developmental shift in ionic gradients takes place at which time activation of the same transmitter receptors allow chloride influx and inhibition. The functional

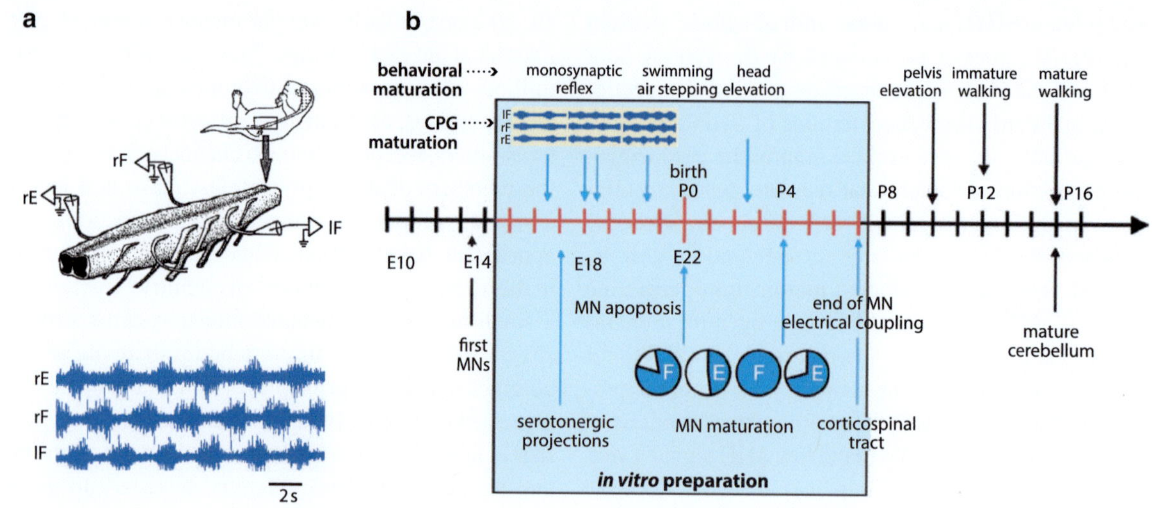


**Fig. 23.25 Neonatal rat *in vitro* preparation and development of motor behavior**. (**a**) The isolated lumbar spinal cord of the rat neonate *in vitro* generates, in the presence of pharmacological excitants, alternating bursts of flexor and extensor motoneuron activity (extensor, right (rE); flexor, left (lF) and right (rF)). (**b**) Correlation of ages suitable for *in vitro* recording of fictive locomotion (*boxed area*) with the development of the lumbar spinal circuitry and motor behaviors between embryonic day (E)10 and postnatal day (P)16. Shortly after lumbar motoneurons (MNs) first appear (~E14) they contribute to both spinal reflexes and rhythmic CPG output (*inset*, CPG maturation; *dark blue* in *circles* represent % of mature MNs), which is initially synchronous, but then switches to flexor/extensor and left/right alternation. Prior to birth some MNs are lost through apoptosis, but the remaining F and E MNs undergo maturation and electrical coupling between them reduces. Inputs from higher brain centers (e.g., serotonergic projections, corticospinal tract) appear perinatally. Quadrupedal walking supporting the body above the ground does not appear until P12. Mature walking is correlated with maturation of the cerebellum ((**a**) Adapted from [29] with permission; original drawing by Dave McLean. (**b**) Adapted from Clarac et al. [9] with permission)

consequences of the initially depolarizing excitatory signaling by transmitters that will later become inhibitory are not fully understood but the resulting opening of voltage-dependent calcium channels will increase intracellular calcium concentrations and calcium is an important signal in neuronal maturation and synaptogenesis. Another feature of developing locomotor networks is that their maturation follows a rostrocaudal path with more rostral loci containing more mature circuitry and with caudal centers lagging behind. For example, neonatal rodents first acquire the ability to crawl over their mother's fur in search of a nipple in order to feed, dragging their hind paws behind them. Only after about 2 weeks of postnatal development can the hindlimbs produce unsupported locomotion (Fig. 23.25b). *In vitro* experiments suggest that the inhibitory GABA system is again involved because when pharmacological excitants are applied to activate the rostral locomotor CPGs of neonates, the rostral forelimb CPGs generate a rhythm but the caudal lumbar CPGs only become active when a GABA receptor antagonists is applied. In addition to changes in the properties of synaptic connections there are also changes in the electrical properties of neurons in the CPG that contribute to the development of locomotor output. In general this is manifest by an increase in firing behavior as the complement of ion channels incorporated into the membranes of CPG neurons changes with time [9].

Across a range of vertebrates it is also noteworthy that whilst the spinal CPGs are present and functional immediately after birth or hatching they have yet to receive a full complement of modulatory inputs from sources in the brainstem which arrive in the spinal cord early in postnatal or larval life. These inputs, which include, for example, the serotonergic projections of the raphe nuclei, and noradrenergic fibers from the locus coeruleus, modify the properties of neurons in the spinal networks in such a way as to confer an increased flexibility on the network output.

In many tetrapods, walking involves the alternation of movements between the limbs on the left and right sides like the axial systems from which limb CPGs presumably evolved. This reciprocity is thought to involve reciprocal glycinergic inhibition and the excitation, which is normally restricted to the same side, involves glutamate-mediated transmission. During the assembly of the walking CPG it is important that certain excitatory interneurons do not stray to the wrong

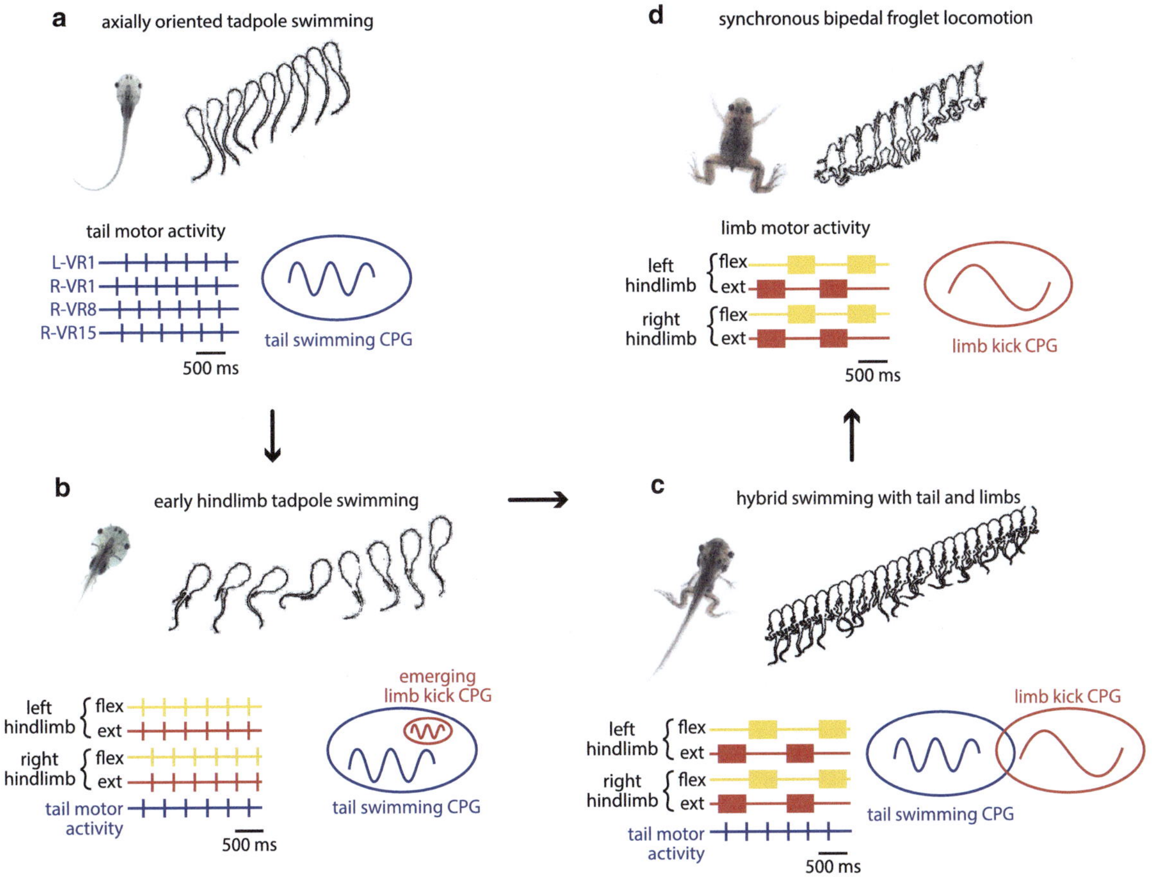


**Fig. 23.26** **CPG development during amphibian metamorphosis**. The development of axial and limb CPGs during metamorphosis in *Xenopus,* from tadpole tail swimming (**a**) with left-right alternation, to limb kicking in postmetamorphic froglet (**d**) with left-right synchrony and flexor-extensor alternation. Initially (**b**), the hindlimbs appear and become functional, but during swimming they are held against the body during swimming. Limb CPG network output shows that extensors (*red*) and flexors (*yellow*) are coactive and burst in time with the tail rhythm (*blue*), as if the emerging CPG for limb kicking emerges from within the preexisting tail network in the spinal cord (*inset schematic*). (**c**) Later during metamorphosis all four limbs are present and functional along with the tail, but now extensors and flexors of each limb alternate. Activity across the body is synchronous so that the limb CPGs now produce a kick rhythm typical of the juvenile frog (**d**). Note tail and limb CPGs have different intrinsic frequencies and although active together now form largely separate networks (*inset*) (Adapted from Sillar et al. [58] with permission. Schematics in (**a**) and (**d**) originally drawn by Denis Combes, and in (**b**) and (**d**) by John Oltoff)

side of the nervous system and in fact the processes of excitatory interneurons are prevented from crossing due to the influence of signaling molecules expressed by the midline. These "**ephrins**" have a repelling effect on the growth of any neurons that express the appropriate EphA4 receptor and since this is selectively expressed by specific glutamate interneurons, these are restricted to mediating ipsilateral excitation. In mice that lack either ephrin, or the EphA4 receptor, during development the processes of excitatory interneurons are able to grow across to the contralateral side and the result is that the rhythmic bursting resulting from CPG activity occurs synchronously across the body. Instead of alternation during locomotion these mice generate a hopping gait [33], as occurs normally in some vertebrate species like rabbits and kangaroos.

### 23.6.8 The Ontogeny of Amphibians as a "Model Case Study" for Different Locomotor Patterns and Their Transitions

A curious feature of amphibian development is the remarkable transition during metamorphosis from tail-based swimming to limb-based locomotion (Fig. 23.26). How does the emerging CPG for the limbs become integrated into the preexisting axial motor system? *In vitro* preparations of the brainstem and spinal cord are capable of spontaneously generating motor rhythms in the limb and/or tail ventral roots appropriate to drive the movements of the host organism's developmental stage (reviewed in [57, 58]). Early in its formation the newly emerging limb network generates a rhythmic output with the same frequency and coordination as the primary tail CPG (left/right alternation), but with flexor/extensor coactivation (as in early stages of chick and mammal development). This indicates that the latter forms a functional scaffolding for the emergence of the former. Only later, once the limbs are fully functional, is the limb network able to "break free" and produce left-right synchrony of limb motoneuron bursting. The two systems coexist and are both functional, often simultaneously, in animals at hybrid stages of development that possess both a functional tail and limbs.

### 23.6.9 Ontogeny of Motor Behavior in Invertebrates

Insects are interesting experimental animals for walking studies as they possess two fundamentally different ontogenetic types: they either hatch from the eggs as a nymph already expressing many morphological features of the adult insect except genital organs and wings (hemimetabolous insects, e.g., grasshoppers, locusts, crickets, and cockroaches), or they hatch as a larva with its own behavior and then undergo complete metamorphosis in a pupal stage where the adult animal is formed (holometabolous insects, e.g., butterflies, moths, beetles, and flies). In the hemimetabolous locust the CPG for flight is already wired in the embryo. Freshly hatched first instars have only minuscule wing buds and tiny wing muscles. They are capable both of walking and hopping, and if they are held in front of a wind tunnel or if the modulator octopamine is injected into the metathoracic ganglion they produce a flight-like motor pattern [61]. Only after the final molt do locusts possess fully functional wings and express the appropriate motor pattern. Nevertheless, in the postembryonic stages, the flight circuitry has to be modified, as, for example, the number of sensory receptors increases enormously between the first and fifth instar.

In holometabolous insects (Fig. 23.27) the larva, or "caterpillar", crawls, feeds, and molts, whereas the adult moth flies around, feeds (often on different food), and reproduces. The flight motor pattern can be induced by the modulator octopamine only in the final pupal stages. The CPG for flight is wired only during the second half of the pupal stage, where a full flight pattern can be elicited 2 days after emergence of the adult animal. In the holometabolous moth *Manduca*, many neurons die at metamorphosis but some persist and among them are motoneurons which innervate slow muscles in the larva and fast flight muscles in the adult. The transition can be followed in individual neurons and involves retraction of the axon from the larval muscle and a widespread retraction of dendritic branches within the ganglion. Exposure to the hormone ecdysone and most likely to other chemical signals induces new dendritic outgrowth and the expression of new genes in the pupa, for example calcium and potassium channels which alter the electrical properties of the neurons. Thus, the mechanisms that take place when a motoneuron changes from slow to fast properties can be revealed.

### 23.6.10 Insect Ecdysis: Cascades of Different Motor Patterns Induced by Neurohormones

Molting is a complex motor behavior displayed by insects involving shedding the old cuticle and emerging as either a larva (larval ecdysis) or a pupa (pupal ecdysis) before metamorphosing as an adult insect (Figs. 23.27b–d and 23.28). Thus, caterpillars and maggots metamorphose into butterflies and moths or adult flies. Most of our knowledge is based on studies of the silk moth (*Bombyx mori*), the tobacco hawkmoth (*Manduca sexta*), and of the fruit fly (*Drosophila melanogaster*). If the hemolymph titers of the two hormones **ecdysone** and **juvenile hormone** are both high, a larval molt follows but if the juvenile hormone titer is very low (or zero), a pupal molt is the result. The process of ecdysis has been separated into three different motor patterns: a preecdysis pattern; an ecdysis pattern; and a postecdysis pattern, each with distinct features and phase relationships of the contracting muscles. A number of neurohormones and neuromodulators, all peptidergic in nature, released from glands or neurons, act on neurons or tissues that bear the corresponding receptor mechanisms and, thus, are activated in a particular sequence giving rise to the respective motor patterns. A hormone called ecdysis-triggering hormone (ETH) is released from epitracheal glands and acts on all neurons bearing ETH receptors, among them neurons that release kinin, FMRFamide, eclosion hormone (EH),

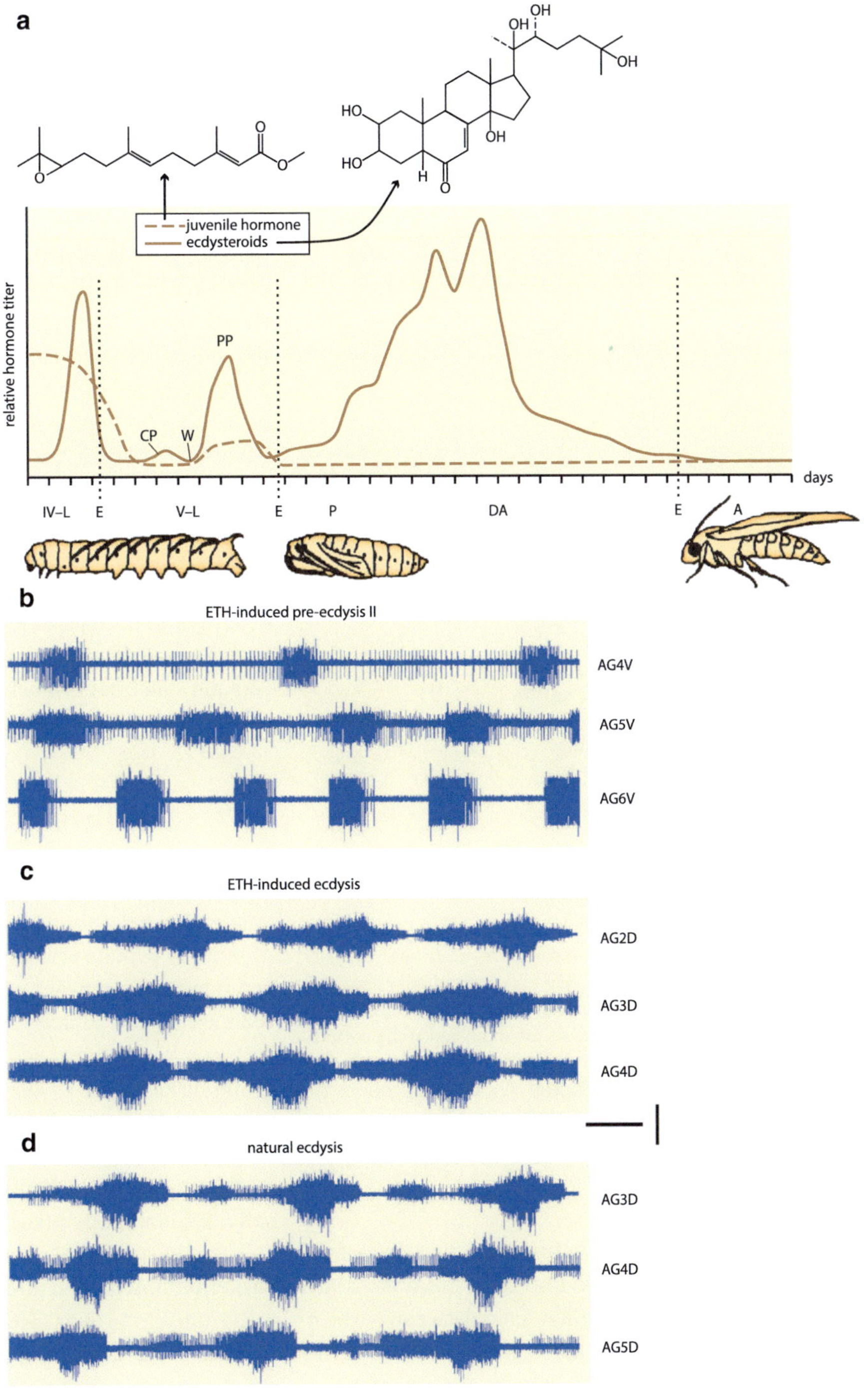


**Fig. 23.27 Metamorphosis of insects.** (**a**) The titers of juvenile hormone and ecdysteroids during metamorphosis of the tobacco hawkmoth, *Manduca sexta*. If the titers of both hormones are high a larval ecdysis (molt) is induced, a decreased titer of juvenile hormone leads to pupal ecdysis. Before this, a small peak in ecdysteroid release (CP, commitment peak) induces the wandering behavior (W) of the last larval instar. The following large peak in ecdysteroid release leads to pupal ecdysis (PP, pupal peak) (After Weeks [65] with permission). (**b–d**) Different motor patterns during larval ecdysis induced by the peptidergic ecdysis-triggering hormone (ETH) and recorded from abdominal nerves. A proecdysis pattern is followed by the true ecdysis pattern. Note how similar the hormone-induced ecdysis pattern is to the naturally recorded ecdysis pattern [71]

crustacean cardioactive peptide (CCAP), myoinhibitory peptides (MIPs), and bursicon, which themselves are all peptides (see Chap. 11). Thus, accurately timed cascades of peptide neurohormones which involve positive and negative feedback regulation are involved in eliciting the various motor patterns and, thus, in the "orchestration of behavior". For example, ETH acts on a pair of brain neurons that run through the whole nerve cord and possess central and peripheral release sites of their peptide transmitter, eclosion hormone (EH). EH has a positive effect on the ETH-releasing epitracheal glands which now release all their ETH.

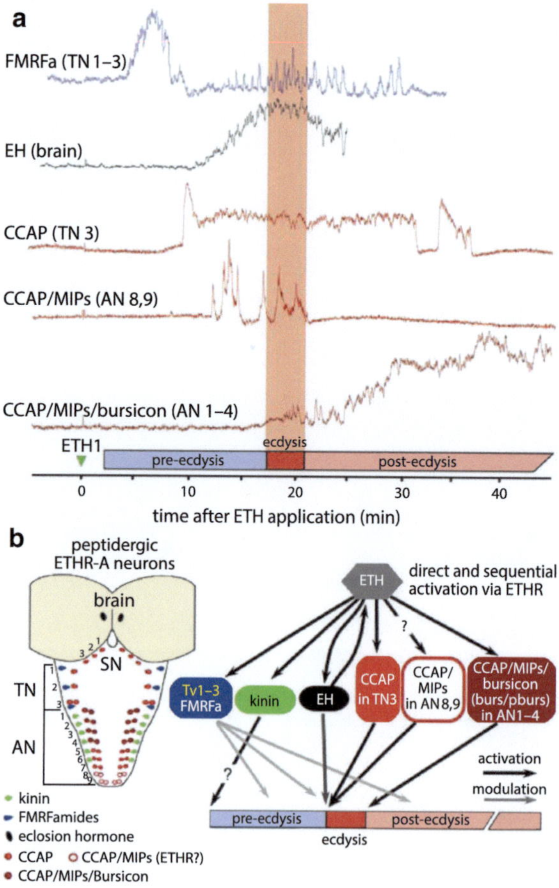


**Fig. 23.28** ***Drosophila* metamorphosis.** (**a**) Optically recorded $Ca^{2+}$ responses in peptidergic neurons which all bear ecdysis-tiggering hormone receptors (ETHR) are sequentially activated by ETH injection into a puparium. The different phases of proecdysis, ecdysis, and postecdysis are indicated at the bottom. (**b**) A model of peptidergic regulation of ecdysis behavior sequences in *Drosophila* (From Kim et al. [30] with permission). Abbreviations: *AN* abdominal neuromeres of CNS, *CCAP* crustacean cardioactive peptide, *EH* eclosion hormone, *ETH* ecdysis-triggering hormone, *ETHR* ecdysis-triggering hormone receptor, *MIP* myoinhibitory peptide, *TN* thoracic neuromeres of CNS

Important contributions to understanding the ontogeny of motor behaviors have also come from the study of the stomatogastric ganglion in crustaceans which controls mastication in the adult. It appears that in the embryo of a lobster the STG already consists of about 30 neurons; therefore, during metamorphosis no new neurons are added. The lobster larva is a planktonic creature which filter-feeds and the embryonic STG produces a swallowing rhythm in which all neurons fire synchronously. Only during metamorphosis, when the planktonic larva changes into a small lobster living a benthic lifestyle, do the adult patterns gradually appear. Thus, the different adult motor patterns evolve from a unique embryonic rhythm [7]. What changes can be observed? First, the number of electrical synapses (gap junctions) decreases. It is likely that this is a prerequisite to the development of the flexible adult rhythms because electrical synapses synchronize neuronal activity. Second, neuromodulators such as GABA and histamine play a role in this process. These substances have been shown to modulate electrical synapses – for example, the gain of the electrical synapse is dramatically reduced in the presence of the modulator. This is a particularly interesting finding, as it shows that not only chemical but also electrical synapses can undergo chemical modulation. If the synthesis of GABA is prevented in an adult STG, the different adult rhythms surprisingly disappear and the unique embryonic rhythm reappears [13]. If GABA synthesis is reinstated, then the embryonic rhythm disappears and the different adult rhythms appear again. Thus, the embryonic rhythm may be the "default state" but is blocked by GABA, and this in turn allows expression of the different adult rhythms.

## 23.7 Summary

Adult motor behavior is sophisticated and flexible. Most animals have the ability to select from a wide repertoire of options which serve a particular behavioral context and transitions between consecutive sequences emerge seamlessly. The underlying neuronal systems must be sufficiently complex and adaptable to accommodate changing demands of the environment as well during ontogeny. It is remarkable that similar design features recur in motor control systems in animals spanning major phylogenetic boundaries and possessing radically different body plans and ecological niches.

Usually a CPG network is capable of producing basic rhythmicity and alternating activation of antagonistic muscle groups. However, this CPG lies at the interface between internal motivational and/or developmental states and the prevailing sensory world. The output generated by CPGs depends not only on the connectivity and properties of its component neurons, but also on the prevailing neuromodulatory and neurohormonal conditions (reflecting the internal state) and changes in the multimodal sensory systems (reflecting the external state).

Whilst more and more is known about the production of motor behavior and its underlying neural control mechanisms, less is known about the decision making process or mechanisms of behavioral selection, two features that may be critical for survival.

## References


1. Ballantyne D, Rathmayer W (1981) On the function of the common inhibitory neurone in the walking legs of the crab, *Eriphia spinifrons*. J Comp Physiol 143:111–122
2. Baro DJ, Cole CL, Harris-Warrick RM (1996) RT-PCR analysis of *shaker*, *shab*, *shaw*, and *shal* gene expression in single neurons and glial cells. Receptor Channel 4: 149–159
3. Boothby KM, Roberts A (1995) Effects of site of tactile stimulation on the escape swimming responses of hatchling *Xenopus laevis* embryos. J Zool 235:113–125
4. Büschges A (2005) Sensory control and organization of neural networks mediating coordination of multisegmental organs for locomotion. J Neurophysiol 93:1127–1135
5. Burrows M (1996) The neurobiology of an insect brain. Oxford University Press, Oxford
6. Burrows M, Hoyle G (1973) Neural mechanism underlying behavior in the locust *Schistocerca gregaria*. III. Topography of limb motorneurons in the metathoracic ganglion. J Neurobiol 4:167–186
7. Casasnovas B, Meyrand P (1995) Functional differentiation of adult neural circuits from a single embryonic network. J Neurosci 15:5703–5718
8. Clarac F, Libersat F, Pflüger HJ, Rathmayer W (1987) Motor pattern analysis in the shore crab (*Carcinus maenas*) walking freely in water and on land. J Exp Biol 133:395–414
9. Clarac F, Pearlstein E, Pflieger JF, Vinay L (2004) The in vitro neonatal rat spinal cord preparation: a new insight into mammalian locomotor mechanisms. J Comp Physiol A 190:343–357
10. Cruse H (1990) What mechanisms coordinate leg movement in walking arthropods? Trends Neurosci 13:15–21
11. Duch C, Pflüger HJ (1995) Motor patterns for horizontal and upside-down walking and vertical climbing. J Exp Biol 198:1963–1976
12. Duch C, Bayline RJ, Levine RB (2000) Postembryonic development of the dorsal longitudinal flight muscle and its innervation in *Manduca sexta*. J Comp Neurol 422:1–17
13. Ducret E, Le Feuvre Y, Meyrand P, Fénelon VS (2007) Removal of GABA within adult modulatory systems alters electrical coupling and allows expression of an embryonic-like network. J Neurosci 27:3626–3638
14. Dumont JPC, Robertson RM (1986) Neuronal circuits: an evolutionary perspective. Science 233:849–853
15. Edwards DH, Heitler WJ, Krasne FB (1999) Fifty years of a command neuron: the neurobiology of escape behavior in the crayfish. Trends Neurosci 22:153–161
16. Evans PD, O'Shea M (1977) An octopaminergic neuron modulates neuromuscular transmission in the locust. Nature 270:257–259
17. Friesen WO, Kristan WB (2007) Leech locomotion: swimming, crawling, and decisions. Curr Opin Neurobiol 17:704–711
18. Furshpan EJ, Potter DD (1959) Transmission at the giant motor synapses of the crayfish. J Physiol 145:289–325
19. Getting PA (1983) Neural control of swimming in *Tritonia*. Symp Soc Exp Biol 37:89–128
20. Grillner S (2006) Biological pattern generation: the cellular and computational logic of networks in motion. Neuron 52: 751–756
21. Grillner S, Deliagina T, Ekeberg O, El Manira A, Hill RH, Lansner A, Orlovsky GN, Wallén P (1995) Networks that co-ordinate locomotion and body orientation in lamprey. Trends Neurosci 18:270–279
22. Heitler WJ, Burrows M (1977) The locust jump. I. The motor programme. J Exp Biol 66:203–219
23. Heitler WJ, Burrows M (1977) The locust jump. II. Neural circuits of the motor programme. J Exp Biol 66: 221–241
24. Hiebert GW, Whelan PJ, Prochazka A, Pearson KG (1996) Contribution of hindlimb flexor afferents to the timing of the phase transitions in the cat step cycle. J Neurophysiol 75: 1126–1137
25. Hoyle G, Burrows M (1973) Neural mechanisms underlying behavior in the locust *Schistocerca gregaria*. I. Physiology of identified motorneurons in the metathoracic ganglion. J Neurobiol 4:3–41
26. Johnston IA (1980) Specialization of fish muscle. In: Goldspink DF (ed) Development and specialization of skeletal muscle. Society for Experimental Biology seminar series, vol 7. p 123–148
27. Kahn JA, Roberts A (1982) The neuromuscular basis of rhythmic struggling movements in embryos of *Xenopus laevis* J Exp Biol 99:197–205
28. Katz PS (1998) Neuromodulation intrinsic to the central pattern generator for escape swimming in *Tritonia*. Ann N Y Acad Sci 860:181–188
29. Kiehn O, Sillar KT, Kjaerulff O, McDearmid JR (1999) Effects of noradrenaline on locomotor rhythm generating networks in the isolated neonatal rat spinal cord. J Neurophysiol 82:741–746
30. Kim YJ, Zitnan D, Galizia GC, Cho KH, Adams ME (2006) A command chemical triggers an innate behavior by sequential activation of multiple peptidergic ensembles. Curr Biol 16:1395–1407
31. Kriellaars DJ, Brownstone RM, Noga BR, Jordan LM (1994) Mechanical entrainment of fictive locomotion in the decerebrate cat. J Neurophysiol 71:2074–2086
32. Kristan WB (1983) The neurobiology of swimming in the leech. Trends Neurosci 6:84–88
33. Kullander K, Butt SJB, Lebret JM, Lundfald L, Restrepo CE, Rydstrom A, Klein R, Kiehn O (2003) Role of EphA4 and EphrinB3 in local neuronal circuits that control walking. Science 299:1889–1892
34. Levine RB, Morton DB, Restifo LL (1995) Remodeling of the insect nervous system. Curr Opin Neurobiol 5:28–35
35. Li WC, Soffe SR, Wolf E, Roberts A (2006) Persistent Responses to brief stimuli: feedback excitation among brainstem neurons J Neurosci 26:4026–4035

36. Li WC, Sautois B, Roberts A, Soffe SR (2007) Reconfiguration of a vertebrate motor network: specific neuron recruitment and context-dependent synaptic plasticity J Neurosci 27:12267–12276
37. Liao C (2004) Neuromuscular control of trout swimming in a vortex street: implications for energy economy during the Karman gait. J Exp Biol 207:3495–3506
38. Maier L, Rathmayer W, Pette D (1984) pH lability of myosin ATPase activity permits discrimination of different muscle fibre types in crustaceans. Histochemistry 81: 75–77
39. Maier L, Pette D, Rathmayer W (1986) Enzyme activities in single electrophysiologically identified crab muscle fibres. J Physiol 371:191–199
40. Marder E (2002) Non-mammalian models for studying neural development and function. Nature 417:318–321
41. Marder E, Bucher D (2001) Central pattern generators and the control of rhythmic movement. Curr Biol 11: R986–R996
42. Marder E, Bucher D (2007) Understanding circuit dynamics using the stomatogastric nervous system of lobsters and crabs. Annu Rev Physiol 69:291–316
43. Marden JH, Kramer MG (1997) Locomotor performance of insects with rudimentary wings. Nature 377:332–334
44. McLean DL, Fetcho JR (2008) Using imaging and genetics in zebrafish to study developing spinal circuits in vivo. Dev Neurobiol 68:817–834
45. Mentel T, Duch C, Stypa H, Wegener G, Müller U, Pflüger HJ (2003) Central modulatory neurons control fuel selection in flight muscle of migratory locust. J Neurosci 23: 1109–1113
46. Mesce KA (2002) Metamodulation of the biogenic amines: second-order modulation by steroid hormones and amine cocktails. Brain Behav Evol 60:339–349
47. Möhl B (1988) Short-term learning during flight control in *Locusta migratoria*. J Comp Physiol A 163:803–812
48. Müller KJ, Nicholls JG, Stent GS (eds) (1981) Neurobiology of the leech. Cold Spring Harbor Laboratory Press, Cold Spring Harbor, pp 113–146
49. Nachtigall W (1989) Mechanics and aerodynamics of flight. In: Goldsworthy GH, Wheeler CH (eds) Insect flight. CRC Press, Boca Raton, pp 1–29
50. Pflüger HJ, Duch C (2011) Dynamic neural control of muscle metabolism related to motor behavior. Physiology (Bethesda) 26:293–303
51. Reichert H (1992) Introduction to Neurobiology. Georg Thieme Verlag, Stuttgart and New York
52. Reichert H, Simeone A (2001) Developmental genetic evidence for a monophyletic origin of the bilaterian brain. Philos Trans R Soc Lond B Biol Sci 356:1533–1544
53. Richter DW, Spyer KM (2001) Studying rhythmogenesis of breathing: comparison of in vivo and in vitro models. Trends Neurosci 24:464–472
54. Richerson GB (2004) Serotonergic neurons as carbon dioxide sensors that maintain pH homeostasis. Nat Rev Neurosci 5:449–461
55. Roberts A, Li WC, Soffe SR (2010) How neurons generate behavior in a hatchling amphibian tadpole: an outline. Front Behav Neurosci 4:1–11
56. Sillar KT (2009) Mauthner cells. Curr Biol 19:R353–R355
57. Sillar KT, Li WC (2010) Tadpole swimming circuit. In: Shepherd GM, Grillner S (eds) Handbook of brain microcircuits. Oxford University Press, Oxford/New York
58. Sillar KT, Combes D, Ramanathan S, Molinari M, Simmers AJ (2008) Neuromodulation and developmental plasticity in the locomotor system of anuran amphibians during metamorphosis. Special issue on "networks in motion". Brain Res Rev 57:94–102
59. Skiebe P (1999) Allatostatin-like immunoreactivity in the stomatogastric nervous system and the pericardial organs of the crab *Cancer pagurus*, the lobster *Homarus americanus*, and the crayfish *Cherax destructor* and *Procambarus clarkii*. J Comp Neurol 403:85–105
60. Sprecher SG, Reichert H (2003) The urbilaterian brain: developmental insights into the evolutionary origin of the brain in insects and vertebrates. Arthropod Struct Dev 32: 141–156
61. Stevenson PA, Kutsch W (1988) Demonstration of functional connectivity of the flight motor system in all stages of the locust. J Comp Physiol A 162:247–259
62. Strauss R (2002) The central complex and the genetic dissection of locomotor behaviour. Curr Opin Neurobiol 12: 633–638
63. Stuart DG, Hultborn H (2008) Thomas Graham Brown (1882–1965), Anders Lundberg (1920–2009), and the neural control of stepping. Brain Res Rev 59:74–95
64. Watkins BL, Burrows M, Siegler MVS (1985) The structure of locust nonspiking interneurones in relation to their segmental ganglion. J Comp Neurol 240:233–255
65. Weeks JC (2003) Thinking globally, acting locally: steroid hormone regulation of the dendritic architecture, synaptic connectivity, and death of an individual neuron. Prog Neurobiol 70:421–442
66. Wiens TJ, Wolf H (1993) The inhibitory motoneurons of crayfish thoracic limbs: identification, structures, and homology with insect common inhibitors. J Comp Neurol 336: 261–278
67. Wilson DM (1961) The central nervous control of locust flight. J Exp Biol 38:472–490
68. Wolf H (1990) Activity patterns of inhibitory motoneurons and their impact on leg movement in tethered walking locusts. J Exp Biol 152:281–304
69. Wolf H, Pearson KG (1988) Proprioceptive input patterns elevator activity in the locust flight system. J Neurophysiol 59:1831–1853
70. Yeh SR, Fricke RA, Edwards DH (1996) The effect of social experience on serotonergic modulation of the escape circuit of crayfish. Science 271:366–369
71. Zitnan D, Hollar L, Spalovska I, Takac P, Zitnanova I, Gill SS, Adams ME (2002) Molecular cloning and function of ecdysis-triggering hormones in the silkworm *Bombyx mori*. J Exp Biol 205:3459–3473

# The Neural Bases of Emotions

# 24

Tamara B. Franklin and Isabelle M. Mansuy

The term "emotion" is understood intuitively but is difficult to define, and currently there is no real consensus in the literature as to its meaning. It is commonly described as a mental state, associated with bodily changes, which arise spontaneously but are consciously felt. Thus, emotions can encompass a wide range of personal states accompanied by a number of observable behaviors and physiological changes. Great progress in the understanding of the neural bases for emotion has been made in the past decades. However, several fundamental questions related to emotions and the processing of emotional information remain unanswered. The field of **affective neuroscience** addresses these questions by investigating how emotions and mood are represented in the brain. Currently research interests lie in delineating which neural structures and networks are required for emotional responses, and further identifying the molecular mechanisms acting in these brain areas. Other important questions include how emotional events are learned and stored, and how changes in the functioning of the mechanisms and networks engaged during emotional processing lead to mood disorders such as depression and anxiety disorders. This chapter will address past and present theories on the question of "What is emotion?" and will outline our current understanding of the neural mechanisms involved in emotional processing.

T.B. Franklin(✉)
European Molecular Biology Laboratory Monterotondo,
Via Ramarini 32, Monterotondo, IT-00015 Rome, Italy
e-mail: tamara.franklin@embl.it

I.M. Mansuy
Brain Research Institute,
Medical Faculty of the University Zürich,
Zürich, Switzerland

Department of Health Sciences and
Technology of the ETH Zürich, Neuroscience Center Zürich,
Winterthurerstrasse 190, CH-8057 Zürich, Switzerland

## 24.1 Theories of Emotion

### 24.1.1 Charles Darwin

Thirteen years after publishing *On the Origin of Species* in 1859, Charles Darwin (1809–1882) published another seminal book entitled *The Expression of Emotions in Man and Animals*. In this publication, he presented three major principles for the origin of emotions [15]. First, the *Principle of Serviceable Associated Habits* proposed that some emotions help to deal with emotional stimuli and, thus, are beneficial to the organism. Second, the *Principle of Antithesis* proposed that some emotions are just the opposite emotions to the aforementioned beneficial emotions, and they, themselves, have no beneficial attributes. Last, the *Principle of the Direct Action of the Nervous System* proposed that physiological changes that occur as the result of emotional stimuli make up the final part of emotional states.

He further developed two additional theories that have become an integral part of how we think about and research emotions today [13]. Importantly, he drew parallels between animal and human emotions, and explored cross-species similarities in emotional states. This concept has justified the use of animals in research aimed at understanding human emotions. A second significant claim was that there are a certain number of basic emotions, including anger, fear,

C.G. Galizia, P.-M. Lledo (eds.), *Neurosciences - From Molecule to Behavior: A University Textbook*,
DOI 10.1007/978-3-642-10769-6_24, 

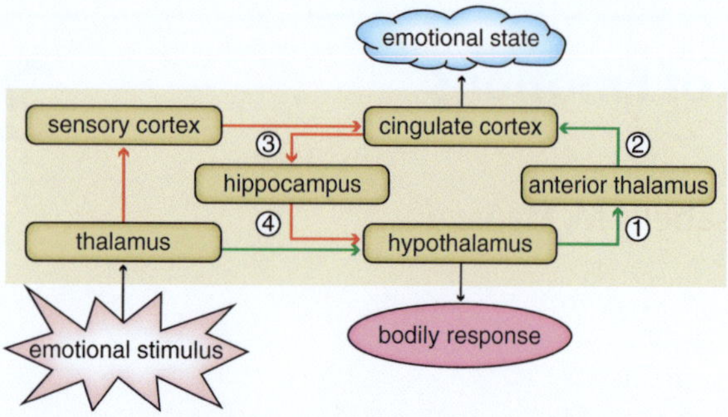


**Fig. 24.1** **The Papez circuit.** The Papez circuit theory of the functional neuroanatomy of emotion consists of two separate streams: "thought" and "feeling". Sensory messages containing information about emotional stimuli are processed by the thalamus and then transmitted to the cortex (stream of thinking, *red arrows*) and the hypothalamus (stream of feeling, *green arrows*). The stream of feeling carries emotional information through pathways from the hypothalamus to the anterior thalamus (*1*) and then to the cingulate cortex (*2*). The stream of thought corresponds to *top-down* cortical control of emotional processes; output pathways from the cingulate cortex to the hippocampus (*3*) and then to the hypothalamus (*4*). Feelings or emotional experiences are proposed to be the result of the integration of signals from the hypothalamus with information from the sensory cortex. While this theory is now outdated, many of the pathways proposed do exist and are key components of the emotional response (Adapted from Dalgleish [13] with permission)

surprise, and sadness, that are present across species and culture. This claim formed the basis for today's belief that different basic emotions have distinct underlying brain circuits and mechanisms.

### 24.1.2 The James-Lange Theory

Soon after Darwin's publication containing his concepts of emotion, a pioneering essay by William James (1842–1910) entitled *What is an Emotion?* proposed that emotion is the experience of **physiological changes** in the body which occur on exposure to emotional stimuli in our external environment [14, 25]. That is, when we see a snake, our muscles contract and our breathing rate increases in preparation for flight. According to James, it is these instinctive bodily changes experienced when seeing a snake that we perceive as fear of the snake. Consequently, it is not because we feel afraid that we prepare to run, but rather that we prepare to run, and so we feel afraid. Similarly, it is not that we feel sad so we cry, but rather we cry so we feel sad. This concept was further developed by Carl Lange (1834–1900), who claimed that different emotions are encoded by different patterns of physiological changes [13]. Thus, the concept that sets of physiological changes are the reason for emotional states is now termed the James-Lange theory of emotions. One of the major caveats for this theory is that it cannot explain the difference between similar but differing emotions such as shame and embarrassment, or fear and anxiety. Additionally, this theory does not explain why emotion often persists after the physiological response has subsided. However, this theory was the first to compile scientific knowledge of the time into a comprehensive theory of emotion.

### 24.1.3 The Cannon-Bard Theory

Walter B. Cannon (1871–1945) criticized the James-Lange theory for several reasons. One criticism was that several studies undertaken in light of James' theory demonstrated that the same set of physiological changes can be associated with both different emotional and non-emotional states [10, 13]. A collaboration between Cannon and Philip Bard (1898–1977) further opposed the James-Lange theory [7, 13]. Bard studied the effect of brain lesions on emotional behaviors in cats, and found inappropriate anger responses in decorticated cats, i.e., cats where the neocortex had been removed. Cannon and Bard argued that this was not in accordance with the James-Lange theory, as sensory and motor cortices would need to be intact if emotions were indeed the perception of bodily changes [13]. Instead, they argued that the **hypothalamus elicits emotional responses**, and that emotional responses are inhibited by neocortical regions. This would explain why decorticated animals display a lack of behavioral control and irrational emotional responses. Thus, Cannon and Bard were the first to propose a specific **brain mechanism for emotion** and, importantly, advocated the use of surgical lesion studies as a means to understand brain processes.

### 24.1.4 The Papez Circuit

The Papez circuit was first suggested in 1937 [13] (Fig. 24.1). Papez (1883–1958) believed that the brain has two separate streams: "thought" and "feeling". In his theory, sensory input of an emotional stimulus is processed by the thalamus, and then transmitted to the sensory cortices, to either the **cingulate cortex** (the thought stream) or the **mamillary bodies** of the hypothalamus (the feeling stream). The thought stream continues from the cingulate cortex through the cingulum pathway to the hippocampus, through the fornix, to the mamillary bodies of the hypothalamus, and back to the anterior thalamus through the mamillothalamic tract. The feeling stream continues from the mamillary bodies, through the anterior thalamus up to the cingulate cortex. Thus, the cingulate cortex acts to integrate information from the hypothalamus and the sensory cortex to produce emotional states. Additionally, Papez's theory allowed for cortical regulation of emotions as a result of projections from the cingulate cortex to the hypothalamus. While it is now known that not all regions named in Papez's circuit are required for the activation of emotional states, many of the proposed pathways do exist, and several regions including the hypothalamus and cingulate cortex are indeed key contributors to emotional processing.

### 24.1.5 The Limbic System

In 1949, Paul MacLean (1913–2007) proposed an anatomical model of the brain regions involved in emotional processing. This model took components of both the Papez circuit and Cannon-Bard theory and integrated them with the findings of Kluver and Bucy published in 1939 which demonstrated that bilateral removal of the temporal lobes in monkeys resulted in a particular set of behaviors, including reduced emotional reactivity, increased exploratory behavior, and hypersexuality [27]. Based on these findings, structures within the **temporal lobe** became an important component in MacLean's theory of the limbic system [13].

MacLean viewed the brain in three parts. The first consists of the evolutionary primitive reptilian brain and includes the basal ganglia, a group of nuclei composed of the striatum, the globus pallidus, the substantia nigra, and the subthalamic nucleus. He suggested that these regions are responsible for primitive emotions like fear and aggression. The second part consists of the visceral brain, and includes several parts of the Papez circuit such as the thalamus, hypothalamus, hippocampus, and cingulate cortex, as well as the amygdala and prefrontal cortex. He suggested that these regions were responsible for augmenting the primitive responses controlled by the ancient reptilian brain, and for social emotions. The third part consists of the new mammalian brain, which mostly includes the neocortex. He proposed that these regions act as an interface between emotions and cognition, and are responsible for administering top-down control (control by brain structures responsible for higher level functions) over all emotional responses.

Like James, MacLean proposed that external events lead to **physiological changes** which, when recognized by the brain and **integrated with perceived events in the outside world**, result in emotional experience [13]. In his view, this integration occurred within the **visceral brain**, which he later termed "the **limbic system**". While the concept of the limbic system has since dominated theories on the neural basis of emotion, it has come under criticism and some have even suggested to drop the term altogether. One of the reasons is that some regions of the limbic system may be less important for emotional experiences than MacLean supposed. For instance, MacLean suggested the hippocampus as a key region in the integration of stimuli that generate emotional experience. However, damage to the hippocampus does not alter emotional responsivity, or result in clinical changes in emotional processes. Instead, the hippocampus is now believed to play a much larger role in cognitive aspects of emotional tasks. Further, regions suggested to play a larger cognitive rather than emotional role are now known to be key in mediating emotional processes. One significant example of this is the **amygdala**, a region now known to be a major component in emotional processing (see Sect. 24.2). More generally, no criteria have been established for inclusion of a particular structure in the limbic system. Thus, while the limbic system is arguably one of the most well-known neural systems of the twentieth century, its relevance in current neuroscience is under scrutiny.

### 24.1.6 Modern Theories of Emotion

#### 24.1.6.1 Damasio and the Somatic Marker Hypothesis

Antonio R. Damasio, a modern neurobiologist with a distinctly pro-Jamesian theory of emotion, suggests that an emotion is the collection of **bodily changes in response to visual or auditory mental images** [14]. He further distinguishes between an emotion, and the feeling of an emotion. For Damasio, feeling an emotion is a cognitive response to the image or thought that stimulated the emotion, combined with the realization of this causal relationship between the thought and the resulting bodily changes.

Damasio further theorizes that emotions can guide decision-making. He suggests that a physiological reaction tags previous events with emotionally relevant stimuli, and has termed this phenomenon, a **somatic marker**. A negative somatic marker acts as an alarm bell, while a positive somatic marker is an incentive. The somatic marker hypothesis suggests that somatic responses to thought(s) increase the accuracy and efficiency of decision-making, and implies that decision processing is not only the result of logical reasoning, but also includes a "gut reaction" which is essential for appropriate and rational behavior.

#### 24.1.6.2 The Schachter-Singer Theory

The Schachter-Singer theory of emotion (sometimes called the **two-component theory of emotion**) proposes that there are two main parts in emotion: a part composed of Jamesian bodily factors, and a cognitive component that allows us to properly label and dissociate emotions [45]. This theory was based on a series of experiments that manipulated both the level of physiological arousal through injections of epinephrine, and the cognitive state of the individual by alteration of social environments. These experiments demonstrated that people do identify emotions based on bodily cues stemming from arousal levels, and cognitive determinants resulting from the circumstance in which they are placed.

#### 24.1.6.3 Edmund Rolls and Emotion

Another two-system approach to emotion is taken by Rolls, who proposed that emotions are elicited by either positive/rewarding, or negative/punishing **instrumental reinforcers** [9, 43] (Fig. 24.2). Emotional states are produced by **delivery** of a reward or punishment, or **omission** or **termination** of a **reward or punishment**. Thus, it is due to the fact that instrumental reinforcers have a goal-related aspect (delivery of a reward/termination or omission of a punishment) that they elicit emotions. For example, sweet taste when

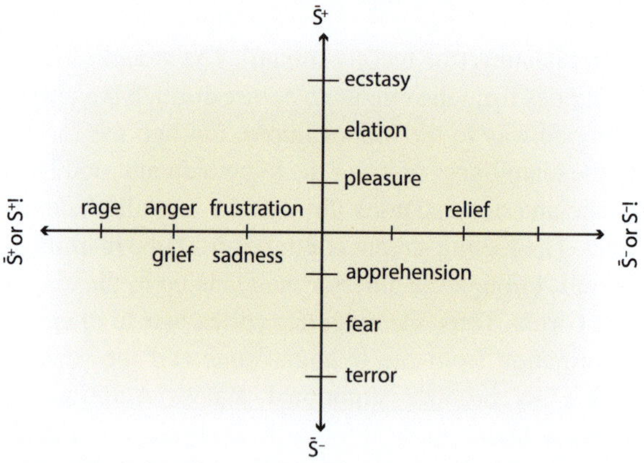


**Fig. 24.2** **Rolls' theory of emotion.** According to Rolls, emotions are states resulting from positive (rewarding) and negative (punishing) instrumental reinforcers. The vertical axis of the graph is emotions elicited by presentation of a reward (S+) or punishment (S−). Termination (or omission) of a reward (S+or S+!) or punishment (S− or S−!) is placed on the horizontal axis. The intensity of the emotion increases as distance from the midpoint increases. The possibility of an active or passive behavioral response is also accounted for. An example of this is that a termination (or omission) of a positive reinforcer may result in anger (an active response), but if this is not possible, in sadness/grief (a passive response) (Reproduced from Calder [9] with permission)

hungry is an instrumental reinforcer that will produce physiological responses, such as salivation. However, the neural processes required for producing salivation can be distinct from those specifying this stimulus as a goal for action, and inducing goal-directed instrumental learning. Thus, Rolls' theory greatly differs from the Damasio/Jamesian view by minimizing the role of physiological responses, and instead defining emotion as states induced by instrumental reinforcers.

## 24.2 Neuroanatomy of Emotion

Findings from human imaging and animal studies have identified the **amygdala**, **prefrontal cortex**, **anterior cingulate cortex**, and **hypothalamus** (Fig. 24.3), as four main structures involved in emotional processing. The following is a brief discussion of these brain areas and their known function in terms of emotions [13].

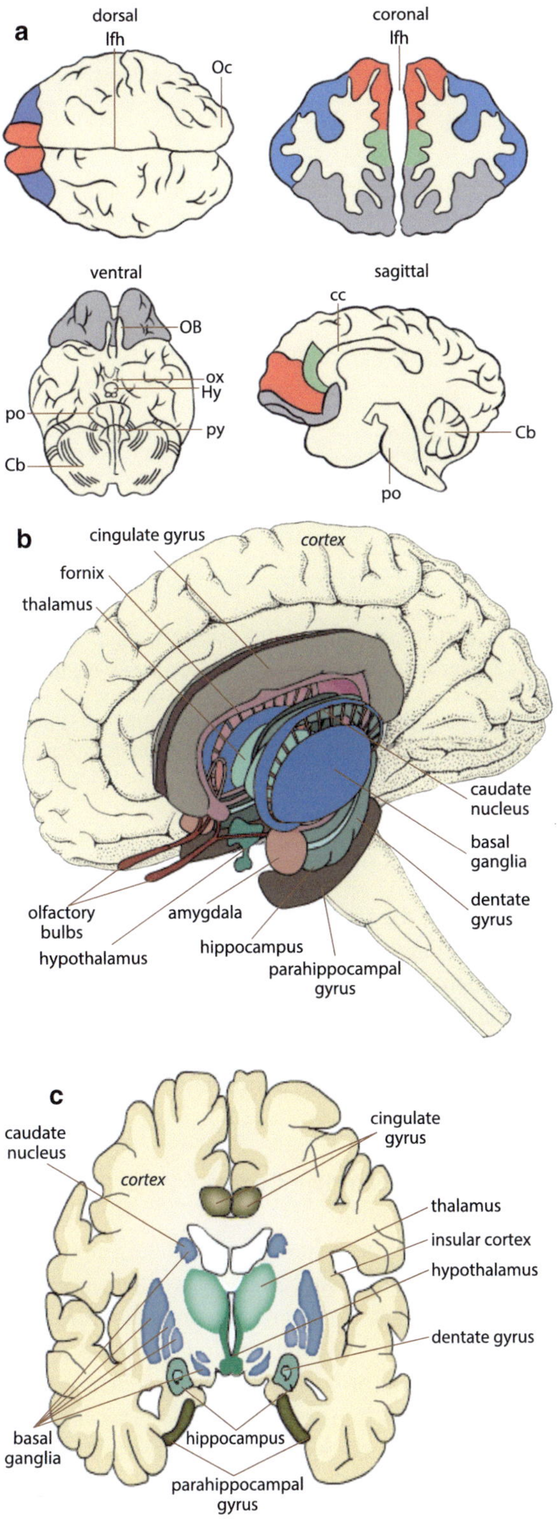


### 24.2.1 The Amygdala

The amygdala is an almond-shaped structure in the medial temporal lobe important for the processing of emotional information. It is a heterogeneous area of the brain, containing several distinct subnuclei with differing cell types and density, neurochemical composition, and functional connectivity. Although it was first characterized as a brain region already in the early nineteenth century, there is continuing debate about the appropriate divisions of subnuclei due to the heterogeneity of the structure. Currently, the most commonly used division is based on the evolutionary origins of two different regions of the amygdala: the primitive division, associated with the olfactory system (the cortico-medial region, including the cortical, medial, and central nuclei), and an evolutionarily newer division associated with the neocortex (the basolateral region, including the lateral, basal, and basal accessory).

The **lateral amygdala**, often referred to as the **gatekeeper** of the amygdala, receives input from sensory systems, including the visual, auditory, somatosensory, olfactory, and taste systems [32, 42]. These sensory inputs generally terminate in the dorsal subnucleus of the lateral amygdala, which in turn is connected with the ventrolateral and medial areas of the lateral amygdala. The **central nucleus** is known as the **output region** for the expression of innate emotional and physiological responses. Connections from the central nucleus to the brainstem act to control both behaviors and physiological responses. While the connections between lateral amygdala and the central nucleus are minimal, there are many connections between the

**Fig. 24.3 Key structures involved in emotional processing in the human brain.** Location of key structures involved in emotional processing, including the amygdala, prefrontal cortex, hypothalamus, and cingulate cortex. The involvement of these regions in aspects of emotional processing is discussed in Sect. 24.2 (**a** Reproduced from Perry et al. [39] with permission, (**b**, **c**) From http://uhpsych100.wikispaces.com/chapter+three). Abbreviations: (*OB*) olfactory bulb; (*Cb*) cerebellum; (*lfh*) longitudinal fissure of hemisphere; (*Oc*) occipital cortex; (*cc*) corpus callosum; (*ox*) optic chiasm; (*Hy*) hypothalamus; (*py*) pyramidal tract; (*po*) pons

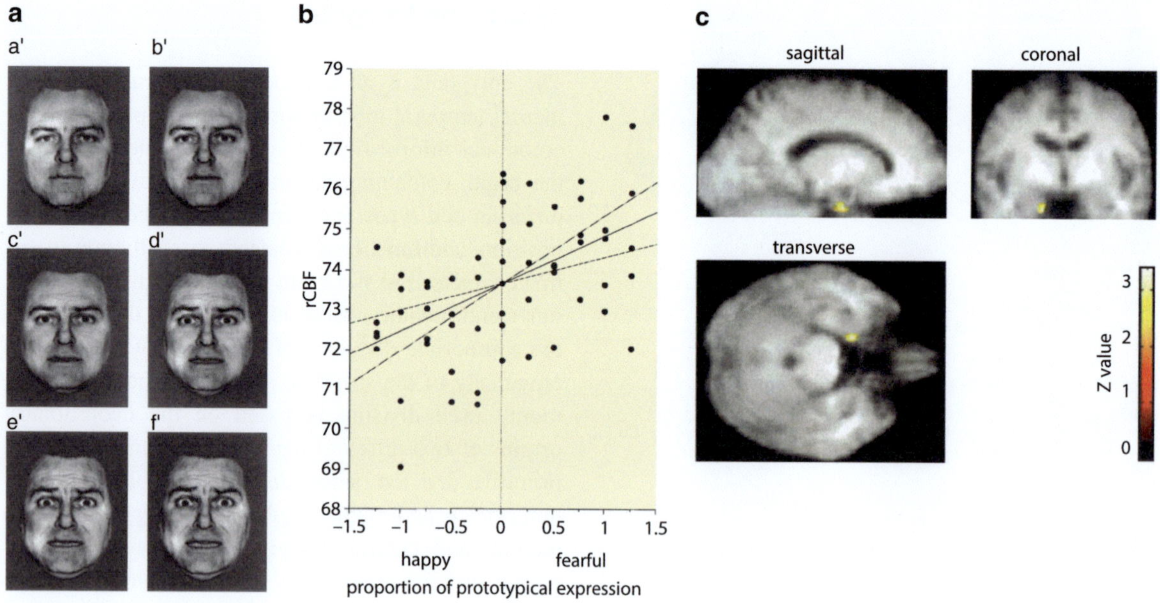


**Fig. 24.4** **Differential response to fearful and happy facial expressions in the left amygdala of human subjects.** (**a**) Prototypical neutral and fearful facial expressions are shown in faces *a* and *f*, respectively. Faces *b*–*e* are facial expressions falling within this continuum. (**b**) Regional cerebral blood flow (*rCBF*) as measured by positron-emission tomography (PET) values. (**c**) Statistical parametric maps (SPM) showing activation in only the left amygdala is significantly greater (*yellow*) when viewing fearful pictures as opposed to those showing happy expressions. SPM is representative of this contrast in rCBF values and shows significant difference in activation in the left amygdala. The z-score is representative of the deviation of activation elicited from fearful faces compared to the mean activation elicited from happy faces (Reproduced from Morris et al. [37] with permission)

medial part of the lateral nucleus to other amygdala subnuclei, that in turn connect with the central nucleus. Additionally, output connections from the basal nucleus to striatal areas mediate the expression of instrumental behaviors, such as running to safety in response to a predator.

The amygdala is an important region for emotional processing, known to be involved in the processing of social signals of emotion, emotional learning, and in the consolidation of emotional memory. Much of the research on the role of the amygdala is associated with **fear**, but the amygdala is activated by multiple **negative stimuli**. For example, fearful faces activate the left amygdala, and increased amygdalar response significantly correlates with images of increased intensity of fearfulness, and decreased response with increasing level of happiness (Fig. 24.4). Increased intensity of sad facial expressions have also been demonstrated to increase neuronal activity in the left amygdala, confirming its role in appraisal of negative facial expressions. In addition to imaging studies, patients with bilateral amygdala damage have demonstrated the importance of the amygdala in fear recognition, as recognition of fear in fearful faces is severely impaired in these patients. Along with its role in emotional processing, the amygdala has also been suggested to be involved in **reward processing**, and the use of reward signals in decision-making processes.

### 24.2.2 The Prefrontal Cortex

The participation of the prefrontal cortex in emotional processing was first suggested by a tragic accident occurring in 1848 to a construction site foreman, Phineas Gage [13]. While tamping gunpowder in a blast hole with an iron rod, the gunpowder exploded and propelled the iron rod through his head. The rod entered under his left eyebrow and exited through the top of the skull, causing major damage and a significant loss of brain tissue in the prefrontal cortex. Gage recovered from his injury, but his personality was greatly altered. Before the accident, Gage was personable and good-natured, but following the lesion he became impatient and quick to anger. These changes in emotional behavior were the first to suggest that the prefrontal cortex is important for establishing appropriate emotional behavior.

The prefrontal cortex is thought to influence emotion by recognizing the emotional and motivational **saliency of stimuli**. In particular, the orbitofrontal region of the prefrontal cortex has been suggested to work in concert with the amygdala to learn and represent associations between new stimuli and primary or innate reinforcers such as food, drink and sex [13]. It has been proposed that the prefrontal cortex is important for changing the **reward value** attached to learned stimuli, and thereby altering associated behavioral responses to these stimuli [43]. The ventromedial prefrontal cortex has been suggested to process the code of somatic markers important in decision processing, as described by Damasio (see 24.1.6.1). In support of this, Damasio et al. [14] described a patient with damage to the ventromedial prefrontal cortex who had difficulties when the application of subtle emotional values to multiple stimuli, and not logical reasoning alone, were required to make appropriate decisions. The involvement of the ventromedial prefrontal cortex in the development of somatic markers was further confirmed in studies with patients with ventromedial prefrontal cortex damage performing a card-sorting task. In this task, patients and control subjects were asked to play a card game in which they could win or lose a cash amount. Participants were allowed to choose 100 cards from any of 4 decks, but were not told how many cards they will be allowed to select ahead of time. Two of the decks contained cards that gave large rewards, but also large losses, while two of the decks gave smaller rewards, but smaller losses. In the long-term these latter two decks were advantageous. Thus, to win this task it was required that immediate rewards were ignored in order for delayed rewards to be provided. Control subjects performed this task based on intuitive decisions, and were unaware of any particular strategy. However, skin conductance response was increased in anticipation of poor choices within the game. In contrast, patients with ventromedial prefrontal cortex lesions performed poorly in the task, and did not exhibit enhanced skin conductance response in anticipation of poor decisions. This suggests that while control subjects developed somatic markers enabling appropriate decision-making, patients lacking ventromedial prefrontal cortex were unable to do this.

### 24.2.3 The Anterior Cingulate Cortex

First suggested to be involved in conscious emotional experience by Papez, the current concept of the anterior cingulate cortex is that it is divided into a dorsal cognitive subdivision, and a rostral, ventral affective subdivision [13]. The anterior cingulate cortex is thought to incorporate visceral, attentional, and emotional information and be a key brain area involved in the **top-down regulation of emotion**. The current hypothesis is that the anterior cingulate cortex monitors for any differences between the ongoing functional state of the organism and new incoming information with emotional or motivational importance. If a conflict between the current state and additional information is identified, then information concerning this discrepancy is projected to areas of the prefrontal cortex, within which appropriate response options are weighed (see Sect. 24.2.2 on prefrontal cortex above).

The anterior cingulate cortex is activated by a variety of emotional stimuli, as demonstrated by human imaging studies. In particular, the anterior cingulate cortex has been implicated in emotional tasks requiring a cognitive aspect, and in emotional recall or imagery. For example, increased neuronal responses to increasing intensity of angry facial expression in the anterior cingulate cortex have been observed.

### 24.2.4 The Hypothalamus

The role of the hypothalamus in emotional response was first clearly suggested by Walter Hess' experiments in the 1920s [13]. Hess implanted electrodes into the hypothalamus of cats and demonstrated that electrical stimulation in one part of the hypothalamus results in an "affective defense reaction". This defense reaction included increased heart rate, heightened levels of alertness, and enhanced likelihood to attack. In the 1950s, James Olds and Peter Milner performed similar electrical stimulation studies in rats [13]. Here, the hypothalamus was observed to be involved in the processing of reward signals, such that a rat continues to press a lever to deliver an electrical self-stimulation to the hypothalamus. The drive to perform such behavior is so strong

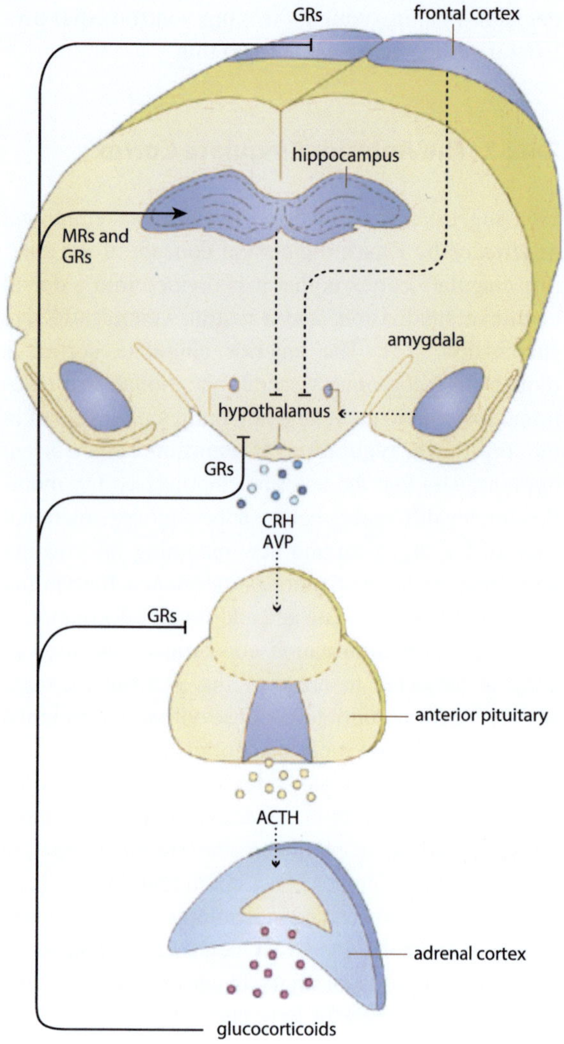


**Fig. 24.5 The hypothalamic-pituitary-adrenal axis.** On activation of the stress response, corticotrophin-releasing hormone (*CRH*) and arginine vasopressin (*AVP*) are released from the medial parvocellular region of the paraventricular nucleus of the hypothalamus. This leads to the release of adrenocortropic hormone (*ACTH*) from the pituitary gland, which in turn, leads to the production of glucocorticoids by the adrenal cortex. Binding of glucorticoids to the glucocorticoid receptor (*GR*) and the mineralocorticoid receptor (*MR*) regulate the responsiveness of the HPA axis to stressful stimuli by regulating ACTH and CRH release. Following withdrawal of the stressor, feedback loops from the adrenal gland to the hypothalamus, hippocampus, and frontal cortex shut down the HPA axis and returns it to baseline activity. The amygdala initiates the stress response by activating the HPA axis (Modified from Lupien et al. [35] with permission)

that a rat will self-stimulate continuously for 75 % of the time, up to 4 h/day. Robert G. Heath further demonstrated the role of the hypothalamus in **reward signaling** by observing self-stimulation of the hypothalamus through electrodes in humans. The hypothalamus, along with the prefrontal cortex, amygdala, and ventral striatum, has now been identified as a key region in the reward pathway in the brain. It has also been observed to be involved in the motivation for basic behaviors like **sex** and **food-seeking** (see Sect. 24.6).

## 24.3 Stress

The subjective state of perceiving adverse changes in the environment, whether potential or actual, is termed "stress" [26]. It is a precipitating factor for the development of psychiatric disorders like **depression** and **anxiety disorders**. In addition to stress, genetic variation may also predispose individuals to mental illness. Stressful stimuli cause the brain to activate several stress mediators, including neurotransmitters, peptides, and steroid hormones. The following is a discussion of the mechanisms associated with and underlying the processing of stressful stimuli.

### 24.3.1 The Hypothalamic-Pituitary-Adrenal Axis

One of the main systems activated when perceiving a stressor is the hypothalamic-pituitary-adrenal (HPA) axis, which regulates the **hormonal aspects of the stress response** [16, 35]. Activation of this axis results in the release of **glucocorticoids** from the **adrenal gland** through several steps (Fig. 24.5). On activation of the stress response, neurons in the medial parvocellular region of the paraventricular nucleus of the hypothalamus release corticotropin-releasing hormone (CRH, also known as corticotropin-releasing factor, CRF) and arginine vasopressin (AVP) into the portal vessel system, which activates the synthesis of pro-opiomelanocortin (POMC) in the anterior pituitary. POMC is then processed into adrenocorticotropic hormone (ACTH), and opioid and melanocortin peptides. ACTH released from the pituitary gland in turn stimulates the release of glucocorticoids (cortisol in humans, and corticosterone in humans, rats, and mice) from the adrenal cortex. Additionally, the adrenal medulla releases catecholamines (adrenaline and noradrenaline). These hormones can produce multiple effects including increased blood pressure and heart rate, reduced digestive capacity, and discharge of stored energy to facilitate muscle use.

The amygdala can act directly to activate the stress response. This is thought to be predominantly due to disinhibition of the PVN as a result of sequential GABA synapses (GABAergic neurons in the amygdala projecting to GABAergic neurons in PVN-projecting brain areas). The HPA axis is also under negative feedback control. On removal of the stressful stimulus, the HPA axis returns to baseline via several negative feedback loops. Circulating glucocorticoids negatively regulate ACTH and CRH release by binding glucocorticoid receptors (GRs) and mineralocorticoid receptors (MRs) in several brain areas including the hippocampus, the prefrontal cortex, and the hypothalamus. This occurs as a result of direct inhibitory influence on the hypothalamus, and indirect inhibitory influence of the hippocampus and prefrontal cortex as a result of excitation of inhibitory PVN-projecting brain areas. Thus, **effective stress coping** occurs when the **stress response is activated on exposure** to a stressful stimulus, but is **quickly turned off** on removal of the stimulus. An alteration of these circuits leading to an inadequate stress response, or an extreme or prolonged response to stressful stimuli is thought to underlie most stress-related mental disorders, such as depression or anxiety disorders.

GRs are activated by cortisol, corticosterone, and other glucocorticoids, and are expressed ubiquitously throughout the body, with very high expression levels in stress-related brain regions. MRs are activated by aldosterone, deoxycorticosterone, as well as glucocorticoids and progestins, and are predominantly located in limbic structures, including the hippocampus, amygdala, and prefrontal cortex. When homo- or heterodimeric, these receptors act as transcriptional regulators by interacting with glucocorticoid response elements (GREs), and recruiting corepressors or coactivators, thereby mediating the responsiveness of the system [16]. GR monomers interact with stress-induced transcription factors (TFs), such as nuclear factor-κB (NF-κB) and activator protein 1 (AP1) to reduce transcriptional activity.

Perhaps due to the importance of appropriate corticosterone signaling, many different mechanisms exist to mediate its availability. In the blood, corticosteroid-binding globulin can act as a sink to bind corticosterone, making it unavailable as a signaling molecule [16]. Additionally, the multidrug resistance (MDR) P-glycoprotein, a protein present in the blood–brain barrier (BBB), reduces the levels of glucocorticoids entering into the brain. As well, factors like heat shock proteins, can bind corticosterone to form a multimeric receptor-protein complex, consequently changing the conformation of the receptor. Steroid metabolism is another important factor in mediating corticosteroid signaling, for instance by converting the bioactive cortisol to its inactive metabolite, 11-dehydrocortisol (cortisone), or in contrast, by regenerating bioactive cortisol in the brain. Signaling is further regulated by interaction of GRs with transcription factors, and interactions of MRs and GRs with coregulators, which can act to repress or activate gene transcription [16].

### 24.3.2 The Binary Organization of the Stress System

The stress system is binary in nature. It consists of a **fast response**, involving the CRH-dependent sympathetic **fight-flight response**, and a **slow response** associated with **adaptation and recovery**, involving the urocortin-dependent parasympathetic response (Fig. 24.6) [16]. The sympathetic system is mediated through CRHR1 activation, while the parasympathetic system depends on CRHR2 activation. Transgenic mice deficient in CRHR1 are less anxious and display an impaired stress response, while mice deficient in CRHR2 display increased anxiety, and a more rapid stress response. However, pharmacological studies administering either CRHR2 agonists or antagonists into specific brain areas have suggested that CRHR2 may have both an anxiolytic and anxiogenic role [5].

Both CRHR1 and CRHR2 are seven-transmembrane G protein-coupled receptors, which signal mainly by coupling to Gs, thereby activating adenylyl cyclase and protein kinase A [5]. There are four known ligands for CRHR1 and CRHR2: CRF, urocortin I, urocortin II, also known as stresscopin-related peptide, and urocortin III, also known as stresscopin. CRH has a much higher affinity for CRHR1 than CRHR2, urocortin I has equal affinity for both receptors, and urocortin II and III appear to be selective for CRHR2. CRHR1 is widely expressed throughout the brain with high levels in the cerebral cortex, cerebellum, amygdala, hippocampus, and olfactory bulb [51]. CRHR2 appears to be more discretely expressed in the brain, with highest expression in the striatum (lateral septal nucleus), pallidum (bed nucleus of stria terminalis), ventromedial hypothalamus, olfactory bulb, and midbrain raphe nuclei.

The binary nature of the stress response can also be illustrated by the theorized role of GRs and MRs. The **GR:MR balance hypothesis** is linked to the recent discovery of membrane MRs. Corticosterone binds to nuclear MRs with a ten-fold higher affinity than to GRs. However, membrane MRs have a much lower affinity for corticosterone than nuclear MRs. Therefore,

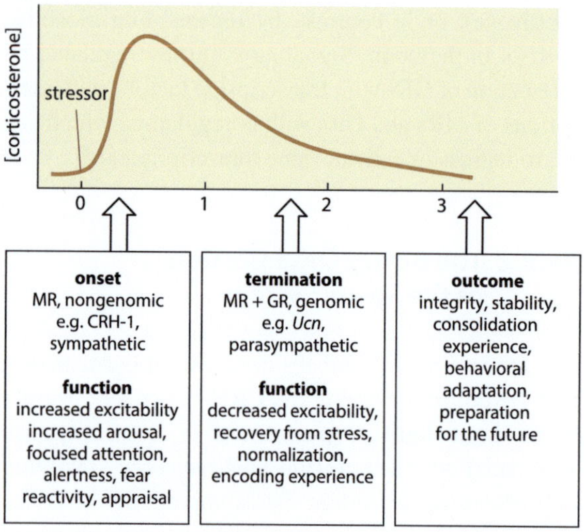


**Fig. 24.6** **Phases following exposure to a stressor.** Both fast nongenomic and slow genomic actions are required for encoding the emotional experience, behavioral adaptation, and preparation for future events, in response to stressful stimuli. On activation of the HPA response by a stressor, corticosterone levels rise. Increased levels of corticosterone can result in nongenomic actions on the excitability and activation of neurons that can enhance the action of the initial stress mediators. In order to prevent stress reactions from overshooting, gene-mediated pathways are activated that lessen the initial stress response. Structural and functional changes promoting homeostatic recovery, replenishment of depleted energy resources, and emotional memory also occur (Reproduced from de Kloet et al. [17] with permission)

GRs and membrane MRs are only occupied at pulsatile peaks of basal corticosteroid secretion resulting from stressful conditions. Thus, while nuclear MRs can regulate gene expression, resulting in slow and persistent changes in the conductance of the plasma membrane, membrane MRs can have rapid effects on the cell. They promote presynaptic glutamate release, and thereby increase both pre- and postsynaptic excitatory transmission (see Chap. 8). Thus, low-affinity membrane MRs are thought to increase excitability and facilitate a fast corticosteroid feed-forward response, thereby enhancing attention, vigilance, and risk-assessment responses. Nongenomic and genomic actions of GRs subsequently stop the initial stress reaction by suppressing the enhancement in excitability. Activation of GRs is suggested to promote mechanisms that prepare for future situations by not only facilitating recovery from the stress reaction, but also promoting memory storage, and mobilizing energy stores.

### 24.3.3 The Effects of Chronic Stress

Chronic stress has detrimental consequences on the brain and can severely alter neuronal structures, but differently depending on the brain structure. It has region-specific effects on dendritic organization in the CA3 subregion of the hippocampus, prefrontal cortex, and amygdala [52]. Animals exposed to chronic stress have decreased dendritic branching, shorter CA3 pyramidal apical dendrites, and fewer synaptic contacts on these neurons. Similarly, chronic corticosterone administration or chronic stress results in decreased dendritic length and spine density in the layer II/III neurons of the prefrontal cortex. In the prefrontal cortex, mushroom-shaped spines, thought to be important in the establishment of synaptic connections, are most susceptible to the effects of chronic stress. Contrary to effects seen in the hippocampus and prefrontal cortex, chronic immobilization stress increases dendritic arborization, spine formation, and synaptic connectivity in the basolateral amygdala [52], but not in the central nucleus. While both, acute and chronic stress increase spine density in basolateral amygdala, only chronic stress increases dendritic arborization. A period without stress reverses the stress-induced atrophy of hippocampal and prefrontal cortex neurons but fails to reverse the **hypertrophy of pyramidal neurons in the amygdala** [12, 40, 53].

Both chronic overexposure to stress hormones and chronic stress treatments also have a negative impact on cell proliferation and neurogenesis in the adult brain. Overexposure to corticosteroids reduces the proliferation and survival of progenitor cells in the subgranular zone of the dentate gyrus. However, this effect is not permanent, and the reduction in progenitor cell proliferation and survival is partially corrected after several weeks of recovery. Two models that parallel high levels of corticosteroids and reliably induce depression-like behaviors in rodents are unpredictable chronic mild stress and chronic social defeat stress. Both models also affect neurogenesis and proliferation; chronic mild stress reduces survival of new brain cells in the hippocampus and subventricular zone, and chronic social defeat stress reduces cell proliferation in the medial prefrontal cortex and neurogenesis in the dentate gyrus. Additionally, chronic, but not acute, restraint stress reduces neurogenesis in both the dentate gyrus and the dorsal vagal complex, an integrative area of the brainstem that relays the autonomic neural responses to stress. Interestingly, both antidepressant administration and treatment with CRH or vasopressin antagonists can prevent the reduction in proliferation of progenitor cells induced by chronic exposure to high levels of corticosteroid or exposure to a chronic stress situation. Thus, the deficit in neurogenesis resulting from chronic stress conditions is **not irreversible** further emphasizing the ability of the hippocampus to be plastic even in the adult.

## 24.4 Disorders of Emotional State: Depression and Anxiety Disorders

Mental illnesses can be classified into **thought disorders**, which result from a disturbance in cognitive functions, or **mood disorders**, reflecting a perturbed emotional state [29]. While normal emotional responses such as euphoria, depression, and anxiety can have a beneficial outcome for an individual, these emotions can be persistently disordered, thereby developing into a disease state.

### 24.4.1 Depression

Depression is a common chronic mental disease affecting approximately 5 % of the U.S. population [49]. It is very difficult to treat, and currently only half of depressed patients show complete remission. Presently, depression can be classified into two main types: unipolar and bipolar. **Unipolar depression** is characterized by dysphoria (a state of feeling discontent or unhappy) most of the day, the inability to experience pleasure (anhedonia), and reduced motivation [4]. Additionally, at least three of the following symptoms are also present: disturbed sleep, diminished appetite and weight loss, decreased energy, decreased sex drive, restlessness, reduced speed of thoughts and actions, difficulty concentrating, indecisiveness, feelings of worthlessness, guilt, pessimistic thoughts, and thoughts about dying and suicide. An episode of depression typically lasts 4–12 months if untreated.

**Bipolar depression** occurs in patients who demonstrate both depressive and manic episodes. About a quarter of all patients with major depression will experience a manic episode at some point [2]. Depressive episodes in patients with bipolar disorder are similar to that seen in unipolar depression. The **manic episodes** are characterized by a heightened or irritable mood lasting for a minimum of 1 week, expressed with a combination of the following symptoms: hyperactivity, over-talkativeness, social intrusiveness, increased energy and libido, flight of ideas, grandiosity, distractibility, reduced need for sleep, and reckless spending [4]. Delusions and hallucinations are also observed in patients with severe cases.

Both post-mortem and neuroimaging studies have reported decreased gray-matter volume, small cell bodies, and glial density in the prefrontal cortex and hippocampus in depressed patients compared to control subjects [24, 30]. These brain regions are critical because they are thought to be involved in cognitive aspects of depression, including feelings of worthlessness and guilt. However, due to comorbid diagnoses and medication history of patients, their involvement has been difficult to prove, and obvious cause-effect relationship between the pathology and the diagnosis of depression is difficult to observe.

Functional studies using fMRI or PET have also demonstrated that depressed patients have chronically increased activity in the amygdala, a brain area where transient sadness results in increased activation in healthy subjects, and chronically decreased activity in the cingulate cortex. Following successful treatment of depression, normal activity in these brain regions was restored. Additionally, deep brain stimulation of the cingulate cortex (implantation of a device that sends electrical impulses to proximal brain areas) was successfully used to treat depressive symptoms in a selected cohort of treatment-resistant patients. It also successfully ameliorated depressive symptoms when applied to the nucleus accumbens, a subregion of the striatum important in the reward pathway and thought to be responsible for anhedonic behaviors often exhibited by depressed patients.

Although the neurobiological basis of depression is still not fully understood, two hypotheses have been proposed: a monoamine hypothesis and a neurotrophin hypothesis [30]. The **monoamine hypothesis** is based on early findings that two drugs developed for unrelated conditions, iproniazid and imipramine, have antidepressant effects in humans. Both drugs were also shown to enhance neurotransmission of the monoamines serotonin or noradrenaline. More selective agents have now been designed to increase monoamine transmission, through two main mechanisms: inhibiting neuronal uptake of a monoamine (e.g., selective serotonin reuptake inhibitors (SSRIs)) or inhibiting the breakdown of monoamines (e.g., monoamine oxidase inhibitors (MAOIs)). While SSRIs and MAOIs have immediate effects on neurotransmission, their antidepressant properties are not observed until several weeks of treatment. It is therefore now hypothesized that antidepressant medications initially increase the level of synaptic monoamines, then induce secondary effects. These secondary effects are thought to take time to develop, because they involve transcriptional and translational changes.

The **neurotrophin hypothesis** emerged from evidence that brain regions involved in emotional processing often have decreased volume in depressed patients compared to healthy controls [30]. However, imaging studies of depressed patients have shown that such reduction in volume can be reversed at least in the hippocampus, by antidepressant treatment. The mechanisms underlying such reduction are suggested to involve an alteration in the level of neurotrophic factors, a group of proteins known as growth factors involved in synaptic plasticity in the adult brain. Much of the research has focused on **brain-derived neurotrophic factor** (**BDNF**), a neurotrophin highly expressed in adult brain areas involved in emotional processing. BDNF-mediated signaling is reduced by stress in the hippocampus, but increased by antidepressant treatment [30]. Additionally, lower BDNF expression has been observed in the brain of depressed patients post-mortem. Because neurotrophins such as BDNF are involved in adult neurogenesis, it was theorized that their reduction in depressed patients could contribute to reduced hippocampal volume by altering neurogenesis. Thus, in the neurotrophin hypothesis the mechanism of action of some antidepressants is postulated to be increased neurotrophin signaling and neurogenesis in depressed patients.

Because current antidepressants have a slow onset of action and patients have a high rate of remission, new targets for the pharmacological treatment of depression have been sought. Recently, a **subset of depression that is associated with weight gain**, has suggested a link between pathways involved in mood and those involved in feeding and metabolism. One of these pathways involves melanin-concentrating hormone (MCH), a hormone secreted from neurons projecting from the lateral hypothalamus to several brain regions involved in emotion and motivation [30]. Antidepressant-like effects have been demonstrated by decreasing MCH-mediated signaling either globally or locally within the nucleus accumbens. Additionally, orexin and ghrelin, two peptides known to promote food intake, have been proposed to have an antidepressant effect, particularly during periods of caloric restriction [30]. Thus, the interaction between metabolic pathways and depression may provide new targets for antidepressant medication.

Another mechanism potentially underlying the slow onset of treatment and amelioration of symptoms is **epigenetic dysregulation**. Posttranslational modifications (PTMs) of histone proteins and DNA methylation, mechanisms known to modulate gene transcription, have been suggested as targets of antidepressant treatment. Recent studies in animals have examined the link between histone PTMs and depression using the chronic social defeat model. **Chronic social defeat** is a rodent model of chronic stress that induces behavioral abnormalities that can be reversed by chronic, but not acute antidepressant treatment. In mice, chronic social defeat alters gene expression and histone (H) methylation. It decreases the expression of two splice variants of *BDNF* (*BDNF III* and *BDNF IV*) in the hippocampus, and the dimethylation of H3 lysine (K) 27, a mark of transcriptional repression (Fig. 24.7). The behavioral abnormalities induced by chronic social defeat can be reversed by chronic antidepressant treatment, but H3K27 dimethylation is not affected. However, two marks of transcriptional activation, H3 acetylation and H3K4 methylation, increase at the same promoters after antidepressant treatment, suggesting a potential mechanism for the action of chronic antidepressant treatment.

DNA methylation has also been shown to be a target of antidepressant treatment. Chronic treatment increases methyl CpG binding protein 2 (MeCP2) and methyl binding domain protein 1 (MBD1), two proteins that bind methylated CpG dinucleotides and act as either transcriptional activators or repressors in the rodent brain. The increase in MeCP2 is specific to $\gamma$-aminobutyric acid (GABA)ergic interneurons, which is particularly significant because altered GABAergic transmission has been linked to major depression and suicide. Further, depressed patients who committed suicide have higher levels of methylation in the GABA-A $\alpha 1$ receptor subunit promoter in the prefrontal cortex when compared to control individuals who died of other causes. Thus, antidepressant treatments may specifically target factors affecting epigenetic modifications in cell types important in depression.

### 24.4.2 Anxiety Disorders

Anxiety disorders are extremely common in the general population, and their lifetime prevalence is currently almost 30 % [46]. Anxiety disorders thus represent a significant problem for our communities. They are characterized by excessive fear, and are expressed by avoidance of situations, people, or objects that are usually no threat to the individual. These disorders have been investigated in several functional neuroimaging studies, which are summarized in Table 24.1.

#### 24.4.2.1 Specific Phobias

Specific phobia is a relatively common disorder, with a lifetime prevalence of 7–11 % [46]. Patients with specific phobia have excessive and persistent fear and avoidance of specific situations or objects such as flying, small animals, enclosed places, or height. These situations or objects cause marked distress and deficits in social, occupational, and academic performance. Anatomically, specific phobias are associated with the amygdala and insular cortex. Increased responses in the amygdala, and enhanced functional connectivity between the right amygdala and periamygdaloid area, fusiform gyrus and motor cortex has been observed when patients were presented with images of the phobia. Such images also result in decreased activity in the prefrontal, orbitofrontal, and ventromedial cortex. This suggests that phobic reactions deactivate areas of the prefrontal cortex which are involved in top-down control over emotion-related areas like the amygdala, which results in preparation of motor responsiveness for fight-or-flight behaviors. Increased activation of the prefrontal, insular, and posterior cingulate cortex on presentation of phobia-related versus phobia-unrelated words was also observed in patients with specific phobia compared to control subjects.

#### 24.4.2.2 Social Anxiety Disorder

Similar to specific phobia, patients with social anxiety disorder, also known as **social phobia**, have a marked and persistent fear of social situations or performance that involve the judgement of others [46]. This fear of embarrassment causes significant distress and deficits in social, occupational, and academic performance. Patients with social anxiety disorder display a recognition bias towards negative voices; they are better at identifying sad or fearful prosodies, but show impairments in identifying happy prosodies versus control subjects.

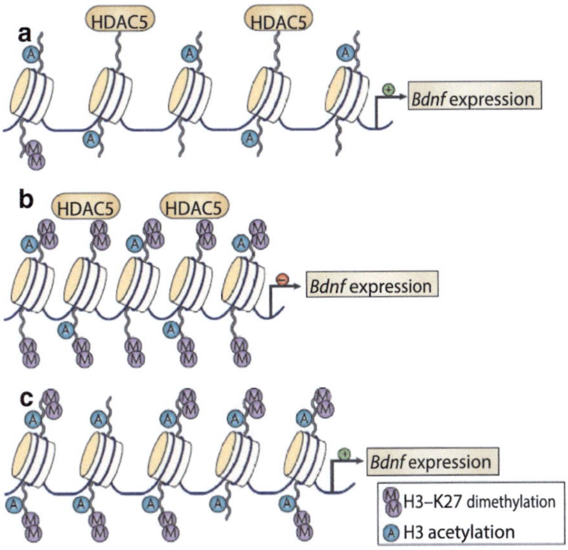


**Fig. 24.7 The role of posttranslational histone modifications in a rodent model of depression.** (**a**) Under basal conditions, the promoter region of the *BDNF* gene has moderate levels of H3 acetylation and H3K27 dimethylation, and is bound by the histone deacetylase HDAC5. (**b**) Following chronic social defeat stress, H3K27 dimethylation is increased. This results in increased condensation of the *BDNF* promoter, which leads to decreased *BDNF* gene expression. (**c**) Following chronic antidepressant (imipramine) treatment, HDAC5 levels are reduced, which leads to increased H3 acetylation. However, H3K27 dimethylation is not changed. Despite this, the increase in H3 acetylation is sufficient to reinstate normal levels of *BDNF* gene expression. *A* acetyl, *BDNF* brain-derived neurotrophic factor, *H* histone, *HDAC* histone deacetylase, *K* lysine, *M* methyl (Reproduced from Tsankova et al. [49] with permission)

Increased amygdalar response in patients with social anxiety disorder is observed during and in anticipation of public speaking, and on presentation of negative comments or emotional facial expressions. This exaggerated amygdalar response during public speaking is decreased in patients successfully treated for social phobia. Similarly, the insular cortex displays heightened activation in response to emotional facial expressions, and in anticipation of public speaking in patients with social anxiety disorder [34].

There are conflicting reports about the role of the dorsal and anterior cingulate cortex in social anxiety disorder. Increased rostral anterior cingulate cortex activation was elicited by facial expressions of fear and disgust in patients with social anxiety disorder. Additionally, orbitofrontal cortex activation increases

**Table 24.1** Summary of altered activity in functional neuroimaging studies of anxiety disorders.

| | Amygdala | rACC | dACC | Hippocampus | Insular cortex |
|---|---|---|---|---|---|
| Posttraumatic stress disorder | ↑ | ↓ | ↑* | ↑↓ | ↑↓ |
| Panic disorder | ↑↓* | ↑* | – | ↑↓ | – |
| Social phobia | ↑ | ↑↓* | ↑↓ | – | ↑ |
| Specific phobia | ↑ | ↑↓* | ↑ | – | ↑ |
| Generalized anxiety disorder | ↑↓* | ↑* | ↑* | – | – |

Reproduced with permission from Shin and Liberzon [46]
rACC rostral anterior cingulated cortex, dACC dorsal anterior cingulated cortex
↑ increased function in the disorder (relative to control groups)
↓ decreased function in the disorder (relative to control groups)
↑↓ mixed findings
* based on a very small number of studies
– too little information available

on presentation of angry versus neutral voices. However, decreased activation on provocation by a social anxiety-inducing stimulus [50], and decreased rate of glucose metabolism were demonstrated in the ventromedial prefrontal cortex of patients with social anxiety disorder. Increased dorsal anterior cingulate activation has been demonstrated as a result of negative comments and harsh or disgusted facial expressions. In contrast, decreased dorsal anterior cingulate cortex activation has been reported in response to angry faces, in anticipation of public speaking, and decreased glucose metabolism has been demonstrated in patients with social anxiety disorder [20]. Thus, the role of the cingulate cortex in social anxiety disorder still requires further clarification.

Social anxiety disorder has been associated with abnormal serotonergic and dopaminergic signaling. Patients with social anxiety disorder display reduced serotonin 1a receptor binding in the amygdala and insular cortex. Additionally, reduced D2 receptor (a dopamine receptor) binding and dopamine transporter densities have been found in the striatum of patients with social anxiety disorder, although one study has failed to replicate this.

#### 24.4.2.3 Panic Disorder

Patients with panic disorder experience unexpected panic attacks, accompanied by a persistent worry about having future attacks or concern for the implications of the attacks [46]. These panic attacks recur over a period of at least several months. **Panic attacks** themselves tend to be brief periods of terror (usually 10–15 min but lasting up to an hour in rare circumstances) during occasions which should not normally induce a fear response or, additionally, during times of social stress when the patients feel as if they were the focus of attention. A key feature of panic disorder is that panic attacks are unexpected and unpredictable. Panic attacks are accompanied by overactivation of the sympathetic nervous system, and result in increased heart rate, shortness of breath, dizziness, and chest pain. Panic disorders are often first diagnosed during adolescence.

Panic disorder has been associated with enhanced sensitivity of brain structures involved in fear pathways, including the amygdala, hippocampus, thalamus, and brain stem [46]. Additionally, failure of the frontal cortex to provide inhibitory input to the amygdala, thereby enhancing amygdalar output, has been suggested as an underlying mechanism. Increased gray matter volume in the midbrain and pons and reduced serotonin transporter and receptor binding, have been observed in patients with panic disorder [22]. The periaqueductal gray (PAG) is a key midbrain structure involved in the defensive reaction that is also affected. Both electrical and chemical stimulations of the PAG cause behavioral reactions including freezing, fight or flight, similar to responses elicited by a proximal threat, such as a predator. Thus, it has been suggested that structural changes and alterations in serotonergic signaling in the PAG are a key component in the neurobiology of panic disorder.

#### 24.4.2.4 Generalized Anxiety Disorder

Generalized anxiety disorder (GAD) is characterized by uncontrollable and excessive anxiety and worry [4]. Symptoms of GAD include restlessness, fatigue, irritability, muscle tension, and sleep and concentration difficulties. The lifetime occurrence rate for GAD is approximately 5 % [3]. It is rare in children and adolescents and, in contrast to the majority of anxiety disorders, it is unlikely to occur prior to the age of 25 years old. However, it is twice as likely to occur in women than men, and is increasingly likely as women age; in women over 45, the incidence rate is approximately 10 %, while in men it remains at approximately 3–4 %.

GAD has not been well studied up to this point, perhaps due to its relatively new classification; it was first classified as a distinct disorder in 1980 and was fully considered independent from panic disorder in 1994 [3]. Currently, both the amygdala and medial prefrontal cortex have been suggested to play a role in GAD by neuroimaging studies in human patients [46]. Patients with GAD display enhanced amygdalar and prefrontal activation in response to fearful and angry facial expressions, and enhanced amygdalar activation in anticipation of aversive photographs. Amygdala activation has also been positively correlated with GAD symptom severity. At present, such neuroimaging studies have been limited and it is still unclear exactly to what extent these structures contribute to the pathophysiology of GAD.

#### 24.4.2.5 Posttraumatic Stress Disorder

Posttraumatic stress disorder (PTSD) occurs as a consequence of a stressful or traumatic event, making it one of the few mental illnesses that require a stressor for its occurrence [4]. In order for PTSD to be diagnosed, a traumatic event, defined as **actual or threatened death or serious injury**, **or a threat to the physical integrity of self or others**, must occur, **followed by intense fear**, **helplessness**, **or horror**. Following this, the traumatic event is persistently reexperienced, and there is continuous avoidance of stimuli associated with the trauma. Traumatic events include combat situations experienced by soldiers, childhood abuse, rape, assault, and serious accidents. However, while the large majority of individuals are exposed to a traumatic event in their lifetime, only a minority of individuals are unable to cope with the stressor, and exhibit prolonged, abnormal behavioral and bodily responses due to the traumatic experience; approximately 7–12 % of the U.S. population will develop PTSD.

Dysregulation of the HPA axis has been implicated in the neurobiological mechanism underlying this disorder. A hallmark endocrinological marker of PTSD is **hypocortisolism**, distinguishing it endocrinologically from depression, a comorbid but distinct disorder. The neuroendocrinology suggests that patients with PTSD have increased sensitivity of the HPA axis to negative glucocorticoid feedback, as a result of increased GR binding and function. Additionally, in PTSD patients, there is a downregulation of pituitary CRF receptors that result in a blunted ACTH response to CRF stimulation. The downregulation of pituitary CRF receptors is the result of increased hypothalamic CRF activity resulting from increased CRF concentrations, found present in cerebrospinal fluid of PTSD patients. PTSD patients also have structural changes in the brain, including reduced hippocampal volume similar to that

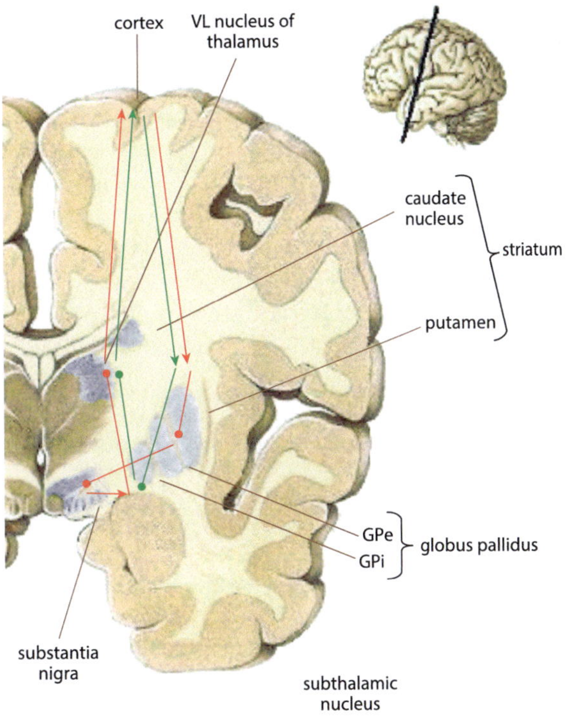


**Fig. 24.8 Direct and indirect pathways within cortico-basal-ganglia-thalamo-cortical loops.** The direct (excitatory) pathway (*green*) leads from the cortex to the striatum, then to the internal segment of the globus pallidus (*GPi*), and substantia nigra pars reticulata (*SNr*), to the thalamus, and back to the cortex. The direct pathway, referred to as a positive feedback loop, contains two inhibitory GABAergic connections (*filled circles*) making its net effect from the cortex back to the cortex excitatory. The indirect (inhibitory) pathway (*red*), a negative feedback loop, leads from the cortex to the striatum, to the external segment of the globus pallidus (*GPe*), the subthalamic nucleus, the GPi/SNr, the thalamus, then back to the cortex. It contains three inhibitory GABAergic connections, making its net effect from the cortex back to the cortex inhibitory. Excitatory connections are represented by *arrows* and inhibitory GABAergic connections are represented by *filled circles*

seen in depressed patients. Thus, PTSD is thought to result in hypersensitivity of the HPA axis to stressful stimuli. Several polymorphisms have been identified as possible risk factors for the development of PTSD, including a genetic variation of the glucocorticoid receptor cochaperone FKBP5 that reduces the risk of developing PTSD in the case of childhood abuse, but not in adulthood trauma. Hypocortisolism is suggested to be a preexisting risk factor for the development of PTSD, as low cortisol levels at the time of exposure to a particular trauma predicts the development of PTSD. This is supported by studies demonstrating that hydrocortisone treatment immediately following exposure to a psychological trauma can effectively treat PTSD.

#### 24.4.2.6 Obsessive-Compulsive Disorder

Obsessive-compulsive disorder (OCD) is a condition characterized by recurrent obsessions and compulsions [4]. **Obsessions** are persistent, inappropriate, and intrusive thoughts, impulses, or images that cause anxiety and distress. To reduce this anxiety or distress, patients do repetitive behaviors or mental acts termed compulsions. These **compulsions** are excessive, inflexible, and are not realistically linked with the obsessions they are intended to prevent or counteract (i.e., neutralizing an obsession of germs by counting). The most common compulsions are cleaning and decontamination rituals, checking, counting, repeating actions, ordering, hoarding, confessing, and praying. Diagnosis of OCD occurs when obsessions or compulsions cause evident distress, consume more than 1 h/day for a month or more, or significantly interferes with life's normal routines. Symptoms of OCD include fears of contamination, worries about harm to self or others, the need for symmetry or order, and religious preoccupations.

OCD has a bimodal distribution and can thus be differentiated as childhood-onset and adult-onset. It is **one of the most common psychiatric disorders affecting children and adolescents** with an average age of onset of pediatric OCD at 7.5–12.5 years of age. In adults the age of onset generally occurs between 22 and 35 years of age. Childhood-onset OCD differs from adult-onset OCD, in that it is more familial and is also more highly associated with tic disorders. Additionally, OCD with a prepubertal onset is more likely to affect males with a ratio of 3:2, but this predominance in males is no longer present in adolescence. OCD affects 1–4 % of children and adolescents and 1–3 % of adults. Approximately 40 % of patients with OCD onset during childhood still struggle with OCD in adulthood.

The neurobiological basis of OCD has yet to be elucidated. However, neuroimaging studies have consistently identified hyperactivity of the orbitofrontal cortex, anterior cingulate cortex, and the head of the caudate nucleus in patients that have not been treated for OCD, that is not present in OCD patients who have received treatment. OCD has been hypothesized to be the result of an imbalance between the "direct" and "indirect" pathways through the basal ganglia (Fig. 24.8). The "direct" pathway is excitatory and the "indirect" pathway is inhibitory. These opposing pathways are thought to be involved in both OCD, and in hyperkinetic and hypokinetic movement disorders. In movement disorders, it has been hypothesized that when the balance between these two pathways shifts towards "excitatory", cortical motor programs are disinhibited resulting in hyperkinetic symptoms, such as is observed in Huntington's disease. When the balance shifts to the indirect pathway this inhibits cortical motor programs, resulting in hypokinetic symptoms, such as is observed in Parkinson's disease. While these symptoms are motor related, it has been suggested that imbalance in these two pathways in nonmotor loops may result in OCD. Increased relative activity in the direct pathway in the OFC/ACC loops may result in a positive feedback loop that results in obsessive thoughts becoming "trapped". This is consistent with findings that the symptoms of obsessive-compulsive disorder are greater in patients with Huntington's disease than in the regular population.

### 24.4.3 Stress Is a Precipitating Factor for Mental Illness

An adverse early environment, or traumatic events occurring during early development, may alter developmental pathways in such a way that it predisposes the individual to abnormal stress responses and stress coping in later life. Increased vulnerability to stress and other forms of pathological stress-coping behaviors is strongly influenced by early life experiences during both pre- and postnatal periods.

Both acute and chronic exposure to stress or glucocorticoids during pregnancy increase maternal glucocorticoid secretion [35]. Since the placenta allows a certain proportion of these glucocorticoids to pass through to the fetus, this results in increased fetal HPA axis activity and altered fetal brain development. This increase in HPA activity is persistent into adulthood in rodents. Additionally, prenatal stress results in a persistent decrease in levels of MRs and GRs in the hippocampus, as well as decreased dendritic spine density in the anterior cingulate gyrus and orbitofrontal cortex in adult animals [35]. Behavioral effects of prenatal stress include learning impairments, increased sensitivity to drugs of abuse, and increased anxiety- and depression-like behaviors in adults. These behavioral effects are thought to be the result of developmental changes in the brain. Specifically, learning impairments are thought to be due to persistent alterations in hippocampal function, anxiety, and depression-like behaviors, the result of effects on amygdalar function, and increased sensitivity to drugs, the result of alterations in the developing dopaminergic system, a system known to be involved in reward-seeking behaviors (see also Sect. 24.6).

In mammals, the quality of early life is primarily defined by nutrition and maternal care. In rodents, maternal care is expressed by **arched-back nursing** (**ABN**) and **licking and grooming** (**LG**) **behaviors**. These two behavioral traits influence the offspring's overall anxiety and stress-related behaviors. Anxiety and stress-related behaviors are in part regulated via glucocorticoids and GRs, such that high levels of GR in forebrain areas such as the hippocampus provide a negative feedback to the brain to reduce the production of glucocorticoids and thereby dampen the stress response (Fig. 24.5) Offspring of high-LG-ABN mothers show increased GR expression and reduced reactivity to stress, whereas offspring of low-LG-ABN mothers have decreased GR expression and increased stress reactivity (Fig. 24.9). In addition to reduced stress responsivity, mice provided with high levels of maternal care have greater dendritic length and spine density in neurons in the CA1 region of the hippocampus and increased dendritic arborization and spine density in the dentate gyrus, as well as increased synaptic plasticity compared to mice raised with low levels of maternal care [35]. Interestingly, if corticosterone is present, mice receiving poor maternal care show enhanced synaptic plasticity. Similarly, in a hippocampal-dependent fear learning task, mice raised with poor levels of maternal care perform better than mice raised with high levels of maternal care. This suggests that certain forms of adverse early environments while detrimental at basal conditions, may better prepare the animal to perform under stressful conditions as an adult.

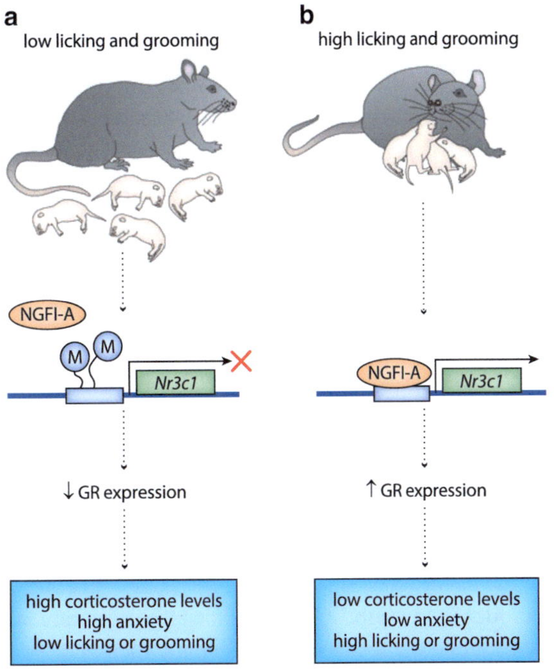


**Fig. 24.9 Poor early environment can cause persistent alterations in stress responsiveness.** Female rat dams demonstrate a range of maternal behaviors, with some dams providing low levels of licking and grooming and some dams providing high levels. (**a**) Low levels of licking and grooming during postnatal development result in increased methylation of the promoter region of the glucocorticoid receptor (*GR*) in the hippocampus of the adult brain. This is associated with reduced binding of the transcription factor nerve growth factor-inducible protein A (*NGFI-A*) and decreased expression of GR. Lower levels of hippocampal GR result in higher levels of baseline and corticosterone secretion following a stressor, increased anxiety-like behavior, and reduced maternal behaviors in females. (**b**) High levels of licking and grooming during postnatal development result in reduced methylation of the GR gene, increased binding of NGFI-A to the GR promoter, and increased GR expression in the adult hippocampus. This is associated with reduced levels of baseline corticosteroid, increased corticosteroid following a stressor, reduced anxiety-like behavior and increased maternal care in females when adult (Reproduced from Feder et al. [21] with permission)

Abnormal levels of circulating corticosteroids have been linked to a variety of disorders in the human population [16]. **Hypercortisolemia**, or the hypersecretion of corticosteroids, is a risk factor for a variety of conditions including depression, obesity, osteoporosis, and cardiovascular conditions. This can occur during chronic stress conditions and results in an inability to coordinate adaptation on release of neuropeptides like CRF. In contrast, hypocortisolemia has been strongly linked with PTSD.

Several hypotheses have been put forward in regard to the persistent effects of stress exposure and its link with mental illness. The **neurotoxicity hypothesis**, formerly known as the **glucocorticoid cascade hypothesis**, suggests that chronic overexposure to glucocorticoids makes neurons unable to withstand insults occurring either during toxic challenges, or normally across development [35]. This results in cell death that manifests itself in the reduction in hippocampal volume often seen following chronic exposure to stress, depression or PTSD. The major caveat to this hypothesis is that it cannot account for the hyposecretion of glucocorticoids present in PTSD patients.

In contrast to the neurotoxicity hypothesis, the **vulnerability hypothesis** claims that reduced hippocampal volume is not the result of chronic stress conditions, but rather is a predisposing factor for the development of stress-related disorders [35]. This hypothesis includes the possibility that reduced hippocampal volume can be the result of either genetic factors or early life stress.

A third hypothesis, called the **life cycle model of stress**, takes aspects of both the neurotoxicity and vulnerability hypotheses, and suggests that there are windows of vulnerability when the developing brain is more or less sensitive to environmental factors causing neurotoxicity [35]. Thus, stressful conditions during the time of hippocampal development could lead to different emotional disorders than those resulting from stressful events during the time of frontal cortex development (Fig. 24.10). In line with this hypothesis, it has been shown that women who experience trauma before the age of 12 have increased risk of major depression, while those who experience trauma between 12 and 18, are at increased risk of PTSD. Similarly, repeated sexual abuse occurring before the age of 12 is associated with reduced hippocampal volume, but the same trauma during adolescence, was associated with reduced prefrontal cortex volume [47].

### 24.4.4 Genetic Variation May Predispose Individuals for Mental Illness

Many of the genetic variants linked with emotional processing are in the serotonergic system. Here we will focus on three genes encoding for enzymes, as examples of how genetic variation can predispose individuals for psychological disorders: **catechol**-O-**methyltransferase gene** (**COMT**), the **serotonin transporter gene** (**5-HTT**), and the **tryptophan hydroxylase 2 gene** (**TPH2**).

The gene for **COMT**, an enzyme that degrades catecholamine neurotransmitters (dopamine, epinephrine, norepinephrine), contains a functional polymorphism that results in a Met–Val substitution at codon 158 [1]. The methionine (Met) allele is associated with low enzymatic activity, and the valine (Val) allele is associated with high enzymatic activity. The Met allele has been associated with increased levels of tonic dopamine and decreased phasic dopamine activity. This polymorphism has been suggested to play a role in a wide range of emotional and cognitive processing. While the Met allele is beneficial for working memory and attentional processing, the Val allele is beneficial for emotional recognition of negative stimuli; healthy Val/Val carriers recognize negative facial expressions faster and better than healthy Met/Met carriers. Met/Met carriers also exhibit increased reactivity to emotional stimuli in brain areas associated with emotional processing, including the hippocampus, amygdala, and thalamus, as well as areas of the prefrontal cortex, compared to Val/Val carriers. Thus, it has been suggested that overactivation of these contribute to emotional dysregulation present in Met/Met individuals. Additionally, COMT Val$^{158}$Met genotype may contribute to the emergence of particular personality traits. **Alexithymia** is a personality trait that is characterized by difficulties in identifying or verbalizing ones feelings [1]. Val/Val carriers have higher alexithymia scores than Met/Met or Met/Val carriers, further suggesting a role for COMT in affective processing.

**5-HTT** is important in serotonergic signaling, as it removes serotonin released into the synaptic cleft [1]. The 5-HTTLPR (serotonin-transporter-linked polymorphic region) is a relatively common polymorphism present in the promoter region of the 5-HTT gene (SLC6A4). It consists of a 22-base-pair repeat with a short (S) and long (L) version that results in differential 5-HTT expression; the short produces less 5-HTT mRNA and protein than the long variant, resulting in increased serotonin in the synaptic cleft. The S allele has been associated with increased amygdala reactivity, neuroticism, social anxiety disorder, and the increased risk for the development of PTSD, especially

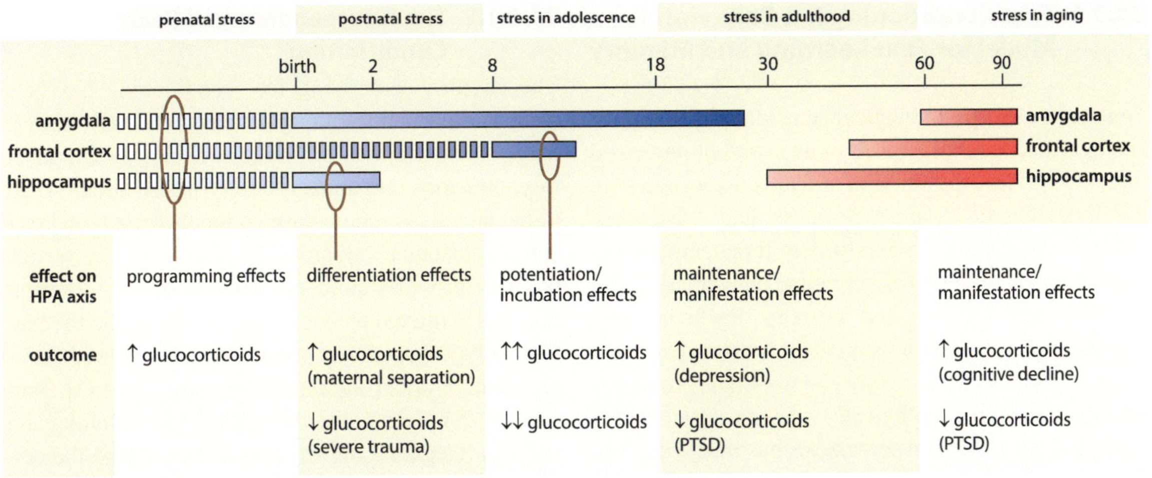


**Fig. 24.10** **The life cycle model of stress.** The effects of an exposure to stress, either acute or chronic, are dependent on the stage of life and the stage of brain development during which it occurs. Exposure to a stressor during the prenatal period affects the development of the hippocampus, frontal cortex, and the amygdala, brain areas involved in the HPA axis (programming effects). Exposure to a stressor during postnatal development has variable effects, including increased secretion of glucocorticoids, as in the case of exposure to maternal separation during childhood, or decreased levels of glucocorticoids, in the case of severe abuse. Thus, the environment during childhood induces differential glucocorticoid production (differentiation effects). After the prenatal period, stress hormones can affect all developing brain areas (*broken blue bars*). Some brain areas may be more susceptible to the effects of stress hormones during periods when they undergo rapid growth (*solid blue bars*). For example, the hippocampus grows rapidly between birth and 2 years of age, and therefore may be more vulnerable to the effects of stressful stimuli during this time. This differs from the amygdala, which continues to grow from birth until the late 20 s, and therefore may be susceptible to the effects of stressful stimuli occurring during childhood. The frontal cortex undergoes a major period of growth during adolescence. A prolonged glucocorticoid response to stress occurring during adolescence may persist into adulthood (potentiation/incubation effects). Brain areas that degenerate most quickly due to aging (*red bars*) are more vulnerable to the effects of stress hormone and this can result in the manifestation of incubated effects of earlier adversity (manifestation effects) or to maintenance of the effects of stressful stimuli (maintenance effects). *PTSD* posttraumatic stress disorder (Reproduced from Lupien et al. [35] with permission)

when social support was limited. However, these associations remain controversial, as many studies have failed to replicate these findings.

Tyrptophan hydroxylase is the rate-limiting enzyme for the synthesis of serotonin. A single-nucleotide polymorphism (SNP) in the **TPH2 gene** (−703 G/T SNP, rs4570625) has been associated with altered serotonergic function and emotional processing. A significantly higher proportion of T-allele carriers are present in two separate cohorts of patients with anxiety-related personality disorders compared to control subjects. Additionally, in healthy subjects, the genetic variant of TPH2 is significantly associated with anxiety-related traits defined by the TPQ Harm Avoidance test, including emotional instability. Brain imaging of T-allele carriers also displays heightened responsiveness to emotional stimuli in the amygdala. Thus, evidence suggests that T-allele carriers are at a higher risk than G-allele carriers for the presence of anxiety-related personality traits, and for the development of anxiety-related personality disorders.

## 24.5 Fear Learning and Memory

**Emotional memory** is the acquisition and storage of information concerning emotionally salient events [32]. It recruits several brain areas depending on the nature of the memory, but always requires the amygdala. The hippocampus and associated cortical areas can also be involved, in particular in the storage of explicit or conscious memory that has an emotional component. Thus, emotional events result in the recruitment of both amygdala-based emotional systems and hippocampal-based nonemotional systems. Here, we will focus on fear learning and memory as a basis for understanding emotional memory processes.

### 24.5.1 Fear Conditioning Is a Behavioral Model for Fear Learning and Memory

**Fear** refers to a psychological state and physiological changes that occur as a result of an actual or perceived threat [42]. Thus, the emotion of fear is associated with the induction of a stress response and subsequent defensive behaviors, on exposure to threatening situations. Fear conditioning is a common behavioral model for emotional learning and memory. Pavlovian fear conditioning is based on Ivan Pavlov's discovery that a neutral stimulus can acquire emotional properties when paired with a biologically relevant stimulus [38] (see Chap. 26). Since every species has innate defense mechanisms that help cope with threat, fear conditioning is highly conserved across species [42].

**Fear conditioning** paradigms involve one or several pairings of a neutral stimulus (the conditioned stimulus; CS) such as a tone, light, or place (context) with an intrinsically aversive stimulus (the unconditioned stimulus or US), most often a mild foot shock. Following conditioning, animals are once again exposed to the CS, in the absence of the US. In animals that have learned and remember the CS-US association, this elicits measurable behavioral, autonomic, and endocrine fear responses such as freezing or increased blood pressure (Fig. 24.11). Repeated exposure of the CS without the US leads to extinction of the conditioned response (CR) [42]. Cue conditioning (i.e., association of the US with a tone or light) is amygdala-dependent, while contextual conditioning (i.e., association of the US with the environment) involves both the amygdala and the hippocampus. Accordingly, the hippocampus is not thought to be involved specifically in forming the association between the CS and the US, but rather in placing these associations within the appropriate contexts.

In general, chronic stress enhances fear acquisition and memory. Both cued and contextual fear conditioning can be enhanced by exposure to either acute or chronic stress prior to training [44]. Pre-training stress also increases the sensitivity of the animal to the stressor, such that the same CRs can be elicited from a less extreme stressful stimulus. This has been suggested to mimic PTSD patients, who demonstrate abnormally strong reactions to relatively mild stimuli. Acute stress following training has also been demonstrated to enhance long-term memory in a cued fear conditioning paradigm.

#### 24.5.1.1 The Neuroanatomy of Fear Conditioning

The circuitry underlying fear conditioning has been well studied and is thought to be relatively simple. Sensory information, including auditory, visual, gustatory, olfactory, and somatosensory information, is sent to the lateral amygdala through inputs from both cortical and thalamic regions [32]. The cortico-amygdala pathway provides slow and detailed sensory information that is filtered by conscious control, while the thalamo-amygdala pathway provides rapid, but less detailed information about the threatening stimulus [42]. Both CS and US information converge in the lateral amygdala, and this integrated information is then sent to the central nucleus, either directly or indirectly via the basal, accessory basal, and intercalated nuclei. The central nucleus has many projections out of the amygdala. From the central amygdala, there are projections to the brainstem and hypothalamus, and it is these connections which control the expression of defensive behaviors, hormonal secretions, and autonomic responses which are part of the behavioral fear response [32] (Fig. 24.11). For example, projections to the hypothalamus can mediate autonomic responses like increased blood pressure and pupil dilation [19]. Additionally, projections to midbrain nuclei can mediate behavioral responses such as freezing and vigilance.

There are direct connections between the lateral amygdala and central amygdala, but these connections are quite minimal [42]. However, there are several indirect connections through other subnuclei of the amygdala. For instance, both the lateral amygdala and basal amygdala project to an inhibitory network of cells, the intercalated region that, in turn, projects to the central nucleus. Since the majority of neurons in the central amygdala are also inhibitory, activation of the neurons projecting from the intercalated region to the central amygdala may act to reduce the inhibitory actions of the central amygdala, and thereby increase the fear-related behaviors.

In mice, neurons in the lateral amygdala are preferentially activated by the expression of fear memories [23]. When these neurons are deleted following acquisition of a fear memory, the expression of this fear memory is blocked. This memory loss is persistent, suggesting that some or all of the fear memory can be permanently erased by ablation of a subpopulation of specific neurons within the lateral amygdala. Importantly, following the erasure of the fear memory

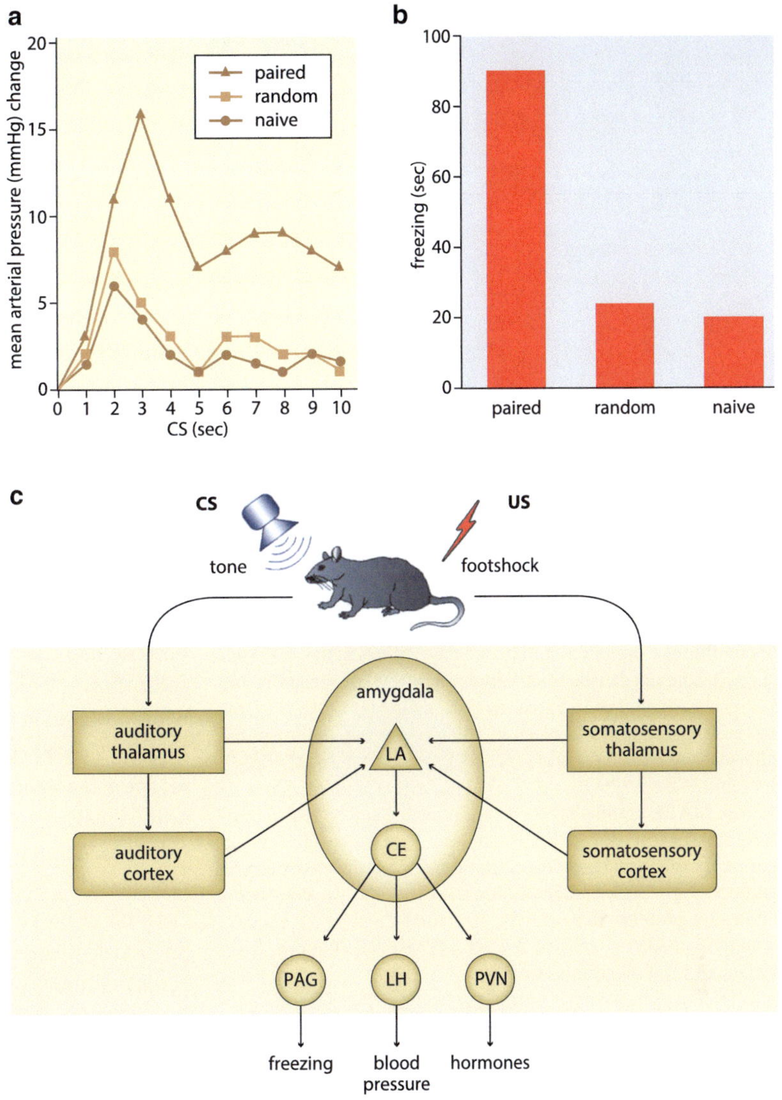


**Fig. 24.11 Autonomic and behavioral responses following auditory fear conditioning, and the brain pathways involved.** Fear conditioning occurs when an animal associates an initially neutral conditioned stimulus (*CS*), such as a tone, with an aversive unconditioned stimulus (*US*), such as a foot shock, after the CS and US are paired in time. (**a**, **b**) Following classical auditory fear conditioning, a tone (CS) elicits both autonomic and behavioral responses. The conditioned group experienced 10-s CS associated with a 0.5-s foot shock (US), the random group was exposed to both the CS and US but this exposure was random, and the naïve group never experienced the CS or the US. Twenty-four hours following conditioning, the testing phase occurs. Here, three CS trials, in the absence of the US, were presented, and arterial pressure responses were measured (a). Additionally, one 120-s CS alone trial was presented and freezing was measured (b). (**c**) Auditory inputs, associated with the CS, and somatosensory inputs, associated with the US, converge onto single neurons within the lateral nucleus of the amygdala. The LA projects, both directly and indirectly, to the central nucleus of the amygdala. Projections from the central amygdala to the brainstem and hypothalamus regulate the fear response, including behaviors like freezing, autonomic responses including increased blood pressure and heart rate, and endocrine responses, including increased pituitary-adrenal hormones. *PAG* periaqueductal gray, *LH* lateral hypothalamus, *PVN* paraventricular nucleus (Reproduced from LeDoux [32] with permission)

by such selective ablation, a new fear memory trace can be produced by retraining the animal.

There is some evidence that overtraining results in fear conditioning which does not rely on the fear circuits involving the lateral amygdala, but rather on weak connections to the central amygdala which are normally not recruited [6]. The concept that the central amygdala is involved in a conditioning independent of the lateral amygdala has been theorized mainly as a result of appetitive conditioning findings; its presence in aversive conditioning requires further study [42].

#### 24.5.1.2 The Molecular Mechanisms Underlying Fear Acquisition and Consolidation

A model of fear conditioning at the molecular level is shown in Fig. 24.12 [33]. Glutamate released as a result of CS inputs, binds to glutamate receptors, including AMPA receptors, NMDA receptors, and metabotropic glutamate receptors, on lateral amygdala cells [41]. At resting membrane potentials, NMDA receptors are gated by magnesium. However, when glutamate is bound to the receptor and the cell is depolarized, the channel is opened and calcium is

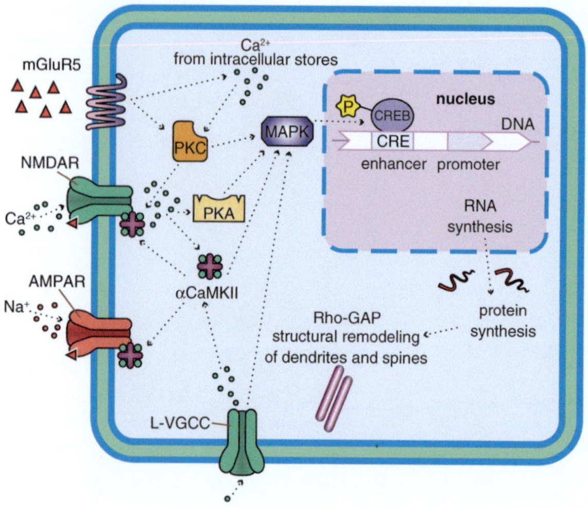


**Fig. 24.12 Molecular mechanisms of fear learning and memory in the lateral amygdala.** On presentation of the auditory conditioned stimulus, glutamate is released from presynaptic terminals in the lateral amygdala. Glutamate binds postsynaptic NMDA, AMPA, and mGluR5 receptors. If a strong unconditioned stimulus simultaneously activates the cell (not shown), calcium can enter the cell through NMDA receptors and L-type voltage-gated calciums channels (*L-VGCC*). This increase in intracellular calcium leads to the activation (phosphorylation) of MAP kinase (*MAPK*), protein kinase C (*PKC*), protein kinase A (*PKA*), and CaM kinase II (*CamKII*). On activation, MAPK is translocated to the nucleus, and activates transcription factors (i.e., CREB), thereby leading to changes in gene expression and protein translation. The new proteins can be additional glutamate receptors for insertion into the cell membrane, or proteins required for structural changes, a process which is mediated in part by Rho-GAP (Reproduced from LeDoux [33] with permission)

allowed to flow into the cell. Thus, NMDA receptors are coincidence detectors, and are key mediators of experience-dependent synaptic modification (see Chaps. 6 and 26). Consequently, when the US depolarizes cells within the lateral amygdala while glutamate released by CS inputs is bound to NMDA receptors, calcium can enter into the cell. This increase in intracellular calcium leads to the activation of protein kinase second messenger cascades that are necessary for memory formation. The role of cyclic AMP-dependent protein kinase (PKA) and the mitogen-activated protein kinase (MAPK) have been studied extensively in relation to fear conditioning and the formation of long-term memory. On activation of PKA and MAPK, MAPK is translocated into the nucleus where it acts on gene transcription factors and thereby mediates changes in gene expression and, ultimately, protein levels. New proteins include additional glutamate receptors for insertion into the cell membrane, and proteins that alter actin and other cytoskeletal functions, thereby contributing to structural changes required for synaptic connectivity.

Calcium entry through voltage-gated calcium channels during training is also necessary for memory formation [41]. It is believed that calcium entry via L-type voltage-gated calcium channels, occurring as a result of strong depolarization, is necessary not for acquisition, but for consolidation of long-term memory. Blockade of L-type voltage-gated calcium channels prior to training results in specific impairment of long-term memory, but leaves short-term memory intact.

#### 24.5.1.3 Reconsolidation and Extinction of Fear Memories

After consolidation, fear memories can once again become labile as a result of reactivation, termed **reconsolidation**. A summary of the molecular signaling pathways involved in memory reconsolidation is provided in Fig. 24.13 [48]. In rodents, infusion of a β-adrenergic receptor antagonist, propanolol, into the amygdala immediately after reactivation of a previously learned fear association impairs subsequent recall. The effect of propanolol is also seen in humans given an oral administration of propanolol before memory reactivation of a previously learned fear response. This treatment prevents the future behavioral expression of fear providing a possible therapeutic option for those suffering with PTSD. Similarly, animal studies have demonstrated that glucocorticoid administration immediately after reactivation of a contextual fear memory reduces subsequent recall, suggesting that glucocorticoids can act to enhance fear memory extinction. This concept of enhanced extinction resulting from glucocorticoid administration has now been successfully used in humans for the treatment of PTSD and anxiety disorders [18].

Extinction of the conditioned response through repeated exposure of the CS in the absence of US is not thought to be the result of an erasure of the original CS-US association, but rather the induction of a stronger association between the CS and the absence of the US [8]. A proposed mechanism for extinction involves the downregulation of amygdala output by infralimbic neurons in the medial prefrontal cortex. These neurons activate GABAergic neurons within the basolateral

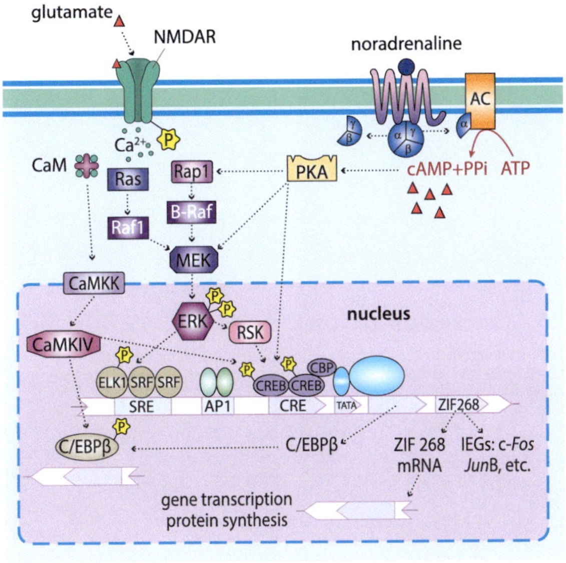


**Fig. 24.13 Molecular mechanisms of memory reconsolidation.** Activation of NMDA receptors result in an influx of calcium that, in turn, activates small GTPases, like Ras, Raf, and Rap. These activated GTPases result in the activation of the extracellular signal-regulated kinase pathway (*ERK*). Additionally, activation of adrenergic receptors result in increased levels of cyclic AMP (*cAMP*), resulting in the activation of protein kinase A (*PKA*). PKA can act directly or indirectly via ERK and ribosomal protein S6 kinase (*RSK*) to activate transcription factors including camp response element-binding protein (*CREB*), zinc finger 268 (*ZIF268*) and ELK1, thereby initiating gene transcription. Memory reconsolidation activates the immediate-early genes c-Fos and JunB. In addition, the CCAAT-enhancing binding protein-b (*C/EBP*) is necessary for memory reconsolidation. Although not yet studied, NMDA receptor activation suggests that the calcium/calmodulin (*CaM*)-CaM-dependent protein kinase kinase (*CamKK*)-CaMKIV pathway is also involved in memory reconsolidation. *AP1* activator protein complex 1 (a complex of c-Fos and c-JUN), *CBP* CREB-binding protein, *MEK* mitogen-activated protein kinase/ERK kinase, *SRE* serine response element, *SRF* serum response factor, TATA box required for transcription (Reproduced from Tronson and Taylor [48] with permission)

amygdala or intercalated cells that, in turn, act to inhibit output from the central nucleus (Fig. 24.14).

NMDA receptors play a role in extinction learning. When NMDA antagonists are infused in the amygdala, extinction is blocked, while administration of an NMDA partial agonist, D-cycloserine, facilitates extinction, whether administered systemically or directly into the amygdala. These findings led to the use of D-cycloserine in phobic patients. Patients with a fear of heights (acrophobia) were administered D-cycloserine while undergoing behavioral exposure therapy. Administration of this NMDA agonist resulted in a faster reduction in acrophobia symptoms when compared to those taking placebo. This reduction in symptoms included both decreases in posttreatment skin conductance responses (SCRs), and better scores in scales measuring acrophobia symptoms, suggesting that D-cycloserine also results in a facilitation of extinction learning in humans.

### 24.5.1.4 Fear Conditioning in Human Studies

Imaging studies have demonstrated an important role for the amygdala in fear learning in humans. In a fear conditioning model similar to that used in animals, neutral stimuli are paired with mild shocks. During acquisition, subjects demonstrate increased bilateral activation of the amygdala to the association between CS and US. Additionally, in human studies, skin conductance response (SCR) is often used as the CR elicited by exposure to the CS. In the above-mentioned study, the increase in amygdala activation to the US was further correlated with the magnitude of SCR, demonstrating a positive correlation between amygdala activation and the conditioned fear response. Similarly, when using white noise as the US, bilateral amygdala

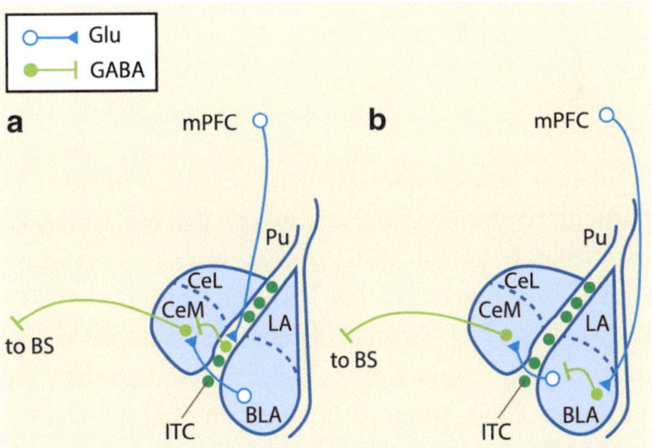


**Fig. 24.14 Proposed mechanisms for the downregulation of amygdala output by the medial prefrontal cortex (*mPFC*).** (**a**) In rat, infralimbic (*IL*) neurons in mPFC inhibit central medial nucleus (*CeM*) projection neurons indirectly via GABAergic intercalated cells (*ITC*). This reduces CeM responses to inputs from the basolateral amygdala (*BLA*) or cortex. (**b**) A second hypothesis is that the mPFC inhibits CeM output through excitatory connections to BLA GABAergic interneurons. *LA* lateral nucleus of the amygdala, *CeL* central lateral nucleus of the amygdala, *BS* brain stem, *Pu* putamen, *Glu* excitatory glutamatergic projections, *GABA* inhibitory GABAergic projections (Adapted from Bishop [8] with permission)

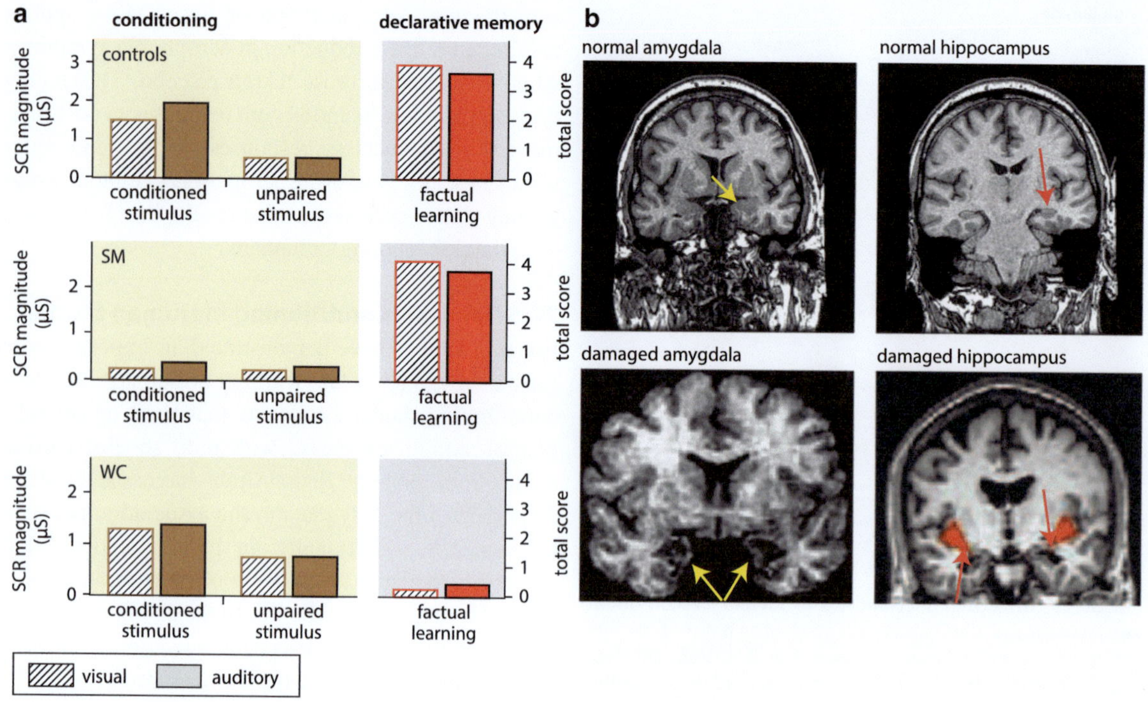


**Fig. 24.15 Dissociable role of amygdala and hippocampus in conditioned fear learning in human subjects.** (**a**) Healthy control subjects have factual knowledge and exhibit a conditioned response, increased skin conductance responses (*SCR*) to conditioned fear stimuli (visual or auditory cues) paired with a loud noise, relative to unpaired control stimuli (*top*). A patient with selective amygdala damage (*SM*) has factual knowledge about the conditioned association but does not acquire the appropriate conditioned response (*middle*). An amnesic patient with selective hippocampal damage (*WC*) has no factual knowledge but has intact conditioned fear responses (*bottom*). (**b**) Structural MRIs displaying normal amygdala (*top left*, *yellow arrow*) and normal hippocampus (*top right*, *red arrow*) compared to selective amygdala damage (*bottom left*, *yellow arrows*) and selective hippocampal damage (*bottom right*, *red arrows*) (Adapted from LaBar and Cabeza [31] with permission)

activation was demonstrated in response to the CS, particularly during initial learning phases but decreased over time. In human fMRI studies, it was also demonstrated that repeated CS-US associations are not required for fear memory to occur, but rather verbal instructions explaining CS-US associations are sufficient. Thus, when participants are told that the CS may be associated with an aversive shock, then activation of the left amygdala and associated SCRs are displayed. Similarly, patients with lesions in the left, but not right, amygdala demonstrate deficits in the fear response to a stimulus that had previously been associated with an aversive stimulus through a verbal description.

The use of patients with specific lesions to the hippocampus or amygdala has allowed researchers to study the differential role of these two brain regions in fear acquisition. Amnesic patients with damage to the hippocampus, but not amygdala, are unable to verbally report the relationship between CS and US, but still display the appropriate conditioned response (increased SCR) to the appropriate CS [31]. The opposite is seen in patients with damage to the amygdala, but not hippocampus; they can verbally report the association between CS and US, but do not display increased SCR to the CS (Fig. 24.15). Further, increased SCR is observed even when the CS is masked to prevent conscious awareness.

### 24.5.2 Fear Learning and Memory Is the Result of Changes in Synaptic Strength in the Amygdala

Learning and memory are thought to be the result of long-term changes in synaptic strength. **Long-term**

**potentiation (LTP)**, a model for the persistent increase in synaptic strength resulting from coincident pre- and postsynaptic activity, is a mechanism for fear learning (see also Chap. 26). Fear conditioning is thought to be the result of LTP-like synaptic changes at sensory inputs to the lateral amygdala. It is suggested that prior to conditioning, the CS inputs are not strong enough to evoke fear responses. US inputs are stronger, and therefore are capable of inducing responses in lateral amygdala neurons. Since the lateral amygdala receives inputs from both CS and US during fear conditioning, the US induces strong postsynaptic depolarization while the CS inputs are also active. This results in increased synaptic strength between CS inputs and lateral amygdala neurons, which, in turn, can affect downstream structures that regulate fear responses.

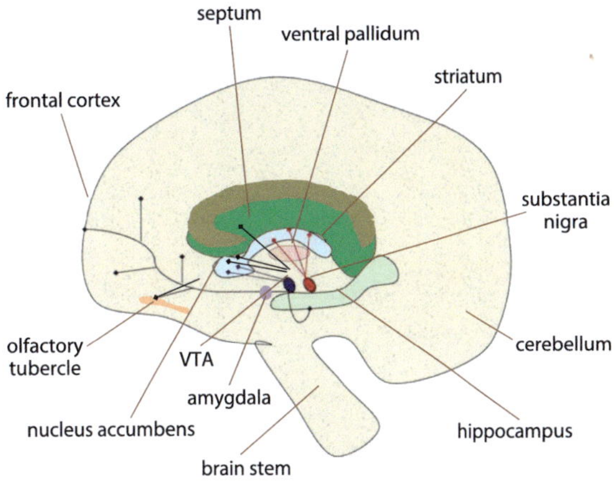


**Fig. 24.16 Dopaminergic pathways in the brain.** Dopaminergic cells present in the substantia nigra project to the striatum (caudate-putamen) and are involved primarily in motor function. The mesocorticolimbic dopamine pathway arises from the ventral tegmental area (*VTA*) and projects to the nucleus accumbens, amygdala, and hippocampus, as well as to frontal cortical areas (Adapted from Knab and Lightfoot [28] with permission)

## 24.6 Motivation and Emotion

It is impossible to know whether animals experience emotions in a similar way as humans. However, humans experience the strongest emotions when they are anticipating activation of the motivational systems, or when this activation of the motivational systems is completed. **Motivational systems** can be defined as systems governing goal-relevant, reflexive behaviors. Thus, **motivation** can be defined as **the drive to obtain a reward or to avoid a punishment** and **motivated behavior is the result of appetitive and aversive cues**. These motivated behaviors are conserved across species due to their importance in the promotion of survival [11]. Thus, motivation may provide an important means of understanding emotion in less complex animals.

As the majority of research on motivational systems has focused on the role of the dopaminergic system, a brief description of dopaminergic projections is provided here [11, 54]. The mesolimbic and mesocortical dopamine systems, arising from the ventral tegmental area (VTA), are thought to be predominantly involved in motivational function, whereas dopaminergic cells arising from the substantia nigra (SNc) are thought to be predominantly involved in motor function [54]. The substantia nigra projects mainly to the caudate putamen. The **mesolimbic dopamine system** refers to the dopaminergic projections from the VTA that mainly innervate the nucleus accumbens and olfactory tubercle, but also the septum, amygdala, and hippocampus. The **mesocortical dopamine system** refers to the dopaminergic projections from the VTA to the medial prefrontal, cingulate, and perirhinal cortices. Currently, the two systems are often collectively referred to as the **mesocorticolimbic dopamine system**, due to the substantial overlap between mesocortical and mesolimbic sytems (Fig. 24.16).

Both primary rewards (like food, water, and sex), and reward-associated stimuli have a drive-like effect [54]. These stimuli cause an energizing effect or motivational arousal, accompanied with an increased likelihood of responses in anticipation of a reward that has yet to be attained. For example, after a reward signal, there is not only reinforcement of the rewarding behavior, but also arousal of the animal before and during the subsequent reward-seeking behavior. This energizing or motivating effect of the presence of a reward, prior to a reward-driven act is known as priming. **Priming** has been suggested as a mechanism underlying the increased likelihood of depressed patients to make negative associations, and as a reason why patients with anxiety disorders respond more quickly to threatening cues. The priming effect decays rather quickly, while the **reinforcing effect** is stored in long-term memory. **Dopamine** is involved in both the priming and reinforcing effect of rewarding stimuli and has been

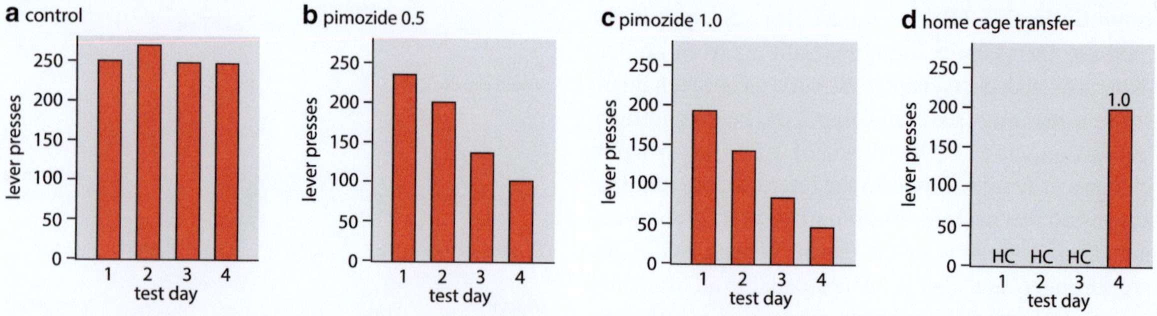


**Fig. 24.17** **Dopamine blockade, via administration of pimozide, reduces reward-seeking behaviors.** Rats were trained for 2–3 weeks to press a lever for food under food deprivation conditions prior to testing. (**a–c**) Testing of control and pimozide-treated rats occurred on 4 days, separated by 2 days of retraining. On the first day, nonrewarded and pimozide-treated animals (0.5 or 1.0 mg/kg 4 h prior to testing) responded similarly. Thus, pimozide-treatment does not affect performance capacity in this test. On subsequent testing sessions, responding decreased in pimozide-treated groups, but not control. (**d**) This reduction in response to reward on the fourth test day in pimozide-treated animals was not due to accumulation of the drug, since rats given the first three pimozide injections without the opportunity to receive food reward (home cage (*HC*) transfer condition) had similar response levels after their fourth injection as animals given their first injection in the test box (Reproduced from Wise [54] with permission)

demonstrated to be important in brain self-stimulation as well as food-, cocaine-, and heroin-seeking behaviors (Fig. 24.17).

Learning the association between an emotional value and a context, stimulus, or event is influenced by motivational state [54]. **Incentive motivation** refers mainly to the priming effects or the drive to encounter a neutral stimulus with enhanced motivational importance as the result of previous pairing with a primary reward. Thus, a previously neutral stimulus can gain incentive value through its association with a reward. Again, dopamine is the neurotransmitter that has been most often speculated to be important for both the establishment and maintenance of incentive motivation.

Recent findings have suggested that a subgroup of dopaminergic neurons encode both reward- and punishment-related signals; in monkeys, certain dopaminergic neurons were both excited by reward-predicting stimuli and inhibited by stimuli predicting an aversive outcome [36]. However, the majority of dopaminergic neurons were excited by stimuli predicting both appetitive and aversive outcomes. There was also a neuroanatomical distribution of these neurons, with those excited by the stimuli predicting an aversive outcome located more dorsolaterally in the substantia nigra pars compacta, but those inhibited by these stimuli located more ventromedially, with a portion of these in the VTA [36] (Fig. 24.18).

## 24.7 Summary and Outlook

There are still many unanswered questions concerning the neural basis of emotions. However, an explosion of research over recent years has contributed to a greater understanding of the neural circuits underlying emotional processing. From these findings, the **hypothalamus**, **amygdala**, and **prefrontal cortex** have emerged as key structures in the regulation of emotions. The persistent effects of stressful events on these, and other brain areas involved in emotion, are thought to increase the risk of developing psychiatric illnesses like depression, posttraumatic stress disorder, and obsessive-compulsive disorder. In recent years, there have been a growing number of treatments for these disorders. However, due to the complexity of these illnesses, they have been particularly difficult to treat and current pharmacological interventions need to be improved. Since it is not known to what extent emotions are conserved across species, **motivational systems** may provide a means for understanding emotion-related processes in less complex animals.

Insect models for emotion related conditions are currently not well studied, and have not been covered in this chapter. However, recent attempts have been made to find models of emotional behaviors particularly in *Drosophila*, where the role of genetic and epigenetic effects are easier to study. Importantly, *Drosophila* express some behaviors that may be related

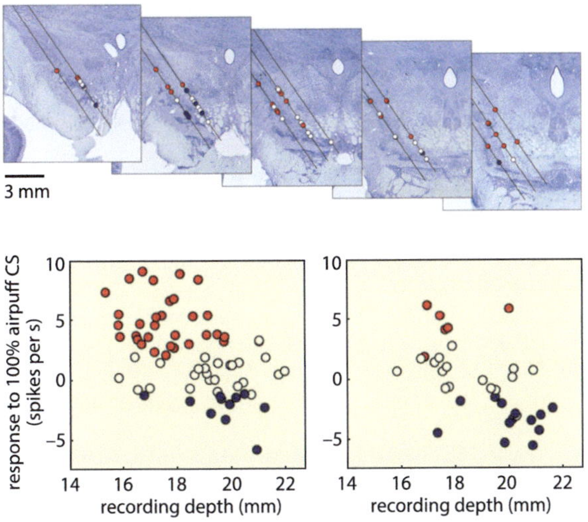


**Fig. 24.18 Location of dopamine neurons in relation to their responses to punishment-predicting conditioned stimuli in the monkey.** Five coronal sections, shown rostrocaudally from *left* to *right* (interval, 1 mm), showing recording sites of 68 dopamine neurons in the substantia nigra and ventral tegmental area of a single monkey. Neurons significantly excited by all punishment-related CS are shown in *red*, neurons significantly inhibited by all punishment-related CS are shown in *blue*, and nonresponsive neurons are shown in *white*. Electrode penetration tracks are indicated by *black lines*. (**b**, **c**) *Red*, *blue*, and *white circles* indicating neurons excited, inhibited, and nonresponsive to punishment-related CS, respectively, in relation to recording depth for two individual monkeys (*right* and *left*) (Reproduced from Matsumoto and Hikosaka [36] with permission)

to emotions in mammals, including avoidance of aversive areas and fear conditioning.

Finally, the critical question of the potential heritability of defects in emotional processing induced by stress has recently gained importance. Its study is expected to reveal novel insight into the contribution of epigenetic processes in the future [55].

## References


1. Aleman A, Swart M, van Rijn S (2008) Brain imaging, genetics and emotion. Biol Psychol 79:58–69
2. Akiskal HS, Bourgeois ML, Angst J, Post R, Möller H, Hirschfeld R (2000) Re-evaluating the prevalence of and diagnostic composition within the broad clinical spectrum of bipolar disorders. J Affect Disord 59:S5–S30
3. Allgulander C (2006) Generalized anxiety disorder: what are we missing? Eur Neuropsychopharmacol 16:S101–S108
4. APA (2000) Diagnostic and statistical manual of mental disorders: DSM-IV-TR. American Psychiatric Association, Washington, DC
5. Bale TL, Vale WW (2004) CRF and CRF receptors: role in stress responsivity and other behaviors. Annu Rev Pharmacol Toxicol 44:525–557
6. Balleine BW, Killcross S (2006) Parallel incentive processing: an integrated view of amygdala function. Trends Neurosci 29:272–279
7. Bard PA (1928) A diencephalic mechanism for the expression of rage with special reference to the central nervous system. Am J Physiol 84:490–513
8. Bishop SJ (2007) Neurocognitive mechanisms of anxiety: an integrative account. Trends Cogn Sci 11:307–316
9. Calder AJ, Lawrence AD, Young AW (2001) Neuropsychology of fear and loathing. Nat Rev Neurosci 2: 352–363
10. Cannon WB (1927/1987) The James-Lange theory of emotions: a critical examination and an alternative theory. Am J Psychol 39:106–124 (reprinted: 100: 567–586)
11. Cardinal RN, Parkinson JA, Hall J, Everitt BJ (2002) Emotion and motivation: the role of the amygdala, ventral striatum, and prefrontal cortex. Neurosci Biobehav Rev 26:321–352
12. Conrad CD, Magariños AM, LeDoux JE, McEwen BS (1999) Repeated restraint stress facilitates fear conditioning independently of causing hippocampal CA3 dendritic atrophy. Behav Neurosci 113:902–913
13. Dalgleish T (2004) The emotional brain. Nat Rev Neurosci 5:583–589
14. Damasio AR (2004) William James and the modern neurobiology of emotion. In: Evans D, Cruse P (eds) Emotion, evolution, and rationality. Oxford University Press, Oxford/New York, pp 3–14
15. Darwin C (1872) The expression of the emotions in man and animals. J. Murray, London
16. de Kloet ER, Joels M, Holsboer F (2005) Stress and the brain: from adaptation to disease. Nat Rev Neurosci 6:463–475
17. de Kloet ER, de Jong IE, Oitzl MS (2008) Neuropharmacology of glucocorticoids: focus on emotion, cognition and cocaine. Eur J Pharmacol 585:473–482
18. de Quervain DJ, Margraf J (2008) Glucocorticoids for the treatment of post-traumatic stress disorder and phobias: a novel therapeutic approach. Eur J Pharmacol 583:365–371
19. Delgado MR, Olsson A, Phelps EA (2006) Extending animal models of fear conditioning to humans. Biol Psychol 73:39–48
20. Evans KC, Wright CI, Wedig MM, Gold AL, Pollack MH, Rauch SL (2008) A functional MRI study of amygdala responses to angry schematic faces in social anxiety disorder. Depress Anxiety 25:496–505
21. Feder A, Nestler EJ, Charney DS (2009) Psychobiology and molecular genetics of resilience. Nat Rev Neurosci 10: 446–457
22. Graeff FG, Del-Ben CM (2008) Neurobiology of panic disorder: from animal models to brain neuroimaging. Neurosci Biobehav Rev 32:1326–1335
23. Han JH, Kushner SA, Yiu AP, Hsiang HL, Buch T, Waisman A, Bontempi B, Neve RL, Frankland PW, Josselyn SA (2009) Selective erasure of a fear memory. Science 323:1492–1496
24. Harrison PJ (2002) The neuropathology of primary mood disorder. Brain 125:1428–1449
25. James W (1884) What is an emotion? Mind 9:188–205

26. Joels M, Baram TZ (2009) The neuro-symphony of stress. Nat Rev Neurosci 10:459–466
27. Klüver H, Bucy PC (1997) Preliminary analysis of functions of the temporal lobes in monkeys. J Neuropsychiatry Clin Neurosci 9:606–620 (reprinted from: Arch Neuro Psych 42: 979–1000, 1939)
28. Knab AM, Lightfoot JT (2010) Does the difference between physically active and couch potato lie in the dopamine system? Int J Biol Sci 6:133–150
29. Kraepelin E, Diefendorf AR (1907) Clinical psychiatry. Macmillan, New York/London
30. Krishnan V, Nestler EJ (2008) The molecular neurobiology of depression. Nature 455:894–902
31. LaBar KS, Cabeza R (2006) Cognitive neuroscience of emotional memory. Nat Rev Neurosci 7:54–64
32. LeDoux JE (1993) Emotional memory systems in the brain. Behav Brain Res 58:69–79
33. LeDoux JE (2007) The amygdala. Curr Biol 17: R868–R874
34. Lorberbaum JP, Kose S, Johnson MR et al. (2004) Neural correlates of speech anticipatory anxiety in generalized social phobia. Neuroreport 15:2701–2705
35. Lupien SJ, McEwen BS, Gunnar MR, Heim C (2009) Effects of stress throughout the lifespan on the brain, behaviour and cognition. Nat Rev Neurosci 10:434–445
36. Matsumoto M, Hikosaka O (2009) Two types of dopamine neuron distinctly convey positive and negative motivational signals. Nature 459:837–841
37. Morris JS, Frith CD, Perrett DI, Rowland D, Young AW, Calder AJ, Dolan RJ (1996) A differential neural response in the human amygdala to fearful and happy facial expressions. Nature 383:812–815
38. Pavlov IP (1927) Conditioned reflexes: an investigation of the physiological activity of the cerebral cortex (translated by Anrep GV). Oxford University Press/Humphrey Milford, Oxford/London
39. Perry JL, Joseph JE, Jiang Y, Zimmerman RS, Kelly TH, Darna M, Huettl P, Dwoskin LP, Bardo MT (2011) Prefrontal cortex and drug abuse vulnerability: translaton to prevention and treatment interventions. Brain Res Rev 65:124–149
40. Radley JJ, Rocher AB, Janssen WGM, Hof PR, McEwen BS, Morrison JH (2005) Reversibility of apical dendritic retraction in the rat medial prefrontal cortex following repeated stress. Exp Neurol 196:199–203
41. Rodrigues SM, Schafe GE, LeDoux JE (2004) Molecular mechanisms underlying emotional learning and memory in the lateral amygdala. Neuron 44:75–91
42. Rodrigues SM, LeDoux JE, Sapolsky RM (2009) The influence of stress hormones on fear circuitry. Annu Rev Neurosci 32:289–313
43. Rolls ET, Grabenhorst F (2008) The orbitofrontal cortex and beyond: from affect to decision-making. Prog Neurobiol 86:216–244
44. Sandi C, Merino JJ, Cordero MI, Touyarot K, Venero C (2001) Effects of chronic stress on contextual fear conditioning and the hippocampal expression of the neural cell adhesion molecule, its polysialylation, and L1. Neuroscience 102:329–339
45. Schachter S, Singer JE (1962) Cognitive, social, and physiological determinants of emotional state. Psychol Rev 69:379–399
46. Shin LM, Liberzon I (2010) The neurocircuitry of fear, stress, and anxiety disorders. Neuropsychopharmacology 35:169–191
47. Teicher MH, Tomoda A, Andersen SL (2006) Neurobiological consequences of early stress and childhood maltreatment: are results from human and animal studies comparable? Ann N Y Acad Sci 1071:313–323
48. Tronson NC, Taylor JR (2007) Molecular mechanisms of memory reconsolidation. Nat Rev Neurosci 8:262–275
49. Tsankova NM, Renthal W, Kumar A, Nestler EJ (2007) Epigenetic regulation in psychiatric disorders. Nat Rev Neurosci 8:355–367
50. Van Ameringen M, Mancini C, Szechtman H et al. (2004) A PET provocation study of generalized social phobia. Psychiatry Res 132:13–18
51. Van Pett K, Viau V, Bittencourt JC, Chan RKW, Li HY, Arias C, Prins GS, Perrin M, Vaie W, Sawchenko PE (2000) Distribution of mRNAs encoding CRF receptors in brain and pituitary of rat and mouse. J Comp Neurol 428:191–212
52. Vyas A, Mitra R, Shankaranarayana Rao BS, Chattarji S (2002) Chronic stress induces contrasting patterns of dendritic remodeling in hippocampal and amygdaloid neurons. J Neurosci 22:6810–6818
53. Vyas A, Pillai AG, Chattarji S (2004) Recovery after chronic stress fails to reverse amygdaloid neuronal hypertrophy and enhanced anxiety-like behavior. Neuroscience 128: 667–673
54. Wise RA (2004) Dopamine, learning and motivation. Nat Rev Neurosci 5:483–494
55. Bohacek J, Gapp K, Saab BJ, Mansuy IM (2012) Transgenerational epigenetic effects on brain functions. Biol Psy 73:313–320

# 25 Experience-Dependent Plasticity in the Central Nervous System

José Fernando Maya-Vetencourt and Matteo Caleo

Environmental influences play a key role in shaping the central nervous system architecture. The interaction of a living organism with the external environment is critical for the survival of the species diversity as it allows an individual to perceive and respond properly to changing environmental conditions. The capacity of the nervous system to change in response to environmental stimuli, referred to as **plasticity**, underlies experience-dependent modifications of brain functions. This feature is fundamental for neural circuitries to be sculpted by external signals and is of particular relevance in processes of learning and memory. How does sensory experience modify synaptic circuitries in the brain? This will be one of the topics we shall concentrate on along this chapter. The nervous system translates information from the external world through signals generated by the electrical activity associated to sensory inputs, and orchestrates adaptive responses to changing environmental conditions. This, in turn, depends upon adaptations of the brain that gathers and processes information to increase the probability of an individual to survive and reproduce. A bird and a bat, for instance, may share a given ecosystem but how perception of the environment occurs in these animals largely differs, as in the former the sense of vision predominates whereas in the latter it is a world of echolocation.

Because experience-dependent changes of brain functions depend, at least partially, on the expression of genes that have evolved to meet specific environmental demands, we shall consider cellular and molecular mechanisms underlying neuronal plasticity in different invertebrate and vertebrate species, with particular emphasis on the mammalian central nervous system. The structure and function of the *BDNF* gene is a particularly clear example of molecular mechanisms that mediate experience-dependent plasticity. BDNF promoter areas are regulated by distinct neurotransmitter systems the levels of which vary in response to environmental inputs. Therefore, BDNF expression in the brain is regulated by sensory experience in a spatiotemporal-dependent manner. This neurotrophin is essential for neuronal survival during development and mediates experience-dependent modifications of synaptic transmission throughout life. Likewise, postsynaptic neurotransmitter receptors such as the (*N*-methyl-D-aspartate) NMDA-type glutamate receptor, which upon high or low neuronal activity modify the efficacy of synaptic transmission, epitomize how the nervous system mediates fast adaptive responses to changing environmental conditions.

We shall also address how intracellular signal transduction pathways associated to experience regulate changes of chromatin structure that underlie phenomena of plasticity in the brain. Epigenetic mechanisms that exert a long-lasting control of gene expression by altering chromatin structure rather than changing the DNA sequence itself, are a common theme in long-term plastic changes from vertebrates to invertebrates. Changes in the pattern of DNA methylation and posttranslational modifications of histones such as

J.F. Maya-Vetencourt (✉)
Centre for Nanotechnology Innovation,
Piazza San Silvestro 12,
56127 Pisa, Italy

Centre for Neuroscience and Cognitive Systems,
Italian Institute of Technology, Corso Bettini 31,
38068 Rovereto, Trento, Italy
e-mail: maya.vetencourt@iit.it

M. Caleo
CNR Neuroscience Institute,
Via G. Moruzzi 1, 56124 Pisa, Italy
e-mail: caleo@in.cnr.it

C.G. Galizia, P.-M. Lledo (eds.), *Neurosciences - From Molecule to Behavior: A University Textbook*,
DOI 10.1007/978-3-642-10769-6_25, © Springer-Verlag Berlin Heidelberg 2013

acetylation, methylation, and phosphorylation have recently emerged as a conserved process by which the nervous system accomplishes the induction of plasticity.

## 25.1 Genetic Factors and Sensory Experience Sculpt the Nervous System Architecture

### 25.1.1 Electrical Activity Modifies Synaptic Circuitries in the Brain

The formation of neural circuitries in the nervous system relies on a tight interaction between genes and environment. Although intrinsic factors mediate the initial assembly of neuronal circuitries in the brain, experience during early postnatal life shapes immature circuitries into the organized connectivity that subserve adult brain function. Spontaneous electrical activity as well as activity caused by sensory experience drives neural circuitry formation, promoting the establishment, elimination, and rearrangement of synaptic connections [24].

In the visual system, for instance, intrinsic signals guide the projection of retinal inputs to different subcortical and cortical structures, creating the neuronal substrate for a complete retinotopic map of the visual field. In different species (e.g., monkeys and cats) the segregation of retinal projections into eye-specific patches at the level of the lateral geniculate nucleus occurs during embryonic life. This anatomical segregation is initially driven by spontaneously generated synchronous firing of action potentials in the mammalian retina before eye opening. Experimental evidence for such spontaneous patterns of electrical activity, in fact, has been reported in rodents, ferrets and monkeys (e.g., [18]). Other functional characteristics in the visual system such as rough orientation selectivity of cortical neurons are also present before visual experience, indicating that the initial anatomical and functional organization of the visual cortex relies, at least partially, on innate processes. However, once basic patterns of neural circuitries are formed sensory experience plays a key role in the refinement of synaptic connectivity in the brain, this phenomenon being particularly prominent during early stages of brain development.

Strabismic or anisometric children that underwent no clinical treatment during early development clearly illustrate this point. In these children, proper visual experience is altered and this, in consequence, causes a marked and permanent impairment of normal visual functions (e.g., visual acuity and contrast sensitivity) later in life. Children with unilateral congenital cataracts, i.e., opacities of the lenses that interfere with the flow of visual inputs face the same problem. If cataracts are not surgically removed in early life, normal visual development is irreversibly impaired. In summary, proper sensory experience is necessary for normal perception to occur and the influence of the environment is particularly evident at early stages of brain development during short-time windows known as "critical periods" (see Sect. 25.2.1). Neural activity associated to sensory experience thus provides a mechanism by which the environment influences brain structure and function.

### 25.1.2 Intrinsic Factors and Sensory Experience Determine Behavior

A variety of sophisticated innate behaviors rely on the interaction between genetic and environmental factors. The behavioral repertoire in most animals including, for instance, foraging and mating strategies in insects, birds, and mammals, consist of an intrinsically predetermined system of adaptive responses that ensures survival and reproduction of a given species. Innate behaviors are subject to genetic changes through natural selection and reflect adaptive responses to selective pressures of an individual's environment. Early postnatal life is a period of particular sensitivity to environmental influences on animal behavior. It seems to be a brief period of time during which the animal acquires an indelible memory of relevant stimuli in the environment, thus ensuring proper development of perception and/or behavior. The groundbreaking work by Konrad Lorenz in the 1960s with graylag geese led him to develop the concept of "imprinting" to explain predetermined behaviors (see also Chap. 26). Based upon the remarkable observation that goslings recognize the first large, moving object they see and hear during the first day of life as a natural tutor – normally their mother – the term **imprinting** was coined to refer to cases of "**programmed learning**" during the critical period (Fig. 25.1). Lorenz found that this period of particular sensitivity to experience in geese is limited to a few hours soon after birth and that once imprinting occurs it is irreversible. Such built-in behavior and the neural substrates underlying it, is likely to give newborns a better chance to survive and reproduce.

Imprinting also plays a key role in the establishment of social behavior and sexual preferences in zebra finches. During sexual imprinting young animals learn the characteristics of their future partners in two stages: an initial very short acquisition period in which

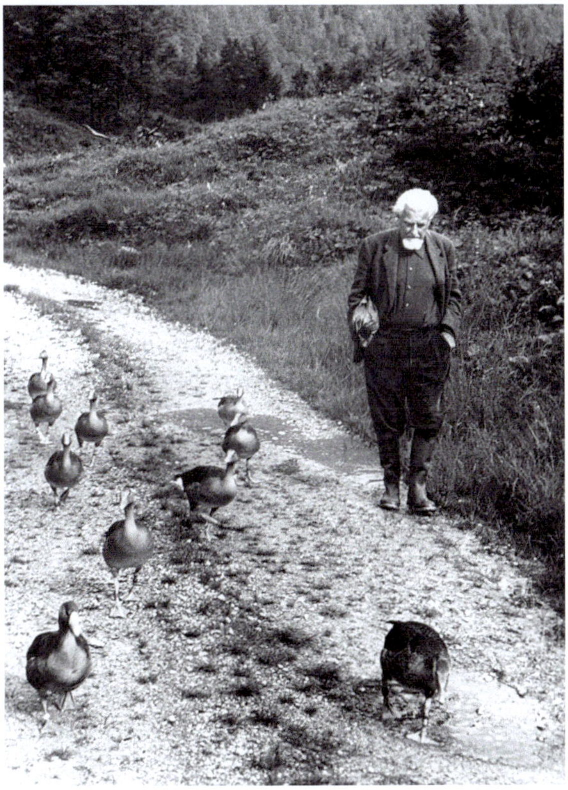

Fig. 25.1 **Konrad Lorenz followed by imprinted geese**

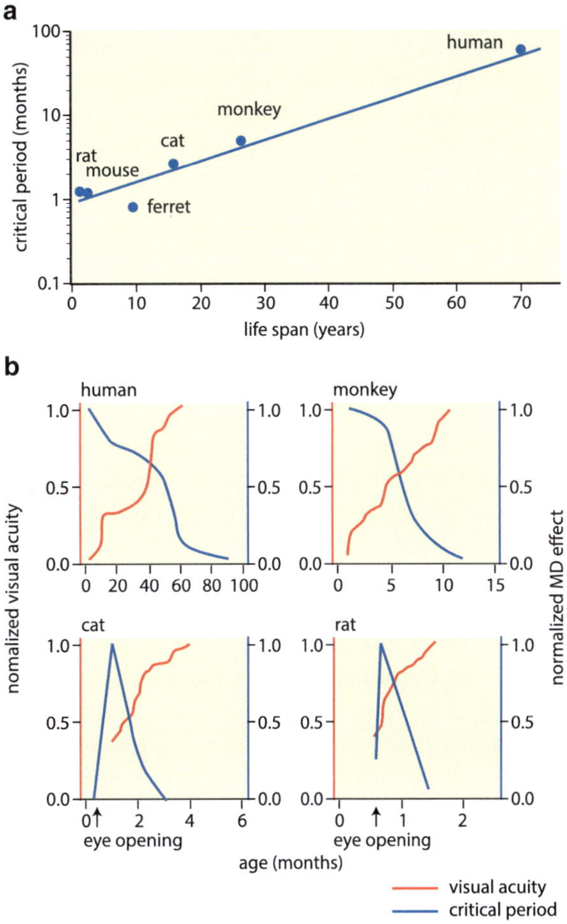


Fig. 25.2 **Critical period, life span, and development of sensory functions**. (**a**) Duration of the critical period for the effects of unilateral eye-lid suture (monocular deprivation) as a function of life span in mammals. The longer the life span of a given species, the longer the critical period. (**b**) Development of visual acuity in *humans*, *monkeys*, *cats* and *rats* as a function of age and compared with the critical period for monocular deprivation. Full development of visual acuity is completed towards the end of the critical period for monocular deprivation (Modified with permission from Berardi et al. [3])

features of the social environment are learnt and a second stabilization stage where a preference for a sexual partner is established upon the previously acquired social information [5].

Most mammals possess a poorly developed visual system at birth, and maternal imprinting relies on olfactory cues (see Chap. 13). During the first week of postnatal life in rodents, the offspring display odor-preference learning that ensures attachment to the mother's odor, thus facilitating huddling and grooming behavior both of which ensure pups survival [37].

## 25.2 Plasticity of the Nervous System is High During Critical Periods but Decreases Thereafter

### 25.2.1 Critical Periods During Sensory Development

The **critical period** is defined as a time frame during development in which neuronal networks are highly sensitive to experience [3] (Fig. 25.2). Once this period ends, the brain is less affected by environmental influences. A failure to be exposed to normal experience during the critical period permanently impairs sensory functions and/or behavior. The existence of critical periods has been reported in different modalities of sensory experience in all species tested so far. Their duration is normally evaluated in terms of either the induction of sensory deprivation effects or the recovery from sensory deprivation [3]. Interestingly, critical periods in response to enhanced sensory experience have been reported as well. An expansion of auditory cortical representations, for instance, is seen in musicians who learnt to play musical instruments before the age of nine but not later.

### 25.2.2 Birdsong and Human Language

Song in birds and language in humans are two compelling examples of critical periods during early development. In these two groups the quality of early sensory experience is the main determinant of perceptual and behavioral capabilities later in life. An initial learning phase is necessary for both birdsong and speech to develop properly: birds do not learn to sing normally, nor infants to speak, if they are not exposed to appropriate acoustic signals from adult tutors [12]. Congenitally deaf children, for instance, do not acquire spoken language, although they can learn sign language if exposed to it during early development. Likewise, songbirds still sing when deafened young but they produce abnormal indistinct series of sounds that are much less songlike than normal song structures.

### 25.2.3 Mechanosensory Deprivation in *Caenorhabditis elegans*

An experience-dependent critical period has also been described in nematodes. The effects of sensory deprivation on synaptic transmission and behavior of *C. elegans* have been studied by comparing animals raised in isolation with animals raised in colonies in the **tap withdrawal paradigm**. These experiments revealed that colony-reared worms display an accelerated development and have larger responses to mechanosensory stimuli. Presynaptic vesicle markers and the expression of AMPA glutamate receptor in *C. elegans* sensory neurons were higher in colony-raised worms when compared with isolated ones, suggesting that synaptic transmission is increased in these synapses. In the absence of conspecific interactions during early development, synapses show a weak development in sensory neurons that cause a decrease in responsiveness to mechanical stimulation. In contrast, brief mechanosensory stimulation during the third larval stage of development rescues the effects of isolation in the tap withdrawal response model and leads to the expression of glutamate receptors and presynaptic vesicle markers [41].

These findings indicate that, just as in birds and mammals, there is a sensitive period in *C. elegans*' life cycle (immediately after hatching) in which high levels of mechanosensory stimulation can rescue the behavioral effects of sensory deprivation. This period ends by the third larval stage. Noteworthy, the delivery of an increased pattern of mechanosensory stimulation in the first larval stage of *C. elegans* causes a significant enhancement of behavioral responses to mechanosensory stimulation (tap withdrawal model) in adult animals. This phenomenon also follows a critical period since the same mechanosensory stimulation at late larval stages does not cause similar effects.

### 25.2.4 Binocularity in *Xenopus laevis*

The *Xenopus* binocular visual system displays activity-dependent synaptic modifications that follow a well-defined critical period during development. In frogs, each lobe of the optic tectum receives visual information from both eyes: contralateral information is transmitted by direct retinotectal projections whereas ipsilateral information is relayed via the nucleus isthmi. The ipsilateral pathway appears in late larval animals and is subject to experience-dependent changes in response to experience during early development. Abnormal visual experience (caused for instance by surgical rotation of one eye in larval stage) causes a dramatic rearrangement of isthmotectal projections that requires NMDA-type glutamate receptor activation. Adult *Xenopus*, in contrast, do not display a structural rearrangement of isthmotectal axons in response to abnormal visual inputs [25]. Of note, the ability for isthmotectal projections to undergo experience-dependent modifications in adulthood can be rescued by exogenous administration of NMDA.

## 25.3 Experience-Dependent Forms of Plasticity in the Visual System

### 25.3.1 Anatomical and Functional Organization of Visual Pathways in Mammals

The extent to which the functional and structural development of the brain depends on early sensory experience has been extensively studied in the developing mammalian visual system. Early electrophysiological studies revealed that activity-dependent reorganization of eye-specific inputs during development is a major mechanism by which neuronal connectivity is established in the primary visual cortex.

Visual perception begins in the retina (see Chap. 18). Retinal ganglion cells project to diverse subcortical structures. Most retinal ganglion cells project to the

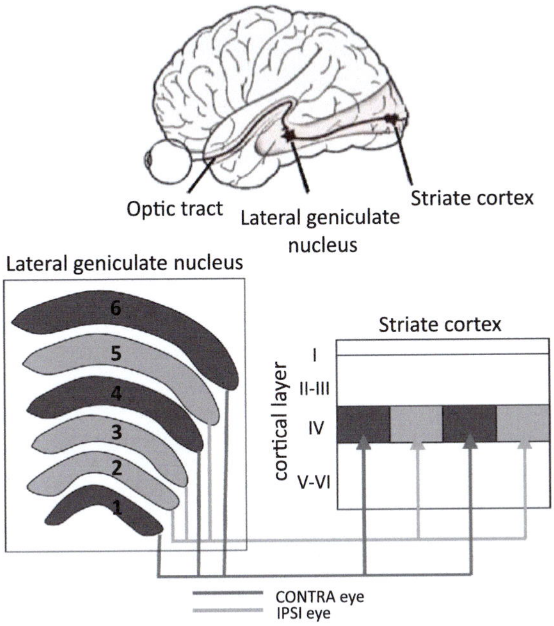


**Fig. 25.3** **Columnar organization of the striate cortex**. The structural and functional organization of the visual system is such that inputs from both eyes remain segregated in separate layers at the level of the lateral geniculate nucleus (pattern of alternating *dark gray* and *light gray stripes*). In many species, including most primates and carnivores, these afferent projections remain segregated in ocular dominance columns of layer IV, the primary cortical target of lateral geniculate neurons. Layer IV neurons send their axons to other cortical layers. It is at this stage that the information from the two eyes converges onto individual neurons

lateral geniculate nucleus in the thalamus and to the superior colliculus in the midbrain. From the thalamus geniculocortical afferents project to the visual cortex and segregate into a pattern of alternating stripes called ocular dominance columns, which are driven by stimulation of one or the other eye [31] (Fig. 25.3). All cells in a column perpendicular to visual cortical layers show a preferred stimulus orientation, and as one goes across adjacent columns a gradual shift of orientation preference occurs, an observation from which the concept of functional "**orientation columns**" arises. Columnar organization in the primary visual cortex was initially evidenced anatomically in monkeys and cats, by injecting one eye of the animal with a radioactive tracer, which was carried by trans-neuronal transport to geniculate neurons and up to cortical layer IV of striate cortex. Afferents related to each eye were evidenced by autoradiography as alternate ocular dominance patches from each eye distributed throughout the binocular area of the primary visual cortex.

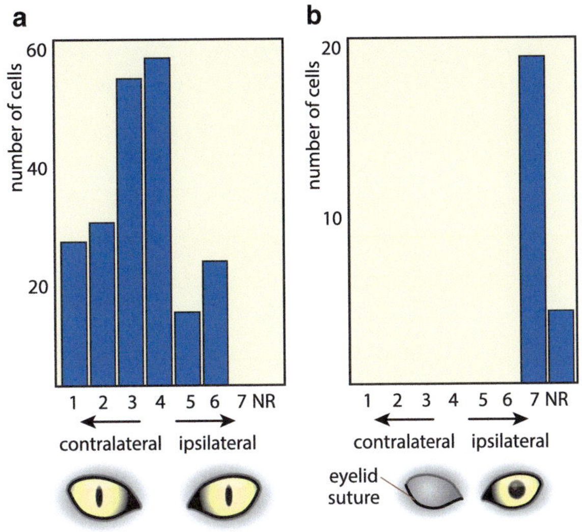


**Fig. 25.4** **Histograms of ocular dominance distributions in striate cortex of normal and monocularly deprived kittens**. Cells in the ocular dominance groups 1 and 7 are exclusively activated by the contralateral and ipsilateral eye, respectively. Class 2/3 cells show a marked dominance by the contralateral eye whereas class 5/6 have a marked ipsilateral dominance. Class 4 cells equally driven by both eyes. (**a**) In the visual cortex of normally reared cats, the contralateral eye drives most visual cortical neurons. (**b**) In kittens reared with monocular deprivation cells stop responding to the contralateral, deprived eye, and become dominated by the ipsilateral eye. Monocular deprivation in adult cats does not alter the ocular dominance distribution in striate cortex although it reduces overall activity. *NR* not responsive cells

### 25.3.2 Unilateral Eyelid Suture During the Critical Period Impairs Visual Function

The groundbreaking work of Hubel and Wiesel in the early 1960s provided the first evidence for structural and functional modifications of the visual system in response to sensory deprivation [49]. In an extraordinary influential set of experiments performed in cats and monkeys Hubel and Wiesel observed that the ocular dominance distribution in the binocular area of the primary visual cortex dramatically changes in response to unilateral eyelid suture (**monocular deprivation**) during the critical period. The number of cells responding to the deprived eye is reduced while the number of neurons activated by the open eye is increased (Fig. 25.4a, b). At the anatomical level, the ocular dominance columns related to the open eye are greatly expanded in layer IV while the columns for the closed eye are shrunken [31]. Accordingly, the arborization of single geniculocortical terminals serving the deprived eye is reduced while spread of terminals serving the open eye is increased [1]. In addition, the deprived eye becomes

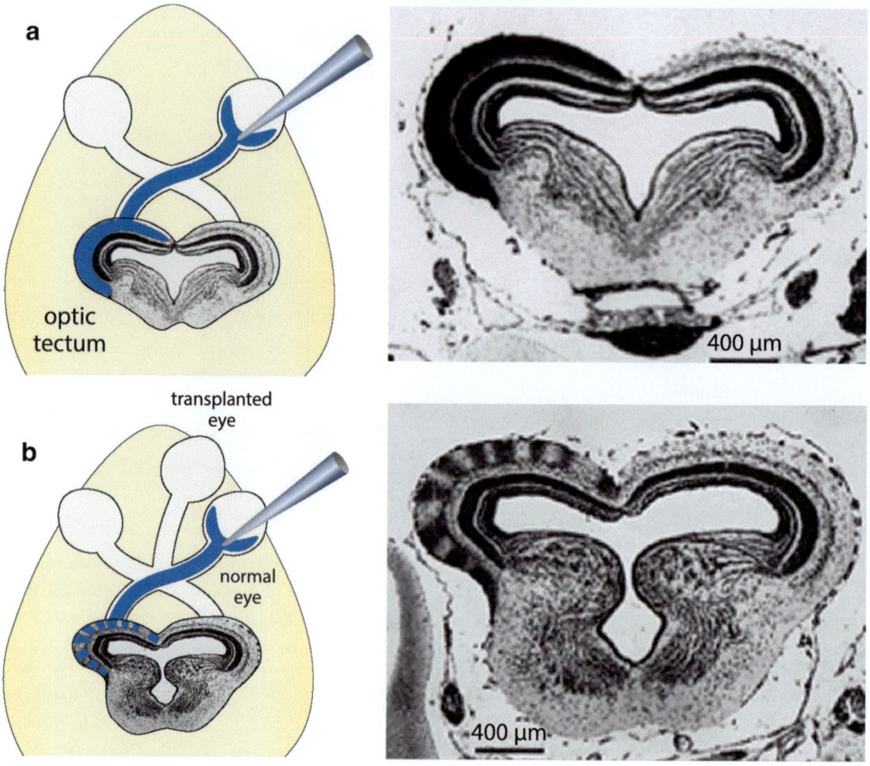


**Fig. 25.5** **The transplant of a third eye in frogs induces the formation of alternating ocular dominance columns in the optic tectum**. The figure depicts coronal sections of the midbrain of the frog (Reproduced from Constantine-Paton and Law [10] with permission) following injection of a labeled amino acid into the contralateral eye. The amino acid is transported into the contralateral tectum. (**a**) A continuous synaptic zone is innervated by the contralateral eye in normal frogs. (**b**) After transplantation of a third eye next to the normal contralateral eye, the injection of a radiolabel into one eye shows alternating stripes in the contralateral optic tectum. The normally continuous synaptic area of the contralateral eye becomes divided into alternating columns due to the activity-dependent competition between retinotectal projections for postsynaptic cells in the tectum

amblyopic: its visual acuity and contrast sensitivity are markedly impaired. Because monocular deprivation does not cause amblyopia in adulthood, this period of early life in which neural circuitries are highly susceptible to experience is a typical example of a critical period.

Most cortical cells remain responsive to both eyes (and there is no shrinkage of ocular dominance columns) following a period of binocular deprivation in juvenile age. Thus, afferents from the two eyes compete for cortical territory, and the relative rather than the absolute amount of activity of the two eyes determines the outcome of this competitive process. The notion of a competition between afferent fibers for postsynaptic space is confirmed by classical experiments performed in frogs by Constantine-Paton and Law [10]. As previously described, in frogs, axons from the retina project only to the contralateral tectum. In order to generate competition for postsynaptic cells in the tectum, a third eye was transplanted into the frog's head near one of the normal eyes early in larval development. Retinal axons from the transplanted eye projected to the contralateral tectum, where they formed periodic, eye-specific bands that alternated with those serving the normal, nearby eye (Fig. 25.5). Thus, competition between two sets of inputs generates alternating domains that are occupied by the terminals of each afferent structure.

### 25.3.3 Sensory Deprivation by Dark Rearing Affects Cortical Development

Another classical paradigm used to assess the impact of experience on the brain is rearing animals in total darkness from birth (**dark rearing**). Total absence of visual inputs leads to a delay in the maturation of striate cortex as indicated by a reduced spatial resolution (acuity) and longer latency of response to visual stimuli [14]. Furthermore, visual cortical neurons have larger receptive field sizes, indicating an immature state of the visual cortex. Moreover, the size and density of dendritic spines are reduced in dark-reared animals with respect to normally reared ones. Of note, dark rearing also extends the critical period and prolongs neuronal plasticity beyond its normal limits [3, 14].

It is worth mentioning that while visual deprivation delays maturation of the visual cortex, raising animals under enriched environmental conditions (with increased motor activity, and enhanced sensory and social stimulation) appears to accelerate visual system development. Enriched animals actually show an acceleration of visual acuity maturation and a precocious closure of the critical period for ocular dominance plasticity [8].

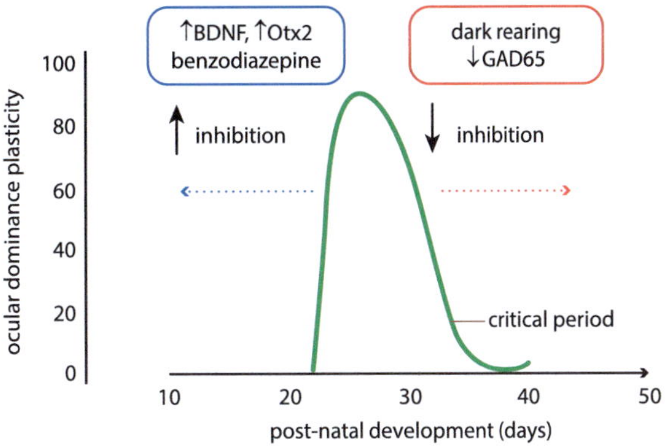


**Fig. 25.6 Inhibitory processes control the time-course of the critical period.** Sensitivity to monocular deprivation in mice is restricted to the critical period that begins about 1 week after eye opening and peaks nearly 1 month after birth. The onset of plasticity can be delayed by preventing the maturation of inhibition (e.g., by dark rearing or by deletion of *GAD65*) (*red arrow*). Conversely, enhancing GABA transmission (e.g., by benzodiazepine treatment, BDNF or Otx2) results in a precocious onset of the critical period (*blue arrow*)

## 25.4 Structural and Functional Mechanisms that Control Critical Period Plasticity in the Visual System

Recent experimental research has shed light into the mechanisms that regulate experience-dependent plasticity in the mammalian visual cortex. In the following, we will cover a number of different signaling molecules including, neurotrophic factors, extracellular matrix molecules, NMDA-type glutamate receptors, GABA-mediated inhibitory transmission and neuromodulatory systems, which all have been recognized as important regulators of visual cortical plasticity [32].

### 25.4.1 Inhibitory Processes Regulate the Time Course of the Critical Period

Intracortical inhibitory processes play a key role in the regulation of visual cortex plasticity [21] (Fig. 25.6). The maturation of GABAergic inhibition sets the threshold for the start of the critical period for experience-dependent plasticity by enabling visual cortical neurons to detect differences in the activity between competing retinal inputs. Actually, transgenic mice with reduced levels of intracortical inhibition, due to the absence of one isoform of the GABA synthesizing enzyme (glutamate decarboxylase of 65 kDa – GAD65), show no susceptibility to monocular deprivation during their entire lifespan. The impairment of experience-dependent plasticity in these animals is rescued by enhancing inhibitory transmission by means of benzodiazepine treatment (agonist of $GABA_A$ receptors). Thus, a reduction of GABA during early postnatal life halts the onset of the critical period, an effect that is rescued by enhancing inhibition.

A second inhibitory threshold that causes the closure of the critical period is reached over postnatal development as well. Transgenic animals that show an accelerated maturation of intracortical inhibitory circuitries, due to overexpression of BDNF in forebrain regions, actually display a precocious end of the critical period for ocular dominance plasticity [23]. In summary, an initial threshold of inhibition triggers the developmental period in which neuronal networks in the visual system are highly susceptible to experience, whereas a second inhibitory threshold signals the end of this phase of enhanced plasticity (Fig. 25.6).

Inhibition triggers plasticity through $GABA_A$ receptors containing the alpha-1 subunit. These receptors are enriched at somatic synapses on pyramidal neurons made by large basket cells (a class of parvalbumin-positive GABAergic interneurons that extend horizontally across ocular dominance columns). Recent studies indicate that visual experience controls the time-course of the critical period by promoting the transfer of the homeoprotein Otx2 from the retina to the visual cortex, where it seems to trigger the maturation of parvalbumin-positive GABAergic interneurons [44].

### 25.4.2 Plasticity Involves the Remodeling of Extracellular Matrix Components

An emerging view in the field of neuronal plasticity is that the effects caused by early sensory experience in the remodeling of visual cortical circuitries are actively preserved throughout life by the late appearance of molecular factors in the extracellular milieu that restrict plasticity. Some components of the extracellular matrix in the central nervous system, in fact, need to be removed for experience-dependent plasticity to occur. The proteolytic activity of tissue plasminogen activator (tPA), for instance, increases after monocular deprivation during the critical period. When tPA is inhibited pharmacologically, the shift of ocular dominance following unilateral eyelid suture is reduced, showing that tPA plays a permissive role for plasticity. Experience-dependent plasticity during the critical period is impaired in tPA-knockout mice and this impairment is rescued by exogenous tPA administration. Changes of binocularity in response to monocular deprivation are accompanied by rapid structural changes of the dendritic spines in deep and superficial layers of the primary visual cortex. This effect is mimicked by exogenous tPA administration suggesting that its proteolytic activity is involved in the structural plasticity of dendrites.

Proteins in the extracellular matrix play a key role in exerting the inhibitory function that terminates the critical period for cortical plasticity. In rodents, the glycoproteins chondroitin sulfate proteoglycans (CSPGs) are inhibitory for experience-dependent plasticity. Degrading sugar chains of CSPGs by exogenous administration of the bacterial enzyme chondroitinase-ABC reactivates ocular dominance plasticity in the adult rat [38]. Similarly, chondroitinase-ABC treatment promotes structural and functional recovery from visual deficits in long-term deprived rats, although the same treatment seems not to be as much effective in adult amblyopic cats [47]. It is important to remark that degradation of CSPGs may alter the ratio of inhibitory/excitatory transmission in the visual cortex, as these glycoproteins condense in perineuronal nets (PNNs) mainly around parvalbumin-positive GABAergic interneurons. Alternatively, or in addition, CSPGs may influence structural plasticity by modifying dendritic spine dynamics and associated functional modifications in the visual cortex.

### 25.4.3 The Maturation of Intracortical Myelination Down-Regulates Plasticity

Myelin-associated proteins are glia-derived proteins that have an important role in controlling neurite outgrowth. Not surprisingly, these proteins are also involved in the regulation of the critical period. In particular, proteins of the myelin sheath including Nogo-A and myelin-associated glycoprotein (MAG), act as molecules that are inhibitory for plasticity by binding to the Nogo receptor (NgR). NgRs are expressed on cortical neurons, and their activation triggers intracellular signaling cascades that inhibit neurite growth. Moreover, the maturation of intracortical myelination correlates with the end of the critical period. In NgR knockout mice ocular dominance plasticity actually persists well into adulthood [36]. Adult transgenic animals lacking the NgR ligand Nogo-A display a similar susceptibility to monocular deprivation thus confirming that NgR-dependent mechanisms participate directly in restricting experience-dependent plasticity in the developing visual system. The paired immunoglobulin-like receptor B (PirB) also has a high affinity for Nogo-A and MAG. When activated, PirB restricts ocular dominance plasticity in the visual cortex [45].

### 25.4.4 Neurotrophins Mediate Experience-Dependent Forms of Plasticity

Neurotrophins are growth factors released by neuronal and glial cells that promote the growth of target neurons, i.e., the development of their axonal and/or dendritic arborizations. "Nerve growth factor" (NGF), "brain-derived neurotrophic factor" (**BDNF**), and "neurotrophin-4" (NT-4), are three neurotrophins involved in various forms of synaptic plasticity. Neurotrophins act via specific receptors: NGF binds to TrkA receptors while BDNF and NT-4 bind to TrkB receptors. These receptors act via several intracellular second messengers, including PKA, ERK1/2, CaMKII, ultimately controlling transcription factors such as CREB, which activate the CRE promoter (see Chap. 6). Typically, neurotrophins function in an activity-dependent manner. Actually, release of neurotrophins only activates those synapses that were electrically active (see Chap. 26). This activity-dependent growth stimulant

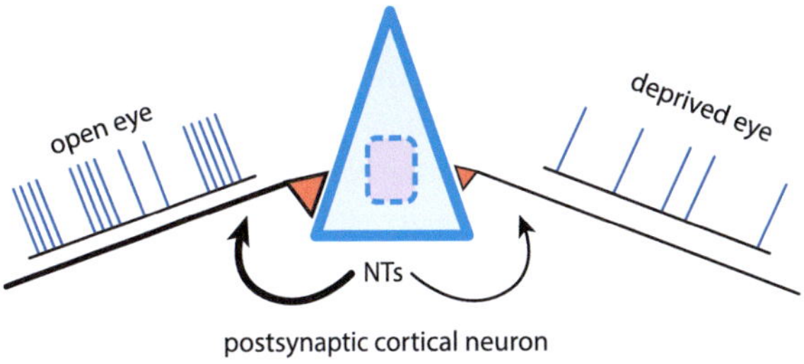


**Fig. 25.7 Potential mechanism of action of neurotrophins on visual cortical plasticity: the hypothesis of activity dependent reward.** LGN fibers driven by the nondeprived eye display a rich electrical discharge and receive a strong neurotrophic support (*thick arrow*) from the postsynaptic cell. In consequence, synaptic inputs are strengthened. In contrast, LGN fibers driven by the deprived eye, having a poor discharge, get a weak neurotrophic support (*thin arrow*). This causes synaptic connections to weaken

is ideally suited for developmental plasticity: active afferents compete for a limited amount of neurotrophins, and those afferents that are electrically active benefit most [7]. The notion that neurotrophins mediate different forms of plasticity initially came from the observation that exogenous NGF administration prevents the loss of visual acuity and the shift of ocular dominance caused by monocular deprivation in the rat visual system. This finding suggests that ocular dominance plasticity involves the competition between geniculocortical projections from either eye for a neurotrophic factor, which is produced and released by visual cortical neurons in an activity-dependent manner (Fig. 25.7). Consistent with this view, infusion of either BDNF or NT-4 into kitten visual cortex desegregates ocular dominance columns [6].

The NGF-induced regulation of ocular dominance plasticity depends on afferent electrical activity. The inactivation of NGF signaling by specific antibodies not only impairs the anatomical and functional development of the rat visual system but also prolongs the critical period into adult life. In kittens, intraventricular, but not intracortical, infusion of NGF attenuates the effects of eyelid suture, indicating that NGF does not act directly on visual cortical neurons but is likely to activate subcortical structures bearing NGF receptors such as afferents from the cholinergic system in the basal forebrain. The primary NGF receptors TrkA, indeed, is mostly but not exclusively expressed in cholinergic fibers.

The action of neurotrophins is site-specific. For instance, the arborization of basal dendrites in ferret visual cortex is stimulated by TrkB ligands BDNF and NT-4, whereas growth of apical dendrites is stimulated by all neurotrophins. BDNF acts on those dendrites and terminals that are activated by presynaptic transmission, providing a mechanism for selective stabilization of active synapses. Actually, inhibition of spontaneous activity abolishes the effects of BDNF on dendritic arborization.

In rodents, BDNF is expressed in the retina and anterogradely transported along the optic nerve to the geniculate nucleus and to the superior colliculus. Levels of BDNF in the retina influence not only development of the lateral geniculate nucleus and superior colliculus, but also ocular dominance plasticity in the cortex. In particular, monocular deprivation reduces BDNF levels in the deprived retina and microinjection of BDNF into the deprived eye counteracts the shift of ocular dominance following eyelid suture.

The neurotrophin BDNF appears to be a critical mediator of the effects of early sensory experience in structural and functional maturation of the visual cortex. In this scenario, visual afferent activity drives the production and release of BDNF, that then promotes the development of intracortical inhibitory circuitries that underlie normal visual functions. In keeping with this, BDNF overexpression in the forebrain allows for normal development of basic physiological properties of the visual cortex even in the total absence of visual experience [19].

### 25.4.5 Long-Distance Diffusely Projecting Systems Modulate Visual Cortical Plasticity

The major modulatory systems in the brain (i.e., noradrenaline, dopamine, histamine, acetylcholine, serotonin) regulate complex functions of the central nervous system such as cognition, behavior, and mediate different forms of neuronal plasticity. Experience-dependent modifications of cortical circuitries are not determined solely by local correlations of electrical activity but are also influenced by attentional mechanisms [43]. Sensory signals, for instance, promote marked modifications of neural circuitries only when animals attend to sensory input and use this information for the control of behavior. Early studies performed in cats demonstrated that changes of visual cortical circuitries in response to experience during early life, are lessened when noradrenergic, cholinergic, and serotonergic projections to the cortex are inactivated. Additionally, there is evidence that these neuromodulatory systems mediate forms of plasticity in the adult visual system of cats and rodents (e.g., [33, 34]).

## 25.5 Hebbian Plasticity and NMDA-Type Glutamate Receptors

The amount and pattern of neural activity specify synaptic modifications [24]. As we have seen in the neurotrophins section, the simultaneous activation of the pre- and postsynaptic neuron leads to an increase in the synaptic strength between those cells. This is summarized by Hebb's principle: "**neurons that fire together wire together**", whereas neurons that fire out of synchrony lose their link. **NMDA receptors** (NMDARs) play a key role in mediating these processes as they function as coincidence detectors in synaptic plasticity (see Chap. 26). NMDARs mediate excitatory synaptic transmission and open only when the presynaptic release of glutamate is coupled with the postsynaptic depolarization, thus implementing Hebb's rule at the molecular level. Under resting conditions, NMDARs are blocked by $Mg^{2+}$, which is removed from the ion channel by intense depolarization, thus allowing $Ca^{2+}$ neuronal inflow that eventually activates intracellular signal transduction pathways. This phenomenon leads to lasting modifications of synaptic efficacy known as long-term potentiation (see Chap. 26). Interestingly, NMDARs are not all alike. NMDARs are heterotetramers composed of different subunits, two obligatory NR1 subunits and two regulatory subunits which come in four types: NR2A to NR2D. Composition of NMDARs determines the functional properties and amount of $Ca^{2+}$ influx through them. The NR1/NR2A heteromeric channel exhibits higher open probability and faster deactivation than the NR1/NR2B channel, therefore NR1/NR2A channels tend to open and close earlier than NR1/NR2B channels. In the visual system of cats and rodents, NMDAR subunit composition is regulated by age and experience. Early in development receptors containing the NR2B subunit predominate but later there is a progressive inclusion of the subunit NR2A. Accordingly, NMDA currents are progressively shortened thus enabling NMDARs to work as high fidelity coincidence detectors of pre- and postsynaptic activity.

Blocking NMDARs inhibits the effects of monocular deprivation in kittens showing that NMDARs are important for visual cortical plasticity. Interestingly, knockout mice with deletion of the NR2A subunit have a normal sensitivity to deprivation by eyelid suture and the effects of monocular deprivation are restricted to early stages of development, suggesting that the switch from NR2B to NR2A may not be crucial for the termination of the critical period. Nevertheless, NMDAR subunit composition influences the outcome of developmental plasticity. In wild-type mice, eyelid suture during the critical period induces two different processes: an initial input-specific weakening of synapses from the deprived eye and a later strengthening of synapses from the open eye [17]. Interestingly, the rapid depression of inputs from the deprived eye is absent in transgenic animals that lack the NR2A subunit. Instead, these mice exhibit a precocious potentiation of open eye inputs.

## 25.6 Long-Term Potentiation (LTP) and Long-Term Depression (LTD)

Modifications of synaptic transmission in response to sensory experience have long been studied in terms of LTP and LTD (see Chap. 26). In the visual system, LTP and LTD can be induced by different patterns of electrical stimulation. Brief and strong episodes of high-frequency stimulation promote LTP while prolonged low-frequency stimulation yields LTD. The Bienenstock-Cooper-Munro (BCM) theory was developed to explain features of experience-dependent plasticity in the kitten visual cortex and describes the extent of LTP and LTD occurrence as a function of spiking activity in postsynaptic cells [4]. Synaptic input at high frequency that activates postsynaptic cells above a certain threshold, results in an increase in synaptic strength (LTP), whereas low-frequency input that produces only low levels of postsynaptic firing results in a decreased synaptic efficacy (LTD). An important feature of the model is that the threshold is not fixed but is itself a function of postsynaptic activity. It does slide as to make potentiation more likely whenever average activity is low, and less likely when it is high (Fig. 25.8). What role do these phenomena play in OD plasticity? The initial weakening of visual inputs from the deprived eye that occurs after brief monocular deprivation may be mediated by homosynaptic LTD.

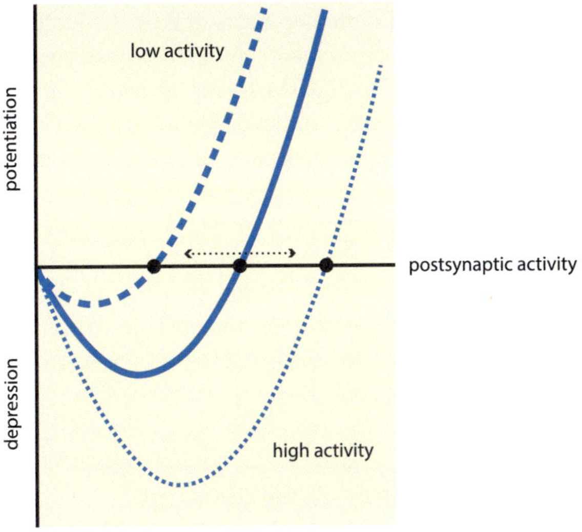


**Fig. 25.8** **Schematic representation of the BCM model**. The *solid curve* at center represents the starting condition: when postsynaptic activity is *below* (or *above*) a threshold (*x-intercept*), presynaptic activity produces depression (or potentiation). If average postsynaptic activity drops to lower values, the entire curve shifts to the left (*dashed line*), making potentiation more likely. If, instead, it rises to higher values, the curve shifts to the right (*dotted line*), making depression more likely

Indeed *in vivo* occurring LTD occludes LTD induction ex vivo: this means that when a cortical slice is taken from an eyelid sutured young animal, LTD can no longer be induced, because LTD is already saturated. LTD is induced by decreased afferent activity from the deprived eye, which causes a reduced activation of NMDARs and thus a low-amplitude rise in $Ca^{2+}$ concentration. This activates postsynaptic protein phosphatases (see Chap. 8), which dephosphorylate and internalize AMPA receptors (AMPARs). Less AMPARs mean a decreased sensitivity to glutamate, and thus a depression of synaptic transmission from the deprived eye. The BCM theory also predicts that open-eye potentiation following monocular deprivation is mediated by homosynaptic LTP. This potentiation could be favored by the shift of the modification threshold to the left due to a reduced afferent input from the deprived eye that decreases average activity. Specific mechanisms remain unclear, but in analogy to LTP in other systems (see Chap. 26), correlated synaptic inputs from the not-deprived eye may activate NMDARs, thus increasing $Ca^{2+}$ levels. $Ca^{2+}$-activated protein kinases eventually increase AMPAR insertion into the postsynaptic membrane enhancing sensitivity to glutamate (see Chap. 8).

Support to this notion comes from the occurrence of NMDAR-dependent forms of LTP in the rat visual system and from the fact that OD plasticity is impaired in mice lacking the protein kinase CaMKII, a key player in LTP expression.

LTP differs across the layers of the visual cortex. Remember that the thalamocortical fibers synapse onto pyramidal cells in layer IV, and these in turn synapse onto cells in other layers, notably layer II/III. High frequency stimulation of layer IV in visual cortical slices of both young and adult rats induces reliable LTP of synaptic responses in layer II/III (Fig. 25.9). In contrast, stimulation of thalamocortical afferents in the white matter (WM-LTP) induces LTP in layer II/III only in young animals. WM-LTP can be rescued in adult life when $GABA_A$ receptor antagonists are applied to adult visual cortex slices. Therefore, the maturation of inhibitory circuitries in layer IV might be one of the mechanisms responsible for the downregulation of WM-LTP over development. Notably, susceptibility to WM-LTP roughly coincides with the critical period for OD plasticity and, as the critical period, can be prolonged by dark rearing [28]. Accordingly, BDNF-overexpressing mice, which show an accelerated maturation of intracortical inhibition, display an accelerated developmental decline of WM-LTP in the visual cortex [23].

A role for BDNF in controlling developmental plasticity of the visual system via LTP/LTD *in vivo* has been shown in the developing retinotectal system of *Xenopus laevis*. Application of BDNF into the tectum leads to a persistent potentiation of retinotectal synapses. This potentiation spreads retrogradely to the retina, with a significant enhancement of synaptic inputs at the dendrites of retinal ganglion cells [13]. In developing neural circuits, the induction of retrograde modifications of neurotransmission can be important for allowing coordinated adjustments of synaptic inputs in neuronal networks.

## 25.7 Homeostatic Plasticity

Hebbian mechanisms, based on the idea that correlated pre- and postsynaptic activity strengthens synapses, provide an important framework for the interpretation of experience-dependent changes in the brain. However, due to positive feedback, Hebbian plasticity would lead to rapid saturation of synaptic

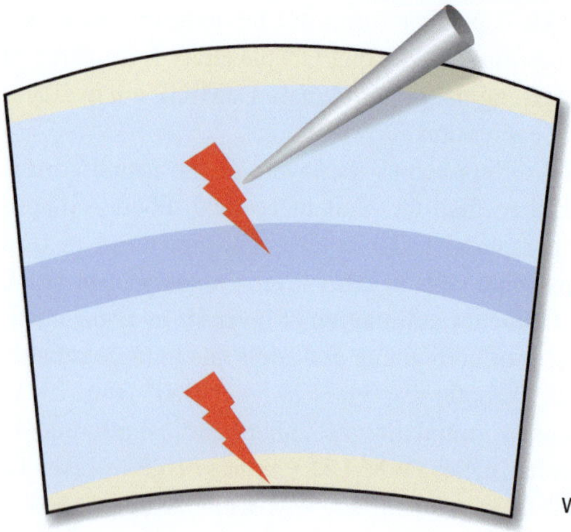


**Fig. 25.9** **Sketch of long-term potentiation (LTP) experiments in visual cortical slices.** High-frequency stimulation of layer IV in visual cortical slices of both young and adult rats induces reliable LTP of synaptic responses in layer II–III. Conversely, electrical stimulation of the white matter yields LTP in layer II–III only in young animals. This form of synaptic plasticity (WM-LTP) follows the time-course of the critical period for ocular dominance plasticity and it can be restored in adult slices only if GABAergic inhibition is reduced (e.g., by exogenous administration of $GABA_A$ receptor antagonists). Stimulation sites are indicated by *lightning bolts*

strength in the absence of proper constraints. Several lines of evidence indicate the presence of homeostatic plasticity, i.e., regulatory adjustments that work to maintain the stability and functionality of neural network subject to Hebbian changes [46]. The BCM model is an example of such homeostatic plasticity, since the crossover point between potentiation and depression moves as to make potentiation less likely when average postsynaptic activity is high (Fig. 25.8). Other mechanisms for homeostatic plasticity include synaptic scaling (scaling of the strength of excitatory synapses depending on average activity of the postsynaptic neuron) and regulation of intrinsic excitability (i.e., changing the way in which postsynaptic neurons integrate synaptic inputs and fire action potentials).

Several experiments have shown that cortical neurons engage in homeostatic adaptations in response to changes in afferent activity, for instance, following sensory deprivation. Specifically, a few days of visual deprivation can scale up excitatory synaptic strength in visual cortex, as measured by the increase in the amplitude of miniature excitatory postsynaptic currents (mEPSCs). Measurements of activity using calcium-sensitive dyes and two-photon imaging have demonstrated that homeostatic response regulation contributes to changes of eye-specific responsiveness after monocular deprivation in mouse visual cortex. Similar adaptations in excitatory synaptic transmission are also observed in auditory cortex following hearing loss, suggesting that homeostatic regulation of cortical synapses by sensory experience is a general phenomenon.

## 25.8 Structural Underpinnings of Experience-Dependent Plasticity

Experience-dependent functional changes of neural circuitries are accompanied by extensive structural rearrangements in neurons. Dendritic spines, which are the sites of excitatory synaptic input to pyramidal cells, are particularly sensitive to experience. A total lack of visual experience early in life (dark rearing), for instance, induces profound changes in spine morphology and density, which are partially reversible by subsequent light exposure. Similarly, monocular deprivation influences motility, turnover, number, and morphology of dendritic spines in the visual cortex. A crucial question is whether and to what extent these forms of structural plasticity contribute to experience-dependent changes in neural connectivity. Longitudinal two-photon imaging experiments in mouse neocortex show that spine dynamics is maximal during early stages of development and decreases thereafter, in parallel to declining plasticity in adulthood [22]. Structural changes of dendritic spines outlast the original experience, and thereby provide a morphological basis for long-term information storage. Visual experience-dependent structural alterations have also been demonstrated in the optic tectum of amphibians and flies. In particular, light stimulation of *Xenopus leavis* tadpoles triggers dendritic arbor growth in tectal cells. In the laboratory fruitfly *Drosophila melanogaster*, alterations in visual experience (monocular deprivation or rearing in darkness) modify the size of the optic lobes and the terminals of photoreceptor cell axons.

## 25.9 Short Noncoding RNAs and the Regulation of Experience-Dependent Plasticity

Recent experimental evidence supports a role for short noncoding RNAs (microRNAs), which interact with and control translation of mRNA targets, in the regulation of experience-dependent plasticity. MicroRNAs are powerful regulators of gene expression and act by binding to the 3′-untranslated region (3′-UTR) of the target mRNA, making it possible for a single microRNA to control expression of multiple genes that posses the same sequence in this region of the mRNA.

A role of microRNAs in synaptic plasticity was first indicated by the observation that a brain-specific microRNA, miRNA-134, localizes to the synapto-dendritic compartment mediating the effects of neurotrophins such as BDNF on structural plasticity of dendritic spines in rat hippocampal neurons [42]. In the visual cortex, another microRNA, miRNA-132, is rapidly up-regulated after eye opening and its expression may control dendritic spine rearrangements associated with functional plasticity triggered by monocular deprivation in early life. Thus, miRNA-132 appears to be a molecular transducer of the action of visual experience on developing visual circuitries, possibly acting through modulation of dendritic spine plasticity.

## 25.10 The Process of Plasticity Reactivation in the Adult Visual System

As previously mentioned, the developmental maturation of intracortical inhibitory circuitries causes the end of plasticity in the visual system [21]. In line with this notion, it is possible to restore plasticity in adult life by reducing levels of inhibition. A direct demonstration that GABAergic transmission is a crucial brake limiting visual cortex plasticity derives from the observation that a pharmacological reduction of inhibitory transmission effectively restores ocular dominance (OD) plasticity in adulthood [20]. This is consistent with the fact that experimental paradigms such as dark exposure, environmental enrichment, food restriction, long-term fluoxetine treatment (a selective serotonin reuptake inhibitor, SSRI), and exogenous IGF-I (insulin-like growth factor 1) administration, all promote plasticity late in life by shifting the intracortical inhibitory/excitatory (I/E) ratio in favor of excitation (e.g., [33]). This has prompted the search for endogenous factors with the potential to enhance plasticity in adult life by modulating the intracortical I/E balance. The process of plasticity reactivation in adulthood is a multifactorial event that comprises the action of different cellular and molecular mechanisms, working in parallel or in series, the sum of which results in the activation of intracellular signal transduction pathways regulating the expression of plasticity genes. In rodents, experimental paradigms based upon the enhancement of environmental stimulation levels, genetic manipulations, and pharmacological treatments have revealed that either the enhanced action of long-distance projection systems (e.g., serotonergic and cholinergic transmission) or IGF-I signaling modulate the intracortical inhibitory/excitatory balance in favor of excitation, which, in turn, sets in motion intracellular events that eventually mediate the expression of genes associated with structural and functional modifications of neural circuitries in the adult visual system.

Experience-dependent modifications of chromatin structure that control gene expression are, recruited as targets of plasticity-associated processes. The plastic outcome caused by fluoxetine in adult rats, is actually mediated by a transitory expression of the neuron-specific transcription factor NPAS4 through a mechanism that involves a reduction in the methylation status at the NPAS4 promoter region [35], this epigenetic modification being normally associated to gene transcription activation. Furthermore, NPAS4 overexpression in the visual cortex of adult naive animals restores ocular dominance plasticity whereas NPAS4 down-regulation during fluoxetine treatment effectively prevents the process of plasticity reactivation [35]. Of note, NPAS4 mediates BDNF expression and drives the formation of inhibitory synapses on excitatory neurons in the rodent visual cortex. Hence, NPAS4 seems to turn on a transcriptional program that regulates the expression of plasticity genes while facilitating, in parallel or in series, a functional reorganization of inhibitory circuitries that might contribute to the homeostasis of cortical excitability during this phase of enhanced plasticity. The role of this transcription factor in mediating phenomena of brain plasticity, is further highlighted by the observation that it is critically involved in processes of memory formation and consolidation, at least, in rodents.

In summary, histone posttranslational modifications seem to be, at least, one of the epigenetic mechanisms (see Chap. 6) underlying the expression of genes that act as downstream effectors of plasticity in the nervous system (Fig. 25.10). The reinstatement of adult visual cortex plasticity caused by intracortical infusion of serotonin, for instance, is partially mediated by 5-HT1A receptors signaling and accompanied by a transitory increment of BDNF expression. This is paralleled by an enhanced histone acetylation status at the activity-dependently regulated *BDNF* promoter regions and by a decreased expression of histone deacetylase (HDACs) enzymes [34]. Accordingly, environmental enrichment, long-term fluoxetine treatment, and food restriction, all increase acetylation of histones in the hippocampus and cortex in adult life. Moreover, a developmental down-regulation of histone posttranslational modifications controls critical period plasticity in the mouse visual system [40] whereas pharmacologically increasing acetylation of histones by treatment with HDACs inhibitors effectively reactivates susceptibility to monocular deprivation and promotes a complete recovery of visual functions in adult amblyopic animals [34, 40].

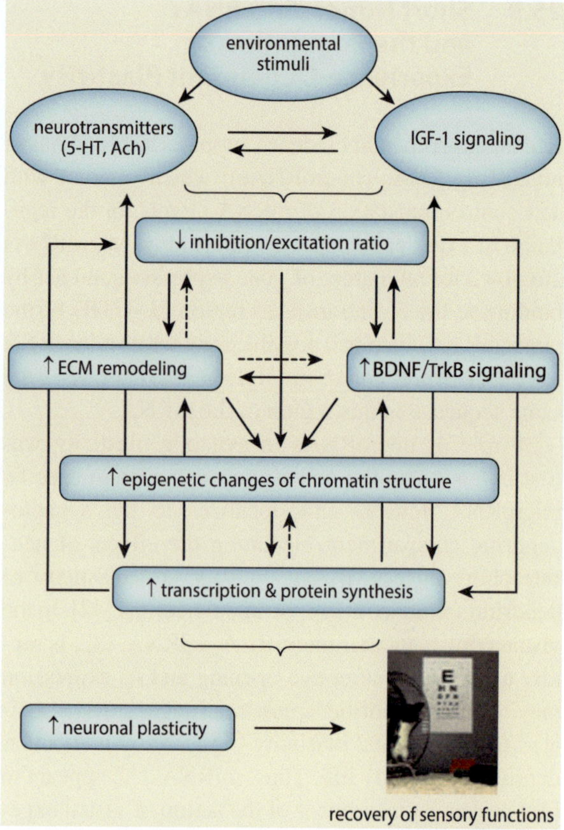


**Fig. 25.10** **The reinstatement of plasticity in adulthood** is associated with signal transduction pathways that involve the enhanced action of either neuromodulatory projection systems (e.g., serotonin and acetylcholine) or IGF-1 signaling, which all set in motion physiological processes that modulate the inhibitory/excitatory ratio. A reduction of the inhibitory/excitatory balance in the visual cortex may directly activate intracellular mechanisms that eventually promote epigenetic modifications of chromatin structure (e.g., posttranslational modifications of histones), which in turn allow for the expression of genes that act as down-stream effectors of plastic phenomena in adult life. A pharmacological reduction of intracortical inhibition enhances plasticity while promoting the activity-dependent BDNF expression and degradation of extracellular matrix (ECM) components that are inhibitory for plasticity. BDNF/TrkB signaling might upregulate the expression of additional genes associated with functional modifications in the visual cortex. Degradation of ECM components may modify the inhibition/excitation ratio in the visual system. The interaction between BDNF/TrkB signaling and extracellular matrix reorganization has yet to be explored. *Continuous arrows* represent established interactions between the molecular and cellular processes mentioned (*boxes*). *Dashed lines* represent interactions that remain to be ascertained. The potential for the reactivation of plasticity to promote the recovery of sensory functions after long-term sensory deprivation in adult life, has been reported using amblyopia as a paradigmatic model

## 25.11 Early Experience Influences Rodents' Behavior by Modifications of Chromatin Structure

In rodents, some dams give more care (licking/grooming) to their offspring than others. Later in life, "high-licked/groomed" pups are less anxious than their "low-licked/groomed" counterparts and they perform better in tests of learning and memory. This effect emerges over the first week of postnatal life and is reversed with cross fostering: if "high licking/grooming" dams rear the biological offspring of "low licking/grooming" mothers, the offspring will behave as "high-licked/groomed" pups [48]. These changes are induced by epigenetic regulation of gene expression, i.e., chromatin structure changes that control gene transcription, including DNA methylation and histone posttranslational modifications (see Chap. 6).

Maternal care increases the expression of the glucocorticoid receptor as well as that of the transcription factor *zif-268*, and leads to a decreased DNA methylation pattern at the glucocorticoid receptor gene promoter in the hippocampus [48]. In addition, changes at the level of chromatin structure in "high-licked/groomed" pups include higher levels of acetylation at lysine-9 on histone-3 (H3K9): a marker of gene transcription activation. Increasing H3K9 acetylation by central infusion of the histone deacetylase inhibitor trichostatin-A in rats born to "low-licking/grooming" mothers causes the methylation pattern to change to that of animals brought up by "high-licking/grooming" dams. Behaviorally, "low-licked/groomed" pups treated with the HDAC inhibitor trichostatin-A are less anxious than vehicle-treated "low-licked/groomed" counterparts and display no difference at behavioral level when compared to "high-licked/groomed" pups.

## 25.12 Experience-Dependent Plasticity in the Auditory System

Considerable plasticity exists in neural circuits that process auditory information. Auditory information ascends via the lateral lemniscus to the midbrain (inferior and superior colliculus) and is then relayed to the medial geniculate nucleus (nucleus ovoidalis in birds). The geniculate neurons project in turn to the primary auditory cortex (A1; field L in birds) (see Chap. 17). Neurophysiological studies in animals raised with abnormal sensory inputs have shown that the function of these auditory circuits is potently shaped by environmental influences. As described above for the visual system, the effects of experience are particularly prominent during critical periods in early development. However, auditory circuits retain the capacity for experience-dependent plasticity throughout adulthood, even if the conditions for plasticity to occur are much more stringent in adults than in juveniles [26]. The following sections describe examples of plasticity in the neural circuits that subserve sound localization and the analysis of sound frequency.

### 25.12.1 Plasticity of Sound Localization

To localize sounds, the central nervous system uses several cues including interaural time differences, interaural volume differences, and the frequency spectrum of the sound at each ear. One important function of auditory localization is the guidance of reflexive orienting responses that shift attention towards unexpected sounds in the environment [27]. This is accomplished at the level of the midbrain (inferior and superior colliculus). In particular the deep layers of the superior colliculus/optic tectum integrate several cues to build a map of auditory space that is aligned with a map of visual space. Indeed, the discharge of superior colliculus neurons is enhanced when auditory and visual stimuli are presented in close spatial and temporal proximity.

Several paradigms have been adopted to study the influence of an altered sensory experience on the sound localization pathway in the midbrain. In the guinea pig, the normal map of auditory space appears in the deep layers of the superior colliculus at 32 days after birth (P32). Formation of this map is prevented by rearing animals from birth in an environment of omnidirectional white noise (i.e., an auditory stimulus that contains every frequency within the audible range). Collicular responses from such noise-reared animals reveal large auditory spatial receptive fields, and the representation of auditory space in the colliculus shows no topographic order. Importantly, exposure to omnidirectional white noise after P30 (postnatal day 30) produces no effect on the collicular map, pointing to a well-defined critical period for map development. Thus, establishment of topographic order in the representation of auditory space in the colliculus requires normal sensory experience during a sensitive phase of development.

One second type of manipulation of auditory experience has involved raising animals with unnatural binaural cues produced by plugging one of the ears, reminiscent of unilateral eyelid suture in the visual system. These experiments, performed in ferrets and barn owls, reveal remarkable adaptive changes in sound localization performance, especially in young animals [27, 29]. Initially, monaurally occluded animals mislocalize sounds towards the side of the open ear [29]. However, after several weeks of monaural occlusion, young animals can perform a sound localization task just as accurately as normal controls, indicating that they progressively compensate for the presence of the earplug [27, 29]. Compensation is much less efficient after a comparable period of monaural occlusion in adulthood [27], again pointing to the existence of a critical period for circuit rearrangement. In keeping with the behavioral data showing greater compensation in juvenile than in adult age, neurophysiological examination reveals an adaptive change in the auditory space map in the superior colliculus when ear occlusion is performed in young animals, but not in adulthood. Indeed, the correspondence between the auditory and the visual map in the superior colliculus of young ferrets with an earplug is similar to that of normally reared ferrets, while there is no significant realignment of the two maps in ferrets that are subjected to a similar period of monaural occlusion as adults [27].

Monaural occlusion experiments performed in young barn owls have yielded comparable results. Young owls subjected to monaural occlusion are capable of recovering accurate sound localization responses. In this species, the earplug induces a compensatory shift in the tuning of auditory neurons towards the abnormal interaural time and level differences.

Another experimental protocol consists in perturbing the normal alignment of the auditory and visual map. Barn owls can be raised with displacing prisms mounted in spectacle frames in front of the eyes; these prismatic spectacles cause a horizontal shift of the visual field of about 20° (Fig. 25.11a–c). Over the course of several weeks, the animals wearing prisms adjust their sound localization mechanisms to match their shifted visual worlds. Basically, prism-reared owls challenged with a sound stimulus orient to the side of the source by an amount that corresponds to the prismatic displacement of the visual field. Thus, vision has a dramatic impact on auditory space processing and dominates it. This learning is adaptive because the owls bring their auditory and visual worlds back into mutual alignment (Fig. 25.11b–d). A critical period exists for these adaptive rearrangements, as adult owls cannot adjust sound localization in response to a large prismatic displacement of the visual field. However, when the prismatic shift is experienced in small increments, maps of sound localization do change also in adults.

Where does this plasticity take place, and how does it occur? The external nucleus of the inferior colliculus (ICX) has been shown to be a site of large-scale plasticity during experience-driven modification of the auditory space map. The ICX is located in the midbrain of the barn owl and contains an ordered map of auditory space. It receives information from the central nucleus of the superior colliculus (ICC) and projects its auditory space map to the optic tectum (OT), thus contributing to auditory orienting behaviors; the ICX also receives visual information via a feedback projection from the tectum (Fig. 25.11a) [29]. In juvenile owls reared with prisms, response to sound localization cues (such as interaural time difference, ITD) in the ICX and optic tectum change to match the visual field displacement (Fig. 25.11b–d). These new, learned responses become progressively dominant, while responses to the normal ITD range (normal responses recorded before prism experience) disappear over several weeks.

This remarkable plasticity includes both anatomical and functional rearrangements. At the morphological level, there is a robust axonal remodeling in the projection from the ICC to the ICX. The ICX receives auditory input from the central nucleus of the inferior colliculus (ICC) via topographic axonal connections; in prism-reared owls, these projections are broader than normal, and axon terminals are located both in their normal target territory and in an abnormal zone where they can support the new, learned responses (Fig. 25.12a). Thus, the formation of an adaptive neural circuit is mediated, at least in part, by axonal sprouting; importantly, the naïve circuit also persists, indicating that these normal projections must be somehow silenced at the completion of learning (see below).

The expression of new, learned responses in the ICX depends on the activation of NMDARs. After a few weeks of prism experience, ICX neurons show both naïve and new responses; focal application into the ICX of a selective blocker of the NMDA receptor wipes out learned responses, while normal responses are reduced only by about 50 % [29]. In contrast, blockade of AMPA glutamate receptors tends to affect normal responses more than new responses. These pharmacological data indicate that synapses mediating adaptive plasticity in the ICX have a high NMDA/AMPA ratio (Fig. 25.12b); since the NMDA receptor is gated by both glutamate and voltage, synapses that are newly formed in the ICX during the process of plasticity may faithfully transmit information only when the postsynaptic cell is depolarized. Such depolarization is provided by an instructive signal from the visually responsive neurons in the optic tectum to the ICX; retinotopic activity in the optic tectum could serve as the template that instructs the auditory space map. A small lesion in the optic tectum eliminates plasticity in the corresponding portion of the auditory space map in the ICX. Thus, excitatory projections

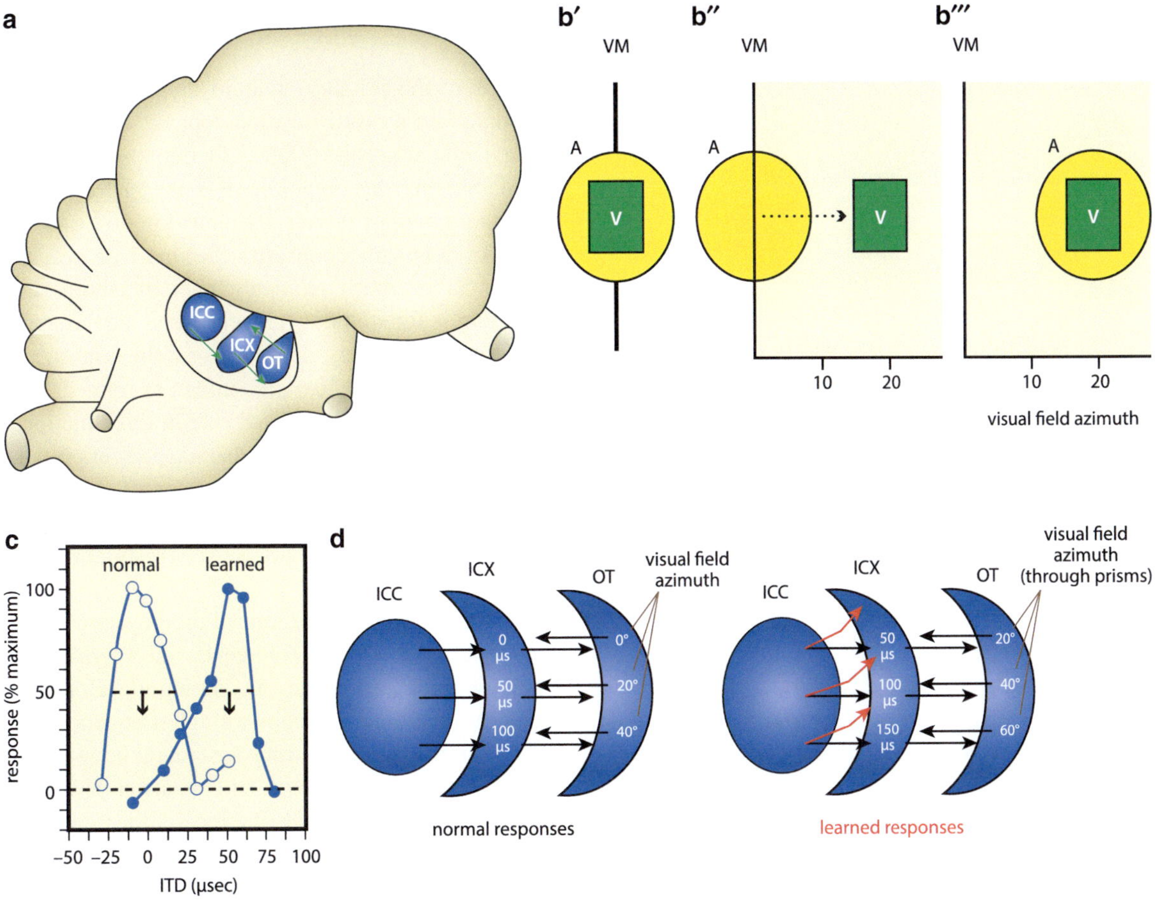


**Fig. 25.11** **Plasticity of the midbrain pathway that mediates sound localization in the barn owl.** (**a**) Schematic view of the barn owl brain showing the position of ICC, ICX, and OT. Information is transferred from the central nucleus of the inferior colliculus (*ICC*) to the external nucleus of the inferior colliculus (*ICX*), where a map of auditory space is created. The ICX projects to the optic tectum (*OT*), where the auditory map merges with a visual map of space. In turn, the OT sends a feedback projection to the ICX. *Green arrows* indicate excitatory synaptic connections between areas. (**b**) Effects of prism rearing on the auditory spatial receptive fields of neurons in the midbrain. Visual (*V*) and auditory (*A*) receptive fields are normally in close correspondence (*b′*). Wearing prisms shifts the visual receptive field by 20° and disrupts the alignment of visual and auditory receptive fields (*b″*). Over several weeks of prism experience, the auditory receptive field shifts to align with the new position of the visual receptive field (*b‴*). *VM* vertical meridian. (**c**) Plasticity of auditory tuning in the optic tectum of a juvenile barn owl after prismatic displacement of the visual field is accompanied by a corresponding change in the response to sound localization cues (*ITD* interaural time differences). The graph shows the response of cells in the optic tectum, before (*open symbols*) and after (*filled symbols*) several weeks of prism experience (learned responses). The neuron normally responds maximally to 0 *ITD* and has a visual receptive field on the vertical meridian (0° azimuth). After wearing prisms for 8 weeks, the cell responds to a value of *ITD* (50 µs) that exactly corresponds to the amount of visual field displacement (20°). *L* left ear leading, *R* right ear leading (Adapted from Keuroghlian and Knudsen [26], with permission). (**d**) The shift in the projection of information from the ICC to the ICX (*black* and *red arrows*) that results from early prism experience. *Left* normal responses, *right* after several weeks of prism experience. Note that units that normally respond to 0 ITD (receptive field at 0° azimuth, i.e., *vertical meridian*), following prism experience are tuned to 50 µs ITD (corresponding to 20° from the *vertical meridian*) (Adapted from Keuroghlian and Knudsen [26], with permission)

from the optic tectum could promote plasticity in the ICX by stabilizing auditory synapses that contribute to activity patterns that match those evoked by the visually based template signal [29] (Fig. 25.12b).

Another key player in the plasticity mechanisms induced by prism rearing is GABAergic inhibition. Iontophoretic application of a $GABA_A$ antagonist into the ICX of prism-reared owls at the completion of the

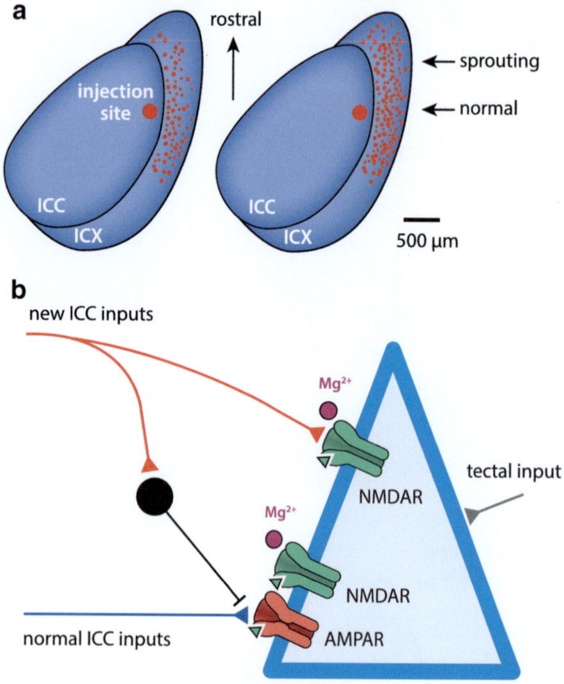


**Fig. 25.12 Mechanisms of plasticity in the barn owl following prism experience.** (**a**) Plasticity of the ICC-ICX projection. *Left* normal anterograde labeling, *right* anterograde labeling in owls wearing prisms for 8 weeks. Note that axons from the ICC terminate in a broader zone of the ICX following prism experience, thus supporting new, learned responses (Modified from Knudsen [29], with permission). (**b**) Schematics of the pattern of neuronal connectivity in the ICX following prism experience. A pyramidal neuron (*triangle*) and an inhibitory, GABAergic neuron (*circle*) are shown. *On the left* are normal (*blue*) as well as sprouting inputs (*red*) from the ICC; *on the right*, inputs from the optic tectum are depicted. During the first phases of plasticity, sprouting axons from the ICC signal to the pyramidal cell mainly via NMDA receptors (*green channels*, NMDARs) and therefore require a coincident input from the optic tectum to reliably activate the postsynaptic neuron. The new set of inputs from the ICC apparently suppress expression of the naïve map by $GABA_A$-mediated inhibition

learning process (i.e., in animals in which the newly learned ITD responses are fully mature) causes the immediate appearance of normal responses in ICX neurons. This indicates that in the ICX of prism-reared owls, excitatory connections supporting normal and learned responses coexist (in agreement with the anatomical data showing intact normal ICC-ICX projections after prism rearing), but responses to normal inputs are masked by GABAergic inhibition (Fig. 25.12b). Thus, the functional selection of the behaviorally appropriate map is accomplished by $GABA_A$-mediated inhibition of the inappropriate, naïve map.

### 25.12.2 Plasticity of Frequency Tuning

Neurons in the primary auditory area (A1) of adult mammals are arrayed in a tonotopic map according to their characteristic (preferred) frequencies of the sound stimulus; this organization is not predetermined but rather emerges during a sensitive period of postnatal development. Soon after the onset of hearing, neurons in A1 have relatively high thresholds and broad tuning for stimulus frequency, and the tonotopic organization is poorly defined. Over the subsequent days, these physiological properties are refined; sensory experience plays a crucial role in this functional maturation. This is similar to what happens in primary visual cortex whose functional properties are poorly developed at the onset of vision, and undergo a progressive, visual experience-dependent maturation during a critical period.

Several reports have examined the impact of auditory experience on the emergence of a systematic tonotopic organization in A1 during development. When rat pups are exposed to a single frequency tone for several hours a day during the critical period (second postnatal week) there is an expansion of the A1 area devoted to the representation of the presented frequency, and a corresponding reduction in the area tuned to other frequencies (Fig. 25.13). Thus, A1 comes to over-represent the experienced frequency, and in this sense becomes customized to the acoustic environment experienced by the subject. The tonotopic modifications induced by early exposure to a single frequency are long-lasting and persist unaltered throughout adulthood [11, 50]. No significant changes of the primary auditory cortex are detected following monotonic stimulation in adult rats.

Another experiment has shown that when rat pups are chronically exposed to white noise during the critical period, the functional properties of A1 neurons do not properly mature and remain very similar to those recorded in very young infant control pups [9]. In particular, while naïve A1 neurons are characterized by high selectivity for sound frequency, cells of animals exposed to the degrading white noise rather exhibit a broad tuning. The exposure to masking noise not only disturbs the functional development of A1 but also leaves cortical circuits more malleable to sensory experience, thus extending the critical period for plasticity. In rats raised under white noise until P50 and then exposed to persistent stimulation with a single frequency, A1 becomes dominated by the representation of the experienced frequency [9]. It is worth recalling that such chronic monotone stimulation has absolutely no effect on the naïve P50 cortex. Thus, as in other sensory modalities, degradation of the normal sensory experience causes the sensitive period to be extended.

Under certain conditions, however, auditory circuits, including A1, retain the ability for experience-dependent plasticity throughout adulthood. Stimuli must be behaviorally relevant in order to induce significant plasticity in adulthood [26]. For instance, animals must be conditioned to attend to (i.e., respond to) the frequency, usually accomplished via positive or negative reinforcement for altering frequency representation and tuning in the adult A1. In one experiment, rats were presented with tones of different frequencies and volume levels, and were trained to detect either the frequency or level of the sounds [39]. A subset of animals were rewarded for responding to a particular frequency at any sound level, whereas other rats were rewarded for responding to a particular level at any frequency; overall, however, rats in both groups were exposed to the same stimuli. In animals rewarded for detecting a particular frequency, the proportion of neurons responding to that frequency increased by a factor of two in primary auditory cortex. This expansion did not occur in animals trained to detect sound level, indicating the specificity of the effect (Fig. 25.14). In another experiment, monkeys were trained to discriminate a specific frequency. They received a juice reward for moving the head immediately after hearing the target frequency tone; learning of this task was accompanied by a significant sharpening of frequency tuning curves that was selective for the target frequency. Collectively, these data demonstrate that plasticity can be triggered in the adult A1 by stimuli that are behaviorally relevant; in addition, the plasticity that is induced appears to be specific for the acoustic feature to which the animal has attended.

The critical importance of attention and reward for triggering plasticity in the adult A1 suggests the involvement of activating, diffuse neuromodulatory systems. One of such neuromodulators is acetylcholine, which is released in the cortex by axonal terminals of cholinergic neurons located in the basal forebrain

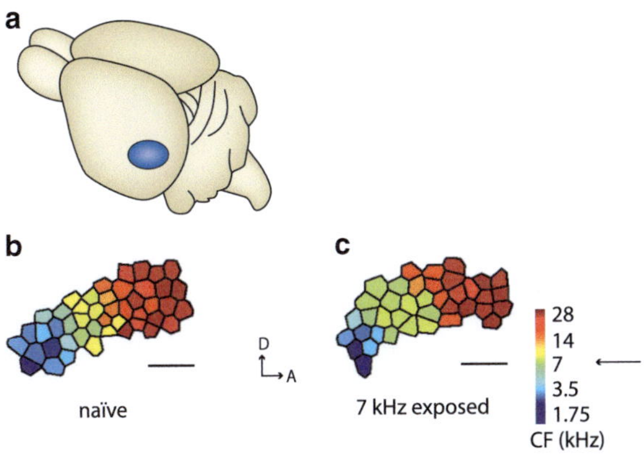


**Fig. 25.13 Effect of early exposure to a single frequency on the tonotopic organization in primary auditory cortex.** (**a**) Schematics of the rat brain showing location of the primary auditory cortex (*blue spot*). (**b**, **c**) Plasticity of the tonotopic map. The color of each polygon indicates the frequency at which neurons in that site are most sensitive (*CF* characteristic frequency, in kHz). Panel (b) shows the CF map from a naïve P18 rat, while panel (c) represents the map of a P18 rat exposed to 7 kHz pure tones. Note that the experienced and nearby frequencies (*yellow* and *green polygons*) are over-represented in the cortex of the stimulated animal with respect to the naïve cortex. *D* dorsal, *A* anterior. Scale bars = 0.5 mm. The *arrow* indicates the frequency of stimulation (Data reproduced from de Villers-Sidani et al. [11], with permission)

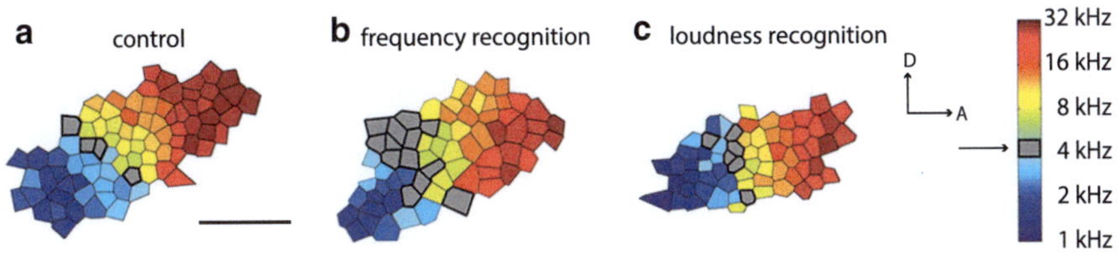


**Fig. 25.14 Task-specific tonotopic reorganization of the adult primary auditory cortex.** Representative tonotopic maps from A1 in adult control rats (**a**), rats trained to respond to a frequency of 5 kHz (**b**), and rats trained to detect sound levels (**c**). The color of each polygon indicates the frequency at which neurons in that site are most sensitive (*CF* characteristic frequency, in kHz); *gray-shaded polygons* indicate recording sites with CF values close to 5 kHz. Note expansion of the representation of the 5 KHz frequency in adult rats trained to recognize that frequency. The *arrows* indicate dorsal (*D*) and anterior (*A*) orientations. Scale bar = 0.5 mm (Data reproduced from Polley et al. [39], with permission)

(nucleus basalis). These cholinergic neurons are activated as a function of the behavioral significance of stimuli and provide the cortex with such information. Pairing a specific tone with nucleus basalis stimulation results in a robust remodeling of cortical area A1, with a clear expansion of the region of the cortex that represents the paired frequency. This cortical reorganization is remarkably similar to that obtained in animals trained to attend to a specific frequency. Thus, behavioral relevance of the sensory stimulus, signaled by an increase in neuromodulators such as acetylcholine, enhances the effectiveness of plasticity mechanisms in adult cortical networks, enabling them to overcome the restrictions on plasticity that exist in mature circuits [26].

## 25.13 Experience-Dependent Plasticity in the Somatosensory System

Experience-dependent plasticity has been amply documented in the primary somatosensory cortex. One of the most successful model systems for analyzing plasticity is the rodent barrel cortex (Fig. 25.15a). Whiskers are the main tactile organs in rodents, and the barrel cortex is the area of somatosensory cortex that contains the representation of the whiskers. The neurons in layer IV are arranged in discrete groups called barrels; each barrel receives inputs from the thalamus carrying information from a single whisker (called the principal whisker); the barrels are arranged in a pattern that corresponds to the topography of the whiskers on the animal's snout (Fig. 25.15a).

The microcircuitry of the barrel cortex corresponds to other cortical areas: thalamic inputs terminate on pyramidal neurons in layer IV, in each barrel these project to neurons in superficial layers (layers II–III) within the same radial column. This feed-forward pathway drives strong responses to each column's principal whisker, but there is also a spread of excitation to nearby columns, mainly mediated by horizontal connections (Fig. 25.15b). Spread of excitation within and across functional columns is precisely regulated by the action of several classes of inhibitory interneurons.

Plasticity can be readily induced in barrel cortex by whisker deprivation (i.e., by trimming some of the whiskers) during adolescence and in adulthood [15]. This plasticity involves rearrangements of layers II–III circuitry, while plasticity of thalamocortical connections can generally be induced only by manipulations

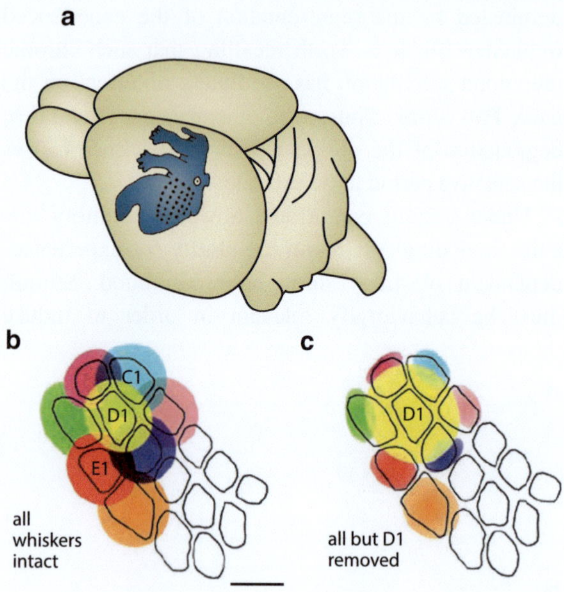


**Fig. 25.15 Plasticity in the rodent barrel cortex.** (**a**) Schematics of the mouse brain showing the somatosensory cortex and the representation of the different body parts. Note the wide area devoted to whisker representation (barrel cortex). (**b**, **c**) Schematized tangential view of layers II–III of barrel cortex showing functional whisker maps in normal animals (**b**) and animals with all whiskers trimmed apart from whisker D1 (**c**). *Colored* regions represent cortical areas responding to different whiskers. In normal rodents (**b**), each whisker activates a cortical area slightly larger than the cortical column defined by its layer IV barrel (barrel outlines are shown in *black*). Removing all but the D1 whisker in adolescent rodents causes expansion of the spared, D1 whisker and a depression of deprived, nearby whiskers within the map (**c**). Scale bar = 0.5 mm (Modified from Feldman and Brecht [15], with permission)

very early in development (within the first days after birth). For instance, trimming all whiskers during a narrow critical period during the second postnatal week (at the peak of layers II–III synaptic development) causes neurons in superficial layers to adopt broad, unfocused receptive fields and a disordered whisker map, while the layer-IV map remains normal. These cortical effects are reminiscent of the deficits in synaptic maturation and map formation observed by deprivation of experience in other sensory modalities, e.g., vision (see above).

Experiments with trimming of some whiskers in young animals (age around P30–P60) have shown two different processes: down-regulation of the deprived input (depression) and up-regulation of the spared input (potentiation). These processes have different kinetics, which implies two mechanistically distinct processes for plasticity. Depression and potentiation combine to produce an expansion of the cortical representation of spared whiskers into the representation of adjacent deprived whiskers (Fig. 25.15b–c). Interestingly, in adulthood (after 6 months of age) plasticity only depends on potentiation of spared vibrissae input in layers II–III.

Weakening of evoked responses to deflection of the trimmed whiskers is the first plastic event induced by sensory deprivation in the adolescent barrel cortex, and precedes synaptic potentiation. Depriving all of the whiskers causes depression, but the reduction of the response is much greater if only a few vibrissae are trimmed. These results imply that one component of depression certainly resides in the reduction of input activity: leaving only spontaneous activity in the afferent pathway from the whisker follicles to the cortex is capable of inducing depression in the cortex. However, superimposed on this effect is the suppressive action of the remaining whiskers, which causes competitive depression. This result is conceptually similar to that obtained in the visual cortex, where deprivation of input from both eyes (via either dark rearing or binocular deprivation) causes some depression, but monocular occlusion causes far greater weakening of the closed eye pathway.

Depression of responses evoked by trimmed whiskers is primarily caused by a reduction in the efficacy of the excitatory synaptic connections between layer IV and layers II–III. Experiments using local glutamate uncaging to activate neurons in specific layers show that layer II–III pyramidal neurons no longer receive a strong input from layer IV following whisker trimming. This depression involves a presynaptic reduction in neurotransmitter release probability, and is likely generated by a Hebbian mechanism of spike timing-dependent plasticity (STDP) (see Chap. 26). In the intracolumnar layer IV-layer II–III pathway of the barrel cortex, LTP is induced when the layer-IV cell fires 0–20 ms before layers II–III cells, and LTD is induced when firing order is reversed, for spiking delays of 0 to 50–100 ms. These observations provide a possible mechanism for response depression in the deprived barrels. In the deprived pathways, whisker trimming abolishes the close correlation between pre- and postsynaptic activity, thus rendering LTP unlikely; moreover, any spontaneous, random presynaptic activity tends to cause depression rather than potentiation, since the time window for depression is greater (50–100 ms) than that for potentiation (20 ms). This mechanism of STDP also accounts for the additional depression observed by sparing some whiskers. Spared vibrissae inputs can drive neurons in nearby deprived columns via horizontal connections, and this increases the number of presynaptic depolarizations occurring out of synchrony with the random presynaptic activity, thus promoting LTD.

Plasticity observed following whisker trimming also involves potentiation of the responses to the spared whisker. Potentiation is seen both in adolescent and adult rats, and involves both intracolumnar and extracolumnar circuitry. First, when a single whisker is spared, there is a potentiation of synaptic transmission from layer IV to layers II–III within the spared barrel column. This is seen electrophysiologically as an increased responsiveness of layers II–III neurons to spared whisker stimulation. This enhancement of response is likely due to the removal of phasic inhibition exerted by the nearby cortical barrel columns. Second, potentiation of spared whisker response also occurs in the surrounding deprived barrels over a period of several days [16]. The possibility of interaction between several active barrel columns has also been tested by imposing on the animals a "chessboard" pattern of deprivation, in which every other whisker is trimmed. In this case, potentiation of spared responses occurs faster than with a single spared whisker. This implies that there is a cooperative interaction in the deprived barrel columns between the input activities generated by the spared whiskers.

Structural changes at the level of axons, dendrite branches, and dendrite spines underlie some of the plastic changes induced by whisker trimming in the barrel cortex. In particular, there is a remodeling of the axonal arbor of excitatory connections in layers II–III. Experiments with longitudinal two-photon imaging have also provided clear support for a role of growth and retraction of dendritic spines in experience-dependent plasticity of the barrel cortex. Spines are the main site of excitatory input and are highly dynamic structures. Induction of plasticity stabilizes new spines (13–15 %, compared with 5 % under baseline conditions) on pyramidal neurons over 2–3 weeks after whisker trimming. These data support the idea that spine growth provides a structural basis for response potentiation, as seen in the visual cortex (see above).

There are several other examples of how experience alters the somatosensory maps in the cortex. One crucial experiment has assessed the effect of training on the representation of the fingers in the somatosensory cortex of the monkey. In this experiment, monkeys are trained to touch a rotating disk with the tips of their middle three fingers to obtain food. After several months of training, the cortical representation of the tips of the stimulated fingers is substantially enlarged. This occurs at the expense of the representation of the adjacent proximal phalanges. Examples of large-scale reorganizations of somatosensory maps are also available in humans, specifically in musicians. In particular, in string players the cortical representation of the digits of the left hand (the fingering hand) is larger than in controls. This cortical reorganization is more pronounced in musicians who have begun their musical training at an early age. This is again consistent with the idea that experience-dependent cortical rearrangements, although available in adulthood to support lifelong learning, are particularly prominent during early stages of development.

## 25.14 Plasticity in the Olfactory System

The olfactory system is particularly plastic since it exhibits lifelong turnover of both peripheral sensory neurons and bulbar interneurons. In the olfactory epithelium, cell renewal persists throughout adult life to replace olfactory sensory neurons; the newly generated sensory neurons must extend their axons and contact the proper targets in the bulb. The turnover of sensory neurons in the olfactory epithelium is tightly regulated by environmental factors. Indeed, neurogenesis is enhanced by ablation of the olfactory bulb, while blocking airflow through one side of the nasal cavity causes an ipsilateral reduction in cell proliferation (see Chap. 13).

Another site of plasticity is represented by newly generated bulbar interneurons [30]. Indeed, the number of newborn interneurons influences olfactory memory. Exposure of mice to a complex olfactory environment (olfactory enrichment) leads to increased newborn cell survival and improvements in olfactory memory. Furthermore, in the female mouse olfactory system, cell proliferation increases after mating and during gestation and lactation, suggesting that adult neurogenesis is important for the high olfactory perceptual and memory demands associated with reproduction. Synaptic properties of newly generated interneurons in the bulb differ from those of pre-existing neurons. Glutamatergic synapses show long-term potentiation (LTP) in an early stage of interneuron maturation (shortly after cell arrival in the bulb). LTP fades as the newborn neurons mature. Thus, neurogenesis provides a basis for lifelong plasticity in the adult olfactory bulb circuit.

Another system in which neurogenesis plays a key role in experience-dependent plasticity is the olfactory pathway of decapods crustaceans. In these animals adult neurogenesis occurs in the olfactory lobe: the first synaptic relay in the brain that receives information from olfactory receptor neurons. As previously described in vertebrates, plasticity in the crustaceans' olfactory lobe is regulated by sensory experience. Unilateral ablation of the olfactory sensilla (containing olfactory receptor neurons) down-regulates the generation and survival of newborn cells in the crayfish olfactory lobe. Impoverished (individuals isolated in small spaces) and enriched (animals living together in large areas) rearing conditions also influence neurogenesis by altering the proliferation of precursor cells and survival of newly generated neurons. Adult neurogenesis may contribute to increase the resolution of odorant decoding and allow adaptation of the olfactory system in crustaceans to ever changing odor environments. Thus regulation of neurogenesis by environmental influences and its role in olfactory system function and plasticity appear to be conserved themes from invertebrates to vertebrates.

## 25.15 Cross-Modal Developmental Plasticity

Several reports have shown that sensory deprivation in one modality (especially when it occurs early in development) can have striking effects on the development of the remaining modalities. This phenomenon is referred to as cross-modal plasticity. Deaf or blind humans have provided convincing behavioral, electrophysiological, and neuroimaging evidence of increased capabilities and compensatory expansion in their remaining modalities [2]. For instance, congenitally blind subjects display better sound localization abilities than sighted individuals and have better two-point tactile discrimination skills. What is the basis for these enhanced abilities? Functional imaging studies of people who are blind from an early age reveal that Braille reading and other tactile discrimination tasks activate their primary visual cortex. Using transcranial magnetic stimulation to disrupt the function of visual cortex in blind people during Braille reading induces errors in Braille reading and distorts the tactile perceptions of blind subjects. Thus, blindness from an early age recruits the visual cortex to process somatosensory information in a functionally relevant way.

Is there a critical period for cross-modal plasticity? This question has been tackled by examining activation of visual cortical areas by Braille reading in subjects who became blind after 14 years of age, following a period of normal vision (late-onset blind). Functional neuroimaging indicates that the visual cortex is not activated in the late-onset blind group; moreover, repetitive transcranial magnetic stimulation applied to inactivate visual cortical areas does not disrupt their Braille reading. These data indicate that a critical period does exist for the visual cortex to be recruited to a role in somatosensory processing, and this critical period does not extend beyond 14 years of age in humans.

One explanation how cross-modal plasticity is accomplished involves the stabilization of long-range cortico-cortical connections between sensory modalities. These connections have been described in immature cats and hamsters. In adult monkeys there is a direct connection between primary auditory cortex and primary visual cortex that could provide a basis for cross-modal plasticity [2].

## 25.16 Summary

In this chapter we have described the key role of environmental influences in shaping the functional and structural architecture of the central nervous system. Although intrinsic, genetic factors mediate the initial formation of neuronal networks, adult patterns of neural connectivity are determined by experience. Through experience-dependent plasticity, the brain remodels its connections in order to adjust the response of the organism to changing conditions. The brain shows greatest plasticity during early life. Early developmental phases during which neural circuits are particularly sensitive to environmental influences are known as “critical periods”. A failure to be exposed to normal experience during the critical period permanently impairs sensory functions and/or behavior. Sensory experience modulates brain structure and function via the activation of different structural and functional factors such as inhibitory transmission, neurotrophins and extracellular matrix molecules. Signal transduction cascades driving gene expression changes ultimately result in long-lasting alterations in synaptic structure and function.

## References


1. Antonini A, Stryker MP (1993) Rapid remodeling of axonal arbors in the visual cortex. Science 260:1819–1821
2. Bavelier D, Neville HJ (2002) Cross-modal plasticity: where and how? Nat Rev Neurosci 3:443–452
3. Berardi N, Pizzorusso T, Maffei L (2000) Critical periods during sensory development. Curr Opin Neurobiol 10:138–145
4. Bienenstock EL, Cooper LN, Munro PW (1982) Theory for the development of neuron selectivity: orientation specificity and binocular interaction in visual cortex. J Neurosci 2: 32–48
5. Bischof HJ, Rollenhagen A (1999) Behavioural and neurophysiological aspects of sexual imprinting in zebra finches. Behav Brain Res 98:267–276
6. Cabelli RJ, Hohn A, Shatz CJ (1995) Inhibition of ocular dominance column formation by infusion of NT-4/5 or BDNF. Science 267:1662–1666
7. Caleo M, Maffei L (2002) Neurotrophins and plasticity in the visual cortex. Neuroscientist 8:52–61
8. Cancedda L, Putignano E, Sale A, Viegi A, Berardi N, Maffei L (2004) Acceleration of visual system development by environmental enrichment. J Neurosci 24:4840–4848
9. Chang EF, Merzenich MM (2003) Environmental noise retards auditory cortical development. Science 300: 498–502
10. Constantine-Paton M, Law MI (1978) Eye-specific termination bands in tecta of three-eyed frogs. Science 202: 639–641
11. de Villers-Sidani E, Chang EF, Bao S et al. (2007) Critical period window for spectral tuning defined in the primary auditory cortex (A1) in the rat. J Neurosci 27:180–189
12. Doupe AJ, Kuhl PK (1999) Birdsong and human speech: common themes and mechanisms. Annu Rev Neurosci 22: 567–631
13. Du JL, Poo MM (2004) Rapid BDNF-induced retrograde synaptic modification in a developing retinotectal system. Nature 429:878–883
14. Fagiolini M, Pizzorusso T, Berardi N et al. (1994) Functional postnatal development of the rat primary visual cortex and the role of visual experience: dark rearing and monocular deprivation. Vision Res 34:709–720
15. Feldman DE, Brecht M (2005) Map plasticity in somatosensory cortex. Science 310:810–815
16. Fox K (2002) Anatomical pathways and molecular mechanisms for plasticity in the barrel cortex. Neuroscience 111: 799–814
17. Frenkel MY, Bear MF (2004) How monocular deprivation shifts ocular dominance in visual cortex of young mice. Neuron 44:917–923
18. Galli L, Maffei L (1988) Spontaneous impulse activity of rat retinal ganglion cells in prenatal life. Science 242:90–91

19. Gianfranceschi L, Siciliano R, Walls J et al. (2003) Visual cortex is rescued from the effects of dark rearing by overexpression of BDNF. Proc Natl Acad Sci USA 100: 12486–12491
20. Harauzov A, Spolidoro M, DiCristo G et al. (2010) Reducing intracortical inhibition in the adult visual cortex promotes ocular dominance plasticity. J Neurosci 30:361–371
21. Hensch TK (2005) Critical period plasticity in local cortical circuits. Nat Rev Neurosci 6:877–888
22. Holtmaat A, Svoboda K (2009) Experience-dependent structural synaptic plasticity in the mammalian brain. Nat Rev Neurosci 10:647–658
23. Huang ZJ, Kirkwood A, Pizzorusso T et al. (1999) BDNF regulates the maturation of inhibition and the critical period of plasticity in mouse visual cortex. Cell 98:739–755
24. Katz LC, Shatz CJ (1996) Synaptic activity and the construction of cortical circuits. Science 274:1133–1138
25. Keating MJ, Grant S (1992) The critical period for experience-dependent plasticity in a system of binocular visual connections in *Xenopus laevis*: its temporal profile and relation to normal development requirements. Eur J Neurosci 4:27–36
26. Keuroghlian AS, Knudsen EI (2007) Adaptive auditory plasticity in developing and adult animals. Prog Neurobiol 82:109–121
27. King AJ, Parsons CH, Moore DR (2000) Plasticity in the neural coding of auditory space in the mammalian brain. Proc Natl Acad Sci USA 97:11821–11828
28. Kirkwood A, Lee HK, Baer MF (1995) Co-regulation of long-term potentiation and experience-dependent synaptic plasticity in visual cortex by age and experience. Nature 375:328–331
29. Knudsen EI (2002) Instructed learning in the auditory localization pathway of the barn owl. Nature 417:322–328
30. Lazarini F, Lledo PM (2011) Is adult neurogenesis essential for olfaction. Trends Neurosci 34:20–30
31. LeVay S, Wiesel TN, Hubel DH (1980) The development of ocular dominance columns in normal and visually deprived monkeys. J Comp Neurol 191:1–51
32. Levelt CN, Hubener M (2012) Critical period plasticity in the visual cortex. Annu Rev Neurosci 35:309–330
33. Maya-Vetencourt JF, Sale A, Viegi A et al. (2008) The antidepressant fluoxetine restores plasticity in the adult visual cortex. Science 320:385–388
34. Maya-Vetencourt JF, Tiraboschi E, Spolidoro M, Castrén E, Maffei L (2011) Serotonin triggers a transitory epigenetic mechanisms that promotes adult visual cortex plasticity. Eur J Neurosci 33:49–57
35. Maya-Vetencourt JF, Tiraboschi E, Greco D et al. (2012) Experience-dependent expression of NPAS4 regulates plasticity in adult visual cortex. J Physiol 590:4777–4787
36. McGee AW, Yang Y, Fischer QS, Daw NW, Strittmatter SM (2005) Experience-driven plasticity of visual cortex limited by myelin and Nogo receptor. Science 309:2222–2226
37. Moriceau S, Sullivan RM (2006) Maternal presence serves as a switch between learning fear and attraction in infancy. Nat Neurosci 9:1004–1006
38. Pizzorusso T, Medini P, Berardi N, Chierzi S, Fawcett JW, Maffei L (2002) Reactivation of ocular dominance plasticity in the adult visual cortex. Science 298:1248–1251
39. Polley DB, Steinberg EE, Merzenich MM (2006) Perceptual learning directs auditory cortical map reorganization through top-down influences. J Neurosci 26:4970–4982
40. Putignano E, Lonetti G, Cancedda L et al. (2007) Developmental downregulation of histone posttranslational modifications regulates visual cortical plasticity. Neuron 53: 747–759
41. Rai S, Rankin CH (2007) Critical and sensitive periods for reversing the effects of mechanosensory deprivation on behavior, nervous system, and development in *Caenorhabditis elegans*. Dev Neurobiol 67:1443–1456
42. Schratt GM, Tuebing F, Nigh EA et al. (2006) A brain specific microRNA regulates dendritic spine development. Nature 439:283–289
43. Singer W (1995) Development and plasticity of cortical processing architectures. Science 270:758–764
44. Sugiyama S, Di Nardo AA, Aizawa S et al. (2008) Experience-dependent transfer of Otx2 homeoprotein into the visual cortex activates postnatal plasticity. Cell 134:508–520
45. Syken J, Grandpre T, Kanold PO, Shatz CJ (2006) PirB restricts ocular-dominance plasticity in visual cortex. Science 313:1795–1800
46. Turrigiano GG (2008) The self-tuning neuron: synaptic scaling of excitatory synapses. Cell 135:422–435
47. Vorobyov V, Kwok JC, Fawcett JW, Sengpiel F (2013) Effects of digesting chondroitin sulfate proteoglycans on plasticity in cat primary visual cortex. J Neurosci 33:234–243
48. Weaver IC, Cervoni N, Champagne FA et al. (2004) Epigenetic programming by maternal behavior. Nat Neurosci 7:847–854
49. Wiesel TN, Hubel DH (1963) Single-cell responses in striate cortex of kittens deprived of vision in one eye. J Neurophysiol 26:1003–1017
50. Zhang LI, Bao S, Merzenich MM (2001) Persistent and specific influences of early acoustic environments on primary auditory cortex. Nat Neurosci 4:1123–1130

# Cellular Correlates of Learning and Memory

# 26

Martin Korte

*There is nothing like future and past (. . .). There is only the presence of the past, the presence of the presence, and the presence of the future. These three I see in the soul, but I cannot see them independent of it: present is the memory of the past, present is the perception of the presence, and present is the expectation of the future.*

*– (Augustine, Confessions, Chap. 20 in Book 11)*

Learning is a process of information input and processing as well as storage, and, on the other hand, it is a product which includes changes in the behavior of an animal due to experience.

Learning and memory are two sides of the same coin: both deal with the fact that the brain changes its processing algorithms as a reaction to a stimulus. While learning is the process by which experiences can change the brain, memory processes are concerned with how these changes can be stored. One might also ask, what is memory conceptionally and how is it implemented on the cellular level in different animals? One answer would be, that memory is an internal representation of a past experience that is reflected in thought or behavior. It depends at the cellular level on changes in the connectivity of neuronal circuits. Learning would then simply be the acquisition of information during the lifetime of an animal.

Processes of learning and memory can be studied in different animals, depending on the intended level of analysis. Typical model organisms are *Drosophila* and mice (or rats), and in addition the sea slug *Aplysia* and the honeybee were instrumental and will be described in this chapter in detail. Learning and memory can be studied at different levels of complexity, from molecules to behavioral observations. This chapter will focus on the cellular events that take place when an animal learns, stores, or retrieves information.

To make a long story short, one can say that the past is present in either changes in synaptic weight (synaptic strength, which is measured by the size of the postsynaptic potential) or in how neurons change their input/output characteristics. The term activity-dependent synaptic plasticity is instrumental in this context. It refers to the observation that synapses change their strength (increase or decrease) and/or their structure due to changes in neuronal activity. In order to incorporate and process new information, a neuronal network needs to be modifiable, or in the technical term: "plastic". A primary focus of current research on how plastic alterations can be implemented in neuronal circuits is the connection between individual neurons – the chemical synapse. Plasticity events can be implemented in different forms, permanent and nonpermanent, depending on the strength and timing of neural activity. Alterations comprise interdependent functional and structural changes in neuronal connectivity. Synaptic plasticity is a phenomenon that can be observed in all animals, vertebrates and invertebrates alike.

Research into the cellular foundation of memory processes follows this logic: First, one needs to identify a brain area of interest that is crucial for specific forms of learning. It could be the mushroom bodies in *Drosophila* or the honeybee; the hippocampus or cerebellum in mice and rats or simply a set of neurons organized in a ganglion in *Aplysia*. If neuronal circuits in these brain areas are believed to be **necessary** to store memories (in other words to be the place for a memory trace or an engram), one has to prove that actually neuronal plasticity happens in these cells. Finally, one has to check whether these defined

M. Korte
Zoological Institute, Division of Cellular Neurobiology,
TU Brauschweig, Spielmannstr. 6,
D-38106 Braunschweig, Germany
e-mail: m.korte@tu-bs.de

C.G. Galizia, P.-M. Lledo (eds.), *Neurosciences - From Molecule to Behavior: A University Textbook*,
DOI 10.1007/978-3-642-10769-6_26, © Springer-Verlag Berlin Heidelberg 2013

neuronal circuits are **sufficient** for information storage and whether they are actually necessary. In the end, learning and memory are system properties which are put into operation by the concerted action of different levels in the nervous system. Therefore, research on the processes and mechanisms of learning and memory requires a multilevel approach.

## 26.1 Forms of Learning

Three forms of learning are conceptually and molecularly different: nonassociative learning, associative learning, and complex learning.

### 26.1.1 Nonassociative Learning

Habituation, sensitization, and dishabituation are nonassociative events that can occur at single synapses.

Nonassociative learning is defined as a change in the behavior of an animal that is the reaction of this animal to a single stimulus type over time: habituation, sensitization, and dishabituation.

**Habituation** – Usually if an animal is confronted with a new stimulus it reacts to it with an orienting response. If the stimulus, which could be a loud noise or a touch on the skin, is neither beneficial nor harmful it will be ignored in the future (Fig. 26.1). This simplest form of learning, in which an animal learns about the irrelevance of a stimulus, is called habituation. Habituation was first investigated neurobiologically by Charles Sherrington (1857–1952). He observed that certain reflexes become weaker after repeated stimulation. Later it could be shown that habituation of motor reflexes in vertebrates is partly due to a weakening of the synaptic connection between excitatory interneurons and motor neurons in the spinal cord (whereas the sensory part of the reflex is unaffected). Habituation is different from sensory adaptation or fatigue, which occurs in the receptor cells. While habituation can be instantaneously reversed by a dishabituating stimulus, adaptation always needs time to recover.

**Sensitization** – A defensive reflex to a stimulus (e.g., withdrawal or escape) becomes stronger after exposure to a different harmful or threatening stimulus. Thus, the response of an animal to a potentially harmful stimulus can be enhanced in the future, even for stimuli that did not lead to a visible response originally (Fig. 26.1).

**Dishabituation** – means that a habituated response returns back to a normal level due to a very strong or painful stimulus (generally of a different modality) and that the original response in its strength can reoccur. This is related to sensitization, but the history of events is different.

### 26.1.2 Associative Learning

Associative learning comprises the combination of two events or stimuli; the brain learns the correlation between events or between its own action and an event.

**Classical Conditioning** – Here, a stimulus (unconditioned stimulus, US) that normally leads to a reaction by the animal is associated with a stimulus (conditioned stimulus, CS) that normally does not lead to a response. The Russian physiologist Ivan Pavlov (1849–1936) studied this type of learning extensively and is most famously associated with the scientific exploration into the mechanism of classical conditioning. In his best-known experiment he measured the salivation of a dog when presented a piece of meat (US). The CS was a tone (sound of a bell) and was given right before the US (the tone always anticipated the food). By repeating this procedure many times, no sight of food was necessary anymore in order to trigger the salivation response (Fig. 26.1). The dog had learned that the tone was a good predictor for the food. It is important that the CS always precedes the US, in Pavlov's case, the tone has to come before the food is visible. Also the time difference between the two stimuli is important; the US has to follow the CS after a few seconds, otherwise the association between the two stimuli will not be learned or the learned response to the CS will be significantly weaker.

**Operant Conditioning** – Whereas classical conditioning associates two external events to each other, operant conditioning is the association of an animal's own behavior to an external stimulus. Operant conditioning deals with the modification of a "voluntary

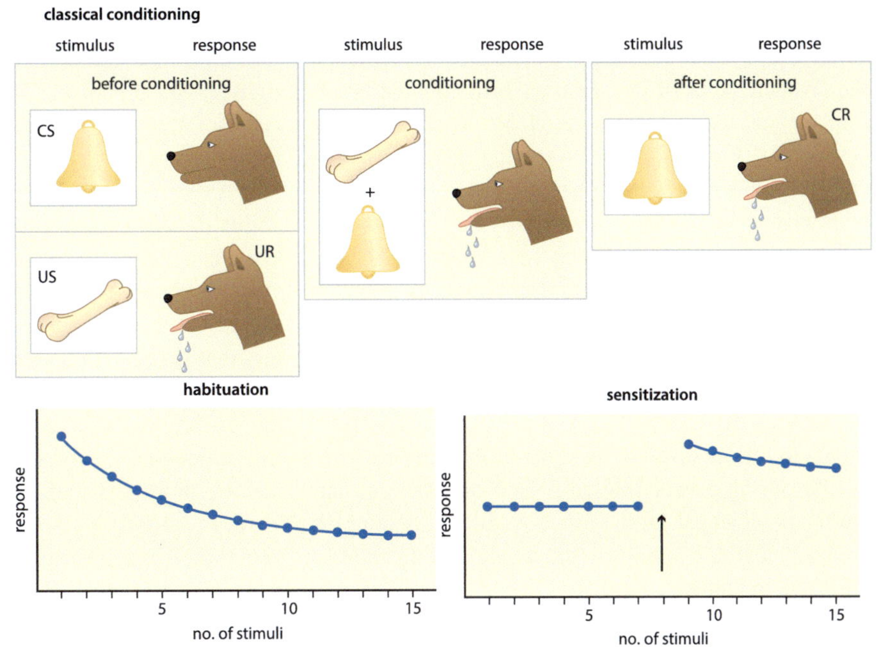


**Fig. 26.1 Classical forms of learning**. Pavlovian associative conditioning of the salivary response of a dog. Repeated pairings of an auditory cue (*CS*), which comes first, with food (*US*) causes the animal to learn the predictive value of bell sound in order to get food. This is measured by objective criteria – in this case the amount of saliva found before and after training. Habituation: the measured response to a stimulus decreases after repeated stimulation with a nonaversive stimulus. Sensitization: the response increases after an aversive stimulus

behavior" or operant behavior. The term "operant" is used, because with this type of conditioning a certain behavior by the animal produces consequences upon the environment, so the animal has to act (operate) in order to associate two events.

**Food (taste) aversion** – Food aversion is a special and unusual form of classical conditioning. In this form of learning, the US comes after the CS. Animals vomit or feel sick after they have eaten something rotten or harmful. The CS (taste of the food) and the US (feeling sick) can be separated by many hours. Afterwards these animals will avoid the taste of this type of nutrition, sometimes for the rest of their life and after a one-trial learning period. This type of learning specifically works for taste associated with feeling sick, but not if, for example, an electric shock is combined with a food taste. This type of learning depends on a prewiring of the nervous systems that allows the association of a particular behavior (rejection) with a special type of stimulus (taste).

**Extinction** – When the CS is presented repeatedly after associative conditioning without pairing to the US, the response will decline. This form of learning is called extinction, and is fundamentally different from forgetting. If a bell rings for a dog that has previously learned that this predicts food, and that bell does not lead to food any more, the salivation of that dog will decrease over time. An animal that would respond to a stimulus that has lost its predictive value would be maladapted.

**Reconsolidation** – Presenting the CS alone does not necessarily lead to extinction, but can induce reconsolidation: during the activation of an engram (memory trace) this trace is changed due to current experience. It is a distinct process that serves to maintain, strengthen, and modify memories which were already formed. As a consequence, once memories are stored in long-term memory, they are not entirely stable, because the retrieval of their memory trace can reprogram it and change it. This process can be blocked by pharmacological agents that block protein synthesis. As Yadin Dudai termed it: "Retrieving a memory involves mingling the internal representations of the past with the percepts of the present" [6, 7].

### 26.1.3 Complex Learning

Imprinting, observational learning, and priming are forms of complex learning.

**Imprinting** – describes a process of phase-sensitive and species-specific learning which occurs only during a particular time in the life span of an animal, most often in very young age. It was first used to describe situations in which an animal learns the characteristics of a certain stimulus (which is then "imprinted" into the animals brain). The best-known form of imprinting is the one in which a young animal learns to recognize the characteristics of its parent to form parental bonding. It was extensively studied by Konrad Lorenz (1903–1989) working with greylag geese. Lorenz could demonstrate how incubator-hatched geese would be imprinted on the first suitable moving stimulus they encountered – even this being Lorenz himself – and restricted to a critical period of a few hours after hatching (see Chap. 25).

Imprinting is a form of prepared learning, in which the development and maturation of the brain is modified by sensory experience in a critical period. The neuronal circuits involved in this type of learning are only plastic during a very limited period of time and can hardly be changed later in life. Additional examples are vocal learning in birds or the learning of the mother language in humans.

**Observational learning** (**social learning**) – This type of behavior can be defined as learning by imitation, which is especially developed in humans. Humans can extract three types of information simultaneously from observing other humans: the demonstrator's goals, actions, and environmental results. Children are so predisposed to learn from observing others that a more appropriate name for the human species would be *Homo imitans* (man who imitates, see Chap. 29).

**Priming** – occurs when an earlier stimulus influences response to a later stimulus. For example in humans, when a person sees the word *table*, and is later asked to complete a word starting with *tab*, the chance that the subject will answer *table* is higher than for a nonprimed person (associative priming). Another example of priming involves people being shown an incomplete sketch which they are unable to identify. They are then progressively shown more details until they recognize the picture. Later they will recognize the sketch at an earlier stage (perceptual priming). In priming, the first stimulus activates parts of the neuronal network in a specific brain region just before carrying out an action or task. The network is already activated when the second stimulus is encountered, thus improving performance of the task. It is part of the implicit (procedural) memory system.

## 26.2 Model Systems

All the above-mentioned types of learning paradigms have been studied in different animal species. The success of biological experiments depends on splendid ideas and concepts as much as it depends on technological advances and last but not least, on choosing the right model organism. The most widely used organisms for studying the cellular foundations of learning and memory processes are mice, *Drosophila*, the sea slug *Aplysia*, and honeybees. Also birds, especial their ability of song learning and several other species (monkeys, cats, ants, the snail *Hermissenda*) are intensely studied, but are not covered in this chapter. The paradigmatic use of just a few species is justified by the fact that many cellular and molecular building blocks of learning and memory processes are conserved during evolution. Indeed, the overall outcome of these experimental (mostly simple) model systems is that cellular processes that mediate learning and memory show an astonishing amount of conservation between vertebrates and invertebrates. Nature has produced a common set of molecules and signaling events that can be used by different animals and in different brain regions in order to store information.

## 26.3 Nonassociative Learning in Simple Organisms

What is the cellular basis of the storage capabilities of neurons in the brain of animals? For addressing the cellular foundations of learning and memory *Aplysia californica* was instrumental. Why *Aplysia*? *Aplysia* does not really show a great variation of behavior and it is indeed not a particularly fast learner. However, it can learn and it has only 20,000 neurons, approx. 2,000 in each of its ten ganglia. Most importantly these neurons have large somata (sometimes gigantic, up to 1 mm in diameter as compared to vertebrate neuronal somata with a diameter of 10–20 μm). In addition, neurons are stereotypical across individuals, and can be identified on the basis of their size, pigmentation, and position in the nervous system. Hundreds of neurons have been reliably linked to specific behaviors. In addition, neuronal circuits can be identified and single neurons can be isolated and cultured *in vitro* and *in vivo* in order to form circuits, which can be explored at the molecular, biochemical, and cellular level. As a result, using *Aplysia* guarantees access to molecular mechanisms of basic neuronal functions and the possibility of studying these mechanisms in real physiological time with single-neuron resolution. The genome of *Aplysia* is rather small and already single cells can provide enough mRNA to perform gene expression

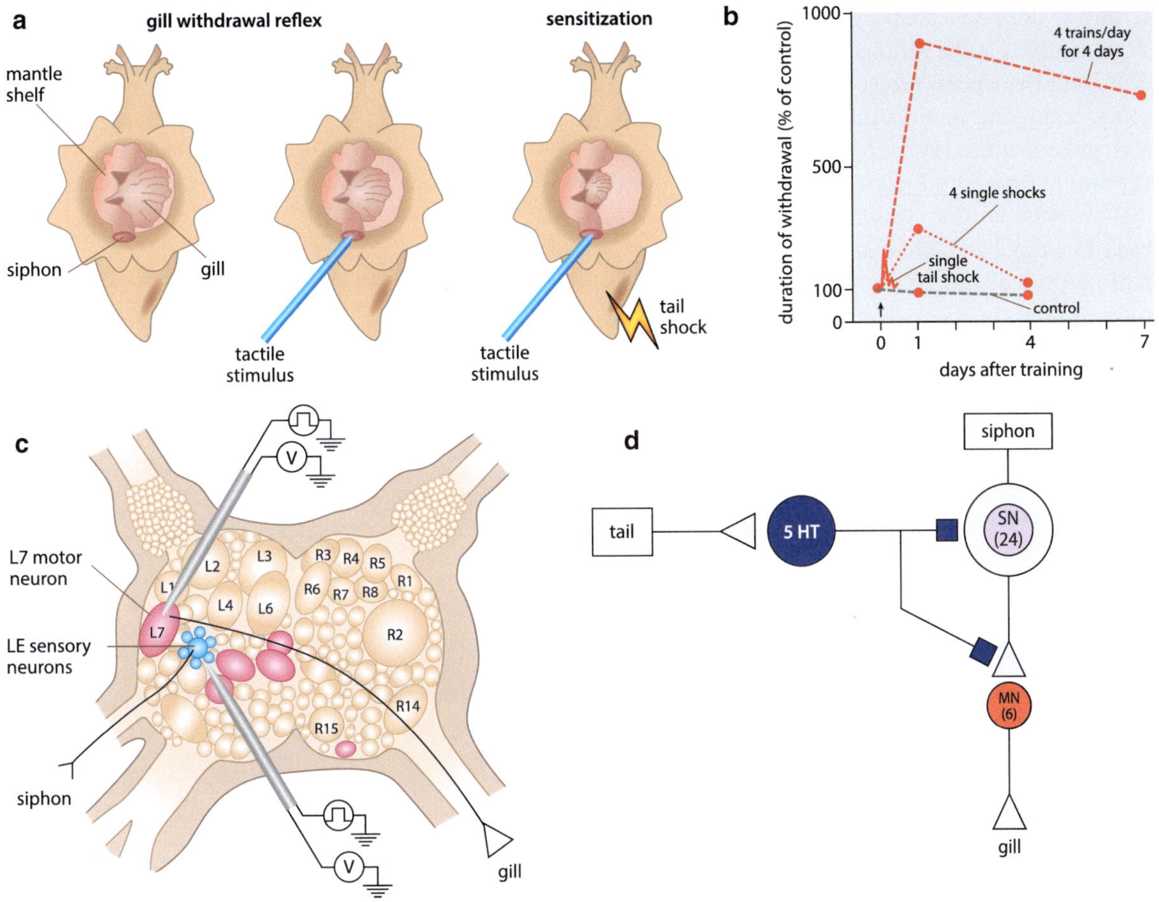


**Fig. 26.2** **Nonassociative learning in *Aplysia*.** (**a**) Cartoon of *Aplysia* (dorsal view). The gill, *Aplysia*'s respiratory organ, is highlighted in *pink*. A light tactile stimulus (like touching or spraying water) at the siphon causes it to contract (gill-withdrawal reflex). Sensitization of the gill-withdrawal reflex is invoked by applying a noxious stimulus at the tail (or the head, not shown here), which leads to enhancing the withdrawal reflex of both siphon and gill. (**b**) Repetitive stimulation converts short-term into long-term memory in *Aplysia*. Before sensitization, a weak touch to the siphon causes only a weak, brief siphon- and gill-withdrawal reflex. Following a single noxious, sensitizing, shock to the tail, that same weak touch produces a much larger response, an enhancement that lasts about 1 h. More tail shocks increase the size and duration of the response. (**c**) View of the abdominal ganglion and the neurons within, some of which mediate the gill-withdrawal reflex. In the example shown, an electrode is recording signals from the L7 motor neurons. These neurons receive monosynaptic sensory input directly from the siphon (LE sensory neurons). (**d**) The simplified connectivity of the gill-withdrawal reflex. *5 HT* serotonergic neuron (in *blue*), *SN* sensory neuron, *MN* motoneuron (Adapted from Kandel [13] with permission)

studies for analyzing how gene expression changes in response to environmental stimuli.

The most frequently observed behaviors in a sea slug like *Aplysia* are, as you might expect, feeding and reproduction (copulation and egg laying). But in the context of learning and memory important behaviors of *Aplysia* are the siphon withdrawal reflex and the gill withdrawal reflex (Fig. 26.2). Both are defensive mechanisms occurring as the animal pulls its vulnerable parts back into the mantle cavity when disturbed. The siphon is normally used by *Aplysia* to expel seawater and waste. These reflexes can be modified by experience and experimentally by an electric shock. With such simple behaviors it was possible to decipher the molecular, cellular, and circuit nature of habituation, sensitization, and even classical conditioning.

### 26.3.1 Habituation

The most simple of the three forms of learning is habituation of the gill-withdrawal reflex, which can be easily studied by a tactile stimulus to the siphon.

If this is done repeatedly the reflex will habituate. What happens at the cellular level? Sensory information from the siphon projects to the abdominal ganglion, where the input is distributed to interneurons and motoneurons (Fig. 26.2). One of these motor neurons is named L7 which innervates the muscle that retracts the gill. This monosynaptic reflex was used to study habituation at the cellular level. After repeated stimulation of the sensory neuron, less neurotransmitter is released by the sensory neuron, whereas the sensitivity of the postsynaptic side of the motor neuron is not changed. Therefore it can be concluded that habituation is a presynaptic phenomenon. Voltage-gated $Ca^{2+}$ channels become less sensitive to incoming action potentials at the presynapse, and with less $Ca^{2+}$ in the presynapse, less vesicles fuse with the presynaptic membrane and less neurotransmitter is released, resulting in a smaller EPSP (excitatory postsynaptic potential) in the motor neuron. At the end point of this process the motor neuron will generate less action potentials and thus release less neurotransmitter to activate the gill withdrawal muscle. So far, the molecular mechanisms are not entirely clear. One hypothesis is that high amounts of glutamate release might inhibit the activity of adenylyl cyclase presynaptically and therefore reduce the production of cyclic AMP (cAMP).

## 26.3.2 Sensitization

In *Aplysia*, sensitization can be achieved by giving a noxious stimulus to the head of the animal (not as with habituation directly to the siphon). If later the siphon is stimulated, the gill-withdrawal reflex is stronger (Fig. 26.2). In addition to the already introduced sensory neuron from the siphon to the abdominal ganglion and to the gill motor neuron (L7), a sensory neuron from the head (named L29) has to be added to the circuitry. This sensory neuron releases serotonin (5-HT) onto the presynaptic side of the siphon sensory neuron. The release of serotonin triggers a cascade of events, which eventually lead to an increase in $Ca^{2+}$ entry into the presynapse of the siphon sensory neuron, in turn increasing neurotransmitter release. The motor neuron EPSP increases and the muscle produces a stronger gill withdrawal. Specifically, the serotonin receptor (a GPCR) activates an adenylyl cyclase which produces

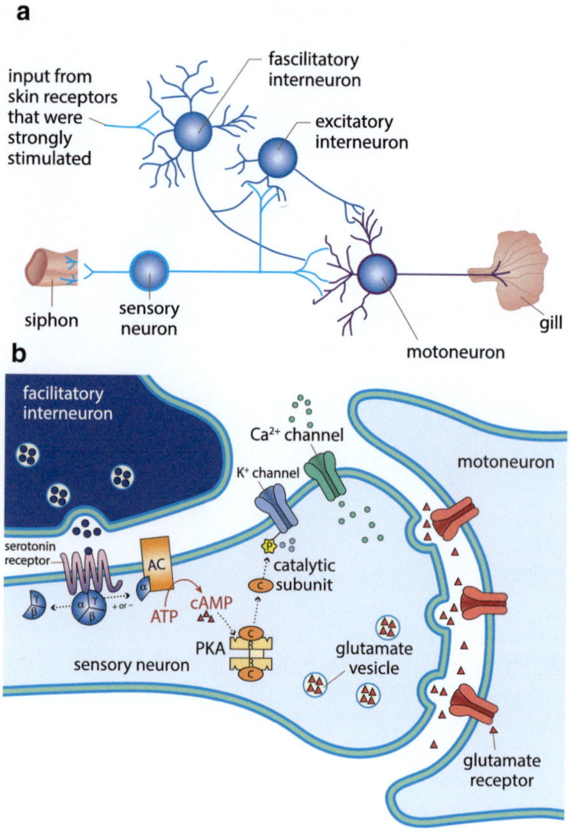


**Fig. 26.3** **Sensitization in *Aplysia*.** (**a**) Connectivity of sensitization in *Aplysia*. Sensitization can be induced by giving a noxious stimulus to the tail or the head of the animal, which activates a facilitatory interneuron that terminates on the presynaptic release sites of the sensory neuron from the siphon onto the gill motoneuron. (**b**) Molecular details on how sensitization is achieved in *Aplysia*. Serotonin is released from the facilitatory neuron and activates serotonin receptors in the sensory neuron. These activate an adenylyl cyclase (*AC*) which produces cAMP. The rising cAMP level leads to the activation of the protein kinase A (*PKA*), which phosphorylates $K^+$ channels. Phosphorylated $K^+$ channels have a reduced open probability, and this increases the duration of action potentials. As a consequence, $Ca^{2+}$ channels at axon terminals remain open for longer durations and more neurotransmitter is released. Thus, motoneurons of the gill receive a stronger excitatory signal

cAMP [5], which activates PKA (see Chap. 6). PKA phosphorylates a potassium channel, shifting it towards the closed state. This leads to presynaptic action potentials to repolarize more slowly, resulting in longer activation of voltage-dependent $Ca^{2+}$ channels, and thus more neurotransmitter release (Fig. 26.3). It is possible that in addition to these presynaptic changes, postsynaptic changes might take place in the motor neuron, by adding glutamate receptors into the postsynaptic synapse (glutamate is the neurotransmitter between the sensory neuron and the motor neuron).

## 26.4 Associativity and Classical Conditioning in Simple Organisms

### 26.4.1 Aversive Conditioning in Aplysia

Animals – and humans – need to learn to predict "what will happen next". If event 'A' regularly precedes event 'B', one can assume that when 'A' happens, also 'B' will happen. Most famous is the Pavlovian conditioning of a dog, where the tone became a good predictor "of having food soon". In classical conditioning, an animal learns that two events in the exterior world have a very high probability to appear in sequence.

> Adenylyl cyclase can act as a coincidence detector of presynaptic and postsynaptic activity.

In an experiment in *Aplysia*, a weak touch to the siphon (the conditioned stimulus, CS, which does at the beginning not lead to much of a response) is followed by a strong tail shock (the unconditioned stimulus, US). The animal can learn to associate the noxious tail stimulus with the touch to the siphon, and retracts its gill, siphon, and tail when the siphon is touched after conditioning [9]. A touch to the siphon alone (CS) leads to a much greater response after conditioning, also in comparison to sensitization, because the animal predicts a strong electric shock. The CS has to precede the US (Figs. 26.4 and 26.5) during training. At the cellular level the US causes the release of serotonin (5-HT) from the sensory neuron L29 and activates an adenylyl cyclase in the sensory neuron axon terminal, as we have already seen for sensitization. The CS causes the arrival of axon potentials in the sensory nerve terminal, leading to $Ca^{2+}$ influx. What is the difference in comparison to pure sensitization, where the adenylyl cyclase was activated as well? Adenylyl cyclase is significantly more active when higher $Ca^{2+}$ levels are present in the presynaptic terminal. $Ca^{2+}$ binds to calmodulin and the $Ca^{2+}$/calmodulin-complex activates adenylyl cyclase more efficiently than the G protein activated by 5-HT alone. This makes the adenylyl cyclase a **coincidence detector** of the CS-US relationship: Learning depends on the coincidence of a $Ca^{2+}$ influx with the activation of adenylyl cyclase by a G protein. The "storage" of this memory is the phosphorylated $K^+$ channel, which leads to a prolonged depolarization of the membrane, resulting in increased release of neurotransmitter molecules. Thus, after learning, a weak activation of the sensory neuron alone (CS) leads to a motor response (CR).

### 26.4.2 Differential Conditioning in Drosophila

A population of flies can be taught to avoid specific odors which they have experienced in conjunction with electric shock. This is generally done with a number of flies, e.g., a group of 50 flies will be sequestered in a closed chamber and trained by exposing them sequentially to two odors in air currents. Flies will then receive electric shock pulses in the presence of the first odor (CS+) but not in the presence of the second odor (CS–). Flies are then tested in a T-maze, between converging currents of the two odors. If they learned the task, the majority of trained flies will avoid the shock-associated odor (CS+). This procedure is called differential conditioning.

Habituation and sensitization can also be tested in *Drosophila*. Indeed, mutations which affect sensitization, also affect classical conditioning. The genes involved can be studied using reverse genetics. Reverse genetics seeks to find what phenotypes arise as a result of mutations in particular genes (whereas forward genetics addresses the genetic basis of a particular phenotype). For example, when a gene is placed under the control of a heat-inducible promoter, this gene can be turned off or on by heating (or cooling). The advantage of this method is that the gene can be turned off in the adult fly and any developmental or long-term compensatory effect is avoided. By switching off the PKA gene in the adult fly, the ability of the fly to learn is disabled, showing that PKA is necessary for learning.

The brain areas that are most important for olfactory learning in the fly are the mushroom bodies (Fig. 26.6 for the honey bee see Fig. 26.8). A simplified cellular network model is shown in Fig. 26.7. Different odors

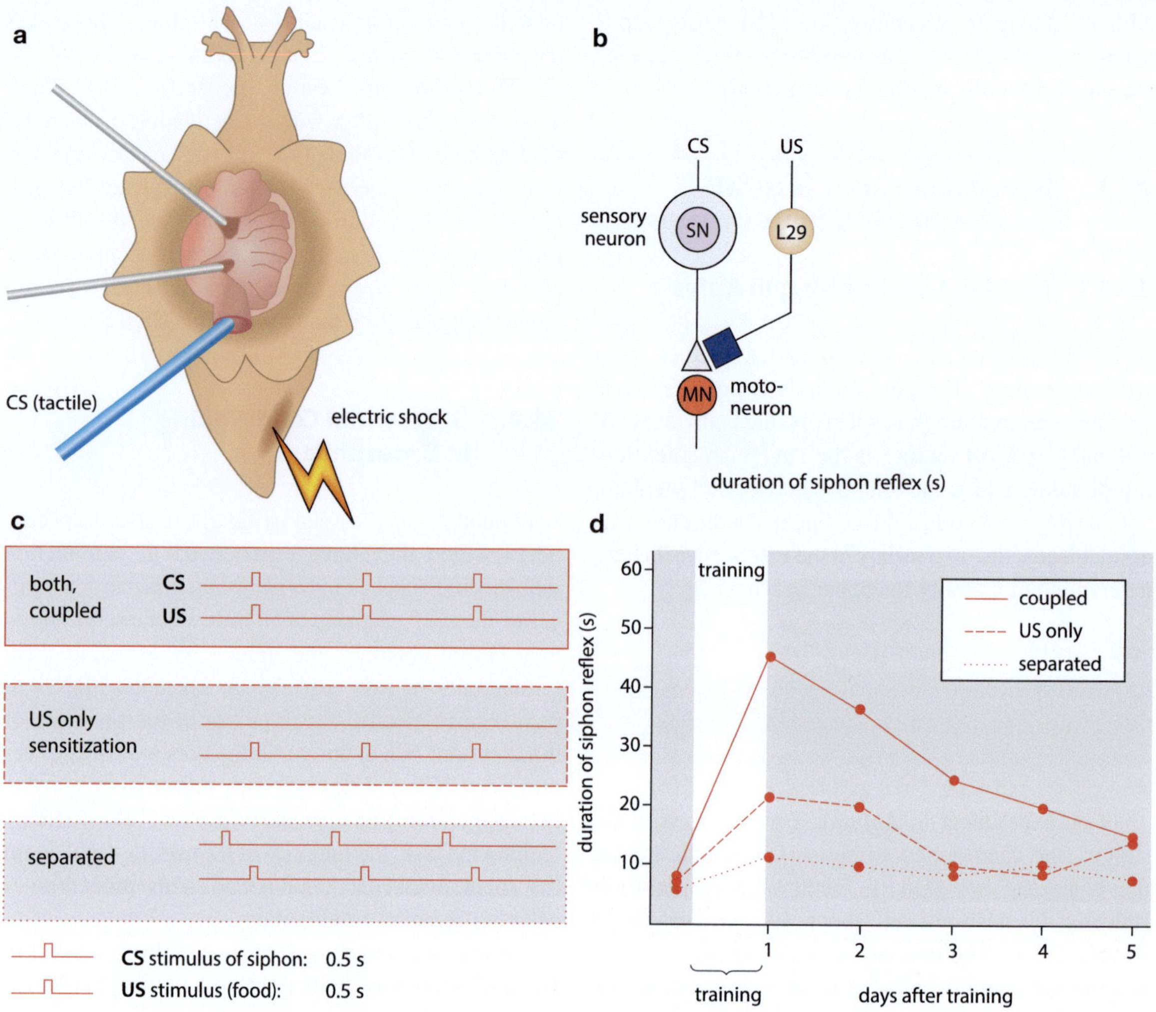


**Fig. 26.4** **Classical conditioning in *Aplysia*.** (**a**) An electric shock at the tail of the animal is paired with a touch to the siphon. (**b**) Schematic drawing of the neuronal connections needed for classical conditioning in *Aplysia*. Shown are the two sensory neurons that mediate the CS und the US. The US neuron projects to the presynaptic terminal of the CS neuron. The CS neuron forms a synapse onto a motor neuron (*MN*). (**c**) Diagram of the pairing procedure. The US needs to narrowly precede the CS (or to be given at the same time). No learning occurs if the US and the CS are separated by a longer time period. (**d**) Duration of the siphon reflex plotted against time. The siphon reflex is increased after CS-US pairing. Giving only the US leads to sensitization, separate delivery of US and CS do not increase the syphon reflex

are represented by distinct subsets of active projection neurons from the antennal lobe, which generate a distinct combinational pattern of active Kenyon cells (see Chap. 13). Modulatory neurons convey the information about an unconditioned stimulus (US) to the mushroom bodies during the learning phase when a particular odor is paired with danger (electric shock). The conditioned response (CR) is mediated by an extrinsic mushroom-body output neuron (CR neuron), receiving synapses from all Kenyon cells. Now let us consider the following situation: before the conditioning procedure has started, these synapses would all be silent and an odor would not elicit a CR. If one odor (the CS+) is paired with an electric shock, those Kenyon cells that are activated by the odor will strengthen their output synapses. This would make the CR neuron responsive to the odor alone, and after conditioning, the CR neuron would act as a specific sensor reporting the presence of the learned odor as an alerting signal for the expected US [11].

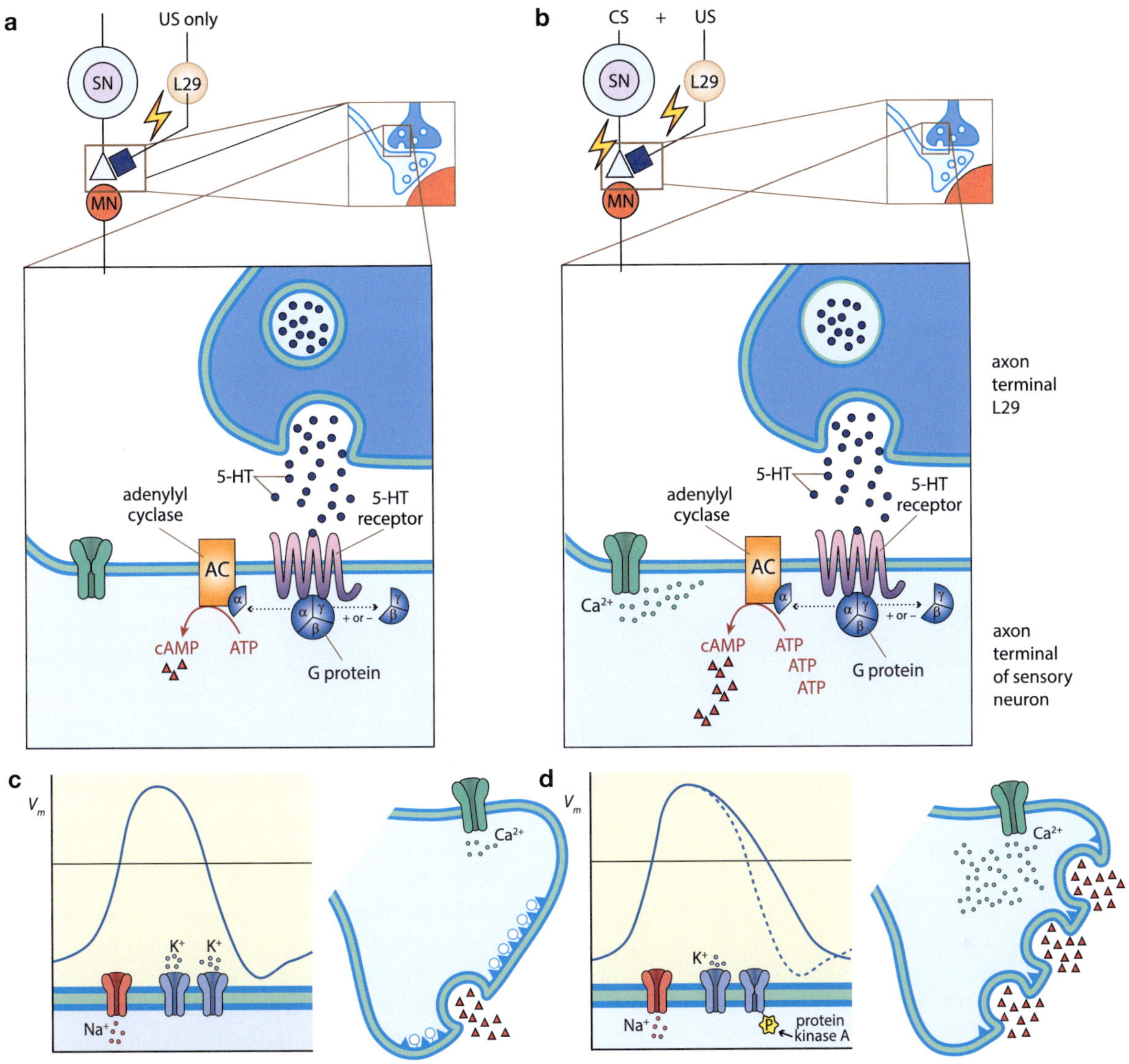


**Fig. 26.5 Molecular basis of classical conditioning in *Aplysia*.** (**a**) Sensitation procedure (US only) leads to a moderate activation of AC which produces a low concentration of cAMP. (**b**) Pairing the CS and the US causes a much greater activation of the AC, because the presynaptic membrane of the CS neuron is depolarized by the CS. More $Ca^{2+}$ increases the affinity of AC to G proteins, leading to greater cAMP production. cAMP activates PKA, which phosphorylates $K^+$ channels, (**c**) under baseline conditions, $K^+$ channels quickly repolarize the membrane, and a limited amount of transmitter is released, (**d**) with phosphorylated $K^+$ channels, repolarization is delayed, leading to more $Ca^{2+}$ influx, and thus more transmitter release

It is one thing to show that MBs are necessary for learning, but another to show that they are indeed the site where memory traces are stored. To tackle this point, changes on the molecular level have to be shown in the MBs neuronal circuit. As in *Aplysia*, the cAMP-signaling system is crucial. The $Ca^{2+}$/CaM-dependent adenylyl cyclase (rutabaga, Rut-AC) is considered to be the **coincidence detector** of the US and CS pathways in associative learning. Very similar to classical conditioning in *Aplysia*, Rut-AC is stimulated by the US through a GPCR, and by the CS through $Ca^{2+}$/CaM. Coincident activation of both pathways is required for the most effective stimulation of cAMP synthesis (Fig. 26.7). This model postulates that Rut-AC is needed exclusively in the Kenyon neurons at the output synapses. Indeed, the

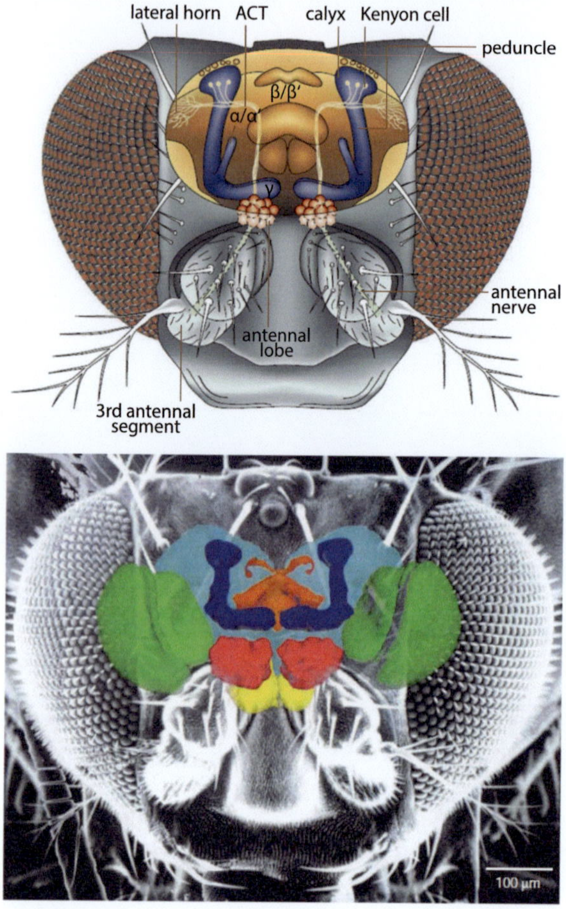


**Fig. 26.6** **The *Drosophila* brain**. *Upper panel*: head view of *Drosophila*. Olfactory receptor cells are on the third antennal segments. In the antennal lobe receptor fibers are organized according to their chemosensitivitry in olfactory glomeruli. From here, projections in the antennocerebral tract (*ACT*) lead to the dorsolateral protocerebrum (lateral horn) and the calyx of the mushroom body. α/α′, β/β′, and γ mark the three mushroom body lobes. *Lower panel*: Color-coded brain areas in the *Drosophila* head capsule, *green* optic lobes, *yellow* subesophageal ganglion, *red* antennal lobes, *blue* mushroom bodies, *orange* central complex (From Heisenberg [11] with permission)

expression of Rut-AC restricted to Kenyon cells in a fly that otherwise did not express Rut-AC was enough to completely restore odor discrimination learning, showing that Kenyon cells are not only **necessary**, but also **sufficient** for learning. This picture is a simplified one: many more neurons and molecules are involved and add to learning under different circumstances. The mutant *dunce* fails to degrade cAMP, because a specific phosphodiesterase is dysfunctional. In the mutant with the telling name *amnesiac* a peptide transmitter is affected which regularly would activate the adenylyl cyclase. The mutant PKA-R1 has a mutation in the *PKA* gene. All of these mutants have learning deficits, each in a particular way. Furthermore, memory mechanisms for short-term memory (STM), middle-term memory (MTM), early long-term memory (eLTM), and late-term memory (lLTM) all have their specific peculiarities. In addition, LTM is possible in the absence of STM, showing that these are not sequential, but parallel systems.

### 26.4.3 Learning and Memory in Honeybees

A classical experimental paradigm with honeybees depends on their superior ability to learn which odor belongs to which nectar (flower) – something a bee must be able to do in the meadow full of different flowers. When a bee touches sugar water with her antenna, she will extend the proboscis (proboscis extension reflex, abbreviated as PER). This reflex can be conditioned: If an odor (CS+) is given repeatedly right before the US (sugar water), then the PER will later be elicited whenever the smell is given alone (Fig. 26.9). The analysis of PER conditioning can be combined with electrophysiological, pharmacological, or imaging approaches that allow the study of the bee brain areas involved in learning *in vivo* [8]. As the bee is immobilized in such an experiment, it is possible to expose its central nervous system through a small window cut in the cuticle of the head. And with this approach neuronal correlates of the different forms of olfactory conditioning can be found at different levels of the nervous system, ranging from single identified neurons to neuronal networks. Three areas in the bee brain are crucial for PER conditioning: On the sensory side, the antennal lobes are involved, then the mushroom bodies, and finally the mushroom body extrinsic neurons which are involved in premotor functions. For example, $Ca^{2+}$ imaging in the antennal lobe shows that both habituation and associative conditioning occur in parallel in the network, leading to odor representations that are altered after conditioning, possibly to increase the signal-to-noise ratio.

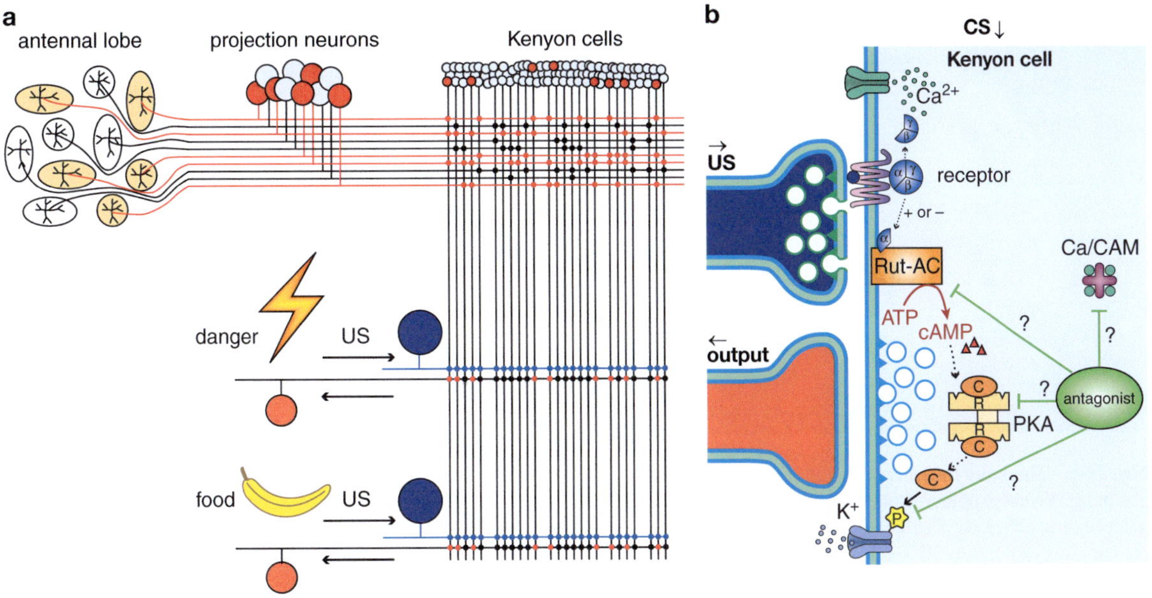


**Fig. 26.7 Circuit model of odor memory**. (**a**) Odors are represented in the mushroom bodies by sets of active Kenyon cells. Patterns of Kenyon cells synapse onto output neurons signalizing, e.g., danger or food. These output neurons are accompanied by modulatory input neurons (*blue*) presenting the unconditioned stimulus to the Kenyon cells. Simultaneous arrival of the conditioned (CS, odor) and unconditioned stimulus (US, danger or food) strengthens the respective synapses from Kenyon cells to output neurons. (**b**) Simultaneous arrival of the conditioned stimulus (*CS*) and the unconditioned stimulus (*US*) in the Kenyon cells activates AC, which in turn increases cAMP synthesis. Elevated levels of cAMP activate protein kinase A (*PKA*), which might phosphorylate target proteins at the synapse. To explain extinction, an antagonist is postulated that could interfere with any of the steps in the cAMP signaling pathway. Local, independently modulated synaptic domains for different USs might reside in the same Kenyon cells. *Ca/CAM* calcium/calmodulin, *Rut-AC* $Ca^{2+}$/CAM-dependent AC (From Heisenberg [11] with permission)

The "US neuron" in honeybees has been studied in detail – the VUMmx1 (ventral unpaired median neuron of the maxillary neuromere 1). This neuron is the neural correlate of the US. It receives gustatory inputs in the subesophageal ganglion, and is extensively connected with the antennal lobes, the lateral protocerebrum, and the olfactory part of the mushroom bodies. The transmitter it uses is octopamine, a biogenic amine that is closely related to the mammalian noradrenaline (norepinephrine). The importance of the VUM neurons was proven by replacing the sugar water in a conditioning experiment with electrical stimulation of the VUMmx1 neuron or with the application of octopamine. This was sufficient to condition the bees to a certain olfactory cue. After, but not before, such a conditioning electrophysiological recordings show that the VUMmx1 neuron responds to the conditioned olfactory stimulus (Fig. 26.7, see also Chap. 28).

Different forms of memory can be distinguished: First, there is the early short-term memory (eSTM) in which the association between an olfactory stimulus and a reward can be disrupted easily. This phase is only a few seconds long and can be correlated with a higher acetylcholine receptor activity in the antennal lobes, in the mushroom bodies, and the VUMmx1 neuron. Second, there is a late form of short-term memory (lSTM) which establishes itself after repetition of the learning situation and consolidates the response. This phase is the prerequisite for the transition from short-term to long-term memory. On the molecular level it depends on the production of nitric oxide and long-lasting PKA activity in neurons of the antennal lobes. Long-term memory (LTM) needs at least three repeats of the conditioning experiments to be established (Fig. 26.9). In the antennal lobe a long-lasting activation of PKA via cAMP of at least 25 min is necessary in order to form long-term

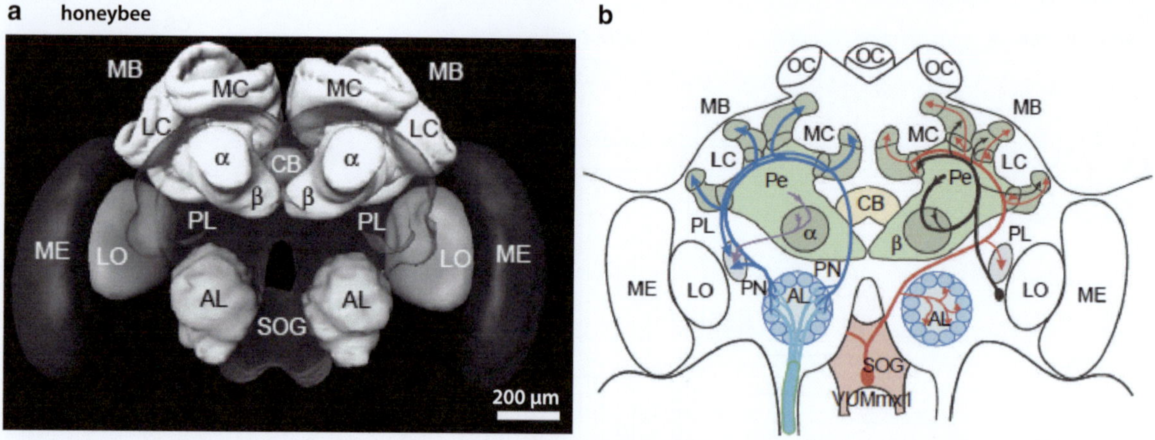


**Fig. 26.8** **The honeybee brain**. (**a**) Plastic view of the honeybee brain. Visual ganglia include medulla (*ME*) and lobula (*LO*), the primary olfactory neuropil is the antennal lobe (*AL*). The primary sensory neuropils send projections to the lateral (*LC*) and medial (*MC*) calyces of the mushroom bodies (*MB*). The alpha-lobes ($\alpha$) and beta-lobes ($\beta$) are the output areas of the MBs. *PL* protocerebral lobe, *SOG* subesophageal ganglion, *CB* central body. (**b**) Information processing pathway in the honeybee brain in accordance with the brain structures shown to the *left*. *PN* projection neurons, *Pe* peduncle of the MB, *OC* ocellus, *VUMmx1* a neuron representing the appetitive US in the brain (From Menzel et al. [20] with permission)

memories. If after one conditioning experiment the cAMP level is artificially increased (e.g., uncaging of cAMP with a light source), long-term memories can be formed, whereas without this treatment the response quickly returns to baseline levels. Long-term memories also come in two forms (Fig. 26.10): (*i*) an early LTM which does not depend on protein synthesis and can last up to 3 days and (*ii*) a late LTM which depends on new protein synthesis and kicks in 3 days after the original conditioning experiments were performed. It can be prevented by protein synthesis inhibitors.

## 26.5 Learning and Memory in the Vertebrate Brain

The cellular mechanisms of learning and memory in vertebrates resemble those occurring in invertebrates, but only recently attempts have been made to elucidate these powerful mechanisms in vertebrates.

One of best-studied mammals is the mouse. An important tool for studying cellular aspects of learning and memory processes in this model organism has been the use of gene targeting methods in order to study the genes involved in learning. Gene targeting methods are quite diverse and can be very sophisticated.

Synapses can change their strength as a result of changes in neuronal activity. Important determinants for this process are the amount of neurotransmitter released at the presynaptic (sending) side in response to action potentials and the amount and opening probability of receptors at the postsynaptic (receiving) side, mainly of receptor-coupled ion channels or Gprotein-coupled receptors (metabotropic receptors). Recently also astrocytes (whose podocytes terminate directly on synapses) have been shown to actively participate in synaptic transmission and plasticity by taking up glutamate (see Chap. 9). This uptake of glutamate might be regulated and the effect of this regulation would change the strength of the affected synapse, e.g., when glutamate is taken up at a slower rate, postsynaptic receptors are activated for longer times, and the synaptic connection is stronger.

### 26.5.1 Hippocampus

Although synaptic plasticity has been found in all regions in the mammalian brain, most of the research on activity-dependent synaptic plasticity has been performed on a brain region called the hippocampus.

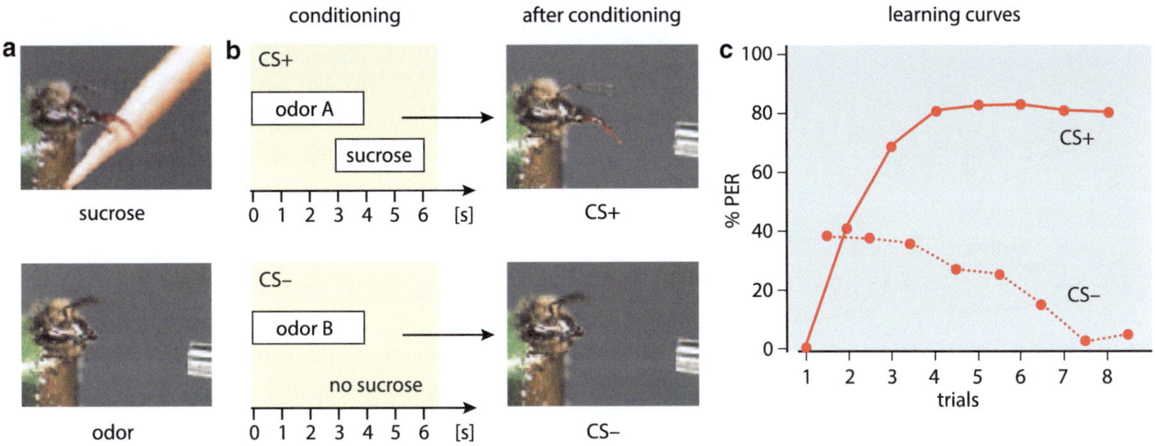


**Fig. 26.9** **The proboscis extension reflex (PER).** (**a**) A bee licks sugar water from a toothpick, an odor is presented to a bee. (**b**) Presenting an odor (CS+) immediately before sucrose solution (US) leads to classical conditioning (forward pairing), giving another odor without reward (CS–) leads to differential conditioning. (**c**) The acquisition curves for a typical differential conditioning experiment (PER: proboscis extension reflex). The bees learn to respond to the CS+ and not to the CS–. The physiological correlates of olfactory conditioning can be studied at different levels, ranging from the molecular and biochemical levels to that of single identified neurons and neuronal ensembles (From Menzel and Giurfa [19] with permission)

The hippocampus is an evolutionarily old part of the cerebral cortex, lying bilaterally in the depth of the cerebral hemispheres (Fig. 26.11). In humans, the hippocampus plays a central role in the processing of declarative memory, semantic memory, and autobiographical memories (see Chaps. 28 and 29). In animals, it is especially important for place memories and to remember the relationship between objects. If the hippocampus is removed in mice or rats, they lose the ability to learn the position of a hidden platform in a water basin, the so-called Morris watermaze test. In this test, the rodent has to swim in milky water and it has to find a platform hidden in the water. The first time it finds the platform is pure chance, but if the animal is put into the water a couple of times, it will memorize the location of the platform by using external visual cues around the water basin. If the hippocampus is removed after the animals have learned the position of the platform, they still can find it. Thus, the hippocampus is not the place in the brain for long-term storage of information. If the hippocampus is removed before the animals are trained, they will not learn the platform position.

The hippocampal formation is an ideal model for the investigation of specific synaptic contacts between groups of neurons, also because it is organized in transverse lamellas where axons are running and pyramidal cells are aligned in the same plane. In order to understand the experimental design of many studies that have used the hippocampus as a cellular model system to elucidate cellular mechanisms of learning and memory it is important to understand its structure: The hippocampus comprises the dentate gyrus (DG) and the ammon's horn (cornu ammonis, CA). The hippocampal formation includes three subregions, forming a trisynaptic circuit between distinct populations of excitatory neurons (Fig. 26.11). Information enters the hippocampus from the entorhinal cortex, terminating on granular cells of the dentate gyrus. Axons from these cells (so-called *mossy fibers*) project to pyramidal neurons of the area CA3. The axons from these cells (termed Schaffer collaterals) connect to CA1 pyramidal neurons. Axons from these cells then leave the hippocampus towards the contralateral hippocampus and these axons project back to the entorhinal cortex. Besides excitatory neurons, several highly diverse populations of inhibitory interneurons are present in all subregions of the hippocampus. They mediate feedback and feedforward inhibition within an area, and shape rhythms of activity by grouped discharge predisposing the network for alterations. In all

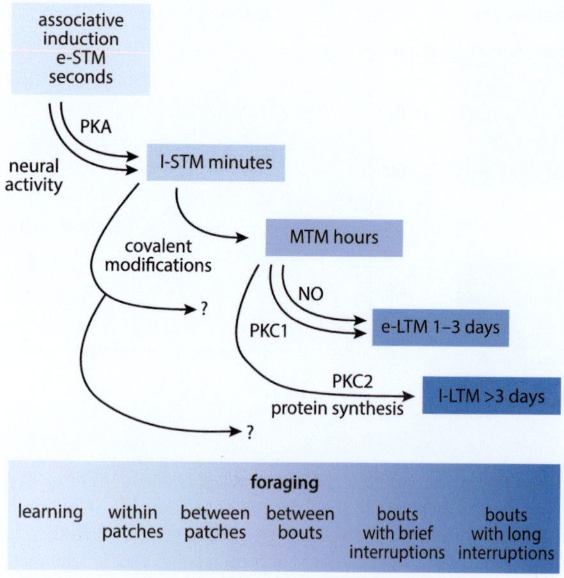


**Fig. 26.10** **Memory stages**. Five memory phases are distinguished on the basis of retention scores, dependence on single or multiple learning trials, susceptibility to retrograde amnesic treatment, participation of particular cellular reaction cascades (prolonged and enhanced activity of the protein kinases PKA and PKC, activity of NO synthase), and sensitivity to protein synthesis inhibition during the acquisition process and the time period immediately following it. Early and late short-term (*e-* and *l-STM*) and middle-term memory (*MTM*) are initiated by a single learning trial; early and late long-term memory (*e-* and *l-LTM*) require multiple learning trials. l-LTM lasts for a lifetime even after only three learning trials. Consolidation during l-STM leads to enhanced retention scores and resistance to amnesic treatment. e- and l-LTM differ in the fact that e-LTM is not sensitive to protein synthesis inhibition, but l-LTM is. These different phases corresponds to different ecological requirements for a bee (*bottom*) (From Menzel [18] with permission)

subregions synaptic connections can be altered, but the hippocampal CA3-CA1 synapse is one of the best-specified synapses in the mammalian brain in terms of synaptic plasticity. Many modulatory neurotransmitter systems project into the hippocampus. The most prominent fibers release acetylcholine, dopamine, serotonin, and histamine onto neurons of the hippocampal formation.

### 26.5.2 Hebb's Postulate

A fundamental new insight into the cellular correlates of learning and memory processes was the concept proposed by the psychologist Donald O. Hebb (1904–1985) in 1949. He postulated that memories are stored in networks of nerve cells and not in single cells and that the information in a network and thus also the *locus* of alteration is in the contacts between nerve cells:

> When an axon of cell A … [excites] cell B and repeatedly or persistently takes part in firing it, some growth process or metabolic change takes place in one or both cells so that A's efficiency as one of the cells firing B is increased [10].

Hebb's statement predicted that associative neuronal activity is instrumental in order to induce alterations in the input–output characteristics of a neuronal network. These changes can happen at synapses and are therefore called activity-dependent synaptic plasticity. Hebb's postulate also includes an important rule, predicting that synaptic weights change when the pre- and postsynaptic neurons show coincident activity (associativity).

As it turned out, not all forms of learning depend on it (as we have already learned from invertebrates), but in order to study the cellular correlate of information storage, Hebb's postulate was instrumental, especially at the CA3-CA1 synapse in the hippocampus. In addition, synaptic plasticity has to be bidirectional, because if synapses were only potentiated, the information would ultimately be lost once all synapses had reached maximal strength. Therefore, equally important is the ability for a long-lasting reduction of synaptic strength. Furthermore, alterations in the strengthening or weakening of synapses vary not only in size and direction, but also in temporal maintenance. Various forms of synaptic plasticity last for seconds to less than 1 h and are regarded as forms of short-term plasticity. Changes, however, can also last for many hours, days, or years.

While this chapter focuses on synaptic plasticity, other processes of information storage are also conceivable. Information stored in neuronal circuits is also changed if action potential thresholds or latencies are changed or if rhythmic/oscillatory activity in whole networks are modified due to learning experiences.

### 26.5.3 Short-Term Synaptic Plasticity

Changes in the input–output characteristics of neurons due to changes in the stimulus pattern can be short

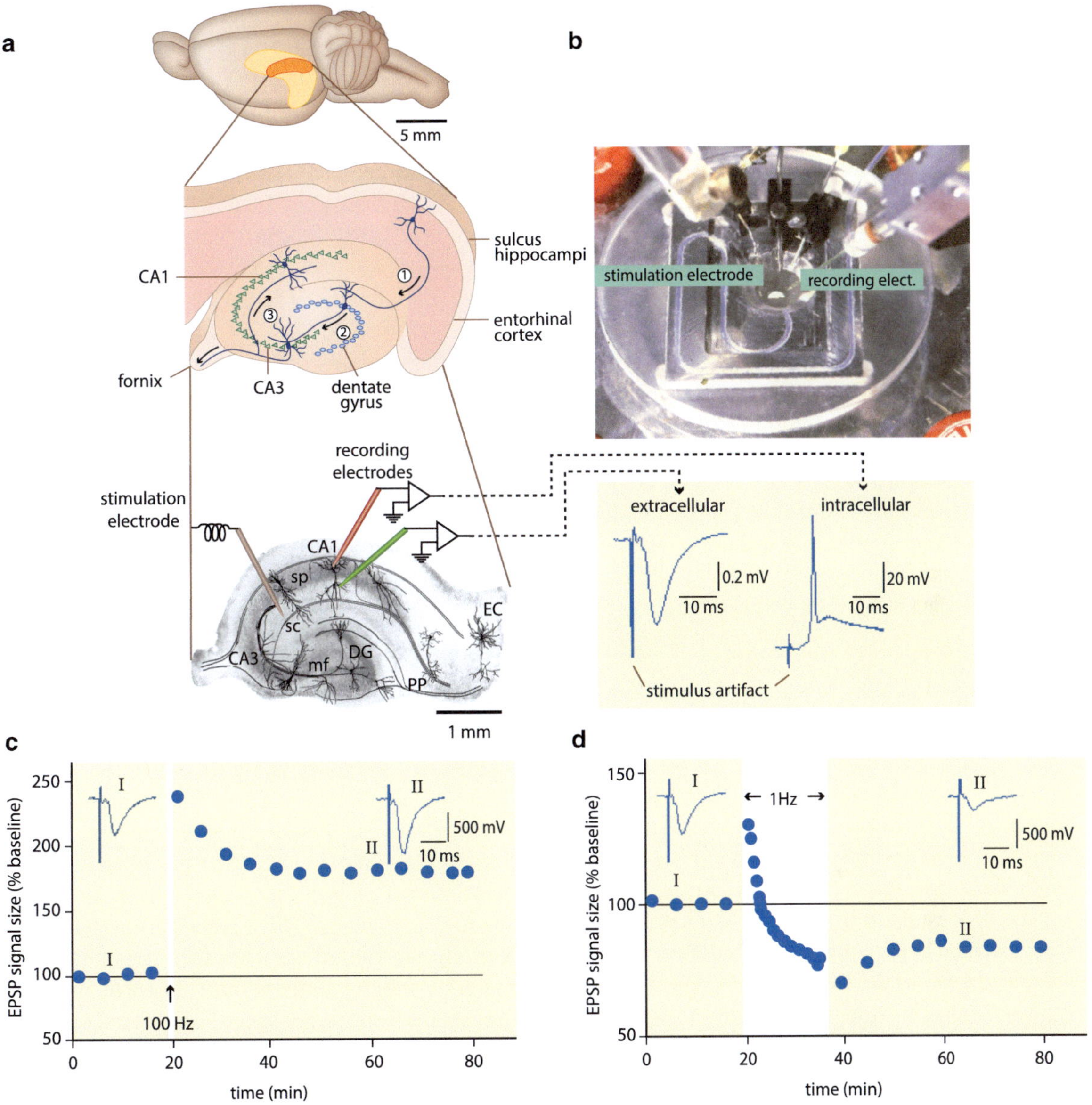


**Fig. 26.11** **The rodent hippocampus.** (**a**) A rodent brain and the position of the hippocampus are shown. In the *lower panel* a single, transverse slice of the hippocampus is drawn, as it is usually used for electrophysiological or imaging experiments. Input into the hippocampus comes from the entorhinal cortex (*EC*) via the perforant path, *PP* (1). The fibers of the PP terminate on granule cells in the dentate gyrus (*DG*). These neurons project to the CA3 pyramidal cell region via mossy fibers (*mf*, 2). The CA3 neurons project to CA1 pyramidal neurons via the Schaffer collaterals (*sc*, 3). All three projections are excitatory. *CA* cornus ammonis, *sp* stratum pyramidale. (**b**) A hippocampal slice is positioned in a recording chamber for electrophysiological experiments. In this case, the stimulus electrode is positioned in the CA3 region and the recording electrode in the CA1 region. Extracellular and intracellular recording examples are shown. (**c**) LTP and (**d**) LTD experiments are shown with original extracellular recording traces from the CA1 region. In the LTP experiment a high-frequency tetanus stimulation (100 Hz) was used; in the LTD experiment a low-frequency stimulus (1 Hz) was used for 15 min. LTP leads to a lasting potentiation of the EPSPs when tested with single stimulations, LTD leads to a depression of EPSPs

**Genetic Methods in Mice**

The function of a gene is studied by its removal/inactivation from the genome. In order to do so the gene of interest has to be cloned. The gene is then mutated and genetically modified by a selectable marker, like a resistant gene from bacteria, usually *neo* (codes for a protein that makes the cells resistant to the antibacterial drug neomycin). When the DNA fragment is transfected into embryonic stem cells, it can bind to the original DNA sequence and via random homologous recombination is inserted into a few cells where the original gene will be substituted by the mutated one. The gene-targeted cells can be selected by their resistance to neomycin and this clone of cells can be cultured. In the end, the genetically manipulated cells are injected into a mouse embryo (blastocyste state with approx. 100 cells). These blastocytes are inserted into a pregnant mother. Now luck is needed: Since embryonic stem cells are pluripotent, they can develop into all kinds of tissues. But they can also become part of the germline, and that is what is wanted here – and this happens by chance. Now the next question is, how to find out which of the offspring mice contains the mutated gene in their progeny? The simplest trick is to use stem cells, let's say from a white mouse and implant the cells in a blastocyst from a pure black mouse. If the fur color of the offspring is mixed (chimeric), these mice can be bred with a black mouse again. If the mutated gene made it into the germline, the progeny of this breeding should have gray color (heterozygous mice). These mice can now be crossed to each other and ¼ of the offspring will be homozygous for the gene, leading to a dysfunctional or absent protein. The phenotype of these mice can now be studied.

There are now even more sophisticated methods of gene targeting technologies, and these are mandatory, because one disadvantage of the method described above is that the gene of interest is missing also during the development of the mouse, and it is missing in all body tissues. This might lead to the death of the animal, making it impossible to use it to address the cellular mechanisms of learning processes. In addition, it might also lead to a dysfunctional nervous system, e.g., synaptic transmission or the anatomy of a certain brain area may be compromised. Therefore it is advantageous to have mice in which the gene of interest is deleted only in a certain brain area, let's say the hippocampus, but not in the cerebellum, the brain stem, and the peripheral nervous system or other tissues of the mouse body. And it would even be more advantageous if the gene is removed when the nervous system if fully developed, let's say 2 or 3 weeks after birth. And precisely this has been done by using the *CRE-loxP* system. Here the gene of interest in one strain of mice is flanked by *lox-P*, which usually does not alter gene expression and does not affect the gene function itself. A second mouse strain is also needed carrying a *CRE* sequence inserted into its genome under the control of a certain promoter. If, for example, the promoter for the enzyme CaMKIIalpha is used, CRE will only be expressed in the forebrain and the expression will start 3 weeks after birth, because this is where and when the promoter is usually turned on – in all other brain areas this promoter stays silent. In these CRE-line mice, CRE will have no effect, since there are no *loxP* sides on any genes in these mice. However, once the CRE mice are crossed to the loxP mice, the gene of interest is removed from the genome, because the CRE enzyme cuts every DNA sequence out of the genome that is between two flanking *loxP*s. Now the gene of interest is removed from the genome in an area- and time-restricted fashion. These mice can be probed for their learning behavior and for their cellular phenotype, e.g., in neurons of the hippocampus, amygdala, or cerebellum.

lasting. This process can be experimentally studied by inducing paired-pulse facilitation (PPF) in hippocampal neurons. PPF occurs at the CA3-CA1 synapse and consists in the gain in postsynaptic responses to the second of two successive pulses within the range of tens of milliseconds (see Chap. 8). Within this time range, transmitter release is determined by the average vesicle release probability, the stimulus-dependent accumulation of $Ca^{2+}$ in the presynaptic terminal and the speed at which $Ca^{2+}$ is cleared from the terminal. The response to the second stimulus is higher, because residual, intracellular $Ca^{2+}$ from the first depolarization increases the amount of transmitter release. The increase depends on the amount and time scale of $Ca^{2+}$ influx through voltage-dependent $Ca^{2+}$ channels.

Paired-pulse facilitation (PPF), together with paired-pulse depression (PPD, a reduced response to the second pulse) and posttetanic potentiation (PTP, an increased response to an isolated pulse after a titanic train) are forms of short-term plasticity. PPF and PPD can be induced if two stimuli are given with a short interstimulus-interval (10–200 ms), PTP is visible after a train of 10–100 stimuli with an interstimulus interval of 10–30 ms. PPF, PPD, and PTP are typically caused presynaptically by changes of the intracellular $Ca^{2+}$ levels in the presynaptic terminal. They can be observed at many synapses in the CNS and PNS, e.g., also at the neuromuscular synapse.

### 26.5.4 Long-Term Potentiation

The hippocampus links neural activity representing aspects of information to neuronal assemblies in the way D.O. Hebb had postulated. If these circuits are activated together again (and only then), the memory engram can be reactivated (i.e., remembered). These neuronal assemblies are stabilized if their connections are strengthened. Mandatory for this process is that the neural response to stimulation of a given intensity is enhanced. In 1973, such an enhancement was observed by Timothy Bliss and Terje Lømo. Applying a short train of high-frequency pulses *in vivo* to the perforant pathway projecting into the hippocampus of anesthetized rabbits, they could observe a long-lasting enhancement in the response properties of the target neurons, a phenomenon that is now termed long-term potentiation (LTP). LTP is usually defined by an increase in synaptic strength (potentiation) that lasts for at least 1 h (long-lasting). It consists of an induction phase (time period directly after a high-frequency stimulus) and the expression or maintenance of LTP thereafter (Figs. 26.11 and 26.12). LTP is characterized by different phases: at 1–3 h it is independent of transcription and translation (early or E-LTP), while if it lasts longer than 3 h, it is generally dependent on altered gene expression and referred to as late-LTP (L-LTP, Fig. 26.18) (Bliss and Lomo 1973, reviewed in [3]).

#### 26.5.4.1 NMDA Receptor-Dependent LTP

Hebb's postulate for synaptic strengthening asks for coincident neural, i.e., pre-and postsynaptic activity and a subcellular mechanism of detection. In 1986, a cellular implementation of this requirement was found. Wigström and Gustafsson showed a strong potentiation of synaptic responses in a rat hippocampal slice preparation by simultaneously activating the Schaffer collaterals and a postsynaptic neuron in the CA1 region. This LTP depends on the activation of NMDA receptors, an ionotropic glutamate receptor (see Chap. 8). If this receptor is pharmacologically blocked, no LTP in the CA3-CA1 synapse can be induced, and hippocampus-dependent spatial learning is blocked [21]. This was surprising, at first, since the regular (basal) synaptic transmission at CA3-CA1 hippocampal synapses is mediated predominantly by ionotropic AMPA receptors mediating $Na^{+}$ influx in response to glutamate binding.

The NMDA receptor is a ligand-coupled ion channel, permeable to $Na^{+}$ and $Ca^{2+}$, which in the resting state is occluded by $Mg^{2+}$ ions. Depolarization of the membrane releases the $Mg^{2+}$ block and upon glutamate binding allows $Ca^{2+}$ influx which does not normally enter the postsynaptic cell as a result of baseline stimulation. Elevation of the $Ca^{2+}$ concentration then triggers several intracellular reactions (Figs. 26.12 and 26.18). The requirement of postsynaptic depolarization during glutamate binding demands coordinated pre- and postsynaptic activity. Therefore the NMDA receptor was termed molecular **coincidence detector**: only when the postsynaptic terminal is already or still active and thus the $Mg^{2+}$ block removed while presynaptic glutamate is released, $Ca^{2+}$ can enter the postsynaptic cell and act as a second messenger. This requirement of coincidence determines a key feature of LTP: **associativity**. Another crucial characteristic of LTP is **input specificity**, meaning that the modifications of the synaptic network affect only the participating synapses – or close-by neighboring synapses. Input specificity seems not to be confined to a single synapse, but to restricted areas around the activated synapse, making it likely that the activated postsynaptic synapse releases a diffusible messenger that in turn might act on presynaptic terminals (retrograde messenger) and on neighboring synapses in close

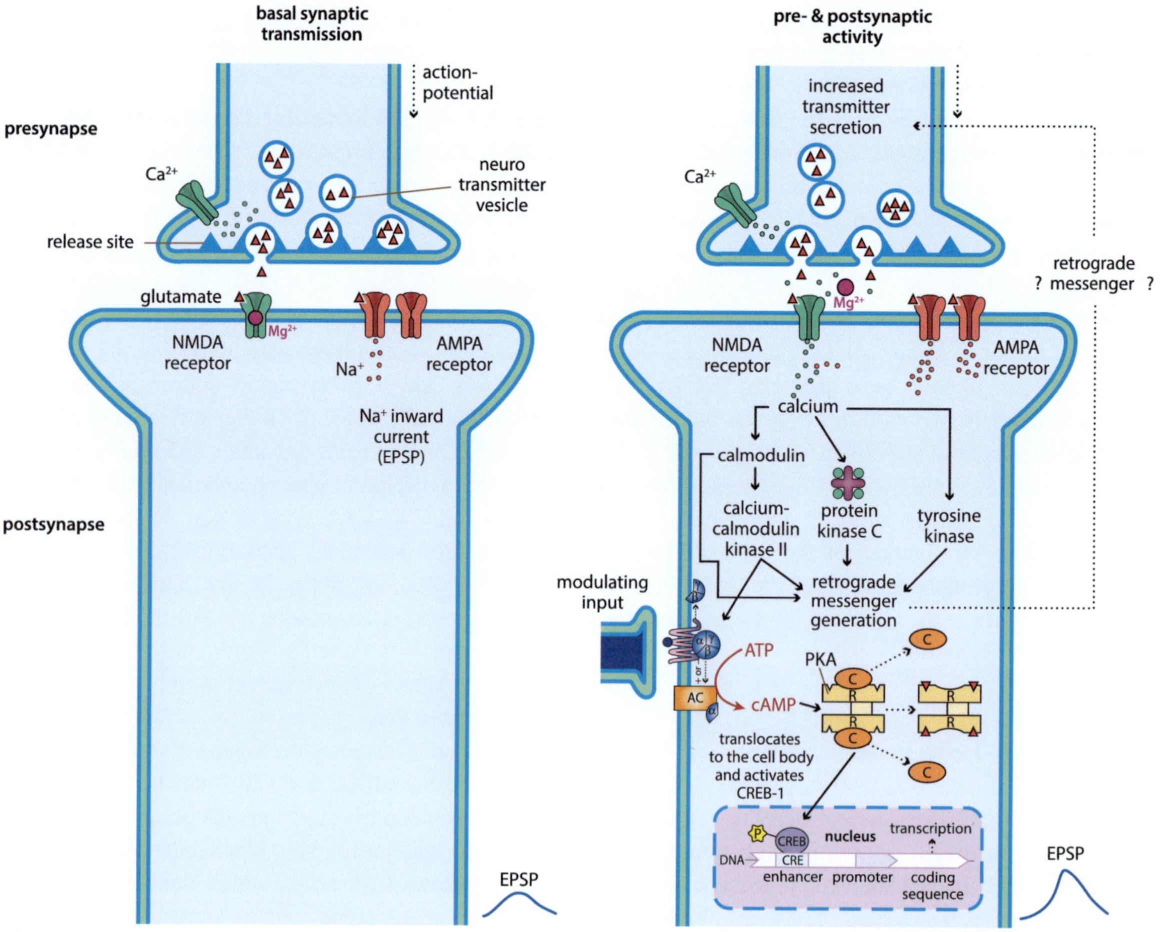


**Fig. 26.12 Induction mechanism of LTP**. The *left panel* shows synaptic transmission via baseline stimulation. The neurotransmitter glutamate only activates AMPA receptors. This induces an excitatory postsynaptic potential (*EPSP*, shown *bottom right*). NMDA receptor channels are blocked by $Mg^{2+}$ ions. In the *right panel*, the pre- and postsynaptic neurons are active within the same time window (associativity). Postsynaptic depolarization leads to the removal of the $Mg^{2+}$ ions, so in addition to AMPA receptors, NMDA receptors open. The resulting influx of $Ca^{2+}$ activates a series of signaling events, some shown in the figure, that strengthen the synaptic connection (LTP). A modulating input can further increase, e.g., the level of cAMP, which might lead to a signaling event inducing gene transcription, and leading to long-lasting changes

proximity. The retrograde messenger is able to enter the presynaptic side and increases the release of neurotransmitter by, e.g., increasing the local $Ca^{2+}$ concentration.

Not only pre- and postsynaptic coactivation leads to synaptic strengthening. It is also possible to induce LTP by high-frequency stimulation of a bundle of axons via tetanic stimulations or a burst of stimuli, usually in the range of 100 Hz. Here, the postsynaptic depolarization, caused by repeated transmitter release, adds up to exceed the voltage threshold for activation of the NMDA receptor. This mechanism ensures the third important characteristic of LTP: **cooperativity**. In order to confine plastic alterations to stimuli representing either simultaneous heterosynaptic information or strong (homosynaptic) inputs, these have to pass a certain intensity threshold. Furthermore, alterations are largely confined to the active connection. At glutamatergic synapses, the postsynaptic elements are located on small membrane protrusions with bottleneck connections to the dendrite – the dendritic spines. Their geometry ensures that electric and ional alterations, especially local alterations of cytosolic $Ca^{2+}$, stay largely confined within the spine, thanks to the fast kinetics of the AMPA receptors and local extrusion by $Na^+$-$Ca^{2+}$ pumps. The specificity, however, is not abso-

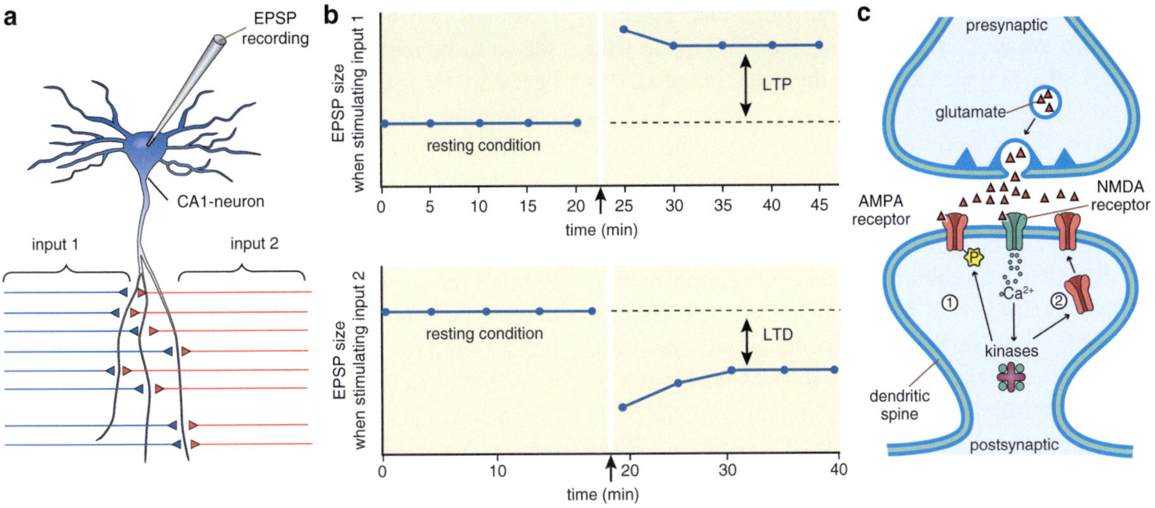


**Fig. 26.13** **LTP and LTD in the same neuron**. (**a**) One recording electrode monitors two independent inputs to a CA1 hippocampal neuron. (**b**) *Input 1* induces LTP (*upper panel*), whereas *input 2* induces LTD by low frequency stimulation (*lower panel*). (**c**) Receptors are constantly recycled from the postsynaptic density (PSD) membrane. Under baseline conditions (steady state) receptor density remains equal. AMPA receptors can be phosophorylated by kinases in order to change opening probabilities (*1*) and under LTP conditions the steady state is shifted, with more new AMPA receptors inserting into the postsynaptic membrane (*2*) enlarging it. Under LTD conditions the number of AMPA receptors is reduced, the PSD becomes smaller and in a new steady state condition fewer receptors reduce the synaptic strength (Adapted from Bear et al. [2] with permission)

lute; part of the depolarization spreads to nearby parts of the dendrite. This contributes to the phenomenon of cooperativity, permitting that inputs converging in close spatial vicinity can be potentiated if they are active simultaneously, but it also allows associativity. Also a weak input can be potentiated if it coincides with a stronger one, because the depolarization from nearby active synapses helps to surpass the threshold of the NMDA receptor or of voltage-dependent $Ca^{2+}$ channels. So, new connections between previously unrelated inputs can be formed if one of them surpasses the induction threshold within a narrow time window. This may well be the underlying mechanism of classical conditioning in vertebrates, as it was described by Pavlov.

Taken together, the $Ca^{2+}$ signal created upon opening of the NMDA receptor triggers a cascade of events leading to the increase in synaptic transmission by the activation of kinases (see below). Therefore, the induction of LTP is a postsynaptic event at the CA3-CA1 synapse of the hippocampus. $Ca^{2+}$ triggers additional influx of $Ca^{2+}$ into the cytosol from intracellular stores. The enhanced elevation of cytosolic $Ca^{2+}$ levels can then activate different effector kinases (enzymes that transfer phosphate groups from one molecule to another, e.g., from ATP to AMPA receptors). Of particular importance for LTP induction is the calcium-calmodulin-kinase II (CaMKII). If CaMKII is inhibited, LTP induction is prevented. CaMKII phosphorylates amino acid residues on AMPA receptors prolonging their open probability. In addition, it also activates further signaling molecules leading to the insertion of additional AMPA receptors into the postsynaptic membrane (Fig. 26.13). This latter process is of special importance because there are synapses in the hippocampus that show no regular transmission unless LTP in induced. These synapses are so-called silent synapses – not a very intuitive term, because you would expect that the transmitter release is missing, but instead these synapses are deaf – meaning that only NMDA receptors are inserted into the postsynaptic membrane. Therefore, under baseline conditions, these synapses do not respond to the release of glutamate (AMPA-receptors are missing in the postsynaptic membrane and NMDA-receptors are blocked by $Mg^{2+}$ ions). When NMDA receptor channels are opened due to glutamate release and postsynaptic depolarization in a narrow time window, then AMPA receptors are inserted into the postsynapse and silent synapses can also be activated under baseline conditions – leading to a stronger connection between the two nerve cells.

At some synapses, LTP induces changes at the presynaptic side. At the CA3-CA1 synapse, most likely both pre- and postsynaptic alterations take place to enhance synaptic efficacy. How does that agree with the fact that at theses synapses the induction of LTP is a postsynaptic event? At synapses where LTP is induced postsynaptically, retrograde signaling to the presynaptic side boosts neurotransmitter release in order to maintain these synaptic changes. Retrograde signaling can take the form of postsynaptic secretion of diffusible transmitters, namely NO (nitric oxide), BDNF (brain-derived neurotrophic factor), and endocannabinoids (see below). Exogenous NO has been shown to lead to LTP if coupled to weak tetanic stimulation. In the hippocampus NO is produced in the postsynapse by a $Ca^{2+}$-dependent NO synthase. After synthesis, NO quickly travels through membranes and activates the soluble guanylyl cyclase (sGC)-coupled receptors. It also stays local because it quickly reacts as a radical with oxygen. Alternatively, BDNF, the major ligand of the TrkB receptor tyrosine kinase, has been shown to be an important component controlling activity-dependent synaptic plasticity possibly via retrograde signaling [22]. Genetic approaches have shown that homozygous as well as heterozygous BDNF knockout mice exhibit a marked reduction in LTP, which can be rescued by adenovirus-mediated local overexpression of BDNF or by application of recombinant BDNF. On the other hand, direct application of the recombinant protein to normal hippocampal slices enhances synaptic strength and the probability of LTP induction. Moreover, the local application of BDNF together with a weak burst of presynaptic activity can induce LTP. Conversely, blockade of BDNF/TrkB interaction strongly reduces hippocampal LTP. Experiments employing conditional TrkB knockout mice in which TrkB is removed postnatally from the forebrain by a Cre-recombinase under the control of the promoter for the CaMKIIalpha enzyme have confirmed that BDNF is mediating the effect on synaptic plasticity via activation of TrkB. Binding of BDNF leads to dimerization of TrkB receptors and to autophosphorylation of its tyrosine residues (see Chap. 6). This activates a second messenger pathway and the resulting cytosolic $Ca^{2+}$ increase augments the likelihood of LTP induction and maintenance.

Another form of retrograde signaling could be achieved by membrane-spanning molecules that are able to bridge the synaptic cleft. Among the possible candidates are the tyrosine kinases named EphrinBs that are enriched in the postsynaptic density (see Chap. 8). They bind to AMPA receptors and have been shown to be required for synaptic plasticity. They are necessary for structural plasticity, ensuring optimal signal transmission due to correct apposition of the pre- and postsynapse by binding to synaptic scaffolding proteins and to receptors. Also neuroligins bind to synaptic scaffolding proteins and have been shown to trans-synaptically modulate transmitter release. Both EphrinBs and neuroligins are regulated by phosphorylation and therefore modified by the alterations in cytosolic $Ca^{2+}$ which activates kinases, as is the case during the induction of LTP. In addition, also *N*-cadherin is a key transmembrane, cell-adhesion molecule with important roles in synaptic plasticity and memory formation.

The mechanisms described here refer to the CA3-CA1 synapse [17]. Also, in the hippocampus there are "non-Hebbian" synapses, at which LTP is induced presynaptically and without NMDA receptor activation, e.g., at the mossy synapse of the mossy fibers with CA3 pyramidal neurons.

#### 26.5.4.2 Long-Term Depression (LTD)

Synaptic depression constitutes a necessary counterpart of LTP. By reducing synaptic efficacy it establishes the necessary difference in strength between synapses, and at the same time it helps to prevent excessive synaptic activity. An earlier notion was that LTD (long-term depression) is a correlate of forgetting by weakening of unused connections, but it is now undisputed that also activity-dependent reduction of synaptic strength is directly involved in learning. One theory holds that whereas LTP establishes connectivity within a network, LTD accomplishes the fine-tuning of connections.

Prolonged low-frequency stimulation induces a long-lasting reduction of synaptic responses (Fig. 26.11) – at the same synapses capable to show LTP. Even more surprisingly the same molecular players which mediate LTP are also involved in LTD. However, they act in opposite directions: AMPA receptors are dephosphorylated and internalized during the induction of LTD. Some forms of LTD are initiated when NMDA receptors are activated by prolonged low-frequency stimulation (1 Hz for 15 min is a typical protocol), resulting in a modest increase of cytosolic $Ca^{2+}$ levels (whereas $Ca^{2+}$ levels after tetanic stimulation increase strongly). Instead of activation of kinases a modest increase in $Ca^{2+}$ levels leads predominantly to activation of phosphatases, due to their higher affinity for $Ca^{2+}$ in comparison to kinases.

Especially the phosphatase calcineurin/PP2B is involved here.

NMDA receptor-dependent LTD is most easily inducible in the hippocampus during development. It decreases with age, concomitantly with a switch between NMDA receptor subunits with different open kinetics. Expression of the NR2B subunit (which helps to mediate LTD) decreases in favor of the NR2A subunits (which helps to mediate LTP).

In addition, one has identified a form of LTD that is independent from NMDA receptor activation, but relies on activation of metabotropic glutamate receptors (mGluRs). This third type of glutamate receptors is not linked to a ion channel but instead to G proteins. The mechanism of mGluR-dependent LTD is much less clear. As it is blocked by the postsynaptic injection of $Ca^{2+}$ chelators, its induction seems to rely on postsynaptic mechanisms. However, it results in decreased frequency of miniEPSPs (postsynaptic potentials resulting from single transmitter vesicle release), not amplitudes, suggesting a presynaptic locus of expression. Occlusion studies indeed indicate that NMDAR-LTD can be induced even after saturation of mGluR-LTD, confirming that both forms of LTD (one NMDAR-dependent, one NMDAR-independent) rely on independent signaling mechanisms. NMDA receptor-dependent LTD is readily induced in juvenile, but not in adult mice suggesting that its main task is to refine the young neuronal network by targeted pruning of less significant contacts during late phases of development. mGluR-dependent LTD is still inducible in the adult system and thus seems to be the main mechanism needed to keep the balance of synaptic weights in the adult hippocampus [17].

### 26.5.5 Modulators of Synaptic Plasticity

Several other transmitter systems have been found to be involved in synaptic plasticity in the hippocampus. These include GABAergic signaling and several neuromodulating factors, acetylcholine, dopamine (especially at corticostriatal synapses), and endocannabinoids.

#### 26.5.5.1 Inhibitory GABAergic Signaling Gates LTP

So far we only looked at pre- and postsynaptic events at the CA3-CA1 synapse. However, different subtypes of inhibitory neurons in the hippocampus also critically determine hippocampal activity-dependent synaptic plasticity. Firstly, locally projecting inhibitory interneurons releasing GABA temporally coordinate the action potentials of pyramidal neurons and thereby synchronize information transmission between hippocampal areas. Secondly, these inhibitory interneurons also facilitate plastic rearrangements. In area CA1, Schaffer collateral axons coming from CA3 pyramidal neurons activate GABAergic interneurons simultaneously with pyramidal neurons. Although blockade of GABAergic activity can facilitate NMDA receptor dependent LTP, in its natural form it is a key feature of LTP induction. During regular synaptic transmission, a delicate balance between excitation and inhibition is crucial. This needs to be transiently shifted towards less inhibitory activity to allow the induction of LTP. This is accomplished by the two different receptor systems for GABA: GABA acts on ionotropic $GABA_A$- and metabotropic $GABA_B$ receptors. $GABA_A$ receptors are predominantly located on the postsynaptic side (pyramidal cells in the hippocampal CA3 and CA1 region) and mediate quick hyperpolarization of the postsynaptic neuron via $Cl^-$ influx. On the other hand, $GABA_B$ receptors are located pre- and postsynaptically, and in addition on the GABAergic terminals themselves. These extrasynaptic receptors are only activated after spillover of GABA in response to a strong stimulation. They have longer latencies and inhibit voltage-dependent $Ca^{2+}$ and $K^+$ channels. Therefore, $GABA_A$ receptors shape the temporal properties of regular synaptic transmission, whereas $GABA_B$ receptors selectively act on synaptic plasticity during rhythmic stimulation. This can be explained as follows: The slow opening constant of the NMDA receptor results in slow rise time of NMDA receptor-mediated EPSPs. Therefore, at low rates of synaptic transmission, $GABA_A$ receptor-mediated hyperpolarization curtails the AMPA receptor-mediated EPSP and intensifies the $Mg^{2+}$ block of NMDA receptors, fine-tuning the temporal specificity of the NMDA receptor for detection of simultaneous inputs. During high-frequency stimulation, GABA accumulates and thereby also stimulates extrasynaptic, metabotropic $GABA_B$ autoreceptors on GABAergic presynaptic terminals, thus reducing the release of GABA as stimulation continues. This mechanism is particularly effective during intermittent stimulation, because $GABA_B$ receptors have a peak of activity after 200 ms. This tunes the system to suppress GABAergic inhibition during θ-type rhythmic activity (theta rhythm, rate of 5–7 Hz spike activity), which is predominant during explorative activity and mnemonic processing and facilitates learning by activity-dependent synaptic plasticity under these circumstance.

#### 26.5.5.2 Endocannabinoids Modify Inhibitory Synaptic Plasticity

The mechanisms that underlie processes of synaptic plasticity are more diverse than previously expected. A particularly good example for this statement is a group of molecules that were discovered recently: endocannabinoids (endogenous cannabinoids, eCB). Endocannabinoid signaling influences the propensity of the synaptic network to react plastically by modifying GABAergic inhibitory contributions, thereby realizing a form of metaplasticity. eCBs are the ligands of CB1 receptors. As hydrophobic molecules, eCBs do not need to be packed into vesicles and can be released directly from cell membranes if the necessary enzymes are activated.

In a number of brain regions, among them the hippocampus, eCBs are retrograde messengers released by postsynaptic neurons in response to strong depolarization and/or activation of GPCRs (e.g., mGluRs or muscarinic receptors) and function to inhibit neurotransmitter release at either excitatory or inhibitory synapses. CB1 receptors are among the most abundant GPCRs in the brain and are mostly found on axon terminals. In the hippocampus, eCBs mediate a form of LTD at inhibitory, but not excitatory, synapses, thereby restricting LTP. The best-described model here is eCB-mediated-heterosynaptic LTD. "Heterosynaptic" means that the activity pattern of one synapse affects synaptic plasticity at another synapse. Repetitive glutamate release from Schaffer collateral terminals activates mGluRs of the mGlu1/5 type leading to activation of PLC (phospholipase C) and the formation of DAG (diacylglycerol). DAG can then be converted to 2-AG (arachidonoylglycerol), which is a ligand to the presynaptic CB1 receptor on GABAergic neurons, and induces long-lasting depression of evoked GABAergic release. The role of CB1 receptors is a mechanism that can restrict or enhance synaptic strength depending on the synapse (inhibitory or excitatory) it acts on.

### 26.5.6 Spike-Timing-Dependent Plasticity (STDP)

Another plasticity phenomenon that can be observed in the vertebrate brain is spike-timing-dependent plasticity (STDP). It is a process that adjusts the connection strengths between neurons based on the relative timing of a particular neuron's outputs and inputs. It also offers an explanation for the mechanisms of LTP and LTD, taking into account the opening time of the NMDA receptor. Hebb postulated, as referred to above, that coincident activation represents information that demands strengthening of the connections of the involved neurons, whereas noncoincident activation leads to weakening of the connectivity. STDP can be experimentally induced by double-patching and triggering precisely-timed spikes in pre- and postsynaptic neurons. In the presynaptic neuron this causes transmitter release, while in the postsynaptic neuron it generates a back-propagating action potential running back into the relevant dendrite. If presynaptic activity precedes the postsynaptic spike by 5–15 ms, representing a situation in which presynaptic neurons contribute to drive the postsynaptic neuron, the strength of the active synapses is enhanced. Similarly, if the postsynaptic spike occurs shortly before transmitter release, uncorrelated activity is mimicked and synaptic strength is reduced (LTD). As a consequence, according to this model, inputs on a given neuron compete for grouped activity to form stronger connections, whereas less precisely timed inputs are weakened. Taken together, synaptic connections increase their efficacy if the presynaptic neuron is activated right before the postsynaptic neuron is activated and synapses in which the presynapse fired before the postsynaptic neuron get stronger; in the inverse situation, the synapse gets weaker. As a result, timing becomes equally or even more important than overall activity, and neural networks become rhythmically controlled.

### 26.5.7 Metaplasticity

The activity impinging on a neuron is changing all the time and a particular neuron might be involved in different neuronal assemblies that store information of different content. Neurons therefore change their ability to

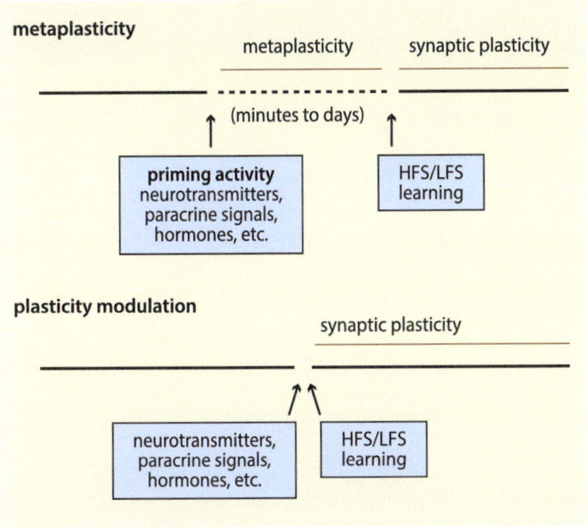


**Fig. 26.14 Processes of metaplasticity**. The standard paradigm for studying metaplasticity is to have an episode of priming activity at one point in time and then a subsequent plasticity-inducing event, such as low-frequency stimulation (*LFS*), high-frequency stimulation (*HFS*), or learning that evokes synaptic plasticity such as long-term potentiation (*LTP*) or long-term depression (*LTD*). The priming signal can include electrical stimulation of neural activity, pharmacological activation of specific transmitter receptors, or behavioral events that might cause hormone release in addition to neural activity. An essential aspect of this model is that there must be a change in neural function as a result of the priming that persists after the termination or washout of the priming stimulus and that alters the response to a subsequent plasticity-inducing event (*upper panel*). This distinguishes metaplasticity from conventional modulation of plasticity (*lower panel*), where the modulation occurs together with the induction of plasticity (Adapted from Abraham [1] with permission)

change, a process termed metaplasticity. Metaplasticity is the plasticity of the plasticity rules of a neuron. So, depending on its history, a neuron might be less or more prone to change its synapses [1]. For example, LTP or LTD might be easier or less easy to induce (Fig. 26.13). These mechanisms depend on the current synaptic "state", as set by ongoing extrinsic influences such as the level of synaptic inhibition and the activity of modulatory neurotransmitters on a subset of synapses. Recent data suggest that individual synapses might not strengthen or weaken on a sliding scale. Instead, there are possibly discrete states into which synapses can move to. These states are active, silent, recently-silent, potentiated and depressed. Thus, the future state is determined by the state gained by previous activity. Not much is known about the molecular nature of this metaplasticity, but it is of great theoretical importance in brain and cognitive science and it adds another layer of complexity to the biology of the synapse (Fig. 26.14).

### 26.5.8 Learning in the Cerebellum

Another brain area, intensively studied by scientists who study synaptic plasticity, is the cerebellum. It is involved in motor control and body balance. In addition it is also necessary for several types of motor learning, most notably learning to adjust to changes in sensorimotor relationships. The learning events where the cerebellum does play a role are those in which it is necessary to make fine adjustments to the way a motor action is performed. It is believed that the cerebellum always corrects the motor output until it fits the expectation and for this process the cerebellum needs to be plastic (see Chap. 23). In the cerebellum each Purkinje cell receives two different types of input: (i) thousands of inputs from parallel fibers, each individually very weak coming from a vast number of cerebellar granule cells, which in turn receive their input from various region of the brain stem; (ii) one single climbing fiber, originating in the inferior olive, which collects proprioceptive input from muscles. Each climbing fiber is so strong that a single climbing fiber action potential will reliably cause a target Purkinje cell to fire a burst of action potentials. The climbing fiber serves as a teaching signal, which induces a long-lasting change in the strength of synchronously activated parallel fiber inputs (Fig. 26.15).

The best-studied form of synaptic plasticity occurring in the cerebellum is LTD, which decreases the efficacy of parallel fiber input to Purkinje cells. Both parallel fibers and climbing fibers must be simultaneously activated for LTD to occur, but LTD will only be observed at parallel fiber synapses to Purkinje cells dendrites. Because the output of Purkinje cells is inhibitory, a decrease in synaptic transmission of Purkinje cells as the only output cells of the cerebellum actually means an increased excitation (Fig. 26.17).

At the molecular level, the parallel fiber terminals release glutamate and activate ionic AMPA receptors and metabotropic glutamate receptors in the postsynaptic Purkinje cell (Fig. 26.16). AMPA receptors depolarize the membrane so strongly that voltage-gated $Ca^{2+}$ channels open as well. mGluRs activate the second messenger diacylglycerol (DAG), which activates the protein kinase C.

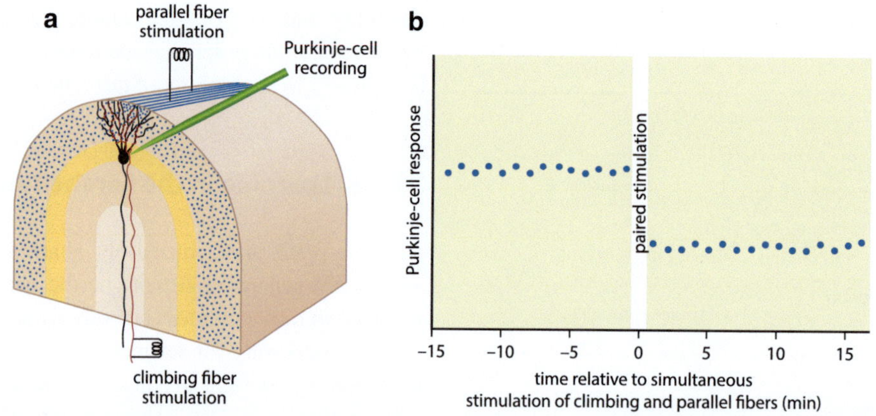


**Fig. 26.15** **Cerebellar learning.** (**a**) Scheme of the cerebellar cortex with the cellular organization of granule cells (*blue*), Purkinje cells (*black*), and climbing fibers (*red*). Parallel fibers from granule cells are the main input onto Purkinje cells. Climbing fibers project from the inferior olive in the brain stem into the cerebellum. Mossy fibers that project from the potine nuclei innervate granule cells. A recording electrode (*green*) inside the Purkinje cell monitors the response to parallel fiber stimulation. (**b**) LTD experiments in which parallel fiber stimulation is paired with stimulation of the climbing fibers. After pairing, stimulating parallel fibers elicit a significantly lower response in Purjinje cells (Adapted from Bear et al. [2] with permission)

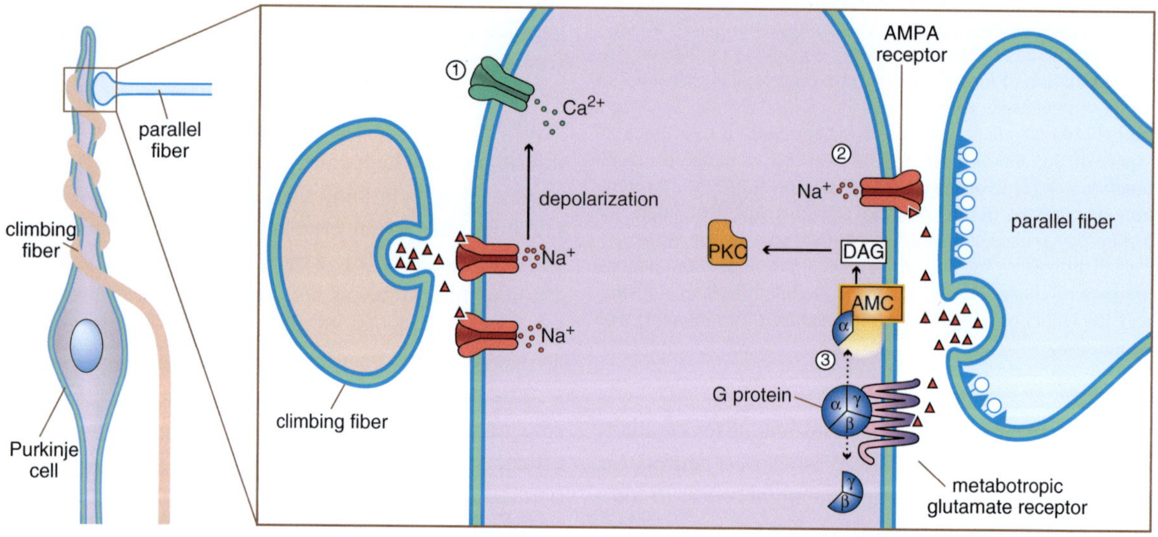


**Fig. 26.16** **LTD in the cerebellum.** Climbing fiber activation strongly depolarizes Purkinje cell dendrites, which leads to the activation of voltage-gated $Ca^{2+}$-channels (*1*). Parallel fiber activation (*2*) leads to $Na^+$ entry through AMPA receptors and the generation of DAG via the stimulation of metabotropic glutamate receptors (*3*). DAG then activates the protein kinase C (*PKC*), which leads to LTD (Adapted from Bear et al. [2] with permission)

PKC phosphorylates AMPA receptors, causing receptor internalization – similar to hippocampal LTD. With the loss of AMPA receptors, the postsynaptic Purkinje cell response to glutamate release from parallel fibers is depressed and this depression can last for hours [12].

The experimental learning paradigm very often used to study cerebellar learning is the eyeblink conditioning test. Here a neutral conditioned stimulus (CS) such as a tone or a light (represented by the climbing fibers) is repeatedly paired with an unconditioned stimulus (US), e.g., air puff, represented by the parallel fibers, which elicits a blink response. After such repeated presentations of the CS and US, the CS will eventually elicit a blink before the US, a conditioned response or CR mediated by the Purkinje cells which project to various motor centers of the brain.

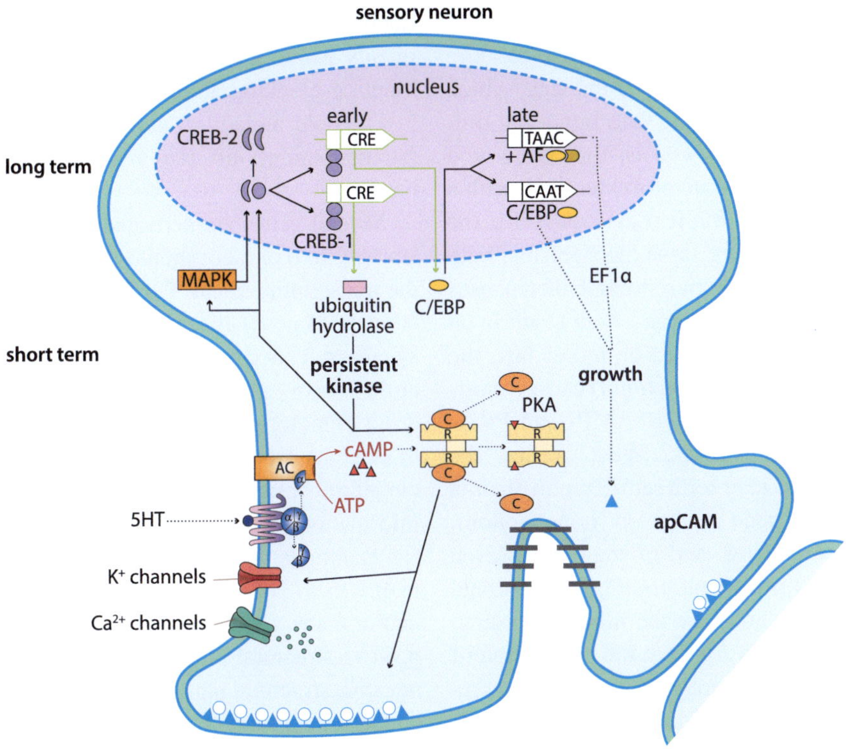


**Fig. 26.17** **Cellular modules for long-term memory**. Effects of short- and long-term sensitization on the monosynaptic component of the gill-withdrawal reflex of *Aplysia*. In short-term sensitization (lasting minutes to hours) a single tail shock causes a transient release of serotonin (*5-HT*) that leads to covalent modification of preexisting proteins. 5-HT acts on the metabotropic serotonin receptor to activate AC, which converts ATP to the second messenger cAMP. In long-term sensitization, repeated stimulation causes the level of cAMP to rise and persist for several minutes. cAMP activates PKA, with the catalytic unit (*C*) dissociating from the regulatory unit (*R*). The catalytic subunits can translocate to the nucleus, and recruit the mitogen-activated protein kinase (*MAPK*). In the nucleus, PKA and MAPK phosphorylate and activate the cAMP response element-binding (*CREB-1*) protein and remove the repressive action of CREB-2, an inhibitor of CREB-1. CREB-1 in turn activates several immediate-response genes, including an ubiquitin hydrolase necessary for regulated proteolysis of the regulatory subunit of PKA. Cleavage of the (inhibitory) regulatory subunit results in persistent activity of PKA, leading to persistent phosphorylation of the substrate proteins of PKA. A second immediate-response gene activated by CREB-1 is C/EBP, which acts both as a homodimer and as a heterodimer with activating factor (*AF*) to activate downstream genes, including elongation factor 1a (EF1a), that lead to the growth of new synaptic connections (Adapted from Kandel [13] with permission)

## 26.6 Long-Term Maintenance of Synaptic Plasticity

> In order to ensure long-term storage of information, gene expression, protein synthesis, and structural changes at synapses are necessary.

How are the early events at synapses consolidated in order to produce a stable and long-lasting memory trace (engram)? To incorporate alterations permanently into a neural network, additional functional and structural changes are necessary. Therefore cellular mechanisms of information storage should distinguish between processes of short-term and long-term memory. The more often we practice something, the better we become and the items learned will be remembered much longer. Indeed repeated learning can consolidate memory contents by converting short-term memory into long-term storage.

### 26.6.1 Long-Lasting Memory Storage in Aplysia

Sensitization in *Aplysia* only lasts for minutes after a one-trial learning session (or in a dish, after a one-time application of serotonin to the sensory neuron). However, if the same procedure is done five or more times, long-term sensitization can be induced, which can last for days or even weeks. One important outcome of these studies in *Aplysia* is, that short- and long-term memory are separate processes each with a specific molecular nature. But it is also clear that the two phases of memory share some mechanisms: both involve changes in the connection strength of synapses between sensory and motor neurons; both result in an increase in neurotransmitter release; and at last, the same neurotransmitter (5-HT, serotonin) causes short- and long-term sensitization via a cAMP- and PKA-dependent molecular pathway. What makes the processes of short- and long-term sensitization distinct is the involvement of gene expression (transcription), translation of new proteins, and in some cases even structural changes (growth or pruning of neuronal structures). This was shown by the blocking protein synthesis during or within 1 h from the application of 5-HT which does not interfere with short-term memory, but with long-lasting memories. This leads to the question about which genes and proteins are involved in converting short- into long-lasting memory and how gene expression can be switched on or off by an external stimulus. For *Aplysia* this switch is the PKA, which has four domains, two regulatory ones and two catalytic ones. With the one-time application of 5-HT, the regulatory domains are removed from the catalytic ones after binding cAMP for a short time period. This enables the catalytic domains of the PKA to phosphorylate $K^+$ channels, but does not give them enough time to travel to the nucleus. As soon as the level of cAMP returns back to normal, the regulatory subunit re-unites with the catalytic one (Fig. 26.17). If after repeated stimulation the cAMP level is high for a long time period, the PKA (catalytic domain) is activated for a long time. In addition, the PKA can recruit the mitogen-activated protein kinase (MAP kinase). These 2 s messenger kinases now translocate to the nucleus of the sensory neurons and change the expression pattern of a whole set of genes by phosphorylating and thereby activating transcription factors (Fig. 26.17). The best-known one is a transcription factor called CREB-1 (cAMP response element-binding protein 1). Phosphorylated CREB-1 now can bind to promoters which have a CRE element (cAMP response element) and activate them. In addition, MAP-kinase and PKA switch off the repressor CREB-2 which under resting conditions reduces the expression of genes which are important for long-term storage (Fig. 26.17). In experiments where CREB-2 was inactivated (by specific function-blocking antibodies) a single application of 5-HT could already produce long-lasting facilitation (strengthening) of the response of the sensory neuron.

Several genes are activated for long-term memory storage. Two of them should be mentioned here: one is the transcription factor C/EBP, which itself activates a cascade of genes that are necessary for creating new synapses. This can lead to new presynaptic terminals at the sensory neuron and to growth processes at dendrites of motor neurons. These changes in the structure of neurons are especially long-lasting and might be one of the basic principles how long-lasting storage of information can be executed, and could also be shown for mammals. Interestingly, in long-term habituation in *Aplysia* 1/3 of synaptic connections between sensory neurons and motor neurons are lost indicating that positive structural plasticity (growth of synapses) and negative structural plasticity (pruning of synapses) can happen at the same neuronal connections.

The other gene activated for long-term storage is a gene which codes for an enzyme to degrade proteins. This enzyme – ubiquitin hydrolase – activates the proteasome in order to degrade protein components that usually inhibit the transfer of short- to long-term memory storage, which is after all a highly regulated process in the nervous system. In particular, it degrades about 25 % of the available regulatory subunits of PKA in one cell. This leads to an increase in the basal phosphorylation of proteins by PKA, including CREB-1, even a long time after the cAMP level has returned to its baseline levels. By this mechanism PKA remains active for at least 24 h in *Aplysia* [13].

### 26.6.2 Long-Lasting Memory Storage in Mammals

*Aplysia* set the stage for the analysis of cellular processes in other animals, and indeed many parallels were found: Also phenomena like hippocampal LTP in vertebrates consist of a protein synthesis-independent early phase (early-LTP) which decays to baseline within 2–3 h after its induction, and a long-lasting phase (late-LTP), lasting hours (or even longer), requiring protein synthesis. Accordingly, both transcriptional and translational mechanisms of gene expression are regulated by plasticity-inducing events all over the

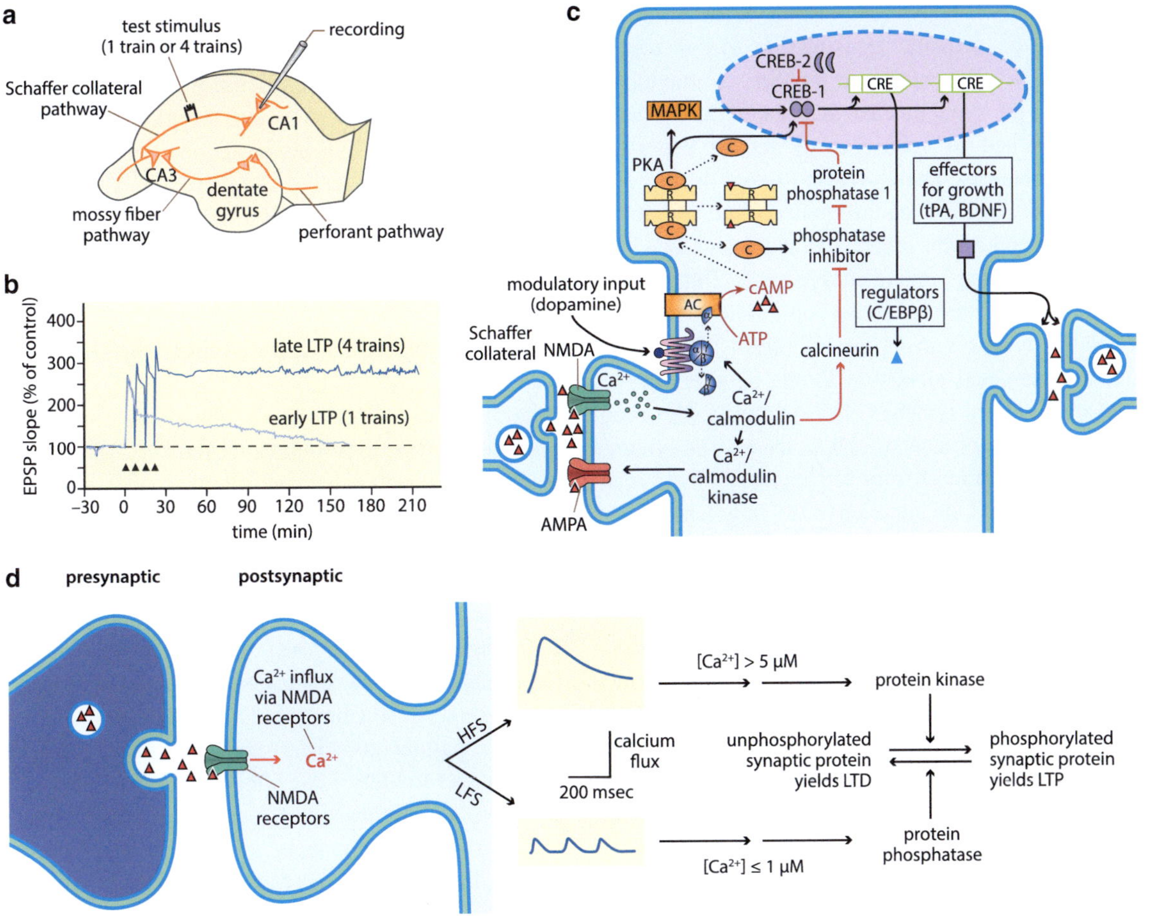


**Fig. 26.18 Long-lasting information storage in the hippocampus.** (**a**) The *Schaffer collateral pathway* connects the pyramidal cells of the CA3 region with the pyramidal cells in the CA1 region of the hippocampus. If this pathway is stimulated with one train of high stimuli, early LTP is induced; if four trains (100 Hz) are used, late LTP (l-LTP) is induced. (**b**) Early LTP lasts about 2 h. l-LTP lasts for more than 3 h and is protein synthesis dependent. (**c**) A model for l-LTP in the CA3-CA1 pathway. A single train of action potentials initiates early LTP by activating NMDA receptors, $Ca^{2+}$ influx into the postsynaptic cell, and the activation of a set of second messengers. With repeated trains of action potentials the $Ca^{2+}$ influx also recruits an adenylyl cyclase (*AC*), which activates the cAMP-dependent protein kinase PKA. Its catalytic subunit is transported to the nucleus where it phosphorylates CREB-1. CREB-1 in turn activates targets (C/EBPβ, tPA, BDNF) that are thought to lead to structural changes. Mutations in mice that block tPA or CREB reduce or eliminate l-LTP. AC can also be modulated by dopamine signals and perhaps other modulatory inputs. In addition, there are constraints (in *red*) that inhibit l-LTP and memory storage. Removal of these constraints lowers the threshold for l-LTP and enhances memory storage. (**d**) A model for how $Ca^{2+}$ can trigger both LTP and LTD in the CA3-CA1 synapse of the hippocampus. A tetanus (*HFS* high-frequency stimulation) leads to LTP by causing a large elevation of $Ca^{2+}$ (which activates kinase). Low-frequency stimulation (*LFS*) yields to LTD causing a small elevation of $Ca^{2+}$, which activates phosphatases (a–c, Adapted from Kandel [13] with permission. d, Adapted from Malenka and Bear [17] with permission)

animal kingdom. Most likely the $Ca^{2+}$ influx together with the activity of modulatory inputs (e.g., in the hippocampus by dopamine) leads to an increase in cAMP, which activates PKA. This kinase then travels to the nucleus and activates the transcription factor CREB-1 (cAMP-response element-binding protein 1). This leads to the expression of genes that are required to maintain the increase in synaptic strength for days, month, or even years. Conversely, the activation of the phosphatase calcineurin after induction of LTD in the mammalian hippocampus via a cascade of phosphatases inhibits the activation of CREB (Fig. 26.18). Also protein translation itself is targeted after induction of synaptic plasticity, a phenomenon that can be observed in snails as well as in mammals. At the activated synapse, protein synthesis was shown to occur and to be necessary for long-lasting potentiation and depression of synaptic transmission (local protein synthesis). Newly synthesized proteins are thought to consolidate the synaptic change initiated during early-phase LTP and to contribute to activity-dependent structural changes. Therefore, on the one hand, long-term memory storage requires general transcription in the nucleus and, on the other hand, it depends on a local specificity for the stimulated synapses. But how do gene products generated in the soma 'know' to which of all the neuron's synapses (10,000–100,000) they have to be targeted, since they should only be incorporated at stimulated synapses? One solution to this problem is the tagging (or capture) hypothesis. The synaptic tag is a molecular marker for activity-dependent synaptic plasticity and its function is to target plasticity-related proteins (PRPs) only to previously activated synapses. One possible mechanism would be to transport generally available proteins into the activated synapses only after stimulation. Alternatively, the synaptic tag could be implemented by spatially restricted new protein synthesis (translation) at stimulated synapses. Synaptic tagging is also a mechanism explaining how newly synthesized proteins from a strongly tetanized input can transform a weakly tetanized input to a long-lasting form of plasticity in an independent but nearby neuronal population (heterosynaptic associativity). According to this model, activated synapses are 'tagged' so that plasticity factors from a strong synaptic input (late-LTP) can transform a transient early-LTP to a late-LTP in a nearby input. It can be concluded that associativity between different synaptic inputs can transform short-lasting forms of synaptic plasticity (<3 h) to long-lasting ones. Synaptic tagging and capture (STC) might be able to explain this heterosynaptic support because it distinguishes between local mechanisms of synaptic tags and cell-wide mechanisms responsible for the synthesis of plasticity-related proteins (PRPs). According to this model, STC initiates storage processes only when the strength of the synaptic tag and the local concentration of essential proteins are above a certain plasticity threshold.

The involvement of local protein synthesis and transcription of new gene products are two mechanisms for memory consolidation. Consolidation is defined as a progressive stabilization of an engram – it can take a rather long time (days to weeks) before long-term memories are formed. In addition, recent findings suggest that many of the molecules that are activated during behavioral training stay active well beyond those typically invoked for memory consolidation processes. So, in addition to local translation, posttranslational modification of proteins already synthesized and present within the synapse can serve as a substrate for the consolidation of long-lasting memory (probably for a time period of up to 24 h).

### 26.6.3 Consolidation by Structural Plasticity

In addition to maintaining the functional changes at synapses in order to mediate long-term changes at synapses, the connection between neurons can be strengthened by building more synaptic contacts between these neurons. Functional changes (LTP, LTD) can lead to structural changes as well (Fig. 26.19). In hippocampal organotypic cultures 30 min after the induction of LTP (functional changes), existing spines became larger and even new spines were formed (structural changes) (see also Chap. 8). These spines have a presynaptic partner and therefore a functional synapse forms approximately 1 day after a new spine is formed. Local protein synthesis is necessary in order to provide, e.g., cytoskeletal elements to increase the size or change the shape of spines. When blocking protein synthesis, lasting structural changes are prevented, without affecting immediate, functional changes. In addition, it was shown that the neurotrophin BDNF is necessary and sufficient for the induction of long-lasting structural changes at spines, as shown by the application of BDNF itself or blockade of BDNF signaling via TrkB. BDNF and the TrkB receptor have been shown before to be important modulators of activity-dependent synaptic plasticity (see above), especially for the induction of LTP and the maintenance of synaptic plasticity. It is interesting to speculate that the BDNF/TrkB system might be an important translator of functional into structural plasticity (Fig. 26.20). Other experiments suggest that the action of BDNF in mediating activity-dependent changes might occur on both sides of the synaptic cleft – a new postsynaptic side must have a presynaptic partner in order to be a functional synapse.

After a first period in which the emphasis of research into structural plasticity in the CNS has been on positive structural plasticity, new evidence now

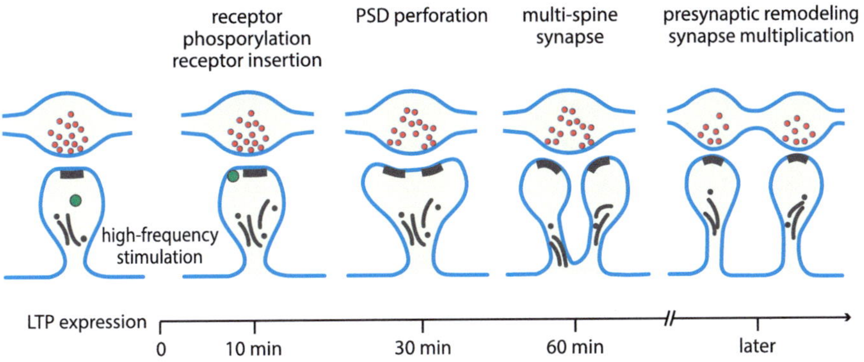


**Fig. 26.19** **From functional to structural plasticity**. Within 10 min after the induction of LTP, activation of $Ca^{2+}$-dependent signal-transduction pathways results in phosphorylation of AMPA receptors and an increase in their single channel conductance. In addition, the size of the spine apparatus increases, and AMPA receptors are delivered to the postsynaptic membrane by exocytosis of coated vesicles (*green*). This membrane insertion also leads to an increase in the size of the postsynaptic density (*PSD*) and, eventually, to the production of perforated synapses within the first 30 min. At 1 h, through an unknown mechanism, some synapses use the expanded membrane area to generate multispine synapses (where two or more spines contact the same presynaptic bouton). Concomitant retrograde communication, possibly through cell-adhesion molecules, may trigger appropriate presynaptic structural changes, eventually increasing the total number of synapses

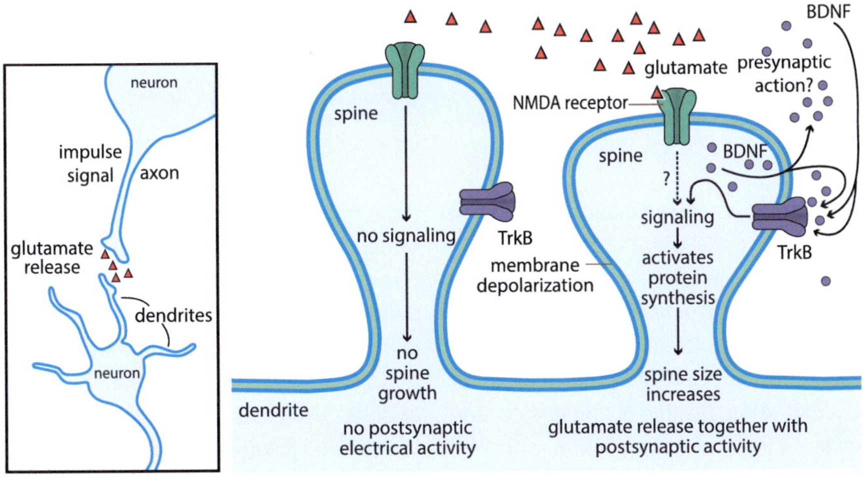


**Fig. 26.20** **Spine growth and memory**. When glutamate binds to postsynaptic *N*-methyl-D-aspartate (*NMDA*) receptors in the dendritic spine membrane, and postsynaptic electrical signals are induced, there is a coincidence of pre- and postsynaptic activity (right spine). Only then is BDNF released in high amounts, activating TrkB receptors in the spine. This initiates local protein synthesis necessary for structural changes in the spine, which may be linked to the formation of long-lasting memories in the hippocampus. Just how BDNF is released is not clear, it might be a presynaptic release mechanism (Adapted from Korte [14] with permission)

indicates that negative functional plasticity can lead to negative structural changes (negative structural plasticity) as well. These structural modifications therefore mirror, in some respects, the functional changes. LTP and LTD are associated not only with opposite morphological effects, but these morphological effects are reversible. These findings are not only of interest for basic research: Malformation of dendritic spines and reduced spine density are one of the hallmarks of many neurological conditions in humans and might be linked to a missing – or abnormal – activity-dependent structural plasticity at synapses. In

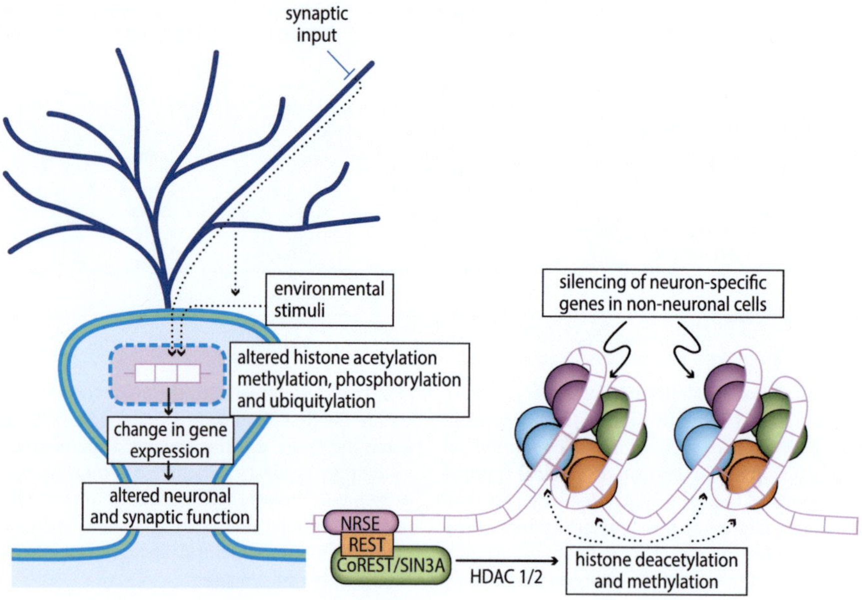


**Fig. 26.21** **Epigenetics in the adult nervous system**. *Left*: Regulation of the epigenetic state of the genome in adult neurons occurs in response to synaptic inputs and/or other environmental stimuli. These external stimuli result in changes in the transcriptional profile of the neuron and, ultimately, neural function. *Right*: The RE1-silencing transcription factor (*REST*)/REST co-repressor (*CoREST*) system. The neuron-restrictive silencer element (*NRSE*) is upstream of genes to be silenced in non-neuronal cells and recruits REST as a mediator of transcriptional repression. SIN3A, CoREST, and REST function with additional factors such as histone deacetylase 1 (*HDAC1*) and HDAC2 to lead to chromatin condensation and gene silencing (From Levenson and Sweatt [16] with permission)

all diseases with cognitive impairments changes in spine shape and density can be observed. This is paradigmatically evident in Alzheimer's disease. It cannot be excluded that the loss of structural plasticity is ultimately related to aging. In addition, many forms of mental retardation show a reduced spine density on pyramidal cells, including fragile X syndrome and Down's syndrome.

## 26.6.4 Epigenetics

Epigenetic mechanisms (see Chap. 6) are important for long-lasting information storage. Classically the term epigenetic was used for the heritable and self-perpetuating modifications to DNA and DNA-associated proteins, like histones, without changes in the DNA sequence itself. These epigenetic modifications include acetylation, methylation, or phoshorylation. The epigenetic footprint on the DNA or histones can lead to a specific pattern of gene expression and can be considered as a persistent form of a cellular memory, which can even be transmitted from mothers to their offspring, e.g., if the mother is under stress during pregnancy this might lead to a different methylation of the DNA bases or acetylation of histones in ovaries and can still be traced in the offspring. The most compelling example of epigenetic changes in the developing cell is the process of cellular differentiation. Totipotent stem cells become the various pluripotent cell lines in the embryo which in turn become fully differentiated cells. In other words, a single fertilized egg cell changes into the many cell types of the animal, including neurons, muscle cells, epithelium, blood vessels etc. as it continues to divide. It does so by activating some genes while inhibiting others using epigenetic gene regulation. The adult nervous system has adopted this developmental program to use it for long-term memory storage (Fig. 26.21). This means that the long-term expression of specific genes can be up- or downregulated epigenetically.

Here is an example for the epigenetic control of the *BDNF* gene, which is a mediator of activity-dependent synaptic plasticity. The expression of the *BDNF* gene is induced in the hippocampus after contextual and spatial learning. The *BDNF* gene consists of nine 5′ noncoding exons each linked to individual promoter regions, and a 3′ coding exon (IX) which codes for the pre-protein amino acid sequence. There are activity-dependent regions within the promoter of the *BDNF* gene and these dictate the spatial and temporal expression of specific *BDNF* transcript isoforms.

Infusion of zebularine (an inhibitor of the DNA methyltransferase NDMT) changes the *BDNF* DNA methylation in the hippocampus leading to different levels of exon-specific *BDNF* mRNAs and indeed also to a different ability to learn. This shows that altered DNA methylation is sufficient to drive differential *BDNF* transcript regulation. That this is linked to activity-dependent synaptic plasticity was shown by the application of an NMDA receptor blocker which prevented activity-associated alterations in *BDNF* DNA methylation.

DNMTs (DNA methyltransferases) detect so-called CpG island sites within DNA – and there are plenty in the *BDNF* gene. "CpG" means "C–phosphate–G", that is, cytosine and guanine linked by a phosphate. The "CpG" abbreviation is used in order to prevent confusion with base-pairing of cytosine and guanine. The methylation of these CpG sites can direct the transcription of genes by altering local chromatin structure. CpG islands are regions of DNA that can be found in 40 % of mammalian gene promoters.

It is currently believed that DNA methylation represents an epigenetic mechanism for potentially contributing to long-lasting changes in the expression pattern of a series of activity-regulated genes, one of them being the *BDNF* gene. These changes in transcription are likely to be important for long-lasting memory storage.

## 26.7 Summary

Some of the signaling pathways used for information storage contribute to a single stage of memory formation, whereas others impact on multiple stages of memory, from short- to long-lasting (Fig. 26.22). Overall memory is not monolithic – there are many different types of learning and memory, most of which are subserved by different signaling systems in different anatomical areas of the animals' brain. Distinct cellular pathways may also act in series or in parallel

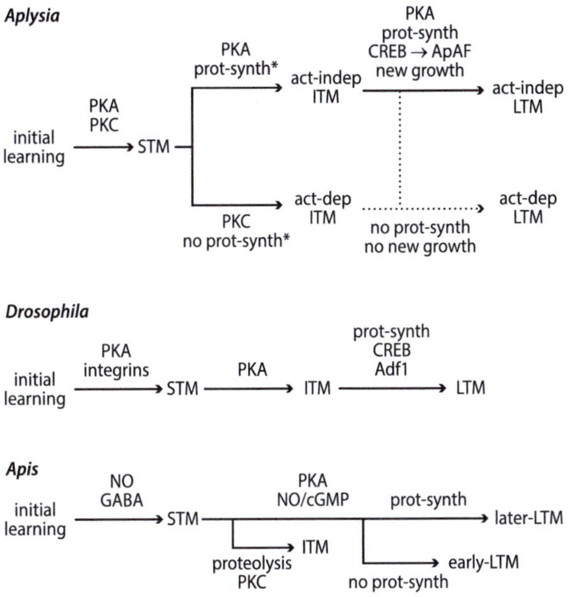


**Fig. 26.22** **Learning in simple systems**. Stages of memory formation in different invertebrate species. *Asterisks* (*) indicate protein synthesis that is RNA synthesis-independent. In *Aplysia*, short-term memory formation (*STM*) is mediated by PKA- and PKC-dependent processes in parallel. Two forms of intermediate-term memory (*ITM*) may be produced depending on the training conditions: activity-independent (act-indep) ITM that is mediated by PKA and is protein synthesis (prot-synth)-dependent; or activity-dependent (act-dep) ITM that is mediated by PKC and is not protein synthesis-dependent. Long-term memory (*LTM*) also appears to have an activity-independent form that is mediated by PKA-induced activation of CREB and ApAF transcription activators and depends on the growth of new synapses, and an activity-dependent form that does not involve protein synthesis or new synaptic growth. It is not known whether activity-dependent ITM necessarily leads to the formation of activity-dependent LTM or whether the activity-dependent and -independent forms of ITM/LTM are mutually exclusive or if they interact (*broken lines*). In *Drosophila*, initial learning requires PKC (not shown) and STM involves PKA- and integrin-mediated processes. PKA activity is critical for ITM formation. LTM requires activation of two separate transcription activators, CREB and Adf1 (the latter encoded by the *nalyot* gene). In honeybees (*Apis mellifera*), STM formation requires NO- and GABA-mediated processes. The precise cellular mechanism for ITM formation is not known, but it is known to involve proteolysis, and blocking ITM development does not affect the subsequent formation of LTM. There are two forms of LTM in the honeybee, an early form that is protein synthesis-independent, and a late form that is protein synthesis-dependent (From Burrell and Sahley [4] with permission, see also [15] for mammals)

during various stages of memory formation. It is also surprising to see that on the molecular level the same signaling pathways are used in different animals in order to learn and memorize. Taken together, there is a lot to learn from simple model systems in order to understand the cellular basis of human learning and memory processes.

## References


1. Abraham WC (2008) Metaplasticity: tuning synapses and networks for plasticity. Nat Rev Neurosci 9:387
2. Bear M, Connors BW, Paradiso MA (2006) Neuroscience: exploring the brain, 3rd edn. Lippincott Williams & Wilkins, New York
3. Bliss TV, Collingridge GL (1993) A synaptic model of memory: long-term potentiation in the hippocampus. Nature 361:31–39
4. Burrell BD, Sahley CL (2001) Learning in simple systems. Curr Opin Neurobiol 11:757–764
5. Cedar H, Kandel ER, Schwartz JH (1972) Cyclic adenosine monophosphate in the nervous system of *Aplysia californica*. I. Increased synthesis in response to synaptic stimulation. J Gen Physiol 60:558–569
6. Dudai Y (2004) Memory from A to Z. Oxford University Press, Oxford/New York
7. Eichenbaum H (2002) The cognitive neuroscience of memory. Oxford University Press, New York
8. Giurfa M (2003) Cognitive neuroethology: dissecting non-elemental learning in a honeybee brain. Curr Opin Neurobiol 13:726–735
9. Hawkins RD, Abrams TW, Carew TJ, Kandel ER (1983) A cellular mechanism of classical conditioning in *Aplysia*: activity-dependent amplification of presynaptic facilitation. Science 219:400–405
10. Hebb DO (1949) The organization of behavior. A neuropsychological theory. Wiley, New York
11. Heisenberg M (2003) Mushroom body memoir: from maps to models. Nat Rev Neurosci 4:266–275
12. Ito M (2002) The molecular organization of cerebellar long-term depression. Nat Rev Neurosci 3:896–902
13. Kandel ER (2001) The molecular biology of memory storage: a dialogue between genes and synapses. Science 294:1030–1038
14. Korte M (2008) Neuroscience. A protoplasmic kiss to remember. Science 319:1627–1628
15. LeDoux JE (2001) Synaptic self: how our brains become who we are. Viking, New York
16. Levenson JM, Sweatt JD (2005) Epigenetic mechanisms in memory formation. Nat Rev Neurosci 6:108–118
17. Malenka RC, Bear MF (2004) LTP and LTD: an embarrassment of riches. Neuron 44:5–21
18. Menzel R (1999) Memory dynamics in the honeybee. J Comp Physiol A 185:323–340
19. Menzel R, Giurfa M (2001) Cognitive architecture of a mini-brain: the honeybee. Trends Cogn Sci 5:62–71
20. Menzel R, Leboulle G, Eisenhardt D (2006) Small brains, bright minds. Cell 124:237–239
21. Morris RGM, Anderson EG, Lynch GS, Baudry M (1986) Selective impairment of learning and blockade of long-term potentiation by an *N*-methyl-D-aspartate receptor antagonist, AP5. Nature 319:774–776
22. Park H, Poo MM (2012) Neurotrophin regulation of neural circuit development and function. Nat Rev Neurosci 14:7–23

# Circadian Timing

# 27

François Rouyer

Living organisms are submitted to day-night cycles that are imposed by the Earth's rotation within a 24-h period. Biological clocks running with a period of about 24 h are named circadian clocks (from Latin: *circa* (about); *diem* (day)). A circadian clock is present in most living organisms, including prokaryotes, and temporally controls a wide range of physiological and behavioral functions. Circadian clocks have three important properties: in constant environmental conditions they usually tick with a period of about 24 h (free running), they keep a near-constant period over the range of physiological temperatures (temperature compensation), and they synchronize to environmental cycles (entrainment). A clock provides the organism with the capacity to anticipate environmental daily changes, such as those affecting light, temperature, or humidity. It also provides a temporal reference to compensate for the changing position of the Sun used for orientation and navigation. The internal timing system also allows the temporal organization of physiological functions.

In metazoans, clocks are present in most tissues and are able to tick autonomously in individual cells. The clock that controls sleep-wake and body-temperature rhythms in mammals is located in the suprachiasmatic nuclei (SCN) of the hypothalamus. The SCN clock synchronizes to day-night cycles by receiving photic inputs from the retina. Circadian photoreception uses rods and cones photoreceptors as well as a subset of retinal ganglion cells that act as photoreceptors thanks to the presence of the melanopsin photopigment. Cellular clocks outside the SCN do not sense light directly but are entrained by day-night cycles through the SCN, which acts as a master clock of the body. Body temperature and humoral signals act as synchronizing cues between the SCN and peripheral clocks. Many insects use a more decentralized clock organization, with peripheral clocks directly sensing light through the blue light-sensitive molecule cryptochrome that is expressed in all clock cells. In addition to cryptochrome, the brain clock neurons use retinal inputs to synchronize rest-activity rhythms to day-night cycles.

Cellular circadian clocks rely on molecular oscillations of clock proteins that result from transcriptional feedback loops. In both insects and mammals, transcription factors activate the expression of transcriptional repressors, which accumulate with a delay to finally antagonize the transcriptional activators. The delay is mostly due to posttranslational modifications that tightly control the stability of the repressors, to tightly control the 24-h period of the molecular feedback loop. This 24-h transcriptional cycle controls a large body of clock-regulated genes, which represent about 5 % of the genome and largely differ among cell types. Genetic polymorphism in the circadian gene network defines molecular chronotypes that shape our individual physiology and behavior. The control of sleep-wake cycles results from the interaction between the circadian clock and a homeostatic process, the sleep pressure, which builds up during wake. Sleep will thus occur when the circadianly driven arousal is low and sleep pressure is high. The cellular and molecular mechanisms underlying sleep remain poorly understood. Recent studies have uncovered the

F. Rouyer
Institut de Neurobiologie Alfred Fessard,
CNRS Gif-sur-Yvette, Paris, France
e-mail: francois.rouyer@inaf.cnrs-gif.fr

C.G. Galizia, P.-M. Lledo (eds.), *Neurosciences - From Molecule to Behavior: A University Textbook*,
DOI 10.1007/978-3-642-10769-6_27, © Springer-Verlag Berlin Heidelberg 2013

existence of sleep-related states in insects and begin to provide molecular insights into sleep mechanisms.

Circadian and sleep-wake cycles are the best-studied examples for neurally controlled rhythms, but others also exist: lunar, seasonal, circannual, and endogenous cycles to name but a few, controlling humoral, behavioral, cognitive, and developmental aspects of neural organization. For the sake of simplicity we focus on circadian rhythms and sleep in this chapter.

## 27.1 Circadian Clocks

The first experimental evidence for a circadian clock came from the French astronomer Jean-Jacques d'Ortous de Mairans who reported his observation to the "Académie Royale des Sciences" in 1729 [1]. Mimosa plants open their leaves during daytime in day-night cycles and still show a 24-h rhythm of leaf movement when placed in complete darkness conditions. They thus need to have some time-keeping mechanism to control the day-night cycle in the absence of light cues. During the following 200 years, similar experiments were carried out with different rhythms in various plant and animal species. In particular, the observation of circadian periods slightly deviating from the usual 24 h supported the existence of an internal pacemaker rather than a response to a physical factor linked to the rotation of the Earth. However, the possible contribution of an uncharacterized external factor fueled the debate up to the 1960s. The extensive characterization of circadian rhythms by Erwin Bünning (1906–1990) in plants, by Colin Pittendrigh (1918–1996) in *Drosophila*, and by Jürgen Aschoff (1913–1998) in rodents and humans set the ground for a biological clock being a key component of the physiological organization of organisms living on our 24-h-rotating planet.

### 27.1.1 Properties of Circadian Clocks

The **circadian period** illustrates the first property of circadian clocks: they persist running (they free-run) in constant conditions with a period of about 24 h. Experimental conditions are constant temperature and lighting, the later being either constant darkness (also named DD for Dark:Dark) or constant light (named LL for Light:Light), as opposed to Light-Dark (LD) or temperature cycling (Fig. 27.1).

A second property is **temperature compensation**. The free-running period of circadian clocks is largely independent of the external temperature, in the range of physiological temperatures. This holds not only for animals that maintain a constant body temperature (homeotherms) but also for those who do not (poikilotherms), such as reptiles, amphibians, and all invertebrates. A clock with a pace that would change with temperature would act as a thermometer rather than providing time information. However, the rate of biochemical reactions strongly increases with rising temperature ($Q_{10}$ is usually around 2, meaning that reaction rates increase by about two-fold for a 10 °C temperature increase). Circadian clocks thus need to include a compensating mechanism in order to maintain a constant free-running period over a wide range of temperatures. This was nicely demonstrated by C. Pittendrigh in the 1950s using the emergence of the young adults of the *Drosophila pseudoobscura* from their pupal cases, at the end of the metamorphosis process (Fig. 27.2). Between 16 and 26 °C the circadian period of emergence slightly decreased from 24.5 to 24.0 h, indicating a strong although not perfect temperature compensation.

A third property of circadian clocks is their ability to synchronize to the day-night cycles that are imposed by the rotation of the Earth. In contrast to their endogenously defined period in constant conditions, the phase of circadian rhythms is determined by the day-night cycles or the LD cycles in laboratory conditions. In other words, environmental cues entrain the circadian clock. Such cues have thus been named **Zeitgebers** by J. Aschoff. In addition to the light and temperature changes that are associated with day-night cycles, environmental changes such as food availability or social cues also act as Zeitgebers for the entrainment of animal circadian rhythms. Light and to a lesser extent temperature are the most potent Zeitgebers for behavioral rhythms and have been used in the majority of the experiments aimed at understanding circadian entrainment. Since endogenous circadian periods are often not exactly 24 h, clocks would be quickly out of phase with day-night cycles if not entrained. If one takes sleep-wake cycles as an example (see Fig. 27.1), a 25-h clock would

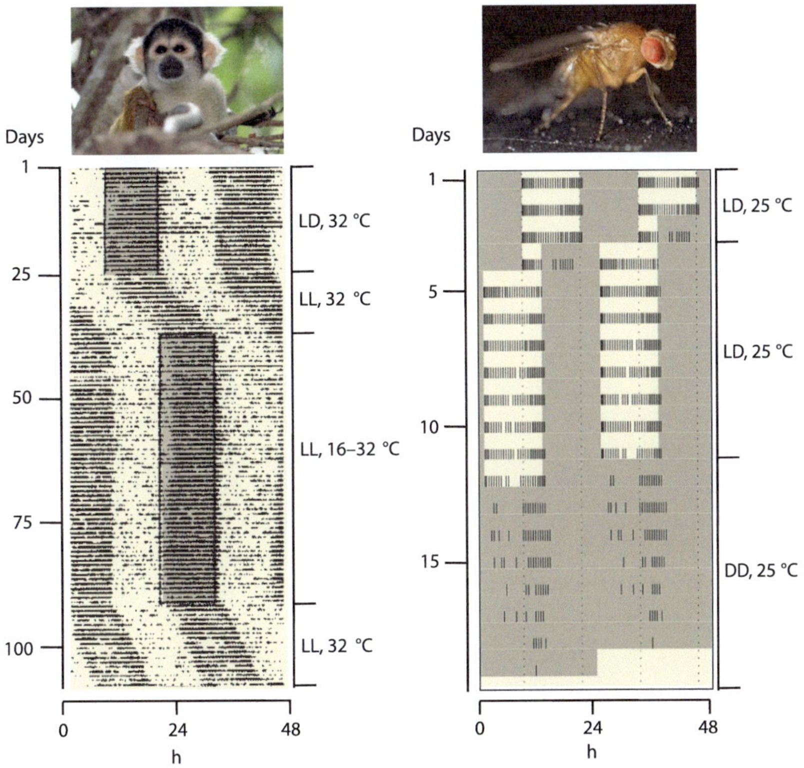


**Fig. 27.1 Sleep-wake cycles in squirrel monkeys and fruit flies.** Sleep distribution is shown by activity graphs or actograms. *Vertical black bars* of the actograms are proportional to animals' activity, which is recorded by an infrared photocell. The activity of each day is plotted twice (i.e., day 2 activity is shown on the *right part* (24–48 h) of row 1 and on the *left part* (0–24 h) of row 2, etc.). Squirrel monkey *(Saimiri sciureus)*: activity in light-dark cycles (lights-ON is *boxed*), constant light (*LL*), or constant light with temperature cycles (high temperature is *boxed*). Drosophila *(Drosophila melanogaster)*: activity in light-dark cycles (lights-ON in *white*) and constant darkness. The temporal distribution of activity shows that sleep-wake cycles can be synchronized by light or temperature. Their persistence in constant conditions indicate that they are driven by an endogenous clock (Squirrel monkey data from Aschoff et al. [31] with permission)

results in diurnal animals falling asleep 1 h later everyday, thus going from nighttime sleep to daytime sleep (12-h shift) every 12 days. The entrainment of the clock allows the animals to keep a stable phase angle between the sleep-wake rhythm and the environmental day-night cycle. To adapt the sleep-wake rhythms to the 24-h period of the day-night cycles, light needs to delay fast (period <24 h) clocks and to advance slow (period >24 h) clocks. This is the consequence of circadian clocks responding differently when hit by light at different phases of their cycle. This property is best described by phase-response curves obtained from experiments where animals are exposed to a short light pulse a particular time of the day (hence of the circadian cycle) in constant darkness (Fig. 27.3). Light pulses given in the middle of the subjective day elicit no responses, pulses given in the early night induce phase delays, and late-night pulses induce phase advances. A short-period clock exposed to a 24-h LD cycle will enter its nighttime phase when light is still ON outside and will thus be delayed to better fit the LD cycle. In contrast, a long-period clock will see outside light before ending its nighttime phase and will be advanced to resynchronize to the LD cycle. These phase-response curves are similar in all species, as expected for such an important adaptive process [2]. The entrainment capacities show limits, which can be defined by recording the sleep-wake rhythms in light-dark cycles of various lengths. When the difference between the free-running period of the animal and the length of the day-night cycles exceed a few hours, the clock will start free-running. This can be used to estimate the free-running period of animal or human subjects, using a so-called forced

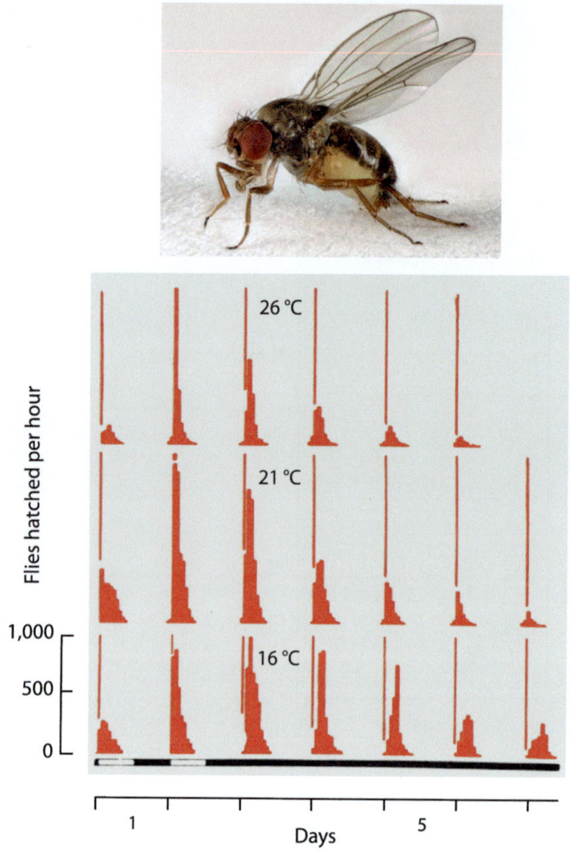


**Fig. 27.2 Eclosion rhythms of fruit flies at different temperatures.** *Drosophila pseudoobscura* cultures are kept in LD cycles and then released in constant darkness. The number of adult individuals emerging from pupal cases is plotted as a function of time. The very small change of period through a wide range of temperatures shows that the circadian clock is temperature-compensated (Adapted from Pittendrigh [32] with permission)

desynchrony protocol with, for example, a 6-h:6-h LD cycle.

Besides laboratory experiments, circadian clocks do not experience phase shifts of the LD cycle or light pulses in natural conditions. However, they are submitted to the seasonal variation of day length, which advances or delays morning and evening light-dark transitions. Within the 24-h cycle, day length is 12 h all year round at the Equator but increasingly varies from the Equator to the Poles between the June solstice (North Pole maximally tilted towards the Sun) and the December solstice (South Pole maximally tilted towards the Sun). For example, at 50 °C North of latitude (Brussels), day length varies from about 16 h (June 21) to about 8 h (December 21). Animals active during the early day or early night would miss dawn or dusk during summer if keeping the winter phase, but the daily resetting of the clock by light allows them to track dawn or dusk from solstice to solstice.

### 27.1.2 What Is a Circadian Clock for?

The presence of a circadian clock in most organisms indicates its strong adaptive value for living on Earth. Living organisms need to adapt their physiology and behavior to these daily environmental changes, in particular light and temperature variations. Having an internal clock allows to anticipate and thus prepare the organism for these changes, rather than simply responding to them. For example, young adult *Drosophila* have a very thin and permeable cuticle (protective layer all over the insect body) when emerging from their pupal case at the end of metamorphosis and are thus very susceptible to desiccation. The circadian clock sets the time of emergence in the morning when humidity is higher and prevents them from being exposed to the hot Sun of the afternoon, which would have deleterious effects especially on wing expansion [3]. Since the emergence process takes a few hours, it has to be started during the night, and the circadian clock allows the fly to anticipate sunrise. It is also important to be in time for available food sources, as nicely illustrated by the synchrony between flowers and their pollinators. Some flowers open each day during a few hours only, with different species opening at different times, as illustrated by the famous flower clock of Linné (1707–1778). This time specialization increases the chance to get pollen from the same species and is controlled by the circadian clock. On the other side, bees can learn where to go at a given time to find an open flower and collect nectar. Famous experiments initiated by Karl von Frisch (1886–1982) have shown that they can do so in the absence of sky light and thus use their internal clock to associate time with the learned location of the food source. The clock thus allows to predict remote events that the bee cannot see or smell (see Chap. 28). Time and space are even more related if one considers navigation. The apparent course of the Sun in the sky throughout the day requires migratory birds and insects to time-compensate their orientation.

Indeed, G. Kramer (1910–1959) studied migratory birds in the laboratory and showed that their attempts to fly during migration periods were orientated in a

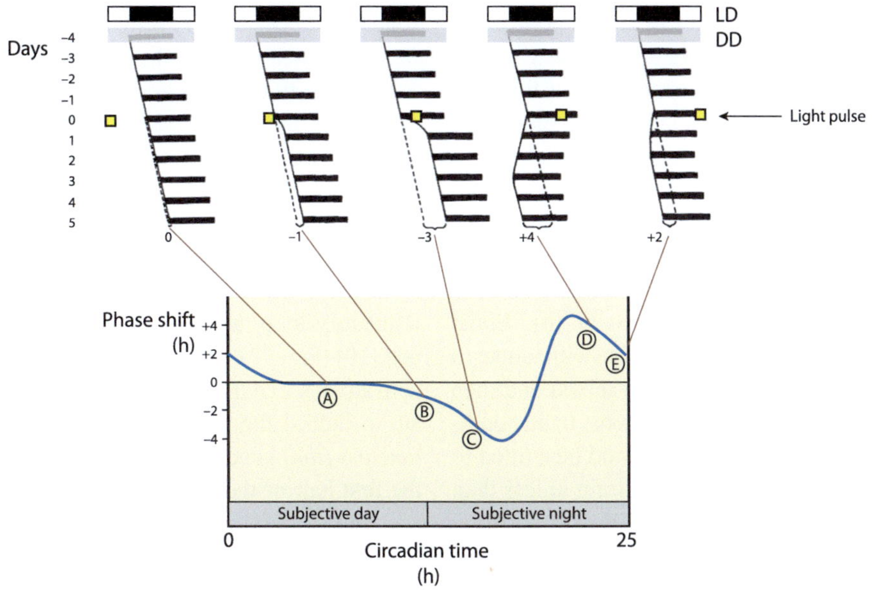


**Fig. 27.3 Photic entrainment of sleep-wake rhythms: the phase-response curve.** Five experiments (**a–d**) are done with individual rats, whose night locomotor activity is reported as a *black bar*. Each animal is entrained in 12-h–12-h light-dark cycles (LD, *white*: light, *black*: dark) and released in constant darkness (DD, *gray*). From day −4 to day −1, the animals display free running rhythms in DD with a circadian period of 25 h. On day 0, rats are submitted to a short (≤1 h) light pulse (*yellow square*), which is applied at a different time of the 25-h circadian cycle for each individual. The animals are then left in DD from day 1 to day 5. Depending of the time at which the pulse was given, the rat will show either no change or a phase shift of its activity rhythm after the pulse. Pulses applied during the subjective day (**a**) produce no response, pulses applied at the beginning of the subjective night (**b**, **c**) produce phase delays, and pulses applied at the end of the subjective night (**d**, **e**) produce phase advances. These phase shifts are plotted against circadian time to derive the phase response curve that is shown at the bottom (Modified from Moore-Ede et al. [1] with permission)

manner that depends on the time of the day. Imposing a phase-shifting to the clock (see below) altered the orientation with a predictable angle. Similar experiments were conducted with monarch butterflies, which migrate over ~3,000 km from northeast America to central Mexico during the fall. When submitted to light conditions that abolish the function of their brain circadian clock, they lose time-dependent orientation and orient randomly in the flight simulator. A circadian clock thus allows animals to keep a constant direction as the Sun moves in the sky from east to west.

The circadian system may stop timing some clock outputs in the absence of cycling environmental conditions. Reindeers living far above the Arctic Circle show robust sleep-activity rhythms in spring and autumn but they lose behavioral rhythms during either summer days when the Sun does not set or winter days when it does not rise [4]. In some cases, evolution has profoundly changed the properties of the circadian system to adapt it to specific ecological niches. Some organisms such as fish species living in subterranean caves never see daylight.

As opposed to its surface counterparts, the eyeless Somalian cavefish *Phreatychtis andruzzii* does not show activity rhythms in light-dark cycles [5]. This appears to be the consequence of mutations that have accumulated in the genes encoding the photopigments involved in the synchronization of the clock by light. Providing food at the same time everyday, induces activity rhythms that persist when the animals are starved, indicating that a food-entrainable internal clock is running but has lost the ability to synchronize to light-dark cycles. Having a clock not only allows to anticipate environmental cycles, but also provides a temporal reference for the internal synchronization between physiological processes. A well-known example in humans is the opposite phasing of sleep and urine production, which is very low at night. Also critical is the circadian phasing of nitrogen fixation and photosynthesis in some prokaryotic cyanobacteria. Photosynthesis produces oxygen during the day, which inhibits the nitrogenase that reduces atmospheric nitrogen to ammonia. The circadian clock of these photosynthetic bacteria restricts the expression of the gene encoding the nitrogenase to the night, thus temporally separating two incompatible biochemical pathways.

The fitness value of circadian clocks was experimentally tested in a few studies. Spectacular results have been obtained in cyanobacteria, whose cell division time (about 24 h in LD 12:12 cycles) allows competition studies to be performed easily [6]. Equal numbers of wild-type or mutant cells with either a short or a long circadian period were mixed and then grown in LD cycles of different durations. In all conditions, the strain with the circadian period best fitted to the LD cycle outcompeted the other strain in less than 20 generations. A circadian clock thus confers a strong fitness advantage in cycling environmental conditions. In an attempt to estimate the adaptive value of the clock in mammals, the survival in the wild of either squirrels or chipmunks in which the SCN were surgically ablated was compared to sham-operated controls. The mortality by predation of SCN-lesioned animals was higher, suggesting that the circadian system provides an advantage, likely by restricting activity at times where predators were less actively seeking a meal [7].

## 27.2 Molecular Mechanisms of Circadian Oscillators

### 27.2.1 Finding the Clock in *Drosophila*

The first genetic mutations affecting sleep-wake cycles were described in fruit flies, in a pioneer genetic screen led by Konopka and Benzer [8]. *Drosophila* cultures issued from mutagenized individual flies were tested for their eclosion rhythms and mutants either arrhythmic or rhythmic with an altered circadian period were obtained. The mutants were then tested for their rest-activity rhythms. This can be measured by placing individual adult flies in small food-containing glass tubes that are crossed by an infrared beam. When the fly is moving, it intercepts the beam and its activity is defined by the number of times the beam is crossed during a time interval of usually 5–30 min. The period of the circadian clock is determined by measuring rest-activity rhythms in constant darkness, and the clock mutations identified by Konopka and Benzer were found to affect similarly eclosion and activity rhythms, indicating that they are driven by the same molecular clock. The three mutations were found to affect the same gene, named *period*. After the molecular cloning of the *period* gene in 1984, the first molecular studies unraveled the negative feedback loop mechanism as a core engine of the circadian oscillator [9]. The power of *Drosophila* genetics soon led to the discovery of the small set of genes that participate in the negative autoregulatory loop taking place in each individual clock cell [10] (Fig. 27.4). In the evening, two transcriptional activators, CLOCK (CLK) and CYCLE (CYC) associate to induce the expression of the *period (per)* and *timeless (tim)* genes, whose transcripts reach a peak in the first half of the night. The PER and TIM proteins accumulate during the night, and reach a peak in the early morning, a few hours after the mRNA peak. This delay is a key feature of the circadian oscillator. Indeed, PER and TIM proteins interact in the cytoplasm and enter the nucleus in the middle of the night, where they induce the repression of the CLK-CYC transcriptional activity. As a consequence, *per* and *tim* mRNA levels drop to low levels. This defines a transcriptional negative feedback loop. The delayed protein accumulation thus creates a transcriptional oscillation, with active transcription in the early night and repression in the late night.

The delayed accumulation of PER and TIM is a consequence of posttranslational mechanisms, which finely control the stability, the subcellular localization and the activity of the clock proteins through their phosphorylation, ubiquitylation, and degradation by the proteasome. Various genetic screens have identified key enzymes involved in these steps, which include kinases such as Casein kinase 1 [11] and 2, phosphatases, and ubiquitin ligases that determine the speed of the molecular oscillator, hence the period of the sleep-wake cycles. In addition to the *per* and *tim* genes, the transcriptional feedback loop controls two other transcription factors which regulate the expression of the *clk* gene itself, thus building a secondary loop. How do these cycling molecules transmit cyclic information within the cell? Large-scale transcriptional studies have revealed that about 5 % of the *Drosophila* transcripts display circadianly regulated levels in the fly head and all of them stop cycling in the absence of either CLK or PER. It thus appears that, in addition to the clock genes themselves, the transcriptional feedback loop controls hundreds of downstream

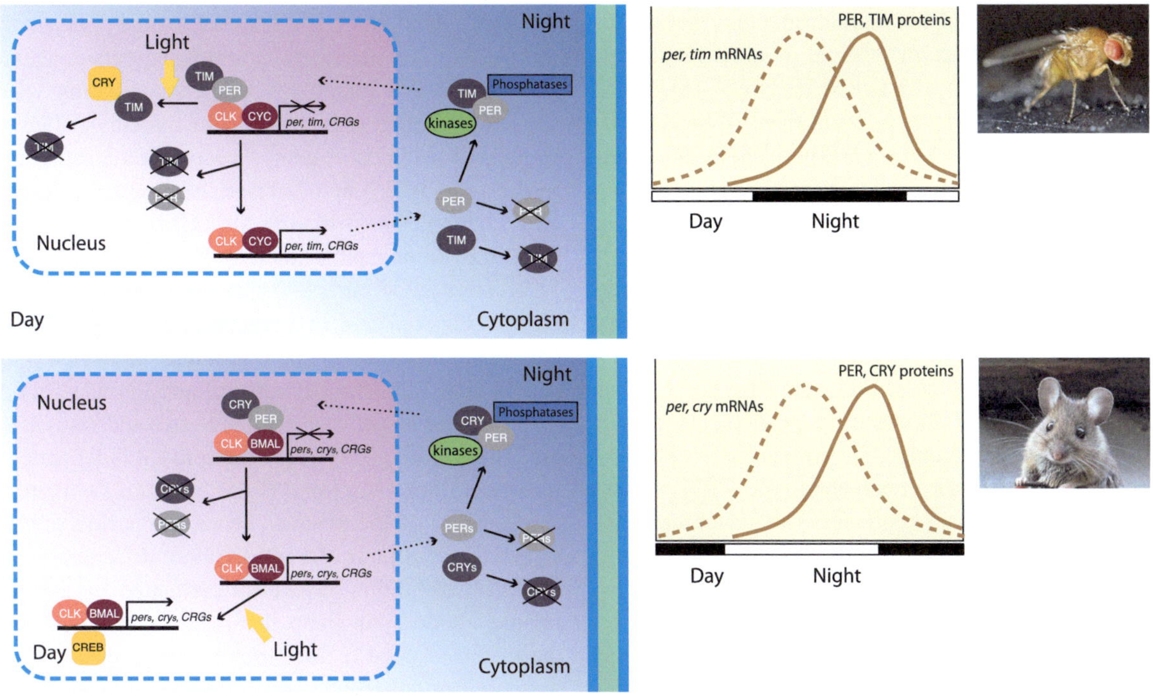


**Fig. 27.4 The molecular feedback loop model of the circadian clock.** In mammals and insects, major clock genes are involved in a negative transcriptional feedback loop. In *Drosophila*, the *per* and *tim* genes are activated by the CLK and CYC proteins at the end of the day. Posttranslational modifications destabilize the PER and TIM proteins and delay their night accumulation. PER and TIM proteins repress the activation of the *per* and *tim* genes in the morning. Degradation of the PER and TIM repressors during the day allows a new cycle to resume. A similar feedback loop occurs in the mouse, where multiple PER and CRY proteins repress the expression of the three *per* genes and the two *cry* genes that is induced by the CLK and BMAL activators. The feedback loop also controls a large number of downstream circadianly-regulated genes (*CRGs*) that are involved in various aspects of circadian physiology. In *Drosophila*, CRY acts as a photoreceptor to induce the light-dependent degradation of the TIM protein. In mammals, light resets the clock by inducing a CREB-dependent increase of *per* genes' transcription. Crosses indicate protein degradation or transcriptional repression

genes that are involved in many aspects of cell physiology. Importantly, blocking electrical activity of clock cells appears to abolish molecular cycling, suggesting that neuronal activity feeds back to the molecular oscillator [12].

### 27.2.2 Molecules in the Mammalian Clock

The first identified mammalian clock genes were isolated in 1997, on the basis of their similarity with the *Drosophila* genes *(per)* or through a sleep-wake cycle behavioral screen using a running wheel as an activity monitor [13]. The architecture of the molecular clockwork is very similar in flies and mammals and the two clocks share several components (Fig. 27.4). The two transcription factors, CLK (or a similar protein named NPAS2) and BMAL (the *bmal1* gene is the mammalian ortholog of *cyc*) activate the expression of the genes encoding the repressors of their activity. There are three *per* genes in mammals, and two of them act as repressors in the transcriptional feedback loop, whereas the role of the third one remains elusive. In contrast to *Drosophila*, the mammalian PER partner does not seem to be TIM but CRYPTOCHROME (CRY). CRY was first characterized as a blue light-sensitive photoreceptive molecule in plants and plays a role in *Drosophila* circadian photoreception (see below). There are two *cry* genes in mammals, and surprisingly, mice devoid of the two *cry* genes are behaviorally arrhythmic but are not impaired in any type of photoreception tested so far, indicating that CRY acts as a bona-fide clock protein and not as a photoreceptor. Similarly to *Drosophila* PER/TIM complexes, PER/CRY complexes enter the nucleus and repress CLK/BMAL-dependent transcription.

The repressing complex recruits enzymes that modify the chromatin-structural histone proteins in the promoter region of the CLK-BMAL target genes, thus changing their ability to be transcribed. Why is CRY a photoreceptor in the *Drosophila* brain clock and a transcriptional repressor in the mammalian clock? The availability of the genome sequence in several insect species has shed light on this evolutionary mystery. In some insects such as butterflies, two *cry* genes were found, with one more similar to the *Drosophila cry*, whereas the other one is closer to the mammalian variant. Other insects such as bees only have the mammalian-type *cry* gene. Biochemical experiments indicate that the *Drosophila* type CRY proteins act as photoreceptors, whereas the mammalian-type CRY proteins act as transcriptional repressors. Sequence comparisons suggest an evolutionary scenario where gene duplications followed by divergence generated two *cry* gene families, with some lineages keeping either the photoreceptor-encoding gene or the transcriptional repressor-encoding gene, and other lineages keeping both. Intriguingly, the absence of a *tim* gene ortholog in bees suggests that their molecular clock machinery might be closer to the mammalian one than to *Drosophila* [14]. Insects devoid of photoreceptive cryptochrome are likely to use their compound eyes as the only light input or yet unknown photoreceptive pathways.

In contrast to flies, the mammalian secondary feedback loop mostly controls *bmal1* expression and relies on a different pair of transcription factors, which belong to the nuclear receptor family. The pace of the two feedback loops depends on posttranslational modifications of the transcription activators and repressors by kinases, phosphatases, and ubiquitin ligases, with several of them being identical in flies and mammals. Such modifications are also sensitive to the cell metabolism and a growing number of studies show that the cell metabolism strongly influences the feedback loop mechanism, thus fine tuning the speed of the clock.

The alteration of the core clock genes either abolishes sleep-wake cycles or alters their period. In LD conditions, where the clock is reset every cycle, a change in the period of the circadian clock will have consequences similar to a phase defect, namely sleep onset and/or offset will be advanced (short period) or delayed (long period). The familial advanced sleep phase syndrome (FASPS) is an autosomal dominant inherited trait that induces an early (up to 4 h advance) sleep onset and offset. Linkage analysis in a large family with several affected patients revealed that the FASPS phenotype was cosegregating with the chromosomal region where the human *per2* gene is located. Molecular analysis identified a mutation in the gene, which induced a serine to glycine amino acid change, thus affecting the ability of the PER2 protein to be phosphorylated by the CK1delta kinase [15]. Transgenic mice carrying the human mutation showed short period rhythms, strongly supporting a causal link between the *per2* mutation and FASPS in this family. In another family, a mutation was found in the *ck1delta* gene of a FASPS-affected individuals. These studies show that multiple CK1-dependent phosphorylation sites on PER2, differently influence the stability of the protein, hence the period of the circadian oscillator. Interindividual differences in human populations have been estimated by asking people about their sleep habits. In a survey more than 55,000 people filled out a questionnaire that was analyzed to calculate individual chronotypes (half-way time between sleep onset and end of sleep) (Fig. 27.5). A wide distribution of clock-dependent chronotypes was observed, with influences of age and sex. Importantly, the interaction between chronotype and social habits strongly influences sleep duration. Early chronotypes sleep less on free days since they go to bed later, thus missing the first part of their night sleep, whereas late chronotypes sleep less on work days since they have to wake up early and miss the end of their night sleep [16]. The survey also revealed that chronotypes follow solar time, especially those leaving outside of large cities: inhabitants on the eastern border of Germany wake up half an hour before those on the western border, even though official time and working hours are identical [17].

In addition to the core clock genes, hundreds of mammalian genes show circadian oscillations of their expression in mammals, and do so in a tissue-specific fashion. Many of these genes are involved in different metabolic pathways, thus generating oscillations of various metabolites. The physiological consequences of these oscillations are very important. For example, changes in amino acid levels affect the synthesis of neurotransmitters and are thus likely to modulate brain activity. A study has shown that the circadian clock controls the expression of the gene encoding monoamine oxidase A (MAO-A), which is involved in dopamine catabolism [18]. *per2* mutant mice not only

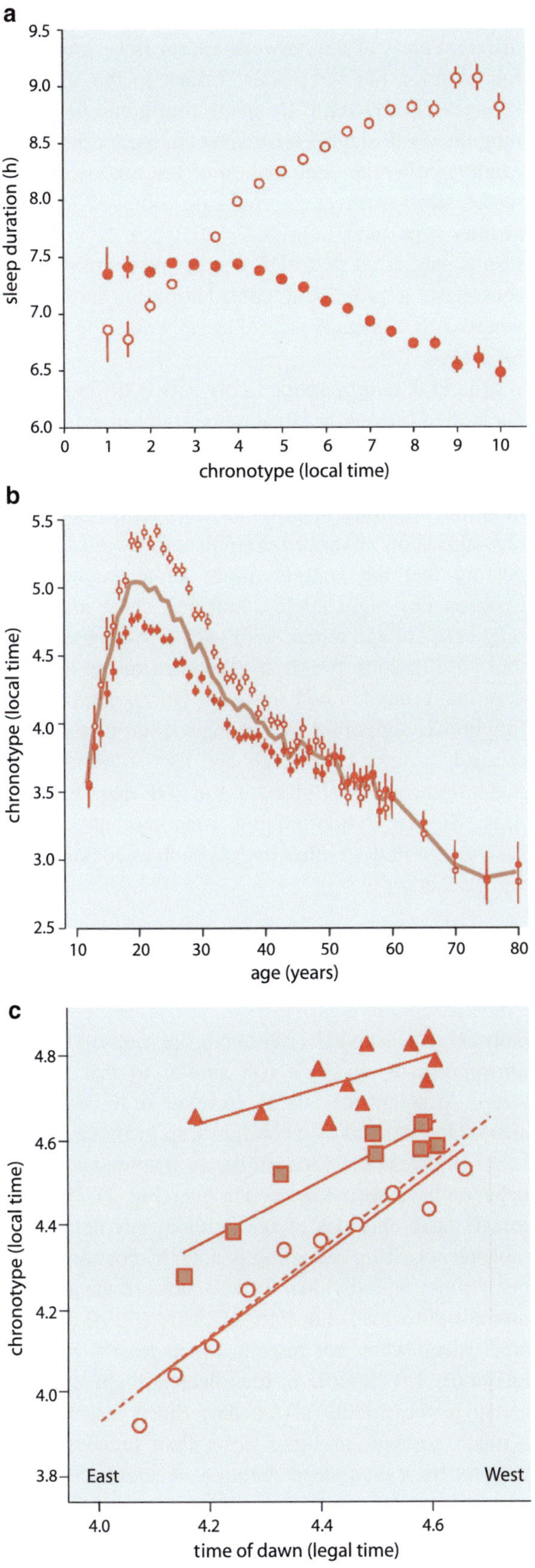


are altered for sleep-wake rhythms but show high levels of dopamine and altered neuronal activity in the striatum. These animals behave abnormally in despair-based behavioral tests, which model human mood disorders. This suggests that genetic or environmental conditions (such as shift work) that perturb the circadian system may have significant consequences on mood regulation.

## 27.3 Neural Organization of the Fly Clock

### 27.3.1 A Network of Clock Neurons Controls Activity Rhythms in *Drosophila*

In order to locate the clock that controls *Drosophila* sleep-wake cycles, the brains of flies carrying a $per^S$ allele (displaying 20-h behavioral rhythms) were introduced in the abdomen of $per^0$ null mutants (no PER function). Flies with a $per^S$ mutation have strong short period (20-h) sleep wake cycles, whereas $per^0$

**Fig. 27.5 Inter-individual variability of circadian clocks: human chronotypes.** (**a**) Chronotype affects sleep duration in real life. Chronotype is defined by mid-sleep in free days (MSF), which is the half-way between sleep-onset and end of sleep. Sleep duration is reported for work days (*filled circles*) and free days (*open circles*). Early chronotypes sleep less on free days since they go to bed later. In contrast, late chronotypes sleep less on workdays since they wake-up earlier. People who sleep between 11:30 p.m. and 6:30 a.m. (MSF at 3:00 a.m.) show no difference in sleep duration between work days and free days. (**b**) Chronotype depends on age and sex. Children are early chronotypes and progressively delay their body clock with age. The maximum delay is reached in the early 20s, then sleep times become earlier with age. Women (*filled circles*) reach their later chronotype around 19.5 years of age, while men (*open circles*) delay until 21. Men are thus, on average, later chronotypes than women but the sex difference disappears around 50. (**c**) Chronotype depends on solar time rather than on social time. Average chronotypes of the German population follow dawn when going from East to West. Dawn time changes by 36 min between the eastern and western borders of the country. Chronotypes of people living in areas with up to 300,000 inhabitants (82 % of the population) are shown as *circles*, while those living in cities with 300,000–500,000 inhabitants and more than 500,000 inhabitants are shown as *squares* and *triangles*, respectively (a, b, From Roenneberg et al. [16]. c, From Foster and Roenneberg [17] with permission)

are behaviorally arrhythmic. The surviving transplanted individuals displayed short-period sleep-wake rhythms indicating that the brain carries the circadian pacemaker for sleep-wake cycles and that this pacemaker does not need synaptic connections to drive circadian rhythms [19]. Antibodies directed against the clock proteins identified about 150 brain neurons that show circadian oscillations of the clock proteins. These neurons are distributed in half a dozen clusters located in the lateral and dorsal regions of the brain (Fig. 27.6). When submitted to light-dark cycles in the laboratory, *Drosophila* flies display morning and evening activity peaks anticipating lights-on and lights-off. This bimodal pattern is often observed and defines a so-called **crepuscular behavior**. Activity records of rodents or insects in short (winter) or long (summer) days show that the morning and evening activity bouts partially follow dawn and dusk, suggesting that they reflect the behavior of somehow independent morning and evening circadian oscillators. The anatomically distinct groups of clock neurons in the *Drosophila* brain provide a cellular substrate for these oscillators. About 16 lateral neurons express a neuropeptide, the pigment-dispersing factor (PDF), which was previously identified in crustaceans where it controls the light-induced changes of screening pigment distribution in the cuticular chromatophores and in the retina (see Chap. 11). With the sophisticated *Drosophila* genetic tools, transgenic animals can be produced where a specific subset of clock cells is killed or synaptically silenced or is the only one running a functional clock in an otherwise clockless genetic background. Such cellular manipulations have shown that flies with a molecular clock restricted to the PDF cells display only morning behavior in light-dark cycles, whereas flies with a clock restricted to the PDF negative subset of lateral neurons display only evening behavior [20, 21]. Manipulations of some of the dorsal groups show that light and temperature conditions define whether or not the dorsal neurons will contribute to these activities. Light also controls the contribution of the lateral neurons with the morning cells driving sustained sleep-wake rhythms in constant darkness whereas the evening cells can do so in constant light. The adaptation of the activity pattern to various environmental conditions thus appears to result from light and temperature modulating the contribution of different subsets of the circadian neuronal network to the behavioral output. The opposite effects of light on different parts of the network appear to be important for adapting the sleep-wake cycles to the seasonal changes of daylength. By using transgenic flies running the clock at a different speed in the morning and evening cells, the contribution of the two oscillators to the sleep-wake cycle could be followed in short (winter type) and long (summer type) days. These results suggested that PDF-expressing morning neurons have a prominent contribution in short days whereas PDF-negative evening cells take the lead in long days.

The PDF neuropeptide likely acts at different levels in the *Drosophila* circadian system. In agreement with the expression of peptide in the morning cells, *pdf*$^{0}$ mutants lose robust rhythms in constant darkness and morning activity in LD cycles. However, they also show advanced evening activity in LD, suggesting that the peptide might be a synchronizer between day-night cycles, morning cells, and evening cells. In agreement with such a role, clock protein oscillations tend to desynchronize among neuronal groups in *pdf*$^{0}$ mutants. This synchronizing function is supported by the expression of the PDF receptor in most clock cells and its requirement for maintaining a proper phase in the PDF-negative neurons. Similar synchronizing functions have been assigned to PDF in other insects such as cockroaches and crickets.

### 27.3.2 Sleep-Wake Cycles in *Drosophila*

Robust rest-activity rhythms are observed in *Drosophila*. Is the fly's rest similar to mammalian sleep? Although there is no correlate to the different sleep states defined by electrophysiology in mammals, behavioral features very similar to mammalian sleep behavior have been described in flies (Fig. 27.7). First, insects have episodes of rest without any detectable movements. Sleep recording is done by counting how many times an individual fly will move in the activity monitor during a 1-min interval. A fly will be considered asleep when not moving for at least 5 min. In laboratory LD conditions, flies sleep at night and take a nap in the middle of the day. Since males show a much stronger mid-day siesta than females, they accumulate a total sleep duration of about 15 h/day, whereas females only sleep for about 10 h/day. Second,

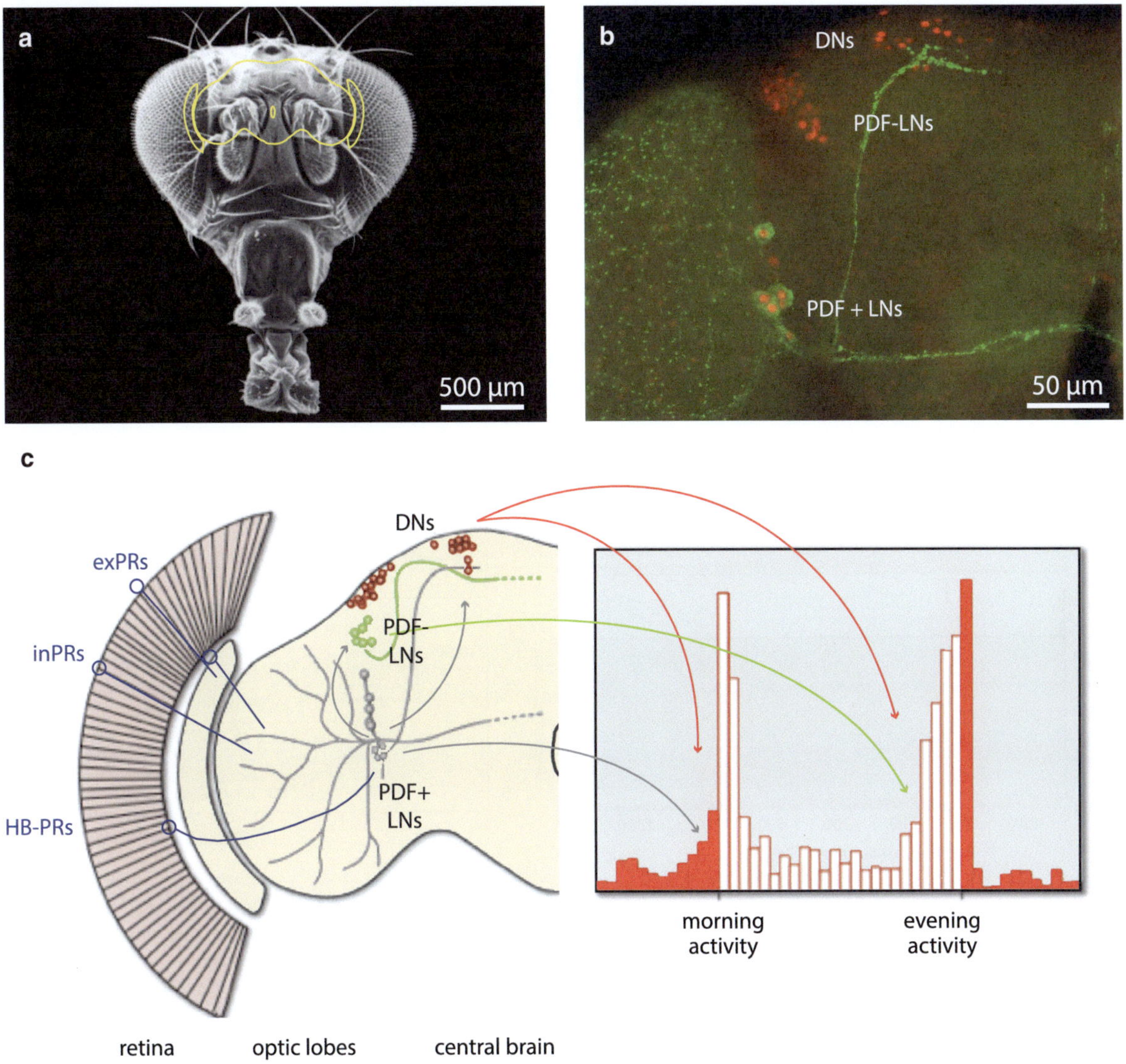


**Fig. 27.6 Several groups of clock neurons control sleep-wake rhythms in *Drosophila*.** (**a**) The *Drosophila* brain contains about 150 neurons that show clock protein cycling. (**b**) A brain hemisphere collected in the morning is labeled for the PER protein (*red*, all clock neurons including lateral (LNs) and dorsal neurons (DNs)) and the PDF neuropeptide (*green*, subset of lateral neurons). (**c**) The behavioral function of several subsets is partly identified. LNs that express PDF (*gray*) are key players in the generation of morning activity whereas PDF-negative lateral neurons (*green*) play a major role in the control of evening activity. Morning cells also provide time information to evening cells. Additional neurons located in the dorsal brain (dorsal neurons, *red*) regulate morning and evening activities, as a function of light and temperature. The brain clock neurons receive light inputs from the visual system through different types of photoreceptor cells (*blue*): internal (*inPRs*) and external (*exPRs*) photoreceptors in the retina, as well as extra-retinal Hofbauer-Buchner photoreceptors (*HB-PRs*)

responses to mechanical stimuli are reduced during sleep episodes. Third, flies respond to sleep deprivation by increasing their time of inactivity in the following hours, indicating the homeostatic nature of sleep control in insects as it is in mammals.

The possibility of doing large genetic screens in *Drosophila* has been exploited to search for genes involved in sleep control. One of the first isolated sleep mutation was *minisleep*, which reduces sleep duration to about 4 h in females and 5 h in males and also reduces lifespan. The *minisleep* mutation affects the *shaker* gene, which encodes a voltage-dependent potassium channel involved in synaptic transmission [22]. Further studies have revealed

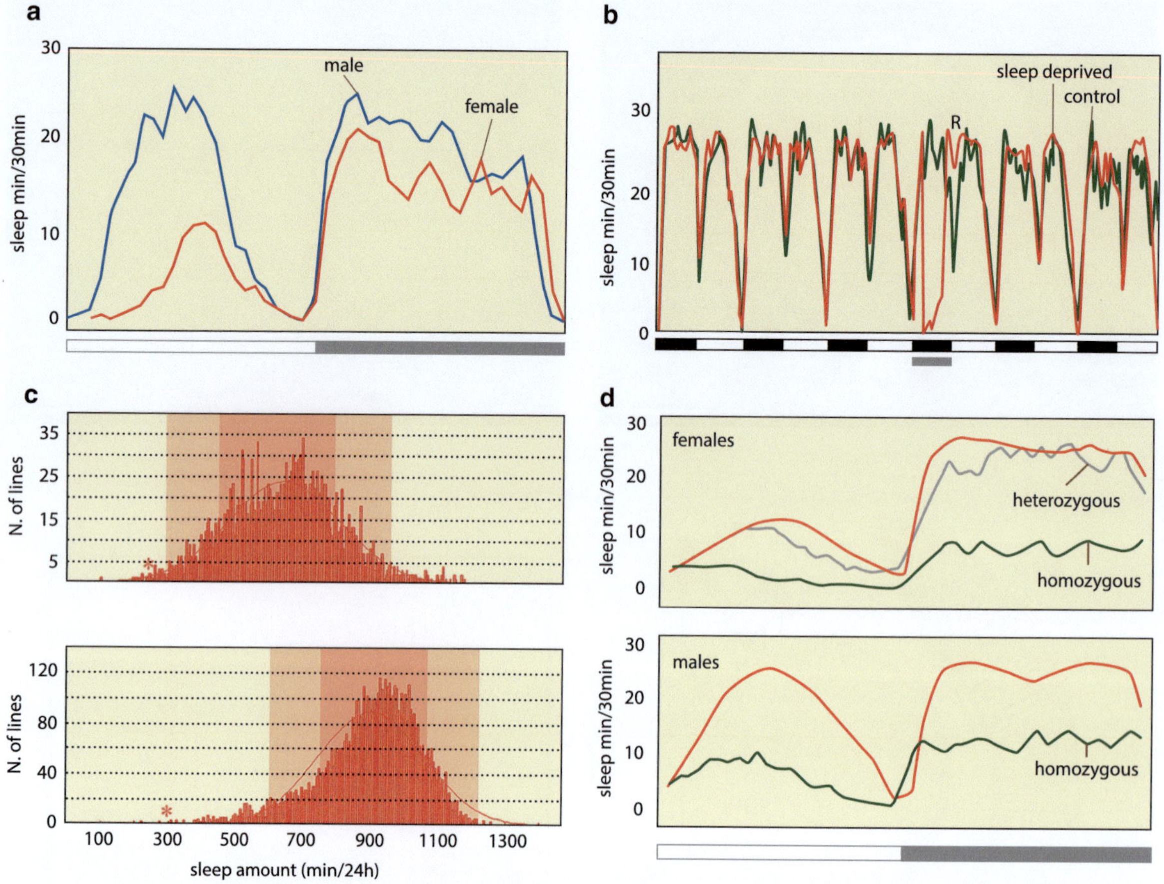


**Fig. 27.7** **Defining sleep in *Drosophila*.** (**a**) Temporal distribution of sleep in wild-type *Drosophila melanogaster*. In 12 h–12 h light-dark laboratory conditions, sleep follows a bimodal distribution with peaks at night and during the middle of the day. Females (*red*) show much less day sleep than males (*blue*). (**b**) When sleep-deprived at night (*gray bar*) fruit flies sleep more (*red line*) than nondeprived controls (*green*) the following day. *R* indicates the sleep recovery period in the sleep-deprived animals. (**c**) Sleep mutants can be identified by genetic screens. Sleep amounts in 9,000 *Drosophila* male (*bottom*) and female (*top*) mutant lines. Mean daily sleep duration is about 900 min in males and 600 min in females. *Shaded areas* show one and two standard deviations from the mean. *Asterisks* show the *minisleep* mutant. (**d**) Daily distribution of sleep amount in wild-type (*red*) and *minisleep* flies (heterozygous in *gray* and homozygous in *green*). *White* and *gray bars* indicate light and dark periods respectively (Modified from Cirelli et al. [22] with permission)

that modulators of Shaker channel activity also affect sleep. Other pathways include dopamine and serotonin receptor signaling in the mushroom bodies, a brain structure involved in learning and memory (see Chap. 26). Circadian and sleep pathways intersect in a particular subset of PDF neurons, which increase action-potential firing and behavioral arousal in response to light. This PDF-induced arousal is inhibited by GABAergic inputs indicating that light and GABA compete to control PDF-dependent arousal in flies.

### 27.3.3 Light Entrainment of the Fly Clock

How do day-night cycles synchronize *Drosophila* sleep-wake cycles? A temporal shift of the LD cycle (for example, an 8-h delay would mimic a transatlantic flight from Moscow to New York) will induce a resetting of the sleep-wake cycle that will take place within a day in wild-type flies (Fig. 27.8). A mutation that prevents the differentiation of all photoreceptor cells slows the resetting, but these blind flies nevertheless synchronize to the new LD cycle within a few days.

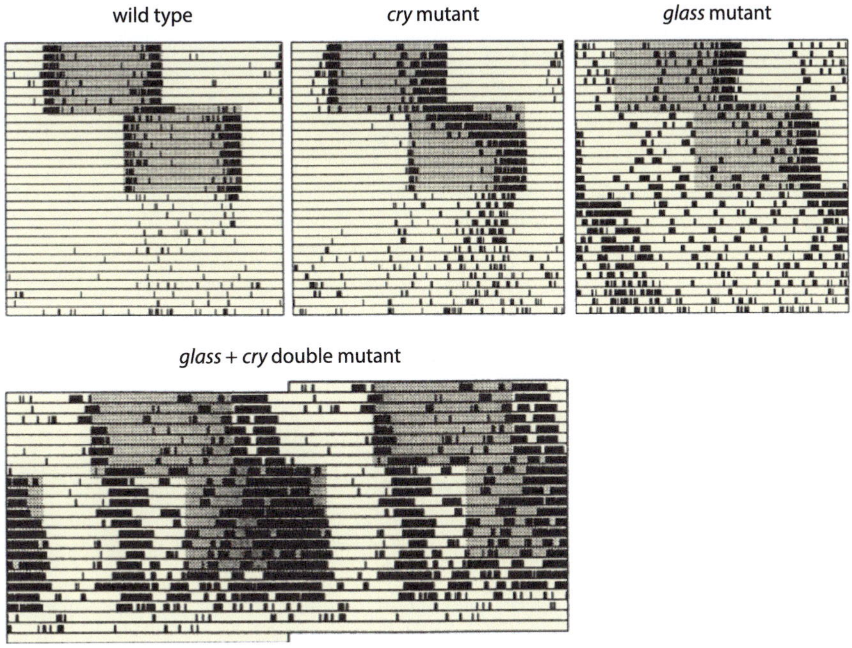


**Fig. 27.8** **Visual system and cryptochrome synchronize the *Drosophila* clock.** Individual actograms of wild-type and mutant flies submitted to an 8 h delay of the environmental light-dark cycle (*dark gray* is lights-ON). A wild type fly synchronizes its sleep-wake rhythm to the new light regime within a day, whereas mutant flies with no CRY protein (*cry* mutant) or no visual system (*glass* mutant) need a few days to complete the resetting of their clock. In contrast, a fly devoid of both the visual system and CRY (double mutant) is not able to synchronize its sleep-wake rhythm to the new LD cycle (Modified from Helfrich-Förster et al. [23] with permission)

Mutant flies devoid of the cryptochrome protein show a similarly slow resetting. In contrast, double mutants without CRY nor photoreceptors do not seem to synchronize: they appear to be circadianly blind to light-dark cycles [23]. The blue-light-sensitive CRY protein is expressed in most clock neurons. Light-activated CRY binds TIM protein, which then recruits the JETLAG ubiquitin ligase. This will lead to the fast proteasome-dependent degradation of TIM, thus resetting the molecular oscillations. Recent experiments indicate that this cell autonomous mechanism may not be the only way for CRY to reset the clock driving the sleep-wake rhythm, but its nonautonomous function has not yet been characterized. CRY is also expressed in clock cells outside the brain (peripheral clocks), where it acts as a cell autonomous circadian photoreceptor.

Body parts from transgenic *Drosophila* carrying a luciferase reporter gene driven by a *per* gene regulatory region were cultured separately for several days in LD cycles, and showed robust bioluminescence rhythms (bioluminescence is produced by luciferase in the presence of luciferin in the culture medium). In DD, the oscillations of the peripheral clocks quickly dampen, but applying new LD cycles restarted the oscillations in phase with the new LD cycle [24]. CRY thus allows insect peripheral clocks to perceive light inputs and synchronize to LD cycles independently of the brain and the visual system.

Photoreceptors in insects are distributed in several organs. The *Drosophila* compound eye is the main visual structure and contains about 350 ommatidia that each include six external photoreceptors and two internal ones, in addition to support cells (see Chap. 18). The external photoreceptors express one type of rhodopsin, whereas the internal ones express four different types that are believed to confer color vision. The eye photoreceptors send axonal projections to the optic lobes of the brain, which are in part innervated by the arborization of the PDF neurons. Flies also have a tiny photoreceptive organ, the eyelet, with four photoreceptor cells expressing only one type of rhodopsin. The eyelet is located underneath the retina and sends axons to the optic lobe area where the PDF neurons spread their dendritic arborization.

Finally, three ocelli on the top of the head are involved in horizon detection during flight and express a sixth type of rhodopsin. Experiments using different types of light quality and intensity indicate that the compound eye and the eyelet both contribute to the synchronization of sleep-wake rhythms. Invertebrate photoreceptors use histamine as neurotransmitter, but how the histaminergic signals reset the molecular oscillations in the clock neurons is unknown. The complexity of the *Drosophila* photoreception system suggests that the different cellular oscillators of the circadian network may receive different types of photic information, possibly related with the changes of light quality during the course of the day.

Similarly to light, temperature cycles efficiently reset insect circadian clocks [3]. Although the entrainment pathways remain uncharacterized, recent data indicate that the chordotonal organs (known as mechanical sensory organs) act as sensory organs for temperature entrainment of the behavioral clock. Two groups of clock neurons respond better to temperature entrainment whereas others are more responsive to light. How temperature and light inputs are integrated by the clock is not known yet.

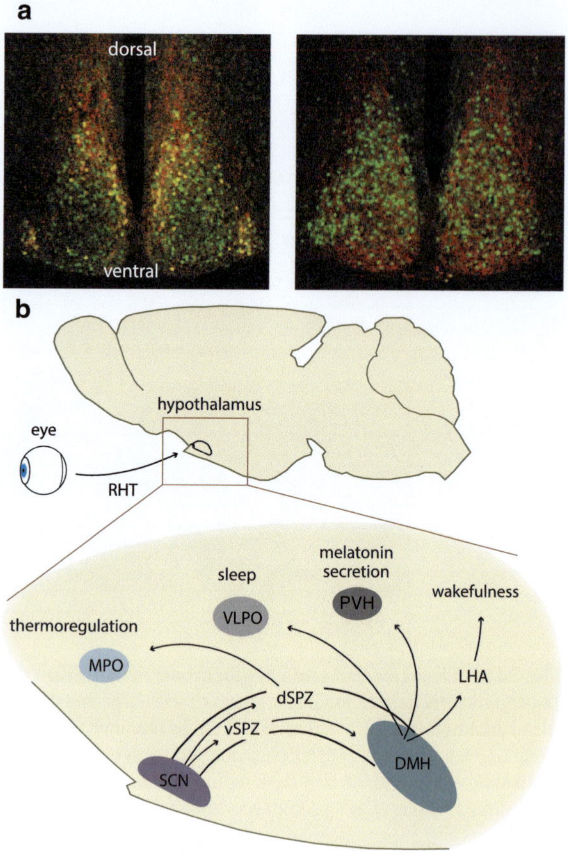


**Fig. 27.9 The suprachiasmatic nuclei (SCN) control sleep-wake rhythms in mammals.** The SCN are two paired nuclei of the hypothalamus located above the optic chiasm. (**a**) The mouse SCN each comprise about 10,000 neurons. Different subsets of SCN neurons can be identified according to the expression of clock genes as well as several neuropeptides. *Green color* identifies neurons that express the CLK protein. *Red* identifies neurons that express either the arginine vasopressin (*AVP*, *left*) or the vasoactive intestinal polypeptide (*VIP*, *right*). *Yellow* indicates co-expression of CLK and the neuropeptide. VIP is more abundant in the ventrolateral SCN, which receives afferents from the retinohypothalamic tract (*RHT*). AVP is prominent in the dorsomedial SCN, which sends efferent fibers. (**b**) SCN outputs mostly target the ventral (*vSPZ*) and dorsal (*dSPZ*) subparaventricular zone. The dSPZ neurons are required for body temperature rhythms through projections to the medial preoptic area (*MPO*). Outputs from the vSPZ are integrated in the dorsomedial nucleus of the hypothalamus (DMH), which send projections to the ventrolateral preoptic area (*VLPO*) to drive sleep-wake cycles. The DMH also targets the paraventricular nucleus (*PVH*) to control melatonin secretion by the pineal gland through an indirect pathway, and send projections to the lateral hypothalamic (*LHA*) neurons to control wakefulness (orexin neurons) and feeding cycles (SCN photograph drawn from Mohawk and Takahashi [33] with permission; drawings adapted from Saper et al. [34] with permission)

## 27.4 Neural Organization of the Mammalian Clock

### 27.4.1 A Master Clock Resides in the Suprachiasmatic Nuclei

In mammals, the circadian clock that controls sleep-wake rhythms resides in the suprachiasmatic nuclei (SCN) of the hypothalamus. The SCN is a bilateral structure that contains about 10,000 neurons in each side of the brain. The SCN is located above the optic chiasma close to the third ventricle of the brain, and receives retinal inputs though the retinohypothalamic tract (RHT) (Fig. 27.9). Lesioning this small region in rodents disrupts the temporal pattern of sleep-wake cycles, as well as body temperature rhythms [25, 26]. Furthermore, the fortuitous finding of a *tau* mutant in hamsters which displays a 20-h circadian period provided a nice tool to perform SCN transplantation studies between animals with different behavioral periods. Grafting an SCN from a short-period *tau* hamster to a lesioned wild-type animal (24-h period before lesion) restored a 20-h sleep-wake rhythm to the arrhythmic host [27]. Similarly, a wild-type mouse SCN graft transplanted into a genetically arrhythmic animal restores 24-h rhythms. A clock in the SCN is thus sufficient to drive sleep-wake cycles in the absence of any other clock-running tissue and to define the behavioral period over all the other clocks of the organism. SCN grafts embedded into semipermeable capsules that prevent neural outgrowth also restored sleep-wake rhythms, indicating that diffusible factors could drive rhythmic behavior [28]. Most SCN neurons are GABAergic neurons, but neurochemical and anatomical analyses have divided the nuclei in two regions. The ventral-lateral region (often named core) receives most of the afferents (in particular those from the visual system), whereas most of the efferents originate from the dorsomedial region (often named shell). The two regions also show differences in their neuropeptide content. Arginine-vasopressin peptide (AVP) is expressed in many shell neurons, whereas the core is enriched in gastrin-releasing peptide (GRP) and vasoactive intestinal peptide (VIP) neurons, with a small region expressing calbindin (CalB). How molecular oscillations control electrical activity in SCN neurons is not clear but circadian regulation of several ion channels at the transcriptional or post-translational level has been observed, suggesting a possible link. Electrical activity also feeds back to the molecular machinery, as reported in *Drosophila* clock neurons. Electrophysiological recordings indicate that SCN neurons have high electrical activity during the day and low electrical activity at night, in both diurnal and nocturnal animals. Similarly, the SCN shows the same phase of molecular oscillations in diurnal and nocturnal species, although the link between clock protein oscillations and neuronal firing oscillations is unknown. High SCN activity thus corresponds to behavioral activity in diurnal species and to resting in nocturnal species, indicating that the rest/activity control is defined downstream of the clock.

Isolated SCN slices show firing rhythms *in vitro*, indicating that it behaves as an autonomous oscillator. The finding of circadian clocks in cyanobacteria or unicellular eukaryotes indicates that a single cell is able to run a molecular clock. Are brain clocks networks of individual oscillators? The answer is yes, **brain clocks contain multiple oscillators that are coupled** (synchronizing each other) to generate synchronous oscillations of neuronal activity.

First evidence for a neuronal clock being made of individual oscillators came from the eye of the sea snail *Bulla gouldiana*, where a circadian clock controls the firing frequency of a small population of retinal neurons. Recording isolated neurons at different times of the day, indicated that each individual neuron displayed a circadian rhythm of membrane electrical conductance, which was higher at night than during the day. Similarly, isolated SCN neurons express rhythms of neuronal firing and keep doing so in the absence of synaptic transmission between cells, indicating a cell autonomous clock property (Fig. 27.10). When isolated, SCN neurons show variable periods and phases of their firing rhythms, indicating that some coupling occurs within the tissue to keep cells in synchrony. Behavioral recording of mice defective for the genes encoding either VIP or the VIP receptor VPAC2 show strongly altered sleep-wake rhythms. The analysis of individual SCN neurons in such animals indicates that some neurons stop oscillating whereas other cells still cycle, but not in synchrony. VIP/VPAC2 signaling is thus a key component for keeping the clock running in some neuronal populations and coupling between the individual oscillators in others. Recent SCN imaging experiments indeed show that in wild-type animals, not all SCN neurons have the same phase, revealing that regional subnetworks may play different roles in the SCN circadian function.

### 27.4.2 The SCN Controls Sleep-Wake Cycles in Mammals

The output pathways through which the SCN controls sleep-wake rhythms remain poorly understood. SCN efferent projections reach different areas of the hypothalamus where they modulate many neuroendocrine and autonomic pathways. SCN neurons densely project to the area located dorsocaudal to the SCN (Fig. 27.9). This region contains the subparaventricular zone (SPZ) and the dorsomedial hypothalamic (DMH) nucleus, the latter receiving direct innervation from the SCN as well as indirect SCN outputs through

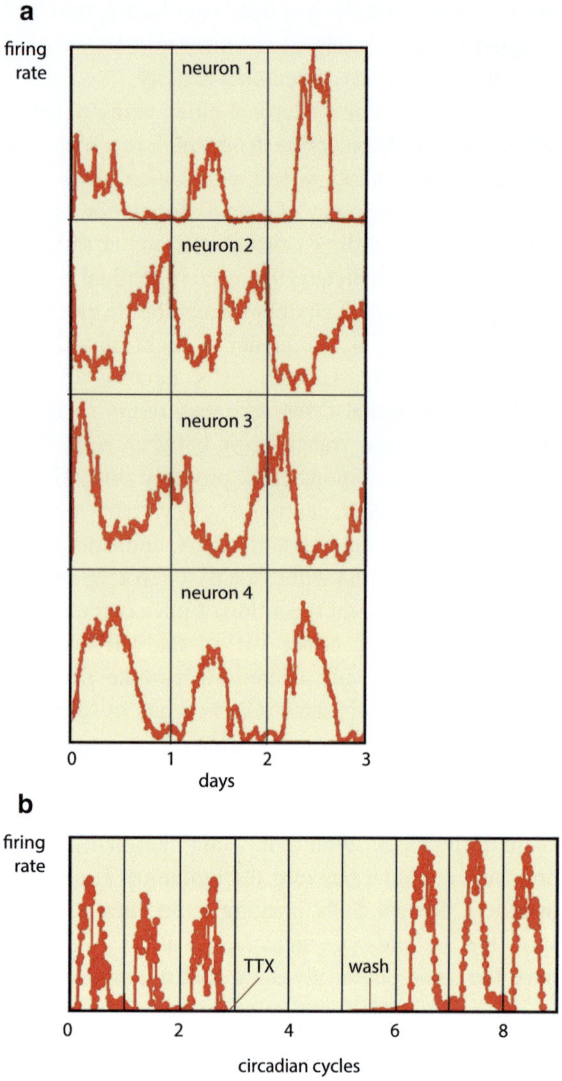


**Fig. 27.10 The SCN is made of coupled individual circadian oscillators.** (**a**) Recording dissociated SCN neurons in culture shows that individual cells display circadian firing rhythms with different periods and phases. (**b**) After a reversible blockage of neuronal firing with tetrodoxin (*TTX*), isolated SCN neurons resume their firing rhythm with the same phase, showing that the circadian clock keeps ticking in each individual neuron in the absence of firing and synaptic transmission (Modified from Welsh et al. [35] with permission)

the ventral SPZ. Lesions of the dorsal SPZ specifically disrupts body temperature rhythms. Lesioning ventral SPZ or DMH strongly alters the temporal control of sleep-wake rhythms with affected animals showing several sleep-wake cycles per day (ultradian rhythms). The DMH sends inhibitory GABAergic projections to the ventrolateral preoptic area (VLPO) and excitatory glutamatergic projections to the lateral hypothalamic area (LHA). The VLPO promotes sleep while the orexin-secreting neurons of the LHA promote wakefulness. Since the disruption of the SCN or its hypothalamic targets reduces wakefulness, it suggests that the circadian system promotes wakefulness during the active period [29].

Sleep-wake cycles result from the interactions between two independent processes. First, a homeostatic process, which controls the rise of sleep propensity during waking and its dissipation during sleep. Second, a circadian process, which defines a cycling level of the arousal state. At each time of the day, the behavioral state will depend on the level of these two parameters. Increasing awake time will increase sleep pressure, but sleep onset requires that the circadianly regulated arousal signal is sufficiently low (Fig. 27.11).

The SCN indirectly controls the nocturnal synthesis of melatonin in the pineal gland, with synaptic relays from the paraventricular nucleus of the hypothalamus to the thoracic spinal cord and superior cervical ganglia. A seasonal variation of neuronal firing has been observed in the SCN, with a compressed phase distribution during short days and a broad distribution during long days. The encoding of day length by the SCN thus relies on a variation of the phase distribution of individual neuron firing oscillations, suggesting that day length modulates the coupling between individual SCN neurons. After receiving a signal from the SCN, the pineal gland releases melatonin in the bloodstream, where the hormone reaches receptors in different brain areas, in particular in the SCN itself. Although melatonin has a clear sleep-promoting effect, the mechanisms of its action remain unclear. Furthermore, laboratory mice with virtually no melatonin do not show altered sleep, suggesting that other pathways play a more important role in the control of sleep-wake cycles. The clearest function of melatonin is controlling seasonal reproductive rhythms in many vertebrate species. Here, melatonin acts as a night hormone to interpret the day length (photoperiod) and induce physiological changes, in particular sexual rest or activity. In mammals, this control occurs through melatonin receptors that are located in the pars tuberalis of the pituitary gland.

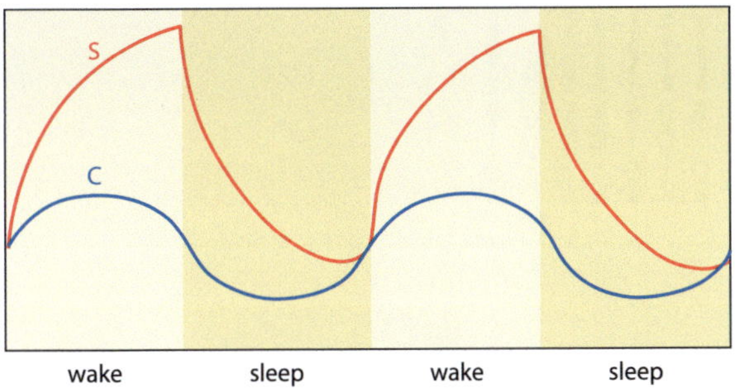


**Fig. 27.11** **The Borbely sleep regulation model.** Sleep results from an interaction between the homeostatic sleep pressure *S* and the circadian process *C*. The sleep pressure builds up during wakefulness and quickly decreases during sleep. The circadian process drives alertness and is linked to the time of the day but does not depend on sleep duration. S and C oppose each other during the day. Sleep occurs when S is high and C is low (Adapted from Daan [36])

### 27.4.3 Light Entrainment of the Mammalian Clock

As in flies, light is the strongest Zeitgeber for the mammalian clock. Compared to image-forming visual photoreception, circadian photoreception requires higher irradiance and longer stimuli. The SCN gets inputs from the retina through the RHT. However, mice devoid of rods and cones keep synchronizing with LD cycles, suggesting that other photoreceptors can mediate light-driven circadian entrainment. The discovery that a subset of retinal ganglion cells express a new photopigment, **melanopsin**, and directly respond to light provided a candidate photoreceptor for circadian entrainment (Fig. 27.12a) [30].

Retinal ganglion cells receive inputs from rod and cone photoreceptors, which control their firing, and send the visual information to the brain by forming the optic nerve with their axons. The melanopsin-expressing intrinsically photoreceptive retina ganglion cells (ipRGC) represent a small fraction (<5 %) of the retinal ganglion cells which reach the SCN through the RHT. Mice devoid of rods and cones, on one hand, and melanopsin mutants, on the other hand, both synchronize to LD cycles, indicating that each of the two photoreceptive pathways can mediate entrainment of the SCN clock (Fig. 27.12b). In contrast to melanopsin mutants, mice genetically ablated for the ipRGCs do not synchronize anymore. This demonstrates the central role of ipRGCs in circadian photoreception: they receive inputs from rods and cones and have an intrinsic photoreceptive capacity. Different subtypes of ipRGCs have been characterized. Melanopsin is also involved in other non-image forming photoreceptive responses such as pupillary constriction and melatonin suppression. The RHT projects in other retino-recipient brain areas, providing the anatomical substrate for these responses. Recent results also indicate that melanopsin may participate in visual photoreception.

Melanopsin is a blue-light photoreceptor that is structurally and functionally related to invertebrate opsins. The phototransduction pathway that takes place downstream opsins in rods and cones involves the activation of the G protein transducin, which activates a phosphodiesterase, thus inducing a decrease of cyclic GMP levels. This triggers the closure of cyclic nucleotide-gated cation channels, hence the hyperpolarization of the photoreceptor cell. In contrast, invertebrate rhabdomeric opsins such as the *Drosophila* rhodopsins, activate a phospholipase C, which increases levels of the second messenger diacylglycerol. This leads to the opening of transient receptor potential (TRP) calcium channels, hence the depolarization of the photoreceptor cell. Although the melanopsin transduction pathway is not fully characterized, it clearly involves the opening of TRP-like channels that lead to depolarization. Glutamate is the main neurotransmitter of ipRGCs and triggers a cAMP pathway that activates the CREB transcription factor in the ventrolateral SCN neurons. CREB activates the transcription of the *per1* and *per2* genes, leading to a rapid increase in PER protein levels. This transient increase will alter PER oscillations, hence inducing a phase shift of the molecular oscillator.

As indicated above, circadian clocks are present in most cells, which are called **peripheral clocks** by comparison to the SCN. These peripheral clocks do not directly see light and thus require the SCN to be synchronized to LD cycles. How the SCN synchronizes body clocks is not fully understood but a large body of experimental work suggests that humoral factors are involved, in particular glucocorticoids. SCN outputs reach the autonomic nervous system, whose target organs thus receive indirect synchronizing signals from the light-entrained SCN. Recent work has identified body temperature to be a strong resetting signal driven

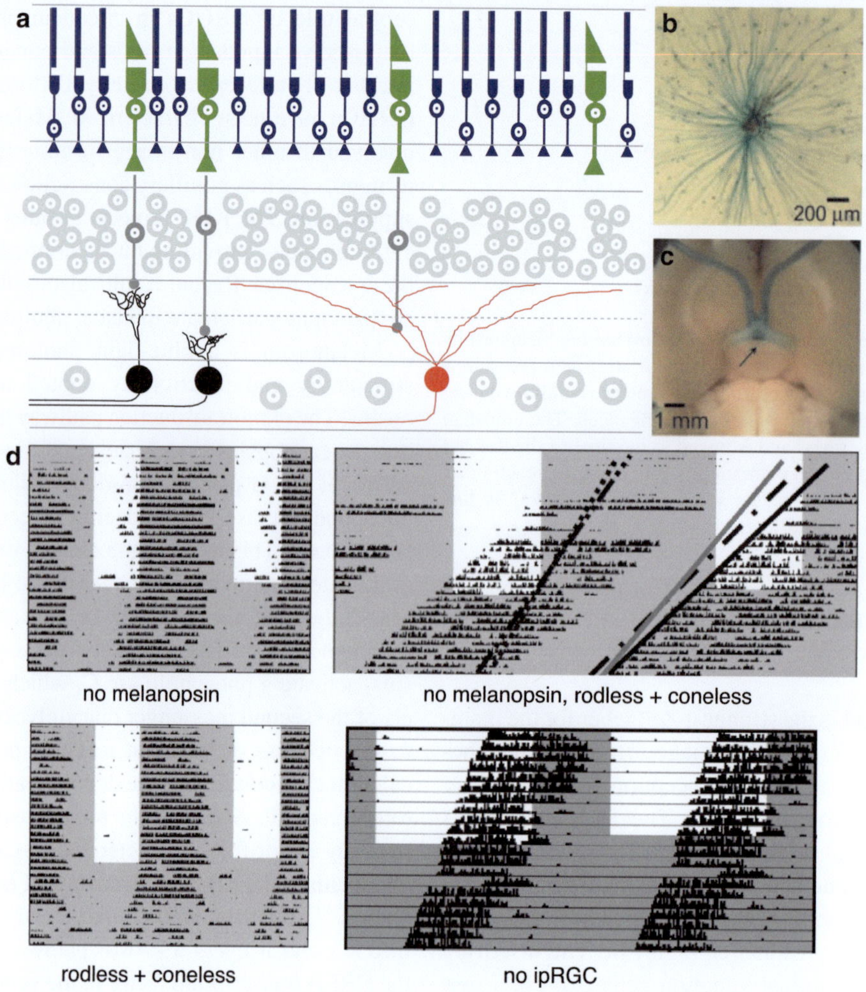


**Fig. 27.12 Different types of retinal cells are involved in the synchronization of sleep-wake rhythms in mammals.** (**a**) Light goes through the ganglion cells (*black*) and bipolar cells (*gray*) before reaching the rod (*blue*) and cone (*green*) photoreceptors. The rods and cones are connected to the bipolar cells, which then connect the ganglion cells that send their axons in the optic nerve (for better clarity, only cone connections are shown here). A small fraction of the retinal ganglion cells, the ipRGCs (*red*), express melanopsin and act as circadian photoreceptors. (**b**, **c**) Targeting a reporter gene in the melanopsin locus allows to observe ipRGCs and their projections. The ipRGCs (labeled in *blue*) axons gather in the center of the retina (**b**) and join the optic nerve to reach the SCN in the brain (**c**) (From Berson [30] and Hattar et al. [37] with permission). (**d**) Synchronization of sleep-wake rhythms to light-dark cycles in mutant mice devoid of melanopsin, rods and cones or ipRGCs. Opsins of the rods and cones, on one hand, and melanopsin of the ipRGCs, on the other hand, are sufficient to synchronize sleep-activity rhythms. Only animals devoid of the two pathways lose synchronization. However, light inputs going through rods and cones require ipRGCs since mice with no ipRGCs are unable to synchronize their rhythms (From Panda et al. [38] and Güler et al. [39] with permission)

by the SCN to synchronize peripheral oscillators. Some of the transcriptional components that translate the synchronizing signals into changes of the molecular oscillator in various peripheral tissues appear to be shared by several pathways, allowing the integration of various cues to orchestrate circadian timing in the body.

## References


1. Moore-Ede MC, Sulzman FM, Fuller CA (1982) The clocks that time us. Physiology of the circadian timing system. Harvard University Press, Cambridge
2. Dunlap JC, Loros JJ, DeCoursey PJ (eds) (2004) Chronobiology – biological timekeeping. Sinauer, Sunderland
3. Saunders DS (2002) Insect clocks. Elsevier, Amsterdam

4. van Oort BEH, Tyler NJC, Gerkema MP, Folkow L, Blix AS, Stokkan KA (2005) Circadian organization in reindeer. Nature 438:1095–1096
5. Cavallari N, Frigato E, Vallone D et al. (2011) A blind circadian clock in cavefish reveals that opsins mediate peripheral clock photoreception. PLoS Biol 9:e1001142
6. Johnson CH, Mori T, Xu YA (2008) Cyanobacterial circadian clockwork. Curr Biol 18:R816–R825
7. DeCoursey PJ, Walker JK, Smith SA (2000) A circadian pacemaker in free-living chipmunks: essential for survival? J Comp Physiol A 186:169–180
8. Konopka RJ, Benzer S (1971) Clock mutants in *Drosophila melanogaster*. Proc Natl Acad Sci USA 68:2112–2116
9. Hardin PE, Hall JC, Rosbash M (1990) Feedback of the *Drosophila* period gene product on circadian cycling of its messenger RNA levels. Nature 343:536–540
10. Hardin PE (2011) Molecular genetic analysis of circadian timekeeping in *Drosophila*. Adv Genet 74:141–173
11. Price JL, Blau J, Rothenfluh A, Abodeely M, Kloss B, Young MW (1998) *double-time* is a novel *Drosophila* clock gene that regulates PERIOD protein accumulation. Cell 94:83–95
12. Nitabach MN, Blau J, Holmes TC (2002) Electrical silencing of *Drosophila* pacemaker neurons stops the free-running circadian clock. Cell 109:485–495
13. Antoch MP, Song EJ, Chang AM et al. (1997) Functional identification of the mouse circadian clock gene by transgenic BAC rescue. Cell 89:655–667
14. Rubin EB, Shemesh Y, Cohen M et al. (2006) Molecular and phylogenetic analyses reveal mammalian-like clockwork in the honey bee (*Apis mellifera*) and shed new light on the molecular evolution of the circadian clock. Genome Res 16:1352–1365
15. Toh KL, Jones CR, He Y, Eide EJ, Hinz WA, Virshup DM, Ptacek LJ, Fu YH (2001) An h*Per2* phosphorylation site mutation in familial advanced sleep-phase syndrome. Science 291:1040–1043
16. Roenneberg T, Kuehnle T, Juda M, Kantermann T, Allebrandt K, Gordijn M, Merrow M (2007) Epidemiology of the human circadian clock. Sleep Med Rev 11:429–438
17. Foster RG, Roenneberg T (2008) Human responses to the geophysical daily, annual and lunar cycles. Curr Biol 18: R784–R794
18. Hampp G, Ripperger JA, Houben T et al. (2008) Regulation of monoamine oxidase a by circadian-clock components implies clock influence on mood. Curr Biol 18:678–683
19. Handler AM, Konopka RJ (1979) Transplantation of a circadian pacemaker in *Drosophila*. Nature 279:236–238
20. Grima B, Chélot E, Xia R, Rouyer F (2004) Morning and evening peaks of activity rely on different clock neurons of the *Drosophila* brain. Nature 431:869–873
21. Stoleru D, Peng P, Agosto J, Rosbash M (2004) Coupled oscillators control morning and evening locomotor behaviour of *Drosophila*. Nature 431:862–868
22. Cirelli C, Bushey D, Hill S, Huber R, Kreber R, Ganetzky B, Tononi G (2005) Reduced sleep in *Drosophila Shaker* mutants. Nature 434:1087–1092
23. Helfrich-Förster C, Winter C, Hofbauer A, Hall JC, Stanewsky R (2001) The circadian clock of fruit flies is blind after elimination of all known photoreceptors. Neuron 30:249–261
24. Plautz JD, Kaneko M, Hall JC, Kay SA (1997) Independent photoreceptive circadian clocks throughout *Drosophila*. Science 278:1632–1635
25. Stephan FK, Zucker I (1972) Circadian rhythms in drinking behaviour and locomotor activity of rats are eliminated by hypothalamic lesions. Proc Natl Acad Sci USA 69:1583–1586
26. Moore RY, Eichler VB (1972) Loss of a circadian adrenal corticosterone rhythm following suprachiasmatic lesions in the rat. Brain Res 42:201–206
27. Ralph MR, Foster RG, Davis FC, Menaker M (1990) Transplanted suprachiasmatic nucleus determines circadian period. Science 247:975–978
28. Silver R, Lesauter J, Tresco PA, Lehman MN (1996) A diffusible coupling signal from the transplanted suprachiasmatic nucleus controlling circadian locomotor rhythms. Nature 382:810–813
29. Saper CB, Fuller PM, Pedersen NP, Lu J, Scammell TE (2010) Sleep state switching. Neuron 68:1023–1042
30. Berson DM (2003) Strange vision: ganglion cells as circadian photoreceptors. Trends Neurosci 26:314–320
31. Aschoff J, Daan S, Honma K (1982) Zeitgebers, entrainment, and masking: some unsettled questions. In: Aschoff J, Daan S, Groos GA (eds) Vertebrate circadian systems: structure and physiology. Springer, New York/Berlin/Heidelberg, pp 13–24
32. Pittendrigh CS (1954) On temperature independence in the clock system controlling emergence time in *Drosophila*. Proc Natl Acad Sci USA 40:1018–1029
33. Mohawk JA, Takahashi JS (2011) Cell autonomy and synchrony of suprachiasmatic nucleus circadian oscillators. Trends Neurosci 34:349–358
34. Saper CB, Scammell TE, Lu J (2005) Hypothalamic regulation of sleep and circadian rhythms. Nature 437: 1257–1263
35. Welsh DK, Logothetis DE, Meister M, Reppert SM (1995) Individual neurons dissociated from rat suprachiasmatic nucleus express independently phased circadian firing rhythms. Neuron 14:697–706
36. Daan S, Beersma DG, Borbely AA (1984) Timing of human sleep: recovery process gated by a circadian pacemaker. Am J Physiol 246:R161–R183
37. Hattar S, Liao HW, Takao M, Berson DM, Yau KW (2002) Melanopsin-containing retinal ganglion cells: architecture, projections, and intrinsic photosensitivity. Science 295: 1065–1170
38. Panda S, Provencio I, Tu DC et al. (2003) Melanopsin is required for non-image-forming photic responses in blind mice. Science 301:525–527
39. Güler AD, Ecker JL, Lall GS et al. (2008) Melanopsin cells are the principal conduits for rod-cone input to non-image-forming vision. Nature 453:102–105

# 28 Learning, Memory, and Cognition: Animal Perspectives

Randolf Menzel

## 28.1 Cognition: Definition

**Cognition** is the integrating process that utilizes many different forms of memory, creates internal representations of the experienced world and provides a reference for expecting the future of the animal's own actions. It thus allows the animal to decide between different options in reference to the expected outcome of its potential actions. All these processes occur as intrinsic operations of the nervous system, and provide an implicit form of knowledge for controlling behavior. None of these processes need to – and certainly will not – become explicit within the nervous systems of many animal species (in particular invertebrates and lower vertebrates), but such processes must be assumed to also exist in these animal species. It is the goal of comparative animal cognition to relate the complexity of the nervous system to the level of internal processing.

Neural integration processes are manifold and span a large range of possibilities all of which can be viewed from an evolutionary perspective as adaptations to the specific demands posed by the environment to the particular species. At one extreme one can find organisms dominated by their inherited information (**phylogenetic memory**) developing only minimal experience-based adaptation. At the other extreme, phylogenetic memory merely provides a broad framework, and **experience-based memory** dominates. Phylogenetic memory controls behavior in rather tight stimulus-response connections requiring little if any internal processing other than sensory coding and generation of motor programs. The factors determining the specific combination and weights of inherited and experience-dependent memories in an individual are not yet well understood. A short individual lifetime, few environmental changes during a lifetime, and highly specialized living conditions will favor the dominance of inherited information; a longer individual lifetime, less adaptation to particular environmental niches and rapid environmental changes relative to the lifespan reduce the value of phylogenetic memory and increase the role of **individual learning**. Social living style also seems important. Here the species' genome must equip the individuals for acting under much more variable environmental conditions because of the society's longer lifetime, and because the communicative processes within the society demand a larger range of cognitive processes.

The complexity and size of the nervous system may be related to the dominance of inherited or experience-dependent memories, in the sense that individual learning demands a larger nervous system having greater complexity. However, the primary parameter determining the size of the nervous system is body size, and secondary parameters like richness of the sensory world, abundance of motor patterns and cognitive capacities, are difficult to relate to brain size, because such parameters cannot be adequately measured and thus a comparison based on them is practically impossible between animals adapted to different environments. Nevertheless, it appears obvious that animals differ with respect to their sensory, motor, and cognitive capacities. Individual learning within the

R. Menzel
Freie Universität Berlin, Institut Biologie-Neurobiologie, Königin Luisestr. 28/30, 14195 Berlin, Germany
e-mail: menzel@neurobiologie.fu-berlin.de

C.G. Galizia, P.-M. Lledo (eds.), *Neurosciences - From Molecule to Behavior: A University Textbook*, DOI 10.1007/978-3-642-10769-6_28, 

species-specific sensory and motor domains will lead to more flexible behavior, and thus to more advanced cognitive functions. Predicting the future will therefore be less constrained, and more options will enrich the animal's present state.

Cognitive components of behavior are characterized by the following faculties: (i) rich and cross-linked forms of sensory and motor processing; (ii) flexibility and experience-dependent plasticity in choice performance; and (iii) long-term (on the timescale of the respective animal's lifespan) adaptation of behavioral routines. These three features allow the creation of novel behavior through different forms of learning and memory processing. Among them, we can cite: (i) rule learning and causal reasoning; (ii) observatory learning during communication, imitation, and navigation, and (iii) recognition of individuals in a society and self-recognition. All these characteristics are based on implicit forms of knowledge and do not require any explicit (or conscious) processing. However, internal processing at the level of working memory (or representation) as an indication of rudimentary forms of explicit processing may exist in invertebrates and lower vertebrates within the context of observatory learning and social communication.

## 28.2 Innate and Learned Behavior

Innate and learned behaviors are intimately connected leading to the concept of preparedness for learning. Mice associate nausea induced by injection of LiCl or radioactive irradiation with novel taste and smell but not with light or sound. Song birds are prepared to learn the species-specific song, and only some species may be more open to aberrant songs. The idea that anything can be learned if associativity rules are followed as put forward by Pavlov (1849–1936) and the behaviorists like B.F. Skinner (1904–1990) is not substantiated, and many examples have been described for species-specific constraints in learning.

Often similar behaviors are performed by closely related species, but in one species it involves learning, in the other it is solely controlled innately. The two species of braconid wasps *Cotesia glomerata* and *Cotesia flavipes* are stem-boring parasitoids. While *C. flavipes* exhibits innate preference for its host's odors – the larvae of *Pieris brassica* (Lepidoptera) – the closely-related *C. glomerata* learns the varying odor profiles of its *Pieris* host larvae, which depend on the plants it feeds on. No other differences in behavior between these two species were found, indicating that experience-dependent adjustment and innate stereotypy are two close strategies and are not related to any great differences between the neural systems involved. It will be interesting to search for structures in the brain that differ in these two species and may be related to these two strategies.

A particularly close connection between innate and learned behavior is **imprinting** (see Chap. 25), the programmed forms of learning described in great detail by ethologists like Konrad Lorenz (1903–1989) for birds, but fast and stable learning early on in ontogeny is a phenomenon in all animal species. Slave-making ants have colonies in which two species of social insects coexist, one of which parasitizes on the other. Slave-making ants invade colonies of other ant species and transport the pupae back to their own nest. Adults emerging from these pupae react and work for the slave-making species as if it were its own species. The basis for this phenomenon may be olfactory imprinting by which the slave ants learn to recognize the slave-makers as members of their own species.

The mechanistic basis of olfactory imprinting has been studied in the fruit fly *Drosophila melanogaster*. Synaptogenesis in the antennal lobe, the primary olfactory neuropile in the insect brain (see Chap. 13), starts in late pupae and continues during the first days of adult life, at the same time as the behavioral response to odors matures. The antennal lobe is made up of functional units, the glomeruli. The glomeruli DM6, DM2, and V display specific growth patterns between days 1 and 12 of adult life. The modifications associated with olfactory imprinting take place at the critical age. Exposure to benzaldehyde at days 2–5 of adult life, but not at 8–11, causes behavioral adaptation as well as structural changes in DM2 and V glomeruli.

These examples show that (i) animals often exhibit innate preferences for signals allowing to rapidly and efficiently detect biologically relevant stimuli in their first encounters with them; (ii) such preferences can but may not always be modified by the animals' experience. It is still unknown how these preferences are hardwired in the naïve nervous system, but since they have been selected through the species' evolutionary history and thus belong to its phylogenetic memory it must be assumed that they are programmed by developmental processes.

**Table 28.1** Categories of learning

| Learning – descriptive categories | Relation of the stimuli | Evaluation of stimuli | What is learned |
|---|---|---|---|
| **Nonassociative learning** | | | |
| Habituation | Stimulus repetition | No evaluation | Stimulus has no meaning |
| Sensitization | Strong stimulus | Causes attention | General arousal |
| **Associative learning** | | | |
| Classical (Pavlovian) conditioning | Neutral stimulus CS co-occurs with meaningful stimulus | US is a reinforcer | Association of CS with US |
| Operant or instrumental learning | Own actions lead to reinforcement | US is a reinforcer | Hierarchy of associations |
| High-level learning (observation, navigation, play, learning by insight) | Directed attention in the course of self-produced behavior | Internal rather than external evaluation | Association between stimuli and behavior |
| Imprinting | Stimuli occurring during active behavior | Developmental preparedness | Stimuli and developmental program |

## 28.3 Learning: Elemental Forms of Associative Learning

Learning is the capacity to change behavior as the result of individual experience in such a way that the new behavior is better adapted to the changed conditions of the environment. Learning can be grouped into three broad categories: simple nonassociative learning like habituation and sensitization, associative learning including classical conditioning and instrumental (operant) learning, and higher forms of associative learning characterized by the lack of an obvious external reinforcing stimulus and by directed attention of the animal to the outcome of self-generated behavior as in observatory learning and learning during playing (Table 28.1).

### 28.3.1 Nonassociative Learning

Stimulus repetition without any consequences leads to a stimulus-specific decrease of stimulus-induced responses (habituation), a lower sensitivity to the stimulus and less attention. **Habituation** is characterized by stimulus specificity, spontaneous recovery and dishabituation, a phenomenon that results from strong, sensitizing stimuli. These properties exclude the possibilities that habituation is based on sensory adaptation or motor fatigue. **Sensitization** results from a strong and unexpected stimulus that induces a state of general arousal, higher sensory sensitivity, and alerted motor responses. Repetition of sensitizing stimuli leads to fast habituation, and single stimulations may induce only short-lasting arousal. Sensitizing stimuli often carry an aversive innate meaning relating these stimuli to unconditioned stimuli in associative learning (see below).

The cellular and neural correlates of nonassociative learning are covered in Chap. 26. Eric Kandel and his coworkers conceptualized a cellular alphabet of neural plasticity leading to a hierarchy of brain mechanisms of learning and memory formation [10].

### 28.3.2 Classical Conditioning

Associative connections between stimuli, events, and actions are the source of information that animals use to extract causal relations in the environment. Figure 28.1 gives examples of basic paradigms of classical conditioning and Table 28.2 lists additional paradigms.

A hungry honeybee responds to the stimulation of the sucrose receptors on the antennae by an extension of its proboscis (tongue) and sucks the sucrose solution. This stimulus arouses the animal, induces directed searching responses, releases an innate response (the proboscis extension response, which represents the unconditioned responses, UR), and acts as an **unconditioned rewarding stimulus** (**US**) for **neutral stimuli** (**CSs**) like odors experienced shortly before the US. These CSs can either precede the US which leads to an association with the US (denoted as CS+), or they follow the US at an interval (backward pairing), or are not paired with the US at all. In these cases they are not associated with the US, thus called CS–. The probability of a group of animals to extend their probosces during the CS+ in expectation of the US (**conditioned responses**, **CR**) increases with the number of forward pairing trials (CS+/US acquisition Fig. 28.1a) and does not change for backward pairing of

US/CS (or may decrease over trials if CS– is first responded to due to generalization, see Fig. 28.1d). Pavlov called the first form of conditioning **excitatory learning**, the latter one **inhibitory learning**. No change of behavior occurs if the animals experience stimulations of the CS alone or the US alone, but multiple exposures to the CS alone may retard acquisition of this CS in later forward-pairing trials (CS+/US). Conditioned animals lose their CR to the CS+ if the CS is presented multiple times without the US, an indication of extinction learning (Fig. 28.1a, right graph). This form of inhibitory learning can depend on the context conditions in which the animal experiences the loss of predictive power of the CS for US. It is, therefore, concluded that the memory of the CS-US connection is not lost but rather a new memory is formed, namely that now the CS predicts the absence of the UC. This conclusion is supported by the finding that at a later time the CS will partially gain its predictive power for the US (spontaneous recovery after extinction learning, Fig. 28.1a, right graph). In classical

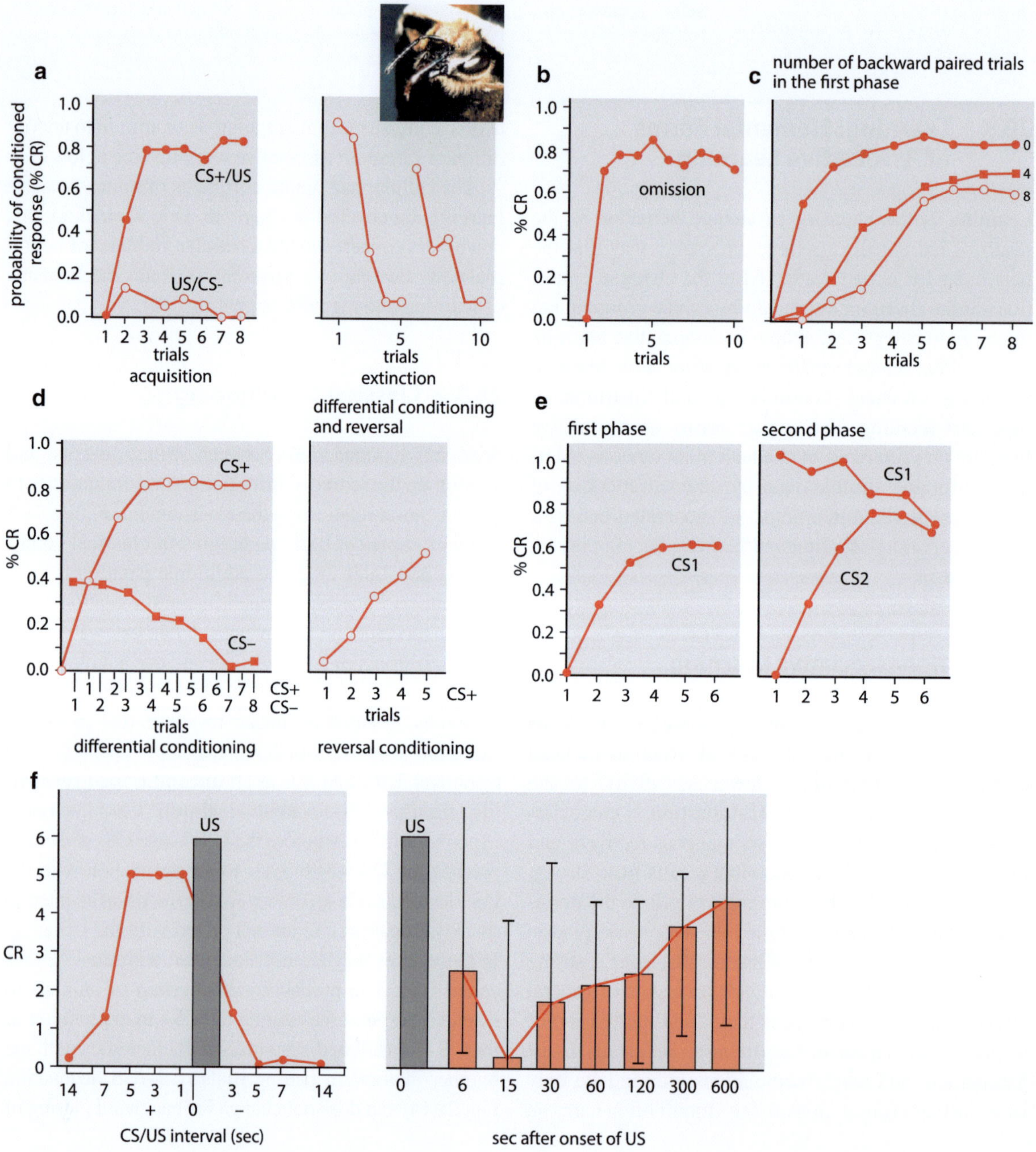

**Table 28.2** A selection of associative learning paradigms

| Paradigm | Phase 1 | Phase 2 | Test |
|---|---|---|---|
| 1. Sensory preconditioning | A/B | A/US | B tested |
| 2. Differential conditioning | A/US | B– | A and B tested |
| 3. Inhibitory conditioning | A– | A/US | A tested |
| 4. Second-order conditioning | A/US | B/A | B tested |
| 5. Within compound association | A/X, B/Y | A/US | A and B tested |
| 6. Blocking | A/US | A+B/US | B tested |
| 7. Massed and spaced trials on acquisition and retention | | | |

*A*, *B*, *X*, *Y* denote neutral stimuli, *US* reinforcer; the sign/indicates paired presentation of the stimuli, *A–* or *B–* presentation of the respective stimulus alone, *A+B* both stimuli are presented simultaneously, *A tested* or *B tested* means that the respective stimuli are presented without the reinforcer in order to evaluate what the animal has learned about A and B

conditioning, typically the establishment of a CS-CR connection does not require any particular behavior of the animal. In an omission paradigm (Fig. 28.1b) the animal is rewarded when it does **not** respond to the CS but not rewarded when it responds. Still honeybees under these conditions learn equally well indicating that an operant component is not involved. Often learning is not directly reflected in a behavioral change and can be uncovered only later when animals are exposed to forward-pairing trials. In honeybees backward pairing of US and CS in a first phase of conditioning will lead to lower acquisition to the same CS in a second phase (Fig. 28.1c). The acquisition function to forward pairing of the same CS with the US is depressed as compared to an acquisition function without or with a lower number of such preexposures. This form of learning (latent learning) appears to involve associations of the stimuli with those emanating from the context, because the resistance to acquisition effect can be context dependent, indicating that stimulus associations are established without the contribution of an US. Inhibitory learning of CS– in a first phase of differential conditioning is also seen when the CS– becomes the CS+ in a second phase (**reversal conditioning**, Fig. 28.1d). Acquisition in the second phase is retarded. The reinforcing property of the US can also be transferred to a CS (**second-order conditioning**, Fig. 28.1e). CS1 is forward paired with US in a first phase. In a second phase a novel CS (CS2) is forward paired with CS1 (CS2/CS1 pairing). CS2 is learned (rise of CR probability to CS2) and responses to CS1

**Fig. 28.1 Elemental forms of classical conditioning exemplified for odor reward conditioning in honey bees.** (**a**) The inset shows how a bee is harnessed in a tube. The hungry bee extends its proboscis (tongue) reflexively when its antennae taste sucrose, the unconditioned stimulus, US. Odors received by the olfactory receptors on the antennae serve as conditioned stimuli, CS. The *left graph* shows how the probability of the conditioned responses in a group of animals (proboscis extension response, PER) increases with the number of forward pairing trials (CS/US). Backward paired trials (US/CS) do not lead to conditioned PER. The *right graph* gives the result of an extinction experiment. Repeated exposures to the CS without the US leads to reduction of PER probability. After an interruption of 1–2 h the conditioned PER recovers partially, an effect known as spontaneous recovery from extinction. (**b**) Classical conditioning can be distinguished from operant conditioning by an omission procedure. The US is only given if the animal did not show the conditioned response (omission). This omission acquisition function of the PER shows pure classical conditioning, because acquisition is not reduced (compare with **a**). (**c**) If animals experience unpaired CS und US presentations, subsequent acquisition to paired CS/US trials is retarded. The *upper curve* (0 unpaired trials) provides the control group (normal acquisition after no exposure to unpaired trials). The *other two curves* show that retardation of acquisition depends on the number of previously experienced unpaired trials. (**d**) Differential conditioning involves two CSs, CS+ is forward-paired with US, CS– is backward-paired. Conditioned PER rises for the CS+. CS– is initially responded to more strongly because of generalization between CS+ and CS–. Further CS– unpaired trials lead to a reduction of PER to CS–. In a second phase CS– becomes the CS+. Now acquisition is retarded because of the unpaired trials in the first phase. (**e**) Second-order conditioning. CS1 is forwards paired with US in the first phase. In the second phase a new odor (CS2) is forward-paired with CS1 (first CS2 then CS1). The reinforcing capacity of CS1 which it gained in the first phase is transmitted to CS2. Concurrently the conditioned PER is reduced for CS1. (**f**) The effect of CS/US interval. The *left graph* shows how conditioned PER depends on the interval between CS and US (*gray* vertical bar extending from 0 to 2 s on the interval scale). The CS is presented for 2 s either before (*left side*: +) or after US (*right side*: –). Optimal conditioning is found for forward-pairing. The *right graph* shows the effect of backward-pairing. Since PER is zero to the CS when it follows the US, the hidden inhibitory component of backward-pairing was uncovered by exposing the animal in a second phase to forward-pairing to the same CS. The inhibitory backward-pairing effect is strongest for a UC–CS interval of 15 s

extinguish. A crucial parameter in classical conditioning is the contiguity between CS and US. Figure 28.1f shows how excitatory and inhibitory learning depend on the timing of the CS with the US. A preceding CS (CS+) gains its predictive power for the US most effectively for short intervals before the onset of US. Excitatory learning is reflected directly in the probability of CR induced by CS+. Inhibitory learning is often not directly reflected in a behavioral change. In that case it needs to be uncovered in a second phase of conditioning, e.g., a forward pairing of the same CS+ with US (as shown in Fig. 28.1c). Stronger inhibitory learning will lead to stronger resistance to acquisition in the second phase. The timing of US and CS– in inhibitory learning can be different from that of excitatory learning. In the honeybee optimal intervals between US and CS– lie between 5 and 25 s after US onset.

Pavlov's terms excitatory and inhibitory conditioning do not refer to the strength or probability of behavior controlled by learning but to the connections developed between CS and US. If the US is an aversive stimulus, excitatory conditioning will lead to less CR, and inhibitory learning to more CR. It is an interesting but unresolved question how excitatory and inhibitory conditioning in reward and punishment learning are related. Does inhibitory reward learning resemble aspects of excitatory punishment learning? In other words, does backward conditioning to an aversive stimulus induce some rewarding potential ("release from punishment") of the CS?

The paradigms of classical conditioning emphasize the importance of CS-US contiguity, however the latent learning phenomenon (Fig. 28.1c) indicates already that this cannot be the only parameter controlling learning. Other paradigms strengthen this conclusion. In **sensory preconditioning** (Table 28.2) two CSs (CS1 and CS2) are first presented together without any US. In a second training phase, one of them (CS1) is paired with the US. In the test phase, CS2 is tested alone and it is found that also CS2 induced CR although CS2 was never paired with the US. In blocking experiments (Table 28.2) a first training phase consists of CS1-US pairings. In the second training phase, CS2 is added, so that the compound CS1CS2 is paired with the US. Surprisingly, it is found that CS2 is less well or not at all learned although it is paired with the US. Learning of CS2 is somehow blocked by the experience of CS1-US pairing in the first phase. These and other paradigms of classical conditioning document the limitations of a simple contiguity effect and call for other explanations. These will be described below.

The rules of **associative learning** have been worked out in great detail by Pavlov and the American behaviorists. These rules are of heuristic value if applied with adequate care and if one considers the restriction that species-specific constraints and environmental conditions may lead to exceptions. In particular, they provide a frame for the design of experiments aiming to extract the crucial components in associative behavioral change. Rules of learning emphasize the role of the contiguity of events (their temporal relations) and their contingencies (the probabilities of co-occurrence of events). Because contiguity and contingency of events are more important than sensory modality or the motor pattern involved, forms of learning can be compared across species, environmental conditions, sensory modalities, and behavioral acts. In the course of learning, neutral environmental stimuli and actions of the animal lead to meaningful outcomes and thus become predictors for that outcome. When a hungry animal finds food, the own actions and the signals associated with this meaningful outcome are stored for future behavior. The meaningful component plays a decisive role because it surprises the animal, it has an innate or learned value with respect to the need of the animal, and it has the potential to reinforce the stimuli and actions with its value. Such a meaningful stimulus can either be (i) an external stimulus (in classical conditioning it is called the unconditioned stimulus, US, with the value of acting as a reward or a punishment), (ii) a successful or failing action (as in instrumental learning, see below), or (iii) an internally generated value function as in observatory representation related to meaning and value (see below).

The most powerful rule of associative learning has been formalized by Rescorla and Wagner [21] (see Box 28.1). Many phenomena of learning are well captured by this concept called the **delta rule**, which states that animals learn when an event is not expected and therefore surprises the animal. The rule defines whether and how much is learned depending on how surprising the association of CS and US is, and surprise is quantified by the difference between expected and actual event. The theory also states that all CSs involved in a learning trial compete for the limited capacity of this difference (**capacity of attention**). A number of learning phenomena are well captured by the theory (Box 28.1). **Expectation** and **surprise** are concepts of a cognitive interpretation for these simple forms of associative learning, and although the authors did not relate their theory to cognitive concepts the success of the theory also lies in the cognitive dimension of its key parameter, expectation.

**Box 28.1 The Delta Rule of Classical Conditioning**

Rescorla and Wagner [21] realized that the simple co-occurrence of CS and US in associative learning does not capture the results found in several conditioning experiments. They argued associative learning occurs not because two events co-occur but because that co-occurrence is unanticipated on the basis of current associative strength. The Rescorla-Wagner model attempts to capture that idea in a more formal way.

The Rescorla-Wagner model is based on the assumption that an associative change in each stimulus depends not only on its own state (how well it can be learned, how often it had been associated with the reinforcer) but also on the states of other stimuli concurrently present. Take the associative strength for a stimulus $X$ to be $V_X$, and for a stimulus A to be $V_A$. Then, the associative strength to the compound stimulus ($V_{AX}$) will be:

$$V_{AX} = V_A + V_X$$

On a learning trial in which a compound stimulus, AX , is followed by $US_1$, the changes in associative strength ($\Delta V$) for A and X are:

$$\Delta V_A = [\alpha_A \beta_1](\lambda_1 - V_{AX})$$

and

$$\Delta V_X = [\alpha_X \beta_1](\lambda 1 - V_{AX})$$

$\lambda_1$ is the maximum conditioning that $US_1$ can produce, and therefore represents the limit of learning. The $\alpha$ and $\beta$ are rate parameters dependent, respectively, on the CS and US. These parameters are viewed as having fixed values based on the physical properties of the particular CS and US. On any given trial the current associative strength $V_{AX}$ is compared with $\lambda$ and the difference is treated like an error to be corrected. This happens by producing a change in associative strength ($\Delta V$) accordingly. Consequently, this is an error-correction model.

This model predicts a number of previously unknown results particularly in blocking and conditioned inhibition experiments. In a blocking experiment one stimulus A has been paired with the US, then the compound stimulus AX is paired with US. Little or nothing is learned by the stimulus X. The model describes the effect as follows: A pairing with the US makes X ineffective to US pairing because the first conditioning of A results in $V_A$ being close to $\lambda$; then on an AX trials, $V_{AX}$ is close to $\lambda$ because $V_X$ is zero resulting in an error term ($\lambda - V_{AX}$) that is close to zero; hence $\Delta V_X$ is close to zero and there is little resulting change in $V_X$.

The neural substrates of associations as established in classical conditioning are thought to be related to a rule of neural plasticity as formulated by Hebb in 1949 [11], and which can be summarized by the catchy sentence: "Wire together what fires together" (see Chap. 25). Indeed phenomena like associative LTP and the molecular properties of coincidence detectors such as the NMDA receptor or the adenylyl cyclase and other proteins (see Chap. 26) illustrate molecular mechanisms of associative plasticity in the nervous system. However, there are several problems with these ideas: (1) The timing of CS and US in both excitatory and inhibitory conditioning (usually several to many seconds) are very different from the timing of spikes in spike timing plasticity (usually in the range of a few ms). The discrepancy becomes even more drastic in learning phenomena like nausea-induced learning in which the interval between CS and US can be hours. (2) Latent learning, sensory preconditioning and blocking indicate that contiguity of stimuli is not the only and possibly not even the decisive parameter in associative learning. Rather properties like expectation, deviation from expectation (error signal), and attention need to be considered for which individual molecular and cellular properties are not sufficient but network properties need to be considered.

### 28.3.3 Instrumental or Operant Conditioning

As opposed to classical conditioning, instrumental or operant conditioning requires an active involvement of the animal. A spontaneously generated behavior leads to an event (a value signal V). For example, a hungry animal searches for food (it produces actions A) in a particular sensory environment (S) and finding it induces a rewarding signal (V). Under these conditions the animal learns the relations between its own actions A, the external

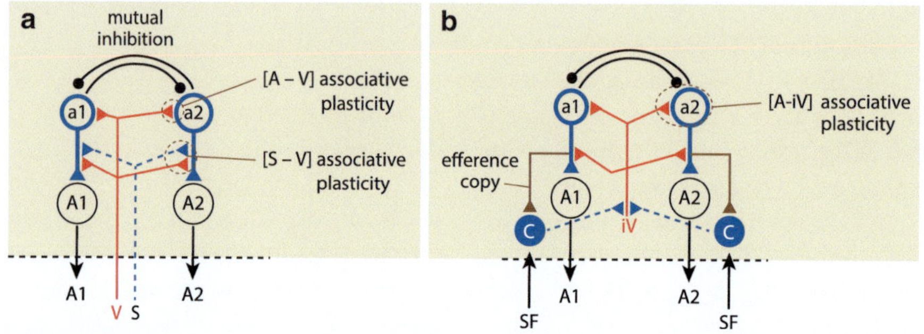


**Fig. 28.2 Conceptual model of neural operations potentially underlying operant learning**. Two motor patterns (actions A1 and A2) are spontaneously generated by pattern generators (a1, a2), whose execution is mutually exclusive by reciprocal inhibition. (**a**) In the case of operant learning with an external value signal either action A1 or A2 leads to a desirable (or avoided) effect as signaled by reward or punishment (*red line*: value signal V). The neural correlate of V strengthens or weakens the respective motor patterns by associatively altering their corresponding neural activities a1 or a2. These associative alterations may also depend on the sensory signals S (*dotted line*) connected with the conditions under which the value signal occurs. The two forms of associations (sensory–value association [S–V]) and (action–outcome association [A–V]) can be related to classical and operant conditioning effects, respectively (see text). (**b**) In the case of operant learning without an external value signal, the sensory feedback SF resulting from the execution of motor pattern A1 or A2 is compared with the corollary discharge (*brown lines*) of the respective motor patterns a1 or a2 (C: comparator). A mismatch leads to an error signal (*blue dotted line*) that activates an internal value system (iV, *red line*) leading to associative alterations in the activation of a1 or a2

conditions signaling the outcome (S) and the evaluating stimulus (V). Thus, there are three associations to be formed: that between action A and stimulus (S) [A–S], the action-value association [A–V], and the stimulus-value association [S–V]. Figure 28.2a presents a conceptual model of neural operations potentially underlying **operant learning**. Two motor patterns (actions A1 and A2) are spontaneously generated by pattern generators (see Chap. 23) a1, a2, whose execution is made mutually exclusive by reciprocal inhibition. Action A1 or A2 lead to a desirable (or avoided) effect as signaled by reward or punishment (value signal V). The neural correlate of V strengthens or weakens the respective motor pattern by associatively altering their corresponding neural activities a1 or a2. These associative alterations may also depend on the sensory signals S connected with the conditions under which the value signal occurs. The three associations ([A–S], [S–V], [A–V]) are partly independent, because the same [A–V] association can be formed for different S, or different actions can lead to the same outcome. [S–V] associations are of the Pavlovian kind (**classical conditioning**) and appear to be established independently and in different neurons than the other two associations: [A–V] associations are related to **goal-directed behavior** (or operant behavior in a strict sense); [A–S] associations are thought to lead to **habit formation**, the development of stereotypical motor patterns under particular stimulus conditions. Behavioral test procedures allow to at least partially separate between these three forms of learning. For example, if the animal is exposed to highly variable contingencies between own actions and outcome [A–V] associations are down-graded and habit formation [A–S] becomes the dominant behavior adaptation. In the extreme case when own actions are fully independent of the occurrence of the value signal animals become passive and produce no actions any more, a situation called **learned helplessness**.

Different brain regions are involved in habitual [A–S] and goal directed [A–V] learning in mammals. [A–S] learning can be mediated at many locations within the nervous system, including the spinal cord, the basal ganglia and the striatum, whereas goal directed [A–V] learning is mediated by cortical structures such as the prelimbic area and the insular cortex, and by neurons in the basal ganglia. In fast sequences of operant learning trials (seconds), neurons of the prefrontal cortex and the caudate nucleus (a structure of the basal ganglia in the mammalian brain) code outcome-reward relations in their sustained activity: correct responses in the last trial lead to lasting activity, wrong responses to reduced activity. This finding indicates that the network of neurons involved in operant learning stores the neural correlates of outcome-reward relations for some extended time. This might have two reasons: to keep a transient memory trace for the next decisions to be made and to facilitate long-lasting storage of the memory trace in stable altered synaptic strengths which requires protein synthesis (see Chap. 26).

Studies on operant learning in *Aplysia* and *Drosophila* allow tracing some of these associations to particular identified neurons and cellular pathways. The motor neuron B52 in the feeding pattern generator of *Aplysia* receives input from both mechanosensory neurons (representing the CS) and the value-representing dopamine neurons. The dopamine neurons also synapse onto presynaptic terminals of the mechanosensory neurons upstream to B52. Coincident activity of dopamine neurons and B52 leads to [A–V] associations of operant behavior by enhancing the excitability of B52 (threshold for spiking is reduced and input resistance enhanced). Coincident activity of dopamine neurons and mechanosensory neurons strengthens their presynaptic activity (classical conditioning, [S–V] association). Interestingly, when [S–V] associations are induced alone B52 excitability is reduced indicating a (as yet unknown) link between classical and operant conditioning in this circuit. In *Drosophila*, operant learning (e.g., a stationary flying fly in an arena in which the animal controls the appearance of stimuli by its behavior and is heated up when is steers towards a particular stimulus) involves both classical [S–V] and operant [A–V] conditioning. The first one develops fast, the latter slowly. The transition from the fast to the slow learning effect requires the mushroom body. The cellular pathways underlying the two forms of plasticity differ. Classical conditioning requires the rutabaga gene-related adenylyl cyclase, operant conditioning a protein kinase C-dependent pathway (see Chap. 26).

Some forms of operant learning lead to improvement of motor performance just by the repetition of motor program without an obvious evaluating signal. In many animal species movements are not perfect when performed the first time. Running, swimming, flying, singing, and other forms of communication may become better, faster, and less energy consuming with practice. New motor patterns in manipulating objects (e.g., pollinating insects extracting nectar and pollen from flowers, birds building a nest and using tools, or mammals preparing food for ingestion) improve with exercise. If the execution of a motor pattern is disturbed (for example, by injury to a limb) changes of the movement pattern can adjust for the damage. The concepts accounting for these forms of learning (Fig. 28.2b) assume the comparison between a sensory feedback (SF) resulting from the execution of movement A1, A2 (the outcome) and a neural template (an **efference copy**) of the neural program initiating the movement. The efference copy is also called a **corollary discharge**, because it accompanies the neural activity leading to the motor pattern and runs in parallel to it (see Chap. 23). The comparison between efference copy and sensory feedback in a neural comparator (C) leads to an error signal (*blue line* in Fig. 28.2b) that activates an internal neural value system leading to associative alterations in the activation of a1 or a2. The corollary discharge can be considered as a neural correlate of expectation, a pattern of activity that precedes conditions in the external world. Thus, on a formal ground the neural operation of comparison between corollary discharge and sensory feedback is equivalent to the deviation from expectation $\Delta V_A$ as derived in the delta rule in classical conditioning (see Box 28.1), and $\Delta V_A$ can be considered to be equivalent to a neural error signal.

The corollary discharge (efference copy) was postulated already by Helmholtz (1821–1894) and has been conceptualized to be forwarded from the motor system to the sensory system providing inhibitory input to the incoming sensory signals. However, it has been difficult to identify such neural pathways. A single multisegmental interneuron (Fig. 28.3, corollary discharge interneuron CDI) was found in the cricket which provides presynaptic inhibition to auditory afferents and postsynaptic inhibition to auditory interneurons when the animal produces its own song but not when it hears songs from other animals. When the animal sings without sound (fictive song) CDI is excited and inhibits the coding of played songs. The authors [19] managed to stimulate CDI intracellularly, resulting in inhibited auditory encoding. They also found that excitation of CDI is specific for self-generated songs and not for other motor patterns, demonstrating that CDI is both necessary and sufficient for the blocking of sensory input expected to be received from own song production. It will be interesting to see whether a mismatch between the expected song pattern and the received song produces an error signal that may be used to fine-tune own song production, e.g., after some disturbance of the song production by the wings.

The concept of an **error signal** driving associative learning has also a strong impact on neural studies of learning related plasticity in the nervous system. The dopamine neurons of the ventral tegmentum of the mammalian brain, for example, change their response properties to the CS according to a modified delta rule [22]. A similar effect was found in an identified neuron in the honeybee brain [9]. This neuron known as VUMmx1 (ventral unpaired median neuron 1 in the maxillary neuromere) codes for the reinforcing property of the US sucrose in olfactory learning, and thus appears to have similar properties as the dopamine neurons in the mammalian brain. As Fig. 28.4 shows, VUMmx1 changes its response properties in the course of learning as do the dopamine neurons in primates (Fig. 28.5). During differential conditioning VUMmx1 develops responses to the CS+ and stops responding to

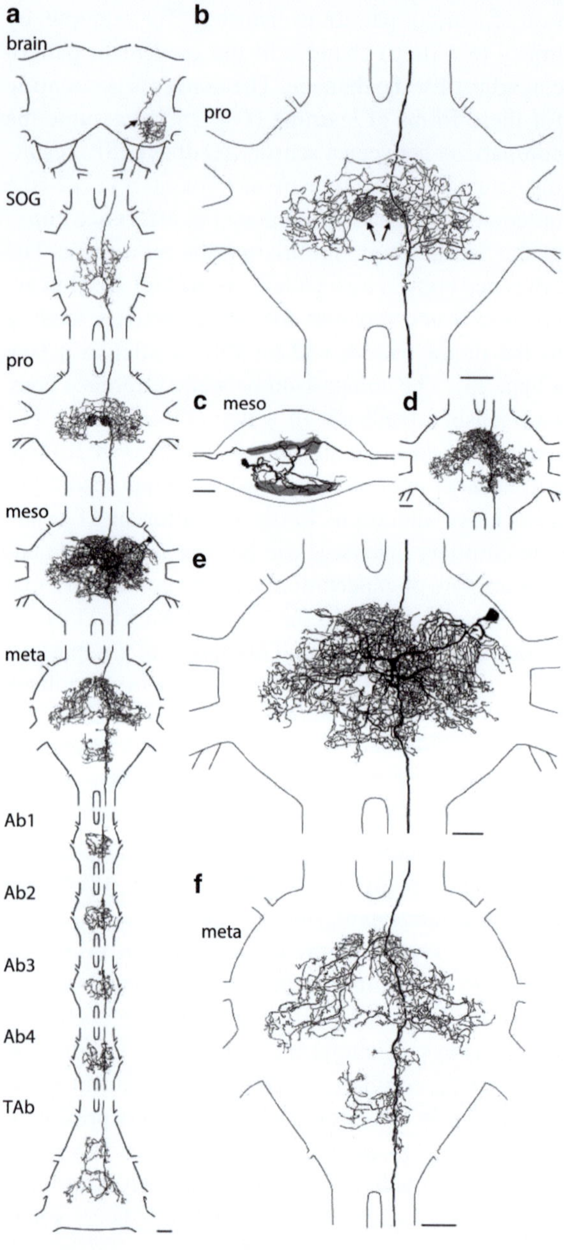


**Fig. 28.3 Morphology of a single, multisegmental interneuron** responsible for pre- and postsynaptic inhibition of auditory neurons in the singing cricket (*Gryllus bimaculatus* ) representing a corollary discharge neuron (CDI). (**a**) A whole-mount staining of the CDI in the CNS of an adult male cricket in ventral view. The soma and dendrites are located in the mesothoracic ganglion, and two axons project throughout the whole CNS with extensive varicose arborizations that are bilateral in every ganglion except the brain. The *arrow* in brain indicates the anterior branch of CDI. (**b**) Axonal arborizations in the prothoracic ganglion; *arrows* indicate overlap with the auditory neuropils. (**c**) Lateral view of CDI in mesothoracic ganglion. The soma is positioned medially near the dorsolateral edge of the ganglion. From the soma the primary neurite extends in a loop toward the middle of the ganglion and gives off a widespread bilateral array of smooth branches typical of insect dendrites. Two axons originate centrally in the ganglion and extend both anteriorly and posteriorly. (**d**) Ventral axonal arborizations in the mesothoracic ganglion. (**e**) Dendritic (dorsal) and axonal (ventral) arborizations of CDI in the mesothoracic ganglion. (**f**) Axonal arborizations of CDI in the metathoracic ganglion have a similar morphology to those in the mesothoracic ganglion. Abbreviations: *SOG* subesophageal ganglion, *Pro* prothoracic ganglion, *Meso* mesothoracic ganglion, *Meta* metathoracic ganglion, *Ab1 to Ab4* abdominal ganglia 1–4, *TAb* terminal abdominal ganglion. Scale bars: 100 μm (After [19] with permission from AAAS)

CS–. If now the US is given after the CS+ one finds no responses to the US anymore, but a US after CS– is well responded to. Thus, an expected US (after CS+) is ineffective, whereas an unexpected US (after CS–) is highly effective. The delta rule and the concept of comparison between corollary discharge and sensory feedback postulate that learning occurs only if $\Delta V_A > 0$. In classical conditioning these properties were related to blocking and second-order conditioning (see Box 28.1 and Table 28.2). In operant conditioning no learning occurs when the error signal is zero. It is conceivable that the acquired responses of the value neurons (the activity of dopamine neurons or of VUMmx1) implement the neural error signal and act as neural reinforcement. However, it has not yet been proven that the neural error signals resulting from a mismatch between the corollary discharge and the sensory feedback really drive the internal value system.

## 28.4 Nonelemental Forms of Associative Learning

### 28.4.1 Definition and Standard Paradigms

Nonelemental forms of learning were developed to reach a more detailed analysis of learning. Here, the associative strength of a stimulus, event, or action is

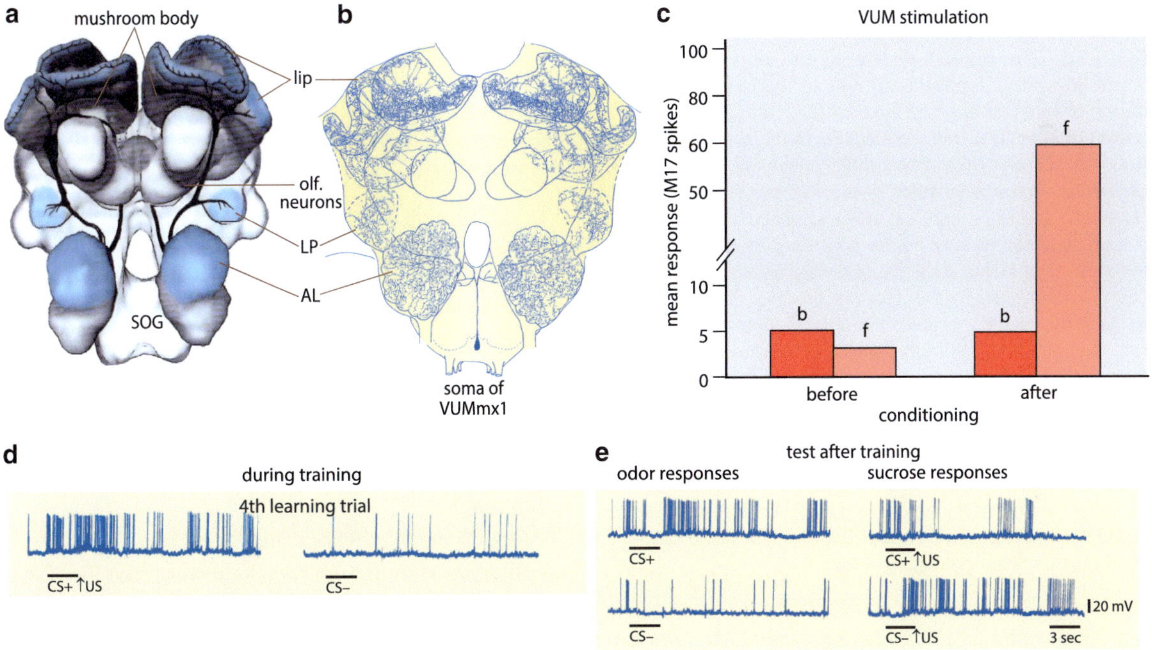


**Fig. 28.4** **Properties of a single identified neuron** (*VUMmx1*, ventral unpaired median neuron 1 of the maxillary neuromere) in the bee brain that represents the reward in olfactory learning. (**a**) The honeybee brain. The olfactory neuropils are marked in *blue*, the antennal lobe (*AL*), the lip region of the mushroom body, the lateral protocerebrum (*LP*). The tracts of olfactory interneurons (*olf. neurons*) connect the AL with the lip and the lateral protocerebrum. (**b**) The VUMmx1. Dendritic branches converging with the olfactory neuropils are marked in *blue*. The soma of VUMmx1 is located in the ventral midline of the subesophageal ganglion (*SOG*). (**c**) Intracellular stimulation of the VUMmx1 replaces the sucrose reward in olfactory conditioning. Before conditioning, the low level of potentials in a muscle involved in the extension of the proboscis (ordinate) indicates that the animals did not respond to the odor. After conditioning, the responses are high in animals experiencing forward pairing of odor and VUMmx1 excitation f but did not change after backward-pairing b. (**d**) VUMmx1 learns about the CS. In differential conditioning the response increases for CS+ and decreases for CS–. Note that a CS+ trial includes the stimulation with US (sucrose) to which VUMmx1 shows a strong response. (**e**) Tests after differential conditioning show an enhanced response to CS+ and a reduced response to CS–. If an expected US follows CS+ the US response is blocked, whereas an unexpected US after CS– is strongly responded to (After [9] with permission from Macmillan Publishers Ltd.: [Nature], © (1993) and [17], © (2001), with permission from Elsevier)

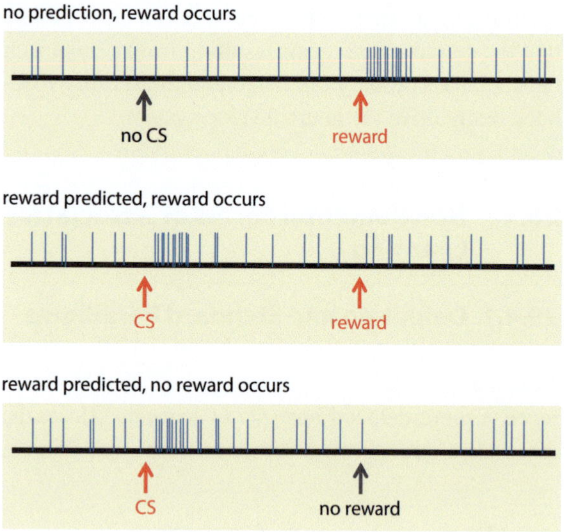


**Fig. 28.5 The reward system in the mammalian brain.** Recordings from dopamine neurons of the ventral tegmentum during reward learning in a monkey [23]. *Top*: Before learning, a drop of appetitive fruit juice occurs in the absence of prediction – hence a positive error in the prediction of reward. The dopamine neuron is activated by this unpredicted occurrence of juice. *Middle*: After learning, the conditioned stimulus predicts reward, and the reward occurs according to the prediction – hence no error in the prediction of reward is seen. The dopamine neuron is activated by the reward-predicting stimulus but fails to be activated by the predicted reward. *Bottom*: After learning, the conditioned stimulus predicts a reward, but the reward fails to occur because of a mistake in the behavioral response of the monkey. The activity of the dopamine neuron is depressed exactly at the time when the reward would have occurred. The depression occurs more than 1 s after the conditioned stimulus without any intervening stimuli, revealing an internal representation of the time of the predicted reward. Each panel shows the time histogram and raster of impulses from the same neuron *CS* conditioned reward-predicting stimulus

ambiguous and therefore cannot predict obvious ways of solving a problem. For example: Stimulus A is rewarded as often as not rewarded, but it is rewarded whenever it is presented together with a different stimulus B, whereas it is nonrewarded when presented together with a third stimulus C (AB+ vs. AC–, + indicating reward, – no reward). The animal cannot rely on the pure associative strength of A, but must include the context B or C.

Standard paradigms for nonelemental learning include: negative patterning (A+, B+, AB–), biconditional discrimination (AB+, CD+, AC–, BD–), and feature-neutral discrimination (B+, AC+, C–, AB–). In all of these, each stimulus appears rewarded as often as not. In **negative patterning** the animal learns to respond to the single stimuli A and B but not to their compound AB. This problem does not admit elemental solutions, since the animals learn that AB has to be different from the linear sum of A and B. In **biconditional discrimination**, the animal learns to respond to the compounds AB and CD and not to the compounds AC and BD. Each element A, B, C, or D, appears rewarded as often as not such that it is impossible to rely on the associative strength of a given stimulus to solve the task. In **feature-neutral discrimination**, the animal learns to respond to B and to the compound AC but not to C and the compound AB. In this case, each element is again ambiguous such that the animal learns the predictive value of the compounds AB and AC, independent of their composing elements. Other less formalized paradigms are contextual learning and rule learning (see below). The more formalized problems appear to be closer to neural interpretations, and have been assigned to cortical and hippocampal circuits in mammals as opposed to elemental forms of associative learning which may not require these circuits.

Two behavioral theories have been proposed for explaining negative patterning and biconditional discrimination: the **configural theory**, which proposes that a compound AB creates an entity different from its components ($AB = X \neq A + B$), and the **unique-cue theory**, which proposes that a compound is processed as the sum of its components plus a stimulus (u) that is unique to the joint presentation of the elements in the mixture ($AB = A + B + u$). In the latter case, the unique cue supports an inhibitory strength assigned to the compound.

Free-flying honeybees and fixed bees conditioned to olfactory stimuli with sucrose reward (see Fig. 28.1) solve a biconditional discrimination (AB+, CD+, AC–, BD–) and negative patterning task (A+, B+, AB–). Thus, bees base their discrimination on separate neural processing of the compound AB. It is argued that a high-order integration center of the insect brain, the mushroom body (see Chap. 13) is involved. The data are in line with the unique cue theory, and it is concluded that the unique cue is created by convergent neural pathways in the mushroom body. Lobsters placed in an aquarium learn to avoid an olfactory stimulus delivered in water with a mechanosensory disturbance. When they are trained to an olfactory compound AX lobsters stop searching AX but still search when presented A alone, X alone, or a novel odor Y. Similarly, a novel compound AY does not inhibit searching behavior. This result is consistent with learning the compound AX as an entity different from its components A and X.

### 28.4.2 Selective Attention, Discrimination, and Generalization

Animals attend to stimuli depending on their motivations and needs. Hungry animals are more sensitive to food-related stimuli, sexually motivated animals to sexual stimuli coming from a potential partner, frightened animals to stimuli that signal protection and shelter. Selective attention consists in the ability to focus perceptually on a particular stimulus and to ignore nonrelevant stimuli. It implies that the representation of the stimulus has been filtered or modified, presumably so that it can be processed or responded to more efficiently. Ethologists illustrate selective attention with "search images" (innate or learned perceptual mechanisms that promote the behavior in question), sensory physiologists point out that selective attention leads to higher sensitivity and more accuracy in perceiving attended stimuli, and learning theorists notice that selective attention can be induced and modified by particular procedures of training. We are dealing here with the latter. **Discriminative learning** is the traditional approach. The animal gradually learns to attend a discriminative stimulus. For example, a rabbit is trained in an eye blink paradigm to respond to a sound of 1,200 Hz (Fig. 28.6a). In one situation (T1) only the 1,200 Hz pulse appears shortly before the air puff against the eye (absolute training), in a different situation (T1–T2) a sound of 2,400 Hz (T2) is intermixed with T1 and is not followed by an air puff (differential training), in a third situation (T1-L) a light bulb is switched on which is also not followed by an air puff (differential training with another modality). Sound discrimination is best after differential training, but the light stimulus leads also to better sound discrimination. The attention-inducing effect of a stimulus of other modality can be very strong (Fig. 28.6b). A pigeon learns to peck an illumined key when a sound S+ of the frequency 1,000 Hz appears (upper curve: frequency discrimination is very low). If the illumination of the key is switched on and off from time to time without the sound but food pellets appear only when both the sound rings and the key light is on, sound frequency discrimination is much better (lower function). A discriminative signal of the same modality can also shift stimulus discrimination. Figure 28.6c shows the result of training a horse to a circle of 60 mm diameter (S+) who had to discriminate it from circles with smaller or larger diameters. The generalization profile for different diameters of circles was rather symmetrical. When the same horse was exposed to differential training with a 38-mm circle, the generalization function was quite asymmetrical shifting best discrimination to even larger circles than S+. These experiments document that animals generalize less after differential conditioning, probably because excitatory and inhibitory learning interact (see above). Furthermore, generalization decreases when the animal is more attentive to the stimulus.

Attention also changes in long series of training as indicated in a rather paradoxical but well-documented phenomenon, **overtraining reversal effect**. In such a situation animals are trained to, e.g., dual choice discrimination (S1+, S2−); then the schedule is reversed (S1−, S2+). Animals are found to be more prepared to reversal after longer training. It is argued that overtraining leads to a loss of attention, and the reversal makes the animals attentive again by the surprise effect. This interpretation postulates that the animal develops an expectation about the outcome of their (trained) behavior. Evidence for this interpretation comes from experiments in which two stimulus conditions were trained, one (S1) associated with a particular reward (dry food), another one (S2) with water. When the animals were made either thirsty or hungry and exposed to both S1 and S2 they chose S2 when thirsty and S1 when hungry (differential outcome effects).

Do insects have selective attention? *Drosophila* flying stationary in a circular arena switch their visual tracking between two vertical bars thus demonstrating selection of two possible targets. Local field potentials recorded in the central brain (possibly originating in the mushroom bodies) show that activity in the 20–30 Hz range increases as a response to selecting a bar. The local field potentials increase with the novelty and the salience of the stimuli, are anticipatory and are reduced when the fly is in a sleep-like state. These results suggest that selective attention underlies visual tracking in flies. Honeybees discriminate colors and patterns depending on the kind of training (absolute, differential). Overtraining leads to better reversal learning as it does in mammals. Honeybees learn different stimulus-reward associations for different contexts indicating that they pay attention to context-relevant stimuli. However, the effects of attention-inducing stimuli on discrimination learning and generalization as well as differential outcome effects were not systematically tested in any invertebrate yet.

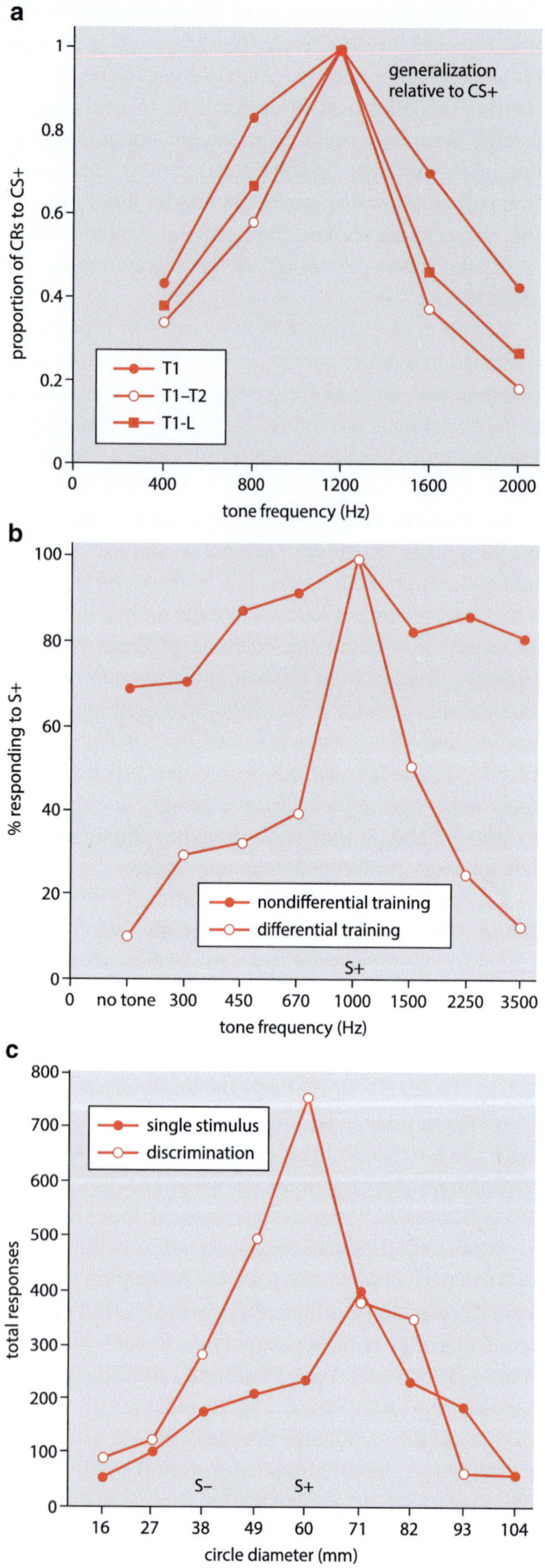


**Fig. 28.6 Discrimination and generalization.** (**a**) The eyeblink response of rabbits was trained to a tone of a frequency of 1,200 Hz either alone (*T1*, absolute training) or differentially to two tones, T1 (CS+: 1,200 Hz) forward-paired, T2 (CS–: 2,400 Hz) backward-paired, or to a tone T1 (CS+: 1,200 Hz) forward-paired and a light stimulus (CS–: L) backward-paired. Notice that the generalization gradient is narrower after differential conditioning even if CS– is of a different modality. (**b**) Differential conditioning leads to sharp generalization functions. Pigeons were trained to peck a key when a tone of 1,000 Hz was presented. In the case of absolute training, pigeons generalized to tones of all frequencies tested, after differential training, the generalization gradient becomes sharp and centered around the trained tone of 1,000 Hz. (**c**) Differential operant conditioning can lead to a peak shift of generalization. Initially a horse was rewarded for pressing a bar in a 60-mm-diameter circle after 60 s. Testing different circles of different diameters (abscissa) shows a symmetrical generalization function centering around the trained stimulus (*open circles*, single stimulus). Afterwards a CS– of 38 mm diameter was introduced and again different diameters were tested (*filled circles*, discrimination). The generalization function shifts to larger diameters (After [14], © (2008), with permission from Elsevier)

### 28.4.3 Rule Learning, Rational Reasoning, and Insight

#### 28.4.3.1 Delayed Matching to Sample

Rules need to be extracted from multiple exposures to a perceptual and/or a performance task when single exposures do not provide the information about the underlying rule.

A rather simple problem for rule-learning is **delayed matching to sample or non-sample (DMTS, DMTNS)**. In such a problem the animal has to choose (or to not choose) the target which resembles the one it was exposed to just before. Successful extraction of the rule implies that the animal performs correctly to novel stimuli. Imagine a honeybee is trained to fly into a maze in which she has to make a decision to enter one or the other of two arms that are arranged like a Y (therefore this arrangement is called a Y-maze). The bee learns to choose the visual targets (colors, patterns) at the end of each arm depending on what it has seen when entering the maze. DMTS and DMTNS tasks are well learned by the bee under such conditions, and the bee transfers the rule to a novel task. For example, if she had learned to fly to the blue target when blue was seen at the entrance, and to the yellow target when yellow was seen, the bee transfers this rule to a novel visual task: if she is exposed for the first time to two patterns of vertical or horizontal stripes she will apply the same rule. She even transfers the rule from learned visual targets to an olfactory task [8]. In mammals it was found that extended delays between the signals to be compared in DMTS and DMTNS requires the hippocampus. A more complicated problem is **transitive inference**. In this problem, animals have to learn a transitivity rule, i.e., if A>B and B>C, then A>C. Preference for A over C in this context can be explained by two strategies: either a form of deductive reasoning in which the experimental subjects construct and manipulate a unitary and linear representation of the implicit hierarchy A>B>C; or a form of associative responding as a function of reinforced and not reinforced experiences, in which case animals choose among stimuli based on their associative strength. The latter explanation can be ruled out by careful experimental design, e.g., by training an animal to five different visual stimuli A, B, C, D, and E in a multiple discrimination task A+ vs. B–, B+ vs. C–, C+ vs. D–, D+ vs. E–. Such training involves overlapping of adjacent premise pairs (A>B, B>C, C>D, D>E), which underlie a linear hierarchy A>B>C>D>E. After training, animals are tested with B vs. D, a non-adjacent pair of stimuli that was never explicitly trained. In theory, B and D have equivalent associative strengths because they were associated with reinforcement or absence of it equally often. Thus, if the choice of the animal is guided by the stimulus' associative strength, it should choose randomly between B and D. If, however, it applies a transitivity rule, it should prefer B over D. Many mammals extract the transitivity rule, honeybees appear not to learn it.

#### 28.4.3.2 Occasion Setting

The so-called **occasion setting** problem requires also the extraction of a rule although nonelemental forms of associations may be sufficient. In this problem, a given stimulus, the occasion setter, informs the animal about the task. This basic form of conditional learning admits different variants depending on the number of occasion setters and discriminations involved, which have received different names. An example involving two occasion setters is the so-called **transwitching problem**. In this problem, an animal is trained differentially with two stimuli, A and B, and with two different occasion setters C1 and C2. With C1 stimulus A is rewarded while stimulus B is not (A+ vs. B–), with C2 it is the opposite (A– vs. B+). Focusing on the elements alone does not allow solving the problem as each element (A, B) appears equally as rewarded and nonrewarded. Each occasion setter (C1, C2) is also rewarded and nonrewarded, depending on its occurrence with A or B. Animals have therefore to learn that C1 and C2 define the valid contingency. The transwitching problem is considered a form of contextual learning because the occasion setters C1 and C2 can be viewed as contexts determining the appropriateness of each choice. Note that biconditional discrimination (AB+, CD+, BC–, AD-see above) is also a transwitching problem, and thus an occasion setting problem, if one considers A and C as occasion setters for B and D (i.e., given A, B+ vs. D–, and given C, B– vs. D+). All of these problems are forms of conditional learning in which a stimulus can have different associates depending on the conditions in which it is presented.

#### 28.4.3.3 Categorization

When animals categorize objects, they apply both the rule of similarity and of difference. Some objects are treated as belonging to the same category, and others to a different category. Pigeons learn hundreds of pictures of natural objects and categorize them differently (houses, humans, flowers, cars, etc). When exposed to new exemplars they group them accordingly. Honeybees can extract the feature as symmetrical vs. asymmetrical from multiple instances and transfer this categorization to new exemplars. Both pigeons and honeybees learn a reversal of their behavior to the category much faster than establishing the rule at the very first instance.

In all these experiments it is crucial to control that some low-level feature (e.g., overall brightness or color, overall spatial frequency distribution, a common particular key feature) may not explain the behavioral categorization effect. Even so, it is not clear what it means that animals categorize objects. Do they create an abstract concept of a category as humans do, e.g., that of symmetry, of houses, of trees? Since little is known about the neural correlates of such concepts, the question cannot yet be answered. Possibly there is one exception – the concept of number.

#### 28.4.3.4 Counting

Can animals *count*? One of the difficulties in answering this question lies in the enormous variety of behaviors that can be controlled by numerical attributes of stimuli. An organism may be trained to select the larger (or the smaller) of two arrays of items, with the experimenter controlling the non-numerical attributes of the stimuli (e.g., area or density), so that only the number of items in the array can reliably predict reinforcement. The concept of number is abstract and should allow the animal to transfer across different sensory modalities and across different test procedures. True counting requires the presence of cardinality, the one-to-one assignment of a numerical tag to an array, and the presence of ordinality, the ability to order these numerical tags. Furthermore, the animal needs to be able to transfer to new numbers. For example, a rat trained to press a lever twice after two light flashes and four times after four light flashes ought to be able to spontaneously press a lever three times after three light flashes with no additional training. This strict definition of counting has been met in very few experiments. Rats were trained to press the right lever when two sounds were presented and to press the left lever when four sounds were presented. The non-numerical features of the stimuli – such as the duration of each sound, the interval between sounds, and the total duration of the sound sequence – was controlled so that a reliable discrimination could be based only on the number of the sounds in a sequence. After rats learned this discrimination, the sounds were replaced by light flashes. The rats followed the previously learned rule. In several cases the transfer between items has been well documented, e.g., in ravens, a gray parrot, monkeys, and apes. The chimpanzee Ai was found to be able to perform a three-unit ordering task which included different behaviors depending on whether a higher or a lower algebraic number (up to three) was expected. It thus seems clear that primates and birds can think in at least simple terms about how many objects they perceive. Thinking about numbers in animals may seem to be a matter that would seldom have been useful enough in the past for natural selection to have favored it. Yet when it becomes important to think in this way in order to get food, ravens and few other birds, as well as rats, monkeys, and apes learn to do so, apparently employing a general ability to learn simple concepts.

A less strict definition of counting includes the capacity of animals to judge about the approximate number of items (numerosity) and the sequential experience of items in navigation (precounting). Discrimination between numbers of items up to seven is well documented in birds and mammals, but less well in insects. Sequentially experienced signals have numerical attributes, and animals may use this for navigation (see below). Bees trained to fly in a tunnel experienced up to four visual signals at varying distances. The feeding place was located at a constant relative position with respect to the sequential signals. In a test situation bees searched accurately between the first and second signal if trained to such a relative position, less accurately between the second and third position when trained to that position, and behaved randomly when trained to the position between the third and fourth position. This result indicates that bees might be capable of some form of precounting up to a number of three.

#### 28.4.3.5 Causal Reasoning

Do animals understand that their actions lead to particular consequences? Causal reasoning in the strict sense has been considered as a key cognitive faculty that divides humans from animals. Animals accordingly have been thought to approximate causal learning by associative processes. It has been difficult to ask whether animals understand that their actions cause an outcome rather than just learning about the correlation between stimuli, actions, and outcome. Figure 28.7 describes a series of experiments showing that rats have a much deeper understanding of the causal nature of their actions. One group of rats observed that presentations of a light (L) was followed by a tone (T) and by food (F). In an operant conditioning paradigm they then learned to expect food after pressing a lever that causes a tone. Another group of animals learned that food is predicted by a noisy tone N indicating a direct cause of N for food. If animals of group one caused the appearance of the tone T (intervening situation) they searched less for food than animals of group 2 after they caused N. Obviously, causing the tone by their own action led to a different expectation of the outcome in the first group than causing the tone in the second group. The kind of "thought" about the physical world which the animal may have implicitly applied could be: "I did not cause the tone, therefore the light must be predictive, and thus I expect food" (for the observing situation). For the intervening situation it could be: "I caused the tone, therefore there should be no light, and thus I do not expect food". Obviously the rats derived predictions of the outcomes of interventions after passive observational learning. These competences cannot be explained by associative theories and require the assumption that rats are capable of causal reasoning [3].

### 28.4.3.6 Insight

Selecting and constructing **tools** is frequently seen in animals ranging from rather stereotypical and innately programmed behavior (as e.g., in weaver ants that use their larvae for knitting together bent leaves to construct a nest) to highly flexible and learned behavior (such as tool use in primates). Animals manipulate material giving the impression that they have an insight into the physical conditions of the world. Ravens spontaneously pull a string with a piece of meat at its end upwards by stepwise catching the string with the pick, lifting it, and stepping on the string with one leg (Fig. 28.8b). The most compelling evidence for the understanding of causal properties of physical objects comes from corvids. A New Caledonian crow spontaneously bent a piece of ineffective straight wire into an effective hook tool for retrieving food (Fig. 28.8a) [2].

### 28.4.3.7 Individual Recognition and Self-Recognition

A cognitive component of **self** may be related to individuality as recognized by others and by the animal itself. Cricket males perform rivalry songs, defend their territories, and fight against each other. Winners and losers appear to learn to recognize each other on an individual basis. The yellow-black patterns of the faces and the abdomen of the paper wasp *Polistes fuscatus* vary considerably, making it possible that individual animals in these small colonies might recognize each other. Altering these facial and/or abdominal color patterns induces aggression against such animals, irrespective of whether their patterns were made to signal higher or lower ranking, arguing that this altered aggressiveness indicates individual recognition [24].

Queens of small ant colonies (*Formica fusca*) are individually recognized by their offspring [6], but how about the workers of insect societies? Insect societies are highly structured in groups of animals performing particular behavior (brood care, cleaning, defense, foraging). Members of some of these groups may differ in body morphology (e.g., soldier ants) and stay with the group for their entire life, other group assignments are age dependent and highly adaptive to the colony needs (as is the case in a honeybee colony). Stable or temporal group membership is mutually recognized most likely by odor profiles. E.O. Wilson states on the final page of "The Insect Societies" (1971): "The insect societies are, for the most part, impersonal. The small, relatively primitive colonies of bumble bees and Polistes wasps are based on dominance hierarchies, and individuals appear to recognize one another to a limited extent. In other kinds of social insects, however, personalized relationships play little or no role. The sheer size of the colonies and the short life of the members make it inefficient, if not impossible, to establish individual bonds." However, the sheer unlimited capacity of

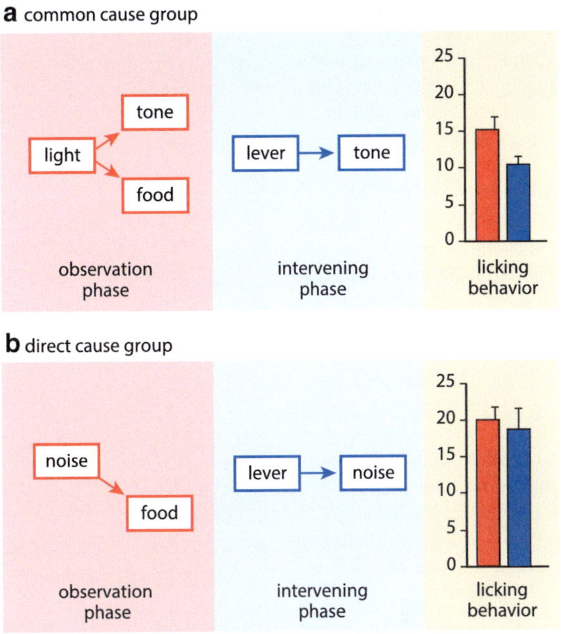


**Fig. 28.7** **Causal reasoning in rats.** (**a**) *Top*: Group one experienced L (light) as a predictor of T (tone) and F (food). In group two a noisy tone N predicted F (observing situation). *Bottom*: Later, in an intervening situation animals of group one could elicit T and those of group 2 could elicit N by pressing a bar (*arrow* to T and N). If the animals had learned the logic relations between light (L), tone (T) and food (F) (group one) or between noise (N) and food (F) (group two) the two groups should behave differently in an intervening situation because eliciting T alone by animals of group one should not lead them to expect F. (**b**) Experimental results. Mean licking behavior was used to measure expected food rewards. Animals of group one expect the food significantly less often than those of group two (Modified from [3] with permission from AAAS)

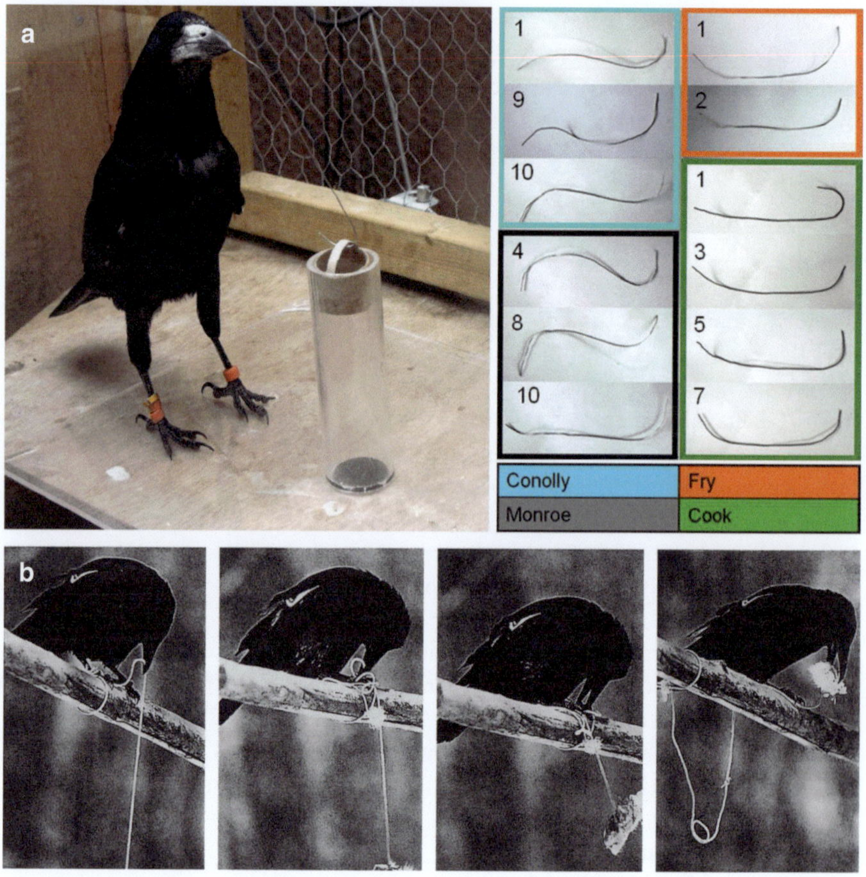


**Fig. 28.8** **Two examples of insight in birds**. (**a**) The Caledonian crow bends a wire to a hook such that it can be used to pull up a container with food deposited in a transparent container. The hooks produced by four different birds are shown on the *right side* (After [2] © (2009), National Academy of Sciences, U.S.A., with permission). (**b**) A raven pulls up a piece of meat hanging on a string. The behavior is not learned but performed at once after the bird has inspected the situation from the distance. Pulling procedure requires sequences of catching the string with the beak, pulling it up, and stepping on it (After [12], with permission)

insects to discriminate odors (see Chap. 13) provides the potential for discrimination of a very large number of group constellations, potentially even to the level of individual recognition.

Does an animal know about itself? **Self recognition** in animals involves at least four different levels: intentionality, individual recognition in a social context, response to a mirror image of the own body, and episodic-like memory. Operant learning includes an intentional component: self-generated behavior creates an expectation via a parallel pathway (efference copy) which allows the distinction between self-produced causal events and environmental events (intentional aspect of behavior). One may even assume that the internal representation of actions (efference copy) developed in evolution for self–nonself discrimination. The identification of the nervous system with its body is probably the most basic component of this distinction. Thus, any animal can be considered as an agent that causes things to happen and recognize these things as caused by itself. Operant behavior is, furthermore, goal directed. The self-generated causes will eventually be more favorable to the animal than the environmental causes. At this level of argumentation any animal will experience "self" as different from "nonself". Such a body-self could be considered to indicate a low level of cognition.

Feeling **pain** may be taken as an indication of body-self experience. Do invertebrates experience pain, a form of self-recognition that includes an emotional and a warning component that points to the future (see Chap. 21)? Locusts and crabs cast off body appendages when attacked. Do they experience different forms of sensory input when they perform these actions themselves or when the same appendages are removed? When honeybees lose their stinger the abdomen is damaged so much that the animal will die. It has been observed that alarm pheromone, which usually triggers an attack flight, induces stress analgesia via an opioid system in the honeybee, potentially indicating that a preparatory response of the nervous system leads to a reduction of the strong sensory input from the body distraction. Opioids, which are usually associated with stress-induced analgesia, have been found in other invertebrates such as crickets and the praying mantis, thus suggesting that their presence may serve to counteract the effect of nociceptive stimuli as in vertebrates.

A common test of self-recognition uses a mirror. Octopus react aggressively to both the own mirror image and a conspecific making it unlikely that they recognize themselves (but they may well be recognized by others as an individual). Primates and some birds pass the mirror test. A magpie spontaneously tries to clean a white paint mark on its head when detecting its mirror image [20]. Based on the different evolutionary history of the bird and primate brain, it is argued that the neural mechanisms of self-recognition in these two groups of animals are convergent traits. Practically nothing is known about their neural basis.

Humans remember episodes of the past, recollect them consciously, and relate them to their own memory in space and time (**episodic** or **autobiographic memory** also called **autonoetic consciousness**). Operant learning provides continuous information about the self-induced causes and distinguishes them from environmental causes (see above). This allows the animal to test whether its expectations about the physics of the world are met, a form of exploring the world. Because the relevant conditions of the world change with the changing requirements of the own body, the distinction between self-related memory and memory about environmental conditions is highly important. It allows to store the own needs and desires together with the actions leading to their satisfaction, which is, in essence, an episodic memory. Since we do not know the conscious state of the animal, such memories have been called episodic-like. Episodic-like memory provides the strongest hint of self-recognition in animals because they include the experience that "I can control the world by my actions". It is this cognitive "I" which defines the self much more than own body recognition in the mirror test.

## 28.4.4 Learning by Observation

### 28.4.4.1 What Is Observation?

Animals improve their perceptual and motor performances and develop new behaviors by observation. Imprinting and many forms of social learning (imitation, acoustic and visual communication, traditions) are based on observation without obvious external reinforcing stimuli rather than on associative learning. How do animals know when and what to learn? Most likely several neural mechanisms are involved given the large range and conditions under which observation learning occurs. A rather low-level mechanism has been mentioned already above in the context of improvement of motor performance by an error signal. It is possible that in a way akin to motor learning (see above, Fig. 28.2b), an internal value coding pathway is generated. In this case, observation learning could function mechanistically in a way similar to associative learning, if the conditions under which it happens activate an internal value system. In imprinting, this could be just the innate program of a sign stimulus. The emotions involved in the social context could form such a value system. In other conditions, e.g., exploration in space, tool manipulation, and play, selective attention could be the source for the activation of a value system – however, little is known about the neural mechanisms. The apparent lack of an obvious external value signal may also result from the replacement of a primary reinforcing stimulus by a learned (second-order) reinforcing stimulus. For example, the positive feeling of social coherence could result from primary rewarding signals (feeding, care taking) in the context of social embedding. Learning in the social context may then be second-order associative conditioning.

These arguments support the notion that the mechanisms of observation learning are of associative nature. Although multiple observations indicate an activation of modulatory systems under conditions of enhanced and directed attention as it happens during exploration, social communication, and playful tool manipulation, it cannot be excluded that observation learning may also involve nonassociative mechanisms possibly based on the sheer sequence of events.

### 28.4.4.2 Navigation

Animals and humans know where they are and where they are going next. The question is, how do they know and what does "knowing" actually mean? In migrating animals (insects, fish, turtles, birds, mammals) long-distance vectors are innately determined. Learning the route may play a role in cases of multiple migrations of the same individual but is not a requirement for successful migration. Median-range navigation involves multiple starts and returns from and to a central point, e.g., a nesting site. Learning is an essential strategy under these conditions. We humans experience our ability to orient and navigate in space and time as a set of functions to which we can relate our attention separately and specifically. We identify an object, take a bearing, and approach it when we sense it from a distance. When planning a route we retrieve from memory a sequence of views, and coordinate our navigational task accordingly. Our introspection of a mental map is experienced as a type of frame for localizing ourselves and the geometric relations of objects. This mental map corresponds to some degree to the physical map that results from measurements of distances and angles or from a bird's-eye view [25]. At the same time we have the ability to communicate locations, how to reach them, and what to expect from them.

Obviously, navigation involves multiple perceptual and computational mechanisms at peripheral and central levels of neural integration. Objects are identified, picture memories (not only in the visual domain) and their sequences are formed, motor performances along traveled routes are learned relative to the own body (**egocentric navigation**) and to the spatial relations to and between objects (**allocentric navigation**). All of these multiple cognitive faculties may be partially integrated into a coherent spatial representation, a cognitive (or mental) map. Studies in humans and animals tell us that cognitive maps are not the only possible reference system – for example, path integration or picture memories may be used instead or in addition.

**Path integration** (Fig. 28.9a) requires the computation of the rotatory and translatory components of movement. Body rotation can be measured with respect to external information (e.g., sun compass, far distant cues) or to internal information (kinestetics, e.g., by vestibular system, movements of body parts, see Chap. 16). The translatory component requires an odometer (distance measure) that may gain its information visually (as in flying insects by visual flow) or from some form of step counting (as shown in ants). The neural mechanisms of path integration are unknown. In mammals, modeling studies suggest that the path integrator resides in the entorhinal cortex integrating the signals from spatially tuned principle cells (grid cells) and that of head direction cells, which then is communicated to the hippocampus forming the properties of the place cells [5, 16] (Fig. 28.10). Head direction cells, grid cells, and place cells provide the animal with information about its location relative to local and further distant landmarks. Place cells in particular code not only spatial relations but also local cues like the odor and the sequence of experiences made along the way towards the location. Spatial coding in the hippocampus resides in multiple spatial representations (neural maps) and is highly dynamic. Spatial coding changes (is remapped) when the geometry of the environment changes. These properties make it likely that the multiple and adaptive neural maps in the hippocampus provide the substrate of a cognitive map (see below).

Path integration allows the animal to return to the point of origin along a straight path. The precision of the inbound path decreases with the length of curved outbound path because of error accumulation in the integration process. Path integration is an egocentric mechanism as long as the animal does not learn anything about the spatial relations of the objects experienced during inbound and outbound movements. If it does (this is the case, e.g., in the honeybee and most likely in other animals), then it is a component of an allocentric mechanism.

**Goal-directed vectors** (Fig. 28.9b): Animals steer towards goals without access to any signals emanating from the goal. This is evident in migratory movements, but occurs also in close range navigation. In that case, vector information relative to a compass (Sun, Moon, stars, Earth magnetic field, steady winds, far distant cues) and an odometer is derived from former learning (e.g., the straight return path in path integration, or in the case of the honeybee the vector communicated in the waggle dance). Vectors provide sufficient information for reaching a goal but also for communicating goals among individuals if the angular and distance components are related to commonly agreed references. The honeybees' waggle dance is such a case. Frequently used routes along memorized vectors lead to learning about the spatial relations of the landmarks along the route, and thus convert an initially egocentric navigation into an allocentric one.

**Picture memories** (Fig. 28.9c): Animals learn the visual appearance of the environment around a particularly important location (e.g., the nest) often when viewing from a rather stable vantage point. This behavior is particularly well known in insects. The simple model as shown in Fig. 28.9c assumes a retinotopically stable visual memory and a search strategy that minimizes the angular deviations between the memory and the actually experienced image. Although the mechanisms assumed in this heuristic model do certainly not apply to navigating animals they still propose a minimal concept of navigation according to picture memories. Animals may store several to many picture memories, and thus may navigate from one vantage point to the next.

**Cognitive map** (Fig. 28.9d): Animals are particularly attentive to landmark features when exploring the environment. It is this situation of observatory learning that may allow to determine and memorize spatial relations between objects in a general sense. It is possible that particularly salient features of the landscape (like boarder lines between areas, slope of the landscape, panorama, rivers etc. which are usually referred to as **gradients**) may establish a geometrically organized but spatially coarse "gradient map". Such a gradient map could include islands of fine-grain picture memories such that animals traveling according combined maps have a patchwork of information about where they are and where to go next.

Evidence that animals form cognitive maps comes from the following observations: (i) Tolman who coined the term cognitive map observed that animals (rats, mice) chose the shortest distance to a desired place if they had explored the area (e.g., a maze) before. Since these seminal studies **novel shortcuts** became the signature for a map structure of spatial memory. (ii) When animals are not yet fully trained in a maze they may hesitate at the choice points and perform movements in the direction of the intended goal. (iii) When animals are trained in a complex maze with multiple routes to the goal and one path is blocked they tend to decide for the nearest and shortest open path. (iv) Rats swimming in a milky water learn the location of a safe platform under water with respect to the geometry of the surrounding marks (Morris' water maze). All these behaviors require a functioning hippocampus.

In mammals the hippocampus (together with the entorhinal cortex) communicates with the prefrontal cortex (Fig. 28.10). Functional imaging studies in humans support the view that the hippocampus complex represents locations (grid and place cells), computes shortcuts by path integration together with neurons coding head direction and yet unknown signals from an odometer (the neural distance measuring device), and supports the learning of places from particular views. Action-based representations have been linked to the dorsal striatum, and observer-independent cognitive maps appear to depend critically on retrospinal cortex together with the hippocampus. The essential role of the hippocampus for navigation is also known in fish, reptiles, and birds, and the volume

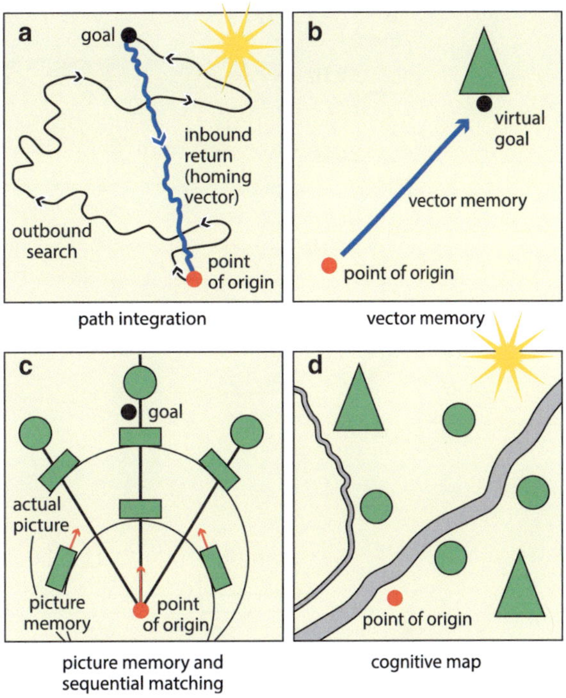


**Fig. 28.9 Mechanisms of navigation.** (**a**) In path integration animals are able to return to the point of origin (e.g., a nest) along a straight path after reaching the return point (goal) along curved movement. The rotatory component of their movements may be measured relative to a far distance source (e.g., the sun), and the translatory movement by some form of an odometer (distance measure). If the animals do not relate its outbound and inbound movements to the geometry of landmarks but to body centred measures path integration is an egocentric form of navigation. (**b**) A vector memory provides the animal with the possibility to reach a distant goal without access to stimuli emanating from the goal (virtual goal). The directional component may be read from a compass (e.g., sun compass), the distance component from an odometer. Both components may be innate as in the case of migration over very long distances or learned (for navigation in the close surrounding). (**c**) A picture memory allows the animal to localize itself relative to the geometry of landmarks as seen from a vantage point (point of origin). Finding this vantage point may involve sequential matching procedures in which the deviation of the actually experienced view from the picture memory is used to reduce the deviation which will bring the animal closer and closer to the vantage point by trial and error. (**d**) The memory structure of a cognitive map relates to the geometry of landmarks relative to a compass system. Such a cognitive map could either be a complete representation of the spatial relations of local landmarks or it could store predominantly those relations between long ranging landmarks (gradient map) into which local view based memories are embedded. In the first case an animal will be able to reach the point of origin from any location within the map, in the latter case the gradient map would be used first to reach an estimated location and then to create a homeward flight by multiple matching procedures of multiple picture memories

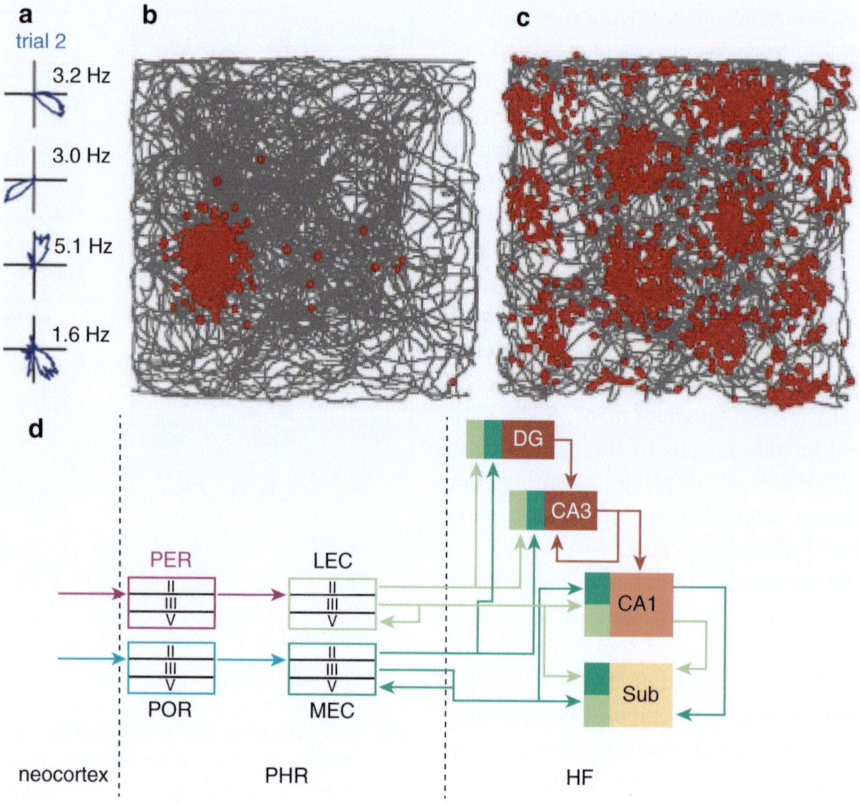


**Fig. 28.10 Examples of head-direction cells, a place cell, and a grid cell recorded from a rat hippocampus.** (**a**) Response characteristics of four head-direction cells. The polar plot describes the neural activity of the respective cell when the animal is looking into the direction indicated by the angular orientation of the *blue line*. The strength of the neural response is given by the deviation from the center (After [15] with permission from AAAS). (**b**) Place cell. The *gray lines* give the running path (trajectory) of a rat exploring a square environment. The *red color* marks the location when the place cell is firing. The population of place cells with different spatial firing properties defines the location of the animal. (**c**) A grid cell is characterized by multiple firing locations arranged in a hexagonal pattern. Grid cells differ with respect to the spatial separation and size of the subfields. (**d**) Circuitry of the hippocampus. The neocortex is connected to the hippocampus mainly via two pathways through the parahippocampal cortex (*PHR*). One projects through the perirhinal cortex (*PER*) and the lateral entorhinal cortex (*LEC*); the other projects through the postrhinal cortex (*POR*) and the medial entorhinal cortex (*MEC*). Cells that carry information about the position of the animal, such as grid cells, head-direction cells, and boarder cells (not shown in the upper diagram) are found in the MEC but not in the LEC. MEC and LEC project to the same regions in the hippocampus, both via direct projections to each hippocampal subfield and via the indirect trisynaptic circuit through dentate gyrus (*DG*) and CA3. Place cells are pyramidal cells of the hippocampal formation (*HF*) (After [5], © (2010), with permission from Elsevier)

of the hippocampus does not only increase in London taxi drivers but also in pigeons with more navigational experience.

Nothing is known about the neural substrate of navigation in any insect. The multisensory convergence in the mushroom body of insects makes it likely that this integration center is involved in navigation. In addition, the central complex may well be involved because it contains neural nets that code the sun compass-related polarization pattern of the sky and the movement of the animal relative to objects [13, 18].

## 28.5 Working Memory: Planning and Decision Making

### 28.5.1 Working Memory: A Definition

Learning leads to a change in behavior. The information necessary to control new and better adapted behavior resides in the nervous system at many levels and is used to control behavior in the future. The entirety of all neural changes induced by learning represents a **memory trace**. Three different components of the memory trace are to be distinguished: consolidation, retrieval, and execution. Learning does not produce the final memory trace immediately. Time- and event-dependent processes form the trace, and are conceptualized as **consolidation processes**. Short-term memory is transformed into mid-term and long-term memory, and the molecular, cellular, neural, and systems-related processes are one of the most intensively studied questions in neuroscience today (see Chap. 26). Stable memory traces need to be moved from a silent into an activated state by **retrieval processes**. Internal conditions of the animal, external cues, and a neural search process (see below) shift a silent memory into an active memory. The **expression** of the active memory may undergo *selection processes* before its content is expressed. Animals need to decide between different options as they reside in memory, and the decision process requires access to the expected outcomes. The expected outcomes are stored in memory, too, and only when the respective memory contents are retrieved they will be accessible to selection processes. This network of interactions between retrieval, selection, and execution is conceptualized in a particular form of memory – working memory.

The concept of **working memory** has been derived from psychological studies in humans, particularly children, which examined the interference of two or more tasks to be performed simultaneously [7]. For example, a subject may be asked to do a simple arithmetic (counting backwards) and at the same time keeping an item in memory. It was found that the capacity of working memory is limited, grows with age of the children, and can be assigned to subcomponents (called the phonological loop, the visuospatial sketchpad, and the episodic buffer) which all interact and converge with a central executive [1]. In animal studies, working memory is often related to a particular form of short-term memory as it is tested in delayed matching to sample or delayed matching to nonsample tasks (DMTS, DMTNS, see above). The memory span is also limited, depends heavily on the task, is sensitive to interference from distracting signals, and depends on the hippocampus. If the DMTS and DMTNS task requires the application of a rule learned in multiple trails, animals have to recruit the memory for the rule and decide whether the rule applies or not.

A basic form of working memory was already mentioned in the model of operant learning (Fig. 28.2b). The comparator C receives input both from the efference copy (which can be considered as the readout of a memory trace) and the sensory feedback accompanying the execution of the action SF. In this elementary form, a decision has to be made on the basis of the match/mismatch between efference copy and SF. Under more complex conditions the animal will find itself exposed to internal body conditions (S1 in Fig. 28.11) and environmental signals that retrieve multiple memory traces which lead to different outcomes if applied (S2, S3 in Fig. 28.11). These signals together will retrieve from stable reference memory several potentially relevant memories (e.g., sensory memories, motor performance memories, value memories) that are shifted into working memory and constitute the active conditions of working memory. The central executive processes receiving such input from working memory will produce the respective patterns of corollary discharge representing the expected outcome of the potential motor patterns. Multiple rounds between working memory and central executive are thought to lead to a decision process that finally will initiate actions.

The conceptual model in Fig. 28.11 does not require any conscious recollection but captures the processing of implicit knowledge as it is available to any nervous system that needs to decide between different behaviors. Any motor command produces an expectation of its outcome (the corollary discharge or efference copy) which is available to the working memory for internal processing leading to an evaluation of whether the expected outcome is desirable on the basis of former experience and body conditions. As all forms of memory, working memory is a process of global neural nets rather than a localized function in any specialized area. In the mammalian brain the striatum, premotor cortex, and inferior parietal cortex will be more involved in the evaluation of potential motor performances, the hippocampus more in those of spatial and sequential navigation tasks, the dopamine system of the ventral tegmentum more on that of the expected reward.

The decision-making process involves components which are well studied in cortical sensory systems. Lower signal-to-noise ratios (equivalent to less strong neural representations, e.g., of a memory readout) are outcompeted by neural activations with higher signal-to-noise ratio. The basic mechanism of neural nets to settle in well-defined representations is mutual inhibition as experienced in visual illusion flip images. Thus, decisions between options that can switch indicate rivalling neural processes. Decision between more

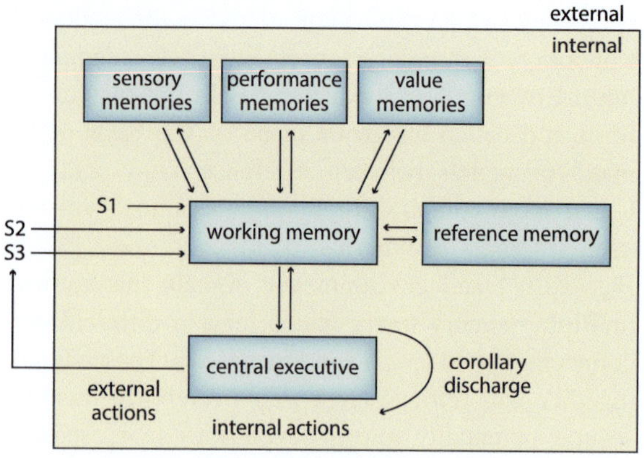


**Fig. 28.11 Conceptual model of processes underlying working memory as it may apply to implicit knowledge of any nervous system that needs to decide between several behaviors.** Working memory receives sensory input from body signals (*S1*) and external signals (*S2*, *S3*). Using these signals it retrieves from stable reference memory relevant memory contents and transfers it into an active state. These active memory contents may belong to sensory memory, performance memory, value memory, or others. Each of these multiple memories are related to particular behaviors that are executed centrally (central executive) leading to the activation of internal actions as represented by their corollary discharge or efference copy. The reciprocal interaction between working memory and central executive will lead to a decision between competing outcomes and finally execute external actions

close neural representations requires longer processing time, as well as decision between more options. A major yet open question relates to those neural mechanisms which connect and orchestrate the multiple parallel processes characterizing working memory function. One proposal is that synchronous spiking activity may be the requirement and the signature for neural decision processes.

## 28.6 Animal Thinking: The Basics

Studies in animal behavior and its neural basis developed into a science by rejecting anthropomorphic terminology and strictly applying descriptive terms. Behavioral processes not directly measurable were either ignored or not accepted as topics for scientific endeavor, as e.g., memory, spontaneity, and creativity of the brain. Behaviorism and ethology, although differing in many respects, developed rather similar strategies in understanding behavior as reflection of input/output properties. As a consequence, spontaneity and creativity of the nervous system was not in the focus of mainstream comparative behavioral biology and neuroscience for most of the last century. In the wake of the cognitive revolution, the conceptual move from black box attitudes to the recognition of the brain as a creative system, research on animal cognition has begun to ask what kind of knowledge animals use to find their way around, how they make decisions between options, and how they represent the social relationships of others around them.

Nervous systems vary in size and architecture, and thus animals come with different adaptations to similar problems. It is sometimes assumed that "simple" nervous systems like those of arthropods and mollusks solve the problems by radically different mechanisms relying on innate routines and elementary forms of associative learning. However, constructing a great divide between simple and advanced nervous systems will lead us astray because the basic logical structure of the processes underlying spontaneity, decision making, planning, and communication are more or less the same. It is a more productive position to envisage the differences in quantitative terms rather than qualitative terms providing us with a wealth of "model systems" to elucidate the essence of the basic processes.

Thinking about the basic design of a brain that subserves cognitive functions, one recognizes a structure of essential modules and their interconnectivity (Fig. 28.12). This architecture of modules appears to be shared by a large range of animal species and may even apply to the worm-like creature at the basis of the evolutionary divide between protostomes and deuterostomes, these two largest evolutionary streams of bilateral animals with a centralized nervous system [4]. Although there are multiples of each of the modules depicted in Fig. 28.11 (multiple perceptual systems, multiple belief-generating systems, multiple desire-generating systems, multiple action-planning systems, multiple motor control systems), the basic idea put forward in this scheme is that perceptual systems feed to three downstream systems arranged both serially and in parallel that converge on the action planning system. Thus, perceptual systems can reach the action-planning systems directly, and, in addition, the desire- and belief-generating systems receiving the same perceptual information will act in parallel onto action-planning, as well.

Given the similarity in the basic design of nervous systems we may ask: Are animals aware of themselves, of what they are doing, of what they are expecting, and

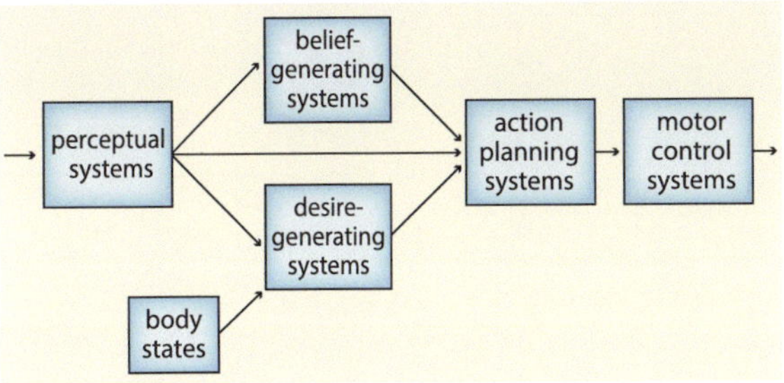


**Fig. 28.12 The cognitive structure of brains composed of modules for perception, desire, and memory stores (belief-generating systems).** Action planning can either be generated by direct inputs from the perceptual systems or result from processes that are generated in parallel pathways weighting the perceptual inputs with respect to expected outcomes (belief-generating systems) and the motivational conditions of the animals body (desire-generating systems). These modules can either be rather basic (as in more simple nervous systems) or highly complex, but in any case the basic structure particularly with respect to direct and indirect pathways and necessity of operations between the models may apply to any nervous system (After [4] with permission from Oxford University Press)

what they intend? This question touches on an epistemological problem, and we would have to ask what is meant by "**awareness**". Certainly we do not mean human awareness, as it is accessible to us by introspection. We certainly do also not imply that all animal species with their enormously different nervous systems have the same kind of "animal awareness". This means that terms like self-awareness, expectation, planning, creativity, and even learning and memory have different meanings for different animals. However, judging from all that we know so far, the difference relates to the level of complexity and not to fundamental differences. It will be the goal of future comparative studies to understand how quantitative differences in the structure and operation of nervous systems lead to the large range of animal cognition around us.

## References


1. Baddeley AD (2001) Is working memory still working? Am Psychol 56:851–864
2. Bird CD, Emery NJ (2009) Insightful problem solving and creative tool modification by captive nontool-using rooks. Proc Natl Acad Sci USA 106:10370–10375
3. Blaisdell AP, Sawa K, Leising KJ, Waldmann MR (2006) Causal reasoning in rats. Science 311:1020–1022
4. Carruthers P (2006) The architecture of the mind. Clarendon Press, Oxford, p 462
5. Derdikman D, Moser E (2010) A manifold of spatial maps in the brain. Trends Cogn Sci 14:561–569
6. El-Showk S, Van Zweden JS, D'Ettorre P, Sundström L (2010) Are you my mother? Kin recognition in the ant *Formica fusca*. J Evol Biol 23:397–406
7. Gathercole SE (2008) Working memory. In: Roediger HL, Byrne JH (eds) Learning and memory – a comprehensive reference, vol 2. Academic Press/Elsevier, Amsterdam, pp 33–52
8. Giurfa M, Zhang SW, Jenett A, Menzel R, Srinivasan MV (2001) The concepts of 'sameness' and 'difference' in an insect. Nature 410:930–933
9. Hammer M (1993) An identified neuron mediates the unconditioned stimulus in associative olfactory learning in honeybees. Nature 366:59–63
10. Hawkins RD, Kandel ER (1984) Is there a cell-biological alphabet for simple forms of learning? Psychol Rev 91: 375–391
11. Hebb DO (1949) The organization of behaviour. Wiley, New York
12. Heinrich B (1995) An experimental investigation of insight in common ravens (*Corvus corax*). The Auk 112:994–1003
13. Homberg U, Heinze S, Pfeiffer K, Kinoshita M, el Jundi B (2011) Central neural coding of sky polarization in insects. Philos Trans R Soc Lond B Biol Sci 366:680–687
14. Kehoe EJ (2008) Discrimination and generalization. In: Byrne JH (ed) Learning and memory: a comprehensive reference. Elsevier/Academic Press, Amsterdam, pp 123–149
15. Langston RF, Ainge JA, Couey JJ et al. (2010) Development of the spatial representation system in the rat. Science 328: 1576–1580
16. McNaughton BL, Battaglia FP, Jensen O, Moser EI, Moser MB (2006) Path integration and the neural basis of the 'cognitive map' 1. Nat Rev Neurosci 7:663–678
17. Menzel R, Giurfa M (2001) Cognitive architecture of a mini-brain: the honeybee. Trends Cogn Sci 5:62–71
18. Neuser K, Triphan T, Mronz M, Poeck B, Strauss R (2008) Analysis of a spatial orientation memory in *Drosophila*. Nature 453:1244–1247
19. Poulet JF, Hedwig B (2006) The cellular basis of a corollary discharge. Science 311:518–522
20. Prior H, Schwarz A, Güntürkün O (2008) Mirror-induced behaviour in the magpie (*Pica pica*): evidence of self-recognition. PLoS Biol 6:e202. doi:10.1371/journal.pbio.0060202
21. Rescorla RA, Wagner AR (1972) A theory of classical conditioning: variations in the effectiveness of reinforcement and non-reinforcement. In: Black AH, Prokasy WF (eds) Classical conditioning II: current research and theory. Appleton-Century-Crofts, New York, pp 64–99
22. Schultz W (1998) Predictive reward signal of dopamine neurons. J Neurophysiol 80:1–27
23. Schultz W, Dayan P, Montague PR (1997) A neural substrate of prediction and reward. Science 275:1593–1599
24. Tibbetts EA (2002) Visual signals of individual identity in the wasp *Polistes fuscatus*. Proc Biol Sci 269:1423–1428
25. Wolbers T, Hegarty M (2010) What determines our navigational abilities? Trends Cogn Sci 14:138–146

# Primate Social Intelligence

# 29

Julia Fischer

Nonhuman primates play an important role for elucidating the foundations of human cognition and social behavior. Indeed, the current predominant hypothesis regarding the evolution of large brains posits that the challenges of living in complex social groups drives brain evolution – the so-called "Social Brain" hypothesis. After a brief introduction into the issue of brain size evolution, this chapter will provide a review over some key abilities that have been suggested to play a role in mastering the complexities of social life in primates, namely the ability to learn from others (social learning), the knowledge about the relationships between others and the processing of social cues (social knowledge), the attribution of mental states to others and oneself ("theory of mind"), including the question of whether or not nonhuman primates know what they know (metacognition), and the use of and responses to signals to regulate social relationships (communication). This chapter will conclude with a brief discussion of the strengths and limits of using nonhuman primates as models for human behavior.

## 29.1 Brain Size Evolution

> Brain size increases may be due to frugivory, to social living, or to both.

One striking feature within the primate order is a disproportionate increase in relative brain size from lemurs to monkeys, apes and humans. In particular, the **neocortex** has experienced considerable expansion. The neocortex is the outer layer of the cerebral hemispheres, and consists of six layers, labeled I to VI (with VI being the innermost and I being the outermost). Together with phylogenetically older structures of the limbic system, the archicortex and paleocortex, it makes up the cortex. The neocortex is important for sensory perception, generation of motor commands, and higher cognition. Figure 29.1 illustrates the enlargement during primate evolution, but also highlights that primates are not the only taxon with an enlarged brain. In the 1980s, the most prominent hypothesis was that the increase in brain size in primates was related to their reliance on fruit in their diet (frugivory), and the resulting need to find food that is patchily distributed in space and time [2]. In recent years, the focus has shifted towards the challenges of living in large and complex groups [3]. Because of the close link between the ecological conditions and social organization, however, the two hypotheses are not mutually exclusive. Several studies have stressed the correlation between neocortex ratio (the ratio between the size of the neocortex relative to the size of the entire brain) and group size (Fig. 29.2); others have called this link into question [5]. A number of scholars have aimed to derive more specific links between particular brain areas and cognitive performance. For instance, it has been proposed that the neocortex is particularly important for problem solving and executive control. Others, however, have pointed out that attempts to link brain size to function is fraught with problems, including the choice of the variables entered

J. Fischer
Cognitive Ethology Lab, German Primate Center,
Kellnerweg 4, 37077 Göttingen, Germany
e-mail: jfischer@dpz.eu

C.G. Galizia, P.-M. Lledo (eds.), *Neurosciences - From Molecule to Behavior: A University Textbook*,
DOI 10.1007/978-3-642-10769-6_29, 

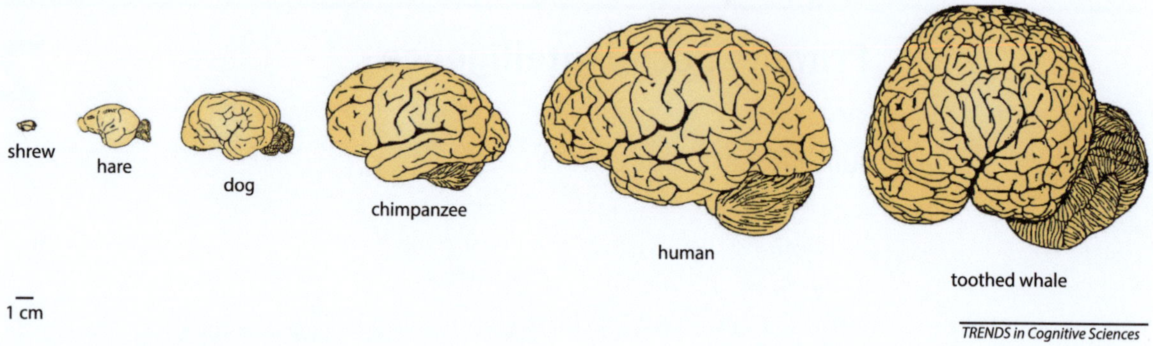


**Fig. 29.1 Expansion of the cortex and variation in brain size in different mammal species**. Humans do not have the largest brain in absolute terms and are exceeded in size by many cetaceans (whales, dolphins, porpoises) and the elephants. Within the primate order, however, humans reveal a significant increase in relative brain size and neocortex ratio (Reprinted from Roth and Dicke [1] with permission from Elsevier)

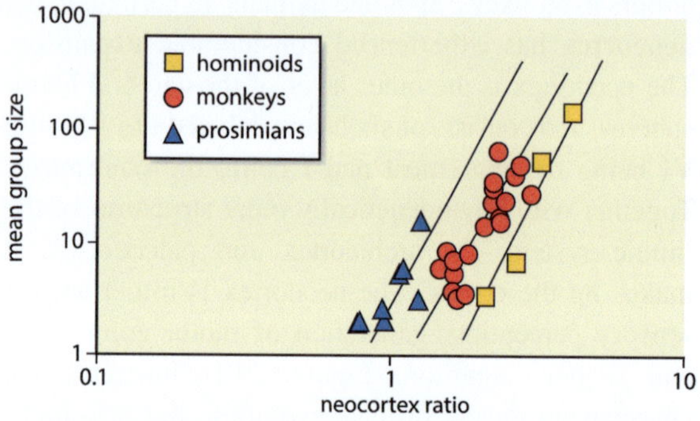


**Fig. 29.2 Neocortex ratio vs. group size**. This graph plots a measure of social complexity (in this case, group size) against a measure of relative brain size (in this case, the ratio of neocortex volume to the volume of the rest of the brain). Hominoids (apes and humans, *orange squares*) require more neural capacity to support a given group size than monkeys (*pink circles*) do, who in turn require more than prosimians (*blue triangles*). Each datapoint is a single genus. Humans are represented by the *top right-hand square* (Reprinted from Barrett et al. [4] with permission from Elsevier)

in the analyses, and the issues associated with multiple correlations [6]. Further, size per se might not be decisive, while the modularity and interconnectedness of different brain areas may confer advantages in processing capacities [7].

> It is still a matter of debate whether brain size per se is crucial, and can be linked to specific cognitive functions such as problem solving and executive control.

## 29.2 Primate Social Relationships

Social bonds are of particular importance for primates, and they use various affiliative behaviors to foster and maintain positive social relationships. Stable social relationships enhance the reproductive success and longevity in female monkeys [8]. Inevitably, however, group living animals also compete over resources such as food or mating partners, leading to sometimes fierce aggression, particularly among males. Such conflicts can seriously disturb relationships. Perhaps not surprisingly, nonhuman primates are said to manage their social relationships, in the sense that they often reconcile with former opponents, depending on the value of the relationship as well as the value of the resource [9].

> Group living is characterized by cooperation and competition.

### 29.2.1 Affiliation

In mammals, the primary bond is the one between mother and infant. In nonhuman primates, positive relationships also occur between other members of a matriline (i.e., female relatives), between males and females, as well as between unrelated members of the same sex, depending on the species. Affiliative relationships are characterized by extended "grooming" sessions between partners, where they comb through the fur of the partner and appear to clean the skin (Fig. 29.3). Although it is acknowledged that such grooming contributes to the health and well-being of the animal, the primary function appears to be in the social domain. Two hormones, oxytocin (OT) and vasopressin (AVP), have been implied in positive social interactions; see Chap. 24. OT had previously mainly been implied in parturition and lactation in mammals, and induces contractions of the smooth muscles. AVP was best known for its role in osmoregulation. More recently, it was acknowledged that OT and AVP also function as central neurotransmitters, which are produced in the parvocellular and supraoptic neurons of the hypothalamus. Central OT affects the establishment of the bond between mother and infant, but is also essential for normal interactions between adult conspecifics and plays an important role in formation of the pair bond [10]. In a study of cotton-top tamarins, affiliative behaviors such as contact sitting, grooming, and sex were correlated with OT levels. Contact and grooming explained most of the variance in female OT levels, whereas sexual behavior explained most of the variance in male OT levels. The initiation of contact by males and solicitation of sex by females were related to increased levels of OT in both [11].

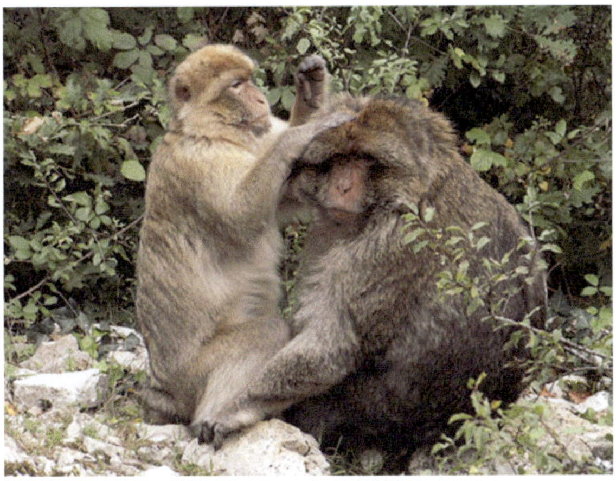

**Fig. 29.3 Grooming in Barbary macaques.** A female Barbary macaque grooming a male group member. Nonhuman primates spend considerable time grooming each other. Grooming patterns are frequently used to determine the strength of affiliative relationships between different individuals. In the short term, grooming may also vary in relation to conflicts or the availability of different resources (Photograph by Julia Fischer)

### 29.2.2 Aggression

Aggression appears to serve a variety of functions, as for instance, the challenge and defense of resources such as food, territories, or mating partners. Aggressive behavior is, however, also accompanied by several disadvantages, for instance the risk of injuries, the consumption of time and energy, or the severance of individual relationships. Aggressiveness is the *propensity* of animals to exert aggressive behaviors. The costs and benefits associated with aggressive behavior vary with the life-history stage of an animal and its evolutionary history.

The androgen testosterone (T) appears to be associated with the effort of males in dominance-seeking behaviors. The increase in production of T when males reach puberty and begin to seek out mating opportunities is at the same time associated with costs, such as reduced immune function [12]. Testosterone levels have been related to the respective life history patterns in reproductive effort in chacma baboons; levels increase with the onset of puberty when males augment their reproductive effort, are maintained relatively high during several years of intense reproductive activity, and then decrease when males change their strategy towards parenting behavior or stop breeding [13] (Fig. 29.4).

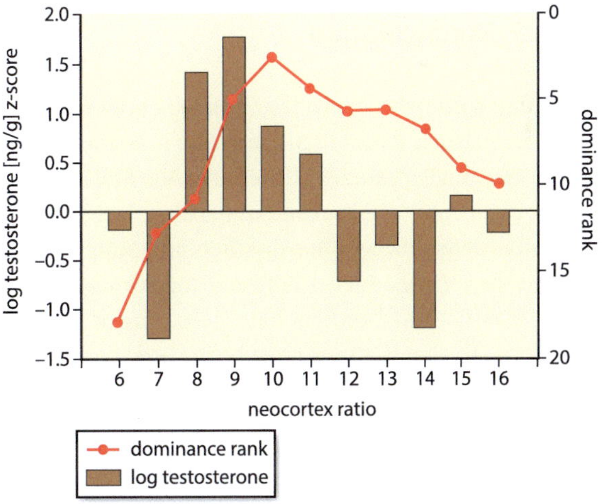


**Fig. 29.4 Rank and T levels in chacma baboons.** Testosterone z-scores (*bars*) and ordinal dominance ranks (*dots*) across 1-year age categories for wild chacma baboon males. Data were collected from $N=41$ males, and dominance ranks were calculated based on agonistic interactions and reflect the rank hierarchy in this group (Redrawn from Beehner et al. [13] with permission from Elsevier. Original data were kindly made available by J. Beehner and T. Bergman)

In humans, nonhuman primates, and other animals variation in aggressiveness between individuals is in part genetically determined and therefore heritable. The key neurotransmitter implied in regulating aggressive behavior is the monoamine serotonin (5-hydroxytryptamine = 5-HT). 5-HT appears to play an inhibitory role in aggressive behaviors. For instance, monkeys with high levels of the 5-HT metabolite 5-hydroxyindoleacetaldehyde (5-HIAA), an indicator for serotonergic activity, are less aggressive than their low-level counterparts. 5-HT levels are influenced by tryptophan hydroxylase (TPH), 5-HT transporter (5-HTT), and monoamine oxidase A (MAOA). TPH is the rate-limiting enzyme in the synthesis of 5-HT, while 5-HTT is responsible for the reuptake of 5-HT into serotonergic neurons. The blocking of the transporter using specific serotonin reuptake inhibitors (SSRIs) increases 5-HT concentrations in the synaptic cleft and decreases aggressiveness, both in humans and nonhuman primates. MAOA degrades 5-HT to 5-HIAA within serotonergic neurons, and is also responsible for the degradation of norepinephrine and dopamine. Polymorphisms of the genes involved in the serotonergic system are thought to affect levels of aggressiveness, both in humans and nonhuman primates.

### 29.2.3 Reconciliation

Nonhuman primates, as well as other socially interacting species such as spotted hyenas or bottleneck dolphins show higher rates of affiliation in the first minutes following a conflict compared to baseline levels. Such affiliative interactions after conflicts are supposed to restore the tolerance level between former opponents after a conflict and therefore have been termed "reconciliation". Reconciliations have been shown to reduce stress and self-directed behaviors, lower the probability of renewed aggression, and permit tolerance around desirable resources [9]. Levels of postconflict affiliation may vary between species in relation to their degree of despotism as well as in relation to the specific relationship between two subjects. The valuable relationship hypothesis predicts that animals that support each other frequently or maintain close bonds will have a higher tendency to reconcile than animals whose relationship is less valuable. While it is sometimes taken for granted that friendly interactions between former opponents will diminish renewed aggression, a study in Barbary macaques (*Macaca sylvanus*) showed that this is not always the case: after 'reconciliation' following conflicts over food, the former victim was frequently aggressed again [14].

## 29.3 Social Knowledge

Given the importance of social relationships among primates, it seems highly likely that evolution has put a premium on the refinement of socio-cognitive abilities. One precondition to maneuver in a complex social setting is the ability to keep track of others' social relationships. There is both observational as well as experimental evidence that monkeys have an exquisite understanding of the different types of relationships among their group members, such as kin, dominance, and mating partners. Long-tailed macaques (*M. fascicularis*) tested in a match-to-sample (see Chap. 28) or paired comparison task to pick the mother-offspring pair, extended their knowledge to other mother-offspring pairs in their group. Based on these results, it was suggested that monkeys have a social concept [15]. Vervet and rhesus monkeys that had been attacked by the member of a specific matriline redirected their aggression preferentially against kin of the former aggressor. At a higher level of complexity, kin of the former victim have been noted to attack members of the clan of the former aggressor [16] (Fig. 29.5).

Monkeys distinguish between different matrilines as well as between different members within a matriline [18]. Moreover, they keep track of short-term changes in relationships, as evidenced by the responses of chacma baboon males in an ingenious experimental design. Chacma baboon males secure exclusive access to a fertile female by closely following her during her receptive period ("consortship"). In the experiment, males were presented with the mating call of a given consort female and the grunt of her consort male, which was played from the opposite direction, indicating that the two individuals whose calls were played were not together. In these situations, males quickly approached the direction from which the female's calls were played, because they apparently inferred that she was no longer in the consort [19].

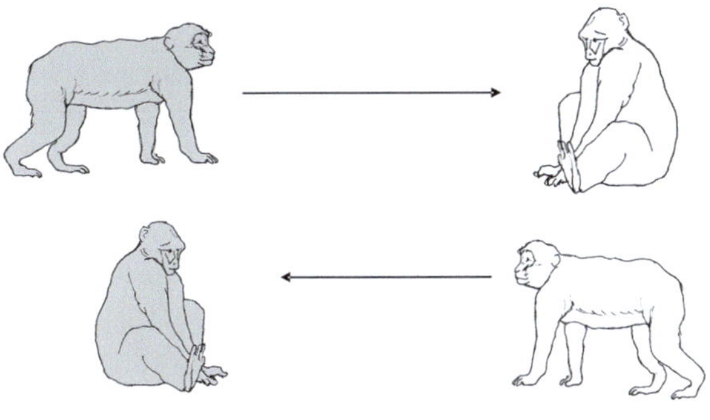

**Fig. 29.5 Nonhuman primates reveal an intricate understanding of their social relationships**. A member of matriline A (depicted in *gray*) attacks a member of matriline B (*white*). Subsequently, another member of B redirects the aggression towards a previously uninvolved member of matriline A ("complex redirected aggression"). Such observations suggest that the animals know the relationships among third parties (Original drawings by A. Skrzeszewska; reprinted from Fischer [17] with permission from Springer)

## 29.4 Social Learning

### 29.4.1 Social Facilitation, Stimulus Enhancement, and Local Enhancement

Animals pay attention to what others are doing, and where.

**Social facilitation** is invoked when an individual's learning is affected by the activity of another animal. Social animals typically pay a lot of attention to what others, particularly their group mates, are doing. This may lead to **stimulus enhancement**, i.e., an increase in salience of stimuli others are paying attention to, as well as **local enhancement**, i.e., the subjects learn something about the contingencies of a specific local situation simply because it is near an individual who does something particular. Social facilitation and stimulus as well as local enhancement have been found in a broad array of species, including guppies and cleaner fish. Agent-based models have shown that these mechanisms are sufficient to create and maintain stable traditions [20].

### 29.4.2 The 'Means-End' Structure of Social Learning

One way to classify different forms of social learning is to consider whether or not the pupil shows an understanding of the means and the ends of a given action. **Emulation** has been defined as a form of social learning when the subject learns something about the consequences of a certain action through observation. Emulation does not require that the animal understands or follows the exact procedure (the means) of the action, nor does it entail an understanding of the other subject's intention or the causal relationship of action and outcome.

Although the terminology is unfortunately not always used consistently, the terms imitation, copying, and emulation refer to different ways an animal learns a behavior from other individuals in a social group.

Within the framework of the means-ends structure, **copying** would be the adoption of some other individuals' behavior with no understanding of what it is good for. In this sense, one follows the exact procedure of the model, but also follows any dysfunctional (or accidental) aspects of it. Copying has been invoked in a large array of situations, including mate choice where females may prefer males that have already mated successfully, bird song learning where songsters dutifully produce exact renditions of song models heard earlier, or the acquisition of food preferences.

**Imitation** can be viewed as a relatively complex ability that requires an understanding of the means as well as the end of an action. One classic experiment on imitation involved captive chimpanzees and human children. Subjects watched a human demonstrator first stab a stick into a hole at the top of an opaque box, then remove the stick and insert it into a second hole at the front panel to obtain a food reward (Fig. 29.6). Both chimpanzees (aged 2–6 years) and children (aged about 4 years) followed this sequence of actions. However, when presented with a transparent box which revealed that sticking the tool into the top hole was inefficient, the chimpanzees switched from copying to a more goal-oriented emulation and simply used the

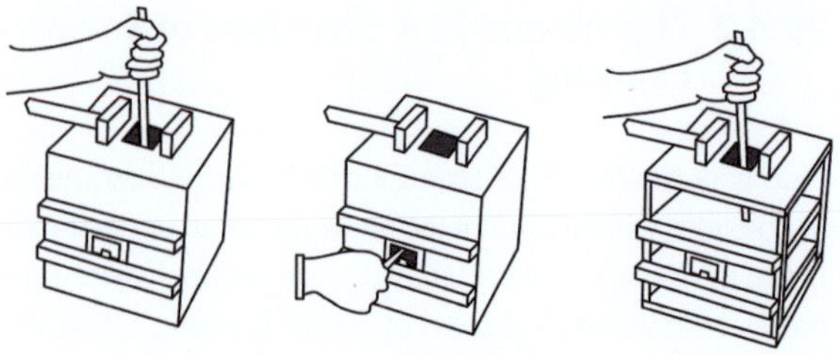

**Fig. 29.6 Experimental setup designed to test social learning in chimpanzees.** Opaque and transparent boxes used in experiments with chimpanzees and children. Subjects saw a familiar human first stab a stick tool into the top and then into the front hole to recover food. In a second condition, the transparent box revealed that the first move (stabbing from the top) had no effect. In this second condition, chimpanzees switched from copying the sequence to emulation, while children continued to follow the human model (Modified from Whiten et al. [21] with permission)

stick to gather the reward. In contrast, the children continued to follow the demonstrated procedure of first inserting the stick into the top and then into the front hole. The result was taken as evidence for the strong **conformity bias** of our own species [21].

### 29.4.3 Teaching

Commonly, the following criteria are used to diagnose the occurrence of **teaching** among animals: "An individual actor A can be said to teach if it modifies its behavior only in the presence of a naïve observer, B, at some cost or at least without obtaining an immediate benefit for itself. A's behavior thereby encourages or punishes B's behavior, or provides B with experience, or sets an example for B. As a result, B acquires knowledge or learns a skill earlier in life or more rapidly or efficiently than it might otherwise do, or that it would not learn at all" [22]. This rather broad definition does not assume instruction, higher order intentionality, or attribution of mental states. This definition therefore remains at the functional level, without regarding the psychological mechanisms underpinning this behavior.

> There is no evidence for active instruction or punishment in nonhumans.

Several studies have shown that older animals adjust their behavior depending on the capability of young, for instance by pre-processing food or by holding young that are too weak to cling to the fur themselves. Notably though, there is no evidence for active punishment or encouragement. There is also no indication that nonhuman primates correct or help youngsters to acquire a certain skill, e.g., nut cracking. Therefore, it seems more plausible that the acquisition of nut cracking can be explained through a combination of social facilitation, emulation, and trial-and-error learning. Obviously, however, socially interacting animals do provide each other with opportunities that they might not have if they lived alone, and thus, the acquisition of skills generally takes on different dynamics when social learning plays a role compared to situations where everything has to be learnt individually.

### 29.4.4 Do Animals Have Culture?

There have been countless attempts to define human culture. One of the prominent definitions goes "that complex whole which includes knowledge, belief, art, law, morals, custom, and any other capabilities and habits acquired by man as a member of society" [23]. A number of animal researchers found this and similar definitions that start with the derived features of human culture of little use, and thus, they focused on those features that are more likely to be found in animals. One standard definition today would be that animal culture consists of "group-typical behavior patterns, shared by members of animal communities, that are to some degree reliant on socially learned and transmitted information" [24]. It should be noted that this change in definition is not without problems when it comes to comparing human and animal culture.

> Social learning can lead to animal traditions, but human cultural systems are characterized by a rich symbolic structure.

There is now ample evidence that different primate populations reveal group-specific behaviors that cannot simply be explained by ecological differences. West African chimpanzees, for instance, crack nuts using stones as hammers and other stones or roots as anvils. East African chimpanzees, in contrast, have never been observed to crack nuts, despite the fact that nuts occur in their habitat. A comprehensive survey of the different variants of tool use and social behavior identified 39 different behavioral patterns which were either observed in only one or a few of the sites taken into account in this study [25]. While the initial analysis assumed that ecological conditions or genetic variation among subjects could be ruled out as an explanation, a more recent study suggested that genetic variation cannot be ruled out as an explanation for behavioral diversity [26]. Thus, the observation of behavioral variation could be well substantiated. However, there is no convincing evidence that animals show cumulative culture in the sense that one innovation is based on a previous one [27].

Human speech plays a key role in the evolution of cumulative culture.

Social learning is clearly a prerequisite for the development of human culture, but it is not sufficient to explain its complexity. One of the hallmarks of human culture is its rich symbolic structure. Moreover, cumulative cultural evolution [28] requires a system to represent, store, and transmit knowledge. One such system is human speech: speech encompasses external reference as well as temporal and spatial transcendence. Once such a symbolic communication system is in place, it is possible to transmit information and knowledge without forcing the individuals to re-enact each action every time information needs to be conveyed. Speech generally and writing in particular can be used to compress information transmission, store acquired knowledge and drive technological development. In conclusion, while both animals and humans share a number of social learning mechanisms, humans in addition are apt imitators, attribute knowledge to others, and are in command of a complex communication system. All of these contribute to the rich symbolic structure that characterizes human cultures.

## 29.5 Theory of Mind

Monkeys and apes track the social relationships of others – but what do they know about what others know or want?

The umbrella term of 'Theory of Mind' research covers a field that investigates the attribution of beliefs, desires, and knowledge to others and oneself. Studies on Theory of Mind in animals have investigated what animals know about the link between seeing and knowing, whether they understand another animal's intentions, knowledge or belief states, and whether they know what they know themselves (**metacognition**). It has become clear that the psychological mechanisms underlying the understanding of goals and intentions may not be the same as the ones underlying the attribution of beliefs and desires, and it is therefore necessary to distinguish between these different aspects. Whether or not animals attribute mental states to each other (in the broad sense) is hotly contested. The same result might be taken as evidence for an understanding of perception-goal psychology, or, alternatively, as an outcome of operant conditioning. For humans, we know that they spend much time reasoning about the knowledge and belief states of others, as well as their intentions. The ability to distinguish one's own mental life from that of others appears to develop around the age of about 4 years. Functional magnetic resonance imaging in humans has revealed a network of brain structures that appears to be involved in the perception of oneself and other individuals as cognitive entities and, more specifically, in the determination of mental states ("ToM network" for "Theory of Mind"). This network consists of the medial prefrontal cortex (PFC), the temporal poles, and the posterior superior temporal sulcus (STS) extending into the temporoparietal junction (TPJ) (Fig. 29.7).

### 29.5.1 The Mirror System

Seeing or hearing other people's actions recruits the same neurons (in monkey studies) and brain regions (in human studies) that are involved in performing

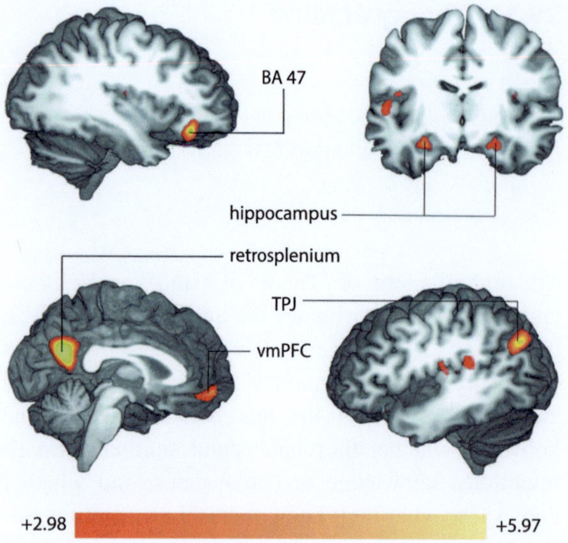


**Fig. 29.7 Brain activation during different emotion recognition tasks.** Brain activation when subjects were asked to judge the authenticity of the verbal expression of an emotion (authentic vs. play-acted) compared to when they were asked to determine the type of emotion. These areas are part of the so-called "Theory of Mind" network. *Top left*: left view sagittal section through temporal lobe. *Top right*: posterior view coronal section through thalamus. *Bottom left*: right view through corpus collosum. *Bottom right*: Anterior view coronal section through prefrontal cortex. *BA47* Brodman Areal 47, Retrosplenium, *TPJ* Temporoparietal junction, *vmPFC* ventro-medial prefrontal cortex (Reprinted from Drolet et al. [29] with permission from Springer)

such or similar actions [30]. These neurons or areas are found in the premotor and posterior parietal cortices. This link between perceiving and performing a particular action has led to the idea that understanding the inner state of other individuals relies on implicit motor simulation – that is, the activation of motor programs that we would use to perform similar actions. In monkeys, neurons that were found to fire both when the subject sees an action or performs that action itself, have been dubbed "mirror neurons". This metaphor proved to be very powerful, and there are numerous debates whether the existence of the mirror system is the basis of empathy and attribution of mental states. The monkey data suggest that the mirror system is a necessary condition for such phenomena, but not a sufficient one. Moreover, such close links between perception and action have now been found in other taxa such as birds as well [31], and appear to be a fundamental organizing principle in various animals. In humans, indirect measures of neural activity support the existence of sensory-motor mirroring mechanisms in homologous frontal and parietal areas and other motor regions. A recent study in patients undergoing surgery because of pharmacologically intractable epilepsy examined firing patterns in cells in medial frontal and temporal cortices while patients performed or observed hand-grasping actions and facial emotional expressions. A significant proportion of neurons in supplementary motor area and hippocampus responded to both observation and performance of these actions, and a subset of these neurons demonstrated excitation during action-performance and inhibition during action-observation [32].

### 29.5.2 Gaze Cues and Gaze Following

Just like humans, nonhuman primates are highly attentive to other individuals' eyes (Fig. 29.8). The ability to track other individuals' gaze to distant objects is one of the most important sociocognitive skills. Gaze following is of adaptive value because where another individual is looking can be an important cue for information about the environment such as food or predators. In addition, gaze can provide information about others' intentions, emotions, and desires.

> Gaze following is based on an involuntary, reflexive response, but may be altered through experience and top-down modulation.

Gaze following occurs regularly in humans and has been extensively studied in the field of cognitive neuroscience, child development and sociocognitive disorders. Comparative studies revealed that gaze following occurs in a large number of animal taxa, ranging from birds to goats. The largest body of evidence has been accumulated for nonhuman primates. Particularly in young animals, gaze following is enhanced when the model shows a facial expression [34]. Apes and corvids may even follow gaze geometrically behind barriers by repositioning themselves if a barrier disallows them to see what someone else is looking at. **Gaze cueing** (i.e., the shift of attention towards the direction where someone is looking) and gaze following appear to be based on an involuntary, reflexive response, but may be altered through experience and can be modulated by top-down cognitive control. A study in humans revealed that the attribution of mental states to others, in particular whether another person is actually able to see something, has perceptible influence on the likelihood of following that other person's gaze. This study challenged the traditional view that social perception is just a bottom-up process in which the human brain uses social signals to make inferences about another person's mental state. Instead, it appears that even simple perceptual processes like gaze cueing can be modulated by mental-state attribution [35].

### 29.5.3 Emotion Recognition

Akin to the link between perception and action, research in humans has also suggested that brain areas involved in emotion processing, including the anterior insula and the rostral cingulate cortex (rCC), are active not only when we experience positive and negative emotions but also when we witness such emotions in others. Possibly, we perform an 'emotional simulation' of other individuals' experience when we process the emotion expression in the faces and voices of others. Whether or not this is also true for nonhuman primates remains an issue for further investigation. What is certainly true is that monkeys are sensitive to the facial and vocal expressions of emotions in conspecifics.

### 29.5.4 Attribution of Knowledge

The studies on gaze following have shown that monkeys are highly attentive to available social cues such as gazes and emotion expressions. But what do they know about what others have seen? One classic study examined the attribution of knowledge in Japanese and rhesus macaque mother-offspring pairs in a captive setting. Mother and offspring were separated, and the mother watched how either a 'predator', in this case a Ph.D. student disguised as a veterinarian, or food was placed in an experimental arena. Next, the offspring was released into the arena and the question was whether the mother would in some way alert the offspring to the presence of the food or the danger. Mothers did not alter their behavior, and it was concluded that they did not understand the mental state of their offspring [16]. Likewise, a number of studies in which monkeys or apes were tested for their willingness to provide information to others yielded negative results. In contrast, competitive paradigms in

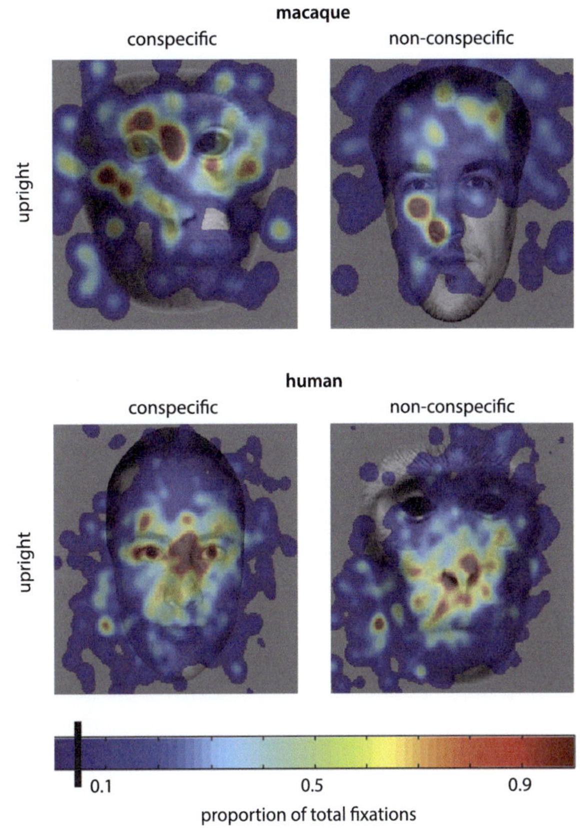


**Fig. 29.8 Monkeys and humans use similar face-processing strategies.** Total fixation densities for both participants (macaque and human) while observing conspecific and nonconspecific faces. Eye fixations are higher for conspecific faces than for nonconspecific faces (Reprinted from Dahl et al. [33] with permission from Elsevier)

which the behavior of the subject was tested when it competed with a knowledgeable or ignorant competitor yielded larger variation in behavior. One of the most highly cited papers in this field examined how subdominant animals adjusted their behavior in various situations in which they competed with a dominant over food (Fig. 29.9). Two food items were placed such that they were physically accessible and visible to both parties or to just one of the animals. The animals watched the placement of the food from behind sliding doors and were then released. In those cases where the food was placed in the middle, or one piece was placed on the dominant's side, the dominant virtually got all the food. In the situation where the subordinate had exclusive visual access to one of the food batches, it was able to secure this food. In a second set of experiments, food was hidden behind occluders, in order to minimize the effect of blocked

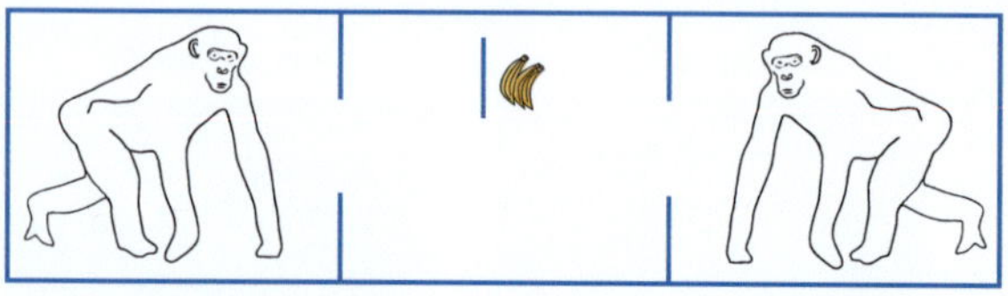

**Fig. 29.9 Experimental setup to test the link between seeing and knowing in chimpanzees.** A competitive paradigm for testing what chimpanzees understand when they see conspecifics. A subordinate and a dominant chimpanzee face each other across a room. Between them is a highly valued piece of food that both can see. Off to the side is the same type of food hidden behind a barrier so that only the subordinate chimpanzee can see it. If the subordinate chimpanzee understands what the dominant chimpanzee can and cannot see, then he should always go after the hidden food to avoid conflict (Modified from Hare et al. [36] with permission from Elsevier)

physical access. There were three experimental conditions: one, in which both animals had visual access (visible-visible), two, in which one piece of food was visible to the subordinate but not to the dominant while another was visible to both (hidden-visible), and a third in which both food items were visible to the subordinate animal but hidden from the dominant (hidden-hidden). In the critical 'hidden-visible' condition, the subordinate obtained almost 50 % of the hidden food items, compared to about 30 % of the visible items. The authors concluded that the apes have some understanding of the visual perspective of others. Two objections have been put forward, namely the so-called 'new evil eye' hypothesis, which posits that subordinates may have learned to avoid any food that has been seen by a dominant. Alternatively, the subordinates might have been able to watch the dominant's behavior, and maybe the dominant was already looking at the food. Thus, the mental operation needed to explain the behavior of the subordinates would not be visual perspective taking but simply the ability to infer (through learning) that others will go where they are looking at (behavioral rule based on gaze following) [36].

In an experiment it is difficult to differentiate whether an animal is "seeing" and has learned a response from previous events, or whether it "knows" about the intentions of an interaction partner.

### 29.5.5 Metacognition

Early research into metacognitive abilities mainly dealt with the question of how humans monitored their progress while acquiring knowledge, and how people retrieved stored knowledge. Whether or not also animals have metacognitive abilities is a matter of debate. Because we cannot directly ask the animals if they know the correct answer, we have to rely on indirect evidence. Most metacognition studies put the animal in a test situation and provide it with an 'opt out' choice [37]. A comparison of the performance levels when animals freely choose the test compared to when they are forced to make a choice is assumed to provide insight into the question of whether animals monitor their own knowledge state. In the so-called **'uncertain paradigm'**, for instance, a subject is asked to solve a discrimination task such as judging the density of pixels in a given square (high vs. low) [38]. Both humans and monkeys decline to take the test significantly more frequently when the difference between two stimuli falls below the noticeable difference than when the stimuli are clearly distinct. However, this behavior could also be explained by the simple rule to pick one button when the density is high, another when it is low, and to opt out when it falls in between. Thus, simple recognition would suffice to explain the behavior [37]. In a more elaborate experiment, monkeys were asked to judge their own memory. They were shown an item to remember and after variable delays, they could either decline or accept the test, which consisted of a presentation of four items from which they had to pick the sample that had been shown before. Animals were rewarded with a small and unattractive food item when declining the test, and a highly desirable food item when they mastered the test. Firstly, the performance of the monkeys was significantly better in those trials in which they chose to take the test than in forced experiments. Second, the longer the delays between the presentation of the sample and the test, the more likely the monkeys were to decline. Both results were taken as evidence that the subjects knew what they remembered [37]. However, it is also conceivable that those situations in which the monkeys did not remember the sample caused aversion, which in turn led the monkeys to opt out of the experiment (Fig. 29.10).

Metacognition means that an animal does not only know the answer to a problem, but also that it knows that it has this knowledge.

Whatever the underlying psychological mechanisms, there is at least (slowly) increasing evidence that there is some modulation of behavior in relation to the knowledge state of the animal. Two rhesus monkeys were trained to "gamble" after taking a discrimination test. After making their choice, they could indicate their confidence by choosing to win or lose three tokens, or gain just one. They chose the three-token option more frequently when they were correct than when they had failed in the discrimination task [39]. Again, however, the result might be explained by an emotional response instead of the assumption that the monkeys knew whether or not they had made the correct choice.

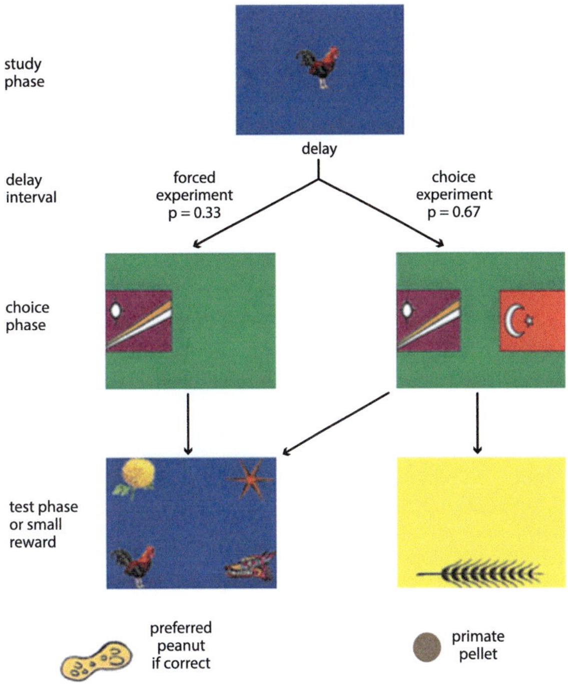


**Fig. 29.10 Experimental design to test whether monkeys know what they know**. Each colored panel represents what monkeys saw on a touch-sensitive computer monitor at a given stage in a trial. At the start of each trial, monkeys studied a randomly selected image. A delay period followed over which monkeys often forgot the studied image. In two-thirds of trials, animals chose between taking a memory test (*right*, left-hand stimulus) and declining the test (*right*, right-hand stimulus). If they declined, they saw another image and received a small reward. In one-third of trials, monkeys were forced to take the test (*left*). They were then shown four images, and received a large reward when choosing the right one. Better accuracy on chosen than on forced tests indicates that monkeys know when they remember and decline tests when they have forgotten, if given the option (Modified from Hampton [37] with permission)

## 29.6 Primate Communication

Nonhuman primates keep track of who is doing what to whom, they pay attention to the activities of others, establish local traditions, and are exquisite behavior readers. But how does this apparent social intelligence translate into their communicative abilities? Strikingly, and somewhat counterintuitively, the communicative abilities of monkeys and apes appear rather limited.

### 29.6.1 Vocal Communication

Studies of the ontogeny of vocal production as well as the neurobiological foundations of vocal control in nonhuman primates suggest that the structure of primate vocalizations is largely innate. Unlike in most songbirds, exposure to species-specific calls is not a prerequisite for the proper development of the vocal repertoire. Nevertheless, developmental modifications occur. These can be mainly related to growth in body size and the influence of hormones during adolescence. As a corollary, learning plays little to no role in the development of vocalizations, and hence, nonhuman primates are unable to produce anything resembling a spoken language, which rests on conventionalization and learning. In contrast to vocal production, comprehension of, and correct responses to calls, is strongly affected by experience [40]. This asymmetry in vocal production vs. comprehension is shared with most other terrestrial mammals, and taken to some extreme in the example of Rico, the domestic dog who was shown to be able to remember the names of over 200 toys [41]. A more recent study featured another border collie, who knew the names of more than 1,000 items, as well as the names of several people, locations, and activities. The study on Rico revealed that he was able

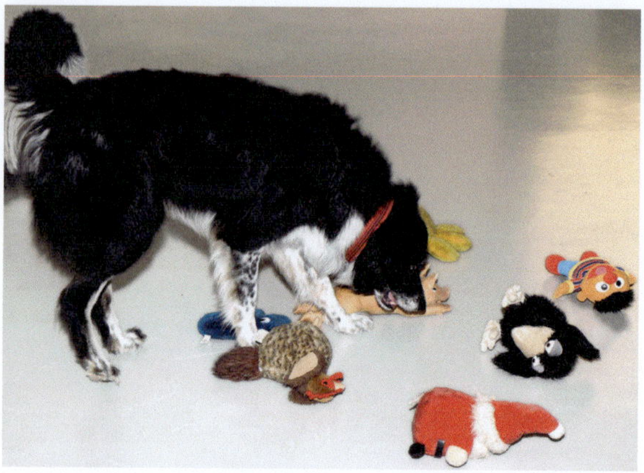

**Fig. 29.11 Rico, the border collie who remembered the names of over 200 toys.** Rico remembered not only the names of over 200 toys, but he was also able to link a new name to a novel toy based on exclusion. Moreover, in some cases, he remembered these new names over a period of 4 weeks after single exposure, in a way akin to the "Fast Mapping" that plays an important part in the word acquisition of human toddlers (Photograph by Renate Ritzenhoff, reprinted with permission)

to infer the name of a new toy by exclusion. This was tested by placing familiar items in a room, and adding a new item. After he was asked to fetch one or two of the familiar items, he was requested to fetch the item with the novel name, which he had never heard before. In most cases, Rico picked the novel toy. In addition, he was able to keep that link in memory, even for several weeks during which he had no access to the novel toy [41] (Fig. 29.11). Despite these exquisite abilities in the domain of comprehension, his vocal utterances were restricted to some barks and growls.

> Language, vocal communication, and comprehension involve innate and learned components, as well as brain lateralization, in all species studied so far.

In nonhuman primates, the neurobiological substrate for the control of vocalizations has been well studied. Current evidence suggests that the anterior cingulate cortex serves to control the initiation of vocalizations, facilitating voluntary control over call emission and onset. The periaqueductal gray (PAG) appears to serve as a relay station. Electrical stimulation of the PAG yields natural-sounding, species-specific vocalizations. In a study with squirrel monkeys, most of the neurons in the PAG fired only before the start of vocalizations, and did not show any vocalization correlated activity. Retrograde tracing studies revealed that vocalization-eliciting sites of the PAG receive a widespread input from cortical and subcortical areas of the forebrain, large parts of the midbrain, as well as the pons and medulla oblongata. The actual motor patterns appear to be generated in a discrete area in the reticular formation just before the olivary complex. Neurons in this area increased their activity just before and during vocalizations, and showed significant correlations with the syllable structure of these vocalizations [42] (Fig. 29.12).

### 29.6.2 Lateralization

Early studies on patients with difficulties in speech perception and production already were used to localize cognitive processes to specific areas of the brain. Brain lesions detected by post-mortem autopsies implied a central role of the left hemisphere in speech processing. Almost 150 years later, research in the field of lateralized perception and production of speech is still thriving. A number of studies involving neuroanatomical, physiological, and behavioral methods revealed hemispheric lateralization in the processing of species-specific sounds in nonhuman primates analogous to the processing of speech stimuli in humans [43].

Japanese macaques, *M. fuscata*, for instance, revealed a right ear advantage – corresponding to a left hemispheric dominance – when they were trained to discriminate communicatively important features in their species-specific sounds. After lesioning of the left superior temporal gyrus (STG), members of the same species showed a stronger postoperative decrement in the discrimination of species-specific calls than after an equivalent lesioning of the right STG. Similarly, a PET study revealed greater local cerebral metabolic activity in the left temporal pole in response to species-specific calls as opposed to any other sounds in rhesus monkeys (*M. mulatta*) [43]. However, monkeys do not reveal handedness at the population level. There is some indication that with increasing age, animals become increasingly more lateralized to one side or the other when performing simple as well as more complex manual actions [44], but there is typically the

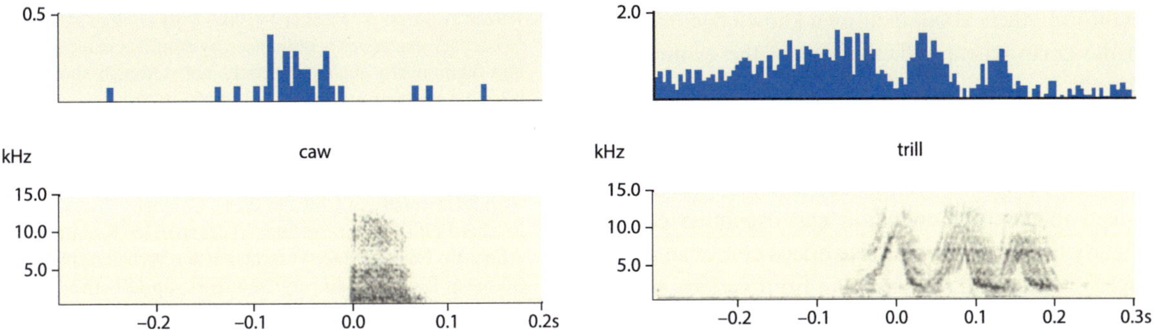


**Fig. 29.12 Spike trains correlating with vocalizations.** Examples of vocalization-correlated activity in single reticular neurons of squirrel monkeys. The *upper trace* shows the averaged neuronal spiking activity of a single neuron in area during the production of repeated renditions of calls of one and the same call type. The *lower trace* represents a superposition of the sonograms of the calls underlying the neuronal activity shown in the upper trace. *Left*: neuronal activity and sonographical displays were triggered to vocal onset ($N=13$ renditions); *right*: neuronal activity and sonographical display were triggered to the first maximum of the fundamental frequency ($N=10$ renditions). In both diagrams the same neuron is shown, but during the production of different call types (Reprinted from Jürgens [42] with permission from Elsevier)

same number of left- and right-handed animals in a given population.

### 29.6.3 Gestural Communication

Possibly as a consequence of the frustration with the lack of elaboration in the vocal production of nonhuman primates, researchers have become more interested in the gestural communication of apes, and recently also monkeys. The reasoning is that while primates lack voluntary control over their vocal output, they do have excellent voluntary control over their hands. Therefore, it has been hypothesized that human speech possibly evolved via a proto-sign language [45]. In addition, the brain area in monkeys (area F5) that controls manual movements is supposed to be homologous to the human Broca's area which is involved in speech production, further fuelling the interest in this modality of communication.

In this context, it is important to distinguish between enculturated subjects raised by humans, captive and wild animals. Some enculturated apes like the chimpanzee Lana have been taught to use certain gestures like 'open' and 'more', which she used in a range of different contexts [46]. Detailed studies of the signaling behavior of chimpanzees, bonobos (*Pan paniscus*), gorillas (*Gorilla gorilla*), and siamangs (*Symphalangus syndactylus*) in captivity, in contrast, did not provide any evidence that gestures referred to certain objects or events in the subjects' environment [47]. The majority of flexibly used gestures occurred in the play context – a context that by its definition is characterized by highly variable behaviors. The evidence for gestural communication among wild subjects is still somewhat sketchy, and no study to date reported that the animals signaled about objects or events in their surroundings. This finding suggests that a lack of voluntary control over the vocal apparatus is only one of the constraints that prevent a more elaborate communication among our closest living relatives.

## 29.7 Summary

Nonhuman primates constitute important models to elucidate the evolutionary origins of human-specific traits such as language, culture, and full-blown mental state attribution. Research on captive and wild animals shows that a number of traits are shared with our closest living relatives, such as lateralized processing of conspecific sounds, the enormous interest in the activities and interactions of others, and the importance of social relationships. Social learning and the attendance to other subjects' behavior is important in monkeys and apes; in addition, there is some evidence that great apes also have an understanding of what other subjects can see and know. There is, however, a striking dichotomy between their cognitive and communicative abilities: while nonhuman primates are apt interpreters of signals and other behaviors, their communicative repertoires are largely innate and they have no intentions to inform others about their own knowledge or events in the environment. Whether or not the evolution of large brains is causally related to life in complex groups remains an open issue, and it is also a matter of debate whether and in which way brain size or other measures such as complexity or neocortex ratio are directly linked to complex cognition and executive control. Clearly though, more complex brains confer an advantage for integrating information from various sources and may allow animals to choose responses more flexibly, thereby gaining a selective advantage. The close integration of perception and action allows for the simulation of other organisms' states, and appears to be important in mental state attribution in humans.

## References


1. Roth G, Dicke U (2005) Evolution of the brain and intelligence. Trends Cogn Sci 9:250–257
2. Harvey PH, Clutton-Brock TH, Mace GM (1980) Brain size and ecology in small mammals and primates. Proc Natl Acad Sci USA 77:4387–4389
3. Dunbar R (2003) Evolution of the social brain. Science 302:1160–1161
4. Barrett L, Henzi P, Dunbar R (2003) Primate cognition: from 'what now?' to 'what if?'. Trends Cogn Sci 7:494–497
5. Holekamp KE (2007) Questioning the social intelligence hypothesis. Trends Cogn Sci 11:65–69
6. Healy SD, Rowe CA (2007) Critique of comparative studies of brain size. Proc Biol Sci 274:453–464 [key paper]
7. Chittka L, Niven J (2009) Are bigger brains better? Curr Biol 19:R995–R1008
8. Silk JB, Beehner JC, Bergman TJ, Crockford C, Engh AL, Moscovice LR, Wittig RM, Seyfarth RM, Cheney DL (2010) Strong and consistent social bonds enhance the longevity of female baboons. Curr Biol 20:1359–1361
9. Thierry B, Aureli F, Nunn CL, Petit O, Abegg C, De Waal FBM (2008) A comparative study of conflict resolution in macaques: insights into the nature of trait covariation. Anim Behav 75:847–860
10. Bales KL, Carter CS (2003) Sex differences and developmental effects of oxytocin on aggression and social behavior in prairie voles (*Microtus ochrogaster*). Horm Behav 44: 178–184
11. Snowdon CT, Pieper BA, Boe CY, Cronin KA, Kurian AV, Ziegler TE (2010) Variation in oxytocin is related to variation in affiliative behavior in monogamous, pairbonded tamarins. Horm Behav 58:614–618
12. McGlothlin JW, Jawor JM, Ketterson ED (2007) Natural variation in a testosterone-mediated trade-off between mating effort and parental effort. Am Nat 170:864–875
13. Beehner JC, Gesquiere L, Seyfarth RM, Cheney DL, Alberts SC, Altmann J (2009) Testosterone related to age and life-history stages in male baboons and geladas. Horm Behav 56:472–480
14. Patzelt A, Pirow R, Fischer J (2009) Post-conflict affiliation in Barbary macaques is influenced by conflict characteristics and relationship quality, but does not diminish short-term renewed aggression. Ethology 115:658–670
15. Dasser VA (1988) Social concept in java monkeys. Anim Behav 36:225–230
16. Cheney DL, Seyfarth RM (1990) How monkeys see the world. University of Chicago Press, Chicago
17. Fischer J (2006) Untersuchungen der sozialen Kognition bei Affen. In: Naguib M (ed) Methoden der Verhaltensbiologie. Springer, Berlin/Heidelberg/New York, pp 157–162
18. Bergman TJ, Beehner JC, Cheney DL, Seyfarth RM (2003) Hierarchical classification by rank and kinship in baboons. Science 302:1234–1236 [key paper]
19. Crockford C, Wittig RM, Seyfarth RM, Cheney DL (2007) Baboons eavesdrop to deduce mating opportunities. Anim Behav 73:885–890
20. Franz M, Nunn CL (2010) Investigating the impact of observation errors on the statistical performance of network-based diffusion analysis. Learn Behav 38:235–242
21. Whiten A (2005) The second inheritance system of chimpanzees and humans. Nature 437:52–55
22. Caro TM, Hauser MD (1992) Is there teaching in nonhuman animals? Q Rev Biol 67:151–174
23. Tylor EB (1871) Primitive culture: researches into the development of mythology, philosophy, religion, language, art, and custom. Murray, London
24. Laland KN, Janik VM (2006) The animal culture debate. Trends Ecol Evol 21:542–547 [key paper]
25. Whiten A, Goodall J, McGrew WC, Nishida T, Reynolds V, Sugiyama Y, Tutin CEG, Boesch C (1999) Cultures in chimpanzees. Nature 399:682–685
26. Langergraber KE, Boesch C, Inoue E, Inoue-Murayama M, Mitani JC, Nishida T, Pusey A, Reynolds V, Schubert G, Wrangham RW et al. (2011) Genetic and 'cultural' similarity in wild chimpanzees. Proc Biol Sci 278:408–416
27. Tomasello M (1999) The cultural origins of human cognition. Harvard University Press, Cambridge
28. Boyd R, Richerson PJ (2005) The origin and evolution of cultures. Oxford University Press, Oxford/New York
29. Drolet M, Schubotz R, Fischer J (2012) Authenticity affects the recognition of emotions in speech: behavioral and fMRI evidence. Cogn Affect Behav Neurosci 12:140–150
30. Rizzolatti G, Sinigaglia C (2010) The functional role of the parieto-frontal mirror circuit: interpretations and misinterpretations. Nat Rev Neurosci 11:264–274
31. Prather JF, Peters S, Nowicki S, Mooney R (2008) Precise auditory-vocal mirroring in neurons for learned vocal communication. Nature 451:305–310
32. Mukamel R, Ekstrom AD, Kaplan J, Iacoboni M, Fried I (2010) Single-neuron responses in humans during execution and observation of actions. Curr Biol 20:750–756
33. Dahl CD, Wallraven C, Bülthoff HH, Logothetis NK (2009) Humans and macaques employ similar face-processing strategies. Curr Biol 19:509–513
34. Teufel C, Gutmann A, Pirow R, Fischer J (2010) Facial expressions modulate the ontogenetic trajectory of gaze-following among monkeys. Dev Sci 13:913–922
35. Teufel C, Fletcher PC, Davis G (2010) Seeing other minds: attributed mental states influence perception. Trends Cogn Sci 14:376–382 [key paper]

36. Hare B, Call J, Agnetta B, Tomasello M (2000) Chimpanzees know what conspecifics do and do not see. Anim Behav 59:771–785 [key paper]
37. Hampton RR (2001) Rhesus monkey know when they remember. Proc Natl Acad Sci USA 98:5359–5362
38. Smith JD, Washburn DA (2005) Uncertainty monitoring and metacognition by animals. Curr Dir Psychol Sci 14:19–24
39. Kornell N, Son LK, Terrace HS (2007) Transfer of metacognitive skills and hint seeking in monkeys. Psychol Sci 18:64–71
40. Fischer J (2002) Developmental modifications in the vocal behaviour of nonhuman primates. In: Ghazanfar AA (ed) Primate audition. CRC Press, Boca Raton, pp 109–125
41. Kaminski J, Call J, Fischer J (2004) Word learning in a domestic dog: evidence for 'fast mapping'. Science 304:1682–1683 [key paper]
42. Jürgens U (2009) The neural control of vocalization in mammals: a review. J Voice 23:1–10
43. Teufel C, Ghazanfar AA, Fischer J (2010) On the relationship between lateralized brain function and orienting asymmetries. Behav Neurosci 124:437–445
44. Schmitt V, Melchisedech S, Hammerschmidt K, Fischer J (2008) Hand preferences in Barbary macaques (*Macaca sylvanus*). Laterality 13:143–157
45. Corballis MC (2002) From hand to mouth. Princeton University Press, Philadelphia
46. Kellogg WN (1968) Communication and language in home-raised chimpanzee – gestures, words, and behavioral signals of home-raised apes are critically examined. Science 162:423–427
47. Call J, Tomasello M (2006) The gestural communication of apes and monkeys. Lawrence Earlbaum Associates, Mahwah

# Computational Neuroscience: Capturing the Essence

# 30

Shaul Druckmann, Albert Gidon, and Idan Segev

The nervous system faces a most challenging task – to receive information from the outside world, process it, to change adaptively, and to generate an output – the appropriate behavior of the organism in a complex world. The research agenda of **computational neuroscience** (**CN**) is to use theoretical tools in order to understand how the different elements composing the nervous system: membrane ion channels, synapses, neurons, networks, and the systems they form, address this demanding challenge. CN deals with theoretical questions at both the cellular and subcellular levels, as well as at the networks, system, and behavioral levels. It focuses both on extracting basic biophysical principles (e.g., the rules governing the input-output relationship in single neurons) as well as on high-level rules governing the computational functions of a whole system, e.g., "How is a spot of light moving in the visual field encoded in the retina?" Or "how do networks of interconnected neurons represent and retain memories?" Ultimately, CN aims to understand, via mathematical theory, how do high-level phenomena such as cognition, emotions, creativity, and imagination, as well as brain disorders such as autism and schizophrenia, emerge from elementary brain-mechanisms. Here we highlight a few theoretical approaches used in CN and provide the respective fundamental insights that were gained. We start with biophysical models of single neurons and end with examples for models at the network level.

## 30.1 Neurons: Input-Output Plastic Devices

A quantitative understanding of how neurons compute means to link, via theory and modeling, their biophysical and anatomical properties to their input-output function(s).

Probably the most striking feature of neurons is the branching structures of their dendritic and axonal tree; each neuron type with its unique characteristic. One such example is provided in Fig. 30.1 for a cat layer 5b pyramidal cell: the dendrite tree is shown in *red* and the ramifying axon in *white*, with its many varicosities (presynaptic release sites) that are marked by *yellow spots*. The total length of the dendritic tree (which compose the gray matter) of a single neuron may reach 10 mm and its membrane surface may go from a few hundred μm$^2$ for small interneurons up to 750,000 μm$^2$ in the case of cerebellar Purkinje cells. Axons of single neurons may extend either locally, or sometimes many centimeters and even meters; each axon may contact many thousands of postsynaptic target neurons, mostly their dendrites.

S. Druckmann • I. Segev (✉)
Department of Neurobiology,
Interdisciplinary Center for Neural Computation,
and Edmond/Lily Safra Center for Brain Sciences,
Hebrew University of Jerusalem,
Givat Ram, Jerusalem 91904, Israel
email: drucks@lobster.ls.huji.ac.il; idan@lobster.ls.huji.ac.il

A. Gidon
Alexander Silberman Institute of Life Sciences,
The Hebrew University of Jerusalem,
Givat Ram, Jerusalem 91904, Israel
email: agidon20@gmail.com

C.G. Galizia, P.-M. Lledo (eds.), *Neurosciences - From Molecule to Behavior: A University Textbook*,
DOI 10.1007/978-3-642-10769-6_30, © Springer-Verlag Berlin Heidelberg 2013

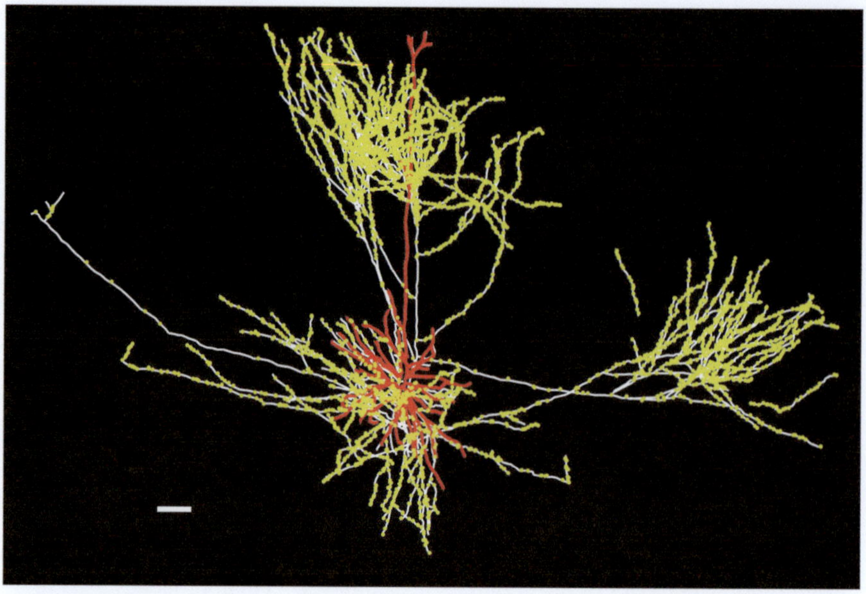

**Fig. 30.1** **The unique branching structure of neurons**. A layer 5b pyramidal cell from cat visual cortex. Dendrites are depicted in *red*, the axon in *white* with its varicosities marked by *yellow dots*. Scale bar 100 μm (Courtesy of Kevan Martin and Tom Binzegger, INI, ETH, Zurich)

Dendrites are the major receptive region of neurons in the mammalian central nervous system (CNS). Typically, a few thousand excitatory and inhibitory synapses contact a single dendritic tree; each of these contacts transmits information about the spiking activity of the presynaptic (input) cell. With the refined anatomical, optical, and physiological techniques available today (e.g., the "brainbow" and "connectomics" methods [26, 46]) it became increasingly clear that groups of axons, originating from specific input sources, target distinct dendritic domains; some contact distal dendritic regions, some more proximal regions, and others may contact the initial segment of the axon (for further reading see [23]). What could be the functional meaning of such domain-specific "division of labor" among the input sources for both the computations and the plastic processes in dendrites?

Furthermore, synapses are plastic rather than static devices; they may change their efficacy (the amount of postsynaptic current that they produce per presynaptic spike) in an activity-dependent manner and at a time scale ranging from few milliseconds to days and even to a life-time. In recent years a huge effort has been focused on the mechanisms underlying synaptic plasticity, as it is believed that learning and memory processes rely on synaptic plasticity. Understanding the interplay between dendritic synapses and their plastic properties (the input), the passive and active electrical properties of dendrites (local integration), and the output in terms of the rate and temporal structure of spike trains in the axonal tree (the output), is a central challenge for neuroscience and in particular for computational neuroscience. In other words, understanding the neuron as an input-output plastic device is a fundamental requirement for understanding how information is encoded and adaptively processed in nervous systems.

This challenge was the primary drive for the development of a "cable theory for dendrites" by Wilfrid Rall [35], which serves since that time as a foundation for understanding neurons as computational and plastic input-output devices.

> Rall's cable theory for dendrites highlights the key parameters that determine the spread of synaptic current from its dendritic site of origin to the soma/axon region. It provides fundamental intuition for signal processing in dendrites.

### 30.1.1 Cable Theory: Current Flow in Dendritic Trees

When a given synapse is activated – triggered by an action potential in the presynaptic axon – a local conductance change is produced (specific transmitter-gated membrane ion channels are opened) in the dendritic membrane, through which ion current may transiently flow and a local synaptic potential is generated. This local transmembrane synaptic current spreads in the dendritic tree where it encounters ion currents and membrane conductance changes resulting from the activation of other synaptic inputs. A certain portion of the synaptic current reaches the soma and axon membrane where, if sufficiently depolarized, an action potential (AP) or a train of them is discharged. These APs travel along the axon to activate the thousands of presynaptic release sites that are distributed along the axon.

We need to understand the effect of many converging excitatory and inhibitory dendritic synapses. Quantifying this input-output relationship requires a rigorous model of all biophysical parameters that determine signal processing in dendrites to the spike generation mechanism in the axon. These biophysical parameters include the spatial distribution of the activated synapses over the dendritic surface, their activation times, and the biophysical properties (time course and magnitude of conductance change, etc.) of the synapses, as well as the electrical properties and detailed morphology of the dendritic tree. Rall's **passive cable theory** [22, 35] provides the *skeleton upon which later more complicated nonlinear cases were considered*; example complications are that the membrane resistance is voltage-dependent and synaptic inputs are transient conductance changes, rather than linear current sources. Typically, the nonlinear cable equation cannot be solved analytically (but see a counterexample below) and numerical methods are needed [36].

The one-dimensional cable theory assumes that current flows in a core-conductor, either longitudinally along the cable ($x$) or through the membrane. The longitudinal current, $i_a$, encounters the cytoplasm (axial) resistance, $r_a$ (per unit length, in Ω/cm) thus $i_a = (1/r_a)(\mathrm{d}V/\mathrm{d}x)$, (Ohm's law) where $V$ is the transmembrane voltage. The membrane is electrically modeled by an equivalent R-C circuit; thus current can either cross the membrane via the passive membrane ion channels, represented by $r_m$ for unit length (in $\Omega \times \mathrm{cm}$), or charge (and discharge) the membrane capacitance $c_m$ (per unit length, in F/cm). At any point $x$ along the cable, the *change* ($\mathrm{d}i_a/\mathrm{d}x$) of the longitudinal current (per unit length) becomes the membrane current, $i_m$, per unit length (Kirchhoff's law). Thus, $i_m = (1/r_a) \times (dV^2/dx^2)$ and because $i_m = V/r_m + c_m dV/dt$ we get the passive cable equation,

$$\frac{r_m}{r_a}\frac{d^2V}{dx^2} = V + r_m c_m \frac{dV}{dt}. \qquad (30.1)$$

or:

$$\frac{d^2V}{dX^2} = V + \frac{dV}{dT}, \qquad (30.2)$$

where $X = x/\lambda$, $\lambda = \sqrt{r_m / r_a}$ is defined as the space constant and $T = t/\tau_m$, where $\tau_m = r_m c_m$ is the membrane time constant. Rall [35] has shown that Eqs. 30.1 and 30.2 can be solved analytically for arbitrary passive dendritic trees receiving current inputs at any location $x$. This allowed calculation of the membrane potential $V$ at any dendritic location at any time $t$ (assuming that the dendritic membrane is passive). The specific solution depends on the morphology of the tree, the electrical properties of the membrane ($r_m$, $c_m$) and cytoplasm ($r_a$), as well as on the boundary condition at the end of the cable segment towards which the current flows [40].

#### 30.1.1.1 Basic Biophysical Insights from Passive Cable Theory

Several key parameters control current flow in dendrites. One is the membrane time constant which, to a first approximation, sets the relevant time-window for synaptic integration. Synaptic inputs that arrive within this time-window will interact (summate) with each other (see discussion regarding the implications of the membrane time-constant in [25]). The average electrotonic cable length of the dendritic tree, $L$ (= $l/\lambda$; unitless), is another key parameter; a short (electrically compact) dendritic tree implies that a significant portion of the charge injected by the synapse will reach the soma and the spike generation mechanism at the axon. A key parameter that affects current flow in branched dendrites is the "geometrical ratio" ($GR$) between the daughters (with diameter $d_1$, $d_2$) and the parent branch, $d_p$, whereby $GR = (d_1^{3/2} + d_2^{3/2})/d_p^{3/2}$. Voltage attenuation from the parent branch towards the daughters is steeper when $GR$ increases [39].

The solution to the cable equation implies that the time-course of the voltage response to a step current injection (voltage build-up and decay) is governed by a sum over an infinite set of decaying exponents, $C_0 e^{-t/\tau_0} + C_1 e^{-t/\tau_1} + \ldots + C_n e^{-t/\tau_n} + \ldots$ with time constants $\tau_1, \tau_2, \ldots \tau_n, \ldots$ all faster than the membrane time constant, $\tau_m (= \tau_0)$. These time constants are also called "equalizing" time constants because they determine how fast the potential equalizes along the entire dendritic cable at the end of the current injection. Rall showed that the values of $\tau_i$ ($i = 1, 2, \ldots, n, \ldots$)

depend on the electrotonic length, $L$, of the dendritic tree; he showed that $L$ can be estimated directly from the ratio $\tau_m/\tau_1$,

$$L = \frac{\pi}{\sqrt{\tau_m/\tau_1 - 1}}, \quad (30.3)$$

where $\tau_1$ is the first (second largest after $\tau_m$) equalizing time constant. Both, $\tau_m$ and $\tau_1$ can be recovered directly from the experimental transient voltage response to current injection to the soma by "peeling" the exponents (first $\tau_m$ then $\tau_1$, etc.) [38].

Based on this theoretical understanding (Eq. 30.3 and using the "peeling" method), the passive membrane time constant, $\tau_m$, of many central neuron types was shown to range between 5 and 100 ms implying that $R_m$, the specific membrane resistivity ranges between $5{,}000$ and $100{,}000\ \Omega \times \text{cm}^2$, assuming specific capacitance, $C_m$ close to $1\ \mu\text{F}/\text{cm}^2$. The average electrotonic length, $L$, of dendrites of different neuron types was estimated to range between 0.2 and 2. For electrically compact dendritic trees (such as cerebellar Purkinje cells or spiny stellate cells) in the mammalian cortex $L$ ranges between 0.3 and 0.5, whereas more extended dendritic trees may be 1.5–2 long such as those of cat $\alpha$-motoneurons in the cat spinal cord, or the large apical tree of layer 5 pyramidal cells in the neocortex; see reviews in [44]. This means that dendrites are electrically rather compact and that (in sharp contrast with what was commonly assumed) charge transfer to the soma is effective even from distal excitatory dendritic synapses in highly branched dendritic trees [40]. Nonetheless, because of the "leaky" boundary conditions imposed on thin distal dendritic arbors by its thicker parent branches ($GR > 1$ for current flowing from the daughter branch towards the thicker parent branch), the peak voltage is expected to attenuate very steeply from the distal excitatory dendritic synapse towards the soma. In other words, the peak synaptic potential near the synaptic input site is significantly larger (10–100-fold) than that observed at the soma. Many of these early theoretical results can now be tested directly with pair electrode recordings from the soma and dendrites of the same cell, using infrared DIC video microscopy. These include assessing the degree of peak voltage attenuation and degree of charge transfer from dendrites to soma.

Another powerful method based on cable theory enabled the estimation of the distance, $X$, ($= x/\lambda$; unitless) between an activated synapse and the soma. Rall [37] showed that the time-course (rise-time and half-width, defined as "shape indices") of the synaptic potential at the soma increases as a function of the distance of the synapse from the soma. The excitatory postsynaptic potential (EPSP) at the soma, for synapses impinging on proximal dendritic sites, have a faster rise time and are narrower compared to the somatic EPSP originating from distal synapses. Additionally, Rall [37] showed that the shape indices of the somatic EPSP can be used to estimate the time course of the synaptic current (see also [22]). This theoretical insight was used for the first time by Rall [37] to estimate the distance, $X$, of the excitatory Ia synapses in spinal motoneurons from the soma. With a new staining method for dendrites and their synapses, Walmsley et al. [48] confirmed that this estimation indeed matched the actual distance of the Ia synapses from the motoneuron soma – a triumph for Rall's cable theory.

> Recent extensions of cable theory address a variety of anatomical and electrical complexities that were not included in Rall's early studies, including the case of dendrites with nonuniform membrane properties, dendrites covered with spines and dendrites receiving multiple synaptic inputs.

### 30.1.2 Recent Analytic Extensions of Passive Cable Theory

For simplicity, early implementation of the analytic cable theory assumed certain morphological idealizations as well as uniformity of membrane and cytoplasm properties. The consequences of relaxing these constraints were considered in later analytical studies by Rall and by others.

This is important, because the membrane properties of dendrites are not uniform and in some neuron types the membrane is more leaky for the distal dendritic membrane. Therefore, cable theory was generalized to account for the effects of nonuniform passive membrane properties on current flow in dendrites and to account for fine-scale (order of micrometers) spatial fluctuations in the dendritic membrane conductance. These fluctuations may result from a large number of dendritic spines and varicosities.

Another extension of passive cable theory enabled to analytically explore an important question – how long does it take for the synaptic potential to travel from its dendritic site of origin to the soma (the "dendritic delay")? Towards this end, the dendritic voltage transients were characterized by their various moments, rather than by the more standard measures such as "time to peak" or "half width". The solution to the cable equation for the various moments of the signal (e.g., the EPSP) provided, among other insights, a new understanding of the time-delay that dendrites impose on their synaptic potentials.

Briefly, the *i*th moment of a transient signal function *f*(*t*) is,

$$m_{i,f} = \int_{-\infty}^{\infty} t^i \cdot f(t)dt \qquad (30.4)$$

The 0th moment is the area (time integral) of *f*(*t*); the ratio between the 1st and the 0th moment is the "center of gravity", or centroid, of *f*(*t*), which is one possible (and mathematically convenient) measure of the characteristic time of the transient. The 2nd moment provides a measure for the "width" of *f*(*t*). Replacing the voltage, *V*, in Eq. 30.2 with $m_{i,f}$, a linear ordinary differential equation for each of the *i*th moments is obtained. The solution describes the behavior of the *i*th moments in arbitrary passive trees. Using the moments to compute the net dendritic time-delay (*TD*) for the propagation of EPSPs between the synaptic site and any other dendritic location, in particular the cell body, shows that TD may range between a few milliseconds to tens of milliseconds [3]. In cortical pyramidal neurons, for example, this delay is expected to range between 0 (for somatic inputs) and $\tau_m$ which is on the order of 5–20 ms (for distal synapses on the apical dendrite of pyramidal neurons).

One can also compute the local delay, *LD*, the time difference between the centroid of the input current and the centroid of the resultant voltage transient at the input site. The *LD* can serve as a measure for the time-window for input synchronization (the time-window in which local synaptic potentials can integrate locally with each other). *LD* at distal dendritic arbors may be as small as $0.1\tau_m$, which is about 10-times shorter compared to *LD* at the soma. Namely, in a passive system, a more precise temporal synchronization is required for local summation of synaptic potentials in the dendrites as compared to the soma. However, this can be "reversed" for nonlinear dendrites when slower nonlinear processes (e.g., the NMDA currents) are more pronounced at distal sites [8].

> The strategic placement of inhibition in dendrites strongly affects the input-output properties of neurons. Distal inhibitory conductance change spreads poorly distally but it effectively dampens proximal dendritic nonlinearities ("hotspots"); multiple inhibitory synapses surrounding a dendritic sub-domain effectively control the excitatory activity within this subdomain.

Below we provide the results of yet another theoretical challenge: to solve the cable equation when the dendritic tree receives multiple inhibitory conductance perturbations [17]. As highlighted below, this study has provided several surprising new results about the principles that govern the operation of synaptic inhibition in dendrites.

#### 30.1.2.1 Dendrites with Multiple Inhibitory Synapses

In all chemical synapses, the transmitter released to the synaptic cleft binds to the postsynaptic receptors and triggers a local conductance change, $g_i$. The synaptic current, *I*, that is generated follows Ohm's law and is

$$I = g_i \times (E - V_{rest}) \qquad (30.5)$$

where *E* is the synaptic battery and $V_{rest}$ and is the resting potential of the membrane; ($E - V_{rest}$) is the synaptic driving (electromotive) force. Excitatory synapses are characterized by a synaptic battery, ranging from 50 to 80 mV *more positive* than the resting potential. Therefore, excitatory synapses generate negative (inward) currents (excitatory postsynaptic currents, EPSCs), which cause membrane depolarization. In contrast, inhibitory synapses have smaller driving force, with a battery that is typically around 0–20 mV *more negative* than the resting potential (namely *E* is close to $V_{rest}$ for inhibitory synapses). Consequently, because $E \approx V_{rest}$, the inhibitory postsynaptic potential (IPSP) is typically small and may even be zero when $E = V_{rest}$. Yet, as for all synapses, inhibitory synapses induce a local conductance change causing a reduction in the input resistance ($R_{in}$) at the vicinity of the synaptic location (sometimes named "shunt" or "shunting inhibition").

Initially, the behavior of synaptic potential in the dendrite was the focus of studies that followed from Rall's cable theory, while the effect of the synaptic shunt (which functionally perturbs the membrane properties) gained less attention. What would be the impact of such perturbation on the processing of synaptic potentials in dendrites?

We start by defining a functional parameter for measuring the impact of the inhibitory conductance change, which we term the shunt level (*SL*). For this analysis we assume that $g_i$ is constant (steady state). *SL* is simply the relative drop in input resistance at any dendritic location *d*, following the activation of one (or several) steady-state synaptic conductance change $g_i$ in the dendritic tree,

$$SL_d = \Delta R_d / R_d. \tag{30.6}$$

$\Delta R_d$ is the change in input resistance at location *d* and $R_d$ is the input resistance before the activation of $g_i$. $SL_d$ depends on the particular dendritic distribution and values of the various $g_i$'s in the tree, and ranges from 0 (no shunt) to 1 (infinite shunt); e.g., $SL_d$=0.2 implies that the inhibitory synapse(s) reduced the input resistance at location *d* by 20 %. *SL* is thus a natural and straightforward measure for the functional impact of inhibitory synapses but is also applicable for excitatory synapses, which, as any synaptic input, also exert a local membrane conductance change when activated. When only a single conductance change, $g_i$, is active at location *i*, it is easy to show that *SL* at any location *d* is,

$$SL_d = \left[\frac{g_i R_i}{1 + g_i R}\right] A_{i,d} \times A_{d,i} \tag{30.7}$$

where $R_i$ is the input resistance at location *i* before the activation of the synapse, and $A_{i,d}$ ($A_{d,i}$) is the steady voltage attenuation from *i* to *d* (and vice versa). Here we define the voltage attenuation $A_{i,d}$ for steady current injected at location *i* as the ratio between the voltage at location *i* and that in location *d*; $V_d / V_i$ (note that typically $A_{i,d} \neq A_{d,i}$). Equation 30.7 shows that the amplitude of *SL* at the input site (in square brackets) depends on both $g_i$ and $R_i$. Interestingly, the attenuation of *SL* depends on the voltage attenuation in both the "forward" direction (from synaptic locus, *i*, to the measured locus *d*) as well as on the "backward" direction (i.e., $A_{i,d} \times A_{d,i}$). Therefore, it is affected by the boundary conditions in *both* ends of the cable. One straightforward outcome of this is that the attenuation of *SL* (e.g., from a proximal synaptic site) towards the terminals depends on the size of the soma. This is in marked contrast to the behavior of the voltage, *V*, whereby the attenuation (from soma to dendrites) does not depend on the soma size (i.e., on the current sink provided by the soma).

Figure 30.2 compares the attenuation of *SL* and of *V* from a distal synaptic input site in an idealized dendritic tree. The voltage attenuation is steep in the

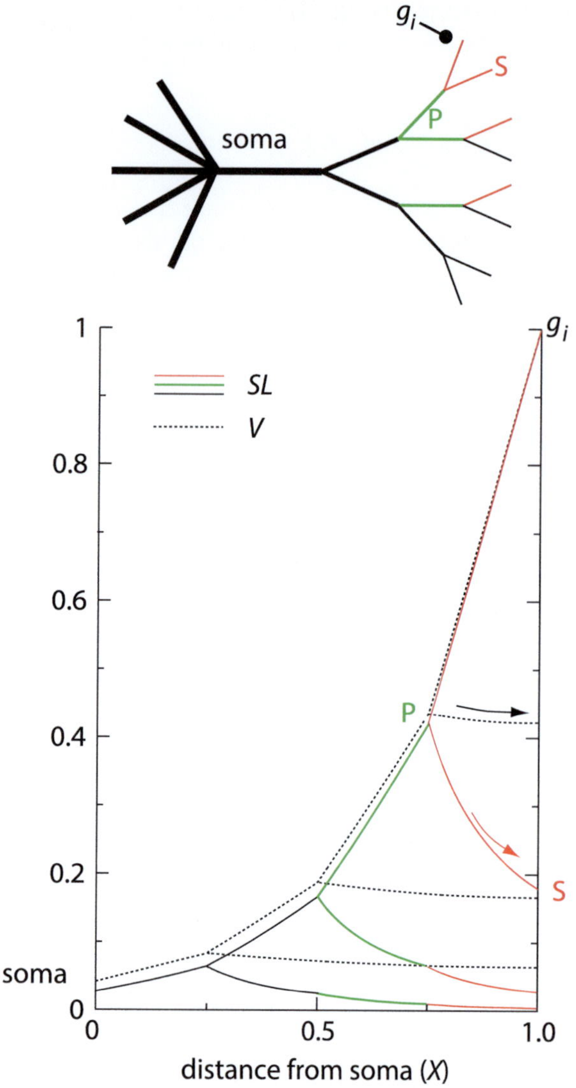


**Fig. 30.2 The spread of synaptic shunt in dendrites receiving distal synaptic conductance perturbation**. *Top*: Idealized symmetrically branched dendritic modeled tree [40]. $g_i$ is located at a single terminal end. *Bottom*: *SL* attenuation (*continuous line*) and steady voltage attenuation (*dotted line*) from the distal dendritic terminal end to the soma. Note that *SL* attenuates steeply towards the dendritic terminals (*red arrow*) compared to voltage which attenuates only slightly (*black arrow*)

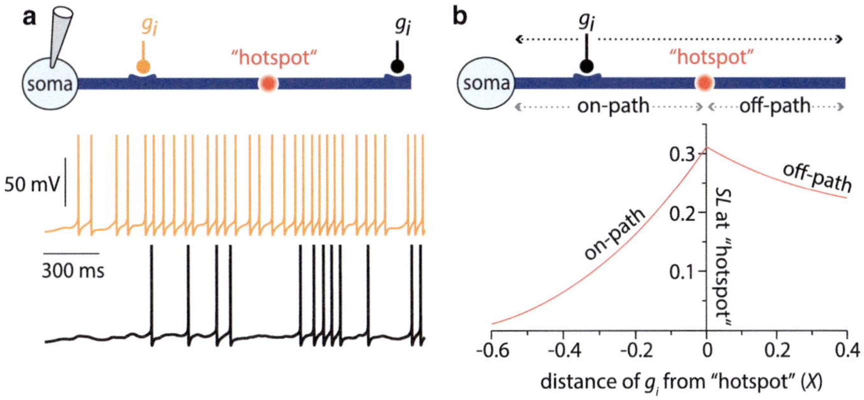


**Fig. 30.3 Distal inhibition is more effective than proximal inhibition in dampening the local dendritic "hotspot".** (**a**) A model of a cylindrical cable (sealed end at $L = 1$) coupled to an excitable soma. Twenty NMDA synapses are clustered at the "hotspot." Each synapse is randomly activated at 20 Hz. Inhibition of the somatic spikes is more effective when placed distally to the hotspot (*black* synapse and corresponding somatic spikes) than proximally (*orange* synapse and corresponding somatic spikes). Both synapses are placed at the same electrotonic distance ($X = 0.4$) from the hotspot with $g_i = 1$ nS. (**b**) Cylindrical model as in (a), but with a passive soma. *SL* is analytically computed at the hotspot as a function of the distance of $g_i$ from the hotspot. *SL* is maximal when $g_i$ impinges directly on the hotspot. *SL* at the hotspot diminishes more steeply when $g_i$ is placed on the path between the hotspot and the soma ("on-path" condition) than when placed distally to the hotspot ("off-path" condition)

centripetal direction (as *GR* is large in this direction, see above) whereas it is shallow in the centrifugal direction ($GR = 1$ in this direction, for the idealized tree modeled). *SL*, however, attenuates steeply in the thin distal dendritic branches, in both the centrifugal and the centripetal direction (*red arrow* in top scheme). This implies that the impact of a single conductance perturbation diminishes rapidly with distance in such thin dendritic branches.

Next, still considering the case of a single inhibitory conductance perturbation, one may ask: "what is the strategic placement of an inhibitory synapse for maximally dampening local dendritic excitable hotspots"? By "hotspot" we refer to a dendritic region containing high density of voltage-dependent ("active") ion channels, e.g., NMDA-receptors or voltage-gated $Ca^{2+}$ channels.

Classically, it was shown both computationally and analytically that the optimal locus for an inhibitory synapse to maximally reduce the somatic depolarization is when it is placed between the excitatory synapse and the soma ("on-the-path" inhibition) [22, 24, 37]. Surprisingly, this rule is typically "reversed" (Fig. 30.3a), i.e., inhibitory synapses are more effective in dampening the active current generated at the hotspot when placed distally to the hotspot ("off-the-path") rather than "on path" at a corresponding location. This is due to the steeper attenuation of *SL* from the proximal site to the hotspot as compared to the corresponding distal site (Fig. 30.3b) [17]. This strong effect of distal inhibition on the more proximal dendritic hotspots may explain why inhibitory synapses are found also at the very distal dendritic regions.

In the mammalian central nervous system, a single inhibitory axon typically forms multiple, sometimes up to 20, contacts per axon on the target dendritic tree (for further reading see [30]). Consequently, when these inhibitory axons fire an action potential, the dendritic tree of the postsynaptic neuron will be shunted at multiple sites (almost) simultaneously. What are the functional implications of such multiple inhibitory contacts?

The steep attenuation of the *SL* with a single inhibitory synapse supports the prevailing view that inhibition acts highly locally and that it is always maximal at the synaptic site itself. However, with *multiple* inhibitory synapses inhibition acts globally, spreading its impact hundreds of micrometers away from the synaptic contacts. Even more surprising is the finding that, with multiple inhibitory contacts, inhibition in regions lacking inhibitory synapses might be stronger than inhibition at the synaptic sites themselves. Both these results are depicted in Fig. 30.4 for a model of

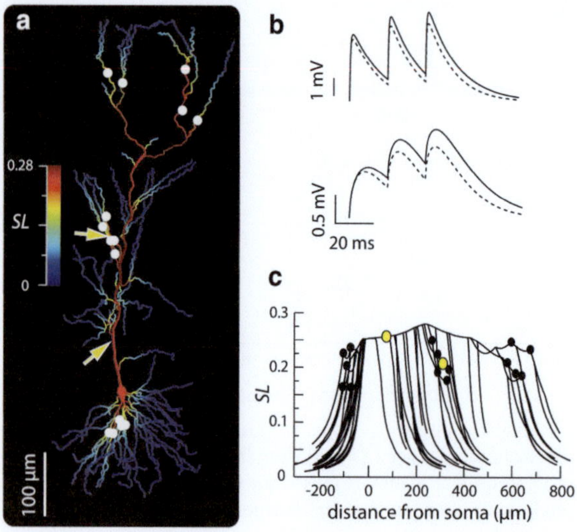


**Fig. 30.4 Spread of SL in dendrites receiving multiple inhibitory synapses.** (**a**) *SL* (color-coded) in a model of a reconstructed CA1 pyramidal neuron receiving a total of 15 inhibitory synapses (*white dots*; each synapse exerts a steady conductance change, $g_i = 0.5$ nS ) targeting distinct dendritic subdomains (basal, oblique, and apical). Note the spread of the inhibition (*red*) hundreds of µm from the synapses. (**b**) AMPA-like excitatory postsynaptic potentials (EPSPs) at the sites (denoted by *yellow arrows* in a) before (*continuous line*) and following (*dashed line*) the activation of the 15 inhibitory synapses. The top traces are for the distal site, whereas lower traces are for the proximal site. The distal excitatory synapse is colocalized with one of the inhibitory synapse; the EPSP generated by this synapse is less inhibited ($SL = 0.2$) than the EPSP generated by the proximal excitatory synapse ($SL = 0.25$), far from any one of the inhibitory synapses. (**c**) *SL* as a function of distance from the soma for the model shown in (a). *Black dots* denote the inhibitory synapses at the three dendritic subdomains. *Yellow dots* denote the location of the two excitatory synapses in (**a**)

CA1 neuron receiving 15 synaptic contacts in three separate dendritic domains (Fig. 30.4a). These results indicate that relatively few inhibitory synapses may "cover" (effectively inhibit) a large dendritic region; perhaps explaining why in most central neurons only about 20 % of the total number of dendritic synapses are inhibitory.

The case of multiple conductance perturbation in arbitrarily passive dendrites can be solved recursively. Based on Rall's method [35] the input resistance at any dendritic location *d* can be computed twice; once prior and once following the activation of the synaptic inhibition. *SL* at any given location *d* is the difference between the two solutions ($\Delta R_d$) divided by $R_d$, the input resistance without inhibition (see Eq. 30.6).

Dendrites, with their unique anatomy, nonlinear membrane properties and the large number of excitatory and inhibitory synapses, could potentially implement a variety of sophisticated computations, including multiplication-like ("AND-NOT") operations, coincidence detection, and input classification. Recent experimental work provided direct evidence that the nervous system does exploit the computational capabilities of dendrites.

### 30.1.3 Computing with Dendrites and Their Synapses

The models described above focused on understanding how the neuron's substrate, dendrites, synapses, and membrane ion channels effect the processing of synaptic inputs in neurons. At a higher, more abstract, level, one may ask: "what are the elementary computations that could be implemented due to the unique structure and biophysics of dendrites?" In particular, it was argued that the electrically distributed system provided by dendrites endows neurons with the potential for performing several elementary computations, beyond those that a structureless "point" neuron could implement. Indeed, one may view the dendritic tree as composed of a system of semi-independent functional subunits. In each of these functional units, local plastic processes as well as local computations may take place almost independently from other subunits. The result of this local nonlinear computation may (or may not) be then delivered to the soma-axon region [6].

For example, suppose that a distal dendritic branch is endowed with a local nonlinear "hotspot" which is activated by a local excitatory synapse, *E*, and that a distal inhibition, *I*, in this branch is not active. In this case (*E* AND NOT *I*), the excitatory synapse will be boosted by the "hotspot", thus delivering a significant amount of excitatory current to the soma. In another similar branch, *I* is active when *E* is active (*E* AND *I*). In this case (if *SL* provided by *I* is large), this branch will not contribute an excitatory charge to the soma and, thus, will not affect the output of that neuron.

Some of the key computations that can be performed with dendrites are:

1. *Detection of motion direction.* This operation is of particular importance for the survival of any living/

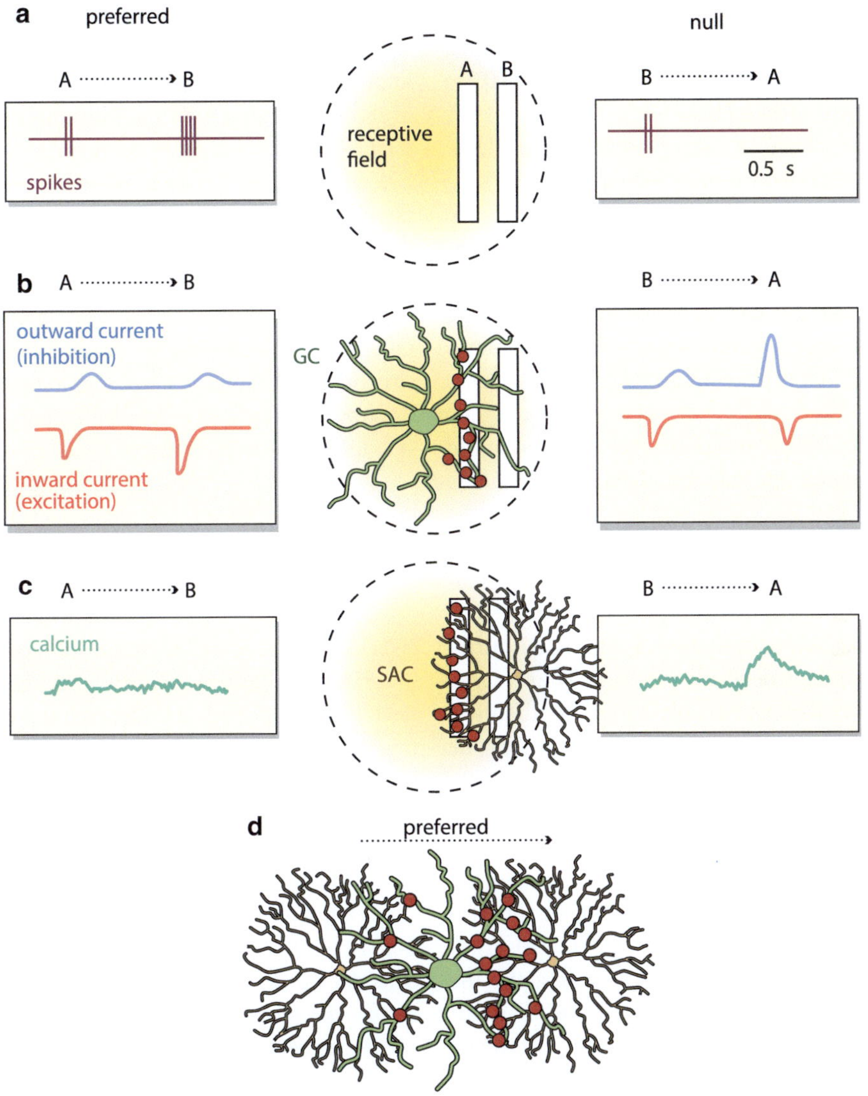


**Fig. 30.5 Computing the direction of visual motion in ganglion cells in the retina.** (**a**) Experimental demonstration that some retina ganglion cells (GCs) are directional selective. Visual motion is simulated by flashing slits of light bars on the retina in successive times (*A* then *B* or *B* then *A*) and recording the resultant spikes from a retinal GC. The preferred sequence (*A* then *B*) evoked greater firing than the null sequence (*B* then *A*). (**b**) For the preferred sequence, the excitatory input to the GC (*red traces*) starts earlier and is larger than the (later) inhibitory input (*blue traces*). In contrast, in the null sequence, the inhibitory input starts earlier and is larger than the excitatory input. (**c**) Calcium influx is enhanced in the dendrites of the starburst amacrine cells (*SAC*) only for motion in the null direction. Calcium influx leads to increased release of the inhibitory transmitter (GABA) onto the postsynaptic GCs via dendro-dendritic synapses (*red dots*). This demonstrates that direction selectivity is computed locally in SACs' dendritic branches at a stage before ganglion cells. (**d**) The larger number of inhibitory synapses on one side of the GC's dendrites (more *red dots* on right side of GC dendrites) serves as the anatomical basis for the GC's directional selectivity. The asymmetric inhibition is achieved by a developmental program that specifically strengthens the inhibition from SACs located on the null side (Adapted from Sterling [47] with permission; copyright 2012, Macmillan Publishers Ltd: Nature.)

moving creature (e.g., detecting whether a car is traveling towards you or in the opposite direction). Indeed, directional selective neurons are found already in the retina of both vertebrates and invertebrates. One possible mechanism for implementing this computation is highlighted in Fig. 30.5 [24, 36].

2. *Collision avoidance*. For instance in the lobula giant movement detector (LGMD) of locusts two inputs

(one excitatory, one inhibitory) converge onto LGMD dendritic trees, where they are multiplied. This operation underlies the response of LGMD to looming visual stimuli [16].

3. *Storage and classification of multiple input features.* Models show that mapping of excitatory synapses onto specific dendritic branches combined with local nonlinearities in these branches, enhances the input classification capacity of neurons [28].
4. *Calculation and memory of position variables.* This requires a temporal integration of velocity signals – an operation that is essential for, e.g., maintaining a particular posture and for navigation. It is argued that time-dependent position representation could be implemented by calcium dynamics in single neurons [27].
5. *Recovering input signals in the presence of strong noise.* The mechanism suggested and experimentally tested in the optic neuropile of the blowfly is that temporal summation of corrupted signals arriving from different dendritic branches is averaging out local fluctuations, thus giving rise to a faithful (smooth) representation in the axon (output) of the visual input [45].
6. *Enhancing temporal resolution for coincidence detection.* The short effective time constant (short local delay, *LD*, see above) in dendrites and the strong voltage saturation of the synaptic input in dendrites (due to high local input resistance), combined with dendritic nonlinearities could equip the neuron with a temporal resolution in the submillisecond range – which is faster than its membrane time-constant or the typical time constants of its synapses [2].
7. *Allowing different temporal coding in the same neurons. In vitro* experiments in layer 2/3 cortical pyramidal cell dendrites have shown that different coding schemes could be implemented in different parts of the same dendrite. Due to stronger activation of the slow NMDA receptors at distal sites, the summation of synaptic inputs in these sites is less sensitive to the exact input timing ("rate code"). Inputs to more proximal sites (with less activation of NMDA current) are more sensitive to temporal precision of the input ("temporal code") [8].

Thanks to the advance of new optical and anatomical technologies, the direct involvement of dendrites in specific computation was experimentally demonstrated *in vivo* [45]. Probably the most complete example to date is the computation of direction of visual motion in retinal ganglion cells (GC). This computation requires some source of asymmetry whereby the synaptic input impinging on the GC for one (e.g., preferred) direction of the visual motion differs from the synaptic input to the GCs for movement in the opposite (null) direction. Indeed, theoretical ideas regarding computation of directional selectivity in dendrites were put forwards some 50 years ago [24, 36] and *in vitro* experiments have confirmed some of these predictions. With the recent convergence of several new technologies (optical imaging from dendrites *in vivo* and large-scale reconstruction of synaptic connections onto the GCs using electron microscopy) a detailed solution for this long-standing problem has been provided (Fig. 30.5) [9, 47].

### 30.1.4 Diversity of Neurons and Their Computation

Ever since the intricate shape of neurons was first observed, similarities and differences between neurons have been described, starting with the beautiful drawings by Ramón y Cajal. A fundamental question that immediately strikes the viewer upon seeing these illustrations is whether the fine shape of each neuron is unique, like snowflakes, to that particular neuron in this particular animal or whether neurons come in a small number of neuronal types, or classes. This question is hard to resolve (see below). However, when considering invertebrate systems one can clearly recognize the particular anatomical shapes of individual cell types repeatedly from animal to animal and the concept of an "identified neuron" is quite clear [31]. To what degree this holds for vertebrate systems is still an open question, though some examples of identifiable neurons are known.

> Neurons may be classified into "classes" based on a variety of attributes – their particular morphology, their electrical properties, their axonal targets, or their genetic expression profile. How many classes of neurons there are in a particular brain region and what is the functional significance of having such classes are still open questions.

Beyond the empirical question of the similarity of neurons, one can ask the more fundamental question of whether the computing elements – the neurons – are stereotypical for a case where we know that a neuronal circuit performs a stereotypical computation. The most thorough and elegant exploration of this question has been the study of the lobster stomatogastric ganglion, in the laboratory of Eve Marder. The ganglion is responsible for, roughly speaking, controlling the processing of further breaking up swallowed food which is not effectively chewed by the mouth parts. This system contains approximately 30 neurons, and it generates a simple, stereotypic output – a rhythmic pattern and continues to provide its output also when removed from the animal (see Chap. 23). One of the most interesting findings of this work has been that underlying the same macroscopic electrical behavior ("phenotype") of a given cell type, one finds *different* microscopic elements that subserve this particular electrical behavior, i.e., different combinations of types and densities of ion channels provide the same macroscopic behavior [29, 34].

#### 30.1.4.1 The Diversity of Cortical Inhibitory Interneurons

Is microscopic diversity of neurons with similar macroscopic electrical behavior unique to invertebrates? Do vertebrates have a subset of macroscopic neural classes, despite neurons being less stereotypical? Different classifications of the variety of neurons have been suggested according to various criteria such as structure of the axon, extent of the dendritic arbor, measures of electrical activity, and other criteria [7]. Yet the most striking dichotomy is that known as "Dale's law" – the principle that an individual neuron will have either only an excitatory or an inhibitory effect on *all* its postsynaptic cells. This principle divides neurons of the central nervous system into excitatory and inhibitory neurons.

Inhibitory interneurons compose approximately 20 % of the neurons in the neocortex and are found within all its layers. The role interneurons play in network dynamics is still a subject of open research with many suggestions proposed. Of course, the most basic role is to counteract (or balance) excitation. More subtle roles for inhibition have also been suggested, from increasing the separation between similar and competing representations, e.g., "winner takes all" algorithms, through increasing temporal fidelity of neurons, to enabling oscillations in different frequencies in the network activity.

Attempts to classify inhibitory neurons have brought forth a number of intriguing findings regarding the specific nature of their diversity. For example, the response of inhibitory neurons, in term of action potential (AP) firing, to applied step depolarizing current varies widely between different cells (Fig. 30.6). Yet at the same time, several general classes of firing responses can be observed. Some respond to step current stimuli with orderly trains of APs ("regular spiking" neurons), some emit high-frequency bursts followed by a quiescent period ("stuttering" neurons), others respond with a series of APs with pauses that increase over the span of the stimulation ("accommodating" neurons). Intriguingly, these different patterns of AP firing match quite well with patterns observed in measures of completely different qualities, such as the morphology of the neuron, the profile of different genes expressed in these different electrical cell-types, or the territory of postsynaptic innervation by these neurons (e.g., close to the postsynaptic cell's soma, or distally in the dendrites of the target cell [19]).

Such specificity within diversity tantalizingly suggests it could be explained in terms of a set of distinct classes, building blocks, which compose the inhibitory circuitry of cortical neural networks. Modern techniques, most notably genetic analysis, are increasingly used [32]. Notably, interneuron diversity is not solely a cortical phenomenon and has also been observed for instance in the hippocampus. Moreover, the question of neuronal cell types is relevant for all nervous systems, especially those with a small number of neurons, such as *C. elegans*, where the entire neuronal population may eventually be described.

#### 30.1.4.2 Classifying Cortical Inhibitory Interneurons

The existence of electrophysiological classes of neurons, and in particular the rich firing repertoire of cortical interneurons, their macroscopic electrical behavior, has been examined both *in vivo* and *in vitro* [30]. Typically, the difference in electrical classes of neurons has been observed and discussed in terms of the response to *depolarizing step currents*. Indeed, the naming conventions of these classes mostly arise from the response to such current clamp stimulations ("regular firing" "adapting", "bursters", "stutterers" etc.). Recently, a meeting between members of different prominent labs worldwide that study the diversity of cortical interneurons took place in Petilla de Aragon, the birthplace of Ramón y Cajal. This group, "the Petilla Interneuron Nomenclature Group (PING)" met in an attempt to unify ideas, intuitions, and methods in order to arrive at a uniform classification of neuronal diversity. The final result [7] lists six main types of

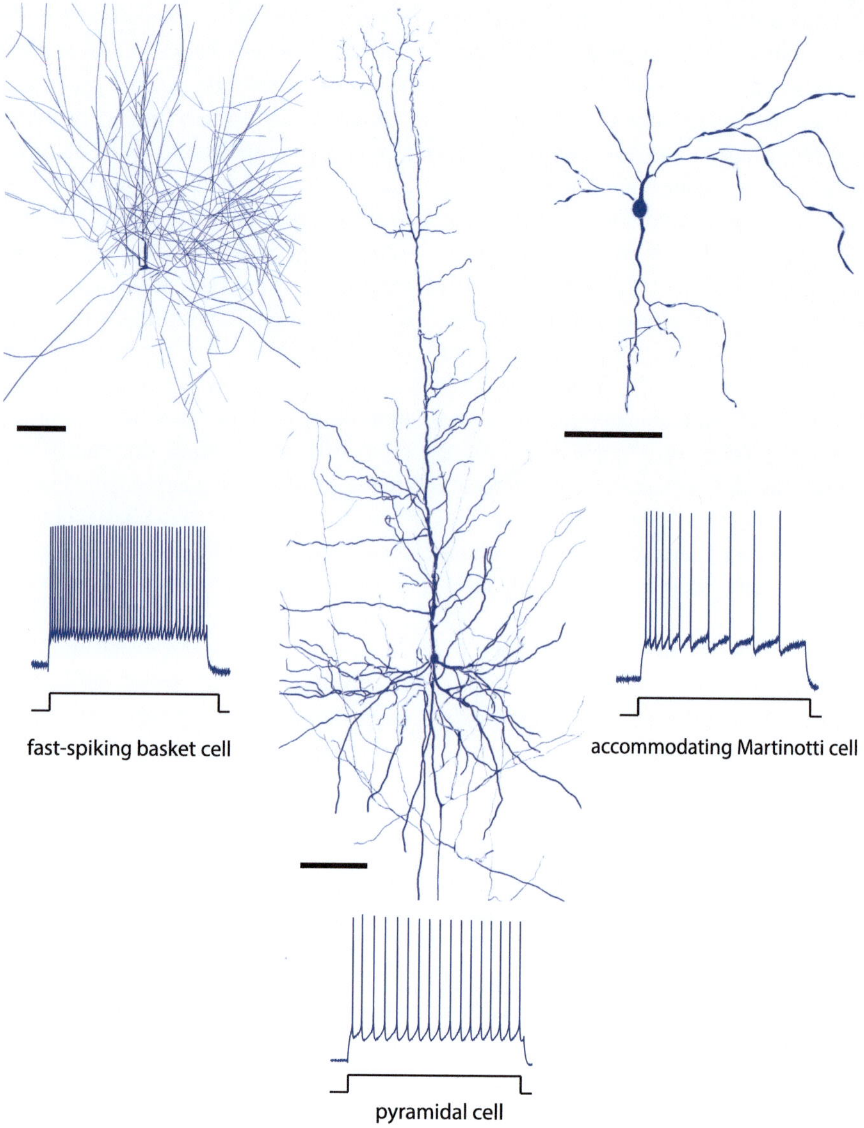


**Fig. 30.6 Examples for neocortical cell types**. Three typical cell types of the neocortex, fast spiking basket cell (*left*), regular firing pyramidal cell (*middle*), and accommodating Martinotti cell (*right*) are shown with their morphology (*top*) and typical response to intracellular step current injection 2 seconds long (*bottom*). Note that the axon of the Martinotti cell is not shown and neurons have different scale. Scale bars indicate 100 μm

firing patterns: fast spiking, adapting, nonadapting non-fast spiking, accelerating, irregular spiking, and intrinsic burst firing (Fig. 30.7).

#### 30.1.4.3 Conductance-Based Models Capturing Diversity of Electrical Classes

Unlike the lobster stomatogastric system mentioned above, direct measurement of all the different ionic conductances underlying an electrical class of cortical inhibitory interneurons is not yet technically feasible. An alternative approach to explore the microscopic basis of the macroscopic electrical classes is to mathematically model their electrical properties, using conductance-based models and the Hodgkin-Huxley formalism (see Chap. 7) for describing the kinetics of

> Conductance-based models aim to describe in detail the microscopic basis of a neuron's electrical behavior, i.e., the electrical current that flows through the different specific ion channels embedded in the neuron's membrane.

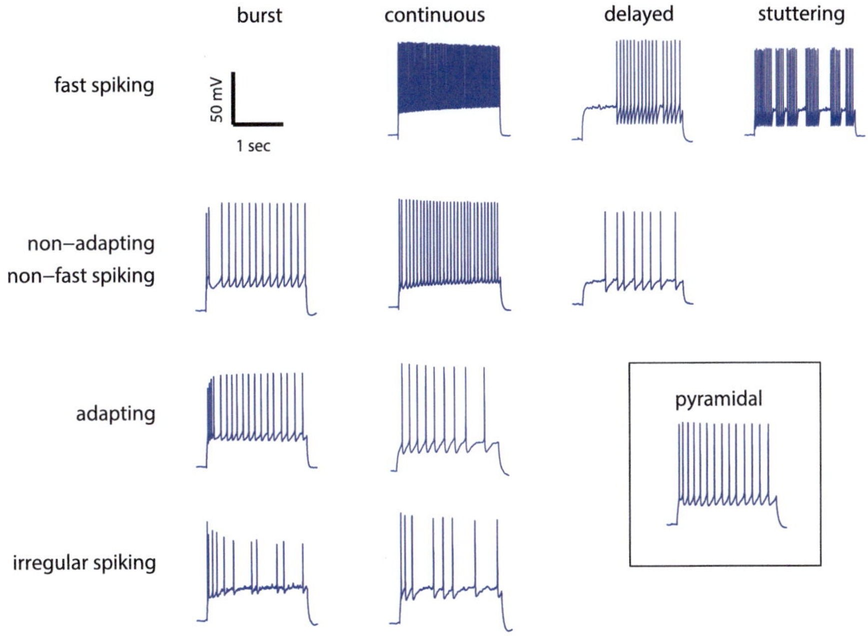


**Fig. 30.7 Petilla interneuron nomenclature for cortical interneurons.** Examples of different types of electrical behavior as presented in an attempt to standardize nomenclature between experimental groups. *Rows* indicate steady state behavior while *columns* indicate initial transient behavior. This is a partial list of all neuron types in [7]. Namely, bursting and accelerating interneurons are not shown (Adapted from Ascoli et al. [7] with permission; copyright 2012, Macmillan Publishers Ltd: Nature Reviews Neuroscience.)

the underlying membrane ion channels. Moreover, such neuron models allow to perform detailed simulations of cortical network dynamics, in which *in silico* manipulations of the nature and details of inhibitory interneuron diversity will serve to tease apart the effect of the different classes of electrical diversity on the overall dynamics of the network. Since the density, distribution and even the type of ion channels involved are typically left as free parameters of the model due to the aforementioned technical difficulties in measuring them, the parameter constraining procedure is a crucial part of the generation of conductance-based models for these cells. As there are at least 13 classes of cortical inhibitory interneurons and as many morphological classes (Fig. 30.7), it is clear that manual parameter tuning, which requires a vast amount of time, is inappropriate for the scale of this effort. Instead, an algorithmic automated approach is required (Fig. 30.8).

#### 30.1.4.4 Feature-Based Distance Functions for Spiking Neurons

The first step in an algorithmic approach to capturing the firing repertoire of neurons is to develop distance functions that indicate how close is a modeled neuronal firing pattern to the experimental firing (target) behavior: when is the electrical behavior of the model close enough to be considered a valid match for the electrical behavior of interest? Traditional approaches attempted to bring the model voltage trace into as close as possible with a single target voltage trace. However, due to the intrinsic variability inherent to neurons, responses to the exact same stimulus in the same neuron can vary considerably, making the approach of trying to bring the model in agreement with a single voltage trace problematic. Rather than using the whole voltage (spiking) trace as a target for fitting, certain attributes, or features, of the neuron's response to a repeated stimulus can serve as measures of similarity between model and experiments. For instance, the features could be: the rate of the APs discharge, the height of APs, their width, etc. These features are considerably well conserved between repetitions, and their experimental mean and variability are easy to define.

These features of the firing of interneurons can be used both to classify the diversity of inhibitory interneurons and to model them [14]. Often multiple features will be required to faithfully capture the differences in firing repertoire of different electrical classes. Multiple Objective Optimization (MOO [13]) is a technique for combined optimization of multiple

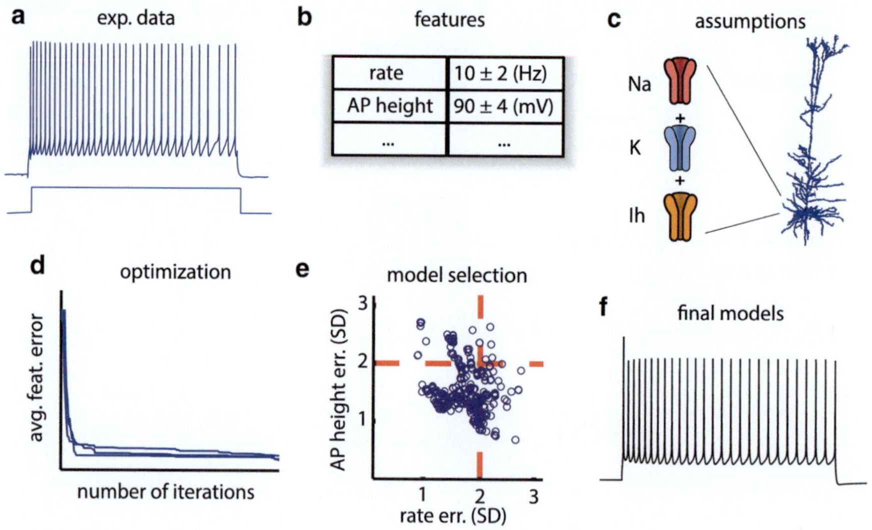


**Fig. 30.8 From experimental data to acceptable conductance-based neuron model.** (**a**) Data is collected from voltage responses to a set of intracellular current injections (steps in the example shown, but could also be ramp current, noise currents, etc.) recorded from single cells' somata. (**b**) The voltage traces are characterized using a set of features (e.g., firing rate, height of action potentials). For each feature both the experimental mean and standard deviation (*SD*) are obtained. (**c**) The generic form of a model to be constrained consists of a reconstructed morphology and an assumed set of membrane ion channels (including their kinetics but not their densities). (**d**) A multiple objective, genetic algorithm-based process of stochastic optimization is applied in order to obtain values for $g_i$ that minimize the distance between the experimentally-measured set of features and those of the model. The convergence of the average error is shown by the *blue curves*. (**e**) For the many possible solutions at the final iteration, a selection criterion of two experimental SDs in each feature is used for choosing a subset of solutions (sets of $g_i$ values); these are considered acceptable models. (**f**) An example of the response of one successful model to a step current input as in (a)

incommensurate and possibly conflicting measures of success (termed objectives).

Applying the MOO procedure to the diversity of interneurons it is possible to obtain faithful conductance-based models of different firing behaviors (Fig. 30.9). The same approach can of course be extended to other neuron types.

As was the case with the lobster stomatogastric neurons, for each interneuron class modeled, multiple combinations of the densities of ion channels are acceptable solutions. Thus, we find that the notion of many microscopic solutions for generating the same macroscopic behavior holds also for the very diverse firing repertoire of cortical inhibitory interneurons [14].

On a final note regarding this issue, any component of a biological system will show some variability. The discussion above implicitly assumed that we have a way of defining how large of a change from the recorded macroscopic behavior is still within the accepted variability. The guideline used for the inhibitory neurons was that the type of changes found within repeated representation of the same stimulus to the same neuron must be within the accepted variability. More generally, on the conceptual level the type and amount of variability that is reasonable to expect from a single neuron would be that variability which still results in a functioning neural circuit, e.g., variability that does not interfere with the function of downstream neurons. But how do we know what interferes with the function of downstream neurons? Perhaps something that does not interfere with their own downstream neurons? Clearly we run the risk of an argument that continues *ad infitum*, or at least until we can record from each neuron in the brain. Fortunately, one solution that allows breaking this infinite chain can be found if we can connect the function of the neural circuit to a defined behavior. If that behavior is still viable with a certain variability then that variability can be deemed reasonable. For instance, if the neural components still produce the correct rhythmic patterns in the case of the lobster stomatogastric ganglion [29]. Thus, we must ultimately strive to think of neurons and the circuits they compose in terms of the behavior of the organism as a whole.

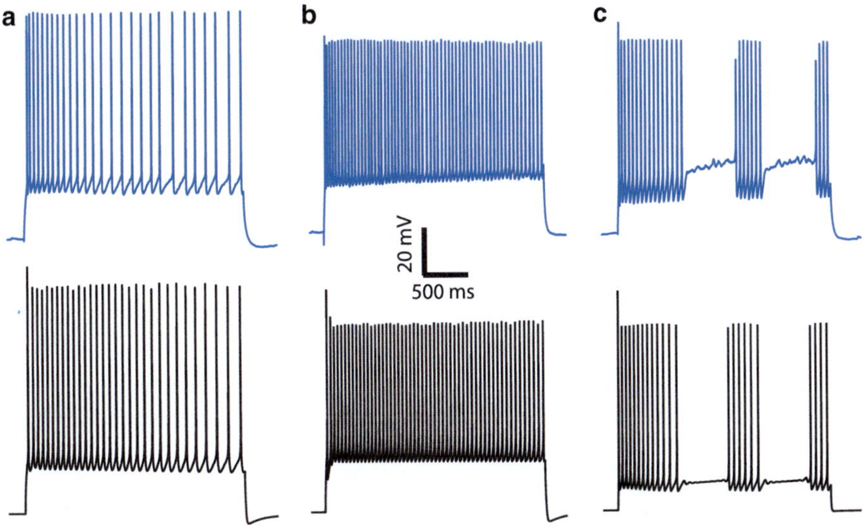


**Fig. 30.9** **Modeling the diversity of interneuron behavior**. For each of three interneuron types, an experimental voltage trace of the response of an interneuron to a step current injection is shown (*top*) along with a corresponding model trace (*bottom*). The conductance-based model of that interneuron was obtained using the MOO procedure. Shown are (**a**) accommodating, (**b**) fast-spiking, (**c**) stuttering interneurons. The models clearly differ in terms of the density of ion channels. For instance, the adaptation in **a** is mediated by calcium-dependent potassium channels which are not present in the solution for case **b** [14]

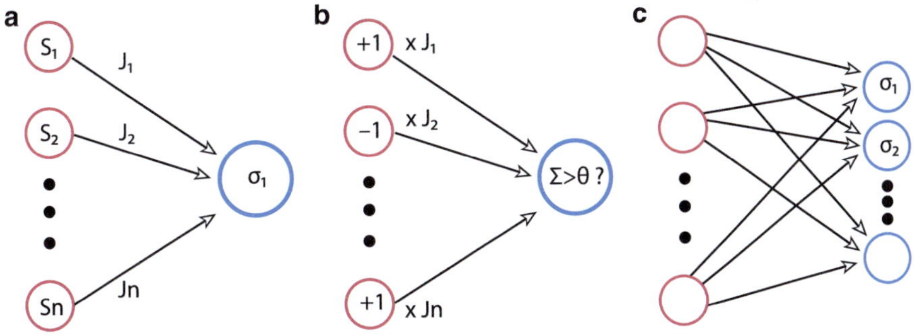


**Fig. 30.10** **Perceptrons**. (**a**) Schematic of a simple perceptron in which a set of input neurons $S_1, S_2, \ldots S_n$ (*red*) are connected to an output neuron, sigma (*blue*). (**b**) Given a particular input pattern of "1"s and "−1"s, the activation of the perceptron is found by multiplying the inputs by their weights, $J_i$. This activation is then compared to the threshold, θ (**c**) Multiple perceptrons can be connected in parallel to the same input neurons to perform different computations

## 30.2 Computation in Neuronal Networks

### 30.2.1 Perceptron

The perceptron [42] is perhaps the simplest model of network computation, yet it is a cornerstone for many theories of neural computation and for the field of machine learning. The basic form of the model is a single neuron that receives inputs from a group of neurons (Fig. 30.10). It is convenient to think of these neurons belonging to stages, or layers, of processing. The first layer is the input layer and the second the output layer. Each of the input neurons drives the output neuron with a certain weight $j$, which can be different for each neuron; conceptually, the weight corresponds to the strength of an input synapse. A positive weight indicates that activity in the input (presynaptic) neuron excites the output (postsynaptic) neuron, making it more likely to discharge an action potential. Conversely, a negative weight indicates that activity in the input neuron inhibits the output neuron making it less likely to discharge an action potential. To find the activity of the perceptron, we simply add up all the inputs coming to the neuron, multiplied by their weight. In other words, we make the simplifying assumption that the neuron linearly integrates across its presynaptic partners. We assume that a perceptron has a simple threshold; if the input is greater than threshold the perceptron responds with a positive (spiking) response, if it is under threshold, with a negative (silent) response. One can make an additional simplifying assumption that the activity of the neurons is binary, being equal to plus one (spiking) or minus one (silent).

The perceptron is a simple model for learning in neural networks. In this "supervised learning" model, the synaptic weights of a simple neural network are repeatedly adjusted in response to stimuli, in order to produce a desired output.

There are numerous interpretations of the computation performed by the perceptron model. For instance, one can describe the computation as the weighing of different input data in order to make a decision. In this view, the activity of the output neuron represents a decision that the organism needs to make, e.g., "Should I run away very quickly?" while each of the input neurons indicates the truth or falseness of a certain piece of information regarding the world, e.g., "I see a tiger in front of me". The synaptic weight between the input neuron and the output neurons measures how strongly and in which direction (positive or negative) that piece of information should influence the decision. In this case the "tiger" (input) neuron would probably be connected with a high weight to the output "I should run" neuron.

Mathematically, the perceptron is a "dichotomizer", i.e., it embodies a dichotomous decision all input patterns are assigned one of two values: true or false. Though it might seem natural to use 1 to designate true (or spiking) and 0 to designate false (silent) it turns out to be more mathematically convenient to designate 1 for true and −1 for false. Each input can take two values and there are $n$ such input neurons, for a total of $2^n$ different possible patterns. The perceptron is defined by its weight vector $J$, which has as many entries as there are neurons in the input layer and by its threshold, θ. Multiplying each input pattern by the weight vector and summing, we have a given total input current to the perceptron for each input pattern. For some patterns this input current will be greater than θ, these are the patterns for which the perceptron gives a positive response whereas for the other patterns the perceptron will yield a negative response.

The perceptron can be trained to give the correct responses for a set of predefined input patterns. Namely, the output of the perceptron is required to be +1 for some (arbitrary) set of patterns and −1 for other patterns. Consider the example of a perceptron receiving two inputs. There are $2^2=4$ possible input patterns ([+1, +1], [+1, −1], [−1, +1], [−1, −1]). We can require different outputs for these input patterns. Some of these input-output functions can be correctly performed by a perceptron (the "AND" output, Fig. 30.11a), while others cannot (the "XOR" output, Fig. 30.11b) since a perceptron can only perform linear classification. There is a simple geometrical intuition for the success of a perceptron classification. If a straight line that separates the input patterns so that all those that should be classified with +1 are on one side and all those that should be classified as −1 are on the other, the particular set of patterns can be classified by a perceptron (compare Fig. 30.11a, b). The weight of the perceptron will be a vector that is perpendicular to the line and pointing in the direction of the positive examples (Fig. 30.11a).

Training is performed by "supervised learning", the type of learning where the student is provided with both the input pattern to be classified and the correct response (+1 or −1) for each pattern. When the predefined responses can in principle be correctly performed with a perceptron, i.e., when there exists a fixed weight vector, $J$, by which the input patterns are multiplied and summed over, such that all patterns that should have a positive response (+1) sum up to a number greater than the threshold, and the sum for all other patterns sum up to a number lesser than threshold, this classification task is called linearly separable. It can be proven that a simple learning rule which updates the perceptron weights a little bit at every time step, will converge after presentation of a reasonable number of examples to a solution that correctly classifies the patterns. The perceptron learning rule is as follows:

$$J_i^{new} = J_i^{old} + \eta \times \Delta J_i \tag{30.7}$$

where

$$\Delta J_i = \begin{cases} 2x_i^k y^k & y^k \neq \tilde{y}^k \\ 0 & otherwise \end{cases} \tag{30.8}$$

Where $J$ denotes the weight vector and $i$ its $i$th component, $x_i^k$ denotes the $i$th component of the $k$th input pattern, $y^k$ denotes the output of the perceptron (plus or minus one) for the $k$th input pattern and $\tilde{y}^k$ the correct output for the $k$th pattern. If the perceptron gives the correct output its weights will not be changed. If the answer is false, the perceptron's weights are adjusted in the direction of the $k$th input pattern that was falsely mislabeled, with a sign according to the sign of the mistake. For instance, in Fig. 30.11c the input pattern in the top left corner, $x^1$, is incorrectly classified as negative, since it is on the negative side of the separating, classification line (if we would have considered a higher dimensional example this line would be a plane). Following the perceptron learning rule, a component proportional to $x^1$ is added to the weight vector. Geometrically, adding vectors amounts to joining the vectors end-to-tail (Fig. 30.11c, right). After learning, following the change in the weight vector, this pattern is correctly classified, since it is now on the positive side of the separating line (Fig. 30.11c, right). Note that the learning is not complete yet since the pattern in the top right corner is on the positive side of the separating line even though its classification should be negative. In this fashion, by iterating through the different input patterns, examining whether the perceptron correctly classifies it, and if not pushing the weights in the direction of the input pattern (with the appropriate sign) it becomes more likely that the patterns will yield the correct response. If the patterns can be correctly classified by a perceptron then this approach will converge to a correct solution in a finite number of steps. The full convergence proof can be found in many textbooks [11, 20].

One can take the simple perceptron model and elaborate it in several ways. For instance, one can have multiple perceptrons reading the same input and performing different classifications on them (Fig. 30.10c). To continue our above example regarding the interpretation of a visual stimulus and decide whether we need to run away, one could have an additional perceptron looking at the same inputs and deciding whether the object should be eaten. Naturally, the different perceptrons will have different synaptic weights. Additionally, multiple perceptrons can be chained one after the other, the decision of the previous perceptron becoming the input to the next one, e.g., the running away neuron might be an input to a neuron representing dropping whatever is in one's hands before running.

The simple perceptron cannot implement every computation. For instance, in Fig. 30.11b we see that the perceptron cannot implement the XOR function. One solution to this problem is to add another layer of units between the input and the output. Since this layer is neither the input nor the output of the system it is often referred to as a "hidden" layer. It can be shown that perceptron networks with hidden layers can solve more demanding computational tasks [11, 20].

Up until now we only discussed differentiating between activation patterns of the presynaptic (input) neurons and have not mentioned at all the timing of the activation of the neurons. In an elegant extension of the perceptron, Gutig and Sompolinsky recently suggested a model that is able to distinguish input patterns based on the timing of the inputs and not just whether or not the inputs are active – the "tempotron" [18]. The tempotron framework provides a supervised synaptic learning rule that enables neurons to extract the information that is embedded in the spatiotemporal structure of spike patterns.

In summary, the perceptron is a basic model that demonstrates how a simple neuronal circuit, composed of neurons that sum up their (weighted) inputs and compare this sum to a threshold can accomplish, via supervised learning, a classification task over its inputs. Despite being highly simplified, the perceptron served as an important conceptual model of how learning may take place in neural networks.

### 30.2.2 Hopfield Model of Associative Memory

The above sections dealt with models at the single neuron level, and discussed some of the computations that might be implemented by single neurons and their synapses. Now we describe several fundamental ideas regarding the computations that networks of neurons could perform. In these models, individual neurons are represented in a reduced and simplified manner, sometimes called "point neurons" to remind us that we ignore the neurons' complicated morphology. The focus here is on the collective computations that large, highly interconnected networks of simple neurons can perform as a whole, while neglecting any additional computational power that might emerge from the specific morphology, synaptic distribution, and particular firing pattern of the individual neurons.

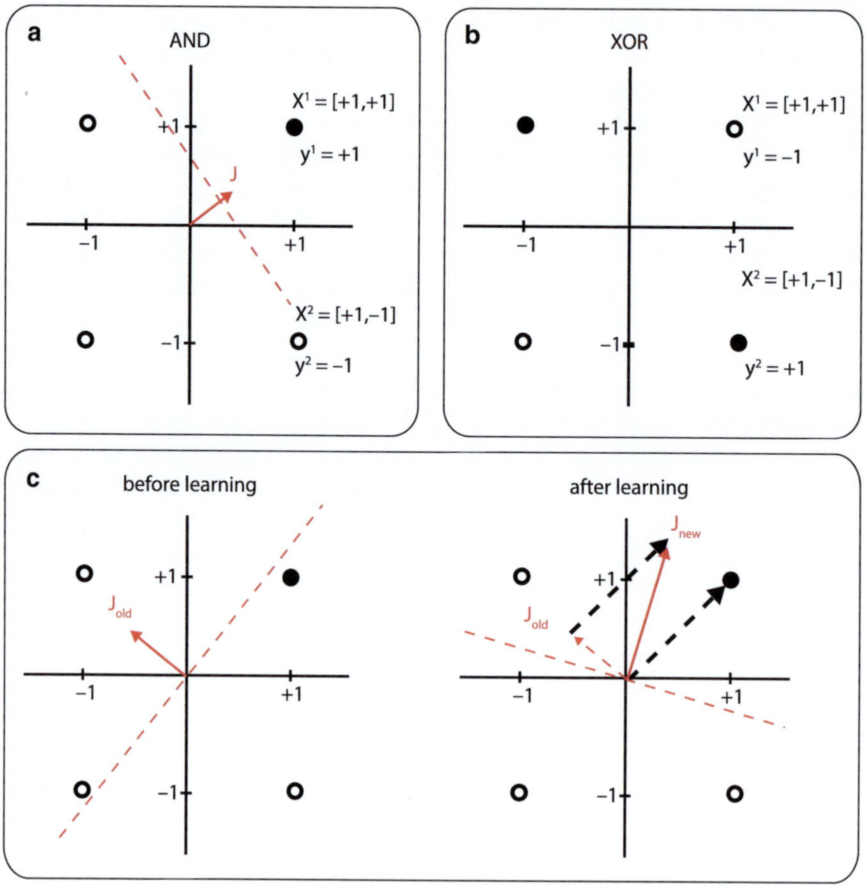


**Fig. 30.11** **Perceptron classification and learning**. (**a**) The "AND" pattern where the classification of all input patterns is negative (*empty circles*) except the pattern for which both input patterns are positive, which is to be classified positively (*full circle*). This pattern can be correctly classified by a perceptron, e.g., one with the weight vector $J$ shown in *red*. The *red dashed line* corresponds to the line perpendicular to $J$. Patterns on the side of $J$ (the side to which $J$'s tip is pointing) will provide positive input and will elicit activity in a perceptron with zero threshold. Patterns on the other side will provide negative input and will not elicit activity. The displacement of the line from the origin is due to the threshold. (**b**) The "XOR" pattern where patterns in which one and only one input is positive should be classified as positive. This output cannot be correctly classified by a perceptron since no line can divide the plane in a way that the positive output patterns are on one side and the negative output on the other. (**c**) *Left*: Before learning the positive pattern, $x^1$ , (superscripts indicate pattern number while subscripts indicate neuron number) is incorrectly classified by the weight vector $J_{\text{old}}$ since it is on the negative side of the line. (**c**) *Right*: According to the perceptron rule, Eqs. 30.7 and 30.8 in the text, a component matching the misclassified point is added to the weight vector to arrive at the new vector $J_{\text{new}}$ . Note that the learning rule operates by making very small changes each time a mistake is made. In this plot we display a large change, which would accumulate over many corrections, in one single step for the purposes of illustration. As can be seen, now the pattern that drove the change, $x^1$ , is on the positive side of the classification line. One more pattern is still on the wrong side and will be corrected in one of the next iterations of learning

One of the most renowned ideas regarding "neural network computation" is that introduced by John Hopfield some 30 years ago [21]. It would not be an exaggeration to claim that these studies ushered in a new era in computational neuroscience. The Hopfield model is a conceptual model that captures one of the most striking differences between the brain and current-day machines – the capacity for associative memory. Associative memory is the ability to recall a full mental image (memory) of a previously encountered stimulus from a fragmentary or noisy version of it. One can come up with dozens of examples of this phenomenon: our ability to recognize a whole musical tune from a few bars, a familiar face from a distorted image, or a complete text from its beginning ("*One small step for man…*"). In contrast, in order for a typical computer to retrieve a missing piece of information, its location in memory (its address), rather than the content of the piece must be known.

The Hopfield model is a conceptual model for associative memory. It shows how a pattern of neural activity representing a given memory can be made an "attractor state" to which similar patterns of activity will converge, allowing recovery of the full memory from a partial version.

Hopfield showed how neuronal-like networks might be constructed so that a specific set of memories can be retrieved in an associative manner. Namely, starting from a certain pattern of activity, the network dynamics passes through different stages, neurons switching on and off, until the network settles into a fixed state in which the neurons' activity remains constant (each neuron remains in its either "on" or "off" state). Many different initial configurations eventually settle into (are attracted into) a given, special state. Each such state represents a single memory that the network supports. If the network is initialized with a pattern that is similar to a particular memory state, corresponding to partial or distorted representations of that memory, the deviations from the memory state gradually fade away as the network activity evolves, until the network dynamics terminates in the attractor state. Thus, the network has converged to the original "clean" memory represented by that attractor state. Figure 30.12 shows a graphical illustration of associative memory performed by a Hopfield network.

We shall begin by describing the deterministic, binary Hopfield model. In this formulation the state of each neuron can take one of two values: plus one (representing spiking) or minus one (silent). The state of the neuron at any given point in time is determined simply by comparing the value of its synaptic input to its threshold. If the input is higher than the threshold the neuron switches to the spiking state. If the input is lower than threshold the neuron switches to the silent state. For the sake of convenience we will assume in this treatment that the threshold of all the neurons is identical and set it at zero. Thus, if the sum of the excitatory and inhibitory inputs that a neuron receives is positive it will be in the spiking state; otherwise it will be in the silent one.

The Hopfield model proposes a specific choice for the connectivity of the network that will allow the network to serve as an associative memory network for a set of predefined patterns (memories to be stored). How can we confirm whether this specific connectivity will indeed allow the network to perform its function? We would like the dynamics of the network, when faced with an incomplete version of one of the memories, to cause the network to change its state and ultimately stabilize on the clean version of the memory. A minimal requirement is that the memory patterns are stable, i.e., if the network reaches the pattern in the course of its dynamics it will then remain at that state (until a new input is presented) indicating that the memory has been recalled.

In order to examine the stability of the network patterns we introduce the following notations. Let us note the number of neurons in the network by $N$. The state of the $i$th neuron will be represented by $S_i; i = 1,2,3\ldots N$. Each neuron can take the value +1 or −1. Thus, there are $2^N$ possible patterns of activity. Assume that we now choose $P$ specific activity patterns to represent memories that the network should be able to recall. Note each memory by $\xi^{u}; \mu = 1,2,3\ldots P$ and the state of the $j^{\text{th}}$ neuron for that pattern by $\xi_j^{\mu}$, e.g., in a network of three neurons one might have: $\xi^1 = \left[\xi_1^1, \xi_2^1, \xi_3^1\right] = [+1,+1,-1]$. We shall denote the connections of the network by a matrix $J$ whose elements represent $J_{ij}$ the strength of connection from neuron $j$ to neuron $i$. The Hopfield model proposes the following rule for the strength of connections: $J_{ij} = \frac{1}{N}\sum_{\mu}\xi_i^{\mu}\xi_j^{\mu}$. In words, for each of the $N$ neurons in the network, the strength of the connection between neuron $j$ and neuron $i$ is determined by the sign of the $j$th neuron times that of the $i$th neuron in each memory pattern, summed over all memory patterns.

In order to examine the logic of this choice let us consider the case of a network that is supposed to store only one memory $\xi^1$. Then the synaptic connectivity matrix will be simply: $J_{ij} = \frac{1}{N}\xi_i^1\xi_j^1$ the sum having been reduced to only one memory pattern. Accordingly, the input that the $i$th neuron receives, denoted by $h_i$, is equal to $h_i = \sum_{j=1}^{N} J_{ij}S_j = \frac{1}{N}\sum_{j=1}^{N}\xi_i^1\xi_j^1 S_j = \frac{1}{N}\xi_i^1\sum_{j=1}^{N}\xi_j^1 S_j$ in

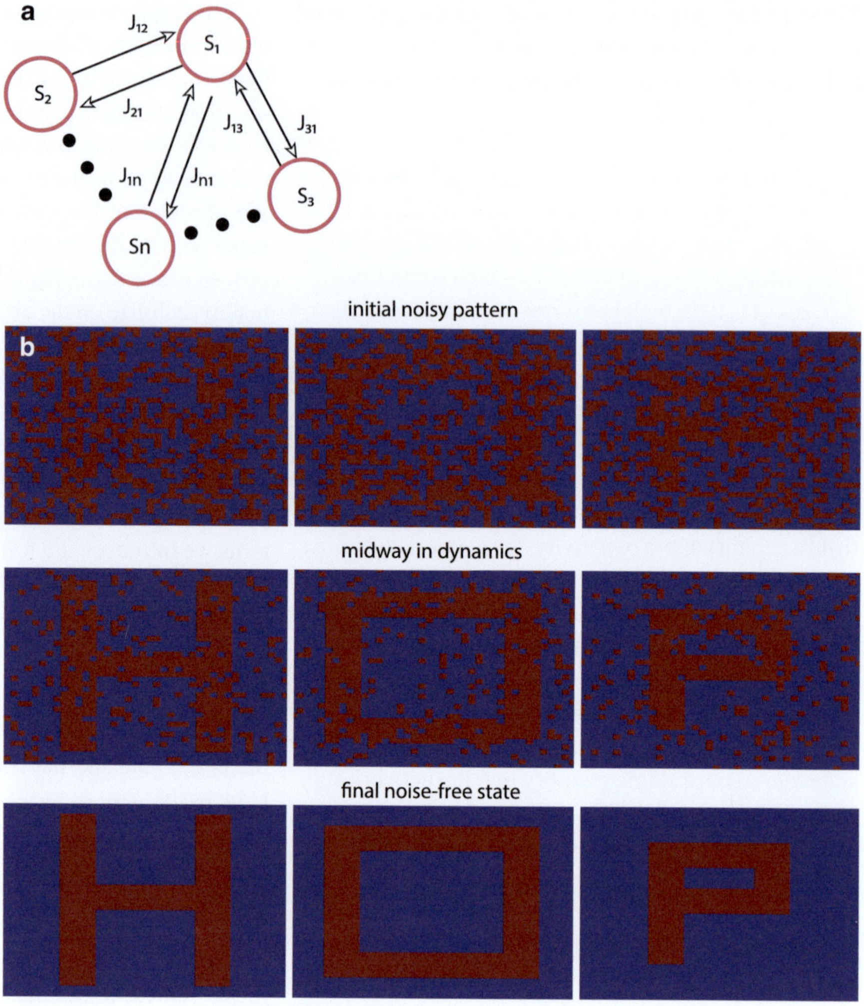


**Fig. 30.12** **Memory recall in Hopfield model**. (**a**) Schematic of a network and its connectivity matrix *J*. (**b**) State of a network that has the letters "h", "o", "p" embedded as memory patterns. Neurons that are on are shown in red and neurons that are off in blue. Top row shows the initial pattern of activation that is a noisy version of the letters. *Columns* indicate three repetitions of the dynamics, once starting with a noisy "h" as the initial state (*left column*), then "o" (*middle column*), then "p" (*right column*). The dynamics of the Hopfield network serve to recall the stored patterns and the state of the network becomes closer and closer (*middle row*) to the stored patterns until the state of the network matches these original memory patterns (*bottom row*)

the last step we took $\xi_i^1$ out of the sum since the sum is over *j*, not *i*.

Let us now assume that at a given moment the network state is equal to the memory pattern, $S_i = \xi_i^1$ for all *i*. Let us now plug in the fact that we know that the network state is equal to that of the memory pattern $h_i = \frac{1}{N}\xi_i^1 \sum_{j=1}^{N} \xi_j^1 S_j = \frac{1}{N}\xi_i^1 \sum_{j=1}^{N} \xi_j^1 \xi_j^1 = \frac{1}{N}\xi_i^1 \sum_{j=1}^{N} \left(\xi_j^1\right)^2$. Now recalling that $\xi_j^1$ can be either plus or minus one, we note that the square of that value $\left(\xi_j^1\right)^2$ will always be equal to one. So we find $h_i = \frac{1}{N}\xi_i^1 \sum_{j=1}^{N} 1 = \frac{1}{N}\xi_i^1 N = \xi_i^1$

where in the next to last step we just computed the sum, 1 summed over *N* times being equal to *N*.

In summary, we find that if the network is at some point in time at the special state that equals the memory pattern, $S_i = \xi_i^1$ then for each neuron the input is equal to the current state, $h_i = \xi_i^1$. In order for a neuron to change its state, it must receive in one time point input that brings it above threshold and then in the next time point a current that is under the threshold (or vice versa). However, as we found that the input is identical to the state, then if it was previously positive it will remain positive, and if it was negative it will remain negative (recall we decided on a zero threshold, making positive input above threshold and negative input below). Thus, no neuron will switch its state from firing to silent or vice versa and this pattern will be stable for the next time step. If it is stable for the next time step, we will again have the same pattern of activity at the next step and again the state will be stable.

Thus, the pattern will remain stable for all time or until reset by some external input. Thus, the memory pattern will be a stable point of the dynamics of the network, just as we required it to be in order to serve as an associative memory network.

In order for a network to serve as an associative memory we not only need the memory patterns to be stable as we have shown but must also require that if we start from a partial, or noise corrupted version of the memory pattern, the dynamics of the network will bring the network state to that of the full, uncorrupted pattern. In order to show that the network dynamics indeed converge to the memory patterns, let us pick up the previous discussion where we calculated the input to a given neuron: $h_i = \frac{1}{N}\xi_i^1 \sum_{j=1}^{N} \xi_j^1 S_j$. Now we assume that the network state is not exactly equal to the memory pattern: $S_i = \xi_i^1$ for all $i$, but rather that the network is in a corrupted version of the network state. In other words, $S_i = \xi_i^1$ holds true not for all $i$, but rather for most $i$. Namely, that a given set of neurons are in the wrong state (either firing when they are supposed to be silent or vice versa). For the sake of simplicity, let us start by assuming that only one neuron is off. Namely, $S_i = \xi_i^1$ for all $i$ but one for $i=k$, where $S_k = -\xi_k^1$. Let us then write down the expression for the input: $h_i = \frac{1}{N}\xi_i^1 \sum_{j=1}^{N} \xi_j^1 S_j = \frac{1}{N}\xi_i^1 \sum_{j=1, j\neq k}^{N} \left(\xi_j^1\right)^2 + \left(-\frac{1}{N}\xi_i^1 \left(\xi_k^1\right)^2\right)$.

Again the square of the binary variable is equal to one, let us plug that into our expression: $h_i = \frac{1}{N}\xi_i^1 \sum_{j=1, j\neq k}^{N} 1 - \frac{1}{N}\xi_i^1 = \xi_i^1\left(\frac{N-1}{N} - \frac{1}{N}\right) = \xi_i^1\left(\frac{N-2}{N}\right)$.

The number we get, especially if $N$ is very large, is almost one times $\xi_i^1$. Thus for all $i$ which are not equal to $k$, the network state was equal to $\xi_i^1$ and will remain that way in the next time point. For the $k$th neuron the state at the previous time point was $-\xi_i^1$, the wrong or corrupted state for that memory pattern, but the input now is equal to almost one times $\xi_i^1$. Thus, at the next time step the neuron will change its state from $-\xi_i^1$ to $\xi_i^1$ changing from the corrupted state to the correct state (Fig. 30.12).

We showed the calculation when there is only one neuron in the wrong state, but we could perform the same calculation for multiple neurons being in the wrong state and, as long as the majority of neurons will be in the correct state, the dynamics of the network will guide the neurons in the corrupted state to converge to the right state as time evolves, thus restoring the complete memory.

If there is more than one memory pattern embedded in the network, then the expression for the connectivity matrix $J$ becomes more complicated and we have: $h_i = \sum_{j=1}^{N} J_{ij} S_j = \frac{1}{N}\sum_{j=1}^{N}\sum_{\mu=1}^{P} \xi_i^\mu \xi_j^\mu S_j$. Again let us consider the stability of the memory pattern, by assuming that at a certain point in time the network reaches a state equal to that memory pattern, $S_i = \xi_i^1$, and examining the stability of the network. Plugging in the fact that the network state is at the first memory pattern and the expression for the connectivity when there are multiple memories we get: $h_i = \frac{1}{N}\sum_{j=1}^{N}\sum_{\mu=1}^{P} \xi_i^\mu \xi_j^\mu \xi_j^1$. Notice that now one can break the sum over $\mu$ to two parts, one in which $\mu = 1$ and one in which $\mu \neq 1$. Writing that down we get:

$$h_i = \frac{1}{N}\sum_{j=1}^{N}\sum_{\mu=1}^{P} \xi_i^\mu \xi_j^\mu \xi_j^1 = \frac{1}{N}\sum_{j=1}^{N}\left[\xi_i^1 \xi_j^1 \xi_j^1 + \sum_{\mu=2}^{P} \xi_i^\mu \xi_j^\mu \xi_j^1\right]$$
$$= \frac{1}{N}\sum_{j=1}^{N} \xi_i^1 \xi_j^1 \xi_j^1 + \frac{1}{N}\sum_{j=1}^{N}\sum_{\mu=2}^{P} \xi_i^\mu \xi_j^\mu \xi_j^1.$$

The first expression is exactly the one we had when there was only one memory pattern and we have just shown that it (by itself) will make the pattern stable, and that it is equal to one times $\xi_i^1$. Plugging that in we obtain:

$$h_i = \xi_i^1 + \frac{1}{N}\sum_{j=1}^{N}\sum_{\mu=2}^{P} \xi_i^\mu \xi_j^\mu \xi_j^1.$$

With the first term alone, the memory pattern will be stable as we require. However, the second term may end up dominating the first term causing the network to be unstable. How would this occur? For instance if, for a certain neuron, we have $\xi_i^1 = 1$, then the first term is equal to one. Now if the second term were equal to negative one (or even more negative) then the overall input to that neuron, the sum of the two terms, would be negative. This will cause that neuron to change its state at the next time point. Thus, the value that was equal to the memory pattern will become corrupted, indicating that the memory pattern is unstable.

We shall not go through the calculation here, but the relation between the size of the first term (the stabilizing, or signal, term) and the second (destabilizing, or noise term) can be used to derive a limit for the number of memories that can be stored in the network without causing the memories to become unstable. Examining the expression above, it is intuitively reasonable that adding more memories will cause the network to become less stable since the destabilizing term contains a sum over $P$ elements. Thus, the larger $P$ gets the more contributions there are to this term. This analysis is referred to as the signal-to-noise analysis in the Hopfield model and yields the limit for the number of memories $P$. It can be shown [21] that it is equal to:

$$P = \frac{N}{4\ln(N)} \tag{30.9}$$

This is a strong result since for networks with a large number of neurons $N$ (and cortical networks indeed have numerous neurons) we can store numerous memories. If for instance we want to store 10,000 memories, we need about 500,000 neurons, which is roughly the amount of neurons in 5 $mm^3$ of the mammalian neocortex. Note that the natural log of a number is a function that grows very slowly. Thus, a ratio of $N$ over $\ln(N)$ grows rapidly with $N$. For instance, for $N$ equal to 350,000 the log is about 13 (12.77), and the ratio is in the tens of thousands. Hence, for every time we double $N$ we are able to store almost twice as many memories per neuron since the numerator doubles and the denominator increases only by a small amount. This result confirms the ability to store numerous memories in large neural networks.

The Hopfield model was of additional crucial importance since it brought into sharp focus the extraordinary similarity between neural networks and a highly developed field in physics – spin glass theory. Indeed, the powerful mathematical techniques developed in this field of physics allowed the macroscopic dynamics of the Hopfield model to be solved analytically, both in the case of deterministic neurons and in the face of noise in the underlying neuronal dynamics [5]. This rigorous analysis confirms the main results of the Hopfield model as well as allowing more detailed analysis of its properties.

Although very elegant conceptually, the Hopfield model suffers from some mismatches between its assumptions and what we know about the connectivity and activity of cortical networks. For instance, the connectivity in the Hopfield model is assumed to be symmetric, the connection between neuron $j$ and neuron $i$ being identical in value to that between neuron $i$ and neuron $j$, while this is clearly not the case in cortical networks. Many of the assumptions of the Hopfield model can be gradually lifted with the qualitative results remaining the same (though others are more problematic, see [15]). These efforts are beyond the scope of this chapter and are described in [4]. Indeed, the Hopfield model remains a crucial conceptual model for how associative memory can be implemented in recurrent neural networks.

## 30.3 Summary

Neuronal networks are highly interconnected, nonlinear dynamical systems. As such, their study and mathematical treatment ranges across many fields and approaches, only some of which were addressed in this chapter. We briefly outlined several classic as well as new theoretical insights at the single neuron level, and we sketched two classical network computation models – the Perceptron and the Hopfield model. These early models have been key in inspiring modern thinking about the operation of neural networks and have been extensively followed up in many directions [1, 4].

Numerous elegant studies could not be mentioned in this brief chapter: studies focusing on understanding the anatomical structure, not only the dynamics, of neurons and neuronal circuits [10], models describing higher-level phenomena such as language processing or decision making [12], as well as models for neurological diseases.

Another important branch of computational neuroscience that fell beyond the scope of this chapter is that of information theory. In brief, this approach focuses less on the aspect of neurons as computational units performing operations on inputs but rather on the important role of neurons in representing the stimulus, serving as a channel for information regarding the outside world. Using the classical tools stemming from the seminal work of Claude Shannon, a rigorous quantitative treatment of the ability and limits of neurons to pass forth information has been applied with much success in diverse studies in neuroscience [41].

The field of computational neuroscience is entering a very exciting era [1]. Modern experimental techniques continue to elaborate and enrich our understanding of the fascinating biophysical underpinnings of the dynamics of neurons, networks and large-scale systems. The ability to record from increasingly large sets of neurons both electrically and by imaging allows neuronal network models to be compared to, and help interpreting, their experimental counterparts in ever increasing detail. Perhaps the most difficult challenge in modern computational neuroscience is to distill from our vast and rapidly-increasing knowledge of the richness of neurons and neuronal networks those key elements that explain the essence of the amazing computing devices embedded in our minds and brains. Their performance, which allows us to explore, interact, and adapt to our natural and social environments, outstrips by leaps and bounds that of any manmade machine. It is this essential interaction between biological knowledge and conceptual ideas based on firm theoretical foundations, which will allow us to understand "how the brain works".

Understanding how the nervous system processes and codes for real-life information so successfully will have a myriad of applications. Deciphering sensory and motor computations allows us to connect artificial systems, e.g., robotic limbs, to nervous systems [33], a field known as Brain Machine (or Brain-Computer) Interface (BMI, BCI) and to build brain-inspired robots [43]. Unraveling the theoretical principles for memory storage and retrieval in the brain could open the door to develop methods for memory repair and perhaps to improvement of memory and other mental capacities (Augmenting Cognition – AugCog). Finally, the parallel-distributed method of computation used by neural circuits can serve to inspire new generations of biomimetic computers.

**Acknowledgments** This work was supported by a grant from the Blue Brain Project and by the Gatsby Charitable Foundations.

## References

1. Abbott LF (2008) Theoretical neuroscience rising. Neuron 60:489–495
2. Agmon-Snir H, Carr CE, Rinzel J (1998) The role of dendrites in auditory coincidence detection. Nature 393:268–272
3. Agmon-Snir H, Segev I (1993) Signal delay and input synchronization in passive dendritic structures. J Neurophysiol 70:2066–2085
4. Amit DJ (1992) Modeling brain function: the world of attractor neural networks. Cambridge University Press, Cambridge/New York
5. Amit DJ, Gutfreund H, Sompolinsky H (1985) Storing infinite numbers of patterns in a spin-glass model of neural networks. Phys Rev Lett 55:1530–1533
6. Archie KA, Mel BW (2000) A model for intradendritic computation of binocular disparity. Nat Neurosci 3:54–63
7. Ascoli GA, Alonso-Nanclares L, Anderson SA et al. (2008) Petilla terminology: nomenclature of features of GABAergic interneurons of the cerebral cortex. Nat Rev Neurosci 9:557–568
8. Branco T, Hausser M (2011) Synaptic integration gradients in single cortical pyramidal cell dendrites. Neuron 69: 885–892
9. Briggman KL, Helmstaedter M, Denk W (2011) Wiring specificity in the direction-selectivity circuit of the retina. Nature 471:183–188
10. Chen BL, Hall DH, Chklovskii DB (2006) Wiring optimization can relate neuronal structure and function. Proc Natl Acad Sci USA 103:4723–4728
11. Dayan P, Abbott LF (2001) Theoretical neuroscience. MIT Press, Cambridge
12. Dayan P, Daw ND (2008) Decision theory, reinforcement learning, and the brain. Cogn Affect Behav Neurosci 8:429–453
13. Deb K (2001) Multi-objective optimization using evolutionary algorithms. Wiley, New York
14. Druckmann S, Banitt Y, Gidon A et al. (2007) A novel multiple objective optimization framework for constraining conductance-based neuron models by experimental data. Front Neurosci 1:7–18
15. Fusi S, Abbott L (2007) Limits on the memory storage capacity of bounded synapses. Nat Neurosci 10:485–493
16. Gabbiani F, Krapp HG, Koch C, Laurent G (2002) Multiplicative computation in a visual neuron sensitive to looming. Nature 420:320–324
17. Gidon A, Segev I (2012) Principles governing the operation of synaptic inhibition in dendrites. Neuron 75:330–341
18. Gutig R, Sompolinsky H (2006) The tempotron: a neuron that learns spike timing-based decisions. Nat Neurosci 9:420–428
19. Helmstaedter M, Sakmann B, Feldmeyer D (2009) L2/3 interneuron groups defined by multiparameter analysis of axonal projection, dendritic geometry, and electrical excitability. Cereb Cortex 19:951–962
20. Hertz J, Krogh A, Palmer RG (1991) Introduction to the theory of neural computation. Westview Press, Boulder
21. Hopfield JJ (1982) Neural networks and physical systems with emergent collective computational abilities. Proc Natl Acad Sci USA 79:2554–2558
22. Jack JJB, Noble D, Tsien RW (1975) Electric current flow in excitable cells. Clarendon, Oxford
23. Klausberger T, Somogyi P (2008) Neuronal diversity and temporal dynamics: the unity of hippocampal circuit operations. Science 321:53–57
24. Koch C, Poggio T, Torre V (1983) Nonlinear interactions in a dendritic tree: localization, timing, and role in information processing. Proc Natl Acad Sci USA 80:2799–2802
25. Koch C, Rapp M, Segev I (1996) A brief history of time (constants). Cereb Cortex 6:93–101

26. Livet J, Weissman TA, Kang H et al (2007) Transgenic strategies for combinatorial expression of fluorescent proteins in the nervous system. Nature 450:56–62
27. Loewenstein Y, Sompolinsky H (2003) Temporal integration by calcium dynamics in a model neuron. Nat Neurosci 6:961–967
28. Losonczy A, Makara JK, Magee JC (2008) Compartmentalized dendritic plasticity and input feature storage in neurons. Nature 452:436–441
29. Marder E, Taylor AL (2011) Multiple models to capture the variability in biological neurons and networks. Nat Neurosci 14:133–138
30. Markram H, Toledo-Rodriguez M, Wang Y et al (2004) Interneurons of the neocortical inhibitory system. Nat Rev Neurosci 5:793–807
31. Meinertzhagen IA (2010) The organisation of invertebrate brains: cells, synapses and circuits. Acta Zoologica 91:64–71
32. Nelson SB, Hempel C, Sugino K (2006) Probing the transcriptome of neuronal cell types. Curr Opin Neurobiol 16:571–576
33. Nicolelis MAL (2001) Actions from thoughts. Nature 409: 403–407
34. Prinz AA, Bucher D, Marder E (2004) Similar network activity from disparate circuit parameters. Nat Neurosci 7:1345–1352
35. Rall W (1959) Branching dendritic trees and motoneuron membrane resistivity. Exp Neurol 1:491–527
36. Rall W (1964) Theoretical significance of dendritic trees for neuronal input-output relations. In: Reiss RF (ed) Neural theory and modelling. Stanford University Press, Stanford, pp 73–97
37. Rall W (1967) Distinguishing theoretical synaptic potentials computed for different soma-dendritic distributions of synaptic input. J Neurophysiol 30:1138–1168
38. Rall W (1969) Time constants and electrotonic length of membrane cylinders and neurons. Biophys J 9:1483–1508
39. Rall W, Agmon-Snir H (1998) Cable theory for dendritic neurons. In: Koch C, Segev I (eds) Methods in neuronal modeling: from ions to networks. MIT Press, Cambridge, pp 27–92
40. Rall W, Rinzel J (1973) Branch input resistance and steady attenuation for input to one branch of a dendritic neuron model. Biophys J 13:648–687
41. Rieke F (1999) Spikes: exploring the neural code. MIT Press, Cambridge
42. Rosenblatt F (1958) The perceptron. Psychol Rev 65:386–408
43. Rucci M, Bullock D, Santini F (2007) Integrating robotics and neuroscience: brains for robots, bodies for brains. Adv Robot 21:1115–1129
44. Segev I (1995) Dendritic processing. In: Arbib MA (ed) The handbook of brain theory and neuronal networks. MIT Press, Cambridge
45. Single S, Borst A (1998) Dendritic integration and its role in computing image velocity. Science 281:1848–1850
46. Sporns O, Tononi G, Kötter R (2005) The human connectome: a structural description of the human brain. PLoS Comput Biol 1:e42
47. Sterling P (2002) Neuroscience: how neurons compute direction. Nature 420:375–376
48. Walmsley B, Graham B, Nicol MJ (1995) Serial E-M and simulation study of presynaptic inhibition along a group Ia collateral in the spinal cord. J Neurophysiol 74:616–623

# Subject Index

Page numbers set in **boldface** refer to instances of particular emphasis or definitions, page numbers set in *italics* refer to entries found within figures and figure legends



C.G. Galizia, P.-M. Lledo (eds.), *Neurosciences - From Molecule to Behavior: A University Textbook*,
DOI 10.1007/978-3-642-10769-6, © Springer-Verlag Berlin Heidelberg 2013

## C

## D

## E

**G**

## M

MIX
Papier aus verantwortungsvollen Quellen
Paper from responsible sources
FSC® C105338

If you have any concerns about our products,
you can contact us on
**ProductSafety@springernature.com**

In case Publisher is established outside the EU,
the EU authorized representative is:
**Springer Nature Customer Service Center GmbH**
**Europaplatz 3, 69115 Heidelberg, Germany**

Printed by Libri Plureos GmbH
in Hamburg, Germany